Günter Zschornack

Handbook of X-Ray Data

Günter Zschornack

Handbook
of X-Ray Data

With 113 Figures and 161 Tables

 Springer

Ass.-Prof. Dr. rer. nat. habil. Günter Zschornack

Technische Universität Dresden
Institut für Angewandte Physik
Mommsenstraße 13
01062 Dresden

g.zschornack@fz-rossendorf.de

Library of Congress Control Number: 2006937001

ISBN 978-3-540-28618-9 Springer Berlin Heidelberg New York

Springer is a part of Springer Science+Business Media

springer.com

© Springer-Verlag Berlin Heidelberg 2007

Typesetting: PTP-Berlin GmbH, Berlin
Production: LE-TEX Jelonek, Schmidt & Vöckler GbR, Leipzig
Cover: WMXDesign, Heidelberg

Printed on acid-free paper   57/3100/YL - 5 4 3 2 1 0

mn 5·21·2015d

Preface

Today, with energy-dispersive and wavelength-dispersive techniques, modern methods in X-ray analysis are used in a wide range of applications, as for example in X-ray fluorescence analysis, electron microbeam analysis, X-ray fluorescence analysis with charged particles, and so on. In many applications – for instance in metallurgy, mining, microelectronics, medicine, biology, environmental protection, chemistry, archeology, X-ray astronomy, and so on – fast and effective information about the probes under investigation can be obtained by simultaneous multielement analysis. Therefore, it is of outstanding importance for every analyst to have a carefully edited collection of basic atomic data. This is also true in the age of the communication society, where a large quantity of data is available via the Internet. Practical experience shows that it is still important to have data sets in the form of printed matter. This is even more important, because much of the data available in the Internet are given without correct citation of data sources and without any evaluation.

Accurate sets of atomic data are necessary for the calibration of energy- and wavelength-dispersive X-ray spectrometers and serve as basic data for applications in atomic physics, nuclear physics and astrophysics, as well as data for diagnostics and modelling of different plasmas, as for instance in fusion research and ion source physics.

The present book characterizes, in a compact and informative form, the most important processes and facts connected with the emission of X-rays. Beginning with the description of the atomic shell structure, classification of atomic electron transitions is given, and important details about these processes are discussed. Subsequently, an overview of the basic processes of the interaction of X-rays with matter and of the most common detection systems for the detection of X-rays are summarized.

Individual sets of experimental atomic data, known from the literature, are compared between themselves and with theoretical results. In this way it becomes possible to reach conclusions about the consistency of the data sets to be used. In describing the procedures for energy and intensity calibration of energy- and wavelength-dispersive X-ray spectrometers, guidance in the application of atomic data for the calibration of the detection systems to be used and for transition line identification is given.

The present book fills a gap in the existing scientific literature in a field important for commercial and technological applications, as well as for basic and applied research. It is offered to all scientists and engineers using X-rays in research, technology, process control, and in other fields. The spectrum of working areas includes spheres of activity such as X-ray and solid-state physics, atomic and nuclear physics, plasma physics, astrophysics, physical chemistry, as well as special applications in such fields as metallurgy, microelectronics, geology, silicate techniques, environment protection, medicine, and so on.

The tables in the present book contain a wide range of data contributing to the value of the book by their easy availability. From my own investigations I know how costly the search for data can be. During the writing of this book I have made an effort to summarize all the data in a transparent way. When, for a certain quantity, different data are known in the literature, as a rule I have given all known information to characterize the actual knowledge in a definite objective way. Here, readers should judge for themselves how successful I have been. I would be grateful to receive comments from users of the book, in order to help me improve its conception and content for further editions.

During the preparation of the first edition published in 1989 by Verlag für Grundstoffindustrie, Leipzig and in the same year as licence edition by the Springer publishing house, my colleagues Dr. S. Fritzsche, Dr. I. Reiche, Dipl.-Phys. K. Mädler, Dipl.-Phys. R. Paul and Dipl.-Phys. J.-U. Uhlenbrok supported me in different ways. I express my gratitude to all of them. During the preparation of the new edi-

tion the support given by Dipl.-Phys. G. Beulich, who updated the extensive data set, was of extraordinary value. Dr. D. Küchler and Dr. A. El-Shemi, as well as Dipl.-Phys. T. Werner and Dr. F. Ullmann, gave me a lot of valuable tips for the improvement of the first edition. Furthermore, the comprehensive technical assistance from Mrs. K. Arndt is acknowledged.

Last but not least, my gratitude is directed not only to those actively contributing to the success of the present book, but also to those contributing, via appreciation and passive tolerance, to the finishing of the present book – my family.

Dresden, October 2006

Günter Zschornack

Contents

Part I

Atomic Structure, X-ray Physics and Radiation Detection

1 X-Ray Physics and Practice

1.1 Historical Development of X-Ray Physics

In antiquity, the Greek philosophers Leukipp[1] and Demokrit[2] developed *atomism* as the science of the existence of the smallest particles. This science says that matter consists of small indivisible particles, the so-called atoms. Only at the beginning of the nineteenth century was this thesis experimentally proved by Dalton[3], Gay-Lussac[4] and Avogadro[5] and the conception of the existence of atoms and molecules was developed. A description of the Dalton atomic theory is given by Roscoe [459].

In the second half of the nineteenth century the conception of the existence of atoms and molecules was applied successfully in the kinetic theory of gases. Important contributions to the new establishment of the kinetic theory came from Clausius[6], Maxwell[7], Boltzmann[8] and Gibbs[9]. With the determination of the mass and size of atoms the concept of the corpuscular structure of matter became more and more important.

In the framework of the kinetic theory of gases the atom was assumed to be a structureless, homogenous small sphere. At the end of the nineteenth century experiments gave the first indications for an inner structure to the atom. In 1881, investigations of gas discharges by Helmholtz[10] gave experimental confirmation that an elementary unit of electricity must exist. In 1897 Thomson[11] demonstrated that a negative electrical elementary charge is associated with a light (in comparison to the atom) particle, the electron. The magnitude of the elementary charge and the electron mass were determined, and electrons were now understood as constituents of the atom because they could be split from the atom. At the beginning of the twentieth century experiments carried out by Lenard [322] and Rutherford [467] on the transmission of electrons and α-particles through thin metallic foils gave a deeper understanding of the atomic structure.

In the evaluation of these and other experiments the *Rutherford atomic model* was developed:

The atom consists of a massive positively charged nucleus with an extension of about 10^{-14}–10^{-15} m. Far from the nucleus, move massless negatively charged electrons. The mass of the atomic nucleus is about 2000–4000 times greater than the mass of the electron shell. With a diameter in the order of 10^{-10} m the space occupied by an atom is almost empty.

---

[1] about 480–420 before Christ
[2] about 460–370 before Christ
[3] John Dalton: British chemist and physicist (1766–1844)
[4] Louis Joseph Gay-Lussac: French physicist and chemist (1778–1850)
[5] Amedeo Avogadro: Italian naturalist (1776–1856)
[6] Rudolf Clausius: German physicist (1822–1888)
[7] James Clerk Maxwell: British physicist (1831–1879)
[8] Ludwig Eduard Boltzmann: Austrian physicist (1844–1906)
[9] Josiah Willard Gibbs: American physicist (1839–1903)
[10] Hermann Helmholtz: German physicist and physiologist (1821–1894)
[11] Sir Joseph John Thomson: British physicist (1856–1940)

From the outside, atoms appear as electrically neutral. Within the atom the entire negative charge of the electrons is equal to the positive charge concentrated in the nucleus. Somewhat later in the year 1920, Chadwick[12] determined the atomic number Z of atomic nuclei. He found that the nuclear charge Z of an atom is equal to the atomic number in the periodic table. Because atoms are electrically neutral it follows that

atomic number = nuclear charge Z
= electron number of
the neutral atom.

In this way, an important relationship between the atomic number and the atomic structure was formulated and it was deduced that the arrangement of the elements in the periodic table must be closely connected with the atomic structure.

Attempts to formulate, on the basis of the available knowledge, a joint theory of atomic structure with the aid of the laws of classical mechanics were unsatisfactory. First in 1925–1926 was quantum mechanics formulated by Heisenberg[13] and Schrödinger.[14] Since that date a relevant description of microscopic systems is given. Thus, in the present book all statements will be formulated on a quantum mechanical basis.

A pioneering paper was published in the year 1895 by Röntgen [456] in the *Sitzungsberichte der Würzburger Physikalischen und Medizinischen Gesellschaft* with the title *"Über eine neue Art von Strahlen"*. In this paper Röntgen described a form of radiation, which he called X-rays[15] . Röntgen discovered these X-rays during experiments with a gas-discharge tube. The rays discovered by Röntgen, and called after him, were an important highlight of physical research at this time. In 1896, more than 1000 papers tracing the investigations of X-rays were published in scientific journals. Today X-rays are of fundamental importance for basic and applied research as well as for applications in production and in production control techniques. Examples of the wide field of applications were given in Sect. 1.2.

The first spectroscopic investigations of light emission and absorption of atoms were made about 110 years ago. These studies had shown that light emitted by an atom has wavelengths characteristic of each individual atom. Based on this understanding, it was deduced that the appearance of characteristic wavelengths must be connected with the internal atomic structure.

The discovery of important regularities in the wavelengths of hydrogen and in other comparably simple spectra (for instance [28,83,332,370,419,428]) in 1913 was the first success of the interpretation of the hydrogen spectrum by the Bohr theory. Because atoms are stable and emit line spectra under certain conditions, Bohr [77] had formulated postulates, the so-called Bohr postulates:

1. *In the atom there exist discrete orbits where, in contradiction to classical electrodynamics, the electrons can move without emitting radiation. Each of these quantum orbits corresponds to a certain energy state which is determined by an integer number (the main quantum number n). The hydrogen atom is in the ground state (the state with the smallest total energy) if the electron rotates in an orbit with the smallest allowed radius. Outer localized orbits correspond to energetically higher states (excited states) of the atom.*
2. *To excite the electron from a stationary state (electron orbit) to a higher state, excitation energy must be transferred to the electron. After a short characteristic time the electron is transferred back from the energetically higher orbit to the orbit with the lower energy. The energy difference between the initial and final orbit is emitted in the form of a photon.*

The application of quantum mechanics to many-electron atoms has allowed us to understand and to describe quite complicated spectra. For our purposes, it is not necessary here to retrace the complicated path to the present state of knowledge. For

---

[12] Sir James Chadwick: British physicist (1891–1974)
[13] Werner Karl Heisenberg: German physicist (1901–1976)
[14] Erwin Schrödinger: Austrian physicist (1887–1961)
[15] in German: X-Strahlen

example, the interested reader can find adequate overviews in [236,373,510,539].

A detailed discussion of the fundamental problems of the understanding of atomic spectra was given, for example, in the overviews of Sobelman [509], Cowan [126] and Shevelko [494].

Closely connected with the development of the conception of the atomic structure is the use of X-rays from atoms and ions for the purposes of basic and applied research, such as their use in the control of technological processes and for medical applications. After the discovery of X-rays, the diagnostic possibilities of this radiation were understood very rapidly.

In 1912, with the papers of Laue[16], Friedrich[17], Knipping, the formulation of the Bragg equation [82] and investigations by Moseley [370], important prerequisites for the use of spectroscopic methods for the analysis of elements were developed. The use of these new diagnostic possibilities led, for example, to the discovery of the elements helium, cesium, gallium, indium and thallium. A deeper understanding of X-ray emission analysis started with the work of Hevesy [235].

Principally, information on the atomic or ionic electron shell structure can be obtained[18] in different ways. Thus data from the elastic or inelastic scattering of electrons, ions or X-rays and the energies from photoelectrons can help to clarify the atomic structure. An important source of very precise information is spectroscopic studies of the quanta emitted or absorbed by atoms. X-ray analytical investigations require the precise knowledge of such atomic data, as for instance X-ray transition energies, the relative intensities of different X-ray transitions, the energies of absorption edges and fluorescence yields.

This is why the data in the present book is, after the discussion of important for X-ray analysis termini, principles and facts to give a comprehensive description of atomic data on the basis of an extensive evaluation of the existing X-ray literature. Here we add information from calculated data if experimental data are not available.

1.2 Significance of Atomic Data for Practice

In basic research X-ray spectroscopy is widely used and often contributes substantially to increasing knowledge in different scientific disciplines. To illustrate this circumstance we give in the following some examples underlying the importance of X-ray physics:

1. Considering the dependence, known from Moseley's law, between the atomic number Z of a given element and the corresponding transition frequencies of the X-rays, the elements calcium ($Z = 20$) and zinc ($Z = 30$) were placed into Mendeleev's periodic table on the basis of their measured X-ray transition energies. Likewise, the positions in the periodic table of the as yet undetermined elements technetium, promethium, astatine and francium could be determined. These elements were found only with great difficulty because the elements astatine and francium have only radioactive short-lived isotopes and the half-life times of technetium and promethium are shorter than the age of the Earth.

2. The elements hafnium ($Z = 72$) and rhenium ($Z = 75$) were discovered by the detection of their X-ray emission.

3. The spectrometry of X-rays from element 104 confirms the production of this element [56].

4. Atomic states, as they are produced for example in ion–atom collisions and in the cascade decay of inner-shell vacancies, can be determined by X-ray spectroscopy. Furthermore, the physical properties of X-ray-emitting matter, such as density, temperature or composition of high-temperature plasmas, can be analyzed by X-ray spectroscopy.

---

[16] Max von Laue: German physicist (1879–1960)

[17] Walter Friedrich: German physicist (1883–1968)

[18] Usually the term "atom" characterizes a neutral atom, i.e. a nucleus with charge $+Z$ around which revolves N electrons, where $N = Z$. By "ions" we understand positively or negatively charged systems with $N < Z$ or $N > Z$.

5. X-rays from electrons leaving the conduction band of metals contain information on the wavefunction of these electrons [169,418].

6. The binding energy of electrons bound in the atomic orbitals weakly reflects the nuclear volume. Because in heavy atoms the lifetimes of inner-shell vacancies are much smaller than the lifetime of many nuclei, the radii of excited nuclei can be determined by the spectrometry of X-rays emitted during β-decay (electron capture).

7. X-ray spectrometry allows one to get information about the excitation process during the production of X-rays. For example, X-ray transition lines from electron capture (weak outer-shell excitation) and from photoionization processes (intense satellite lines) have been investigated.

8. By X-ray spectrometry Coster–Kronig transition probabilities [574] as well as K and L transition probabilities can be determined.

9. To determine nuclear lifetimes in the region of 10^{-16} s X-rays have been used in experiments with delayed particles.

10. X-rays are applied in the structure analysis of huge molecules. For example, the structure of insulin, hemoglobin and of DNA were clarified by the help of X-rays.

Without attempting completeness we will give a list of applications of different X-ray techniques (e.g. XIXE, RFA, PIXE[19]) where precise atomic data are prerequisites for successful results:

- *Medicine and biomedical research*
    - investigations of tissue probes, cells, etc.,
    - analysis of trace elements in blood and glands,
    - investigations of the assimilation of nonradioactive iodine,
    - analysis of clinical X-ray sources;
- *Metallurgy*
    - analysis of binding structures, diffusion profiles, etc.,
    - qualitative and quantitative analysis of alloys,
    - studies on the lattice structures at high pressures,

- phase transitions at high temperatures and other transition phenomena;
- *Geophysical research and source exploration*
    - seafloor analysis, history of the seafloor,
    - classification of mineral probes,
    - mineral and precious metal exploration;
- *Solid-state and semiconductor research*
    - fault analysis of semiconductor elements,
    - investigation of the spatial impurity distribution in electronic devices and basic materials,
    - classification of minerals in lunar probes, etc. to get information on the temperature, chemical composition and cooling rates of different probes;
- *Space research*
    - X-ray astronomy (identification of characteristic lines in the X-ray continuum),
    - analysis of meteors, asteroids and planets,
    - space exploration;
- *Forensic science*
    - identification of art and coin forging,
    - authenticity of credit cards, passports, etc.,
    - customs examination according to precious metals,
    - identification of poisons of heavy elements,
    - missile abrasion, gunpowder tracks;
- *Industry*
    - quality control and error analysis,
    - on-line process control in steelworks, metal refining, etc.,
    - control of machine wear by fluorescence analysis of the lubricating oil,
    - calibration of materials thickness,
    - identification and arranging of unknown materials,
    - investigation of catalytic after-burning processes in the car industry (Pb, S),
    - food chemistry investigations;
- *Archeology and art*
    - authenticity and classification of precious archeological objects, determination of the age, origin, etc.,
    - identification and authenticity of old works of art, coins, etc.;
- *Environment analysis and protection*

---

[19] PIXE: Particle Induced X-ray Emission

– analysis of environmental pollution (gaseous and powdery pollutants) and air filter probes,
– trace analysis of toxic heavy elements.

Depending on the area of responsibility and on the detection techniques (energy-dispersive or wavelength-dispersive spectrometers) the demands on atomic data are different, depending on the actual problem. Here we will note that not in every case is the precision of the available atomic data the limiting factor, because also quantities such as the effect-to-background ratio, the energy resolution of the applied spectrometer or the chemical composition of the investigated probes can influence the analysis results substantially. Thus for solving a concrete problem a detailed analysis considering special details and circumstances must be done.

2 Physical Fundamentals

2.1 Electronic Configurations and Atomic Ground States

2.1.1 Occupation of Electronic Levels

An atom consisting of electrons and a nucleus as a many-body system can be described correctly only as a whole ensemble. Here the description of the electron motion is approximated in a central symmetric effective potential produced by the nucleus and all other electrons in the atom. Then in a many-electron atom the effective central potential is the sum of the screened Coulomb potential of the nucleus and of the centrifugal potential. The field generated in this way is a so-called self-consistent field.

In the nonrelativistic approximation the electron motion in the central field is completely described by the orbital angular momentum $l = 0, 1, 2, 3, \ldots$, the projection of this momentum on a certain direction $l_z = m = l, l - 1, \ldots, -l$ and by the electron energy connected with the principal quantum number n. The dependence between the principal quantum number n and the energetic states characterized by n has, for the hydrogen atom, the form

$$E_n = -\frac{Z^2 e^4 m_e}{8 h^2 \varepsilon_0^2} \frac{1}{n^2} \tag{2.1}$$

where Z is the atomic number, e is the elementary charge, m_e is the electron rest mass, h is Planck's constant and ε_0 is the dielectric constant.

Often the *main subshells* in the atom are characterized, beside the principal quantum number, with the capital letters

$n =$	1	2	3	4	5	6	7	...

shell	K	L	M	N	O	P	Q	...

The orbital angular momentum l is an integer quantum number characterizing the angular momentum state of the particle and corresponds, for an electron on a stationary orbit, to a multiple of $\hbar^1$.

The magnetic quantum number m is also an integer number, describing the component of the orbital angular quantum momentum according to the quantization axis (as a rule the z-axis). There are $2l + 1$ possible orientations of the orbital angular momentum l.

Generally, in quantum mechanics the declaration of angular momenta occurs in units of Planck's constant $\hbar$. But in most cases this unit is omitted for simplicity. The characterization of states with different orbital angular momenta l is based on symbols using letters of the Latin alphabet:

orbital angular momentum l [$\hbar$]	0	1	2	3	4	5	6	7	8	9	10	11	12	13	14

designation	s	p	d	f	g	h	i	k	l	m	n	o	q	r	t
	S	P	D	F	G	H	I	K	L	M	N	O	Q	R	T

Capital letters are used as notation for the whole atomic state or for a significant finite population of electrons. Electron states of a given orbital angular momentum l are, according to increasing energy, described by the principal quantum number n, where n can take the values $l + 1, l + 2, \ldots$.

Single electron states with different quantum numbers n and l are characterized by a notation where the leading number declares the principal quantum number n and the following letter gives the

---

[1] $\hbar = h/2\pi$

value of the orbital momentum l. If a certain state is occupied by some electrons with the same values of n and l then the number of these electrons is given by a superscript to the symbol for the orbital angular momentum. For instance the occupation of the electronic shells in the neutral oxygen atom is then designated as

$$O\,(1s^2\,2s^2\,2p^4)$$

This means that two electrons occupy the state with $n = 0$ and $l = 0$, two electrons the state with $n = 1$ and $l = 0$ and four electrons the state with $n = 1$ and $l = 1$. The electron distribution over the different atomic substates with different n and l is called the *electron configuration*.

Because electrons are fermions the electron configuration of a many-electron atom is determined by the Pauli exclusion principle [421]:

All many-electron systems are described by wavefunctions antisymmetrized according to electron exchange. For the occupation of the electron shells this means that each quantum state characterized by a set of certain quantum numbers can be occupied only by one electron, i.e. electrons in a many-electron atom must differ in at least one quantum number.

For simplicity and clarity it is standard notation to give the electron configuration in boxed brackets for that atom which is the latest before the considered one that has closed subshells with the same n and l. Then only the occupation of nonclosed subshells is given explicitly. In the example considered this means for the oxygen atom:

$$[\text{Be}]\,2p^4 \quad \text{with} \quad \text{Be}\,(1s^2\,2s^2)\,.$$

Frequently, in the case that only one electron populates a state with given n and l the superscript will be neglected, so for example in the case of the sodium ($Z = 11$) electron configuration we have

$$[\text{Ne}]\,3s\,.$$

Furthermore, besides the described quantum numbers the state of an electron in the central field is characterized by the *parity P*, depending only on the orbital angular momentum and describing the properties of a wavefunction $\Psi(r)$ in the case of reflection at the origin of the system of coordinates ($r \rightarrow -r$). Then

$$\Psi(-r) = +\Psi(r)\,,\ P = +1\,,\ \text{even parity}\,;$$
$$\Psi(-r) = -\Psi(r)\,,\ P = -1\,,\ \text{odd parity}\,.$$

The overall scheme with the orbital angular momentum has the form

$$P = (-1)^l\,. \tag{2.2}$$

States with even l were declared as states with even (positive) parity. In this case the electron wavefunction does not change sign under sign reversal of all particle coordinates. The opposite case (odd l) is declared as odd (negative) parity.

Therefore

orbital angular momentum l = 0,2,4,... :
  states with positive parity, $P = 1$
orbital angular momentum l = 1,3,5,... :
  states with negative parity, $P = -1$.

Thus, the state of a system with N electrons is even according to

$$P = (-1)\exp\left(\sum_{i=1}^{N} l_i\right) \tag{2.3}$$

if the sum of the orbital angular momenta is even. Odd terms are characterized by an index "o" so for instance

$$^2S_{1/2},\,^2P^o_{1/2}{}^2$$

The requirement to classify electron states by their parity follows because of the need to draw up selection rules for radiative electron transitions.

Finally, to every electron there is associated a *spin*[3]. The spin s and its projection s_z on a selected direction are not connected with the motion of the electron and do not have a classical analogy. For one electron it yields

$$s = 1/2 \qquad s_z = \pm 1/2 \qquad \text{(in units of } \hbar\text{)}.$$

---

[2]   read: doublet S one-half, doublet P odd one-half
[3]   The spin follows without any additional assumptions from momentum conservation in the Dirac theory of the hydrogen atom.

Therefore the state of one electron is completely characterized by the four quantum numbers n, l, l_z and s_z.

For given quantum numbers n and l there exist $2(2l + 1)$ electron states with different l_z and s_z (degenerate[4] levels nl). Considering the Pauli exclusion principle, in each quantum state with quantum numbers n, l, l_z and s_z there can be only one electron, i.e. in an atom maximally $2(2l + 1)$ electrons can have the same values n and l. Here electrons with the same principal quantum number are called *equivalent electrons* and electrons with different principal quantum numbers are referred to as *nonequivalent electrons*. In the case of equivalent electrons the Pauli principle reduces the number of possible states considerably.

The population of all electrons with the same n and l forms a *closed subshell* (subshell occupied by the maximum number of electrons) or an *open subshell* whereas the population of all electrons with the same principal quantum number n forms an *atomic main shell*.

Considering the above-mentioned relations, a closed l-subshell can have the following occupation:

$l = 0$	s-subshell	2 electrons,
$l = 1$	p-subshell	6 electrons,
$l = 2$	d-subshell	10 electrons,
$l = 3$	f-subshell	14 electrons and so on.

Starting from this assumption, the occupation of different main shells yields

K-shell	$1s^2$	2 electrons,
L-shell	$2s^2\,2p^6$	8 electrons,
M-shell	$3s^2\,3p^6\,3d^{10}$	18 electrons,
N-shell	$4s^2\,4p^6\,4d^{10}\,4f^{14}$	32 electrons and so on.

Considering the magnetic interaction between the orbital angular momentum and the electron spin and of the coupling of the magnetic moments connected with this, the total angular momentum results

$$j = l + s \tag{2.4}$$

respectively for absolute values

$$j = |l \pm s|$$

and for a single electron system

$$j = |l \pm \frac{1}{2}| \,.$$

The consequence is that each electron energy level nl must be described additionally by the *total angular momentum* of the electron (fine structure level splitting). In the standard nomenclature this angular momentum is given as a subscript on the right side of the spectroscopic destination of l:

$ns_{1/2}$, $np_{1/2}$, $np_{3/2}$, $nd_{3/2}$, $nd_{5/2}$, etc.

To each electron energy level nlj there exist $(2l + 1)$ sublevels, only distinguished by the value of the projections of the total angular momentum

$$m_j = l_z + s_z \,.$$

2.1.2 Systematics of Electron Energy Levels

LS Coupling

In the central field approximation the energy of an electron is completely described by the specification of the electron configuration nl. If noncentral electrostatic interactions take place between the electrons and the spin–orbit interaction the energy levels n_1l_1, n_2l_2, ... split into many different sublevels. The systematics of these sublevels occurs on the basis of two borderline cases if one of the mentioned interactions is small in comparison to the other one.

Experimental results show that with increasing atomic number a more or less continuous transition occurs from the case where the electrostatic interaction dominates compared with the spin–orbit interaction (LS coupling), to the opposite case of jj coupling. In pure form, jj coupling is rare; quite often an intermediate coupling is characteristic.

Strictly speaking, the naturally observed cases correspond to neither of the described couplings exactly. So it may be that in the case of electron excitation for the description of the atomic core, Russell–Saunders (LS) coupling is favored, and for the de-

---

[4] Degeneration: Because of certain symmetries of the system different eigenfunctions (wavefunctions) belong to each energetic state.

scription of the excited electron, jj coupling must be used.

Quantum numbers L and S. In the case of *LS* coupling, also called *Russell–Saunders coupling* [464, 466], the electrostatic interaction leads to a level splitting of a given electronic configuration into a series of sublevels characterized by different values of the total orbital angular momentum L

$$L = \sum_i l_i ; \qquad |L| = \hbar \sqrt{L(L+1)} \qquad (2.5)$$

and of the total spin S

$$S = \sum_i s_i ; \qquad |S| = \hbar \sqrt{S(S+1)} . \qquad (2.6)$$

For the total angular momentum it then follows that

$$J = L + S ; \qquad |J| = \hbar \sqrt{J(J+1)} . \qquad (2.7)$$

According to equation (2.5) the individual orbital angular momenta couple first to the total orbital angular momentum and the individual spins according to (2.6) to the total spin. Afterwards the total orbital angular momentum and the total spin couple to the total angular momentum J (equation 2.7)).

The dependence of the energetic L splitting can be explained by a simple analogy: Different values of L correspond to different electron orbits. That is the reason that electrons with different values of L are localized at different distances from each other. This leads to deviating values of the electrostatic repulsion between the electrons, i.e. to different energetic substates. The energy dependence of S is also of electrostatic origin but does not have a classical analogy because it is of a quantum mechanical nature.

The interaction energy of the electrons with the nucleus and the interaction energy of the electrons among each other are of different sign. Thus the electrostatic electron interaction produces an energetic level shift towards higher energies, i.e. the absolute energy values decrease.

Based on empirical facts for the description of the energy level structure of configurations containing more than two electrons, Hund [254] proposed relations applicable for *LS* coupling conditions, and these are referred in the literature as *Hund's rules*:

- *Complete occupied shells or subshells do not contribute to the total orbital angular momentum **L** and to the total spin **S**.*
- *Equivalent electrons (with the same l) are inserted into the ground state so that a maximum total spin S[5] results. This is the reason that levels with the highest multiplicity are the energetically deepest localized ones.*
- *In a complex configuration electrons are inserted so that the term with maximum S has the largest value of L. This means the terms with the largest value of **L** are the lowest-energy terms in the case of the same multiplicity $S(S+1)$.*
- *For less than half-filled shells the total angular momentum J has the value*

$$J = L - S$$

and for more than half-filled shells it yields

$$J = L + S .$$

At first Hund's rules were formulated on the basis of the then known experimental results in too general a form. Certainly, Hund's rules can be applied to configurations with a single open subshell or with one subshell plus an s-electron and then only in restricted form [126]:

- *The lowest energy term of a configuration l^n or $l^n s^1$ is that term of maximum S with the largest value of L.*

This restricted form of Hund's rules has been well verified experimentally.

Level fine structure. If we assume that the spin–orbit coupling is small in comparison to the Coulomb interaction then this results in a splitting of the *LS*

---

5   The claim to maximum S can be explained as follows [307]: In a system of two electrons the spin can be $S = 0$ or $S = 1$ whereby corresponding to $S = 1$ there is an antisymmetric local part of the wavefunction $\Psi(r_1, r_2)$. For $r_1 = r_2$ this function vanishes, i.e. the probability of both electrons being localized close to each other is small for the case $S = 1$. As a result the electrostatic repulsion is small and the electrons have a smaller energy. For a system of several electrons the *most antisymmetric* local part of the wavefunction corresponds to the greatest spin.

terms corresponding to the different values of the total orbital angular momentum of the electrons

$$J = L + S \ . \tag{2.8}$$

This splitting is called the *fine-structure* or *multiplet* splitting.

According to the general quantum mechanical rules for the addition of angular momenta, the total angular momentum J of an atom can take values of

$$L + S \geq J \geq |L - S| \ . \tag{2.9}$$

Here we distinguish between two cases:

$L \geq S$: there are $2S + 1$ different values of J, i.e. the term is split into $2S + 1$ components.

$L \leq S$: In this case the number of components is $2L + 1$, but with the addition that the designation of multiplicity remains according to the value of $2S + 1$

The quantity $2S + 1$ describing here the number of term components is called the *multiplicity* of the terms. The designation of the terms is the following:

$2S + 1 = 1$ singlet;
$2S + 1 = 2$ doublet;
$2S + 1 = 3$ triplet;
$2S + 1 = 4$ quartet, etc.

The notation of atomic electronic levels is governed by

multiplicity $\quad\quad\quad\quad\quad\quad\quad\quad$ parity
$\quad$ total orbital angular momentum $_{\text{total angular}}$
$\quad\quad\quad\quad\quad\quad\quad\quad\quad\quad\quad\quad\quad\quad_{\text{momentum}}$

Here on the left upper side multiplicity $2S + 1$ and on the right lower side the total angular momentum J of the electrons are indicated. For odd parity on the right upper term side the index "o" (for odd) is added:

$$^{2S+1}L_J^{(o)} \ .$$

In the case of an absent index the term is of even parity. So for instance the term $^2D^o_{3/2}$ describes an electron level with odd parity with $D = 2$, $S = 1/2$ and $J = 3/2$ and the term $^3D_1$ stands for a level with even parity with $L = 2$, $S = 1$ and $J = 1$.

The specification of L and S describes a *term*; L, S and J describe a *level* and L, S, J and M a *state*.

An *LS*-term is associated with

$$(2L + 1)(2S + 1)$$

states differing in the z-components of the orbital angular momentum M_L and of the spin M_S. Furthermore, each J-component is $2J + 1$ times degenerated because each level with the quantum numbers L, S, J is degenerated according to the orientation of the vector J. The reason is that the energy of an isolated atom does not depend on the orientation of the total angular momentum in space. Thus $2J + 1$ electronic states correspond to the different possible values of the z-component of the total angular momentum J with the same energy.

A situation typical of LS coupling is shown for a sp-configuration in Fig. 2.1. The starting point is here the mean central field energy of a spatial symmetric atom split by contributions from the electron–electron interaction and from the Coulomb exchange interaction. A further energy splitting occurs by the spin–orbit interaction. This is also the situation verified experimentally for the field-free case. The presence of an external magnetic field causes an additional splitting into M components of the levels resulting from the spin–orbit interaction (cancelling of degeneration to M). The total splitting

$$\sum_J (2J + 1) = (2L + 1)(2S + 1) \tag{2.10}$$

in the considered system into $2J + 1$ sublevels for each state ^SL_J can be used to determine the value of j for each level.

If the level splitting is small in comparison to the level separation caused by more stronger interactions then the center of gravity of each group of split levels is identical with the position of the nonsplit parent levels[6].

Landé's interval rule. The multiplet splitting follows Landé's interval rule:

The energetic difference between two terms of a fine-structure multiplet with quantum numbers

---

[6] For a proof see [126], p. 357

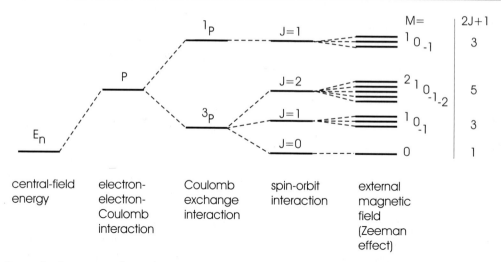

Fig. 2.1. Energy level structure of a sp-electron configuration in *LS* coupling. For clarity the splitting for the spin–orbit interaction and that by an external magnetic field is represented increased and not in a true scale

J + 1 and J of the total quantum momentum is proportional to the greater of both total angular momenta of the terms.

This yields

$$\Delta E_{J+1} - \Delta E_J = A_J (J + 1) . \tag{2.11}$$

The so-called *interval factor* A_J is a constant with different values for different terms. A_J can be positive or negative and the following apply:

- For $A_J > 0$ the multiplet component has the lowest energy such that $J = L - S$. Such kinds of multiplets are called *regular*.
- For $A_J < 0$ the multiplet component has the lowest energy having the greatest possible value of $J = L + S$. These multiplets are called *irregular multiplets*.

For $n = 2l + 1$ a multiplet splitting is absent. The interval rule is valid only for *LS* coupling. This fact allows it to prove the existence of *LS* coupling in fine-structure multiplets by the application of the interval rule.

A situation typical of *LS* coupling is shown for a Ne^{3+} configuration in Fig. 2.2. The distance between *LS* terms of a certain configuration is much smaller than the distance between terms of different electron configurations. Each term – excluding singlet and S terms – has a fine structure whereby the energetic difference between the components of this structure is much smaller than the difference between different terms.

Term determination for many-electron configurations under *LS* coupling conditions. Possible terms for configurations of nonequivalent electrons can be determined using the rules for the quantum mechanical addition of orbital angular momenta. For two angular momenta L_1 and L_2 for the quantum number L it follows that

$$L = L_1 + L_2, \ L_1 + L_2 - 1, \ldots, |L_1 - L_2|$$

and for the resulting spin quantum number is

$$S = S_1 + S_2, \ S_1 + S_2 - 1, \ldots, |S_1 - S_2| .$$

If the system consists of more than two electrons, then at first quantum momenta for two electrons are summarized. As the next step the quantum momenta of a third electron are added, then for a fourth, and so on. This procedure should be demonstrated in the following:

Configuration np n'd: $l = 1, 2, 3$; this results in the terms $^1P, ^1D, ^1F, ^3P, ^3D, ^3F$.

If to this configuration a further electron $n''p$ is added, then it yields:

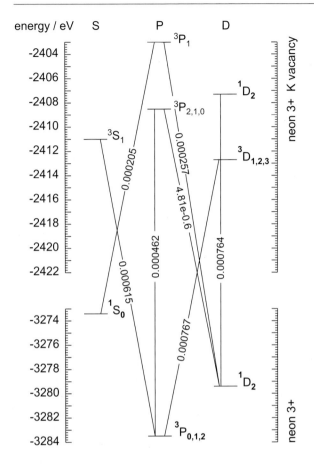

Fig. 2.2. Energy levels of a Ne^{3+} electron configuration under *LS* coupling conditions. Emission rates are given for selected possible X-ray transitions

Addition of a *p*-electron

to the term ^{1}P results in ^{2}S, ^{2}P, ^{2}D;
to the term ^{1}D results in ^{2}P, ^{2}D, ^{2}F;
to the term ^{1}F results in ^{2}D, ^{2}F, ^{2}G;
to the term ^{3}P results in ^{2}S, ^{2}P, ^{2}D, ^{4}S, ^{4}P, ^{4}D;
to the term ^{3}D results in ^{2}P, ^{2}D, ^{2}F, ^{4}P, ^{4}D, ^{4}F;
to the term ^{3}F results in ^{2}D, ^{2}F, ^{2}G, ^{4}D, ^{4}F, ^{4}G .

As a whole we have

2 terms ^{2}S	1 term ^{4}S
4 terms ^{2}P	2 terms ^{4}P
6 terms ^{2}D	3 terms ^{4}D
4 terms ^{2}F	2 terms ^{4}F
2 terms ^{2}G	1 term ^{4}G.

Such an entirety is usually summarized as

$$^2SPDFG \; ^4SPDFG$$
$$2\;4\;6\;4\;2 \quad 1\;2\;3\;2\;1$$

The ascertainment of the resulting terms can also be realized with a scheme where the initial term $np\, n'd$ is given in square brackets: $[np\, n'd]$. The specification of the initial term is also called the *genealogy*.

The presence of one electron in an electron shell leads to a doublet, two electrons result in a singlet and a triplet, three electrons result in a doublet and a quartet and so on. Adding an electron to a term of given multiplicity, results in terms with multiplicities of one step higher and lower than the initial multiplicity, respectively.

In the course of increasing the number of electrons in a given shell for the different terms, multiplicities result, as given in Table 2.1. Here multiplic-

Table 2.1. Term multiplicities during the filling of an electron shell

number of electrons	1	2	3	4	5	6	7	8
multiplicity		2	1, 3	2, 4	1, 3, 5	2, 4	1, 3	2	1, 3

ities with even and odd numbers change with each other. In the region near to a noble gas shell the possible multiplicities decrease again.

The ascertainment of the terms of a given multiplet resulting from a LS-term after addition of a further electron is called *polyades*. For the example given above the two terms

$np\,n'd\ p\ [{}^3F]$ ${}^2D, {}^2F, {}^2G$ and
$np\,n'd\ p\ [{}^3F]$ ${}^4D, {}^4F, {}^4G$

form two different polyades.

The determination of possible terms will be more complicated if it concerns equivalent electrons. Here the Pauli exclusion principle can lead to the situation that for the values of L and S, derived from the common rules of angular momentum addition, there are terms not allowed by the exclusion principle. Possible terms of the configuration p^n, d^n and f^n are summarized in Table 2.2. In the last column of Table 2.2 the statistical weight $\sum (2L+1)(2S+1)$ of the configuration is given, i.e. the total number of states of the given configuration.

If the electron configuration contains equivalent as well as nonequivalent electrons, then the terms of equivalent electrons are determined first. After that the remaining electrons are added according to the rules of adding angular momenta. Examples for LS-terms of nonequivalent electrons are given in Table 2.3.

If a configuration contains two groups of equivalent electrons, then first the terms for each group are determined and added than for the total configuration according to the addition rules of angular momenta. Then in the framework of the vector model of angular momenta the vector sums are

$$L = [(L_1 + L_2) + L_3] + \dots$$
$$S = [(S_1 + S_2) + S_3] + \dots \qquad (2.12)$$
$$J = L + S.$$

Here the L_i, S_i describe terms for the i-th nonclosed subshell given in Table 2.2.

Closed shells must not be considered because the corresponding values of L_i and S_i add to zero.

In Tables 2.4 and 2.5 a summary of common notations for angular momenta in one- and many-electron atoms is given according to [126].

jj-Coupling

Coupling schemata. The term scheme of many atoms is different from that expected from the scheme based on LS coupling conditions. Therefore from the analysis of many experimental results it follows that there is only a restricted region of application for the LS coupling.

Thus we will here discuss the opposite case to the LS coupling, where the spin–orbit interaction is much stronger than the electrostatic interaction. Here the orbital angular momentum and the spin are not conserved independent of each other but are characterized by a *total angular momentum j*. For a many-electron atom the different total angular momenta j can be added to give a total angular momentum J of the atom. In the literature this case is called *jj-coupling*, because with a dominant spin–orbit interaction the introduction of single spins and orbital angular momenta no longer describe the situation correctly. Only the total angular momentum J of the electron is conserved, i.e. the resulting total angular momentum of a N-electron system follows by the addition of the total angular momenta of the single electrons

$$J = \sum_{i=1}^{N} j_i = \sum_{i=1}^{N} (l_i + s_i)\,; \qquad |J| = \hbar\,\sqrt{J(J+1)}\,.$$
$$(2.13)$$

Undistorted jj-coupling appears only in some atomic spectra. Nevertheless, as a rule spectra of heavy elements show structures near to those expected from

Table 2.2. Terms of the configuration l^n of equivalent electrons under LS coupling conditions (see [126,198,307,465,509]). $g(l^n)$ is the statistical weight

configuration	terms	number of terms	$g(l^n)$
s^1	^2S	1	2
s^2	^1S	1	1
$p^1,\ p^5$	$^2P^0$	1	6
$p^2,\ p^4$	$^1(SD)\ ^3P$	3	15
p^3	$^2(PD)^o\ ^4S^0$	3	20
$d^1,\ d^9$	^2D	1	10
$d^2,\ d^8$	$^1(SDG)\ ^3(PF)$	5	45
$d^3,\ d^7$	$^2(PDFGH)_{2}\ ^4(PF)$	8	120
$d^4,\ d^6$	$^1(SDFGI)_{2\,2\ \,2}\ ^3(PDFGH)_{4\ \,2}\ ^5D$	16	210
d^5	$^2(SPDFGHI)_{\ \,3\,2\,2}\ ^4(PDFG)\ ^6S$	16	252
$f^1,\ f^{13}$	$^2F^0$	1	14
$f^2,\ f^{12}$	$^1(SDGI)\ ^3(PFH)$	7	91
$f^3,\ f^{11}$	$^2(PDFGHIKL)^0_{\ \,2\,2\,2\,2}\ ^4(SDFGI)^0$	17	364
$f^4,\ f^{10}$	$^1(SDFGHIKLN)_{2\,4\ \,4\,2\,3\ \,2}\ ^3(PDFGHIKLM)_{3\,2\,4\,3\,4\,2\,2}$ $^5(SDFGI)$	47	1001
$f^5,\ f^9$	$^2(PDFGHIKLMNO)^0_{4\,5\,7\,6\,7\,5\,5\,3\,2}$ $^4(SPDFGHIKLM)^0_{\ \,2\,3\,4\,4\,3\,3\,2}\ ^6(PFH)^0$	73	2002
$f^6,\ f^8$	$^1(SPDFGHIKLMNQ)_{4\ \,6\,4\,8\,4\,7\,3\,4\,2\,2}$ $^3(PDFGHIKLMNO)_{6\,5\,9\,7\,9\,6\,6\,3\,3}\ ^5(SPDFGHIKL)_{\ \,3\,2\,3\,2\,2}$	119	3003
f^7	$^7F\ ^2(SPDFGHIKLMNOQ)^0_{2\,5\,7\,10\,10\,9\,9\,7\,5\,4\ \,2}$ $^4(SPDFGHIKLMN)^0_{\ \,2\,2\,6\,5\,7\,5\,5\,3\,3}\ ^6(PDFGHI)^0\ ^8S^0$	119	3432

Table 2.3. *LS*-terms of nonequivalent electrons (after [534])

configuration	terms
ss	$^1S,\ ^3S$
sp	$^1P,\ ^3P$
sd	$^1D,\ ^3D$
pp	$^1S,\ ^1P,\ ^1D,\ ^3S,\ ^3P,\ ^3D$
pd	$^1P,\ ^1D,\ ^1F,\ ^3P,\ ^3D,\ ^3F$
dd	$^1S,\ ^1P,\ ^1D,\ ^1F,\ ^1G,\ ^3S,\ ^3P,\ ^3D,\ ^3F,\ ^3G$

Table 2.4. Angular momenta for one-electron systems (after [126])

	generally	angular momentum for one electron		
		orbital	spin	total
vector	J	l	s	$j = l + s$
operator	J^2	l^2	s^2	j^2
eigenvalue	$j(j+1)\hbar^2$	$l(l+1)\hbar^2$	$s(s+1)\hbar^2$	$j(j+1)\hbar^2$
possible values of the quantum numbers	$j = 0, 1, 2, \ldots$ or $j = \frac{1}{2}, \frac{3}{2}, \frac{5}{2}, \ldots$	$l = 0, 1, 2, 3, \ldots$	only $s = \frac{1}{2}$	$j = l - \frac{1}{2}\,(l > 0)$ and $j = l + \frac{1}{2}$
operator	J_z	l_z	s_z	j_z
eigenvalue	$m\hbar$ (or $m_j\hbar$)	$m_l\hbar$	$m_s\hbar$	$m\hbar$
possible values of the quantum numbers	$m = -j,$ $-j+1, \ldots$ $j-1, j$	$m_l = -l,$ $-l+1, \ldots,$ $l-1, l$	$m_s = \pm\frac{1}{2}$	$m = -j,$ $-j+1, \ldots$ $j-1, j$

jj-coupling conditions. Generally it yields that in the transition from light elements to heavier ones a more or less continuous transition from *LS*-coupling to *jj*-coupling occurs, i.e. there appears an intermediate coupling type.

The ratio between the electrostatic and the spin–orbit interaction often differs for different levels of the same atom. For light and not too heavy atoms the atomic ground state levels could be described in *LS*-coupling, but this is no longer true for strongly excited states. In this case the electrostatic interaction from electrons of the atomic core with outer electrons is small in comparison to the spin–orbit interaction.

With some exclusions all real spectra can be described in the framework of *LS*- or *jj*-coupling. Often this is also possible for cases if neither *LS*- nor *jj*-coupling can be applied exactly. Furthermore, later

Table 2.5. Angular momenta for many-electron systems (after [126])

	generally	angular momenta of a N-electron system		
		orbital	spin	total
vector	J	$L = \sum_{i=1}^{N} l_i$	$S = \sum_{i=1}^{N} s_i$	$J = \sum_{i=1}^{N} (l + s)$
operator	J^2	L^2	S^2	J^2
eigenvalue	$J(J+1)\hbar^2$	$L(L+1)\hbar^2$	$S(S+1)\hbar^2$	$J(J+1)\hbar^2$
possible values of the quantum numbers	$J = 0, 1, 2, \ldots$ $J = \frac{1}{2}, \frac{3}{2}, \frac{5}{2}, \ldots$	L integer	S integer (N even) or S half-integral (N odd)	J integer (N even) or J half-integral (N odd)
operator	J_z	l_z	s_z	j_z
eigenvalue	$M\hbar$ (or $M_L\hbar$)	$M_L\hbar$	$M_S\hbar$	$M\hbar$
possible values of the quantum numbers	$M = -J,$ $-J+1, \ldots$ $J-1, J$	$M_L = -L,$ $-L+1, \ldots,$ $L-1, L$	$M_S = \pm\frac{1}{2}$	$M = -J,$ $-J+1, \ldots$ $J-1, J$

we will discuss other coupling schemes such as pair coupling and jK-coupling too.

Systemization of electron states for jj-coupling conditions. If the conditions for jj-coupling are satisfied, then every state of an electron can be described by four quantum numbers $nljm_j$, where for a given l the value of j result to

$$l = j \pm \frac{1}{2}.$$

One of the possible values of l is even and the other is odd. Thus, if the value of j and the parity of the state are given then the value of l is determined.

If the electrostatic interaction is neglected, the energy of the electron does not depend on the spatial orientation of the total angular momentum j in the space, i.e. the energy is only dependent on the quantum numbers nlj. In this case each state for a given j is $(2j+1)$-times degenerated. This means for m:

each state with $j = l + 1/2$ contains $2l + 2$ levels with different m;

each state with $j = l - 1/2$ contains $2l$ levels with different m.

In Fig. 2.3 a characteristic level structure for jj-coupling conditions is shown schematically (distances are not proportional to the level energies).

Usually the value of the total angular momentum j is given as a subscript on the right side of the orbital angular momentum l. Possible states are then

$s_{1/2}$, $p_{1/2}$, $p_{3/2}$, $d_{3/2}$, $d_{5/2}$, $f_{5/2}$, $f_{7/2}$, $g_{7/2}$, $g_{9/2}$, $h_{9/2}$, $h_{11/2}$, $\ldots$

Due to the spin–orbit interaction the states $j = l+1/2$ and $j = l - 1/2$ correspond different energy levels.

For a many-electron system under jj-coupling conditions, coupling of the angular momenta occurs, so that first the spin and the orbital angular momentum of each electron couple to a value j_i and then the resulting j_i couple in an arbitrary sequence to a total angular momentum J. For a two-electron system it yields

$$\left[(l_1, s_1)j_1, (l_2, s_2)j_2\right] JM$$

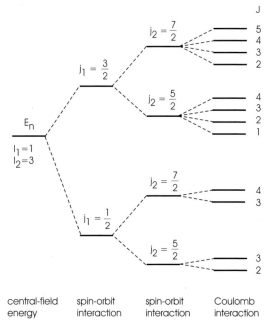

central-field   spin-orbit   spin-orbit   Coulomb
energy   interaction   interaction   interaction

Fig. 2.3. Formation of a pf energy level structure for jj-coupling conditions (after [126])

In jj-coupling the notation of energy levels has the form

$$(j_1, j_2)_J .$$

Here for jj-coupling the exclusion principle means that two electrons must be different at least in one of the quantum numbers nlj.

Considering the electrostatic interaction, a level described by the quantum numbers nlj splits into levels characterized by the total angular momentum J. Possible values of J can be found here analogous to the determination of the possible values under LS-coupling conditions. For nonequivalent electrons the allowed values of J can be determined with the common rules of addition of angular momentum.

As an example we will consider a $nd^1 nf^1$ configuration: Here for the d-electron, $j = 3/2, 5/2$, and for the f-electron, $j = 5/2, 7/2$. The possible values of J are listed in Table 2.7. Here we must note that the total level number for a given value of J should be the same in LS- as well as in jj-coupling.

As in LS-coupling for equivalent electrons the Pauli exclusion principle must be considered too. In

Table 2.7. Terms for a $nd^1 nf^1$-configuration under jj-coupling conditions

j_1	j_2	$|j_1 - j_2| \leq J \leq j_1 + j_2$	terms
$\frac{3}{2}$	$\frac{5}{2}$	1,2,3,4	$\left(\dfrac{3}{2}, \dfrac{5}{2}\right)_{1,2,3,4}$
$\frac{3}{2}$	$\frac{7}{2}$	2,3,4,5	$\left(\dfrac{3}{2}, \dfrac{7}{2}\right)_{2,3,4,5}$
$\frac{5}{2}$	$\frac{5}{2}$	0,1,2,3,4,5	$\left(\dfrac{5}{2}, \dfrac{5}{2}\right)_{0,1,2,3,4,5}$
$\frac{5}{2}$	$\frac{7}{2}$	1,2,3,4,5,6	$\left(\dfrac{5}{2}, \dfrac{7}{2}\right)_{1,2,3,4,5,6}$

Table 2.6 allowed levels of j^n configurations are given for $j \leq 9/2$.

In nuclear physics jj-coupling is widely used for equivalent particles. Otherwise, for atomic subshells ln the coupling conditions are often much closer to the LS scheme than to the jj coupling, if we refrain from some very heavy elements. In the same way the coupling between different subshells or between excited electrons and the atomic core is much closer to jj-coupling.

Furthermore, we will refer to the fact that LS or jj coupling lead to a different number of atomic states if in the case of LS coupling the spin–orbit interaction or for the jj coupling the electrostatic interaction is neglected. This means, if closely located transitions could not be resolved in the case of jj coupling, then there are far fewer transition lines than in the case of LS coupling. This circumstance allows us to deduce the coupling scheme from an experimentally observed spectrum. This is also possible if the observed transition lines are so wide that it is impossible to resolve adjacent spectral lines.

Additional Coupling Types

Beside LS- and the jj-coupling there exits further coupling types much more common than the jj-coupling but much less well known.

Paircoupling. Paircoupling appears in ions for highly excited states. Paircoupling conditions are observed for excited configurations, where the orbital energy only weakly depends from the spin s of the excited

Table 2.6. Allowed states of the configuration j^n for jj-coupling conditions (after [214]; see also [126,509])

l	j	n	J
s,p	$\frac{1}{2}$	0,2	0
		1	$\frac{1}{2}$
p,d	$\frac{3}{2}$	0,4	0
		1,3	$\frac{3}{2}$
		2	0,2
d,f	$\frac{5}{2}$	0,6	0
		1,5	$\frac{5}{2}$
		2,4	0,2,4
		3	$\frac{3}{2},\frac{5}{2},\frac{9}{2}$
f,g	$\frac{7}{2}$	0,8	0
		1,7	$\frac{7}{2}$
		2,6	0,2,4,6
		3,5	$\frac{3}{2},\frac{5}{2},\frac{7}{2},\frac{9}{2},\frac{11}{2},\frac{15}{2}$
		4	0,2,2,4,4,5,6,8
g,h	$\frac{9}{2}$	0,10	0
		1,9	$\frac{9}{2}$
		2,8	0,2,4,6,8
		3,7	$\frac{3}{2},\frac{5}{2},\frac{7}{2},\frac{9}{2},\frac{9}{2},\frac{11}{2},\frac{13}{2},\frac{15}{2},\frac{17}{2},\frac{21}{2}$
		4,6	0,0,2,2,3,4,4,5,6,6,6,7,8,8,9,10,12
		5	$\frac{1}{2},\frac{3}{2},\frac{5}{2},\frac{5}{2},\frac{7}{2},\frac{7}{2},\frac{9}{2},\frac{9}{2},\frac{9}{2},\frac{11}{2},\frac{11}{2},\frac{13}{2},\frac{13}{2},\frac{15}{2},\frac{15}{2},\frac{17}{2},\frac{17}{2},\frac{19}{2},\frac{21}{2},\frac{25}{2}$

electron. This is characteristic of states with high orbital angular momenta, because such electrons disturb the atomic core only slightly and the strength of the spin–orbital interaction is proportional to l^{-3}. Here the energetic level structure depends on the possible values of J, resulting from the addition of s to the resulting vector K of all other angular momenta.

For paircoupling, the following coupling types are known as borderline cases:

1. **jK-coupling.** This is known also as jl-coupling and arises if the spin–orbital interaction of the core electrons is greater than the electrostatic interaction of these electrons with the excited electron.

For two electrons the following coupling scheme results:

$$l_1 + s_1 = j_1$$
$$j_1 + l_2 = K \qquad (2.14)$$
$$K + s_2 = J$$

with the notation

$$j_1[K]_J .$$

2. *LK*-**coupling.** This is also known as *Ls*-coupling and is the alternative borderline case to *jK*-coupling. In two-electron configurations this corresponds to the case if the electron–electron Coulomb interaction is greater than the spin–orbital interaction of each electron. Then the coupling scheme has the form

$$l_1 + l_2 = L$$
$$L + s_1 = K \qquad (2.15)$$
$$K + s_2 = J .$$

In this way the individual states states are characterized by

$$L[K]_J .$$

Overview to the coupling types. For two nonequivalent electrons with the orbital angular momenta $l_1 s_1$ and $l_2 s_2$:

LS-coupling $l_1 + l_2 = L$, $s_1 + s_2 = S$, $L + S = J$
jj-coupling $l_1 + s_1 = j_1$, $l_2 + s_2 = j_2$, $j_1 + j_2 = J$
LK-coupling $l_1 + l_2 = L$, $L + s_1 = K$, $K + s_2 = J$
jK-coupling $l_1 + s_1 = j_1$, $j_1 + l_2 = K$, $K + s_2 = j$,

According to [494] for a two-electron configuration $npn'p$ for the individual coupling types the possible states can be written as:

LS-coupling:

$$^1S_0, \ ^1P_1, \ ^1D_2, \ ^3S_1, \ ^3P_{0,1,2}, \ ^3D_{1,2,3} \ ; .$$

jj-coupling:

$$\left[\frac{1}{2}\,\frac{1}{2}\right]_{0,1}, \ \left[\frac{1}{2}\,\frac{3}{2}\right]_{1,2}, \ \left[\frac{3}{2}\,\frac{1}{2}\right]_{1,2}, \ \left[\frac{3}{2}\,\frac{3}{2}\right]_{0,1,2,3} \ ; .$$

LK-coupling:

$$S\left[\frac{1}{2}\right]_{0,1}, \ P\left[\frac{1}{2}\right]_{0,1}, \ P\left[\frac{3}{2}\right]_{1,2},$$
$$D\left[\frac{3}{2}\right]_{1,2}, \ D\left[\frac{5}{2}\right]_{2,3} \ ;$$

jK-coupling:

$$\frac{1}{2}\left[\frac{1}{2}\right]_{0,1}, \ \frac{3}{2}\left[\frac{1}{2}\right]_{0,1}, \ \frac{1}{2}\left[\frac{3}{2}\right]_{1,2},$$
$$\frac{3}{2}\left[\frac{3}{2}\right]_{1,2}, \ \frac{3}{2}\left[\frac{5}{2}\right]_{2,3} .$$

2.1.3 Electron Shell Structure

Atomic ground states and excited states. The fact that photons are emitted from the atomic shell only with certain discrete wavelengths is connected with the discrete atomic structure, i.e. the atom can exist only in states with discrete energies. The first scientific proof of the existence of discrete atomic energy levels was obtained by the bombardment of atoms with monoenergetic photons. As a result photoelectrons with discrete kinetic energies were observed.

For X-ray analytic purposes important inner-shell electron binding energies can be determined by

- photoelectron spectroscopy after the absorption of monoenergetic and sufficiently energetic photons,
- the analysis of X-ray emission and absorption spectra and
- by absorption spectroscopy with synchrotron radiation.

The possible discrete electron energies of an atom are called *energy levels*. Each quantum state representing the lowest possible energy of an atom is called the *ground state*; all other states are excited states and the corresponding quantum states are named *excited states*.

Energy units of atomic states and X-ray transitions. In atomic and X-ray physics energies are commonly declared in electronvolts (eV) or in the corresponding wavelengths in nanometers (nm). Energy and wavelength are connected by

$$E = h\nu = \frac{hc}{\lambda} \ . \tag{2.16}$$

Here h is Planck's constant and c is the vacuum speed of light. With $hc = (1.23984191 \pm +0.00000011) \, 10^{-4} \, \text{eV cm}$ [157] we find

$$E[\text{keV}] = \frac{1.23984191}{\lambda[\text{nm}]} \quad \text{and} \quad E[\text{Ry}] = \frac{911.27}{\lambda[\text{nm}]} \ . \tag{2.17}$$

Here Ry is the Rydberg unit[7] often used in the theory. It yields

$$1 \, \text{Ry} = \frac{2\pi^2 m e^4}{h} = \frac{e^2}{2a_0} = 13.6056923 \, \text{eV} \tag{2.18}$$

with $a_0 = h^2/4\pi^2 m e^2 = 0,529177 \cdot 10^{-8}$ cm as the first Bohr radius of the hydrogen atom. Also used is the Hartree

$$1 \, \text{Hartree} = \frac{e^2}{a_0} = 2 \, \text{Ry} \ . \tag{2.19}$$

For the sake of completeness here we also give some other non-SI units used up to now in different parts of atomic and X-ray physics:

- **Ångström.** Especially in wavelength-dispersive X-ray spectroscopy the non-SI unit Ångström[8] is widely used:

$$1 \, \text{Å} = 1 \times 10^{-10} \, \text{m} = 1 \times 10^{-1} \, \text{nm} = 1 \times 10^{-8} \, \text{cm} \ .$$

Likewise the International Ångström is also applied , defined by the red line of cadmium ($\lambda_{\text{Cd}} = 6438, 48698$ Å):

$$1 \, \text{I.Å} = 0.000155316413 \, \lambda_{\text{Cd}} \ .$$

- **Kayser.** For levels corresponding to weak bound valence-electrons often the unit *Kayser*[9] (abbreviated notation K) is used as a pseudoenergy unit. Here a spectral line with the wavenumber k is characterized by the number of vacuum wavelengths λ per length unit (usually per centimeter):

$$k = \frac{1}{\lambda} = \frac{E}{hc} \ .$$

For λ in nm:

$$k = \frac{10^9}{\lambda} \, \text{cm}^{-1} \ .$$

The kilokayser is applied to the XUV spectral range and the millikayser characterizes effects such as the isotope shift and hyperfine structure splitting in optical spectra.

- **Relations between energy and pseudoenergy units.** The following relations exist between energy and pseudoenergy units:

  1 Ry = 109737.31568525 cm^{-1}
  1 eV = 8065.544 cm^{-1}

  These relations are valid for the Rydberg constant R_∞ for infinitely large nuclear mass.

Energy level diagrams. The possible energy levels of an atom can be expressed by an energy level diagram (Grotrian diagram). Here each level is represented by a horizontal line along a vertical energy scale. Every level is placed in columns characterizing certain quantum states. The individual levels are characterized by the quantum numbers n (main quantum number) and l (orbital angular momentum). Known are expressions whose end levels are placed in one of several columns, according to certain angular momenta that correspond to that level. Furthermore there exist expressions where in each column the notation to characterize the quantum state is given in the complete description of *LS*-coupling.

In each column with fixed l, the levels form a sequence of energies known as *Rydberg series*. For high n every series converges to an energy, the so-called *series limit*. This value is also known as the *ionization limit* and describes the physical state for which the electron is ionized, i.e. where the electron is infinitely far from the atom and has zero kinetic energy.

Generally there exist ionization limits corresponding to different energetic states of the remain-

---

[7] according to the Swedish physicist J.R. Rydberg (1854–1919)

[8] Spectroscopic length unit named after the Swedish physicist and astronomer Anders Jonas Ångström. Officially, since January 1, 1980 the Ångström hast lost its validity but is still used nowadays in practice.

[9] Named after the German physicist H.G.J. Kayser (1853–1940), the unit is not accepted by the IUPAP, but in practice it is often used instead of cm^{-1}.

ing ion. The lowest lying ionization limit relates to the ion ground state and is known as the *first ionization limit* (or likewise as the ionization limit).

Ionization energy. The energy difference between the atom or ion ground state and that of the next higher charged ion is called the *ionization energy* of the atom or ion, i.e. it characterizes the energy necessary to transform an atom or ion to the next higher charge state. The energy needed for the production of a positive charged atom X^{1+} is known as the *first ionization energy*, for higher charge states yield corresponding designations.

The often used term *ionization potential* is understood to refer to a potential in volts with the same numerical value as the ionization energy in eV.

A summary of experimental ionization energies for all neutral atoms up to uranium is given in Table 2.8. Here the values tabulated in [126] contain results from [367], [99], and [338].

As a rule, ionization occurs by removing an electron from the outermost orbital (most weakly bound electron). It should be mentioned that not for every case is the ionization equivalent to a simple removal of an electron from a certain orbital. For some cases in the following the ionization process can additionally lead to excited states in the atom to be ionized.

Electron affinity. Some neutral atoms and molecules are able to catch an electron to form a negative ion in a stable state. In this way the *electron affinity* characterizes the energy released by the attachment of an electron by a neutral atom or molecule. The electron affinity can be calculated by the Born–Haber circular process.

For example, a summary of electron affinities is given in [494] and [441]. The values of the electron affinity alternate through the periodic systems of elements between energies of less than 100 meV (for instance C^- (^2D)) to about 3 eV (for instance I^- (^1S)).

Electronic structure and ground state terms. In Table 2.9 the electronic structure and the ground state terms for the neutral atoms of the periodic systems of the elements and for the first two ion ground states are given.

Notation of ion charge states. To characterize the ion charge state for a q-fold positive or negative charged

ion, it is common to write $Z^{q\pm}$ or to add to the element symbol a roman number describing the charge state of the positive charged ion. In this form we understand by Ge I the neutral germanium atom, by Ge II the singly ionized germanium ion, by Ge III the twofold ionized germanium ion, and so on.

2.2 Characteristic X-Rays

2.2.1 Classification

Origin of X-rays. Characteristic X-rays originate from electron transitions between different energetic atomic or ionic substates after creating an inner-shell vacancy, i.e. after the ionization of a strongly bound electron. The X-ray properties were basically determined by the atomic number and the number and distributions of the electrons according to the different possible quantum states. Here we define the energy E_i of a state taking part in the X-ray transition as the energy of an electron state, where the atom has one vacancy in the i-th subshell. Neglecting nonradiative electron transitions, then the transition $i \rightarrow f$ from an initial state i into a final state f of lower energy leads to the emission of an X-ray quantum with energy $E_i - E_f$. This quantum corresponds to the emission of radiation with a frequency f, a vacuum wavelength λ and a wavenumber k. Here it yields

$$E_i - E_f = hf = \frac{hc}{\lambda} = hck \qquad (2.20)$$

or, if we give the energy of the individual levels in pseudoenergy units E/hc

$$\frac{E_i}{hc} - \frac{E_f}{hc} = \frac{1}{\lambda} = k \ . \qquad (2.21)$$

On the other hand, if for the atom there exists a state f and if the atom is localized in a radiation field of frequency

$$f' = \frac{E_i - E_f}{h}$$

then it can be excited in a state i by the absorption of a photon with energy $E_i - E_f$.

Table 2.8. Experimental ionization energies for neutral atoms ($q = 0$) and for the first ion ground states q (in eV) for all elements up to $Z = 102$ (after [126,441,494,528]).

Z	Atom	ionization stage q					
		q=0	q=1	q=2	q=3	q=4	q=5
1	H	13.598					
2	He	24.587	54.418				
3	Li	5.392	75.641	122.455			
4	Be	9.322	18.211	153.896	217.720		
5	B	8.298	25.155	37.931	259.374	340.228	
6	C	11.260	24.384	47.888	64.494	392.091	490.00
7	N	14.534	29.602	47.450	77.474	97.891	552.12
8	O	13.618	35.118	54.936	77.414	113.900	138.12
9	F	17.422	34.971	62.709	87.141	114.244	157.17
10	Ne	21.564	40.963	63.46	97.12	126.22	157.93
11	Na	5.139	47.287	71.621	98.92	139.39	172.15
12	Mg	7.646	15.035	80.144	109.266	141.27	186.51
13	Al	5.986	18.829	28.488	119.994	153.72	190.48
14	Si	8.151	16.346	33.493	45.142	166.796	205.06
15	P	10.486	19.726	30.203	51.444	65.026	220.43
16	S	10.360	23.33	34.83	47.31	72.68	88.054
17	Cl	12.967	23.814	39.61	53.47	67.8	97.03
18	Ar	15.759	27.630	40.74	59.81	75.02	91.01
19	K	4.341	31.626	45.73	60.91	82.66	100.0
20	Ca	6.113	11.872	50.914	67.10	84.41	108.78
21	Sc	6.54	12.80	24.757	73.669	91.66	111.1
22	Ti	6.82	13.58	27.492	43.267	99.30	119.5
23	V	6.74	14.66	29.311	46.709	65.282	128.13
24	Cr	6.766	16.50	30.96	49.1	69.46	90.64
25	Mn	7.435	15.640	33.668	51.2	72.4	95
26	Fe	7.870	16.183	30.652	54.8	75.0	99.1
27	Co	7.86	17.083	33.50	51.3	79.5	102
28	Ni	7.635	18.169	35.17	54.9	75.5	108
29	Cu	7.726	20.293	36.84	55.2	79.9	103
30	Zn	9.394	17.965	39.724	59.4	82.6	108
31	Ga	5.999	20.51	30.71	64	83	110
32	Ge	7.899	15.935	34.22	45.71	93.5	113
33	As	9.81	18.589	28.352	50.14	62.63	127.6
34	Se	9.752	21.19	30.821	42.945	68.3	81.81
35	Br	11.814	21.8	36	47.3	59.7	88.6
36	Kr	13.999	24.360	36.95	52.5	64.7	78.5
37	Rb	4.177	26.050	39.02	52.6	71.0	84.4
38	Sr	5.695	11.030	42.884	56.28	71.6	90.8
39	Y	6.38	12.24	20.525	60.60	75.0	89.26
40	Zr	6.84	13.13	22.99	34.412	80.35	94
41	Nb	6.88	14.32	25.05	38.3	50.55	102.06
42	Mo	7.099	16.16	27.17	46.4	61.2	68

Table 2.8. (cont.)

Z	Atom	ionization stage q					
		q=0	q=1	q=2	q=3	q=4	q=5
43	Tc	7.28	15.26	29.55	43	59	76
44	Ru	7.37	16.76	28.47	46.5	63	81
45	Rh	7.46	18.08	31.06	45.6	67	85
46	Pd	8.34	19.43	32.93	48.8	66	90
47	Ag	7.576	21.484	34.83	52	70	89
48	Cd	8.993	16.908	37.48	55	73	94
49	In	5.786	18.870	28.044	54	77	98
50	Sn	7.344	14.632	30.503	40.735	72.28	103
51	Sb	8.641	16.53	25.3	44.2	56	99
52	Te	9.009	18.6	27.96	37.42	58.76	70.7
53	I	10.451	19.131	33	41.7	71	75.76
54	Xe	12.130	21.21	32.1	46.7	59.7	71.8
55	Cs	3.894	23.14	33.38	45.5	62	74
56	Ba	5.212	10.004	35.844	47.1	62	80
57	La	5.577	11.06	19.177	49.95	61.6	80
58	Ce	5.539	10.85	20.198	36.76	65.55	77.6
59	Pr	5.473	10.55	21.624	38.98	57.53	81
60	Nd	5.525	10.73	22.1	40.41	60.00	80
61	Pm	5.582	10.90	22.3	41.1	61.69	82
62	Sm	5.644	11.07	23.4	41.4	62.66	79
63	Eu	5.670	11.24	24.92	42.6	63.23	82
64	Gd	6.150	12.09	20.63	44.0	64.76	85
65	Tb	5.864	11.52	21.91	39.79	66.46	89
66	Dy	5.939	11.67	22.8	41.47	62.08	84
67	Ho	6.022	11.80	22.84	42.5	63.93	86
68	Er	6.108	11.93	22.74	42.65	65.10	88
69	Tm	6.184	12.05	23.68	42.69	65.42	89
70	Yb	6.254	12.176	25.05	43.74	65.58	89
71	Lu	5.426	13.9	20.955	45.250	66.47	93
72	Hf	6.65	14.9	23.3	33.37	68.38	95
73	Ta	7.89	16.2	22.3	33.1	48.27	99
74	W	7.98	17.7	24.1	35.4	48	61
75	Re	7.88	16.6	26	37.7	51	64
76	Os	8.7	17	25	40	54	68
77	Ir	9.1	17.0	27	39	57	72
78	Pt	9.0	18.563	28.5	41.1	55	75
79	Au	9.225	20.5	30.5	43.5	58	73
80	Hg	10.437	18.756	34.2	46	61	77
81	Tl	6.108	20.428	29.83	50.7	64	81
82	Pb	7.416	15.032	31.938	42.32	68.8	84
83	Bi	7.289	16.69	25.56	45.3	56.0	88.3
84	Po	8.42	19.4	27.3	38	61	73
85	At	9.0	20.1	29.3	41	51	78
86	Rn	10.748	21.4	29.4	43.8	55	67

Table 2.8. (cont.)

Z	Atom	q=0	q=1	q=2	q=3	q=4	q=5
				ionization stage q			
87	Fr	3.98	22.5	33.5	43	59	71
88	Ra	5.279	10.147	34.3	46.4	58.5	76
89	Ac	5.170	12.1	16	49	62	76
90	Th	6.080	11.5	20.0	28.8	65	69
91	Pa	5.890	14	22	26	34	84
92	U	6.050	14	24	33	36	44
93	Np	6.190					
94	Pu	6.062					
95	Am	5.993					
96	Cm	6.021					
97	Bk	6.229					
98	Cf	6.298					
99	Es	6.422					
100	Fm	6.500					
101	Md	6.580					
102	No	6.650					

Ritz combination principle. The appearance of characteristic X-ray lines can be explained by the *Ritz combination principle* [454]:

The wavenumbers (frequencies, energies) from lines of characteristic X-rays can be expressed as differences between two atomic substates. The addition or the subtraction of the wavenumbers of two spectral lines of an atom give the wavenumber of a further transition line.

Corresponding to this mechanism for the origin of X-rays, it means that any two levels can combine and produce in this way a line in the X-ray spectrum. This is in principle true, but most level combinations can be neglected because of their low intensity as a result of different selection rules which can reduce the intensity of transitions significantly.

The practical relevance of the combination principle consists in the circumstance that the observation of known X-ray lines and series can be used to predict new, as yet unknown spectral series or spectral lines by combining observed terms or lines. In the past the application of the Ritz combination principle has been used to predict the Paschen, Lyman, Brackett and Pfund series of the hydrogen atom, quite earlier before their experimental scientific proof [236].

Term energies and X-ray emission. X-ray emission occurs if in an atomic inner-shell there exists an initial vacancy, i.e. if at least one electron is absent in the state to be considered. This is equivalent to the situation where only one electron occupies the considered subshell. Thus, the energy levels involved in an X-ray emission process, i.e. the energy levels with one inner-shell vacancy, can be described in a first approximation as optical energy levels of a hydrogen-like ion, if the influence of the electrons on the atomic field is considered by the constants of the total screening σ_t and of the inner screening σ_i, respectively [4]. Then for the energy of the electron orbitals:

$$E(nlj) = Rhc\,\frac{M}{M+m_e}\left[\frac{(Z-\sigma_t)^2}{n^2}+\frac{\alpha^2(Z-\sigma_i)^4}{n^4}\right.$$
$$\left.\left(\frac{n}{j+0.5}-\frac{3}{4}\right)\right]. \qquad (2.22)$$

In this description M is the atomic mass, m_e the electron mass, R the Rydberg constant and the quantum

Table 2.9. Electronic structure and ground state terms for neutral atoms and ions of charge $q = 1$ and 2 for all elements with $Z \leq 102$. For the electronic configuration, experimentally determined LS terms are given for the ground states of free atoms and ions (after [126,367,441]). For some heavy elements for the coupling of the outermost electron the jj-coupling description is more appropriate and is given according to [441].

Z	S	$q = 0$ occupation	term	$q = 1$ occupation	term	$q = 2$ occupation	term
1	H	$1s^1$	$^2S_{1/2}$	–	–	–	–
2	He	$1s^2$	$^1S^0$	$1s^1$	$^2S_{1/2}$	–	–
3	Li	[He] $2s^1$	$^2S_{1/2}$	[He]	$^1S_0$	$1s^1$	$^2S_{1/2}$
4	Be	[He] $2s^2$	$^1S_0$	[He] $2s^1$	$^2S_{1/2}$	[He]	$^1S_0$
5	B	[Be] $2p^1$	$^2P^0_{1/2}$	[Be]	$^1S_0$	[He] $2s^1$	$^2S_{1/2}$
6	C	[Be] $2p^2$	$^3P_0$	[Be] $2p^1$	$^2P^0_{1/2}$	[Be]	$^1S_0$
7	N	[Be] $2p^3$	$^4S^0_{3/2}$	[Be] $2p^2$	$^3P_0$	[Be] $2p^1$	$^2P^0_{1/2}$
8	O	[Be] $2p^4$	$^3P_2$	[Be] $2p^3$	$^4S^0_{3/2}$	[Be] $2p^2$	$^3P_0$
9	F	[Be] $2p^5$	$^2P^0_{3/2}$	[Be] $2p^4$	$3P_2$	[Be] $2p^3$	$^4S^0_{3/2}$
10	Ne	[Be] $2p^6$	$^1S_0$	[Be] $2p^5$	$^2P^0_{3/2}$	[Be] $2p^4$	$^3P_2$
11	Na	[Ne] $3s^1$	$^2S_{1/2}$	[Ne]	$^1S_0$	[Be] $2p^5$	$^2P^0_{3/2}$
12	Mg	[Ne] $3s^2$	$^1S_0$	[Ne] $3s^1$	$^2S_{1/2}$	[Ne]	$^1S_0$
13	Al	[Ne] $3s^2\,3p^1$	$^2P^0_{1/2}$	[Ne] $3s^2$	$^1S_0$	[Ne] $3s^1$	$^2S_{1/2}$
14	Si	[Ne] $3s^2\,3p^2$	$^3P_0$	[Ne] $3s^2\,3p^1$	$^2P^0_{1/2}$	[Ne] $3s^2$	$^1S_0$
15	P	[Ne] $3s^2\,3p^3$	$^4S^0_{3/2}$	[Ne] $3s^2\,3p^2$	$^3P_0$	[Ne] $3s^2\,3p^1$	$^2P^0_{1/2}$
16	S	[Ne] $3s^2\,3p^4$	$^3P_2$	[Ne] $3s^2\,3p^3$	$^4S^0_{3/2}$	[Ne] $3s^2\,3p^2$	$^3P_0$
17	Cl	[Ne] $3s^2\,3p^5$	$^2P^0_{3/2}$	[Ne] $3s^2\,3p^4$	$^3P_2$	[Ne] $3s^2\,3p^3$	$^4S^0_{3/2}$
18	Ar	[Ne] $3s^2\,3p^6$	$^1S_0$	[Ne] $3s^2\,3p^5$	$^2P^0_{3/2}$	[Ne] $3s^2\,3p^4$	$^3P_2$
19	K	[Ar] $4s^1$	$^2S_{1/2}$	[Ar]	$^1S_0$	[Ne] $3s^2\,3p^5$	$^2P^0_{3/2}$
20	Ca	[Ar] $4s^2$	$^1S_0$	[Ar] $4s^1$	$^2S_{1/2}$	[Ar]	$^1S_0$
21	Sc	[Ar] $3d^1\,4s^2$	$^2D_{3/2}$	[Ar] $3d^1\,4s^1$	$^3D_1$	[Ar] $3d^1$	$^2D_{3/2}$
22	Ti	[Ar] $3d^2\,4s^2$	$^3F_2$	[Ar] $3d^2\,4s^1$	$^4F_{3/2}$	[Ar] $3d^2$	$^3F_2$
23	V	[Ar] $3d^3\,4s^2$	$^4F_{3/2}$	[Ar] $3d^4$	$^5D_0$	[Ar] $3d^3$	$^4F_{3/2}$
24	Cr	[Ar] $3d^5\,4s^1$	$^7S_3$	[Ar] $3d^5$	$^6S_{5/2}$	[Ar] $3d^4$	$^5D_0$
25	Mn	[Ar] $3d^5\,4s^2$	$^6S_{5/2}$	[Ar] $3d^5\,4s^1$	$^7S_3$	[Ar] $3d^5$	$^6S_{5/2}$
26	Fe	[Ar] $3d^6\,4s^2$	$^5D_4$	[Ar] $3d^6\,4s^1$	$^6D_{9/2}$	[Ar] $3d^6$	$^5D_4$
27	Co	[Ar] $3d^7\,4s^2$	$^4F_{9/2}$	[Ar] $3d^8$	$^3F_4$	[Ar] $3d^7$	$^4F_{9/2}$
28	Ni	[Ar] $3d^8\,4s^2$	$^3F_4$	[Ar] $3d^9$	$^2D_{5/2}$	[Ar] $^3d^8$	$^3F_4$

Table 2.9. (cont.)

Z	S	$q = 0$		$q = 1$		$q = 2$	
		occupation	term	occupation	term	occupation	term
29	Cu	$[\text{Ar}]\,3d^{10}\,4s^1$	$^2S_{1/2}$	$[\text{Ar}]\,3d^{10}$	$^1S_0$	$[\text{Ar}]\,3d^9$	$^2D_{5/2}$
30	Zn	$[\text{Ar}]\,3d^{10}\,4s^2$	$^1S_0$	$[\text{Ar}]\,3d^{10}\,4s^1$	$^2S_{1/2}$	$[\text{Ar}]\,3d^{10}$	$^1S_0$
31	Ga	$[\text{Zn}]\,4p^1$	$^2P^0_{1/2}$	$[\text{Zn}]$	$^1S_0$	$[\text{Ar}]\,3d^{10}\,4s^1$	$^2S_{1/2}$
32	Ge	$[\text{Zn}]\,4p^2$	$^3P_0$	$[\text{Zn}]\,4p^1$	$^2P^0_{1/2}$	$[\text{Zn}]$	$^1S_0$
33	As	$[\text{Zn}]\,4p^3$	$^4S^0_{3/2}$	$[\text{Zn}]\,4p^2$	$^3P_0$	$[\text{Zn}]\,4p^1$	$^2P^0_{1/2}$
34	Se	$[\text{Zn}]\,4p^4$	$^3P_2$	$[\text{Zn}]\,4p^3$	$^4S^0_{3/2}$	$[\text{Zn}]\,4p^2$	$^3P_0$
35	Br	$[\text{Zn}]\,4p^5$	$^2P^0_{3/2}$	$[\text{Zn}]\,4p^4$	$^3P_2$	$[\text{Zn}]\,4p^3$	$^4S^0_{3/2}$
36	Kr	$[\text{Zn}]\,4p^6$	$^1S_0$	$[\text{Zn}]\,4p^5$	$^2P^0_{3/2}$	$[\text{Zn}]\,4p^4$	$^3P_2$
37	Rb	$[\text{Kr}]\,5s^1$	$^2S_{1/2}$	$[\text{Kr}]$	$^1S_0$	$[\text{Zn}]\,4p^5$	$^2P^0_{3/2}$
38	Sr	$[\text{Kr}]\,5s^2$	$^1S_0$	$[\text{Kr}]\,5s^1$	$^2S_{1/2}$	$[\text{Kr}]$	$^1S_0$
39	Y	$[\text{Kr}]\,4d1\,5s^2$	$^2D_{3/2}$	$[\text{Kr}]\,5s^2$	$^1S_0$	$[\text{Kr}]\,5s^1$	
40	Zr	$[\text{Kr}]\,4d^2\,5s^2$	$^3F_2$	$[\text{Kr}]\,4d^2\,5s^1$	$^4F_{3/2}$	$[\text{Kr}]\,4d^2$	$^3F_2$
41	Nb	$[\text{Kr}]\,4d^4\,5s^1$	$^6D^0_{1/2}$	$[\text{Kr}]\,4d^4$	$^5D_0$	$[\text{Kr}]\,4d^3$	$^4F_{3/2}$
42	Mo	$[\text{Kr}]\,4d^5\,5s^1$	$^7S_3$	$[\text{Kr}]\,4d^5$	$^6S_{5/2}$	$[\text{Kr}]\,4d^4$	$^5D_0$
43	Tc	$[\text{Kr}]\,4d^5\,5s^2$	$^6S_{5/2}$	$[\text{Kr}]\,4d^5\,5s^1$	$^7S_3$	$[\text{Kr}]\,4d^5$	$^6S_{5/2}$
44	Ru	$[\text{Kr}]\,4d^7\,5s^1$	$^5F_5$	$[\text{Kr}]\,4d^7$	$^4F_{9/2}$	$[\text{Kr}]\,4d^6$	$^5D_4$
45	Rh	$[\text{Kr}]\,4d^8\,5s^1$	$^4F_{9/2}$	$[\text{Kr}]\,4d^8$	$^3F_4$	$[\text{Kr}]\,4d^7$	$^4F_{9/2}$
46	Pa	$[\text{Kr}]\,4d^{10}$	$^1S_0$	$[\text{Kr}]\,4d^9$	$^2D_{5/2}$	$[\text{Kr}]\,4d^8$	$^3F_4$
47	Ag	$[\text{Kr}]\,4d^{10}\,5s^1$	$^2S_{1/2}$	$[\text{Kr}]\,4d^{10}$	$^1S_0$	$[\text{Kr}]\,4d^9$	$^2D_{5/2}$
48	Cd	$[\text{Kr}]\,4d^{10}\,5s^2$	$^1S_0$	$[\text{Kr}]\,4d^{10}\,5s^1$	$^2S_{1/2}$	$[\text{Kr}]\,4d^{10}$	$^1S_0$
49	In	$[\text{Cd}]\,5p^1$	$^2P^0_{1/2}$	$[\text{Cd}]$	$^1S_0$	$[\text{Kr}]\,4d^{10}\,5s^1$	$^2S_{1/2}$
50	Sn	$[\text{Cd}]\,5p^2$	$^3P_0$	$[\text{Cd}]\,5p^1$	$^2P^0_{1/2}$	$[\text{Cd}]$	$^1S_0$
51	Sb	$[\text{Cd}]\,5p^3$	$^4S^0_{3/2}$	$[\text{Cd}]\,5p^2$	$^3P_0$	$[\text{Cd}]\,5p^1$	$^2P^0_{1/2}$
52	Te	$[\text{Cd}]\,5p^4$	$^3P_2$	$[\text{Cd}]\,5p^3$	$^4S^0_{3/2}$	$[\text{Cd}]\,5p^2$	$^3P_0$
53	I	$[\text{Cd}]\,5p^5$	$^2P^0_{3/2}$	$[\text{Cd}]\,5p^4$	$^3P_2$	$[\text{Cd}]\,5p^3$	$^4S^0_{3/2}$
54	Xe	$[\text{Cd}]\,5p^6$	$^1S_0$	$[\text{Cd}]\,5p^5$	$^2P^0_{3/2}$	$[\text{Cd}]\,5p^4$	$^3P_2$
55	Cs	$[\text{Xe}]\,6s^1$	$^2S_{1/2}$	$[\text{Xe}]$	$^1S_0$	$[\text{Cd}]\,5p^5$	–
56	Ba	$[\text{Xe}]\,6s^2$	$^1S_0$	$[\text{Xe}]\,6s^1$	$^2S_{1/2}$	$[\text{Xe}]$	$^1S_0$
57	La	$[\text{Xe}]\,5d^1\,6s^2$	$^2D_{3/1}$	$[\text{Xe}]\,5d^2$	$^3F_2$	$[\text{Xe}]\,5d^1$	$^2D_{3/2}$
58	Ce	$[\text{La}]\,4f^1$	$^1G^0_4$	$[\text{Xe}]\,4d^1\,5d^2$	$^4H^0_{7/2}$	$[\text{Xe}]\,4f^2$	$^3H_4$
59	Pr	$[\text{Xe}]\,4f^3\,6s^2$	$^4I^0_{9/2}$	$[\text{Xe}]\,4f^3\,6s^1$	$^5I^0_4$	$[\text{Xe}]\,4f^3$	$^4I^0_{9/2}$

Table 2.9. (cont.)

Z	S	$q=0$ occupation	term	$q=1$ occupation	term	$q=2$ occupation	term
60	Nd	$[\text{Xe}]\,4f^4\,6s^2$	$^5I_4$	$[\text{Xe}]\,4f^4\,6s^1$	$^6I_{7/2}$	$[\text{Xe}]\,4f^4$	$^5I_4$
61	Pm	$[\text{Xe}]\,4f^5\,6s^2$	$^6H^0_{5/2}$	$[\text{Xe}]\,4f^5\,6s^1$	$^7H^0_2$	$[\text{Xe}]\,4f^5$	$(^6H^0_{5/2})$
62	Sm	$[\text{Xe}]\,4f^6\,6s^2$	$^7F_0$	$[\text{Xe}]\,4f^6\,6s^1$	$^8F_{1/2}$	$[\text{Xe}]\,4f^6$	$^7F_0$
63	Eu	$[\text{Xe}]\,4f^7\,6s^2$	$^8S^0_{7/2}$	$[\text{Xe}]\,4f^7\,6s^1$	$^9S^0_4$	$[\text{Xe}]\,4f^7$	$^8S^0_{7/2}$
64	Gd	$[\text{Eu}]\,5d^1$	$^9D^0_2$	$[\text{Cs}]\,4f^7\,5d^1$	$^{10}D^0_{5/2}$	$[\text{Xe}]\,4f^7\,5d^1$	$^9D^0_2$
65	Tb	$[\text{Xe}]\,4f^9\,6s^2$	$^6H^0_{17/2}$	$[\text{Xe}]\,4f^9\,6s^1$	$^7H^0_8$	$[\text{Xe}]\,4f^9$	$^6H^0_{15/2}$
66	Dy	$[\text{Xe}]\,4f^{10}\,6s^2$	$^5I_8$	$[\text{Xe}]\,4f^{10}\,6s^1$	$^6I_{17/2}$	$[\text{Xe}]\,4f^{10}$	$(^5I_8)$
67	Ho	$[\text{Xe}]\,4f^{11}\,6s^2$	$^4I^0_{15/2}$	$[\text{Xe}]\,4f^{11}\,6s^1$	$^5I^0_8$	$[\text{Xe}]\,4f^{11}$	$^4I^0_{15/2}$
68	Er	$[\text{Xe}]\,4f^{12}\,6s^2$	$^3H_6$	$[\text{Xe}]\,4f^{12}\,6s^1$	$^4H_{13/2}$	$[\text{Xe}]\,4f^{12}$	$^3H_6$
69	Tm	$[\text{Xe}]\,4f^{13}\,6s^2$	$^2F^0_{7/2}$	$[\text{Xe}]\,4f^{13}\,6s^1$	$^3F^0_4$	$[\text{Xe}]\,4f^{13}$	$^2F^0_{7/2}$
70	Yb	$[\text{Xe}]\,4f^{14}\,6s^2$	$^1S_0$	$[\text{Xe}]\,4f^{14}\,6s^1$	$^2S_{1/2}$	$[\text{Xe}]\,4f^{14}$	$^1S_0$
71	Lu	$[\text{Yb}]\,5d^1$	$^2D_{3/2}$	$[\text{Yb}]$	$^1S_0$	$[\text{Xe}]\,4f^{14}\,6s^1$	$^2S_{1/2}$
72	Hf	$[\text{Yb}]\,5d^2$	$^3F_2$	$[\text{Yb}]\,5d^1$	$^2D_{3/2}$	$[\text{Xe}]\,4f^{14}\,5d^2$	$^3F_2$
73	Ta	$[\text{Yb}]\,5d^3$	$^4F_{3/2}$	$[\text{Cs}]\,4f^{14}\,5d^3$	$^5F_2$	$[\text{Xe}]\,4f^{14}\,5d^3$	–
74	W	$[\text{Yb}]\,5d^4$	$^5D_0$	$[\text{Cs}]\,4f^{14}\,5d^4$	$^6D_{1/2}$	$[\text{Xe}]\,4f^{14}\,5d^4$	–
75	Re	$[\text{Yb}]\,5d^5$	$^6S_{5/2}$	$[\text{Cs}]\,4f^{14}\,5d^5$	$^7S_3$	$[\text{Xe}]\,4f^{14}\,5d^5$	–
76	Os	$[\text{Yb}]\,5d^6$	$^5D_4$	$[\text{Cs}]\,4f^{14}\,5d^6$	$^6D_{9/2}$	$[\text{Xe}]\,4f^{14}\,5d^6$	–
77	Ir	$[\text{Yb}]\,5d^7$	$^4F_{9/2}$	$[\text{Cs}]\,4f^{14}\,5d^7$	–	$[\text{Xe}]\,4f^{14}\,5d^7$	–
78	Pt	$[\text{Cs}]\,4f^{14}\,5d^9$	$^3D_3$	$[\text{Xe}]\,4f^{14}\,5d^9$	$^2D_{5/2}$	$[\text{Xe}]\,4f^{14}\,5d^8$	–
79	Au	$[\text{Cs}]\,4f^{14}\,5d^{10}$	$^2S_{1/2}$	$[\text{Xe}]\,4f^{14}\,5d^{10}$	$^1S_0$	$[\text{Xe}]\,4f^{14}\,5d^9$	–
80	Hg	$[\text{Yb}]\,5d^{10}$	$^1S_0$	$[\text{Cs}]\,4f^{14}\,5d^{10}$	$^2S_{1/2}$	$[\text{Xe}]\,4f^{14}\,5d^{10}$	–
81	Tl	$[\text{Hg}]\,6p^1$	$^2P^0_{1/2}$	$[\text{Hg}]$	$^1S_0$	$[\text{Cs}]\,4f^{14}\,5d^{10}$	$^2S_{1/2}$
82	Pb	$[\text{Hg}]\,6p^2$	$^2P_0$	$[\text{Hg}]\,6p^1$	$^2P^0_{1/2}$	$[\text{Hg}]$	$^1S_0$
83	Bi	$[\text{Hg}]\,6p^3$	$^4S^0_{3/2}$	$[\text{Hg}]\,6p^2$	$^2P_0$	$[\text{Hg}]\,6p^1$	$^2P^0_{1/2}$
84	Po	$[\text{Hg}]\,6p^4$	$^3P_2$	$[\text{Hg}]\,6p^3$	–	$[\text{Hg}]\,6p^2$	–
85	At	$[\text{Hg}]\,6p^5$	$^2P_{3/2}$	$[\text{Hg}]\,6p^4$	–	$[\text{Hg}]\,6p^3$	–
86	Rn	$[\text{Hg}]\,6p^6$	$^1S_0$	$[\text{Hg}]\,5p^5$	–	$[\text{Hg}]\,6p^4$	–
87	Fr	$[\text{Rn}]\,7s^1$	$^2S_{1/2}$	$[\text{Rn}]$	–	$[\text{Hg}]\,6p^5$	–
88	Ra	$[\text{Rn}]\,7s^2$	$^1S_0$	$[\text{Rn}]\,7s^1$	$^2S_{1/2}$	$[\text{Rn}]$	$^1S_0$
89	Ac	$[\text{Rn}]\,6d^1\,7s^2$	$^2D_{3/2}$	$[\text{Rn}]\,7s^2$	$^1S_0$	$[\text{Rn}]\,7s^1$	$^2S_{1/2}$

Table 2.9. (cont.)

Z	S	q = 0 occupation	term	q = 1 occupation	term	q = 2 occupation	term
90	Th	$[\mathrm{Rn}]\,6d^2\,7s^2$ mixed with	$^3F_2$	$[\mathrm{Rn}]\,6d^1\,7s^2$ $[\mathrm{Rn}]\,6d^2\,7s^1$	$''D_{3/2}$ $^4F_{3/2}$	$[\mathrm{Rn}]\,5f^1\,6d^1$	$^3H_4^0$
91	Pa	$[\mathrm{Rn}]\,5f^2$ $6d^1\,7s^2$	$^3H_4$ $(4,\tfrac{3}{2})_{\frac{11}{2}}$	$[\mathrm{Rn}]\,5f^2\,7s^2$	$^3H_4$	$[\mathrm{Rn}]\,5f^2\,7s^1$	–
92	U	$[\mathrm{Ra}]\,5f^3$ $6d^1\,7s^2$	$^4I_{9/2}^0$ $(\tfrac{9}{2},\tfrac{3}{2})_6^0$	$[\mathrm{Rn}]\,5f^3\,7s^2$	$^4I_{9/2}^0$	$[\mathrm{Rn}]\,5f^3\,7s^1$	–
93	Np	$[\mathrm{Ra}]\,5f^4$ $6d^1\,7s^2$	$^5I_4$ $(4,\tfrac{3}{2})_{\frac{11}{2}}$	$[\mathrm{Rn}]\,5f^4\,7s^2$	–	$[\mathrm{Rn}]\,5f^4\,7s^1$	–
94	Pu	$[\mathrm{Rn}]\,5f^6\,7s^2$	$^7F_0$	$[\mathrm{Rn}]\,5f^6\,7s^1$	$^8F_{1/2}$	$[\mathrm{Rn}]\,5f^6$	–
95	Am	$[\mathrm{Rn}]\,5f^7\,7s^2$	$^8S_{7/2}^0$	$[\mathrm{Rn}]\,5f^7\,7s^1$	$^9S_4^0$	$[\mathrm{Rn}]\,5f^7$	$^8S_{7/2}^0$
96	Cm	$[\mathrm{Ra}]\,5f^7$ $6d^1\,7s^2$	$^8S_{7/2}^0$ $(\tfrac{7}{2},\tfrac{3}{2})_2$	$[\mathrm{Rn}]\,5f^7\,7s^2$	$(S_{7/2}^0$	$[\mathrm{Rn}]\,5f^7\,7s^1$	–
97	Bk	$[\mathrm{Rn}]\,5f^9\,7s^2$	$^6H_{15/2}^0$	$[\mathrm{Rn}]\,5f^9\,7s^1$	$^7H_8^0$	$[\mathrm{Rn}]\,5f^9$	–
98	Cf	$[\mathrm{Rn}]\,5f^{10}\,7s^2$	$^5I_8$	$[\mathrm{Rn}]\,5f^{10}$ $7s^1$	$^5I_8$ $(8,\tfrac{1}{2})_{\frac{17}{2}}$	$[\mathrm{Rn}]\,5f^{10}$	–
99	Es	$[\mathrm{Rn}]\,5f^{11}\,7s^2$	$^4I_{15/2}^0$	$[\mathrm{Rn}]\,5f^{11}$ $7s^1$	$^4I_{15/2}^0$ $(\tfrac{15}{2},\tfrac{1}{2})_8^0$	$[\mathrm{Rn}]\,5f^{11}$	$^4I_{15/2}^0$
100	Fm	$[\mathrm{Rn}]\,5f^{12}\,7s^2$	$^3H_6$	$[\mathrm{Rn}]\,5f^{12}\,7s^1$	–	$[\mathrm{Rn}]\,5f^{12}$	–
101	Md	$[\mathrm{Rn}]\,5f^{13}\,7s^2$	$^2F_{7/2}^0$	$[\mathrm{Rn}]\,5f^{13}\,7s^1$	–	$[\mathrm{Rn}]\,5f^{13}$	–
102	No	$[\mathrm{Rn}]\,5f^{14}\,7s^2$	$^1S_0$	$[\mathrm{Rn}]\,5f^{14}\,7s^1$	$^2S_{1/2}$	$[\mathrm{Rn}]\,5f^{14}$	$^1S_0$

numbers nlj characterize the main quantum number, the orbital angular momentum and the total angular momentum, respectively. The first term in equation (2.22), the sum of the kinetic and potential energies of the electrons, contain the main part of the energy. If we neglect the second term in equation (2.22) which describes the interaction between the magnetic moment of the electron and that of the other remaining electrons, then it follows that the square root of the level energies $E(nlj)$ depends linearly from the atomic number Z. This means that the square root

of the X-ray transition energy between two energy levels depends linearly on the atomic number:

$$\frac{1}{\lambda} = C\,(Z - \sigma)^2 \qquad (2.23)$$

where λ is the wavelength of an X-ray transition line emitted from an atom of atomic number Z; C is a constant for the considered X-ray series and σ is the screening constant. In the literature this dependence is known as *Moseley's law*[10]. The statement of Moseley's law [371] in 1914 firstly allows us to formulate a physically based systematization for the localization

---

[10] Henry Gwyn Jeffreys Moseley (1887–1915)

Table 2.10. Classification of electron levels according to the quantum numbers nlj.

n	l	j	nl_j		nl
1	0	1/2	$1s_{1/2}$	K	$1s$
2	0	1/2	$2s_{1/2}$	L_1	$2s$
2	1	1/2	$2p_{1/2}$	L_2	$2p$
2	1	3/2	$2p_{3/2}$	L_3	$2p$
3	0	1/2	$3s_{1/2}$	M_1	$3s$
3	1	1/2	$3p_{1/2}$	M_2	$3p$
3	1	3/2	$3p_{3/2}$	M_3	$3p$
3	2	3/2	$3d_{3/2}$	M_4	$3d$
3	2	5/2	$3d_{5/2}$	M_5	$3d$
4	0	1/2	$4s_{1/2}$	N_1	$4s$
4	1	1/2	$4p_{1/2}$	N_2	$4p$
4	1	3/2	$4p_{3/2}$	N_3	$4p$
4	2	3/2	$4d_{3/2}$	N_4	$4d$
4	2	5/2	$4d_{5/2}$	N_5	$4d$
4	3	5/2	$4f_{5/2}$	N_6	$4f$
4	3	7/2	$4f_{7/2}$	N_7	$4f$
⋮	⋮	⋮	⋮	⋮	⋮

Table 2.11. Regular doublets of X-ray K and L series.

K series			L series		
	initial state	final state		initial state	final state
$\alpha_1\alpha_2$	K	L_3L_2	$l\eta$	L_3L_2	M_1
$\beta_1\beta_3$	K	M_3M_2	$\alpha_2\beta_1$	L_3L_2	M_4
			$\beta_6\gamma_5$	L_3L_2	N_1
			$\beta_{15}\gamma_1$	L_3L_2	N_4
			$\beta_5\gamma_6$	L_3L_2	O_4
			$\beta_4\beta_3$	L_1	M_2M_3
			$\alpha_1\alpha_2$	L_3	M_5M_4
			$\gamma_2\gamma_3$	L_1	N_2N_3
			$\beta_{15}\beta_2$	L_3	N_4N_5

of elements in the periodic table of elements. According to equation (2.22) the energy of the electron orbitals depends on the quantum numbers nlj. In Table 2.10 a list of possible subshells is given, where for the subshells the X-ray term notation K, L_1, L_2 …as well as the optical term designation $s, p, d, \ldots$ are summarized. Here the classification of the X-ray terms is so constructed, that the index 1 describes a subshell containing an s-electron, the indices 2 and 3 describe subshells with p-electrons containing subshells with $j = 1/2$ and $j = 3/2$, and so on[11].

Every subshell can contain at maximum of $2j + 1$ electrons. This *multiplicity* $2j + 1$ of the state is founded on the fact that the total angular momentum vector can, corresponding to the rules of spatial quantization, be orientated in different directions. To describe this behavior an additional quantum number m_j is introduced. Then a complete subshell contains electrons with all possible values of m_j.

Atoms in magnetic fields. If the atom is localized in an outer magnetic field, a further splitting of the electron energy levels known from optical spectroscopy as the *Zeeman effect* can be observed. This splitting is not observed in X-ray spectra, but the level multiplicity influences the observed line intensity as a result of the dependence between multiplicity and electron transmission probabilities of the involved states.

Regular doublets. Two levels characterized by the same quantum numbers n and l, but different in j, form *regular doublets* (also called *relativistic* or *spin doublets*). The energetic difference in the doublet depends on the difference in the interaction energy of the magnetic moments. As a result X-ray line doublets can be observed if the initial and final levels of the transition form a spin doublet. The energy difference between both lines of a regular doublet is given by equation (2.22) for two terms differing in $\Delta j = 1$:

$$E = R\,h\,c\,\alpha^2 \frac{(Z - \sigma_i)^4}{n^3 l(l + 1)} . \qquad (2.24)$$

A summary of regular doublets for the X-ray K- and L-series is given in Table 2.11.

An analysis of the relativistic spin screening constant σ_i leads to the conclusion that for a considered

---

[11] Instead of Roman numbers often Arabic numbers are used.

Table 2.12. Relativistic spin screening constants σ_i (after [4])

L_1	$L_{2,3}$	M_1	$M_{2,3}$	$M_{4,5}$	N_1	$N_{2,3}$	$N_{4,5}$	$N_{6,7}$
2,0	3,5	6,8	8,5	13,0	14,0	17,0	24	34

doublet this constant is independent of the atomic number. In Table 2.12 values for σ_i given in [4] are tabulated.

Irregular doublets. An additional doublet structure arises also for level pairs with the same values of the quantum numbers n and j but different in the orbital angular momentum l. Such doublets ($\Delta l = 1, \Delta j = 0$) are called *irregular doublets* (or *screening doublets*).

Irregular doublets are formed as a result of transitions from or to level pairs such as $L_1 L_2$, $M_1 M_2$, $M_3 M_4$ ... and are connected with σ_t. The selection rule $\Delta l = \pm 1$ for dipole transitions is not valid for one line of the doublet, i.e. one line is a forbidden transition, as a rule a quadrupole transition. σ_t depends on Z in a characteristic manner and the values of σ_t are substantially greater than these ones from σ_i. Values of the screening constant σ_t are given in [516].

Internal and outer screening. Theoretical calculations of screening constants show good agreement with experimental values. Partially the screening depends on charges, localized at radii smaller than the subshell radius of the screened electronic shell and in some degree on charges localized on radii greater than that of the actual subshell. The first effect (*internal screening*) depends only, to first order, on the total charge inside a sphere with the radius of the screened shell, where the second effect (*outer screening*) depends on the charge distribution from a shell with greater radius. Here the contribution from a shell with greater radius gives a smaller effect than the contribution from inner shells. The potential at the radius r_0 of the considered shell can be written as

$$V(r_0) = \frac{Ze}{r_0} - \frac{1}{r_0} \int_0^{r_0} \varrho(r) 4\pi r^2 \, dr - \int_{r_0}^{\infty} \varrho(r) 4\pi r^2 \, dr$$

$$(2.25)$$

with $\varrho(r)$ electron charge density at the radius r.

Representation of electron levels for the description of X-ray transitions. As a rule by the interpretation of optical and X-ray spectra it is assumed that the observed lines correspond to electron transitions between two electron orbitals. For the representation of X-ray transitions energy level diagrams can be constructed in two alternative ways where each combination of n, l and j lead to different electronic levels.

In Fig. 2.4 a diagram is given where the levels with the strongest bound electrons are localized on the bottom of the representation. If an electron is ionized from a subshell considering actual selection rules the vacancy is filled by electrons of outer subshells. Such an electron transition from an outer to an inner subshell leads, in accordance with nonradiative electron transitions, to the emission of a characteristic X-ray quantum.

Alternative to the representation shown in Fig. 2.4 the energy levels of an atom can be given as presented in Fig. 2.5 in the so-called *Grotrian diagram*. The ionization of a strong bound electron from an inner-shell $n_1 l_1$ leads as a result of an X-ray transition to a vacancy in a higher subshell $n_2 l_2$, i.e. the emission of an X-ray quantum then corresponds to an electron transition from a higher level to a lower one. In the course of such a transition the vacancy $n_1 l_1$ is fulfilled by a transition $n_2 l_2 \to n_1 l_1$ leading to the localization of the vacancy in the weaker bound

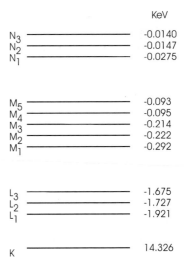

Fig. 2.4. Energy level diagram of krypton

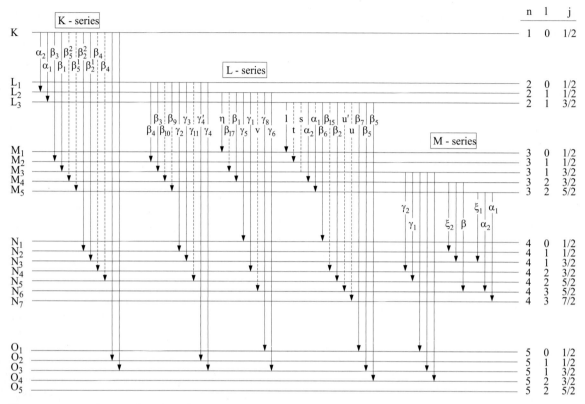

Fig. 2.5. Grotrian diagram for selected X-ray transitions of the K-, L- and M series. The solid lines characterize dipole transitions, dashed lines X-ray transitions of higher multipolarity

subshell 2. This secondary vacancy can then be filled by another radiative transition $n_3l_3 \rightarrow n_2l_2$ from a weaker bound subshell 3. X-ray transitions registered in an energy level diagram as shown in Fig. 2.4 are known as a *Kossel diagram* of X-ray transitions [422].

If we neglect molecular and solid-state effects and the interaction with partially filled subshells the individual levels in Fig. 2.5 are characterized by the quantum numbers nlj. Usually in X-ray spectroscopy the vacancies $1s_{1/2}$, $2s_{1/2}$, $2p_{1/2}$, $2p_{3/2}$, $3s_{1/2}$, etc. are described by K, L_1, L_2, L_3, M_1, etc.[12]

X-ray transition in atoms with one initial vacancy.
X-ray transitions are described by a combination of the symbols of both the levels participating in the transition or by an empirical symbol which consists of the designation of the shell where the initial va-

cancy is localized and a single or double index, as for instance

$$K_{\alpha_1} = KL_3 = 1s_{1/2} - 2p_{3/2}$$
$$K_{\alpha_2} = KL_2 = 1s_{1/2} - 2p_{1/2}$$

K_α = unresolved doublet (transitions $n=2 \rightarrow n=1$)
K_β = unresolved doublet (transitions $n=3 \rightarrow n=1$).

Beside the given nomenclature, often summarizing designations for lines of the K series are used:

K'_{β_1} all KM transitions (especially $K_{\beta_1} + K_{\beta_3} + K_{\beta_5}$)
K'_{β_2} all KN transitions and transitions from higher shells (especially $K_{\beta_2} + K_{\beta_4}$).

An overview of Siegbahn's nomenclature of X-ray transitions as it is used today in X-ray physics is

---

[12] Roman indices are often used instead of Arabian indices.

Table 2.13. Siegbahn's nomenclature for X-ray transitions of the K, L and M series. Here the series is determined by the vacancy in the upper level. Note: In some publications the $L_{\beta_4^{II}}$ transition is also denoted $L_{\beta_{4x}}$ and the $L_{\beta_4^{I}}$ transition is denoted L_{β_4}

lower level		upper level								
level		K	L_1	L_2	L_3	M_1	M_2	M_3	M_4	M_5
L_1	$2s_{1/2}$	α_3								
L_2	$2p_{1/2}$	α_2								
L_3	$2p_{3/2}$	α_1								
M_1	$3s_{1/2}$			η	l					
M_2	$3p_{1/2}$	β_3	β_4		t					
M_3	$3p_{3/2}$	β_1	β_3	β_{17}	s					
M_4	$3d_{3/2}$	β_5^{II}	β_{10}	β_1	α_2					
M_5	$3d_{5/2}$	β_5^{I}	β_9		α_1					
N_1	$4s_{1/2}$			γ_5	β_6					
N_2	$4p_{1/2}$	β_2^{II}	γ_2						ξ_2	
N_3	$4p_{3/2}$	β_2^{I}	γ_3					δ	ξ_1	
N_4	$4d_{3/2}$	β_4^{II}		γ_1	β_{15}			γ_2		
N_5	$4d_{5/2}$	β_4^{I}		γ_{11}	β_2			γ_1		
N_6	$4f_{5/2}$			ν	u'				β	α_2
				γ_8'	β_7'					
N_7	$4f_{7/2}$			γ_8'	u					α_1
					β_7'					
O_1	$5s_{1/2}$			γ_8	β_7					
O_2	$5p_{1/2}$		$\gamma_{4'}$						η	
O_3	$5p_{3/2}$		γ_4							
O_4	$5d_{3/2}$			γ_6	β_5					
O_5	$5d_{5/2}$				β_5					
$P_{2,3}$	$6p_{1/2,3/2}$		γ_{13}							

given in Table 2.13. For corresponding summaries see also [50, 126, 337, 356], etc.

A new nomenclature considering the atomic structure was proposed by the IUPAC[13]. Contrary to Siegbahn's notation based on the intensities of the observed X-ray transitions the *IUPAC notation* is based on the designation of the energy levels and understand form the transition by declaring the sub-shells involved in the transition [264]. For instance this means:

Siegbahn's notation	IUPAC notation
K_{α_1}	K - L_3
K_{α_2}	K - L_2
L_{α_1}	L_3 - M_5
M_{α_1}	M_5 - N_7 etc.

Selection rules. Although according to the Ritz combination principle transitions between all occupied electronic levels into the state of the primary vacancy are possible, the transitions are truncated by selection rules for the quantum numbers of the states involved in the transition.

For dipole transitions (the most intense X-ray lines, also called *main transition* or *main transition lines*) yields:

$|\Delta l| = 1$

$|\Delta j| = 0, 1$ ($\Delta j = 0$ only for $j \neq 0$)

$\Delta n \neq 0$.

With reduced intensity we observe

- electric quadrupole transitions with $|\Delta l| = 0, 2$; $|\Delta j| = 0, 1, 2$ and
- magnetic dipole transitions with $|\Delta l| = 0$; $|\Delta j| = 0, 1$.

X-ray series. Lines emitted by an atom with one K shell vacancy, i.e. all transitions finishing in the K shell, belong to the so-called K series. Analogously all transitions, arising due to the filling of a L shell vacancy, belong to the L series, etc. If an orbital contains more than one subshell a differentiation with respect to the exact localization of the initial vacancy can be made. If, for example, the initial vacancy is localized in the L_1 subshell all transitions filling this vacancy form the L_1 subseries. All transitions formed by the filling of an inner-shell vacancy in the otherwise neutral atom are called *diagram lines*.

X-ray satellites. Beside the diagram lines, transition lines with strongly reduced intensity exist, the so-called *X-ray satellite lines*. The transition energy of these lines can be higher or lower than the transition energy of the diagram lines and cannot be

---

[13] IUPAC: International Union of Pure and Applied Chemistry

explained in the framework of usually applied energy level schemes. Satellite line are formed by radiative electron transitions in multiply ionized atoms, by multielectron processes, by multiplet splitting of the X-ray terms, by the influence of the chemical surroundings of the emitting atom, and by other effects.

X-ray transitions in ions. As a rule with increasing ionization the X-ray transition energies shift to higher energies [587]. An energy shift to the low-energy region occurs, if electrons are ionized from orbitals with high orbital angular momentum (d, f electrons). The nomenclature of the X-ray lines emitted by ions follows the notation given in Table 2.13 until extreme few-electron states are reached. As a rule the notation for the most important spectral lines in hydrogen- and helium-like ions as well as for higher isoelectronic sequences is given in the notation given in [194]. The notation of the most important spectral lines in hydrogen- and helium-like ions is summarized in Table 2.14.

Table 2.14. Notation of the most important X-ray spectral lines in hydrogen- and helium-like ions

isoelectronic sequence	notation	transition
H	Ly_{α_1}	$2p^1\,2p_{3/2} \to 1s^1\,1s_{1/2}$
H	Ly_{α_2}	$2p^1\,2p_{1/2} \to 1s^1\,1s_{1/2}$
He	w	$1s^1 2p^1\,{}^1P_1 \to 1s^2\,{}^1S_0$
He	x	$1s^1 2p^1\,{}^3P_2 \to 1s^2\,{}^1S_0$
He	y	$1s^1 2p^1\,{}^3P_1 \to 1s^2\,{}^1S_0$
He	z	$1s^1 2p^1\,{}^3S_1 \to 1s^2\,{}^1S_0$

Band spectrum. In each series the components with the shortest wavelengths arise from transitions of valence electrons. For the case of bound atoms these transitions are widened, have a structure and are known as emission bands. The energetic width of the emission bands is determined by the characteristics of the discrete energy levels that could be occupied by valence electrons. The course of the intensity distribution of individual bands in the solid-state can be determined by the electron theory of solid states

and for the case of molecular compounds or gases by molecular orbital (MO) theory. For instance, examples can be found in [356].

2.2.2 Line Profile and Line Broadening

In practice it is often assumed that the energy of all levels is strongly fixed, i.e. that no uncertainty in the orbital energy exists. Then the energy of an X-ray quantum is also strongly defined and the X-ray transition is monochromatic.

In reality X-ray transition lines are never strongly monochromatic and their frequency is distributed over a certain wavelength region. The main reasons for this broadening are:

- the natural linewidth of the atomic energy levels caused by their finite lifetime and the from it following uncertainty of the energy levels,
- the Doppler width as a result of the Doppler broadening (Doppler effect) modifying the radiation frequency from atoms moving relative to the radiation detector and
- the collision and pressure broadening resulting from modifications of the radiation by collisions between radiation emitting atoms and their neighbors.

A systemization of line broadening distinguishes between two classes of broadening effects:

- *Inhomogenous line broadening.* Inhomogenous line broadening is caused by the Doppler effect. This designation comes from the fact that a spectral line broadened by the Doppler effect arises from many spectral lines each of them smaller in their linewidth and corresponding to atoms, molecules or ions with different velocities.
- *Homogenous line broadening.* Spectral lines are characterized as homogenously broadened if the Doppler effect does not act. Many effects contribute to homogenous line broadening, such as for instance the natural linewidth, collision broadening and interaction broadening.

Natural Linewidth

From quantum mechanics we know that as a result of the Heisenberg uncertainty principle the energy of an excited atomic level with a finite lifetime τ cannot be better known than

$$\Delta E \approx \frac{\hbar}{\tau} . \qquad (2.26)$$

Thus, the finite lifetime of an excited atomic state results in unsharp energy levels. The decay probability per time interval is then

$$P = \frac{1}{\tau} = \frac{\Delta E}{\hbar} . \qquad (2.27)$$

Level decay probabilities (or level widths divided by $\hbar$) are commonly declared in eV/$\hbar$ or in atomic units (a.u.) and *level widths* in eV or in atomic energy units. For these units:

- transition probabilities:
  1 a.u. = 4.1341×10^{16} s^{-1} = 27.2113845 eV/$\hbar$;
- widths:
  1 a.u. = 27.2113845 eV.

Experimental information on atomic energy level widths can be obtained by the measurement of linewidths from X-ray emission lines, absorption edges or from absorption lines.

The theory of natural linewidths was based on the Dirac theory of radiation formulated in [562]. Corresponding to the quantum mechanical result a radiative electron transition between the initial state i and the final state f has the spectral distribution

$$I(\omega)\, d\omega = \frac{\Gamma(i) + \Gamma(f)}{h} \frac{d\omega}{\left(\omega_{if} - \omega\right)^2 + \frac{\pi}{h}\left(\Gamma(i) + \Gamma(f)\right)} \qquad (2.28)$$

with $\omega_{if} = (E_i - E_f)/\hbar$. From (2.28) it follows that the width Γ of a spectral line is the sum of the widths of the initial and final states. The Lorentz distribution described by (2.28) characterizes a resonance process and the resulting line is symmetric according to the intensity maximum of the spectral line.

From (2.26) it follows that the natural linewidth is determined by the lifetime of a vacancy in the considered level. The natural linewidth of an X-ray transition connecting two electron levels is influenced by different effects, acting on the vacancy lifetime. In the following we describe the most important effects and consequences of it.

According to [301] all these effects can be subdivided into two main groups: (i) lifetime effects and (ii) other effects not connected with the lifetime. For the analysis in both groups two Z regions are distinguished:

- a region where the levels are inner-shell core levels (commonly in the penultimate occupied subshell); and
- a region where the levels are inner-shell core levels separated from the peripheral shell at least by one fully occupied shell.

The lifetime of a vacancy is influenced by vacancies in other atomic subhells. This effect is important if the vacancy is localized near to other peripheral levels, as for instance in the case in light atoms. For deep core levels this effect can be neglected if the additional vacancies are localized in outer subshells and if between the vacancy and the core levels there are enough electrons contributing to the screening of the outer-shell electrons. The same situation is observed for chemical bond atoms in a solid-state environment because with the chemical bonding the addition or removal of outer-shell electrons is connected.

If the peripheral electron configuration is disturbed by multiple ionization or chemical bonding all level energies were changed and channels for nonradiative electron transitions (Coster–Kronig transitions) could be closed. According to (2.27) this effect can substantially influence the level and transition widths because of

$$P = \frac{1}{\tau} = \frac{\Gamma}{\hbar} \qquad (2.29)$$

with

$$P = P_R + P_A + P_{CK} ; \qquad \Gamma = \Gamma_R + \Gamma_A + \Gamma_{CK} . \qquad (2.30)$$

Here the indices R, A and CK describe radiative, Auger and Coster–Kronig transitions.

An overview of partial and total widths of electronic levels and corresponding vacancy lifetimes is given in Fig. 2.6. Extensive tables of calculated lifetime dominated level widths are given in [346–349] and in [113–115]. Calculated level widths for all elements and all subshells beginning with the the K up to the Q$_1$ subshell are known from [423].

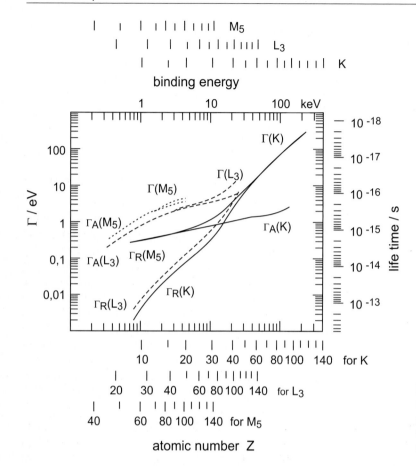

Fig. 2.6. Dependence of partial X-ray (Γ_R) and Auger widths (Γ_A) as well as of energetic total widths (Γ) and of the corresponding vacancy lifetimes on the atomic number Z

In [301] semiempirical tabulations of K and L shell level widths are given. An overview of known energetic widths of the K to N_7 subshells is presented in [93, 94]. Here papers describing individual subshells and Z-regions are cited. Examples of the determination of level widths for L and M subshells by the analysis of measured with high spectral resolution K_α and and K_β X-ray lines are given in [148]. Details of the measurements and of data analysis can be found for example in [147,229].

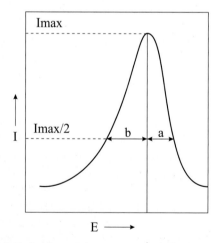

Fig. 2.7. Definition of the asymmetry index. I_{max} is the maximum line intensity

Multiplet Splitting of X-Ray Lines

In spectroscopic practice, X-ray lines with asymmetric profiles are observed. Quantitatively this asymmetry is described by the introduction of a so-called

Z	Symbol	K_{α_1}	K_{α_2}		Z	Symbol	K_{α_1}	K_{α_2}
17	Cl	0,9	1,0		28	Ni	1,2	1,3
19	K	1,0	1,0		29	Cu	1,2	1,3
21	Sc	1,0	1,0		30	Zn	1,1	1,3
22	Ti	1,2	1,0		31	Ga	1,0	1,2
23	Va	1,2	1,1		32	Ga	1,0	1,1
24	Cr	1,4	1,0		35	Br	1,0	1,1
25	Mn	1,5	1,3		38	Sr	1,0	1,1
26	Fe	1,6	1,3		40	Zr	1,0	1,1
27	Co	1,4	1,3		42	Mo	1,0	1,0

Table 2.15. Asymmetry indices for K_{α_1} and K_{α_2} X-ray lines of elements with $Z \leq 42$ (from [279])

asymmetry index

$$\alpha = \frac{a}{b} . \tag{2.31}$$

The quantities a and b characterize the short- and long-wavelength part of the linewidth (see Fig. 2.7). For the K_{α_1} and K_{α_2} lines α increases from 1 for $Z = 20$ (calcium) to 1.6 for $Z = 26$ (iron) and is again 1 at $Z = 42$ (molybdenum).

In Table 2.15 asymmetry indices for the K_{α_1} and K_{α_2} lines up to molybdenum are given in [279]. Reference [241] gives a graphical overview of asymmetry indices for the K_{α_1} and K_{α_2} lines of elements from chromium ($Z = 24$) up to copper ($Z = 29$) – see Fig. 2.8. Results from [241] and from other authors are listed below:

K_{α_1} asymmetry index	K_{α_2} asymmetry index
Parratt 1936 [416]	Parratt 1936 [416]
Edamoto 1950 [162]	Edamoto 1950 [162]
Meisel 1961 [355]	Meisel 1961 [355]
Blochin 1964 [73]	Tsutsumi 1968 [541]
Tsutsumi 1968 [541]	Tsutsumi 1973 [542]
Nigavekar 1969 [395]	Lee 1974 [317]
Pessa 1973 [425]	Sorum 1987 [519]
Tsutsumi 1973 [542]	Maskil 1988 [339]
Lee 1974 [317]	
Onouel 1978 [406]	
Deutsch 1982 [146]	
Sorum 1987 [519]	

The observed line asymmetry can be explained by the spin–spin interaction of 2p electrons with elec-

trons from the partially filled 3d subshells [556]. This interaction leads to an asymmetric splitting according to the initial 2p level. Because this splitting is smaller than the total width of each sublevel there arises an asymmetric line shape. Analogous line asymmetries are observed in chemical compounds and alloys. Here the character of the chemical bond is the reason for the line form deviation from the expected symmetric intensity distribution.

Especially in light atoms, multiplet splitting is spectroscopically easily detectable. For example, we quote the known multiplet splitting of the L_2 and L_3 levels of 3d elements in the region $22 \leq Z \leq 28$, leading to an observable broadening of X-ray as well as of Auger transition lines resulting in relatively high values of the antisymmetry index. This effect is known for bound as well as for free atoms. Because the exchange interaction decreases with increasing separation of partially filled subshells the energetic widths of the K and L levels and the widths of transitions between these orbitals are influenced only weakly.

As a rule the calculation of X-ray satellite spectra is complicated because of the multiplicity of the initial states which can decay in a multiplicity of final states. Often in a small spectral region hundreds of transition lines can be present for which the transition energies, transition rates and linewidths must be determined. Therefore in the literature only less detailed analyses are known for selected X-ray transitions. As an example we refer to the description of the K_β X-ray emission spectrum of argon in [159].

Line Broadening by Multiple Vacancies

X-ray line brodening and line deformation can also be connected with the appearance of multiple vacancies in the atom. If besides an inner-shell vacancy there exist additional vacancies in outer shells, satellite lines within the energetic width of the diagram line are observed because as a rule outer-shell vacancies result in small energetic shifts of the X-ray transition lines [587]. This leads to a broadening of the emission lines what is especially reflected in light atoms because in atoms with higher Z the energetic shift caused by one or some additional outer-shell vacancies is small in comparison to the natural linewidth of the transition. An overview on partial and total widths of individual electron levels and on corresponding lifetimes is given in Fig. 2.6.

In Fig. 2.9 a fit of the $K_{\beta1,3}$ X-ray transition line for electron impact excitation is shown [187]. An exclusive consideration of direct contributions of the K_{β_1} and K_{β_3} X-ray transitions would not lead to a satisfactory approximation of the measured transition lines. Only the consideration of additional vacancies in the $3p$ and $3d$ subshells allow a sufficiently correct approximation. For example, for cobalt and copper the influence of additional M shell vacancies is analyzed in [147].

X-ray transition energy shifts for selected X-ray transitions calculated in the framework of single-configuration Dirac–Fock calculations for the case of one outer-shell vacancy are shown in Figs. 2.10–2.12 (see [587]). In these calculations the ion ground state of a singly ionized positive ion is assumed.

Doppler Broadening

If the X-ray emitted atoms take part in a thermal motion or have velocity components to or from the spectrometer, then the Doppler effect causes a line broadening. In contrast to the natural linewidth the resulting broadening is an *inhomogenous broadening* where individual atoms contribute to different parts of the spectral line.

For the frequency f resulting from the relative motion v of the emitter and the detection system it yields

$$f = f_0 \frac{\sqrt{1 - \left(\frac{v}{c}\right)^2}}{1 + \frac{v}{c}\cos\vartheta} . \qquad (2.32)$$

Here ϑ is the angle between the observation direction and the direction of the source moving relative to the observer.

The *Doppler broadening* Δf of a spectral line of frequency f_0 on the basis of the thermal motion of atoms in a gas of absolute temperature T is determined for the resulting FWHM as

$$\Delta f = 2\frac{f_0}{c}\sqrt{\frac{2kT}{m}\ln 2} . \qquad (2.33)$$

Here m is the mass of the radiation emitting atom, k the Boltzmann constant and c is the velocity of light in vacuum. From equation (2.33) it follows that the Doppler broadening increases with increasing temperature and has the greatest value for light atoms. The resulting distribution is a Gauss distribution. If all constants are summarized in a numerical factor then for the FWHM of a Doppler broadened line follows

$$\Delta f = 7,16 \times 10^{-7} f_0 \sqrt{\frac{T}{m}} \; s^{-1} . \qquad (2.34)$$

The principal situation for line broadening is shown in Fig. 2.13 if we assume a Maxwellian velocity distribution in a radiation-emitting gas.

Besides the traditional investigations of line broadening effects in the optical spectral range in the last time the consideration of Doppler broadening phenomena is of increasing importance so for example in X-ray spectroscopic investigations of ion–atom collisions, in astrophysical studies and for plasma diagnostics. A detailed discussion of the Doppler effect can be found in [199].

Collision Broadening

Impact excitation and deexcitation by electrons, protons, α-particles or heavy ions likewise can lead to a dramatic decrease of the lifetime of excited atoms or ions. This is connected with a level broadening of atomic states and in this course with a broadening of spectral lines. In classical consideration, collisions can be understood as an interruption of the wave of

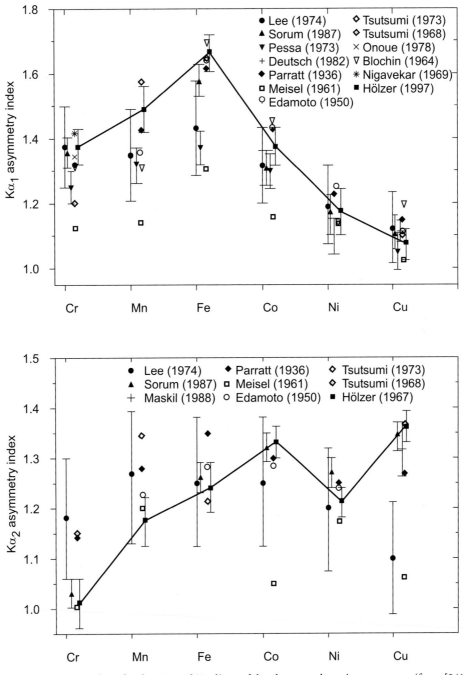

Fig. 2.8. Asymmetry indices for the K_{α_1} and K_{α_2} lines of the elements chromium to copper (from [241])

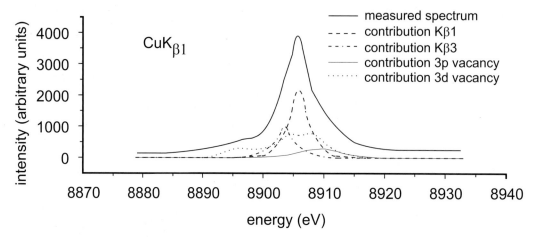

Fig. 2.9. Contributions of 3p and 3d vacancies to the form of the $K_{\beta_{1,3}}$ X-ray transition line in copper for electron impact excitation (from [187])

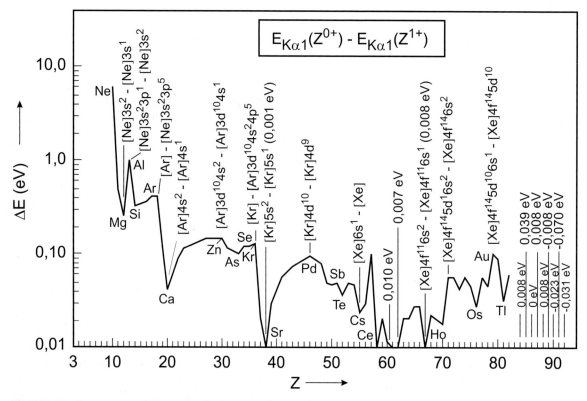

Fig. 2.10. K_{α_1} X-ray energy shifts ΔE for the ion ground state of ions with one vacancy of an outer shell as a function of the atomic number Z. On characteristic points the electron configuration of the atom ground state and of the ground state of the singly ionized atom is given [587]

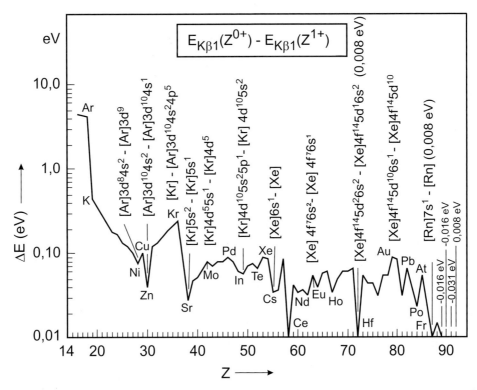

Fig. 2.11. K_{β_1} X-ray energy shifts ΔE for the ion ground state of ions with one vacancy of an outer-shell as a function of the atomic number Z. On characteristic points the electron configuration of the atom ground state and of the ground state of the singly ionized atom is given [587]

the emitted radiation, leading to a broadening of the corresponding Fourier spectrum. The collisions lead a Lorentz function in the spectrum where the resulting FWHM is proportional to the particle density of the colliding atoms.

For the description of spectral line broadening in gases, Weisskopf [563] distinguishes three contributions to the collision broadening[14]:

- **Collision attenuation.** The *collision attenuation* can be described as phase jumps if the emitting or absorbing atoms collides with other gas atoms. In this context the radiant atom is considered as an electron oscillator. The inneratomic correlation which is responsible for the phase jump is treated according to the quantum theory. The emitted radiation at the collision is disregarded.

- **Static approximation.** The static approximation considers all emitted radiation but no Fourier analysis is accomplished. The approximation can be applied if the resulting frequency shift $\Delta f = f - f_0$ is greater than the collision broadening of the line. The frequency shift at the interaction between the colliding atoms can be calculated under the assumption of the validity of the Franck–Condon principle. The consideration of van der Waals interactions allows a description of the line profile at great distances from the collision center. An approximate formula for the resulting FWHM, applicable to hydrogen and similar atomic states, has the form [396]

$$\Delta\lambda^{W}_{\text{FWHM}} \cong 3.0 \times 10^{16} \lambda^2 \, C^{2/5} \left(\frac{T}{\mu}\right)^{3/10} N \quad (2.35)$$

---

[14] for details see [236,237]

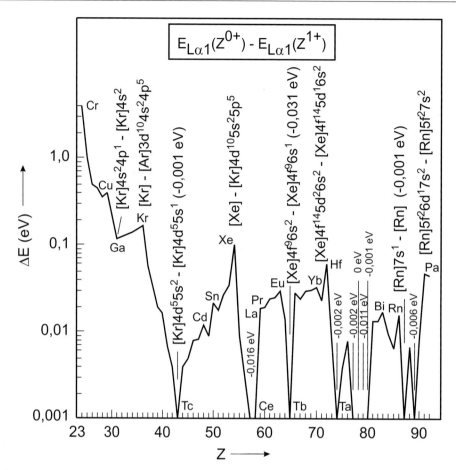

Fig. 2.12. L_{α_1} X-ray energy shifts ΔE for the ion ground state of ions with one vacancy of an outer-shell as a function of the atomic number Z. On characteristic points the electron configuration of the atom ground state and of the ground state of the singly ionized atom is given [587]

with μ the reduced mass of the atom and of the incoming projectile, N the density of incoming projectiles and T the temperature of the emitters. The quantity C can be estimated as $C = C_u - C_l$ with

$$C_{l,u} = 9.8 \times 10^{10} \, \alpha_P \, R^2_{l,u} \; .$$

For the mean atomic polarizability it yields

$$\alpha_P \approx 6.7 \times 10^{-25} \left(\frac{3 \, I_H}{4 \, E^*} \right)^2 \quad \text{cm}^3$$

where I_H is the ionization energy of hydrogen and E^* the energy of the first excited level of the incoming projectile. For the quantity R we have

$$R^2_{l,u} \approx 2.5 \left(\frac{I_H}{I - E_{l,u}} \right)^2$$

with I the ionization energy of the emitting atom.

- **Resonance broadening.** *Resonance broadening* or *coupling broadening* occurs as a result of the exchange of excitation energy between excited and nonexcited atoms of the same element and causes a decrease of the lifetime of both states and so a line broadening of the spectral lines. The line profile is a Lorentz function. The FWHM may be estimated as [156]

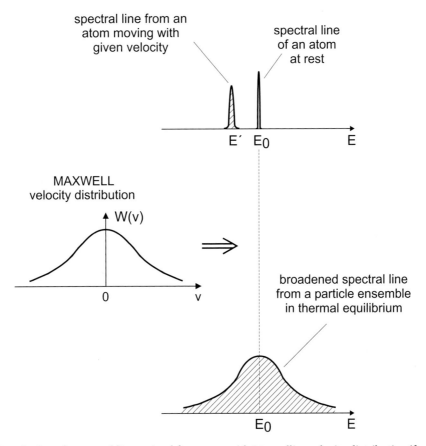

Fig. 2.13. Broadening of a spectral line emitted from a gas with Maxwellian velocity distribution (from [324])

$$\Delta\lambda_{\mathrm{FWHM}}^{R} \cong 8.6 \times 10^{-30} \left(\frac{g_l}{g_u}\right)^{1/2} \lambda^2 \, \lambda_r \, f_r \, N_i$$

$$(2.36)$$

where λ is the wavelength of the observed line, g_u and g_l are the statistical weights of the upper and lower levels, λ_r is the wavelength of the resonance line, f_r is the oscillator strength of the resonance line and N_i is the ground state density.

Interaction and Intensity Broadening

For the sake of completeness we also mention here *interaction broadening* and *intensity broadening*. These types of line broadening are of importance for laser spectrometry but of less importance in X-ray physics. The reason of interaction broadening is the finite interaction time between the light passing an atom or molecule and the passed object. The intensity broadening is connected with the increasing of the transition rate between individual levels in a field of light. A more detailed description of these broadening processes can be found in [324] or [57].

Voigt Profile

Although most of the considered broadenings lead to a Lorentz line profile the intensity profile of measured transition lines results from a mathematical convolution of Gauss and Lorentz functions because all transition lines at first have a Gauss distribution caused by the Doppler effect. The resulting intensity profile is called the Voigt profile [552]. The actual form of the profile is determined by the balance between the FWHMs of the individual processes.

If we assume that a monoenergetic line can be described as a Gauss function then a measured spectrum $M(x)$ can be understood as the convolution of an existing spectral distribution with a Gauss function [302]:

$$G(x) = A \exp\left(-\frac{x^2}{2\sigma^2}\right) \qquad (2.37)$$

$$M(x) = \int_{-\infty}^{\infty} S(x')\, G(x - x')\, dx' . \qquad (2.38)$$

A convolution with a delta function reproduces the Gauss function (i.e. the monoenergetic line) and a convolution with a Lorentz function

$$L(x) = \frac{A}{1 + 4\dfrac{x^2}{\Gamma^2}} \qquad (2.39)$$

with the substitution

$$t = \frac{x - x'}{\sigma}$$

gives the Voigt profile

$$V(x) = \frac{A\,\Gamma^2}{\sigma} \int_{-\infty}^{\infty} \frac{e^{-t^2/2}\, dt}{\left(\dfrac{\Gamma}{\sigma}\right)^2 + 4\left(\dfrac{x}{\sigma} - t\right)^2} . \qquad (2.40)$$

Here A is the amplitude of the considered function, Γ the natural linewidth of the Lorentz function and σ the width of the Gauss function. Equation (2.40) cannot be solved analytically and can be approximated by a Taylor series or by a Gauss–Hermite integration.

Line Broadening by External Fields

The existence of external electric or magnetic fields can cause a small increase or decrease of the total energy of atoms or ions in these fields. Here each atomic state splits into sublevels where the amplitude of the splitting depends on the strength of the external field. For the presence of electric fields this is known as the Stark effect [521] and for magnetic fields as the Zeeman effect [584]. As a rule the strength of the cited effects (level splitting) is for most X-ray standard applications of less or negligible importance.

Hyperfine Structure and Isotope Shifts

Additional to the level splitting by external fields each orbital can split into many substates by the hyperfine structure interaction. This result is due to the magnetic interaction of the electrons with the magnetic moment of the nucleus. In heavy elements these splittings often are big enough to distinguish individual components in spectroscopic investigations. In light elements the level splitting is small and spectral lines overlap in a form that the effect can be detected only as a common asymmetric broadening of the spectral line. The atomic energy levels not only depend on the nuclear spin, but also on the mass of the nucleus and his form (quadrupole effects). These effects also produce small energy shifts of the spectral lines of different isotopes.

Instrumental Effects

Besides the effects discussed above methodical features can influence the form and width of spectral lines. These are especially:

- finite diaphragm widths,
- diffraction effects,
- geometrical aberrations,
- error sources and disturbing effects during electrical signal registration, and
- the finite resolution of the applied spectrometer.

All these circumstances can produce deformations, shifts and broadening effects in different spectrometers and will be discussed later.

2.2.3 X-Ray Emission Rates

Einstein Transition Probabilities

Interaction of X-rays with an atom. Besides knowledge of X-ray transition energies in applications of X-ray fluorescence analysis, it is important to know the transition probabilities in an atom between individual atomic substates. In principle an atom can interact with X-rays in three different ways (see also Fig. 2.14):

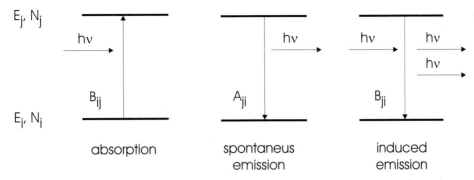

Fig. 2.14. Absorption, spontaneous and induced emission of radiation of the energy $h\nu$ between the energy levels E_1 and E_2 with the occupation numbers N_i and N_j. The transitions occur with the corresponding probabilities B_{ij}, A_{ji} and B_{ji}

1. **Spontaneous emission.** Here under emission of a photon an atom from an excited state j of energy E_j goes to a state i of lower energy E_i. In this way the energy of the emitted photon is equal to the energy difference of the electron states involved in the transition

$$h\,\nu_{ji} = E_j - E_i \,, \tag{2.41}$$

corresponding to a spectral line with wavenumber

$$k_{ji} = \frac{1}{\lambda_{ji}} = \frac{E_j - E_i}{hc} \,. \tag{2.42}$$

The transition probability a_{ji} per time unit for an atom from the state j into the state i considers that for an isolated field free atom in a state with the total angular momentum J_i there exist

$$g_i = 2J_i + 1$$

degenerate quantum states of energy E_i corresponding to $2J_i + 1$ possible values of the magnetic quantum number M_i. The Einstein spontaneous transition probability [165] is defined as the total probability per time unit that an atom from the state j is transferred to a state g_i of the energy level i:

$$A_{ji} = \sum_{M_i} a_{ji} \,. \tag{2.43}$$

A_{ji} is independent of M_j, i.e. the transition probability is not dependent on an arbitrary choice of the coordinate axes in space.

2. **Absorption.** The absorption of an X-ray quantum occurs by excitation of an electron from a energetically stronger bound state (energetic lower state) to a weaker bound state.

If atoms are irradiated with radiation of frequency ν_{ji} and radiation density $u(\nu)$ then these atoms were transferred into excited states j with a probability $B_{ji}u(\nu)$ under absorption of a quantum of energy $h\nu_{ji}$ [165]. Here:

$$B_{ij} = \frac{c^3}{8\pi h\nu^3} A_{ji} \,, \tag{2.44}$$

i.e. the probabilities for spontaneous emission and absorption are proportional to one another.

3. **Induced emission.** Einstein [165] further showed that excited atoms with the same probability B_{ji} can also be stimulated to go into the ground state under emission of a quantum of energy $h\nu_{ji}$. This process is called *induced emission*.

Then the total transition probability for a transition from a state j to i is the sum of the probability of spontaneous emission A_{ji} and of the probability of induced emission $u(\nu)B_{ji}$.

The Einstein A-coefficient (spontaneous emission) is of importance for all fluorescent processes and the B-coefficient (induced emission) plays an important role for laser devices, because here we have induced emission.

Line intensity and lifetime of excited states. In classical understanding the *lifetime* is those time periods needed for the atom, as a radiating dipole, to emit its excitation energy. From classical considerations it follows that the mean lifetime becomes shorter if the radiation amplitude and therefore the line intensity increase. This is also conserved in the sense of the correspondence principle of quantum mechanics. According to this principle for a certain excited atom there exists a probability to emit its energy in a certain time interval in the form of a photon. If there is no other transition possibility, this emission take place "in mean" after the so-called *mean lifetime* τ. For this reason the number of atoms per second emitted photons and in this course also the intensity of the considered spectral line is indirect proportional to the mean lifetime if the deexcitation is restricted only to the considered deexcitation channel.

If $N_j(t)$ is the number of excited atoms at time t then the total transition rate as a result of spontaneous transitions is

$$\frac{dN_j[t]}{dt} = -N_j(t) \sum_i A_{ji} . \tag{2.45}$$

Here the summation is taken over all atomic states with an energy lower than E_j. If no other excitation or deexcitation process exists then

$$N_j(t) = N_j(0) \exp\left(-\frac{t}{\tau_j}\right) \tag{2.46}$$

with

$$\tau_j = \left(\sum_i A_{ji}\right)^{-1} \tag{2.47}$$

as the natural lifetime of the atom in each of the states of the level j. For a finite lifetime, from the uncertainty principle (2.26), there is a finite width for the level j and a corresponding natural width of the spectral line.

Oscillator strength. A further quantity to characterize the transition probability is the oscillator strength f_{ji} which is connected with the lifetime τ_j by the relation

$$\tau_j = \frac{f_{ji}}{A_{ji}} . \tag{2.48}$$

Furthermore,

$$f_{ji} = \frac{A_{ji}}{\sum_i A_{ji}} \tag{2.49}$$

or [4]

$$f_{ji} = \frac{\pi \varepsilon_0 m_e c^3 \hbar^2 A_{ji}}{e^2 E_{ji}^2} \tag{2.50}$$

with E_{ji} the transition energy.

The oscillator strength is dimensionless and has the physical meaning of an effective number of electrons as classical oscillators emitting radiation with the same strength as the considered atom. For intense spectral lines the oscillator strength is in the order of 1 and is exactly 1 for the case of only one possible transition. A detailed theoretical description of oscillator strengths and transition probabilities can be found for instance in [126, 482, 508] and in [494].

Calculation of Relativistic X-Ray Emission Rates

To get a complete and systematic overview of X-ray emission rates and of X-ray intensity ratios between individual transition lines (*coupling factors*) it is often insufficient to analyze the experimental data published in the literature because there are different gaps for individual transition ratios. A more complete picture can be derived by calculations of X-ray intensity ratios with different calculation (approximation) methods.

Calculations of X-ray intensity ratios are reported in many papers. This problem is coupled with the common problem of the calculation of atomic transition probabilities for radiative transitions, reviewed in a series of former papers, for instance in [124], [231], [507], [496], [360], [509], and others. Short overviews are given in [195], [313], [170], [171] [130], [561], [196], [38], [131], [504], [489] and [116]; these papers must be understood as a selection of known work only.

More modern calculations consider all multipole contributions for the interaction between the electromagnetic field and each electric or magnetic multipole moment of the atom as a whole make it possible to the transition rate. The different contributions are described with E1, E2, E3, ... for the electric dipole, quadrupole, octupole ... moments and with M1, M2, M3, ... for the corresponding magnetic moments. Selection rules for multipole moments dominating X-ray transitions are given in Table 2.16.

Table 2.16. Selection rules for dominating electron transitions by the emission of characteristic X-rays

radiation	multipole	$|\Delta l|$	$|\Delta j|$	limitation for j
electric	dipole	1	0,1	$0 \leftrightarrow 0$
	quadrupole	0,2	0,1,2	$\frac{1}{2} \leftrightarrow \frac{1}{2}$
	octupole	1,3	0,1,2,3	$\frac{1}{2} \leftrightarrow \frac{1}{2}$
magnetic	dipole	0,2	0,1	$0 \leftrightarrow 0$
	quadrupole	1	0,1,2	$\frac{1}{2} \leftrightarrow \frac{1}{2}$

Massey [341] has calculated relativistic transition rates under neglect of retardation effects. The first correct relativistic formulation for the emission from single-particle states was given in [29] (see also [30,31]).

The emission rate for photons with energy $\hbar\omega$ and impulse $\hbar k$ in the angle $d\Omega$ with a polarization vector $\boldsymbol{\varepsilon}$ from an atom, which passes from the initial state j into the final state i, results in the first order of perturbation [489]:

$$\Gamma_{ji} = \frac{\alpha\omega}{2\pi} \left| \langle \Psi_1 | \sum_j \alpha\,\boldsymbol{\varepsilon}\, e^{ikr} | \Psi_2 \rangle \right|^2 d\Omega_k . \quad (2.51)$$

According to equation (2.51) for the calculation of X-ray emission rates the energy of the emitted photons and the atomic wavefunctions Ψ of the initial and of the final state of the electrons taking part in the transition must be known. For many-electron atoms this can be realized only approximately. For instance, overviews about different approximation methods are given in [448,489,588].

For a one-electron atom an analytical expression for the wavefunction can be given in the nonrelativistic approximation. This is impossible for a many-electron atom. Here simple calculations can be done under neglect of the electron–electron interaction with Coulomb wavefunctions equal for all involved orbitals. A variety of approximate methods is known to take into consideration effects of the electron–electron interaction in many-electron atoms. A simple approximation is to replace the actual nuclear charge by an effective charge considering the screen-ing of the nuclear charge by electrons localized in the near vicinity of the nucleus [341].

More realistic wavefunctions can be calculated with iterative procedures in the independent-particle model. Methods are known that optimize the evolution coefficients of the one-electron wavefunctions by analytical procedure [457,458] or after parameters of analytical functions [88]. The most today used methods are based on an approach of one-electron wavefunctions in the central-field approximation whereby the radial wavefunctions were given as many numerical values on a radial mesh.

Numerical values for the description of the atomic structure under consideration of the interaction between the electrons of a multielectron atom can be obtained on an iterative way, where the standard approximation is the *Hartree–Fock method* [174,175,190,228]. Here the atomic wavefunctions are assumed as an antisymmetrized sum of products of one-electron wavefunctions. The one-electron wavefunctions at the same time satisfy a self-consistency criterion where they are eigenvalue solutions for the potential and for the exchange terms which are determined by the same wavefunctions.

Generally there exist different methods based on self-consistent field calculations different in the treatment of the exchange term of the two-electron interaction:

- *Hartree model* with neglect of the exchange term;
- *Hartree–Fock–Slater model* with approximation of the exchange term by an easy to calculate local exchange potential [506]; and
- *Hartree–Fock model* with correct treatment of electron exchange (nonlocal potential).

Both the relativistic (Dirac–Fock or Dirac–Fock–Slater) and the nonrelativistic method (Hartree–Fock or Hartree–Fock–Slater) are widely used for the calculation of X-ray emission rates. As formulated in [448] a simplified comparison of relativistic and nonrelativistic formulations for atomic structure calculations is given in Table 2.17.

For instance extensive nonrelativistic calculations are known from McGuire [346–350] and Bhalla [68]. Results from relativistic calculations were published at about the same time in [64–66,460,486] and [329].

Table 2.17. Comparison of relativistic and nonrelativistic formulations for atomic structure calculations

atomic structure calculations	
nonrelativistic	relativistic
Schrödinger equation	Dirac equation
Hamilton operator:	Hamilton operator:

$$-\frac{1}{1}\sum_{i=1}^{N}\left(\nabla_i^2 + \frac{2Z}{r_i}\right) + \sum_{i<j}\frac{1}{r_{ij}}$$

$$\sum_{i=1}^{N}\left(c\boldsymbol{\alpha}_i \times \boldsymbol{p}_1 + (\beta - 1)c^2 + \frac{Z}{r_i}\right)$$

$$+\sum_{i<j}\frac{1}{r_{ij}}$$

nonrelativistic	relativistic
in atomic units	in atomic units
wavefunction:	wavefunctions:
one-component spinor	two-component spinor
relativistic corrections in perturbation theory:	Breit operator and most important quantum electrodynamic corrections in perturbation theory:
• relativistic mass correction	• retardation of the electrostatic interaction as a result of the finite velocity of light
• one-particle and two-particle Darwin terms (retardation of the electrostatic interaction)	
• spin–orbit coupling	• interaction between the magnetic momenta of the orbital angular momenta and of the spins
• spin–other-orbit coupling	
• magnetic interaction between spin and orbital angular momenta of different electrons	• electron self-energy (fluctuation in the electron position during the interaction of electrons with the radiation field)
	• vacuum polarization (screening of the nuclear potential by virtual electron-positron pairs)

All cited calculations are based on the assumption that the electrons see the same potential in their initial and final states. Relativistic Hartree–Fock calculations were discussed in [139,213] and [334].

The fact that the initial and final state wavefunctions are different from each other causes exchange effects described in [32] and [487] using different one-particle wavefunctions for the initial and final states.

Higher precision in the description of the electron–electron interaction in comparison to single configuration Hartree–Fock calculations is possible by using different techniques:

- multiconfigurational Hartree–Fock method;
- variation calculations with correlated wavefunctions;
- $1/Z$ development techniques;
- pair correlation technique; and
- diagram perturbation calculations.

To calculate X-ray data such as transition energies, emission rates and intensity ratios in X-ray physics the multiconfigurational Dirac–Fock (MCDF) program GRASP given in [214,351] (GRASP-Code) with extensions from [160] and [448] is often used. This MCDF program solves the Dirac equation and calculates atomic wavefunctions and energy eigenvalues with the method of the self-consistent field approach. For the calculation of orbital energies two-particle corrections and QED contributions are included. The theoretical fundamentals of the method can be found in [213] and [214].

The GRASP code with its subprogram OSCL allows us to calculate X-ray emission rates in the frozen core approximation. This subprogram is described in [160]. Reiche [448] describes the subprogram RATEN, which enables an over the momentum coupling averaged calculation of transition rates from radiative electron transitions in frozen core approximation under the assumption that initial and final state can be desribed by a single Slater determinant. The program NONORTHO [448] allows to calculate emission rates from radiative electron transitions between any states characterized by MCDF wavefunctions in the adiabatic approximation.

As a new development the RATIP code for accurate calculations of atomic properties and structures became available in the last few years [188,189]. The RATIP code extends the widely applied GRASP code to calculate a diversity of atomic transition and ionization properties within a relativistic framework. Differences and further developments in comparison to GRASP92 code [415] have been discussed in [188]. Of special interest here is the capacity to obtain reliable results for open-shell atoms, especially also for case studies of d- and f-shell elements.

Polarization and Angle Correlations

The X-ray radiation emitted from a statistical ensemble of free atoms is isotropic and nonpolarized. However, if in the preparation process of atomic initial states a direction is distinguished then the problem characterizing the density matrix is diagonal with respect to the magnetic quantum numbers with the distinguished direction as polar axis and the emitted

radiation is distributed axially symmetric around the distinguished direction. Generally it yields that for a nondiagonal density matrix or if states with different magnetic quantum numbers are not occupied equally, as a rule the radiation is polarized and the radiation intensity varies with the direction. For instance, a description of the density matrix formalism in connection with polarization processes as given in [184] and [44].

While for standard applications in X-ray spectral analysis polarization is of rather less importance, in special applications the use of polarized X-rays as exciting radiation allows us to increase the signal-to-noise ratio and to lower the detection limit. For the production of polarized X-rays Barkla scattering [42] or Bragg diffraction polarization is used [6]. An overview of both methods can be found in [469].

Barkla scattering arrangements can be realized in three different ways [469]:

1. **Orthogonal scattering systems.** Here a low divergent and good collimated X-ray beam is scattered under a scattering angle of $\pi/2$, where the electric field vector in the x-z scattering plane is eliminated. If the beam is scattered a second time in a suitable manner under an angle of $\pi/2$ relative to the first scattering plane then in the x-z plane the orthogonal field vector can be eliminated.

2. **Multilayer scattering.** Multilayer scattering is applied if some elements are to be be analyzed simultaneously with a polychromatic source. Actual layer systems are assembled from a material of low atomic number with a support of a high-Z material or a heavy molecular compound [582].

3. **Zylindrical scattering systems** To increase the number of registered quanta the scattering material is here formed cylindrically. A more detailed descriptions of this kind of arrangements can be found in [469].

The Bragg diffraction polarization can be realized in orthogonal assemblies close to Barkla scattering for plane multilayer scattering as well as for higher X-ray energies with plane or bent (Johann or Johansson geometry) crystals to increase the scattered intensity.

X-Ray Emission Rates in Ions

The existence of additional vacancies in different atomic subshells influences the emission probabilities of radiative electron transitions. This can be observed as intensity variations in X-ray transitions and in this way in different atomic fluorescence yields.

To describe this effect in ions a simple statistical weighting procedure was proposed in [312]. This procedure allows us to describe fluorescence yields and oscillator strengths in multiply ionized atoms on the basis of single-vacancy transition rates and oscillator strengths. If a subshell, which can contain at maximum n_0 electrons, is occupied with n electrons the transition rate for the analyzed subshell is reduced in comparison to the fully occupied shell by the ratio n/n_0 for single electron transitions and by $n(n-1)/n_0(n_0-1)$ if the transition contains two electrons from the partially filled shell. The oscillator strength is then reduced by n/n_0.

Radiative transition rates Γ are proportional to the product of the oscillator strength and the square root of the transition energy. If the binding energy difference between the electron states involved in the considered transition is known the influence of transition energy shifts on the transition rates can be considered. To give an overview of changes in X-ray emission rates for increasing outer-shell ionization in the following we show X-ray emission rates of selected transitions in argon, krypton and lead.

In Figs. 2.15 and 2.16 X-ray emission rates of selected K-series transitions in argon ions for different ion charge states are shown. During the ionization of electrons from substates with $n = 3$ the emission rates change only slightly. With the ionization of $2p_{3/2}$ electrons the K_{α_2} rate changes only weakly, but the K_{α_1} rate and the total emission rate decrease noticeably. In the same way the K_{β_3} rate increases during the ionization of $3p_{3/2}$ electrons at decreasing K_{β_1} rate. This behavior is also reflected in Fig. 2.17 which shows the intensity ratios. In Figs. 2.18 and 2.19 X-ray emission rates for krypton ions of the

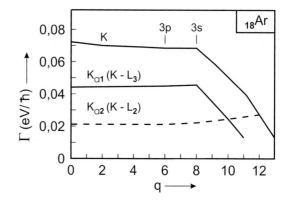

Fig. 2.15. Argon K_{α_1} and K_{α_2} X-ray emission rates and total K shell emission rates Γ as a function of the ionization stage q of the ground state ion. In the upper part fully ionized subshells are indicated

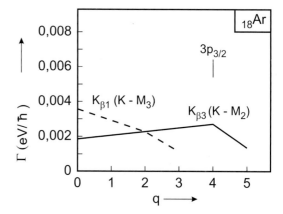

Fig. 2.16. Argon K_{β_1} and K_{β_3} X-ray emission rates as a function of the ionization stage q of the ground state ion. In the upper part the fully ionized $3p_{3/2}$ subshell is indicated

K, L_1, L_2 and L_3 series and from some intense X-ray transitions are shown. Strong changes in the X-ray emission rates are observed if electrons are ionized from levels involved in the examined transition. The characteristic decrease in the L X-ray emission rates for the L_1 series is explained by the closing of the L_{β_9} and $L_{\beta_{10}}$ channels, for the L_2 series by closing of L_{β_1} and L_{γ_5} channels and for the L_3 series by closing of the L_{β_6}, L_{α_1}, L_{α_2} and L_s channels as a result of the ionization of the considered subshell.

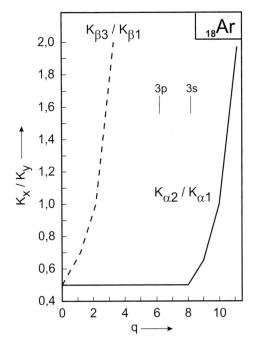

Fig. 2.17. Intensity ratios of argon K X-ray transitions as a function of the ionization stage q of the ground state ion. In the upper part fully ionized subshells are indicated

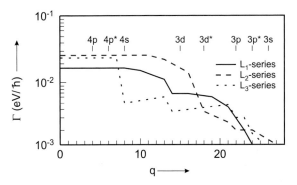

Fig. 2.19. Krypton X-ray emission rates Γ for the L_1, L_2 and L_3 series as a function of the number of outer shell vacancies q. In the upper part fully ionized subshells are indicated. $l^*: j = l - 0,5; l: j = l + 0,5$

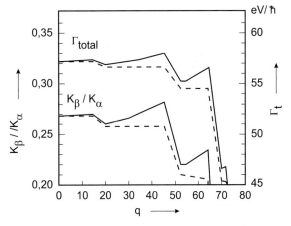

Fig. 2.20. Total X-ray emission rates Γ_t and $K_\beta/K\alpha$ intensity ratios for lead as a function of the number of outer shell vacancies q. The dashed line gives the result of the Larkin statistical weighting procedure [312]

In Fig. 2.20 the total K X-ray emission rate and the K_β/K_α intensity ratio as a function of the outer-shell ionization in lead is shown [17]. In Fig. 2.20 the quantum mechanical calculated transitions rates contain contributions from the atomic reorganization after the ionization. The reorganization effects appear as clear deviations from the results derived with the statistical weighting procedure [312] and are characteristic of a wide range where ionization of d and f electrons take place. These electrons give only small contributions to the radiative deexcitation of K-shell vacancies but more intensively contribute to

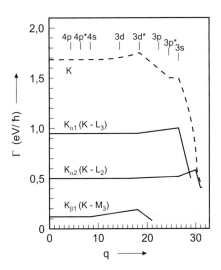

Fig. 2.18. Krypton X-ray emission rates Γ for K_{α_1}, K_{α_2} and K_{β_1} transitions and for the total K X-ray emission rate as a function of the ionization stage q of the ground state ion. In the upper part fully ionized subshells are indicated. $l^*: j = l - 0,5; l: j = l + 0,5$

the screening of p-electrons which are more relevant for K X-ray emission. Generally we state the decrease of the total X-ray emission rates from the Larkin procedure in comparison to the quantum mechanical results. In the same way the K_β/K_α ratio is calculated to be too small over the complete ionization region. The reason is that in the Larkin procedure only the decrease of the configuration weights of levels involved in the analyzed transition are considered, i.e. reorganization effects are fully neglected.

2.2.4 X-Ray Energy Shifts

Physical Fundamentals

Under certain conditions X-ray energy shifts relative to the X-ray diagram lines can be observed. X-ray energy shifts appear if the potential field wherein the electrons move is changed. Because the effective atomic potential is the sum of contributions from the field of the nucleus and from different interactions of electrons among each other, changes of the nuclear potential as well as in the occupation of electron shells influence the X-ray transition energies. Sources for changes of the effective atomic potential are known:

- processes connected with nuclear effects such as
  - isotope effects,
  - hyperfine interactions,
  - electron capture,
  - internal conversion;
- and interactions in the atomic shell.

Alterations in the atomic potential can occur by excitation processes, by ionization after the interaction with ionizing radiation (photons, electrons, protons, α-particles or heavy ions), by capture processes of exotic particles such as muons or kaons or by modifying the atomic valence state during chemical bounding. The magnitude of the energy shifts varies between meV and some hundreds of eV. Here, how the effective potential is modified in comparison to the potential characteristic for the appearance of diagram lines is decisive.

For the first time in [502] X-ray satellite lines were observed at the beginning of the twentieth century in the K emission spectra of the elements sodium to zinc ($11 \leq Z \leq 30$). The measured transitions were identified as satellite lines of the corresponding K_{α_1} transitions. Coster [125] first reported satellite lines in the L X-ray emission spectra and Stenstrom [522] reported satellite lines in the M X-ray emission spectra. These results were a proof that under certain conditions X-ray lines of the K, L and M series can have energetic shifts in comparison to the transition energies of the diagram lines.

Satellite lines in ion collisions were always observed [120] in experiments with different targets bombarded with mercury ions. While doing so the appearance of lines was observed to be different in their energy from the energies of parent diagram lines of the beam atoms as well as different from the energies of the target materials. This phenomenon was explained by satellite lines.

The chemical surroundings of atoms and ions also influence the parameters of characteristic X-rays. Chemically caused X-ray energy shifts were first investigated systematically in [327]. Extensive work for determining chemical X-ray energy shifts was done in [531]. Here detailed measurements on compounds of molybdenum, tin, silver, antimony, tungsten and of rare earth elements with wavelength-dispersive X-ray spectroscopy are known. Somewhat later measurements of other authors on tin [203] and on ruthenium, praseodymium and ytterbium [316] were published.

The study of other sources for nondiagram lines occur basically in the last few decades, conditioned by the progress in accelerator techniques, plasma physics, ion source development, measurement techniques and by the availability of powerful computers.

Properties of characteristic X-rays such as transition energies, emission rates, energies of absorption edges or energetic widths of X-ray lines can be calculated with the self-consistent-field method. To study X-ray energy shifts, experiments have been done by many different authors with high-resolving semiconductor detectors. Although it has significantly less luminous intensity, high-resolving wavelength-dispersive X-ray spectroscopy is an excellent tool to study X-ray energy shifts. Here X-ray energy shifts were measured with a precision up to 0.1 percent of the natural linewidth of the measured X-ray transition line.

Isotope Effects

Atoms of the same element, but with different neutron numbers (isotope[15]) emit characteristic X-rays different from isotope to isotope in a characteristic manner. This behavior is caused by small energetic differences in the the binding energies of atomic levels as a result of the co-motion and the finite size of the atomic nuclei. An overview of the influence of isotope effects on electron binding energies and X-ray transition energies can be found for instance in [76] and [227].

By means of the addition of neutrons to the nucleus, for each nucleus the characteristic charge distribution $\varrho(r)$ is transformed into a radial broadened distribution for the nucleus A'. This situation is shown in Fig. 2.21. From spectroscopic investigations of mesonic atoms[16], electron scattering experiments etc., the nuclear radius depends on the number of nucleons A in the nucleus and can be estimated as

$$R = r_0 A^{1/3}$$

with R is the mean nuclear radius and with the values of r_0 deduced from different experiments:

$$r_0 = (1, 1 \ldots 1, 4) \times 10^{-15} \text{ m} .$$

The nuclear charge density and the radial wavefunction are shown in Fig. 2.21 for the case that one neutron is added to the nucleus A. As shown in Fig. 2.21 the overlap of the electron wavefunction with the nuclear charge distribution decreases and the 1s-electron for the atom with the nucleus A' is more weakly bound by the quantity ΔE. Isotope effects are the strongest for the 1s level, because here as a result of the near localization of the 1s wavefunction to the nucleus the overlap with the nuclear wavefunctions is more intense than for other electron states.

The changes of electron binding energies as a result of isotope effects for in an X-ray transition involved levels can be observed as X-ray energy shifts. The energetic level shift between two isotopes is described as

$$\delta E^{AA'} = \delta E_{Coul}^{AA'} + \delta E_{Mas}^{AA'} + \delta E_{Pol}^{AA'} . \quad (2.52)$$

The first and dominant term on the right side of equation (2.52) describes the energy shift connected with a change of the Coulomb field if a certain number N of neutrons is added to the nucleus. The second term contains a correction for the nuclear mass. Nuclear polarization effects are described by the third term of equation (2.52).

Another in the literature known reading is to present isotope shifts by two contributions [283]:

- the mass shift $\delta E_{Mas}^{AA'}$ and

- the field shift $\delta E_{Feld}^{AA'}$.

As an example of optical spectra the separation procedure of mass and field isotope shift is described for titanium atoms in [193].

Coulomb energy shifts $\delta E_{Coul}^{AA'}$. The Coulomb energy shift results from the finite dimension of the nucleus causing deviations from the Coulomb field of a point nucleus and from the deviation of the nucleus from an ideally spherical form. The resulting term shifts are proportional to the probability of finding the electrons in the nuclear region, i.e. s-electrons are much more strongly influenced than p-electrons or electrons with higher angular momenta. The Coulomb energy shift can be written in terms of the nuclear charge parameters C:

$$\delta E_{Coul} = C_1 \delta \langle r^2 \rangle + C_2 \langle r^4 \rangle + C_2 \langle r^6 \rangle + \ldots \quad (2.53)$$

with

$$\delta \langle r^{2N} \rangle = \frac{\int \delta \varrho \, r^{2N} \, dV}{\int \varrho \, dV} . \quad (2.54)$$

Here ϱ is the nuclear charge density. The coefficients C_n can be calculated with the self-consistent-field

---

[15] Isotopes: nuclei with the same atomic number but with different neutron numbers and therefore with different atomic masses. Isotopy was discovered in 1910 by Soddy for the radioactive decay series. The natural isotopy of chemical elements was demonstrated by mass spectroscopy in 1919 by Aston.

[16] In such atoms an electron is replaced by a pion or muon.

Table 2.18. Coefficients of the charge moments $\delta\langle r^2\rangle$, $\delta\langle r^4\rangle$, $\delta\langle r^6\rangle$, known also as *Seltzer coefficients* (from [491]). If the units are given in $\mathrm{fm}^2, \mathrm{fm}^4, \mathrm{fm}^6$ then the energy shifts result in meV (the order of magnitude is given in brackets).

Z	C_1	C_2	C_3	Z	C_1	C_2	C_3
30	0.683 (1)	−0.232 (−2)	0.950 (−5)	67	0.468 (3)	−0.412	0.114 (−2)
31	0.795 (1)	−0.277 (−2)	0.109 (−4)	68	0.515 (3)	−0.463	0.128 (−2)
32	0.918 (1)	−0.331 (−2)	0.128 (−4)	69	0.565 (3)	−0.518	0.143 (−2)
33	0.106 (2)	−0.398 (−2)	0.152 (−4)	70	0.623 (3)	−0.579	0.158 (−2)
34	0.122 (2)	−0.471 (−2)	0.174 (−4)	71	0.683 (3)	−0.648	0.177 (−2)
35	0.140 (2)	−0.565 (−2)	0.208 (−4)	72	0.750 (3)	−0.722	0.196 (−2)
36	0.161 (2)	−0.662 (−2)	0.237 (−4)	73	0.825 (3)	−0.807	0.219 (−2)
37	0.183 (2)	−0.788 (−2)	0.282 (−4)	74	0.904 (3)	−0.899	0.242 (−2)
38	0.208 (2)	−0.926 (−2)	0.325 (−4)	75	0.991 (3)	−0.100 (1)	0.271 (−2)
39	0.237 (2)	−0.110 (−1)	0.385 (−4)	76	0.109 (4)	−0.111 (1)	0.299 (−2)
40	0.267 (2)	−0.128 (−1)	0.445 (−4)	77	0.119 (4)	−0.124 (1)	0.333 (−2)
41	0.303 (2)	−0.150 (−1)	0.517 (−4)	78	0.131 (4)	−0.139 (1)	0.370 (−2)
42	0.341 (2)	−0.173 (−1)	0.588 (−4)	79	0.144 (4)	−0.155 (1)	0.414 (−2)
43	0.384 (2)	−0.203 (−1)	0.687 (−4)	80	0.157 (4)	−0.171 (1)	0.455 (−2)
44	0.432 (2)	−0.233 (−1)	0.770 (−4)	81	0.172 (4)	−0.191 (1)	0.505 (−2)
45	0.485 (2)	−0.269 (−1)	0.885 (−4)	82	0.188 (4)	−0.211 (1)	0.559 (−2)
46	0.543 (2)	−0.309 (−1)	0.100 (−3)	83	0.207 (4)	−0.236 (1)	0.625 (−2)
47	0.609 (2)	−0.356 (−1)	0.115 (−3)	84	0.228 (4)	−0.266 (1)	0.709 (−2)
48	0.680 (2)	−0.405 (−1)	0.128 (−3)	85	0.249 (4)	−0.297 (1)	0.795 (−2)
49	0.758 (2)	−0.462 (−1)	0.144 (−3)	86	0.272 (4)	−0.320 (1)	0.831 (−2)
50	0.845 (2)	−0.523 (−1)	0.161 (−3)	87	0.297 (4)	−0.356 (1)	0.931 (−2)
51	0.939 (2)	−0.594 (−1)	0.181 (−3)	88	0.326 (4)	−0.396 (1)	0.103 (−1)
52	0.104 (3)	−0.662 (−1)	0.196 (−3)	89	0.357 (4)	−0.443 (1)	0.116 (−1)
53	0.117 (3)	−0.775 (−1)	0.232 (−3)	90	0.392 (4)	−0.489 (1)	0.127 (−1)
54	0.129 (3)	−0.872 (−1)	0.258 (−3)	91	0.429 (4)	−0.549 (1)	0.144 (−1)
55	0.144 (3)	−0.995 (−1)	0.293 (−3)	92	0.470 (4)	−0.601 (1)	0.156 (−1)
56	0.158 (3)	−0.111	0.322 (−3)	93	0.516 (4)	−0.676 (1)	0.178 (−1)
57	0.177 (3)	−0.127	0.368 (−3)	94	0.563 (4)	−0.740 (1)	0.192 (−1)
58	0.195 (3)	−0.144	0.419 (−3)	95	0.618 (4)	−0.830 (1)	0.218 (−1)
59	0.214 (3)	−0.163	0.474 (−3)	96	0.678 (4)	−0.918 (1)	0.240 (−1)
60	0.238 (3)	−0.185	0.534 (−3)	97	0.744 (4)	−0.103 (2)	0.271 (−1)
61	0.263 (3)	−0.210	0.607 (−3)	98	0.814 (4)	−0.114 (2)	0.299 (−1)
62	0.291 (3)	−0.234	0.667 (−3)	99	0.892 (4)	−0.126 (2)	0.331 (−1)
63	0.320 (3)	−0.264	0.750 (−3)	100	0.980 (4)	−0.142 (2)	0.376 (−1)
64	0.351 (3)	−0.292	0.817 (−3)	101	0.107 (5)	−0.157 (2)	0.417 (−1)
65	0.388 (3)	−0.330	0.921 (−3)	102	0.118 (5)	−0.178 (2)	0.479 (−1)
66	0.427 (3)	−0.368	0.102 (−2)	103	0.130 (5)	−0.197 (2)	0.530 (−1)

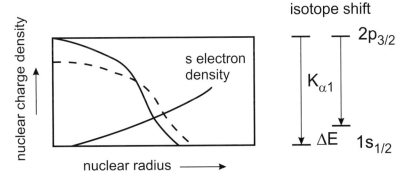

Fig. 2.21. K_{α_1} isotope shift of characteristic X-rays (after [76]). The addition of neutrons increases the width of the radial nuclear charge distribution (dashed line) in comparison to the charge distribution of the initial nucleus (solid line). This leads to a change in the overlap between nucleus and the 1s wavefunction

method and are tabulated in Table 2.18 for mean mass numbers. In particular the Coulomb energy shift is important in heavier elements, where the heavier isotope has the higher energy.

Nuclear mass correction $\delta E_{\text{Mas}}^{\text{AA}'}$. The nuclear mass correction can be divided into

- the normal nuclear mass correction, and
- the specific mass correction.

Thus, we have

$$\delta E_{\text{mas}}^{\text{AA}'} = \delta E_{\text{norm}}^{\text{AA}'} + \delta E_{\text{spez}}^{\text{AA}'} . \qquad (2.55)$$

Because the nucleus has a finite mass the *normal mass correction* is caused by the nuclear motion. For one-electron spectra the wavenumber difference yields

$$\Delta k = m_e \frac{A - A'}{AA'} k$$

with m_e as the electron mass.

The *specific mass correction* is caused by correlations between the outer electrons of the radiation emitting system.

The mass effect is clearly pronounced for light atoms and less important for heavier atoms. Here the heavier isotope is associated with the lower energy state. The contribution of the mass effect to the energy shift of the K_{α_1} X-ray transitions is approximated in [532]

$$\delta E_{\text{mas}}^{A+\Delta A} \approx \frac{2}{3} \frac{\Delta A}{1836 A^2} E(K_{\alpha_1}) \qquad (2.56)$$

with ΔA as the mass increase after the addition of one neutron.

Isotope shifts and nuclear parameters. The isotope shift δE measured for transition lines of characteristic K X-rays and corrected for the mass shift $\delta E_{\text{mas}}^{\text{AA}'}$ directly give the value of the coefficient $\delta E_{\text{Coul}}/C_1$ equal to the nuclear parameter λ:

$$\lambda = \delta\langle r^2 \rangle + \frac{C_2}{C_1} \delta\langle r^4 \rangle + \frac{C_3}{C_1} \delta\langle r^6 \rangle + \dots \qquad (2.57)$$

The K_α isotope shift essentially is determined by the energetic shift of the 1s level; the contributions of the L_2 and L_3 levels are only in the order of $< 10^{-2}$ and 10^{-5} of the 1s energy level shift. As a rule the nuclear parameter λ deviate by less than 10 percent from $\delta\langle r^2 \rangle$ [227]. In the case of titanium we have for instance [193]

$$\lambda = 0.995(4) \times \delta\langle r^2 \rangle .$$

In the case of the K shell the isotope shift of the binding energy can be estimated as [63]

$$\Delta E \approx 2 \times 10^{-9} A^{5/3} \Delta A .$$

An overview of the experimentally observed Coulomb shifts ΔE_{Coul} and of variations of the nuclear charge radius $\delta\langle r^2 \rangle$ is given in Table 2.19.

Isotope shifts and hyperfine structure effects are of the same order of magnitude. The most pronounced isotope shifts can be observed in molecular spectra. For instance more detailed discussions can be found in [237] and [422].

Table 2.19. K X-ray isotope shifts δE_{Coul} and changes in the radius of the charge distribution $\delta\langle r^2 \rangle$ for selected isotope pairs (after [76]) as determined by different authors. Corresponding citations can be found in the original sources [76]

isotope	ΔE_{Coul}/meV	λ/fm^2
$^{92-100}$Mo	35.0 ± 5.0	1.030 ± 0.150
$^{94-100}$Mo	31.0 ± 8.0	0.910 ± 0.230
$^{116-124}$Sn	35.0 ± 1.3	0.414 ± 0.016
$^{134-135}$Ba	-6.3 ± 2.0	-0.040 ± 0.013
$^{134-136}$Ba	-3.4 ± 2.0	-0.022 ± 0.013
$^{136-137}$Ba	-0.2 ± 2.0	-0.001 ± 0.013
$^{140-142}$Ce	51.5 ± 1.9	0.274 ± 0.010
$^{144-145}$Nd	29.8 ± 4.2	0.126 ± 0.018
$^{144-146}$Nd	55.4 ± 7.7	0.233 ± 0.033
$^{146-148}$Nd	65.4 ± 8.0	0.275 ± 0.034
$^{148-150}$Sm	88.3 ± 3.0	0.303 ± 0.010
$^{149-150}$Sm	65.1 ± 3.0	0.224 ± 0.010
$^{150-152}$Sm	119.5 ± 3.5	0.411 ± 0.012
$^{154-155}$Gd	39.2 ± 8.1	0.112 ± 0.024
$^{156-157}$Gd	10.4 ± 4.4	0.030 ± 0.013
$^{156-158}$Gd	50.6 ± 3.6	0.144 ± 0.010
$^{157-158}$Gd	40.6 ± 3.7	0.116 ± 0.011
$^{158-160}$Gd	54.0 ± 3.4	0.154 ± 0.010
$^{161-162}$Dy	39.4 ± 5.1	0.092 ± 0.012
$^{162-163}$Dy	4.2 ± 3.3	0.010 ± 0.008
$^{162-164}$Dy	55.6 ± 3.7	0.130 ± 0.009
$^{163-164}$Dy	51.4 ± 4.7	0.120 ± 0.011
$^{166-168}$Er	69.5 ± 4.5	0.135 ± 0.009
$^{168-170}$Er	80.0 ± 6.1	0.155 ± 0.012
$^{170-171}$Yb	48.0 ± 20.2	0.077 ± 0.032
$^{170-172}$Yb	101.4 ± 11.6	0.163 ± 0.019
$^{171-172}$Yb	53.4 ± 8.6	0.086 ± 0.014
$^{172-173}$Yb	31.2 ± 16.9	0.050 ± 0.027
$^{172-174}$Yb	88.0 ± 8.4	0.141 ± 0.013
$^{173-174}$Yb	56.8 ± 20.6	0.091 ± 0.033
$^{174-176}$Yb	65.5 ± 7.4	0.103 ± 0.012
$^{178-180}$Hf	77.4 ± 5.3	0.103 ± 0.007
$^{182-184}$W	92.3 ± 10.5	0.102 ± 0.012
$^{184-186}$W	60.0 ± 8.0	0.066 ± 0.009
$^{200-204}$Hg	254.0 ± 37.0	0.162 ± 0.024
$^{204-206}$Pb	200.0 ± 38.0	0.106 ± 0.020
$^{206-207}$Pb	50.0 ± 20.0	0.027 ± 0.011
$^{206-208}$Pb	186.0 ± 18.0	0.099 ± 0.010
$^{207-208}$Pb	136.0 ± 25.0	0.072 ± 0.013
$^{235-238}$U	1800.0 ± 200.0	0.383 ± 0.044

Nuclear polarization effects. A further contribution to the isotope shift can arise from nuclear polarization effects. Estimations show that in comparison to other contributions this contribution can be neglected [491].

Odd–even staggering. The energy shift for an isotope with odd atomic mass number is not localized in the middle of the energy shifts for neighboring isotopes with even mass numbers but somewhat closer to the energy shift of the isotope with lower mass number.

Meanwhile the combination of results from isotope shifts from muonic and electronic X-ray spectra and from optical data allow us to get very comprehensive information on nuclear radii. This also concerns isotopes far away from the line of nuclear stability.

Chemical Shifts

Chemical shifts modify X-ray emission spectra by effects connected with the presence of valence electrons. If a valence electron is removed from an atom[17] a reduction in the screening of the nuclear potential in relation to the other electrons takes place and for the remaining electrons a certain increase in the binding energies can be observed (see Fig. 2.22). In atoms of intermediate atomic number the energy shift can be in the order of up to 10 eV and as a rule has about the same value for ls, 2p, 3p and 4p electron levels. For similar reasons the difference in X-ray transition energies is typically two orders of magnitude lower, i.e. about 0,1 eV.

Because chemical shifts are only in the order of fractions of a percent of the natural linewidth of the investigated X-ray line, in the past chemical X-ray energy shifts were measured predominantly with high-resolving wavelength-dispersive methods. Thus, for instance Sumbaev [533] used a Cauchois-type crystal diffraction spectrometer (Bragg reflection in transmission geometry) and Gohshi [208] a two-crystal diffraction spectrometer with two plane crystals (Bragg reflection on the crystal surface). Meanwhile a variety of papers is known where with the comparatively low energy resolution of semicon-

---

[17] This is especially significant for s-electrons because of their localization near to the nucleus.

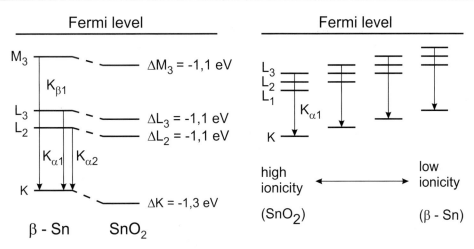

Fig. 2.22. Electron levels and chemical shifts in Sn and SnO_2 (after [161])

ductor detectors chemical influences on X-ray lines (for instance see [271] and [447]) were measured.

Alder et al. [8] discuss the influence of the orbital angular momentum of the valence electron on the expected K_α and K_β X-ray energy shifts. They demonstrate that the total electron charge density decreases in the nuclear region if s-electrons are removed. By removing p, d or f valence electrons the charge density increases as a result of decreasing screening. While the screening effect for p valence electrons is comparatively small the removal of d or f electrons leads to significantly greater changes in the total electron density.

The study of chemical shifts makes it possible to get qualitative results for the donor–acceptor behavior of ligands, of the charge of the atom and of specific features of phase transitions [546]. Furthermore, it is possible to extract information on the redistribution of the electron density in the atom as a result of the chemical bonding. Depending on the system there arise clear effects depending on the chemical bonding, such as for instance differences in the relaxation processes of diamagnetic and paramagnetic compounds of 3d transition elements and of free molecules or molecules absorbed on metallic surfaces [217].

For atomic inner-shells these effects lead to the strongest observed contributions. This is the reason that in most papers handling chemical shifts the X-

ray K series is investigated. Typically the energy shift of the K-shell is the greatest one and is much lower for outer shells. Hence the highest terms of an X-ray series show the greatest energy shift. The removal of an electron as a result of a chemical bond must be incomplete, i.e. the size of the X-ray energy shift can be connected with the ionicity of the compound.

For light elements chemical shifts have the greatest values, because the valence electrons are localized in the L shell and so strongly influence the K shell electrons. For a transition between K and L levels the energy shift can be expressed as

$$\Delta E_{KL} = mi\left[p_K\,\Delta E_{K(Z,Z-1)} - p_L\,\Delta E_{L(Z,Z-1)}\right] \quad (2.58)$$

where the quantities ΔE describe the energetic level shifts of K and L orbitals as they result for the complete ionization of an outer-shell electron. Furthermore, i is the ionicity, m the valence and p the time that for each valence electron is localized in the K or L shell. The results of X-ray measurements for transitions which include different electron orbitals characterize the different sensitivities of electron orbitals according to chemical valence states resulting in different strengths of chemical shifts. In Table 2.20 chemical shifts are given as summarized in [76] from different literature sources. Additionally included are results from [480]. The strongest chemical shifts are measured for elements of the rare earth region. These shifts are explained by strong changes in the screening of inner electrons during the addition or ionization of 4f electrons.

With the availability of sophisticated atomic structure programs and of powerful computing techniques it is possible to get, from systematic investigations of chemical shifts, information about the valence structure in different regions of the periodic system of elements. Such investigations are restricted so that as a rule the calculations are done for free atoms, i.e. they do not exactly represent the chemical bonding but are extrema for strong ionic bonds.

Hyperfine Structure

In the late nineteenth century additional splittings in optical spectra of atoms and molecules which could not be explained by fine structure effects and which were three orders of magnitude smaller than those from fine structure splittings were observed. A qualitative explanation of the hyperfine structure was given by Pauly [420]. He understood the observed level splitting as resulting from the interaction of electrons from the atomic shell with the magnetic dipole moment of the nucleus. Classical experiments as those of Terenin and Dobrezov in the year 1927 demonstrate that the sodium D lines[18] show a further splitting. This situation is shown in Fig. 2.23. A quantitative description of the observed effects was given in [210] and [172].

The level splitting caused by the interaction of the nuclear moments with the electrons of the atomic shell can be distinguished into

- *magnetic hyperfine structure*, caused by the nuclear magnetic dipole moments and
- *electric hyperfine structure*, caused by the nuclear electric quadrupole moments. As a rule this interaction is small in comparison to the magnetic hyperfine structure.

We will mentioned here that the isotope shift (see Sect. 2.2.4) is often included in the generic term *hyperfine structure splitting*.

Magnetic hyperfine structure. Hyperfine structure splittings arise for all spectral lines of atoms with nuclei with odd mass number[19]. The main reason for the hyperfine structure is the interaction of the magnetic moments of the electrons and of the nucleus, respectively. The nuclear magnetic moment is formed as the sum of the magnetic moments of the nucleons and is in the order of the nuclear magneton μ_K. Because of the smallness of the nuclear magneton in comparison to the Bohr magneton (about three orders of magnitude smaller by the ratio of the electron mass to the proton mass) the hyperfine structure splitting is very small and is in the order of about 100 meV to 1 eV for atomic K levels.

The magnetic interaction energy between the nuclear magnetic moment and the magnetic field produced from the electrons at the nucleus leads to a mutual orientation of the intrinsic nuclear angular momenta I (or nuclear spin) and the atomic shell momenta J. As a rule the angular momenta I of the nucleus are integral or half-integral between 0 and 9/2. Connected with this are nuclear angular momenta between -2 and $+6$ nuclear magnetons. Because the interaction energy is small in comparison to the coupling energy of the electrons in the atomic shell and of the nucleons in the nucleus, the hyperfine interaction does not disturb the inner couplings of the atomic shell and of the nucleus, i.e. the hyperfine interaction between J and I results only in the coupling of both angular momenta to the total angular momentum F of the atom:

$$F = I + J . \tag{2.59}$$

Because for the addition of two angular momenta the triangle relation $|I - J| \leq F \leq I + J$ must be fulfilled, for F we have

- $2I + 1$ values for $I < J$ or
- $2J + 1$ values for $J < I$.

The magnetic interaction energy between shell and nuclear moments has the value

$$E_{\mathrm{HFS}} = -\mu_I B_J . \tag{2.60}$$

---

[18] Sodium D lines: D_1: transition $3\,^2P_{1/2} \rightarrow 3\,^2S_{1/2}$; D_2: transition $3\,^2P_{3/2} \rightarrow 3\,^2S_{1/2}$
[19] In addition, also for the isotopes ^2D and ^{14}N.

Table 2.20. Experimental determined chemical energy shifts. All shifts are given in meV. A and B chemical as X-ray energies for the compounds A or B. The given in parenthesis values indicate the oxydation degree of the compound

Z	A	B	$\Delta E_{K_{\alpha_1}}$	$\Delta E_{K_{\alpha_2}}$	$\Delta E_{K_{\beta_1}}$	$\Delta E_{K_{\beta_2}}$	$\Delta E_{K_{\beta_{1,3}}}$
24	Cr	Cr_2O_3 (3+)	50	380			
		$LaCrO_3$ (3+)	40	350			
		CrO_2 (4+)	60	360			
		Ba_2CrO_4 (4+)	50	370			
		$Cr_{0.6}Ru_{0.4}O_2$ (5+)	320	500			
		CrO_3 (6+)	860	960			
		K_2CrO_4 (6+)	820	900			
		$NiCrO_4$ (6+)	830	870			
32	Ge	GeO_2	244±20				
		GeS_2	123±13				
		GeS	110±11				
33	As	As_2O_3	151±6				
37	Rb	RbCl	21±4		28±10	131±22	
38	Sr	SrO	−30±4				
39	Y	Y_2O_3	−146±10				
40	Zr	ZrO_2	−299±5		−149±8	480±15	
41	Nb	Nb_2O_3	−260±5				
42	Mo	MoO_3	−199±5		137±5	224±15	
44	Ru	RuF_3	−94±3			−50±25	
		RuO_2	−42±4			−24±25	
47	Ag	Ag_2S	51±4		59±10	−125±15	
		AgCl	122±5		129±9	−104±16	
48	Cd	CdO	115±6				
		CdSe	82±13				
49	In	InO_3	112±8				
50	Sn_α	Sn_β	37±10				
		SnO	108±12				
		SnO_2	204±11		193±16	101±30	
	Sn_β	SnO	81±5		122±11	89±21	
51	Sb	Sb_2O_3	121±17				
		Sb_2O_4	172±10				
51		Sb_2O_5	200±15				
52	Te	TeO_2	176±5				
		TeO_3	269±5				
56	Ba	BaO	42±20				
57	La	La_2O_3	−3±10				
58	Ce	CeO_2	−457±15				
59	Pr	Pr_2O_3	−20±15				
		PrO_2	−416±5				
	PrF_3	Pr	−45±4			−48±29	−55±7
		Pr_2O_3	5±8	48±6		31±p	11±10
		PrO_2	371±3	320±5		416±18	1018±4

Table 2.20. (cont.)

Z	A	B	$\Delta E_{K_{\alpha_1}}$	$\Delta E_{K_{\alpha_2}}$	$\Delta E_{K_{\beta_1}}$	$\Delta E_{K_{\beta_2}}$	$\Delta E_{K_{\beta_{1,3}}}$
62	Sm	$SmCl_2$	606±14	448±19	1455±40	415±50	K_{β_3}: 1360±50
63	Eu	EuF_3	−644±12				
		EuI_3	−618±21				
	EuF_2	EuF_3	644±11	582±15	1450±40	300±65	K_{β_3}: 1730±50
64	Gd	GdF_3	−13±15				
65	Tb	$TbO_{1.66}$	−266±13				
		$TbO_{1.72}$	394± 18				
69	Tm	TmF_3	46±26				
70	Yb	Yb_2S_3	−520±25				
		YbF_3	−579±26	570±114		586±114	1402±43
71	Lu	LuF_3	−48±38				
		Lu_2O_3	3±20				
72	Hf	HfO_3	6±30				
73	Ta	Ta_2O_5	113±30				
74	W	WO_3	110±33				

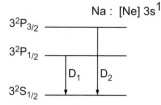

fine structure splitting

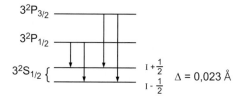

hyperfine structure splitting

Fig. 2.23. Fine structure and hyperfine structure splitting in sodium

The vector $\boldsymbol{\mu}_I$ describes the magnetic nuclear moment and $\boldsymbol{B}_J$ the magnetic field produced by the electrons of the atomic shell on the nucleus. Because the magnetic field produced by the electrons is practically constant over the nuclear sphere we can assume here a mean field $\langle B_0 \rangle$. Averaging over the precision motion of spin and orbital angular momenta this field is oriented in the direction of the orbital angular momentum vector J

$$\boldsymbol{B_0} = \langle B_0 \rangle \frac{J}{J\hbar} \; . \tag{2.61}$$

The magnetic moment $\boldsymbol{\mu}_I$ of the nucleus should be oriented in the direction of J:

$$\boldsymbol{\mu}_I = g_K\, \mu_K\, \frac{\boldsymbol{I}}{\hbar} = \frac{\mu_I\, \mu_K}{\hbar}\frac{\boldsymbol{I}}{I} \quad \text{with} \quad g_K = \frac{\mu_i}{I} \; . \tag{2.62}$$

Here g_K describes the nuclear g-factor. Then each orientation of the nuclear spin in the field B_J corresponds a certain energy state of the system. Starting from (2.60) we have

$$E_{\text{HFS}} = -\frac{\mu_I}{\hbar^2}\frac{\mu_K}{I \times J}\,(\boldsymbol{IJ}) \; . \tag{2.63}$$

To calculate the perturbation energy for the expectation value

$$\boldsymbol{IJ} = \frac{1}{2}\left(\boldsymbol{F}^2 - \boldsymbol{I}^2 - \boldsymbol{J}^2\right)$$

the actual operators are replaced by their eigenvalues

$$E_{\mathrm{HFS}} = \frac{A}{2} \left[F(F+1) - I(I+1) - J(J+1) \right] \quad (2.64)$$

with

$$A = -\frac{\mu_I \, \mu_K \, \langle B_0 \rangle}{2 \, I \times J} \quad \text{(interval factor)} . \quad (2.65)$$

The quantity A is known as the *interval factor* or as the *hyperfine structure constant*. For positive nuclear momenta the hyperfine structure constant is positive. This means that the hyperfine structure splitting is greater if F increases.

The relative term difference is described by the so-called *interval rule*

$$\Delta E_{F+1} - \Delta E_F = A(F+1) . \quad (2.66)$$

This rule reads:

- *The difference of two terms in a hyperfine structure multiplet is proportional to the greater value of both F-values. The splittings in a multiplet are related as*

$$F : F - 1 : F - 2$$

and so on.

The absolute energetic spacing of hyperfine structure components is given by the *interval factor A*, which contain a product of the nuclear momentum μ and the atomic field $\langle B_0 \rangle$ (see (2.65)). If one of these values is known the other one can be determined with the observed level splitting. Only for some trivial cases is it possible to calculate $\langle B_0 \rangle$ from the electron configuration exactly. On the other hand, many nuclear moments are well known because they can be measured by the use of external fields. For a known nuclear moment $\langle B_0 \rangle$ can be derived from the measurement.

In spectral transitions the intensities are proportional to $(2F+1)$ because the terms are $(2F+1)$-times degenerate relative to m_F and for transitions this degeneration influences the state density of the final state. The analysis of the level splitting is complicated by the fact that normally both orbitals involved in the transition are the subject of hyperfine structure splitting.

Electric hyperfine structure. Nuclei can be deformed. In this way the deformation is described by the *electric quadrupole moment*. According to [534] the contribution of the electric hyperfine structure is given by

$$\Delta E_{\mathrm{HFS}} = B \, \frac{\frac{3}{4} (C^2 + C) - I(I+1) J(J+1)}{2I \, (2I-1) \, (2J-1)} \quad (2.67)$$

with

$$C = F(F+1) - I(I+1) - J(J+1) .$$

The *constant of electric quadrupole interaction B* contains the product of the electric quadrupole moment and of a quantity proportional to the electric field gradient. The quantity B is only different from zero if the form of the nucleus deviates from spherical symmetry.

The contribution of the electric hyperfine structure is principally somewhat smaller than the magnetic contribution, but of the same order of magnitude [534].

Total energy of the hyperfine structure interaction. The total energy can be calculated as follows

$$E_{\mathrm{HFS}} = E_{\mathrm{FS}} + \frac{A}{2} C + B \, \frac{\frac{3}{4} (C^2 + C) - I(I+1) J(J+1)}{2I \, (2I-1) \, (2J-1)} \quad (2.68)$$

where E_{FS} is the contribution from the fine structure interaction with respect to the total angular momentum J of the atomic shell.

Importance of hyperfine structure. Above all, the investigation of the hyperfine structure interaction is of importance for studies in nuclear physics. From spectroscopic measurements of the hyperfine structure splitting the quantities A and B can be determined from which information on nuclear properties (nuclear spin, magnetic dipole moment, electric quadrupole moment) can be deduced.

Numerical values of the hyperfine structure splitting. Extensive tabulations of the hyperfine structure splitting and of the parameters of the energy levels of stable isotopes are given in [441], [442] and [494].

Atomic Structure Shifts for Transitions with $\Delta Z = 1$

Atomic structure shifts were observed between different elements in similar chemical states. In *electron capture*, an electron from an inner atomic orbital is captured by the nucleus and absorbed by the process

$$p + e^- \rightarrow n + \nu_e$$

i.e. an element of the atomic number Z is transformed into an element with $Z - 1$. The X-rays connected with this process are emitted from an atom with an outer-shell structure of the atom with next higher atomic number. Comparing the characteristic X-ray energies with energies observed in processes with $\Delta Z = 0$ (photoionization, internal conversion) X-ray energy shifts can be determined corresponding to the difference in the atomic structure of neighboring elements [227]. In Fig. 2.24 this situation is shown for elements of the rare earth group.

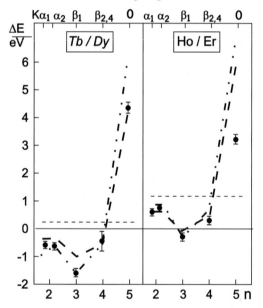

Fig. 2.24. X-ray energy shifts ΔE (E(photoionization), np$\rightarrow$1s - E(electron capture, np$\rightarrow$1s) for rare earth elements (after [227]). Experimental results are given for Tb/Dy from [226] and for Ho/Er from [79]. The dashed horizontal lines give calculated contributions to the energy shift by hyperfine structure effects. The dotted-dashed lines characterize calculated energy shifts for the electron configuration of trivalent metals. The dashed line give results from calculations with divalent metals

In electron capture the electrons are meanwhile in a stable state. Under ionization of the same atom by photoionization an additional electron can be bound but the vacancy lifetime for the K shell is too short for electron relaxation processes (10^{-15} to 10^{-17} s in heavy elements; see Fig. 2.6). Thus it is possible, by the measurement of the atomic structure shift, to get a "snapshot" of both electron configurations.

Dynamical Shifts

The production mechanism of inner-shell vacancies is connected with outer-shell excitations that temporarily modify the atomic screening and so produce small energy shifts. These processes were extensively discussed in [227].

As a rule for low-energetic nuclear transitions and photoionization processes, effects from additional inner-shell vacancies can be detected. These secondary effects are weak but detectable. Outer-shell ionization occurs more frequently but the magnitude of these effects is small (often inside the natural linewidth). On the other hand many cases are known where asymmetric or complex X-ray line profiles were observed. For example: Measured with high spectral resolution $K_{\alpha_{1,2}}$ lines of copper ($Z = 29$) show a complex line structure that can be explained by additional $3d^{-1}$ and $3p^{-1}$ vacancies [479]. For the line broadening we refer also to Sect. 2.2.2 and [187].

Because of the small vacancy production probability for electron capture, dynamical shifts are negligible in comparison to contributions from photoionization processes. The order of magnitude for dynamical shifts of the K X-ray series is shown in Fig. 2.25.

X-Ray Outer-Shell Satellites

For example ions with *outer-shell vacancies* play a dominant role in plasmas, in collective particle ensembles and in heavy ion sources. The existence of outer-shell vacancies influences the course and the intensity of interaction processes and structure effects in the atomic shell and of atom–nucleus interactions. Up to now in the literature systematic experimental investigations of X-ray energy shifts by the presence of an increasing number of outer-shell

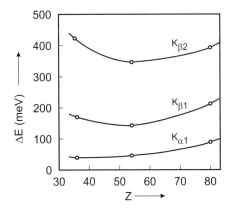

Fig. 2.25. Dynamical shifts ΔE of K X-rays for photoionization processes as a function of the atomic number Z (after [227])

vacancies are not known, but otherwise many individual experimental results have been published.

Energetic shifts of individual X-ray lines depend on the outer-shell ionization and can be estimated with self-consistent-field calculations. For example, for intense X-ray dipole transitions of all elements up to uranium this is done in [587]. In the same way there exist many papers describing properties of isoelectronic sequences.

In Table 2.21 experimentally determined energies of neon satellite lines are compared with calculated energy shifts. The case of neon allows us to compare the intrinsic influence of inner-shell vacancies on X-ray transition energies, because nonradiative for additional L shell ionization processes as it is

Table 2.21. Influence of L shell vacancies on K_α X-ray transition energies for neon ($Z=10$). Experimental values E_{exp} from [272] are compared with calculated values E_{th} from [587]. ΔE is the energy difference compared with the parent diagram line; ΔE_{th} is the calculated energy difference

Satellite	E_{exp}/eV	$\Delta E/eV$	$\Delta E_{th}/eV$
KL^0	848±2		
KL^{-1}	855±2	7±4	6,1
KL^{-2}	863±2	15±4	14,5
KL^{-3}	873±2	25±4	25,4
KL^{-4}	882±2	34±4	31,7
KL^{-5}	895±2	47±4	45,3
KL^{-6}	907±2	59±4	56,1

typical for L shell ionization in more complex atoms.

A comparison of experimental and calculated X-ray energy shifts gives good agreement within the error margins of the experiment. For the transition to heavier elements the situation loses its unambiguous character, because it can no longer be assumed that only inner-shell or outer-shell vacancies exist alone. Furthermore the energy shift per outer-shell vacancy decreases per additional outer-shell vacancy due to the increasing number of electrons in the atomic shell and increasing in this way the electronic screening.

As a result of the ionization process a variety of vacancy configurations is possible. Here such rearrangement processes should be considered where the ground state for the positive ion of the corresponding degree of ionization is reached, i.e. the state with a minimum of total energy of the multielectron system. To reach minimal total energy the *ion ground state configurations* can show significant differences to the electron configurations of neutral atoms with the same number of electrons.

As a rule the localization of vacancies in the ion ground states begins with the 7s and 6d orbitals [441] and continues with the sequence 5f, 6p, 6s, 5d, 4f, 5p, 5s, 4d, 4p, 4s, 3d, 3p, 3s, 2p, 2s and 1s.

In Figs. 2.26–2.30 energetic X-ray shifts for selected transitions in argon ($Z = 18$) and thorium ($Z = 90$) as a function of increasing ionization are shown for the ion ground states.

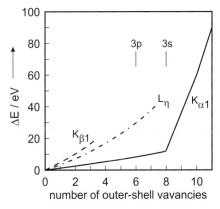

Fig. 2.26. Argon X-ray energy shifts ΔE of selected transitions as a function of the number of outer-shell vacancies. Completely ionized subshells are indicated

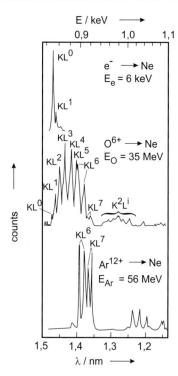

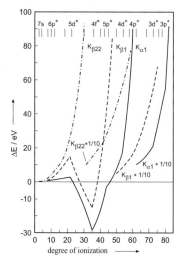

Fig. 2.27. Neon K_α X-ray emission spectrum for excitation by electrons and heavy ions (from [61])

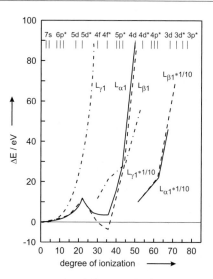

Fig. 2.29. X-ray energy shifts ΔE of selected L transitions in thorium for ion ground states with different degree of ionization. Completely ionized subshells are indicated. $l^*:j = l - 0.5; l:j = l - 0.5$

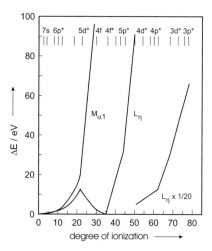

Fig. 2.30. X-ray energy shifts ΔE of the L_η and $M_{\alpha 1}$ transitions in thorium for ion ground states with different degree of ionization. Completely ionized subshells are indicated. $l^*:j = l - 0.5; l:j = l - 0.5$

Figures 2.31 and 2.32 give the tendency of X-ray energy shifts for increasing degree of ionization and increasing atomic number. In Figs. 2.33–2.35 for K_{α_1}, L_{α_1} and M_{α_1} transitions X-ray energy shifts for all elements up to uranium for selected ionization stages are summarized.

Fig. 2.28. X-ray energy shifts ΔE of selected K transitions in thorium for ion ground states with different degree of ionization. Completely ionized subshells are indicated. $l^*:j = l - 0.5; l:j = l - 0.5$

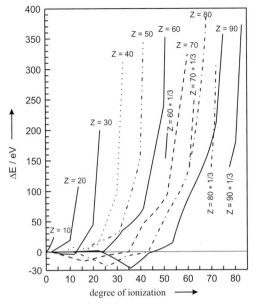

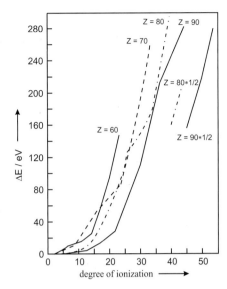

Fig. 2.32. L_{γ_1} X-ray transition energy shifts ΔE for selected elements as a function of the ionization stage of the ion ground state

Fig. 2.31. K_{α_1} X-ray transition energy shifts ΔE for selected elements as a function of the ionization stage of the ion ground state

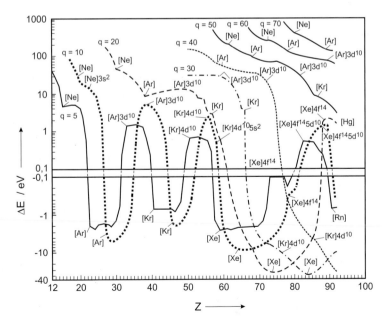

Fig. 2.33. K_{α_1} X-ray energy shifts ΔE for selected ionization stages q as a function of the atomic number Z. The ground state electron configuration is given at characteristic points

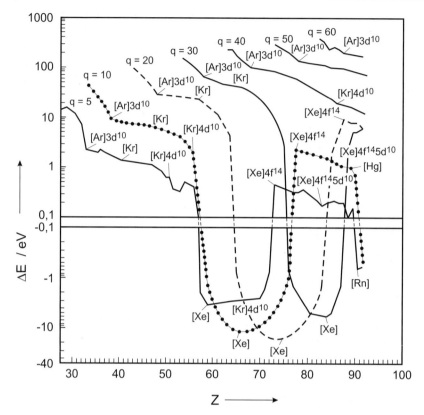

Fig. 2.34. L$_{\alpha_1}$ X-ray energy shifts ΔE for selected ionization stages q as a function of the atomic number Z. The ground state electron configuration is given at characteristic points

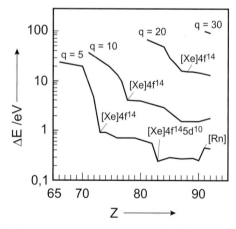

Fig. 2.35. M$_{\gamma_1}$ X-ray energy shifts ΔE for selected ionization stages q as a function of the atomic number Z. The ground state electron configuration is given at characteristic points

From these graphical dependencies we see that the energy shifts are greatest for X-ray series where the initial level has the greatest main quantum number. With increasing degree of ionization the energy shifts have at different ionization regions significant different energy shift gradients. While for elements with occupied 4f subshells for K and L transitions decreasing energy shifts are characteristic, for lines of the M-series this behavior could not be observed. This circumstance can be used for the determination of ionization stages of ions by X-ray diagnostics to clear up ambiguities in X-ray spectra of the K and L series by measuring lines of the M series and of the energy gradient in K and L spectra for the case of developing ionization.

The influence of outer-shell vacancies on the X-ray transition energies depends on the localization of the electrons involved in the considered X-ray transition. For the K series the $K_{\beta_{21}}$ transitions have the greatest energy gradient for increasing outer-shell ionization. Generally it yields that the X-ray energy shifts are greatest for the highest difference in the main quantum numbers of electrons involved in the transition.

For different degrees of ionization the X-ray energy shifts of K and L transitions have clear local minima. For the K_{α_1} transition a decreasing X-ray energy shift for an increasing occupation of s and p electron orbitals and an increasing shift during the occupation of d and f levels is characteristic. For the L_{α_1} transition line for a fixed ionization state we find a permanent decrease of the X-ray energy shifts over a wide Z-region. Only after reaching the xenon electron configuration do the energy shifts increase again.

The comparison of results for the K_{α_1} ($1s_{1/2}$-$2p_{3/2}$), K_{β_1} ($1s_{1/2}$-$3p_{3/2}$) and L_{α_1} ($2p_{3/2}$-$3d_{5/2}$) transitions show that outer-shell vacancies cause different periodicities in the crossing of level energies for the orbitals involved in the X-ray transition. The crossing between the $1s_{1/2}$ and the $2p_{3/2}$ energy shifts has periodic maxima for low ionization degrees if the electron configuration reaches the noble gas configurations of argon, krypton or xenon. For the level pair $1s_{1/2}$-$3p_{3/2}$ and $2p_{3/2}$-$3d_{5/2}$ the same behavior does not appear results until reaching the xenon electron configuration. The reason for the observed phenomena is based on the different interactions of d and f electrons with the electrons involved in the transition.

The dependencies shown in Figs. 2.26–2.35 can be used for the identification of ionization degrees of atoms emitting characteristic X-rays by the energy shift relative to transition energy of the parent diagram lines. For a wide Z-region X-ray transition lines can be found where for a certain element the X-ray energy shifts for different ion charge states do not cross each other, i.e. the energy shifts can be classified unambiguously to different ion charge states. If this is not the case the measurement loses its unambiguous character and further physically independent quantities must be analyzed.

X-Ray Inner-Shell Satellites

Multiple vacancy states in atomic inner-shells can be induced by photon, electron or heavy ion impact or result from atomic rearrangement processes. For electron and photon excitation it is characteristic that diagram lines resulting from an isolated primary vacancy (see Fig. 2.27) will be mainly excited. Here only weak contributions from multiple inner-shell vacancies (KL^{-1} and KL^{-2}) can be observed. On the other hand, light ions can excite the whole satellite spectrum. Here the excitation in outer shells as well as the intensity of hypersatellite lines $K^{-2}L^{-i}$ is small. For heavy projectiles the ionization probability increases and *few-electron states* can be observed. For both light projectiles excited *few-vacancy states* and few-electron states a simplification of the spectra (i.e. small spectral overlap) is typical.

Beside satellites on the short-wavelength side of the diagram lines, also lines on the long-wavelength side of the spectrum can be observed. Long-wavelength satellites can occur from electron transitions from outer atomic levels to the K orbital in crystals (intersection transitions) as well as from two-electron processes, where the X-ray transition energy is decreased by the simultaneous excitation of an outer-shell electron (*radiative Auger process*).

For the case of sufficiently strong interaction between incompletely filled subshells and the X-ray vacancy, the X-ray multiplet structure can be appear as level splitting. Thus for instance X-ray satellites on the long-wavelength side of the parent diagram lines from atoms with incompletely filled d and f orbitals can be attributed to multiplet splitting.

Calculations with atomic structure codes for atoms with closed shells, with partially filled shells, molecules and solid states allow the interpretation of diagram lines and of spectral structures as a result of multiplet splitting, crossing transitions and other effects. Because of the small lifetime of an X-ray state (see Fig. 2.6) from processes leading in the atom to multiple vacancy configurations (successive ionization by some electrons, multiple ionization by nonradiative electron transitions, multiple

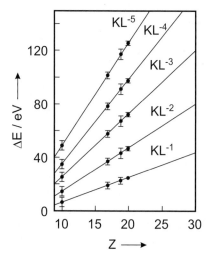

Fig. 2.36. X-ray K_α satellite energy shifts ΔE in light atoms as a function of additional L-shell vacancies and of the atomic number Z (with data from Kauffman [272] and Watson [559]).

ionization by impact ionization) only multiple ionization and direct multiple ionization (*shake-off processes*) are probable. Of principal importance here are Coster–Kronig processes $X_1 \rightarrow X_j Y$ where the vacancy states X_i and Y_j belong to different subshells of the same main shell. The fundamentals of the shake-off theory and corresponding numerical values for different configurations and elements can be found in [1,2,97,100].

While for the ionization of inner-shell electrons by photon or electron impact states with a small number of inner-shell vacancies are characteristic, the situation for collisions with heavy particles will be more complex. Here configurations with a high number of inner-shell vacancies can be observed and as a result one-, two- or three-electron ions can appear. With high-resolving X-ray spectrometry various characteristic spectra from ion–atom collisions were observed. Together with reaction models and self-consistent field calculations the analysis of these spectra allows us to get conclusions about the interaction mechanisms during the collision process. Satellite and hypersatellite lines can be classified by

their energy or wavelength. This allows us to get conclusions about the vacancy distribution in the atomic shell after the interaction process.

In Fig. 2.36 the energetic shift ΔE of K_α X-ray satellite lines in comparison to the transition energies of the parent diagram lines is shown as a function of the atomic number Z. Table 2.22 gives the order of magnitude of X-ray energy shifts per additional L and M vacancy for transition lines of the K and L series. It should be mentioned that the values given in Table 2.22 should be understood only as orientation but for exact values atomic structure calculations should be done.

Table 2.22. Satellite energy shifts for additional vacancies in the L and M subshells as given by Mokler [363]. Z_L and Z_M are screened charges: $Z_L = Z - 4, 15$; $Z_{M_{1,2,3}} = Z - 11, 25$; $Z_{M_{4,5}} = Z - 21, 15$

transition	initial vacancy	final vacancy	shift per L vacancy/eV	shift per M vacancy/eV
K_α	K	$L_{2,3}$	$1,66 \times Z_L$	$0,06 \times Z_M$
$K_{\beta_{1,3}}$	K	$M_{2,3}$	$4,38 \times Z_L$	$0,91 \times Z_M$
L_α	L_3	$M_{4,5}$	$2,24 \times Z_L$	$0,56 \times Z_M$
L_{β_1}	L_2	M_4	$2,24 \times Z_L$	$0,56 \times Z_M$
L_{β_2}	L_3	N_5	$3,71 \times Z_L$	$1,72 \times Z_M$

X-ray Emission from Highly Charged Ions

For many investigations connected with the spectroscopy of characteristic X-rays the appearance of X-ray lines stemming from ionized atoms is characteristic. With the development of such fields of investigations as plasma physics, X-ray astronomy, ion accelerators and sources of highly charged ions processes become more and more important, where the emission of X-rays typically deviates significantly from the standard signature as known from X-ray standard applications. Of main importance are here:

- Direct Excitation (DE)[20];
- Radiative Recombination (RR)[21] and
- Dielectronic Recombination (DR)[22].

The itemized processes are summarized in Fig. 2.37.

---

[20] DE: Direct Excitation
[21] RR: Radiative Recombination
[22] DR: Dielectronic Recombination

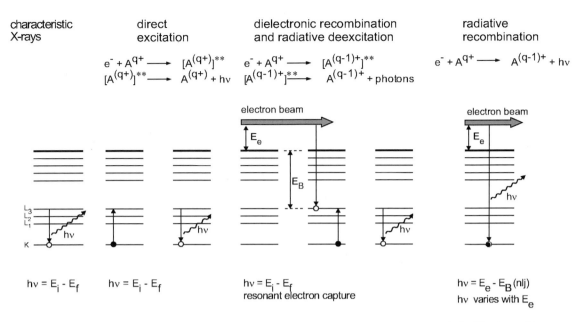

Fig. 2.37. Origin of X-ray diagram lines and the most important X-ray emission processes in highly charged ions

Direct Excitation

A direct excitation of electrons

$$e^- + A^{q+} \rightarrow A^{(q+)**}$$

occurs if the energy of the exciting electron is greater than the binding energy difference between the sublevels involved in the excitation process (see Fig. 2.37). The excited state so produced can deexcite by photon emission

$$A^{(q+)**} \rightarrow A^{q+} + h\nu \ .$$

For this process the typical multiplet splitting can produce a complex spectrum with many transition lines, resolvable only with wavelength-dispersive X-ray spectroscopy.

A simple estimation of excitation cross-sections for the electron excitation from the state i into the state j can be derived from the Van-Regemorter formula (see for instance [173]):

$$\sigma_{ij}(E) = \frac{8\pi^2}{\sqrt{3}} \, a_0^2 f_{ij} \, \frac{I_H^2}{E_e I_{ij}} \, G\left(\frac{E_e}{E_{ij}}\right) \qquad (2.69)$$

with a_0 the Bohr radius, f_{ij} the oscillator strength, E_{ij} the transition energy and E_e the electron energy.

The Gaunt factor G for $\Delta n = 0$ can be calculated as

$$G(x) = (0,33 - 0,3\,x^{-1} + 0,08\,x^{-2}) \, \ln x$$

and for $\Delta n > 0$ as

$$G(x) = (0,276 - 0,18\,x^{-1}) \, \ln x \ .$$

For $x < 1$ the Gaunt factor becomes zero. The calculation formula for the oscillator strength f_{ij} is given by (2.50).

In Fig. 2.38 an iridium spectrum is shown as it was measured the Dresden EBIT at 15 keV electron energy. DE lines for the L as well as for the M series are clearly expressed in the shown spectrum. The observed peaks are ensembles of overlapping transition lines from X-rays emitted by ions of adjacent ion charge states as is characteristic for electron impact ion sources. For the L excitation, lines can be identified for electron excitation processes into shells with $n = 3$ and 4 and for M excitation transitions for electron excitation into $n = 4$ and 5 states.

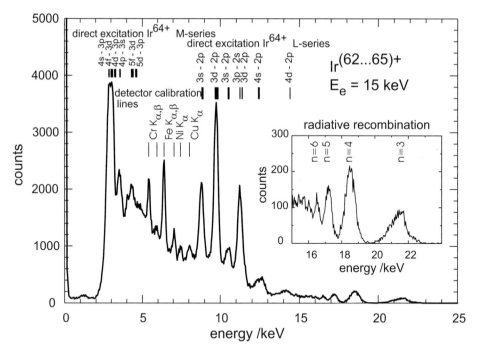

Fig. 2.38. Measured X-ray spectrum with a Si(Li) solid-state detector on the Dresden EBIT (see [407]) obtained from highly charged iridium ions excited with an electron energy of $E_e = 15$ keV. Shown are direct excitation lines of the L and M series and in the inset lines from radiative recombination processes

Radiative Recombination

Radiative recombination (RR; see also Fig. 2.37) is a process

$$A^{q+} + e^- \rightarrow A^{(q-1)+} + h\nu$$

where a free electron is captured into an ionized atomic state with emission of an X-ray quantum with energy

$$E_{RR} = E_e - E_B(nlj) . \quad (2.70)$$

$E_B(nlj)$ is here the (negative) binding energy for the considered subshell (nlj) (see Fig. 2.39) and E_e the kinetic energy of the captured electron. The consequence is that X-rays can be observed with a transition energy higher than the kinetic energy of the electrons involved in the RR process. This behavior can be seen in the inset of Fig. 2.38.

A calculation of the RR cross-section can be done as described in [282]:

$$\sigma_{RR}^q(E_e) = \frac{8\pi}{3\sqrt{3}} \, \alpha \, \lambda_e^2 \, \chi_q(E_e) \, \ln\left(1 + \frac{\chi_q(E_e)}{2\hat{n}^2}\right) \quad (2.71)$$

with

$$\chi_q(E_e) = (Z+q)^2 \frac{I_H}{4E_e}$$

and

$$\hat{n} = n + (1 - W_n) - 0{,}3 ,$$

where n is the main quantum number of the valence shell and W_n the ratio of the unoccupied states of the valence shell to the total number of possible states in this shell. α is the Sommerfeld fine-structure constant ($\alpha = 1/137.0360$), $\lambda_e = 3.861 \times 10^{-11}$ cm the electronic Compton wavelength and $I_H = 13.6$ eV the ionization potential of the hydrogen atom.

Dielectronic Recombination

Dielectronic Recombination (DE) [340] (see also Fig. 2.37) is a resonant recombination process of a free electron with an ion X^{q+} with subsequent radia-

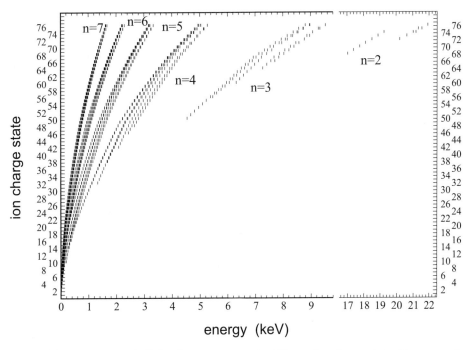

Fig. 2.39. Iridium binding energies $E_B(nlj)$ for electron capture in different ion charge states. X-ray energies calculates than corresponding to (2.70)

tive stabilization. In this way in a first step a free electron is nonradiatively transferred in a bound state of the ion under excitation of a further core electron. As a result a doubly-excited state $[X^{(q-1)+}]^{**}$ appears. Because of the discrete level structure the kinetic energy of the free electron must fulfill the resonance condition

$$E_r = E_d - E_i \qquad (2.72)$$

where E_d is the total binding energy of the dielectronic intermediate state and E_i is the energy of the initial state. In a second step the doubly-excited ion stabilizes by an electronic transition in a lower lying level or in the ground state

$$X^{q+} + e^- \rightarrow X^{(q-1)+^{**}} \rightarrow \begin{cases} X^{(q-1)+^*} + \hbar\omega & \text{(a)} \\ \text{or} \\ X^{(q-1)+} + \hbar\omega & \text{(b)} \end{cases}.$$
$$(2.73)$$

The X-rays produced in the process (a) are the so-called *dielectronic satellites*.

The DE cross-section is calculated according to [62] as

$$\sigma_{DR}(E_e) = \frac{\pi \hbar^2}{p_e^2} \frac{g_d}{2 g_i} \frac{\Gamma_A(d \rightarrow i)\,\Gamma_r(d \rightarrow f)}{(E_e - E_r)^2 + \Gamma_d/4} \qquad (2.74)$$

with p_e the initial electron momentum, g_d, g_i the statistical weights for the states d and i, Γ_A the Auger width, Γ_r the radiative width, Γ_d the total width of the doubly-excited state and E_r the resonance energy. The identification of the states in (2.74) occurs according to

$$e^- + |i\rangle \rightarrow |d\rangle \rightarrow |f\rangle + \hbar\omega .$$

If the natural linewidth Γ_d is small in comparison to the experimental linewidth, for the theoretic line profile a delta-function can be assumed:

$$\sigma_{DR}(E_e) = \frac{2\pi^2\hbar^2}{p_e^2} \frac{g_d}{2 g_i} \frac{\Gamma_A(d \rightarrow i)\,\Gamma_r(d \rightarrow f)}{\Gamma_d} \times \delta(E_e - E_r) .$$
$$(2.75)$$

---

[23] Here in Sect. 4.3 Lisitsa discusses the dielectronic recombination as a resonance process of equivalent photons.

The dielectronic recombination is discussed in more detail by Lisitsa [311][23] and Beyer et al. [62], among others.

2.3 Fluorescence Yields

2.3.1 Implications for Basic Research and Practice

A detailed knowledge of nonradiative electron transition characteristics is important for the interpretation of many X-ray measurements in atomic physics, nuclear physics, solid-state physics and other areas of research. For instance we refer here, as an application in nuclear physics, to nuclear decay via electron capture where the transition energies as well as the multipolarities of the internal converted γ-transitions are determined by the measurement of relative X-ray ratios. Here X-ray fluorescence yields are needed for the determination of the initial vacancy distribution in the atom. In the same way the internal ionization or emission of electrons from β-decay investigated by the measurement of X-rays in coincidence with β-particles requires an exact knowledge of the properties of atomic transitions. Because the fluorescence yields of most atoms were determined from electron or photon ionization, atomic ionization cross-sections can be derived from the measurement of the total X-ray or from the total Auger electron production cross-sections. The ionization cross-section σ_i^I for a given state i can be characterized with the X-ray production cross-section σ_i^R or with the cross-section for the emission of Auger electrons σ_i^A in the following form:

$$\sigma_i I = \frac{\sigma_i^R}{\omega_i} = \frac{\sigma_i^A}{1 - \omega_i} . \tag{2.76}$$

To describe Auger cascades numerical values for fluorescence yields are of importance for a wide field in applied physics. In the course of cascade deexcitation an Auger transition starting from an inner-shell vacancy produces a twofold ionized atom and the successive filling of the vacancy transformed in a higher subshell can lead to multiple ionized atoms.

Beside investigations of basic phenomena exact fluorescence yields are also of importance for a wide field of applied physics. For instance such data are required in X-ray fluorescence analysis (for example for medical investigations, trace element analysis, *in situ* probe analysis in geological applications and others), Auger electron spectroscopy, low-energy electron diffraction (LEED) and in the description of photon transport processes.

2.3.2 Notations and Definitions[24]

The fluorescence yield of an atomic shell or subshell is defined as the probability that a vacancy in the considered shell will be filled by a radiative electron transition. According to (2.30) the total probability for the filling of an inner-shell vacancy includes contributions from Röntgen, Auger and Coster–Kronig transitions. The rates corresponding to the individual processes can be calculated as

$$\omega = \frac{\Gamma_R}{\Gamma} \quad \text{fluorescence yield} \tag{2.77}$$

$$a = \frac{\Gamma_A}{\Gamma} \quad \text{Auger yield} \tag{2.78}$$

$$f = \frac{\Gamma_{CK}}{\Gamma} \quad \text{Coster–Kronig yield.} \tag{2.79}$$

In detail Bambynek et al. [38] discussed the role of fluorescence yields very completely. Thus following [38] we give here in this subsection a summary of the most important relations.

The sum of the yields from radiative electron transitions ω and from nonradiative transitions a and f is normalized to

$$\omega + a + f = 1 . \tag{2.80}$$

For a probe containing many atoms the fluorescence yield of an individual subshell is equal to the number of emitted X-ray quanta during the filling of the vacancy divided by the number of initial vacancies.

From this definition the K-shell fluorescence yield is

$$\omega_K = \frac{I_K}{n_K} . \tag{2.81}$$

---

[24] Reprinted with permission from Bambynek et al. [38]. Copyright 1972 by the American Physical Society

Here I_K describes the total number of emitted K X-ray quanta and n_K the number of initial vacancies.

For higher shells the definition of fluorescence yields is much more complicated. The reasons are:

1. Main shells above the K-shell have more than one subshell and the electrons in these subshells can have different orbital angular momenta. That is why the mean fluorescence yield depends on the ionization mechanism of the considered shells because different ionization methods lead to different distributions of initial vacancies.
2. In the same shell Coster–Kronig transitions can transform initial vacancies from one to another subshell before the vacancy is filled by another electron transition.

These circumstances require us to check measured values very carefully with regard to their consistency with the introduced definitions. From the existence or absence of Coster–Kronig transitions two definitions for the *mean fluorescence yields* can be introduced:

Mean fluorescence yields in the absence of Coster–Kronig transitions. In most experiments for studying fluorescence yields of individual atomic shells vacancies are produced simultaneously in different subshells. If ω_{nlj} describes the subshell fluorescence yield of subshell nlj then according to (2.81) for a fixed n (i.e. into a main shell K, L, M, …):

$$\omega_{nlj} = \frac{I_{nlj}}{n_{nlj}} \, . \qquad (2.82)$$

Then the mean fluorescence yield of a given main shell has the form

$$\bar{\omega}_X = \sum_{i=1}^{k} N_i^X \, \omega_i^X \qquad (2.83)$$

where X describes the corresponding main shell and i the quantum numbers (nlj) of k individual subshells. Here the quantity N_i^X characterizes the relative number of initial vacancies in the subshell i of the X-th main shell:

$$N_i^X = \frac{n_i^X}{\sum n_i^X} \, ; \qquad \sum_{i=1}^{k} N_i^X = 1 \, , \qquad (2.84)$$

with summation over all k subshells of the main shell X.

If the total vacancy number in all subshells of the X-th shell is given by n_X then

$$n_X = \sum_{i=1}^{k} n_i^X \qquad (2.85)$$

and analogously to (2.81) the mean X-shell fluorescence yield $\bar{\omega}_X$ is written as

$$\bar{\omega}_X = \frac{I_X}{n_X} \, . \qquad (2.86)$$

Here I_X is the number of emitted X-ray quanta from the shell X.

The definition of $\bar{\omega}_X$ is applicable in accordance with (2.83) and the initial vacancy distribution must be unchanged until the vacancies are filled by electron transitions from higher main shells, i.e. Coster–Kronig transitions should not arise.

Then it follows that the mean fluorescence yield for a main shell X in any one experiment depends on the initial vacancy distribution. This means that two experiments can give different values in $\bar{\omega}_X$ if the chosen ionization methods produce different vacancy distributions. Thus $\bar{\omega}_X$ is not a fundamental constant for any atom but depends on the atomic subshell yields ω_i^X and on the parameters N_i^X of the individual experiment.

To determine the atomic quantities ω_i^X for all k subshells of the main shell X we need k experiments leading to different known initial vacancy distributions. These experiments result in a set of mean fluorescence yields $(\bar{\omega}_X)_1 \ldots (\bar{\omega}_X)_k$

$$(\bar{\omega}_X)_1 = \sum_{i=1}^{k} \left(N_i^X \right)_1 \, \omega_i^X$$

$$(\bar{\omega}_X)_2 = \sum_{i=1}^{k} \left(N_i^X \right)_2 \, \omega_i^X$$

$$\vdots \quad \vdots \qquad (2.87)$$

$$(\bar{\omega}_X)_k = \sum_{i=1}^{k} \left(N_i^X \right)_k \, \omega_i^X .$$

Then the given set of k equations is solved for the k subshell fluorescence yields ω_i^X.

Mean fluorescence yield in the presence of Coster–Kronig transitions. Considering Coster–Kronig transitions, two alternative assumptions can be made:

1. The mean fluorescence yield $\bar{\omega}_X$ can be assumed to be a linear combination of the subshell fluorescence yields ω_i^X with a vacancy distribution v_i^X which is changed by Coster–Kronig transitions. This method results in equations containing the subshell fluorescence yield ω_i^X for the initial situation and which correspond to the real experimental situation.
2. Mathematically the mean fluorescence yields $\bar{\omega}_X$ can be represented as linear combinations of the initial vacancy distribution N_i^X with a set of coefficients v_i^X. In this way the selection of the coefficients v_i^X must consider the possible Coster–Kronig transitions.

In the literature both approximations were used. This is the reason that mistakes in the distinction between data sets can lead to misunderstandings in the data interpretation. Therefore following [38] we formulate, for each approximation, equations and transformation equations valid between both alternative approximations.

A. Description in terms of a changed vacancy distribution. The mean fluorescence yield of the main shell X is described as a linear combination of the subshell fluorescence yields ω_i^X:

$$\bar{\omega}_X = \sum_{i=1}^{k} v_i^X \, \omega_i^X . \qquad (2.88)$$

In contrast to the initial vacancy distribution from (2.83) the coefficients v_i^X here describe the relative number of vacancies in the subshells X_i including the vacancies transformed to individual subshells by Coster–Kronig transitions. For the quantities v_i^X this yields

$$\sum_{i=1}^{k} v_i^X > 1 . \qquad (2.89)$$

We note that we have $v_i^X > 1$. The reason is that some vacancies produced in subshells below X_i are accounted for several times because they shift to higher shells by Coster–Kronig transitions. We declare the Coster–Kronig transition probability for the transformation of a vacancy from a subshell X_i to a higher subshell X_j with f_{ij}^X. Then the quantities v_i^X can be expressed in terms of the relative number N_i^X of initial vacancies:

$$
\begin{aligned}
v_1^X &= N_1^X \\
v_2^X &= N_2^X + f_{12}^X N_1^X \\
v_3^X &= N_3^X + f_{23}^X N_2^X + (f_{13}^X + f_{12}^X f_{23}^X) N_1^X
\end{aligned}
\qquad (2.90)
$$

$$\vdots \quad \vdots$$

$$v_k^X = N_k^X + f_{k-1,k}^X N_{k-1}^X + (f_{k-2,k-1}^X f_{k-1,k}^X) N_{k-2}^X \qquad (2.91)$$

$$+ \ldots + (f_{1k}^X + f_{12}^X f_{2k}^X + f_{12}^X f_{23}^X f_{3k}^X + \ldots) N_1^X \qquad (2.92)$$

B. Description in terms of the initial vacancy distribution N_i^X. In this approximation the mean fluorescence yield is written as

$$\bar{\omega}_X = \sum N_i^X \, v_i^X . \qquad (2.93)$$

The definition of the coefficients v_i^X is stated in such a way that their consistency with (2.90) is ensured. The choice of the coefficients v_i^X guarantees that they characterize the total number of X-rays produced after creating an initial vacancy in the subshell X_i. This definition is different from the definition of the subshell fluorescence yields ω_i^X. Here we note that the products $v_i^X \omega_i^X$ and $n_i^X v_i^X$ are not identical. Only the sum of this products as given in (2.88) and (2.93) are the same for the mean fluorescence yield ω_i^X. The physical definition of $v_i^X \omega_i^X$ contains the number of radiative electron transitions from higher shells to the i-th subshell per created vacancy. The quantities $N_i^X v_i^X$ correspond to the number of X-ray quanta emitted in transitions to all subshells of the X-th main shell per vacancy in the i-th subshell.

C. Transformation equations. The transformation equations between the coefficients v_i^X and the subshell fluorescence yields ω_i^X follow from (2.88), (2.90) and (2.93):

$$v_1^x = \omega_1^X + f_{12}^X \omega_2^X + (f_{13}^X + f_{12}^X f_{23}^X)\, \omega_3^X + \dots \quad (2.94)$$

$$(f_{1k}^X + f_{12}^X f_{2k}^X + f_{13}^X f_{3k}^X + \dots + f_{1,k-1}^X f_{k-1,k}^X$$

$$+ \text{ products of } 3, 4, \dots, (k-1)\, f_{ij}^X, \text{ ordered for}$$

the transition of the vacancy

from the subshell 1 to the subshell k) $\times\; \omega_k^X$

$$(2.95)$$

$$\vdots \quad \vdots$$

$$v_{k-1}^X = \omega_{k-1}^X + f_{k-1}^X\, \omega_k^X$$

$$v_k^X = \omega_k^X.$$

We demonstrate the application of these equations in detail for the quantities N_i^X, v_i^X and $v_i^X \omega_i^X$ for the L and M shells. For these shells initial and final vacancy distributions are connected as following:

| L-shell |

$$v_1^L = N_1^L$$

$$v_2^L = N_2^L + f_{12}^L N_1^L \qquad (2.96)$$

$$v_3^L = N_3^L + f_{23}^L N_2^L + (f_{13}^L + f_{12}^L f_{23}^L)\, N_1^L$$

| M-shell |

$$v_1^M = N_1^M$$

$$v_2^M = N_2^M + f_{12}^M N_1^M$$

$$v_3^M = N_3^M + f_{23}^M N_2^M + (f_{13}^M + f_{12}^M f_{23}^M)\, N_1^M$$

$$v_4^M = N_4^M + f_{34}^M N_3^M + (f_{24}^M + f_{23}^M f_{34}^M)\, N_2^M \qquad (2.97)$$

$$+ (f_{14}^M + f_{13}^M f_{34}^M + f_{12}^M f_{24}^M + f_{12}^M f_{23}^M f_{34}^M)\, N_1^M$$

$$v_5^M = N_5^M + f_{45}^M N_4^M + (f_{35}^M + f_{34}^M f_{45}^M)\, N_3^M$$

$$+ (f_{25}^M + f_{24}^M f_{45}^M + f_{23}^M f_{35}^M + f_{23}^M f_{34}^M f_{45}^M)\, N_2^M$$

$$+ (f_{15}^M + f_{14}^M f_{45}^M + f_{13}^M f_{35}^M + f_{12}^M f_{25}^M + f_{13}^M f_{34}^M f_{45}^M$$

$$f_{12}^M f_{24}^M f_{45}^M + f_{12}^M f_{23}^M f_{35}^M + f_{12}^M f_{23}^M f_{34}^M f_{45}^M)\, N_1^M.$$

The coefficients v_i^X and the subshell fluorescence yields ω_i^X are combined in the following way in the L shell:

$$v_1^L = \omega_1^L + f_{12}^L \omega_2^L + (f_{13}^L + f_{12}^L f_{23}^L + f_{13}^{L'})\, \omega_3^L$$

$$v_2^L = \omega_2^L + f_{23}^L \omega_3^L \qquad (2.98)$$

$$v_3^L = \omega_3^L.$$

The quantity $f_{13}^{L'}$ describes the radiative intrashell rate for the transition $L_1 \rightarrow L_3$. In the literature often $f_{13}^{L'}$ is not considered because it yields $f_{13}^L \gg f_{13}^{L'}$ (for example according to [300] we have $f_{13}^L : f_{13}^{L'}$ for krypton $0.52 : 4.1 \times 10^{-5}$ and for uranium $0.57 : 0.0097$).

Neglecting the intrashell rates for radiative electron transitions for the M-shell we have:

$$v_1^M = \omega_1^M + f_{12}^M \omega_2^M$$

$$+ (f_{13}^M + f_{12}^M f_{23}^M)\, \omega_3^M + (f_{14}^M + f_{13}^M f_{34}^M$$

$$+ f_{12}^M f_{24}^M + f_{12}^M f_{23}^M f_{34}^M)\, \omega_4^M + (f_{15}^M + f_{14}^M f_{45}^M$$

$$+ f_{13}^M f_{35}^M + f_{12}^M f_{25}^M + f_{13}^M f_{34}^M f_{45}^M + f_{12}^M f_{24}^M f_{45}^M$$

$$+ f_{12}^M f_{23}^M f_{35}^M + f_{12}^M f_{23}^M f_{34}^M f_{45}^M)\, \omega_5^M \qquad (2.99)$$

$$v_2^M = \omega_2^M + f_{23}^M \omega_3^M + (f_{24}^M + f_{23}^M f_{34}^M)\, \omega_4^M$$

$$+ (f_{25}^M + f_{24}^M f_{45}^M + f_{23}^M f_{35}^M + f_{23}^M f_{34}^M f_{45}^M)\, \omega_5^M$$

$$v_3^M = \omega_3^M + f_{34}^M \omega_4^M + (f_{35}^M + f_{34}^M f_{45}^M)\, \omega_5^M$$

$$v_4^M = \omega_4^M + f_{45}^M \omega_5^M$$

$$v_5^M = \omega_5^M.$$

From the cited equations it follows that the determination of individual subshell fluorescence yields ω_i^X from experiment can be very complicated. Additionally, with the measurement of the mean main shell fluorescence yield $\bar{\omega}_X$ for a sufficient number of different initial vacancy distributions, it is also necessary to know the corresponding Coster–Kronig transition probabilities. In some cases the situation is simplified by the circumstance that the fluorescence yields for the subshells with the most tightly bound electrons (for example ω_3^L and ω_5^M) can be determined directly and that in some regions of the periodic table of the elements some Coster–Kronig transition probabilities vanish. In this case the above equations were simplified.

Table 2.23. Regions of atomic numbers where strong Coster–Kronig transitions are energetically possible (after [38])

transition	Z-region	transition	Z-region
L_1-$L_2O(P,...)$	all Z, where O-,...	L_1 - L_2M_4	$21 \leq Z \leq 40$
	levels are occupied	L_1 - L_2M_5	$26 \leq Z \leq 41$
L_1 - L_2N_1	$19 \leq Z \leq 70$	L_1 - $L_3N\,(O, P,...)$	all Z where N-,...
L_1 - L_2N_2	$31 \leq Z \leq 76$		levels are occupied
L_1 - L_2N_3	$33 \leq Z \leq 81$	L_1 - L_3M_1	$11 \leq Z \leq 31$
L_1 - L_2N_4	$39 \leq Z \leq 91$	L_1 - L_3M_2	$13 \leq Z \leq 35$
L_1 - L_2N_5	$Z \geq 42$	L_1 - L_3M_3	$15 \leq Z \leq 36$
L_1 - L_2N_6	$Z \geq 58$	L_1 - L_3M_4	$21 \leq Z \leq 49, Z \geq 77$
L_1 - L_2N_7	$Z \geq 63$	L_1 - L_3M_5	$26 \leq Z \leq 50, Z \geq 74$
L_1 - L_2M_1	$11 \leq Z \leq 29$	L_2 - $L_3N(O,...)$	all Z where N-,...
L_1 - L_2M_2	$13 \leq Z \leq 32$		levels are occupied
L_1 - L_2M_3	$15 \leq Z \leq 33$	L_2 - L_3M_4	$21 \leq Z \leq 30$
		L_2 - L_3M_5	$26 \leq Z \leq 30, Z \geq 91$

To explain this situation in Table 2.23 the appearance of Coster–Kronig transitions in different Z-regions is given. The Auger yield a_i^X introduced in (2.80) represents the probability that a vacancy in the i-th subshell is filled by a nonradiative electron transition from a higher main shell. Here one must consider that Coster–Kronig transitions are not immanent in the definition of the Auger yield. The Coster–Kronig yield f_{ij}^X is the probability that a vacancy in the subshell X_i is filled by an electron transition from the subshell X_j of the same main shell X. The emitted electron stems here from the same or a higher main shell.

With the given definition between the fluorescence yield, the Auger yield and the Coster–Kronig yield follows:

$$\omega_i^X + a_i^X + \sum_{j=i+1}^{k} f_{ij}^X = 1 . \qquad (2.100)$$

Analogously to the definition of the mean fluorescence yield (2.88) the mean Auger yield is described as

$$a^X = \sum_{i=1}^{k} v_i^X a_i^X , \qquad (2.101)$$

with v_i^X as the modified relative vacancy numbers (2.90). The sum of the mean fluorescence yield and the mean Auger yield for a given shell is, for the same initial vacancy distribution, equal to one:

$$\bar{a}_X + \bar{\omega}_X = 1 . \qquad (2.102)$$

An overview of the dependence of the K and L fluorescence yields, the Auger and Coster–Kronig yields from the atomic number Z is shown in Figs. 2.40–2.43.

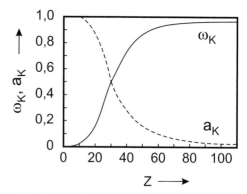

Fig. 2.40. K-shell fluorescence yield ω_K and K-shell Auger yield a_K as a function of the atomic number Z

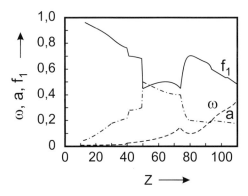

Fig. 2.41. Fluorescence yield ω, Auger yield a and Coster–Kronig yield f_1 for the L_1 subshell as a function of the atomic number Z

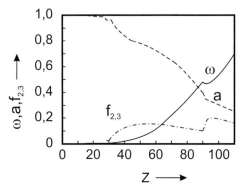

Fig. 2.42. Fluorescence yield ω, Auger yield a and Coster–Kronig yield f_{23} for the L_2 subshell as a function of the atomic number Z

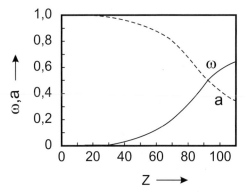

Fig. 2.43. Fluorescence yield ω and Auger yield a for the L_3 subshell as a function of the atomic number Z

2.3.3 Natural Level Widths and Fluorescence Yields

The energetic natural level width Γ_i of a subshell i is calculated from the emission rate of radiative electron transitions (X-ray emission rate) Γ_i^R and from the fluorescence yield ω_i of the subshell where the initial vacancy is localized

$$\Gamma_i = \frac{\Gamma_i^R}{\omega_i} \, . \qquad (2.103)$$

Consequently the natural width for an X-ray transition results from the sum of the level widths of the orbitals involved in the X-ray transition.

For example for the energetic width of the K_{α_1} transition we have

$$\Gamma(K_{\alpha_1}) = \Gamma(K) + \Gamma(L_3) \, .$$

2.4 Ionization

Inner-shell ionization processes can happen as a result of different excitation mechanisms. For X-ray physics electron impact ionization and photoionization are of outstanding importance. Nevertheless processes such as ionization during proton or heavy ion impact are also of importance for different applications. Thus in the following we will give a short overview of these processes.

2.4.1 Electron Impact ionization

Electron impact ionization is the basic process for producing characteristic X-rays in X-ray tubes, is important for electron beam microprobes and is a fundamental process in all plasma types, in astrophysics and in laboratory plasmas. A lot of plasma properties, as for instance the emitted radiation, are dependent on the ionization of the plasma. Here the ions are also produced by electron impact ionization. Furthermore, for the interpretation of spectroscopic observations and for the modelling of X-ray production processes X-ray production cross-sections must be available.

Electron impact ionization can occur in direct and in indirect processes. *Direct ionization* is connected with the emission of an electron from an inner- or outer-shell as a result of direct electron collisions. *Indirect ionization* takes place as a result of an Auger decay of an intermediate autoionizing state. The balance between direct and indirect processes essentially depends on the energy of the incident electron and on the atomic structure of the target.

Direct ionization. Direct ionization occurs if the energy of the incident electron is greater than the ionization potential of the target atom or the considered substate. Then the energy of the emitted electron is equal to the difference between the kinetic energy of the incident electron and that of the binding energy of the ionized electron.

Although the ionization of atoms by electron impact was first investigated experimentally in 1894 by Lenard [321] and in 1912 by Bloch [71] and theoretically in 1912 by Thompson [538], it was Dolden [151] who determined the first reliable electron impact ionization cross-sections for He^{1+} with the crossed-beam method. For the study of ionization processes different experimental methods were developed, as for instance crossed-beam experiments, spectroscopy on stored ions (electron beam ion traps, electron-ion storage rings) and plasma spectroscopy.

As result of the application of these methods a sufficiently satisfying data base for ionization cross-sections of neutral and low-charged ions (see for example [536][25]) exists. Precise data for higher than twofold ionized ions have been available since 1978, but most data are known here only for comparatively low ion charges states.

The theoretical description of electron impact ionization processes is connected with different difficulties. Besides the fact that a multibody problem is given here, many ionization mechanisms are possible. The most important ionization mechanisms are [475]:

$$X^{n+} + e^- \rightarrow X^{(n+1)} + 2\,e^- \tag{2.104}$$

$$X^{n+} + e^- \rightarrow X^{(n+m)+} + (m+1)\,e^- \tag{2.105}$$

$$X^{n+} + e^- \rightarrow X^{*n+} + e^- \tag{2.106}$$
$$\quad\quad \hookrightarrow \quad A^{(n+1)+} + e^-$$

$$A^{n+} + e^- \rightarrow A^{*(n+1)+} + 2\,e^- \tag{2.107}$$
$$\quad\quad \hookrightarrow \quad A^{(n+m)+} + (m-1)\,e^-.$$

Equations (2.104) and (2.105) describe direct impact ionization, where the energy of the incident electron is transferred to one or more electrons and thereby exceeds, for each individual bound electron, its binding energy. The excitation of an autoionizing state is characterized by (2.106). Equation (2.107) stands for another two-step process, an Auger transition after inner-shell ionization, where first an inner-shell electron is ionized and then some electrons are emitted.

Usually it is stated that in electron–ion collisions the process (2.104) occurs with the highest probability. But the experimental experience of the last few years shows that also the process (2.106) give an important contribution and in some cases can be the dominant process. An analogous situation exists for multi-ionization processes described by (2.105) and (2.107).

Because of the fundamental importance of precise electron impact ionization cross-sections, much work has been done in the theoretical interpretation of this process. A complete resolution of the long-range three-body problem with two continuum electrons in the final state is impossible and the precision of approximations based on classical or semiclassical descriptions or on quantum mechanics is often problematic.

Thompson [538] described electron impact ionization processes for neutral atoms in the framework of the classical theory. Assuming that the energy transfer occurs by Coulomb interaction from the incident electron of energy E to an atomic electron at rest, the total ionization cross-section has the form

$$\sigma(E) = 4\,\pi\,a_o^2\,E_H \sum_j \frac{n_j}{I_j\,E}\left(1 - \frac{I_j}{E}\right) \tag{2.108}$$

$$= 6.5 \times 10^{-14} \sum_j \frac{n_j}{I_j\,E}\left(1 - \frac{I_j}{E}\right) \text{cm}^2$$

---

[25] Experimental values for electron impact ionization for atoms and ions from hydrogen up to uranium

with $a_0 = 0.529 \times 10^{-10}$ m as the Bohr radius, $E_H = 13.6$ eV the ionization energy of the hydrogen atom and n_j the number of electrons in the subshell j bounded with the energy I_j (in eV).

Besides many modifications of the Thomson formula (2.109) in the framework of the classical two-center theory, the formula given by Gryzinski [216] is often applied successfully:

$$\sigma_j I_j^2 = 6.51 \times 10^{-14} \, n_j \, g(U_j) \; [\text{cm}^2 \, \text{eV}^2] \qquad (2.109)$$

with

$$g(U_j) = \frac{1}{U_j} \left(\frac{U_j - 1}{U_j + 1} \right)^{3/2}$$
$$\left\{ 1 + \frac{2}{3} \left(1 - \frac{1}{2\,U_j} \right) \ln \left[2.7 + (U_j - 1)^{1/2} \right] \right\}$$

and

$$U_j = \frac{E}{I_j} \, .$$

The most important defect of Thomson's theory is the incorrect high-energy behavior of (2.109). Bethe [58] shows that the ionization cross-section at high energies must behave as

$$\sigma_{U_j} \sim A \, \frac{\ln U_j}{U_j} + \frac{B}{U_j} \, , \qquad (2.110)$$

leading to a corresponding modification of the theory. Thus the limitation that the atomic electron during the collision must be at rest was removed. Many formulations with different initial velocities of the electron are known (for instance see the early works of Gryzinski [216], Kingston [278], Friens [186] and Bell [55]). A little later quantum properties as electron exchange and interference were taken into account by Burges [89] and Kumar [303]).

Another classical approximation is based on the impact parameter method [7,89,490] which considers many-body effects and which reproduces the correct high-energy behavior of the cross-sections (see 2.110). The advantages of both methods are summarized in the "classical exchange impact parameter method" [89] which reproduces the correct ionization cross-section behavior for low electron energies as well as for high energy ones.

All classical and semiclassical approximations describe direct Coulomb ionization only. Autoionization processes after inner-shell ionization or excitation processes must be considered here by independent calculations. Here for example Salop [474] adds the approximated ionization cross-sections for the individual processes in a simple way to the total ionization cross-section.

The first quantum mechanical calculation of ionization cross-sections was done by Bethe [58] in the Born approximation. The most common quantum mechanical calculation method for electron impact ionization is the Coulomb–Born approximation where the incident electron as well as the emitted electron are described with Coulomb wavefunctions. Different versions of this approximation vary in the treatment or neglect of exchange effects. Modern methods for the quantum mechanical calculation of ionization cross-sections for instance are known from Younger [581] (consideration of screening effects) and from Jakubowicz and Moores [261] and Moores [368] (consideration of direct and indirect contributions). Overviews of the many calculation methods are given by Powell [434], Crandall [127] and Salzborn [475]. Because of the high complexity of derivative calculations and considering the fact that it is often sufficient to know simple approximated cross-sections different semiempirical formulas were developed. Especially successful is here the formula derived from Lotz [328]:

$$\sigma(E) = \sum_{j=1}^{N} a_j \, n_j \, \frac{\ln(E/I_j)}{E \, I_j} \left\{ 1 - b_j \exp \left(-c_j \, \frac{E}{I_j - 1} \right) \right\} \, .$$
$$(2.111)$$

The coefficients a_j, b_j and c_j are tabulated constants derived from the approximation of experimental data. The estimation error is here in the region from +40 to −30%. The indicated error region is characteristic of most calculations of ionization cross-sections. Only in some special cases precisions of some percentage can be derived. In the case of electron impact ionization the typical precision of experimental results also percentage-wise.

Details of the treatment of electron impact ionization are discussed, for instance, by Maerk and Dunn [333].

Indirect ionization. In many-electron atoms or ions indirect processes can lead to important contributions to the ionization cross-section. Because it is not the aim of the present book we will characterize these processes only briefly:

- *Excitation-Autoionization (EA).* For this process after an electronic excitation an autoionization process follows:

$$X^{q+} + e^- \rightarrow [X]^{**} + e^- \qquad (2.112)$$
$$\rightarrow X^{(q+1)+} + 2\,e^-.$$

The decay of the twofold excited state can, according to (2.112), result in the emission of an additional electron and in this way in an increase of the ionization cross-section (up to a factor two or more). Alternatively the doubly excited state can undergo radiative stabilization

$$X^{**} \rightarrow X^{q+} + \hbar\,\omega\,. \qquad (2.113)$$

The balance between additional electron emission and radiative stabilization is characterized by the *branching ratio.*

- *Resonant ionization (RI) via electron capture.* RI processes are characteristic only for certain resonant energies of the incident electron and produce resonance structures in the ionization cross-section. Processes contributing up to some 10% of the total ionization cross-section are:
  - Resonant-Excitation-Auto-Double Ionization (READI) and
  - Resonant Excitation-Double Autoionization (REDA).

2.4.2 Ionization in Ion–Atom and Ion–Ion Collisions

Ionization processes

$$Z_1^{q_1+} + Z_2^{q_2+} \rightarrow Z_1^{q_1+} + Z_2^{(q_2+1)+} + e^- \qquad (2.114)$$

are discussed in ion–atom collisions for instance in [363] and in [85]. The range of validity of corresponding approximations is investigated in detail by Aberg [2].

For $Z_1 \cong Z_2$ (Z_1 is the atomic number of the projectile; Z_2 the atomic number of the target atom) in comparison to the electron orbital velocity v_e and low projectile velocities v_p for the description of inner-shell ionization processes the *quasimolecular electron promotion model* is used. The basic idea of this model is that an inner-shell electron as a result of an ion–atom collision of overlapping molecular orbitals changes to a higher state. This process is only valid for projectile–target combinations where the orbital binding energies are approximately equal or overlapping. For slow collisions ($v_e \gg v_p$) the reaction cross-section from the MO model is some orders of magnitude greater than the contribution from direct Coulomb processes. The contribution of the direct Coulomb process derives its maximum at $v_p \approx v_e$. The most important approximation methods for Coulomb ionization processes are:

- the plane wave Born approximation;
- the semiclassical approximation; and
- the classical two-center approximation.

The order of magnitude of ionization cross-sections for ionization by heavy ion bombardment is 10^4–10^7 barn and varies with the atomic number and with the energy of the projectile. For protons and α-particles the ionization cross-section decreases depending on the incident energy to orders of magnitude from 10^1 barn to 10^4 barn.

2.4.3 Multiple Ionization Processes

Multiple ionization processes occur as a result of collision processes of energetic particles with atoms or ions or after reorganization processes in the atomic shell after the filling of an initial inner-shell vacancy. During a collision the direct production of a multiple vacancy state as well as the formation of an initial vacancy as starting point for nonradiative reorganization processes such as Coster–Kronig transitions and Auger cascades are probable. The occurrence of multiple vacancy states is of importance for precise element analysis methods, for X-ray diagnostics of plasma states and for the interpretation of different classes of interaction processes between ionizing radiation and matter. The proof for the existence of multiple vacancy states was given by Krause [299].

If vacancies in corresponding electron states arise, groups of X-ray satellite lines are observed where each transition line corresponds to a certain number of vacancies. Analogously to X-ray satellites, Auger electron satellites were observed on the low-energy side of the parent diagram lines. In comparison to X-ray satellites these satellites have a more complex structure. The reason is that Auger satellites have a higher possible number of final states. This is especially pronounced if the transition goes in an outer shell. Because of the after the ionization following nonradiative reorganization processes the ionization degree of the atom can increase significantly. Because of the small vacancy lifetime for the development of multiple vacancy configurations after photon or electron impact only the indirect multiple ionization by nonradiative electron transitions and the direct multiple ionization (shake-off) are probable.

Of special importance are Coster–Kronig processes of the type $X_i \rightarrow X_j Y$ where the vacancy states X_i and X_j belong to the same main shell.

The basic ideas of the theory and numerical values for selected elements are among others given by Carlson et al. [96,97,100] and by Aberg [1]. Shake-off processes are characterized by the assumption that the excitation of an electron into a bound state or in the continuum is connected with a sudden change in the atomic potential. The result is that another weakly bound electron is suddenly localized above the ionization border.

Such a situation is seen if another electron is rejected from the atomic shell in a time short in comparison to the orbital period of the excited electron. Then the excited electron performs a monopole transition ($\Delta l = 0$) while for the other electrons result in a change in the orbital angular momentum $\Delta l = 1$. This selection rule holds for photon ionization cross-sections in the dipole approximation. Additionally, for electron impact ionization processes multipole transitions are also possible.

The most tightly bound electron has the maximum likelihood for the ionization of a second electron. For a certain subshell the ionization probability decreases with Z_{eff}^2 (Z_{eff} is the effective nuclear charge). This situation is shown in Fig. 2.44 for a change of the nuclear charge $\Delta Z = 1$ corresponding

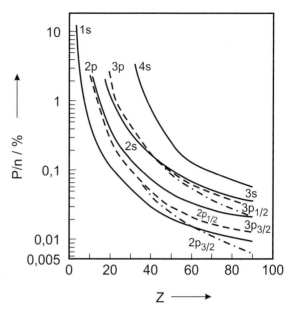

Fig. 2.44. Calculated probabilities P (in %) for electron shake-off processes in s- and p-shells divided by the number n of electrons in the observed subshell as a function of the atomic number Z (after [97])

approximately to a change in Z_{eff} for K-shell ionization.

To estimate the intensity of shake-off processes the following rules can be formulated in accordance with Fig. 2.44:

1. For a given subshell shake-off processes decrease with increasing Z.
2. For a given atom and a certain angular orbital momentum the intensity increases with increasing main quantum number.
3. For a given main quantum number the contribution of shake-off processes per electron for low Z is higher than for high orbital angular momenta; for high Z an inverse tendency is characteristic.
4. The total contribution of shake-off processes summed over all subshells is approximately independent of Z.

Up to now extensive experimental information about the appearance of K and L vacancies for elements up to $Z = 32$ has accumulated. Comparatively less systematic work exists for XY inner-shell vacancy states where the vacancies were localized others than in the K and L shells.

For the interpretation of X-ray and Auger satellite spectra some general properties for the production and the decay of XY vacancies can be summarized:

1. Ionization due to shake-off processes has the lowest probability for the most tightly bound electrons.
2. Coster–Kronig transitions are important for preferentially strong bound electrons.
3. Shake-off processes can be reduced by a proper selection of the energy of the exciting photons or electrons if their energy is only less different from the binding energy of the X-shell electrons.
4. In the case of photoionization Coster–Kronig processes can be eliminated by a suitable selection of the excitation energy.

An inner-shell vacancy can be filled up either by a radiative electron transition or by an Auger transition. The vacancies created during these processes can be filled again by further electron transitions. This process continues up to the moment when all vacancies reach the outermost shell or the ion ground state. With the exception of the K and L shells Auger processes are more probably than radiative electron transitions. Because in each Auger process an electron is emitted, a series of such processes leads to a vacancy cascade, i.e. to highly ionized atoms.

In Fig. 2.45 the relative ratios of ions are shown produced by vacancy multiplication processes (cascades) after creation of a vacancy in a certain subshell.

Starting from the situation as shown in Fig. 2.45 the following predictions can be made. For a given charge spectrum as a rule there exists a peaked distribution because the intensities of the different ion charge states for heavy ions are approximately distributed symmetrically. In the K-shell spectra of argon and xenon we found small saddle-points at low ionization charge states referring to competing vacancy cascades. The heavier the atom and lower lying the shell in which the initial vacancy is produced the more complex the ion charge spectrum becomes. Exceptions to this rule are observed in the K-shell of neon and krypton. Here we found a smaller mean ion

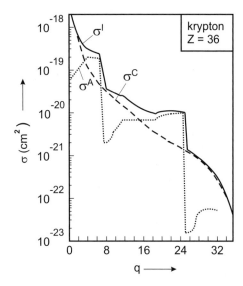

Fig. 2.46. Ionization cross-sections for the ionization of krypton as a function of the ion charge state q by electrons of an energy of 20 MeV (after [474]). σ^I is the total ionization cross-section, σ^C is the continuum ionization cross-section, and σ^A is the Auger ionization cross-section

charge state than in the L_1 subshell although the K-shell charge spectra show contributions from higher charged ions. This situation is caused by radiative electron transitions if the initial vacancy is transformed to a higher shell without further ionization processes.

Ionization cross-sections for all ionization stages of carbon, nitrogen, oxygen, neon and argon were determined experimentally by Donets and Ovsyannikov [152,153] with the electron beam method. Selected values from these experiments are given in Table 2.24.

In Fig. 2.46 the total ionization cross-section σ_I for krypton and its ions is given for the electron impact ionization process at electron energies of 20 MeV for all ion ground states [474]. Here the ionization cross-section is the sum of contributions from continuum ionization cross-sections σ_C and of multiple ionization cross-sections σ_A with consideration of nonradiative electron transitions.

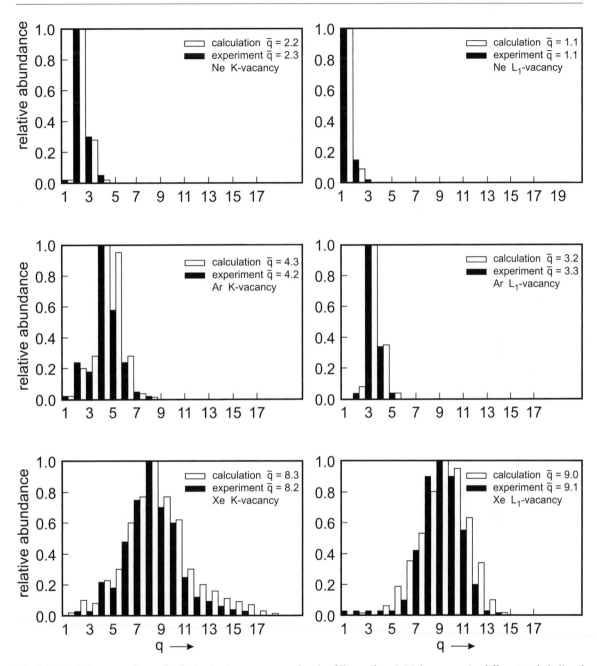

Fig. 2.45. Relative rates for multiple ionization processes for the filling of an initial vacancy in different subshells of selected atoms. The experimental values are from Carlson [96]

$q \rightarrow q+1$	carbon 4.85 keV	nitrogen 5.45 keV	oxygen 5.65 keV	neon 8.30 keV	argon 11 keV
$1 \rightarrow 2$	560 ± 110	670 ± 90	780 ± 100		
$2 \rightarrow 3$	162 ± 12	255 ± 45	330 ± 40	315 ± 30	
$3 \rightarrow 4$	78 ± 8	100 ± 11	145 ± 15	180 ± 20	188 ± 38
$4 \rightarrow 5$	$12{,}2 \pm 0.8$	43 ± 3	60 ± 7	88 ± 20	95 ± 25
$5 \rightarrow 6$	4.5 ± 0.4	7.5 ± 0.5	30 ± 4	44 ± 4.5	74 ± 15
$6 \rightarrow 7$		4.8 ± 0.4	43 ± 0.4	20.5 ± 2.4	23 ± 7
$7 \rightarrow 8$			1.7 ± 0.1	8.8 ± 0.6	39 ± 9
$8 \rightarrow 9$				1.8 ± 0.2	25 ± 5
$9 \rightarrow 10$				0.86 ± 0.12	17 ± 2.4

Table 2.24. Ionization cross-sections for the ionization of ions of the charge state q (in 10^{-20} cm^2) by electron impact ionization at different excitation energies E_e as given in the table head as parameter (after [152])

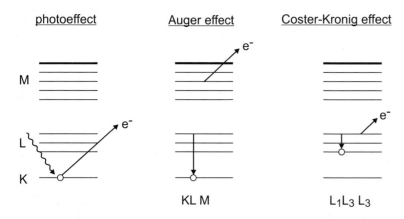

Fig. 2.47. Electron emission following photoionization, Auger and Coster–Kronig transitions

2.5 X-Ray Spectra and the Auger-Effect

2.5.1 Auger Electrons and Inner-Shell Ionization

Meitner [357] and Robinson [455] found that atoms after inner-shell ionization emit monoenergetic electrons with energies independent of the ionization method. For these electrons Auger [26] [22–26] found the following coherences:

1. Under quantum irradiation the photoelectrons and accompanying Auger electrons were emitted by the same atom.
2. Auger electrons were emitted in arbitrary directions independent of the emission direction of the photoelectrons.

3. The energy of Auger electrons does not depend on the energy of the ionizing photon radiation.
4. The energy of Auger electrons increases with the atomic number Z of the target atoms.
5. Not every photoelectron emission is accompanied by Auger electron emission.

If for example a K-shell electron is ionized then two processes dominate:

1. Radiative electron transitions where an electron from a higher shell (for instance from the L-shell) fills up the K-shell vacancy under X-ray emission.
2. Nonradiative electron transitions where instead of X-ray emission an additional electron (for in-

---

[26] The by Auger during to studies with a Wilson chamber observed slow monoenergetic electrons are after the discoverer called as Auger electrons.

stance from the L or M shell) is ionized. This situation is shown in Fig. 2.47.

The notation for Auger transitions is

$$X\text{-}YZ$$

(sometimes also written as X YZ) and means that the initial vacancy is localized in the X shell and the final vacancies are in the Y and Z shells. Here a nonradiative transition occurs from the X to the Y shell and an Auger electron is emitted from the Z shell. The Auger effect has important consequences for X ray spectra [4]:

- In deexcitation processes the Auger process is in concurrence with radiative electron transitions, i.e. the influence on the energetic widths of X-ray emission lines.
- In the Auger process vacancies are transformed from one shell to another. This influences the intensities of X-ray emission lines.
- The Auger process leads to double-vacancy states in the atom and is a reason for the appearance of X-ray satellite lines characteristic of atoms with inner-shell double vacancies.

For instance, overviews of the Auger effect were published by Bambynek et al. [38] and by Burhop and Asaad [91].

2.5.2 Auger Effect and Energetic Width of X-Ray Emission Lines and Absorption Edges

According to (2.28) in the frequency range between ν and $\nu + d\nu$ the irradiated energy $E_{if}(\nu)$ for a transition $i \to f$ of a photon of frequency

$$\nu_{fi} = \frac{E_i - E_f}{\hbar}$$

is given as [4]

$$E_{if}(\nu) = \frac{1}{(2\pi)^2} \; \frac{(\Gamma_i + \Gamma_f)\,\nu}{(\nu_{fi} - \nu)^2 + \left(\Gamma_i + \dfrac{\Gamma_i + \Gamma_f}{2\hbar}\right)^2} \; . \tag{2.115}$$

Here Γ_i and Γ_f are the sums over the widths of the initial and final states.

If p_i^R and P_i^A are the probabilities that an atom is deexcited by radiative or nonradiative (Auger) electron transitions then the mean lifetime of the state i is

$$\tau_i = \frac{1}{p_i^R + p_i^A} \; . \tag{2.116}$$

The Heisenberg uncertainty principle gives, for the width of the initial state,

$$\Gamma_i = \frac{\hbar}{\tau_i} = \hbar\,(p_i^R + p_i^A) = \Gamma_i^R + \Gamma_i^A \; , \tag{2.117}$$

with $\Gamma_i^R = \hbar\,p_i^R$ as the width of radiative transitions and $\Gamma_i^A = \hbar\,p_i^A$ as the width for Auger transitions. Thus the total width of the initial state consists of two partial widths. This scheme also described absorption edges.

2.5.3 Auger Effect and Intensities of X-Ray Emission Lines

The connection between the Auger effect and intensities of X-ray emission lines is described in [4]. If n_i is the number of atoms ionized in a certain time interval in an inner-shell i then for equilibrium conditions this is equal to the deexcitation rate of the state i for all possible processes (radiative and Auger transitions).

The total number n_{if} of X-ray quanta emitted at the transition $i \to f$ per unit of time is equal to

$$n_{if} = \frac{n_i\,\Gamma_{if}^R}{\Gamma_i^R + \Gamma_i^A} \; . \tag{2.118}$$

This means that the absolute intensity of the transition is influenced by the Auger transition probability.

The relative intensity of two emission lines

$$\frac{I_{if}}{I_{ab}} = \frac{\nu_{fi}\,n_i\,\Gamma_{if}^R\,(\Gamma_a^R + \Gamma_a^A)}{\nu_{ba}\,n_a\,\Gamma_{ab}^R\,(\Gamma_i^R + \Gamma_i^A)} \tag{2.119}$$

is then influenced by the Auger process.

For the case that both lines have an identical initial state we have

$$\frac{I_{if}}{I_{ib}} = \frac{\nu_{fi}\,\Gamma_{if}^R}{\nu_{bi}\,\Gamma_{ib}^R} \; , \tag{2.120}$$

i.e. the result is independent of Auger transitions from the initial state.

2.6 X-Ray Atomic Scattering Factors

In the wavelength-dispersive X-ray spectral analysis for the estimation of the reflecting power of individual analyzer crystals at different photon energies and crystal curvatures it is necessary to determine the *structure factor* for the crystal to be used [103].

An introduction to atomic scattering factors is given for example by Cromer and Waber [129]. The coherent X-ray scattering of atoms with energy high in comparison to the binding energy of all electrons of the considered atom is described by the *atomic scattering factor*. According to the classical theory the atomic scattering factor characterizes the ratio between the amplitude of the radiation scattered by the atom to the amplitude of the radiation scattered on a single electron under identical conditions. The calculation of atomic scattering factors is given in detail by James [262].

In the nonrelativistic approximation the scattering factor is expressed by

$$f(s) = \int \Psi_f^*(\boldsymbol{r}) \, e^{i\boldsymbol{s}\boldsymbol{r}} \, \Psi_i(\boldsymbol{r}) \, dV \qquad (2.121)$$

where Ψ is the total wavefunction of the atom, and where the indices i and f characterize the initial and final states of the atom. In the case of coherent scattering the initial and final states are identically. The vector s has the magnitude

$$|\boldsymbol{s}| = 4\,\pi\,\sin\frac{\vartheta}{\lambda}\,,$$

with ϑ as the Bragg angle and λ as the wavelength of the incident wave. Because $\Psi^*\,\Psi$ define the charge distribution, (2.121) defines f as the Fourier transform of the atomic charge distribution.

If for a given atom the nl subshells contain $2(2l+1)$ electrons then the corresponding atom is spherically symmetric. The same symmetry is also characteristic of s electrons. For a spherically symmetric atom the direction of s must not be considered.

For the spherical case (2.121) is reduced to

$$f(s) = \int \frac{\varrho(r)\,\sin sr}{sr}\,dr \qquad (2.122)$$

with

$$\varrho(r) = [r\,P(r)]^2\,.$$

Here $P(r)$ describes the radial part of $\Psi(\boldsymbol{r})$. If relativistic wavefunctions were used then $\varrho(r)$ becomes $r^2\,[F^2(r) + G^2(r)]$. $F(r)$ and $G(r)$ are here the large and small components of the radial wavefunction.

In Chap. 8 scattering factors for all free atoms and of the chemically most significant ions are listed as given by Cromer and Waber [129]. The intervals between the quantities expressed in Chap. 8 were chosen in such a way that a linear interpolation between the listed values is possible with sufficient accuracy. Apart from hydrogen all calculations were done with relativistic wavefunctions.

Chapter 9 summarizes the coefficients a_i, b_i and c for a fit of the scattering factors given in Chap. 8 based on the approximation

$$f\left(\frac{\sin\vartheta}{\lambda}\right) = \sum_{i=1}^{4} a_i\,\exp\left(\frac{-b_i\,\sin^2\vartheta}{\lambda^2}\right) + c. \tag{2.123}$$

To characterize the precision of the analytical approximation the value and the position of the maximum and minimum deviation from the true curves and the mean absolute deviation is given.

2.7 X-Ray Absorption

When a beam of X-rays traverses matter the individual X-ray quanta are absorbed in a single interaction or lost from the beam by scattering processes in a single interaction. Among others Heitler [231] and Agarwal [4] describe the attenuation process in detail and we follow these descriptions here in essential parts.

Linear attenuation coefficient. The intensity of an X-ray beam is reduced along a certain direction dx by an intensity dI in comparison to the initial intensity I. With μ_l as *linear attenuation coefficient* this can be described as

$$\frac{dI}{I} = -\mu_l\,dx\,. \tag{2.124}$$

The negative sign in (2.124) means that I decreases with increasing x. After integration of (2.124) it follows that

$$\ln I = -\mu_l\,x + C \tag{2.125}$$

with C as integration constant. With the initial conditions $I = I_0$ at $x = 0$ and $C = \ln I_0$ this yields

$$I = I_0 \, e^{-\mu_l x} \, . \tag{2.126}$$

The dimension of μ_l is cm^{-1}.

Mass attenuation coefficient. In the from (2.126) derived equation

$$\mu_l = \frac{1}{x} \, \ln \frac{I_0}{I}; \tag{2.127}$$

all quantities on the right side are observables. Experiments have shown that the value of μ_l is dependent on the state of aggregation of the absorber. Thus it seems practical to introduce a *mass attenuation coefficient* μ_m independent of the physical state of the material.

Consider an X-ray beam of intensity I for a given attenuation cross-section. With ϱ as linear material density the beam interacts along the direction dx with a mass

$$dm = \varrho \, dx \, .$$

Then the relative change of intensity dI/I along the direction dx is proportional to the mass m along dx. This yields

$$\frac{dI}{I} = -\mu_m \, dm = -\mu_m \, \varrho \, dx \tag{2.128}$$

and

$$I = I_0 \, e^{-\mu_m m} = -I_0 \, e^{-\mu_m \varrho \, dx} \, . \tag{2.129}$$

Here the product $m = \varrho \, x$ describes the *mass surface density* (in g/cm^2).

Atomic attenuation coefficient. For the characterization of the radiation attenuation per atom of the absorber material it is assumed that the number of atoms in dm is given by

$$dn = \frac{dm}{A} \, L = \frac{\varrho \, dx}{A} \, L \tag{2.130}$$

with A the atomic weight and L Avogadro's number. Then it follows that

$$\frac{dI}{I} = -\mu_a \, dn = -\frac{\varrho \, L}{A} \, \mu_a \, dx \tag{2.131}$$

and

$$I = I_0 \, e^{\mu_a n} \tag{2.132}$$

where μ_a is the *atomic attenuation coefficient*. Since dn describes the number of atoms per square centimeter, of a material layer with the dimension of μ_a, it follows that the unit of area is cm^2. The atomic attenuation coefficient interrelates the linear and the mass attenuation coefficients as

$$\mu_a = \frac{A}{\varrho L} \, \mu_l = \frac{A}{L} \, \mu_m. \tag{2.133}$$

The dimension of μ_a suggests that this quantity can be understand as the atomic cross-section of the interaction. The concept of the atomic attenuation coefficient makes it possible to calculate molecular attenuation coefficients of any molecules over the addition of the contributions from all involved atoms. For example we get

$$\mu_a(CaF_2) = \mu_a(Ca) + 2\,\mu_a(F) \, .$$

Electronic attenuation coefficient. The number of electrons per unit of area of absorber material is $dn_e = Z \, dn$. Thus from (2.131) it follows that

$$\frac{dI}{I} = -\frac{\mu_a}{Z} \, dn_e \tag{2.134}$$

and

$$I = I_0 \, e^{-\mu_e n_e} \, . \tag{2.135}$$

μ_e is the so-called *electronic attenuation coefficient* with dimension cm^2 and is equal to μ_a/Z.

Comparison of the attenuation coefficients. Summarizing the details for the attenuation coefficients we have:

coefficient	dimension
atomic attenuation coefficient μ_a	l^2
linear attenuation coefficient $\mu_l = \mu_a \dfrac{L}{A} \varrho$	l^{-1}
mass attenuation coefficient $\mu_m = \mu_a \dfrac{L}{A}$	$l^2 \, m^{-1}$
electronic attenuation coefficient $\mu_e = \dfrac{\mu_a}{Z}$	l^2

Attenuation coefficients and atomic structure. If X-rays penetrate matter their intensity is attenuated by absorption processes or atomic scattering. Photoelectric absorption takes place if an inner-shell electron is removed. Scattering processes dominate at the interaction of X-rays with tightly bound outer-shell electrons. Assuming independent processes the atomic attenuation coefficient μ_a can be expressed as a sum of the atomic photoabsorption coefficient τ_a and the scattering coefficient σ_a

$$\mu_a = \tau_a + \sigma_a . \tag{2.136}$$

Furthermore,

$$\mu_l = \tau_l + \sigma_l \tag{2.137}$$

$$\mu_m = \tau_m + \sigma_m \tag{2.138}$$

$$\mu_e = \tau_e + \sigma_e \tag{2.139}$$

with

$$\tau_e = \frac{\tau_a}{Z} = \frac{A}{LZ} \, \tau_m = \frac{A}{LZ\varrho} \, \tau . \tag{2.140}$$

The electronic photoabsorption absorption coefficient τ_e considers the interaction of the radiation field with all electrons of the atom. Then the range of validity of (2.139) is restricted by the relation

$$\lambda < \lambda_K \quad \text{or} \quad E > E_K$$

where λ_K, E_K is the wavelength corresponding to the K-shell absorption edge or corresponding energy.

If σ_m is constant for increasing wavelength and increasing atomic number Z, then τ_m and τ_a decrease rapidly. For long wavelengths and heavy elements we have approximately

$$\begin{array}{l} \text{atomic attenuation coefficient } \mu_a \approx \\ \text{atomic photoabsorption coefficient } \tau_a \end{array} \tag{2.141}$$

and

$$\begin{array}{l} \text{mass attenuation coefficient } \mu_m \approx \\ \text{mass related photoabsorption coefficient } \tau_m. \end{array} \tag{2.142}$$

If the impinging photon radiation is sufficiently energetic and overcomes the binding energy of the

---

[27] 1 barn = 1×10^{-24} cm^2

K shell then the photoabsorption coefficients can be calculated as the sum of the absorption coefficients of all atomic subshells of the atom. If the photon energy is not high enough to ionize all electrons of the atom, then only subshells which are capable of ionization can summarize contributions.

Between the absorption edges the change of τ_a with λ and Z shows the dependence

$$\tau \sim Z^4 \lambda^3 . \tag{2.143}$$

From (2.143) it follows that energetic X-rays are highly penetrating and that is why they are called "hard" radiation. Low-energetic X-rays are strongly absorbed and are called "soft" X-rays. A description of the quantum theory of photoabsorption including the calculation of τ_{nlj} is given by Stobbe [524] and Heitler [231]. Details for the consideration of the electronic screening in multielectron atoms for the calculation of τ are discussed by Pinsker [429].

Total cross-section for X-ray attenuation. The total cross-section for the attenuation of X-ray intensity for the transmission of X-rays through matter contains the following contributions:

$$\sigma_{\text{tot}} = \mu_a = \underbrace{\tau_a}_{\text{photoeffect}} + \underbrace{\sigma_{\text{koh}}}_{\text{Rayleigh scattering}} +$$
$$\underbrace{\sigma_{\text{inkoh}}}_{\text{Compton scattering}} + \underbrace{\sigma_{\text{pair}}}_{\text{pair production}} . \tag{2.144}$$

Here σ_{tot} is the total interaction cross-section per atom[27] in barn/cm^2, τ_a the atomic photoelectric cross-section, σ_{koh} the cross-section for coherent scattering (Rayleigh scattering), σ_{inkoh} the cross-section for incoherent scattering (Compton scattering) and σ_{pair} the interaction cross-section for the production of electron–positron pairs (pair production cross-section).

The mass attenuation coefficient μ_m relates to σ_{tot} as

$$\frac{\mu_l}{\varrho} = \mu_m = \sigma_{\text{tot}} \frac{L}{A} . \tag{2.145}$$

Frequently used conversion factors for the conversion of σ_{tot} to μ_m are given in Table 2.25. Here

$$\sigma \left[\frac{\text{barn}}{\text{Atom}} \right] = K \times \mu \left[\frac{\text{cm}^2}{\text{g}} \right] .$$

The introduced in (2.144) of the pair production cross-section is relevant only for photon energies above 1 MeV ($E > 2\, m_e\, c^2$), i.e. in the X-ray energy region we neglect this process.

Absorption-jump ratio. Because of the behavior of the attenuation coefficients at the absorption edges for each absorption edge there are two values for the attenuation coefficient at a fixed wavelength λ_{nlj} or energy E_{nlj}. If the mass attenuation coefficient on the short-wavelength side of λ_{nlj} is given as $\tau_m(\lambda_{nlj})$ and on the low-wavelength side with $\tau'_m(\lambda_{nlj})$ than it follows that $\tau_m(\lambda_{nlj}) > \tau'_m(\lambda_{nlj})$. The ratio

$$r_{nlj} = \frac{\tau_m\,(\lambda_{nlj})}{\tau'_m\,(\lambda_{nlj})} = \frac{\tau_a\,(\lambda_{nlj})}{\tau'_a\,(\lambda_{nlj})} > 1 \qquad (2.146)$$

is the *absorption edge jump ratio* for the level (*nlj*).

An overview of empirically determined ratios r_{nlj} is given in [4].

In the wavelength region $\lambda \leq \lambda_K$ the photoelectric absorption occurs preferentially on K-shell electrons. Then the ratio

$$\frac{r_k - 1}{r_K} = \frac{\tau_K}{\tau_a}$$

describes the part of the emitted photoelectrons coming from the K shell. Analogously the jump ratios for the absorption edges L_1, L_2, L_3, M_1, etc. can be determined.

Energy transfer by X-rays. In some cases for the interaction of X-rays with matter also the energy transfer into the target is of interest. To describe this process a *mass energy transfer coefficient*

$$\frac{\mu_{\text{kin}}}{\varrho} = \frac{1}{E\,\varrho}\,\frac{d\,E_{\text{kin}}}{dx} \qquad (2.147)$$

is introduced. Here $d\,E_{\text{kin}}$ is the sum of all kinetic energies from all particles released in the target and E the sum of the energies of all quanta impinging on a layer of thickness dx. If we consider that the charged particles emit part of their energy as Bremsstrahlung the real energy absorbed in the target along a path

dx will be described by the *mass energy absorption coefficient*

$$\mu_E = (1 - \eta_B)\,\mu_{\text{kin}} \qquad (2.148)$$

where η_B is part of the energy of the charged secondary particles lost as Bremsstrahlung. The mass energy transfer coefficient can be determined from the cross-sections of the photo-, Compton and pair production effects.

X-Ray Transmission. According to (2.126) the transmission I/I_0 of X-rays through matter (air, spectroscopic windows, shielding, ...) depends on the actual X-ray energy and on the nature of the medium acting as attenuator. For instance, in Table 2.26 we give the transmission of X-rays with different energies through beryllium.

Table 2.26. X-ray transmission through beryllium ($Z = 4$) in percent

transition line energy/keV	thickness / μm			
	20	30	40	50
NaKα 1.04	13.8	5	1	–
Mg K$_\alpha$ 1.26	30.4	16.7	9.0	5.0
Cr K$_{\alpha_1}$ 5.41	98.0	97.7	96.3	95.4
Cu K$_{\alpha_1}$ 8.04	100	99.5	99.0	98.4

2.8 Continuous X-Rays (Electron Bremsstrahlung)

Besides the emission of characteristic X-rays the characteristic line spectra for this type of radiation often are superimposed with continuous radiation. Continuous radiation arise if sufficiently energetic light charged particles are decelerated in the Coulomb fields of nuclei or other charged particles. The continuous radiation emitted in this process is termed *Bremsstrahlung* and can be explained in the framework of classical electrodynamics. Bremsstrahlung can arise as disturbing background radiation or, as in the case of X-ray tubes, as well-aimed produced radiation. For the issues of X-ray physics *electron Bremsstrahlung* is of special interest and is discussed here in more detail.

Table 2.25. Conversion factors K for the conversion from barn/atom into cm^2/g (after Henke [234])

Z	element	K	Z	element	K	Z	element	K
1	H	1.674	33	As	124.4	65	Tb	263.9
2	He	6.646	34	Se	131.1	66	Dy	269.8
3	Li	11.52	35	Br	132.7	67	Ho	273.8
4	Be	14.96	36	Kr	139.1	68	Er	277.7
5	B	17.95	37	Rb	141.9	69	Tm	280.5
6	C	19.94	38	Sr	145.5	70	Yb	287.3
7	N	23.26	39	Y	147.6	71	Lu	290.5
8	O	26.56	40	Zr	151.5	72	Hf	296.3
9	F	31.54	41	Nb	154.3	73	Ta	300.4
10	Ne	33.50	42	Mo	159.3	74	W	305.2
11	Na	38.17	43	Tc	164.2	75	Re	309.1
12	Mg	40.35	44	Ru	167.8	76	Os	315.8
13	Al	44.80	45	Rh	170.9	77	Ir	319.1
14	Si	46.63	46	Pd	176.7	78	Pt	323.9
15	P	51.43	47	Ag	179.1	79	Au	327.0
16	S	53.24	48	Cd	186.6	80	Hg	333.0
17	Cl	58.86	49	In	190.6	81	Tl	339.3
18	Ar	66.33	50	Sn	197.1	82	Pb	344.0
19	K	64.90	51	Sb	202.1	83	Bi	347.0
20	Ca	66.54	52	Te	211.9	84	Po	348.7
21	Sc	74.64	53	I	210.7	85	At	348.7
22	Ti	79.53	54	Xe	218.0	86	Rn	368.6
23	V	84.58	55	Cs	220.7	87	Fr	370.2
24	Cr	86.33	56	Ba	228.0	88	Ra	375.2
25	Mn	91.21	57	La	230.6	89	Ac	376.9
26	Fe	92.72	58	Ce	232.6	90	Th	385.3
27	Co	97.85	59	Pr	233.9	91	Pa	383.5
28	Ni	97.48	60	Nd	239.5	92	U	395.2
29	Cu	105.5	61	Pm	240.7	93	Np	393.6
30	Zn	108.6	62	Sm	249.6	94	Pu	405.1
31	Ga	115.8	63	Eu	252.3			
32	Ge	120.5	64	Gd	261.1			

2.8.1 Properties of Electron Bremsstrahlung

For the production of electron Bremsstrahlung two mechanisms of the electron–atom interaction are of importance:

- *Ordinary Bremsstrahlung*: Electron Bremsstrahlung as a result of photon emission of a charged particle decelerated in the field of a target electron or nucleus.
- *Polarizational Bremsstrahlung* (atomic Bremsstrahlung) as a result of the dynamical polarization of the atom by the incident particle (photon emission of the target electrons virtually excited by the projectile).

Because of its practical importance, in the following we will consider in detail ordinary Bremsstrahlung in comparison to polarization Bremsstrahlung. Ordinary Bremsstrahlung is much more important in most cases.

The production of electron Bremsstrahlung is characterized by a change of the momentum p of the impinging electron with simultaneous emission of electromagnetic radiation (Bremsstrahlung):

$$A + e^-(p_0) \rightarrow A + e^-(p_1) + \hbar\omega . \qquad (2.149)$$

In this way the inner state of the atom remains unchanged.

For electron Bremsstrahlung the spectrum is characteristic: quanta can be emitted up to a limiting energy of

$$E_{\max} = \frac{hc}{E_e}. \qquad (2.150)$$

The probability that a charged particle emits Bremsstrahlung is proportional to

$$\frac{q^2 Z^2 E}{m^2}$$

with q the charge of the particle (in units of the electron charge), E the particle energy and m the mass of the particle. This proportionality is the reason that especially light particles such as electrons can intensively emit Bremsstrahlung. On the other hand, in comparison to electrons, protons generate Bremsstrahlung with a probability six orders of magnitude lower.

For the case of the deceleration of an electron in the field of a nucleus the basic situation for the production of Bremsstrahlung is shown in Fig. 2.48.

2.8.2 Bremsstrahlung Cross-Sections

Fundamental calculations of the energy loss of electrons by Bremsstrahlung in the fields of atomic nuclei and atomic electrons were done by Bethe [59] and Heitler [231], among others. An overview on cross-sections for Bremsstrahlung production processes can be found by Koch and Motz [287]. Furthermore, a summary of the treatment of Bremsstrahlung photon emission in the framework of many-body perturbation theory was given by Amusia [13]. Basic formulas for Bremsstrahlung production and corresponding numerical material can also be found by Beyer et al. [62].

The energy loss $(-dE/dx)_{rad}$ of electrons during Bremsstrahlung emission can be explained in the framework of quantum electrodynamics using the Thomas–Fermi model

$$\left(-\frac{dE}{dx}\right) \approx 4\,\alpha\,r_e^2 N_0 E Z^2 \ln \frac{183}{Z^{1/3}} . \qquad (2.151)$$

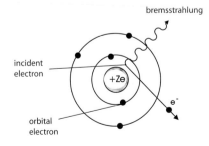

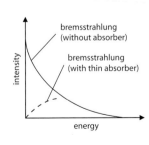

Fig. 2.48. Production of electron Bremsstrahlung by deceleration of electrons in the field of a nucleus. On the ride side the influence of a thin absorber on the undisturbed spectrum is shown.

Here the quantity $(-dE/dx)_{rad}$ describes the energy loss per electron and path length. Furthermore, it yields E, the electron energy, N_0, the atomic density of a medium with atomic number Z, $\alpha = e^2/\hbar c$, the fine-structure constant and $r_e = e^2/(4\pi \varepsilon_0 mc^2) = \alpha^2 a_0 = 2.818 \times 10^{-13}$ cm the classical electron radius.

2.8.3 Radiation Length and Radiation Density

The electron Bremsstrahlung loss increases with increasing electron energy. This will be transparent if we write (2.151) in the form

$$\left(-\frac{dE}{dx}\right) = \frac{1}{x_0} E . \tag{2.152}$$

Here the quantity x_0 describes the so-called *radiation length*, which characterizes the target thickness for which the particle energy is reduced to the e-th part of the impact energy:

$$\frac{1}{x_0} = 4 \alpha \, r_e^2 N_0 Z^2 \, \ln \frac{183}{Z^{173}} . \tag{2.153}$$

Furthermore, the product of the radiation length and the density ϱ of the medium is also applied for the characterization of the Bremsstrahlung and is known as the *radiation thickness x_r*:

$$x_r = \varrho \, x_0 = \frac{A}{4 \alpha \, r_e^2 N_A Z^2 \, \ln \dfrac{183}{Z^{1/3}}} \tag{2.154}$$

with A the atomic number of the target atoms and N_A the Avogadro constant.

2.8.4 Electron Bremsstrahlung from X-Ray Tubes

The photon spectrum from an X-ray tube contains contributions from the electron Bremsstrahlung of accelerated electrons impinging on the thick (in the most cases) anode and lines of X-ray transitions from the atoms of the anode material. Here we will tract in more detail the Bremsstrahlung part of the tube spectrum.

An X-ray tube working with an acceleration voltage U emits Bremsstrahlung quanta up to a threshold wavelength

$$\lambda_{max} = \frac{hc}{eU} \qquad (\text{Duane-Hunt limit}) . \tag{2.155}$$

After considering the values for the natural constants we get

$$\lambda_{max}[\text{nm}] = 1.239 \times 10^9 \frac{1}{U \, [\text{kV}]} . \tag{2.156}$$

For the emitted spectrum the intensity $I(\lambda)$ can be described by the Kramer rule [296]:

$$I(\lambda) \, d\lambda = \frac{KiZ}{\lambda^2} \left(\frac{\lambda}{\lambda_{max}} - 1\right) \tag{2.157}$$

where K is the Kramer constant, i is the tube current and Z is the atomic number of the anode material. $I(\lambda)$ describes the intensity density and $I(\lambda) \, d\lambda$ the spectral intensity. The total radiation intensity emitted by an X-ray tube results as the integral of the spectral intensity integrated over all wavelengths of the spectrum. The units of the intensity density are $\text{s}^{-1}\text{nm}^{-1}$ or $\text{s}^{-1}\text{eV}^{-1}$ if the quantity is declared as $I(\lambda)$ or as $I(E)$.

For the X-ray intensity at a tube voltage U and C as a constant it follows that

$$I = CiZU^2 . \tag{2.158}$$

The maximum wavelength for the intensity distribution of the Bremsstrahlung spectrum can be estimated from the critical wavelength λ_{max}

$$\bar{\lambda} = \frac{3}{2} \lambda_{max} \tag{2.159}$$

and depends on the tube voltage, the characteristics of the applied voltage and on the anode material (atomic number).

For the total absorption of an electron beam of current i and acceleration voltage U in a material of atomic number Z the Bremsstrahlung power P_X is

$$P_X[\text{W}] = 1.5 \times 10^{-9} Z \, i[\text{A}] \, U^2[\text{V}] . \tag{2.160}$$

In Fig. 2.49 the spectral distribution emitted by an X-ray tube is presented. For real tubes the long-wave part of the emitted radiation is attenuated by the tube window (as a rule beryllium or aluminum).

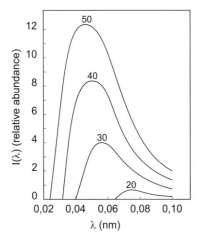

Fig. 2.49. Spectral intensity $I(\lambda)$ of the electron Bremsstrahlung in an X-ray tube with thick tungsten anode as a function of the acceleration voltage (after [72]); as parameter the acceleration voltage is given in keV.

For an X-ray tube working with a sinusoidal alternating-current voltage for the intensity it follows [284] that

$$I_\sim(\lambda) = \frac{1}{\pi}\, CUI \left[\sqrt{\lambda^2 - \lambda^2_{max}} - \lambda_{max}\, arccos\, \frac{\lambda_{max}}{\lambda}\right]\frac{1}{\lambda^3}$$

$$(2.161)$$

with C a constant.

The ratio of the spectral densities of the emitted Bremsstrahlung of an X-ray tube working with an alternating-current voltage to a tube working with direct-current voltage can be calculated in the following way [284]

$$\frac{I_\sim(\lambda)}{I(\lambda)} = \frac{1}{\pi}\left(\frac{1}{1 - \dfrac{\lambda_{max}}{\lambda}}\right) \times$$

$$\left[\sqrt{1 - \left(\frac{\lambda_{max}}{\lambda}\right)^2} - \frac{\lambda_{max}}{\lambda}\, arccos\, \frac{\lambda_{max}}{\lambda}\right].$$

$$(2.162)$$

2.8.5 Bremsstrahlung Intensity Distribution

For Bremsstrahlung the differential scattering cross-section for the deflection of an electron in an atomic or nuclear field was calculated in the Born approximation by Bethe [60].

A characteristic azimuthal spatial intensity distribution of Bremsstrahlung emission from an X-ray tube is given in Fig. 2.50. Here the Bremsstrahlung emission is peaked in the forward direction. The mean angle of the emitted Bremsstrahlung quanta can be approximated by

$$\bar{\vartheta} \approx \frac{m_e c^2}{E_e}$$

$$(2.163)$$

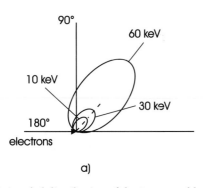

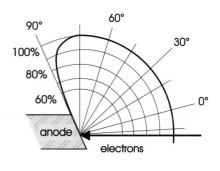

a)   b)

Fig. 2.50. Azimuthal distribution of the Bremsstrahlung intensity from an X-ray tube (after Kluev [284]). Part (**a**) of the figure characterizes the theoretical distribution for different anode voltages in the case of a thin anode and part (**b**) shows the experimental measured emission from a thick tungsten anode at an acceleration voltage of 100 keV

with E_e as the electron energy. According to [72] from Sommerfeld considerations [514, 515] for the theoretical angle distribution we have for the Bremsstrahlung emitted from a massive anode

$$I(\varphi) \sim \frac{\sin^2 \varphi}{\cos \varphi} \left[\frac{1}{\left(1 - \dfrac{v_e}{c} \cos \varphi\right)^4} - 1 \right] . \quad (2.164)$$

The angle φ describes the angle between the acceleration vector of the electron and the radius vector directed to the point where the field strength components B and H were measured.

For the short-wave limit of the Bremsstrahlung spectrum we have the exact relation

$$I(\varphi) = \text{const.} \frac{\sin^2 \varphi}{\left(1 - \dfrac{v_e}{c} \cos \varphi\right)^6} . \quad (2.165)$$

2.8.6 Polarizational Bremsstrahlung

Polarizational Bremsstrahlung is of special interest in ion–atom and atom–atom collisions because, as a result of the mass of the impinging particles, here the ordinary Bremsstrahlung is clearly truncated. This special process is described for example by Amusia et al. [11, 12].

For a description of polarizational Bremsstrahlung many-electron correlations must be considered. In the literature this is described on different levels. For example more information can be found in [289, 290] and the citations therein. Bremsstrahlung production in more complicated systems such as solid states, molecules and clusters can be

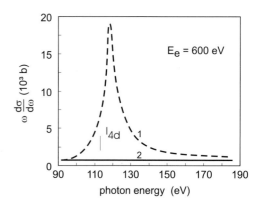

Fig. 2.51. Bremsstrahlung cross-sections of the ordinary (curve 2) and polarizational Bremsstrahlung (curve 1) in the region of the 4d ionization potential I_{4d} of lanthanum (after Korpl et al. [291]) for an electron energy of $E_e = 600$ eV. The ionization potential I_{4d} for the 4d subshell is shown

described semiempirically under consideration of the actual atomic structure (ionization threshold in many-electron systems) by using photoabsorption data [292].

The consideration of polarizational Bremsstrahlung leads to a clear increase of the ordinary Bremsstrahlung cross-section in the vicinity of the ionization potentials of the individual atomic substates. In Fig. 2.51 this is shown for Bremsstrahlung cross-sections in the vicinity of the 4d ionization potential of lanthanum. Above the 4d ionization potential the Bremsstrahlung cross-sections show a maximum and decrease again with increasing energy of the emitted photons.

2.8.7 Bremsstrahlung Emission from Plasmas

For astrophysical investigations, ion source physics and other applications it is of interest to calculate Bremsstrahlung spectra in plasmas produced by the deceleration of electrons on the fields of ions. This is of special importance because Bremsstrahlung in hot plasmas give essential contributions to energy loss processes and can be used for plasma temperature diagnostics.

The Bremsstrahlung spectrum of ions in the nonrelativistic Elwert–Born energy region of the incidence electrons was calculated in a model potential approximation by Avdonina and Pratt [27]. Analytical results for a hot plasma, assuming an ionic radial potential, are given by Lamoreux and Avdonina [306].

Bremsstrahlung processes can lead to substantial power losses from the plasma. The power P_{Br} emitted from a plasma by Bremsstrahlung emission can be described quantum mechanically [231]. For the power density of the Bremsstrahlung emission it follows that

$$P_{Br} \left[\frac{W}{m^3} \right] = 1.5 \times 10^{-38} \, q^2 \sqrt{kT_e[eV]} \, n_e[m^{-3}] \, n_i[m^{-3}]$$

(2.166)

with q the ion charge, k the Boltzmann constant, T_e the electron temperature, n_e the electron density and n_i the ion density. If in the plasma there exist ions with different charge states q then the product $q^2 n_i$ must be replaced by the sum $\sum_j q_i^2 \, n_i^j$ over all ion species j.

3 Energy and Intensity Measurements

3.1 Calibration Normals for Electromagnetic Radiation

3.1.1 Classification of the Calibration Normals

The hierarchy of calibration normals begins with the fact that for the characterization of electromagnetic radiation, beginning with infrared radiation ($\lambda \approx 10^{-6}$ m) to short-wavelength γ-radiation ($\lambda \approx 10^{-13}$ m), a part that includes many orders of magnitude comprising wavelength range must be considered. For the use of a transition line as a calibration normal it must be required that its wavelength should be localized in a well-accessible area.

Therefore because of the wide wavelength range it becomes necessary to introduce some additional local normals. In this framework the wavelengths are distinguished from each other mostly by a multiple and the error produced in this way should be sufficiently small.

Local calibration normals must be related also to the main normal. Because the normals can differ by three or four orders of magnitude this can be connected with considerable difficulties. To avoid this for the apportionment of the wavelength areas additional normals are introduced. Choosing such normals we must adopt a compromise:
Too many normals lead to decreasing accuracy for the normals related to each other, and too few normals can result in strongly differing wavelengths where the accuracy of the comparison is also affected.

To avoid possible confusion in the selection of local normals the normals should be classified. In the optical spectral region such a classification was done some 50 years ago. As first-order calibration normal, the red cadmium line has been chosen and several transition lines from neon and iron were selected as second-order calibration normals.

A classification of the spectral range of X-ray and γ-radiation is given by Dselepov and Schestopalova [157].

The corresponding classification scheme has the following structure:

First-order normal. As first-order normal a transition is used from ^{86}Kr with a vacuum wavelength of 605,780211 nm.

Second-order normals. Second-order normals were selected by the following rules:

(a) The wavelength should be determined by a comparison with the wavelength from the first-order normal.
(b) The comparison is made with a precision in the order of the best known values from actual data. In this way the relative error in the wavelength comparison should not exceed $(2-8) \times 10^{-7}$.
(c) The comparison must be based on a method founded on completely known physical laws. For the application of these laws unprovable assumptions are not allowed.
(d) A second-order normal must be comparable with the first-order normal and with the normals of lower priority. It must be well and easily reproducible.
(e) The spectroscopic purity of the normals should be beyond doubt. The transition used as normal must be a well separated line and should be so far from other transition lines so that their background contributions were negligible.

(f) A comparison between second-order normals should be consistent with the comparison of the wavelength of the first-order normal.

For the determination of second-order normals all conditions (a)–(f) must be fulfilled. On the basis of conditions (a)–(c) the γ-lines from ^{169}Yb, ^{170}Tm, ^{192}Ir, ^{198}Hg and the X-ray emission line W K_{α_1} are proper transitions. The condition (e) excludes ^{169}Yb and ^{192}Ir as well as ^{170}Tm. Thus only the W K_{α_1} and ^{198}Hg lines are suitable as second-order normals.

Third-order normals. These normals can be chosen by using the following rules:

(a) Wavelengths from third-order normals must be determined by the direct comparison with wavelengths from first- and second-order normals, where the comparison should occur on an up-to-date precision level and based on well-known physical laws.
(b) New third normals can be introduced on the basis of properly determined third-order normals by the Ritz combination principle $E_1 + E_2 = E_3$.
(c) As upper error margins ΔE and $\Delta E/E$ we have here:

$\Delta E \quad < 10$ eV   for E < 200 keV
$\Delta E/E < 5 \times 10^{-5}$ for E > 200 keV.

In Table 3.1. γ lines as third-order normals are listed with energies in the X-ray transition energy region of the stable elements of Mendeleev's periodic table.

Fourth-order normals. Fourth-order normals are lines where the transition energy was determined by comparison with second- and third-order normals. Here:

(a) In comparison to those from third-order normals the upper error margins are increased by a factor of two:

$\Delta E \quad < 20$ eV   for E < 200 keV
$\Delta E/E < 1 \times 10^{-4}$ for E > 200 keV.

(b) An unconditional comparison with second- and third-order normals is not demanded. Methods are possible where the measuring device is calibrated with second- and third-order normals and the line energy is derived from the corresponding calibration curve.

3.1.2 X-Ray Calibration Normals

To join the X-ray and optical spectral region, wavelength measurements were first made with diffraction gratings (Compton and Duane in 1925 [123], Bearden in 1931 [46, 47], Hanins and Bearden in 1964 [223], and others). With this techniques Henins in 1971 [232] measured the Al $K_{\alpha_{1,2}}$ line in six diffraction orders with a relative error of 9×10^{-6}. Because at shorter wavelengths the experimental effort increases dramatically the method is restricted to the determination of wavelengths of light elements.

The most common method to the wavelength determination of X-rays is Bragg's diffraction law. For a simple cubic crystal lattice this yields

$$n\lambda = 2d \sin \vartheta \qquad (3.1)$$

with ϑ as the Bragg angle, d as the lattice constant and n as the diffraction order. Experimentally the wavelength is determined by the measurement of the angle where the diffraction shows an intensity maximum. In this way the lattice constant of the crystal is assumed to be known from previous experiments.

For precision measurements, (3.1) can be written as

$$\lambda = \frac{2d \sin \vartheta}{n\,(h^2 + k^2 + l^2)^{1/2}} \left(1 - \frac{1 - \mu}{\sin^2 \vartheta}\right) . \qquad (3.2)$$

Here h, k, l describe the Miller crystallographic index for the crystallographic plane and μ is the index of refraction of the analyzed wavelength λ. Therefore for a precise determination of the wavelength λ it is required to measure ϑ and d exactly because μ is almost equal to one $(1 - \mu < 1 \times 10^{-5})$. For example the Bragg angle ϑ can be determined with a goniometer at $10°$ to 0, 1$''$ $(\Delta\vartheta/\vartheta = 3 \times 10^{-6})$ and with interferometric methods to $\leq 0.01''$.

For the determination of the lattice constant d the relation yields

$$d = \left(\frac{f A}{\varrho L}\right)^{1/3} . \qquad (3.3)$$

Here f describes the number of atoms in an elementary cell ($f = 1$ for a cubic lattice), A is the mean

Table 3.1. Energies E_γ from γ emitters, having as third-order normals an energy uncertainty of $\Delta E < 10$ eV (after [157]). $T_{1/2}$ is the half-time; Kr, W, Hg denote the determination with the krypton normal or with lines from second-order normals; and ΔE_γ is the error of the energy determination

initial nuclide	E_γ [keV]	ΔE_γ [eV]	$T_{1/2}$	method of determination
^{171}Lu	9.149	1	8.22d	Hg
^{57}Co	14.408	5	269.80d	
^{171}Lu	19.384	2	8.22d	Hg
^{241}Am	26.345	1	432.80a	
^{171}Lu	27.126	3	8.22d	Hg
^{182}Ta	31.7376	0.5	115d	W,Hg
^{183}Ta	40.9765	1	5.10d	W
^{182}Ta	42.7151	0.7	115d	W,Hg
^{183}Ta	46.48501	0.20	5.10	Hg
^{171}Lu	46.516	4	8.22d	Hg
^{199}Au	49.82655	0.18	3.13d	Hg
^{183}Ta	52.59648	0.18	5.10d	Hg
^{171}Lu	55.677	6	8.22d	Hg
^{241}Am	59.5370	1	432.80a	W
^{169}Yb	63.12080	0.17	30.70d	Hg
	63.12077	0.17	30.70d	Kr
^{182}Ta	65.7225	0.4	115d	Hg
	65.7230	0.9	115d	W,Hg
	65.7218	2	115d	W
^{171}Lu	66.720	7	8.22d	Hg
^{182}Ta	67.74998	0.22	115d	Hg
^{153}Sm	69.67340	0.21	46.44h	Hg
^{187}W	71.995	4	23.90h	W
	72.004	4	23.90h	Hg
^{171}Lu	72.366	7	8.22d	Hg
^{193}Os	73.046	3	31.50h	Hg
^{153}Sm	75.42256	0.26	46.44h	Hg
^{171}Lu	75.876	8	8.22d	Hg
^{197}Hg	77.345	8	64.10h	W
^{166}Ho	80.557	4	27h	W
^{133}Ba	80.998	8	10.90a	
^{183}Ta	82.989	2	5.10d	W
^{153}Sm	83.36764	0.26	46.44h	Hg
^{170}Tm	84.262	4	128.6d	W
	84.25523	0.23	128.6d	Kr
	84.25478	0.26	128.6d	Hg
^{182}Ta	84.6822	0.9	115d	W,Hg
^{183}Ta	84.7123	2	5.10d	W
^{160}Tb	86.786	2	72.3d	
^{176}Lu	88.361	9	3.6×10^{10}a	W
^{153}Sm	89.48646	0.28	46.44d	Hg
^{147}Nd	91.1073	1.6	10.98d	Hg

Table 3.1. (cont.)

initial nuclide	E_γ [keV]	ΔE_γ [eV]	$T_{1/2}$	method of determination
^{169}Yb	93.61496	0.27	30.70d	Hg
	93.61514	0.27	30.70d	Kr
^{165}Dy	94.697	5	2.334h	W
	94.694	3	2.334h	Hg
^{153}Sm	97.43155	0.30	46.44d	Hg
^{183}Ta	99.08182	0.27	5.10d	Hg
	99.0806	2	5.10d	W
^{182}Ta	100.1033	2	115d	W
	100.10652	0.27	115d	Hg
^{183}Ta	101.9360	2	5.10d	W
	102.483	3	5.10d	W
	103.149	5	5.10d	W
^{153}Sm	103.175	4	46.44d	W
	103.18072	0.30	46.44d	Hg
^{193}Os	107.019	8	31.50h	Hg
^{183}Ta	107.9329	2	5.10d	W
	107.9337	0.3	5.10d	Hg
^{183}Ta	109.728	3	5.10d	W
^{160}Yb	109.77987	0.30	30.70d	Hg
	109.77987	0.30	30.70d	Kr
^{171}Er	111.624	4	7.52h	Hg
^{182}Ta	113.673	3	115d	W
	113.6724	0.4	115d	Hg
^{187}W	113.749	8	23.90h	Hg
^{182}Ta	116.416	4	115d	W
	116.4171	1	115d	W,Hg
^{171}Er	116.659	6	7.52h	Hg
^{160}Yb	118.1900	0.4	30.70d	Hg
	118.1902	0.4	30.7d	Kr
^{165}Dy	119.493	8	2.334h	Hg
^{183}Ta	120.375	3	5.10d	W

atomic mass, ϱ the crystal density and L the Avogadro number. With this method Hanins and Bearden (1964) [223] reached a relative error in the determination of the lattice constant d of 5×10^{-6}.

More precise measurements became possible through the use of an X-ray interferometer [142]. Therefore the relative error in the determination of d could be reduced to $1, 5 \times 10^{-7}$. Such precise determinations of d are impossible on the basis of (3.3), because the values for ϱ, A and L have typical relative errors of $\Delta\varrho_{Si}/\varrho_{Si} = 4.6 \times 10^{-6}, \Delta A_{Si}/A_{Si} = 3 \times 10^{-6}$ and $\Delta L/L = 1 \times 10^{-6}$.

In the utilization of X-ray lines as calibration normals a range of circumstances are considered that can influence the measurement:

Energetic widths of X-ray lines. The most common X-ray lines correspond to electron transitions of the E1 type. E1 transitions are very fast and therefore have a considerable energetic width (see Chap. 5). Small energetic widths extend only for transitions of types M1, E2, ...

Asymmetry of X-ray transition lines. Various X-ray lines have a measurable asymmetry (see Table 2.15).

This and the energetic width of the X-ray transition lines limits the accuracy in the analysis of experimental lines, while it is often unclear which point of the diffraction profile corresponds to the Bragg angle.

Isotope effects. The X-ray wavelengths of one and the same element differ from one another in the meV region as a result of isotope effects (see Sect. 2.2.4 and Table 2.19).

Chemical effects. If an atom builds up a chemical bond, then the transition energies change by a small amount (see Sect. 2.2.4 and Table 2.20).

In contrast to these disadvantages of X-ray transition lines these difficulties do not extend to nuclear calibration normals, because the lifetime of nuclear states is essentially greater than those from atomic electronic states. Therefore the energetic widths of emitted γ lines are some orders of magnitude smaller. The reasons for this are:

1. In comparison to the atomic size the extent of the nucleus is very small. Because the probabilities for electric dipole or quadrupole radiation are proportional to x^2 or x^4 (x is the linear dimension of the emitting system), a considerable difference arises as a result of the atomic dimension ($x \approx 10^{-10}$ m) and the dimension of the nucleus ($x \approx 10^{-14}$ m);
2. The most intense atomic transitions are dipole transitions. Most nuclear transitions are quadrupole transitions which are substantially slower than atomic transitions.

Intensities of X-ray lines as they are used for the calibration of the detection efficiency for X-ray transition lines for various spectrometers are summarized in Chap. 5. It contains intensity ratios calculated by Scofield [489].

Intensities for γ lines of all isotopes were tabulated by Gusev and Dmitriev [219].

3.2 Energy and Intensity Calibration

3.2.1 Calibration of Wavelength-Dispersive Spectrometers

The determination of unknown energies or wavelengths of measured X-ray lines can be realized on the basis of the well-known dependence between the experimentally determined Bragg angle and the corresponding wavelength. If the measurement is related to a calibration line of wavelength λ_c in the diffraction order n_c, which is located at the position $\sin \vartheta_c$, wavelength, the diffraction order and line position of an unknown line (index x) are related by

$$\frac{n_c \, \lambda_c}{n_x \, \lambda_x} = \frac{\sin \vartheta_c}{\sin \vartheta_x} \, . \tag{3.4}$$

In (3.4) we consider that for a single measurement in diffraction order n a line is analyzed whose position is determined as $P_i(n) \pm \Delta P_i(n)$. If m single measurements in different diffraction orders n are considered after reduction to the first order of diffraction for the mean position of the analyzed line, we find

$$P = \frac{1}{m} \sum_{i=1}^{m} \frac{P_i(n)}{n} \, . \tag{3.5}$$

The calculation of the uncertainty ΔP for the determination of the line position can be done according to Reidy [449] in two ways. As the standard deviation of the weighted average of P we have

$$\Delta P = \left(\frac{\sum_{i=1}^{m} [P - (P_i(n)/n]^2}{m(m-1)} \right)^{1/2} \tag{3.6}$$

and as the quadratic summation of the error calculated from the angle deviation we have

$$\Delta P = \left(\frac{\sum_{i=1}^{m} (\Delta P_i(n)/n)}{m(m-1)} \right)^{1/2} . \tag{3.7}$$

In practice for ΔP the greatest value resulting from (3.6) and (3.7) is assumed.

For the ratio of the calibration line and of the line to be measured we have

$$\frac{\lambda_c}{\lambda_x} = \frac{P_c \pm \Delta P_c}{P_x \pm \Delta P_x} = R \pm \Delta R \qquad (3.8)$$

where the wavelength ratio is characterized by R and the error of this ratio is given by ΔR. In this way we get

$$R = \frac{P_c}{P_x} \qquad (3.9)$$

and

$$\Delta R = \left[\left(\frac{\Delta P}{P} \right)^2 + \left(\frac{\Delta P_c}{P_x} \right)^2 \right]^{1/2} R . \qquad (3.10)$$

For the energy of the analyzed line it follows that

$$E_x = R E_c \qquad (3.11)$$

and for the uncertainty in the determination of E_x

$$\Delta E_x = \left[\left(\frac{\Delta R}{R} \right)^2 + \left(\frac{\Delta E_c}{E_c} \right)^2 \right]^{1/2} E_x . \qquad (3.12)$$

If the possible sources of mistakes are not independent of one another, correlations between different parameters must be considered.

3.2.2 Calibration of Energy-Dispersive Spectrometers

Problem Definition

The method of measurement for the energy and the relative intensities by means of energy-dispersive spectrometers is based on the comparison of apparatus spectra from calibration sources and from samples to be examined. At the same time there is the task of finding the peak maxima P in the analyzed spectra to get an assignment to the energy E and to determine from the peak area S a corresponding intensity I. Because of the outstanding spectroscopic properties of high-resolving silicon or germanium based semiconductor detectors it becomes possible to correct results without repetition of the experiment if data of the used calibration normals were changed.

The attainable precision of the results is limited by a range of processes accompanying the detection process. This should be explained by the effects which could affect the form of the apparatus spectrum during the application of semiconductor detectors [555]:

1. scattering of X-rays on the target holder;
2. influence of background radiation (e.g. electrons);
3. X-ray scattering in the source;
4. excitation of X-rays in the target holder or in the source;
5. energy dependence of the photo- and Compton-effect;
6. X-ray emission out of the sensitive volume of the semiconductor detector;
7. scattering in the detector entrance window;
8. efficiency of the charge carrier collection in the sensitive detector volume;
9. radiation scattering in the entrance window of the vacuum chamber and in absorption filters;
10. excitation of X-rays in the construction materials of the detector and of the radiation source;
11. scattering of X-ray quanta in the materials surrounding the detector;
12. influence of outer electron Bremsstrahlung;
13. dependence of the maximum of the spectral line from the geometry of the measurement;
14. influence of the natural radioactive background;
15. summation of impulses.

In the following the determination of the nonlinearity of spectrometers as well as the calibration of the energy and registration efficiency will be described by the example of semiconductor detectors according to Vylov et al. [555]. The described methods can also be used for other energy-dispersive spectrometers.

Nonlinearity of Spectrometers

In the examination of the nonlinearity of the spectrometer we start from two arbitrary peaks of different energies E_i at positions P_i. From these peaks it is supposed that both peak energies lie on a straight line which can be approximated with the coefficients

b_1 and b_2

$$E = b_1 + b_2 P \tag{3.13}$$

with

$$b_1 = \frac{E_1 P_2 - E_2 P_1}{P_2 - P_1} \quad \text{und} \quad b_2 = \frac{E_1 - E_2}{P_1 - P_2} . \tag{3.14}$$

The accuracy in the determination of the nonlinearity depends on the suitable choice of the difference $(P_2 - P_1)$ and on the corresponding errors of the energy measurement. With equation (3.13) the positions P_j^{th} of the maxima of the chosen peaks of known energy and the differences $\Delta_j = P_j - P_j^{th}$ are determined.

The quantities Δ_j can be approximated by the least squares method with the polynomials

$$\Delta_j = \sum_{i=1}^{m} a_i P_j^{i-1} . \tag{3.15}$$

As a rule for the quantity m we have $2 \leq m \leq 5$. The selection of $m > 5$ can lead to oscillations between the experimental points. The choice of m and of the approximation region result by an optimization of the minimally attainable error in the observed experiment.

$$\bar{F} = \frac{1}{n} \sum_{j=1}^{n} \left(\Delta_j - \Delta_j^{exp} \right) \tag{3.16}$$

with n as the number of experimental values to be used. Usually $\bar{F}$ has a value of ≤ 0.05 channels [555], whereby the attainable value of $\bar{F}$ is determined by the quality of the spectrometer and by the choice and consistency of the calibration normals.

Intensity Calibration

Depending of the measurement problem in the efficiency calibration we distinguish between the absolute and relative registration efficiency.

Absolute registration efficiency. The calibration of the absolute registration efficiency can be done with intensity normals under standard conditions and under consideration of effects which can affect the form of the apparatus spectrum (see Sect. 2.2.2). The experimental conditions must be selected here in a way that allows us to minimize errors which are connected with the geometric reproduction of the source position, the total dead time of the spectrometer, the measurement time and the error in the half-lifetime of radionuclides to be used. It yields

$$\varepsilon_{abs} = \frac{S/\tau}{I \, \exp(-\lambda t)} . \tag{3.17}$$

S describes the area under a peak for a line of given energy, τ the measurement time, I the number of monoenergetic quanta emitted per second from the used source into an angle of 4π, λ the decay constant of the used nuclide and t the time between the date of attestation and the moment of measurement. The approximation of the detector efficiency is done by the least squares method for a polynomial

$$\ln \varepsilon_j = \sum_{i=1}^{m} c_i \left(\ln E_j \right)^{i-1} \tag{3.18}$$

with $2 \leq m \leq 5$. Based on the calibration lines, which are not, as a rule, distributed equidistantly over the observed energy region and the complicated dependency of the registration efficiency on the energy of the X-ray quanta the energy region of interest is often subdivided into several approximation areas and adapted with an equation (3.18). At the same time the optimal value of m can be reached from the minimization of

$$F = \left. \frac{\varepsilon_{abs} - \varepsilon_{calc}}{\varepsilon_{abs}} \right|_{E_j} . \tag{3.19}$$

Control of the relative error in the determination of the absolute registration efficiency is possible with

$$\frac{\Delta \varepsilon}{\varepsilon} = \left[\left(\frac{\Delta I}{I} \right)^2 + \left(\frac{\Delta S}{S} \right)^2 + \left(\frac{0.693 \, \Delta t}{T_{1/2}} \right)^2 + \right.$$
$$\left. \left(\frac{0.693 \, t \, \Delta T_{1/2}}{T_{1/2}^2} \right)^2 + \left(\frac{\Delta \tau}{\tau} \right)^2 \right]^{1/2} \tag{3.20}$$

where $T_{1/2}$ is the half-lifetime of the radionuclide.

Relative registration efficiency. For a range of tasks it is sufficient to know the relative registration efficiency of the spectrometer relative to an intensity normal. At the same time the relative registration efficiency is

$$\varepsilon_{\text{rel}} = \frac{S}{I} . \qquad (3.21)$$

The conversion of ε_{rel} into absolute values can be done by means of a standardized multiplier $\bar{u}$. In this way the quantity $\bar{u}$ can be found from parabolic interpolation between adjacent points of ε_{abs} where the values of $\bar{u}_i$ are related to $\bar{u}_i = \varepsilon_{\text{abs}}^i / \varepsilon_{\text{rel}}^i$. With an averaging of the calculated $\bar{u}_i$ we have

$$\bar{u} = \frac{1}{n} \sum_{i=1}^{n} \bar{u}_i , \qquad (3.22)$$

with n as the number of experimental points to be used. After ordering the efficiency values $\varepsilon(E)$ according to increasing quanta energy E, an approximation of the efficiency curve over the whole energy region of interest is obtained.

Line intensity. For analyzed lines the intensities and their errors can be calculated with

$$I_i = \frac{S_i}{\varepsilon_i} \frac{I_N}{S_N / \varepsilon_N} ; \quad \Delta I_i = I_i \left[\left(\frac{\Delta S_i}{S_i} \right)^2 + \left(\frac{\Delta S_N}{S_N} \right)^2 \right]^{1/2} . \qquad (3.23)$$

Here ε_i is the registration efficiency of the spectrometer for quanta of energy E_i. I_N and S_N are the intensity and the peak area of the calibration peak and S_i the peak area of the peak to be calibrated.

Energy Calibration

After the consideration of the spectrometer nonlinearity coefficients of the calibration curve are calculated with the least squares method as follows

$$E_i = b_1 + b_2 P_i . \qquad (3.24)$$

The error of a single measurement is

$$\Delta E_i = \left[(\Delta b_1)^2 + (\Delta b_2)^2 P_i^2 + r P_i + (b_2 \Delta P_i)^2 \right]^{1/2} . \qquad (3.25)$$

The first three terms in (3.25) characterize the error region of the calibration curve with $r P_i$ as correlation term. The fourth term characterizes the error in the determination of the position of the peak maximum.

The quality of the experiment can be estimated by the analysis of the expression

$$\Delta E_i = E_i - E_i^{\text{th}} . \qquad (3.26)$$

Here the E_i^{th} are the result of the energy values derived from the calibration for the used transition line peaks and the E_i are the input energy values.

3.3 X-Ray Detectors

3.3.1 Energy- and Wavelength-Dispersive X-Ray Spectroscopy

Principally, X-ray spectrometry can be done

- energy-dispersive or
- wavelength-dispersive.

The physical processes for energy-dispersive detectors are:

semiconductor detectors, gas ionization detectors: production of charge carriers with following proportional to the energy amplification of the generated charge or current pulses

scintillation detectors: excitation of scintillation processes with following production of photoelectrons by the scintillation light and amplification proportional to the energy of the electron current in a photomultiplier

cryogenic detectors: calorimetric detection of the quantum energy deposited into the detector volume

Wavelength-dispersive detection methods are based on the diffraction of X-rays on crystallographic lattices according to the Bragg law (3.1)

$$n \lambda = 2d \sin \vartheta$$

with n the order of diffraction, λ the wavelength of the X-rays to be analyzed, d the lattice plane spacing of the used crystallographic plane and ϑ the diffraction angle.

Table 3.2. Energy- and wavelength-dispersive X-ray spectroscopy

	energy-dispersive	wavelength-dispersive
detection method	electric signals proportional to the X-ray energies	Bragg diffraction on the crystal lattice
resolution	$E/\Delta E \leq 50$	$\lambda/\Delta\lambda \leq 30000$ for n=1
registration efficiency	high low for cryogenic detectors	low
typical detectors	semiconductor detectors gas ionization detectors szintillation detectors cryogenic detectors	crystal diffraction spectrometer: Bragg spectrometer (Bragg diffraction on the crystal surface) Laue spectrometer (diffraction on the crystallographic lattice during the radiation transmission through the crystal)
measurement methods	simultaneous multichannel measurement over the whole energetic region of interest	sequential measurement or simultaneous position-sensitive measurement over a registricted energy region

3.3.2 Gas Ionization Detectors

Apart from Geiger–Müller counters gas ionization detectors register X-rays energy-dispersively. A detailed description of the principles of gas ionization detectors and of the basic physics for X-ray detection is given for example by Kment and Kuhn [285], Wiesemann [569] and Delaney and Finch [133].

Physics of X-Ray Detection

Gas ionization detectors as well as semiconductor detectors consist basically of two opposite electrodes between which gases (or semiconductor crystals) are located as detector medium. If a voltage is applied at the electrodes between them an electrical field arises which separates the charge carriers produced in the detector volume and which causes the collection of the charge carriers at the electrodes. This principle is shown in Fig. 3.1. If sufficiently energetic X-rays enter the detector volume in the gas ionization detector *electron–ion pairs* will be produced.

Each entering X-ray quantum generates

$$N = \frac{E_{abs}}{w} \qquad (3.27)$$

charge carriers. Here E_{abs} is the energy absorbed by

Table 3.3. Mean energy for the creation of an electron–ion pair.

element	He	Ne	Ar	Kr	Xe
energy/eV	27.8	27.4	24.4	22.8	20.8

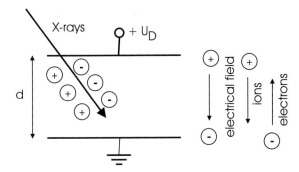

Fig. 3.1. Principle of ionization detectors. d is the electrode spacing

the detector medium and w is the mean energy necessary for the creation of a charge carrier pair (see also Table 3.3) . The corresponding primary charge is then

$$Q = e\,\frac{E_{abs}}{w} . \qquad (3.28)$$

Corresponding to their electric charge, electrons and ions move then to the cathode or to the anode and are registered there. As electric information the mean current or pulses produced by the charge carriers are available. If the potential difference at the electrodes is small (small electric field strength) then the particle movement results from the overlapping of thermal and of directed motion resulting from the electric field. For high field strengths the particle motion is determined by the direction of the electric forces

$$F = eE .$$

Therefore the electric field causes charge separation and collection of electrons and ions on the electrodes, respectively.

Mean ion velocity in field direction. The mean ion velocity in the field direction (we assume a field distribution along the x-axis) can be calculated as

$$v = \mu^+ \frac{p_0}{p} E_x \qquad (3.29)$$

with μ^+ the ion mobility as mean ion velocity for a field strength of 1 V/cm and at normal pressure ($p_0 = 760$ torr). This yields

$$[\mu] = \frac{cm^2}{V s} .$$

Selected values for the ion mobility in different gases are given in Table 3.4.

Table 3.4. Ion mobility of gases at normal pressure

gas	H_2	He	N_2	O_2	Ar	air
$\mu^+ / \dfrac{cm^2}{V s}$	13.0	10.2	1.3	2.2	1.7	1.4

For detectors filled with gases composed of several elements the mobility μ^+ of ions of the component i is

$$\frac{1}{\mu_i^+} = \sum_{j=1}^{n} \frac{c_j}{\mu_{ij}} . \qquad (3.30)$$

Here n describes the number of gas components, c_j the volume concentration of the gas j and μ_{ij}^+ the mobility of ions of the gas i in the gas j.

Electron mobility. The electron mobility μ^- is up to three orders of magnitude higher than those of positive ions. In Table 3.5 some electron mobilities for pure electropositive gases are given.

Table 3.5. Electron mobilities of pure gases at normal pressure

gas	H_2	He	N_2	Ar
$\mu^- / \dfrac{cm^2}{V s}$	7900	500	145	206

The electron mobility does not depend linearly from the field strength E and is sensible to gas composition. For

$$\frac{E}{p} \leq 10^3 \frac{V}{cm\ atm} \quad \text{yields the proportionality } v \sim \sqrt{\frac{E}{p}} .$$

For greater E/p the mean ion velocity increases slowly and can reach values up to $(10^4 \ldots 10^5)$ m/s.

For the complete understanding of gas ionization detectors further processes must be considered:

- drift and mobility of charge carriers,
- recombination of charge carriers,
- electron capture processes and
- charge multiplication processes.

Drift and mobility. In a gas ionization detector in the presence of an electric field the produced charge carriers are accelerated towards the anode or cathode. The charge carrier flow to the detector electrodes is disturbed by collisions with the gas molecules, i.e. the pressure of the working gas of the detector influences the properties of charge carrier collection.

The charge carrier mobility μ characterizes the described circumstance and satisfies the relation

$$\mu = \frac{v}{E} . \qquad (3.31)$$

In a classic approach the charge carrier mobility is connected with the diffusion constant D by

$$\frac{D}{\mu} = \frac{k T}{e} \qquad (3.32)$$

where k is the BOLTZMANN constant and e the elementary electric charge.

Recombination of charge carriers. Because the number of produced charge carriers is decisive for the energy resolution as well as for the registration efficiency of the detector, charge carrier loss processes influence the detector properties substantially.

Positively charged ions can recombine under emission of a photon:

$$A^+ + e^- \rightarrow A + h\nu \ .$$

Molecular ions recombine under emission of an electron:

$$A^- + B^+ \rightarrow AB + e^- \ .$$

For a given detector the characteristic recombination rate dn depends on the densities n_+ and n_- of the positive and negative charged ions. With r as recombination coefficient depending on the chosen gas filling the detector we have

$$-\frac{dn_+}{dt} = -\frac{dn_-}{dt} = r\,n_+\,n_- \ . \qquad (3.33)$$

The influence of recombination processes decreases with increasing detector voltage (increasing field strength E).

Electron capture processes. Electron attachment is based on the capture of free electrons by electronegative atoms when forming negative ions

$$e^- + A \rightarrow A^- + h\nu \ .$$

Therefore electronegative gases (for example O_2, H_2O, NO, NH_3, H_2S, SO_2, halogens) decrease the collection efficiency of electrons and ions before these reach the collection electrodes. This means that electronegative gases influence the function of radiation detectors even then, if it arrives on the speed of the charge carrier transport (pulse detectors).

Electropositive gases (for example noble gases, N_2, H_2, CH_4, CO_2) cannot catch electrons. Therefore it is of importance for gas ionization detectors to use pure gases as the detector medium.

Charge multiplication processes. Multiplication processes in gas detectors occur if the electrons from primary ionization processes gain sufficient energy from the accelerating electric field to ionize gas molecules. The secondary electrons so produced produce further electrons, and so on. This process leads to the formation of an *avalanching process*.

Electron impact ionization produces ion–electron pairs according to

$$e^- + A \rightarrow A^+ + 2\,e^- \ .$$

Furthermore excitation processes

$$e^- + A \rightarrow A^* + e^- \ ,$$

are possible where the photons are emitted as a result of deexcitation processes.

The ionization process can be characterized by the *first Townsend gas amplification coefficient α_T*, which characterizes the number of ionizations that an accelerated electron causes in an electric field. α_T depends on the electric field in the detector and on the gas pressure in the detector volume:

$$\alpha_T = \alpha_T \left(\frac{E}{p}\right) \ .$$

Between two collisions the electrons are accelerated in such a way that they on their part can also cause ionization and therefore the number of negative charge carriers is increased to

$$dN = \alpha_T\,E N\,dx \ . \qquad (3.34)$$

For the number of negative charge carriers the integration of equation (3.34) gives, for varying field strength and initially existing N_0 charge carriers,

$$N = N_0\,\exp\left(\int_0^x \alpha_t(x')\,E(x')\,dx'\right) = N_0\,A \qquad (3.35)$$

with A as the *gas multiplication factor*. The unit of α_T in the chosen form is $1/V$.

Therefore the gas multiplication factor has the form

$$A = \frac{N}{N_0} = \exp\left(\int_0^x \alpha_t(x')\,E(x')\,dx'\right) \ . \qquad (3.36)$$

For a constant electric field equation (3.35) simplifies to

$$N = N_0\,\exp(\alpha_T\,E\,x) \ . \qquad (3.37)$$

Because the gas multiplication factor (3.36) can increase practically to infinity there exist some limitations:

$$\left.\begin{array}{l} A < 10^8 \\ \alpha_T E x < 20 \end{array}\right\} \quad \text{Raether limit.}$$

After exceeding this limit in the gas ionization detector a charge carrier avalanche develops to a spark. Typically an avalanche develops in about 100 ns and a streamer mechanism develops in 10 ns.

Secondary electrons. Primarily, in the detector the produced electrons drift under the influence of the electric field to the anode and gain kinetic energy along the field gradient. This allows them on their part to ionize or excite again. At the same time the produced secondary electrons can ionize again, etc. Therefore each primary electron releases an avalanche of n electrons. Additionally excited atoms and ions with a probability f can produce photoelectrons in the working gas and in the cathode. The quantity f describes the probability that one photoelectron per electron emerges. Then for a primary electron we have

$$1 \text{ electron} \longrightarrow n \text{ electrons} + nf \text{ photoelectrons.} \tag{3.38}$$

The produced electrons generate a new avalanche and we have:

2. step : $(nf)n$ electrons $+ (nf)(nf)$ photoelectrons

3. step : $n^3 f^2$ electrons $+ n^3 f^3$ photoelectrons.

Finally per primary electron,

$$A = n + fn^2 + f^2 n^3 + \ldots \tag{3.39}$$

electrons reach the anode. Thus for $|fn| < 1$ we have

$$A = \frac{n}{1 - fn} . \tag{3.40}$$

In practice, this process causes a distorted rise of the pulse, because the electrical charge corresponding to a registered event will be, after each successive ionization time, increased during the actual photoelectron contribution.

A distorted pulse rise can be avoided when the gas filling an organic gas is added (quenching gas) ,

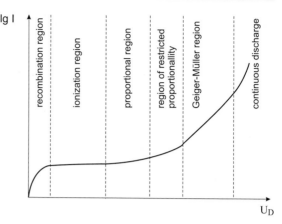

Fig. 3.2. Current–voltage characteristics for the description of physical processes in gas ionization detectors for radiation detection. I is the ionization current; U_D is the detector voltage

which absorbs the emitted photons and in this way prevents the production of photoelectrons. As photon absorbing gases for example CH_4, CO_2, CO or C_2H_5OH are used. Propane, butane, ethylene and propylene also absorb photons, but their dissociation products act also as quenching gas.

Working characteristics of gas ionization detectors. The physical processes in a gas discharge could be described with the current–voltage characteristics for the ionization current I depending on the voltage U_D applied to the detector electrodes for a constant intensity of the impinging X-rays (see Fig. 3.2). A characterization of the region shown in Fig. 3.2 is given in Table 3.6.

Geometry of gas ionization detectors. Generally gas ionization detectors may be constructed in any of three basic geometries:

- parallel plate detectors,
- cylindrical geometry, or
- spherical geometry

The geometries of gas-filled detectors are summarized in Fig. 3.3.

Geiger–Müller Counters

Geiger–Müller counters are gas ionization detectors working in the so-called Geiger–Müller region (see

Table 3.6. Working regions of gas ionization detectors as shown in Fig. 3.2. Q_0 is the total charge, Q_{01} is the charge of primary produced charge carriers, I is the ionization current, and I_S is the saturation current

destination	process	working region
recombination region	Here $U = R \times I$ is valid; not all produced ions arrive at the detector electrodes. Recombination and diffusion processes cause charge carrier losses. $Q_0 < Q_{01}$	not practically used for radiation detection
ionization region	The ionization current is approximately constant. A slow current rise caused by recombination and secondary effects is possible. All primary produced charge carriers become collected. $Q_0 = Q_{01}$	ionization chambers
proportional region	Particles become strongly accelerated, so that they cause secondary ionization $\rightarrow$ avalanche-like charge carrier multiplication arise (gas amplification $k = I/I_S$ up to 10^4). Characteristic is the proportionality of the amplified current to the primary ionization. $Q_0 = k\,Q_{01}$	proportional counter
region of restricted proportionalty	The increasing density of slow positive ions produces space charges which partially screen the detector electrodes. The formation of charge carrier avalanches is restricted by the space charge. Recombinations take effect and yield impinging particles independent of the detector $k = f(U_D)$ and $Q_0 = k\,Q_{01}$ (nonlinear)	rarely used in practice
Geiger–Müller region	The discharge hits the whole electrode geometry; the gas amplification increases so strongly that the current stays independent of the primary ionization. Each produced charge carrier pair can trigger an avalanche. $Q_0 = Q_0(U_D) \rightarrow Q_{01}$	Geiger–Müller counter removal counters spark counter
continuous discharge	Start of a self-sustained gas discharge without influence of ionizing radiation. Continuous discharge or electric breakdown	not useable for particle detection

Fig. 3.2) where they generate a signal that is not proportional to energy independent of the impinging particles. Such pulse counters are suited for applications where only the registered pulse rate is of interest. Because they are applied only as pure pulse counters the price and the demands on the registration electronics are comparatively low.

Counter build-up. Constructions of Geiger–Müller counters include

- cylindrical lateral counters,
- bell counters, and

- conical counters.

The fundamental construction of a cylindrical counter is shown in Fig. 3.4. An outer cylindrical cathode encloses a thick axial wire or a cylindrical anode. The radiation to be analyzed enters through a front-side detector window or along the cylindric geometry into the working volume of the counter and interacts there with the detector gas.

Mode of operation. Through the choice of a suitable detector voltage near the anode wire there emerges an electric field strength high enough that a detector

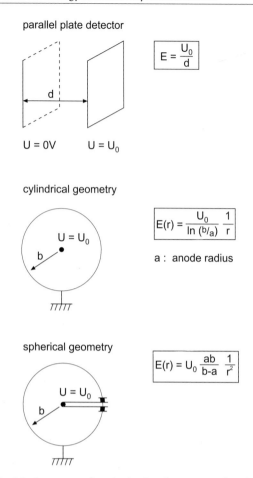

parallel plate detector

$$E = \frac{U_0}{d}$$

U = 0V   U = U_0

cylindrical geometry

$$E(r) = \frac{U_0}{\ln\,(b/a)}\,\frac{1}{r}$$

a : anode radius

U = U_0

b

spherical geometry

$$E(r) = U_0\,\frac{ab}{b-a}\,\frac{1}{r^2}$$

U = U_0

b

Fig. 3.3. Geometry of gas ionization detectors and analytical expressions for the acting electrical field

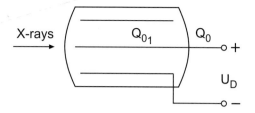

X-rays Q_{01} Q_0 +

U_D

Fig. 3.4. Fundamental construction of a cylindrical Geiger–Müller counter

The produced electrical charge Q_0 is independent of the primary charge Q_{01}, i.e. the counter has no energy dispersion. Independently of the primary ionization all pulses have the same height with an amplitude much higher than the counter noise ($U_{pulse} \approx 1\,V$).

To apply Geiger–Müller counters for the detection of consecutive events the initiated discharge process must be interrupted. The reason is that from the interaction with photons or ions secondary electrons cause an intermittent discharge which must be quenched in order to reproduce the counter functionality again.

The quenching of the discharge can be realized by

- external quenching, or with
- self-quenching counters.

Counters with external quenching are filled with monatomic or diatomic gases which undergo no photoabsorption. Here the quenching of the discharge occurs with on outer electronic quenching circuit. In this way after discharge ignition the counter voltage is lowered below the voltage necessary for charge carrier multiplication up to the moment when all positive ions have reached the cathode. This can be realized with

- a high-ohmic dynamical plate resistance ($\approx 10^9\,\Omega$), or with
- a specially designed electronic circuit.

In *self-quenching counters* a *quenching gas* is added to the working gas to interrupt the discharge without outer influence. As quenching gas organic vapors or gases such as methane, ethanol, halogens (bromine, chlorine) are used. Here the following effects are of importance:

regime in the Geiger–Müller region is formed. Here gas amplifications of up to 10^8 are characteristic. As a result of the high field strengths during the development of the charge carrier avalanches increasingly encourage highly excited gas atoms which on their part can emit photons. In this way in the detector gas and in the cathode, secondary electrons can be produced triggering new charge carrier avalanches which spread over the whole counter region. The produced electrons are collected in about 10^{-8} s and the slowly movable ions on the other hand form a cylindrical space charge surface around the counter wire. This surface acts as a virtual anode and reduces the field strength in the region of the counter wire so much that the avalanche development is interrupted.

1. Quenching gas molecules absorb photons emitted from excited noble gas ions so that these do not arrive at the cathode and initiate there photoelectrons.
2. Positive noble gas ions transfer their charge to quenching gas molecules because their ionization energy is lower than that of the counter gas atoms. Therefore instead of the counter gas ions the quenching gas molecules are transferred to the counter wall. As a rule the excitation energy originating from neutralization processes on the cathode leads to a dissociation of the molecules so that the production of secondary ions does not take place. The discharge stops and starts not before a new photon impinges into the counter and ionizes the working gas.
3. Ion neutralization prevents further photoemission.

Time resolution. Because Geiger–Müller counters do not measure energy-dispersively it is possible to detect the entrance of single photons time-resolved into the counter. In this way the time resolution is determined by the counter dead-time, which is for self-quenching counters in the region of 100 μs up to 300 μs.

Counting response. The counting response describes the measured counting rate as a function of the applied voltage for a constant flux of photons. A dependence typical for Geiger–Müller counters is shown in Fig. 3.5. Beginning with a certain cutoff voltage U_E the counter begins to work. After a sharp increase to the plateau voltage U_P the count rate changes insignificantly up to the breakdown voltage U_D. The operation in this region or at higher voltages leads to a destruction of the counter. For counters with organic quenching gases cutoff voltages of (800 ... 1000) V and for counters with halogen fillings voltages of (200 ... 400) V are characteristic.

The plateau length ($U_D - U_P$) amounts to some hundreds V. For the running voltage U_A should be chosen as

$$U_A = U_P + 100\,\text{V}\,. \qquad (3.41)$$

In this way the slope S of the plateau

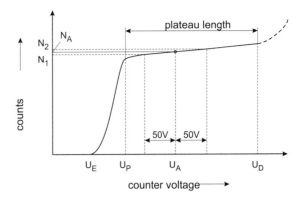

Fig. 3.5. Characteristics of a Geiger–Müller counter (after [527]). U_E is the cutoff voltage, U_P is the plateau voltage, U_A is the running voltage, and U_D is the breakdown voltage

$$S = \frac{N_2 - N_1}{N_M \times 100\,\text{V}}\,100\,\% \qquad (3.42)$$

should not exceed 5%/V.

Registration efficiency. The efficiency for the registration of X-rays depends on the photon energy and the connected radiation transmission of the used detector window and on the type and the thickness of the cathode material. Transmission curves for different materials for instance can be found at the address http://www-cxro.lbl.gov/optical_constants/. Depending on the experimental conditions and the counter to be used the registration efficiency is between 1% and 50%.

Lifetime. Organic quenched counters have a lifetime of about 10^9 impulses and the lifetime of halogen filled counters is in the order of $> 10^{10}$ impulses. As criteria when the counter lifetime is reached, the following items are accounted for:

- the plateau length falls short of 100 V, and
- the plateau slope exceeds 10%/100 V.

Proportional Counters

Build-up. The geometry typical of a proportional counter is the cylindrical geometry, but other geometries are also known (see Fig. 3.3). Next to sealed counters in which the working gas is included in a closed volume, flow proportional counters are also applied where the working gas is permanently exchanged by a suitable gas flow.

Operating mode. Proportional counters can be applied in two different regimes:

energy   : proportional region
detection (high working gas pressure)

counting : region of restricted proportionality
regime $\rightarrow$ high gas amplification is reached
        at low working gas pressure

The mechanism of gas amplification starts at a critical electric field strength that is sufficient to accelerate the electrons to such energies exceeding the ionization energy of the gas molecules. The critical field strength depends on the applied gas and on the gas pressure in the detector and is in the order of $E > 10\,$kV/cm. In cylindrical counters this value is reached next to the central thin (20 μm to 100 μm) anode wire. Therefore the amplification factor k is independent of the position of the ionization track in the gas volume, i.e. independent of the localization of the primary ionization

$$k(U_D) = \exp\left(C\sqrt{U_D\,a\,p}\left[\left(\frac{U_0}{U_{\min}}\right)^{1/2} - 1\right]\right) \approx 2^m \tag{3.43}$$

with C the detector constant (must be determined experimentally), p the gas pressure, m the collision number in the gas amplification region, and $U_{\min}$ the the minimal necessary detector voltage for triggering gas amplification, i.e. $k(U_{\min}) = 1$. The quantity a describes the anode radius and is explained in Fig. 3.3.

In Fig. 3.6 for a proportional counter a simplified circuit in connection with a classic measuring arrangement is shown. In modern measuring devices the impulse hight analyzer is an analog-to-digital converter, which digitizes the energy-proportional signal produced by the amplifier and therefore transforms it into a form possible to store it in a PC or in special electronic devices.

Each registered photon produces at the RC-element of the amplifier a voltage impulse of the form

$$U(t) = \frac{Q}{2C\,\ln(b/a)}\ln\left(1 + \frac{b^2}{a^2}\frac{t}{t_+}\right) \tag{3.44}$$

with t_+ as the time needed by the ions to reach the cathode:

$$t_+ = \frac{p\,\ln(b/a)}{2\,U_D\,\mu_+}\,(b^2 - a^2)\,. \tag{3.45}$$

For the collection time T_- of the electron component, the same expression as (3.45) holds, but the quantity μ_+ is replaced by μ_-.

The course of voltage pulses for different differentiation times RC is shown in Fig. 3.7. Two cases are shown:

- $t_+ \ll$ RC
  At the beginning the voltage pulse rises very steeply as a result of the short electron transit time T_-, then more slowly through the contribution of the ion component up to an amplitude

$$U_{\max} = k\,\frac{e}{C}\,\frac{E_{\text{abs}}}{w}\,.$$

- $t_- \ll$ RC $\ll t_+$
  The pulse height is here also proportional to E_{abs}/w, however the pulse duration is far smaller. This allows count rates up to 10^6 pulses/s without counting losses appearing.

Energy resolution. In the X-ray region the energy resolution of gas proportional counters is at 6 keV typically in the order of 15% to 20%. Counting rates up to 10^6 counts per second are possible.

Time resolution. Based on the short impulse form at suitable differentiation constants (see Fig. 3.7) for proportional counters, a time resolution up to about 10^{-6} s is possible. In this way for high impulse rates only the electron component is analyzed.

Counting response. As for Geiger–Müller counters the working point is placed into the plateau of the counting-rate characteristics (see also Sect. 3.3.2). The form of the counting response depends on the kind of radiation to be detected, the gain of the linear amplifier and on the response limit of the subsequent electronics.

Proportional counters allow a separate measurement of radiation components with different specific ionization. Strongly ionizing particles produce

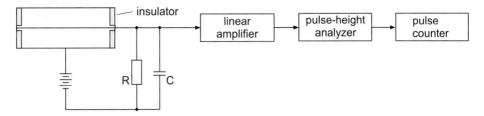

Fig. 3.6. Measuring arrangement for energy-dispersive measurements with proportional counters

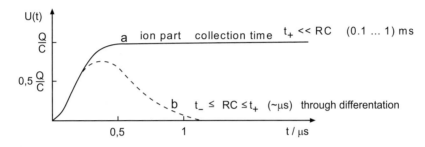

Fig. 3.7. Course of voltage pulses in gas proportional counters (after [527]).

greater pulses and can be detected at lower counter voltages, weakly ionizing radiation needs for the registration higher detector voltages.

Registration efficiency. The registration efficiency depends on the detector gas, its pressure and on the material and thickness of the applied detector windows. In Fig. 3.8 the course of the registration efficiency for a xenon–methane mixture is shown at different working gas pressures. The clearly pronounced edge-structure above 30 keV is caused by the K absorption edge of xenon. Furthermore for the low-energy region the transmission of selected window materials is shown.

Gas composition. The choice of the working gas must fulfill different demands such as for instance proportionality between the energy of the radiation and of the detector signal, a high gas amplification, a lower detector working voltage and the detection of high counting rates. As a rule noble gases are inserted as the detector filling because these gases need the lowest detector voltages for avalanche formation.

Pure argon makes possible gas amplifications in the order of $10^3 \ldots 10^4$. Through the addition of anorganic gases such as CO_2, CH_4 or BF_3 the amplification factor can be increased up to 10^6. The supple-

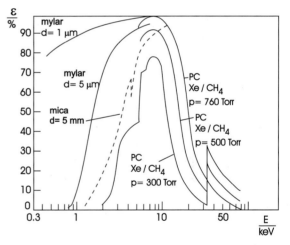

Fig. 3.8. Registration efficiency of a Xe/CH_4 proportional counter as a function of the X-ray energy E. In the low-energy region the transmission of some mylar and mica windows is shown also

ment of electronegative gases such as freon allows us to increase the amplification further to 10^7. Details are discussed in [452], for example.

Lifetime. Closed proportional counters with working gas/quenching gas mixtures show, depending on the gas composition, the gas pressure and on the energy of the detected photons after about $\geq 10^{10}$ counts, a deterioration of the detector characteristics. This problem does not appear if gas flow detectors are inserted.

Position-sensitive proportional detectors

Position-sensitive proportional counters are applied for X-ray detection as

- one-dimensional resolving position-sensitive proportional counters, and
- two-dimensional resolving position-sensitive proportional detectors (multiwire detectors or microstrip detectors).

One-dimensional resolving position-sensitive proportional counters. The following detector types are used as one-dimensional resolving proportional counters:

1. **Detectors with a high-impedance anode wire.** Detectors with a high-impedance anode wire, of whose ends both voltages U_1 and U_2 could be registered and where the position information is obtained from the voltage ratio.
2. **Detectors with integrated delay line.** In detectors with integrated delay line tappings from the delay line are connected to the detector cathode segments. The time difference between the voltage signals U_1 and U_2 includes the position information and the integral of both signals is proportional to the transferred photon energy.
3. **Detectors with wedge-like double cathode.** Detectors with wedge-like double cathodes are based on the principle that the charge distribution produced by the impinging radiation on two wedge-like cathode segments is analyzed. The ratio U_1/U_2 contains the position information and the sum $U_1 + U_2$ is proportional to the absorbed energy.
   The distance between the individual cathode segments can amount to 1 mm to 1.5 mm and allows position resolutions of about 200 μm.

Two-dimensional resolving position-sensitive proportional counters. Two-dimensional resolving proportional counters are known as:

1. **Multiwire proportional counters.** Multiwire proportional counters can be applied for energy spectroscopy as well as for position detection. These detectors have been known for more than thirty years [105, 106].
   Multiwire proportional counters are built up with a lattice of very thin anode wires of which each wire works as an independent proportional counter. The lattice itself is centered between two cathode planes with a typical cathode–anode distance of 7 mm to 8 mm. The distance between the single anode wires is in the order of 2 mm. The cathodes themselves consist of a certain number of parallel wires that are selected by a delay line or by amplifiers arranged in decade groups. As a rule two cathode wire fields crossed at 90° are used, so that a two-dimensional registration of the signal position becomes possible.
   For the measurement of high counting rates or in coincidence experiments also expensive solutions are used, with which to each wire a separate amplifier is assigned. The attainable time resolution of multiwire proportional detectors is essentially determined by the electron drift time.
   For a wire distance s the position resolution is in the order of $s/3$. In general the attainable energy resolution is somewhat smaller than in single-wire proportional counters, because for a good energy resolution an extremely equally spaced wire distance is required. Here small deviations cause a deterioration of the energy resolution.
2. **Microstrip proportional detectors.** Microstrip proportional detectors use a structured anode plane. The position resolution for this detector type can be 30 μm.

3.3.3 Scintillation detectors

If in gas ionization detectors the formation of free charge carriers is used for the detection of X-rays, in scintillation detectors luminescence flashes (scintillations) emitted from excited atoms or molecules of solid, liquid or gaseous scintillators are used to detect the impinging radiation.

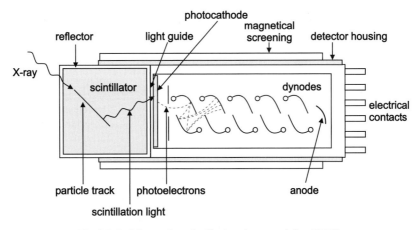

Fig. 3.9. Build-up of a scintillation detector (after [200])

Build-up of scintillation detectors. The scheme of a scintillation detector is given in Fig. 3.9 and the build-up can be characterized as

scintillator $\rightarrow$ light guide $\rightarrow$ photomultiplier $\rightarrow$ pulse processing .

In the scintillator luminescence light is produced by impinging photons, which by a light guide are transferred on the photocathode of a photomultiplier. The photoelectrons initiated from the photocathode were focussed on to so-called dynodes, at which appears a step by step multiplication of the secondary electrons. The electrical pulses so produced contain information on the energy of the registered quanta and can also be used for time spectroscopy.

Operational mode. For example, a description of the operational mode of a scintillation detector is given in [526]: The energy E_X of the in the incoming photons can be absorbed in the scintillator by ionization and excitation processes partially $E_{abs} < E_X$ or complete $E_{abs} = E_X$. In this way the *scintillation response* η_S describes the part of the absorbed energy E_{abs} that is transferred into the energy of the luminescence light (scintillations with a decay constant τ).

Is E_m the mean energy of a photon then

$$N_l = \eta_s \, \frac{E_{abs}}{E_m}. \qquad (3.46)$$

Luminescence quanta will be emitted isotropically. Because as much as possible of the produced light quanta should arrive at the photocathode, the scintillator is often enveloped with a light reflecting layer and the optical contact between scintillator and photocathode is realized by a light guide, which is additionally optical coupled to the scintillator and photocathode by light-guiding pastes. The quality of the optical coupling is characterized by the *light transfer factor* k_l, which describes the number of photons reaching the photocathode in relation to the number of primary produced quanta.

Therefore $N_l \, k_l$ photons reach the photocathode and produce

$$N_c = N_l \, k_l \, \eta_l \qquad (3.47)$$

photoelectrons. The *quantum efficiency of the photocathode* η_q describes the ratio between the number of emitted photoelectrons to the number of quanta reaching the cathode.

Through an electron-optical focussing system integrated into the photomultiplier, a fraction k_e (*electron transfer factor*) of the photoelectrons reach the first dynode of the photomultiplier and produces Δ secondary electrons. Towards the anode an increasing positive electrical potential is formed for the dynodes by a potential divider, so that the electrons are guided from dynode to dynode up to the anode. When impinging on each dynode a further Δ secondary electrons are initiated so that an

scintillator	Z_{eff}	ϱ	λ_m	λ_c	τ	I_r	H
NaI(Tl)	46.5	3.67	415	320	230	100	yes
CsI(Tl)	54	4.52	565	330	980	45	slightly
CsI(Na)	54	4.52	420	300	630	85	yes
CsF	49	4.64	390	220	5	6	yes
$Bi_4Ge_3O_{12}$ (BGO)	62,5	7.13	480	350	300	12	no
$CdWO_4$	54	7.87	540	450	800	40	no
CaF_2(Eu)	14.6	3.18	435	405	940	50	no

Table 3.7. Physical characteristics of selected scintillation materials. Z_{eff} is the effective atomic number, ϱ is the density in g/cm², λ_m is the wavelength of maximum emission in nm, λ_c is the wavelength of the scintillation cutoff wavelength in the crystal, τ is the decay constant in ns, I_r is the light output relative to NaI(Tl), and H is hygroscopic

electron amplification of Δ^n

results for n dynodes. Therefore

$$N_e = N_c\, k_e\, \Delta^n \qquad (3.48)$$

electrons arrive at the anode of the photomultiplier. This corresponds to a charge of $Q = e\, N_e$.

If the time constant RC on the photomultiplier output increases in comparison to the time constant for the described elementary processes, then the charging of the initial capacitor C leads to a voltage impulse with the quantity

$$U = \frac{Q}{C} = \frac{e}{C}\, \eta_q\, \frac{E_{\max}}{E_m}\, k_e\, \eta_q\, \Delta^n \,. \qquad (3.49)$$

Scintillators. The most important scintillator materials are

1. anorganic crystals,
2. organic crystals,
3. organic liquid scintillators, and
4. plastic scintillators in polymerized synthetic materials.

In addition doped glasses or gases can also be applied. A detailed discussion of single scintillator materials and of their characteristics is given for example in [200, 323].

Some characteristic values of scintillators used in X-ray spectroscopy are summarized in Table 3.7.

Often λ_m does not agree with the maximum of the spectral sensitivity of the photocathode. This can be improved through the addition of a second luminescent substance, a so-called *frequency convertor*.

In Table 3.7 luminescence decay times of selected scintillators are given for room temperature. For the optimal use of scintillation detectors short luminescence decay times are necessary. Note that because of their composition a row of scintillators has different decay times τ_i. Therefore the measured form of a scintillation pulse $n(t)$ consists of the overlapping of different components $n_{0i}(t)$:

$$n(t) = \sum_i n_{0i}\, \exp(-t/\tau_i) \qquad (3.50)$$

with n_{0i} total number of photons emitted.

For the X-ray energy region the photoeffect is dominant in photon detection processes. Therefore as a rule scintillator thicknesses of up to 10 mm are sufficient.

Light reflectors and light guides. Because the light emission in the scintillator is isotropic only a part of the produced photons impinges on the photocathode. To increase the number of photons falling on the photocathode the scintillator can be equipped with a reflection layer. Often reflectors such as MgO powder ($\approx 97\%$ reflection) or aluminum evaporated polyester ($\approx 93\%$ reflection) are used. The quality of the reflector determines the number of secondary electrons N_c emitted per registered event from the cathode. Ideally N_c should be independent of the origin of the light in the scintillator.

In certain cases (for example big scintillators and small photocathodes) the application of *light guides* is helpful. Organic glasses are used as light guides. In this way light transport occurs through total reflection. In longer light guides the resulting transmission coefficient must be considered.

Table 3.8. Operational parameters for commercial photomultipliers

parameter	value
photocathode materials	multialkali coating: Na_2KSb activated with caesium bialkali coating: K_2CsSb activated with oxygen and caesium
quantum output of the photocathode	$0.1 \ldots 0.4$
gain of a single dynode	$2 \ldots 20$
inter-dynode voltages	$50\,V \ldots 300\,V$
dynode materials	beryllium or magnesium oxydes Cs_3Sb gallium phosphide
total gain	$10^4 \ldots 10^8$
charge per pulse	$10^{-12} \ldots 10^{-8}\,As$
cross current in the voltage divider for spectroscopic photomultipliers for short-time photomultipliers	$0.1 \ldots 1\,mA$ $1 \ldots 10\,mA$

Air gaps between scintillator, light guide and photomultiplier must be avoided because otherwise light losses through total reflection can appear.

Photomultiplier head. The build-up of a scintillation detector is shown in Fig. 3.9. Typical operation parameters for commercial photomultipliers are given in Table 3.8.

The basic processes in photomultiplier tubes can be summarized as follows:

1. Light quanta that arrive at the photocathode produce photoelectrons.
2. The photoelectrons are focussed electrostatically on the first dynode. There electron multiplication takes place, i.e. a gain of the primary electron current occur.
3. The electron amplification increases from dynode to dynode.
4. The amplified signal is collected at the anode and then reaches the subsequent electronics.

The generation of the dynode potentials takes place by means of a voltage divider as shown in Fig. 3.10. As a rule the ratio between the different R_i are given by the manufacturer. Then the absolute

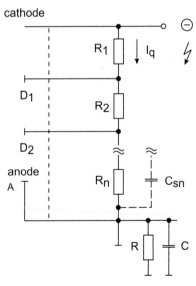

Fig. 3.10. Voltage divider for the generation of the dynode potentials. D_i is the dynodes, F is the focussing electrodes, K is the photocathode, P is the anode, and R_i are the resistances of the voltage divider

values of the resistances arise from the actual cross-currents (see Table 3.8). In addition, supporting capacitors on the last dynodes serve to hold the dynode potentials constant during a pulsative current load.

The *amplification stability* substantially depends on the voltage U put onto the photomultiplier. For the amplification k we have

$$\frac{\Delta k}{k} \approx \frac{\Delta U}{U} \times n \qquad (3.51)$$

with n the number of dynodes. For the application of scintillation detectors in energy-dispersive X-ray spectroscopy the detector voltage should be stabilized to about 0,1%.

Of special design are so-called *short-time photomultipliers* for applications in time spectroscopy. Here very high output pulses appear, i.e. the anode signal no longer depends linearly on the photon energy. On the other hand, for energy spectroscopy a dynode output is used.

For the use of photomultipliers it must be considered that a magnetic induction of μT can clearly influence the measuring result. Therefore photomultipliers are often shielded against outer magnetic fields (among others also against the Earth's magnetic field) with iron or μ-metal.

Registration electronics. The pulse amplitude generated on the dynodes or on the anode is comparatively small and must be amplified. Common preamplifiers include:

1 **Charge amplifiers.** A charge amplifier converts the arriving signals into a voltage proportional to the detected electrical charge (see Fig. 3.11). For the time constant counts

$$\tau = RC \; > \; \text{pulse width at the amplifier input}$$

i.e. τ is chosen greater than the width of the impulses to be amplified. To warrant low noise, field-effect transistors are used as operational amplifiers. The disadvantage of a charge amplifier consists in its missing suitability for high counting rates. On the other hand, its advantage is the good signal dispersion.

2 **Voltage amplifier.** In a voltage amplifier (see Fig. 3.12) the initial current is converted into a

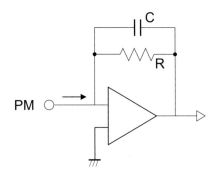

Fig. 3.11. Principal build-up of a charge amplifier

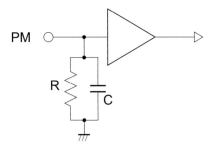

Fig. 3.12. Principal build-up of a voltage amplifier

voltage. For the choice of the time constant we have

$$\tau = RC \; < \; \text{pulse width on the amplifier input} \; .$$

The advantages of a voltage amplifier are its high count-rate processing and the circumstance that here the internal noise is not of importance.

Energy resolution. For scintillation detectors energy and time resolution are substantially determined by inhomogeneities of the scintillator crystal and of the photocathode as well as by the statistics of the detection process.

For X-ray detection only a small part of the photon energy is converted into photoelectrons because it counts

$$\eta_s \, \eta_q \ll 1 \; .$$

The root-mean-square fluctuation of the output signal calculated for the case of the detection of monoenergetic radiation satisfies

$$\Delta_{\text{output}} = \frac{1}{N_c} + \frac{1}{N_c \, k_d} + \frac{1}{N_c \, k_d^2} + \ldots \qquad (3.52)$$

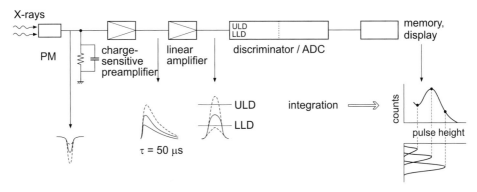

Fig. 3.13. Measuring arrangement for X-ray spectrometry with a scintillation detector

with k_d the gain of a single dynode. Therefore for the root-square fluctuation of the output signal we have

$$\frac{\Delta Q_A}{Q_A} = \Delta^2_{\text{output}} = \frac{1}{E_{\text{abs}}} \frac{E_m}{\eta_S} \frac{1}{\eta_q k_l} \frac{k_d}{k_d - 1} \,. \quad (3.53)$$

Scintillation detectors have a comparatively low energetic resolution. This is caused by the relatively small number of produced photons and electrons as well as through a Fano factor of $F \approx 1$. Typically a relative resolution of 8% up to 15% is reached. At low energies values up to 40% can appear.

A typical measuring arrangement is shown in Fig. 3.13. The pulses are amplified and formed by a preamplifier and are further amplified in a linear amplifier and evaluated in an ADC or discriminator. In the subsequent storage unit (as a rule a PC with interface/processing card) a multichannel spectrum of the measured radiation field is registered.

The structure of the measured spectrum is determined for a certain energy resolution by the interaction processes of the detected photons with the scintillator atoms. For radiation detection in the X-ray energy region, the following effects are of importance:

1. **Photoeffect.** Through the absorption of an X-ray of energy E_X a photoelectron with energy

$$E_e = E_X - E_B$$

is produced. E_B is the binding energy of the emitted photoelectron. The photoeffect dominates, especially in scintillator materials with high atomic number Z and at low X-ray energies E_X:

$$\sigma_{\text{photo}} \sim \frac{Z^5}{E_X^{7/2}} \,.$$

2. **Compton effect.** By the Compton effect only a part of the X-ray quantum energy is transferred to a slightly bound electron and electrons up to a certain edge-energy (Compton edge) are emitted. The Z- and E_X-dependence for Compton scattering has the form

$$\sigma_{\text{Compton}} \sim \frac{Z}{E_X} \,.$$

The photon spectrum corresponding to individual elementary processes is shown in Fig. 3.14. At X-ray energies below the threshold for pair-production processes (1.02 MeV) the photo- and the Compton effect dominate. For the photoeffect a sharp line is characteristic, which corresponds to the registered X-ray energy. On the other hand, for Compton scattering a continuum, the so-called *Compton continuum*, is observed, which delivers contributions to the spectrum up to a certain limiting energy. The resulting spectrum is represented on the right side of Fig. 3.14. To characterize the balance of the elementary processes given in Table 3.9 the linear attenuation coefficients for photon radiation in a NaI(Tl) scintillator are indicated.

From Table 3.9 one can see that for a photon energy of up to 200 keV the photoeffect dominates. Therefore in this energy region it is sufficient to

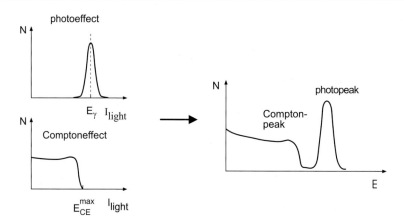

Fig. 3.14. Photon spectrum arising from the individual elementary processes and resulting total spectrum

Table 3.9. Linear absorption coefficients μ for the photo- and the Compton effect in a NaI(Tl) crystal

E_γ/kev	μ_{photo}/cm^{-1}	$\mu_{Compton}$/cm^{-1}	μ_{total}/cm^{-1}
50	20	0.5	20.5
100	5.5	0.45	5.95
200	0.8	0.3	1.1
500	0.07	0.28	0.35

use, for example, NaI(Tl) crystals with a thickness of about 1 cm. Only for the detection of photons with energies higher than about 1 MeV are greater scintillator volumes required, because then the total absorption peak preferably arises from multiple interaction processes.

Time resolution. Real-time fluctuations δt_{PM} in the photomultiplier are in the order of

- 10 ns to 100 ns for spectroscopic photomultipliers, and
- about 0.1 ns to some ns for short-time photomultipliers.

This means that counting rates up to 10^7 pulses per second can be registered without impulse losses.

The form of the output pulse results then from the convolution of the pulse form for the scintillator light with the pulse resulting from the charge carrier collection in the photomultplier ($2.35 \times \delta t_{PM}$). Therefore statistical fluctuations are connected, leading to

a certain trigger unsharpness of the output current, which on the other hand leads to time fluctuations in the registration of time signals. Depending on the scintillator, the photomultiplier and on the absorbed photon energy, fluctuations in the order of 0.1 ns up to some ns are possible.

For a sufficient time resolution the following combination is advantageous:

- photomultiplier with low transit time fluctuations, and
- scintillator with short luminescence decay time and high light output.

Registration efficiency. The determination of the registration efficiency results from the evaluation of the full-energy peak from the photoeffect. For this purpose X-ray sources with known source strength are used. One of the possible definitions of the registration efficiency is

$$\varepsilon = \frac{\text{total number of in the full-energy peak registered counts}}{\text{from the source emitted quanta in the dead-time corrected measurement time}} .$$

(3.54)

For a NaI(Tl) scintillation detector a characteristic registration efficiency as a function of the registered X-ray energy is given in Fig. 3.15.

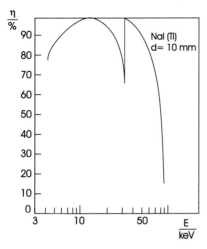

Fig. 3.15. Registration efficiency η of a NaI(Tl) scintillation detector of thickness 10 mm

3.3.4 Semiconductor Detectors

For more than 50 years semiconductors have been successfully applied with an energy resolution clearly higher than those from other detectors such as gas ionization detectors or scintillation detectors and with comparatively good registration efficiency. On the basis of the wide application of semiconductor detectors in different scientific areas and commercial applications there exist a multitude of excellent monographs on this type of detector (see, for example, [331]), to which readers are referred.

Build-up. Semiconductor detectors work as solid-state ionization chambers. They use properties of semiconductor materials with p- or n-conduction produced by doping pure semiconductor materials with donors or acceptors:

- **p-conduction** means an excess on *acceptors*, i.e. mobile positive charge carriers (holes) produced by doping of the semiconductor material with trivalent atoms (B, Al, Ga, In).
- **n-conduction** means an excess on *donors*, i.e. negative charge carriers produced by doping with pentavalent atoms (P, As, Sb, Bi).

If n- and p-conducting materials are brought into contact then electrons and holes diffuse in the adjacent area with opposite types of charge. In this way the electric charges partially compensate so that in the n-region positive and in the p-region negative excess charges arise. Between both areas a potential difference ($U_{pn} \approx 0.6\,\mathrm{V}$) originates, which prevents further diffusion of charge carriers.

The hole and electron concentration is different by many orders of magnitude in the p- and n-region of the semiconductor and produce a junction space-charge region that, because of the depletion of free-movable charge carriers, acts as a barrier layer (Fig. 3.16: region from -a to b). If in the reverse direction at the p-n contact an outer voltage is applied (negative polarity at the p-region), then in the reverse direction a high electrical field strength appear

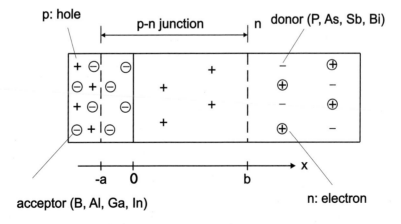

Fig. 3.16. Basic structure of a p-n contact

material	Si	Ge
atomic number	14	32
density [g cm^{-2}]	2.3283	5.3234
dielectric constant (relative) ε_r	12	16
forbidden energy gap (300 K) [eV]	1.115	0.665
forbidden energy gap (0 K) [eV]	1.21	0.785
electron mobility (300 K) [cm^2 V^{-1} s^{-1}]	1350	3900
hole mobility (300 K) [cm^2 V^{-1} s^{-1}]	480	1900
electron mobility (L N$_2$; 77 K) [cm^2 V^{-1} s^{-1}]	2.1×10^4	3.6×10^4
hole mobility (L N$_2$; 77 K) [cm^2 V^{-1} s^{-1}]	1.1×10^4	4.2×10^4
mean energy per ion–electron pair (77 K) [eV]	3.76	2.96
specific resistance (300 K) [Ω m]	2300	0.47
specific resistance (77 K) [Ω m]		500
Fano factor	0.1	0.1

Table 3.10. Properties of semiconductor materials

which allows fast charge carrier separation, if a collision of ionizing radiation generates ion–electron pairs.

Operational mode. During the entrance of an X-ray quantum into the semiconductor material, ion–electron pairs are formed by ionization processes. In the barrier zone electrons and ions are quickly separated by the electric field and accelerated towards the field boundaries (electrons move towards the n-region, ions towards the p-region). In the outer circuit this causes a short-time current on the detector and a spurious capacity which can be measured. The registered pulse amplitude increases with increasing detector voltage, which on the other hand, is limited by the increasing noise of the reverse current.

The layer thickness of the depletion depth of the n-region x_n (region 0 to b in Fig. 3.16) and of the p-region x_p (region -a to 0 in Fig. 3.16) can be calculated as

$$x = \sqrt{2\,\varepsilon_r\,\varepsilon_0\,\varrho\,\mu\,(U_D + U_A)} \qquad (3.55)$$

with ε_r the dielectric constant of the base material, ε_0 the electric field constant, μ the charge carrier mobility, ϱ the specific resistance of the base material, U_A the applied outer blocking bias and U_D the diffusion voltage. Some properties of semiconductor materials important for the physics of the detector material are summarized in Table 3.10. Basic physical prop-

erties of the elemental semiconductors germanium and silicon are also reviewed in detail by Haller [220].

The total width of the space–charge region

$$x_D = x_n + x_p \qquad (3.56)$$

describes the thickness of the sensitive detector volume. Then the barrier-layer capacity can be calculated with

$$C_d = \varepsilon_0\,\varepsilon_r\,\frac{F}{x_D} \,. \qquad (3.57)$$

The quantity F describes the area of the barrier layer.

Already, through the diffusion voltage, a thin barrier-layer arises. If the detector voltage is switched on, in the high-resistive space-charge region the complete blocking voltage releases, so that a zone of high field strength is formed. Barrier-layer thickness, detector capacity and also the pulse high (because of $U = Q/C$) depend on the blocking voltage U_A.

Important advantages in the application of semiconductor detectors are that:

- semiconductor detectors such as solid-state devices have a high density, i.e. it is possible to reach a comparatively high registration efficiency;
- the small energy extent, in comparison to other detector types, for the production of charge carrier pairs allows the production of more charge carriers per impinging X-ray and therefore a better energy resolution;

Table 3.11. Properties of semiconductors from binary compounds

parameter	HgI_2	GaAs	CdTe
atomic number	80/53	31/33	48/34
density [g cm^{-2}]	6.30	5.35	6.06
energy gap (300 K) [eV]	2.13	1.43	1.45
mean energy per electron–ion pair [eV]	4.22	4.51	4.42
electron mobility [cm^{-2} V^{-1} s^{-1}]	100	8500	1050
hole mobility [cm^{-2} V^{-1} s^{-1}]	4	420	80

- the signal produced in the detector is independent of the place of origin in the detector volume which favors semiconductors for spectrometric measurements; and
- the collection time for charge carriers is in the order of 10 ns up to 100 ns, i.e. good time resolution is possible.

Disadvantages are that

- semiconductor detectors are relative expensive;
- for most semiconductor detectors, cooling with liquid nitrogen is necessary; and
- semiconductor detectors under the influence of ionizing radiation can be damaged, i.e. for intense radiation fields the lifetime of the detector can be restricted.

Detector materials. To realize ideal spectroscopic conditions certain demands must be addressed for the detector materials:

1. the energy for the creation of a charge carrier pair must be as low as possible;
2. high charge carrier mobility by small losses or a high charge carrier lifetime (pure crystals without dislocations);
3. good insulation properties to avoid a small residual current flowing at high applied voltages. This is connected with the demand to have a low concentration of thermally produced charge carriers, i.e. a wide band gap and detector operation at low temperatures is favorable;
4. high resistance to get large active detector volumes;

5. high Z-materials, because the photoeffect is proportional to Z^5 (large absorption coefficient);
6. the material must allow good contact so that the negative electrode cannot inject electrons and the positive electrode cannot inject positive charge carriers;
7. the material must be available in reasonable amounts at reasonable costs.

In semiconductor detectors most frequently silicon or germanium monocrystals are used, because these crystals allow the realization of large detector volumes at the best energy resolution. Because of the small band gap in the case of germanium, cooling with liquid nitrogen (77 K) is required. Silicon detectors must be cooled if highest energy resolution should be preserved. A compilation of material parameters is given in Table 3.10.

Because of their wide band gap semiconductor detectors from binary compounds such as HgI_2, GaAs, CdTe and others, can be stored and operated at room-temperatures. In Table 3.11 selected parameters for some binary compound semiconductors are summarized. The high atomic number of these materials (larger absorption coefficients) make them attractive for X-ray spectroscopy. For example, 2 mm of CdTe for X-ray absorption is equivalent to 10 mm of germanium.

Packaging. To achieve good energy resolution as a rule semiconductor detectors are equipped with a Dewar vacuum flask to minimize contributions from detector noise. A typical detector configuration with cryostat is shown in Fig. 3.17. There are different de-

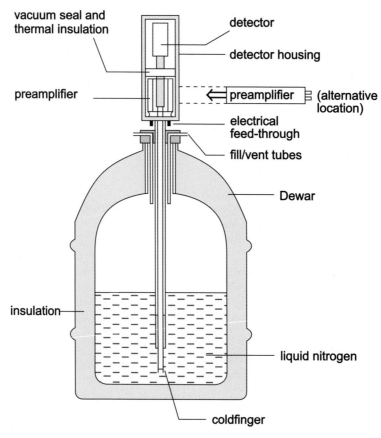

vacuum seal and
thermal insulation

detector

detector housing

preamplifier

preamplifier ⇐ preamplifier (alternative
location)

electrical
feed-through

fill/vent tubes

Dewar

insulation

liquid nitrogen

coldfinger

Fig. 3.17. Typical packaging of semiconductor detectors for the application at LN_2 temperatures (after [200])

tector versions where the detector finger is directed upwards (Fig. 3.17), is mounted rectangular above the Dewar, or rectangular at the Dewar plate.

Detector types. In energy-dispersive X-ray spectroscopy different types of semiconductor detectors are in use. The most common detector types are:

- **Barrier-layer detectors.** These are Si detectors that use the p-n transition as a radiation detector. In this way, to prevent the current flow, the detector operates in the reverse direction. The sensitive region (barrier-layer) depends on the material properties and on the applied voltage. The thickness of the barrier-layer can be calculated according to equation 3.55.

At room temperature for silicon we have [354]:

p-conducting material:

$$d[\mu m] = 0.3 \sqrt{\varrho_p[\Omega\,cm]\,(U_D[V] + U_A[V]}$$

n-conducting material:

$$d[\mu m] = 0.5 \sqrt{\varrho_n[\Omega\,cm]\,(U_D[V] + U_A[V])}\,.$$
$$(3.58)$$

The capacitance of the depletion layer is given by equation (3.57) and is approximately:

$$Si: C[pF] \approx 1.05\,\frac{F[cm^2]}{d[cm]} \qquad (3.59)$$

$$Ge: C[pF] \approx 1.40\,\frac{F[cm^2]}{d[cm]}\,.$$

In most cases the p-n transition is produced through diffusion of phosphor in p-conducting silicon (Si(P) detectors). If a detector basis material with high specific resistance is used detectors with depletion zones of 1 mm to 3 mm can be achieved.

- **p-i-n detectors.** To achieve good registration efficiency for energetic photon radiation, lithium drifted detectors are used which are designated depending on the basic material, such as Si(Li) or Ge(Li) detectors.

  With the lithium drift technique thick intrinsic barrier layers can be produced. Here lithium atoms are drifted into p-conducting germanium or silicon at high temperatures. Then the drifted lithium atoms get deposited on interstitial sites forming a p-n junction. When a normal barrier layer emerges a blocking voltage is applied and lithium ions drift into the p-zone. Through charge compensation between the n- and p-zones a high-impedance intrinsic conducting zone, the so-called i-zone (intrinsic layer: p-i-n structure) arises, which shows a practically constant field strength. Depending of the drift and contacting technologies to be used, a detector thickness of up to some cm can be realized.

  For a temperature of 77 K the reverse current of common Ge(Li) detectors amounts to about 10–100 pA, corresponding to a noise level of 100–300 eV respectively. To realize highest energy resolutions, detectors operate principally at liquid nitrogen temperatures. To avoid contamination of the detector surface the detector is mounted in vacuum because such contaminations can lead to field inhomogeneities and leakage currents.

- **HP Ge detectors.** Germanium detectors can be produced with a broad intrinsic conducting zone without lithium drift, if pure materials with $10^9 \ldots 10^{11}$ impurity atoms per cm^3 are used. These detectors are so-called HP[1] detectors. They can be stored at room temperatures, but for reaching highest detector resolution operation at 77 K is necessary (LN_2 temperature).

- **Semiconductor detectors with binary compounds** are applied for X-ray detection at room temperature or with thermoelectric cooling in detectors with comparatively small detector volumes.

- **Position-sensitive detectors.** Silicon or germanium detectors are applied as energy-dispersive position-sensitive X-ray detectors. Common realizations are:

  - *Continuous detector structures with resistive layer.* The principle of such detectors is based on the charge dividing principle, in which the from of the registration position x-dependent charge signal is callipered from the two contacts on the ends of the resistive layer. The ratio of the signal amplitudes gives the radiation impact point with a measurement accuracy of some micrometers. The addition of both signals then gives the energy information.

  - *Strip detectors.* Detectors produced in planar technology and consisting of many individual strip detectors of strip widths of $\geq 10\ \mu m$ and strip distances of $\geq 10\ \mu m$ also allow the position-sensitive detection of X-ray radiation.

  - *Honeycomb detectors.* Here the detector consists of a field of independent detectors, and each detector is equipped with its own spectroscopic electronics.

  - *CCD detectors.* Fully depleted back-illuminated CCDs have a good registration efficiency in the range from eV up to 10 kev or higher, and can in this energy region be applied to position-sensitive X-ray detection. For example, the soft X-ray performance of back-illuminated CCDs is discussed by Bailey et al. [40].

Energy resolution. The spectrometer energy resolution ΔE is determined by contributions stemming from the detector $\Delta E_{det.}$ and from the electronics $\Delta E_{el.}$:

$$\Delta E = \sqrt{\Delta E_{det.}^2 + \Delta E_{el.}^2} \ . \tag{3.60}$$

The ultimate detector resolution from charge carrier collection is

---

[1] HP: High Purity

$$\Delta E = \sqrt{w\,E\,F} \qquad (3.61)$$

where E is the X-ray energy and F is the Fano factor.

If the detected transition is represented by a Gaussian shape the FWHM[2] is determined as

$$\text{FWHM} = 2.35 \times \Delta E\,. \qquad (3.62)$$

Furthermore, contributions must be added from caught charge carriers on trapping sites (lattice dislocations) and from noise through leakage currents.

The Fano factor is defined as

$$F = \frac{\text{(Standard deviation of the produced charge carrier pairs)}^2}{\text{number of the produced charge carrier pairs}}\,. \qquad (3.63)$$

In the case of Poisson statistics this yields

$$\Delta E = \sqrt{\frac{F\,E}{w}}\,. \qquad (3.64)$$

The value of the Fano factor lies between zero and unity. This means:

$F = 0$: no statistical fluctuations
(the whole energy is used for the production of charge carrier pairs)

$F = 1$: the number of charge carriers is determined by Poisson statistics

$F < 1$: the production of charge carrier pairs does not exactly following Poisson statistics
(because Poisson statistics assume independent events; $F < 1$ means that the individual ionization acts are correlated to one another)

The electronic part of the noise contains statistical fluctuations in the current flow leading to errors in the amplitude measurement. Important noise parts are:

- *Parallel noise.* This contribution depends on the detector current $I_{\text{det.}}$ and on the dynamical resistance R_w. The equivalent noise reads as

$$\Delta E_{\text{pn}} \sim I_{\text{det.}} \times \tau \quad \text{or} \quad \frac{\tau}{R_w} \qquad (3.65)$$

---

² FWHM: Full With at Half Maximum

with τ the charge carrier collection time. This means that for a small noise contribution, $I_{\text{det.}}$ must be small and R_w should be great.

- *Series noise.* In the amplifier circuit the series noise is determined by the input noise of the first FET:

$$\Delta E_{\text{sn}}^2 \sim \frac{C_{\text{det.}}^2}{\tau\,S_{\text{FET}}} \qquad (3.66)$$

with S_{FET} as the dynamical slope of the FET. Therefore we must aim to increase S_{FET} and to reduce the detector capacitance $C_{\text{det.}}$.

- *Sparkle noise.* This noise is generated by surface effects and depends on the electronic circuits and on the individual detector, but is independent of τ.

The noise parts discussed here are added quadratically.

For an optimal energy resolution the time constant must fulfill the following condition:

$$\tau_{\text{opt.}} = \tau_{\text{int.}} = \tau_{\text{diff.}}\,. $$

$\tau_{\text{int.}}$ and $\tau_{\text{diff.}}$ describe the integration and the differentiation time constant, respectivelly. It counts

$$\tau > \tau_{\text{opt.}} \Rightarrow \text{parallel noise dominates,}$$
$$\tau < \tau_{\text{opt.}} \Rightarrow \text{series noice dominates.}$$

For the case of Si(Li) and Ge X-ray detectors for photon energies of 6 keV, energy resolutions lie in the order of 130 eV up to 150 eV, respectively.

Time resolution. For semiconductor detectors typical impulse rise times are in the order of 10^{-7} s to 10^{-9} s. With suitable electronic methods a time resolution of up to < 1 ns can be reached.

Registration efficiency. In Fig. 3.18 the registration efficiency for typical X-ray semiconductor detectors is shown. Because, in the low-energy region, the registration efficiency is dominated by the radiation

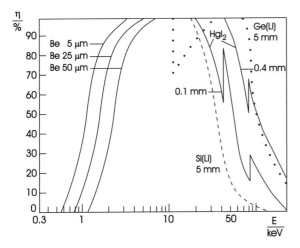

Fig. 3.18. Registration efficiency of semiconductor detectors in the X-ray energy region. Additionally the transmission of X-rays through beryllium is shown for the low-energy limit

transmission through the detector window the transmission of low-energy X-rays through beryllium is also shown in the figure. Furthermore also window-less detectors are available but the operation of these detectors needs special measures. In this case it is possible to detect very low-energetic X-rays with approximately 100% registration efficiency.

Radiation damage. At sufficiently intense exposure ionizing radiation causes anomalous electronic states in the detector crystal through the creation of certain levels in the band structure, which influence the charge carrier lifetime, the energy resolution and the material resistance. A discussion of these circumstances is given for example by Newman [378], Kraner [297] and Bischoff et al. [70].

As a rule in the X-ray energy region radiation damage in semiconductor crystals can be neglected.

3.3.5 Crystal Diffraction Spectrometers

Parameter and Classification

Wavelength-dispersive X-ray spectroscopy uses the diffraction of X-rays on crystallographic planes or on the crystal lattice. The diffraction on crystallographic lattices can be described by the Laue reflection condition for a three-dimensional point lattice

$$n\,\lambda = 2\,d\,\sin\vartheta\;. \tag{3.67}$$

where $d = d_{hkl}$ is the lattice constant (with h, k, l as Miller indices), ϑ is the Bragg angle and n is the diffraction order (with $n = 1, 2, 3, \ldots$).

Equation (3.67) is exactly valid inside of the crystal. For transitions between different media the refraction of radiation must be considered. This can be done if we consider an effective lattice constant

$$d_n = d\,(1 - T)\;. \tag{3.68}$$

The correction factor T is

$$T = \frac{\delta}{\sin^2\vartheta_n}\;. \tag{3.69}$$

Above the region of self-absorption we have

$$\delta \approx 2.7 \times 10^{-8}\,\frac{\varrho\,Z_M}{M}\,\lambda^2 \tag{3.70}$$

with λ the wavelength of the radiation to be analyzed (in nm), ϱ the crystal density (in $g\,cm^2$), Z_M the number of electrons in the crystal molecule and M the molar weight (in g).

For short wavelengths, X-rays can be detected in *open or short-wave spectrometers*, if the energy of the X-ray quanta leads not to a significant reduction of the radiation intensity for the given spectrometer geometry. Here a first estimation of the transmission of radiation through the effective attenuation media give an impression of whether spectrometry is possible under normal conditions. For longer wavelengths where significant transmission losses during the transmission through air are characteristic, the spectrometer can be operated under a helium atmosphere or under vacuum. In this case one speaks about so-called *vacuum- or long-wave spectrometers*.

A systematization of crystal diffraction spectrometers can be undertaken depending on the type of inserted crystals:

1. **single-crystal spectrometers.** Analysator crystals are used as
   - plane crystals or as
   - bent (focussing) crystals.
2. **Double-crystal spectrometers.** With the goal of increasing the spectrometer resolution and to get

Table 3.12. Systematization of crystal diffraction spectrometers depending on the type of diffraction

Bragg spectrometer	Laue spectrometer
diffraction at the crystal surface diffraction on atomic planes oriented perpendicular to the vertical line of the crystal surface	transmission geometry diffraction on atomic planes oriented parallel to the vertical line of the crystal surface
Bragg [82] Johann [267] Johansson [268] von Hamos [554]	Cauchois [101,102] DuMond [158]

a possibly undistorted form of the diffraction reflexes double-crystal spectrometers consist of two closely positioned sequential crystals.

3. **Grating spectrometers.** Grating spectrometers include mechanically produced diffraction lattices for use in the low-energy region (UV, EUV) and are not discussed here.

Crystal diffraction spectrometers can also be distinguished depending on the type of diffraction. A corresponding classification is given in Table 3.12.

Plane-Crystal Spectrometers.

Plane-crystal spectrometers are realized as

- single-crystal spectrometers, or as
- double-crystal spectrometers.

Exceptionally three-crystal spectrometers have been applied.

Single-Crystal Spectrometers
Bragg [82] described a crystal diffraction spectrometer whose geometry is still applied in different modifications. The basic scheme of this spectrometer is represented in Fig. 3.19. Characteristic here is the diffraction at the crystal surface under consideration of the diffraction condition (3.67). Typically for these spectrometers is a simple set-up and a high spectral resolution. X-rays which pass through the diaphragm A are diffracted through the Bragg angle ϑ (the angle between the crystal surface and the incident photon) and detected after entering the diaphragm B. The spectrum of an X-ray source can be

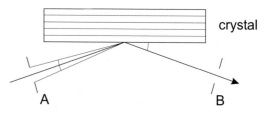

Fig. 3.19. Basic Bragg geometry for wavelength-dispersive X-ray spectroscopy with use of a plane crystal

detected if the crystal is moved in $\Delta\vartheta$ steps and the exit slit (detector) is simultaneously moved in $2\,\Delta\vartheta$ steps. In this set-up the analyzed radiation is quasi-parallel. The aperture of the applied diaphragms and the quality of the crystal to be used determine the possible resolution. Because a small zone of the crystal always contributes to the diffraction process such spectrometers have a low luminosity.

Generally different spectrometer principles are applied for X-ray spectroscopy with different levels of success. Figure 3.20 shows an overview of common methods. Spectrometers with entrance or detector slits are known in different versions. The edge-spectrographs described in the literature give less precise results.

For the simultaneous measurement of a certain wavelength region the measurement set-up can be modified in such a way that a position-sensitive detector is used. Then the detector can measure X-ray reflexes of different wavelengths at the diaphragm B over the whole detector width.

Double-Crystal Spectrometers
In spite of its high spectral resolution and the excellent reproducibility of their results, double-crystal

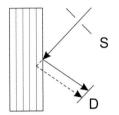 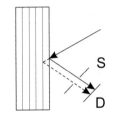 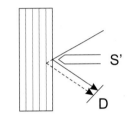

Fig. 3.20. Different types of plane-crystal spectrometers

spectrometers are rarely used because their registration efficiency in comparison to single-crystal spectrometers is some orders of magnitude lower ($10^{-10} - 10^{-11}$). There are double-crystal spectrometers such as Bragg spectrometers as well as Laue spectrometers. A representation of both geometries is given in Fig. 3.21.

For the parallel geometry the beam which exits from the spectrometer is parallel to the beam which enters the set-up, i.e. the angle of deflection is zero. This means that the angular dispersion D vanishes in this case. For the antiparallel case the angular dispersion is equal to the sum of both individual dispersions of each crystal

$$D = D_{\vartheta_1} + D_{\vartheta_2} = \frac{1}{2d}\left(\frac{n_1}{\cos\vartheta_1} + \frac{n_2}{\cos\vartheta_2}\right) \quad (3.71)$$

with $n_{1,2}$ is the order of reflection for the first and second crystal.

Double-crystal spectrometers in antiparallel geometries are already known from Davis and Purks [132], Ehrenberg and Mark [163] and Ehrenberg and

Susich [164]. Additional double-crystal spectrometers have been described by a multitude of authors, as for example by Deslattes [141], Goshi et al. [209] and Förster et al. [176]. The operation of a vertical dispersion mode double-crystal spectrometer is discussed by Renner et al. [450].

Focussing Crystal Diffraction Spectrometers

Focussing crystal diffraction spectrometers are known for Bragg geometry as well as for Laue geometry. In particular, Bragg spectrometers have been applied for wavelength-dispersive X-ray spectral analysis; we will discuss these in more detail later. We note here that for bent crystals the width of the rocking curve can, as a result of the bending process, be wider than for plane crystals.

Bragg Spectrometer
Spectrometer geometries. The most frequently used spectrometer geometries are listed in Table 3.12:

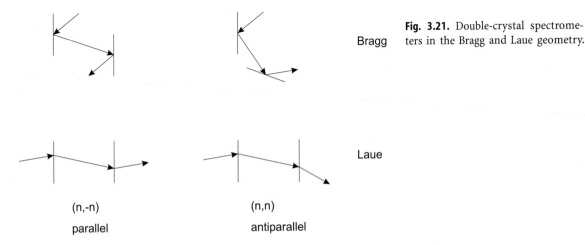

Fig. 3.21. Double-crystal spectrometers in the Bragg and Laue geometry.

Bragg

Laue

(n,-n)
parallel

(n,n)
antiparallel

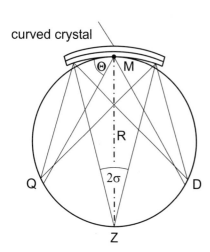

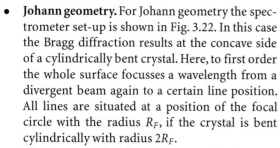

Fig. 3.22. Bragg spectrometer in Johann geometry. Q is the source; D is the detector

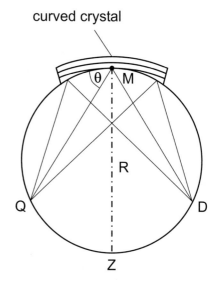

Fig. 3.23. Bragg spectrometer in Johansson geometry. Q is the source; D is the detector

- **Johann geometry.** For Johann geometry the spectrometer set-up is shown in Fig. 3.22. In this case the Bragg diffraction results at the concave side of a cylindrically bent crystal. Here, to first order the whole surface focusses a wavelength from a divergent beam again to a certain line position. All lines are situated at a position of the focal circle with the radius R_F, if the crystal is bent cylindrically with radius $2R_F$.

  The focussing in Johann geometry is not exact; however it is sufficient for many uses if the operation is restricted to small crystal apertures. An essential advantage of this method is that the beam must not be limited by a diaphragm or a cutting edge. Details of the linewith to be expected, of the imaging errors and of the registration efficiency have been discussed by Johann [267].

- **Johansson geometry.** Johansson spectrometers are exactly focussing. For the Johansson geometry a ground and bent crystal is characteristic, whereby the crystal is first bent on the radius $2R_F$ of the focal circle and is then ground on the radius R_F. The corresponding geometry is represented in Fig. 3.23. This diffraction scheme allows precise selective focussing with a high gain

on registration efficiency. Details of the derivation of the focussing condition, of the influence of different factors on the received line sharpness, on the resolution and on the luminous intensity have been discussed by Johansson [268]. Johansson spectrometers work without diaphragms, because all rays satisfying the Bragg condition are selected and after diffraction focussed onto the focal point. In comparison to the Johann spectrometer the advantage here is that much larger crystal apertures can be used.

- **Von Hamos geometry.** The basic idea of this geometry (see Fig. 3.24) is the diffraction on a cylindrically bent crystal [553, 554]. The von Hamos geometry combines the advantage of planar crystal Bragg spectrometers (high resolution) and Johann/Johansson spectrometers (higher luminosity). For example, descriptions of spectrometers in the von Hamos geometry can be found in [54, 243].

- **Doubly curved geometries.** For specific applications new spectrometer concepts based on oriented polycrystalline films deposited on doubly curved substrates have also been introduced. For

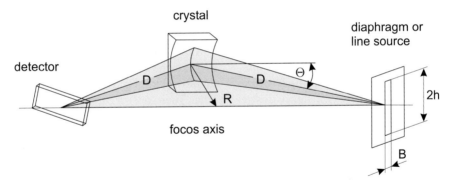

Fig. 3.24. Bragg spectrometer in the von Hamos geometry (after [54])

example, different geometries are discussed by Sparks et al. [520] and Wittry [572].

Laue Spectrometer

Here the most common spectrometers are arrangements named after Cauchois and DuMond, respectively. Because these spectrometers work in transmission geometry, they have been applied especially for the analysis of hard X-rays and of γ-radiation.

With the recovery of semiconductor detectors both spectrometer types have lost importance, because in the hard X-ray region semiconductor detectors show, besides good energy resolution, a registration efficiency some orders higher than that from Laue crystal diffraction spectrometers.

Cauchois Spectrometer. The Cauchois diffraction geometry [101] is shown in Fig. 3.25. In this geometry a

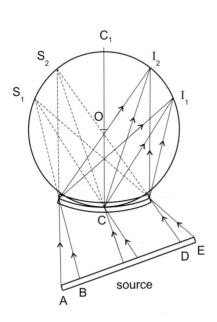

Fig. 3.25. Laue spectrometer in the Cauchois geometry

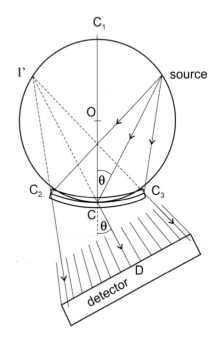

Fig. 3.26. Laue spectrometer in the DuMont geometry

singly bent crystal is used in transmission geometry. An extended source or a line source in the focal plane and lengthwise across the transmission crystal (divergent beam from broad focal spot) is most appropriate to this set-up. In the Cauchois set-up large volumes or areas can be analyzed. For instance this geometry is useful for X-ray fluorescence studies with large-area targets. In the Cauchois geometry all wavelengths are focused simultaneously. The recording of the diffraction images can be done with a slit detector or a position-sensitive detector, whereby different wavelengths are recorded along the detector area. High-resolution measurements can be realized if the slit width is chosen equal to the intrinsic linewidth of the crystal.

Measurements are performed by moving the slit detector or the position-sensitive detector continuously or in a step-by-step regime along the focal circle or with a fixed slit detector with rotating crystal. Both cases allow us to reach the same information.

The focusing error for this arrangement is the same as for the Johann geometry. For low X-ray energies the use of transmission geometries is restricted because the transmission of the beam through the crystal can lead to a signficant loss of intensity.

DuMont Spectrometers. The DuMond geometry [158] is shown in Fig. 3.26. Here the source is ideally a line source and is located on the focal circle. In this geometry the X-rays diverge after Bragg reflection and form a virtual image. For a given source-crystal orientation the spectrometer acts as a monochromator since only one wavelength will be coherently reflected.

The original measurement procedure here is that the diffraction angle is varied by rotating the source and the crystal about the point C. More practically, most instruments work with a fixed source where the crystal and detector rotate around the point C. For example, a description of an in-beam installed DuMond spectrometer can be found in [424].

Characteristics of Analyzer Crystals

In Table 3.13 properties of commonly used crystals important for X-ray spectral analysis are summarized. In this way values for the lattice constant d, the reflex intensity I and for the resolving power $\lambda/\Delta\lambda$ known from the literature are specified.

Although for an optimal measurement the reflectivity of the crystal as well as the resolving power should be high, both quantities are alternative to each other. Therefore the choice of suitable analyzing crystals requires a careful analysis of the single-crystal characteristics. Based on selected crystals and geometries the most important viewpoints of such an analysis are discussed in the following.

Influence of Instrumental Effects on the Image Characteristics of Crystals

Most instrumental effects typical for the operation of crystal diffraction spectrometers can be minimized or excluded by careful construction, adjustment and optimal operation of the spectrometer.

To consider or exclude possible sources of error it is necessary to know which mistakes can appear and in which order of magnitude the mistakes are to be expected. In general different instrumental effects combine with the natural characteristics of X-ray lines from which arise the observed intensity and line form. The resulting instrumental effects either become evoked through technical mistakes or are of a fundamental nature. Here the first class of errors are typical effects that lead to shifts of diffraction reflexes and therefore to a change of the measured wavelength:

1. Misorientation of the crystal with respect to the Rowland circle.
   Because the magnitude of the misorientation Δr is much smaller than the radius R of the Rowland circle for the wavelengths shift we have

$$\Delta\lambda/\lambda \sim \frac{\Delta r^2}{R^2} \; .$$

   $\Delta\lambda/\lambda$ remains negligibly small.
2. Mistakes in the centering or in the axiality of the mechanical parts of the spectrometer.
   In the vertical direction the axis of rotation can be adjusted with high accuracy, but if it does not pass through the neutral axis of the crystal it results in a wavelength shift of

$$\frac{\Delta\lambda}{\lambda} \approx \frac{\Delta a}{2R} \qquad (3.72)$$

with Δa as the deviation of the axis of rotation from the neutral axis of the crystal [449].

3. Deviations of the crystals from cylindrical form.

In focussing Bragg spectrometers technical defects also cause a line broadening [103]. A difference between the radius of curvature r of the crystal and of the effective diameter of the Rowland circle R causes a broadening of

$$b_1 = \frac{t}{R}\,|R - r|\,\sin\vartheta \qquad (3.73)$$

with t as the linear aperture of the crystal. If the reflecting surfaces are not parallel to the axis of the focal cylinder, but rather positioned at an angle φ, then the broadening is

$$b_2 = \frac{h}{2}\,\varphi\,\cos\vartheta \qquad (3.74)$$

with h as the crystal height.

Fundamental effects cannot be removed by the increase of the technical precision. These effects are established in the nature of the diffraction process and cause, through the finite dimensions of source and crystal, a focussing defect. A uniform description of the connected aberrations for focussing Bragg spectrometers is given by Zschornack et al. [585]. In Table 3.14 analytical expressions for the description of the influence of crystal and source dimensions on the wavelength shifts are summarized for the Johann geometry. On the basis of the characteristic R^{-2} dependence for all expressions, a reduction of the influence of geometrical effects on wavelength shifts can be achieved by an increase of the diameter of the Rowland circle.

As well as the itemized reasons that can cause a line shift or broadening of a given diffraction reflex, both the natural linewidth as well as the diffraction width of the crystal causes a noticeable contribution to the linewidth. The width of the reflection curve of a bent crystal depends on the initial physical parameters of the crystal plate, the nature (elastic or plastic) of the crystal, the quality of the bending process and the possible grinding, as well as on the subsequent treatment.

The influence of instrumental effects on the image characteristics of crystals, i.e. properties of crystal diffraction spectrometers can be modelled with different ray-tracing techniques. For instance we refer to Sparks et al. [520], Hubbard and Pantos [250], Morita [369], Lai et al. [308], Wittry and Sun [571], Chukhovski et al. [118] and Sanchez et al. [478].

Spectrometer Characteristics

Crystal diffraction spectrometers are often homemade devices. This makes these devices very variably and permits them to be adapted to concrete requirements. To compare individual spectrometers a comparison of general spectrometer characteristics is required.

Diffraction Line Width
Knowledge of the factors leading to a broadening of diffraction reflexes allows an optimization of the construction of the spectrometer and operation with respect to an optimal wavelength resolution.

Following [286] the angular width ω, as it is measured for the full width at half maximum of a diffraction line, can be represented as a function

$$\omega = \omega(\omega_G, \omega_C, \omega_X) \qquad (3.75)$$

of the geometrical width ω_G, that was caused by the crystal width ω_C and the radiation width ω_X [286].

The geometrical width ω_G is a function of partial widths depending on the geometry of the spectrometer

$$\omega_G = \omega_G(\omega_A, \omega_S, \omega_t, \omega_\varphi\;.) \qquad (3.76)$$

The quantity ω_a is the angle determined by the aperture of a shielding diaphragm and is only of importance for single-crystal spectrometers with plane crystals. ω_S characterizes the angle given by the finite source width and ω_t is a contribution essential only for transmission spectrometers through the finite crystal depth t. ω_φ describes the beam divergence given by the source and crystal geometry.

The angular width determined by the crystal

$$\omega_K = \omega_K(\omega_M, \omega_D, \omega_R) \qquad (3.77)$$

corresponds to the half-width of the intensity distribution of a parallel monochromatic plane wave after

Table 3.13. Overview of the most common crystal analyzers. The numbers in brackets mean the wavelength λ in nm and the numbers before the double point the reflection order. The indicated values for $\lambda/\Delta\lambda$ and for the reflex intensity I carry an orienting character because the values were assembled from a multitude of single papers by [356] and [74], completed with data from [544]. For the considered values different experimental conditions were characteristic. Spacings of α-sodium soap hemihydrates originate from [359].

crystal	plane	d[nm]	I	$\lambda/\Delta\lambda$
quartz	$50\bar{5}2$	0.0812		
SiO_2	$22\bar{4}3$	0.10149		
	$13\bar{4}0$	0.11801	0.2	
	$20\bar{2}3$	0.1375		
	$22\bar{4}0$	0.12255		
	$21\bar{3}1$	0.1541		
	$11\bar{2}1$	0.1818		
	$20\bar{2}0$	0.2123		
	$10\bar{1}2$	0.2282		
	$10\bar{2}0$	0.24565		
	$10\bar{1}1$	0.33432	0.3	
	$10\bar{1}0$	0.42548	0.1 (0.007)	16000 (0.14)
			0.5 (0.69)	2: 48000 (0.14)
				3: 140000 (0.14)
LiF	422	0.0823	0.1 (0.07)	
	420	0.0895	0.2 (0.07)	> 200
	220	0.142		
	200	0.20138	1 (0.1)	
			0.9 (0.07)	
			0.7 (0.05)	
Ge	422	0.11		11000
	220	0.200		
	111	0.32702	0.3	2300 (0.15)
			3: 0.4	
Si	111	0.3135	0.5 (0.23)	17000 bis 27000
				(at 0.07 to 0.3)
	220	0.192016		
topaz	400	0.116		8900 (0.07)
$Al_2(F,OH)_2 SiO_4$	303	0.1356	0.25	70000 (0.07)
	200	0.23246		
	002	0.41957		

Table 3.13. (cont.)

crystal	plane	d[nm]	I	$\lambda/\Delta\lambda$
corundum Al_2O_3	146	0.083		
	030	0.1374		
mica, muscovite	$31\bar{1}$	0.1695		
$K_2O \times 3Al_2O_3 \times 6SiO_2 \times 2H_2O$	$\bar{2}01$	0.2535		
	331	0.150		
	100	0.516		
	001	0.99634	0.05	400 (1.0)
			(0.17 bis 0.83)	1800 (1.76)
			3: 0.35	2700 (1.94)
			4: 0.1	2: 4700 (0.95)
			5: 0.40	
			6: 0.05	
			8: 0.15	
gypsum	200	0.2595		
$CaSO_4 \times 2H_2O$	020	0.76005	0.12 (1.46)	
rock salt NaCl	200	0.282		
rock sugar $C_{12}H_{22}O_{11}$	001	0.756		
	100	0.1006		
calcite	200	0.30355	0.3	11000 (0.14)
$CaCO_3$				8900 (0.07)
				2: 70000 (0.07)
	633	0.101		
	422	0.1517		
sylvite KCl	200	0.314		
fluorite CaF_2	111	0.31515		
	220	0.1841		
graphite	002	0.335	5.5 (0.15)	
C			3 (0.27)	
			2 (0.61)	
ADP	200	0.376		830 (1.0)
$NH_4 \times H_2PO_4$	110	0.532	0.12	1000 (0.98)
			(0.27 to 0.99)	
	112	0.307		

Table 3.13. (cont.)

crystal	plane	d[nm]	I	$\lambda/\Delta\lambda$
PET C(CH$_2$OH)$_4$	002	0.43665		
EDDT C$_2$H$_8$N$_2\cdot$C$_4$H$_6$O$_6$	020	0.44015	0.2 (0.53)	3000 (0.8)
MoS$_2$	001	0.615		
sorbitol hexaacetate (SHA)	110	0.700		
Na β aluminia	0004	0.562	weak – medium	
NaAl$_{11}$O$_{17}$	0002	1.1245	weak	
crystal	plane	d[nm]	I	$\lambda/\Delta\lambda$
KAP	100	1.33171	0.2 (0.53)	2000…4000
KHC$_8$H$_4$O$_4$			0.1 (1.83)	(1.0 bis 2.3)
				670 (1.0)
clinochlor	001	1.4196		1600 (2.74)
(Mg,Fe)$_{19}$Al$_5$				
[Si$_{11}$Al$_5$O$_{40}$] [OH]$_{32}$				
indium antimonide	111	0.37403		
OAO		4.5275	0.2 (9.38)	
lead stearate		5.01	0.1	
beryl	10$\bar{1}$0	0.7977		
THP TlHC$_8$H$_4$O$_4$	100	0.1295		
RAP RbHC$_8$H$_4$O$_4$	100	0.130605		
OHM		0.3175		
CH$_3$(CH$_2$)$_{17}$OOC(CH)$_2$COOH				
α-Sodium soap hemihydrates				
Sodium myristate	001	8.245		
C$_{14}$H$_{27}$O$_2$Na$\cdot\frac{1}{2}$ H$_2$O				
Sodium palmitate	001	9.271		
C$_{16}$H$_{31}$O$_2$Na$\cdot\frac{1}{2}$ H$_2$O				
Sodium stearate	001	10.291		
C$_{18}$H$_{35}$O$_2$Na$\cdot\frac{1}{2}$ H$_2$O				
Sodium arachidate	001	11.323		
C$_{20}$H$_{39}$O$_2$Na$\cdot\frac{1}{2}$ H$_2$O				

Table 3.14. Analytical expressions for the description of the influence of finite crystal and source dimensions on the wavelength shifts $\Delta\lambda/\lambda$ for Bragg crystal diffraction spectrometers in the Johann geometry (after [585]). R is the radius of the Rowland circle

crystal dimension	characteristic wavelength shift $\Delta\lambda/\lambda$	source dimension	characteristic wavelength shift $\Delta\lambda/\lambda$
height h_0	$\dfrac{1}{24}\dfrac{h_0^2}{R^2\sin^2\vartheta}$	heigh z_0	$\dfrac{1}{24}\dfrac{z_0^2}{R^2\sin^2\vartheta}$
width t_0	$-\dfrac{1}{24}\dfrac{t_0^2}{R^2}\cot^2\vartheta$	width y_0	$\dfrac{1}{24}\dfrac{y_0^2}{R^2}\sin^2\vartheta$
depth r_0	$\dfrac{1}{8}\dfrac{r_0^2}{R^2}\cot^2\vartheta$	depth x_0	0

diffraction. ω_M is here the distribution caused by the angular contribution from the microcrystallites of a plane crystal. Depending on the individual crystal structure and on the surface treatment in Bragg spectrometers ω_M can take values between a few and some ten, or rarely, to some hundred arcsec [201,356]. The quantity ω_D describes the diffraction width of an ideal crystal and ω_R the angular width that is connected with the bending of the crystal plate.

The radiation width ω_γ for γ lines is small in comparison with ω_G, but results in a non-negligible contribution to the width of the X-ray transition lines.

Luminous Intensity
In order to obtain good counting statistics for the line to be analyzed, it is required that the luminous intensity L of the spectrometer is as large as possible. The luminous intensity as the registration efficiency of a photon through the relevant spectrometer is [53]

$$L = \frac{\Omega\,\alpha}{4\,\pi}\,P_L\,\eta\,e^{-\mu x} \qquad (3.78)$$

where Ω is the effective solid angle, α is the transmission of the collimator, η is the registration efficiency of the detector used to record the diffracted X-rays and $\exp(-\mu x)$ is the attenuation of the X-ray flux through absorption processes. The reflectivity P_L in the line maximum is

$$P_L = \frac{P_I}{\omega} = \frac{\text{integral reflectivity}}{\text{width of the Rocking curve}}\,. \qquad (3.79)$$

In this way the integral reflectivity is defined as follows:

$$P_I = \frac{\displaystyle\int_{\vartheta-\Delta\vartheta}^{\vartheta+\Delta\vartheta} I(\vartheta)\,d\vartheta}{I_0} \qquad \text{(rad)} . \qquad (3.80)$$

$I(\vartheta)$ describes the beam intensity reflected by the atomic plane and I_0 the intensity of the primary beam. The limits of the integration region were determined through $\pm\Delta\vartheta$, whereby the width of the integration region as a rule is selected as the threefold full width at half maximum of the diffraction line.

Because in the literature only a few inconsistent values for the reflectivity of different crystals at various energies and diffraction orders are known, for the estimation of the reflection power a calculation of these dependencies often becomes necessary. For instance, Cauchois and Bonelle [103] and Pinsker [429] describe the corresponding mathematical apparatus for the calculation of crystal reflectivities. Crystallographic formulas and constants can be found for example in Glocker [201]. Programs for the calculation of the reflectivity are also available (for example see [586]).

It has been shown experimentally that the application of static and alternating electric fields can influence the diffraction properties of individual crystals.

Experimental investigations on (110) quartz planes are described in [154] where an increase of reflectivity is observed.

Luminousity of Johansson spectrometers. Because in wavelength-dispersive X-ray spectrometry spectrometers of the Bragg type are used as a rule, in

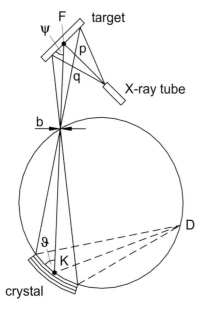

Fig. 3.27. Spectrometer geometry for the derivation of the luminosity of a Johannson spectrometer for X-ray excitation in a target by an X-ray tube (the destinations used in the figure are explained in the text)

the following the luminousity of such a spectrometer in the Johansson geometry is described following the compilation given in [16]. In the derivation of the luminosity we start with equation (3.78). Here in the derivation of the analytical relationship between the luminosity L and the spectrometer geometry we set $\alpha_K = 1$ and contributions from the detector efficiency and from the radiation absorption are not considered. We will assume monochromatic radiation. The spectrometer geometry for the derivation of the luminosity is represented in Fig. 3.27. Then the luminosity for the case that a radiation source is located at a point F yields

$$L_1 = P_L \frac{d_h\, d_v}{4\,\pi} \qquad (3.81)$$

with d_h and d_v as the corresponding horizontal and vertical divergence of the X-ray beam for the diffraction at the crystal K. The horizontal beam divergence is

$$d_h = \frac{b}{q} \quad \text{for} \quad Q > \frac{b\,D}{t}\,. \qquad (3.82)$$

At the same time, q describes the distance between the source point F and the entrance slit. The width b of the entrance slit is determined from the condition that any one ray in the horizontal plane is diffracted on the crystal, i.e.

$$b = D\,\sin\vartheta\,\delta\vartheta\,. \qquad (3.83)$$

$\delta\vartheta$ is the width of the diffraction reflex. Along the focal circle in the Johansson geometry the reflection angle is constant and equal to ϑ. For rays outside the focal circle plane ϑ decreases around the quantity

$$\Delta\vartheta = \left(\frac{d_v}{2}\right)\tan\vartheta\,. \qquad (3.84)$$

Diffraction is possible on the crystal if the condition $\Delta\vartheta \leq \delta\vartheta/2$ is fulfilled. From this the vertical divergence is

$$d_v = 2\,\sqrt{\cot\vartheta\,\delta\vartheta}\,. \qquad (3.85)$$

Then for L_1 it follows that

$$L_1 = \frac{P_L\,D\,(\delta\vartheta)^{3/2}\,\sqrt{2\,\sin 2\,\vartheta}}{4\,\pi\,q}\,. \qquad (3.86)$$

If a target excited through a primary radiation source is located on the point F, then the luminosity of the spectrometer is

$$L = L_1\,L_2 \qquad (3.87)$$

with

$$L_2 = \frac{s}{4\,\pi\,p^2}\,. \qquad (3.88)$$

In this way L_2 describes the rate of the primary X-rays and p is the distance between the point F and the primary radiation source. The effectively useful area s of the sample is limited by the crystal hight h_0 and the crystal width t_0

$$s = \frac{t_0\,h_0\,q}{D\,\sin\psi}\,. \qquad (3.89)$$

Therefore for L_2 and L it follows that:

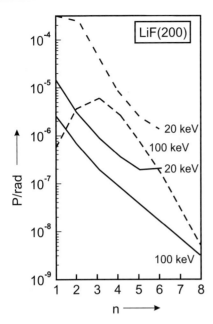

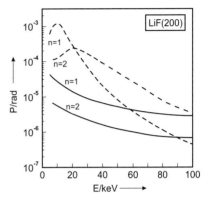

Fig. 3.28. Integral reflectivity P of a plane LiF(200) crystal depending on the diffraction order n at different photon energies E in the kinematical P_{kin} (dashed line) and in the dynamical theory P_{dyn} (solid line)

Fig. 3.29. Integral reflectivity P of a plane LiF(200) crystal as a function of the photon energy E and at different orders of diffraction n in the kinematic P_{kin} (dashed line) and in the dynamic theory P_{dyn} (solid line)

$$L_2 = \frac{t_0\, h_0\, q}{4\,\pi\, p^2\, D\, \sin\psi} \qquad (3.90)$$

$$L = \frac{P_L\, t_0\, h_0\, (\delta\vartheta)^{3/2}\, \sqrt{2}\, \sin 2\vartheta}{16\,\pi^2\, p^2\, \sin\psi}\,. \qquad (3.91)$$

Luminosity for a point source outside of the focal circle. Consider a point source located outside the focal circle at a distance r_B from a slit localized on the focal circle. Then it can be shown that for the luminosity the following relations hold [319]:

plane crystal

$$L = \frac{\Delta\Theta_{\text{hor}}\, \Delta\Theta_{\text{vert}}}{4\,\pi}\, P_{\text{max}}\, \eta_D\, e^{-\mu\,(r_B + 4R\,\sin\vartheta)} \qquad (3.92)$$

Johansson geometry

as equation (3.92) and

von Hamos geometry

$$L = \frac{\Delta\Theta_{\text{hor}}\, \Delta\Theta_{\text{vert}}}{4\,\pi}\, P_{\text{max}}\, \eta_D\, e^{-2\mu\, \frac{R_k}{\sin\vartheta}}\,. \qquad (3.93)$$

Here $\Delta\Theta_{\text{hor}}$ and $\Delta\Theta_{\text{vert}}$ describe the horizontal and vertical diaphragm aperture angles, P_{max} the reflection power in the reflex maximum and η_D the detection efficiency of the diffracted photons. The exponential function with attenuation coefficient μ stands for the radiation attenuation.

Reflectivities for Selected Crystal and Atomic Planes
In order to characterize integral reflectivities for individual crystals and atomic planes at different X-ray energies and for different orders of diffraction, selected values as given by Zschornack et al. [586] are indicated. In Figs. 3.28 and 3.29 the integral reflectivity for ideal monocrystals is examined for a flat LiF(200) crystal. Note the cross-over points between the results from kinematic and dynamic theory for $E = 100$ keV at $n = 2$ in Fig. 3.28 and for $n = 1$ at $E = 55$ keV in Fig. 3.29.

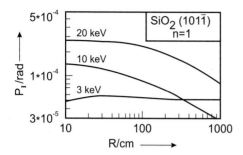

Fig. 3.30. Integral reflectivity P_I of quartz $(10\bar{1}1)$ calculated with the dynamic theory as a function of the crystal bent radius R at different X-ray energies

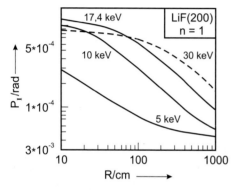

Fig. 3.31. Integral reflectivity P_I of LiF(200) calculated with the dynamic theory as a function of the crystal bent radius R at different X-ray energies

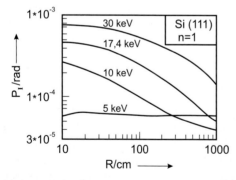

Fig. 3.32. Integral reflectivity P_I of Si(111) calculated with the dynamic theory as a function of the crystal bent radius R at different X-ray energies

In Figs. 3.28–3.32 the integral reflectivity calculated in the framework of the dynamic and the kinematic theory is shown for different radii of curvature R at different energies. The results for quartz $(10\bar{1}1)$,

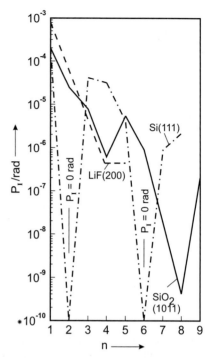

Fig. 3.33. Integral reflectivity P_I of different crystals bent with a radius of $R = 324$ mm depending on the order of diffraction n at a photon energy of $E = 17.4$ keV (Mo K_α)

LiF(200) and Si(111) show that an increase of the radii of curvature by two orders of magnitude lead to a decrease of the integral reflectivity by one order of magnitude. Crystals with bent radii of some ten centimeters have no substantial differences in their reflectivity.

In Fig. 3.33 integral reflectivities for different crystals bent with a radius of $R = 324$ mm are shown. For $n = 1$ LiF(200) has the highest integral reflectivity, however the reflectivity falls away steeply with growing n. On the other hand, Si(111) and quartz$(10\bar{1}1)$ have local maxima for higher n, which for Si(111) at $n = 3, 4$ shows a five and seven times smaller reflectivity.

Angular Dispersion

For the angular dispersion from the differentiation of Braggs law (3.67) it follows that

$$\frac{d\vartheta}{d\lambda} = \frac{n}{2\,d\,\cos\vartheta}. \tag{3.94}$$

Furthermore, from equation (3.94) we have

$$\frac{dE}{d\vartheta} = \frac{E \cos \vartheta}{n} . \qquad (3.95)$$

Inverse Linear Dispersion
For bent crystals with bending radius R we have

$$\frac{d\lambda}{dx} = \frac{2\,d \, \cos \vartheta}{n\,R} . \qquad (3.96)$$

Resolving Power
The resolving power A is defined as

$$A = \frac{\lambda}{\Delta \lambda} \qquad (3.97)$$

with $\Delta\lambda$, the half-width of the apparatus-related distortion function. $\Delta\lambda$ is a function of the crystal properties and additionally depends on the chosen measurement geometry.

Relative Energy Resolution
For the relative energy resolution we can write

$$\frac{|\Delta E|}{E} = \frac{|\Delta \sin \vartheta|}{\sin \vartheta} = \Delta\vartheta \, \cot \vartheta = \frac{2\,d}{h\,c} \Delta\vartheta \, \frac{E}{n} \cos \vartheta . \qquad (3.98)$$

Energy Resolution
To access an optimal energy resolution in precision measurements different points of view must be considered; these result from the Bragg equation for the maximum relative error in the energy determination:

$$\frac{|\Delta E|}{E} = \frac{|\Delta \sin \vartheta|}{\sin \vartheta} = \cot \vartheta \, \Delta\vartheta = \frac{2d}{hc} \Delta\vartheta \, \frac{E}{n} \cos \vartheta . \qquad (3.99)$$

$\Delta\vartheta$ is the angular error resulting from the mosaic width of the crystal and from the uncertainty in the determination of the Bragg angle and ΔE is the uncertainty in the energy calibration. To minimize the relative error of the energy determination we must aim to

- measure at large angles ϑ,
- minimize the angular error $\Delta\vartheta$,
- realize measurements at high orders of diffraction n, and
- use crystals with small lattice spacing d.

Above all, of special interest is the quantity $\Delta\vartheta$. This error consists of the width of the mosaic distribution, the mechanical precision of the spectrometer and the statistical error from the reflex analysis.

At sufficiently high statistics line positions can be determined with an accuracy of 1/100 of the lines half-value or better [53]. Increasingly exact results are achieved if narrow and intense lines are evaluated, i.e. the crystals should have a large reflectivity at small mosaic widths. For the determination of the attainable energy resolution we rewrite equation (3.99) in the form

$$\Delta E = 1.613 \times 10^{-3} \, dE^2 \, \frac{\Delta\vartheta}{n}$$
$$\cos \left[\arcsin \left(\frac{1239.853 \, n}{2\,d\,E} \right) \right] \frac{1}{n} . \qquad (3.100)$$

In equation (3.100) E is given in eV and d in nm. From the literature it is known that because of different mosaic widths different exemplars of the same crystal and of the same atomic plane can give different contributions to the broadening of the reflex profile. Depending from the related crystals and from the X-ray energy for $n = 1$ the mosaic widths to be expected are in the region of some arcsec to some sexagesimal minutes [356], which corresponds to values between about 5×10^{-6} and 2×10^{-3} rad. Therefore for the specification of the attainable energy resolution it is helpful to use the quantity

$$\frac{\Delta E}{\Delta\vartheta} \left[\frac{\text{eV}}{\text{rad}} \right] = 1.613 \times 10^{-3} \, dE^2$$
$$\cos \left[\arcsin \left(\frac{1239.853 \, n}{2\,d\,E} \right) \right] \frac{1}{n} . \qquad (3.101)$$

The quantity $\Delta E / \Delta\vartheta$ characterizes the contribution to the resolving power of the crystal depending on the structure of the atomic crystal planes, without containing a statement about the width of the mosaic distribution that should be also considered to get quantitative values near to reality.

In Fig. 3.34 the dependence of $\Delta E / \Delta\vartheta$ on the energy of the analyzed X-rays is given for $n = 1$ for a row of crystals and atomic planes.

spectrometer	resolving power
plane crystal spectrometer	$\dfrac{E}{\Delta E} = 4\,\dfrac{R}{B}\,\dfrac{h^2\,c^2}{4\,d^2\,E^2\,\sqrt{1 - \dfrac{h^2\,c^2}{4\,d^2\,E^2}}}$
Johansson spectrometer	$\dfrac{E}{\Delta E} = 4\,\dfrac{1}{\vartheta_{\mathrm{FWHM}}}\,\dfrac{h^2\,c^2}{2\,d^2\,E^2\,\sqrt{1 - \dfrac{h^2\,c^2}{4\,d^2\,E^2}}}$
von Hamos spectrometer	$\dfrac{E}{\Delta E} = \dfrac{2\,R}{B\,\sin\vartheta\,\cos\vartheta}$

Table 3.15. Resolving power of different Bragg crystal diffraction spectrometers

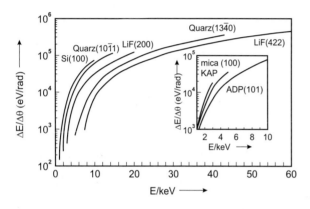

Fig. 3.34. Attainable energy resolutions $\Delta E/\Delta\vartheta$ for different crystals and atomic planes as a function of the photon energy E

Besides the relative energy and wavelength resolution ($\Delta E/E$ and $\Delta\lambda/\lambda$) the resolution can also be described by $E/\Delta E$ or $\lambda/\Delta\lambda$. Table 3.15 summarizes the resolution of different Bragg spectrometers as given by Lehnert [319] and Beiersdorfer et al. [54]. $\vartheta_{\mathrm{FWHM}}$ is here the half-width of the rocking curve.

3.3.6 Cryogenic Detectors

Cryogenic detectors are applied for X-ray detection in some special cases for X-ray fluorescence analysis and for X-ray spectroscopy. The use of these detectors is up to now restricted and their application frequency is rather low in comparison to the use of classical semiconductor detectors. On the other hand, cryogenic detectors can be applied for X-ray detection with an energy resolution about ten times better than that achievable from other energy-dispersive detectors. The cryogenic detectors used in applications of X-ray spectroscopy are:

- superconducting tunnel junctions, and
- hot-electron microcalorimeters.

As in the case of semiconductor or gas ionization detectors here the basic detection mechanism is not the creation of electron–ion pairs by ionization processes in the detector material but low-energy solid-state excitations. These excitations can be

- *quasiparticles* (electron- and hole-like excitations in superconductors),
- *phonons* (quantized crystal lattice vibrations), and
- *hot electrons* in a metal at low temperature.

Characteristic for all these processes is a very low excitation energy of about 1 meV or lower and working temperatures of 1 K or lower, respectively.

Although cryogenic detectors are commercial available, their distribution is restricted because of their price and small active areas. Therefore we will

E_X	ΔE_X	basic material	reference
1 keV	8.9 eV	Nb	leGrand et al. [318]
			Friedrich et al. [185]
1.77 keV	10 eV	Ta	Kraft et al. [295]
5.9 keV	16 eV	Ta/Al	Verhoeve et al. [551]
5.9 keV	29 eV	Nb/Al	Frank et al. [178]
5.9 keV	12 eV	Al with Pb absorber	Angloher et al. [14]

Table 3.16. Energy resolution ΔE_X of superconducting tunnel junction detectors. E_X is the X-ray energy

give here only a short characterization of these detectors and refer for more details to the reviews given by Frank [179] and [166] and the references therein.

Superconducting Tunnel Junctions

In a superconducting tunnel junction detector X-rays are absorbed in a superconducting film and create quasiparticle excitations. The number of excitations is proportional to the absorbed X-ray energy and can reach the order of 10^5–10^6 excitations. Registration of quasiparticle excitations is possible if the tunnelling current is measured from the tunnelling process through a thin insulating barrier. Because the magnitude of the tunnelling current pulse is proportional to the photon energy the detector can be used for energy-dispersive single X-ray counting.

Excellent resolution can also be achieved with alumina superconducting tunnel junction detectors with a lead X-ray absorbing layer [14]. Here the energy of the X-ray photon absorbed in the lead layer is transferred to the detector by phonons.

Superconducting tunnel junction detectors have pulse decay times in the order of 1–30 μs depending on the detector material to be used. This means that count rates of 10 000 pulses per second and more can be registered. The detector size varies from 70 μm × 70 μm to 200 μm × 200 μm. For good registration efficiency measurements in the energy region below 1 keV have been recommended.

Hot-Electron Microcalorimeters

Hot-electron microcalorimeters are true calorimetric detectors. The working principle consists of the measurement of the temperature generated in the detector by the absorption of an X-ray photon after photoelectron creation and following electron–electron scattering processes. During the detection process the measured temperature rise is proportional to the deposited energy. After a single photon detection the temperature decreases back to the initial detector temperature with a time constant typically in the order of tens to hundreds of microseconds. To avoid pile-up effects recommended photon count rates for microcalorimeters are in the order of 1000 counts per second or lower.

Hot-electron microcalorimeters have small detection areas, typically in the order of 100 μm × 100 μm up to 500 μm × 500 μm. The measurement of the temperature in the detector occurs commonly with transition edge sensors. Such sensors consist of a thin film of superconducting material and operate in the transition region between the superconducting and the normal conducting state. For good energy resolution the operation temperature of the microcalorimeter must be in the order of 100 mK or lower. Typical energy resolutions for hot-electron microcalorimeters are summarized in Table 3.17. Up to about 10 keV the detection efficiency is quite high and drops down at lower photon energies.

Table 3.17. Energy resolution ΔE_X of hot-electron microcalorimeters. E_X is the X-ray energy

E_X	ΔE_X	reference
1.5 keV	2 eV	Wollmann et al. [573]
5.9 keV	4.5 eV	Irwin et al. [259]

4 Data Base

4.1 Electron Binding Energies

The comparison of the electron binding energies as given in Chap. 5 shows that binding energies determined by different authors have substantial deviations from each other. As a rule such discrepancies were not based on incorrectly determined values but were determined by the physical nature of the investigated probe.

The values from Bearden and Burr [51] cited in Chap. 5 were determined on the basis of the following experimental techniques:

1. X-ray emission data from which binding energy differences were calculated,
2. X-ray absorption data, and
3. photoelectron data.

The last two methods allow the determination of binding energies in a direct way. Nevertheless here the measured data for insulators differ from other ones because here the Fermi level of the spectrometer is localized between the valence and the conduction band. The data measured with method 1 can be used for the indirect determination of atomic binding energies by a normalization procedure with the data derived from the methods 2 or 3. Sevier [493] extended the data basis by the consideration of

1. Auger emission data (indirect method),
2. autoionization data (indirect method),
3. energy loss data (direct method),
4. data from internal conversion (direct method if the transition energy is known), and
5. data from the measurement of excitation curves (direct method).

Among others, X-ray spectroscopic measurements can be done on large probes of pure elements. Here the sensitivity to surface contaminations as a result of bad vacuum conditions (for instance oxidation effects, oil deposition from vacuum pumps) as a rule is very small. On the other hand the spectroscopy of photoelectrons can lead, in the case of insufficient vacuum conditions, to electron binding energies for oxides or other disturbed states different from the values from the pure element (see also Fig. 4.1). For metals, electron binding energies were often measured for oxides in the most common valence states. The order of magnitude of physically and chemically determined electron binding energy shifts is shown in Fig. 4.1.

A detailed discussion of differences between condensed elementary states and compounds, for conductors in the condensed and in the gas phase and between metals and insulators is given by Sevier [493], Shirley et al. [495], Watson et al. [560], Ley et al. [325] and Williams and Long [570]. To characterize binding energy differences in Table 4.1 atomic electron binding energies for solids and gases were compared.

To characterize the significance of the quantities in Figs. 4.2–4.18, summarized in Chap. 5, the quantities published by Siegbahn [503] and Bearden and Burr [51] were compared with those from Sevier [493]. For atomic inner-shells a consistency of ± 1 eV is characteristic for most data. Only above $Z = 80$ are there greater deviations for all subshells. Deviations for outer shells (N, O) are up to one order of magnitude higher. If ultimate precision from the data is demanded a detailed analysis for the individual element with consideration of the data extraction strategy is imperative.

Table 4.1. Atomic electron binding energies for solids and gases measured with different methods (after [144]). The errors for the last significant positions are given in parentheses. Symbols: XPS – X-ray photoelectron spectroscopie; AES – Auger electron spectroscopy; XAS – X-ray absorption spectroscopy

Z	element	shell	E_{XPS} [eV] gas	E_{XPS} [eV] solid state	E_{AES} [eV] gas	E_{XAS} [eV] solid state
11	Na	1s		1074.0 (1)	1079.1 (3)	1072
		2s	70.84 (10)	65.7 (1)		
		3s	5.14	3.5		
12	Mg	1s	1311.2 (3)	1306.7 (1)	1311.3 (3)	1303
		2s	94.0 (5)	92.25 (10)	96.5 (3)	63
		$2p_{1/2}$	54.8 (5)	52.9 (1)	57.85 (10)	57.54
		$2p_{3/2}$	54.8 (5)	52.9 (1)	57.57 (10)	57.54
21	Sc	2s	503.2 (10)	501.5 (3)		
		$2p_{1/2}$	408.4 (10)	407.08 (10)		
		$2p_{3/2}$	403.9 (10)	402.15 (10)		
		3s	56.4 (10)	54.64 (10)		
		$3p_{1/2,3/2}$	33.6 (10)	31.84 (10)		
26	Fe	2s	857 (2)	6 853 (2)		
		$2p_{1/2}$	733 (1)	725.15 (20)		720.8
		$2p_{3/2}$	720 (1)	712.05 (20)		707.4
		3s	104 (1)	96.06 (20)		
		3p1/2, 3/2	66 (1)	57.54 (20)		54
30	Zn	2s	1203 (1)	1200.46 (25)		
		$2p_{1/2}$	1052.0 (3)	1049.39 (15)		1045
		$2p_{3/2}$	1029.1 (3)	1026.26 (15)		1022
		3s	145.0 (1)	144.18 (15)		
		$3p_{1/2}$		95.6 (15)	98.7 (7)	86
		$3p_{3/2}$		93.00 (15)	96.1 (5)	
82	Pb	$4f_{7/2}$	144.0 (5)	104.5		143.6 (1)

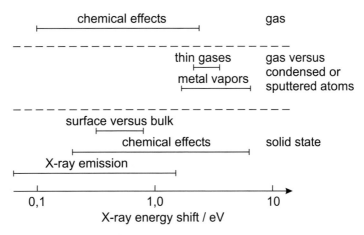

Fig. 4.1. Atomic electron binding energy shifts as a result of the physical and chemical influence of the environment (after [493]). For comparison K X-ray energy shifts as a result of chemical effects are also given

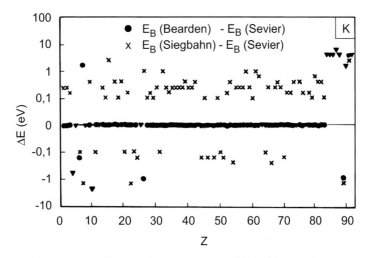

Fig. 4.2. K-shell electron binding energy differences between results published by Bearden and Burr [51] (E_B(BEARDEN)), Siegbahn [503] (E_B(SIEGBAHN)) and Sevier [493] (E_B(SEVIER)). The case that both differences coincide is marked by ▼

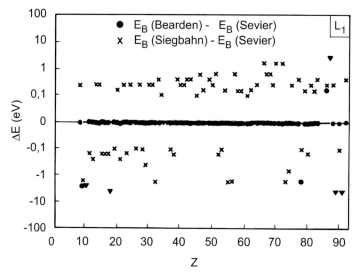

Fig. 4.3. L_1-shell electron binding energy differences between results published by Bearden and Burr [51] (E_B(BEARDEN)), Siegbahn [503] (E_B(SIEGBAHN)) and Sevier [493] (E_B(SEVIER)). The case that both differences coincide is marked by ▼

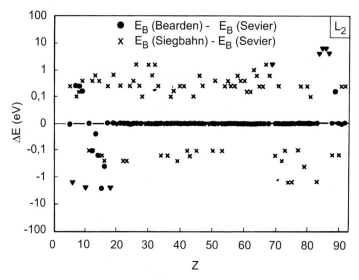

Fig. 4.4. L_2-shell electron binding energy differences between results published by Bearden and Burr [51] (E_B(BEARDEN)), Siegbahn [503] (E_B(SIEGBAHN)) and Sevier [493] (E_B(SEVIER)). The case that both differences coincide is marked by ▼

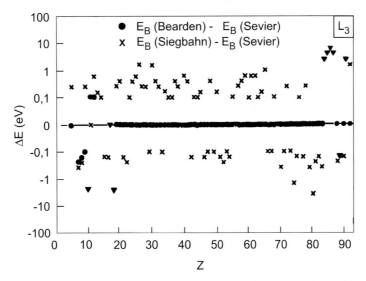

Fig. 4.5. L$_3$-shell electron binding energy differences between results published by Bearden and Burr [51] (E_B(BEARDEN)), Siegbahn [503] (E_B(SIEGBAHN)) and Sevier [493] (E_B(SEVIER)). The case that both differences coincide is marked by ▼

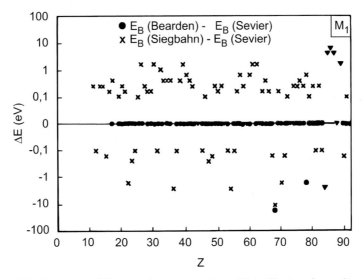

Fig. 4.6. M$_1$-shell electron binding energy differences between results published by Bearden and Burr [51] (E_B(BEARDEN)), Siegbahn [503] (E_B(SIEGBAHN)) and Sevier [493] (E_B(SEVIER)). The case that both differences coincide is marked by ▼

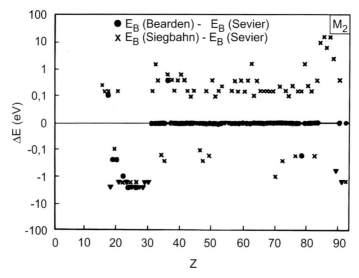

Fig. 4.7. M_2-shell electron binding energy differences between results published by Beardem and Burr [51] ($E_B(\text{BEARDEN})$), Siegbahn [503] ($E_B(\text{SIEGBAHN})$) and Sevier [493] ($E_B(\text{SEVIER})$). The case that both differences coincide is marked by ▼

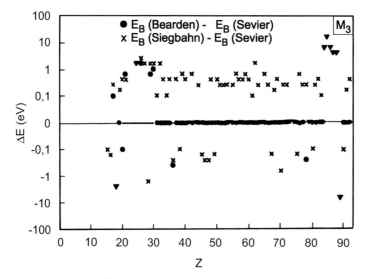

Fig. 4.8. M_3-shell electron binding energy differences between results published by Bearden and Burr [51] ($E_B(\text{BEARDEN})$), Siegbahn [503] ($E_B(\text{SIEGBAHN})$) and Sevier [493] ($E_B(\text{SEVIER})$). The case that both differences coincide is marked by ▼

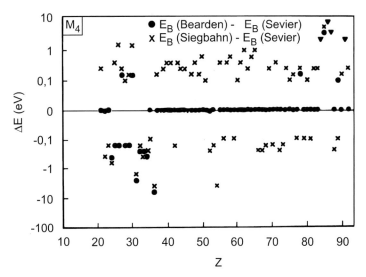

Fig. 4.9. M_4-shell electron binding energy differences between results published by Bearden and Burr [51] (E_B(BEARDEN)), Siegbahn [503] (E_B(SIEGBAHN)) and Sevier [493] (E_B(SEVIER)). The case that both differences coincide is marked by ▼

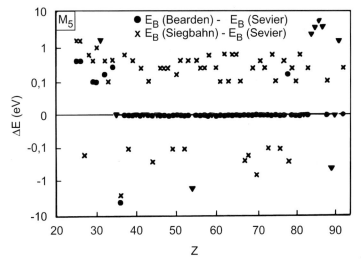

Fig. 4.10. M_5-shell electron binding energy differences between results published by Besrden and Burr [51] (E_B (BEARDEN)), Siegbahn [503] (E_B(SIEGBAHN)) and Sevier [493] (E_B(SEVIER)). The case that both differences coincide is marked by ▼

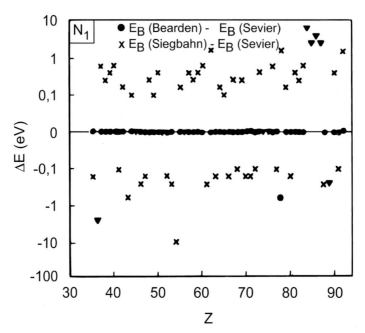

Fig. 4.11. N_1-shell electron binding energy differences between results published by Bearden and Burr [51] (E_B(BEARDEN)), Siegbahn [503] (E_B(SIEGBAHN)) and Sevier [493] (E_B(SEVIER)). The case that both differences coincide is marked by ▼

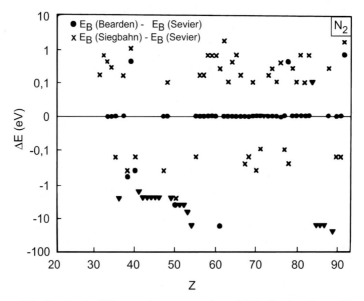

Fig. 4.12. N_2-shell electron binding energy differences between results published by Bearden and burr [51] (E_B(BEARDEN)), Siegbahn [503] (E_B(SIEGBAHN)) and Sevier [493] (E_B(SEVIER)). The case that both differences coincide is marked by ▼

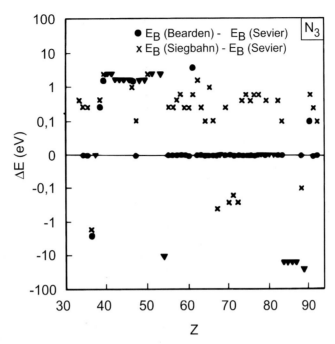

Fig. 4.13. N₃-shell electron binding energy differences between results published by Bearden and Burr [51] (E_B (BEARDEN)), Siegbahn [503] (E_B(SIEGBAHN)) and Sevier [493] (E_B(SEVIER)). The case that both differences coincide is marked by ▼

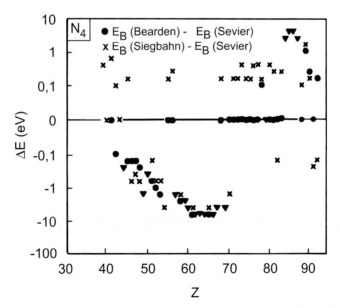

Fig. 4.14. N₄-shell electron binding energy differences between results published by Bearden and Burr [51] (E_B (BEARDEN)), Siegbahn [503] (E_B(SIEGBAHN)) and Sevier [493] (E_B(SEVIER)). The case that both differences coincide is marked by ▼

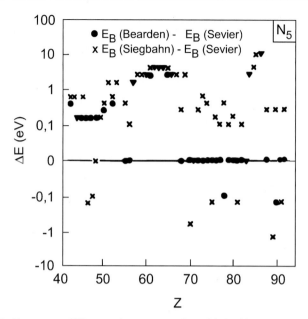

Fig. 4.15. N_5-shell electron binding energy differences between results published by Bearden and Burr [51] (E_B (BEARDEN)), Siegbahn [503] (E_B(SIEGBAHN)) and Sevier [493] (E_B(SEVIER)). The case that both differences coincide is marked by ▼

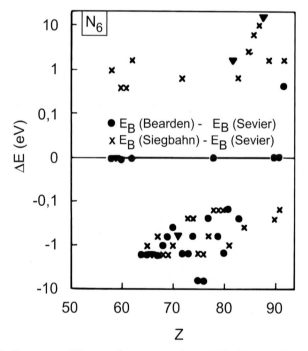

Fig. 4.16. N_6-shell electron binding energy differences between results published by Bearden and Burr [51] (E_B (BEARDEN)), Siegbahn [503] (E_B(SIEGBAHN)) and Sevier [493] (E_B(SEVIER)). The case that both differences coincide is marked by ▼

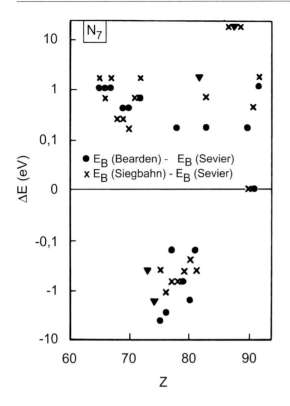

Fig. 4.17. N_7-shell electron binding energy differences between results published by Bearden and Burr [51] (E_B(BEARDEN)), Siegbahn [503] (E_B(SIEGBAHN)) and Sevier [493] (E_B(SEVIER)). The case that both differences coincide is marked by ▼

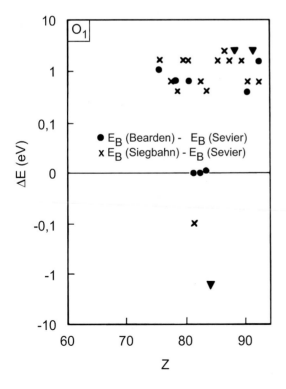

Fig. 4.18. O_1-shell electron binding energy differences between results published by Bearden and Burr [51] (E_B(BEARDEN)), Siegbahn [503] (E_B(SIEGBAHN)) and Sevier [493] (E_B(SEVIER)). The case that both differences coincide is marked by ▼

Basing on photoelectron measurements of absolute energy levels for 81 elements and X-ray absorption wavelengths for eight elements a reevaluation of X-ray atomic energy levels was made by [51]. The energy levels of the remaining elements were interpolated by Moseley's law using energy levels from adjacent elements.

During the evaluation, the circumstance is used that X-ray emission wavelengths are reevaluated and placed on a consistent Å* scale by Bearden [50]. In most cases this results in an overdetermined set of equations for energy level differences which is solved by a least-square procedure. In this way probable errors of energy level differences are also determined. Including the results from the above experimental investigations absolute energies of atomic energy levels were determined.

The electron binding energy data known from the literature were based on the analysis of results from different experimental techniques. This is the reason that uncertainties of conversion factors or of energy standards influence the derived results. In the course of the last few years the precision of calibration normals and of conversion factors has increased. Therefore the uncertainties of calibration normals and conversion factors for the determination of electron binding energies has further decreased. As result we derive, considering results with increased precision, a mean shift of the results from [51] by 13.4 ppm in the direction of increasing energy [493]. Thus the K-shell binding energies from Bearden and Burr [51] shift as

+0.5 eV for Z > 55,
+1.0 eV for Z > 76,
+2.0 eV for Z > 100 .

All other electron binding energies increase by 0.5 eV. Not included in Chap. 5 are the best available theoretical values. Deviations of calculated electron binding energies from experimental results can be explained by the fact that calculations as a rule were done for free atoms although the experimentally determined quantities were derived from solid state or from chemical compounds. Furthermore, not all published calculated data were derived on the highest possible level so that certain contributions to the electron binding energies were neglected.

4.2 X-Ray Transition Energies

4.2.1 Standard Energy Range

The discussed problems and open questions by the determination of electron binding energies are reflected as well at comparisons between X-ray energies determined in different experimental approaches. To characterize the values used in Figs. 4.19–4.21 in Chap. 5 selected X-ray transition lines were mutually compared. As a standard of comparison, X-ray transition energies were determined from electron binding energies derived with high-resolving electron spectroscopy [503]. The typical deviations are, for the K series, in the order of 10 ppm of the transition energy and, for the M series, of about 100 ppm. As a rule activities to increase the precision of X-ray transition energies occur for three classes of experiments:

Measurements with direct indicating instruments. Wavelength-dispersive measurements were realized by absolute angle determinations with inteferometrically calibrated crystals where only an optical wavelength as external reference is used or the measurement is done with precise angle encoders [276, 277]. The errors gained with this method at the determination of K_{α_1} transition energies range from 8 meV for argon to 280 meV for uranium.

Relative measurements to directly measured γ lines. Borchert [78], Hungerford et al. [255] and Barreau and Börner [43] used focussed crystal diffraction spectrometers to measure X-ray transition energies relative to the 411 keV γ line in ^{198}Hg (see Sect. 3.1). In detail this method has been used to determine X-ray transitions in selected heavy elements with $Z > 69$. For instance, for the K_{α_1} transition energy in uranium an error of 500 meV has been obtained.

Relative measurements to directly measured X-ray transitions. There are also the investigations of Bearden et al. [49,50] and Borchert et al. [80]. The obtained precision has been characterized by the errors for the

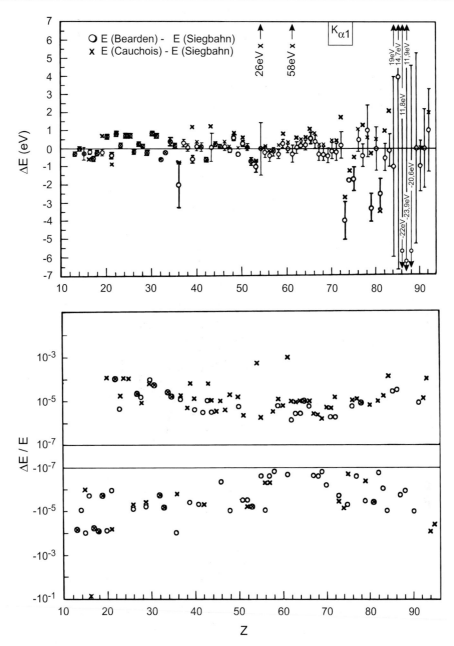

Fig. 4.19. Differences ΔE of experimentally determined K_{α_1} X-ray transition energies E depending on the atomic number Z. Results from Bearden [50] (E (Bearden)), Cauchois and Senemaud [104] (E (Cauchois)) and Siegbahn [503] (E (Siegbahn)) are compared. In the lower part relative deviations in relation to the results derived by Siegbahn [503] are given

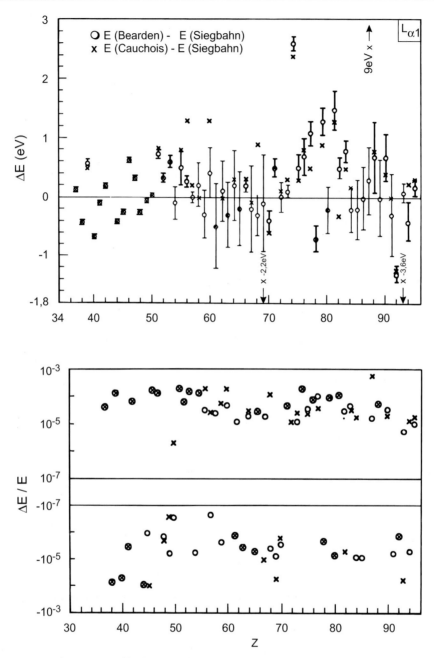

Fig. 4.20. Differences ΔE of experimentally determined L_{α_1} X-ray transition energies E depending on the atomic number Z. Results from Bearden [50] (E (Bearden)), Cauchois and Senemaud [104] (E (Cauchois)) and Siegbahn [503] (E (Siegbahn)) are compared. In the lower part relative deviations in relation to the results derived by Siegbahn [503] were given

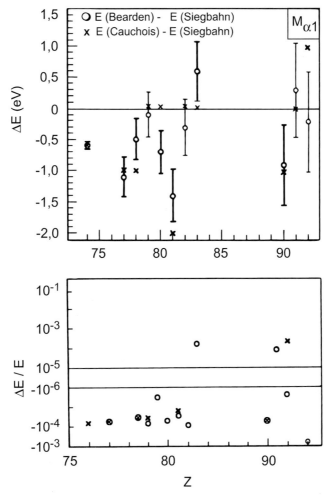

Fig. 4.21. Differences ΔE of experimental determined M_{α_1} X-ray transition energies E in dependence from the atomic number Z. Results from Bearden [50] (E (Bearden)), Cauchois and Senemaud [104] (E (Cauchois)) and Siegbahn [503] (E (Siegbahn)) are compared. In the lower part relative deviations in relation to the results derived by Siegbahn [503] were given

measurements of the K_{α_1} lines of selected elements. For instance, for the Al K_{α_1} line an error of 6 meV and for the Bi K_{α_1} line an error of 1,37 eV was derived. A discussion of individual wavelength ratios derived by different authors can be found in Deslattes and Kessler Jr. [144].

Calculated data for isolated neutral atoms derived from the Livermore Evaluated Atomic Data Library (EADL) were published by Perkins et al. [423] in July

1991. Here K and L shell radiative rates are accurate to about 10%; for outer subshells inaccuracies of up to 30% could be realistic. Transition energies of the M series can deviate from experimental measured values by 10 eV. If cited, these data give a first indication of the transition energy, but high precision, as is necessary for special cases in element X-ray fluorescence analyze (overlap between L and M lines), has not been achieved. If available it is recommended to use other experimental data to first order. The authors mentioned that nonradiative widths for inner-shell vacancies are given better than 15% if Coster–Kronig or super Coster–Kronig transitions do not take place. In this case the nonradiative widths can be overestimated by a factor of two.

In [423] basic subshell data and radiative widths are taken from Scofield, partially published in [486, 488, 489]. Nonradiative widths are from Chen et al. [108–111, 114, 115].

The work of Bearden [50] contains a recomputation of more than 2700 X-ray emission and absorption wavelengths of experimentally determined quantities. As a basis for the reevaluation as an X-ray wavelength standard, the W K_{α_1} line is used to establish an absolute angstrom basis (λ W $K_{\alpha_1} = (0.2090100 \pm 1 \text{ ppm})$). This value was used to define a new unit Å^* in such a way that it yields

$$\lambda \, W \, K_{\alpha_1} = 0.2090100 \text{Å}^*$$

i.e.

$$1\text{Å}^* = 1\text{Å}^* \pm 5\text{ppm} \, .$$

To cover various regions of the X-ray spectrum secondary standards are established with the highest precision available in 1967:

$$\lambda \, \text{Ag} \, K_{\alpha_1} = (0.5594075 \pm 1.1 \text{ or } 5.5 \text{ ppm}) \, \text{Å}^*$$
$$\lambda \, \text{Mo} \, K_{\alpha_1} = (0.709300 \pm 1.3 \text{ or } 5.2 \text{ ppm}) \, \text{Å}^*$$
$$\lambda \, \text{Cu} \, K_{\alpha_1} = (1.540562 \pm 1.3 \text{ or } 5.2 \text{ ppm}) \, \text{Å}^*$$
$$\lambda \, \text{Cr} \, K_{\alpha_1} = (2.293606 \pm 1.3 \text{ or } 5.2 \text{ ppm}) \, \text{Å}^*$$

For the two given uncertainties the first value is related to the W K_{α_1} transition line as primary standard and the second one takes into account the probable error of ± 5 ppm in the conversion factor.

Comparisons of tabulated calculated X-ray transition energies (for instance see Huang et al. [249] or Perkins et al. [423]) with experimental transition energies based on the data from Siegbahn [503] about wide regions shows deviations in the order of eV. This fact can be explained by the neglect of correlation effects and other contributions in the calculations. For instance, the influence of a finite nuclear charge distribution is discussed by Chen et al. [113].

Data preferentially derived due to excitation by electrons or photons are known from Cauchois and Senemaud [104]. In this paper data from wavelength-dispersive X-ray transition and absorption edge measurements available in the literature up to the end of 1976 are evaluated. The conversion factors are

$$\Lambda = \frac{\lambda/\text{Å}}{\lambda/\text{kXu}} = \frac{\lambda/\text{mÅ}}{\lambda/\text{Xu}} = 1.0021017 \, (0.6 \text{ ppm})$$

$$\frac{\lambda}{\text{Å}} \times \frac{E}{\text{eV}} = 12398.52$$

which gives

$$\frac{\lambda}{\text{Xu}} \times \frac{E}{\text{keV}} = 12372.52.$$

For the Rydberg constant it is assumed that

$$R_\infty = 109737.3143(10) \, \text{cm}^{-1} \, .$$

A comparison of the measured most precise X-ray transition energies with calculated results with consideration of contributions from vacuum polarization, velocity-dependent Breit interactions, the nuclear shape and contributions of the electron self-energy under consideration of the electronic screening is given by Deslattes and Kessler Jr. [144]. Here characteristic deviations between experiment and theory were in the order of 100 meV up to 10 eV depending on the atomic number and on the individual transition. For the mutual comparison of precision data the excitation conditions must also be considered. As a rule different excitation processes lead to energy shifts lying in the error region as given in Chapter 5 for the X-ray transition energies listed there.

In [145] K and L X-ray transition energies of all elements for $Z = 10$–100 are tabulated. The summarized X-ray transition energies are the result of a combination of theoretical calculations with selected experimental data. Here for conversions of wavelength transition data from primary measurements to eV the relation

$$E[\text{eV}] = \frac{12398.41857(49)}{\lambda[\text{Å}]}$$

has been applied. Because of the small uncertainty here the authors pointed out that the energy and wavelength representations can be treated as metrologically equivalent.

In [145] the authors give a detailed discussion of the precision of the experimental data and they discuss in detail the theoretical uncertainty of calculated data and in this context also theoretical issues unresolved until now.

4.2.2 Soft X-Ray Energy Region

In the soft X-ray energy range below 1 keV, spectrum analysis can be hindered due to the high density of transition lines, i.e. often some line interference can occur. Additionally, imprecisely known transition energies and line intensity ratios can lead to misinterpretations in the spectrum analysis. To improve this situation different authors give reports on special studies in the low-energy region. Uncertainties in the analysis of M X-ray spectra of rare earth elements were analyzed by Labar and Salter [305]. For the energy region $100 \text{ eV} \leq E_X \leq 700 \text{ eV}$ Assmann and Wendt [18] found differences of up to two orders of magnitude in comparison with other authors. A reinvestigation of L-spectra in the Z-region $24 \leq Z \leq 33$ was undertaken by Assmann et al. [19] and Scheffel et al. [481] report on new results for the iron L-spectrum. X-ray M spectra in the region $39 \leq Z \leq 56$ were studied by Wendt [564]. New data for electron excited M X-ray spectra of the elements $55 \leq Z \leq 58$ can be found in Dellith and Wendt [134].

4.3 X-Ray Fluorescence Yields

4.3.1 K-Shell X-ray fluorescence Yields

Precise available values of fluorescence yields are important for various investigations in areas such as atomic physics, nuclear physics and materials science, as well as for applied physics. Because of the insufficient correspondence between fluorescence yield measurements published in the literature often semiempirical relations were used to describe the fluorescence yields ω as a function of the atomic number Z.

The first relation for the approximation of K shell fluorescence yields as a function of Z was given by Wentzel [566]:

$$\frac{\omega_K}{1 - \omega_K} \approx 10^{-6} Z^4 . \tag{4.1}$$

For the derivation of equation (4.1) Wentzel [566] considers only L electrons under neglect of screening and relativistic effects. After further studies Burhop [90] deduced that a more exact description needs a modification of equation (4.1) in the form that some constants must be entered. An overview of possible approximation formulas for semiempirical fits of the K shell fluorescence yield can be found in Bambynek et al. [38]. The same authors published a three-parameter approximation with the form

$$\left(\frac{\omega_K}{1 - \omega_K} \right)^{1/4} = 0.015 + 0.0327 \cdot Z - 0.64 \cdot 10^{-6} \cdot Z^3 . \tag{4.2}$$

Nevertheless a detailed analysis shows that these recommended empirical values have somewhat systematic deviations from experiment. The deviations have a characteristic Z-dependence since they are dependent on the electronic configuration of the actual outer electron shells. Thus an improved approximation was published by Hanke et al. [224]:

$$\omega_K = 3.3704 \cdot 10^{-1} - 6.0047 \cdot 10^{-2} Z \\ + 3.3133 \cdot 10^{-3} Z^2 - 3.9251 \cdot 10^{-5} Z^3 . \tag{4.3}$$

The range of validity for equation (4.3) was given as $12 \leq Z \leq 42$ [224].

Deviations of K-shell fluorescence yields from Bambynek et al. [38] and Krause [300] in comparison to the values given by Bambynek [39] are shown in Fig. 4.22.

The data sets used for comparison were composed from data derived with different measurement methods as well as by different excitation methods (impact ionization by charged particles, photoionization, internal conversion, electron capture, higher effects in the nuclear decay). Methods known from the literature for the determination of fluorescence yields are summarized in Table 4.2. A detailed description of the techniques used was given by Bambynek et al. [38]. A critical estimation of individual techniques can be found by Langenberg and van Eck [309]. A report of the available experimental and theoretical information on fluorescence yields of K, L and higher shells was summarized by Hubbell in 1989 [252] and somewhat later by Hubbell et al. [253].

The values given by Krause [300] were evaluated by the author and characterized with the errors given in Table 4.3.

4.3.2 L X-Ray Fluorescence Yields

Experimental techniques for the determination of L-shell fluorescence yields were discussed by Bambynek et al. [38].

Discontinuities in Coster–Kronig transition probabilities for Z-regions where the energetic threshold for intense groups of Coster–Kronig transitions were of importance influence the value of corresponding fluorescence yields (see also Table 2.20).

For the L_1 subshell the Coster–Kronig yield is the dominant component of the L_1 deexcitation channel. Thus the quality of their determination has a decisive influence on the class of the determined fluorescence yield. This is of significant importance in regions where directly measured values were not available or are not known precisely enough.

The uncertainties estimated by Krause [300] for the determination of fluorescence yields and Coster–Kronig yields fitted by him between the individual L-subshells are summarized in Table 4.4. In Table 4.4 the greatest uncertainties were observed for light elements. This is connected with the circumstance that the used values as a rule were determined from molecules or theoretical yields derived on the basis of single-particle calculations. This means there it is not considered that in these Z-regions many-body interactions [273] and chemical effects were of importance. The strong influence of many-body effects on M-shell Coster–Kronig transitions was discussed by Ohno and Wendin [399]. These effects also appear in a weaker form in the atomic L shells and cause, in comparison to the values given in [300], a reduction of the Coster–Kronig widths in the vicinity of the discontinuities in the fluorescence yield curves and can lead to uncertainties greater than the given values.

An approximation for the L_3 fluorescence yield in the range $38 \leq Z \leq 79$ is given by Hanke and Wernisch [224]

$$\omega_{L_3} = 4.41 \cdot 10^{-2} - 4.7559 \cdot 10^{-3} Z + 1.1494 \cdot 10^{-4} Z^2$$
$$-1.8594 \cdot 10^{-7} Z^3. \qquad (4.4)$$

Average L shell fluorescence yields were evaluated for elements with $56 \leq Z \leq 92$ by Singh et al. [505]. Tables of calculated L subshell fluorescence yields and Coster–Kronig transition probabilities in the element range $25 \leq Z \leq 96$ are known from Puri et al. [436]. Detailed investigations of Coster–Kronig transition probabilities were done by different authors and listed for instance by Sogut et al. [511, 513]. In detail this concerns papers from Ohno [400], Broll [87], Semmes et al. [492], Rani et al. [445], Zararsiz and Aygün [583], Sorenson [517,518], Jitschin et al. [265], Xu [578], Hubbell et al. [253], Allawadhi et al. [9], Baydas et al. [45] and Kim et al. [281]. The influence of the chemical environment on Koster–Kronig transition probabilities is discussed in [513] for measurements on tungsten and lead.

4.3.3 M X-Ray Fluorescence Yields

To this day the knowledge of M-shell deexcitation is relatively incomplete. Fortunately this information is for many experiments of secondary interest in comparison to information on K and L shell deexcitation. For heavy elements it yields that M-MN Coster–Kronig transitions dominate in relation to the corresponding M-NY and M-XY Auger transitions and to the M-N and M-Y X-ray transitions. Corresponding observations for a lighter element are reported by Mehlhorn [353] for the case of krypton.

For M-shell Coster–Kronig transitions additionally the contribution of super Coster–Kronig transitions must be considered, which appear in certain regions of the periodic table of elements. Here a vacancy in a subshell i leads to two vacancies in the subshell j of the shell with the same principal quantum number. To quantify such processes McGuire [350] calculates a quantity S_{ij} which describes the mean number of M_j vacancies produced as the first step of

Table 4.2. Methods for the determination of K-shell fluorescence yields (after [38,39]). P – given accuracy in percent ; P_{max} – estimated accuracy, G – gas, S – solid state, EC – electron capture, IC – internal conversion, AES – Auger electron spectroscopy

Nr.	method	creation of the primary vacancy by	target	element range	P	P_{max}
1	fluorescence excitation in gas targets	X-rays	G	6 ... 54	0.5 ... 22	2
2	fluorescence excitation in solid states	X-rays	S	3 ... 56	0.2 ... 50	3
3	AES and conversion electron spectroscopy	IC, EC	S	43 ... 93	0.3 ... 9.0	2
4	AES, X-ray- and β-spectroscopy	EC	S	80	1.7	2
5	AES and K X-ray spectroscopy	EC	G	17 ... 54	0.3 ... 6	1
6	AES and K X-ray spectroscopy	EC	S	6 ... 99	0.9 ... 37	5*
7	K X-ray and γ- or conversion electron spectroscopy	EC leading to metastable states	S	27 ... 49	5.9 ... 8.9	5
8	determination of the K X-ray emission and deexcitation rate	EC	S	23 ... 55	0.5 ... 10.0	1
9	K X-ray - γ or K X-ray - K conversion electron coincidences	EC, IC	S	22 ... 97	1.1 ... 9.0	2
10	bubble chamber technique	X-rays	G	8 ... 54	3 ... 75	15
11	jump of the photoionization current at the K-edge	X-rays	G, S	22 ... 53	–	20
12	photoemulsion technique	EC	S	84	5.6	15
13	excitation with	e^-, p, α heavy ions	G, S	6 ... 25	6.4 ... 53	15**

* for high Z;

** at simultaneous detection of X-rays and Auger electrons

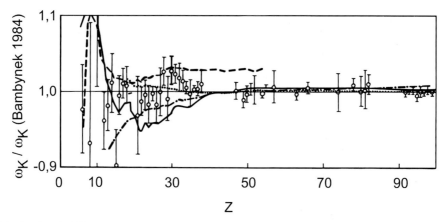

Fig. 4.22. Residuals of K-shell fluorescence yields ω_K regarding the values from Bambynek [39] as a function of the atomic number Z. The semicolon line gives the ratio to the values from Bambynek et al. [38], the solid line those in relation to Krause [300], the dotted line the ratio to the theoretic relativistic values from Chen et al. [111] and the broken line the relation to the theoretic nonrelativistic values from Walters and Bhalla [557]

decay of a vacancy M_i. In Table 4.5 values of S_{ij} and calculated Coster–Kronig probabilities f_{45} are listed.

Because of the small basis of data for experimentally determined M-shell data it is difficult to undertake a comparison with calculated data sets. Thus further experimental investigations are required. A description of different techniques to determine M-shell fluorescence yields can be found in Bambynek et al. [38].

Mean M-shell fluorescence yields $\bar{\omega}_M$ for the element region $29 \leq Z \leq 100$ were calculated by Öz et al. [405]. Total M-shell fluorescence yields for $19 \leq Z \leq 100$ are given by Hubbell et al. [253]. From measurements of $\bar{\omega}_M$ errors of up to 50% are known [446]. Fit values of M subshell fluorescence yields and Coster–Kronig transitions for elements with $20 \leq Z \leq 90$ were published in [512]. Here the basis of the fit were data published by McGuire [350] for selected elements in the region $22 \leq Z \leq 92$.

Table 4.3. Estimated uncertainties (in %) of the K-shell fluorescence yields fitted by Krause [300]

Z	atomic number region									
	5 to 10	10 to 20	20 to 30	30 to 40	40 to 50	50 to 60	60 to 70	70 to 80	80 to 90	90 to 100
$\Delta\omega_K$	10 to 40*	5 to 10	3 to 5	3	2	1 to 2	1	1	1	1

∗) The yields for solid states and molecules could be different in the indicated Z-region than the values given in the table.

Table 4.4. Estimated uncertainties (in %) for L-shell fluorescence yields $\omega_{1,2,3}$ and Koster–Kronig yields f (after [300])

Z	atomic number region								
	10 to 20	20 to 30	30 to 40	40 to 50	50 to 60	60 to 70	70 to 80	80 to 90	90 to 100
ω_1	$\geq 30^*$	30^{**}	30^{**}	20^{**}	15	15	15	15	15
				...	...				...
				30	20				20
ω_2	$\geq 25^*$	$25^{*,**}$	25	10	10	5	5	5	10^{**}
				...		...			...
				25		10			...
ω_3	$\geq 25^*$	25^*	20	10	5	5	3	3	3 ... 5
				...	...		...		...
				20	10		5		...
f_1	3^*	5^*	5	5^{**}	15	10	5^{**}	5	5
			...				...		...
			10				10		10
f_{12}	10^*	15^*	15	20^{**}	20	15	20^{**}	10	10
									...
									50
f_{13}	5^*	10^*	10	10^{**}	15	10	5^*	5	510
				...	...		...		...
				...	...		10		10
f_{23}	-	$\geq 40^*$	20	20	20	15	15	15^{**}	15
			...			...			
			30			20			

*) The yields for solid states and molecules could be different in the indicated Z-region than the values given in the table.

**) In the vicinity of the discontinuities of the given curves the uncertainties can be greater than the given values.

Table 4.5. Calculated mean number S_{ij} of M_j vacancies produced in the first decay step of an M_i vacancy and theoretical M-shell Coster–Kronig probabilities f_{45} (after [350]).

Z	S_{12}	S_{13}	S_{14}	S_{15}	S_{23}	S_{24}	S_{25}	S_{34}	S_{35}	f_{45}
20	0.328	0.655								
22	0.319	0.639	0.314	0.471		1.057	0.672	0.509	1.220	
23	0.315	0.631	0.335	0.503		1.089	0.820	0.558	1.280	
24	0.319	0.638	0.397	0.596		1.123	0.834	0.612	1.342	
25	0.312	0.623	0.357	0.538		1.108	0.797	0.589	1.317	
26	0.311	0.621	0.371	0.556		1.116	0.815	0.600	1.329	
27	0.308	0.616	0.376	0.564		1.120	0.817	0.602	1.335	
28	0.307	0.614	0.381	0.566		1.122	0.827	0.609	1.341	
29	0.304	0.608	0.406	0.610		1.133	0.850	0.623	1.360	
30	0.283	0.566	0.374	0.561		1.107	0.811	0.597	1.320	
32	0.249	0.522	0.273	0.409		1.085	0.786	0.580	1.292	
36	0.270	0.540	0.086	0.127		0.919	0.516	0.395	1.039	
40	0.278	0.475	0.108	0.163	0.032	0.591	0.309	0.252	0.677	
44	0.305	0.457	0.065	0.124	0.067	0.550	0.283	0.236	0.672	
47	0.343	0.461	0.065	0.097	0.073	0.570	0.258	0.223	0.689	
50	0.315	0.475	0.067	0.101	0.016	0.604	0.252	0.213	0.678	
54	0.238	0.505	0.081	0.122	0.031	0.612	0.233	0.206	0.688	
57	0.195	0.506	0.094	0.140	0.034	0.557	0.282	0.198	0.678	
60	0.236	0.489	0.092	0.128	0.057	0.644	0.172	0.174	0.712	0.267
63	0.338	0.485	0.070	0.100	0.062	0514	0.137	0.165	0.720	0.369
67	0.266	0.527	0.061	0.090	0.106	0.667	0.120	0.145	0.751	0.408
70	0.272	0.525	0.056	0.091	0.116	0.680	0.105	0.141	0.761	0.479
73	0.197	0.561	0.065	0.115	0.114	0.674	0.106	0.082	0.810	0.411
76	0.161	0.594	0.067	0.109	0.107	0.684	0.098	0.106	0.764	0.418
79	0.148	0.594	0.067	0.112	0.114	0.673	0.095	0.114	0.782	0.046
83	0.109	0.650	0.065	0.095	0.103	0.662	0.083	0.094	0.750	0.035
86	0.143	0.593	0.069	0.100	0.128	0.610	0.093	0.072	0.768	0.065
90	0.072	0.690	0.063	0.091	0.116	0.623	0.088	0.097	0.725	0.066

4.4 Natural Atomic Level Widths and Widths of X-Ray Transition Lines

Because the natural energetic linewidths of X-ray transitions result as a sum of the energetic level widths of the atomic substates involved in the considered X-ray transition (see Sect. 2.3.3) in the following both quantities are discussed together.

Experimental atomic level widths are not known for all elements. Most information is available for K-shell data. Experimental values for the K_α linewidths and for K-shell level widths are known for example from Allinson [10], Richtmyer and Barnes [453], Ingelstam [258] and Parratt [416]. More presently measurements are published by Gokhale [202], Brogren [86], Watanabe [558], Nelson et al. [374, 377], Svensson et al. [535] and Agren et al. [3].

Table 4.6. Estimated errors (in %) of semiempirical values of natural linewidths and level widths (after [301])

level or line	atomic number region								
	10 to 20	20 to 30	30 to 40	40 to 50	50 to 60	60 to 70	70 to 80	80 to 90	90 to 100
K	10*	5	5	4	4	3	3	3	4
L_1	30*	30*	30**	25**	20	15	15	15	20
L_2	25*	25*	25**	20	10	10	10	10	10**
L_3	25*	25*	20	15	10	10	8	8	8
$K_{\alpha_{1,2}}$	10*	7*	6	5	4	4	3	3	4

*)   Error resulting from the neglect of many-body effects are not included.

**)   In the region of Coster–Kronig discontinuities the errors can exceed the given values.

Experimental values for L-shell level widths derived by measurements of absorption edges and of the form of absorption and emission lines are given by Leisi et al. [320] based on the original data from Richtmyer [453], Parratt [417], Beeman and Friedman [52] and Bearden and Snyder [48]. More modern values for example are known for free atoms from Mehlhorn [353], Wuilleumier [575] and Nordgreen [397].

The semiempirical linewidths published by Krause and Oliver [301] are atomic widths. Generally atomic widths are smaller than observed in experiment. Here the reason is that different effects such as for example multiplet splitting and multiple ionization processes can lead to a widening of levels or emission lines. Apart from light elements these widenings are small in comparison to the natural level width. A detailed discussion of widening effects are given in Sect. 2.2.2.

The errors of the determination of level widths result from a quadratic summation of the errors from the fluorescence yields and of the X-ray emission rates, respectively. Here the dominant part results from the X-ray fluorescence yields [300]; see also Tables 4.3 and 4.4. The error of the K X-ray emission rate is assumed by Krause and Oliver [301] to be 3–5% and the errors of L-subshell rates are 4–7%. These

estimations relate to a comparison of calculated ratios with the following experimental data:

- X-ray intensity ratios for transitions including different main quantum numbers according to [472],
- K_α X-ray linewidths for heavy elements where the radiative width dominates [426,473].

Opposite to the errors assumed in [301] based on a comparison with experimental intensity ratios [489] estimate for the K X-ray emission rate an uncertainty of about 2%. Errors for the determination of the linewidth result from the combined errors of the individual components. The errors derived in this way are summarized in Table 4.6.

The uncertainties for the determination of energetic widths of $K\alpha$ transitions are dominantly influenced by the errors for the determination of the K-shell level width. Large uncertainties are typical for low atomic numbers because here the errors of the fluorescence yields are high because of the small amount of experimental and calculated data (see also Table 4.3 [273]). A graphical representation of this situation is given in Figs. 4.23 and 4.24. Here the behaviour of K-shell level widths in the region $Z > 40$ can be approximated as [320]:

$$\Gamma(K) = 1.73 \cdot 10^{-6} \, Z^{3.93} \quad \text{eV} . \tag{4.5}$$

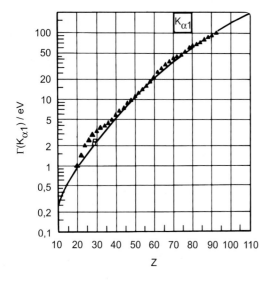

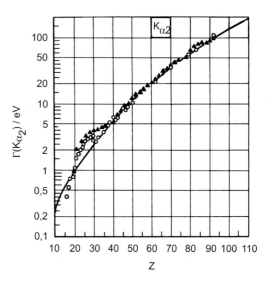

Fig. 4.23. Natural energetic linewidths of K_{α_1} transitions as a function of the atomic number (after [301]). The solid line relates to data from [301]. For comparison fitted data from [473] (▲) and values from [119] (□) are also given.

Fig. 4.24. Natural energetic linewidths of K_{α_2} transitions as a function of the atomic number Z (after [301]). The solid line relates to data from [301]. For comparison fitted data from [473] (▲) and values from [425] (○) are also given.

The semiempirical values from Krause and Oliver [301] shown in Figs. 4.23 and 4.24 show a continuous rise with increasing atomic number, while the fitted data from Pessa et al. [425] and Salem and Lee [473] have the same deviations from the data given in [301]. In the region between $Z = 21$ and $Z = 28$ the increasing widths can be partially explained by the exchange interaction of a 2p hole with the incompletely filled M shell.

In Fig. 4.25 absolute deviations of results derived by several authors in comparison to those given in [301] are shown. Deviations for K_{α_1} and K_{α_2} transition lines in different regions were explained in [301] with imprecise initial data for the approximations done by other authors. The values from [425] in the region $Z < 90$ give a satisfying correspondence to that from [301] and the mean deviation to the results published in [377] does not exceed the 2% boundary.

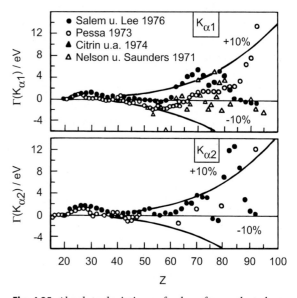

Fig. 4.25. Absolute deviations of values from selected authors in comparison to the energetic widths Γ given in [301] for the K_{α_1} and K_{α_2} X-ray transitions (after [301]). An arbitrary selected boundary for an error region of $\pm$ 10% is shown

4.5 X-Ray Emission Rates

In physical practice the evaluation of X-ray emission rates as a rule occurs by comparison of the inten-

Table 4.7. Comparison of experimental and theoretical $K_{\alpha_2}/K_{\alpha_1}$ intensity ratios. Z – atomic number; A – intensity ratios calculated by Reiche [448] on the basis of relativistic wavefunctions from the GRASP-code published by Grant et al. [214] under consideration of all multipole orders of the radiation field in frozen core approximation; B – intensity ratios calculated by Scofield [487]; E – experiment

Z		A	B	E	reference to E
26	Fe	0.5061	0.5107	0.506 ± 0.01	Salem and Wimmer [470]
				0.507 ± 0.01	McCrary et al. [343]
30	Zn	0.5096	0.5142	0.513 ± 0.01	Salem and Wimmer [470]
38	Sr	0.5163	0.5205	0.521 ± 0.01	Salem and Wimmer [470]
48	Cd	0.5286	0.5317	0.544 ± 0.01	Salem and Wimmer [470]
50	Sn	0.5313	0.5343	0.541 ± 0.01	Salem and Wimmer [470]
				0.531 ± 0.01	McCrary et al. [343]
56	Ba	0.5407	0.5428	0.533 ± 0.01	McCrary et al. [343]
78	Pt	0.5879	0.5850	0.563 ± 0.03	Nelson et al. [374]
80	Hg	0.5926	0.5899	0.595 ± 0.03	Nelson et al. [374]

sity ratios connected with these rates. Intensity ratios are observables in the experiment, and often at the same time are prerequisites for the analysis of measured X-ray spectra. To determine X-ray intensity ratios over the whole Z-region theoretical ratios calculated with self-consistent field models (see for instance Babushkin [29–31], Scofield [486–489], Chen and Craseman [116], Rosner and Bhalla [460], Lu [329], Manson and Kennedy [335], Bhalla [65–67]) or approximations from experimental determined intensity ratios (see for instance Salem et al. [472], Nelson et al. [376], Salem and Schultz [471]) can be used.

The in Sect. 5 calculated values of the X-ray emission rates are given. For the experimentally based approximation of X-ray line intensities a comparison of X-ray emission rates among each other is only permitted if X-ray transitions with the same initial vacancy are considered. If this is not the case than the line intensities change significantly as a result of nonradiative electron transitions.

K-shell. Here a direct comparison of experimentally measured intensity ratios with theoretical emission rates is possible, because all comparisons are related to transitions with the same initial vacancy. For the determination of K-shell values Scofield et al. [488]

give a systematical error of up to 10%. With consideration of exchange corrections Scofield [487] gave total emission rates for light elements increased by 10% and rates for heavy elements increased by 2%. Thus the error for calculated intensity ratios is reduced to about 2% in comparison to experiment. To characterize the reachable precision at the calculation for ratios of intensity in Table 4.7 data by Scofield [487] and Reiche [448] is compared with experimental data. As a rule the derived agreement is here in the order of the error boundaries characteristic of the individual experiments, whereby the calculated values from [487] and [448] differ only by at about 1%. The difference increases if transitions from different main shells to the K-shell were considered (for example K_{β_1}/K_{α_1}: K-M$_3$/K-L$_3$ or K_{β_2}/K_{β_1}: K-N$_{2,3}$/K-M$_3$). Here differences of up to 20% for light and up to 3% for heavy elements are considered.

The absolute size of X-ray intensity ratios is influenced by the gauge used in the calculation (Coulomb or length gauge; see [116]). Calculated with the length gauge, emission rates exceed those calculated in the framework of the Coulomb gauge by about 10% for $Z = 10$ and by 4% for $Z = 30$. For heavy elements an increasing identity of the results derived in the two gauge forms is observed.

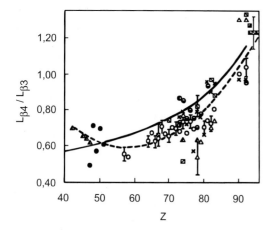

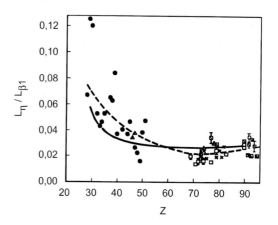

Fig. 4.26. Comparison of calculated ([488]) L_{β_4}/L_{β_3} intensity ratios with theoretical values from [472]. The solid line corresponds to values published by [488] and the dashed line an approximation of experimental values according to [472]. The individual symbols characterize results from different authors

Fig. 4.27. Comparison of calculated ([488]) L_{η}/L_{β_1} intensity ratios with experimental values from [472]. The solid line corresponds to values published by [488] and the dashed line an approximation of experimental results according to [472]. The individual symbols characterize results from different authors

L and M shells. Total L X-ray emission rates in the Coulomb gauge framework coincide with the results published by Scofield [487] to better than 1%. For individual subshells preferentially of light elements deviations of up to a factor of two are known from results derived in the length gauge (for instance the L_2 width for $Z = 18$), but for higher Z the derived values are nearly identical. A comparison of theoretical results with experimental data is given in Figs. 4.26–4.28. Three line intensity ratios are selected and these result from the creation of a primary L_1, L_2 or L_3 vacancy, respectively:

L_{β_4}/L_{β_3} : L_1-M_2 / L_1 - M_3
L_{η}/L_{β_1} : L_2-M_1 / L_2 - M_4
$L_{\beta_{2,15}}/L_{\alpha_1}$: L_3-$N_{4,5}$ - L_3 / M_5.

While for L-series transition lines a wide spectrum of references for experimental and theoretical data is known, the data basis for lines of the M series is much poorer. M-shell X-ray emission for six element rates were calculated in the atomic number region $48 \leq Z \leq 93$ by Bhalla [65] on the basis of relativistic Hartree–Slater wavefunctions and for elements in the region $48 \leq Z \leq 92$ by Chen and Craseman [116]

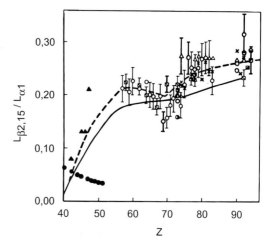

Fig. 4.28. Comparison of calculated ([488]) $L_{\beta_{2,15}}/L_{\alpha_1}$ intensity ratios with experimental values from [472]. The solid line corresponds to values published by [488] and the dashed line an approximation of experimental results according to [472]. The individual symbols characterize results from different authors

on the basis of relativistic Dirac–Fock wavefunctions for the length and Coulomb gauge. Chen et al. [116] state that the uncertainty for calculated M-shell X-ray emission rates calculated in a model of independent particles is in the range between 10% and 40% in the atomic number region $48 \leq Z \leq 80$. A comparison of calculated relativistic and nonrelativistic M-shell X-ray emission rates can, depending on the individual transition, deviate from some 10% up to 500% as a result of the different calculated transition energies and changes in the radial distribution of the single-electron wavefunctions [116]. Calculations in the Coulomb and length gauge give results deviating among each other from a few percent up to some 10%.

With the background of the above-mentioned circumstances M-shell emission rates in Chap. 5 should be understood only as landmarks, as a rule, giving the correct order of magnitude of the emission strength. If for the M_{α_1} and the M_{α_2} lines identical intensities are given it means that the corresponding authors have calculated only the combined X-ray emission rate for the M_α doublet.

Some detailed studies on M spectra of rare earth elements can be found in Wendt [565]. For soft X-rays in the energy region of $100 \text{ eV} \leq E_X \leq 700 \text{ eV}$ Assmann and Wendt [18] discuss actually measured relative intensities of L and M lines differing by up to two orders of magnitude with respect to the accepted literature data.

Because X-ray emission rates can be measured by means of X-ray intensity ratios between different transition lines, we note here that chemical effects [444] and pressure effects [256] can influence the measured results.

4.6 Mass Attenuation Coefficients

The mass attenuation coefficients in the photon energy region below 1 keV given in Chap. 10 relate to the photoabsorption cross-sections published by Henke [234] and above 1 keV to an approximation of experimental data done by Veigele [547].

Because the photoabsorption cross-sections in the photon energy region below 1 keV exceed the cross-sections for coherent and incoherent scattering processes by some orders of magnitude the effective attenuation can be approximated by photoabsorption cross-sections. On the basis of available approximations and data from 140 original papers Henke [234] reevaluates new values of photoabsorption cross-sections in the photon energy region between 30 eV and 300 eV. Data for the energy region between 300 eV and 1 keV were taken from a paper published by Henke and Ebisu [233].

Data above 1 keV given by Veigele [547] are based on about 9000 experimental total photoelectric interaction cross-sections. An extensive bibliography of these data can be found in the original paper [547]. A comparison with the data from Hubbell [251], Storm and Israel [529] and McMaster et al. [352] shows that above a photon energy of 10 keV in most cases these data coincide with the values given in [547] with a precision better than 5%. Below 10 keV the values from [352] are about 20% higher than the results from [547] and from [529]. The deviation between the data from [529] and [547] in the photon energy region below 10 keV are in the order of up to 5%. Uncertainties published in [547] as a result of the evaluation of the used data are tabulated in Table 4.8.

Table 4.8. Uncertainties for the determination of mass attenuation coefficients after [547] (in %). The declaration of the uncertainties occur in regions determined by the electron binding energies for the individual main shells of the element considered

Z	element	>K	K-L	L-M	M-N
1	H	±(2-5)			
2	He	±(2-5)			
3	Li	±(2-5)			
4	Be	±(2-5)			
5	B	±(2-5)			
6	C	±(2-5)			
7	N	±(2-5)			
8	O	±(2-5)			
9	F	±(2-5)			
10	Ne	±(2-5)			
11	Na	±(2-5)	±(2-5)		
12	Mg	±(2-5)	±(2-5)		
13	Al	±(2-5)	±(2-5)		
14	Si	±(2-5)	±(2-5)		
15	P	±(2-5)	±(2-5)		
16	S	±(2-5)	±(2-5)		
17	Cl	±(2-5)	±(2-5)		
18	Ar	±(2-5)	±(2-5)		
19	K	±(2-5)	±(2-5)		
20	Ca	±(2-5)	±(2-5)		
21	Sc	±(2-5)	±(2-5)		
22	Ti	±(2-5)	±(2-5)		
23	V	±(2-5)	±(2-5)		
24	Cr	±(2-5)	±(2-5)		
25	Mn	±(2-5)	±(2-5)		
26	Fe	±(2-5)	±(2-5)		
27	Co	±(2-5)	±(2-5)		
28	Ni	±(2-5)	±(2-5)		
29	Cu	±(2-5)	±(2-5)		
30	Zn	±(2-5)	±(2-5)	±(10-20)	
31	Ga*	±(2-5)	±(2-5)	±(10-20)	
32	Ge	±(2-5)	±(2-5)	±(10-20)	
33	As*	±(2-5)	±(2-5)	±(10-20)	
34	Se	±(2-5)	±(2-5)	±(10-20)	
35	Br	±(2-5)	±(2-5)	±(10-20)	

*) For these elements no experimental data were used.

Table 4.8. (cont.)

Z element	>K	K-L	L-M	M-N
36 Kr	±(2-5)	±(2-5)	±(5-10)	
37 Rb*	±(2-5)	±(2-5)	±(5-10)	
38 Sr*	±(2-5)	±(2-5)	±(5-10)	
39 Y*	±(2-5)	±(2-5)	±(5-10)	
40 Zr	±(2-5)	±(2-5)	±(5-10)	
41 Nb	±(2-5)	±(2-5)	±(5-10)	
42 Mo	±(2-5)	±(2-5)	±(5-10)	
43 Tc*	±(2-5)	±(2-5)	±(5-10)	
44 Ru*	±(2-5)	±(2-5)	±(5-10)	
45 Rh	±(2-5)	±(2-5)	±(5-10)	
46 Pd	±(2-5)	±(2-5)	±(5-10)	
47 Ag	±(2-5)	±(2-5)	±(5-10)	
48 Cd	±(2-5)	±(2-5)	±(5-10)	
49 In	±(2-5)	±(2-5)	±(5-10)	
50 Sn	±(2-5)	±(2-5)	±(5-10)	
51 Sb	±(2-5)	±(2-5)	±(5-10)	
52 Te	±(2-5)	±(2-5)	±(5-10)	
53 I	±(2-5)	±(2-5)	±(5-10)	
54 Xe	±(2-5)	±(2-5)	±(5-10)	
55 Cs*	±(2-5)	±(2-5)	±(5-10)	
56 Ba	±(2-5)	±(2-5)	±(5-10)	
57 La	±(2-5)	±(2-5)	±(5-10)	
58 Ce	±(2-5)	±(2-5)	±(5-10)	
59 Pr	±(2-5)	±(2-5)	±(5-10)	
60 Nd	±(2-5)	±(2-5)	±(5-10)	
61 Pm*	±(2-5)	±(2-5)	±(5-10)	±(10-20)
62 Sm	±(2-5)	±(2-5)	±(5-10)	±(10-20)
63 Eu*	±(2-5)	±(2-5)	±(5-10)	±(10-20)
64 Gd	±(2-5)	±(2-5)	±(5-10)	±(10-20)
65 Tb*	±(5-10)	±(5-10)	±(5-10)	±(10-20)
66 Dy*	±(5-10)	±(5-10)	±(5-10)	±(10-20)
67 Ho	±(5-10)	±(5-10)	±(5-10)	±(10-20)
68 Er	±(5-10)	±(5-10)	±(5-10)	±(10-20)
69 Tm*	±(5-10)	±(5-10)	±(5-10)	±(10-20)
70 Yb	±(5-10)	±(5-10)	±(5-10)	±(10-20)
71 Lu	±(5-10)	±(5-10)	±(5-10)	±(10-20)

*) For these elements no experimental data were used.

Table 4.8. (cont.)

Z element	>K	K-L	L-M	M-N
72 Hf	±(5-10)	±(5-10)	±(5-10)	±(10-20)
73 Ta	±(2-5)	±(2-5)	±(5-10)	±(10-20)
74 W	±(2-5)	±(2-5)	±(5-10)	±(10-20)
75 Re*	±(2-5)	±(2-5)	±(5-10)	±(10-20)
76 Os*	±(2-5)	±(2-5)	±(5-10)	±(10-20)
77 Ir	±(2-5)	±(2-5)	±(5-10)	±(5-10)
78 Pt	±(2-5)	±(2-5)	±(2-5)	±(5-10)
79 Au	±(2-5)	±(2-5)	±(2-5)	±(5-10)
80 Hg	±(2-5)	±(2-5)	±(2-5)	±(5-10)
81 Tl	±(2-5)	±(2-5)	±(2-5)	±(5-10)
82 Pb	±(2-5)	±(2-5)	±(2-5)	±(5-10)
83 Bi	±(5-10)	±(2-5)	±(2-5)	±(5-10)
84 Po*	±(5-10)	±(5-10)	±(10-20)	±(10-20)
85 At*	±(5-10)	±(5-10)	±(10-20)	±(10-20)
86 Rn*	±(5-10)	±(5-10)	±(10-20)	±(10-20)
87 Fr*	±(5-10)	±(5-10)	±(10-20)	±(10-20)
88 Ra*	±(5-10)	±(5-10)	±(10-20)	±(10-20)
89 Ac*	±(5-10)	±(5-10)	±(10-20)	±(10-20)
90 Tb	±(5-10)	±(5-10)	±(5-10)	±(10-20)
91 Pa*	±(5-10)	±(5-10)	±(5-10)	±(10-20)
92 U	±(5-10)	±(5-10)	±(5-10)	±(10-20)
93 Np*	±(5-10)	±(5-10)	±(10-20)	±(10-20)
94 Pu	±(5-10)	±(5-10)	±(10-20)	±(10-20)

*) For these elements no experimental data were used.

Part II

X-Ray Reference Data

5 X-Ray Emission Lines and Atomic Level Characteristics

In the following X-ray transition energies E, intensities I, energetic line widths Γ of X-ray transition lines, electron binding energies E_B, fluorescence yields ω_{nlj}, energetic level widths Γ and absorption edges AE are tabulated. The quantity TPIV (cited from [423]) characterize the transition probability that a certain initial vacancy is filled from an higher electronic level. For example KM_2 means, that there is an initial vacancy in the K subshell and this vacancy is filled from an electron from the M_2 subshell, leaving a vacancy in the M_2 subshell.

Theoretical transition energies from [423] are given for orientation, but they could deviate from experimental measured values up to some tens of eV. In some cases where experimental X-ray transition energies are not known there are also cited transition energies calculated from electron binding energies publicated by [51]. This values give also an orientation but they could deviate from the real experimental values by up to some tens of eV. As latest data of calculated X-ray transition energies and evaluated experimental transition energies data from [145] is included in the presented table. Especially this is the case where other experimental information is not available yet or where not enough other data are available. For K and L X-ray transition data from [50] in the most cases data as corrected by Deslattes et al. [145] for an optically based scale are used.

The head of each table contains

- the chemical symbol and the name of the element;
- the atomic number Z;
- the atomic mass A (the number in parentheses gives the uncertainty in the last digit and an entry in brackets give the mass number of the longest-lived isotope for the case that the element has not stable isotopes. In this case an atomic weight is not exactly defined because of the complete absence in nature or for the case, that there exist a greater number of isotopes.);
- the density ϱ;
- the electron configuration of the atomic ground state.

Please note, that the sequence of citations in the table do not match a rating of the cited values.

Not all positions could be fulfilled in the present table yet. But there is the hope that every year the information density on X-ray and atomic level data rise up. Thus to complete your personally copy of the book you can fill up vacant positions for the case that new data are published after printing the present tables. The author would it also very kindly acknowledge if you inform him on new publications, data, etc.

The in the tables cited values are experimental determined quantities, calculated ones or results from evaluations of experimental and calculated data. Thus, if available we give for a certain physical quantity not only a selected value, but in many cases results from different papers published in the literature. This will be helpful for a first individual evaluation of the physical relevance of the examined values.

For special problems we will recommend to study careful the in the table cited papers to get detailed information on the nature of the derived quantities, possible errors, experimental and theoretical methods etc.

Li   Z = 3

[He] 2s^1

Lithium

$A = 6.941(2)$ [260] $\varrho = 0.533\,\text{g/cm}^3$ [547]

X-ray transitions

line	transition	E/eV		I/eV/$\hbar$	TPIV	Γ_X/eV
K series						
K$_\alpha$	KL	54.30±0.24	[50]			
K$_{\alpha_{2,1}}$	KL$_{2,3}$	54.24	[104]			

level characteristics

level	E_B/eV		ω_{nlj}		Γ/eV	AE/eV	
K	55.0	[503]	2.9(-4)	[39]		54.8	[51]
	54.8±0.1	[493]	1.1(-4)	[183]			
	54.7	[51]					
IP	5.39172	[222]					

Be   Z= 4   [He] 2s^2

Beryllium A = 9.012182(3) [260] ϱ = 1.845 g/cm^3 [547]

X-ray transitions

line	transition	E/eV		I/eV/$\hbar$	TPIV	Γ/eV
K series						
K$_\alpha$	KL	108.50±1.00	[50]			
K$_\alpha$	KL	109.21	[104]			

level characteristics

level	E_B/eV		ω_{nlj}		Γ/eV		AE/eV	
K	111.0	[503]	2.9(-4)	[39]	0.05	[74]	111.0	[51]
	111.0±1.0	[51]	3.6(-4)	[183]				
	111.9±0.4	[493]	3.3(-4)	[310]				
IP	9.3227	[222]						

B	Z= 5	[He] $2s^2\,2p^1$

Boron $A = 10.811(7)$ [260] $\varrho = 2.535\,\text{g/cm}^3$ [547]
$\varrho = 2.36\ \ \text{g/cm}^3$ [483]

X-ray transitions

line	transition	E/eV		I/eV/$\hbar$	TPIV	Γ/eV
K series						
K_α	KL	183.3±0.8	[50]			
K_α	KL	185.2	[104]			

level characteristics

level	E_B/eV		ω_{nlj}		Γ/eV		AE/eV	
K	188.0	[503]	1.41(-3)	[39]	0.06	[74]	188.0	[51]
	188.0±0.4	[51]	1.70(-3)	[301]				
	188.0±0.3	[493]	5.6(-4)	[347]				
			8.0(-4)	[557]				
L_1	14.05	[493]						
L_2	5.0	[503]					4.7	[51]
	4.7±0.9	[493]						
L_3	4.7 ± 0.9	[493]					4.7	[51]
IP	8.29803	[222]						

Carbon $A = 12.0107(8)$ [260] $\varrho = 2.25$ g/cm^3 [547]

[He] 2s^2 2p^2

X-ray transitions

line	transition	E/eV		I/eV/$\hbar$		TPIV	Γ/eV
K series							
K$_\alpha$	KL	277.0$\pm$2.0	[50]	1.03(-4)	[488]		
K$_\alpha$	KL	276.8	[104]				
K$_{\alpha_2}$	KL$_2$	282.02	[423]			5.615(-4)	
K$_{\alpha_1}$	KL$_3$	282.03	[423]			1.121(-3)	

level characteristics

level	E_B/eV		ω_{nlj}		Γ/eV		AE/eV	
K	284.0	[503]	2.58(-3)	[39]	0.06	[74]	283.8	[51]
	283.8$\pm$0.4	[51]	2.80(-3)	[301]	0.061	[423]		
	284.1$\pm$0.3	[493]	2.60(-3)	[347]				
			2.40(-3)	[557]				
			1.68(-3)	[423]				
			1.30(-3)	[183]				
			2.80(-3)	[310]				
L$_1$	19.39	[493]						
L$_2$	7.0	[503]					6.4	[51]
	6.4$\pm$0.4	[51]						
	9.0$\pm$2.1	[493]						
L$_3$	6.4 $\pm$ 0.4	[493]					6.4	[51]
IP	11.26030	[222]						

Nitrogen $A= 14.00674(7)$ [222] $\varrho= 1.165(-3)\,\mathrm{g/cm^3}$ [547]

X-ray transitions

line	transition	E/eV		I/eV/ℏ		TPIV	Γ/eV
K series							
K_α	KL	392.4±0.7	[50]	3.31(-4)	[488]		
K_α	KL	392.73	[104]				
K_α	KL	395.38 (N$_2$)	[493]				
K_{α_2}	KL$_2$	393.35	[423]			1.094(-3)	
K_{α_1}	KL$_3$	393.37	[423]			2.182(-3)	

level characteristics

level	E_B/eV		ω_{nlj}		Γ/eV		AE/eV	
K	399.0	[503]	4.30(-3)	[39]	0.09	[74]	401.6	[51]
	401.06±0.4	[51]	5.20(-3)	[300, 301]	0.093	[423]		
	400.5±1.2	[493]	6.00(-3)	[347]				
			4.70(-3)	[557]				
			3.28(-3)	[423]				
			5.20(-3)	[310]				
L$_1$	25.41	[493]						
L$_2$	9.0	[503]					9.2	[51]
	9.2±0.6	[51]						
	8.9±0.6	[493]						
L$_3$	9.0	[503]					9.2	[51]
	9.2±0.6	[51]						
	9.7±0.6	[493]						
IP	14.53414	[222]						

[He] $2s^2 2p^4$

Oxygen $A = 15.9994(3)$ [260] $\varrho = 1.331(-3)$ g/cm^3 [547]

X-ray transitions

line	transition	E/eV		I/eV/$\hbar$		TPIV	Γ/eV
K series							
K_α	KL	524.9±0.7	[50]	8.23(-4)	[488]		
K_α	KL	526.48	[104]				
K_{α_2}	KL$_2$	523.09	[423]			1.908(-3)	
K_{α_1}	KL$_3$	523.13	[423]			3.800(-3)	

level characteristics

level	E_B/eV		ω_{nlj}		Γ/eV		AE/eV	
K	532.0	[503]	6.90(-3)	[39]	0.13	[74]	532.0	[51]
	532.0±0.4	[493]	8.30(-3)	[301]	0.133	[423]		
			7.70(-3)	[557]				
			9.20(-3)	[423]				
			1.30(-2)	[310]				
			6.70(-3)	[183]				
L$_1$	24.0	[503]					23.7	[51]
	23.7±0.4	[493]						
L$_2$	7.0	[503]					7.1	[51]
	7.1±0.8	[51]						
	6.8±0.8	[493]						
L$_3$	7.0	[503]					7.1	[51]
	7.1±0.8	[51]						
	7.4±0.8	[493]						
IP	13.61806	[72.77]						

Fluorine $A = 18.9984032(5)$ [260] $\varrho = 1.579(-3)$ g/cm^3 [547]

[He] 2s^2 2p^5

X-ray transitions

line	transition	E/eV		I/eV/$\hbar$		TPIV	Γ/eV
K series							
K$_\alpha$	KL	676.8±0.7	[50]	1.758(-3)	[488]		
K$_\alpha$	KL	677.07	[104]				
K$_{\alpha_2}$	KL$_2$	671.32	[423]			3.068(-3)	
K$_{\alpha_1}$	KL$_3$	671.39	[423]			6.107(-3)	

level characteristics

level	E_B/eV		ω_{nlj}		Γ/eV		AE/eV	
K	686.00	[503]	1.30(-2)	[301]	0.18	[74]	685.4	[51]
	685.40 ± 0.40	[493]	1.55(-2)	[557]	0.18	[423]		
			1.05(-2)	[39]				
			9.18(-3)	[39]				
			1.30(-2)	[310]				
L_1	31.00	[503]					31.0	[51]
	31.00	[51]						
	34.00 ± 1.00	[493]						
L_2	9.00	[503]					8.6	[51]
	8.60 ± 0.80	[51]						
	8.40 ± 0.80	[493]						
L_3	9.00	[503]					8.6	[51]
	8.60 ± 0.80	[51]						
	8.70 ± 0.80	[493]						
	17.423	[126]						
IP	17.42282	[222]						

Ne   Z=10

[He] $2s^2\,2p^6$

Neon $A = 20.1797(6)$ [260] $\varrho = 8.391(-4)\ \text{g/cm}^3$ [547]

X-ray transitions

line	transition	E/eV		I/eV/$\hbar$		TPIV	Γ/eV	
K series								
K_{α_3}	KL_1	817.69 ± 0.56	[145]					
$K_{\alpha_{1.2}}$	$KL_{3,2}$	848.61 ± 0.26	[50]	$3.37(-3)$	[488]		0.24	[74]
K_α	KL	848.32	[104]					
K_{α_2}	KL_2	838.10	[423]			$4.643(-3)$		
K_{α_2}	KL_2	849.09 ± 0.54	[145]					
K_{α_1}	KL_3	838.22	[423]			$9.230(-3)$		
K_{α_1}	KL_3	849.17 ± 0.54	[145]					
K_β	KM	857.89 ± 0.44	[50]					
	KM	857.44	[104]					

level characteristics

level	E_B/eV		ω_{nlj}		Γ/eV		AE/eV	
K	867.0	[503]	$1.52(-2)$	[39]	0.24	[301]	866.90	[51]
	866.9 ± 0.3	[51]	$1.80(-2)$	[301]	0.243	[423]	870.73 ± 0.16	[145]
	870.1 ± 0.2	[350]	$1.82(-2)$	[347]	0.24	[94]		
			$1.64(-2)$	[557]				
			$2.04(-2)$	[293]				
			$1.39(-2)$	[423]				
			$1.80(-2)$	[310]				
L_1	45.0	[503]			< 0.1	[301]	45.0	[51]
	48.5	[350]					53.04 ± 0.40	[145]
L_2	18.0	[503]			0.01	[94]	18.3	[51]
	18.3 ± 0.4	[51]					21.63 ± 0.39	[145]
	21.7	[493]						
L_3	18.0	[503]			0.01	[94]	18.3	[51]
	18.3 ± 0.4	[51]					21.55 ± 0.38	[145]
	21.6	[493]						
IP	21.5646	[222]						

Na   Z=11

[Ne] $3s^1$

Sodium $A = 22.989770(2)$ [260] $\varrho = 0.969$ g/cm^3 [547]

X-ray transitions

line	transition	E/eV		I/eV/$\hbar$		TPIV	Γ/eV	
K series								
K_{α_3}	KL_1	1004.99 ± 0.49	[145]				0.81	[423]
K_{α_2}	KL_2	1041.77 ± 0.51	[145]			6.690(-3)		
K_{α_1}	KL_3	1041.94 ± 0.51	[145]			1.330(-2)		
$K_{\alpha_{1,2}}$	$KL_{3,2}$	1040.98 ± 0.12	[50]	5.756(-3)	[488]		0.30	[301]
	$KL_{3,2}$	1041.06	[104]					
K_β	KM	1071.12 ± 0.27	[50]	1.8(-13)	[488]			
	KM	1067.20	[104]					
	KM_1	1073.0 ± 1.2	[145]					
L series								
	$L_1 L_2$					3.803(-5)		
	$L_1 L_{2,3}$	33.00 ± 0.10	[50]	8.84(-6)	[488]			
	$L_1 L_{2,3}$	32.66	[104]					
	$L_1 L_3$	32.7	[104]			7.799(-5)		
	$L_{2,3} M$	30.45 ± 0.04	[50]					
L_η	$L_2 M_1$	31.3 ± 1.4	[145]					
L_l	$L_3 M_1$	31.09 ± 0.37	[145]					

level characteristics

level	E_B/eV		ω_{nlj}		Γ/eV		AE/eV	
K	1072.0	[503]	2.13(-2)	[39]	0.30	[301]	1071.52	[51]
	1072.1±0.4	[493]	2.30(-2)	[301]	0.289	[423]	1080.15 ± 0.15	[145]
			2.24(-2)	[557]	0.28	[94]		
			2.40(-2)	[293]				
			2.00(-2)	[423]				
			2.10(-2)	[253]				
L$_1$	63.0	[503]	1.2(-4)	[423]	0.2	[301]	63.57	[51]
	63.3±0.3	[493]			0.515	[423]	75.16 ± 0.35	[145]
					0.28	[94]		
L$_2$	31.0	[503]			0.02	[94]	30.61	[51]
	31.1±0.4	[493]					38.38 ± 0.37	[145]
L$_3$	31.0	[503]			0.02	[94]	30.61	[51]
	31.0±0.4	[493]					38.21 ± 0.37	[145]
M$_1$	1.0	[503]						
	0.7±0.4	[493]						
IP	5.13908	[222]						

Mg   Z=12

[Ne] $3s^2$

Magnesium $A = 24.3050(6)$ [260] $\varrho = 1.735$ g/cm^3 [547]

X-ray transitions

line	transition	E/eV		I/eV/$\hbar$		TPIV	Γ/eV	
K series								
K_{α_3}	KL_1	1205.04	[423]				1.453	[423]
		1211.54 $\pm$ 0.43	[145]					
$K_{\alpha_{1.2}}$	$KL_{3,2}$	1253.60 $\pm$ 0.03	[50]	9.11(-3)	[488]		0.36	[301]
		1253.60	[104]					
K_{α_2}	KL_2	1254.14 $\pm$ 0.49	[484]			9.273(-3)	0.33	[423]
K_{α_1}	KL_3	1254.39 $\pm$ 0.49	[484]			1.842(-2)	0.33	[423]
K_β	KM	1302.20 $\pm$ 0.40	[50]	1.61(-12)	[488]			
		1297.0	[104]					
L series								
	$L_1L_{2,3}$	39.20 $\pm$ 0.10	[50]	1.04(-5)	[488]			
		39.01	[104]					
	L_1L_2					5.793(-5)	1.39	[423]
	L_1L_3					1.204(-4)		
	L_1M_1	92.0 $\pm$ 1.4	[145]					
	$L_{2,3}M$	49.3 $\pm$ 0.10	[50]	2.30(-6)	[488]			
L_η	L_2M_1	49.66	[423]			3.411(-6)		
		49.4 $\pm$ 1.4	[145]					
L_l	L_3M_1	49.35	[423]			3.180(-6)		
		49.14 $\pm$ 0.36	[145]					

level characteristics

level	E_B/eV		ω_{nlj}		Γ/eV		AE/eV	
K	1305.0	[503]	2.91(-2)	[39]	0.36	[301]	1303.4	[51]
	1305.0±0.4	[493]	3.00(-2)	[301]	0.333	[423]	1312.30±0.14	[145]
			3.36(-2)	[347]	0.33	[94]		
			3.01(-2)	[557]				
			2.77(-2)	[423]				
			2.60(-2)	[253]				
L_1	89.0	[503]	2.9(-5)	[301]	0.41	[301]	92.8	[51]
	89.4±0.4	[493]	1.8(-4)	[423]	1.125	[423]	100.75±0.29	[145]
					0.46	[94]		
L_2	52.0	[503]	1.2(-3)	[301]	0.001	[301]	51.4	[51]
	51.4±0.5	[51]	3.0(-6)	[423]	0.0044	[423]	58.16±0.35	[145]
					0.03	[94]		
L_3	52.0	[503]	1.2(-3)	[301]	0.001	[301]	51.4	[51]
	51.4±0.5	[51]	3.0(-6)	[423]	0.0042	[423]	57.90±0.35	[145]
	51.3±0.5	[493]			0.03	[94]	49.454	[51]
M_1	2.0	[503]						
	2.1±0.4	[493]						
IP	7.64624	[222]						

Al   Z=13

[Ne] $3s^2\,3p^1$

Aluminium $A = 26.981538(2)$ [260] $\varrho = 2.694\ \text{g/cm}^3$ [547]

X-ray transitions

line	transition	E/eV		I/eV/$\hbar$		TPIV	Γ/eV	
K series								
K_{α_3}	KL_1	1430.85	[423]				1.77	[423]
		1438.94 ± 0.38	[145]					
K_{α_2}	KL_2	1486.295±0.10	[50]	4.60(-3)	[488]	1.237(-2)	0.49	[301]
		1486.40	[104]	5.30(-3)	[487]		0.38	[423]
		1486.14±0.03	[143]					
		1486.27±0.03	[277]					
K_{α_1}	KL_3	1486.708±0.10	[50]	9.14(-3)	[488]	2.455(-2)	0.49	[301]
		1486.70	[104]	1.06(-2)	[487]		0.38	[423]
		1486.57±0.01	[143]					
		1486.70±0.01	[277]					
K_{β_3}	KM_2	1561.28 ± 0.96	[145]	7.0(-5)	[487]	7.558(-5)		
K_{β_1}	KM_3	1545.03	[423]	1.4(-4)	[487]	1.500(-4)		
K_β	KM	1557.57±0.58	[50]	8.39(-5)	[488]			
		1553.40	[104]					
L series								
	$L_1L_{2,3}$	42.80±0.15	[50]	1.34(-5)	[488]			
		45.10	[104]					
	L_1L_2					7.900(-5)	1.39	[423]
	L_1M_1	117.7 ± 1.1	[145]					
L_{β_4}	L_1M_2	114.17	[423]			2.077(-6)		
		122.3 ± 1.1	[145]					
L_{β_3}	L_1M_3	114.18	[423]			4.030(-6)		
	$L_{2,3}M$	72.40±0.20	[50]	2.69(-6)	[488]			
L_η	L_2M_1	69.5 ± 1.2	[145]			1.605(-5)		
	L_2M_2	74.2 ± 1.2	[145]					
$L_{\beta_{17}}$	L_2M_3	76.33	[423]			1.756(-10)		
L_l	L_3M_1	69.1 ± 1.2	[145]			1.50(-5)		
L_t	L_3M_2	73.7 ± 1.2	[145]			8.2(-11)		
	L_3M_3	75.86	[423]			8.101(-11)		

level characteristics

line	transition	E/eV			I/eV/$\hbar$		TPIV		Γ/eV	
K	1560.0	[503]	3.87(-2)	[39]	0.42	[301]	1559.6	[51]		
	1559.6±0.4	[493]	3.90(-2)	[301]	0.378	[423]	1569.56 ± 0.13	[145]		
			4.12(-2)	[347]	0.37	[94]				
			3.98(-2)	[557]						
			3.33(-2)	[293]						
			3.81(-2)	[288]						
			3.79(-2)	[41]						
			3.57(-2)	[215]						
			3.87(-2)	[253]						
L_1	118.0	[503]	2.6(-5)	[301]	0.73	[301]	117.7	[51]		
	117.7±0.4	[493]	3.1(-6)	[348]	1.390	[423]	130.62 ± 0.25	[145]		
			3.0(-5)	[423]	0.78	[94]				
L_2	74.0	[503]	7.5(-4)	[301]	0.0057	[423]	73.1	[51]		
	73.1±0.5	[51]	2.0(-5)	[423]	0.04	[94]	72.76	[50]		
	73.2±0.2	[493]					82.44 ± 0.34			
L_3	73.0	[503]	7.5(-4)	[301]	0.004	[301]	73.1	[51]		
	72.7±0.2	[493]	2.4(-3)	[348]	0.0057	[423]	72.76	[50]		
			2.0(-5)	[423]	0.04	[94]	82.03 ± 0.34			
M_1	1.0	[503]								
	0.7±0.4	[493]								
IP	5.98577	[222]								

Si   Z=14

[Ne] $3s^2\,3p^2$

Silicon $A = 28.0855(3)$ [260] $\varrho = 2.32$ g/cm^3 [547]

X-ray transitions

line	transition	E/eV		I/eV/$\hbar$		TPIV	Γ/eV	
K series								
K$_{\alpha_3}$	KL$_1$	1676.95	[423]				2.11	[423]
		1686.54 ± 0.33	[145]					
K$_{\alpha_2}$	KL$_2$	1739.38±0.02	[50]	6.67(-3)	[488]	1.598(-2)	0.49	[301]
		1739.30	[104]	7.60(-3)	[487]		0.44	[423]
		1739.394 ± 0.034	[366]					
K$_{\alpha_1}$	KL$_3$	1739.98±0.02	[50]	0.1323	[488]	3.170(-2)	0.49	[301]
		1740.00	[104]	0.151	[487]		0.44	[423]
		1739.985 ± 0.019	[366]					
K$_\beta$	KM	1835.96±0.40	[50]	3.39(-4)	[488]			
	KM	1836.10	[104]					
	KM$_1$	1834.97 ± 0.84	[145]					
K$_{\beta_3}$	KM$_2$	1841.79 ± 0.72	[145]	2.10(-4)	[487]	2.724(-4)		
K$_{\beta_1}$	KM$_3$	1821.98	[423]	4.20(-4)	[487]	5.404(-4)		
L series								
	L$_1$M$_1$	148.42 ± 0.91	[145]					
L$_{\beta_4}$	L$_1$M$_2$	155.24 ± 0.79	[145]			5.833(-6)		
L$_{\beta_3}$	L$_1$M$_3$	145.03	[423]			1.073(-5)		
	L$_{2,3}$M	91.50±0.30	[50]	4.90(-6)	[488]			
L$_\eta$	L$_2$M$_1$	95.04	[423]			4.720(-5)		
		95.3 ± 1.1	[145]					
	L$_2$M$_2$	102.11 ± 0.93	[145]					
L$_{\beta_{17}}$	L$_2$M$_3$	102.15	[423]			1.828(-9)		
L$_l$	L$_3$M$_1$	93.81 ± 0.93	[145]			4.421(-5)		
L$_t$	L$_3$M$_2$	100.63 ± 0.93	[145]			8.542(-10)		
	L$_3$M$_3$	101.46	[423]			8.431(-10)		

level characteristics

level	E_B/eV		ω_{nlj}		Γ/eV		AE/eV	
K	1839.0	[503]	5.04(-2)	[39]	0.48	[301]	1838.9	[51]
	1838.9±0.4	[493]	5.00(-2)	[301]	0.425	[423]	1840.05	[50]
			5.92(-2)	[347]	0.43	[94]	1850.26 ± 0.13	[145]
			5.14(-2)	[557]				
			4.70(-2)	[38]				
			4.85(-2)	[423]				
			4.82(-2)	[309]				
L₁	149.0	[503]	3.0(-5)	[301]	1.03	[301]	148.7	[51]
	148.7±0.4	[493]	9.77(-6)	[348]	1.684	[423]	163.72 ± 0.20	[145]
			3.4(-4)	[423]	0.9	[94]		
L₂	100.0	[503]	3.7(-4)	[301]	0.015	[301]	99.2	[51]
	99.2±0.5	[51]	5.0(-5)	[423]	0.018	[423]	100.8	[50]
	99.5±0.4	[493]			0.05	[94]	110.59 ± 0.34	[145]
L₃	99.0	[503]	3.8(-4)	[301]	0.004	[301]	99.2	[51]
	98.9±0.4	[493]	1.08(-3)	[348]	0.018	[423]	100.8	[50]
			4.0(-5)	[423]	0.05	[94]	109.10 ± 0.34	[145]
M₁	8.0	[503]						
	7.6±0.5	[493]						
M₂	3.0	[503]						
	3.0±0.4	[493]						
IP	8.15169	[222]						

P   Z=15

[Ne] $3s^2\, 3p^3$

Phosphorus $A = 30.973761(2)$ [222] $\varrho = 1.82$ g/cm^3 [547]

X-ray transitions

line	transition	E/eV		I/eV/ℏ		TPIV	Γ/eV	
K series								
K_{α_3}	KL$_1$	1943.25	[423]				2.50	[423]
		1954.45 ± 0.27	[145]					
K_{α_2}	KL$_2$	2012.70±0.48	[50]	9.35(-3)	[488]	2.012(-2)	0.56	[301]
		2012.90	[104]	1.06(-2)	[487]		0.51	[423]
K_{α_1}	KL$_3$	2013.68±0.48	[50]	1.854(-2)	[488]	3.987(-2)	0.57	[301]
		2013.80	[104]	2.09(-2)	[487]		0.51	[423]
K_β	KM	2139.1±0.11	[50]	8.62(-4)	[488]	4.66		
K_β	KM	2136.1	[104]					
	KM$_1$	2136.1 ± 1.0	[145]					
K_{β_3}	KM$_2$	2143.79 ± 0.58	[145]	4.7	[487]	6.215(-4)		
K_{β_1}	KM$_3$	2138.8	[104]	9.30(-4)	[487]	1.232(-3)		
		2140.8 ± 1.0	[145]					
L series								
	L$_1$L$_2$	48.01	[423]			1.188(-4)	2.07	[423]
	L$_1$L$_3$	48.97	[423]			2.789(-4)	2.07	[423]
	L$_1$M$_1$	181.7 ± 1.0	[145]					
L_{β_4}	L$_1$M$_2$	189.34 ± 0.62	[145]			1.430(-5)		
L_{β_3}	L$_1$M$_3$	186.4 ± 1.0	[145]			2.505(-5)		
	L$_{2,3}$M	119.40±0.46	[50]	7.98(-6)	[488]			
L_η	L$_2$M$_1$	123.0 ± 1.2	[145]			1.073(-4)		
	L$_2$M$_2$	130.65 ± 0.79	[145]					
$L_{\beta_{17}}$	L$_2$M$_3$	127.7 ± 1.2	[145]			9.518(-9)		
L_l	L$_3$M$_1$	121.42 ± 0.78	[145]			1.007(-4)		
L_t	L$_3$M$_2$	129.07 ± 0.78	[145]			4.448(-9)		
L_s	L$_3$M$_3$	126.1 ± 1.2	[145]			4.385(-9)		

level characteristics

level	E_B/eV		ω_{nlj}		Γ/eV		AE/eV	
K	2149.0	[503]	6.42(-2)	[39]	0.53	[301]	2145.5	[51]
	2145.5±0.4	[493]	6.30(-2)	[301]	0.473	[423]	2143.54	[50]
			7.43(-2)	[347]	0.47	[94]	2154.24 ± 0.12	[145]
			6.53(-2)	[557]				
			6.04(-2)	[38]				
			7.10(-2)	[253]				
			7.73(-2)	[423]				
			8.06(-2)	[309]				
L_1	189.0	[503]	3.9(-5)	[301]	1.26	[301]	189.3	[51]
	189.3±0.4	[493]	2.12(-5)	[348]	2.030	[423]	199.79 ± 0.16	[145]
			4.4(-4)	[423]	1.1	[94]		
L_2	136.0	[503]	3.1(-4)	[301]	0.032	[301]	132.2	[51]
	132.2±0.5	[51]	1.1(-4)	[423]	0.0372	[423]	131.9	[50]
	136.2±0.4	[493]			0.07	[94]	141.10 ± 0.33	[145]
L_3	135.0	[503]	4.1(-4)	[348]	0.033	[301]	132.2	[51]
	135.3±0.4	[493]	3.1(-4)	[301]	0.0374	[423]	131.9	[50]
			1.(-4)	[423]	0.07	[94]	139.51 ± 0.32	[145]
M_1	16.0	[503]						
	16.2±0.6	[493]						
M_2	10.0	[503]						
	9.6±0.4	[493]						
M_3	10.0	[503]						
	10.1±0.4	[493]						
IP	10.48669	[222]						

S   Z=16

[Ne] $3s^2\,3p^4$

Sulfur $A = 32.066(6)$ [222] $\varrho = 1.953$ g/cm^3 [547]

X-ray transitions

line	transition	E/eV		I/eV/$\hbar$		TPIV	Γ/eV	
K series								
K_{α_3}	KL_1	2229.93	[423]				2.92	[423]
		2242.56 ± 0.23	[145]					
K_{α_2}	KL_2	2306.70 ± 0.04	[50]	0.01276	[488]	2.478(-2)	0.64	[301]
		2306.70	[104]	0.0143	[487]		0.59	[423]
		2307.01 ± 0.45	[145]					
		2306.700 ± 0.038	[366]					
K_{α_1}	KL_3	2307.89 ± 0.03	[50]	0.0253	[488]	4.906(-2)	0.65	[301]
		2308.00	[104]	0.0283	[487]		0.59	[423]
		2308.80 ± 0.40	[145]					
		2307.885 ± 0.034	[366]					
K_β	KM	2468.10 ± 0.15	[50]					
		2464.90	[104]					
	KM_1	2460.1 ± 1.1	[145]					
K_{β_3}	KM_2	2469.73 ± 0.62	[145]	8.8(-4)	[487]	1.156(-3)		
K_{β_1}	KM_3	2464.07 ± 0.14	[50]	1.18(-3)	[488]	2.289(-3)		
		2467.53 ± 0.72	[145]					
L series								
	L_1L_2	53.24	[423]			1.301(-4)	2.46	[423]
	L_1L_3	54.57	[423]			3.191(-4)	2.46	[423]
	L_1M_1	217.5 ± 1.1	[145]					
L_{β_4}	L_1M_2	227.17 ± 0.60	[145]			3.058(-5)		
L_{β_3}	L_1M_3	224.96 ± 0.72	[145]			5.148(-5)		
L_l,L_η	$L_{3,2}M_1$	148.66 ± 0.79	[50]	2.44(-5)	[488]			
L_l,L_η	$L_{3,2}M_1$	148.70	[104]					
L_η	L_2M_1	153.1 ± 1.4	[145]			2.074(-4)		
	L_2M_2	162.72 ± 0.81	[145]					
$L_{\beta_{17}}$	L_2M_3	160.52 ± 0.93	[145]			3.454(-8)		
L_l	L_3M_1	151.29 ± 0.80	[145]			1.951(-4)		
L_t	L_3M_2	160.93 ± 0.80	[145]			1.614(-8)		
L_s	L_3M_3	158.73 ± 0.92	[145]			1.589(-8)		

level characteristics

level	E_B/eV		ω_{nlj}		Γ/eV		AE/eV	
K	2472.0	[503]	8.64(-2)	[39]	0.59	[301]	2472.0	[51]
	2472.0±0.4	[493]	7.80(-2)	[301]	0.522	[423]	2470.51	[50]
			8.99(-2)	[347]	0.52	[94]	2481.71 ± 0.12	[145]
			8.18(-2)	[557]				
			7.27(-2)	[293]				
			8.20(-2)	[412]				
			7.61(-2)	[38]				
			7.10(-2)	[253]				
			7.73(-2)	[423]				
			8.06(-2)	[309]				
			7.80(-2)	[310]				
			8.20(-2)	[183]				
L_1	229.0	[503]	7.4(-5)	[301]	1.49	[301]	229.2	[51]
	229.2±0.4	[493]	3.63(-5)	[348]	2.395	[423]	239.15 ± 0.11	[145]
			5.3(-4)	[423]	1.3	[94]		
L_2	165.0	[503]	2.6(-4)	[301]	0.054	[301]	164.8	[51]
	164.8±0.7	[51]	2.1(-4)	[423]	0.066	[423]	174.70 ± 0.33	[145]
	165.4±0.4	[493]			0.09	[94]		
L_3	164.0	[503]	2.6(-4)	[301]	0.054	[301]	164.8	[51]
	164.2±0.4	[493]	2.9(-4)	[348]	0.066	[423]	172.91 ± 0.32	[145]
			1.9(-4)	[423]	0.09	]116]		
M_1	16.0	[503]						
	15.8±0.6	[493]						
M_2	8.0	[503]						
	7.8±0.4	[493]						
M_3	8.0	[503]						
	8.2±0.4	[493]						
IP	10.360	[222]						

Cl   Z=17

[Ne] $3s^2\, 3p^5$

Chlorine $A = 35.4527(9)$ [12.77] $\varrho = 2.947(-3)$ g/cm^3 [547]

X-ray transitions

line	transition	E/eV		I/eV/$\hbar$		TPIV	Γ/eV	
K series								
K_{α_3}	KL_1	2536.85	[423]				3.38	[423]
		2250.96 ± 0.26	[145]					
K_{α_2}	KL_2	2620.78 ± 0.05	[50]	0.0170	[488]	2.995(-2)	0.72	[301]
		2620.80	[104]	0.0188	[487]		0.68	[423]
		2620.27 ± 0.55	[366]					
K_{α_1}	KL_3	2622.39 ± 0.05	[50]	0.0337	[488]	5.924(-2)	0.72	[301]
		2622.40	[104]	0.0372	[487]		0.68	[423]
		2623.67 ± 0.43	[366]					
K_β	KM	2815.60 ± 0.28	[50]					
		2815.10	[104]					
	KM_1	2807.0 ± 1.3	[145]					
K_{β_3}	KM_2	2819.04 ± 0.64	[145]	1.52(-3)	[487]	1.909(-3)		
K_{β_1}	KM_3	2817.58 ± 0.49	[145]	1.18(-3)	[488]	3.778(-3)		
L series								
	L_1L_2	58.57	[423]			1.322(-4)	2.91	[423]
	L_1L_3	60.35	[423]			3.387(-4)	2.91	[423]
	L_1M_1	256.0 ± 1.3	[145]					
L_{β_4}	L_1M_2	268.08 ± 0.66	[145]			5.776(-5)		
L_{β_3}	L_1M_3	266.62	[145]			9.346(-5)		
L_η	L_2M_1	184.14 ± 0.36	[50]	1.75(-5)	[488]	3.586(-4)		
		184.0	[104]					
		184.64	[423]					
		185.7 ± 1.5	[145]					
	L_2M_2	197.77 ± 0.84	[145]					
$L_{\beta_{17}}$	L_2M_3	196.31 ± 0.70	[145]			9.980(-8)		
L_l	L_3M_1	182.60 ± 0.36	[50]			3.378(-4)		
		182.40	[104]					
		183.31 ± 0.83	[145]					
L_t	L_3M_2	195.37 ± 0.83	[145]			4.664(-8)		
L_s	L_3M_3	193.91 ± 0.69	[145]			4.585(-8)		

level characteristics

level	E_B/eV		ω_{nlj}		Γ/eV		AE/eV	
K	2823.0	[503]	9.89(-2)	[39]	0.64	[301]	2822.4	[51]
	2822.4±0.3	[493]	9.70(-2)	[301]	0.574	[423]	2819.64	[50]
			0.108	[347]	0.57	[94]	2832.76 ± 0.12	[145]
			0.1004	[557]				
			9.15(-2)	[293]				
			9.42(-2)	[38]				
			0.103	[412]				
			9.50(-2)	[408]				
			9.49(-2)	[423]				
			8.90(-2)	[253]				
L$_1$	270.0	[503]	1.2(-4)	[301]	1.58	[301]	270.2	[51]
	270.2±0.4	[493]	5.60(-5)	[348]	2.803	[423]	281.80 ± 0.15	[145]
			6.2(-4)	[423]	1.5	[94]		
L$_2$	202.0	[503]	2.4(-4)	[301]	0.083	[301]	201.6	[51]
	201.6±0.3	[493]	3.6(-4)	[423]	0.103	[423]	211.49 ± 0.33	[145]
			3.6(-4)	[423]	0.11	[94]		
L$_3$	200.0	[503]	2.4(-4)	[301]	0.087	[301]	200.0	[51]
	200.0±0.3	[493]	2.3(-4)	[348]	0.104	[423]	209.09 ± 0.31	[145]
			3.4(-4)	[423]	0.11	[94]		
M$_1$	18.0	[503]						
	17.5±0.4	[493]						
M$_2$	7.0	[503]						
	6.8±0.4	[51]						
	6.7±0.4	[493]						
M$_3$	7.0	[503]						
	6.8±0.4	[51]						
	6.7±0.4	[493]						
IP	12.96764	[222]						

Ar   Z=18

[Ne] $3s^2\,3p^6$

Argon

$A = 39.948(1)\ [12.77]$ $\varrho = 1.66(-3)\ \text{g/cm}^3\ [547]$

X-ray transitions

line	transition	E/eV		I/eV/$\hbar$		TPIV	Γ/eV	
K series								
K_{α_3}	KL_1	2880.13 ± 0.33	[145]				3.87	[423]
K_{α_2}	KL_2	2955.63 ± 0.03	[50]	0.0222	[488]	3.559(-2)	0.81	[301]
		2955.62	[104]	0.0238(-2)	[487]		0.78	[423]
		2955.69 ± 0.01	[143]					
		2955.566 ± 0.016	[485]					
K_{α_1}	KL_3	2957.70 ± 0.01	[50]	0.0440	[488]	7.033(-2)	0.81	[301]
		2957.69	[104]	0.0472	[487]		0.78	[423]
		2957.81 ± 0.01	[143]					
		2957.682 ± 0.016	[485]					
$K_{\beta_{3,1}}$	$KM_{2,3}$	3190.49 ± 0.24	[50]					
		2815.10	[104]					
	KM_1	3177.4 ± 1.5	[145]				0.78	[423]
K_{β_3}	KM_2	3191.31 ± 0.58	[145]	2.44(-3)	[487]	2.913(-3)		
K_{β_1}	KM_3	3191.47 ± 0.58	[145]	4.80(-3)	[488]	5.756(-3)		
K_{β_2}	$KN_{2,3}$	3203.54 ± 0.10	[84]					
	$KO_{2,3}$	3205.00 ± 0.10	[84]					
	$KP_{2,3}$	3205.51 ± 0.10	[84]					
	$KQ_{2,3}$	3205.77 ± 0.10	[84]					
L series								
	L_1M_1	$297.3 \pm 1.6 \pm 0.23$	[145]					
L_{β_4}	L_1M_2	311.18 ± 0.69	[145]			9.683(-5)		
L_{β_3}	L_1M_3	$311.35 \pm 0.69 \pm 0.23$	[145]			1.527(-4)		
L_η	L_2M_1	221.79 ± 0.59	[50]	7.88(-5)	[488]	5.713(-4)		
		218.70	[104]					
	L_2M_2	335.41 ± 0.78	[145]					
$L_{\beta_{17}}$	L_2M_3	235.58 ± 0.78	[145]			2.460(-7)		
L_l	L_3M_1	220.22 ± 0.58	[50]	2.48(-5)	[488]	5.392(-4)		
		221.00	[104]					
		219.52 ± 0.77	[145]					
L_t	L_3M_2	233.40 ± 0.77	[145]			1.149(-7)		
L_s	L_3M_3	233.57 ± 0.77	[145]			1.128(-7)		

level characteristics

level	E_B/eV		ω_{nlj}		Γ/eV		AE/eV	
K	3203.00	[503]	0.118	[301]	0.68	[301]	3202.9	[51]
	3202.90 ± 0.30	[493]	0.1215	[557]	0.63	[423]	3202.93	[50]
			0.126	[347]	0.66	[94]	3207.44 ± 0.12	[145]
			0.111	[293]				
			0.122	[412]				
			0.121	[412]				
			0.129	[230]				
			0.115	[38]				
			0.1195	[39]				
			0.1146	[423]				
			0.118	[310]				
			0.122	[112]				
L_1	320.00	[503]	1.80(-4)	[301]	1.63	[301]	320.0	[51]
	320.00	[51]	8.58(-5)	[348]	3.24	[423]	327.31 ± 0.23	[145]
	326.00	[493]	7.1(-4)	[423]	1.8	[94]		
			1.1(-4)	[244]				
L_2	247.00	[503]	2.20(-4)	[301]	0.126	[301]	247.3	[51]
	247.30	[51]	5.7(-4)	[423]	0.15	[423]	235.41 ± 0.32	[145]
	250.60	[493]			0.13	[94]		
L_3	245.00	[503]	2.20(-4)	[301]	0.128	[301]	245.2	[51]
	245.20 ± 0.30	[51]	1.90(-4)	[348]	0.15	[423]	249.54 ± 0.31	[145]
	248.50	[493]	5.4(-4)	[423]	0.13	[94]		
			1.8(-4)	[244]				
M_1	25.00	[503]			0.14	[94]	29.3	[51]
	25.30 ± 0.40	[51]						
M_2	12.00	[503]						
	12.40 ± 0.30	[51]						
	15.90	[493]						
M_3	12.00	[503]						
	12.40 ± 0.30	[51]						
	15.80	[493]						
	15.76	[126]						
IP	15.75962	[222]						

K   Z=19

[Ar] 4s^1

Potassium $A = 39.0983(1)$ [12.77] $\varrho = 0.860\,\text{g/cm}^3$ [547]
$\varrho = 0.851\,\text{g/cm}^3$ [483]

X-ray transitions

line	transition	E/eV		I/eV/$\hbar$		TPIV	Γ/eV	
K series								
$K_{\alpha3}$	KL$_1$	3229.98 ± 0.32	[145]				4.36	[423]
$K_{\alpha2}$	KL$_2$	3311.10 ± 0.20	[50]	2.863(-2)	[488]	4.186(-2)	0.89	[301]
		3310.75	[104]	3.05(-2)	[487]		0.86	[423]
		3311.22 ± 0.01	[143]					
		3311.97 ± 0.44	[144]					
$K_{\alpha1}$	KL$_3$	3313.80 ± 0.20	[50]	5.65(-2)	[488]	8.265(-2)	0.89	[301]
		3313.80	[104]	6.03(-2)	[487]		0.86	[423]
		3313.97 ± 0.01	[143]					
		3314.60 ± 0.42	[144]					
	KM$_1$	3375.0 ± 1.4	[145]				0.87	[423]
$K_{\beta3}$	KM$_2$	3591.78 ± 0.42	[145]	3.50(-3)	[487]	3.997(-3)	0.68	[423]
$K_{\beta1}$	KM$_3$	3591.49 ± 0.61	[145]	7.00(-3)	[487]	7.910(-3)	0.68	[423]
$K_{\beta1.3}$	KM$_{2,3}$	3589.63 ± 0.31	[50]	8.14(-3)	[488]			
$K_{\beta5}$	KM$_{4,5}$	3602.78 ± 0.62	[50]					
		3602.40	[104]					
	KN$_1$	3611.14 ± 0.17	[145]					
L series								
	L$_1$M$_1$	345.0 ± 1.5	[145]					
$L_{\beta4}$	L$_1$M$_2$	361.81 ± 0.82	[145]			1.472(-4)		
$L_{\beta3}$	L$_1$M$_3$	361.52 ± 0.71	[145]			2.254(-4)		
	L$_1$N$_1$	381.16 ± 0.27	[145]					
L_η	L$_2$M$_1$	262.45 ± 0.17	[50]	3.46(-5)	[488]	8.354(-4)	0.18	[423]
		262.50	[104]					
	L$_2$M$_2$	279.81 ± 0.93	[145]					
$L_{\beta17}$	L$_2$M$_3$	279.52 ± 0.81	[145]			4.814(-7)		
$L_{\gamma5}$	L$_2$N$_1$	299.17 ± 0.38	[145]			2.039(-5)		
L_l	L$_3$M$_1$	259.703 ± 0.081	[50]	3.54(-5)	[488]	7.895(-4)	0.18	[423]

line	transition	E/eV		I/eV/$\hbar$		TPIV	Γ/eV
		260.30	[104]				
L_t	L_3M_2	277.18 ± 0.91	[145]			2.241(7)-	
L_s	L_3M_3	276.90 ± 0.80	[145]			2.202(-7)	
L_{β_6}	L_3N_1	296.54 ± 0.36	[145]			1.925(-5)	

M series

	$M_{2,3}N_1$	17.90 ± 0.23	[50]	1.66(-7)	[488]		
		18.29	[104]				

level characteristics

level	E_B/eV		ω_{nlj}		Γ/eV		AE/eV	
K	3608.00	[503]	0.140	[301]	0.74	[301]	3607.4	[51]
	3607.40 ± 0.40	[493]	0.1448	[557]	0.69	[423]	3607.81	[50]
			0.149	[347]	0.71	[94]	3616.22 ± 0.12	[145]
			0.132	[293]				
			0.1432	[39]				
			0.138	[38]				
			0.1364	[423]				
			0.140	[310]				
L_1	377.00	[503]	2.40(-4)	[301]	1.92	[301]	377.1	[51]
	377.10 ± 0.40	[493]	1.15(-4)	[348]	3.68	[423]	386.25 ± 0.22	[145]
			8.0(-4)	[423]	2.1	[94]		
L_2	297.00	[503]	2.70(-4)	[301]	0.152	[301]	296.3	[51]
	296.30 ± 0.40	[493]	8.6(-4)	[423]	0.18	[423]	294.5	[50]
					0.18	[94]	304.25 ± 0.33	[145]
L_3	294.00	[503]	2.70(-4)	[301]	0.156	[301]	293.6	[51]
	293.60 ± 0.40	[493]	2.10(-4)	[348]	0.18	[423]	294.5	[50]
			8.1(-4)	[423]	0.18	[94]	301.62 ± 0.31	[145]
M_1	34.00	[503]			0.7	[94]	34.8	[51]
	33.90 ± 0.40	[493]						
M_2	18.00	[503]						
	17.80 ± 0.40	[51]						
	18.10 ± 0.50	[493]						
M_3	18.00	[503]						
	17.80 ± 0.40	[51]						
	17.80 ± 0.50	[493]						
N_1	4.34	[493]						
	4.341	[126]						
IP	4.34066	[222]						

Ca   Z=20   [Ar] 4s²

Calcium $A = 40.078(4)$ [222] $\varrho = 1.55\,\text{g/cm}^3$ [547]

X-ray transitions

line	transition	E/eV		I/eV/$\hbar$		TPIV	Γ/eV	
K series								
K_{α_3}	KL_1	3598.89 ± 0.31	[145]				4.36	[423]
K_{α_2}	KL_2	3687.56 ± 0.43	[50]	3.63(-2)	[488]	4.872(-2)	0.98	[301]
		3688.00	[104]	3.85(-2)	[487]	4.872(-2)	0.951	[423]
K_{α_1}	KL_3	3690.98 ± 0.41	[50]	7.16(-2)	[488]	9.613(-2)	0.98	[301]
		3691.70	[104]	7.61(-2)	[487]	9.613(-2)	1.00	[473]
							0.95	[423]
	KM_1	3993.8 ± 1.4	[145]				0.95	[423]
K_{β_3}	KM_2	4014.32 ± 0.59	[145]	4.80(-3)	[487]	5.178(-3)	0.74	[423]
K_{β_1}	KM_3	4014.68 ± 0.58	[145]	9.60(-3)	[487]	1.025(-2)	0.74	[423]
$K_{\beta_{3,1}}$	$KM_{2,3}$	4012.76 ± 0.38	[50]	1.149(-2)	[488]			
		4012.88	[104]					
K_{β_5}	$KM_{4,5}$	4032.47 ± 0.58	[50]					
		4032.50	[104]					
	KN_1	4043.20 ± 0.18	[145]					
L series								
	L_1M_1	394.9 ± 1.5	[145]					
L_{β_4}	L_1M_2	415.43 ± 0.68	[145]			2.041(-4)		
L_{β_3}	L_1M_3	415.80 ± 0.67	[145]			3.060(-4)		
	L_1N_1	444.32 ± 0.27	[145]					
L_η	L_2M_1	306.43 ± 0.22	[50]	4.78(-5)	[488]	1.196(-3)	0.21	[423]
		306.40	[104]					
	L_2M_2	$326,76 \pm 0.79$	[145]					
$L_{\beta_{17}}$	L_2M_3	327.12 ± 0.78	[145]			8.747(-7)		
L_{β_1}	L_2M_4	344.97 ± 0.28	[50]					
		344.90	[104]					
L_{γ_5}	L_2N_1	355.64 ± 0.39	[145]			2.536(-5)		

line	transition	E/eV		I/eV/$\hbar$		TPIV	Γ/eV	
L_l	L_3M_1	302.69 ± 0.22	[50]	4.90(-5)	[488]	1.131(-3)	0.21	[423]
		302.70	[104]					
L_t	L_3M_2	323.34 ± 0.77	[145]			4.060(-7)		
L_s	L_3M_3	323.71 ± 0.77	[145]			3.988(-7)		
$L_{\alpha_{2,1}}$	$L_3M_{4,5}$	341.27 ± 0.28	[50]					
		341.40	[104]					
L_{β_6}	L_3N_1	352.23 ± 0.37	[145]			2.397(-5)		

M series

	$M_{2,3}N_1$	23.60 ± 0.40	[50]	8.61(-7)	[335]			
		24.80	[104]					

level characteristics

level	E_B/eV		ω_{nlj}		Γ/eV		AE/eV	
K	4038.00	[503]	0.163	[301]	0.81	[301]	4038.1	[51]
	4038.10 ± 0.40	[493]	0.170	[557]	0.74	[423]	4038.12	[50]
			0.177	[347]	0.77	[94]	4049.35 ± 0.12	[145]
			0.155	[293]				
			0.1687	[39]				
			0.163	[38]				
			0.1603	[423]				
			0.163	[310]				
			0.170	[244]				
L_1	438.00	[503]	3.10(-4)	[301]	2.07	[301]	437.8	[51]
	437.80 ± 0.40	[493]	1.56(-4)	[348]	3.95	[423]	450.46 ± 0.20	[145]
			8.8(-4)	[423]	2.5	[94]		
			2.0(-4)	[244]				
L_2	350.00	[503]	3.30(-4)	[301]	0.17	[301]	350.0	[51]
	350.00 ± 0.40	[493]	1.22(-3)	[423]	0.21	[423]	352.92	[50]
					0.21	[94]	361.79 ± 0.39	[145]
L_3	347.00	[503]	3.30(-4)	[301]	0.17	[301]	346.4	[51]
	346.40 ± 0.40	[493]	2.10(-4)	[348]	0.21	[423]	349.34	[50]
			1.16(-3)	[423]	0.21	[94]	358.37 ± 0.30	[145]
			2.88-4)	[244]				

level	E_B/eV		ω_{nlj}		Γ/eV		AE/eV	
M_1	44.00	[503]	8.40(-6)	[38]	0.82	[350]	44.3	[51]
	43.70 ± 0.40	[493]	6.689(-4)	[512]				
					1.1	[94]		
M_2	26.00	[503]	0.062	[38]	0.0003	[350]		
	25.40 ± 0.40	[51]	2.0(-5)	[512]	1.2	[94]		
	25.80 ± 0.60	[493]						
M_3	26.00	[503]			0.001	[74]		
	25.40 ± 0.40	[51]			1.2	[94]		
	25.50 ± 0.60	[493]						
N_1	1.80 ± 0.60	[493]						
IP	6.11316	[222]						

Sc   Z=21

[Ar] $3d^1\,4s^2$

Scandium $A = 44.955910(8)$ [222] $\varrho = 3.00\,\text{g/cm}^3$ [547]
$\varrho = 2.90\,\text{g/cm}^3$ [483]

X-ray transitions

line	transition	E/eV		I/eV/$\hbar$		TPIV	Γ/eV	
	K series							
K_{α_3}	KL_1	3991.57 ± 0.39	[145]				5.39	[423]
$K_{\alpha 2}$	KL_2	4085.95 ± 0.89	[50]	4.55(-2)	[488]	5.638(-2)	1.06	[301]
		4085.50	[104]				1.03	[423]
$K_{\alpha 1}$	KL_3	4090.60 ± 0.15	[50]	8.96(-2)	[488]	1.111(-1)	1.05	[301]
		4090.05	[104]				1.03	[423]
		4090.735 ± 0.019	[145]					
	KM_1	4438.8 ± 1.5	[145]				2.28	[423]
$K_{\beta_{3,1}}$	$KM_{2,3}$	4460.44 ± 0.47	[50]	9.92(-3)	[488]			
		4460.30	[104]					
	KM_2	4461.16 ± 0.80	[145]			6.223(-3)	0.86	[423]
	KM_3	4462.93 ± 0.80	[145]			1.230(-2)	0.85	[423]
$K_{\beta 5}$	$KM_{4,5}$	4486.59 ± 0.72	[50]	3.91(-5)	[488]			
		4486.10	[104]					
$K_{\beta 5}^{II}$	KM_4	4490.79 ± 0.41	[145]			1.961(-7)		
	KM_5					2.890(-7)		
	KN_1	4495.16 ± 0.18	[145]					
	L series							
	L_1M_1	447.3 ± 1.7	[145]					
$L_{\beta_{4,3}}$	$L_1M_{2,3}$	467.90	[104]	7.21(-4)	[488]			
$L_{\beta 4}$	L_1M_2	469.59 ± 0.98	[145]			2.626(-4)	4.64	[423]
$L_{\beta 3}$	L_1M_3	471.36 ± 0.97	[145]			3.888(-4)	4.64	[423]
$L_{\beta 10}$	L_1M_4	499.22 ± 0.59	[145]			1.375(-7)		
	L_1N_1	503.60 ± 0.36	[145]					
L_η	L_2M_1	352.92 ± 0.30	[50]	6.22(-5)	[488]	1.552(-3)	1.71	[423]
		352.90	[104]					

line	transition	E/eV		I/eV/$\hbar$		TPIV	Γ/eV	
	L_2M_2	357.7 $\pm$ 1.0	[145]				0.29	[423]
$L_{\beta_{17}}$	L_2M_3	377.5 $\pm$ 1.0	[145]			1.421(-6)	0.28	[423]
L_{β_1}	L_2M_4	399.69 $\pm$ 0.38	[50]	7.06(-5)	[488]	9.277(-5)	0.23	[423]
		399.80	[104]					
L_{γ_5}	L_2N_1	409.73 $\pm$ 0.40	[145]			3.039(-5)		
L_l	L_3M_1	348.36 $\pm$ 0.43	[50]	6.40(-5)	[488]	1.474(-3)	1.71	[423]
		348.30	[104]					
L_t	L_3M_2	371.01 $\pm$ 0.99	[145]			6.618(-7)	0.29	[423]
L_s	L_3M_3	372.78 $\pm$ 0.99	[145]			6.496(-7)	0.28	[423]
$L_{\alpha_{2,1}}$	$L_3M_{4,5}$	395.48 $\pm$ 0.56	[50]	6.99(-5)	[488]			
		395.70	[104]					
	L_3M_4	400.64 $\pm$ 0.60	[145]			6.522(-6)		
	L_3M_5					7.764(-5)		
L_{β_6}	L_3N_1	405.01 $\pm$ 0.37	[145]			2.878(-5)		

level characteristics

level	E_B/eV		ω_{nlj}		Γ/eV		AE/eV	
K	4493.00	[503]	0.188	[301]	0.86	[301]	4492.8	[51]
	4492.80 $\pm$ 0.40	[493]	0.1991	[557]	0.80	[423]	4488.9	[50]
			0.205	[347]	0.83	[94]	4501.68 $\pm$ 0.12	[145]
			0.189	[293]				
			0.1962	[39]				
			0.190	[38]				
			0.190	[41]				
			0.1860	[423]				
			0.188	[310]				
L_1	500.00	[503]	3.90(-4)	[301]	2.21	[301]	500.4	[51]
	500.40 $\pm$ 0.40	[493]	9.7(-4)	[423]	4.59	[423]		
					3.3	[94]		
L_2	407.00	[503]	8.40(-4)	[301]	0.19	[301]	406.7	[51]
	406.70 $\pm$ 0.40	[493]	1.68(-3)	[423]	0.23	[423]		
					0.36	[94]		
L_3	402.00	[503]	8.40(-4)	[301]	0.19	[301]	402.2	[51]
	402.20 $\pm$ 0.40	[493]	1.59(-3)	[423]	0.23	[423]	411.53 $\pm$ 0.31	[145]
					0.23	[94]		

level	E_B/eV		ω_{nlj}		Γ/eV		AE/eV	
M_1	54.00	[503]	7.282(-4)	[512]	1.7	[423]	51.1	[51]
	53.80 ± 0.40	[493]	2.0(-6)	[423]	1.48	[423]		
M_2	32.00	[503]	2.0(-5)	[512]	1.2	[94]		
	32.30 ± 0.50	[51]			0.06	[423]		
	33.80 ± 0.50	[493]						
M_3	32.00	[503]			0.05	[423]		
	32.30 ± 0.50	[51]			1.2	[423]		
	31.50 ± 0.50	[493]						
M_4	7.00	[503]						
	6.60 ± 0.50	[493]						
N_1	1.70 ± 0.10	[493]						
IP	6.5615	[222]						

Ti   Z=22   [Ar] 3d^2 4s^2

Titanium $A = 47.867(1)$ [222] $\varrho = 4.54\,\text{g/cm}^3$ [547]
$\varrho = 4.6\,\text{g/cm}^3$ [483]

X-ray transitions

line	transition	E/eV		I/eV/$\hbar$		TPIV	Γ/eV	
K series								
$K_{\alpha 3}$	KL_1	4404.59 ± 0.43	[145]				6.13	[423]
$K_{\alpha 2}$	KL_2	4504.92 ± 0.94	[50]	5.63(-2)	[488]	6.460(-2)	1.18	[301]
		4504.80	[104]	5.93(-2)	[487]		2.13	[473]
							1.11	[423]
$K_{\alpha 1}$	KL_3	4510.90 ± 0.42	[50]	0.1107	[488]	1.271(-1)	1.16	[301]
		4510.80	[104]	0.1168	[487]		1.45	[473]
							1.11	[423]
		4931.80	[104]					
	KM_1	4907.1 ± 1.5	[145]				2.98	[423]
$K_{\beta 3}$	KM_2	4930.86 ± 0.85	[145]	7.70(-3)	[487]	7.327(-3)	1.05	[423]
$K_{\beta 1}$	KM_3	4934.46 ± 0.87	[145]	1.52(-2)	[488]	1.448(-2)	1.02	[423]
$K_{\beta 3,1}$	$KM_{2,3}$	4931.827 ± 0.059	[50]	1.898(-2)	[488]			
		4931.80	[104]					
$K_{\beta 5}$	$KM_{4,5}$	4962.27 ± 0.59	[50]	1.26(-6)	[488]			
		4962.90	[104]					
$K_{\beta 5}^{II}$	KM_4	4965.95 ± 0.43	[145]			5.901(-7)		
$K_{\beta 5}^{I}$	KM_5					8.685(-7)		
	KN_1	4971.09 ± 0.18	[145]					
L series								
	$L_1 M_1$	502.5 ± 1.8	[145]				7.38	[423]
$L_{\beta 4}$	$L_1 M_2$	526.3 ± 1.1	[145]			3.223(-4)	5.26	[423]
$L_{\beta 3}$	$L_1 M_3$	529.9 ± 1.1	[145]			4.715(-4)	5.43	[423]
$L_{\beta 10}$	$L_1 M_4$	561.36 ± 0.64	[145]			3.057(-7)		
	$L_1 N_1$	566.50 ± 0.40	[145]					
L_η	$L_2 M_1$	401.37 ± 0.18	[50]	7.93(-5)	[488]	2.020(-3)	2.37	[423]
		401.20	[104]					
	$L_2 M_2$	426.8 ± 1.1	[145]				0.44	[423]

line	transition	E/eV		I/eV/$\hbar$		TPIV	Γ/eV	
$L_{\beta_{17}}$	L_2M_3	430.4 ± 1.1	[145]			2.273(-6)	0.42	[423]
L_{β_1}	$L_2M_{4,5}$	458.35 ± 0.64	[50]	2.02(-4)	[488]			
		457.80	[104]					
L_{β_1}	L_2M_4	461.88 ± 0.64	[145]			3.223(-4)	5.46	[423]
	L_2M_5					4.715(-4)	5.43	[423]
L_{γ_5}	L_2N_1	467.03 ± 0.40	[145]			3.780(-5)		
L_l	L_3M_1	395.35 ± 0.37	[50]	8.18(-4)	[488]	1.916(-3)	2.37	[423]
		395.40 6 [104]						
L_t	L_3M_2	420.5 ± 1.0	[145]			1.077(-6)	0.45	[423]
L_s	L_3M_3	424.1 ± 1.1	[145]			1.056(-6)	0.42	[423]
$L_{\alpha_{2,1}}$	$L_3M_{4,5}$	452.16 ± 0.49	[50]	1.99(-4)	[488]			
		451.70	[104]					
L_{α_2}	L_3M_4	455.67 ± 0.61	[145]			1.242(-5)		
L_{α_1}	L_3M_5					1.460(-4)		
L_{β_6}	L_3N_1	460.72 ± 0.37 [145]				3.605(-5)		

level characteristics

level	E_B/eV		ω_{nlj}		Γ/eV		AE/eV	
K	4965.00	[503]	0.214	[301]	0.94	[301]	4966.4	[51]
	4966.40 ± 0.40	[493]	0.223	[347]	0.86	[423]	4964.58	[50]
			0.2273	[557]	0.89	[94]	4977.92 ± 0.12	[145]
			0.212	[293]				
			0.219	[38]				
			0.221	[41]				
			0.2256	[39]				
			0.2135	[423]				
			0.214	[310]				
L_1	564.00	[503]	4.70(-4)	[301]	2.34	[301]	563.7	[51]
	563.70 ± 0.40	[493]	2.80(-4)	[348]	5.27	[423]	573.33 ± 0.33	[145]
			1.06(-3)	[423]	3.9	[94]		
L_2	461.00	[503]	1.50(-3)	[301]	0.24	[301]	461.5	[51]
	461.50 ± 0.40	[493]	2.23(-3)	[423]	0.25	[423]	454.31	[50]
					0.52	[94]	473.85 ± 0.33	[145]

level	E_B/eV		ω_{nlj}		Γ/eV		AE/eV	
L_3	455.00	[503]	1.50(-3)	[301]	0.22	[301]	455.5	[51]
	455.50 ± 0.40	[493]	1.18(-3)	[348]	0.25	[423]	454.31	[50]
			2.11(-3)	[423]	0.25	[94]	467.55 ± 0.31	[145]
M_1	59.00	[503]	3.20(-6)	[38]	3.24	[350]	58.7	[51]
	60.30 ± 0.40	[493]	7.923(-4)	[512]	2.12	[423]		
			4.0(-6)	[423]	2.1	[94]		
M_2	34.00	[503]	3.40(-5)	[38]	0.21	[350]		
	34.60 ± 0.40	[51]	2.0(-5)	[512]	0.19	[423]		
	35.60 ± 1.80	[493]	3.0(-5)	[423]	1.2	[94]		
M_3	32.20 ± 1.80	[493]			0.16	[423]		
					0.10	[74]		
					1.2	[94]		
M_4	3.00	[503]						
	3.70 ± 0.40	[493]						
N_1	1.60 ± 1.30	[503]						
IP	6.8281	[222]						

V   Z=23   [Ar] 3d³ 4s²

[Ar] $3d^3\ 4s^2$

Vanadium $A = 50.9415(1)$ [12.77] $\varrho = 6.10\,\text{g/cm}^3$ [547]

X-ray transitions

line	transition	E/eV		I/eV/ℏ		TPIV	Γ/eV	
K series								
K_{α_3}	KL_1	4838.50 ± 0.47	[145]				6.79	[423]
$K_{\alpha 2}$	KL_2	4944.59 ± 0.45	[50]	6.89(-2)	[488]	7.328(-2)	1.28	[301]
		4944.60	[104]	7.24(-2)	[487]		1.20	[423]
$K_{\alpha 1}$	KL_3	$4952.21 \pm 0.0.42$	[50]	0.1354	[488]	1.440(-1)	1.26	[301]
		4952.10	[104]	0.1424	[487]		1.20	[423]
	KM_1	5399.3 ± 1.5	[145]				3.56	[423]
$K_{\beta 3}$	KM_2	5424.42 ± 0.89	[145]	9.50(-3)	[487]	8.484(-3)	1.33	[423]
$K_{\beta 1}$	KM_3	5430.00 ± 0.94	[145]	1.88(-2)	[487]	1.674(-2)	1.30	[423]
$K_{\beta_{3,1}}$	$KM_{2,3}$	5427.320 ± 0.071	[50]	2.37(-2)	[488]			
		5427.20	[104]					
$K_{\beta 5}$	$KM_{4,5}$	5462.96 ± 0.21	[50]	2.94(-6)	[488]			
		5462.90	[104]					
$K_{\beta_5^{II}}$	KM_4	5465.33 ± 0.44	[145]			1.265(-6)		
$K_{\beta_5^I}$	KM_5					1.857(-6)		
	KN_1	5471.17 ± 0.19	[145]					
L series								
	L_1M_1	560.5 ± 1.8	[145]				8.55	[423]
$L_{\beta_{4,3}}$	$L_1M_{2,3}$	585.1 ± 3.7	[50]	1.201(-3)	[488]			
L_{β_4}	L_1M_2	585.9 ± 1.1	[145]			3.789(-4)	6.32	[423]
L_{β_3}	L_1M_3	591.5 ± 1.2	[145]			5.539(-4)	6.30	[423]
$L_{\beta_{10}}$	L_1M_4	5399.3 ± 1.5	[145]			5.663(-7)		
	L_1N_1	632.67 ± 0.44	[145]					
L_η	L_2M_1	453.48 ± 0.74	[50]	9.92(-5)	[488]	2.533(-3)	2.92	[423]
		453.80	[104]					
	L_2M_2	480.8 ± 1.1	[145]				0.69	[423]

line	transition	E/eV		I/eV/$\hbar$		TPIV	Γ/eV	
$L_{\beta_{17}}$	L_2M_3	486.4 ± 1.2	[145]			3.475(-6)	0.66	[423]
L_{β_1}	L_2M_4	519.2 ± 1.3	[50]	4.14(-4)	[488]	3.010(-4)	0.28	[423]
		518.80 6	[104]					
L_{γ_5}	L_2N_1	527.58 ± 0.40	[145]			4.681(-5)		
L_l	L_3M_1	446.46 ± 0.24	[50]	1.02(-4)	[488]	2.400(-3)	2.92	[423]
		446.50	[104]					
L_t	L_3M_2	472.6 ± 1.1	[145]			1.659(-6)	0.69	[423]
L_s	L_3M_3	478.2 ± 1.1	[145]			1.625(-6)	0.66	[423]
$L_{\alpha_{2,1}}$	$L_3M_{4,5}$	511.27 ± 0.94	[50]	4.12(-5)	[488]			
		510.94	[104]					
L_{α_2}	L_3M_4	513.54 ± 0.63	[145]			2.282(-5)		
	L_3M_5					2.648(-4)		
	L_3N_1	519.38 ± 0.38	[145]			4.437(-5)		
	$M_{2,3}M_{4,5}$	37.00 ± 1.00	[50]					
		35.70	[104]					

level characteristics

level	E_B/eV		ω_{nlj}		Γ/eV		AE/eV	
K	5465.00	[503]	0.243	[301]	1.01	[301]	5465.1	[51]
	5465.10 ± 0.30	[493]	0.243	[293]	0.92	[423]	5478.28 ± 0.12	[145]
			0.2608	[557]	0.96	[94]		
			0.250	[38]				
			0.250	[41]				
			0.2564	[39]				
			0.2426	[423]				
			0.243	[310]				
L_1	628.00	[503]	5.80(-4)	[301]	2.41	[301]	628.2	[51]
	628.20 ± 0.40	[493]	1.16(-3)	[423]	5.92	[423]	639.78 ± 0.37	[145]
					4.6	[94]		
L_2	520.00	[503]	2.60(-3)	[301]	0.26	[301]	520.5	[51]
	520.50 ± 0.30	[493]	2.88(-3)	[423]	0.28	[423]	534.69 ± 0.33	[145]
					0.78	[94]		
L_3	513.00	[503]	2.60(-3)	[301]	0.24	[301]	512.9	[51]
	512.90 ± 0.30	[493]	2.74(-3)	[423]	0.28	[423]	526.50 ± 0.31	[145]
					0.28	[94]		

level	E_B/eV		ω_{nlj}		Γ/eV		AE/eV	
M_1	66.00	[503]	2.90(-3)	[38]	4.18	[350]	66.3	[51]
	66.50 ± 0.40	[493]	8.613(-4)	[512]	2.64	[423]		
			1.0(-5)	[423]	2.2	[94]		
M_2	38.00	[503]	2.30(-5)	[38]	0.53	[350]		
	37.80 ± 0.30	[51]	2.0(-5)	[512]	0.41	[423]		
	40.60 ± 1.60	[493]	2.0(-5)	[423]	1.2	[94]		
M_3	35.00 ± 1.60	[493]			0.38	[423]		
					1.2	[94]		
M_4	2.00	[503]						
	2.20 ± 0.30	[493]						
N_1	1.70 ± 1.70	[493]						
IP	6.7463	[222]						

Cr   Z=24   [Ar] 3d⁵ 4s¹

Chromium $A = 51.9961(6)$ [12.77] $\varrho = 7.18\,\text{g/cm}^3$ [547]
$\varrho = 7.14\,\text{g/cm}^3$ [483]

X-ray transitions

line	transition	E/eV		I/eV/ℏ		TPIV	Γ/eV	
K series								
K_{α_3}	KL_1	5293.35 ± 0.50	[145]				7.81	[423]
K_{α_2}	KL_2	5405.51 ± 0.01	[50]	8.35(-2)	[488]	8.258(-2)	1.37	[301]
		5405.55	[104]	8.76(-2)	[487]		2.64	[473]
		5405.56 ± 0.01	[143]				1.30	[423]
		5405.57 ± 0.01	[277]					
		5405.5384 ± 0.0071	[241]					
K_{α_1}	KL_3	5414.72 ± 0.05	[50]	0.1640	[488]	1.621(-1)	1.35	[301]
		5414.70	[104]	0.1720	[487]		2.05	[473]
		5414.78 ± 0.07	[143]				1.31	[423]
		5414.79 ± 0.07	[277]					
		5414.8045 ± 0.0071	[241]					
	KM_1	5914.2 ± 1.5	[145]				4.25	[423]
K_{β_3}	KM_2	5940.74 ± 0.92	[145]	1.12(-2)	[488]	9.494(-3)	1.91	[423]
K_{β_1}	KM_3	5947.1 ± 1.0	[145]	2.22(-2)	[488]	1.870(-2)	1.88	[423]
$K_{\beta_{3,1}}$	$KM_{2,3}$	5946.71 ± 0.06	[50]	2.852(-2)	[488]			
		5946.60	[104]					
		5946.823 ± 0.011	[241]					
K_{β_5}	$KM_{4,5}$	5986.97 ± 0.26	[50]	6.30(-6)	[488]			
		5987.10	[104]					
$K_{\beta_5^{II}}$	KM_4	5986.82 ± 0.46	[145]			2.525(-6)		
$K_{\beta_5^{I}}$	KM_5	5990.83 ± 0.12	[145]			3.695(-6)		
	KN_1	5989.03 ± 0.19	[145]					
L series								
	L_1M_1	620.9 ± 1.8	[145]				10.09	[423]
$L_{\beta_{4,3}}$	$L_1M_{2,3}$	653.9 ± 1.0	[50]	1.58(-3)	[488]			
		639.40	[104]					
	L_1M_2	647.4 ± 1.2	[145]			4.334(-4)	7.75	[423]
	L_1M_3	653.8 ± 1.3	[145]			6.328(-4)	7.72	[423]
$L_{\beta_{10}}$	L_1M_4	693.47 ± 0.74	[145]			9.014(-7)		
L_{β_9}	L_1M_5	697.48 ± 0.40	[145]			1.283(-6)		
	L_1N_1	695.68 ± 0.47	[145]					

line	transition	E/eV		I/eV/$\hbar$		TPIV	Γ/eV	
L_η	L_2M_1	510.22 ± 0.93 508.60	[50] [104]	1.20(-4)	[488]	3.063(-3)	3.58	[423]
	L_2M_2	536.7 ± 1.1	[145]				2.83	[423]
$L_{\beta_{17}}$	L_2M_3	543.0 ± 1.2	[145]			5.0168-6)	1.21	[423]
$L_{\beta 1}$	L_2M_4	582.90 ± 0.41 580.60	[50] [104]	8.00(-4)	[488]	5.246(-4)	0.32	[423]
	L_2M_5	586.76 ± 0.33	[145]					
L_{γ_5}	L_2N_1	584.96 ± 0.41	[145]			5.758(-5)		
L_l	L_3M_1	500.33 ± 0.30 499.98	[50] [104]	1.25(-4)	[488]	2.891(-3)	3.58	[423]
L_t	L_3M_2	526.9 ± 1.1	[145]			2.433(-6)	1.25	[423]
L_s	L_3M_3	533.2 ± 1.2	[145]			2.380(-6)	1.21	[423]
$L_{\alpha_{2,1}}$	$L_3M_{4,5}$	572.9 ± 1.2 571.36	[50] [104]	7.88(-4)	[488]			
L_{α_2}	L_3M_4	572.94 ± 0.64	[145]			4.102(-5)		
L_{α_1}	L_3M_5	576.95 ± 0.30	[145]			4.703(-4)		
L_{β_6}	L_3N_1	575.15 ± 0.38	[145]			5.435(-5)		

M series

| | $M_{2,3}M_{4,5}$ | 40 ± 1.20
40.25 | [50]
[104] | | | | | |

level characteristics

level	E_B/eV		ω_{nlj}		Γ/eV		AE/eV	
K	5989.00	[503]	0.275	[301]	1.08	[301]	5989.2	[51]
	5989.20 ± 0.30	[493]	0.276	[293]	0.99	[423]	5989.017 ± 0.040	[50]
			0.2939	[557]	1.02	[94]	5995.656 ± 0.12	[145]
			0.282	[38]				
			0.2885	[39]				
			0.287	[537]				
			0.282	[34]				
			0.280	[427]				
			0.279	[221]				
			0.2729	[423]				
			0.275	[310]				
			0.287	[537]				

level	E_B/eV		ω_{nlj}		Γ/eV		AE/eV	
L_1	695.00	[503]	7.10(-4)	[301]	2.54	[301]	694.6	[51]
	694.60 ± 0.40	[493]	2.97(-4)	[348]	6.83	[423]	742.4 ± 4.4	[50]
			1.26(-3)	[423]	5.2	[94]	702.31 ± 0.40	[145]
L_2	584.00	[503]	3.70(-3)	[301]	0.29	[301]	583.7	[51]
	583.70 ± 0.30	[493]	3.65(-3)	[423]	0.32	[423]	592.60 ± 0.57	[50]
					0.76	[94]	591.60 ± 0.30	[145]
L_3	575.00	[503]	3.70(-3)	[301]	0.27	[301]	574.5	[51]
	574.50 ± 0.30	[493]	3.29(-3)	[348]	0.32	[423]	598.9 ± 4.3	[50]
			3.46(-3)	[423]	0.32	[94]	581,78 ± 0.30	[145]
M_1	74.00	[503]	2.60(-6)	[38]	4.92	[350]	74.1	[51]
	74.10 ± 0.40	[493]	9.352(-4)	[512]	3.26	[423]		
			1.0(-5)	[423]	2.3	[94]		
M_2	43.00	[503]	1.60(-5)	[38]	1.32	[350]		
	42.50 ± 0.30	[51]	2.0(-5)	[512]	0.93	[423]		
	45.90 ± 0.90	[493]	1.0(-6)	[423]	1.2	[94]		
M_3	42.50 ± 0.30	[51]	1.0(-6)	[423]	0.80	[74]		
	39.90 ± 0.90	[493]			0.89	[423]		
					1.2	[94]		
M_4	2.00	[503]						
	2.30 ± 0.40	[51]						
	2.90 ± 0.40	[493]						
M_5	2.20 ± 0.40	[493]						
N_1	1.00 ± 1.70	[493]						
IP	6.7665	[222]						

Mn Z=25

[Ar] $3d^5\,4s^2$

Manganese

$A = 54.93804(9)$ [12.77] $\varrho = 7.30\,\text{g/cm}^3$ [547]

X-ray transitions

line	transition	E/eV		I/eV/$\hbar$		TPIV	Γ/eV	
	K series							
K_{α_3}	KL_1	5769.18 ± 0.54	[145]				7.79	[423]
$K_{\alpha 2}$	KL_2	5887.65 ± 0.06	[50]	0.1003	[488]	9.177(-2)	1.50	[301]
		5887.6859 ± 0.0084	[241]	0.1049	[487]		1.42	[423]
$K_{\alpha 1}$	KL_3	5898.75 ± 0.03	[50]	0.1967	[488]	1.799(-1)	1.48	[301]
		5898.70	[104]	0.2058	[487]		1.43	[423]
		5898.8010 ± 0.0084	[241]					
	KM_1	6455.7 ± 1.5	[145]				4.45	[423]
$K_{\beta 3}$	$KM_2$6485.39 ± 0.96	[145]		1.40(-2)	[488]	1.094(-2)	2.14	[423]
$K_{\beta 1}$	KM_3	6492.7 ± 1.0	[145]	2.76(-2)	[488]	2.152(-2)	2.10	[423]
$K_{\beta_{3,1}}$	$KM_{2,3}$	6490.45 ± 0.07	[50]	3.548(-2)	[488]			
		6490.40	[104]					
		6490.585 ± 0.014	[241]					
$K_{\beta 5}$	$KM_{4,5}$	6535.36 ± 0.51	[50]	1.04(-5)	[488]			
		6535.39	[104]					
$K_{\beta_5^{II}}$	KM_4	6537.52 ± 0.48	[145]			3.855(-6)		
$K_{\beta_5^{I}}$	KM_5	6538.54 ± 0.51	[145]			5.641(-6)		
	KN_1	6544.46 ± 0.20	[145]					
	L series							
	L_1M_1	686.5 ± 1.8	[145]				10.11	[423]
$L_{\beta_{4,3}}$	$L_1M_{2,3}$	721.2 ± 1.2	[50]	2.03(-3)	[488]			
		705.30	[104]					
L_{β_4}	L_1M_2	716.2 ± 1.3	[145]			4.880(-4)	7.80	[423]
L_{β_3}	L_1M_3	723.5 ± 1.4	[145]			7.085(-4)	7.76	[423]
$L_{\beta_{10}}$	L_1M_4	768.34 ± 0.80	[145]			1.261(-6)		
L_{β_9}	L_1M_5	769.36 ± 0.82	[145]			1.762(-6)		
	L_1N_1	775.27 ± 0.51	[145]					
L_η	L_2M_1	567.42 ± 0.77	[50]	1.79(-4)	[488]	3.576(-3)	3.74	[423]
		568.20	[104]					

line	transition	E/eV			I/eV/$\hbar$		TPIV	Γ/eV	
	L_2M_2	599.2 ± 1.2	[145]					1.44	[423]
$L_{\beta_{17}}$	L_2M_3	606.5 ± 1.3	[145]				7.007(-6)	1.39	[423]
L_{β_1}	L_2M_4	648.80 ± 0.70 648.05	[50] [104]	1.80(-3)	[488]		8.777(-4)	0.36	[423]
	L_2M_5	652.34 ± 0.72	[145]						
L_{γ_5}	L_2N_1	658.26 ± 0.41	[145]				6.971(-5)		
L_l	L_3M_1	556.22 ± 0.11 556.60	[50] [104]	1.56(-4)	[488]		3.356(-3)	3.75	[423]
L_t	L_3M_2	587.3 ± 1.1	[145]				3.419(-69)	1.44	[423]
L_s	L_3M_3	594.6 ± 1.2	[145]				3.338(-6)	1.40	[423]
$L_{\alpha_{2,1}}$	$L_3M_{4,5}$	637.44 ± 0.49 637.25	[50] [104]	1.17(-3)	[488]			1.50	[301]
L_{α_2}	L_3M_4	639.42 ± 0.66	[145]				7.066(-5)		
L_{α_1}	L_3M_5	640.44 ± 0.69	[145]				7.980(-4)		
L_{β_6}	L_3N_1	646.36 ± 0.38 [145]					6.557(-5)		

M series

	$M_{2,3}M_{4,5}$	45.00 ± 1.00 45.70	[50] [104]						

level characteristics

level	E_B/eV		ω_{nlj}		Γ/eV		AE/eV	
K	6539.00 6539.00 ± 0.40	[503] [493]	0.308 0.310 0.3276 0.314 0.303 0.3213 0.3042 0.308	[301] [293] [557] [38] [41] [39] [423] [310]	1.16 1.07 1.11	[301] [423] [94]	6539.0 6.537667 6552.12 ± 0.12	[51] [294] [145]
L_1	769.00 769.00 ± 0.40	[503] [493]	8.40(-4) 1.36(-3) 3.3(-4) 3.5(-4)	[301] [423] [436] [244]	6.62 6.72 6.2	[301] [423] [94]	769.0 782.94 ± 0.44	[51] [145]

level	E_B/eV		ω_{nlj}		Γ/eV		AE/eV	
L_2	652.00	[503]	5.00(-3)	[301]	0.34	[301]	651.4	[51]
	651.40 ± 0.40	[493]	4.53(-3)	[423]	0.36	[423]	665.92 ± 0.33	[145]
			3.40(-3)	[310]	0.97	[94]		
			3.40(-3)	[436]				
			3.90(-3)	[244]				
L_3	641.00	[503]	5.00(-3)	[301]	0.32	[301]	640.3	[51]
	640.30 ± 0.40	[493]	4.30(-3)	[423]	0.36	[423]	654.02 ± 0.31	[145]
			3.90(-3)	[436]	0.36	[94]		
			4.57(-3)	[244]				
M_1	84.00	[503]	3.10(-6)	[38]	5.92	[350]	82.3	[51]
	83.90 ± 0.50	[493]	1.0(-5)	[423]	3.39	[423]		
					2.4	[94]		
M_2	49.00	[503]	1.60(-5)	[38]	1.52	[350]		
	48.60 ± 0.40	[51]	2.0(-5)	[512]	1.08	[423]		
	53.10 ± 0.40	[493]	2.0(-6)	[423]	1.2	[94]		
M_3	49.00	[503]	2.0(-6)	[423]	1.04	[423]		
	48.60 ± 0.40	[51]			1.2	[94]		
	46.40 ± 0.40	[493]						
M_4	4.00	[503]						
	3.30 ± 0.50	[51]						
	3.50 ± 0.60	[51]						
M_5	4.00	[503]						
	3.30	[51]						
	2.70 ± 0.50	[493]						
N_1	1.90 ± 0.20	[493]						
IP	7.43402	[222]						

Fe   Z=26   [Ar] 3d⁶ 4s²

Iron $A = 55.845(2)$ [222] $\varrho = 7.86\ \text{g/cm}^3$ [547]

X-ray transitions

line	transition	E/eV		I/eV/$\hbar$		TPIV	Γ/eV	
K series								
$K_{\alpha 3}$	KL_1	6266.07 ± 0.58	[145]				8.53	[423]
$K_{\alpha 2}$	KL_2	6390.84 ± 0.03	[50]	0.1196	[488]	1.014(-1)	1.62	[301]
		6390.80	[104]	0.1248	[487]		3.20	[473]
		6391.0264 ± 0.0099	[241]				1.55	[423]
$K_{\alpha 1}$	KL_3	6403.84 ± 0.03	[50]	0.2341	[488]	1.986(-1)	1.61	[301]
		6403.80	[104]	0.2444	[487]		2.46	[473]
		6404.0062 ± 0.0099	[241]				1.56	[423]
	KM_1	7020.2 ± 1.5	[145]				4.84	[423]
$K_{\beta 3}$	KM_2	7053.2 ± 1.0	[145]	1.67(-2)	[487]	1.221(-2)	2.66	[423]
$K_{\beta 1}$	KM_3 7059.9 ± 1.1	[145]		3.29(-2)	[487]	2.400(-2)	2.60	[423]
$K_{\beta 3,1}$	$KM_{2,3}$	7057.98 ± 0.08	[50]	4.269(-2)	[488]			
		7058.00	[104]					
		7058.175 ± 0.016	[241]					
$K_{\beta 5}$	$KM_{4,5}$	7108.26 ± 0.60	[50]	1.74(-5)	[488]			
		7108.40	[104]					
$K_{\beta_5^{II}}$	KM_4	7110.59 ± 0.50	[145]			6.008(-6)		
$K_{\beta_5^{I}}$	KM_5	7111.50 ± 0.52	[145]			8.772(-6		
	KN_1	7117.93 ± 0.20	[145]					
L series								
	L_1M_1	754.1 ± 1.8	[145]				11.07	[423]
$L_{\beta 4,3}$	$L_1M_{2,3}$	792.2 ± 1.5	[50]	2.525(-3)	[488]			
		790.40	[104]					
$L_{\beta 4}$	L_1M_2	787.2 ± 1.3	[145]			5.335(-4)	8.89	[423]
$L_{\beta 3}$	L_1M_3	793.8 ± 1.5	[145]			3.921(-4)	8.84	[423]
$L_{\beta 10}$	L_1M_4	844.52 ± 0.85	[145]			1.631(-6)		
$L_{\beta 9}$	L_1M_5	845.44 ± 0.87	[145]			2.202(-6)		
	L_1N_1	851.86 ± 0.55	[145]					

line	transition	E/eV		I/eV/ℏ		TPIV	Γ/eV	
L_η	L_2M_1	627.8 ± 1.9	[50]	1.79(-4)	[488]	4.038(-3)	4.10	[423]
		627.75	[104]					
	L_2M_2	663.7 ± 1.2	[145]				1.91	[423]
$L_{\beta_{17}}$	L_2M_3	670.4 ± 1.3	[145]			9.181(-6)	1.86	[423]
L_{β_1}	L_2M_4	718.32 ± 0.62	[50]	1.803(-3)	[488]	1.402(-3)	0.40	[423]
		717.95	[104]					
	L_2M_5	721.99 ± 0.73	[145]					
L_{γ_5}	L_2N_1	728.41 ± 0.41	[145]			8.273(-5)		
L_l	L_3M_1	615.30 ± 0.45	[50]	1.88(-4)	[488]	3.759(-3)	4.11	[423]
		615.50	[104]					
L_t	L_3M_2	650.1 ± 1.2	[145]			4.556(-6)	1.87	[423]
L_s	L_3M_3	656.8 ± 1.3	[145]			1.009(-5)	1.87	[423]
$L_{\alpha_{2,1}}$	$L_3M_{4,5}$	704.8 ± 1.2	[50]	1.77(-3)	[488]			
		704.95	[104]					
L_{α_2}	L_3M_4	707.46 ± 0.67	[145]			1.150(-4)		
L_{α_1}	L_3M_5	708.38 ± 0.70	[145]			1.285(-3)		
L_{β_6}	L_3N_1	714.80 ± 0.38	[145]			7.696(-5)		

M series

	$M_{2,3}M_{4,5}$	51.00 ± 1.00	[50]
		51.15	[104]

level characteristics

level	E_B/eV		ω_{nlj}		Γ/eV		AE/eV	
K	7114.00	[503]	0.340	[301]	1.25	[301]	7112.0	[51]
	7112.00 ± 0.90	[51]	0.364	[347]	1.15	[423]	7110.747 ± 0.020	[294]
	7113.00 ± 0.90	[493]	0.3624	[557]	1.19	[94]	7125.87 ± 0.13	[145]
			0.344	[293]				
			0.342	[35]				
			0.3362	[423]				
			0.340	[310]				
			0.350	[567]				
L_1	846.00	[503]	1.00(-3)	[301]	2.76	[301]	846.1	[51]
	846.10 ± 0.40	[493]	3.84(-4)	[487]	7.38	[423]	859.80 ± 0.48	[145]
			1.47(-3)	[423]	7.05	[94]		

level	E_B/eV		ω_{nlj}		Γ/eV		AE/eV	
			3.80(-3)	[436]				
L_2	723.00	[503]	6.30(-3)	[301]	0.37	[301]	721.1	[51]
	721.10 ± 0.90	[493]	1.43(-3)	[107]	0.40	[423]	720.74 ± 0.31	[50]
			5.33(-3)	[423]	1.14	[94]	736.35 ± 0.34	[145]
			4.80(-3)	[436]				
L_3	710.00	[503]	6.30(-3)	[301]	0.36	[301]	708.1	[51]
	708.10 ± 0.90	[493]	1.49(-3)	[107]	0.41	[423]	707.46 ± 0.30	[50]
			5.25(-3)	[423]	0.41	[94]	722.74 ± 0.31	[145]
			5.40(-3)	[436]				
M_1	95.00	[503]	2.80(-6)	[38]	6.90	[350]	91.3	[51]
	92.90 ± 0.90	[493]	1.097(-3)	[512]	3.70	[423]		
			1.0(-5)	[423]	2.4	[94]		
M_2	56.00	[503]	1.60(-5)	[38]	2.15	[350]		
	54.00 ± 0.90	[51]	2.0(-5)	[512]	1.51	[423]		
	58.10 ± 0.90	[493]	4.0(-6)	[423]	1.23	[94]		
M_3	56.00	[503]	4.0(-6)	[423]	1.40	[74]		
	54.00 ± 0.90	[51]			1.46	[423]		
	52.00 ± 0.90	[493]			1.23	[94]		
M_4	6.00	[503]						
	3.90 ± 0.90	[493]						
M_5	6.00	[503]						
	3.60	[51]						
	3.10 ± 0.90	[493]						
N_1	2.10 ± 2.20	[493]						
IP	7.9024	[222]						

Co   Z=27

[Ar] 3d^7 4s^2

Cobalt

$A = 58.933200(9)$ [222] $\varrho = 8.90\,\text{g/cm}^3$ [547]
$\varrho = 8.71\,\text{g/cm}^3$ [483]

X-ray transitions

line	transition	E/eV		I/eV/$\hbar$		TPIV	Γ/eV	
K series								
K$_{\alpha_3}$	KL$_1$	6784.08 ± 0.61	[145]				9.21	[423]
K$_{\alpha_2}$	KL$_2$	6915.30 ± 0.04	[50]	0.1415	[488]	1.112(-1)	1.76	[301]
		6915.30	[104]				1.69	[423]
		6915.5380 ± 0.0039	[241]					
K$_{\alpha_1}$	KL$_3$	6930.32 ± 0.04	[50]	0.2766	[488]	2.175(-1)	1.76	[301]
		6930.20	[104]				1.70	[423]
		6930.3780 ± 0.0039	[241]					
	KM$_1$	7609.0 ± 1.5	[145]				5.21	[423]
K$_{\beta_{1,3}}$	KM$_{2,3}$	7649.43 ± 0.10	[50]	5.088(-2)	[488]			
		7649.30	[104]					
		7649.445 ± 0.014	[241]					
K$_{\beta_3}$	KM$_2$	7645.5 ± 1.0	[145]			1.350(-2)	3.24	[423]
K$_{\beta_1}$	KM$_3$	7651.5 ± 1.1	[145]			2.651(-2)	3.16	[423]
K$_{\beta_5}$	KM$_{4,5}$	7705.98 ± 0.21	[50]	2.78(-5)	[488]			
		7706.00	[104]					
K$_{\beta_5}^{II}$	KM$_4$	7708.35 ± 0.52	[145]			8.906(-6)		
K$_{\beta_5}^{I}$	KM$_5$	7709.15 ± 0.54	[145]			1.297(-5)		
	KN$_1$	7716.05 ± 0.21	[145]					
L series								
	L$_1$M$_1$	824.9 ± 1.9	[145]				11.94	[423]
L$_{\beta_{4,3}}$	L$_1$M$_{2,3}$	866.4 ± 2.7	[50]	3.10(-3)	[488]			
		864.10	[104]					
L$_{\beta_4}$	L$_1$M$_2$	861.4 ± 1.4	[145]			5.828(-4)	9.98	[423]
L$_{\beta_3}$	L$_1$M$_3$	867.4 ± 1.5	[145]			8.697(-4)	9.90	[423]
L$_{\beta_{10}}$	L$_1$M$_4$	924.27 ± 0.90	[145]			1.956(-6)		
L$_{\beta_9}$	L$_1$M$_5$	925.07 ± 0.92	[145]			2.597(-6)		
	L$_1$N$_1$	931.97 ± 0.59	[145]					

line	transition	E/eV		I/eV/$\hbar$		TPIV	Γ/eV	
L_η	L_2M_1	693.8 ± 1.7	[50]	2.15(-4)	[488]	4.382(-3)	4.43	[423]
		692.25	[104]					
	L_2M_2	731.5 ± 1.2	[145]				2.45	[423]
$L_{\beta_{17}}$	L_2M_3	737.4 ± 1.3	[145]			1.165(-5)	2.38	[423]
L_{β_1}	L_2M_4	791.41 ± 0.60	[50]	2.63(-3)	[488]	2.168(-3)		
		791.30	[104]					
	L_2M_5	795.08 ± 0.74	[145]					
L_{γ_5}	L_2N_1	801.98 ± 0.42	[145]			9.631(-5)		
L_l	L_3M_1	677.80 ± 0.44	[50]	2.82(-4)	[488]	4.070(-3)	4.44	[423]
		677.25	[104]					
L_t	L_3M_2	715.9 ± 1.2	[145]			5.774(-6)	2.47	[423]
L_s	L_3M_3	721.9 ± 0.13	[145]			5.617(-6)	2.40	[423]
$L_{\alpha_{2,1}}$	$L_3M_{4,5}$	776.25 ± 0.43	[50]	2.58(-3)	[488]			
		776.40	[104]					
L_{α_2}	L_3M_4	778.71 ± 0.69	[145]			1.787(-4)		
L_{α_1}	L_3M_5	779.52 ± 0.71	[145]			1.963(-3)		
L_{β_6}	L_3N_1	786.42 ± 0.39	[145]			9.631(-5)		

M series

	$M_{2,3}M_{4,5}$	58.00 ± 1.60	[50]					
		57.70	[104]					

level characteristics

level	E_B/eV		ω_{nlj}		Γ/eV		AE/eV	
K	7709.00	[503]	0.373	[301]	1.33	[301]	7708.9	[51]
	7708.90 ± 0.30	[493]	0.379	[293]	1.24	[423]	7708.776 ± 0.020	[294]
			0.3977	[557]	1.28	[94]	7724.26 ± 0.13	[145]
			0.366	[41]				
			0.381	[38]				
			0.388	[39]				
			0.3687	[423]				
			0.373	[310]				
L_1	926.00	[503]	1.20(-3)	[301]	2.79	[301]	925.6	[51]
	925.60 ± 0.40	[493]	1.58(-3)	[423]	7.97	[423]	940.18 ± 0.51	[145]
			4.48-4)	[436]	7.2	[94]		

level	E_B/eV		ω_{nlj}		Γ/eV		AE/eV	
L_2	794.00	[503]	7.70(-3)	[301]	0.43	[301]	793.8	[51]
	793.60 ± 0.30	[493]	6.66(-3)	[423]	0.46	[423]	793.84 ± 0.38	[50]
			6.308-3)	[436]	1.13	[94]	810.18 ± 0.34	[145]
L_3	779.00	[503]	7.70(-3)	[301]	0.43	[301]	778.6	[51]
	778.60 ± 0.30	[493]	6.31(-3)	[423]	0.47	[423]	779.03 ± 0.36	[50]
			7.10(-3)	[436]	0.47	[94]	794.62 ± 0.31	[145]
M_1	101.00	[503]	2.80(-6)	[38]	7.28	[350]	101.0	[51]
	100.70 ± 0.40	[493]	2.0(-5)	[423]	3.97	[423]		
					2.4	[94]		
M_2	60.00	[503]	1.70(-5)	[38]	2.85	[350]		
	59.50 ± 0.30	[51]	2.0(-5)	[512]	2.00	[423]		
	63.20 ± 0.30	[493]	1.0(-5)	[423]	1.25	[94]		
M_3	60.00	[503]	1.0(-5)	[423]	1.27	[94]		
	57.70 ± 0.30	[493]			1.93	[423]		
M_4	3.00	[503]						
	2.90 ± 0.30	[51]						
	2.70 ± 0.30	[493]						
M_5	3.00	[503]						
	3.30 ± 0.30	[493]						
N_1	1.90 ± 2.40	[493]						
IP	7.8810	[222]						

Ni   Z=28

[Ar] $3d^8\,4s^2$

Nickel $A = 58.6934(2)$ [222] $\varrho = 8.78\,\text{g/cm}^3$ [547]
$\varrho = 8.96\,\text{g/cm}^3$ [483]

X-ray transitions

line	transition	E/eV		I/eV/$\hbar$		TPIV	Γ/eV	
K series								
$K_{\alpha3}$	KL_1	7323.29 ± 0.65	[423]				9.85	[423]
$K_{\alpha2}$	KL_2	7460.89 ± 0.04	[50]	0.1662	[488]	1.211(-1)	1.96	[301]
		7460.80	[104]	0.1730	[487]		3.78	[473]
		7461.0343 ± 00.45	[241]				1.86	[423]
$K_{\alpha1}$	KL_3	7478.15 ± 0.04	[50]	0.325	[488]	2.364(-1)	1.94	[301]
		7478.10	[104]	0.338	[487]		3.07	[473]
		7478.2521 ± 0.0045	[241]				1.86	[423]
	KM_1	8222.2 ± 1.4	[145]				5.58	[423]
$K_{\beta3}$	KM_2	8262.4 ± 1.1	[145]	2.34(-2)	[487]	1.480(-2)	3.87	[423]
$K_{\beta1}$	KM_3	8267.6 ± 1.1	[145]	4.59(-2)	[487]	2.903(-2)	3.78	[423]
$K_{\beta3,1}$	$KM_{2,3}$	8264.66 ± 0.04	[50]	6.023(-2)	[488]			
		8264.52	[104]					
		8264.775 ± 0.017	[241]					
$K_{\beta5}$	$KM_{4,5}$	8328.68 ± 0.33	[50]	4.27(-5)	[488]			
		8328.70	[104]					
$K_{\beta5}^{II}$	KM_4	8330.91 ± 0.54	[145]			1.269(-5)		
$K_{\beta5}^{I}$	KM_5	8331.59 ± 0.56	[145]			1.844(-5)		
	KN_1	8338.94 ± 0.22	[145]					
L series								
	L_1M_1	898.9 ± 1.9	[145]				12.76	[423]
$L_{\beta4,3}$	$L_1M_{2,3}$	940.7 ± 1.1	[50]	3.76(-3)	[488]			
		940.60	[104]					
$L_{\beta4}$	L_1M_2	939.1 ± 1.5	[145]			6.321(-4)	11.05	[423]
$L_{\beta3}$	L_1M_3	944.3 ± 1.6	[145]			9.677(-4)	10.96	[423]
$L_{\beta10}$	L_1M_4	1007.62 ± 0.95	[145]			2.273(-6)		
$L_{\beta9}$	L_1M_5	1008.30 ± 0.97	[145]			2.979(-6)		
	L_1N_1	1015.65 ± 0.63	[145]					
L_η	L_2M_1	762.0 ± 2.1	[50]	2.54(-4)	[488]	4.557(-3)	4.77	[423]

line	transition	E/eV		I/eV/$\hbar$		TPIV	Γ/eV	
		759.70	[104]					
		762.2 ± 1.7	[145]					
	L_2M_2	802.4 ± 1.3	[145]				3.06	[423]
$L_{\beta_{17}}$	L_2M_3	807.6 ± 1.3	[145]			1.394(-5)	2.97	[423]
L_{β_1}	L_2M_4	868.77 ± 0.54	[50]	3.70(-3)	[488]	3.124(-3)		
		868.20	[104]					
	L_2M_5	871.63 ± 0.76	[145]					
L_{γ_5}	L_2N_1	878.98 ± 0.42	[145]					
L_l	L_3M_1	742.72 ± 0.59	[50]	2.69(-4)	[488]	4.261(-3)	4.78	[423]
		742.40	[104]					
L_t	L_3M_2	784.7 ± 1.2	[145]			7.039(-6)	3.07	[423]
L_s	L_3M_3	789.9 ± 1.3	[145]			6.835(-6)	2.97	[423]
$L_{\alpha_{2,1}}$	$L_3M_{4,5}$	851.47 ± 0.21	[50]	3.61(-3)	[488]			
		850.90	[104]					
L_{α_2}	L_3M_4	853.19 ± 0.71	[145]			2.627(-4)		
L_{α_1}	L_3M_5	853.87 ± 0.26	[145]			2.860(-3)		
L_{β_6}	L_3N_1	861.23 ± 0.39	[145]			9.788(-5)		

M series

	$M_{2,3}M_{4,5}$	65.10 ± 0.70	[50]					
		67.00	[104]					

level characteristics

level	E_B/eV		ω_{nlj}		Γ/eV		AE/eV	
K	8333.00	[503]	0.406	[301]	1.44	[301]	8332.8	[51]
	8332.80 ± 0.40	[493]	0.414	[293]	1.33	[423]	8331.486 ± 0.020	[294]
			0.432	[557]	1.39	[94]	8347.42 ± 0.14	[145]
			0.414	[38]				
			0.4212	[39]				
			0.4014	[423]				
			0.406	[310]				
L_1	1008.00	[503]	1.40(-3)	[301]	2.89	[301]	1008.1	[51]
	1008.10 ± 0.40	[493]	4.63(-4)	[348]	8.51	[423]	1024.13 ± 0.55	[145]
			1.72(-3)	[423]	6.4	[94]		
			4.9(-4)	[436]				

level	E_B/eV		ω_{nlj}		Γ/eV		AE/eV	
			1.6(-3)	[21]				
L_2	872.00	[503]	8.60(-3)	[301]	0.52	[301]	871.9	[51]
	871.90 ± 0.40	[493]	2.69(-3)	[107]	0.53	[423]	870.54 ± 0.45	[50]
			7.80(-3)	[423]	0.98	[94]	887.36 ± 0.34	[145]
			7.90(-3)	[436]				
			7.50(-3)	[21]				
L_3	855.00	[503]	9.30(-3)	[301]	0.48	[301]	854.7	[51]
	854.70 ± 0.40	[493]	7.50(-3)	[423]	0.53	[423]	853.58 ± 0.43	[50]
			8.80(-3)	[436]	0.53	[94]	869.70 ± 0.31	[145]
			5.50(-3)	[21]				
M_1	112.00	[503]	3.50(-6)	[38]	7.92	[350]	110.8	[51]
	111.80 ± 0.60	[493]	2.0(-5)	[423]	4.25	[423]		
					2.3	[94]		
M_2	68.00	[503]	1.50(-5)	[38]	3.80	[350]		
	68.10 ± 0.40	[51]	2.0(-5)	[512]	2.54	[423]		
	71.20 ± 0.40	[493]	1.0(-5)	[423]	1.3	[94]		
M_3	68.00	[503]	1.0(-5)	[423]	1.3	[94]		
	69.70 ± 0.40	[493]			2.45	[423]		
M_4	4.00	[503]						
	3.60 ± 0.40	[51]						
	3.90 ± 0.40	[493]						
M_5	4.00	[503]						
	3.30 ± 0.40	[493]						
N_1	2.20 ± 2.50	[493]						
IP	7.6398	[222]						

Cu   Z=29

[Ar] 3d^{10} 4s^1

Copper $A = 63.546(3)$ [12.77] $\varrho = 8.94\,\text{g/cm}^3$ [547]
$\varrho = 8.93\,\text{g/cm}^3$ [483]

X-ray transitions

line	transition	E/eV		I/eV/$\hbar$		TPIV	Γ/eV	
K series								
K_{α_3}	KL$_1$	7884.83 ± 0.68	[145]				10.57	[423]
$K_{\alpha 2}$	KL$_2$	8027.83 ± 0.01	[50]	0.1942	[488]	1.311(-1)	2.17	[301]
		8027.80	[104]	0.2017	[487]		2.04	[423]
		8027.91 ± 0.02	[143]					
		8027.92 ± 0.02	[277]					
		8027.8416 ± 0.0026	[241]					
$K_{\alpha 1}$	KL$_3$	8047.78 ± 0.01	[50]	0.379	[488]	2.557(-1)	2.11	[301]
		8047.70	[104]	0.393	[487]		2.05	[423]
		8047.86 ± 0.01	[143]					
		8047.87 ± 0.01	[277]					
		8047.8227 ± 0.0026	[241]					
	KM$_1$	8859.8 ± 1.4	[145]				6.16	[423]
$K_{\beta 3}$	KM$_2$	8902.90 ± 0.64	[50]	2.354(-2)	[488]	1.589(-2)	5.00	[423]
		8902.80	[104]	2.69(-2)	[487]			
$K_{\beta 1}$	KM$_3$	8905.29 ± 0.64	[50]	4.61(-2)	[488]	3.109(-2)	4.89	[423]
		8905.10	[104]	5.27(-2)	[487]			
$K_{\beta 5}$	KM$_{4,5}$	8977.14 ± 0.29	[50]	6.50(-5)	[488]			
		8974.20	[104]					
$K_{\beta 5}^{II}$	KM$_4$	8977.49 ± 0.56	[145]			1.794(-5)		
$K_{\beta 5}^{I}$	KM$_5$	8977.14 ± 0.29	[145]			2.593(-5)		
$K_{\beta 2}$	KN$_{2,3}$	8977.00	[104]					
	KN$_1$	8980.22 ± 0.23	[145]					
L series								
	L$_1$M$_1$	975.0 ± 1.8	[145]				6.17	[423]
$L_{\beta_{4,3}}$	L$_1$M$_{2,3}$	1022.8 ± 1.0	[50]	4.45(-3)	[488]			
		1023.00	[104]					
L_{β_4}	L$_1$M$_2$	1019.1 ± 1.6	[145]			6.884(-4)	12.69	[423]
L_{β_3}	L$_1$M$_3$	1022.0 ± 1.6	[145]			1.065(-3)	12.59	[423]

line	transition	E/eV		I/eV/$\hbar$		TPIV	Γ/eV	
$L_{\beta_{10}}$	L_1M_4	1092.7 ± 1.0 1.4	[145]			2.601(-6)		
L_{β_9}	L_1M_5	1093.0 ± 1.0	[145]			3.422(-6)		
L_η	L_2M_1	832.1 ± 1.7 830.80	[50] [104]	2.96(-4)	[488]	4.656(-3)	5.32	[423]
	L_2M_2	875.6 ± 1.3	[145]				4.16	[423]
$L_{\beta_{17}}$	L_2M_3	878.5 ± 1.4	[145]			1.625(-5)		
L_{β_1}	L_2M_4	949.84 ± 0.32 949.70	[50] [104]	5.19(-3)	[488]	4.430(-3)	4.60	[423]
	L_2M_5	949.44 ± 0.77	[145]					
L_{γ_5}	L_2N_1	951.84 ± 0.43	[145]					
L_l	L_3M_1	811.08 ± 0.71 810.60	[50] [104]	3.15(-4)	[488]	4.338(-3)	5.34	[423]
L_t	L_3M_2	855.8 ± 1.3	[145]			8.203(-6)	4.17	[423]
L_s	L_3M_3	858.8 ± 1.3	[145]			7.948(-6)	4.07	[423]
$L_{\alpha_{2,1}}$	$L_3M_{4,5}$	929.68 ± 0.31 930.10	[50] [104]	5.06(-3)	[488]			
L_{α_2}	L_3M_4	929.38 ± 0.72	[145]			3.709(-4)		
L_{α_1}	L_3M_5	929.71 ± 0.74	[145]			3.972(-3)		
L_{β_6}	L_3N_1	932.11 ± 0.39	[145]			1.062(-4)		

M series

	$M_{2,3}M_{4,5}$	72.00 ± 1.30 74.30	[50] [104]

level characteristics

level	E_B/eV		ω_{nlj}		Γ/eV		AE/eV	
K	8979.00 8978.90 ± 0.40	[503] [493]	0.440 0.4678 0.445 0.4538 0.451 0.444 0.439 0.4338 0.440 0.443	[301] [557] [38] [39] [537] [221] [35] [423] [310] [33]	1.55 1.44	[301] [423]	8978.9 8980.476 ± 0.020 8987.96 ± 0.15	[51] [294] [145]

level	E_B/eV		ω_{nlj}		Γ/eV		AE/eV	
			0.425	[430]				
L_1	1096.00	[503]	1.60(-3)	[301]	3.06	[301]	1096.1	[51]
	1096.60 ± 0.40	[493]	1.87(-3)	[423]	9.13	[423]	1103.12 ± 0.59	[145]
			5.38-4)	[436]				
			1.708-3)	[21]				
L_2	951.00	[503]	1.00(-2)	[301]	0.62	[301]	951.0	[51]
	951.00 ± 0.40	[493]	3.57(-3)	[107]	0.60	[423]	952.68 ± 0.11	[50]
			9.22(-3)	[423]			959.58 ± 34	[145]
			9.60(-3)	[436]				
			1.40(-2)	[21]				
L_3	931.00	[503]	1.10(-2)	[301]	0.56	[301]	931.1	[51]
	931.10 ± 0.40	[493]	3.83(-3)	[107]	0.61	[423]	933.04 ± 0.10	[50]
			8.80(-3)	[423]			939.85 ± 0.31	[145]
			1.06(-2)	[436]				
			1.20(-2)	[21]				
M_1	120.00	[503]	4.10(-6)	[38]	6.66	[350]	122.5	[51]
	119.80 ± 0.60	[493]	3.0(-5)	[423]	4.73	[423]		
M_2	74.00	[503]	1.6(-5)	[38]	5.22	[350]		
	73.60 ± 0.40	[51]	2.0(-5)	[512]	3.56	[423]		
	75.30 ± 0.40	[493]	2.0(-5)	[423]				
M_3	74.00	[503]	2.0(-5)	[423]	3.46	[423]		
	73.60 ± 0.40	[51]						
	72.80 ± 0.40	[493]						
M_4	2.00	[503]						
	1.60 ± 0.40	[51]						
	1.80 ± 0.40	[493]						
M_5	2.00	[503]						
	1.60 ± 0.40	[51]						
	1.50 ± 0.40	[493]						
N_1	1.20 ± 2.70	[493]						
IP	7.72638	[222]						

Zn   Z=30   [Ar] $3d^{10} 4s^2$

Zinc $A = 65.37(1)$ [222] $\varrho = 7.115\,\mathrm{g/cm^3}$ [547]
$\varrho = 7.13\,\mathrm{g/cm^3}$ [483]

X-ray transitions

line	transition	E/eV		I/eV/$\hbar$		TPIV	Γ/eV	
K series								
K_{α_3}	KL_1	8465.23 ± 0.68	[145]				11.06	[423]
$K_{\alpha 2}$	KL_2	8615.823 ± 0.073	[50]	0.2254	[488]	1.406(-1)	2.39	[301]
		8615.80	[104]	0.2338	[487]		3.96	[473]
							2.23	[423]
$K_{\alpha 1}$	KL_3	8638.906 ± 0.073	[50]	0.439	[488]	2.738(-1)	2.32	[301]
		8638.80	[104]	0.455	[487]		3.40	[473]
							2.24	[423]
	KM_1	9522.7 ± 1.4	[145]				6.32	[423]
$K_{\beta 3}$	KM_2	9570.8 ± 1.2	[145]	3.19(-2)	[487]	1.739(-2)	5.33	[423]
$K_{\beta 1}$	KM_3	9573.6 ± 1.2	[145]	6.24(-2)	[487]	3.400(-2)	5.18	[423]
$K_{\beta 3,1}$	$KM_{2,3}$	9572.03 ± 0.22	[50]	8.237(-2)	[488]			
		9571.80	[104]					
$K_{\beta 5}$	$KM_{4,5}$	9649.9 ± 1.1	[50]	9.22(-5)	[488]			
		9651.60	[104]					
$K_{\beta 5}^{II}$	KM_4	9650.97 ± 0.59	[145]			2.351(-5)		
$K_{\beta 5}^{I}$	KM_5	9651.31 ± 0.59	[145]			3.399(-5)		
	KN_1	9659.54 ± 0.23	[145]					
$K_{\beta 2}$	$KN_{2,3}$	9658.05 ± 0.22	[50]					
		9658.0	[104]					
L series								
	L_1M_1	1057.5 ± 1.9	[145]				14.26	[423]
$L_{\beta 4,3}$	$L_1M_{2,3}$	1107.0 ± 1.0	[50]	5.38(-3)	[488]			
		1122.40	[104]					
$L_{\beta 4}$	L_1M_2	1105.6 ± 1.7	[145]			7.386(-4)	13.27	[423]
$L_{\beta 3}$	L_1M_3	1108.4 ± 1.6	[145]			1.175(-3)	13.12	[423]
$L_{\beta 10}$	L_1M_4	1185.7 ± 1.0	[145]			2.896(-6)		
$L_{\beta 9}$	L_1M_5	1186.1 ± 1.0	[145]			3.916(-6)		

line	transition	E/eV		I/eV/$\hbar$		TPIV	Γ/eV	
	L_1N_1	1194.31 ± 0.66	[145]					
L_η	L_2M_1	906.3 ± 2.0	[50]	3.49(-4)	[488]	4.612(-3)	5.43	[423]
		905.70	[104]					
	L_2M_2	954.6 ± 1.4	[145]				4.44	[423]
$L_{\beta_{17}}$	L_2M_3	957.4 ± 1.4	[145]			1.808(-5)	4.29	[423]
L_{β_1}	L_2M_4	1034.65 ± 0.38	[50]	6.78(-3)	[488]	5.873(-3)		
		1034.50	[104]					
	L_2M_5	1035.10 ± 0.80	[145]					
L_{γ_5}	L_2N_1	1043.33 ± 0.45	[145]			1.304(-4)		
L_l	L_3M_1	884.3 ± 1.9	[50]	3.73(-4)	[488]	4.350(-3)	5.44	[423]
		894.70	[104]					
L_t	L_3M_2	931.7 ± 1.4	[145]			9.215(-6)	4.45	[423]
L_s	L_3M_3	934.5 ± 1.3	[145]			8.910(-6)	4.30	[423]
$L_{\alpha_{2,1}}$	$L_3M_{4,5}$	1011.77 ± 0.37	[50]	6.61(-3)	[488]			
		1011.90	[104]					
L_{α_2}	L_3M_4	1011.87 ± 0.75	[145]			4.963(-4)		
L_{α_1}	L_3M_5	1012.21 ± 0.75	[145]			5.260(-3)		
L_{β_6}	L_3N_1	1020.44 ± 0.40	[145]			1.133(-4)		

M series

| | $M_{2,3}M4, 5$ | 79.00 ± 1.50 | [50] | | | | | |

level characteristics

level	E_B/eV		ω_{nlj}		Γ/eV		AE/eV	
K	9659.00	[503]	0.474	[301]	1.67	[301]	9658.6	[51]
	9658.60 ± 0.60	[493]	0.5014	[557]	1.56	[423]	9660.755 ± 0.030	[294]
			0.449	[347]	1.62	[94]	9666.55 ± 0.15	[145]
			0.482	[293]				
			0.479	[38]				
			0.4857	[39]				
			0.4659	[423]				
			0.470	[310]				
			0.488	[112]				
L_1	1194.00	[503]	1.80(-3)	[301]	3.28	[301]	1193.6	[51]
	1193.60 ± 0.90	[493]	5.23(-4)	[348]	9.50	[423]	1203.31 ± 0.58	[145]

level	E_B/eV		ω_{nlj}		Γ/eV		AE/eV	
			2.02(-3)	[310]	4.8	[94]		
			5.7(-4)	[436]				
			6.0(-4)	[244]				
L_2	1044.00	[503]	1.10(-2)	[301]	0.72	[301]	1042.8	[51]
	1042.80 ± 0.60	[493]	1.06(-2)	[423]	0.67	[423]	1045.21 ± 13	[50]
			1.04(-2)	[436]	1.06	[94]	1052.33 ± 0.36	[145]
			1.14(-2)	[244]				
L_3	1021.00	[503]	1.20(-2)	[301]	0.65	[301]	1019.7	[51]
	1019.70 ± 0.60	[493]	1-08(-2)	[348]	0.68	[423]	1022.03 ± 0.12	[50]
			1.02(-2)	[423]	0.68	[94]	1029.45 ± 0.31	[145]
			1.01(-2)	[436]				
			1.12(-2)	[244]				
M_1	137.00	[503]	4.60(-6)	[38]	5.90	[350]	139.8	[51]
	135.90 ± 1.10	[493]	1.48(-3)	[512]	4.76	[423]		
			3.0(-5)	[423]	2.1	[94]		
M_2	87.00	[503]	2.20(-5)	[38]	4.70	[350]		
	88.60 ± 0.60	[493]	2.0(-5)		3.77	[423]		
			2.0(-5)	[423]	2.1	[94]		
M_3	87.00	[503]	3.0(-5)	[423]	2.10	[74]		
	85.60 ± 0.60	[493]			3.62	[423]		
					2.15	[94]		
M_4	9.00	[503]						
	8.10 ± 0.60	[51]						
	7.90 ± 0.60	[493]						
M_5	9.00	[503]					10.1	[51]
	8.10 ± 0.60	[51]						
	8.00 ± 0.60	[493]						
N_1	1.30 ± 0.30	[493]						
IP	9.3942	[222]						

Ga   Z=31

[Ar] 3d^{10} 4s^2 4p^1

Gallium

$A = 69.723(1)$ [222] $\varrho = 5.918\,\text{g/cm}^3$ [547]
$\varrho = 5.93\,\text{g/cm}^3$ [483]

X-ray transitions

line	transition	E/eV		I/eV/$\hbar$		TPIV	Γ/eV	
K series								
K$_{\alpha_3}$	KL$_1$	9067.89 ± 0.75	[145]				10.40	[423]
K$_{\alpha 2}$	KL$_2$	9224.82 ± 0.07	[50]	0.2603	[488]	1.498(-1)	2.66	[301]
		9224.835 ± 0.027	[366]					
K$_{\alpha 1}$	KL$_3$	9251.74 ± 0.06	[50]	0.506	[488]	2.912(-1)	2.59	[301]
		9251.70	[104]				2.46	[423]
		9251.674 ± 0.066	[366]					
	KM$_1$	10210.4 ± 1.3	[145]				6.15	[423]
K$_{\beta 3}$	KM$_2$	10260.28 ± 0.64	[50]	3.29(-2)	[488]	1.893(-2)	5.62	[423]
		10260.00	[104]					
K$_{\beta 1}$	KM$_3$	10264.19 ± 0.29	[50]	6.43(-2)	[488]	3.699(-2)	5.41	[423]
		10264.00	[104]					
K$_{\beta 5}$	KM$_{4,5}$	10348.2 ± 2.6	[50]	1.39(-4)	[488]			
		10350.40	[104]					
K$_{\beta 5}^{II}$	KM$_4$	10349.92 ± 0.60	[145]			3.009(-5)	1.71	[423]
K$_{\beta 5}^{I}$	KM$_5$	10350.49 ± 0.61	[145]			4.346(-5)		
	KN$_1$	10365.69 ± 0.34	[145]					
K$_{\beta 2}$	KN$_{2,3}$	10366.42 ± 0.26	[50]	5.46(-4)	[488]			
		10366.00	[104]					
K$_{\beta 2}^{II}$	KN$_2$	10370.50 ± 0.21	[145]			1.080(-4)		
K$_{\beta 2}^{I}$	KN$_3$					2.060(-4)		
L series								
	L$_1$M$_1$	1142.5 ± 1.8	[145]				13.16	[423]
L$_{\beta 3,4}$	L$_1$M$_{2,3}$	1196.9 ± 1.4	[50]	6.48(-3)	[488]			
L$_{\beta 4}$	L$_1$M$_2$	1193.9 ± 1.6	[145]			7.968(-4)	12.63	[423]
L$_{\beta 3}$	L$_1$M$_3$	1197.9 ± 1.7	[145]			1.286(-3)	12.42	[423]
L$_{\beta 10}$	L$_1$M$_4$	1282.0 ± 1.1	[145]			3.251(-6)	12.41	[423]
L$_{\beta 9}$	L$_1$M$_5$	1282.6 ± 1.1	[145]			4.519(-6)		

line	transition	E/eV		I/eV/$\hbar$		TPIV	Γ/eV	
	L_1N_1	1297.80 ± 0.83	[145]					
L_{γ_2}	L_1N_2	1302.61 ± 0.70	[145]			4.924(-6)		
L_η	L_2M_1	984.22 ± 0.23	[50]	4.10(-4)	[488]	4.433(-3)	5.21	[423]
		984.20	[104]					
	L_2M_2	1036.7 ± 1.3	[145]				4.68	[423]
$L_{\beta_{17}}$	L_2M_3	1040.7 ± 1.4	[145]			1.963(-5)	4.46	[423]
L_{β_1}	L_2M_4	1124.76 ± 0.30	[50]	8.68(-3)	[488]	7.534(-3)	0.77	[423]
		1124.30	[104]					
	L_2M_5	1125.39 ± 0.80	[145]					
L_{γ_5}	L_2N_1	1140.59 ± 0.54	[145]			1.376(-4)		
	L_2N_2	1145.40 ± 0.40	[145]					
L_l	L_3M_1	957.17 ± 0.22	[50]	4.41(-4)	[488]	4.301(-3)	5.22	[423]
		957.40	[104]					
L_t	L_3M_2	1009.9 ± 1.3	[145]			1.008(-5)	4.68	[423]
L_s	L_3M_3	1013.9 ± 1.3	[145]			9.725(-6)	4.47	[423]
$L_{\alpha_{2,1}}$	$L_3M_{4,5}$	1097.97 ± 0.14	[50]	8.45(-3)	[488]			
		1097.90	[104]					
L_{α_2}	L_3M_4	1098.03 ± 0.75	[145]			6.459(-4)	0.77	[423]
L_{α_1}	L_3M_5	1098.59 ± 0.76	[145]			6.713(-3)		
L_{β_6}	L_3N_1	1113.79 ± 0.49	[145]			1.195(-4)		
	L_3N_2	1118.61 ± 0.36	[145]			6.635(-9)		

level characteristics

level	E_B/eV		ω_{nlj}		Γ/eV		AE/eV	
K	10367.00	[503]	0.507	[301]	1.82	[301]	10367.1	[51]
	10367.10 ± 0.50	[493]	0.514	[293]	1.70	[423]	10368.1 ± 1.3	[50]
			0.5338	[557]	1.76	[94]	10377.76 ± 0.16	[145]
			0.510	[38]				
			0.5166	[39]				
			0.528	[181]				
			0.529	[411]				
			0.4973	[423]				
			0.507	[310]				
			0.5338	[557]				

level	E_B/eV		ω_{nlj}		Γ/eV		AE/eV	
L_1	1298.00	[503]	2.10(-3)	[301]	3.38	[301]	1297.7	[51]
	1297.70 ± 1.10	[493]	2.20(-3)	[423]	8.71	[423]	1302.7 ± 1.0	[50]
			8.90(-4)	[436]	4.1	[94]	1309.87 ± 0.65	[145]
L_2	1143.00	[503]	1.20(-2)	[301]	0.38	[301]	1142.3	[51]
	1142.30 ± 0.50	[493]	1.21(-2)	[423]	0.76	[423]	1145.02 ± 0.78	[50]
			1.22(-2)	[436]	0.77	[94]	1152.66 ± 0.35	[145]
L_3	1116.00	[503]	1.30(-2)	[301]	0.76	[301]	1115.4	[51]
	1115.40 ± 0.50	[493]	1.18(-2)	[423]	0.76	[423]	1116.96 ± 0.15	[50]
			1.18(-2)	[436]	0.77	[94]	1125.86 ± 0.31	[145]
M_1	158.00	[503]	1.588(-3)	[512]	2	[94]	159.5	[51]
	158.10 ± 0.70	[493]	4.0(-5)	[423]	4.46	[423]		
M_2	107.00	[503]	2.0(-5)	[512]	2.25	[94]		
	106.80 ± 0.70	[493]	5.0(-5)	[423]	3.93	[423]		
M_3	103.00	[503]	4.0(-5)	[423]	2.3	[94]		
	102.90 ± 0.50	[493]			3.72	[423]		
M_4	18.00	[503]			0.012	[94]		
	17.40 ± 0.50	[51]			0.01	[423]		
	20.70 ± 0.50	[493]						
M_5	18.00	[503]					18.7	[51]
	17.40 ± 0.50	[51]						
	15.70 ± 0.50	[493]						
N_1	5.60 ± 3.40	[493]						
N_2	1.00	[503]						
	0.80 ± 0.50	[493]						
IP	5.99930	[222]						

Ge   Z=32

[Ar] $3d^{10}$ $4s^2$ $4p^2$

Germanium $A = 72.61(2)$ [222] $\varrho = 5.308\,\text{g/cm}^3$ [547]
$\varrho = 5.32\,\text{g/cm}^3$ [483]

X-ray transitions

line	transition	E/eV		I/eV/$\hbar$		TPIV	Γ/eV	
K series								
K_{α_3}	KL_1	9692.08 ± 0.84	[145]				10.56	[423]
K_{α_2}	KL_2	9855.42 ± 0.10	[50]	0.299	[488]	1.585(-1)	2.92	[301]
		9855.20	[104]	0.309	[487]		4.18	[473]
							2.69	[423]
K_{α_1}	KL_3	9886.52 ± 0.11	[50]	0.580	[488]	3.077(-1)	2.78	[301]
		9886.30	[104]	0.600	[487]		3.75	[473]
							2.69	[423]
	KM_1	10924.0 ± 1.3	[145]				5.98	[423]
K_{β_3}	KM_2	10978.1 ± 1.3	[50]	3.86(-2)	[488]	2.049(-2)	5.84	[423]
		10978.00	[104]	4.40(-2)	[487]			
K_{β_1}	KM_3	10982.19 ± 0.29	[50]	7.55(-2)	[488]	4.004(-2)	5.56	[423]
		10982.00	[104]	8.62(-2)	[487]			
K_{β_5}	$KM_{4,5}$	11074.8 ± 1.5	[50]	1.73(-4)	[488]			
		11075.50	[104]					
$K_{\beta_5}^{II}$	KM_4	11074.59 ± 0.63	[145]			3.763(-5)	1.89	[423]
$K_{\beta_5}^{I}$	KM_5	11075.33 ± 0.63	[145]			5.428(-5)	1.89	[423]
	KN_1	11098.72 ± 0.52	[145]					
$K_{\beta_2^{II}}$	KN_2	11105.84 ± 0.37	[145]	9.70(-4)	[487]	3.342(-4)		
$K_{\beta_2^{I}}$	KN_3			1.87(-3)	[487]	6.409(-4)		
K_{β_2}	$KN_{2,3}$	11100.97 ± 0.29	[50]	1.84(-3)	[488]			
		11101.00	[104]					
L series								
	L_1M_1	1231.9 ± 1.9	[145]				12.83	[423]
L_{β_4}	L_1M_2	1286.12 ± 0.39	[50]	2.76(-3)	[488]	8.459(-4)	12.70	[423]
		1286.10	[104]					
L_{β_3}	L_1M_3	1294.04 ± 0.40	[50]	5.00(-3)	[488]	1.423(-3)	12.42	[423]
		1294.10	[104]					
$L_{\beta_{10}}$	L_1M_4	1382.5 ± 1.2	[145]			3.684(-6)	8.75	[423]

line	transition	E/eV		I/eV/ℏ		TPIV	Γ/eV	
L_{β_9}	L_1M_5	1383.2 ± 1.2	[145]			5.305(-6)	8.74	[423]
	L_1N_1	1406.6 ± 1.1	[145]					
L_{γ_2}	L_1N_2	1413.76 ± 0.93	[145]			1.483(-5)		
L_η	L_2M_1	1067.98 ± 0.27	[50]	4.81(-4)	[488]	4.214(-3)	4.98	[423]
	L_2M_2 1123.7 ± 1.5	1218.60	[104]				4.83	[423]
		1068.00	[104]					
$L_{\beta_{17}}$	L_2M_3	1128.4 ± 1.5	[145]			2.072(-5)	4.55	[423]
	L_2M_2	1123.7 ± 1.5	[145]					
L_{β_1}	L_2M_4	1218.50 ± 0.18	[50]	1.09(-2)	[488]	9.332(-3)	0.89	[423]
		1218.60	[104]					
	L_2M_5	1219.65 ± 0.83	[145]				0.88	[423]
L_{γ_5}	L_2N_1	1242.60	[104]			1.444(-4)		
		1243.05 ± 0.72	[145]					
	L_2N_2	1250.17 ± 0.56	[145]					
L_l	L_3M_1	1036.21 ± 0.51	[50]	5.20(-4)	[488]	4.220(-3)	4.97	[423]
		1037.80	[104]					
L_t	L_3M_2	1092.7 ± 1.4	[145]			1.091(-5)	4.83	[423]
L_s	L_3M_3	1097.4 ± 1.4	[145]					
$L_{\alpha_{2,1}}$	$L_3M_{4,5}$	1188.01 ± 0.13	[50]	1.06(-2)	[488]			
		1188.20	[104]					
L_{α_1}	L_3M_4	1187.91 ± 0.78	[145]			8.057(-4)	0.88	[423]
L_{α_1}	L_3M_5	1188.66 ± 0.13	[145]			8.319(-3)	0.84	[423]
L_{β_6}	L_3N_1	1193.70 ± 2.00	[50]			1.249(-4)		
		1193.70	[104]					
	L_3N_2	1219.17 ± 0.51	[145]			2.107(-8)		

level characteristics

level	E_B/eV		ω_{nlj}		Γ/eV		AE/eV	
K	11104.00	[503]	0.535	[301]	1.96	[301]	11103.1	[51]
	11103.10 ± 0.70	[493]	0.565	[557]	1.85	[423]	11103.76 ± 0.74	[50]
			0.558	[347]	1.92	[94]	11113.82 ± 0.16	[145]
			0.545	[293]				
			0.540	[38]				
			0.5464	[39]				
			0.554	[410]				
			0.5278	[423]				
			0.535	[310]				
			0.538	[430]				

level	E_B/eV		ω_{nlj}		Γ/eV		AE/eV	
L_1	1413.00	[503]	2.40(-3)	[301]	3.53	[301]	1414.3	[51]
	1414.30 ± 0.70	[493]	7.70(-4)	[348]	8.71	[423]	1413.23 ± 0.24	[50]
			2.41(-3)	[423]	3.8	[94]	1421.74 ± 0.73	[145]
			1.05(-3)	[436]				
			2.70(-3)	[21]				
L_2	1249.00	[503]	1.60(-2)	[301]	0.95	[301]	1247.8	[51]
	1247.80 ± 0.70	[493]	7.72(-3)	[107]	0.84	[423]	1249.32 ± 0.19	[50]
			1.37(-2)	[423]	0.86	[94]	1258.15 ± 0.35	[145]
			1.42(-2)	[436]				
			1.10(-2)	[21]				
L_3	1217.00	[503]	1.50(-2)	[301]	0.82	[301]	1216.7	[51]
	1216.70 ± 0.70	[493]	1.44(-2)	[348]	0.84	[423]	1217.06 ± 0.18	[50]
			1.35(-2)	[423]	0.86	[94]	1227.15 ± 0.31	[145]
			1.36(-2)	[436]				
			1.20(-2)	[21]				
M_1	181.00	[503]	9.10(-6)	[38]	4.59	[350]	180.1	[51]
	180.00 ± 0.80	[493]	5.0(-5)	[423]	4.13	[423]		
					2.1	[94]		
M_2	129.00	[503]	2.60(-5)	[38]	5.22	[350]		
	127.90 ± 0.90	[493]	2.0(-5)	[512]	3.99	[423]		
			3.0(-5)	[423]	2.3	[94]		
M_3	122.00	[503]	6.0(-5)	873]	2.00	[74]		
	120.80 ± 0.70	[493]			3.71	[423]		
					2.3	[94]		
M_4	29.00	[503]	5.30(-3)	[26,T]	0.05	[74]		
	28.70 ± 0.70	[51]	2.7(-3)	[512]	0.05	[423]		
	29.20	[493]			0.045	[94]		
M_5	29.00	[503]	2.70(-3)	[26,T]	0.05	[74]	29.2	[51]
	28.70 ± 0.70	[51]			0.04	[423]		
	28.50	[493]			0.044	[94]		
N_1	9.00	[493]						
N_2	3.00	[503]						
	2.30	[493]						
IP	7.8994	[222]						

As   Z=33

$[Ar] 3d^{10} 4s^2 4p^3$

Arsenic $A = 74.92160(2)$ [12.77] $\varrho = 5.72 \text{ g/cm}^3$ [547]
$\varrho = 5.73 \text{ g/cm}^3$ [483]

X-ray transitions

line	transition	E/eV		I/eV/ℏ		TPIV	Γ/eV	
K series								
K_{α_3}	KL_1	10336.70 ± 0.82	[145]				9.86	[423]
K_{α_2}	KL_2	10508.00 ± 0.09	[50]	0.342	[488]	1.669(-1)	3.17	[301]
		10508.00	[104]	0.353	[487]		2.95	[423]
		10507.50 ± 0.15	[366]					
K_{α_1}	KL_3	10543.72 ± 0.09	[50]	0.663	[488]	3.232(-1)	3.08	[301]
		10544.00	[104]	0.685	[487]		2.93	[423]
		$10543.2674 \pm 0.0.0081$	[366]					
	KM_1	11663.0 ± 1.3	[145]				5.97	[423]
K_{β_3}	KM_2	11719.86 ± 0.84	[50]	4.52(-2)	[488]	2.206(-2)	5.96	[423]
		11720.00	[104]	5.14(-2)	[487]			
K_{β_1}	KM_3	11725.73 ± 0.37	[50]	8.82(-2)	[488]	4.309(-2)	5.64	[423]
		11726.00	[104]	0.1005	[487]			
K_{β_5}	$KM_{4,5}$	11821.4 ± 1.7	[50]	2.31(-2)	[488]			
		11822.00	[104]					
$K_{\beta_5}^{II}$	KM_4	11825.05 ± 0.64	[145]			4.614(-5)	2.07	[423]
$K_{\beta_5}^{I}$	KM_5	11825.89 ± 0.65	[145]			6.645(-5)	2.07	[423]
	KN_1	11858.67 ± 0.77	[145]					
$K_{\beta_2}^{II}$	KN_2	11867.94 ± 0.64	[145]	1.93(-3)	[487]	6.722(-4)		
$K_{\beta_2}^{I}$	KN_3	11866.38 ± 0.23	[145]	3.70(-3)	[487]	1.292(-3)		
K_{β_2}	$KN_{2,3}$	11864.34 ± 0.50	[50]	4.03(-3)	[488]			
		11867.00	[104]					
L series								
	L_1M_1	1326.3 ± 1.8	[145]				11.80	[423]
$L_{\beta_{4,3}}$	$L_1M_{2,3}$	1388.54 ± 0.23	[50]	9.25(-3)	[488]			
		1388.30	[104]					
L_{β_4}	L_1M_2	1384.54 ± 0.23	[145]			9.041(-4)	11.80	[423]
L_{β_3}	L_1M_3	1390.4 ± 0.17	[145]			1.552(-3)	11.48	[423]
$L_{\beta_{10}}$	L_1M_4	1488.3 ± 1.2	[145]			4.196(-6)	7.90	[423]
L_{β_9}	L_1M_5	1489.2 ± 1.2	[145]			6.170(-6)	7.90	[423]

line	transition	E/eV		I/eV/$\hbar$		TPIV	Γ/eV	
	L_1N_1	1522.0 ± 1.3	[145]					
L_{γ_2}	L_1N_2	1531.2 ± 1.2	[145]			2.924(-5)		
L_{γ_3}	L_1N_3	1529.68 ± 0.76	[145]			5.227(-5)		
L_η	L_2M_1	1155.04 ± 0.15	[50]	5.61(-4)	[488]	3.972(-3)	4.89	[423]
		1155.10	[104]					
	L_2M_2	1213.8 ± 1.4	[145]				4.88	[423]
$L_{\beta_{17}}$	L_2M_3	1219.3 ± 1.4	[145]			2.161(-5)	4.56	[423]
L_{β_1}	L_2M_4	1316.99 ± 0.17	[50]	1.348(-2)	[488]	1.124(-2)	0.99	[423]
		1316.90	[104]					
	L_2M_5	1318.06 ± 0.82	[145]				0.99	[423]
L_{γ_5}	L_2N_1	1350.84 ± 0.95	[145]					
	L_2N_2	1360.10 ± 0.82	[145]					
	L_2N_3	1358.54 ± 0.41	[145]			9.455(-8)		
L_l	L_3M_1	1119.78 ± 0.15	[50]	6.10(-4)	[488]	4.164(-3)	4.88	[423]
		1120.00	[104]					
L_t	L_3M_2	1178.1 ± 1.3	[145]			1.162(-5)	4.87	[423]
L_s	L_3M_3	1183.7 ± 1.3	[145]			1.117(-5)	4.55	[423]
$L_{\alpha_{2,1}}$	$L_3M_{4,5}$	1282.01 ± 0.16	[50]	1.31(-2)	[488]			
		1281.90	[104]					
L_{α_2}	L_3M_4	1281.56 ± 0.16	[145]			9.902(-4)	0.98	[423]
L_{α_1}	L_3M_5	1282.41 ± 0.79	[145]			1.001(-2)	0.98	[423]
L_{β_6}	L_3N_1	1315.19 ± 0.91	[145]					
	L_3N_2	1324.46 ± 0.78	[145]			4.399(-8)		
	L_3N_3	1322.90 ± 0.37	[145]			4.172(-8)		

level characteristics

level	E_B/eV		ω_{nlj}		Γ/eV		AE/eV	
K	11867.00	[503]	0.562	[301]	2.14	[301]	11866.7	[51]
	11866.70 ± 0.70	[493]	0.574	[293]	2.01	[423]	11864.3 ± 1.7	[50]
			0.594	[557]	2.09	[94]	11876.74 ± 0.18	[145]
			0.567	[38]				
			0.5748	[39]				
			0.5574	[423]				
			0.562	[310]				

level	E_B/eV		ω_{nlj}		Γ/eV		AE/eV	
			0.594	[557]				
L_1	1527.00	[503]	2.80(-3)	[301]	3.79	[301]	1526.5	[51]
	1526.50 ± 0.80	[493]	1.40(-3)	[107]	7.85	[423]	1529.32 ± 0.28	[50]
			2.64(-3)	[423]	3.8	[94]	1540.04 ± 0.70	[145]
			1.23(-3)	[436]				
L_2	1359.00	[503]	1.40(-2)	[301]	1.03	[301]	1358.6	[1369
	1358.60 ± 0.70	[493]	8.85(-3)	[107]	0.93	[423]	1358.71 ± 0.22	[50]
			1.54(-2)	[423]	0.95	[94]	1368.90 ± 0.35	[145]
			1.62(-2)	[436]				
L_3	1323.00	[503]	1.60(-2)	[301]	0.94	[301]	1323.1	[51]
	1323.10 ± 0.70	[493]	9.74(-3)	[107]	0.92	[423]	1323.61 ± 0.21	[50]
			1.53(-2)	[423]	0.94	[94]	1333.26 ± 0.31	[145]
			1.55(-2)	[436]				
M_1	204.00	[503]	1.818(-3)	[512]	2.4	[94]	204.7	[342]
	203.50 ± 0.70	[493]	6.0(-5)	[423]	3.96	[423]		
M_2	147.00	[503]	2.0(-5)	[512]	2.25	[94]		
	146.40 ± 1.20	[493]	9.0(-5)	[423]	3.95	[423]		
M_3	141.00	[503]	8.0(-5)	[423]	2.25	[94]		
	140.50 ± 0.80	[493]			3.63	[423]		
M_4	41.00	[503]	2.7(-3)	[512]	0.058	[94]		
	41.20 ± 0.70	[51]	1.0(-6)	[423]	0.06	[423]		
	41.70 ± 0.70	[493]						
M_5	41.00	[503]			0.06	[94]	41.7	[51]
	40.90 ± 0.70	[493]			0.06	[423]		
N_1	12.50 ± 3.90	[493]						
N_2	3.00	[503]						
	2.50 ± 1.00	[493]						
N_3	3.00	[503]						
	2.50 ± 1.00	[493]						
IP	9.7886	[222]						

Se   Z=34   [Ar] 3d^{10} 4s^2 4p^4

Selenium $A = 78.96(3)$ [12.77] $\varrho = 4.82$ g/cm^3 [547]

X-ray transitions

line	transition	E/eV		I/eV/$\hbar$		TPIV	Γ/eV	
K series								
K$_{\alpha_3}$	KL$_1$	11003.49 $\pm$ 0.84	[145]				10.23	[423]
K$_{\alpha 2}$	KL$_2$	11184.0 $\pm$ 0.20	[50]	0.389	[488]	1.747(-1)	3.46	[301]
		11182.00	[104]	0.401	[487]		4.43	[473]
		11181.53 $\pm$ 0.31	[366]				3.22	[423]
K$_{\alpha 1}$	KL$_3$	11222.40 $\pm$ 0.20	[50]	0.753	[488]	3.379(-1)	3.33	[301]
		11222.40	[104]	0.777	[487]		4.10	[473]
		11222.52 $\pm$ 0.12		[366]			3.20	[423]
	KM$_1$	12427.9 $\pm$ 1.3	[145]				6.43	[423]
K$_{\beta 3}$	KM$_2$	12489.7 $\pm$ 1.0	[50]	5.26(-2)	[488]	2.360(-2)	6.06	[423]
		12490.00	[104]	5.98(-2)	[487]			
K$_{\beta 1}$	KM$_3$	12496.03 $\pm$ 0.67	[50]	0.1027	[488]	4.605(-2)	5.70	[423]
		12496.00	[104]	0.1169	[487]			
K$_{\beta 5}$	KM$_{4,5}$	12596.0 $\pm$ 1.9	[50]	3.02(-4)	[488]			
		12597.00	[104]					
K$_{\beta 5}^{II}$	KM$_4$	12601.47 $\pm$ 0.68	[145]			5.560(-5)	2.26	[423]
K$_{\beta 5}^{I}$	KM$_5$	12602.42 $\pm$ 0.67				7.997(-5)	2.26	[423]
	KN$_1$	12645.7 $\pm$ 1.1	[145]					
K$_{\beta 2}^{II}$	KN$_2$	12656.38 $\pm$ 0.73	[145]	3.30(-2)	[487]	1.119(-3)		
K$_{\beta 2}^{I}$	KN$_3$	12655.11 $\pm$ 0.33	[145]	6.40(-3)	[487]	2.154(-3)		
K$_{\beta 2}$	KN$_{2,3}$	12652.29 $\pm$ 0.96	[50]	7.29(-3)	[488]			
		12653.00	[104]					
L series								
	L$_1$M$_1$	1424.4 $\pm$ 1.8	[145]				12.27	[423]
L$_{\beta 3,4}$	L$_1$M$_{2,3}$	1490.0 $\pm$ 2.4	[50]	1.095(-2)	[488]			
L$_{\beta 4}$	L$_1$M$_2$	1486.8 $\pm$ 1.8	[145]			9.535(-4)	11.90	[423]
L$_{\beta 3}$	L$_1$M$_3$	1493.0 $\pm$ 1.8	[145]			1.688(-3)	11.55	[423]
L$_{\beta 10}$	L$_1$M$_4$	1598.0 $\pm$ 1.2	[145]			4.788(-6)	8.10	[423]
L$_{\beta 9}$	L$_1$M$_5$	1598.9 $\pm$ 1.2	[145]			7.120(-6)	8.11	[423]

line	transition	E/eV		I/eV/$\hbar$		TPIV	Γ/eV	
	L_1N_1	1642.2 ± 1.6	[145]					
L_{γ_2}	L_1N_2	1652.9 ± 1.3	[145]			4.808(-5)		
L_{γ_3}	L_1N_3	1651.62 ± 0.88	[145]			8.573(-5)		
L_η	L_2M_1	1244.55 ± 0.18	[50]	6.52(-4)	[488]	3.747(-3)	5.26	[423]
		1244.80	[104]					
	L_2M_2	1308.4 ± 1.4	[145]			9.535(-4)	4.89	[423]
$L_{\beta_{17}}$	L_2M_3	1314.7 ± 1.4	[145]			2.230(-5)	4.54	[423]
L_{β_1}	L_2M_4	1419.24 ± 0.12	[50]	1.64(-2)	[488]	1.324(-2)	1.09	[423]
		1419.20	[104]					
	L_2M_5	1420.60 ± 0.84	[145]				1.09	[423]
L_{γ_5}	L_2N_1	1463.9 ± 1.3	[145]			1.553(-4)		
	L_2N_2	1474.55 ± 0.89	[145]					
	L_3N_3	1473.29 ± 0.50	[145]			1.620(-7)		
L_l	L_3M_1	1204.41 ± 0.17	[50]	7.13(-4)	[488]	4.089(-3)	5.23	[423]
		1204.50	[104]					
L_t	L_3M_2	1267.7 ± 1.3	[145]			1.224(-5)	4.87	[423]
L_s	L_3M_3	1273.9 ± 1.3	[145]			1.173(-5)	4.52	[423]
$L_{\alpha_{2,1}}$	$L_3M_{4,5}$	1379.11 ± 0.11	[50]	1.60(-2)	[488]			
		1379.00	[104]					
L_{α_2}	L_3M_4	1378.92 ± 0.80	[145]			1.183(-3)	1.07	[423]
L_{α_1}	L_3M_5	1379.88 ± 0.80	[145]			1.185(-2)	1.07	[423]
L_{β_6}	L_3N_1	1423.1 ± 1.2	[145]			1.362(-4)		
	L_3N_2	1433.83 ± 0.86	[145]			7.606(-8)		
	L_3N_3	1432.57 ± 0.46	[145]			7.222(-8)		
M series								
M_{ξ_1}	M_5N_3	53.80 ± 0.50	[50]	2.16(-6)	[335]	1.325(-6)		

level characteristics

level	E_B/eV		ω_{nlj}		Γ/eV		AE/eV	
K	12658.00	[503]	0.589	[301]	2.33	[301]	12657.8	[51]
	12657.80 ± 0.70	[493]	0.602	[293]	2.20	[423]	12654.61 ± 0.19	[50]
			0.623	[557]	2.28	[94]	12666.72 ± 0.19	[145]
			0.596	[38]				
			0.6019	[39]				
			0.5857	[423]				
			0.589	[310]				
			0.623	[557]				

level	E_B/eV		ω_{nlj}		Γ/eV		AE/eV	
L_1	1654.00	[503]	3.20(-3)	[301]	3.94	[301]	1653.9	[51]
	1653.90 ± 3.50	[493]	1.30(-3)	[348]	8.04	[423]	1652.44 ± 0.33	[50]
			2.88(-3)	[423]	3.8	[94]	1663.23 ± 0.73	[145]
			1.45(-3)	[436]				
L_2	1476.00	[503]	1.60(-2)	[301]	1.13	[301]	1476.2	[162]
	1476.20 ± 0.70	[493]	9.94(-3)	[107]	1.03	[423]	1474.72 ± 0.26	[50]
			1.72(-2)	[423]	1.05	[94]	1484.90 ± 0.31	[145]
			1.84(-2)	[436]				
L_3	1436.00	[503]	1.80(-2)	[301]	1.00	[301]	1435.8	[51]
	1435.80 ± 0.70	[493]	1.78(-2)	[348]	1.00	[423]	1433.98 ± 0.25	[50]
			1.73(-2)	[423]	1.02	[94]	1444.18 ± 0.31	[145]
			1.75(-2)	[436]				
M_1	232.00	[503]	1.94(-3)	[512]	4.30	[74]	229.6	[51]
	231.50 ± 0.70	[493]	7.0(-5)	[423]	4.23	[423]		
					2.8	[94]		
M_2	168.00	[503]	2.0(-5)	[512]	2.00	[74]		
	168.20 ± 1.30	[493]	1.2(-4)	[423]	3.86	[423]		
					2.2	[94]		
M_3	162.00	[503]	1.1(-4)	[423]	2.00	[74]		
	161.90 ± 1.00	[493]			3.51	[423]		
					2.2	[94]		
M_4	57.00	[503]	2.7(-3)	[512]	0.06	[74]		
	56.70 ± 0.80	[51]	2.0(-6)	[423]	0.06	[423]		
	57.40 ± 0.80	[493]			0.065	[94]		
M_5	57.00	[503]	1.0(-6)	[423]	0.06	[74]	54.6	[51]
	56.70 ± 0.80	[51]			0.07	[423]		
	56.40 ± 0.80	[493]			0.066	[94]		
N_1	16.20 ± 4.10	[493]						
N_2	6.00	[503]						
	5.60 ± 1.30	[493]						
N_3	6.00	[503]						
	5.60	[51]						
	5.60 ± 1.30	[493]						
IP	9.75238	[222]						

Br   Z=35   [Ar] 3d^{10} 4s^2 4p^5

Krypton $A = 79.904(1)$ [12.77] $\varrho = 3.11\,\text{g/cm}^3$ [547]

X-ray transitions

line	transition	E/eV		I/eV/$\hbar$		TPIV	Γ/eV	
K series								
K$_{\alpha_3}$	KL$_1$	11692.04 ± 0.88	[145]				10.63	[423]
K$_{\alpha 2}$	KL$_2$	11877.75 ± 0.34	[50]	0.442	[488]	1.821(-1)	3.73	[301]
		11878.00	[104]	0.454	[487]		3.52	[423]
K$_{\alpha 1}$	KL$_3$	11924.36 ± 0.34	[50]	0.853	[488]	3.515(-1)	3.60	[301]
		11924.00	[104]	0.876	[487]		3.48	[423]
	KM$_1$	13218.6 ± 1.3	[145]				6.76	[423]
K$_{\beta 3}$	KM$_2$	13284.7 ± 1.1	[50]	6.09(-2)	[488]	2.511(-2)	6.14	[423]
		13285.00	[104]	6.87(-2)	[487]			
K$_{\beta 1}$	KM$_3$	13291.56 ± 0.42	[50]	0.1188	[488]	4.900(-2)	5.74	[423]
		13292.00	[104]	0.1342	[487]			
K$_{\beta 5}$	KM$_{4,5}$	13396.3 ± 2.1	[50]	2.30(-4)	[488]			
		13400.30	[104]					
K$_{\beta_5}^{II}$	KM$_4$	13403.97 ± 0.70	[145]			6.604(-5)	2.46	[423]
K$_{\beta_5}^{I}$	KM$_5$	13405.04 ± 0.70				9.482(-5)	2.47	[423]
	KN$_1$	13459.8 ± 1.5	[145]					
K$_{\beta_2^{II}}$	KN$_2$	13471.50 ± 0.70	[145]	5.20(-3)	[487]	1.674(-3)		
K$_{\beta_2^{I}}$	KN$_3$	13470.84 ± 0.41	[145]	1.01(-2)	[487]	3.222(-3)		
K$_{\beta 2}$	KN$_{2,3}$	13469.60 ± 0.43	[50]	7.82(-3)	[488]			
		13464.50	[104]					
L series								
	L$_1$M$_1$	1526.5 ± 1.8	[145]				12.59	[423]
L$_{\beta 4,3}$	L$_1$M$_{2,3}$	1596.3 ± 2.7	[50]	4.62(-3)	[488]			
L$_{\beta 4}$	L$_1$M$_2$	1592.8 ± 1.8	[145]			1.026(-3)	11.98	[423]
L$_{\beta 3}$	L$_1$M$_3$	1600.0 ± 1.8	[145]			1.834(-3)	11.58	[423]
L$_{\beta 10}$	L$_1$M$_4$	1711.9 ± 1.3	[145]			5.591(-6)	8.30	[423]
L$_{\beta 9}$	L$_1$M$_5$	1713.0 ± 1.3	[145]			8.316(-6)	8.30	[423]
	L$_1$N$_1$	1767.8 ± 2.1	[145]					

line	transition	E/eV		I/eV/$\hbar$		TPIV	Γ/eV	
L_{γ_2}	L_1N_2	1779.5 ± 1.3	[145]			7.286(-5)		
L_{γ_3}	L_1N_3	1778.79 ± 0.97	[145]			1.294(-4)		
L_η	L_2M_1	1339.63 ± 0.21	[50]	7.54(-4)	[488]	3.554(-3)	5.48	[423]
		1339.70	[104]					
	L_2M_2	1407.2 ± 1.4	[145]				4.87	[423]
$L_{\beta_{17}}$	L_2M_3	1414.4 ± 1.4	[145]			2.287(-5)	4.46	[423]
L_{β_1}	L_2M_4	1525.92 ± 0.14	[50]	1.99(-2)	[488]	1.535(-2)	1.19	[423]
		1525.80	[104]					
	L_2M_5	1527.35 ± 0.85	[145]				1.19	[423]
L_{γ_5}	L_2N_1	1582.1 ± 1.7	[145]			1.611(-4)		
	L_2N_2	1593.81 ± 0.87	[145]					
	L_2N_3	1593.15 ± 0.56	[145]			2.520(-7)		
L_l	L_3M_1	1293.50 ± 0.20	[50]	8.30(-4)	[488]	4.074(-3)	5.44	[423]
		1293.70	[104]					
L_t	L_3M_2	1360.8 ± 1.4	[145]			1.305(-5)	4.83	[423]
L_s	L_3M_3	1368.1 ± 1.3	[145]			1.248(-5)	4.42	[423]
$L_{\alpha_{2,1}}$	$L_3M_{4,5}$	1480.46 ± 0.13	[50]	1.732(-2)	[488]			
		1480.30	[104]					
L_{α_2}	L_3M_4	1479.95 ± 0.82	[145]			1.403(-3)	1.15	[423]
L_{α_1}	L_3M_5	1481.02 ± 0.81	[145]			1.374(-2)	1.16	[423]
L_{β_6}	L_3N_1	1535.8 ± 0.6	[145]			1.436(-4)		
	L_3N_2	1547.49 ± 0.83	[145]			1.203(-7)		
	L_3N_3	1546.82 ± 0.52	[145]			1.142(-7)		

M series

	M_1M_2	67.20 ± 0.10	[50]			1.286(-5)	8.10	[423]
		79.40	[104]					
	M_1M_3	75.30 ± 0.14	[50]			3.420(-5)	7.70	[423]
		85.90	[104]					
	M_2M_4	113.30 ± 0.30	[50]			1.330(-4)	3.81	[423]
		113.30	[104]					
	M_2N_1	161.30 ± 0.40	[50]	2.46(-5)	[335]	2.438(-5)		
		161.31	[104]					
	$M_3M_{4,5}$	108.90 ± 0.30	[50]					
		109.00	[104]					
	M_3M_4					1.140(-5)	3.41	[423]
	M_3M_5					1.069(-4)	3.41	[423]
	M_3N_1	155.45 ± 0.60	[50]	2.46(-5)	[335]	2.232(-5)		

line	transition	E/eV		I/eV/$\hbar$		TPIV	Γ/eV
		155.40	[104]				
M$_{\xi_2}$	M$_4$N$_2$	64.88 $\pm$ 0.07	[50]	4.06(-6)	[335]	3.362(-6)	
		64.90	[104]				
M$_\delta$	M$_4$N$_3$	65.40 $\pm$ 0.10	[50]	4.06(-6)	[335]	6.249(-7)	
		66.31	[104]				
M$_{\xi_1}$	M$_5$N$_3$	64.37 $\pm$ 0.07	[50]	4.06(-6)	[335]	3.299(-6)	
		64.38	[104]				

level characteristics

level	E_B/eV		ω_{nlj}		Γ/eV		AE/eV	
K	13474.00	[503]	0.618	[301]	2.52	[301]	13473.7	[51]
	13473.70 $\pm$ 0.40	[493]	0.629	[293]	2.40	[423]	13470.5 $\pm$ 2.2	[50]
			0.6493	[557]	2.49	[94]	13483.86 $\pm$ 0.19	[145]
			0.622	[38]				
			0.6275	[39]				
			0.6128	[423]				
			0.615	[310]				
			0.627	[112]				
L$_1$	1782.00	[503]	3.60(-3)	[301]	4.11	[301]	1782.0	[51]
	1782.00 $\pm$ 0.40	[493]	3.17(-3)	[423]	8.24	[423]	1781.6 $\pm$ 1.9	[50]
			1.65(-3)	[436]	3.8	[94]	1791.81 $\pm$ 0.76	[145]
			1.73(-3)	[244]				
L$_2$	1596.00	[503]	1.80(-2)	[301]	1.21	[301]	1596.0	[51]
	1596.00 $\pm$ 0.40	[493]	1.09(-2)	[107]	1.13	[423]	1599.2 $\pm$ 0.5	[50]
			1.91(-2)	[423]	1.14	[94]	1606.17 $\pm$ 35	[145]
			2.06(-2)	[436]				
L$_3$	1550.00	[503]	2.00(-2)	[301]	1.08	[301]	1549.9	[51]
	1549.90 $\pm$ 0.40	[493]	1.94(-2)	[423]	1.09	[423]	1552.9 $\pm$ 1.4	[50]
			1.94(-2)	[436]	1.1	[94]	1559.84 $\pm$ 0.31	[145]
M$_1$	257.00	[503]	2.07(-3)	[512]	3.2	[94]	257.0	[51]
	256.50 $\pm$ 0.40	[493]	8.0(-5)	[423]	4.36	[423]		

level	E_B/eV		ω_{nlj}		Γ/eV		AE/eV	
M_2	189.00	[503]	2.0(-5)	[512]	2.1	[94]		
	189.30 ± 0.40	[493]	1.6(-4)	[423]	3.74	[423]		
M_3	182.00	[503]	1.4(-4)	[423]	2.15	[94]		
	181.50 ± 0.40	[493]			3.34	[423]		
M_4	70.00	[503]	2.7(-3)	[512]	0.068	[94]		
	70.10 ± 0.40	[493]	4.0(-6)	[423]	0.07	[423]		
M_5	69.00	[503]	3.0(-6)	[423]	0.07	[94]	69.0	[51]
	69.00 ± 0.40	[493]			0.07	[423]		
N_1	27.00	[503]						
	27.30 ± 0.50	[493]						
N_2	5.00	[503]						
	5.20 ± 0.40	[493]						
N_3	5.00	[503]						
	4.60 ± 0.40	[493]						
IP	11.81381	[222]						

Kr   Z=36

[Ar] $3d^{10}\,4s^2\,4p^6$

Krypton

$A = 83.80(1)$ [12.77] $\varrho = 3.484(-3)$ g/cm^3 [547]

X-ray transitions

line	transition	E/eV		I/eV/$\hbar$		TPIV	Γ/eV	
K series								
K_{α_3}	KL_1	12402.57 ± 0.92	[145]				9.79	[423]
$K_{\alpha 2}$	KL_2	12598.00 ± 1.30	[50]	0.499	[488]	1.889(-1)	4.06	[301]
		12598.00	[104]	0.512	[487]		4.62	[473]
		12595.424 ± 0.056	[366]				3.84	[423]
$K_{\alpha 1}$	KL_3	12649.00 ± 1.30	[50]	0.962	[488]	3.642(-1)	3.92	[301]
		12650.00	[104]	0.988	[487]		4.23	[473]
		12648.002 ± 0.052	[366]				3.79	[423]
	KM_1	14034.9 ± 1.2	[145]				7.79	[423]
$K_{\beta 3}$	KM_2	14104.00 ± 1.60	[50]	7.02(-2)	[488]	2.657(-2)	6.16	[423]
		14105.00	[104]	7.76(-2)	[487]			
		14104.96 ± 0.11	[366]					
$K_{\beta 1}$	KM_3	14112.00 ± 1.60	[50]	0.1369	[488]	5.182(-2)	5.70	[423]
		14113.00	[104]	0.1518	[487]			
$K_{\beta 5}$	$KM_{4,5}$	14237.7 ± 4.8	[50]	2.05(-4)	[488]	7.747(-5)	2.69	[423]
		14237.00	[104]					
$K_{\beta 5}^{II}$	KM_4	14232.63 ± 0.72	[145]			7.747(-5)	2.69	[423]
$K_{\beta 5}^{I}$	KM_4	14232.63 ± 0.48	[145]			1.110(-4)	2.69	[423]
	KN_1	14301.2 ± 2.0	[145]					
$K_{\beta_2^{II}}$	KN_2	14313.43 ± 0.59	[145]	7.70(-3)	[487]	2.335(-3)		
$K_{\beta_2^{I}}$	KN_3	14314.10 ± 0.58	[145]	1.50(-2)	[487]	4.492(-3)		
$K_{\beta 2}$	$KN_{2,3}$	14315.0 ± 2.4	[50]	1.187(-2)	[488]			
		14314.00	[104]					
$K_{\beta 4}$	$KN_{4,5}$	14328.2 ± 4.9	[50]					
		14328.90	[104]					
	$KO_{2,3}$	14324.57 ± 0.10	[84]					
	$KP_{2,3}$	14325.86 ± 0.10	[84]					
	$KQ_{2,3}$	14326.45 ± 0.10	[84]					
	$KR_{2,3}$	14326.72 ± 0.10	[84]					
	L_1M_1	1632.4 ± 1.8	[145]				12.35	[423]
$L_{\beta 4}$	L_1M_2	1697.5 ± 1.7	[50]	5.43(-3)	[488]	1.114(-3)	10.71	[423]
		1700.00	[104]					

line	transition	E/eV		I/eV/$\hbar$		TPIV	Γ/eV	
$L_{\beta3}$	L_1M_3	1706.8 ± 1.7	[50]	9.66(-3)	[488]	1.982(-3)	10.25	[423]
		1709.90	[104]					
$L_{\beta10}$	L_1M_4	1830.1 ± 1.3	[145]			6.556(-6)	7.25	[423]
$L_{\beta9}$	L_1M_5	1831.3 ± 1.3	[145]			9.756(-6)	7.25	[423]
	L_1N_1	1898.7 ± 2.5	[145]					
$L_{\gamma2}$	L_1N_2	1910.9 ± 1.2	[145]			1.044(-4)		
$L_{\gamma3}$	L_1N_3	1911.5 ± 1.2	[145]			1.847(-4)		
L_{η}	L_2M_1	1439.4 ± 1.4	[145]			3.38(-3)	6.394	[423]
	L_2M_2	1510.1 ± 1.4	[145]				4.76	[423]
$L_{\beta17}$	L_2M_3	1517.7 ± 1.4	[145]			2.351(-5)		
$L_{\beta1}$	L_2M_4	1636.60 ± 0.65	[50]	2.374(-2)	[488]	1.749(-2)	1.29	[423]
		1639.00	[104]					
		1636.876 ± 0.021	[366]					
	L_2M_5	1638.30 ± 0.87	[145]				1.29	[423]
$L_{\gamma5}$	L_2N_1	1703.3 ± 1.7	[50]	1.08(-4)	[488]	1.675(-4)		
		1706.00	[104]					
	L_2N_2	1717.86 ± 0.75	[145]					
	L_2N_3	1710.00 ± 1.20	[50]	4.11(-7)	[488]	3.662(-7)		
		1713.00	[104]					
		1718.52.52 ± 0.73	[145]					
L_l	L_3M_1	1386.9 ± 1.4	[145]			4.046(-3)	6.34	[423]
L_t	L_3M_2	1457.7 ± 1.4	[145]			1.363(-5)	4.71	[423]
L_s	L_3M_3	1465.3 ± 1.3	[145]			1.301(-5)	4.25	[423]
$L_{\alpha1.2}$	$L_3M_{4,5}$	1586.00 ± 0.60	[50]	2.33(-3)	[488]			
		1589.00	[104]					
		1585.411 ± 0.026	[366]					
$L_{\alpha2}$	L_3M_4	1584.63 ± 0.83	[145]			1.621(-3)	1.24	[423]
$L_{\alpha1}$	L_3M_5	1585.88 ± 0.83	[145]			1.578(-2)	1.24	[423]
$L_{\beta6}$	L_3N_1	1650.9 ± 1.3	[50]	1.19(-4)	[488]	1.508(-4)		
		1654.00	[104]					
	L_3N_2	1665.43 ± 0.70	[145]			1.776(-7)		
	L_3N_3	1666.09 ± 0.68	[145]			1.634(-7)		

line	transition	E/eV			I/eV/$\hbar$	TPIV		Γ/eV
M series								
	M_2N_1	195.00		[104]	3.12(-5)	[335]	3.245(-5)	
	M_3N_1	187.00		[104]	3.12(-5)	[335]	3.231(-5)	

level characteristics

level	E_B/eV			ω_{nlj}		Γ/eV		AE/eV	
K	14326.00		[503]	0.643	[301]	2.75	[301]	14325.6	[51]
	14325.60 ± 0.80		[493]	0.659	[347]	2.62	[423]	14324.61 ± 0.24	[50]
				0.6754	[557]	2.71	[94]	14328.06 ± 0.20	[145]
				0.655	[293]				
				0.646	[38]				
				0.660	[230]				
				0.6517	[39]				
				0.655	[197]				
				0.6385	[423]				
				0.643	[310]				
				0.651	[112]				
L_1	1921.00		[503]	4.10(-3)	[301]	4.28	[301]	1921.0	[51]
	1921.00 ± 0.60		[493]	2.19(-3)	[107]	7.17	[423]	1916.3 ± 4.4	[50]
				1.85(-3)	[348]	3.75	[94]	1925.49 ± 0.79	[145]
				3.50(-3)	[423]				
				2.21(-3)	[244]				
L_2	1727.00		[503]	2.00(-2)	[301]	1.31	[301]	1727.2	[51]
	1727.20 ± 0.50		[493]	1.19(-2)	[107]	1.22	[423]	1729.66 ± 0.36	[50]
				2.20(-2)	[348]	1.25	[94]	1732.49 ± 0.36	[145]
				2.11(-2)	[423]				
				1.99(-2)	[436]				
				2.11(-2)	[244]				
				1.40(-2)	[21]				

level	E_B/eV		ω_{nlj}		Γ/eV		AE/eV	
L_3	1675.00	[503]	2.20(-2)	[301]	1.17	[301]	1674.9	[51]
	1674.90 ± 0.50	[493]	1.23(-2)	[107]	1.17	[423]	1677.25 ± 0.34	[50]
			2.36(-2)	[348]	1.19	[94]	1680.06 ± 0.31	[145]
			2.16(-2)	[423]				
			2.02(-2)	[436]				
			2.16(-2)	[244]				
			1.60(-2)	[20]				
M_1	289.00	[503]	4.9(-5)	[38]	6.11	[350]	292.8	[51]
	292.10 ± 1.00	[493]	2.199(-3)	[512]	5.17	[423]		
			1.0(-4)	[423]	3.5	[94]		
M_2	223.00	[503]	6.0(-5)	[38]	4.14	[350]		
	222.70 ± 1.10	[51]	2.0(-5)	[512]	3.54	[423]		
	222.10	[493]	2.0(-4)	[423]	1.6	[94]		
M_3	214.00	[503]	6.00(-5)	[38]	4.13	[350]		
	213.80 ± 1.10	[51]	1.8(-4)	[423]	3.08	[423]		
	214.40	[493]			1.1	[94]		
M_4	89.00	[503]	2.70(-3)	[38]	0.089	[350]		
	88.00 ± 0.80	[51]	2.70(-3)	[512]	0.07	[423]		
	95.00	[493]	1.0(-5)	[423]	0.07	[94]		
M_5	89.00	[503]	1.0(-5)	[423]	0.09	[74]	93.8	[51]
	88.00 ± 0.80	[51]			0.07	[423]		
	93.80	[493]			0.072	[94]		
N_1	24.00	[503]			0.4	[94]		
	24.00 ± 0.80	[51]						
	27.50	[493]						
N_2	11.00	[503]						
	10.60 ± 1.90	[51]						
	14.70	[493]						
N_3	11.00	[503]						
	10.60 ± 1.90	[51]						
IP	13.99961	[222]						

Rb   Z=37   [Kr] 5s^1

Rubidium $A = 85.4678(3)$ [12.77] $\varrho = 1.53 \, \text{g/cm}^3$ [547]

X-ray transitions

line	transition	E/eV		I/eV/$\hbar$		TPIV [423]	Γ_X/eV	
K series								
K_{α_3}	KL_1	13135.18 ± 0.93	[145]				9.05	[423]
$K_{\alpha 2}$	KL_2	13335.88 ± 0.21	[50]	0.562	[488]	1.953(-1)	4.92	[301]
		13336.00	[104]	0.576	[487]		4.20	[423]
$K_{\alpha 1}$	KL_3	13395.49 ± 0.19	[50]	1.081	[488]	3.757(-1)	4.26	[301]
		13395.00	[104]	1.109	[487]		4.11	[423]
	KM_1	14878.6 ± 1.2	[145]				8.68	[423]
$K_{\beta 3}$	KM_2	14951.86 ± 0.80	[50]	8.05(-2)	[488]	2.800(-2)	6.02	[423]
		14952.00	[104]	8.87(-2)	[487]			
$K_{\beta 1}$	KM_3	14961.42 ± 0.53	[50]	0.1570	[488]	5.459(-2)	5.67	[423]
		14962.00	[104]	0.1735	[487]			
$K_{\beta 5}$	$KM_{4,5}$	15084.8 ± 2.7	[50]	3.70(-4)	[488]			
		15085.30	[104]					
$K_{\beta_5}^{II}$	KM_4	15088.21 ± 0.76	[145]			8.996(-5)	2.92	[423]
$K_{\beta_5}^{I}$	KM_5	15089.65 ± 0.75	[145]			1.286(-4)	2.92	[423]
	KN_1	15166.8 ± 2.0	[145]				3.10	[423]
$K_{\beta_2^{II}}$	KN_2	15185.59 ± 0.61	[145]	1.02(-2)	[487]	3.051(-3)		
$K_{\beta_2^{I}}$	KN_3	15186.55 ± 0.61	[145]	1.88(-2)	[487]	5.880(-3)		
$K_{\beta 2}$	$KN_{2,3}$	15185.54 ± 0.83	[50]	1.628(-2)	[488]			
		15186.00	[104]					
$K_{\beta 4}$	$KN_{4,5}$	15205.1 ± 5.5	[50]					
		15206.00	[104]					
L series								
	$L_1 M_1$	1743.4 ± 1.8	[145]				12.04	[423]
$L_{\beta 4}$	$L_1 M_2$	1817.74 ± 0.12	[50]	6.34(-3)	[488]	1.214(-3)	9.37	[423]
		1817.70	[104]					
$L_{\beta 3}$	$L_1 M_3$	1826.60 ± 0.12	[50]	1.123(-2)	[488]	2.147(-3)	9.03	[423]
		1826.60	[104]					
$L_{\beta 10}$	$L_1 M_4$	1953.0 ± 1.4	[145]			7.673(-6)	6.28	[423]

line	transition	E/eV		I/eV/$\hbar$		TPIV	Γ/eV	
L_{β_9}	L_1M_5	1954.5 ± 1.3	[145]			1.142(-5)	6.28	[423]
	L_1N_1	2031.6 ± 2.5	[145]				6.46	[423]
$L_{\gamma 2.3}$	$L_1N_{2,3}$	2050.72 ± 0.15 2050.70	[50] [104]	1.26(-3)	[488]			
$L_{\gamma 2}$	L_1N_2	2050.4 ± 1.2	[145]			1.352(-4)		
$L_{\gamma 3}$	L_1N_3	2051.4 ± 1.2	[145]			2.400(-4)		
L_η	L_2M_1	1541.78 ± 0.11 1541.80	[50] [104]	9.98(-4)	[488]	3.268(-3)	7.14	[423]
	L_2M_2	1617.0 ± 1.4	[145]				4.47	[423]
$L_{\beta_{17}}$	L_2M_3	1625.9 ± 1.3	[145]			2.440(-5)	4.13	[423]
L_{β_1}	L_2M_4	1752.18 ± 0.11 1752.10	[50] [104]	2.82(-2)	[488]	1.992(-2)	1.38	[423]
	L_2M_5	1753.90 ± 0.88	[145]				1.38	[423]
$L_{\gamma 5}$	L_2N_1	1835.33 ± 0.12 1835.30	[50] [104]	1.36(-4)	[488]	1.762(-4)	1.56	[423]
	L_2N_2	1849.84 ± 0.75	[145]					
	L_2N_3	1850.79 ± 0.74	[145]			4.935(-7)		
L_l	L_3M_1	1482.40 ± 0.10 1482.40	[50] [104]	1.11(-3)	[488]	4.104(-3)	7.08	[423]
L_t	L_3M_2	1557.7 ± 1.4	[145]			1.450(-5)	4.41	[423]
L_s	L_3M_3	1566.6 ± 1.3	[145]			1.381(-5)	4.07	[423]
$L_{\alpha 2}$	L_3M_4	1692.57 ± 0.10 1692.50	[50] [104]	2.75(-3)	[488]	1.886(-3)	1.32	[423]
$L_{\alpha 1}$	L_3M_5	1694.141 ± 0.069 1694.10	[50] [104]	2.448(-2)	[488]	1.784(-2)	1.32	[423]
$L_{\beta 6}$	L_3N_1	1775.18 ± 0.11 1775.20	[50] [104]	1.51(-4)	[488]	1.613(-4)	1.50	[423]
	L_3N_2	1790.54 ± 0.70	[145]			2.399(-7)		
	L_3N_3	1791.50 ± 0.70	[145]			2.293(-7)		

M series

	transition	E/eV		I/eV/$\hbar$		TPIV	Γ/eV	
	M_1M_3	85.90 ± 0.20	[50]			3.666(-5)	8.65	[423]
	M_2M_4	135.50 ± 0.30	[50]			2.053(-4)		[423]
	M_2N_1	217.40 ± 0.80 217.40	[50] [104]	4.07(-5)	[335]	4.318(-5)	3.41	[423]
	$M_3M_{4,5}$	128.20 ± 0.30	[50]					
	M_3M_4					1.711(-5)	2.88	[423]
	M_3M_5					1.620(-4)	2.89	[423]

line	transition	E/eV		I/eV/$\hbar$		TPIV	Γ/eV	
	M_3N_1	208.30 ± 0.70	[50]	4.07(-5)	[335]	4.607(-5)	3.07	[423]
		208.40	[104]					
M_{ξ_2}	M_4N_2	97.00 ± 0.15	[50]	1.05(-5)	[335]	1.320(-5)		
		96.98	[104]					
M_δ	M_4N_3	97.80 ± 0.15	[50]	1.05(-5)	[335]	2.437(-6)		
		97.85	[104]					
M_{ξ_1}	M_5N_3	96.40 ± 0.15	[50]	1.05(-5)	[335]	1.314(-5)		

level characteristics

level	E_B/eV		ω_{nlj}		Γ/eV		AE/eV	
K	15200.00	[503]	0.667	[301]	2.99	[301]	15199.7	[51]
	15199.70 ± 0.30	[493]	0.6779	[293]	2.86	[423]	15202.5 ± 1.4	[50]
			0.6987	[557]	2.96	[94]	15207.74 ± 0.22	[145]
			0.669	[38]				
			0.6744	[39]				
			0.679	[461]				
			0.668	[37]				
			0.6628	[423]				
			0.667	[310]				
L_1	2065.00	[503]	4.60(-3)	[301]	4.44	[301]	2065.1	[51]
	2065.10 ± 0.30	[493]	1.32(-2)	[107]	6.21	[423]	2063.6 ± 2.5	[50]
			3.85(-3)	[423]	3.75	[94]	2072.55 ± 0.81	[145]
			3.40(-3)	[436]				
			2.60(-3)	[576]				
L_2	1864.00	[503]	2.20(-2)	[301]	1.43	[301]	1863.9	[565]
	1863.90 ± 0.30	[493]	2.34(-2)	[126]	1.31	[423]	1866.08 ± 0.42	[50]
			2.23(-2)	[436]	1.34	[94]	1871.98 ± 0.36	[145]
L_3	1805.00	[503]	2.40(-2)	[301]	1.27	[301]	1804.4	[51]
	1804.40 ± 0.30	[493]	2.40(-2)	[423]	1.25	[423]	1806.80 ± 0.39	[50]
			2.26(-2)	[436]	1.27	[423]	1812.69 ± 0.31	[145]
M_1	322.00	[503]	2.336(-3)	[512]	5.83	[423]	326.7	[51]
	322.10 ± 0.30	[493]	1.1(-4)	[423]	4	[94]		
M_2	248.00	[503]	1.0(-5)	[512]	3.16	[423]		

level	E_B/eV		ω_{nlj}		Γ/eV		AE/eV	
	247.40 ± 0.30	[493]	2.5(-4)	[423]	1.9	[94]		
M_3	239.00	[503]	2.2(-4)	[423]	2.82	[423]		
	238.50 ± 0.30	[493]			1.95	[94]		
M_4	112.00	[503]	2.70(-3)	[512]	0.07	[423]		
	111.80 ± 0.30	[493]	2.0(-5)	[423]	0.067	[94]		
M_5	111.00	[503]	1.0(-5)	[423]	0.07	[423]	112,0	[51]
	110.30 ± 0.30	[493]			0.069	[94]		
N_1	30.00	[503]	1.0(-6)	[423]	0.25	[423]		
	29.30 ± 0.30	[493]			1.2	[94]		
N_2	15.00	[503]						
	14.30 ± 0.40	[493]						
N_3	14.00	[503]						
	14.00 ± 0.30	[493]						
IP	4.17713	[222]						

Sr   Z=38   [Kr] 5s^2

Strontium $A = 87.62(1)$ [12.77] $\varrho = 2.60\,\text{g/cm}^3$ [547]

X-ray transitions

line	transition	E/eV		I/eV/$\hbar$		TPIV	Γ/eV	
K series								
K_{α_3}	KL$_1$	13889.75 ± 0.96	[145]				9.37	[423]
K_{α_2}	KL$_2$	14098.03 ± 0.24	[50]	0.630	[488]	2.013(-1)	4.79	[301]
		14098.00	[104]				4.97	[473]
							4.52	[423]
K_{α_1}	KL$_3$	14165.20 ± 0.24	[50]	1.210	[488]	3.865(-1)	4.63	[301]
		14165.00	[104]	1.241	[487]		5.17	[473]
							4.44	[423]
	KM$_1$	15748.1 ± 1.2	[145]				9.68	[423]
K_{β_3}	KM$_2$	15825.17 ± 0.90	[50]	9.20(-2)	[488]	2.937(-2)	5.93	[423]
		15825.00	[104]	0.1010	[487]			
K_{β_1}	KM$_3$	15835.89 ± 0.60	[50]	0.1793	[488]	5.725(-2)	5.59	[423]
		15836.00	[104]	0.1975	[487]			
K_{β_5}	KM$_{4,5}$	15968.9 ± 3.0	[50]	4.62(-4)	[488]			
		15958.70	[104]					
$K_{\beta_5}^{II}$	KM$_4$	15970.20 ± 0.70	[145]			1.035(-4)	3.18	[423]
$K_{\beta_5}^{I}$	KM$_5$	15971.90 ± 0.76	[145]			1.476(-4)	3.18	[423]
	KN$_1$	16068.8 ± 2.1	[145]				3.12	[423]
$K_{\beta_2^{II}}$	KN$_2$	16085.68 ± 0.68	[145]	1.30(-2)	[487]	3.805(-3)	3.12	[423]
$K_{\beta_2^{I}}$	KN$_3$	16086.84 ± 0.41	[145]	2.53(-2)	[487]	7.275(-3)	3.12	[423]
K_{β_2}	KN$_{2,3}$	16084.68 ± 0.93	[50]	2.111(-1)	[488]			
		16085.00	[104]					
K_{β_4}	KN$_{4,5}$	16103.9 ± 1.5	[50]					
		16103.80	[104]					
L series								
	L$_1$M$_1$	1858.4 ± 1.8	[145]				12.82	[423]
L_{β_4}	L$_1$M$_2$	1936.44 ± 0.13	[50]	7.38(-3)	[488]	1.341(-3	9.07	[423]
		1936.30	[104]					
L_{β_3}	L$_1$M$_3$	1947.20 ± 0.14	[50]	1.299(-2)	[488]	2.361(-3)	8.72	[423]

line	transition	E/eV		I/eV/$\hbar$		TPIV	Γ/eV	
		1947.10	[104]					
$L_{\beta_{10}}$	L_1M_4	2080.5 ± 1.4	[145]			9.077(-6)	6.31	[423]
L_{β_9}	L_1M_5	2082.2 ± 1.4	[145]			1.351(-5)	6.31	[423]
	L_1N_1	2179.0 ± 2.7	[145]				6.90	[423]
$L_{\gamma_{2,3}}$	$L_1N_{2,3}$	2196.52 ± 0.17	[50]	1.65(-4)	[488]			
		2196.50	[263]					
L_{γ_2}	L_1N_2	2195.9 ± 1.3	[145]			1.693(-4)	6.25	[423]
L_{γ_3}	L_1N_3	2197.1 ± 1.0	[145]			3.003(-4)	6.25	[423]
L_η	L_2M_1	1649.337 ± 0.097	[50]	1.14(-3)	[488]	3.198(-3)	7.97	[423]
		1649.30	[104]					
	L_2M_2	1728.1 ± 1.4	[145]				4.22	[423]
$L_{\beta_{17}}$	L_2M_3	1738.3 ± 1.3	[145]			2.534(-5)	3.87	[423]
L_{β_1}	L_2M_4	1871.74 ± 0.13	[50]	3.31(-2)	[488]	2.238(-2)	1.46	[423]
		1871.70	[104]					
	L_2M_5	1873.88 ± 0.89	[145]				1.46	[423]
L_{γ_5}	L_2N_1	1969.19 ± 0.14	[50]	1.68(-4)	[488]	1.874(-4)	2.05	[423]
		1969.10	[104]					
	L_2N_2	1987.67 ± 0.82	[145]				1.40	[423]
	L_2N_3	1988.82 ± 0.55	[145]			6.352(-7)	1.40	[423]
L_l	L_3M_1	1582.174 ± 0.090	[50]	1.28(-3)	[488]	4.164(-3)	7.90	[423]
		1582.20	[104]					
L_t	L_3M_2	1661.4 ± 1.4	[145]			1.541(-5)	4.15	[423]
L_s	L_3M_3	1671.5 ± 1.3	[145]			1.465(-5)	3.80	[423]
L_{α_2}	L_3M_4	1804.77 ± 0.12	[50]	3.24(-3)	[488]	2.133(-3)	1.39	[423]
		1804.70	[104]					
L_{α_1}	L_3M_5	1806.585 ± 0.078	[50]	2.88(-2)	[488]	2.006(-2)	1.36	[423]
		1806.60	[263]					
L_{β_6}	L_3N_1	1901.83 ± 0.13	[50]	1.88(-4)	[488]	1.724(-4)	1.98	[423]
		1901.80	[104]					
	L_3N_2	1920.88 ± 0.76	[145]			3.083(-7)	1.33	[423]
	L_3N_3	1922.04 ± 0.50	[145]			2.956(-7)	1.34	[423]

M series

	M_2M_4	144.70 ± 0.30	[50]			2.483(-4)	2.88	[423]
	M_2N_1	241.60 ± 0.50	[50]	5.22(-5)	[335]	5.725(-5)	3.47	[423]
		241.60	[104]					
	$M_3M_{4,5}$	135.70 ± 0.30	[50]					
	M_3M_4					2.066(-5)	2.53	[423]

line	transition	E/eV		I/eV/$\hbar$		TPIV	Γ/eV	
	M_3M_5					1.967(-4)	2.54	[423]
	M_3N_1	231.30 ± 0.40	[50]	5.22(-5)	[335]	6.360(-5)	3.12	[423]
		231.30	[104]					
$M_{\xi 2}$	M_4N_2	114.80 ± 0.20	[50]	1.49(-5)	[335]	2.274(-5)	0.06	[423]
M_δ	M_4N_3	117.73	[423]			4.182(-6)	0.06	[423]
$M_{\xi 1}$	M_5N_3	114.00 ± 0.10	[50]	1.49(-5)	[335]	2.279(-5)	0.07	[423]

level characteristics

level	E_B/eV		ω_{nlj}		Γ/eV		AE/eV	
K	16105.00	[503]	0.690	[301]	3.25	[301]	16104.6	[51]
	16104.60 ± 0.30	[493]	0.702	[293]	3.11	[423]	16107.2 ± 1.5	[50]
			0.691	[38]	3.23	[94]	16115.26 ± 0.23	[145]
			0.6956	[39]				
			0.6857	[423]				
			0.690	[310]				
L_1	2216.00	[503]	9.10(-3)	[301]	4.67	[301]	2216.3	[51]
	2216.30 ± 0.30	[493]	3.00(-3)	[348]	6.25	[423]	2217.1 ± 2.9	[50]
			4.30(-3)	[423]	3.75	[94]	2225.51 ± 0.83	[145]
			4.10(-3)	[436]				
			2.90(-3)	[576]				
			5.50(-3)	[21]				
L_2	2007.00	[503]	2.40(-2)	[301]	1.54	[301]	2006.8	[51]
	2006.80 ± 0.30	[493]	2.24(-2)	[348]	1.40	[423]	2008.46 ± 0.48	[50]
			2.58(-2)	[423]	1.43	[94]	2017.25 ± 0.36	[145]
			2.48(-2)	[436]				
			3.00(-2)	[21]				
L_3	1940.00	[503]	2.60(-2)	[301]	1.39	[301]	1939.6	[51]
	1939.60 ± 0.30	[493]	2.43(-2)	[348]	1.33	[423]	1941.17 ± 0.45	[50]
			2.66(-2)	[423]	1.35	[94]	1950.46 ± 0.31	[145]
			2.51(-2)	[436]				
			3.40(-2)	[21]				
M_1	358.00	[503]	2.478(-3)	[512]	6.20	[74]	358.7	[51]
	357.50 ± 0.30	[493]	1.3(-4)	[423]	6.57	[423]		
					4.4	[94]		

level	E_B/eV		ω_{nlj}		Γ/eV		AE/eV	
M_2	280.00	[503]	7.0(-5)	[512]	2.20	[74]		
	279.80 $\pm$ 0.30	[493]	3.1(-4)	[423]	2.82	[423]		
					1.9	[94]		
M_3	269.00	[503]	2.8(-4)	[423]	2.20	[74]		
	269.10 $\pm$ 0.30	[493]			2.47	[423]		
					1.9	[94]		
M_4	135.00	[503]	2.7(-3)	[512]	0.08	[74]		
	135.00 $\pm$ 0.30	[493]	3.0(-5)	[423]	0.06	[423]		
					0.061	[94]		
M_5	133.00	[503]	2.0(-5)	[423]	0.08	[74]	134.2	[51]
	133.10 $\pm$ 0.30	[493]			0.06	[423]		
					0.064	[94]		
N_1	38.00	[503]	2.0(-6)	[423]	1.00	[74]		
	37.70 $\pm$ 0.30	[493]			0.65	[423]		
					1.6	[94]		
N_2	20.00	[503]			0.0001	[74]		
	19.90 $\pm$ 0.30	[51]			0.4	[94]		
	20.70 $\pm$ 0.40	[493]						
N_3	20.00	[503]						
	19.90 $\pm$ 0.30	[51]			0.0001	[347]		
	19.50 $\pm$ 0.40	[493]						
IP	5.6949	[222]						

Y   Z=39

[Kr] $4d^1\,5s^2$

Yttrium $A = 88.90585(2)$ [12.77] $\varrho = 4.45\,\text{g/cm}^3$ [547]
$\varrho = 4.50\,\text{g/cm}^3$ [483]

X-ray transitions

line	transition	E/eV		I/eV/$\hbar$		TPIV	Γ/eV	
K series								
K_{α_3}	KL_1	14667.13 ± 0.99	[145]				9.43	[423]
K_{α_2}	KL_2	14882.94 ± 0.26	[50]	0.705	[488]	2.068(-1)	5.18	[301]
		14885.00	[104]				4.89	[423]
K_{α_1}	KL_3	14958.54 ± 0.27	[50]	1.351	[488]	3.963(-1)	5.02	[301]
		14960.00	[104]				4.81	[423]
	KM_1	16644.7 ± 1.2	[145]				10.39	[423]
K_{β_3}	KM_2	16725.9 ± 1.0	[50]	0.1047	[488]	3.070(-2)	5.91	[423]
		16726.00	[104]					
K_{β_1}	KM_3	16738.08 ± 0.67	[50]	0.2039	[488]	5.982(-2)	6.07	[423]
		16738.00	[104]					
K_{β_5}	$KM_{4,5}$	16879.8 ± 3.4	[50]	5.72(-4)	[488]			
		16879.00	[104]					
$K_{\beta_5}^{II}$	KM_4	16879.20 ± 0.81	[145]			1.179(-4)	3.46	[423]
$K_{\beta_5}^{I}$	KM_5	16881.39 ± 0.79	[145]			1.679(-4)	3.46	[423]
	KN_1	16994.8 ± 2.3	[145]				5.55	[423]
K_{β_2}	$KN_{2,3}$	17015.6 ± 1.4	[50]	2.589(-2)	[488]			
		17012.30	[104]					
$K_{\beta_2}^{II}$	KN_2	17012.08 ± 0.62	[145]			4.526(-3)	3.46	[423]
$K_{\beta_2}^{I}$	KN_3	17015.30 ± 0.49	[145]			8.714(-3)	3.45	[423]
K_{β_4}	$KN_{4,5}$	17036.2 ± 1.7	[50]	5.72(-6)	[488]			
		17038.00	[104]					
	KN_4					6.957(-7)		
	KN_5					9.831(-7)		
L series								
	L_1M_1	1977.5 ± 1.8	[145]				13.13	[423]
L_{β_4}	L_1M_2	2059.99 ± 0.15	[50]	8.54(-3)	[488]	1.491(-3)	8.56	[423]
		2060.00	[104]					

line	transition	E/eV		I/eV/$\hbar$		TPIV	Γ/eV	
$L_{\beta 3}$	$L_1 M_3$	2072.17 ± 0.15 2072.20	[50] [104]	1.49(-2)	[488]	2.610(-3)	8.71	[423]
$L_{\beta 10}$	$L_1 M_4$	2212.1 ± 1.4	[145]			1.076(-5)	6.10	[423]
$L_{\beta 9}$	$L_1 M_5$	2214.3 ± 1.4	[145]			1.602(-5)	6.11	[423]
	$L_1 N_1$	2327.7 ± 2.9	[145]				8.30	[423]
$L_{\gamma 2.3}$	$L_1 N_{2,3}$	2346.82 ± 0.20 2346.75	[50] [104]	2.06(-3)	[488]			
$L_{\gamma 2}$	$L_1 N_2$	2345.0 ± 1.2	[145]			2.032(-4)	6.10	[423]
$L_{\gamma 3}$	$L_1 N_3$	2348.2 ± 1.1	[145]			3.587(-4)	6.10	[423]
	$L_1 N_4$	2373.8 ± 1.2	[145]			5.058(-8)		
L_η	$L_2 M_1$	1760.96 ± 0.11 1760.90	[50] [104]	1.30(-3)	[488]	3.126(-3)	8.59	[423]
	$L_2 M_2$	1843.8 ± 1.4	[145]				4.01	[423]
$L_{\beta 17}$	$L_2 M_3$	1855.4 ± 1.3	[145]			2.525(-5)	4.17	[423]
$L_{\beta 1}$	$L_2 M_4$	1995.85 ± 0.14 1995.80	[50] [104]	3.87(-2)	[488]	2.486(-2)	1.55	[423]
	$L_2 M_5$	1998.39 ± 0.91	[145]				1.56	[423]
$L_{\gamma 5}$	$L_2 N_1$	2110.19 ± 0.16 2110.20	[50] [104]	2.03(-4)	[488]	1.986(-4)	3.75	[423]
	$L_2 N_2$	2129.08 ± 0.74	[145]				1.55	[423]
	$L_2 N_3$	2132.30 ± 0.61	[145]			7.701(-7)	1.55	[423]
$L_{\gamma 1}$	$L_2 N_4$	2157.90 ± 0.71	[145]			1.343(-4)		
L_l	$L_3 M_1$	1685.39 ± 0.10 1685.40	[50] [104]	1.47(-3)	[488]	4.299(-3)	8.51	[423]
L_t	$L_3 M_2$	1768.6 ± 1.4	[145]			1.664(-5)	3.93	[423]
L_s	$L_3 M_3$	1780.3 ± 1.3	[145]			1.579(-5)	4.09	[423]
$L_{\alpha 2}$	$L_3 M_4$	1920.48 ± 0.13 1920.50	[50] [104]	3.78(-3)	[488]	2.431(-3)	1.47	[423]
$L_{\alpha 1}$	$L_3 M_5$	1922.564 ± 0.088 1922.50	[50] [104]	3.36(-2)	[488]	2.218(-2)	1.48	[423]
$L_{\beta 6}$	$L_3 N_1$	2034.43 ± 0.15 2034.40	[50] [104]	2.27(-4)	[488]	1.873(-4)	3.67	[423]
	$L_3 N_2$	2053.95 ± 0.69	[145]			3.816(-7)	1.48	[423]
	$L_3 N_3$	2057.16 ± 0.56	[145]			3.658(-7)	1.48	[423]
$L_{\beta 15}$	$L_3 N_4$	2082.77 ± 0.66	[145]			1.303(-5)		

M series

	$M_2 M_4$	152.20 ± 0.40	[50]			2.931(-4)	2.58	[423]

line	transition	E/eV		I/eV/$\hbar$		TPIV	Γ/eV	
	M_2N_1	267.00 ± 0.50	[50]	6.46(-5)	[335]	7.405(-5)	4.78	[423]
	$M_3M_{4,5}$	143.40 ± 0.30	[50]					
	M_3M_4					2.379(-5)	2.74	[423]
	M_3M_5					2.278(-4)	2.74	[423]
	M_3N_1	256.00 ± 1.10	[50]	6.46(-5)	[335]	8.369(-5)	4.93	[423]
		256.00	[104]					
M_{γ_2}	M_3N_4	294.86	[423]			5.669(-7)		
M_{γ_1}	M_3N_5	294.79	[423]			4.672(-6)		
	$M_4N_{2,3}$	132.80 ± 0.30	[145]					
	M_4N_2					3.679(-5)	0.13	[423]
	M_4N_3					6.712(-6)	0.12	[423]
M_{ξ_1}	M_5N_3	134.37	[423]			3.709(-5)	0.13	[423]

level characteristics

level	E_B/eV		ω_{nlj}		Γ/eV		AE/eV	
K	17039.00	[503]	0.710	[301]	3.52	[301]	17038.4	[51]
	17038.40 ± 0.30	[493]	0.722	[293]	3.39	[423]	17036.612 ± 0.050	[294]
			0.742	[557]	3.51	[94]	17047.90 ± 0.24	[145]
			0.711	[38]				
			0.7155	[39]				
			0.7071	[423]				
			0.710	[310]				
L_1	2373.00	[503]	5.90(-3)	[301]	4.71	[301]	2372.5	[51]
	2372.50 ± 0.30	[493]	4.79(-3)	[423]	6.04	[423]	2376.5 ± 3.4	[50]
			4.70(-3)	[436]	3.75	[94]	2380.76 ± 0.87	[145]
			6.40(-3)	[576]				
L_2	2155.00	[503]	2.60(-2)	[301]	1.65	[301]	2155.55	[51]
	2155.50 ± 0.30	[493]	2.84(-2)	[423]	1.49	[423]	2153.97 ± 0.55	[50]
			2.76(-2)	[436]	1.53	[94]	2164.89 ± 0.36	[145]
L_3	2080.00	[503]	2.80(-2)	[301]	1.50	[301]	2080.0	[51]
	2080.00 ± 0.30	[493]	2.93(-2)	[423]	1.41	[423]	2079.54 ± 0.52	[50]
			2.79(-2)	[436]	1.43	[94]	2089.76 ± 0.31	[145]

level	E_B/eV		ω_{nlj}		Γ/eV		AE/eV	
M_1	395.00	[503]	2.624(-3)	[512]	4.9	[94]	392.0	[51]
	393.60 ± 0.30	[493]	1.5(-4)	[423]	7.09	[423]		
M_2	313.00	[503]	1.30(-4)	[512]	1.95	[94]		
	312.40 ± 0.30	[493]	3.7(-4)	[423]	2.52	[423]		
M_3	301.00	[503]	3.4(-4)	[423]	1.95	[94]		
	300.30 ± 0.30	[493]			2.68	[423]		
M_4	160.00	[503]	2.7(-3)	[512]	0.062	[94]		
	159.60 ± 0.30	[493]	4.0(-5)	[423]	0.06	[423]		
M_5	158.00	[503]	4.0(-5)	[423]	0.066	[94]	155.8	[51]
	157.40 ± 0.30	[493]			0.07	[423]		
N_1	46.00	[503]	3.0(-6)	[423]	2	[94]		
	45.40 ± 0.30	[493]			2.26	[423]		
N_2	29.00	[503]			0.06	[423]		
	25.60 ± 0.40	[51]			0.8	[94]		
	25.10 ± 2.30	[493]						
N_3	26.00	[503]			0.06	[423]		
	25.60 ± 0.40	[51]			0.3	[94]		
	22.80 ± 2.30	[493]						
N_4	3.00	[503]						
	2.40 ± 1.20	[493]						
IP	6.2171	[222]						

Zr   Z=40   [Kr] $4d^2\,5s^2$

Zirconium $A = 91.224(2)$ [222] $\varrho = 6.44\,\text{g/cm}^3$ [547]

X-ray transitions

line	transition	E/eV		I/eV/$\hbar$		TPIV	Γ/eV	
K series								
K_{α_3}	KL_1	15467.0 ± 1.0	[145]				9.39	[423]
$K_{\alpha 2}$	KL_2	15691.90 ± 0.20	[50]	0.786	[488]	2.119(-1)	5.62	[301]
		15691.00	[104]	0.804	[487]		5.25	[473]
		15690.645 ± 0.050	[366]				5.28	[423]
$K_{\alpha 1}$	KL_3	15775.10 ± 0.20	[50]	1.504	[488]	4.054(-1)	5.40	[301]
		15775.00	[104]	1.539	[487]		5.70	[473]
		15774.914 ± 0.054	[366]				5.19	[423]
	KM_1	17658.0 ± 1.2	[145]				11.29	[423]
$K_{\beta 3}$	KM_2	17654.10 ± 1.00	[50]	0.1186	[488]	3.197(-2)	6.38	[423]
		17654.00	[104]	0.1297	[487]			
		17652.628 ± 0.075	[366]					
$K_{\beta 1}$	KM_3	17667.80 ± 0.80	[50]	0.2310	[488]	6.226(-2)	6.53	[423]
		17668.00	[104]	0.2533	[487]			
		17666.578 ± 0.076	[366]					
$K_{\beta 5}$	$KM_{4,5}$	17816.1 ± 3.8	[50]	7.03(-4)	[488]			
		17816.00	[104]					
$K_{\beta_5}^{II}$	KM_4	17815.21 ± 0.83	[145]			1.333(-4)	3.77	[423]
$K_{\beta_5}^{I}$	KM_5	17817.98 ± 0.81	[145]			1.893(-4)	3.77	[423]
	KN_1	17948.5 ± 2.5	[145]				8.21	[423]
$K_{\beta_2^{II}}$	KN_2	17966.67 ± 0.73	[145]	1.86(-2)	[487]	5.236(-3)	3.96	[423]
$K_{\beta_2^{I}}$	KN_3	17971.47 ± 0.56	[145]	3.62(-2)	[487]	1.006(-2)	3.95	[423]
$K_{\beta 2}$	$KN_{2,3}$	17970.3 ± 1.5	[50]	3.11(-2)	[488]			
		17969.98	[104]					
$K_{\beta 4}$	$KN_{4,5}$	17994.3 ± 1.9	[50]	1.68(-5)	[488]			
		17993.80	[104]					
$K_{\beta_4}^{II}$	KN_4	18000.09 ± 0.75	[145]			1.883(-6)		
$K_{\beta_4^{I}}$	KN_5					2.656(-6)		

line	transition	E/eV		I/eV/$\hbar$		TPIV	Γ/eV	

L series

line	transition	E/eV		I/eV/$\hbar$		TPIV	Γ/eV	
	L_1M_1	2101.0 ± 1.8	[145]				13.28	[423]
$L_{\beta 4}$	L_1M_2	2187.30 ± 0.10	[50]	9.84(-3)	[488]	1.655(-3)	5.60	[473]
		2187.30	[104]				8.38	[423]
		2187.714 ± 0.036	[366]					
$L_{\beta 3}$	L_1M_3	2201.00 ± 0.10	[50]	1.71(-2)	[488]	2.881(-3)	5.60	[473]
		2201.00	[104]				8.53	[423]
		2201.063 ± 0.032	[145]					
$L_{\beta 10}$	L_1M_4	2348.2 ± 1.4	[145]			1.270(-5)	5.76	[423]
$L_{\beta 9}$	L_1M_5	2350.9 ± 1.4	[145]			1.891(-5)	5.76	[423]
	L_1N_1	2481.5 ± 3.1	[145]				10.20	[423]
$L_{\gamma 2,3}$	$L_1N_{2,3}$	2502.87 ± 0.22	[50]	1.42(-2)	[488]			
		2502.80	[104]					
$L_{\gamma 2}$	L_1N_2	2499.6 ± 1.3	[145]			2.395(-4)	5.96	[423]
$L_{\gamma 3}$	L_1N_3	2504.4 ± 1.4	[145]			4.206(-4)	5.94	[423]
	L_1N_4	2533.0 ± 1.4	[145]			1.403(-7)		
L_η	L_2M_1	1876.56 ± 0.13	[50]	1.48(-3)	[488]	3.107(-3)	9.17	[423]
		1875.60	[104]					
	L_2M_2	1963.5 ± 1.4	[145]				4.28	[423]
$L_{\beta 17}$	L_2M_3	1976.8 ± 1.3	[145]			2.759(-5)	4.43	[423]
$L_{\beta 1}$	L_2M_4	2124.40 ± 0.10	[50]	4.50(-2)	[488]	2.729(-2)	1.87	[473]
		2124.40	[104]				1.66	[423]
		2124.394 ± 0.028	[366]					
	L_2M_5	2127.36 ± 0.93	[145]				1.66	[423]
$L_{\gamma 5}$	L_2N_1	2255.17 ± 0.18	[50]	2.40(-4)	[488]	2.134(-4)	6.10	[423]
		2255.10	[104]					
	L_2N_2	2276.06 ± 0.83	[145]				1.86	[423]
	L_2N_3	2280.86 ± 0.66	[145]			9.263(-7)	1.85	[423]
$L_{\gamma 1}$	L_2N_4	2302.66 ± 0.19	[50]	2.52(-4)	[488]	3.429(-4)	3.34	[473]
		2302.60	[104]					
L_l	L_3M_1	1792.01 ± 0.08	[50]	1.68(-3)	[488]	4.431(-3)	9.08	[423]
		1792.00	[104]					
		1792.111 ± 0.023	[366]					
L_t	L_3M_2	1879.2 ± 1.4	[145]			1.775(-5)	4.19	[423]
L_s	L_3M_3	1892.6 ± 1.3	[145]			1.678(-5)	4.34	[423]
$L_{\alpha 2}$	L_3M_4	2039.90 ± 0.10	[50]	4.38(-3)	[488]	2.684(-3)	1.52	[473]
		2039.90	[104]				1.57	[423]
		2040.19 ± 0.16	[366]					
$L_{\alpha 1}$	L_3M_5	2042.36 ± 0.07	[50]	3.89(-2)	[488]	2.443(-2)	1.68	[473]
		2042.30	[104]				1.57	[423]

line	transition	E/eV		I/eV/$\hbar$		TPIV	Γ/eV	
		2042.489 ± 0.027	[366]					
$L_{\beta6}$	L_3N_1	2171.28 ± 0.17	[50]	2.71(-4)	[488]	2.029(-4)	6.01	[423]
		2171.20	[104]					
	L_3N_2	2191.80 ± 0.78	[145]			4.614(-7)	1.77	[423]
	L_3N_3	2196.60 ± 0.62	[145]			4.424(-7)	1.76	[423]
$L_{\beta15}$	L_3N_4	2225.22 ± 0.81	[145]	5.33(-5)	[488]	3.345(-5)		
$L_{\beta2}$	L_3N_5	2219.40	[104]	4.72(-4)	[488]	2.962(-4)	5.13	[473]
$L_{\beta2,15}$	$L_3N_{4,5}$	2219.40 ± 0.18	[50]					
M series								
	M_2M_4	161.70 ± 0.40	[50]			1.560(-5)	2.76	[423]
	M_3M_5	153.30 ± 0.60	[50]			2.559(-4)	2.91	[423]
$M_{\gamma2}$	M_3N_4	324.21	[423]			1.931(-6)		
$M_{\gamma1}$	M_3N_5	324.36	[423]			1.598(-5)		
$M_{\xi1}$	M_5N_3	153.75	[423]			5.737(-5)	0.32	[423]
	$M_{4,5}N_{2,3}$	151.10 ± 0.40	[50]	2.52(-5)	[335]			
	$M_{4,5}O_{2,3}$	177.00 ± 1.00	[50]					

level characteristics

level	E_B/eV		ω_{nlj}		Γ/eV		AE/eV	
K	17998.00	[503]	0.730	[301]	3.84	[301]	17997.6	[51]
	17997.60 ± 0.40	[493]	0.761	[557]	3.70	[423]	17995.872 ± 0.080	[294]
			0.740	[347]	3.83	[94]	18008.15 ± 0.26	[145]
			0.741	[293]				
			0.730	[38]				
			0.734	[39]				
			0.7272	[423]				
			0.730	[310]				
			0.732	[112]				

level	E_B/eV		ω_{nlj}		Γ/eV		AE/eV	
L_1	2532.00	[503]	6.80(-3)	[301]	4.78	[301]	2531.6	[51]
	2531.60 ± 0.30	[493]	3.96(-3)	[107]	5.69	[423]	2541.1 ± 3.9	[50]
			3.97(-3)	[348]	3.75	[94]	2541.10 ± 0.87	[145]
			5.33(-3)	[423]				
			5.30(-3)	[436]				
			5.50(-3)	[244]				
L_2	2307.00	[503]	2.80(-2)	[301]	1.78	[301]	2306.7	[51]
	2306.70 ± 0.30	[493]	1.89(-2)	[107]	1.59	[423]	2305.36 ± 0.63	[50]
			2.94(-2)	[348]	1.63	[94]	2317.53 ± 0.36	[145]
			3.10(-2)	[423]				
			2.92(-2)	[436]				
			3.07(-2)	[244]				
L_3	2233.00	[503]	3.10(-2)	[301]	1.37	[301]	2222.3	[51]
	2222.30 ± 0.30	[493]	2.01(-2)	[107]	1.50	[423]	2222.30 ± 0.59	[50]
			2.95(-2)	[348]	1.51	[94]	2233.28 ± 0.32	[145]
			3.21(-2)	[423]				
			3.04(-2)	[436]				
			3.20(-2)	[244]				
M_1	431.00	[503]	7.00(-5)	[38]	6.47	[350]	430.3	[51]
	430.30 ± 0.30	[493]	2.775(-3)	[512]	7.59	[423]		
			1.7(-4)	[423]	5.4	[94]		
M_2	345.00	[503]	1.40(-4)	[38]	2.43	[350]		
	344.20 ± 0.40	[493]	1.9(-4)	[512]	2.69	[423]		
			4.4(-4)	[423]	2	[94]		
M_3	331.00	[503]	1.50(-4)	[38]	2.40	[350]		
	330.50 ± 0.40	[493]	4.1(-4)	[423]	2.84	[423]		
					2	[94]		
M_4	183.00	[503]	2.70(-3)	[38]	0.073	[350]		
	182.40 ± 0.30	[493]	2.7(-3)	[512]	0.07	[423]		
			7.0(-5)	[423]	0.07	[94]		
M_5	180.00	[503]	6.0(-5)	[423]	0.07	[74]	178.8	[51]
	180.00 ± 0.30	[493]			0.07	[423]		
					0.074	[94]		
N_1	62.00	[503]	4.0(-6)	[423]	5.00	[74]		
	51.30 ± 0.30	[493]			4.51	[423]		
					2.4	[94]		

level	E_B/eV		ω_{nlj}	Γ/eV		AE/eV
N_2	29.00	[503]		0.27	[423]	
	28.70 ± 0.40	[51]		0.01	[74]	
	29.30 ± 2.10	[493]		1.2	[94]	
N_3	29.00	[503]		0.25	[423]	
	28.70 ± 0.40	[51]		0.01	[74]	
	25.70 ± 2.10	[493]		0.6	[94]	
N_4	3.00	[503]				
	3.00 ± 0.30	[493]				
IP	6.63390	[222]				

Nb   Z=41

[Kr] 4d⁴ 5s¹

Niobium $A = 92.90638(2)$ [222] $\varrho = 8.58\,\text{g/cm}^3$ [547]
$\varrho = 8.55\,\text{g/cm}^3$ [483]

X-ray transitions

line	transition	E/eV		I/eV/$\hbar$		TPIV	Γ/eV	
K series								
K_{α_3}	KL_1	16290.1 ± 1.0	[145]				9.92	[423]
$K_{\alpha 2}$	KL_2	16521.28 ± 0.33	[50]	0.874	[488]	2.167(-1)	6.01	[301]
		16521.00	[104]				5.71	[423]
$K_{\alpha 1}$	KL_3	16615.16 ± 0.33	[50]	1.669	[488]	4.139(-1)	5.80	[301]
		16615.00	[104]				5.60	[423]
	KM_1	18518.9 ± 1.2	[145]				12.00	[423]
$K_{\beta 3}$	KM_2	18606.5 ± 1.2	[50]	0.1339	[488]	3.318(-2)	6.81	[423]
		18606.00	[104]					
$K_{\beta 1}$	KM_3	18622.68 ± 0.83	[50]	0.2607	[488]	6.461(-2)	6.95	[423]
		18623.00	[104]					
$K_{\beta 5}$	$KM_{4,5}$	18784.10	[104]	8.55(-4)	[488]			
$K_{\beta_5}^{II}$	KM_4	18778.45 ± 0.87	[145]			1.496(-4)	4.11	[423]
$K_{\beta_5}^{I}$	KM_5	18782.22 ± 0.85	[145]			2.120(-4)	4.11	[423]
	KN_1	18930.5 ± 2.6	[145]				11.98	[423]
$K_{\beta 2}$	$KN_{2,3}$	18952.9 ± 1.7	[50]	3.62(-2)	[488]			
		18953.00	[104]					
$K_{\beta_2}^{II}$	KN_2	18948.34 ± 0.90	[145]			5.866(-3)	5.05	[423]
$K_{\beta_2}^{I}$	KN_3	18955.56 ± 0.63	[145]			1.122(-2)	5.09	[423]
$K_{\beta 4}$	$KN_{4,5}$	18981.3 ± 2.2	[50]	4.15(-5)	[488]			
		18980.00	[104]					
$K_{\beta_4}^{II}$	KN_4	18985.99 ± 0.83	[145]			4.286(-6)		
$K_{\beta_4}^{I}$	KN_5					6.008(-6)		

line	transition	E/eV		I/eV/ℏ		TPIV	Γ/eV	

L series

line	transition	E/eV		I/eV/ℏ		TPIV	Γ/eV	
	L_1M_1	2228.7 ± 1.8	[145]				13.88	[423]
$L_{\beta 4}$	L_1M_2	2319.38 ± 0.19 2319.30 2319.60 ± 1.00	[50] [104] [437]	1.129(-2)	[488]	1.838(-3)	8.69	[423]
$L_{\beta 3}$	L_1M_3	2334.80 ± 0.20 2334.70 2335.20 ± 1.00	[50] [104] [437]	1.953(-2)	[488]	3.182(-3)	8.84	[423]
$L_{\beta 10}$	L_1M_4	2487.00 2488.3 ± 1.5	[104] [145]	9.18(-5)	[488]	1.495(-5)	5.99	[423]
$L_{\beta 9}$	L_1M_5	2492.00 2492.1 ± 1.5	[104] [145]	1.37(-4)	[488]	2.227(-5)	5.99	[423]
	L_1N_1	2640.4 ± 3.3	[145]				13.86	[423]
$L_{\gamma 2,3}$	$L_1N_{2,3}$	2663.88 ± 0.17 2663.80	[50] [104]	2.94(-3)	[488]			
$L_{\gamma 2}$	L_1N_2	2658.2 ± 1.5	[145]			2.749(-4)	6.93	[423]
$L_{\gamma 3}$	L_1N_3	2665.4 ± 1.3	[145]			4.795(-4)	6.97	[423]
	L_1N_4	2695.8 ± 1.4	[145]			3.299(-7)		
L_η	L_2M_1	1996.21 ± 0.14 1996.20	[50] [104]	1.67(-3)	[488]	3.072(-3)	9.67	[423]
	L_2M_2	2087.5 ± 1.5	[145]				4.49	[423]
$L_{\beta 17}$	L_2M_3	2102.7 ± 1.3	[145]			2.883(-5)	4.64	[423]
$L_{\beta 1}$	L_2M_4	2257.38 ± 0.18 2257.40	[50] [104]	5.19(-2)	[488]	2.974(-2)	1.78	[423]
	L_2M_5	2260.88 ± 0.94	[145]				1.80	[423]
$L_{\gamma 5}$	L_2N_1	2406.63 ± 0.21 2406.60	[50] [104]	2.79(-4)	[488]	2.278(-4)	9.65	[423]
	L_2N_2	2426.99 ± 0.98	[145]				2.73	[423]
	L_2N_3	2434.21 ± 0.71	[145]			1.070(-6)	2.77	[423]
$L_{\gamma 1}$	L_2N_4	2461.87 ± 0.22 2461.30	[50] [104]	1.28(-3)	[488]	7.348(-4)		
L_l	L_3M_1	1902.27 ± 0.13 1902.20	[50] [104]	1.91(-3)	[488]	4.613(-3)	9.56	[423]
L_t	L_3M_2	1993.2 ± 1.4	[145]			1.911(-5)	4.37	[423]
L_s	L_3M_3	2008.4 ± 1.3	[145]			1.800(-5)	4.52	[423]
$L_{\alpha 2}$	L_3M_4	2163.02 ± 0.17 2163.00	[50] [104]	5.04(-3)	[488]	2.989(-3)	1.68	[423]
$L_{\alpha 1}$	L_3M_5	2165.89 ± 0.11 2165.90	[50] [104]	4.48(-2)	[488]	2.655(-2)	1.68	[423]

line	transition	E/eV		I/eV/$\hbar$		TPIV	Γ/eV	
$L_{\beta 6}$	L_3N_1	2312.54 ± 0.19	[50]	3.17(-4)	[488]	2.220(-4)	9.55	[423]
		2312.50	[104]					
		2312.90 ± 1.00	[437]					
	L_3N_2	2332.68 ± 0.93	[145]			5.447(-7)	2.61	[423]
	L_3N_3	2339.90 ± 0.66	[145]			5.213(-7)	2.61	[423]
$L_{\beta 15}$	L_3N_4	2370.34 ± 0.86	[145]	1.23(-4)	[488]	7.312(-5)		
$L_{\beta 2}$	L_3N_5	2367.90	[104]	1.09(-3)	[488]	6.450(-4)		
		2366.80 ± 1.00	[437]					
$L_{\beta 15,2}$	$L_3N_{4,5}$	2367.02 ± 0.20	[50]					

M series

line	transition	E/eV		I/eV/$\hbar$		TPIV	Γ/eV	
	$M_1N_{2,3}$	437.00	[104]	4.04(-4)	[335]			
	M_1N_2					5.112(-5)	9.01	[423]
	M_1N_3					7.845(-5)	9.05	[423]
	M_2M_4	171.80 ± 0.70	[50]			3.607(-4)	2.88	[423]
	M_2N_1	323.00 ± 2.50	[50]	9.32(-5)	[335]	1.118(-4)	10.75	[423]
		323.00	[104]					
	M_2N_4	375.00 ± 2.30	[50]	4.52(-5)	[335]	4.144(-5)		
	M_3M_5	158.20 ± 0.40	[50]			2.759(-4)	3.03	[423]
	M_3N_1	305.00 ± 1.50	[50]	9.32(-5)	[335]	1.296(-4)	10.90	[423]
		305.00	[104]					
$M_{\gamma 2,1}$	$M_3N_{4,5}$	356.00 ± 2.00	[50]	4.52(-5)	[335]			
	M_3N_4					5.144(-6)		
	M_3N_5					4.240(-5)		
$M_{\xi 2}$	M_4N_2	174.24	[423]			8.362(-5)	1.12	[423]
M_δ	M_4N_3	176.46	[423]			1.498(-5)	1.16	[423]
$M_{\xi 1}$	M_5N_3	173.45	[423]			8.517(-5)	1.17	[423]
	$M_{4,5}N_{2,3}$	171.70 ± 0.20	[50]					
	$M_{4,5}O_{2,3}$	200.20 ± 0.65	[50]					

level characteristics

level	E_B/eV		ω_{nlj}		Γ/eV		AE/eV	
K	18986.00	[503]	0.747	[301]	4.14	[301]	18985.6	[51]
	18985.60 ± 0.40	[493]	0.759	[293]	4.02	[423]	18982.961 ± 0.040	[294]
			0.7788	[557]	4.16	[94]	18990.67 ± 0.27	[145]
			0.748	[38]				
			0.7512	[39]				
			0.7459	[423]				
			0.74	[310]				
L_1	2698.00	[503]	9.40(-3)	[301]	3.94	[301]	2697.7	[51]
	2697.70 ± 0.30	[493]	5.92(-3)	[423]	5.90	[423]	2710.0 ± 4.4	[50]
			6.10(-3)	[436]	3.8	[94]	2700.53 ± 0.88	[145]
			1.10(-2)	[576]				
L_2	2465.00	[503]	3.10(-2)	[301]	1.87	[301]	2464.7	[51]
	2464.70 ± 0.30	[493]	3.39(-2)	[423]	1.69	[423]	2464.37 ± 0.72	[50]
			3.25(-2)	[436]	1.73	[94]	2469.32 ± 0.35	[145]
L_3	2371.00	[503]	3.40(-2)	[301]	1.66	[301]	2370.5	[51]
	2370.50 ± 0.30	[493]	3.51(-2)	[423]	1.58	[423]	2370.60 ± 0.67	[50]
			3.39(-2)	[436]	1.6	[94]	2375.01 ± 0.30	[145]
M_1	469.00	[503]	2.932(-3)	[512]	5.8	[94]	466.6	[51]
	468.40 ± 0.30	[493]	1.9(-4)	[423]	7.98	[423]		
M_2	379.00	[503]	2.5(-4)	[512]	2.05	[94]		
	378.40 ± 0.40	[493]	5.1(-4)	[423]	2.79	[423]		
M_3	363.00	[503]	4.8(-4)	[423]	2.05	[94]		
	363.00 ± 0.40	[493]			2.94	[423]		
M_4	208.00	[503]	2.7(-3)	[512]	0.092	[94]		
	207.40 ± 0.30	[493]	1.0(-4)	[423]	0.09	[423]		
M_5	205.00	[503]	9.0(-5)	[423]	0.095	[94]	202.3	[51]
	204.60 ± 0.30	[493]			0.10	[423]		
N_1	58.00	[503]	1.0(-5)	[423]	7.96	[423]		
	58.10 ± 0.30	[493]			2.8	[94]		
N_2	34.00	[503]	1.0(-6)	[423]	1.5	[94]		
	33.90 ± 0.40	[51]			1.03	[423]		

level	E_B/eV		ω_{nlj}	Γ/eV		AE/eV
	35.60 ± 2.60	[493]				
N_3	34.00	[503]		1	[94]	
	33.90 ± 0.40	[51]		1.07	[423]	
	29.60 ± 2.60	[493]				
N_4	4.00	[503]				
	3.20 ± 0.30	[493]				
IP	6.75885	[222]				

Mo   Z=42   [Kr] 4d^5 5s^1

Molybdenum A = 95.94(1) [12.77] ϱ = 10.20 g/cm^3 [547]

X-ray transitions

line	transition	E/eV		I/eV/$\hbar$		TPIV	Γ/eV	
K series								
K_{α_3}	KL$_1$	17135.0 ± 1.1	[145]				10.50	[423]
$K_{\alpha 2}$	KL$_2$	17374.30 ± 0.15	[50]	0.970	[488]	2.213(-1)	6.49	[301]
		17375.00	[104]	0.990	[487]		6.80	[473]
		17374.39 ± 0.22	[143]				6.16	[423]
		17374.40 ± 0.22	[277]					
		17374.29 ± 0.29	[144]					
$K_{\alpha 1}$	KL$_3$	17479.35 ± 0.03	[50]	1.848	[488]	4.217(-1)	6.31	[301]
		17480.00	[104]	1.888	[487]		6.82	[473]
		17479.48 ± 0.01	[143]				6.04	[423]
		17479.49 ± 0.01	[277]					
		17479.372 ± 0.010	[144]					
	KM$_1$	19495.8 ± 1.2	[145]				12.87	[423]
$K_{\beta 3}$	KM$_2$	19590.25 ± 0.41	[50]	0.1506	[488]	3.436(-2)	7.31	[423]
		19591.00	[104]	0.1640	[487]			
		19589.00 ± 1.00	[95]					
		19590.37 ± 0.42	[277]					
$K_{\beta 1}$	KM$_3$	19608.34 ± 0.42	[50]	0.2931	[488]	6.689(-2)	7.46	[423]
		19609.00	[104]	0.320	[487]			
		19607.00 ± 1.00	[95]					
		19608.47 ± 0.42	[277]					
$K_{\beta 52}$	KM$_4$	19771.4 ± 2.3	[50]	7.31(-4)	[488]	1.669(-4)	4.48	[423]
		19771.00	[104]					
$K_{\beta 51}$	KM$_5$	19776.4 ± 2.3	[50]	1.03(-3)	[488]	2.360(-4)	4.49	[423]
		19776.00	[104]					
	KN$_1$	19940.3 ± 2.8	[145]				15.20	[423]
$K_{\beta 22}$	KN$_2$	19960.0 ± 1.1	[145]	2.18(-2)	[488]	6.341(-3)	6.34	[423]
		19963.00	[104]	2.48(-2)	[487]			
$K_{\beta 21}$	KN$_3$	19967.16 ± 0.68	[145]	4.23(-2)	[488]	1.223(-2)	6.43	[423]
				4.82(-2)	[487]			
	KN$_{2,3}$	19965.27 ± 0.95	[50]					

line	transition	E/eV		I/eV/$\hbar$		TPIV	Γ/eV	
		19964.50 ± 0.50	[95]					
$K_{\beta_4}^{II}$	KN_4	19990.00	[104]	7.05(-5)	[488]	6.708(-6)		
		20000.73 ± 0.83	[145]					
$K_{\beta_4}^{I}$	KN_5	19997.00	[104]	7.05(-5)	[488]	9.385(-6)		
		19999.84 ± 0.35	[145]					
K_{β_4}	$KN_{4,5}$	19996.8 ± 4.3	[50]					

L series

line	transition	E/eV		I/eV/$\hbar$		TPIV	Γ/eV	
	L_1M_1	2360.9 ± 1.8	[145]				14.63	[423]
L_{β_4}	L_1M_2	2455.68 ± 0.22	[50]	1.29(-2)	[488]	2.066(-3)	5.78	[473]
		2455.60	[104]				9.06	[423]
L_{β_3}	L_1M_3	2473.07 ± 0.22	[50]	2.22(-2)	[488]	3.553(-3)	5.90	[473]
		2473.00	[104]				9.22	[423]
$L_{\beta_{10}}$	L_1M_4	2633.7 ± 1.5	[145]			1.776(-5)	6.24	[423]
L_{β_9}	L_1M_5	2637.8 ± 1.5	[145]			2.646(-5)	6.25	[423]
	L_1N_1	2805.3 ± 3.4	[145]					
$L_{\gamma2,3}$	$L_1N_{2,3}$	2830.65 ± 0.19	[50]	3.49(-3)	[488]			
		2830.70	[104]					
$L_{\gamma2}$	L_1N_2	2825.1 ± 1.7	[145]			3.219(-4)	8.10	[423]
$L_{\gamma3}$	L_1N_3	2832.2 ± 1.3	[145]			5.584(-4)	8.19	[423]
	L_1N_4	2865.8 ± 1.4	[145]			5.410(-7)		
	L_1N_5	2864.87 ± 0.96	[145]			8.042(-7)		
L_η	L_2M_1	2120.26 ± 0.16	[50]	1.88(-3)	[488]	3.128(-3)	10.29	[423]
		2120.30	[104]					
	L_2M_2	2215.7 ± 1.5	[145]				4.72	[423]
$L_{\beta_{17}}$	L_2M_3	2232.9 ± 1.3	[145]			2.042(-5)	4.87	[423]
L_{β_1}	L_2M_4	2394.831 ± 0.055	[50]	5.95(-2)	[488]	3.232(-2)	2.03	[473]
		2394.80	[104]				1.90	[423]
	L_2M_5	2398.63 ± 0.95	[145]				1.91	[423]
$L_{\gamma5}$	L_2N_1	2563.26 ± 0.16	[50]	3.25(-4)	[488]	2.475(-4)	12.62	[423]
		2563.20	[104]					
	L_2N_2	2585.9 ± 1.2	[145]				3.76	[423]
	L_2N_3	2592.98 ± 0.78	[145]			1.275(-6)	3.85	[423]
$L_{\gamma1}$	L_2N_4	2623.52 ± 0.16	[50]	2.05(-3)	[488]	1.111(-3)	3.76	[473]
		2623.50	[104]					
	L_2N_5	2625.66 ± 0.44	[145]					
L_l	L_3M_1	2015.71 ± 0.15	[50]	2.17(-3)	[488]	4.807(-3)	10.17	[423]
		2015.70	[104]					

line	transition	E/eV		I/eV/$\hbar$		TPIV	Γ/eV	
L_t	L_3M_2	2110.8 ± 1.4	[145]			2.042(-5)	4.60	[423]
L_s	L_3M_3	2128.0 ± 1.2	[145]			1.932(-5)	4.75	[423]
$L_{\alpha2}$	L_3M_4	2289.875 ± 0.050	[50]	5.78(-3)	[488]	3.253(-3)	1.80	[473]
		2289.90	[104]				1.78	[423]
$L_{\alpha1}$	L_3M_5	2293.187 ± 0.050	[50]	5.13(-2)	[488]	2.889(-2)	1.86	[473]
		2293.20	[104]				1.79	[423]
$L_{\beta6}$	L_3N_1	2455.68 ± 0.36	[50]	3.72(-4)	[488]	2.439(-4)	12.50	[423]
		2455.60	[104]					
	L_3N_2	2480.9 ± 1.1	[145]			6.497(-7)	3.64	[423]
	L_3N_3	2488.06 ± 0.72	[145]			6.211(-7)	3.73	[423]
$L_{\beta15}$	L_3N_4	2521.63 ± 0.86	[145]	1.97(-4)	[488]	1.107(-4)		
$L_{\beta15,2}$	$L_3N_{4,5}$	2518.33 ± 0.15	[50]					
$L_{\beta2}$	L_3N_5	2518.30	[104]	1.73(-3)	[488]	9.765(-4)	5.30	[473]
		2520.74 ± 0.38	[145]					

M series

line	transition	E/eV		I/eV/$\hbar$		TPIV	Γ/eV	
	$M_1N_{2,3}$	470.00	[104]	5.01(-4)	[335]			
	M_1N_2					5.958(-5)	10.47	[423]
	M_1N_3					9.073(-5)	2.06	[423]
	M_2M_4	179.80 ± 0.50	[50]			3.872(-4)	3.05	[423]
	M_2N_1	351.00 ± 3.00	[50]	1.11(-4)	[335]	1.315(-4)	13.77	[423]
		351.00	[104]					
	M_3M_5	165.60 ± 0.20	[50]			2.925(-4)	3.20	[423]
	M_3N_1	331.00 ± 2.00	[50]	1.11(-4)	[335]	1.541(-4)	13.92	[423]
		332.50	[104]					
$M_{\gamma2}$	M_3N_4	384.63	[423]			9.191(-6)		
$M_{\gamma1}$	M_3N_5	384.86	[423]			7.586(-5)		
$M_{\xi2}$	M_4N_2	195.38	[423]			1.194(-4)	2.08	[423]
M_{δ}	M_4N_3	197.99	[423]			2.120(-5)	2.18	[423]
$M_{\xi1}$	M_5N_3	194.54	[423]			1.222(-4)	2.18	[423]
	$M_{4,5}N_{2,3}$	192.60 ± 0.20	[50]					
	$M_{4,5}O_{2,3}$	226.20 ± 0.80	[50]					

level characteristics

level	E_B/eV		ω_{nlj}		Γ/eV		AE/eV	
K	20000.00	[503]	0.765	[301]	4.52	[301]	19999.5	[51]
	19999.50 ± 0.30	[493]	0.776	[293]	4.37	[423]	20000.351 ± 0.020	[294]
			0.7951	[557]	4.52	[94]	20008.81 ± 0.28	[145]
			0.764	[38]				
			0.7672	[39]				
			0.7633	[423]				
			0.765	[310]				
			0.768	[33]				
			0.765	[112]				
L_1	2866.00	[503]	1.00(-2)	[301]	4.25	[301]	2865.5	[51]
	2865.50 ± 0.30	[493]	3.40(-3)	[107]	6.13	[423]	2880.6 ± 5.0	[50]
			5.75(-3)	[348]	3.8	[94]	2873.84 ± 0.88	[145]
			6.65(-3)	[423]				
			1.10(-2)	[576]				
			1.40(-2)	[20]				
L_2	2625.00	[503]	3.40(-2)	[301]	1.97	[301]	2625.1	[51]
	2625.10 ± 0.30	[493]	2.45(-2)	[107]	1.79	[423]	2627.30 ± 0.82	[50]
			3.50(-2)	[348]	1.83	[94]	2634.63 ± 0.36	[145]
			3.68(-2)	[423]				
			3.60(-2)	[436]				
			3.708-2)	[20]				
L_3	2520.00	[503]	3.70(-2)	[301]	1.78	[301]	2520.2	[51]
	2520.20 ± 0.30	[493]	2.59(-2)	[107]	1.67	[423]	2523.56 ± 0.76	[50]
			3.73(-2)	[348]	1.69	[94]	2529.71 ± 0.31	[145]
			3.83(-2)	[423]				
			4.00(-2)	[20]				
M_1	505.00	[503]	3.093(-3)	[512]	7.00	[74]	506.3	[51]
	504.60 ± 0.30	[493]	2.1(-4)	[423]	8.50	[423]		
					6.3	[94]		
M_2	410.00	[503]	3.1(-4)	[512]	2.70	[74]		
	409.70 ± 0.40	[493]	5.9(-4)	[423]	2.93	[423]		
					2.1	[94]		
M_3	393.00	[503]	5.6(-4)	[423]	2.70	[74]		
	392.30 ± 0.30	[493]			3.08	[423]		
					2.1	[94]		

level	E_B/eV		ω_{nlj}		Γ/eV		AE/eV	
M_4	230.00	[503]	2.7(-3)	[512]	0.10	[74]		
	230.30 $\pm$ 0.30	[493]	1.4(-4)	[423]	0.11	[423]		
					0.22	[94]		
M_5	227.00	[503]	1.2(-4)	[423]	0.10	[74]	227.9	[51]
	227.00 $\pm$ 0.30	[493]			0.12	[423]		
					0.12	[94]		
N_1	62.00	[503]	1.0(-5)	[423]	10.83	[423]		
	61.80 $\pm$ 0.30	[493]			7.10	[74]		
					3.2	[94]		
N_2	35.00	[503]	2.0(-6)	[423]	1.00	[74]		
	34.80 $\pm$ 0.30	[51]			1.97	[423]		
	38.30 $\pm$ 0.70	[493]			2.2	[94]		
N_3	35.00	[503]	1.0(-6)	[423]	1.00	[74]		
	34.80 $\pm$ 0.40	[51]			2.06	[423]		
	32.30 $\pm$ 0.70	[493]			1.6	[94]		
N_4	2.00	[503]						
	1.90 $\pm$ 0.30	[493]						
N_5	2.00	[503]						
	1.80 $\pm$ 0.30	[51]						
	1.20 $\pm$ 0.30	[493]						
IP	7.09243	[222]						

Tc Z=43	[Kr] $4d^5\,5s^2$

Technetium $A = 98$ [222]

$\varrho = 11.50\,\mathrm{g/cm^3}$ [547]
$\varrho = 11.49\,\mathrm{g/cm^3}$ [483]

X-ray transitions

line	transition	E/eV		I/eV/$\hbar$		TPIV [423]	Γ_X/eV	
K series								
K_{α_3}	KL_1	18003.8 ± 1.1	[145]				11.13	[423]
K_{α_2}	KL_2	18250.90 ± 1.2	[50]	1.073	[488]	2.256(-1)	6.99	[301]
		18251.00	[104]				6.64	[423]
K_{α_1}	KL_3	18367.2 ± 1.2	[50]	2.040	[488]	4.288(-1)	6.82	[301]
		18369.00	[104]				6.51	[423]
	KM_1	20501.2 ± 1.2	[145]				13.80	[423]
K_{β_3}	KM_2	20599.2 ± 2.0	[50]	0.1688	[488]	3.547(-2)	7.82	[423]
		20614.00	[104]					
K_{β_1}	KM_3	20619.0 ± 2.0	[50]	0.328	[488]	6.902(-2)	7.98	[423]
$K_{\beta_5}^{II}$	KM_4	20787.80 ± 0.92	[145]			1.850(-4)	4.89	[423]
$K_{\beta_5}^{I}$	KM_5	20792.02 ± 0.89	[145]			2.611(-4)	4.89	[423]
	KN_1	20977.3 ± 3.0	[145]				18.39	[423]
K_{β_2}	$KN_{2,3}$	21005.00 ± 2.00	[50]	2.533(-2)	[488]	1.24		
$K_{\beta_2}^{II}$	KN_2	20999.2 ± 1.4	[145]			6.843(-3)	8.07	[423]
$K_{\beta_2}^{I}$	KN_3	21006.03 ± 0.79	[145]			1.315(-2)	8.24	[423]
$K_{\beta_4}^{II}$	KN_4	21042.72 ± 0.85	[145]			9.789(-6)		
$K_{\beta_4}^{I}$	KN_5	21043.05 ± 0.75	[145]			1.366(-5)		
L series								
	L_1M_1	2497.4 ± 1.8	[145]				13.79	[423]
L_{β_4}	L_1M_2	2595.6 ± 2.0	[145]			2.316(-3)	7.82	[423]
L_{β_3}	L_1M_3	2614.9 ± 1.8	[145]			3.958(-3)	7.97	[423]
$L_{\beta_{10}}$	L_1M_4	2784.0 ± 1.5	[145]			2.101(-5)	4.88	[423]
L_{β_9}	L_1M_5	2788.2 ± 1.5	[145]			3.131(-5)	4.88	[423]
	L_1N_1	2973.4 ± 3.6	[145]				18.38	[423]
L_{γ_2}	L_1N_2	2995.4 ± 2.0	[145]			3.739(-4)	8.07	[423]
L_{γ_3}	L_1N_3	3002.2 ± 1.4	[145]			6.448(-4)	8.24	[423]

line	transition	E/eV		I/eV/ℏ		TPIV [423]	Γ$_X$/eV	
	L$_1$N$_4$	3038.9 ± 1.5	[145]			8.296(-7)		
	L$_1$N$_5$	3039.2 ± 1.4	[145]			1.234(-6)		
L$_\eta$	L$_2$M$_1$	2249.9 ± 1.3	[145]			3.183(-3)	10.94	[423]
	L$_2$M$_2$	2348.2 ± 1.5	[145]				4.96	[423]
L$_{\beta_{17}}$	L$_2$M$_3$	2367.5 ± 1.3	[145]			3.258(-5)	5.12	[423]
L$_{\beta_1}$	L$_2$M$_4$	2536.83 ± 0.61	[50]	6.80(-2)	[488]	3.499(-2)	2.03	[423]
	L$_2$M$_5$	2540.74 ± 0.96	[145]				2.03	[423]
L$_{\gamma_5}$	L$_2$N$_1$	2726.0 ± 3.0	[145]			2.683(-4)	15.53	[423]
	L$_2$N$_2$	2747.9 ± 1.5	[145]				5.21	[423]
	L$_2$N$_3$	2754.75 ± 0.86	[145]			1.503(-6)	5.39	[423]
L$_{\gamma_1}$	L$_2$N$_4$	2791.44 ± 0.91	[145]			1.573(-3)		
	L$_2$N$_5$	2791.77 ± 0.81	[145]					
L$_l$	L$_3$M$_1$	2133.6 ± 1.4	[145]			5.003(-3)	10.81	[423]
L$_t$	L$_3$M$_2$	2231.9 ± 1.4	[145]			2.178(-5)	4.83	[423]
L$_s$	L$_3$M$_3$	2251.2 ± 1.2	[145]			2.065(-5)	4.99	[423]
L$_{\alpha_2}$	L$_3$M$_4$	2420.24 ± 0.93	[145]			3.526(-3)	1.90	[423]
L$_{\alpha_1}$	L$_3$M$_5$	2423.99 ± 0.21	[50]	5.85(-2)	[488]	3.131(-2)	1.90	[423]
L$_{\beta_6}$	L$_3$N$_1$	2609.7 ± 3.0	[145]			2.673(-4)	15.40	[423]
	L$_3$N$_2$	2631.7 ± 1.4	[145]			7.667(-7)	5.09	[423]
	L$_3$N$_3$	2638.46 ± 0.80	[145]			7.321(-7)	5.26	[423]
L$_{\beta_{15}}$	L$_3$N$_4$	2675.15 ± 0.86	[145]			1.566(-4)		
L$_{\beta_2}$	L$_3$N$_5$	2675.49 ± 0.76	[145]			1.382(-3)		

M series

line	transition	E/eV				TPIV [423]	Γ$_X$/eV	
M$_{\gamma_2}$	M$_3$N$_4$	416.88	[423]			1.482(-5)		
M$_{\gamma_1}$	M$_3$N$_5$	417.17	[423]			1.223(-4)		
M$_{\xi_2}$	M$_4$N$_2$	217.41	[423]			1.657(-4)	3.46	[423]
M$_\delta$	M$_4$N$_3$	220.46	[423]			2.915(-5)	3.63	[423]
M$_{\xi_1}$	M$_5$N$_3$	216.52	[423]			1.704(-4)	3.64	[423]

level characteristics

level	E_B/eV		ω_{nlj}		Γ/eV		AE/eV	
K	21044.00	[503]	0.780	[301]	4.91	[301]	21044.0	[51]
	21044.00 ± 0.70	[493]	0.779	[38]	4.75	[423]	21047.49 ± 0.53	[50]
			0.8093	[557]	4.91	[94]	21050.47 ± 0.85	[145]
			0.792	[293]				
			0.7821	[39]				
			0.7794	[423]				
L_1	3043.00	[503]	1.10(-2)	[301]	4.36	[301]	3042.5	[51]
	3042.50 ± 0.40	[493]	7.46(-3)	[423]	6.38	[423]	3055.3 ± 5.6	[50]
			7.60(-3)	[436]	3.8	[94]	3046.63 ± 0.91	[145]
L_2	2793.00	[503]	3.70(-2)	[301]	2.08	[301]	2.7932	[51]
	2793.20 ± 0.40	[493]	4.01(-2)	[423]	1.89	[423]	2794.91 ± 0.93	[50]
			3.96(-2)	[436]	1.93	[94]	2799.19 ± 0.36	
L_3	2677.00	[503]	4.00(-2)	[301]	1.91	[301]	2.6769	[51]
	2676.90 ± 0.40	[493]	4.17(-2)	[423]	1.76	[423]	2677.80 ± 0.86	[50]
			4.10(-2)	[436]	1.78	[94]	2682.91 ± 0.31	[145]
M_1	544.00	[503]	3.259(-3)	[512]	6.7	[94]	544.0	[51]
	544.00 ± 1.00	[493]	2.3(-4)	[423]	9.05	[423]		
M_2	445.00	[503]	3.7(-4)	[512]	2.15	[94]		
	444.90 ± 1.50	[493]	6.8(-4)	[423]	3.08	[423]		
M_3	425.00	[503]	6.5(-4)	[423]	2.15	[94]		
	425.00 ± 1.50	[493]			3.23	[423]		
M_4	257.00	[503]	2.7(-3)	[512]	0.5	[94]		
	256.40 ± 0.50	[493]	1.9(-4)	[423]	0.14	[423]		
M_5	253.00	[503]	1.7(-4)	[423]	0.14	[423]	253.9	[51]
	252.90 ± 0.40	[493]			0.14	[423]		
N_1	68.00	[503]	1.0(-5)	[423]	13.64	[423]		
	68.80 ± 0.80	[493]			3.5	[94]		
N_2	39.00	[503]	3.0(-6)	[423]				
	38.90 ± 1.90	[51]			3.33	[423]		
	42.80 ± 1.90	[493]			2.7	[94]		

level	E_B/eV		ω_{nlj}		Γ/eV		AE/eV
N_3	39.00	[503]	1.0(-6)	[423]	2.2	[94]	
	36.90 ± 1.90	[493]			3.50	[423]	
N_4	2.00	[503]					
	2.00 ± 0.80	[493]					
N_5	2.00	[503]					
	1.20 ± 0.80	[493]					
IP	7.28	[222]					

Ru   Z=44   [Kr] 4d⁷ 5s¹

Ruthenium $A = 101.07(2)$ [222] $\varrho = 12.20\,\text{g/cm}^3$ [547]
$\varrho = 12.10\,\text{g/cm}^3$ [483]

X-ray transitions

line	transition	E/eV		I/eV/ℏ		TPIV	Γ/eV	
K series								
K_{α_3}	KL_1	18895.0 ± 1.283.10	[145]				11.67	[423]
K_{α_2}	KL_2	19150.49 ± 0.18	[50]	1.184	[488]	2.296(-1)	7.56	[301]
		19151.00	[104]				7.96	[473]
		19150.68 ± 0.18	[143]				7.14	[423]
		19150.68 ± 0.18	[277]					
K_{α_1}	KL_3	19279.16 ± 0.18	[50]	2.247	[488]	4.356(-1)	7.33	[301]
		19279.00	[104]				7.41	[473]
		19279.36 ± 0.18	[143]				7.00	[423]
		19279.29 ± 0.18	[277]					
	KM_1	21533.3 ± 1.2	[145]				14.71	[423]
K_{β_3}	KM_2	21634.65 ± 0.16	[50]	0.1885	[488]	3.655(-2)	8.36	[423]
		21635.00	[104]					
		21634.79 ± 0.23	[277]					
K_{β_1}	KM_3	21656.75 ± 0.16	[50]	0.367	[488]	7.109(-2)	8.54	[423]
		21657.00	[104]					
		21656.89 ± 0.23	[277]					
$K_{\beta_{52}}$	KM_4	21828.9 ± 8.00	[50]	1.05(-3)	[488]	2.042(-4)	5.32	[423]
		21828.00	[104]					
$K_{\beta_{51}}$	KM_5	21833.6 ± 5.1	[50]	1.48(-3)	[488]	2.873(-4)	5.33	[423]
		21833.00	[104]					
	KN_1	22046.6 ± 3.1	[145]				21.64	[423]
K_{β_2}	$KN_{2,3}$	22074.3 ± 1.7	[50]	5.65(-2)	[488]			
		22074.00	[104]					
$K_{\beta_2}^{II}$	KN_2	22071.1 ± 1.7	[145]			7.158(-3)	10.27	[423]
$K_{\beta_2}^{I}$	KN_3	22077.45 ± 0.88	[145]			1.378(-2)	10.58	[423]
K_{β_4}	$KN_{4,5}$	22104.6 ± 5.2	[50]	1.68(-4)	[488]			
		22114.00	[104]					
$K_{\beta_4}^{II}$	KN_4	22117.30 ± 0.84	[145]			1.361(-5)		
$K_{\beta_4}^{I}$	KN_5	22117.91 ± 0.83	[145]			1.893(-5)		

line	transition	E/eV		I/eV/$\hbar$		TPIV	Γ/eV	
	L series							
	L_1M_1	2638.1 ± 1.8	[145]				16.07	[423]
$L_{\beta4}$	L_1M_2	2741.15 ± 0.18	[50]	1.664(-2)	[488]	2.612(-3)	5.96	[473]
		2741.10	[104]				9.72	[423]
$L_{\beta3}$	L_1M_3	2763.39 ± 0.27	[50]	2.826(-2)	[488]	4.437(-3)	6.35	[473]
		2763.30	[104]				9.90	[423]
$L_{\beta10}$	L_1M_4	2938.7 ± 1.6	[145]			2.496(-5)	6.68	[423]
$L_{\beta9}$	L_1M_5	2943.6 ± 1.5	[145]			3.720(-5)	6.68	[423]
	L_1N_1	3151.6 ± 3.7	[145]				23.00	[423]
$L_{\gamma2,3}$	$L_1N_{2,3}$	3180.91 ± 0.24	[50]	4.75(-3)	[488]			
		3181.10	[104]					
$L_{\gamma2}$	L_1N_2	3176.1 ± 2.3	[145]			4.349(-4)	11.63	[423]
$L_{\gamma3}$	L_1N_3	3182.91 ± 0.24	[145]			7.456(-4)	11.93	[423]
	L_1N_4	3222.3 ± 1.5	[145]			1.224(-6)		
	L_1N_5	3222.9 ± 1.4	[145]			1.822(-6)		
L_η	L_2M_1	2381.99 ± 0.14	[50]	2.37(-3)	[488]	3.235(-3)	11.55	[423]
		2381.90	[104]					
	L_2M_2	2485.1 ± 1.5	[145]				5.20	[423]
$L_{\beta17}$	L_2M_3	2506.8 ± 1.3	[145]			3.454(-5)	5.37	[423]
$L_{\beta1}$	L_2M_4	2683.263 ± 0.010	[50]	7.73(-2)	[488]	3.769(-2)	2.18	[473]
		2683.30	[104]				2.16	[423]
	L_2M_5	2687.90 ± 0.97	[145]				2.16	[423]
$L_{\gamma5}$	L_2N_1	2891.85 ± 0.20	[50]	4.30(-4)	[488]	2.901(-4)	18.48	[423]
		2891.80	[104]					
	L_2N_2	2920.5 ± 1.8	[145]				7.11	[423]
	L_2N_3	2926.79 ± 0.93	[145]			1.753(-6)	7.41	[423]
$L_{\gamma1}$	L_2N_4	2964.52 ± 0.21	[50]	4.35(-3)	[488]	2.118(-3)	4.15	[473]
		2964.50	[104]					
	L_2N_5	2967.25 ± 0.88	[145]					
L_l	L_3M_1	2252.79 ± 0.18	[50]	2.77(-3)	[488]	5.200(-3)	11.40	[423]
		2252.70	[104]					
L_t	L_3M_2	2356.2 ± 1.4	[145]			2.319(-5)	4.96	[423]
L_s	L_3M_3	2377.9 ± 1.2	[145]			2.194(-5)	5.13	[423]
$L_{\alpha2}$	L_3M_4	2554.330 ± 0.055	[50]	7.48(-3)	[488]	3.810(-3)	1.98	[473]
		2554.40	[104]				2.02	[423]
$L_{\alpha1}$	L_3M_5	2558.579 ± 0.039	[50]	6.64(-2)	[488]	3.381(-2)	2.03	[473]
		2558.60	[104]					[423]

line	transition	E/eV		I/eV/$\hbar$		TPIV	Γ/eV	
$L_{\beta 6}$	L_3N_1	2763.39 ± 0.27	[50]	4.99(-4)	[488]	2.922(-4)	18.34	[423]
		2763.30	[104]					
	L_3N_2	2791.6 ± 1.7	[145]			8.964(-7)	6.87	[423]
	L_3N_3	2797.95 ± 0.87	[145]			8.552(-7)	7.17	[423]
$L_{\beta 15}$	L_3N_4	2837.79 ± 0.83	[145]	4.15(-4)	[488]	2.114(-4)		
$L_{\beta 15,2}$	$L_3N_{4,5}$	2835.96 ± 0.19	[50]					
$L_{\beta 2}$	L_3N_5	2838.40 ± 0.82	[145]			1.865(-3)		

M series

line	transition	E/eV		I/eV/$\hbar$		TPIV	Γ/eV	
	M_2M_4	199.20 ± 0.34	[50]			4.215(-4)		[423]
	M_2N_1	384.00 ± 2.40	[50]	1.51(-4)	[335]	1.679(-4)	19.70	[423]
		384.00	[104]					
	M_2N_4	486.00 ± 2.00	[50]	2.07(-4)	[335]	1.764(-4)		
		486.20	]					
	M_3M_5	181.40 ± 0.30	[50]			3.095(-4)	3.56	[423]
$M_{\gamma 2,1}$	$M_3N_{4,5}$	462.00 ± 2.00	[50]	2.07(-4)	[335]			
$M_{\gamma 2}$	M_3N_4	461.80	[104]			2.208(-5)		
	M_3N_5					1.822(-4)		
$M_{\xi 2}$	M_4N_2	240.34	[423]			2.244(-4)	5.29	[423]
M_δ	M_4N_3	243.87	[423]			3.911(-5)	5.59	[423]
$M_{\xi 1}$	M_5N_3	239.39	[423]			2.318(-4)	5.60	[423]
	$M_{4,5}N_{2,3}$	236.90 ± 0.30	[50]					
	$M_{4,5}O_{2,3}$	276.80 ± 0.60	[50]					

level characteristics

level	E_B/eV		ω_{nlj}		Γ/eV		AE/eV	
K	22117.00	[503]	0.794	[301]	5.33	[301]	22117.2	[51]
	22117.20 ± 0.30	[493]	0.8236	[557]	5.15	[423]	22119.56 ± 0.58	[50]
			0.806	[347]	5.33	[94]	22127.70 ± 0.32	[145]
			0.807	[293]				
			0.793	[38]				
			0.7958	[39]				
			0.7943	[423]				
			0.794	[310]				
L_1	3224.00	[503]	1.20(-2)	[301]	4.58	[301]	3224.0	[51]
	3224.00 ± 0.30	[493]	7.74(-3)	[348]	6.51	[423]	3232.9 ± 6.2	[50]
			8.41(-3)	[423]	3.9	[94]	3232.69 ± 0.93	[145]
			8.30(-3)	[436]				
			1.50(-2)	[576]				
L_2	2967.00	[503]	4.00(-2)	[301]	2.23	[301]	2966.9	[51]
	2966.90 ± 0.30	[493]	4.18(-2)	[348]	1.99	[423]	2966.1 ± 1.1	[50]
			4.34(-2)	[423]	2.03	[94]	2977.03 ± 0.37	[145]
			4.30(-2)	[436]				
L_3	2838.00	[503]	4.30(-2)	[301]	2.00	[301]	2837.9	[51]
	2837.90 ± 0.30	[493]	4.50(-2)	[348]	1.85	[423]	2837.77 ± 0.96	[50]
			4.52(-2)	[423]	1.87	[94]	2848.19 ± 0.32	[145]
			4.50(-2)	[436]				
M_1	585.00	[503]	1.20(-4)	[38]	7.89	[350]	586.1	[51]
	585.00 ± 0.30	[493]	3.429(-3)	[512]	9.56	[423]		
			2.6(-4)	[423]	7.2	[94]		
M_2	483.00	[503]	2.60(-4)	[38]	2.91	[350]		
	482.80 ± 0.30	[493]	3.7(-4)	[512]	3.21	[423]		
			7.7(-4)	[423]	2.2	[94]		
M_3	461.00	[503]	2.30(-4)	[38]	2.98	[350]		
	460.60 ± 0.30	[493]	7.5(-4)	[423]	3.39	[423]		
					2.2	[94]		
M_4	284.00	[503]	2.90(-3)	[38]	0.24	[350]		
	283.60 ± 0.30	[493]	2.7(-3)	[512]	0.17	[423]		
			1.6(-4)	[423]	0.59	[94]		

level	E_B/eV		ω_{nlj}		Γ/eV		AE/eV	
M_5	279.00	[503]	2.3(-4)	[423]	0.22	[74]	280.0	[51]
	279.40 ± 0.30	[493]			0.17	[423]		
					0.17	[94]		
N_1	75.00	[503]	1.0(-5)	[423]	16.49	[423]		
	74.90 ± 0.30	[493]			7.40	[74]		
					3.9	[94]		
N_2	43.00	[503]	1.0(-5)	[423]	3.2	[94]		
	43.10 ± 0.40	[51]			5.12	[423]		
	47.00 ± 0.40	[493]			3.80	[74]		
N_3	43.00	[503]	2.0(-6)	[423]	2.8	[94]		
	43.10 ± 0.40	[51]			5.42	[423]		
	41.20 ± 0.40	[493]			3.80	[74]		
N_4	2.00	[503]						
	2.00 ± 0.30	[51]						
	2.40 ± 0.30	[493]						
N_5	2.00	[503]						
	2.00 ± 0.30	[51]						
	1.80 ± 0.30	[493]						
IP	7.36050	[222]						

Rh   Z=45   [Kr] $4d^8\,5s^1$

Rhodium $A = 102.90550(2)$ [222] $\varrho = 12.39\,\mathrm{g/cm^3}$ [547]
$\varrho = 12.44\,\mathrm{g/cm^3}$ [483]

X-ray transitions

line	transition	E/eV		I/eV/$\hbar$		TPIV	Γ/eV	
K series								
K_{α_3}	KL_1	19809.3 ± 1.1	[145]				12.20	[423]
$K_{\alpha2}$	KL_2	20073.67 ± 0.20	[50]	1.304	[488]	2.333(-1)	7.90	[301]
		20074.00	[104]				9.40	[473]
							7.68	[423]
$K_{\alpha1}$	KL_3	20216.12 ± 0.20	[50]	2.469	[488]	4.417(-1)	7.90	[301]
		20216.00	[104]				9.50	[473]
							7.53	[423]
	KM_1	22586.0 ± 1.2	[145]				15.45	[423]
$K_{\beta3}$	KM_2	22698.83 ± 0.17	[50]	0.2100	[488]	3.756(-2)	8.94	[423]
		22699.00	[104]					
$K_{\beta1}$	KM_3	22723.59 ± 0.17	[50]	0.408	[488]	7.304(-2)	9.13	[423]
		22724.00	[104]					
$K_{\beta52}$	KM_4	22909.6 ± 5.6	[50]	1.25(-3)	[488]	2.241(-4)	5.80	[423]
		22908.00	[104]					
$K_{\beta51}$	KM_5	22916.8 ± 5.6	[50]	1.76(-3)	[488]	3.146(-4)	5.80	[423]
		22916.00	[104]					
	KN_1	23142.4 ± 3.2	[145]				24.80	[423]
$K_{\beta22}$	KN_2	23169.2 ± 2.0	[145]	3.33(-2)	[488]	7.470(-3)	13.00	[423]
$K_{\beta21}$	KN_3	23175.60 ± 0.96	[145]	6.45(-2)	[488]	1.443(-2)	13.49	[423]
$K_{\beta2}$	$KN_{2,3}$	23173.0 ± 1.3	[50]					
		23173.00	[104]					
$K_{\beta4}$	$KN_{4,5}$	23217.2 ± 5.8	[50]	2.43(-4)	[488]			
$K_{\beta_4^{II}}$	KN_4	23206.00	[104]			1.821(-5)		
		23218.69 ± 0.83	[145]					
$K_{\beta_4^{I}}$	KN_5	23216.00	[104]			2.527(-5)		
		23219.47 ± 0.58	[145]					
L series								
	L_1M_1	2776.6 ± 1.8	[145]				16.47	[423]

line	transition	E/eV		I/eV/$\hbar$		TPIV	Γ/eV	
$L_{\beta4}$	L_1M_2	2890.84 ± 0.20	[50]	1.88(-2)	[488]	2.986(-3)	9.96	[423]
		2890.80	[104]					
$L_{\beta3}$	L_1M_3	2915.72 ± 0.20	[50]	3.17(-2)	[488]	5.037(-3)	10.15	[423]
		2915.60	[104]					
$L_{\beta10}$	L_1M_4	3098.5 ± 1.6	[145]			2.998(-5)	6.82	[423]
$L_{\beta9}$	L_1M_5	3103.7 ± 1.5	[145]			4.471(-5)	6.82	[423]
	L_1N_1	3333.0 ± 3.8	[145]				25.82	[423]
$L_{\gamma2.3}$	$L_1N_{2,3}$	3364.06 ± 0.27	[50]	5.48(-3)	[488]			
		3363.90	[104]					
$L_{\gamma2}$	L_1N_2	3360.5 ± 2.7	[145]			5.105(-4)	14.02	[423]
$L_{\gamma3}$	L_1N_3	3366.3 ± 1.6	[145]			8.696(-4)	14.51	[423]
	L_1N_4	3409.4 ± 1.4	[145]			1.769(-6)		
	L_1N_5	3410.1 ± 1.5	[145]			2.635(-6)		
L_{η}	L_2M_1	2519.10 ± 0.15	[50]	2.65(-3)	[488]	3.286(-3)	11.96	[423]
		2519.00	[104]					
	L_2M_2	2626.3 ± 1.5	[145]				5.45	[423]
$L_{\beta17}$	L_2M_3	2650.6 ± 1.3	[145]			3.656(-5)	5.64	[423]
$L_{\beta1}$	L_2M_4	2834.439 ± 0.038	[50]	8.75(-2)	[488]	4.057(-2)	2.30	[423]
		2834.40	[104]					
	L_2M_5	2839.56 ± 0.98	[145]				2.31	[423]
$L_{\gamma5}$	L_2N_1	3065.00 ± 0.22	[50]	4.91(-4)	[488]	3.131(-4)	21.30	[423]
		3065.10	[104]					
	L_2N_2	3096.4 ± 2.1	[145]				9.51	[423]
	L_2N_3	3102.1 ± 1.0	[145]			2.036(-6)	10.00	[423]
$L_{\gamma1}$	L_2N_4	3143.81 ± 0.24	[50]	5.96(-3)	[488]	2.764(-3)		
		3143.80	[104]					
	L_2N_5	3145.99 ± 0.91	[145]					
L_l	L_3M_1	2376.55 ± 0.20	[50]	3.12(-3)	[488]	5.399(-3)	11.80	[423]
		2376.50	[104]					
L_t	L_3M_2	2484.1 ± 1.4	[145]			2.464(-5)	5.44	[423]
L_s	L_3M_3	2508.4 ± 1.2	[145]			2.328(-5)	5.47	[423]
$L_{\alpha2}$	L_3M_4	2692.078 ± 0.078	[50]	8.45(-3)	[488]	4.099(-3)	2.15	[423]
		2692.00	[104]					
$L_{\alpha1}$	L_3M_5	2696.775 ± 0.078	[50]	7.50(-2)	[488]	3.639(-2)	2.15	[423]
		2696.70	[104]					
$L_{\beta6}$	L_3N_1	2922.94 ± 0.20	[50]	5.73(-4)	[488]	3.187(-4)	21.15	[423]
		2922.90	[104]					

line	transition	E/eV		I/eV/$\hbar$		TPIV	Γ/eV	
	L_3N_2	2954.1 ± 2.1	[145]				9.34	[423]
	L_3N_3	2959.86 ± 0.94	[145]				9.83	[423]
$L_{\beta15}$	L_3N_4	3002.94 ± 0.82	[145]	5.68(-4)	[488]	2.755(-4)		
$L_{\beta2}$	L_3N_5	3003.73 ± 0.86	[145]	5.01(-3)	[488]	2.430(-3)		
$L_{\beta15,2}$	$L_3N_{4,5}$	3001.27 ± 0.22	[50]					

M series

line	transition	E/eV		I/eV/$\hbar$		TPIV	Γ/eV	
	M_2M_4	209.00 ± 0.15	[50]			4.304(-4)	3.56	[423]
	M_2N_1	442.00 ± 3.00	[50]	1.75(-4)	[335]	1.838(-4)	22.56	[423]
		443.60	[104]					
	M_3M_4	189.20 ± 0.30	[50]			3.094(-5)	3.75	[423]
	M_3N_1	417.00 ± 1.00	[50]	1.75(-4)	[335]	2.204(-4)	22.75	[423]
		416.20	[104]					
$M_{\gamma2,1}$	$M_3N_{4,5}$	496.00 ± 1.80	[50]	3.06(-4)	[335]			
		495.90	[104]					
$M_{\gamma2}$	M_3N_4					3.078(-5)		
$M_{\gamma1}$	M_3N_5					2.538(-4)		
	$M_{4,5}N_{2,3}$	260.10 ± 0.50	[50]					
$M_{\xi2}$	M_4N_2					2.977(-4)	7.62	[423]
M_δ	M_4N_3					5.136(-5)	8.11	[423]
	M_5N_2						7.62	[423]
$M_{\xi1}$	M_5N_3					3.086(-4)	8.12	[423]
	$M_{4,5}O_{2,3}$	303.00 ± 1.50	[50]					

level characteristics

level	E_B/eV		ω_{nlj}		Γ/eV		AE/eV	
K	23220.00	[503]	0.808	[301]	5.77	[301]	23219.9	[51]
	23219.90 ± 0.30	[493]	0.820	[293]	5.59	[423]	23221.99 ± 0.30	[294]
			0.8367	[557]	5.77	[94]	23230.32 ± 0.30	[145]
			0.807	[38]				
			0.8086	[39]				
			0.8081	[423]				
			0.808	[310]				
			0.808	[112]				
L_1	3412.00	[503]	1.30(-2)	[301]	4.73	[301]	3411.9	[51]
	3411.90 ± 0.30	[493]	9.61(-3)	[423]	6.61	[423]	3416.7 ± 7.0	[50]
			8.90(-3)	[436]	4	[94]	3420.89 ± 0.92	[145]
			9.16(-3)	[244]				
			1.70(-2)	[576]				
L_2	3146.00	[503]	4.30(-2)	[301]	2.35	[301]	3146.1	[51]
	3146.10 ± 0.30	[493]	4.70(-2)	[423]	2.09	[423]	3144.76 ± 0.59	[50]
			4.50(-2)	[436]	2.13	[94]	3156.74 ± 0.37	[145]
			4.78(-2)	[244]				
L_3	3004.00	[503]	4.30(-2)	[301]	2.13	[301]	3003.8	[51]
	3003.80 ± 0.30	[493]	4.90(-2)	[423]	1.94	[423]	3002.07 ± 0.54	[50]
			4.70(-2)	[436]	1.96	[423]	3014.48 ± 0.31	[145]
			5.01(-2)	[244]				
M_1	627.00	[503]	3.429(-3)	[512]	7.6	[94]	628.1	[51]
	627.10 ± 0.30	[493]	2.7(-4)	[423]	9.86	[423]		
M_2	521.00	[503]	4.9(-4)	[512]	2.25	[94]		
	521.00 ± 0.30	[493]	8.6(-4)	[423]	3.35	[423]		
M_3	496.00	[503]	8.5(-4)	[423]	2.25	[94]		
	496.20 ± 0.30	[493]			3.54	[423]		
M_4	312.00	[503]	2.7(-3)	[512]	0.61	[94]		
	311.70 ± 0.30	[493]	3.5(-4)	[423]	0.21	[423]		
M_5	307.00	[503]	3.1(-4)	[423]	0.21	[94]	307.2	[51]
	307.00 ± 0.30	[493]			0.21	[423]		

level	E_B/eV		ω_{nlj}		Γ/eV		AE/eV
N_1	81.00	[503]	2.0(-5)	[423]	19.21	[423]	
	81.00 ± 0.30	[493]			4.2	[94]	
N_2	48.00	[503]	1.0(-5)	[423]	7.41	[423]	
	47.90 ± 0.40	[51]			4.2	[94]	
	51.90 ± 0.50	[493]					
N_3	48.00	[503]	2.0(-6)	[423]	7.90	[423]	
	47.90 ± 0.40	[51]			3.8	[94]	
	46.30 ± 0.50	[493]					
N_4	3.00	[503]					
	2.50 ± 0.40	[51]					
	2.80 ± 0.40	[493]					
N_5	3.00	[503]					
	2.50 ± 0.40	[51]					
	2.20 ± 0.40	[493]					
IP	7.45890	[222]					

Pd   Z=46   [Kr] 4d^{10}

Palladium $A = 106.42(1)$ [222] $\varrho = 12.00\,\text{g/cm}^3$ [547]
$\varrho = 12.20\,\text{g/cm}^3$ [483]

X-ray transitions

line	transition	E/eV		I/eV/$\hbar$		TPIV	Γ/eV	
K series								
K_{α_3}	KL_1	20747.8 ± 1.1	[145]				12.32	[423]
$K_{\alpha 2}$	KL_2	21020.15 ± 0.22	[50]	1.433	[488]	2.367(-1)	8.67	[301]
		21020.00	[104]				9.20	[473]
							8.26	[423]
$K_{\alpha 1}$	KL_3	21177.08 ± 0.17	[50]	2.707	[488]	4.472(-1)	8.49	[301]
		21177.00	[104]				8.80	[473]
							8.09	[423]
	KM_1	23681.7 ± 1.2	[145]				16.51	[423]
$K_{\beta 3}$	KM_2	23791.12 ± 0.19	[50]	0.2332	[488]	3.852(-2)	9.53	[423]
		23791.00	[104]					
		23792.00 ± 2.00	[95]					
$K_{\beta 1}$	KM_3	23818.69 ± 0.19	[50]	0.453	[488]	7.487(-2)	9.74	[423]
		23819.00	[104]					
		23819.00 ± 2.00	[95]					
$K_{\beta 5}$	$KM_{4,5}$	23995.0 ± 6.2	[50]	2.07(-3)	[488]			
		23994.00	[104]					
$K_{\beta_5}^{II}$	KM_4	24011.0 ± 1.0	[145]			2.449(-4)	6.32	[423]
$K_{\beta_5}^{I}$	KM_5	24016.22 ± 0.99	[145]			3.427(-4)	6.32	[423]
	KN_1	24267.0 ± 3.6	[145]				27.84	[423]
$K_{\beta 2}$	$KN_{2,3}$	24299.40 ± 0.28	[50]	7.25(-2)	[488]			
		24299.00	[104]					
		24299.00 ± 1.00	[95]					
$K_{beta_2}^{II}$	KN_2	24297.2 ± 2.2	[145]			7.784(-3)	17.34	[423]
$K_{\beta_2}^{I}$	KN_3	24302.2 ± 2.4	[145]			1.508(-2)	18.05	[423]
$K_{\beta 4}$	$KN_{4,5}$	24346.00 ± 3.90	[50]	3.51(-4)	[488]			
$K_{\beta_4}^{II}$	KN_4	24348.74 ± 0.89	[145]			2.439(-5)		
$K_{\beta_4}^{I}$	KN_5	24349.18 ± 0.88	[145]			3.354(-5)		

line	transition	E/eV		I/eV/$\hbar$		TPIV	Γ/eV	

L series

line	transition	E/eV		I/eV/$\hbar$		TPIV	Γ/eV	
	L_1M_1	2934.0 ± 1.8	[145]				8.18	[423]
$L_{\beta4}$	L_1M_2	3045.43 ± 0.22	[50]	2.12(-2)	[488]	3.594(-3)	6.18	[473]
		3045.40	[104]				9.73	[423]
		3045.60 ± 1.00	[437]					
$L_{\beta3}$	L_1M_3	3072.98 ± 0.23	[50]	3.55(-2)	[488]	6.023(-3)	6.80	[473]
		3073.00	[104]				9.95	[423]
		3074.10 ± 1.00	[437]					
$L_{\beta10}$	L_1M_4	3263.72 ± 0.25	[50]	2.23(-4)	[488]	3.790(-5)	6.52	[423]
		3263.70	[104]					
$L_{\beta9}$	L_1M_5	3269.58 ± 0.26	[50]	3.33(-4)	[488]	5.651(-5)	6.52	[423]
		3269.50	[104]					
	L_1N_1	3519.3 ± 4.1	[145]				28.05	[423]
$L_{\gamma2.3}$	$L_1N_{2,3}$	3553.32 ± 0.30	[50]	6.21(-3)	[488]			
		3553.20	[104]					
$L_{\gamma2}$	L_1N_2	3549.4 ± 2.8	[145]			6.237(-4)	17.55	[423]
$L_{\gamma3}$	L_1N_3	3554.5 ± 3.0	[145]			1.054(-3)	18.26	[423]
	L_1N_4	3601.0 ± 1.5	[145]			2.709(-6)		
	L_1N_5	3601.4 ± 1.5	[145]			4.016(-6)		
L_{η}	L_2M_1	2660.28 ± 0.17	[50]	2.95(-3)	[488]	3.357(-3)	12.67	[423]
		2660.30	[104]					
	L_2M_2	2771.4 ± 1.3	[145]				5.67	[423]
$L_{\beta17}$	L_2M_3	2798.9 ± 1.2	[145]			3.862(-5)	5.89	[423]
$L_{\beta1}$	L_2M_4	2990.250 ± 0.053	[50]	9.87(-2)	[488]	4.345(-2)	2.36	[473]
		2990.30	[104]				2.46	[423]
		2990.10 ± 1.00	[437]					
	L_2M_5	2995.6 ± 1.0	[145]				2.46	[423]
$L_{\gamma5}$	L_2N_1	3243.74 ± 0.25	[50]	5.54(-4)	[488]	3.371(-4)	23.99	[423]
		3243.60	[104]					
	L_2N_2	3276.5 ± 2.2	[145]				13.49	[423]
	L_2N_3	3281.6 ± 2.4	[145]			2.317(-6)	14.20	[423]
$L_{\gamma1}$	L_2N_4	3328.74 ± 0.26	[50]	8.21(-3)	[488]	3.612(-3)	4.50	[473]
		3328.70	[104]					
	L_2N_5	3328.56 ± 0.89	[145]					
L_l	L_3M_1	2503.43 ± 0.22	[50]	3.51(-3)	[488]	5.598(-3)	12.50	[423]
		2503.40	[104]					
L_t	L_3M_2	2614.8 ± 1.3	[145]			2.614(-5)	5.51	[423]
L_s	L_3M_3	2642.3 ± 1.2	[145]			2.465(-5)	5.73	[423]

line	transition	E/eV		I/eV/$\hbar$		TPIV	Γ/eV	
$L_{\alpha2}$	L_3M_4	2833.312 ± 0.067	[50]	9.51(-3)	[488]	4.392(-3)	2.16	[473]
		2833.30	[104]				2.29	[423]
$L_{\alpha1}$	L_3M_5	2838.638 ± 0.048	[50]	8.44(-2)	[488]	3.899(-2)	2.21	[473]
		2838.60	[104]				2.30	[423]
$L_{\beta6}$	L_3N_1	3087.06 ± 0.23	[50]	6.52(-4)	[488]	3.467(-4)	23.82	[423]
		3087.00	[104]					
		3087.50 ± 1.00	[437]					
	L_3N_2	3119.9 ± 2.2	[145]			1-185(-6)	13.33	[423]
	L_3N_3	3125.0 ± 2.4	[145]			1.127(-6)	14.04	[423]
$L_{\beta15}$	L_3N_4	3171.80	[104]	7.79(-4)	[488]	3.598(-4)		
		3171.50 ± 0.85	[145]					
$L_{\beta2}$	L_3N_5	3171.80	[104]	6.83(-4)	[488]	3.154(-3)	5.63	[473]
		3171.93 ± 0.84	[145]					
$L_{\beta15,2}$	$L_3N_{4,5}$	3171.820 ± 0.048	[50]					

M series

line	transition	E/eV		I/eV/$\hbar$		TPIV	Γ/eV	
	$M_1N_{2,3}$	616.00 ± 6.00	[50]	1.04(-3)	[335]			
		617.00	[104]					
	M_1N_2					9.520(-5)	21.75	[423]
	M_1N_3					1.395(-4)	22.46	[423]
	M_2M_4	219.40 ± 0.40	[50]			4.293(-4)	3.74	[423]
	M_2N_1	474.00 ± 4.00	[50]	2.00(-4)	[335]	1.942(-4)	25.26	[423]
		473.00	[104]					
	M_2N_4	560.00 ± 2.50	[50]	4.47(-4)	[335]	3.343(-4)		
		561.00	[104]					
	M_3M_5	197.00 ± 0.30	[50]			3.028(-4)	3.95	[423]
	M_3N_1	445.00 ± 1.60	[50]	2.00(-4)	[335]	2.351(-4)	25.47	[423]
		446.00	[104]					
$M_{\gamma2,1}$	$M_3N_{4,5}$	531.00 ± 2.00	[50]	4.47(-4)	[335]			
$M_{\gamma2}$	M_3N_4					4.183(-5)		
$M_{\gamma1}$	M_3N_5					3.424(-4)		
	$M_{4,5}N_{2,3}$	284.40 ± 0.70	[50]					
$M_{\xi2}$	M_4N_2					3.876(-4)	11.55	[423]
M_{δ}	M_4N_3					6.612(-5)	12.26	[423]
	M_5N_2						11.55	[423]
$M_{\xi1}$	M_5N_3					4.034(-4)	12.26	[423]
	$M_{4,5}O_{2,3}$	332.00 ± 2.00	[50]					

level characteristics

level	E_B/eV		ω_{nlj}		Γ/eV		AE/eV	
K	24350.00	[503]	0.820	[301]	6.24	[301]	24350.3	[51]
	24350.30 ± 0.30	[493]	0.833	[293]	6.05	[423]	24352.59 ± 0.20	[294]
			0.8491	[557]	6.25	[94]	24357.63 ± 0.36	[145]
			0.819	[38]				
			0.8204	[39]				
			0.8208	[423]				
			0.820	[310]				
			0.819	[38]				
L_1	3605.00	[503]	1.40(-2)	[301]	4.93	[301]	3604.3	[51]
	3604.30 ± 0.30	[493]	1.15(-2)	[423]	6.26	[423]	3607.3 ± 1.6	[50]
			1.90(-2)	[576]	3.9	[94]	3609.87 ± 0.93	[145]
L_2	3331.00	[503]	4.40(-2)	[301]	2.43	[301]	3330.3	[51]
	3330.30 ± 0.30	[493]	5.08(-2)	[423]	2.20	[423]	3330.35 ± 0.13	[50]
			5.00(-2)	[436]	2.23	[94]	3337.01 ± 0.37	[145]
L_3	3173.00	[503]	4.90(-2)	[301]	2.25	[301]	3173.01 ± 0.12	[50]
	3173.30 ± 0.30	[493]	5.29(-2)	[423]	2.03	[423]	3173.3	[51]
			5.20(-2)	[436]	2.05	[94]	3180.38 ± 0.32	[145]
M_1	670.00	[503]	3.785(-3)	[512]	9.00	[74]	671.6	[51]
	669.90 ± 0.30	[493]	3.0(-4)	[423]	10.46	[423]		
					8	[94]		
M_2	559.00	[503]	5.5(-4)	[512]	3.70	[74]		
	559.10 ± 0.30	[493]	9.6(-4)	[423]	3.47	[423]		
					2.35	[94]		
M_3	531.00	[503]	9.5(-4)	[423]	3.70	[74]		
	531.50 ± 0.30	[493]			3.69	[423]		
					2.35	[94]		
M_4	340.00	[503]	2.7(-3)	[512]	0.40	[74]		
	340.00 ± 0.30	[493]	4.5(-4)	[423]	0.26	[423]		
					0.26	[94]		
M_5	335.00	[503]	4.0(-4)	[423]	0.40	[74]	335.2	[51]
	334.70 ± 0.30	[493]			0.26	[423]		

level	E_B/eV		ω_{nlj}		Γ/eV		AE/eV
					0.26	[94]	
N_1	86.00	[503]	2.0(-5)	[423]	21.79	[423]	
	86.40 ± 0.30	[493]			7.20	[74]	
					4.35	[94]	
N_2	51.00	[503]	1.0(-5)	[423]	10.00	[74]	
	51.10 ± 0.40	[51]			11.29	[423]	
	54.40 ± 0.60	[493]			6.4	[94]	
N_3	51.00	[503]	4.0(-6)	[423]	10.00	[74]	
	51.10 ± 0.40	[51]			12.00	[423]	
	50.00 ± 0.60	[493]			5.6	[94]	
N_4	1.00	[503]					
	1.50 ± 0.30	[51]					
	1.70 ± 0.30	[493]					
N_5	1.00	[503]					
	1.50 ± 0.30	[51]					
	1.30 ± 0.30	[493]					
IP	8.3369	[222]					

Ag	Z=47

[Kr] 4d^{10} 5s^1

Silver

$A = 107.8682(2)$ [222] $\varrho = 10.48\,\text{g/cm}^3$ [547]
  $\varrho = 10.492\,\text{g/cm}^3$ [483]

X-ray transitions

line	transition	E/eV		I/eV/$\hbar$		TPIV	Γ/eV	
K series								
K_{α_3}	KL$_1$	21709.4 ± 1.2	[145]				12.00	[423]
$K_{\alpha 2}$	KL$_2$	21990.30 ± 0.10	[50]	1.571	[488]	2.400(-1)	9.32	[301]
		21991.00	[104]	1.599	[487]		11.30	[473]
		21990.51 ± 0.18	[143]				8.86	[423]
		21990.44 ± 0.10	[277]					
$K_{\alpha 1}$	KL$_3$	22162.917 ± 0.030	[50]	2.961	[488]	4.523(-1)	9.16	[301]
		22162.99	[104]	3.02	[487]		11.10	[473]
		22163.14 ± 0.04	[143]				8.68	[423]
		22163.06 ± 0.03	[277]					
	KM$_1$	24797.4 ± 1.2	[145]				17.67	[423]
$K_{\beta 3}$	KM$_2$	24911.54 ± 0.30	[50]	0.2582	[488]	3.944(-2)	10.20	[423]
		24911.98	[104]	0.2789	[487]			
		24911.71 ± 0.30	[277]					
$K_{\beta 1}$	KM$_3$	24942.42 ± 0.30	[50]	0.502	[488]	7.663(-2)	10.43	[423]
		24942.60	[104]	0.543	[487]			
		24942.58 ± 0.30	[277]					
$K_{\beta 5}$	KM$_{4,5}$	25145.5 ± 1.5	[50]	2.44(-3)	[488]			
		25145.00	[104]					
$K_{\beta_5}^{II}$	KM$_4$	25141.5 ± 1.5	[145]			2.665(-4)	6.86	[423]
$K_{\beta_5}^{I}$	KM$_5$	25147.6 ± 1.0	[145]			3.719(-4)	6.86	[423]
	KN$_1$	25421.1 ± 3.5	[145]				30.48	[423]
$K_{\beta_2^{II}}$	KN$_2$	25455.6 ± 2.9	[145]	4.77(-2)	[487]	8.016(-3)	20.31	[423]
$K_{\beta_2^{I}}$	KN$_3$	25457.8 ± 2.0	[145]	9.25(-2)	[487]	1.550(-2)	21.26	[423]
$K_{\beta 2}$	KN$_{2,3}$	25456.71 ± 0.31	[50]	8.27(-2)	[488]			
		25457.00	[104]					
$K_{\beta_4}^{II}$	KN$_4$	25490.00	[104]			3.003(-5)		
		25509.99 ± 0.91	[145]					
$K_{\beta_4}^{I}$	KN$_5$	25510.83 ± 0.99	[145]			4.140(-5)		

line	transition	E/eV		I/eV/ℏ		TPIV	Γ/eV	
K_{β_4}	$KN_{4,5}$	25511.8 ± 3.1	[50]	4.68(-4)	[488]			

L series

line	transition	E/eV		I/eV/ℏ		TPIV	Γ/eV	
	L_1M_1	3088.0 ± 1.7	[145]				16.52	[423]
L_{β_4}	L_1M_2	3203.487 ± 0.061	[50]	2.38(-2)	[488]	4.639(-3)	5.90	[279]
		3203.33	[104]				9.04	[423]
L_{β_3}	L_1M_3	3234.49 ± 0.11	[50]	3.96(-2)	[488]	7.719(-3)	6.60	[279]
		3234.30	[104]				9.28	[423]
$L_{\beta_{10}}$	L_1M_4	3432.91 ± 0.13	[50]	2.63(-4)	[488]	5.127(-5)	5.70	[423]
		3432.86	[104]					
L_{β_9}	L_1M_5	3439.21 ± 0.13	[50]	3.92(-4)	[488]	7.647(-5)	5.71	[423]
		3439.00	[104]					
	L_1N_1	3711.7 ± 4.0	[145]				29.33	
L_{γ_2}	L_1N_2	3743.25 ± 0.15	[50]	4.25(-3)	[488]	8.295(-4)	11.00	[279]
							19.16	[423]
L_{γ_3}	L_1N_3	3749.82 ± 0.15	[50]	7.14(-3)	[488]	1.394(-3)	10.20	[279]
							20.11	[423]
$L_{\gamma_{2,3}}$	$L_1N_{2,3}$	3749.70	[104]					
	L_1N_4	3800.6 ± 1.5	[145]			4.091(-6)		
	L_1N_5	3801.4 ± 1.6	[145]			4.103(-6)		
L_η	L_2M_1	2806.11 ± 0.19	[50]	3.28(-3)	[488]	3.437(-3)	13.43	[423]
		2806.10	[104]					
	L_2M_2	2922.1 ± 1.5	[145]				5.95	[423]
$L_{\beta_{17}}$	L_2M_3	2952.4 ± 1.3	[145]			4.073(-5)	6.18	[423]
L_{β_1}	L_2M_4	3150.974 ± 0.036	[50]	0.1109	[488]	4.651(-2)	2.40	[279]
		3151.00	[104]				2.61	[423]
	L_2M_5	3156.9 ± 1.0	[145]				2.62	[423]
L_{γ_5}	L_2N_1	3428.35 ± 0.13	[50]	6.30(-4)	[488]	3.622(-4)	26.23	[423]
		3428.20	[104]					
	L_2N_2	3465.0 ± 2.9	[145]				16.07	[423]
	L_2N_3	3467.1 ± 2.0	[145]			2.684(-6)	17.02	[423]
L_{γ_1}	L_2N_4	3519.625 ± 0.059	[50]	1.04(-2)	[488]	4.347(-3)	3.95	[279]
		3619.60	[104]					
	L_2N_5	3520.17 ± 0.97	[145]					
L_l	L_3M_1	2633.66 ± 0.17	[50]	3.93(-3)	[488]	5.801(-3)	13.25	[423]
		2633.60	[104]					
L_t	L_3M_2	2749.8 ± 1.4	[145]			2.771(-5)	5.76	[423]
L_s	L_3M_3	2780.1 ± 1.2	[145]			2.608(-5)	5.99	[423]

line	transition	E/eV		I/eV/$\hbar$		TPIV	Γ/eV	
$L_{\alpha2}$	L_3M_4	2978.240 ± 0.053 2978.30	[50] [104]	1.07(-2)	[488]	4.713(-3)	2.42	[423]
$L_{\alpha1}$	L_3M_5	2984.340 ± 0.032 2984.30	[50] [104]	9.46(-2)	[488]	4.181(-2)	3.20 2.43	[279] [423]
$L_{\beta6}$	L_3N_1	3256.06 ± 0.11 3255.90	[50] [104]	7.46(-4)	[488]	3.766(-4)	26.05	[423]
	L_3N_2	3292.7 ± 2.8	[145]			1.372(-6)	15.88	[423]
	L_3N_3	3294.8 ± 1.9	[145]			1.306(-6)	16.83	[423]
$L_{\beta15}$	L_3N_4	3347.82 3346.99 ± 0.84	[104] [145]	9.81(-4)	[488]	4.335(-4)		
$L_{\beta2}$	L_3N_5	3347.80 3347.84 ± 0.92	[104] [145]	8.64(-3)	[488]	3.820(-3)	3.72	[279]
$L_{\beta15,2}$	$L_3N_{4,5}$	3347.842 ± 0.040	[50]					

M series

line	transition	E/eV		I/eV/$\hbar$		TPIV	Γ/eV	
	$M_1N_{2,3}$	658.00 ± 7.00 659.50	[50] [104]	1.23(-3)	[335]			
	M_1N_2					1.071(-4)	24.88	[423]
	M_1N_3					1.555(-4)	25.83	[423]
	M_2M_4	229.50 ± 0.40	[50]			4.279(-4)	3.95	[423]
	M_2N_4	600.00 ± 2.00 600.40	[50] [104]	6.06(-4)	[335]	4.159(-4)	7.50	[279]
	M_3M_5	204.80 ± 0.30	[50]			2.956(-4)	4.18	[423]
	M_3N_1	478.00 ± 2.00 477.80	[50] 82]	2.30(-4)	[335]	2.514(-4)	27.80	[423]
$M_{\gamma2,1}$	$M_3N_{4,5}$	568.00 ± 2.00	[50]					
$M_{\gamma2}$	M_3N_4					5.208(-5)		
$M_{\gamma1}$	M_3N_5	568.70	[104]	6.06(-4)	[335]	4.282(-4)	8.60	[279]
	$M_{4,5}N_{2,3}$	311.70 ± 0.60	[50]					
$M_{\xi2}$	M_4N_2					4.965(-4)	14.06	[423]
M_{δ}	M_4N_3					8.390(-5)	15.01	[423]
	M_5N_2						14.06	[423]
$M_{\xi1}$	M_5N_3					5.186(-4)	15.01	[423]
	$M_{4,5}O_{2,3}$	370.00 ± 3.00	[50]					
	M_5N_1	509.00 ± 4.00	[50]				24.23	[423]

level characteristics

level	E_B/eV		ω_{nlj}		Γ/eV		AE/eV	
K	25514.00	[503]	0.831	[301]	6.75	[301]	25515.0	[51]
	25514.00 ± 0.30	[493]	0.8605	[557]	6.56	[423]	25515.59 ± 0.30	[294]
			0.842	[347]	6.76	[94]	25523.71 ± 0.39	[145]
			0.844	[293]				
			0.830	[38]				
			0.8313	[39]				
			0.834	[177]				
			0.8326	[423]				
			0.831	[310]				
			0.836	[112]				
L_1	3806.00	[503]	1.60(-2)	[301]	4.88	[301]	3805.8	[51]
	3805.80 ± 0.30	[493]	1.01(-2)	[107]	5.40	[423]	3807.34 ± 0.17	[50]
			1.02(-2)	[348]	3.8	[94]	3814.27 ± 0.97	[145]
			1.11(-2)	[436]				
			1.14(-2)	[244]				
			2.00(-2)	[577]				
			1.60(-2)	[21]				
L_2	3524.00	[503]	5.10(-2)	[301]	2.57	[301]	3523.7	[51]
	3523.70 ± 0.30	[493]	4.30(-2)	[107]	2.30	[423]	3525.83 ± 0.15	[50]
			5.47(-2)	[348]	2.32	[94]	3533.04 ± 0.38	[145]
			5.47(-2)	[423]				
			5.40(-2)	[436]				
			5.60(-2)	[244]				
			5.50(-2)	[21]				
L_3	3351.00	[503]	5.20(-2)	[301]	2.40	[301]	3351.1	[51]
	3351.10 ± 0.30	[493]	4.49(-2)	[107]	2.12	[423]	3350.96 ± 0.13	[50]
			6.02(-2)	[348]	2.15	[94]	3360.71 ± 0.32	[145]
			5.70(-2)	[423]				
			5.60(-2)	[436]				
			5.77(-2)	[244]				
			5.80(-2)	[21]				
M_1	717.00	[503]	1.70(-4)	[38]	9.62	[350]	719.0	[51]
	717.50 ± 0.30	[493]	3.971(-3)	[512]	11.12	[423]		
			3.3(-4)	[423]	8.4	[94]		
M_2	602.00	[503]	2.90(-4)	[38]	3.74	[350]		

level	E_B/eV		ω_{nlj}		Γ/eV		AE/eV	
	602.40 ± 0.30	[493]	6.1(-4)	[512]	3.64	[423]		
			1.06(-3)	[423]	2.45	[94]		
M_3	571.00	[503]	3.20(-4)	[38]	3.88	[350]		
	571.40 ± 0.30	[493]	1.06(-3)	[423]	3.88	[423]		
			3.3(-4)	[423]	2.55	[94]		
M_4	373.00	[503]	2.70(-3)	[38]	0.44	[350]		
	372.80 ± 0.30	[493]	2.7(-3)	[512]	0.30	[423]		
			5.8(-4)	[423]	0.3	[94]		
M_5	367.00	[503]	5.2(-4)		0.31	[94]	368.0	[51]
	366.70 ± 0.30	[493]			0.31	[423]		
N_1	95.00	[503]	2.0(-5)	[423]	23.93	[423]		
	95.20 ± 0.30	[493]			4.4	[94]		
N_2	62.00	[503]	2.0(-5)	[423]	8.4	[94]		
	62.60 ± 0.30	[493]			13.76	[423]		
N_3	56.00	[503]	1.0(-5)	[423]	8	[94]		
	55.90 ± 0.30	[493]			14.71	[423]		
N_4	3.00	[503]						
	3.30 ± 0.30	[51]						
	3.60 ± 0.30	[493]						
N_5	3.00	[503]						
	3.30 ± 0.30	[51]						
	3.10 ± 0.30	[493]						
IP	7.5763							

Cd	Z=48

[Kr] 4d^{10} 5s^2

Cadmium $A = 112.411(8)$ [222] $\varrho = 8.63\,\text{g/cm}^3$ [547]
$\varrho = 8.65\,\text{g/cm}^3$ [483]

X-ray transitions

line	transition	E/eV		I/eV/$\hbar$		TPIV	Γ/eV	
K series								
K_{α_3}	KL$_1$	22694.5 ± 1.2	[145]				11.37	[423]
K_{α_2}	KL$_2$	22984.05 ± 0.20	[50]	1.719	[488]	2.430(-1)	9.91	[301]
		22984.00	[104]				10.40	[473]
							9.49	[423]
K_{α_1}	KL$_3$	23173.98 ± 0.20	[50]	3.23	[488]	4.568(-1)	9.79	[301]
		23174.00	[104]				9.80	[473]
							9.30	[423]
	KM$_1$	25941.6 ± 1.2	[145]				18.80	[423]
K_{β_3}	KM$_2$	26061.32 ± 0.39	[50]	0.2852	[488]	4.031(-2)	10.93	[423]
		26062.00	[104]					
K_{β_1}	KM$_3$	26095.44 ± 0.39	[50]	0.554	[488]	7.830(-2)	11.17	[423]
		26096.00	[104]					
$K_{\beta_5}^{II}$	KM$_4$	26301.2 ± 1.1	[145]			2.890(-4)	7.43	[423]
$K_{\beta_5}^{I}$	KM$_5$	26307.8 ± 1.2	[145]			4.022(-4)	7.44	[423]
	KN$_1$	26604.9 ± 3.5	[145]				32.43	[423]
K_{β_2}	KN$_{2,3}$	26644.07 ± 0.59	[50]	9.42(-2)	[488]			
		26645.00	[104]					
$K_{\beta_2}^{II}$	KN$_2$	26644.1 ± 3.5	[145]			8.205(-3)	23.56	[423]
$K_{\beta_2}^{I}$	KN$_3$	26643.5 ± 1.6	[145]			1.592(-2)	24.83	[423]
$K_{\beta_4}^{II}$	KN$_4$	26701.15 ± 0.88	[145]			3.607(-5)		
$K_{\beta_4}^{I}$	KN$_5$	26702.5 ± 1.1	[145]			4.969(-5)		
L series								
	L$_1$M$_1$	3247.1 ± 1.7	[145]				15.99	[423]
L_{β_4}	L$_1$M$_2$	3367.23 ± 0.12	[50]	2.66(-2)	[488]	6.458(-3)	6.28	[473]
		3367.30	[104]				8.11	[423]
		3367.50 ± 1.00	[437]					
L_{β_3}	L$_1$M$_3$	3401.48 ± 0.12	[50]	4.40(-2)	[488]	1.057(-2)	7.23	[473]
		3401.50	[104]				8.35	[423]

line	transition	E/eV		I/eV/$\hbar$		TPIV	Γ/eV	
		3401.70 ± 1.00	[437]					
$L_{\beta10}$	L_1M_4	3607.60 ± 0.31	[50]	3.08(-4)	[488]	7.472(-5)	4.62	[423]
		3607.60	]					
		3607.40 ± 1.00	[437]					
$L_{\beta9}$	L_1M_5	3614.49 ± 0.14	[50]	4.59(-4)	[488]	1.115(-4)	4.63	[423]
		3614.40	[104]					
		3614.30 ± 1.00	[437]					
	L_1N_1	3910.4 ± 4.1	[145]				29.62	[423]
$L_{\gamma2}$	L_1N_2	3951.38 ± 0.37	[50]			1.190(-3)	20.75	[423]
$L_{\gamma2.3}$	$L_1N_{2,3}$	3951.40	[104]	8.20(-3)	[488]			
$L_{\gamma3}$	L_1N_3	3949.0 ± 2.2	[145]			1.990(-3)	22.01	[423]
	L_1N_4	4006.6 ± 1.5	[145]			6.517(-6)		
	L_1N_5	4008.0 ± 1.6	[145]			9.757(-6)		
L_{η}	L_2M_1	2956.782 ± 0.094	[50]	3.64(-3)	[488]	3.517(-3)	14.12	[423]
		2956.80	[104]					
	L_2M_2	3076.9 ± 1.5	[145]				6.23	[423]
$L_{\beta17}$	L_2M_3	3110.6 ± 1.3	[145]				6.47	[423]
$L_{\beta1}$	L_2M_4	3316.605 ± 0.053	[50]	0.1242	[488]	4.997(-2)	2.54	[473]
		3316.60	[104]				2.75	[423]
		3316.70 ± 1.00	[437]					
	L_2M_5	3323.3 ± 1.0	[145]				2.74	[423]
$L_{\gamma5}$	L_2N_1	3619.38 ± 0.14	[50]	7.14(-4)	[488]	3.886(-4)	27.74	[423]
		3619.30	[104]					
		3619.20 ± 1.00	[437]					
	L_2N_2	3659.6 ± 3.5	[145]				18.87	[423]
	L_2N_3	3659.0 ± 1.6	[145]			3.119(-6)	20.13	
$L_{\gamma1}$	L_2N_4	3716.898 ± 0.099	[50]	1.279(-2)	[488]	5.142(-3)	4.38	[473]
		3716.80	[104]					
		3717.50 ± 1.00	[437]					
	L_2N_5	3718.0 ± 1.0	[145]					
L_l	L_3M_1	2767.376 ± 0.082	[50]	4.39(-3)	[488]	6.005(-3)	13.93	[423]
		2767.30	[104]					
L_t	L_3M_2	2887.7 ± 1.4	[145]			2.932(-5)	6.04	[423]
L_s	L_3M_3	2921.4 ± 1.2	[145]			2.755(-5)	6.28	
$L_{\alpha2}$	L_3M_4	3126.950 ± 0.070	[50]	1.192(-2)	[488]	5.047(-3)	2.40	[473]
		3126.90	[104]				2.56	[423]
$L_{\alpha1}$	L_3M_5	3133.755 ± 0.047	[50]	0.1057	[488]	4.478(-2)	2.43	[473]
		3133.80	[104]				2.56	[423]

line	transition	E/eV		I/eV/$\hbar$		TPIV	Γ/eV	
$L_{\beta6}$	L_3N_1	3429.98 ± 0.13	[50]	8.52(-4)	[488]	4.083(-4)	27.55	[423]
		3430.00	[104]					
		3430.20 ± 1.00	[437]					
	L_3N_2	3470.4 ± 3.5	[145]			1.587(-6)	18.68	[423]
	L_3N_3	3469.7 ± 1.5	[145]				19.94	[423]
$L_{\beta15,2}$	$L_3N_{4,5}$	3528.159 ± 0.059	[50]	1.21(-3)	[488]			
		3528.20	[104]					
$L_{\beta15}$	L_3N_4	3527.42 ± 0.79	[145]			5.102(-4)		
$L_{\beta2}$	L_3N_5	3527.60 ± 1.00	[437]	1.07(-2)	[488]	4.511(-3)	5.82	[473]
		3528.74 ± 0.97	[145]					

M series

line	transition	E/eV		I/eV/$\hbar$		TPIV	Γ/eV	
	M_2N_1	540.00 ± 5.00	[50]	3.81(-4)	[116]	2.186(-4)	29.18	[423]
		541.40	[104]					
	M_2N_4	238.40 ± 0.90	[50]			5.019(-4)		
	M_2N_4	639.00 ± 2.30	[104]			5.019(-4)		
	M_2N_5	639.00 ± 2.30	[50]	7.95(-4)	[335]			
	M_3M_5	211.10 ± 0.70	[50]			2.866(-4)	4.42	[423]
	M_3N_1	507.00 ± 2.00	[50]	4.19(-4)	[116]	2.667(-4)	29.42	[423]
		507.50	[104]	2.64(-4)	[335]			
M_γ	$M_3N_{4,5}$	606.00 ± 2.00	[50]	7.95(-4)	[335]			
	M_3N_4					7.387(-5)		
$M_{\gamma1}$	M_3N_5	606.00	[104]			5.156(-4)		
$M_{\xi2}$	M_4N_2	341.84	[423]			6.267(-4)	16.81	[423]
M_δ	M_4N_3	347.87	[423]			1.050(-4)	18.08	[423]
M_η	M_4O_2	408.50	[104]					
	$M_4O_{2,3}$	408.00 ± 1.30	[50]					
$M_{\xi1}$	M_5N_3	340.67	[423]			6.569(-4)	18.08	[423]
	M_5O_3	403.00 ± 1.30	[50]					

level characteristics

level	E_B/eV		ω_{nlj}		Γ/eV		AE/eV	
K	26711.00	[503]	0.843	[301]	7.28	[301]	26711.2	[51]
	26711.20 ± 0.30	[493]	0.855	[557]	7.09	[423]	26713.29 ± 0.20	[294]
			0.8707	[347]	7.32	[94]	25720.58 ± 0.41	[145]
			0.840	[38]				
			0.8415	[39]				
			0.8434	[423]				
			0.843	[310]				
			0.853	[33]				
L_1	4018.00	[503]	1.80(-2)	[301]	4.87	[301]	4018.0	[51]
	4018.00 ± 0.30	[493]	2.07(-2)	[423]	4.28	[423]	4019.01 ± 0.19	[50]
			1.22(-2)	[436]	3.5	[94]	4026.07 ± 0.98	[145]
			2.79(-2)	[577]				
L_2	3727.00	[503]	5.60(-2)	[301]	2.62	[301]	3727.0	[51]
	3727.00 ± 0.30	[493]	5.91(-2)	[423]	2.40	[423]	3728.01 ± 0.17	[50]
			5.90(-2)	[436]	2.42	[94]	3736.10 ± 0.39	
L_3	3538.00	[503]	5.60(-2)	[301]	2.50	[301]	3537.5	[51]
	3537.50 ± 0.30	[493]	6.14(-2)	[423]	2.21	[423]	3537.60 ± 0.15	[50]
			6.00(-2)	[436]	2.24	[94]	3546.84 ± 0.32	[145]
M_1	770.00	[503]	4.161(-3)	[512]	10.00	[74]	772.0	[51]
	770.20 ± 0.30	[493]	3.6(-4)	[423]	11.71	[423]		
					8.8	[94]		
M_2	651.00	[503]	6.7(-4)	[512]	3.70	[74]		
	650.70 ± 0.30	[493]	1.16(-3)	[423]	3.83	[423]		
					2.55	[94]		
M_3	617.00	[503]	6.80(-2)	[512]	3.70	[74]		
	616.50 ± 0.30	[493]	1.18(-3)	[423]	4.07	[423]		
					2.8	[94]		
M_4	411.00	[503]	2.70(-3)	[512]	0.48	[74]		
	410.50 ± 0.30	[493]	7.3(-4)	[423]	0.34	[423]		
					0.34	[94]		
M_5	404.00	[503]	6.6(-4)	[423]	0.48	[74]	405.2	[51]
	403.70 ± 0.30	[493]			0.35	[423]		
					0.35	[94]		

level	E_B/eV		ω_{nlj}		Γ/eV		AE/eV
N_1	108.00	[503]	3.0(-5)	[423]	25.34	[423]	
	107.60 ± 0.30	[493]			6.00	[74]	
					4.4	[94]	
N_2	67.00	[503]	3.0(-5)	[423]	14.00	[74]	
	66.90 ± 0.40	[493]			16.47	[423]	
					10.8	[94]	
N_3	67.00	[503]	1.0(-5)	[423]	10.5	[94]	
	66.90 ± 0.40	[51]			14.00	[74]	
	65.00 ± 0.40	[493]			17.73	[423]	
N_4	9.00	[503]					
	9.30 ± 0.30	[51]					
	9.70 ± 0.30	[493]					
N_5	9.00	[503]					
	9.30 ± 0.30	[51]					
	9.00 ± 0.30	[493]					
O_1	2.20 ± 0.50	[493]					
IP	8.9938	[222]					

In   Z=49   [Kr] 4d^{10} 5s^2 5p^1

Indium $A = 114.818(3)$ [222] $\varrho = 7.30\,\text{g/cm}^3$ [547]
$\varrho = 7.28\,\text{g/cm}^3$ [483]

X-ray transitions

line	transition	E/eV		I/eV/$\hbar$		TPIV	Γ/eV	
K series								
K_{α_3}	KL$_1$	23703.5 $\pm$ 1.2	[145]				10.56	[423]
$K_{\alpha 2}$	KL$_2$	24002.03 $\pm$ 0.28	[50]	1.878	[488]	2.457(-1)	10.63	[301]
		24002.00	[104]				10.17	[423]
$K_{\alpha 1}$	KL$_3$	24209.75 $\pm$ 0.22	[50]	3.52	[488]	4.609(-1)	10.56	[301]
		24210.00	[104]				9.97	[423]
	KM$_1$	27114.4 $\pm$ 1.2	[145]				19.49	[423]
$K_{\beta 3}$	KM$_2$	27237.50 $\pm$ 0.25	[50]	0.314	[488]	4.113(-2)	11.69	[423]
		27238.00	[104]					
$K_{\beta 1}$	KM$_3$	27275.55 $\pm$ 0.25	[50]	0.61	[488]	7.985(-2)	11.97	[423]
		27276.00	[104]					
$K_{\beta 52}$	KM$_4$	27491.8 $\pm$ 1.8	[50]	2.38(-2)	[488]	3.123(-4)	8.05	[423]
		27491.00	[104]					
$K_{\beta 51}$	KM$_5$	27499.1 $\pm$ 1.8	[50]	3.31(-3)	[488]	4.333(-4)	8.06	[423]
		27499.00	[104]					
	KN$_1$	27819.7 $\pm$ 3.7	[145]				33.85	[423]
$K_{\beta 2}$	KN$_{2,3}$	27861.20 $\pm$ 0.93	[50]	0.1070	[488]			
		27862.00	[104]					
$K_{\beta_2}^{II}$	KN$_2$	27860.20 $\pm$ 0.93	[145]			8.395(-3)	26.90	[423]
$K_{\beta_2}^{I}$	KN$_3$	27860.9 $\pm$ 1.7	[145]			1.634(-2)	29.13	[423]
$K_{\beta_4}^{II}$	KN$_4$	27899.00	[104]			4.270(-5)	7.70	[423]
		27923.03 $\pm$ 0.99	[145]					
$K_{\beta_4}^{I}$	KN$_5$	27928.00	[104]			5.881(-5)	7.68	[423]
		27924.2 $\pm$ 1.0	[145]					
$K_{\beta 4}$	KN$_{4,5}$	27928.4 $\pm$ 3.7	[50]	7.76(-4)	[488]			
	KO$_{2,3}$	27940.00 $\pm$ 2.00	[50]					
		27940.00	[104]					
	KO$_2$					6.080(-5)		

line	transition	E/eV		I/eV/$\hbar$		TPIV	Γ/eV	
	KO$_3$					1.083(-4)		
L series								
	L$_1$M$_1$	3410.9 $\pm$ 1.7	[145]				14.66	[423]
L$_{\beta4}$	L$_1$M$_2$	3535.31 $\pm$ 0.13	[50]	2.973(-2)	[488]	1.029(-2)	6.92	[423]
		3535.30	[104]					
		3534.40 $\pm$ 1.00	[437]					
L$_{\beta3}$	L$_1$M$_3$	3573.14 $\pm$ 0.14	[50]	4.87(-2)	[488]	1.687(-2)	7.20	[423]
		3573.10	[104]					
		3572.50 $\pm$ 1.00	[437]					
L$_{\beta10}$	L$_1$M$_4$	3786.83 $\pm$ 0.15	[50]	3.60(-4)	[488]	1.244(-4)	3.28	[423]
		3786.70	[104]					
		3787.40 $\pm$ 1.00	[437]					
L$_{\beta9}$	L$_1$M$_5$	3794.26 $\pm$ 0.15	[50]	5.37(-4)	[488]	1.856(-4)	3.28	[423]
		3794.20	[104]					
		3794.90	[437]					
	L$_1$N$_1$	4116.2 $\pm$ 4.3	[145]				29.08	[423]
L$_{\gamma2,3}$	L$_1$N$_{2,3}$	4160.48 $\pm$ 0.41	[50]	9.39(-3)	[488]			
		4160.50	[104]					
L$_{\gamma2}$	L$_1$N$_2$	4157.2 $\pm$ 4.2	[145]			1.953(-3)	22.12	[423]
L$_{\gamma3}$	L$_1$N$_3$	4157.4 $\pm$ 2.2				3.248(-3)	24.35	[423]
	L$_1$N$_4$	4219.5 $\pm$ 1.5	[145]			1.175(-5)	2.93	[423]
	L$_1$N$_5$	4220.6 $\pm$ 1.6	[145]			1.765(-5)	2.91	[423]
L$_{\gamma4',4}$	L$_1$O$_{2,3}$	4236.70 $\pm$ 0.30	[50]	7.34(-5)	[488]			
		4236.70	[104]					
	L$_1$O$_2$					1.654(-5)		
	L$_1$O$_3$					2.539(-5)		
L$_{\eta}$	L$_2$M$_1$	3112.58 $\pm$ 0.10	[50]	4.03(-3)	[488]	3.597(-3)	14.27	[423]
		3112.60	[104]					
	L$_2$M$_2$	3236.3 $\pm$ 1.5	[145]				6.53	[423]
L$_{\beta17}$	L$_2$M$_3$	3273.8 $\pm$ 1.3	[145]			4.516(-5)	6.81	[423]
L$_{\beta1}$	L$_2$M$_4$	3487.244 $\pm$ 0.058	[50]	0.1387	[488]	5.356(-2)	2.89	[423]
		3487.30	[104]					
		3485.90 $\pm$ 1.00	[437]					
	L$_2$M$_5$	3494.8 $\pm$ 1.0	[145]				2.90	[423]
L$_{\gamma5}$	L$_2$N$_1$	3815.93 $\pm$ 0.16	[50]	8.07(-4)	[488]	4.163(-4)	28.69	[423]
		3815.80	[104]					
		3817.10 $\pm$ 1.00	[437]					
	L$_2$N$_2$	3858.4 $\pm$ 3.6	[145]				21.74	[423]

line	transition	E/eV		I/eV/$\hbar$		TPIV	Γ/eV	
	L$_2$N$_3$	3858.6 ± 1.6	[145]			3.612(-6)	23.97	[423]
L$_{\gamma 1}$	L$_2$N$_4$	3920.848 ± 0.073	[50]	1.556(-2)	[488]	6.009(-3)	2.54	[423]
		3920.80	[104]					
	L$_2$N$_5$	3921.84 ± 0.98	[145]				2.53	[423]
L$_l$	L$_3$M$_1$	2904.431 ± 0.091	[50]	4.90(-3)	[488]	6.212(-3)	14.07	[423]
		2904.40	[104]					
L$_t$	L$_3$M$_2$	3028.9 ± 1.4	[145]			3.099(-5)	6.29	[423]
L$_s$	L$_3$M$_3$	3066.4 ± 1.2	[145]			2.908(-5)	6.57	[423]
L$_{\alpha 2}$	L$_3$M$_4$	3279.322 ± 0.079	[50]	1.328(-2)	[488]	5.399(-3)	2.69	[423]
		3279.30	[104]					
L$_{\alpha 1}$	L$_3$M$_5$	3286.982 ± 0.052	[50]	0.1187	[488]	4.789(-2)	2.70	[423]
		3287.00	[104]					
L$_{\beta 6}$	L$_3$N$_1$	3608.27 ± 0.14	[50]	9.71(-4)	[488]	4.419(-4)	28.49	[423]
		3608.20	[104]					
		3607.50 ± 1.00	[437]					
	L$_3$N$_2$	3651.0 ± 3.5	[145]			1.828(-6)	21.50	[423]
	L$_3$N$_3$	3651.1 ± 1.5	[145]			1.743(-6)	23.73	[423]
L$_{\beta 15}$	L$_3$N$_4$	3713.80	[104]	1.46(-3)	[488]	5.942(-4)	2.34	[423]
		3713.24 ± 0.87	[145]					
L$_{\beta 2}$	L$_3$N$_5$	3714.30 ± 1.00	[437]	1.295(-2)	[488]	5.265(-3)	2.32	[423]
		3714.37 ± 0.92	[145]					
L$_{\beta 15,2}$	L$_3$N$_{4,5}$	3713.81 ± 0.03	[50]					
L$_{\beta 7}$	L$_3$O$_1$	3730.00 ± 5.00	[50]	9.61(-5)	[488]	3.905(-5)		

M series

line	transition	E/eV		I/eV/$\hbar$		TPIV	Γ/eV	
M$_{\gamma 2}$	M$_3$N$_4$	635.18	[423]			7.387(-5)	4.34	[423]
M$_{\gamma 1}$	M$_3$N$_5$	636.09	[423]			6.096(-4)	4.32	[423]
M$_{\xi 2}$	M$_4$N$_2$	370.27	[423]			7.739(-4)	19.62	[423]
M$_\delta$	M$_4$N$_3$	377.17	[423]			1.284(-4)	21.85	[423]
M$_\eta$	M$_4$O$_2$	456.13	[423]			6.343(-6)		
M$_{\xi 1}$	M$_5$N$_3$	369.13	[423]			8.151(-4)	21.85	[423]

level characteristics

level	E_B/eV		ω_{nlj}		Γ/eV		AE/eV	
K	27940.00	[503]	0.853	[301]	7.91	[301]	27939.9	[51]
	27939.90 $\pm$ 0.30	[493]	0.865	[293]	7.67	[423]	27940.39 $\pm$ 0.30	[50]
			0.8803	[557]	7.9	[94]	27949.69 $\pm$ 0.44	[145]
			0.850	[38]				
			0.8508	[39]				
			0.8533	[423]				
			0.853	[310]				
L_1	4238.00	[503]	2.00(-2)	[301]	5.00	[301]	4237.5	[51]
	4237.50 $\pm$ 0.30	[493]	3.30(-2)	[423]	2.89	[423]	4237.26 $\pm$ 0.21	[50]
			1.34(-2)	[436]	3	[94]	4246.17 $\pm$ 0.99	[145]
			2.69(-2)	[577]				
L_2	3938.00	[503]	6.10(-2)	[301]	2.72	[301]	3938.0	[51]
	3938.00 $\pm$ 0.30	[493]	6.34(-2)	[423]	2.50	[423]	3939.32 $\pm$ 0.19	[50]
			6.40(-2)	[436]	2.53	[94]	3947.38 $\pm$ 0.39	[145]
L_3	3730.00	[503]	6.00(-2)	[301]	2.65	[301]	3730.1	[51]
	3730.10 $\pm$ 0.30	[493]	6.59(-2)	[126]	2.30	[423]	3730.25 $\pm$ 0.17	[50]
			6.50(-2)	[436]	2.34	[94]	3739.91 $\pm$ 0.31	[145]
M_1	826.00	[503]	4.36(-3)	[512]	11.77	[423]	972.2	[51]
	825.60 $\pm$ 0.30	[493]	4.1(-4)	[423]	9.2	[94]		
M_2	702.00	[503]	7.4(-4)	[512]	2.7	[94]		
	702.20 $\pm$ 0.30	[493]	1.27(-3)	[423]	4.03	[423]		
M_3	664.00	[503]	7.20(-2)	[512]	3.05	[94]		
	664.30 $\pm$ 0.30	[493]	1.30(-3)	[423]	4.31	[423]		
M_4	451.00	[503]	2.7(-3)	[512]	0.38	[94]		
	450.80 $\pm$ 0.30	[493]	9.1(-4)	[423]	0.39	[423]		
M_5	443.00	[503]	8.2(-4)	[423]	0.39	[94]	443.9	[51]
	443.10 $\pm$ 0.30	[493]			0.39	[423]		
N_1	122.00	[503]	4.0(-5)	[423]	26.19	[423]		
	121.90 $\pm$ 0.30	[493]			4.2	[94]		

level	E_B/eV		ω_{nlj}		Γ/eV		AE/eV
N_2	77.00	[503]	3.0(-5)	[423]	13.2	[94]	
	77.40 ± 0.40	[51]			19.23	[423]	
	81.90 ± 0.40	[493]					
N_3	77.00	[503]	1.0(-5)	[423]	14	[94]	
	77.40 ± 0.40	[51]			21.46	[423]	
	75.10 ± 0.40	[493]					
N_4	16.00	[503]					
	16.20 ± 0.30	[51]					
	18.60 ± 0.30	[493]			0.04	[423]	
N_5	16.00	[503]					
	15.80 ± 0.30	[493]			0.02	[423]	
O_1	4.10 ± 0.70	[493]					
O_2	0.75 ± 1.00	[493]					
IP	5.78636	[222]					

Sn   Z=50

[Kr] $4d^{10}\,5s^2\,5p^2$

Tin $A = 118.710(7)$ [222] $\varrho = 7.30\,\text{g/cm}^3$ [547]
$\varrho = 7.29\,\text{g/cm}^3$ [483]

X-ray transitions

line	transition	E/eV		I/eV/$\hbar$		TPIV	Γ/eV	
K series								
K_{α_3}	KL_1	24736.6 ± 1.2	[145]				10.68	[423]
$K_{\alpha 2}$	KL_2	25044.04 ± 0.23	[50]	2.047	[488]	2.482(-1)	11.30	[301]
		25044.00	[104]	2.080	[487]		12.40	[473]
							10.89	[423]
$K_{\alpha 1}$	KL_3	25271.36 ± 0.23	[50]	3.83	[488]	4.643(-1)	11.20	[301]
		25272.00	[104]	3.89	[487]		11.20	[473]
							10.68	[423]
	KM_1	28316.1 ± 1.2	[145]				20.61	[423]
$K_{\beta 3}$	KM_2	28444.43 ± 0.33	[50]	0.346	[488]	4.189(-2)	11.00	[473]
		28445.00	[104]	0.372	[487]		12.49	[423]
$K_{\beta 1}$	KM_3	28486.26 ± 0.33	[50]	0.671	[488]	8.134(-2)	11.80	[473]
		28487.00	[104]	0.722	[487]		12.83	[423]
$K_{\beta_5^{II}}$	KM_4	28710.2 ± 3.00	[50]	2.77(-3)	[488]	3.362(-4)	8.71	[423]
		28709.00	[104]					
$K_{\beta_5^{I}}$	KM_5	28716.2 ± 3.00	[50]	3.84(-3)	[488]	4.651(-4)	8.72	[423]
		28715.00	[104]					
	KN_1	29064.2 ± 3.7	[145]				33.77	[423]
$K_{\beta_2^{II}}$	KN_2	29096.05 ± 0.80	[145]	6.97(-2)	[487]	8.585(-3)	29.94	[423]
$K_{\beta_2^{I}}$	KN_3	29109.2 ± 1.8	[145]	0.1356	[487]	1.677(-2)	33.90	[423]
K_{β_2}	$KN_{2,3}$	29109.64 ± 0.81	[50]	0.1211	[488]			
		29109.00	[104]					
$K_{\beta_4^{II}}$	KN_4	29175.4 ± 1.1	[145]	9.74(-4)	[488]	4.975(-5)	8.36	[423]
$K_{\beta_4^{I}}$	KN_5	29176.00	[104]			6.841(-5)	8.36	[423]
		29176.7 ± 1.1	[145]					
K_{β_4}	$KN_{4,5}$	29175.7 ± 3.00	[50]					
	$KO_{2,3}$	29195.00 ± 2.00	[50]	3.99(-3)	[488]			
		29195.00	[104]	0.006	[487]			
	KO_2					1.719(-4)		
	KO_3					3.120(-4)		

line	transition	E/eV		I/eV/$\hbar$		TPIV	Γ/eV	
	L series							
	L_1M_1	3579.5 ± 1.7	[145]				15.35	[423]
$L_{\beta 4}$	L_1M_2	3708.33 ± 0.15	[50]	3.31(-2)	[488]	1.097(-2)	6.60	[473]
		3708.20	[104]				7.23	[423]
		3708.60 ± 1.00	[437]					
$L_{\beta 3}$	L_1M_3	3750.392 ± 0.050	[50]	5.39(-2)	[488]	1.786(-2)	7.70	[473]
		3750.30	[104]				7.57	[423]
		3750.90 ± 1.00	[437]					
$L_{\beta 10}$	L_1M_4	3971.63 ± 0.17	[50]	4.18(-4)	[488]	1.386(-4)	3.45	[423]
		3971.50	[104]					
$L_{\beta 9}$	L_1M_5	3980.00 ± 0.17	[50]	6.24(-4)	[488]	2.069(-4)	3.45	[423]
		3979.80	[104]					
	L_1N_1	4327.6 ± 4.2	[145]				28.51	[423]
$L_{\gamma 2.3}$	$L_1N_{2,3}$	4376.83 ± 0.46	[50]	1.07(-2)	[488]			
		4376.90	[104]					
$L_{\gamma 2}$	L_1N_2	4359.5 ± 1.3	[145]			2.143(-3)	24.68	[423]
$L_{\gamma 3}$	L_1N_3	4372.6 ± 2.3	[145]			3.547(-3)	28.64	[423]
	L_1N_4	4438.8 ± 1.6	[145]			1.399(-5)	3.10	[423]
	L_1N_5	4440.1 ± 1.6	[145]			2.108(-5)	3.10	[423]
$L_{\gamma 4',4}$	$L_1O_{2,3}$	4463.80 ± 0.30	[50]	2.30(-4)	[488]			
		4463.90	[104]					
	L_1O_2					4.901(-5)		
	L_1O_3					7.628(-5)		
L_η	L_2M_1	3272.37 ± 0.12	[50]	4.45(-3)	[488]	3.676(-3)	14.94	[423]
		3272.30	[104]					
	L_2M_2	3400.4 ± 1.5	[145]				6.82	[423]
$L_{\beta 17}$	L_2M_3	3442.1 ± 1.3	[145]			4.747(-5)	7.16	[423]
$L_{\beta 1}$	L_2M_4	3662.839 ± 0.048	[50]	0.1544	[488]	5.735(-2)	2.75	[473]
		3662.70	[104]				3.04	[423]
		3662.70 ± 1.00	[437]					
	L_2M_5	3671.3 ± 1.0	[145]				3.05	[423]
$L_{\gamma 5}$	L_2N_1	4019.20 ± 0.17	[50]	9.10(-4)	[488]	4.454(-4)	28.09	[423]
		4019.00	[104]					
	L_2N_2	4051.76 ± 0.72	[145]				24.27	[423]
	L_2N_3	4064.9 ± 1.7	[145]			4.170(-6)	28.23	[423]
$L_{\gamma 1}$	L_2N_4	4131.161 ± 0.061	[50]	1.865(-2)	[488]	6.923(-3)	5.23	[473]
		4131.10	[104]				2.69	[423]
	L_2N_5	4132.4 ± 1.0	[145]				2.69	[423]

line	transition	E/eV		I/eV/$\hbar$		TPIV	Γ/eV	
L$_l$	L$_3$M$_1$	3045.01 $\pm$ 0.10	[50]	5.45(-3)	[488]	6.421(-3)	14.73	[423]
		3045.00	[104]					
L$_t$	L$_3$M$_2$	3173.4 $\pm$ 1.4	[145]			3.271(-5)	6.61	[423]
L$_s$	L$_3$M$_3$	3215.0 $\pm$ 1.2	[145]			3.064(-5)	6.95	[423]
L$_{\alpha 2}$	L$_3$M$_4$	3437.356 $\pm$ 0.056	[50]	1.475(-2)	[488]	5.769(-3)	2.62	[473]
		3435.40	[104]				2.83	[423]
L$_{\alpha 1}$	L$_3$M$_5$	3444.011 $\pm$ 0.042	[50]	0.1308	[488]	5.115(-2)	2.62	[473]
		3444.00	[104]				2.83	[423]
L$_{\beta 6}$	L$_3$N$_1$	3792.66 $\pm$ 0.15	[50]	1.10(-3)	[488]	4.774(-4)	27.88	[423]
		3791.90	[104]					
		3792.60 $\pm$ 1.00	[437]					
	L$_3$N$_2$	3824.71 $\pm$ 0.66	[145]			2.100(-6)	24.06	[423]
	L$_3$N$_3$	3837.9 $\pm$ 1.7	[145]			2.005(-6)	28.02	
L$_{\beta 15}$	L$_3$N$_4$	3904.07 $\pm$ 0.95	[145]	1.75(-3)	[488]	6.824(-4)	2.48	[423]
L$_{\beta 2}$	L$_3$N$_5$	3904.90	[104]	1.55(-2)	[488]	6.057(-3)	6.10	[473]
		3905.32 $\pm$ 0.95	[145]				3.2	[423]
							2.48	[423]
L$_{\beta 15,2}$	L$_3$N$_{4,5}$	3904.894 $\pm$ 0.055	[50]					
L$_{\beta 7}$	L$_3$O$_1$	3927.90 $\pm$ 0.40	[50]	1.24(-4)	[488]	4.850(-5)		
		3928.50	[104]					

M series

line	transition	E/eV		I/eV/$\hbar$		TPIV	Γ/eV	
	M$_1$P$_3$	616.80	[263]					
	M$_2$N$_4$	262.10 $\pm$ 0.60	[50]			6.944(-4)		
	M$_2$N$_1$	619.00 $\pm$ 3.00	[50]	3.43(-4)	[335]	2.437(-4)	29.70	[423]
		618.70	[104]					
	M$_2$N$_4$	733.00 $\pm$ 2.00	[50]	1.29(-3)	[335]	6.944(-4)	4.29	[423]
		732.30	[104]					
		733.60	[263]					
	M$_3$N$_5$	228.70 $\pm$ 0.40	[50]			7.095(-4)	4.29	[423]
	M$_3$N$_1$	575.00 $\pm$ 3.00	[50]	3.43(-4)	[335]	2.988(-4)	30.04	[423]
		576.60	[104]					
M$_{\gamma 2}$	M$_3$N$_4$			1.29(-3)	[335]	8.594(-5)	4.63	[423]
M$_{\gamma 1}$	M$_3$N$_5$	691.10	[104]	1.29(-3)	[335]	7.095(-4)	4.63	[423]
M$_{\gamma 2,1}$	M$_3$N$_{4,5}$	962.60 $\pm$ 2.00	[50]					
M$_{\xi 2.\delta}$	M$_{4,5}$N$_{2,3}$	397.00 $\pm$ 1.00	[50]					
	M$_4$N$_2$	491.00 $\pm$ 2.00	[50]			9.378(-4)	22.09	[423]
	M$_4$N$_3$					1.542(-4)	26.05	[423]

line	transition	E/eV		I/eV/ℏ	TPIV	Γ/eV	
M_η	M_4O_2	491.40	[104]		2.068(-5)		
	$M_4O_{2,3}$	491.00 ± 2.00	[50]				
M_{ξ_1}	M_5N_3	398.82	[423]		9.939(-4)	26.06	[423]
	M_5O_3	483.00 ± 2.00	[50]		2.059(-5)		

level characteristics

level	E_B/eV		ω_{nlj}		Γ/eV		AE/eV	
K	29200.00	[503]	0.862	[301]	8.49	[301]	29200.1	[51]
	29200.10 ± 0.40	[493]	0.8889	[557]	8.28	[423]	29200.39 ± 0.20	[294]
			0.871	[347]	8.53	[94]		
			0.874	[293]				
			0.859	[38]				
			0.8595	[39]				
			0.8625	[423]				
			0.862	[310]				
			0.861	[33]				
			0.858	[112]				
L_1	4465.00	[503]	3.70(-2)	[301]	2.97	[301]	4464.7	[51]
	4464.70 ± 0.30	[493]	1.30(-2)	[107]	3.02	[423]	4464.77 ± 0.24	[50]
			1.30(-2)	[348]	2.4	[94]	4473.20 ± 0.99	[145]
			3.53(-2)	[423]				
			3.56(-2)	[436]				
			3.65(-2)	[244]				
			3.04(-2)	[577]				
L_2	4156.00	[503]	6.50(-2)	[301]	2.84	[301]	4156.1	[51]
	4156.10 ± 0.30	[493]	5.65(-2)	[107]	2.61	[423]	4157.27 ± 0.21	[50]
			6.56(-2)	[348]	2.64	[94]	4165.49 ± 0.39	[145]
			6.85(-2)	[423]				
			6.80(-2)	[436]				
			7.00(-2)	[244]				
L_3	3929.00	[503]	6.40(-2)	[301]	2.75	[301]	3928.8	[51]
	3928.80 ± 0.30	[493]	7.37(-2)	[348]	2.40	[423]	3928.84 ± 0.18	[50]
			7.07(-2)	[423]	2.43	[94]	3938.45 ± 0.33	[145]
			7.00(-2)	[244]				

level	E_B/eV		ω_{nlj}		Γ/eV		AE/eV	
M_1	884.00	[503]	2.50(-4)	[38]	12.34	[423]	884.7	[51]
	883.80 ± 0.30	[493]	4.7(-4)	[512]	10.85	[350]		
			4.4(-4)	[423]	9.6	[94]		
M_2	757.00	[503]	7.00(-4)	[38]	3.69	[350]		
	756.40 ± 0.40	[493]	7.9(-4)	[512]	4.21	[423]		
			1.39(-3)	[423]	2.85	[94]		
M_3	715.00	[503]	5.40(-4)	[38]	4.06	[350]		
	714.40 ± 0.30	[493]	1.42(-3)	[423]	4.55	[423]		
					3.3	[94]		
M_4	494.00	[503]	2.70(-3)	[38]	0.52	[350]		
	493.30 ± 0.30	[493]	2.7(-4)	[512]	0.43	[423]		
			1.12(-3)	[423]	0.43	[94]		
M_5	485.00	[503]	1.02(-3)	[423]	0.50	[74]	484.9	[51]
	484.80 ± 0.30	[493]			0.44	[423]		
					0.44	[94]		
N_1	137.00	[503]	4.0(-5)	[423]	25.49	[423]		
	136.50 ± 0.40	[493]			3.60	[74]		
					3.4	[94]		
N_2	89.00	[503]	4.0(-5)	[423]	14.00	[74]		
	88.60 ± 0.40	[51]			21.66	[423]		
	93.90 ± 0.40	[493]			17	[94]		
N_3	89.00	[503]	1.0(-5)	[423]	14.00	[74]		
	86.00 ± 0.40	[493]			25.62	[423]		
					17	[94]		
N_4	24.00	[503]			0.08	[94]		
	23.90 ± 0.30	[51]			0.14	[74]		
	24.60 ± 0.30	[493]			0.08	[423]		
N_5	24.00	[503]			0.14	[74]		
	23.90 ± 0.30	[51]			0.08	[423]		
	23.40 ± 0.30	[493]			0.08	[94]		
O_1	7.00 ± 0.50	[493]						
O_2	1.25 ± 0.30	[493]						
IP	7.3439	[222]						

Sb   Z=51

[Kr] 4d^{10} 5s^2 5p^3

Antimony $A = 121.760(1)$ [222] $\varrho = 6.679\,\text{g/cm}^3$ [547]
$\varrho = 6.62\,\text{g/cm}^3$ [483]

X-ray transitions

line	transition	E/eV		I/eV/$\hbar$		TPIV	Γ/eV	
K series								
K_{α_3}	KL$_1$	25793.9 $\pm$ 1.2	[145]				12.08	[423]
$K_{\alpha 2}$	KL$_2$	26110.78 $\pm$ 0.25	[50]	2.228	[488]	2.504(-1)	12.20	[301]
		26111.00	[104]	2.263	[487]		11.65	[423]
		26111.04 $\pm$ 0.24	[143]					
		26111.95 $\pm$ 0.25	[277]					
$K_{\alpha 1}$	KL$_3$	26358.86 $\pm$ 0.25	[50]	4.16	[488]	4.674(-1)	12.00	[301]
		26359.00	[104]	4.22	[487]		11.43	[423]
		26359.13 $\pm$ 0.25	[143]					
		26359.03 $\pm$ 0.25	[277]					
	KM$_1$	29546.8 $\pm$ 1.2	[145]				21.80	[423]
$K_{\beta 3}$	KM$_2$	29679.20 $\pm$ 0.29	[50]	0.379	[488]	4.261(-2)	13.36	[423]
		29680.00	[104]	0.407	[487]			
		29679.40 $\pm$ 0.42	[277]					
$K_{\beta 1}$	KM$_3$	29725.53 $\pm$ 0.22	[50]	0.735	[488]	8.266(-2)	13.74	[423]
		29726.00	[104]	0.790	[487]			
		29725.72 $\pm$ 0.32	[277]					
$K_{\beta_5^{II}}$	KM$_4$	29956.1 $\pm$ 1.1	[50]	3.21(-3)	[488]	3.609(-4)	9.41	[423]
		29956.00	[104]					
$K_{\beta_5^{I}}$	KM$_5$	29963.3 $\pm$ 1.1	[50]	4.43(-3)	[488]	4.977(-4)	9.41	[423]
		29964.00	[104]					
	KN$_1$	30340.1 $\pm$ 3.9	[145]				31.52	[423]
$K_{\beta_2^{II}}$	KN$_2$	30387.1 $\pm$ 4.1	[145]	7.85(-2)	[487]	8.774(-3)	32.56	[423]
$K_{\beta_2^{I}}$	KN$_3$	30388.0 $\pm$ 1.9		0.1529	[487]	1.719(-2)	39.26	[423]
$K_{\beta 2}$	KN$_{2,3}$	30389.84 $\pm$ 0.55	[50]	0.1365	[488]			
		30390.00	]					
$K_{\beta_4^{II}}$	KN$_4$	30428.00	[104]			5.717(-5)	9.07	[423]
		30458.3 $\pm$ 1.1	[145]					
$K_{\beta_4^{I}}$	KN$_5$	30461.00	[104]			7.845(-5)	9.08	[423]
		30459.7 $\pm$ 1.1	[145]					

line	transition	E/eV		I/eV/$\hbar$		TPIV	Γ/eV	
$K_{\beta 4}$	$KN_{4,5}$	30460.40 ± 0.75	[50]	1.21(-3)	[488]			
	$KO_{2,3}$	30487.50 ± 0.75	[50]	8.17(-3)	[488]			
		30488.00	[104]	1.14(-2)	[487]			
	KO_2					3.239(-4)		
	KO_3					5.951(-4)		

L series

line	transition	E/eV		I/eV/$\hbar$		TPIV	Γ/eV	
	L_1M_1	3752.9 ± 1.7	[145]				16.02	[423]
$L_{\beta 4}$	L_1M_2	3886.42 ± 0.16	[50]	3.68(-2)	[488]	1.169(-2)	7.57	[423]
		3886.30	[104]					
		3886.40 ± 1.00	[437]					
$L_{\beta 3}$	L_1M_3	3932.73 ± 0.17	[50]	5.94(-2)	[488]	1.888(-2)	7.95	[423]
		3932.70	[104]					
		3933.30 ± 1.00	[437]					
$L_{\beta 10}$	L_1M_4	4161.64 ± 0.19	[50]	4.85(-4)	[488]	1.542(-4)	3.62	[423]
		4161.60	[104]					
$L_{\beta 9}$	L_1M_5	4170.82 ± 0.19	[50]	7.24(-4)	[488]	2.301(-4)	3.62	[423]
	L_1N_1	4546.2 ± 4.4	[145]				25.74	[423]
$L_{\gamma 2.3}$	$L_1N_{2,3}$	4599.95 ± 0.51	[50]	1.215(-2)	[488]			
		4600.00	[104]					
$L_{\gamma 2}$	L_1N_2	4593.1 ± 4.8	[145]			2.347(-3)	26.77	[423]
$L_{\gamma 3}$	L_1N_3	4594.0 ± 2.4	[145]			3.862(-3)	33.47	[423]
	L_1N_4	4664.3 ± 1.6	[145]			1.647(-5)	3.29	[423]
	L_1N_5	4665.8 ± 1.6	[145]			2.487(-5)	3.29	[423]
$L_{\gamma 4}$	$L_1O_{2,3}$	4696.70 ± 0.36	[50]	4.77(-4)	[488]			
		4696.70	[104]					
	L_1O_2					9.690(-5)		
	L_1O_3					1.516(-4)		
L_η	L_2M_1	3436.65 ± 0.13	[50]	4.90(-3)	[488]	3.756(-3)	15.58	[423]
		3436.60	[104]					
	L_2M_2	3569.3 ± 1.5	[145]				7.13	[423]
$L_{\beta 17}$	L_2M_3	3615.3 ± 1.3	[145]			4.985(-5)	7.51	[423]
$L_{\beta 1}$	L_2M_4	3843.615 ± 0.071	[50]	0.1717	[488]	6.145(-2)	3.18	[423]
		3846.60	[104]					
		3843.40 ± 1.00	[437]					
	L_2M_5	3853.1 ± 1.0	[145]				3.19	[423]
$L_{\gamma 5}$	L_2N_1	4228.78 ± 0.19	[50]	1.02(-3)	[488]	4.757(-4)	25.30	[423]
		4228.70	[104]					
	L_2N_2	4276.3 ± 4.2	[145]				26.33	[423]

line	transition	E/eV		I/eV/$\hbar$		TPIV	Γ/eV	
	L_2N_3	4277.2 ± 1.8	[145]			4.807(-6)	33.04	[423]
$L_{\gamma 1}$	L_2N_4	4347.831 ± 0.068 4347.80	[50] [104]	2.206(-2)	[488]	7.904(-3)	2.85	[423]
	L_2N_5	4348.99 ± 0.96	[145]				2.85	[423]
L_l	L_3M_1	3188.63 ± 0.11 3188.60	[50] [104]	6.06(-3)	[488]	6.632(-3)	15.36	[423]
L_t	L_3M_2	3321.3 ± 1.4	[145]			3.449(-5)	6.91	[423]
L_s	L_3M_3	3367.3 ± 1.2	[145]			3.226(-5)	7.29	[423]
$L_{\alpha 2}$	L_3M_4	3595.358 ± 0.093 3595.30	[50] [104]	1.634(-2)	[488]	6.155(-3)	12.20 2.97	[473] [423]
$L_{\alpha 1}$	L_3M_5	3604.756 ± 0.062 3604.80	[50] [104]	0.1449	[488]	5.458(-2)	12.00 2.97	[473] [423]
$L_{\beta 6}$	L_3N_1	3980.00 ± 0.17 3979.90 3980.60 ± 1.00	[50] [104] [437]	1.25(-3)	[488]	5.149(-4)	25.08	[423]
	L_3N_2	4028.4 ± 4.1	[145]			2.404(-6)	26.11	[423]
	L_3N_3	4029.3 ± 1.7	[145]			2.298(-6)	32.81	[423]
$L_{\beta 15,2}$	$L_3N_{4,5}$	4100.826 ± 0.060 4100.80	[50] [104]					
$L_{\beta 15}$	L_3N_4	4099.59 ± 0.89	[145]	2.06(-3)	[488]	7.749(-4)	2.63	[423]
$L_{\beta 2}$	L_3N_5	4101.02 ± 0.91	[145]	1.829(-2)	[488]	6.889(-3)	2.64	[423]
$L_{\beta 7}$	L_3O_1	4125.50 ± 0.40 4126.20	[50] [104]	1.54(-4)	[488]	5.813(-5)		

M series

	transition	E/eV		I/eV/$\hbar$		TPIV	Γ/eV	
	M_1P_3	659.50	[263]					
	M_2N_1	658.00 ± 3.50 659.50	[50] [104]	3.90(-4)	[335]	2.556(-4)	27.01	[423]
	M_2N_4	776.00 ± 2.50 775.90 779.80	[50] [104] [263]	1.60(-3)	[335]	7.927(-4)	4.56	[423]
	M_3M_5	237.50 ± 0.45	[50]			2.623(-4)	5.28	[423]
	M_3N_1	612.00 ± 3.00 613.50	[50] [104]	3.90(-4)	[335]	3.151(-4)	27.39	[423]
$M_{\gamma 2,1}$	$M_3N_{4,5}$	733.00 ± 2.00 733.60	[50] [263]					
$M_{\gamma 2}$	M_3N_4			1.60(-3)	[335]	9.816(-5)	4.94	[423]
$M_{\gamma 1}$	M_3N_5	732.80	[104]	1.60(-3)	[335]	8.108(-4)	4.94	[423]
$M_{\xi 2,\delta}$	$M_4N_{2,3}$	429.00 ± 1.00	[50]					

line	transition	E/eV		I/eV/$\hbar$	TPIV	Γ/eV	
	M_4N_2				1.119(-3)	24.09	[423]
	M_4N_3				1.822(-4)	30.80	[423]
M_η	M_4O_2	540.06	[423]		4.443(-5)		
M_{ξ_1}	M_5N_3	429.73	[423]		1.195(-3)	30.80	[423]

level characteristics

level	E_B/eV		ω_{nlj}		Γ/eV		AE/eV	
K	30491.00	[503]	0.870	[301]	9.16	[301]	30491.2	[51]
	30491.20 ± 0.30	[493]	0.867	[38]	8.94	[423]	30490.49 ± 0.20	[50]
			0.8971	[557]	9.2	[94]	30501.27 ± 0.49	[145]
			0.8676	[39]				
			0.8709	[423]				
			0.87	[310]				
L_1	4699.00	[503]	3.90(-2)	[301]	3.13	[301]	4698.3	[51]
	4698.30 ± 0.30	[493]	3.11(-2)	[107]	3.15	[423]	4698.44 ± 0.26	[50]
			3.78(-2)	[423]	2.3	[94]	4707.3 ± 1.0	[145]
			3.82(-2)	[436]				
			4.03(-2)	[577]				
L_2	4381.00	[503]	6.90(-2)	[301]	3.00	[301]	4380.4	[51]
	4380.40 ± 0.30	[493]	6.16(-2)	[107]	2.71	[423]	4381.9 ± 1.1	[50]
			7.37(-2)	[423]	2.74	[94]	4390.54 ± 0.40	[145]
			7.30(-2)	[436]				
L_3	4132.00	[503]	6.90(-2)	[301]	2.87	[301]	4132.2	[51]
	4132.20 ± 0.30	[493]	6.33(-2)	[107]	2.49	[423]	4132.33 ± 0.20	[50]
			7.57(-2)	[423]	2.53	[94]	4142.58 ± 0.33	[145]
			7.50(-2)	[436]				
M_1	944.00	[503]	5.3(-4)	[512]	10	[94]	946.0	[51]
	943.70 ± 0.30	[493]	4.8(-4)	[423]	12.87	[423]		
M_2	812.00	[503]	8.5(-4)	[512]	3	[94]		
	811.90 ± 0.30	[493]	1.50(-3)	[423]	4.42	[423]		
M_3	766.00	[503]	8.30(-2)	[512]	3.6	[94]		
	765.60 ± 0.30	[493]	1.55(-3)	[423]	4.80	[423]		

level	E_B/eV		ω_{nlj}		Γ/eV		AE/eV	
M_4	537.00	[503]	2.70(-3)	[512]	0.47	[94]		
	536.90 ± 0.30	[493]	1.35(-3)	[423]	0.47	[423]		
M_5	528.00	[503]	1.24(-3)	[423]	0.48	[94]	528.2	[51]
	527.50 ± 0.30	[493]			0.48	[423]		
N_1	152.00	[503]	5.0(-5)	[423]	22.59	[423]		
	152.00 ± 0.30	[493]			2.6	[94]		
N_2	99.00	[503]	6.0(-5)	[423]				
	98.40 ± 0.50	[51]			23.62	[423]		
	104.30 ± 0.50	[493]						
N_3	99.00	[503]	2.0(-5)	[423]				
	98.40 ± 0.50	[51]			30.33	[423]		
	95.40 ± 0.50	[493]						
N_4	32.00	[503]			0.14	[94]		
	31.40 ± 0.30	[51]						
	32.20 ± 0.30	[493]			0.14	[423]		
N_5	32.00	[503]			0.14	[94]		
	30.80 ± 0.30	[493]			0.14	[423]		
O_1	4.10 ± 0.40	[493]						
O_2	7.78	[493]						
O_3	0.75 ± 0.10	[493]						
IP	8.6084	[222]						

Te	Z=52

[Kr] $4d^{10}\,5s^2\,5p^4$

Tellurium $A = 127.60(3)$ [12.77] $\varrho = 6.23\,\text{g/cm}^3$ [547]
$\varrho = 6.25\,\text{g/cm}^3$ [483]

X-ray transitions

line	transition	E/eV		I/eV/$\hbar$		TPIV	Γ/eV	
K series								
K_{α_3}	KL_1	26875.7 ± 1.3	[145]				12.91	[423]
$K_{\alpha 2}$	KL_2	27201.99 ± 0.21	[50]	2.420	[488]	2.524(-1)	13.00	[301]
		27202.00	[104]				14.20	[473]
							12.45	[423]
$K_{\alpha 1}$	KL_3	27472.57 ± 0.27	[50]	4.50	[488]	4.699(-1)	12.80	[301]
		27472.00	[104]				12.80	[473]
							12.22	[423]
	KM_1	30806.9 ± 1.2	[145]				22.84	[423]
$K_{\beta 3}$	KM_2	30944.60 ± 0.46	[50]	0.415	[488]	4.327(-2)	12.30	[473]
		30945.00	[104]				14.28	[423]
$K_{\beta 1}$	KM_3	30995.97 ± 0.34	[50]	0.805	[488]	8.392(-2)	13.30	[473]
		30996.00	[104]				14.69	[423]
$K_{\beta_5^{II}}$	KM_4	31231.6 ± 1.2	[145]			3.862(-4)	10.15	[423]
$K_{\beta_5^{I}}$	KM_5	31241.8 ± 1.2	[145]			5.310(-4)	10.16	[423]
	KN_1	31646.5 ± 4.0	[145]				23.08	[423]
$K_{\beta 2}$	$KN_{2,3}$	31700.76 ± 0.72	[50]	0.1534	[488]			
		31701.00	[104]					
$K_{\beta_2^{II}}$	KN_2	31696.4 ± 4.5	[145]			8.965(-3)	34,84	[423]
$K_{\beta_2^{I}}$	KN_3	31698.6 ± 2.2	[145]			1.762(-2)	45,72	[423]
	$KO_{2,3}$	31811.40 ± 0.80	[50]	1.403(-2)	[488]			
		31812.00	[104]					
$K_{\beta_4^{II}}$	KN_4	31772.4 ± 1.1	[145]			6.500(-5)	9.81	[423]
$K_{\beta_4^{I}}$	KN_5	31774.0 ± 1.1	[145]			8.900(-5)	9.83	[423]
	KO_2					5.136(-4)		
	KO_3					9.497(-4)		
L series								
	L_1M_1	3931.2 ± 1.7	[145]				16.47	[423]
$L_{\beta 4}$	L_1M_2	4069.52 ± 0.18	[50]	4.08(-2)	[488]	1.245(-2)	6.82	[473]

line	transition	E/eV		I/eV/$\hbar$		TPIV	Γ/eV	
		4069.40	[104]				7.91	[423]
$L_{\beta3}$	L_1M_3	4120.48 ± 0.18	[50]	6.54(-2)	[488]	1.994(-2)	8.22	[473]
		4120.40	[104]				8.33	[423]
$L_{\beta10}$	L_1M_4	4355.16 ± 0.20	[50]	5.61(-4)	[488]	1.711(-4)	3.79	[423]
		4355.00	[104]					
$L_{\beta9}$	L_1M_5	4367.16 ± 0.20	[50]	8.37(-4)	[488]	2.555(-4)	3.79	[423]
		4367.00	[104]					
	L_1N_1	4770.8 ± 4.5	[145]				16.71	[423]
$L_{\gamma2,3}$	$L_1N_{2,3}$	4829.10 ± 0.56	[50]	1.374(-2)	[488]			
		4829.10	[104]					
$L_{\gamma2}$	L_1N_2	4820.8 ± 2.7	[145]			2.563(-3)	28.47	[423]
$L_{\gamma3}$	L_1N_3	4822.8 ± 2.7	[145]			4.194(-3)	39.35	[423]
	L_1N_4	4896.6 ± 1.6	[145]			1.919(-5)	3.44	[423]
	L_1N_5	4898.2 ± 1.6	[145]			2.904(-5)	3.46	[423]
$L_{\gamma4}$	$L_1O_{2,3}$	4936.90 ± 0.40	[50]	8.26(-4)	[488]			
		4937.00	[104]					
	L_1O_2					1.611(-4)		
	L_1O_3					2.520(-4)		
L_{η}	L_2M_1	3605.90 ± 0.14	[50]	5.39(-3)	[488]	3.835(-3)	16.01	[423]
		3605.80	[104]					
	L_2M_2	3743.0 ± 1.5	[145]				7.45	[423]
$L_{\beta17}$	L_2M_3	3793.7 ± 1.3	[145]			5.229(-5)	7.86	
$L_{\beta1}$	L_2M_4	4029.63 ± 0.12	[50]	0.1899	[488]	6.571(-2)	2.96	[473]
		4029.50	[104]				3.33	[423]
	L_2M_5	4040.1 ± 1.1	[145]				3.32	[423]
$L_{\gamma5}$	L_2N_1	4443.70 ± 0.21	[50]	1.15(-3)	[488]	5.075(-4)	16.25	[423]
		4443.60	[104]					
	L_2N_2	4494.7 ± 4.4	[145]				28.01	[423]
	L_2N_3	4496.8 ± 2.0	[145]			5.520(-6)	38.89	[423]
$L_{\gamma1}$	L_2N_4	4570.93 ± 0.15	[50]	2.583(-2)	[488]	8.938(-3)	5.60	[473]
		4570.80	[263]				2.99	[423]
	L_2N_5	4572.25 ± 0.97	[145]				3.00	[423]
L_l	L_3M_1	3335.58 ± 0.12	[50]	6.72(-3)	[488]	6.844(-3)	15.79	[423]
		3335.50	[104]					
L_t	L_3M_2	3472.6 ± 1.4	[145]			3.633(-5)	7.23	[423]
L_s	L_3M_3	3523.4 ± 1.2	[145]			3.392(-5)	7.64	[423]
$L_{\alpha2}$	L_3M_4	3758.79 ± 0.15	[50]	1.806(-2)	[488]	6.561(-3)	2.88	[473]

line	transition	E/eV		I/eV/ℏ		TPIV	Γ/eV	
		3758.70	[104]				3.10	[423]
$L_{\alpha 1}$	$L_3 M_5$	3769.38 ± 0.10	[50]	0.1601	[488]	5.818(-2)	2.88	[473]
		3769.30	[104]				3.11	[423]
$L_{\beta 6}$	$L_3 N_1$	4173.25 ± 0.19	[50]	1.412(-3)	[488]	5.545(-4)	16.02	[423]
		4173.10	[104]					
	$L_3 N_2$	4224.4 ± 4.4	[145]			2.747(-6)	27.79	[423]
	$L_3 N_3$	4226.5 ± 2.0	[145]			2.627(-6)	38.67	[423]
$L_{\beta 15,2}$	$L_3 N_{4,5}$	4301.70 ± 0.18	[50]					
$L_{\beta 15}$	$L_3 N_4$	4300.31 ± 0.91	[145]	2.40(-3)	[488]	8.719(-4)	2.76	[423]
$L_{\beta 2}$	$L_3 N_5$	4301.60	[104]	2.136(-2)	[488]	7.760(-3)	6.25	[473]
		4301.89 ± 0.91	[145]				2.78	[423]
$L_{\beta 7}$	$L_3 O_1$	4329.80 ± 0.50	[50]	1.88(-4)	[488]	6.832(-5)		
		4330.00	[104]					

M series

line	transition	E/eV		I/eV/ℏ		TPIV	Γ/eV	
	$M_1 P_3$	704.50	[263]					
	$M_2 N_1$	703.00 ± 4.00	[50]	4.41(-4)	[335]	2.656(-4)	18.08	[423]
		702.50	[104]					
	$M_3 M_5$	246.50 ± 0.50	[43]			2.546(-4)	5.57	[423]
	$M_3 N_1$	648.00 ± 3.40	[50]	4.41(-4)	[335]	3.316(-4)	18.49	[423]
		649.10	[104]					
$M_{\gamma 2,1}$	$M_3 N_{4,5}$	778.00 ± 2.00	[50]	1.96(-3)	[335]			
		779.80	[263]					
	$M_3 N_4$					1.106(-4)	5.23	[423]
$M_{\gamma 1}$	$M_3 N_5$	778.30	[104]			9.137(-4)	5.25	[423]
$M_{\xi 2}$	$M_4 N_2$	462.77	[423]			1.321(-3)	25.72	[423]
M_{δ}	$M_4 N_3$	472.82	[423]			2.131(-4)	36.59	[423]
	$M_4 O_1$	581.00 ± 1.00	[50]					
		581.30	[104]					
M_{η}	$M_4 O_{2,3}$	581.00 ± 1.40	[50]					
	$M_4 O_2$					7.967(-5)		
	$M_4 O_3$					1.228(-5)		
$M_{\xi 1}$	$M_5 N_3$	461.82	[263]			1.418(-3)	36.60	[423]
	$M_5 O_3$	569.00 ± 1.30	[50]			8.165(-5)		
		569.80	[104]					

level characteristics

level	E_B/eV		ω_{nlj}		Γ/eV		AE/eV	
K	31814.00	[503]	0.877	[301]	9.89	[301]	31813.8	[51]
	31813.80 ± 0.30	[493]	0.890	[293]	9.64	[423]	31811.5 ± 1.2	[50]
			0.9046	[557]	9.91	[94]	31824.29 ± 0.52	[145]
			0.875	[38]				
			0.8750	[39]				
			0.857	[270]				
			0.8787	[423]				
			0.874	[112]				
			0.875	[38]				
L_1	4939.00	[503]	4.10(-2)	[301]	3.32	[301]	4939.2	[51]
	4939.20 ± 0.30	[493]	4.04(-2)	[423]	3.27	[423]	4939.73 ± 0.29	[50]
			4.10(-2)	[436]	2.2	[94]	4948.5 ± 1.0	[145]
			4.17(-2)	[244]				
			5.01(-2)	[577]				
L_2	4612.00	[503]	7.40(-2)	[301]	3.12	[301]	4612.0	[51]
	4612.00 ± 0.30	[493]	7.91(-2)	[423]	2.81	[423]	4612.56 ± 0.40	[50]
			8.10(-2)	[244]	2.84	[94]	4622.56 ± 0.40	
			0.875	[38]				
L_3	4341.00	[503]	7.40(-2)	[301]	2.95	[301]	4341.4	[51]
	4341.40 ± 0.30	[493]	8.09(-2)	[423]	2.59	[423]	4341.88 ± 0.23	[50]
			8.10(-2)	[436]	2.62	[94]	4352.20 ± 0.34	[145]
			8.29(-2)	[244]				
M_1	1006.00	[503]	5.9(-4)	[512]	11.00	[74]	1006.0	[51]
	1006.10 ± 0.30	[493]	5.3(-4)	[423]	13.20	[423]		
					10.2	[94]		
M_2	870.00	[503]	9.1(-4)	[512]	4.40	[74]		
	869.70 ± 0.30	[493]	1.61(-3)	[423]	4.64	[423]		
					3.2	[94]		
M_3	819.00	[503]	8.90(-2)	[512]	4.40	[74]		
	818.70 ± 0.30	[493]	1.68(-3)	[423]	5.06	[423]		
					3.9	[94]		
M_4	582.00	[503]	2.7(-3)	[512]	0.60	[74]		
	582.50 ± 0.30	[493]	1.63(-3)	[423]	0.51	[423]		
					0.52	[94]		

level	E_B/eV		ω_{nlj}		Γ/eV		AE/eV	
M_5	572.00	[503]	1.50(-3)	[423]	0.60	[74]	573.0	[51]
	572.10 ± 0.30	[493]			0.52	[423]		
					0.52	[94]		
N_1	168.00	[503]	6.0(-5)	[423]	13.44	[423]		
	168.30 ± 0.30	[493]			5.00	[74]		
					2.4	[94]		
N_2	110.00	[503]	7.0(-5)	[423]				
	110.20 ± 0.50	[51]			25.20	[423]		
	116.80 ± 0.50	[493]			7.30	[74]		
N_3	110.00	[503]	2.0(-5)	[423]	7.30	[74]		
	96.90 ± 0.50	[493]			36.08	[423]		
N_4	40.00	[503]			0.17	[94]		
	39.80 ± 0.30	[51]			0.10	[74]		
	40.80 ± 0.30	[493]			0.17	[423]		
N_5	40.00	[503]			0.2	[94]		
	39.80 ± 0.30	[51]			0.10	[74]		
	39.20 ± 0.30	[493]			0.20	[423]		
O_1	12.00	[503]						
	11.60 ± 0.60	[493]						
O_2	2.00	[503]						
	2.30 ± 0.50	[51]						
	2.60 ± 0.50	[493]						
O_3	2.00 ± 0.50	[493]						
IP	9.0096	[222]						

I   Z=53   [Kr] $4d^{10}\,5s^2\,5p^5$

Iodine $A = 126.90447(3)$ [222] $\varrho = 4.92\,\text{g/cm}^3$ [547]

X-ray transitions

line	transition	E/eV		I/eV/$\hbar$		TPIV	Γ/eV	
K series								
K_{α_3}	KL_1	27982.3 ± 1.3	[145]				13.78	[423]
$K_{\alpha 2}$	KL_2	28317.52 ± 0.67	[50]	2.625	[488]	2.543(-1)	13.80	[301]
		28318.00	[104]				13.29	[423]
$K_{\alpha 1}$	KL_3	28612.32 ± 0.49	[50]	4.87	[488]	4.721(-1)	13.70	[301]
		28612.00	[104]				13.06	[423]
	KM_1	32096.7 ± 1.2	[145]				23.38	[423]
$K_{\beta 3}$	KM_2	32239.71 ± 0.50	[50]	0.453	[488]	4.391(-2)	15.27	[423]
		32240.00	[104]					
$K_{\beta 1}$	KM_3	32295.05 ± 0.50	[50]	0.879	[488]	8.515(-2)	15.69	[423]
		32295.00	[104]					
$K_{\beta_5^{II}}$	KM_4	32538.8 ± 1.3	[145]			4.122(-4)	10.94	[423]
$K_{\beta_5^{I}}$	KM_5	32550.1 ± 1.2	[145]			5.651(-4)	10.94	[423]
	KN_1	32984.4 ± 0.40	[145]				16.75	[423]
$K_{\beta 2}$	$KN_{2,3}$	33041.7 ± 2.6	[50]	0.1717	[488]			
		33019.00	[104]					
$K_{\beta_2^{II}}$	KN_2	33037.3 ± 4.8	[145]			9.156(-3)	36.35	[423]
$K_{\beta_2^{I}}$	KN_3	33041.8 ± 2.6	[145]			1.791(-2)	54.36	[423]
$K_{\beta_4^{II}}$	KN_4	33118.2 ± 1.1	[145]			7.322(-5)	10.49	[423]
$K_{\beta_4^{I}}$	KN_5	33120.0 ± 1.1	[145]			1.000(-4)	10.48	[423]
L series								
	$L_1 M_1$	4114.4 ± 1.7	[145]				16.40	[423]
$L_{\beta 4}$	$L_1 M_2$	4257.53 ± 0.19	[50]	4.51(-2)	[488]	1.315(-2)	8.29	[423]
		4257.40	[104]					
$L_{\beta 3}$	$L_1 M_3$	4313.49 ± 0.20	[50]	7.17(-2)	[488]	2.089(-2)	8.71	[423]
		4313.40	[104]					
$L_{\beta 10}$	$L_1 M_4$	4556.43 ± 0.22	[50]	6.46(-4)	[488]	1.883(-4)	3.95	[423]
		4556.40	[104]					

line	transition	E/eV		I/eV/$\hbar$		TPIV	Γ/eV	
$L_{\beta 9}$	$L_1 M_5$	4569.06 ± 0.22	[50]	9.65(-4)	[488]	2.812(-4)	3.96	[423]
		4569.00	[104]					
	$L_1 N_1$	5002.1 ± 4.5	[145]				9.77	[423]
$L_{\gamma 2.3}$	$L_1 N_{2,3}$	5065.67 ± 0.61	[50]	1.548(-2)	[488]			
		5065.70	[104]					
$L_{\gamma 2}$	$L_1 N_2$	5055.0 ± 5.2	[145]			2.774(-3)	29.37	[423]
$L_{\gamma 3}$	$L_1 N_3$	5059.5 ± 3.0	[145]			4.511(-3)	47.38	[423]
	$L_1 N_4$	5135.9 ± 1.6	[145]			2.204(-5)	3.51	[423]
	$L_1 N_5$	5137.7 ± 1.6	[145]			3.343(-5)	3.50	[423]
$L_{\gamma 4}$	$L_1 O_{2,3}$	5184.80 ± 0.40	[50]	1.29(-3)	[488]			
		5184.80	[104]					
	$L_1 O_2$					2.417(-4)		
	$L_1 O_3$					3.771(-4)		
L_η	$L_2 M_1$	3780.19 ± 0.15	[50]	5.92(-3)	[488]	3.924(-3)	15.92	[423]
		3779.90	[104]					
	$L_2 M_2$	3921.6 ± 1.5	[145]				7.80	[423]
$L_{\beta 17}$	$L_2 M_3$	3977.5 ± 1.3	[145]			5.501(-5)	8.23	[423]
$L_{\beta 1}$	$L_2 M_4$	4220.76 ± 0.13	[50]	0.2098	[488]	7.022(-2)	3.47	[423]
		4220.70	[104]					
	$L_2 M_5$	4232.5 ± 1.1	[145]				3.48	[423]
$L_{\gamma 5}$	$L_2 N_1$	4666.08 ± 0.23	[50]	1.28(-3)	[488]	5.405(-4)	9.28	[423]
		4666.00	[104]					
	$L_2 N_2$	4719.6 ± 4.6	[145]				28.89	[423]
	$L_2 N_3$	4724.2 ± 2.4	[145]			6.319(-6)	46.90	[423]
$L_{\gamma 1}$	$L_2 N_4$	4800.98 ± 0.22	[50]	2.997(-2)	[488]	1.004(-2)	3.02	[423]
		4890.90	[104]					
	$L_2 N_5$	4802.33 ± 0.95	[145]				3.03	[423]
L_l	$L_3 M_1$	3485.06 ± 0.13	[50]	7.44(-3)	[488]	7.059(-3)	15.67	[423]
		3485.00	[104]					
L_t	$L_3 M_2$	3627.3 ± 1.4	[145]			3.822(-5)	7.56	[423]
L_s	$L_3 M_3$	3683.2 ± 1.2	[145]			3.564(-5)	7.99	[423]
$L_{\alpha 2}$	$L_3 M_4$	3926.09 ± 0.11	[50]	1.99(-2)	[488]	6.987(-3)	3.24	[423]
		3926.00	[104]					
$L_{\alpha 1}$	$L_3 M_5$	3937.70 ± 0.11	[50]	0.1764	[488]	6.194(-2)	3.24	[423]
		3937.60	[104]					
$L_{\beta 6}$	$L_3 N_1$	4370.62 ± 0.21	[50]	1.59(-3)	[488]	5.963(-4)	9.05	[423]

line	transition	E/eV		I/eV/$\hbar$		TPIV	Γ/eV	
		4370.50	[104]					
	L_2N_2	4425.3 $\pm$ 4.5	[145]			3.118(-6)	28.65	[423]
	L_3N_3	4429.9 $\pm$ 2.4	[145]			2.991(-6)	46.66	[423]
$L_{\beta15}$	L_3N_4	4507.58 $\pm$ 0.19	[50]	2.77(-3)	[488]	9.740(-4)	2.79	[423]
$L_{\beta2}$	L_3N_5	4507.58 $\pm$ 0.19	[50]	2.471(-2)	[488]	8.676(-3)	2.80	[423]
$L_{\beta7}$	L_3O_1	4543.50 $\pm$ 0.50	[50]	2.26(-4)	[488]	7.925(-5)		
		4541.50	[104]					

M series

line	transition	E/eV		I/eV/$\hbar$		TPIV	Γ/eV	
$M_{\gamma2,1}$	$M_3N_{4,5}$	826.60	[263]					
	M_3N_4					1.235(-4)	5.42	[423]
	M_3N_5					1.020(-3)	5.43	[423]
$M_{\xi2}$	M_4N_2	495.93	[423]			1.543(-3)	26.52	[423]
M_δ	M_4N_3	507.25	[423]			2.463(-4)	44.54	[423]
M_η	M_4O_2	630.69	[423]			1.285(-4)		
$M_{\xi1}$	M_5N_3	495.10	[423]			1.665(-3)	44.54	[423]

level characteristics

level	E_B/eV		ω_{nlj}		Γ/eV		AE/eV	
K	33170.00	[503]	0.884	[301]	10.60	[301]	33169.4	[51]
	33169.40 $\pm$ 0.40	[493]	0.882	[38]	10.38	[423]	33167.2 $\pm$ 1.3	[50]
			0.9112	[557]	10.7	[94]	33179.46 $\pm$ 0.54	[145]
			0.8819	[39]				
			0.8858	[423]				
			0.884	[310]				
L_1	5188.00	[503]	4.40(-2)	[301]	3.46	[301]	5188.1	[51]
	5188.10 $\pm$ 0.30	[493]	4.28(-2)	[423]	3.40	[423]	5191.9 $\pm$ 1.6	[50]
			6.00(-2)	[576]	2.1	[94]	5197.2 $\pm$ 1.0	[145]
			6.02(-2)	[577]				
L_2	4852.00	[503]	7.90(-2)	[301]	3.25	[301]	4852.1	[51]
	4852.10 $\pm$ 0.30	[493]	8.49(-2)	[423]	2.91	[423]	4854.1 $\pm$ 1.4	[50]
			8.40(-2)	[436]	2.95	[94]	4861.84 $\pm$ 0.38	
L_3	4557.00	[503]	7.90(-2)	[301]	3.08	[301]	4557.1	[51]

level	E_B/eV		ω_{nlj}		Γ/eV		AE/eV	
	4557.10 ± 0.30	[493]	8.64(-2)	[423]	2.68	[423]	4558.8 ± 1.2	[50]
			8.60(-2)	[436]	2.72	[94]	4567.52 ± 0.32	[145]
M_1	1072.00	[503]	6.5(-4)	[512]	10.4	[94]	1072.1	[51]
	1072.10 ± 0.30	[493]	6.0(-4)	[423]	13.01	[423]		
M_2	931.00	[503]	9.7(-4)	[512]	3.35	[94]		
	930.50 ± 0.30	[493]	1.72(-3)	[423]	4.89	[423]		
M_3	875.00	[503]	9.40(-2)	[512]	4.3	[94]		
	874.60 ± 0.30	[493]	1.81(-3)	[423]	5.31	[423]		
M_4	631.00	[503]	2.7(-3)	[512]	0.56	[94]		
	631.30 ± 0.30	[493]	1.94(-3)	[423]	0.56	[423]		
M_5	620.00	[503]	1.80(-3)	[423]	0.56	[94]	620.0	[51]
	619.40 ± 0.30	[493]			0.56	[423]		
N_1	186.00	[503]	7.0(-5)	[423]	6.37	[423]		
	186.40 ± 0.30	[493]			2.4	[94]		
N_2	123.00	[503]	9.0(-5)	[423]				
	122.70 ± 0.50	[51]			25.97	[423]		
	130.10 ± 0.50	[493]						
N_3	123.00	[503]	3.0(-5)	[423]				
	122.70	[51]			43.98	[423]		
	119.00 ± 0.50	[493]						
N_4	50.00	[503]			0.11	[94]		
	49.60 ± 0.30	[51]			0.11	[423]		
	50.70 ± 0.30	[493]						
N_5	50.00	[503]			0.12	[94]		
	48.90 ± 0.30	[493]			0.12	[423]		
O_1	14.00	[503]						
	13.60 ± 0.60	[493]						
O_2	3.00	[503]						
	3.30 ± 0.50	[51]						
	$3.80 \pm 0\%.50$	[493]						
O_3	3.00	[503]						
	3.30 ± 0.50	[51]						
	2.90 ± 0.50	[493]						
IP	10.45126	[222]						

Xe   Z=54   [Kr] 4d^{10} 5s^2 5p^6

Xenon A = 131.29(2) [222] ϱ = 5.458(-3) g/cm^3 [547]

X-ray transitions

line	transition	E/eV		I/eV/$\hbar$		TPIV	Γ/eV	
K series								
K$_{\alpha 3}$	KL$_1$	29112.8 ± 3.1	[145]				14.45	[423]
		29112.44 ± 0.24	[365]					
K$_{\alpha 2}$	KL$_2$	29458.00 ± 1.50	[50]	2.842	[488]	2.560(-1)	14.80	[301]
		29488.00	[104]	2.883	[487]		15.10	[473]
		29458.44 ± 0.05	[277]				13.96	[423]
		29458.16 ± 0.05	[365]					
		29458.250 ± 0.050	[144]					
K$_{\alpha 1}$	KL$_3$	29779.00 ± 1.50	[50]	5.26	[488]	4.741(-1)	14.60	[301]
		29805.00	[104]	5.34	[487]		14.20	[473]
		29778.97 ± 0.10	[277]				13.73	[423]
		29778.69 ± 0.10	[365]					
		29778.78 ± 0.10	[144]					
	KM$_1$	33416.62 ± 0.06	[365]				22.28	[423]
		33416.0 ± 3.2	[145]					
K$_{\beta 3}$	KM$_2$	33563.20 ± 0.20	[50]	0.494	[488]	4.450(-2)	13.43	[473]
		33563.42 ± 0.12	[277]	0.527	[487]		16.09	[423]
		33563.10 ± 0.12	[365]					
K$_{\beta 1}$	KM$_3$	33624.23 ± 0.12	[50]	0.958	[488]	8.628(-2)	15.30	[473]
		33647.00	[104]	1.022	[487]		16.68	[423]
		33624.45 ± 0.12	[277]					
		33624.13 ± 0.12	[365]					
K$_{\beta_5^{II}}$	KM$_4$	33875.95 ± 0.08	[365]			4.390(-4)	11.70	[423]
K$_{\beta_5^{I}}$	KM$_5$	33888.78 ± 0.10	[365]			5.995(-4)	11.70	[423]
	KN$_1$	34353.4 ± 6.0	[145]				16.31	[423]
K$_{\beta_2^{II}}$	KN$_2$	34408.9 ± 6.9	[145]	0.1073	[487]	9.331(-3)	36.81	[423]
K$_{\beta_2^{I}}$	KN$_3$	34408.0 ± 1.1	[145]	0.2100	[487]	1.820(-2)	43.08	[423]
K$_{\beta 2}$	KN$_{2,3}$	34414.7 ± 4.2	[50]	0.1916	[488]			
		34449.00	[104]					
K$_{\beta_4^{II}}$	KN$_4$	34495.4 ± 3.2	[145]			8.183(-5)	11.19	[423]

line	transition	E/eV		I/eV/ℏ	TPIV	Γ/eV	
$K_{\beta_4^I}$	KN_5	34497.2 ± 3.1	[145]		1.115(-4)	11.19	[423]
L series							
	L_1M_1	4303.2 ± 1.9	[145]			14.51	[423]
L_{β_4}	L_1M_2	4450.328 ± 0.030	[365]		1.388(-2)	8.33	[423]
		4450.36 ± 0.02	[404]				
L_{β_3}	L_1M_3	4512.028 ± 0.030	[365]		2.188(-2)	8.92	[423]
		4511.98 ± 0.04	[404]				
$L_{\beta_{10}}$	L_1M_4	4763.2 ± 2.0	[145]		2.068(-4)	8.91	[423]
L_{β_9}	L_1M_5	4775.8 ± 2.0	[145]		3.088(-4)	8.92	[423]
	L_1N_1	5240.5 ± 4.7	[145]			11.88	[423]
L_{γ_2}	L_2N_2	5296.0 ± 5.5	[145]		2.993(-3)	29.05	[423]
L_{γ_3}	L_1N_3	5306.70 ± 0.20	[404]		4.839(-3)	35.32	[423]
$L_{\gamma_{2,3}}$	$L_1N_{2,3}$	5306.71 ± 0.20	[402]				
	L_1N_4	5382.5 ± 1.8	[145]		2.516(-5)	3.42	[423]
	L_1N_5	5384.3 ± 1.8	[145]		3.821(-5)	3.42	[423]
	$L_1P_{2,3}$	5450.40 ± 0.10	[84]				
	$L_1Q_{2,3}$	5451.67 ± 0.10	[84]				
	$L_1R_{2,3}$	5452.18 ± 0.10	[84]				
	$L_1S_{2,3}$	5452.41 ± 0.10	[84]				
L_η	L_2M_1	3958.368 ± 0.050	[365]		4.029(-3)	14.03	[423]
		3957.80 ± 0.20	[404]				
	L_2M_2	4104.6 ± 1.4	[145]			7.83	[423]
$L_{\beta_{17}}$	L_2M_3	4166.2 ± 1.3	[145]		5.825(-5)	8.42	[423]
L_{β_1}	L_2M_4	4417.668 ± 0.030	[365]		7.491(-2)	3.45	[423]
		4417.66 ± 0.02	[404]				
	L_2M_5	4430.2 ± 1.1	[145]			3.46	[423]
L_{γ_5}	L_2N_1	4895.0 ± 3.9	[145]		5.749(-4)	8.05	[423]
	L_2N_2	4950.5 ± 4.6	[145]			28.56	[423]
	L_2N_3	4949.2 ± 8.9	[145]		7.203(-6)	35.04	[423]
L_{γ_1}	L_2N_4	5037.14 ± 0.10	[404]		1.118(-2)	2.93	[423]
		5036.95 ± 0.93	[145]				
	L_2N_5	5038.78 ± 0.89	[145]			2.93	[423]
	$L_2O_{4,5}$	5104.88 ± 0.10	[84]				
	L_2P_1	5103.09 ± 0.10	[84]				
	$L_2P_{4,5}$	5105.93 ± 0.10	[84]				
	$L_2Q_{4,5}$	5106.45 ± 0.10	[84]				

line	transition	E/eV		I/eV/$\hbar$		TPIV	Γ/eV	
L$_l$	L$_3$M$_1$	3638.008 ± 0.059	[365]			7.276(-3)	13.80	[423]
		3636.90 ± 0.30	[404]					
L$_t$	L$_3$M$_2$	3784.7 ± 1.4	[145]			4.017(-5)	7.61	[423]
L$_s$	L$_3$M$_3$	3846.3 ± 1.2	[145]			3.740(-5)	8.20	[423]
L$_{\alpha_2}$	L$_3$M$_4$	4097.378 ± 0.030	[365]			7.433(-3)	3.22	[423]
		4097.42 ± 0.05	[404]					
L$_{\alpha_1}$	L$_3$M$_5$	4109.9 ± 0.30	[50]	0.1939	[488]	6.588(-2)	3.15	[473]
		4110.088 ± 0.020	[365]				3.22	[423]
		4110.18 ± 0.04	[404]					
L$_{\beta_6}$	L$_3$N$_1$	4575.1 ± 3.8	[145]			6.411(-4)	7.83	[423]
	L$_3$N$_2$	4630.6 ± 4.6	[145]			3.535(-6)	5.20	[423]
	L$_3$N$_3$	4629.3 ± 8.9	[145]			3.392(-6)		
L$_{\beta_{15}}$	L$_3$N$_4$	4717.08 ± 0.89	[145]			1.081(-3)	2.71	[423]
L$_{\beta_2}$	L$_3$N$_5$	4718.86 ± 0.08	[404]			9.640(-3)	2.71	[423]
		4718.90 ± 0.84	[145]					
	L$_3$O$_{4,5}$	4784.23 ± 0.10	[84]					
	L$_3$P$_1$	4782.40 ± 0.10	[84]					
	L$_3$P$_{4,5}$	4785.26 ± 0.10	[84]					
	L$_3$Q$_{4,5}$	4685.70 ± 0.10	[84]					

M series

line	transition	E/eV		I/eV/$\hbar$		TPIV	Γ/eV	
M$_{\gamma_2}$	M$_3$N$_4$	856.72	[423]			1.370(-4)	5.66	[423]
M$_{\gamma_1}$	M$_3$N$_5$	858.78	[423]			1.130(-3)	5.66	[423]
M$_{\gamma_{2,1}}$	M$_3$N$_{4,5}$	873.10	[263]					
M$_{\xi_2}$	M$_4$N$_2$	530.22	[423]			1.782(-3)	26.30	[423]
M$_\delta$	M$_4$N$_3$	542.93	[423]			2.816(-4)	32.57	[423]
M$_\eta$	M$_4$O$_2$	678.56	[423]			1.932(-4)		
M$_{\xi_1}$	M$_5$N$_3$	529.54	[423]			1.936(-3)	32.57	[423]

level characteristics

level	E_B/eV		ω_{nlj}		Γ/eV		AE/eV	
K	34561.00	[503]	0.891	[301]	11.40	[301]	34561.4	[51]
	34561.40 ± 1.10	[493]	0.902	[557]	11.11	[423]	34593.00 ± 0.71	[50]
			0.9176	[347]	11.5	[94]	34566.5 ± 2.6	[145]
			0.889	[38]				
			0.880	[230]				
			0.8883	[39]				
			0.990	[540]				
			0.8924	[423]				
			0.891	[310]				
			0.887	[112]				
L_1	5453.00	[503]	5.84(-2)	[348]	3.64	[301]	5452.8	[51]
	5452.80 ± 0.40	[493]	4.54(-2)	[423]	3.35	[423]	5452.89 ± 0.35	[50]
			4.60(-2)	[436]	2	[94]	5453.7 ± 1.3	[145]
			4.75(-2)	[244]				
			0.889	[38]				
L_2	5104.00	[503]	9.12(-2)	[348]	3.51	[301]	5103.7	[51]
	5103.70 ± 0.40	[493]	9.08(-2)	[423]	2.85	[423]	5103.83 ± 0.31	[50]
			9.00(-2)	[436]	3.05	[94]	5108.10 ± 0.31	[145]
			9.30(-2)	[244]				
L_3	4782.00	[503]	9.70(-2)	[348]	3.25	[301]	4782.2	[51]
	4782.20 ± 0.40	[493]	9.21(-2)	[423]	2.63	[423]	4782.16 ± 0.27	[50]
			9.20(-2)	[436]	2.82	[94]	4788.22 ± 0.32	
			9.42(-2)	[244]				
M_1	1145.00	[503]	4.70(-4)	[38]	10.18	[350]	1144.6	[51]
	1148.40 ± 2.50	[493]	7.1(-4)	[512]	11.17	[423]		
			7.0(-4)	[423]	10.6	[94]		
M_2	999.00	[503]	9.00(-4)	[38]	4.83	[350]		
	999.00 ± 2.10	[493]	1.03(-3)	[512]	4.98	[423]		
			1.87(-3)	[423]	3.5	[94]		
M_3	937.00	[503]	6.80(-4)	[38]	5.48	[350]		
	937.00 ± 2.10	[493]	1.95(-3)	[423]	5.58	[423]		
					4.7	[94]		
M_4	685.00	[503]	2.70(-3)	[38]	0.68	[350]		

level	E_B/eV		ω_{nlj}		Γ/eV		AE/eV	
	690.60 ± 0.70	[493]	2.7(-3)	[512]	0.59	[423]		
			2.29(-3)	[423]	0.6	[94]		
M_5	672.00	[503]	2.14(-3)	[423]	0.6	[94]	676.4	[51]
	672.30 ± 0.50	[51]			0.68	[74]		
	674.70 ± 2.70	[493]			0.60	[423]		
N_1	208.00	[503]	7.0(-5)	[423]	5.20	[423]		
	217.70 ± 5.00	[493]			5.50	[74]		
					2.6	[94]		
N_2	147.00	[503]	1.0(-4)	[423]	3.00	[74]		
	146.70 ± 3.10	[51]			25.71	[423]		
	163.90 ± 3.00	[493]						
N_3	147.00	[503]	3.0(-5)	[423]	3.00	[74]		
	146.70 ± 3.10	[51]			31.97	[423]		
N_4	63.00	[503]	1.0(-6)	[423]	0.09	[74]		
	69.50 ± 0.20	[493]			0.08	[423]		
					0.1	[94]		
N_5	63.00	[503]	1.0(-6)	[423]	0.09	[74]		
	67.60 ± 0.20	[493]			0.08	[423]		
					0.08	[94]		
O_1	18.00	[503]						
	23.40 ± 0.50	[493]						
O_2	7.00	[503]						
	13.40 ± 0.50	[493]						
O_3	12.10 ± 0.50	[493]						
IP	12.1298	[222]						

Cs Z=55	[Xe] 6s^1

Cesium $A = 132.90545(2)$ [222] $\varrho = 1.873 \, \text{g/cm}^3$ [547]

X-ray transitions

line	transition	E/eV		I/eV/$\hbar$		TPIV	Γ/eV	
K series								
K$_{\alpha_3}$	KL$_1$	30270.5 ± 1.3	[145]				15.42	[423]
K$_{\alpha_2}$	KL$_2$	30625.40 ± 0.45	[50]	3.07	[488]	2.576(-1)	15.80	[301]
		30625.00	[104]				14.90	[423]
K$_{\alpha_1}$	KL$_3$	30973.13 ± 0.46	[50]	5.68	[488]	4.757(-1)	15.60	[301]
		30973.00	[104]				14.67	[423]
	KM$_1$	34766.7 ± 1.2	[145]				23.50	[423]
K$_{\beta_3}$	KM$_2$	34919.68 ± 0.58	[50]	0.538	[488]	4.507(-2)	16.89	[423]
		34919.00	[104]					
K$_{\beta_1}$	KM$_3$	34987.3 ± 1.0	[50]	1.042	[488]	8.736(-2)	17.79	[423]
		34987.00	[104]					
K$_{\beta_5^{II}}$	KM$_4$	35244.9 ± 1.4	[145]			4.664(-4)	12.58	[423]
K$_{\beta_5^{I}}$	KM$_5$	35258.6 ± 1.3	[145]			6.352(-4)	12.58	[423]
	KN$_1$	35755.7 ± 4.1	[145]				17.71	[423]
K$_{\beta_2}$	KN$_{2,3}$	35821.7 ± 3.1	[50]	0.2131	[488]			
		35822.00	[104]					
K$_{\beta_2^{II}}$	KN$_2$	35813.2 ± 5.1	[145]			9.483(-3)	35.34	[423]
K$_{\beta_2^{I}}$	KN$_3$	35823.0 ± 3.8	[145]			1.848(-2)	49.22	[423]
K$_{\beta_4^{II}}$	KN$_4$	35905.3 ± 1.3	[145]			9.093(-5)	11.27	[423]
K$_{\beta_4^{I}}$	KN$_5$	35907.4 ± 1.2	[145]			1.236(-4)	11.26	[423]
L series								
	L$_1$M$_1$	4496.1 ± 1.7	[145]				15.02	[423]
L$_{\beta_4}$	L$_1$M$_2$	4649.45 ± 0.52	[50]	5.49(-2)	[488]	1.470(-2)	8.42	[423]
		4649.40	[104]					
L$_{\beta_3}$	L$_1$M$_3$	4716.85 ± 0.53	[50]	8.58(-2)	[488]	2.296(-2)	9.32	[423]
		4717.10	[104]					
L$_{\beta_{10}}$	L$_1$M$_4$	4975.21 ± 0.59	[50]	8.50(-4)	[488]	2.277(-4)	4.10	[423]
		4975.70	[104]					

line	transition	E/eV		I/eV/$\hbar$		TPIV	Γ/eV	
$L_{\beta 9}$	$L_1 M_5$	5002.72 ± 0.60 5003.00	[50] [104]	1.27(-3)	[488]	3.401(-4)	4.11	[423]
	$L_1 N_1$	5485.2 ± 4.6	[145]				9.23	[423]
$L_{\gamma 2}$	$L_1 N_2$	5542.10 ± 0.73 5542.30	[50] [104]	1.209(-2)	[488]	3.237(-3)	26.87	[423]
$L_{\gamma 3}$	$L_1 N_3$	5552.77 ± 0.74 5553.20	[50] [104]	1.943(-2)	[488]	5.202(-3)	40.75	[423]
	$L_1 N_4$	5634.8 ± 1.7	[145]			2.867(-5)	3.55	[423]
	$L_1 N_5$	5636.9 ± 1.7	[145]			4.367(-5)	3.54	[423]
$L_{\gamma 4}$	$L_1 O_{2,3}$	5702.60 ± 0.50 5702.70	[50] [104]	2.52(-3)	[488]			
	$L_1 O_2$					4.289(-4)		
	$L_1 O_3$					6.748(-4)		
L_η	$L_2 M_1$	4142.13 ± 0.41 4142.50	[50] [104]	7.10(-3)	[488]	4.135(-3)	14.50	[423]
	$L_2 M_2$	4293.7 ± 1.5	[145]				7.89	[423]
$L_{\beta 17}$	$L_2 M_3$	4360.9 ± 1.3	[145]			6.161(-5)	8.79	[423]
$L_{\beta 1}$	$L_2 M_4$	4619.83 ± 0.51 4620.10	[50] [104]	0.2543	[488]	7.987(-2)	3.58	[423]
	$L_2 M_5$	4633.8 ± 1.1	[145]				3.59	[423]
$L_{\gamma 5}$	$L_2 N_1$	5128.75 ± 0.63 5129.20	[50] [104]	1.59(-3)	[488]	6.108(-4)	8.72	[423]
	$L_2 N_2$	5188.3 ± 4.8	[145]				26.35	[423]
	$L_2 N_3$	5198.1 ± 3.6	[145]			8.194(-6)	40.22	[423]
$L_{\gamma 1}$	$L_2 N_4$	5280.34 ± 0.67 5280.60	[50] [104]	3.95(-2)	[488]	1.242(-2)	3.03	[423]
	$L_2 N_5$	5282.54 ± 0.98	[145]				3.02	[423]
L_l	$L_3 M_1$	3794.99 ± 0.34 3795.10	[50] [104]	9.07(-3)	[488]	7.495(-3)	14.27	[423]
L_t	$L_3 M_2$	3946.5 ± 1.4	[145]			4.218(-5)	7.66	[423]
L_s	$L_3 M_3$	4013.7 ± 1.2	[145]			3.921(-5)	8.56	[423]
$L_{\alpha 2}$	$L_3 M_4$	4272.31 ± 0.44 4272.60	[50] [104]	2.401(-2)	[488]	7.899(-3)	3.35	[423]
$L_{\alpha 1}$	$L_3 M_5$	4286.49 ± 0.44 4286.80	[50] [104]	0.2127	[488]	6.998(-2)	3.36	[423]
$L_{\beta 6}$	$L_3 N_1$	4781.06 ± 0.55	[50]	2.01(-3)	[488]	6.907(-4)	8.49	[423]

line	transition	E/eV		I/eV/$\hbar$		TPIV	Γ/eV	
		4781.50	[104]					
	L_3N_2	4841.1 ± 4.8	[145]			3.994(-6)	26.12	[423]
	L_3N_3	4851.0 ± 3.5	[145]			3.840(-6)	39.99	[423]
$L_{\beta15,2}$	$L_3N_{4,5}$	4936.00 ± 0.58	[50]					
$L_{\beta15}$	L_3N_4	4933.27 ± 0.98	[145]	3.63(-3)	[488]	1.195(-3)	2.80	[423]
$L_{\beta2}$	L_3N_5	4936.40	[104]	3.24(-2)	[488]	1.066(-2)	2.80	[423]
		4935.36 ± 0.92	[145]					
$L_{\beta7}$	L_3O_1	4989.30 ± 0.40	[50]	3.26(-4)	[488]	1.025(-4)		
		4989.70	[104]					

M series

line	transition	E/eV		I/eV/$\hbar$		TPIV	Γ/eV	
	M_2N_1	836.0	[134]			3.122(-4)	10.705	[423]
	M_2N_4	987.0	[134]			1.244(-3)	5.020	[423]
	M_3N_1	766.0	[134]			3.821(-4)	11.609	[423]
$M_{\gamma2,1}$	$M_3N_{4,5}$	918.40	[263]					
$M_{\gamma2}$	M_3N_4					1.498(-4)	5.92	[423]
$M_{\gamma1}$	M_3N_5					1.235(-3)	5.92	[423]
		922.0	[134]					
$M_{\xi2}$	M_4N_2	565.73	[423]			2.080(-3)	24.03	[423]
M_δ	M_4N_3	579.94	[423]			3.252(-4)	37.90	[423]
		579.0	[134]					
M_η	M_4O_2	729.43	[423]			2.619(-4)		
	$M_4O_{2,3}$	726.0	[134]					
$M_{\xi1}$	M_5N_3	565.22	[423]			2.245(-3)	37.90	[423]
		565.0	[134]					
	M_5O_3	714.0	[134]			2.762(-4)		

N series

line	transition	E/eV		I/eV/$\hbar$		TPIV	Γ/eV	
	N_4O_2	65.74 ± 0.04	[50]			8.083(-7)		
	N_4O_3	67.46 ± 0.04	[50]			1.395(-7)		
	N_5O_3	65.15 ± 0.04	[50]			9.731(-7)		

level characteristics

level	E_B/eV		ω_{nlj}		Γ/eV		AE/eV	
K	35985.00	[503]	0.897	[301]	12.30	[301]	35984.6	[51]
	35984.60 ± 0.40	[493]	0.876	[167]	11.95	[423]	35988.0 ± 1.5	[50]
			0.895	[38]	12.3	[94]	35991.92 ± 0.62	[145]
			0.8942	[39]				
			0.898	[211]				
			0.8984	[423]				
			0.897	[310]				
L_1	5713.00	[503]	4.90(-2)	[301]	3.78	[301]	5714.3	[51]
	5714.30 ± 0.40	[493]	4.83(-2)	[423]	3.47	[423]	5720.6 ± 2.0	[50]
			4.90(-2)	[436]	2	[94]	5721.4 ± 1.0	[145]
			6.30(-2)	[576]				
L_2	5360.00	[503]	9.00(-2)	[301]	3.25	[301]	5359.4	[51]
	5359.40 ± 0.40	[493]	9.72(-2)	[423]	2.95	[423]	5358.15 ± 0.34	[50]
			9.70(-2)	[436]	3.15	[94]	5367.05 ± 0.39	[145]
L_3	5012.00	[503]	9.10(-2)	[301]	3.51	[301]	5011.9	[51]
	5011.90 ± 0.40	[493]	9.81(-2)	[423]	2.72	[423]	5011.31 ± 0.30	[50]
			9.80(-2)	[436]	2.92	[94]	5019.87 ± 0.32	[145]
M_1	1217.00	[503]	7.7(-4)	[512]	10.8	[94]	1217.1	[51]
	1217.10 ± 0.40	[493]	7.5(-4)	[423]	11.55	[423]		
M_2	1065.00	[503]	1.09(-3)	[512]	3.7	[94]	1065.0	[51]
	1065.00 ± 0.50	[493]	2.04(-3)	[423]	4.94	[423]		
M_3	998.00	[503]	0.107	[512]	5	[94]		
	997.60 ± 0.50	[493]	2.08(-3)	[423]	5.85	[423]		
M_4	740.00	[503]	2.7(-3)	[512]	0.63	[94]		
	739.50 ± 0.40	[493]	2.71(-3)	[423]	0.63	[423]		
M_5	726.00	[503]	2.52(-3)	[423]	0.63	[94]	726.6	[51]
	725.50 ± 0.50	[493]			0.63	[423]		
N_1	231.00	[503]	8.0(-5)	[423]	2.8	[94]		
	230.80 ± 0.40	[493]			5.76	[423]		
N_2	172.00	[503]	1.3(-4)	[423]				
	172.30 ± 0.60	[493]			23.40	[423]		

level	E_B/eV		ω_{nlj}		Γ/eV		AE/eV
N_3	162.00	[503]	4.0(-5)	[423]			
	161.60 ± 0.60	[493]			37.27	[423]	
N_4	79.00	[503]	1.0(-6)	[423]	0.08	[94]	
	78.80 ± 0.50	[493]			0.08	[423]	
N_5	77.00	[503]	1.0(-6)	[423]	0.08	[94]	
	76.50 ± 0.50	[493]			0.07	[423]	
O_1	23.00	[503]					
	22.70 ± 0.50	[493]					
O_2	13.00	[503]					
	13.10 ± 0.50	[493]					
O_3	12.00	[503]					
	11.40 ± 0.50	[493]					
IP	3.89390	[222]					

Ba   Z=56   [Xe] 6s^2

Barium $A = 137.327(7)$ [222] $\varrho = 3.50\,\text{g/cm}^3$ [547]

X-ray transitions

line	transition	E/eV		I/eV/$\hbar$		TPIV	Γ/eV	
K series								
K_{α_3}	KL_1	31452.5 ± 1.4	[145]				16.44	[423]
$K_{\alpha 2}$	KL_2	31817.10 ± 0.40	[50]	3.32	[488]	2.591(-1)	16.80	[301]
		31817.00	[104]	3.36	[487]		16.80	[473]
		31816.82 ± 0.06	[277]				15.89	[423]
		31816.615 ± 0.060	[144]					
$K_{\alpha 1}$	KL_3	32193.60 ± 0.30	[50]	6.11	[488]	4.771(-1)	16.50	[301]
		32194.00	[104]	6.20	[487]		16.10	[473]
		32193.47 ± 0.07	[277]				15.66	[423]
		32193.262 ± 0.070	[144]					
	KM_1	36147.3 ± 1.3	[145]				24.71	[423]
$K_{\beta 3}$	KM_2	36304.00 ± 0.40	[50]	0.584	[488]	4.560(-2)	16.00	[473]
		36305.00	[104]	0.621	[487]		17.96	[423]
		36303.58 ± 0.12	[277]					
		36303.35 ± 0.12	[144]					
$K_{\beta 1}$	KM_3	36378.20 ± 0.30	[50]	1.132	[488]	8.835(-2)	18.15	[473]
		36379.00	[104]	1.204	[487]		18.92	[423]
		36377.68 ± 0.08	[277]					
		36377.445 ± 0.080	[144]					
$K_{\beta_5^{II}}$	KM_4	36643.2 ± 3.2	[50]	6.33(-3)	[488]	4.945(-4)	13.49	[423]
		36644.00	[104]					
$K_{\beta_5^I}$	KM_5	36666.0 ± 3.2	[50]	8.60(-3)	[488]	6.712(-4)	13.50	[423]
		36666.00	[104]					
	KN_1	37188.0 ± 4.2	[145]				19.16	[423]
$K_{\beta 2}$	$KN_{2,3}$	37257.7 ± 1.7	[50]	0.2365	[488]			
		37258.00	[104]					
$K_{\beta_2^{II}}$	KN_2	37249.0 ± 1.6	[145]	0.1317	[487]	9.633(-3)	30.25	[423]
$K_{\beta_2^I}$	KN_3	37262.9 ± 1.3	[145]	0.2581	[487]	1.877(-2)	15.32	[423]
$K_{\beta 4}$	$KN_{4,5}$	37311.5 ± 3.3	[50]	3.03(-3)	[488]			
$K_{\beta_4^{II}}$	KN_4	37313.00	[104]			1.005(-4)	12.91	[423]

line	transition	E/eV		I/eV/$\hbar$		TPIV	Γ/eV	
$K_{\beta_4^I}$	KN_5	37349.84 ± 0.86	[145]			1.363(-4)	12.91	[423]
	$KO_{2,3}$	37426.00 ± 2.30	[50]	5.18(-2)	[488]			
		37428.00	[104]	6.22(-2)	[487]			
	KO_2					1.352(-3)		
	KO_3					2.593(-3)		

L series

line	transition	E/eV		I/eV/$\hbar$		TPIV	Γ/eV	
	L_1M_1	4694.8 ± 1.7	[145]				15.49	[423]
L_{β_4}	L_1M_2	4851.97 ± 0.56	[50]	6.04(-2)	[488]	1.557(-2)	7.42	[473]
		4852.50	[104]				8.74	[423]
L_{β_3}	L_1M_3	4926.97 ± 0.58	[50]	9.35(-2)	[488]	2.410(-2)	9.20	[473]
		4927.50	[104]				9.70	[423]
$L_{\beta_{10}}$	L_1M_4	5194.28 ± 0.64	[50]	9.72(-4)	[488]	2.505(-4)	4.26	[423]
		5194.80	[104]					
L_{β_9}	L_1M_5	5217.23 ± 0.65	[50]	1.45(-3)	[488]	3.742(-4)	4.27	[423]
		5217.60	[104]					
	L_1N_1	5735.5 ± 4.6	[145]				9.94	[423]
L_{γ_2}	L_1N_2	5796.90 ± 0.50	[50]	1.357(-2)	[488]	3.499(-3)	21.02	[423]
		5797.50	[104]					
L_{γ_3}	L_1N_3	5809.20 ± 0.50	[50]	2.167(-2)	[488]	5.587(-3)	6.09	[423]
		5809.80	[104]					
$L_{\gamma_{2,3}}$	$L_1N_{2,3}$	5810.11 ± 0.20	[403]					
	L_1N_4	5895.2 ± 5.6	[145]			3.256(-5)	3.69	[423]
	L_1N_5	5897.3 ± 1.3	[145]			4.971(-5)	3.69	[423]
L_{γ_4}	$L_1O_{2,3}$	5973.30 ± 0.90	[50]	3.17(-3)	[488]			
		5974.50	[104]					
	L_1O_2					5.183(-4)		
	L_1O_3					8.163(-4)		
L_η	L_2M_1	4330.96 ± 0.67	[50]	7.75(-3)	[488]	4.241(-3)	14.93	[423]
		4331.80	[104]					
	L_2M_2	4487.4 ± 1.5	[145]				8.17	[423]
$L_{\beta_{17}}$	L_2M_3	4560.9 ± 1.3	[145]			6.511(-5)	9.13	[423]
L_{β_1}	L_2M_4	4827.58 ± 0.14	[50]	0.279	[488]	8.524(-2)	3.45	[473]
		4828.90	[104]				3.71	[423]
	L_2M_5	4842.9 ± 1.1	[145]				3.47	[423]
L_{γ_5}	L_2N_1	5370.7 ± 1.0	[50]	1.77(-3)	[488]	6.482(-4)	9.38	[423]
		5371.20	[104]					
	L_2N_2	5433.0 ± 1.5	[145]				20.46	[423]

line	transition	E/eV		I/eV/ℏ		TPIV	Γ/eV	
	L_2N_3	5446.4 ± 1.1	[145]			9.315(-6)	5.54	[423]
$L_{\gamma 1}$	L_2N_4	5531.22 ± 0.73	[50]	4.51(-2)	[488]	1.377(-2)	6.35	[473]
		5531.60	[104]				3.14	[423]
	L_2N_5	5533.28 ± 0.61	[145]				6.35	[423]
$L_{\gamma 8}$	L_2O_1	5578.00	[104]	3.10(-4)	[488]	8.754(-5)	3.70	[423]
L_l	L_3M_1	3954.15 ± 0.37	[50]	1.00(-2)	[488]	7.717(-3)	14.70	[423]
		3954.50	[104]					
L_t	L_3M_2	4111.1 ± 1.4	[145]			4.425(-5)	7.94	[423]
L_s	L_3M_3	4184.6 ± 1.3	[145]			4.108(-5)	8.90	[423]
$L_{\alpha 2}$	L_3M_4	4450.94 ± 0.12	[50]	2.629(-2)	[488]	8.382(-3)	3.45	[473]
		4451.70	[104]				3.48	[423]
$L_{\alpha 1}$	L_3M_5	4466.30 ± 0.12	[50]	0.2328	[488]	7.425(-2)	3.39	[473]
		4467.30	[104]				3.49	[423]
$L_{\beta 6}$	L_3N_1	4994.05 ± 0.60	[50]	2.25(-3)	[488]	7.433(-4)	9.15	[423]
		4994.40	[104]					
	L_3N_2	5056.0 ± 1.5	[145]			4.499(-6)	20.23	[423]
	L_3N_3	5070.1 ± 1.0	[145]			4.333(-6)	5.31	[423]
$L_{\beta 15,2}$	$L_3N_{4,5}$	5156.58 ± 0.19	[50]					
$L_{\beta 15}$	L_3N_4	5156.50	[104]	4.12(-3)	[488]	1.314(-3)	2.90	[423]
		5154.8 ± 4.9	[145]					
$L_{\beta 2}$	L_3N_5	5156.50	[104]	3.68(-3)	[488]	1.174(-2)	6.70	[473]
		5156.97 ± 0.55	[145]				2.90	[423]
$L_{\beta 7}$	L_3O_1	5207.90 ± 0.40	[50]	3.93(-4)	[488]	1.162(-4)	3.47	[423]
		5208.00	[104]					

M series

line	transition	E/eV		I/eV/ℏ		TPIV	Γ/eV	
	M_2N_1	884.0	[134]			3.236(-4)	18.267	[423]
	M_2N_4	1043.0	[134]			1.452(-3)	5.210	[423]
	M_3N_1	807.0	[134]			4.185(-4)	12.418	[423]
$M_{\gamma 2,1}$	$M_3N_{4,5}$	973.00 ± 2.30	[50]					
		974.70	[263]					
$M_{\gamma 2}$	M_3N_4			3.72(-7)	[116]	1.634(-4)	6.17	[423]
				3.96(-3)	[335]			
$M_{\gamma 1}$	M_3N_5	976.00	[104]	1.31(-5)	[116]	1.347(-3)	6.17	[423]
		971.0	[134]	3.96(-3)	[335]			
		973.0	[50]					
$M_{\xi 2}$	M_4N_2	602.43	[423]			2.418(-3)	18.07	[423]
M_δ	M_4N_3	617.0	[134]			3.740(-4)	3.14	[423]
M_β	M_4N_6	799.0	[134]					

line	transition	E/eV		I/eV/$\hbar$		TPIV	Γ/eV	
M_η	M_4O_2	779.00 ± 2.40	[50]	2.14(-5)	[116]	3.394(-4)		
		779.30	[104]					
		780.80	[263]					
	M_4O_3	789.00 ± 4.50	[50]	3.21(-6)	[116]	5.172(-5)		
		788.70	82]					
		789.70	[263]					
	$M_4O_{2,3}$	779.0	[134]					
$M_{\xi 1}$	M_5N_3	601.00 ± 1.20	[50]	3.78(-4)	[116]	2.594(-3)	3.16	[423]
		599.30	[104]	2.90(-4)	[335]			
		601.90	[263]					
		601.0	[134]					
$M_{\alpha 1}$	M_5N_7	784.0	[134]					
	M_5O_3	765.00 ± 5.00	[50]	2.01(-5)	[116]	3.583(-4)		
		765.30	[104]					
		765.0	[134]					
	N series							
	N_4O_2	75.90 ± 0.10	[50]			1.357(-6)		
	N_4O_3	77.96 ± 0.10	[50]			2.334(-7)		
	N_5O_3	75.30 ± 0.10	[50]			1.627(-6)		

level characteristics

level	E_B/eV		ω_{nlj}		Γ/eV		AE/eV	
K	37441.00	[503]	0.902	[301]	13.20	[301]	37440.6	[51]
	37440.60 ± 0.40	[493]	0.901	[38]	12.83	[423]	37452.4 ± 1.7	[50]
			0.916	[293]	13.2	[94]	37450.23 ± 0.63	[145]
			0.8997	[39]				
			0.9039	[423]				
			0.902	[310]				
			0.899	[112]				
L_1	5987.00	[503]	5.20(-2)	[301]	3.52	[301]	5988.8	[51]
	5988.80 ± 0.40	[493]	5.13(-2)	[423]	3.61	[423]	5995.9 ± 2.1	[50]
			5.30(-2)	[436]	2.1	[94]	5997.7 ± 1.0	
			5.38(-2)	[244]				
			5.70(-2)	[576]				
L_2	5624.00	[503]	9.07(-2)	[107]	3.57	[301]	5623.6	[51]

level	E_B/eV		ω_{nlj}		Γ/eV		AE/eV	
	5623.60 ± 0.30	[493]	9.60(-2)	[301]	3.06	[423]	5623.29 ± 0.28	[50]
			0.1041	[423]	3.25	[94]	5633.67 ± 0.39	[145]
			0.103	[436]				
			0.105	[244]				
			0.916	[293]				
L_3	5247.00	[503]	8.99(-2)	[107]	3.32	[301]	5247.0	[51]
	5247.00 ± 0.30	[493]	9.70(-2)	[301]	2.82	[423]	5247.04 ± 0.33	[50]
			0.104	[436]	3.02	[94]	5257.36 ± 0.32	[145]
			0.107	[244]				
			0.916	[293]				
M_1	1293.00	[503]	8.3(-4)	[512]	10.00	[74]	1292.8	[51]
	1292.80 ± 0.40	[493]	8.1(-4)	[423]	11.88	[423]		
					11.1	[94]		
M_2	1137.00	[503]	1.15(-3)	[512]	5.40	[74]	1136.7	[51]
	1136.70 ± 0.50	[493]	2.15(-3)	[423]	5.13	[423]		
					3.9	[94]		
M_3	1063.00	[503]	0.114	[512]	5.40	[74]	1062.2	[51]
	1062.20 ± 0.50	[493]	2.22(-2)	[423]	6.09	[423]		
					5.4	[94]		
M_4	796.00	[503]	2.7(-3)	[512]	0.75	[74]		
	796.10 ± 0.30	[493]	3.18(-3)	[423]	0.66	[423]		
					0.67	[94]		
M_5	781.00	[503]	2.95(-3)	[423]	0.75	[74]	780.5	[51]
	780.70 ± 0.30	[493]			0.67	[423]		
					0.67	[94]		
N_1	253.00	[503]	9.0(-5)	[423]	6.00	[74]		
	253.00 ± 0.50	[493]			6.33	[423]		
					3.1	[94]		
N_2	192.00	[503]	1.5(-4)	[423]	2.80	[74]		
	191.80 ± 0.70	[493]			17.41	[423]		
					5	[94]		
N_3	179.70 ± 0.60	[493]	5.0(-5)	[423]	2.80	[74]		
					2.49	[423]		
					1.3	[94]		
N_4	93.00	[503]	2.0(-6)	[423]	0.10	[74]		
	92.50 ± 0.50	[493]			0.08	[423]		

level	E_B/eV		ω_{nlj}		Γ/eV		AE/eV
					0.08	[94]	
N_5	90.00	[503]	2.0(-6)	[423]	0.10	[74]	
	89.90 ± 0.50	[493]			0.07	[423]	
					0.08	[94]	
O_1	40.00	[503]					
	39.10 ± 0.60	[493]					
O_2	17.00	[503]					
	16.60 ± 0.50	[493]					
O_3	15.00	[503]					
	14.60 ± 0.50	[493]					
IP	5.21170	[222]					

La   Z=57

[Xe] 5d^1 6s^2

Lanthanum

$A = 138.9055(2)$ [222] $\varrho = 6.166\,\text{g/cm}^3$ [547]
$\varrho = 6.18\,\text{g/cm}^3$ [483]

X-ray transitions

line	transition	E/eV		I/eV/ℏ		TPIV	Γ/eV	
K series								
K_{α_3}	KL$_1$	32660.4 ± 1.4	[145]				17.51	[423]
$K_{\alpha 2}$	KL$_2$	33034.38 ± 0.26	[50]	3.58	[488]	2.605(-1)	17.80	[301]
		33034.00	[104]				16.93	[423]
$K_{\alpha 1}$	KL$_3$	33442.12 ± 0.27	[50]	6.57	[488]	4.784(-1)	17.60	[301]
		33442.00	[104]				16.69	[423]
	KM$_1$	37559.0 ± 1.3	[145]				25.91	[423]
K_{β_3}	KM$_2$	37720.60 ± 0.68	[50]	0.633	[488]	4.611(-2)	19.06	[423]
		37721.00	[104]					
		37721.00 ± 1.00	[95]					
K_{β_1}	KM$_3$	37801.45 ± 0.51	[50]	1.227	[488]	8.932(-2)	19.90	[423]
		37802.00	[104]					
		37802.00 ± 1.00	[95]					
K_{β_5}	KM$_{4,5}$	38092.00 ± 5.00	[95]					
$K_{\beta_5^{II}}$	KM$_4$	38074.6 ± 3.5	[50]	7.19(-3)	[488]	5.233(-4)	14.47	[423]
		38075.4 ± 1.5	[145]					
$K_{\beta_5^I}$	KM$_5$	38094.5 ± 3.5	[50]	9.73(-3)	[488]	7.078(-4)	14.48	[423]
		38095.00	[104]					
	KN$_1$	38654.0 ± 4.2	[145]				20.62	[423]
K_{β_2}	KN$_{2,3}$	38730.3 ± 1.3	[50]					
		38724.00 ± 2.00	[95]					
$K_{\beta_2^{II}}$	KN$_2$	38697.00	[104]	0.1343	[488]	9.784(-3)	19.55	[423]
		38717.0 ± 1.6	[145]					
$K_{\beta_2^I}$	KN$_3$	38731.20	[104]	0.2616	[488]	1.905(-2)	16.56	[423]
		38732.9 ± 1.4	[145]					
K_{β_4}	KN$_{4,5}$	38828.2 ± 3.6	[50]	3.57(-3)	[488]			
$K_{\beta_4^{II}}$	KN$_4$	38784.00	[104]			1.105(-4)	13.87	[423]
		38823.1 ± 5.0	[145]					
$K_{\beta_4^I}$	KN$_5$	38829.00	[104]			1.494(-4)	13.87	[423]
		38825.05 ± 0.90	[145]					
	KO$_{2,3}$	38909.00 ± 2.40	[50]	6.14(-2)	[488]			

line	transition	E/eV		I/eV/$\hbar$		TPIV	Γ/eV	
		38911.00	[104]					
		38916.00 $\pm$ 1.00	[95]					
	KO_2					1.533(-3)		
	KO_3					2.936(-3)		

L series

line	transition	E/eV		I/eV/$\hbar$		TPIV	Γ/eV	
	L_1M_1	4898.6 $\pm$ 1.7	[145]				15.90	[423]
$L_{\beta 4}$	L_1M_2	5061.79 $\pm$ 0.38	[50]	6.63(-2)	[488]	1.653(-2)	9.02	[423]
		5062.80	[104]					
$L_{\beta 3}$	L_1M_3	5143.40 $\pm$ 0.39	[50]	0.1017	[488]	2.536(-2)	9.86	[423]
		5143.90	[104]					
$L_{\beta 10}$	L_1M_4	5416.99 $\pm$ 0.71	[50]	1.11(-3)	[488]	2.761(-4)	4.43	[423]
		5415.00	[104]					
$L_{\beta 9}$	L_1M_5	5435.79 $\pm$ 0.71	[50]	1.65(-3)	[488]	4.127(-4)	4.44	[423]
		5434.00	[104]					
	L_1N_1	5993.6 $\pm$ 4.6	[145]				10.61	[423]
$L_{\gamma 2}$	L_1N_2	6060.73 $\pm$ 0.29	[50]	1.519(-2)	[488]	3.787(-3)	9.51	[423]
		6060.20	[104]					
$L_{\gamma 3}$	L_1N_3	6075.32 $\pm$ 0.29	[50]	2.409(-2)	[488]	6.007(-3)	6.53	[423]
		6075.10	[104]					
	L_1N_4	6162.7 $\pm$ 5.3	[145]			3.691(-5)	3.83	[423]
	L_1N_5	6164.6 $\pm$ 1.3	[145]			5.650(-5)	3.84	[423]
	$L_1N_{4,5}$	6167.10 $\pm$ 0.61	[499]					
	$L_1N_{6,7}$	6264.8 $\pm$ 1.6	[498]					
$L_{\gamma 4}$	$L_1O_{2,3}$	6252.00 $\pm$ 1.30	[50]	3.78(-3)	[488]			
		6252.90	[104]					
	L_1O_2					5.625(-4)		
	L_1O_3					8.778(-4)		
L_{η}	L_2M_1	4524.9 $\pm$ 7.3	[50]	8.45(-3)	[488]	4.352(-3)	15.32	[423]
		4525.00	[104]					
	L_2M_2	4686.2 $\pm$ 1.5	[145]				8.45	[423]
$L_{\beta 17}$	L_2M_3	4766.7 $\pm$ 1.3	[145]			6.879(-5)	9.29	[423]
$L_{\beta 1}$	L_2M_4	5042.17 $\pm$ 0.15	[50]	0.306	[488]	9.069(-2)	3.86	[423]
		5043.20	[104]					
	L_2M_5	5057.8 $\pm$ 1.1	[145]				3.87	[423]
$L_{\gamma 5}$	L_2N_1	5620.13 $\pm$ 0.48	[50]	1.96(-3)	[488]	6.878(-4)	10.03	[423]
		5621.80	[104]					
	L_2N_2	5683 $\pm$ 15	[145]				8.94	[423]
	L_2N_3	5698.6 $\pm$ 1.1	[145]			1.054(-5)	5.95	[423]
		5702.30	[95]					

line	transition	E/eV		I/eV/$\hbar$		TPIV	Γ/eV	
$L_{\gamma 1}$	L_2N_4	5788.30 ± 0.26	[50]	5.11(-2)	[488]	1.515(-2)	3.26	[423]
		5789.10	[104]					
	L_2N_5	5790.83 ± 0.61	[145]				3.26	[423]
L_v	$L_2N_{6,7}$	5887.7 ± 1.4	[498]					
	L_2O_2	5888.00	[104]	3.99(-3)	[488]		1.96	[473]
L_l	L_3M_1	4124.5 ± 6.1	[50]	1.10(-2)	[488]	7.947(-3)	15.08	[423]
		4124.00	[104]					
L_t	L_3M_2	4278.9 ± 1.4	[145]			4.642(-5)	8.21	[423]
L_s	L_3M_3	4359.3 ± 1.3	[145]			4.303(-5)	9.05	[423]
$L_{\alpha 2}$	L_3M_4	4634.26 ± 0.10	[50]	2.872(-2)	[488]	8.887(-3)	17.80	[473]
		4634.40	[104]				3.62	[423]
$L_{\alpha 1}$	L_3M_5	4651.02 ± 0.13	[50]	0.2543	[488]	7.870(-2)	17.60	[473]
		4651.20	[104]				3.63	[423]
$L_{\beta 6}$	L_3N_1	5211.63 ± 0.42	[50]	2.51(-3)	[488]	7.994(-4)	9.79	[423]
		5211.90	[104]					
	L_3N_2	5276 ± 15	[145]			5.051(-6)	8.70	[423]
	L_3N_3	5291.2 ± 1.0	[145]			4.873(-6)	5.71	[423]
$L_{\beta 15,2}$	$L_3N_{4,5}$	5382.87 ± 0.23	[50]	4.65(-3)	[488]	1.439(-3)	3.02	[423]
$L_{\beta 15}$	L_3N_4	5381.4 ± 4.6	[145]			1.439(-3)	3.01	[423]
$L_{\beta 2}$	L_3N_5	5383.43 ± 0.55	[145]	4.16(-2)	[488]	1.287(-2)	3.02	[423]
		5384.00	[104]					
L_u	$L_3N_{6,7}$	5479.1 ± 1.2	[498]					
$L_{\beta 7}$	L_3O_1	5450.00 ± 7.00	[50]	4.62(-4)	[488]	1.295(-4)	3.54	[423]
		5450.00	[104]					
$L_{\beta 5}$	$L_3O_{4,5}$	5478.00	[104]	3.55(-4)	[488]			

line	transition	E/eV		I/eV/$\hbar$		TPIV	Γ/eV	
	M series							
	M_2N_1	931.0	[134]			3.327(-4)	12.155	[423]
	M_2N_4	1101.0	[134]			1.452(-3)	12.155	[423]
M_γ	$M_3N_{4,5}$	1027.00 ± 3.40	[50]	4.62(-3)	[335]			
		1028.00	[104]					
		1028.90	[263]					
M_{γ_2}	M_3N_4					1.777(-4)	6.22	[423]
M_{γ_1}	M_3N_5	1023.0	[134]			1.463(-3)	6.22	[423]
M_{ξ_2}	M_4N_2	640.23	[423]			2.761(-3)	6.48	[423]
M_δ	M_4N_3	654.0	[134]			4.225(-4)	3.49	[423]
M_β	M_4N_6	854.00 ± 3.00	[50]					
		856.20	[263]					
		849.0	[134]					
M_η	M_4O_2	836.63	[423]			4.143(-4)		
M_{ξ_1}	M_5N_3	638.00 ± 1.60	[50]	3.25(-4)	[335]	2.992(-3)	9.10	[279]
		638.10	[104]				8.92	[423]
		639.10	[263]					
		638.0	[134]					
M_{α_1}	M_5N_7	833.0	[134]					
$M_{\alpha1.2}$	$M_5N_{6,7}$	833.00 ± 2.80	[50]					
		834.90	[263]					
	M_5O_3	811.0	[134]			4.421(-4)		
	N series							
	$N_{4,5}O_{2,3}$	81.20 ± 0.30	[50]					

level characteristics

level	E_B/eV		ω_{nlj}		Γ/eV		AE/eV	
K	38925.00	[503]	0.907	[301]	14.10	[301]	38924.6	[51]
	38924.60 ± 0.40	[493]			13.77	[423]	38934.3 ± 9.0	[50]
			0.906	[38]	14.2	[94]	38939.45 ± 0.67	[145]
			0.9049	[39]				
			0.9090	[423]				
			0.907	[310]				
L_1	6267.00	[503]	5.50(-2)	[301]	4.06	[301]	6266.3	[51]
	6266.30 ± 0.50	[493]	5.44(-2)	[423]	3.73	[423]		
			5.70(-2)	[436]	2.2	[94]		
L_2	5891.00	[503]	0.103	[301]	3.63	[301]	5890.6	[51]
	5890.60 ± 0.40	[493]	0.1111	[423]	3.16	[423]		
			0.111	[436]	3.35	[94]		
L_3	5483.00	[503]	0.104	[301]	3.41	[301]	5482.7	[51]
	5482.70 ± 0.40	[493]	0.1109	[423]	2.92	[423]		
			0.112	[436]	3.12	[94]		
M_1	1362.00	[503]	8.40(-4)	[38]	9.30	[350]	1361.3	[51]
	1361.30 ± 0.30	[493]	8.9(-4)	[512]	12.16	[423]		
			8.4(-4)	[423]	11.4	[94]		
M_2	1205.00	[503]	1.10(-3)	[38]	5.76	[350]	1204.4	[51]
	1204.40 ± 0.60	[493]	9.28(-3)	[512]	5.29	[423]		
			8.7(-4)	[423]	4.1	[94]		
M_3	1124.00	[503]	9.90(-4)	[38]	5.41	[350]	1123.4	[51]
	1123.40 ± 0.50	[493]	0.121	[512]	6.12	[423]		
			2.27(-3)	[423]	5.8	[94]		
M_4	849.00	[503]	2.70(-3)	[38]	0.73	[350]		
	848.50 ± 0.40	[493]	2.7(-3)	[512]	0.70	[423]		
			2.70(-3)	[423]	0.7	[94]		
M_5	832.00	[503]	3.43(-3)	[423]			836.0	[51]
	831.70 ± 0.40	[493]			0.70	[423]		
					0.7	[94]		
N_1	271.00	[503]	1.0(-4)	[423]	3.3	[94]		

level	E_B/eV		ω_{nlj}		Γ/eV		AE/eV
	270.40 ± 0.80	[493]			6.87	[423]	
N_2	206.00	[503]	1.7(-4)	[423]	5.03	[94]	
	205.80 ± 1.20	[493]			5.78	[423]	
N_3	192.00	[503]	5.0(-5)	[423]	1.45	[94]	
	191.40 ± 0.90	[493]			2.79	[423]	
N_4	99.00	[503]	3.0(-6)	[423]	0.09	[94]	
	98.90 ± 0.80	[51]					
	100.70 ± 0.80	[493]			0.09	[423]	
N_5	99.00	[503]	3.0(-6)	[423]	0.1	[94]	
	98.90 ± 0.80	[51]					
	97.70 ± 0.80	[493]			0.10	[423]	
O_1	33.00	[503]			0.62	[423]	
	32.30 ± 7.20	[493]					
O_2	15.00	[503]					
	14.40 ± 1.20	[51]					
	16.60 ± 1.20	[493]					
O_3	15.00	[503]					
	13.30 ± 1.20	[493]					
IP	5.5770	[222]					

Ce   Z=58

[Xe] $4f^1\, 5d^1\, 6s^2$

Cerium $A = 140.116(1)$ [222] $\varrho = 6.771\,\text{g/cm}^3$ [547]
$\varrho = 6.79\,\text{g/cm}^3$ [483]

X-ray transitions

line	transition	E/eV		I/eV/$\hbar$		TPIV	Γ/eV	
K series								
K_{α_3}	KL_1	33896.2 ± 1.4	[145]				18.58	[423]
$K_{\alpha 2}$	KL_2	34279.28 ± 0.28	[50]	3.85	[488]	2.618(-1)	18.90	[301]
		34279.00	[104]				19.50	[473]
							17.96	[423]
$K_{\alpha 1}$	KL_3	34720.00 ± 0.29	[50]	7.06	[488]	4.794(-1)	18.60	[301]
		34720.00	[104]				18.60	[473]
							17.74	[423]
	KM_1	39006.9 ± 1.4	[145]				27.68	[423]
$K_{\beta 3}$	KM_2	39170.46 ± 0.73	[50]	0.686	[488]	4.660(-2)	17.95	[473]
		39171.00	[104]				20.43	[423]
$K_{\beta 1}$	KM_3	39257.77 ± 0.37	[50]	1.329	[488]	9.028(-2)	20.60	[473]
		39259.00	[104]				21.31	[423]
$K_{\beta_5^{II}}$	KM_4	39539.0 ± 3.7	[50]	8.16(-3)	[488]	5.540(-4)	15.54	[423]
		39540.00	[104]					
$K_{\beta_5^{I}}$	KM_5	39557.9 ± 3.7	[50]	1.099(-2)	[488]	7.468(-4)	15.46	[423]
		39558.00	[104]					
	KN_1	40155.1 ± 4.3	[145]				22.40	[423]
$K_{\beta 2}$	$KN_{2,3}$	40233.1 ± 1.9	[50]					
$K_{\beta_2^{II}}$	KN_2	40201.00	[104]	0.1454	[488]	9.920(-3)	27.24	[423]
		40220 ± 16	[145]					
$K_{\beta_2^{I}}$	KN_3	40234.10	[104]	0.2827	[488]	1.930(-2)	18.47	[423]
		40237.4 ± 1.4	[145]					
$K_{\beta 4}$	$KN_{4,5}$	40336.5 ± 3.9	[50]	4.02(-3)	[488]			
		40338.00	[104]					
$K_{\beta_4^{II}}$	KN_4	40330.4 ± 4.7	[145]			1.165(-4)	15.35	[423]
$K_{\beta_4^{I}}$	KN_5	40334.22 ± 0.93	[145]			1.568(-4)	15.07	[423]
	$KO_{2,3}$	40427.00 ± 2.60	[50]	6.12(-2)	[488]			
		40429.20	[104]					

line	transition	E/eV		I/eV/$\hbar$		TPIV	Γ/eV	
	KO$_2$					1.673(-3)		
	KO$_3$					3.236(-3)		

L series

line	transition	E/eV		I/eV/$\hbar$		TPIV	Γ/eV	
	L$_1$M$_1$	5110.7 ± 1.7	[145]				16.77	[423]
L$_{\beta 4}$	L$_1$M$_2$	5277.35 ± 0.43	[50]	7.28(-2)	[488]	1.753(-2)	7.82	[473]
		5277.50	[104]				9.53	[423]
L$_{\beta 3}$	L$_1$M$_3$	5365.29 ± 0.42	[50]	0.1106	[488]	2.662(-2)	9.70	[473]
		5365.40	[104]				10.41	[423]
L$_{\beta 10}$	L$_1$M$_4$	5644.98 ± 0.50	[50]	1.26(-3)	[488]	3.036(-4)	4.63	[423]
		5644.90	[104]					
L$_{\beta 9}$	L$_1$M$_5$	5664.63 ± 0.47	[50]	1.88(-3)	[488]	4.538(-4)	4.56	[423]
		5664.60	[104]					
	L$_1$N$_1$	6258.9 ± 4.6	[145]				11.49	[423]
L$_{\gamma 2}$	L$_1$N$_2$	6326.39 ± 0.59	[50]	1.666(-3)	[488]	4.009(-3)	16.34	[423]
		6326.70	[104]					
L$_{\gamma 3}$	L$_1$N$_3$	6342.00 ± 0.59	[50]	2.617(-2)	[488]	6.299(-3)	7.56	[423]
		6342.30	[104]					
L$_{\gamma 11}$	L$_1$N$_{4,5}$	6440.0	[104]	2.55(-4)	[488]			
		6439 ± 0.34	[499]					
	L$_1$N$_4$	6434.2 ± 5.1	[145]			4.018(-5)	4.45	[423]
	L$_1$N$_5$	6438.0 ± 1.3	[145]			6.148(-5)	4.16	[423]
	L$_1$N$_6$	6538.6 ± 1.5	[145]					
	L$_1$O$_1$	6512.90	[104]	1.15(-8)	[488]		4.43	[423]
	L$_1$N$_{6,7}$	6546.60	[501]					
	L$_1$O$_1$	6512.90	[104]				4.44	[423]
L$_{\gamma 4}$	L$_1$O$_{2,3}$	6528.00 ± 1.40	[50]	3.78(-3)	[488]			
		6528.40	[104]					
	L$_1$O$_2$					5.890(-4)		
	L$_1$O$_3$					9.092(-4)		
L$_\eta$	L$_2$M$_1$	4731.6 ± 1.1	[50]	9.21(-3)	[488]	4.455(-3)	16.15	[423]
		4731.90	[104]					
	L$_2$M$_2$	4893.1 ± 1.5	[145]				8.91	[423]
L$_{\beta 17}$	L$_2$M$_3$	4981.4 ± 1.4	[145]			7.248(-5)	9.79	[423]
L$_{\beta 1}$	L$_2$M$_4$	5262.93 ± 0.41	[50]	0.335	[488]	9.746(-2)	3.73	[473]
		5263.10	[104]				4.01	[423]
	L$_2$M$_5$	5281.6 ± 1.1	[145]				3.64	[423]
L$_{\gamma 5}$	L$_2$N$_1$	5874.90 ± 0.51	[50]	2.13(-3)	[488]	7.277(-4)	10.87	[423]

line	transition	E/eV		I/eV/$\hbar$		TPIV	Γ/eV	
		5874.90	[104]					
	L_2N_2	5940.30	[501]				15.72	[423]
		5940 ± 15	[145]					
	L_2N_3	5957.9 ± 1.1	[145]			1.177(-5)	6.94	[423]
	L_2N_3	5960.60	[501]				6.94	[423]
$L_{\gamma1}$	L_2N_4	6052.15 ± 0.29	[50]	5.56(-2)	[488]	1.618(-2)	6.75	[473]
		6052.20	[104]				3.83	[423]
	L_2N_5	6054.66 ± 0.61	[145]					
L_ν	$L_2N_{6,7}$	6161.60	[104]	5.72(-6)	[488]			
	L_2N_6					1.664(-6)		
		6161.57 ± 0.61	[499]					
L_ν	L_2N_6	6155.21 ± 0.84	[145]			1.664(-6)		
$L_{\gamma8}$	L_2O_1	6126.00 ± 1.20	[50]	3.73(-4)	[488]	1.063(-4)	3.81	[423]
		6127.70	[104]					
L_l	L_3M_1	4287.52 ± 0.88	[50]	1.21(-2)	[488]	8.164(-3)	15.93	[423]
		4287.50	[104]					
L_t	L_3M_2	4452.6 ± 1.4	[145]			4.856(-5)	8.69	[423]
L_s	L_3M_3	4540.8 ± 1.3	[145]			4.495(-5)	9.57	[423]
$L_{\alpha2}$	L_3M_4	4823.17 ± 0.34	[50]	3.14(-2)	[488]	9.465(-3)	3.78	[473]
		4823.20	[104]				3.79	[423]
$L_{\alpha1}$	L_3M_5	4840.06 ± 0.31	[50]	0.2777	[488]	8.382(-2)	3.70	[473]
		4840.00	[104]				3.72	[423]
$L_{\beta6}$	L_3N_1	5433.24 ± 0.43	[50]	2.75(-3)	[488]	8.570(-4)	10.65	[423]
		5433.20	[104]					
	L_3N_2	5499.00	[501]			5.587(-6)	15.50	[423]
	L_3N_3	5518.80	[501]			5.386(-6)	6.72	[423]
$L_{\beta15}$	L_3N_4	5610.3 ± 4.3	[145]	5.04(-3)	[488]	1.521(-3)	3.61	[423]
$L_{\beta2}$	L_3N_5	5612.40	[104]	4.50(-2)	[488]	1.359(-2)	6.86	[473]
		5614.13 ± 0.54					3.32	[423]
$L_{\beta15,2}$	$L_3N_{4,5}$	5612.67 ± 0.42	[50]					
$L_{u,u'}$	$L_3N_{6,7}$	5721.10	[104]	4.79(-6)	[488]			
		5721.10 ± 0.53	[205]					
L_u	L_3N_6					2.521(-7)		
	L_3N_7					1.447(-6)		
$L_{\beta7}$	L_3O_1	5713.20 ± 0.50	[50]	4.80(-4)	[488]	1.436(-4)	3.59	[423]
		5682.30	[104]					
	$L_3O_{2,3}$	5699.70	[501]					

line	transition	E/eV		I/eV/$\hbar$		TPIV	Γ/eV	
	L_3O_2					9.834(-7)		
	L_3O_3					8.890(-7)		
M series								
	M_2N_1	982.0	[134]			3.464(-4)	13.341	[423]
	M_2N_4	1159.0	[134]			1.532(-3)	6.301	[423]
$M_{\gamma2,1}$	$M_3N_{4,5}$	1074.90 ± 0.90	[50]	5.13(-3)	[335]			
		1074.80	[104]					
		1077.20	[263]					
$M_{\gamma2}$	M_3N_4					1.914(-4)	7.18	[423]
$M_{\gamma1}$	M_3N_5	1074.0	[134]			1.571(-3)	6.89	[423]
$M_{\xi2}$	M_4N_2					2.937(-3)	13.30	[423]
$M\delta$	M_4N_3	693.0	[134]			4.430(-4)	4.52	[423]
M_β	M_4N_6	902.00 ± 2.60	[50]	7.87(-4)	[335]	2.282(-4)		
		903.70	[263]					
		899.0	[134]					
M_η	M_4O_2					4.625(-4)		
$M_{\xi1}$	M_5N_3	676.00 ± 1.50	[50]	3.53(-4)	[335]	3.225(-3)	10.60	[473]
		674.60	[104]				4.44	[423]
		677.50	[263]					
		675.0	[134]					
$M_{\alpha2,1}$	$M_5N_{6,7}$	883.00 ± 1.30	[50]	7.87(-4)	[335]			
		885.00	[263]					
$M_{\alpha2}$	M_5N_6	879,29	[423]			1.155(-5)		
$M_{\alpha1}$	M_5N_7	880.0	[134]			2.259(-4)		
	$M_5O_{2,3}$	862.00 ± 3.00	[50]					
		862.00	[104]					
	M_5O_3	862.0	[134]			5.062(-4)		
N series								
	$N_{4,5}O_{2,3}$	85.90 ± 0.40	[50]					

level characteristics

level	E_B/eV		ω_{nlj}		Γ/eV		AE/eV	
K	40444.00	[503]	0.912	[301]	15.10	[301]	40443.0	[51]
	40443.00 ± 0.40	[493]	0.911(26)	[38]	14.74	[423]	40453.6 ± 9.8	[50]
			0.926	[293]	15.2	[94]	40446.57 ± 0.71	[145]
			0.9096	[39]				
			0.9137	[423]				
			0.912	[310]				
L_1	6549.00	[503]	5.80(-2)	[301]	4.21	[301]	6548.8	[51]
	6548.80 ± 0.50	[493]	5.74(-2)	[423]	3.84	[423]	6548.1 ± 2.6	[50]
			6.10(-2)	[436]	2.5	[94]	6550.4 ± 1.1	[145]
L_2	6165.00	[503]	0.110	[301]	3.80	[301]	6164.2	[51]
	6164.20 ± 0.40	[493]	0.1190	[423]	3.22	[423]	6160.9 ± 2.3	[50]
			0.119	[436]	3.41	[94]	6167.01 ± 0.31	[145]
L_3	5724.00	[503]	0.111	[301]	3.49	[301]	5723.4	[51]
	5723.40 ± 0.40	[493]	0.1177	[423]	3.00	[423]	5724.0 ± 3.9	[50]
			0.119	[436]	3.19	[94]	5726.47 ± 0.32	[145]
M_1	1435.00	[503]	9.5(-3)	[512]	11.00	[74]	1434.6	[51]
	1434.60 ± 0.60	[493]	9.1(-4)	[423]	12.93	[423]		
					11.6	[94]		
M_2	1273.00	[503]	9.33(-3)	[512]	5.80	[74]	1272.8	[51]
	1272.80 ± 0.60	[493]	2.38(-3)	[423]	5.69	[423]		
					4.3	[94]		
M_3	1186.00	[503]	0.128	[512]	5.80	[74]	1185.4	[51]
	1185.40 ± 0.50	[493]	2.53(-3)	[423]	6.57	[423]		
					6.2	[94]		
M_4	902.00	[503]	2.7(-3)	[512]	0.82	[74]		
	901.30 ± 0.60	[493]	4.14(-3)	[423]	0.79	[423]		
					0.72	[94]		
M_5	884.00	[503]			1.20	[74]	883.8	[51]
	883.30 ± 0.50	[493]	3.97	[423]	0.72	[423]		
					0.72	[94]		
N_1	290.00	[503]	1.1(-4)	[423]	8.10	[74]		
	289.60 ± 0.70	[493]			7.65	[423]		

level	E_B/eV		ω_{nlj}		Γ/eV		AE/eV
					3.5	[94]	
N_2	224.00	[503]	2.0(-4)	[423]	4.50	[74]	
	223.30 ± 1.10	[493]			12.50	[423]	
					5.06	[94]	
N_3	208.00	[503]	6.0(-5)	[423]	4.50	[74]	
	207.20 ± 0.90	[493]			3.72	[423]	
					1.6	[94]	
N_4	111.00	[503]	4.0(-6)	[423]	0.34	[74]	
	113.60 ± 0.60	[493]			0.61	[423]	
					0.61	[94]	
N_5	111.00	[503]	4.0(-6)	[423]	0.34	[74]	
	107.60 ± 0.60	[493]			0.32	[423]	
					0.32	[94]	
N_6	1.00	[503]					
	0.10 ± 1.20	[493]					
O_1	38.00	[503]			0.59	[423]	
	37.80 ± 1.30	[493]					
O_2	20.00	[503]					
	19.80 ± 1.20	[51]					
	21.80 ± 1.20	[493]					
O_3	20.00	[503]					
	18.80 ± 1.20	[493]					
IP	5.5387	[222]					

Pr   Z=59

[Xe] 4f^3 6s^2

Praseodymium A = 140.90765(2) [222] ϱ = 6.772 g/cm^3 [547]
ϱ = 6.71 g/cm^3 [483]

X-ray transitions

line	transition	E/eV		I/eV/$\hbar$		TPIV	Γ/eV	
K series								
K$_{\alpha_3}$	KL$_1$	35157.3 ± 1.4	[145]				19.74	[423]
K$_{\alpha 2}$	KL$_2$	35550.59 ± 0.30	[50]	4.15	[488]	2.630(-1)	20.10	[301]
		35550.00	[104]				19.08	[423]
K$_{\alpha 1}$	KL$_3$	36026.71 ± 0.31	[50]	7.57	[488]	4.802(-1)	19.80	[301]
		36027.00	[104]				18.87	[423]
	KM$_1$	40484.3 ± 1.4	[145]				17.14	[423]
K$_{\beta 3}$	KM$_2$	40653.27 ± 0.99	[50]	0.742	[488]	4.706(-2)	21.79	[423]
		40654.00	[104]					
K$_{\beta 1}$	KM$_3$	40748.67 ± 0.79	[50]	1.436	[488]	9.112(-2)	22.73	[423]
		40749.00	[104]					
K$_{\beta_5^{II}}$	KM$_4$	41039.2 ± 1.5	[145]			5.856(-4)	16.65	[423]
K$_{\beta_5^{I}}$	KM$_5$	41060.5 ± 1.5	[145]			7.851(-4)	16.53	[423]
	KN$_1$	41688.1 ± 4.3	[145]				24.03	[423]
K$_{\beta 2}$	KN$_{2,3}$	41774.4 ± 4.2	[50]					
K$_{\beta_2^{II}}$	KN$_2$	41738.40	[104]	0.1583	[488]	1.006(-2)	27.62	[423]
		41754 ± 16	[145]					
K$_{\beta_2^{I}}$	KN$_3$	41775.40	[104]	0.308	[488]	1.956(-2)	20.13	[423]
		41774.3 ± 1.4	[145]					
K$_{\beta_4^{II}}$	KN$_4$	41872.4 ± 4.5	[145]			1.245(-4)	16.56	[423]
K$_{\beta_4^{I}}$	KN$_5$	41876.35 ± 0.97	[145]			1.670(-4)	16.31	[423]
L series								
	L$_1$M$_1$	5326.9 ± 1.7	[145]				5.32	[423]
L$_{\beta 4}$	L$_1$M$_2$	5498.1 ± 1.4	[50]	7.97(-2)	[488]	1.853(-2)	9.97	[423]
		5497.90	[104]					
L$_{\beta 3}$	L$_1$M$_3$	5591.8 ± 1.1	[50]	0.1199	[488]	2.787(-2)	10.91	[423]
		5593.60	[104]					
L$_{\beta 10}$	L$_1$M$_4$	5884.0 ± 1.7	[50]	1.43(-3)	[488]	3.328(-4)	4.83	[423]

line	transition	E/eV		I/eV/$\hbar$		TPIV	Γ/eV	
		5885.20	[104]					
$L_{\beta9}$	L_1M_5	5902.8 ± 1.7	[50]	2.14(-3)	[488]	4.976(-4)	4.71	[423]
		5904.10	[104]					
	L_1N_1	6530.7 ± 4.6	[145]				12.21	[423]
$L_{\gamma2}$	L_1N_2	6598.0 ± 2.1	[50]	1.837(-2)	[488]	4.269(-3)	15.80	[423]
		6600.40	[104]					
$L_{\gamma3}$	L_1N_3	6615.9 ± 2.1	[50]	2.861(-2)	[488]	6.649(-3)	8.31	[423]
		6618.80	[104]					
	L_1N_4	6715.1 ± 4.8	[145]			4.432(-5)	4.74	[423]
$L_{\gamma11}$	L_1N_5	6719.0 ± 1.3	[145]			6.789(-5)	4.49	[423]
	$L_1N_{4,5}$	6718.70	[104]	2.92(-4)	[488]			
	$L_1N_{6,7}$	6830.40	[501]					
$L_{\gamma4}$	$L_1O_{2,3}$	6815.00 ± 1.50	[50]	4.09(-3)	[488]			
		6811.90	[104]					
	L_1O_2					6.221(-4)		
	L_1O_3					9.503(-4)		
L_η	L_2M_1	4935.6 ± 8.7	[50]	1.002(-2)	[488]	4.563(-3)	16.81	[423]
		4935.00	[104]					
	L_2M_2	5104.0 ± 1.5	[145]				9.31	[423]
$L_{\beta17}$	L_2M_3	5200.4 ± 1.4	[145]			7.667(-5)	10.25	[423]
$L_{\beta1}$	L_2M_4	5488.9 ± 1.1	[50]	0.365	[488]	1.042(-1)	4.17	[423]
		5489.60	[104]					
	L_2M_5	5510.0 ± 1.1	[145]				4.05	[423]
$L_{\gamma5}$	L_2N_1	6136.2 ± 1.8	[50]	2.34(-3)	[488]	7.698(-4)	11.55	[423]
		6135.70	[104]					
	L_2N_2	6204 ± 15	[145]				15.14	[423]
	L_2N_3	6223.8 ± 1.0	[145]			1.317(-5)	7.65	[423]
$L_{\gamma1}$	L_2N_4	6322.1 ± 1.4	[50]	6.13(-2)	[488]	1.747(-2)	4.08	[423]
		6323.50	[104]					
	L_2N_5	6325.84 ± 0.61	[145]				3.83	[423]
L_ν	L_2N_6	6436.70	[104]	1.06(-5)	[488]	3.030(-6)		
		6431.16 ± 0.81	[145]					
	L_2N_7	6436.70	[104]					
$L_{\gamma8}$	L_2O_1	6403.00 ± 1.30	[50]	4.05(-4)	[488]	1.164(-4)	3.87	[423]
		6403.30	[104]					
L_l	L_3M_1	4453.23 ± 0.95	[50]	1.328(-2)	[488]	8.390(-3)	16.59	[423]
		4453.60	[104]					

line	transition	E/eV		I/eV/$\hbar$		TPIV	Γ/eV	
L_t	L_3M_2	4628.2 ± 1.4	[145]			5.080(-5)	9.10	[423]
L_s	L_3M_3	4724.6 ± 1.3	[145]			4.697(-5)	10.04	[423]
$L_{\alpha2}$	L_3M_4	5013.64 ± 0.90	[50]	3.42(-2)	[488]	1.004(-2)	3.96	[423]
		5013.00	[104]					
$L_{\alpha1}$	L_3M_5	5033.79 ± 0.60	[50]	0.302	[488]	8.893(-2)	3.84	[423]
		5033.60	[104]					
$L_{\beta6}$	L_3N_1	5659.7 ± 1.5	[50]	3.04(-3)	[488]	9.185(-4)	11.33	[423]
		5660.20	[104]					
	L_3N_2	5724.20	[501]			6.199(-6)	14.92	[423]
		5728 ± 15	[145]					
	L_2N_3	5748.06 ± 0.97	[145]			5.978(-6)		
$L_{\beta15}$	L_3N_4	5851.30	[104]	5.54(-3)	[488]	1.628(-3)	3.87	[423]
		5846.1 ± 4.1	[145]					
$L_{\beta2}$	L_3N_5	5850.08 ± 0.54	[145]			1.455(-2)	3.62	[423]
$L_{\beta15,2}$	$L_3N_{4,5}$	5849.9 ± 1.6	[50]					
L_u	$L_3N_{6,7}$	5721.10	[104]	8.91(-6)	[488]			
L'_u	L_3N_6	5955.40 ± 0.75	[145]			4.573(-7)		
L_u	L_3N_7					2.621(-6)		
$L_{\beta7}$	L_3O_1	5927.00 ± 1.10	[50]	5.27(-4)	[488]	1.588(-4)	3.66	[423]
		5926.40	[104]					
	$L_3O_{2,3}$	5940.80	[501]					
	L_3O_2					1.168(-6)		
	L_3O_3					1.037(-6)		

M series

line	transition	E/eV		I/eV/$\hbar$		TPIV	Γ/eV	
$M_{\gamma2.1}$	$M_3N_{4,5}$	1127.30 ± 0.90	[50]	5.80(-3)	[335]			
		1127.30	[104]					
		1129.20	[263]					
$M_{\gamma2}$	M_3N_4					2.051(-4)	7.73	[423]
$M_{\gamma1}$	M_3N_5					1.679(-3)	7.48	[423]
$M_{\xi2}$	M_4N_2	715.34	[423]			3.233(-3)	12.71	[423]
M_δ	M_4N_3	736.42	[423]			4.784(-4)	5.22	[423]
M_β	M_4N_6	950.00 ± 1.00	[50]	1.37(-3)	[335]	3.579(-4)		
		951.50	[263]					
M_η	M_4O_2	928.72	[423]			5.254(-4)		
$M_{\xi1}$	M_5N_3	714.00 ± 1.60	[50]			3.580(-3)	5.10	[423]
		716.70	[263]					
$M_{\alpha2,1}$	$M_5N_{6,7}$	929.20 ± 0.35	[50]	1.37(-3)	[335]			

line	transition	E/eV		I/eV/$\hbar$	TPIV	Γ/eV
		930.80	[263]			
M_{α_2}	M_5N_6				1.878(-5)	
M_{α_1}	M_5N_7				3.667(-4)	
N series						
	$N_{4,5}N_{6,7}$	109.50 ± 1.00	[50]		3.667(-4)	
	$N_{4,5}O_{2,3}$	80.80 ± 0.30	[43]		3.667(-4)	

level characteristics

level	E_B/eV		ω_{nlj}		Γ/eV		AE/eV	
K	41991.00	[503]	0.917	[301]	10.20	[301]	41990.6	[51]
	41990.60 ± 0.50	[493]			15.78	[423]	42002 ± 11	[50]
			0.915	[38]	16.2	[94]	41994.11 ± 0.75	[145]
			0.9140	[39]				
			0.9180	[423]				
			0.917	[310]				
			0.923	[33]				
			0.916	[242]				
L_1	6835.00	[503]	6.10(-2)	[301]	4.34	[301]	6834.8	[51]
	6834.80 ± 0.50	[493]	6.04(-2)	[423]	3.96	[423]	6834.4 ± 2.8	[50]
			6.50(-2)	[436]	2.7	[94]	6836.8 ± 1.1	[145]
L_2	6441.00	[503]	0.117	[301]	3.89	[301]	6440.4	[51]
	6440.40 ± 0.50	[493]	0.1272	[423]	3.30	[423]	6439.0 ± 2.5	[50]
			0.128	[436]	3.48	[94]	6443.60 ± 0.39	
L_3	5965.00	[503]	0.118	[301]	3.60	[301]	5964.3	[51]
	5964.30 ± 0.40	[493]	0.1248	[423]	3.09	[423]	5963.3 ± 2.1	[50]
			0.126	[436]	3.27	[94]	5967.84 ± 0.33	[145]
M_1	1511.00	[503]	1.01(-3)		11.8	[94]	1511.0	[51]
	1511.00 ± 0.80	[493]	9.7(-4)	[423]	13.51	[423]		
M_2	1338.00	[503]	9.42(-3)	[512]	4.5	[94]	1337.4	[51]
	1337.40 ± 0.70	[493]	2.49(-3)	[423]	6.01	[423]		
M_3	1243.00	[503]	1.05(-3)	[512]	6.7	[94]	1242.2	[51]

level	E_B/eV		ω_{nlj}		Γ/eV		AE/eV	
	1242.20 ± 0.60	[493]	2.68(-3)	[423]	6.95	[423]		
M_4	951.00	[503]	2.7(-3)	[512]	0.75	[94]		
	951.10 ± 0.60	[493]	4.67(-3)	[423]	0.87	[423]		
M_5	931.00	[503]	4.56(-3)	[423]	0.75	[94]	928.8	[51]
	931.00 ± 0.60	[493]			0.75	[423]		
N_1	305.00	[503]	1.2(-4)	[423]	3.7	[94]		
	304.50 ± 0.90	[493]			8.25	[423]		
N_2	237.00	[503]	2.2(-4)	[423]	5.08	[94]		
	236.30 ± 1.50	[493]			11.84	[423]		
N_3	218.00	[503]	7.0(-5)	[423]	1.75	[94]		
	217.60 ± 1.10	[493]			4.35	[423]		
N_4	114.00	[503]	1.0(-5)	[423]	0.78	[94]		
	113.20 ± 0.70	[51]						
	117.90 ± 0.70	[493]			0.78	[423]		
N_5	114.00	[503]	1.0(-5)	[423]				
	110.10 ± 0.70	[493]			0.53	[423]		
N_6	2.00	[503]			0.53	[94]		
	2.00 ± 0.60	[493]						
O_1	38.00	[503]			0.57	[423]		
	37.40 ± 1.00	[493]						
O_2	23.00	[503]						
	22.30 ± 0.70	[51]						
	24.60 ± 0.70	[493]						
O_3	23.00	[503]						
	21.20 ± 0.70	[493]						
IP	5.464	[222]						

Nd   Z=60   [Xe] $4f^4 6s^2$

Neodymium $A = 144.24(3)$ [12.77] $\varrho = 7.003\,\text{g/cm}^3$ [547]
$\varrho = 6.96\,\text{g/cm}^3$ [483]

X-ray transitions

line	transition	E/eV		I/eV/$\hbar$		TPIV	Γ/eV	
K series								
K_{α_3}	KL_1	36445.1 ± 1.5	[145]				20.95	[423]
$K_{\alpha 2}$	KL_2	36847.40 ± 0.20	[50]	4.45	[488]	2.641(-1)	21.30	[301]
		36848.00	[104]	4.51	[487]		21.50	[473]
		36847.74 ± 0.08	[277]				20.25	[423]
		36847.502 ± 0.080	[144]					
$K_{\alpha 1}$	KL_3	37361.00 ± 0.20	[50]	8.11	[488]	4.810(-1)	20.90	[301]
		37361.00	[104]	8.21	[487]		21.50	[473]
		37360.98 ± 0.07	[277]				20.05	[423]
		37360.739 ± 0.070	[144]					
	KM_1	41993.8 ± 1.5	[145]				30.94	[423]
$K_{\beta 3}$	KM_2	42166.50 ± 0.40	[50]	0.801	[488]	4.751(-2)	21.33	[473]
		42162.30	[104]	0.845	[487]		23.22	[423]
		42166.51 ± 0.74	[277]					
		42166.24 ± 0.57	[144]					
$K_{\beta 1}$	KM_3	42271.30 ± 0.30	[50]	1.550	[488]	9.194(-2)	23.25	[473]
		42269.00	[104]	1.636	[487]		24.16	[423]
		42271.17 ± 0.57	[277]					
		42270.90 ± 0.57	[144]					
$K_{\beta_5^{II}}$	KM_4	42569.0 ± 1.6	[145]			6.161(-4)	17.84	[423]
$K_{\beta_5^I}$	KM_5	42592.8 ± 1.5	[145]			8.241(-4)	17.65	[423]
	KN_1	43254.5 ± 4.3	[145]				25.69	[423]
$K_{\beta 2}$	$KN_{2,3}$	43335 ± 22	[50]					
		43301.40	[104]					
$K_{\beta_2^{II}}$	KN_2	43322 ± 16	[145]	0.1719	[488]	1.020(-2)	28.95	[423]
				0.1857	[487]			
$K_{\beta_2^I}$	KN_3	43344.9 ± 1.4	[145]	0.363	[487]	1.983(-2)	21.82	[423]
$K_{\beta_4^{II}}$	KN_4	43449.2 ± 4.2	[145]			1.327(-4)	17.92	[423]
$K_{\beta_4^I}$	KN_5	43452.3 ± 1.0	[145]			1.772(-4)	17.67	[423]

line	transition	E/eV		I/eV/ℏ		TPIV	Γ/eV	
	KO$_{2,3}$			8.07(-2)	[487]			
L series								
	L$_1$M$_1$	5548.7 ± 1.8	[145]				5.50	[423]
L$_{\beta4}$	L$_1$M$_2$	5721.6 ± 1.2	[50]	8.71(-2)	[488]	1.955(-2)	8.15	[473]
		5722.20	[104]				10.42	[423]
L$_{\beta3}$	L$_1$M$_3$	5829.40 ± 0.50	[50]	0.1298	[488]	2.913(-2)	10.30	[473]
		5830.60	[104]				11.37	[423]
		5827.801 ± 0.052	[366]					
L$_{\beta10}$	L$_1$M$_4$	6124.97 ± 0.41	[50]	1.62(-3)	[488]	3.642(-4)	5.05	[423]
		6126.50	[104]					
L$_{\beta9}$	L$_1$M$_5$	6148.82 ± 0.41	[50]	2.42(-3)	[488]	5.445(-4)	4.86	[423]
		6148.80	[104]					
	L$_1$N$_1$	6809.4 ± 4.6	[145]				12.90	[423]
L$_{\gamma2}$	L$_1$N$_2$	6884.03 ± 0.34	[50]	2.02(-2)	[488]	4.534(-3)	16.15	[423]
		6883.20	[104]					
L$_{\gamma3}$	L$_1$N$_3$	6900.44 ± 0.34	[50]	3.12(-2)	[488]	7.001(-3)	9.23	[423]
		6900.50	[104]					
L$_{\gamma11}$	L$_1$N$_{4,5}$	7007.74 ± 0.36	[206]	3.33(-4)	[488]			
	L$_1$N$_4$	7004.1 ± 4.5	[145]			4.870(-5)	5.13	[423]
	L$_1$N$_5$	7007.2 ± 1.3	[145]			7.470(-5)	4.88	[423]
	L$_1$N$_6$	7117.2 ± 1.5	[145]					
	L$_1$N$_{6,7}$	7122.1 ± 2.0	[500]					
	L$_1$O$_1$	7088.20	[104]	1.75(-8)	[488]		4.63	[423]
L$_{\gamma4,4'}$	L$_1$O$_{2,3}$	7107.00 ± 1.60	[50]					
L$_{\gamma4'}$	L$_1$O$_2$	7106.60	[104]	2.92(-3)	[488]	6.546(-4)		
L$_\eta$	L$_2$M$_1$	5145.70 ± 0.90	[50]	1.088(-2)	[488]	4.672(-3)	17.45	[423]
		5147.10	[104]					
		5145.25 ± 0.17	[366]					
	L$_2$M$_2$	5320.3 ± 1.5	[145]				9.73	[423]
L$_{\beta17}$	L$_2$M$_3$	5424.4 ± 1.2	[500]			8.160(-5)	10.67	[423]
L$_{\beta1}$	L$_2$M$_4$	5721.60 ± 0.50	[50]	0.398	[488]	1.112(-1)	4.00	[473]
		5720.60	[104]				4.35	[423]
		5721.446 ± 0.050	[366]					
	L$_2$M$_5$	5744.6 ± 1.1	[145]				4.16	[423]
L$_{\gamma5}$	L$_2$N$_1$	6405.29 ± 0.33	[50]	2.55(-3)	[488]	8.135(-4)	12.19	[423]
		6406.30	[104]					
	L$_2$N$_2$	6474 ± 15	[145]				15.45	[423]
	L$_2$N$_3$	6496.8 ± 1.0	[145]			1.471(-5)	8.33	[423]

line	transition	E/eV		I/eV/$\hbar$		TPIV	Γ/eV	
$L_{\gamma 1}$	L_2N_4	6601.16 ± 0.24	[50]	6.74(-2)	[488]	1.882(-2)	7.16	[473]
		6600.80	[104]				4.43	[423]
	L_2N_5	6604.15 ± 0.61	[145]				4.18	[423]
L_ν	$L_2N_{6,7}$	6719.80	[104]	1.73(-5)	[488]			
L_ν	L_2N_6	6714.16 ± 0.78	[145]			4.833(-6)		
	$L_2N_{6,7}$	6718.98 ± 0.61	[500]					
$L_{\gamma 8}$	L_2O_1	6683.00 ± 1.80	[50]	4.40(-4)	[488]	1.273(-4)	3.93	[423]
		6681.40	[104]					
	$L_2O_{2,3}$	6702.30	[501]					
L_l	L_3M_1	4633.00 ± 0.70	[50]	1.456(-2)	[488]	8.618(-3)	17.25	[423]
		4633.40	[104]					
		4631.849 ± 0.052	[366]					
L_t	L_3M_2	4807.1 ± 1.5	[145]			5.311(-5)	9.52	[423]
L_s	L_3M_3	4912.2 ± 1.4	[145]			4.903(-5)	10.46	[423]
$L_{\alpha 2}$	L_3M_4	5207.70 ± 0.70	[50]	3.71(-2)	[488]	1.065(-2)	4.08	[473]
		5209.80	[104]				4.15	[423]
		5207.7 ± 1.1	[366]					
$L_{\alpha 1}$	L_3M_5	5230.40 ± 0.40	[50]	0.329	[488]	9.426(-2)	3.93	[473]
		5231.30	[104]				3.96	[423]
		5230.239 ± 0.035	[366]					
$L_{\beta 6}$	L_3N_1	5892.99 ± 0.25	[50]	3.35(-3)	[488]	9.832(-4)	11.99	[423]
		5893.90	[104]					
	L_3N_2	5963.10	[501]			6.858(-6)	18.42	[423]
	L_3N_3	5988.80	[501]			6.618(-6)	8.12	[423]
$L_{\beta 15,2}$	$L_3N_{4,5}$	6091.25 ± 0.26	[50]	5.42(-2)	[488]		7.18	[473]
		6090.60	[104]					
$L_{\beta 15}$	L_3N_4	6087.8 ± 3.8	[145]			1.738(-3)	4.23	[423]
$L_{\beta 2}$	L_3N_5	6090.96 ± 0.55	[145]			1.554(-2)	3.97	[423]
$L_{u',u}$	$L_3N_{6,7}$	6204.90	[104]	1.45(-6)	[488]			
L'_u	L_3N_6	6202.34 ± 0.52	[500]			7.268(-7)		[423]
L_u	L_3N_7					4.159(-6)		[423]
$L_{\beta 7}$	L_3O_1	6170.80 ± 0.90	[50]	5.76(-3)	[488]	1.722(-4)	3.73	[423]
		6169.60	[104]					
	$L_3O_{2,3}$	6184.70	[501]			1.280(-6)		

M series

line	transition	E/eV		I/eV/$\hbar$		TPIV	Γ/eV	
$M_{\gamma 2,1}$	$M_3N_{4,5}$	1180.00 ± 1.00	[50]					
		1183.20	[263]					
$M_{\gamma 2}$	M_3N_4	1158,42	[423]	6.51(-3)	[335]	2.203(-4)	8.34	[423]

line	transition	E/eV		I/eV/$\hbar$		TPIV	Γ/eV	
$M_{\gamma 1}$	M_3N_5	1180.20	[104]	6.15(-3)	[335]	1.800(-3)	8.08	[423]
M_{ξ_2}	M_4N_2	754.54	[423]			3.512(-3)	13.04	[423]
M_δ	M_4N_3	777.58	[423]			5.077(-4)	5.91	[423]
M_β	M_4N_6	997.00 $\pm$ 1.60	[50]	2.10(-3)	[335]	5.627(-4)		
		999.10	[263]					
M_η	M_4O_2	979.18	[423]			5.859(-4)		
$M_{\xi 1}$	M_5N_3	753.00 $\pm$ 1.80	[50]			3.901(-3)	12.23	[423]
		756.00	[263]					
$M_{\alpha_{2,1}}$	$M_5N_{6,7}$	978.00 $\pm$ 1.50	[50]					
		980.10	[263]					
	M_5N_6					3.056(-5)		
	M_5N_7					5.955(-4)		
N series								
	$N_4N_{6,7}$	116.00 $\pm$ 1.10	[50]					
	$N_{4,5}O_{2,3}$	96.20 $\pm$ 0.50	[50]					

level characteristics

level	E_B/eV		ω_{nlj}		Γ/eV		AE/eV	
K	43569.00	[503]	0.921	[301]	17.30	[301]	43568.9	[51]
	43568.90 $\pm$ 0.40	[493]	0.920	[38]	16.87	[423]	43574 $\pm$ 11	[50]
			0.935	[293]	17.4	[94]	43575.27 $\pm$ 0.79	[145]
			0.9181	[39]				
			0.9220	[423]				
			0.921	[310]				
			0.918	[112]				
L_1	7126.00	[503]	6.40(-2)	[301]	4.52	[301]	7126.0	[51]
	7126.00 $\pm$ 0.40	[493]	6.00(-2)	[107]	2.9	[94]		
			7.45(-2)	[348]	4.08	[423]	7129.52 $\pm$ 0.61	[50]
			6.36(-2)	[423]			7130.2 $\pm$ 1.1	[145]
			6.70(-2)	[436]				
			6.78(-2)	[244]				

level	E_B/eV		ω_{nlj}		Γ/eV		AE/eV	
L_2	6722.00	[503]	0.124	[301]	3.97	[301]	6721.5	[51]
	6721.50 ± 0.40	[493]	0.120	[107]	3.55	[94]	6723.55 ± 0.54	[50]
			0.133	[348]	3.38	[423]	6727.09 ± 0.40	[145]
			0.1357	[423]				
			0.136	[436]				
			0.139	[244]				
			0.136	[436]				
			0.139	[244]				
L_3	6208.00	[503]	0.125	[301]	3.65	[301]	6207.9	[51]
	6207.90 ± 0.40	[493]	0.120	[107]	3.36	[94]	6209.36 ± 0.46	[50]
			0.135	[348]	3.18	[423]	6213.90 ± 0.33	
			0.1321	[423]				
			0.134	[436]				
			0.136	[244]				
			0.119	[168]				
M_1	1576.00	[503]	8.10(-4)	[38]	12.87	[350]	1575.3	[51]
	1575.30 ± 0.70	[493]	1.07(-3)	[512]	14.07	[423]		
			1.01(-3)	[423]	12	[94]		
M_2	1403.00	[503]	1.32(-3)	[38]	6.69	[350]	1402.8	[51]
	1402.80 ± 0.60	[493]	9.55(-3)	[512]	6.34	[423]		
			2.61(-3)	[423]	4.7	[94]		
M_3	1298.00	[503]	1.05(-3)	[38]	6.81	[350]	1297.4	[51]
	1297.40 ± 0.50	[493]	1.11(-3)	[512]	7.29	[423]		
			2.85(-3)	[423]	7.3	[94]		
M_4	1000.00	[503]	2.60(-3)	[38]	1.39	[350]		
	999.90 ± 0.60	[493]	2.7(-3)	[512]	0.97	[423]		
			5.25(-3)	[423]	0.78	[94]		
M_5	978.00	[503]	3.20(-3)	[38]	1.00	[350]	980.4	[51]
	977.70 ± 0.60	[493]	1.23(-3)	[512]	0.78	[423]		
			5.21(-3)	[423]	0.78	[94]		
N_1	316.00	[503]	1.3(-4)	[423]	10.00	[74]		
	315.20 ± 0.80	[493]			8.82	[423]		
					4	[94]		
N_2	244.00	[503]	2.5(-4)	[423]	6.00	[74]		
	243.30 ± 1.60	[493]			12.07	[423]		
					5.1	[94]		
N_3	225.00	[503]	8.0(-5)	[423]	6.00	[74]		
	224.60 ± 1.30	[493]			4.95	[423]		
					1.9	[94]		
N_4	118.00	[503]	1.0(-5)	[423]	1.05	[423]		
	117.50 ± 0.70	[51]			0.80	[74]		
	123.40 ± 0.70	[493]			1.05	[94]		
N_5	118.00	[503]	1.0(-5)	[423]	0.80	[423]		
	117.50 ± 0.70	[51]			0.80	[74]		

level	E_B/eV		ω_{nlj}	Γ/eV		AE/eV
	113.50 ± 0.70	[493]		0.8	[94]	
N_6	2.00	[503]				
O_1	1.50 ± 0.90	[493]				
	38.00	[503]		0.55	[423]	
	37.50 ± 0.90	[493]				
O_2	22.00	[503]				
	21.10 ± 0.80	[51]				
	23.60 ± 0.80	[493]				
O_3	22.00	[503]				
	19.80 ± 0.80	[493]				
IP	5.5250	[222]				

Pm   Z=61

[Xe] $4f^5$ $6s^2$

Promethium $A = 144.9127$ [222] $\varrho = 7.22$ g/cm^3 [547]

X-ray transitions

line	transition	E/eV		I/eV/$\hbar$		TPIV	Γ/eV	
K series								
K_{α_3}	KL$_1$	37759.6 ± 1.5	[145]				22.22	[423]
$K_{\alpha 2}$	KL$_2$	38171.55 ± 0.70	[50]	4.78	[488]	2.653(-1)	22.50	[301]
		38225.00	[104]				21.47	[423]
$K_{\alpha 1}$	KL$_3$	38725.11 ± 0.72	[50]	8.67	[488]	4.814(-1)	22.20	[301]
		38783.00	[104]				21.28	[423]
	KM$_1$	43535.6 ± 1.6	[145]				32.60	[423]
$K_{\beta 3}$	KM$_2$	43712.7 ± 9.1	[50]	0.863	[488]	4.792(-2)	24.69	[423]
$K_{\beta 1}$	KM$_3$	43825.5 ± 6.9	[50]	1.670	[488]	9.273(-2)	24.99	[423]
		43874.00	[104]					
$K_{\beta_5^{II}}$	KM$_4$	44131.7 ± 1.6	[145]			6.484(-4)	19.09	[423]
$K_{\beta_5^{I}}$	KM$_5$	44158.1 ± 1.6	[145]			8.644(-4)	18.83	[423]
	KN$_1$	44854.1 ± 4.3	[145]				27.37	[423]
$K_{\beta 2}$	KN$_{2,3}$	44937 ± 24	[50]	0.362	[488]	4.18		
		44986.00	[104]					
$K_{\beta_2^{II}}$	KN$_2$	44923 ± 16	[145]			1.034(-2)	30.14	[423]
$K_{\beta_2^{I}}$	KN$_3$	44948.9 ± 1.5	[145]			2.009(-2)	23.53	[423]
$K_{\beta_4^{II}}$	KN$_4$	45060.8 ± 4.0	[145]			1.411(-4)	19.40	[423]
$K_{\beta_4^{I}}$	KN$_5$	45061.8 ± 1.8	[145]			1.877(-4)	19.13	[423]
L series								
	L$_1$M$_1$	5776.1 ± 1.8	[145]				5.68	[423]
$L_{\beta 4}$	L$_1$M$_2$	5955.2 ± 2.2	[145]			2.061(-2)	10.89	[423]
$L_{\beta 3}$	L$_1$M$_3$	6071.3 ± 1.8	[50]	0.1402	[488]	3.040(-2)	11.18	[423]
		6071.20	[104]					
$L_{\beta 10}$	L$_1$M$_4$	6372.1 ± 1.9	[145]			3.975(-4)	5.28	[423]
$L_{\beta 9}$	L$_1$M$_5$	6398.5 ± 1.8	[145]			5.943(-4)	5.02	[423]
	L$_1$N$_1$	7094.5 ± 4.6	[145]				13.56	[423]
$L_{\gamma 2}$	L$_1$N$_2$	7163 ± 16	[145]			4.805(-3)	16.34	[423]

line	transition	E/eV		I/eV/$\hbar$		TPIV	Γ/eV	
L_{γ_3}	L_1N_3	7189.3 ± 1.7	[145]			7.351(-3)	9.73	[423]
	L_1N_4	7301.2 ± 4.3	[145]			5.335(-5)	5.59	[423]
	L_1N_5	7302.3 ± 1.4	[145]			8.193(-5)	5.32	[423]
	L_1N_6	7416.9 ± 1.5	[145]					
L_η	L_2M_1	5363.0 ± 1.3	[145]			4.780(-3)	18.03	[423]
	L_2M_2	5542.1 ± 1.5	[145]				10.13	[423]
$L_{\beta_{17}}$	L_2M_3	5656.5 ± 1.5	[145]			8.676(-5)	10.43	[423]
L_{β_1}	L_2M_4	5961.5 ± 1.7	[50]	0.433	[488]	1.185(-1)	4.53	[423]
		5961.50	[104]					
	L_2M_5	5985.4 ± 1.1	[145]				4.26	[423]
L_{γ_5}	L_2N_1	6681.4 ± 3.9	[145]			8.589(-4)	12.80	[423]
	L_2N_2	6750 ± 15	[145]				15.58	[423]
	L_2N_3	6776.2 ± 1.0	[145]			1.638(-5)	8.97	[423]
L_{γ_1}	L_2N_4	6892.1 ± 5.1	[50]	7.39(-2)	[488]	2.022(-2)	4.84	[423]
		6892.00	[104]					
	L_2N_5	6889.17 ± 0.65	[145]				4.56	[423]
L_ν	L_2N_6	7003.78 ± 80	[145]			7.143(-6)		
L_l	L_3M_1	4810.0 ± 1.5	[145]			8.849(-3)	17.84	[423]
L_t	L_3M_2	4989.2 ± 1.5	[145]			5.586(-5)	9.94	[423]
L_s	L_3M_3	5103.6 ± 1.4	[145]			5.115(-5)	10.24	[423]
L_{α_2}	L_3M_4	5407.9 ± 1.4	[50]	4.03(-2)	[488]	1.128(-2)	4.34	[423]
		5407.80	[104]					
L_{α_1}	L_3M_5	5432.6 ± 1.1	[50]	0.357	[488]	9.980(-2)	4.08	[423]
		5432.50	[104]					
L_{β_6}	L_3N_1	6128.5 ± 3.8	[145]			1.051(-3)	12.61	[423]
	L_3N_2	6197 ± 15	[145]			7.571(-6)	15.39	[423]
	L_3N_3	6223.31 ± 0.97	[145]			7.309(-6)	8.78	[423]
$L_{\beta_{15,2}}$	$L_3N_{4,5}$	6338.9 ± 2.9	[50]	6.62(-3)	[488]	1.852(-3)	4.65	[423]
$L_{\beta_{15}}$	L_3N_4	6335.2 ± 3.5	[145]			1.852(-3)	4.64	[423]
L_{β_2}	L_3N_5	6336.23 ± 0.59	[145]	5.92(-2)	[488]	1.656(-2)	4.38	[423]
L'_u	L_3N_6	6450.84 ± 0.72	[145]			1.071(-6)		

M series

line	transition	E/eV		I/eV/$\hbar$		TPIV	Γ/eV	
$M_{\gamma_{2,1}}$	$M_3N_{4,5}$	1238.60	[263]					
M_{γ_2}	M_3N_4	1211,85	[423]			2.323(-4)	8.36	[423]
M_{γ_1}	M_3N_5	1276,31	[423]			1.893(-3)	8.09	[423]
M_{ξ_2}	M_4N_2	794.59	[423]			3.740(-3)	13.20	[423]
M_δ	M_4N_3	819.73	[423]			5.295(-4)	6.59	[423]

line	transition	E/eV		I/eV/$\hbar$	TPIV	Γ/eV	
M_β	M_4N_6	1050.70	[263]		8.838(-4)		
M_η	M_4O_2	1030.64	[423]		6.380(-4)		
$M_{\xi 1}$	M_5N_3	794.80	[263]		4.131(-3)	6.33	[423]
$M_{\alpha_{2,1}}$	$M_5N_{6,7}$	1023.80	[263]				
M_{α_2}	M_5N_6	1025,72	[423]		4.990(-5)		
M_{α_1}	M_5N_7	1026.25	[423]		9.705(-4)		

level characteristics

level	E_B/eV		ω_{nlj}		Γ/eV		AE/eV	
K	45185.00	[503]	0.925	[301]	18.50	[301]	45184.0	[51]
	45184.00 $\pm$ 0.70	[493]			18.01	[423]	45198 $\pm$ 12	[50]
			0.924	[38]	18.5	[94]	45189.77 $\pm$ 0.87	[145]
			0.9226	[39]				
			0.9257	[423]				
			0.925	[310]				
L_1	7428.00	[503]	6.40(-2)	[301]	4.67	[301]	7427.9	[51]
	7427.90 $\pm$ 0.80	[493]	6.67(-2)	[423]	4.20	[423]	7435.7 $\pm$ 3.3	[50]
			7.10(-2)	[436]	3.1	[94]	7430.2 $\pm$ 1.1	[145]
L_2	7013.00	[503]	0.124	[301]	4.05	[301]	7012.8	[51]
	7012.80 $\pm$ 0.60	[493]	0.1446	[423]	3.45	[423]	7014.2 $\pm$ 2.9	[50]
			0.145	[436]	3.63	[94]	7017.09 $\pm$ 0.44	[145]
L_3	6460.00	[503]	0.125	[301]	3.75	[301]	6459.3	[51]
	6459.30 $\pm$ 0.60	[493]	0.1397	[423]	3.23	[423]	6460.44 $\pm$ 0.50	[50]
			0.142	[436]	3.44	[94]	6464.15 $\pm$ 0.36	[145]
M_1	1650.00	[503]	1.13(-3)	[512]	12.2	[94]	1646.5	[51]
	1648.60 $\pm$ 1.50	[493]	1.07(-3)	[423]	14.58	[423]		
					4.2	[94]		

level	E_B/eV		ω_{nlj}		Γ/eV		AE/eV	
M_2	1472.00	[503]	9.72(-3)	[512]	5	[94]	1471.4	[51]
	1471.40 ± 6.20	[493]	2.72(-3)	[423]	6.67	[423]		
					5.13	[94]		
M_3	1357.00	[503]	1.17(-3)	[512]	7.8	[94]	1356.9	[51]
	1356.90 ± 1.40	[493]	2.97(-2)	[423]	6.98	[423]		
					2.05	[94]		
M_4	1052.00	[503]	8.42(-3)	[512]	0.82	[94]	1051.5	[51]
	1051.50 ± 0.90	[493]	5.88(-3)	[423]	1.07	[423]		
					1.38	[94]		
M_5	1027.00	[503]	2.67(-3)	[512]	0.82	[94]	1026.9	[51]
	1026.90 ± 1.00	[493]	5.92(-3)	[423]	0.82	[423]		
					1.11	[94]		
N_1	331.00	[503]	1.4(-4)	[423]				
	331.40 ± 2.00	[493]			9.35	[423]		
N_2	255.00	[503]	2.8(-4)	[423]				
	242.00 ± 1.60	[51]			12.13	[423]		
	254.70 ± 3.20	[493]						
N_3	237.00	[503]	9.0(-5)	[423]				
	242.00 ± 1.60	[51]			5.52	[423]		
	236.20 ± 2.50	[493]						
N_4	121.00	[503]	1.0(-5)	[423]				
	120.40 ± 2.00	[51]			1.38	[423]		
	127.60 ± 2.00	[493]						
N_5	121.00	[503]	1.0(-5)	[423]				
	120.40 ± 2.00	[51]			1.11	[423]		
	115.60 ± 2.00	[493]						
N_6	4.00	[503]						
	3.50 ± 2.00	[493]						
O_1	38.00	[503]			0.54	[423]		
	36.00 ± 2.50	[493]						
O_2	22.00	[503]						
	24.50 ± 1.00	[493]						
O_3	22.00	[503]						
	20.10 ± 1.00	[493]						
IP	5.58	[222]						

Sm   Z=62

[Xe] $4f^6 6s^2$

Samarium $A = 150.36(3)$ [222] $\varrho = 7.537$ g/cm^3 [547]
$\varrho = 7.50$ g/cm^3 [483]

X-ray transitions

line	transition	E/eV		I/eV/$\hbar$		TPIV	Γ/eV	
K series								
K_{α_3}	KL_1	39101.2 ± 1.6	[145]				23.55	[423]
K_{α_2}	KL_2	39522.40 ± 0.25	[50]	5.12	[488]	2.663(-1)	23.80	[301]
		39523.00	[104]				24.70	[473]
		39523.64 ± 0.10	[277]				22.76	[423]
		39523.39 ± 0.10	[144]					
K_{α_1}	KL_3	40118.10 ± 0.25	[50]	9.27	[488]	4.819(-1)	23.60	[301]
		40119.00	[104]				26.00	[473]
		40118.74 ± 0.06	[277]				22.58	[423]
		40118.481 ± 0.060	[144]					
	KM_1	45110.3 ± 1.7	[145]				34.30	[423]
K_{β_3}	KM_2	45289.00 ± 3.30	[50]	0.927	[488]	4.829(-2)	24.56	[473]
		45287.00	[104]				26.24	[423]
		45288.6 ± 4.9	[144]					
K_{β_1}	KM_3	45413.00 ± 3.30	[50]	1.797	[488]	9.344(-2)	25.65	[473]
		45412.00	[104]				26.59	[423]
		45413.0 ± 4.9	[144]					
K_{β_5}	$KM_{4,5}$	45731.4 ± 7.5	[50]	1.739	[488]			
		45729.00	[104]					
$K_{\beta_5^{II}}$	KM_4	45728.1 ± 1.8	[145]			6.808(-4)	20.40	[423]
$K_{\beta_5^{I}}$	KM_5	45756.9 ± 1.6	[145]			9.043(-4)	20.08	[423]
	KN_1	46488.6 ± 4.4	[145]				29.09	[423]
K_{β_2}	$KN_{2,3}$	46575 ± 26	[50]	0.391	[488]			
		46576.00	[104]					
$K_{\beta_2^{II}}$	KN_2	46530 ± 16	[145]			1.047(-2)	31.25	[423]
$K_{\beta_2^{I}}$	KN_3	46588.2 ± 1.5	[145]			2.033(-2)	25.29	[423]
$K_{\beta_4^{II}}$	KN_4	46709.1 ± 3.8	[145]			1.495(-4)	21.00	[423]

line	transition	E/eV		I/eV/$\hbar$		TPIV	Γ/eV	
$K_{\beta beta_4^l}$	KN_5	46706.4 ± 1.1	[145]			1.980(-4)	20.70	[423]
	$KO_{2,3}$	46801.00 ± 5.00	[50]	8.11(-2)	[488]			
	KO_2					2.114(-3)		
	KO_3					4.276(-3)		

L series

line	transition	E/eV		I/eV/$\hbar$		TPIV	Γ/eV	
	L_1M_1	6009.1 ± 1.9	[145]				19.42	[423]
$L_{\beta 4}$	L_1M_2	6196.19 ± 0.26	[50]	0.1036	[488]	2.174(-2)	8.60	[473]
		6196.10	[104]				11.35	[423]
$L_{\beta 3}$	L_1M_3	6318.00 ± 0.10	[50]	0.1512	[488]	3.171(-2)	10.80	[473]
		6317.70	[104]				11.69	[423]
		6316.36 ± 0.13	[366]					
$L_{\beta 10}$	L_1M_4	6630.40 ± 0.10	[50]			4.338(-4)	5.49	[423]
		6628.69 ± 0.26	[440]					
$L_{\beta 9}$	L_1M_5	6659.70 ± 0.10	[50]			6.488(-4)	5.17	[423]
		6655.60 ± 0.14	[440]					
	L_1N_1	7387.5 ± 4.7	[145]				14.19	[423]
$L_{\gamma 2}$	L_1N_2	7467.19 ± 0.37	[50]	2.427(-2)	[488]	5.090(-3)	16.35	[423]
		7466.70	[104]					
$L_{\gamma 3}$	L_1N_3	7486.82 ± 0.20	[50]	3.68(-2)	[488]	7.714(-3)	10.39	[423]
		7486.70	[104]					
$L_{\gamma 11}$	$L_1N_{4,5}$	7605.90	[104]	4.28(-4)	[488]			
		7606.24 ± 0.38	[204]					
	L_1N_4	7608.0 ± 4.0	[145]			5.840(-5)	6.10	[423]
$L_{\gamma 11}$	L_1N_5	7605.2 ± 1.4	[145]			8.976(-5)	5.80	[423]
	L_1N_6	7733.00	[104]	8.74(-7)	[488]			
		7724.4 ± 1.5	[145]					
$L_{\gamma 4}$	$L_1O_{2,3}$	7713.70 ± 0.14	[50]	5.07(-3)	[488]			
		7713.60	[104]					
	L_1O_2					7.201(-4)		
	L_1O_3					1.064(-3)		
L_{η}	L_2M_1	5589.20 ± 0.08	[50]	1.279(-2)	[488]	4.911(-3)	18.61	[423]
		5589.10	[104]					
		5585.55 ± 0.91	[366]					
	L_2M_2	5769.7 ± 1.6	[145]				10.55	[423]
$L_{\beta 17}$	L_2M_3	5891.70	[104]	2.79(-4)	[488]	9.214(-5)	10.89	[423]
		5891.6 ± 1.4	[498]					
$L_{\beta 1}$	L_2M_4	6205.10 ± 0.09	[50]	0.471	[488]	1.265(-1)	4.33	[473]
		6206.30	[104]				4.71	[423]
		6204.073 ± 0.093	[366]					
	L_2M_5	6232.6 ± 1.2	[145]				4.38	[423]

line	transition	E/eV		I/eV/$\hbar$		TPIV	Γ/eV	
$L_{\gamma 5}$	L_2N_1	6967.67 ± 0.17	[50]	3.02(-3)	[488]	9.059(-4)	13.40	[423]
	L_2N_2	7006 ± 15	[145]				15.55	[423]
	L_2N_3	7064.27 ± 0.81	[443]			1.825(-5)		
$L_{\gamma 1}$	L_2N_4	7178.09 ± 0.17	[50]	8.09(-2)	[488]	2.171(-2)	7.50	[473]
		7177.90	[104]				5.31	[423]
	L_2N_5	7182.12 ± 0.66	[145]				5.01	[423]
L_{ν}	L_2N_6	7301.27 ± 0.77	[145]			1.007(-5)		
	$L_2N_{6,7}$	7308.00 ± 0.86	[204]					
$L_{\gamma 8}$	L_2O_1	7265.80	[104]	5.14(-4)	[488]	1.486(-4)	4.06	[423]
	L_2O_2	7288.00	[104]	5.06(-9)	[488]			
$L_{\gamma 6}$	L_2O_4	7307.60 ± 0.40	[50]					
L_l	L_3M_1	4994.50 ± 0.80	[50]	1.743(-2)	[104]	9.081(-3)	18.43	[423]
		4993.20	[104]					
		4990.43 ± 0.17	[366]					
L_t	L_3M_2	5174.6 ± 1.5	[145]			5.874(-5)	10.37	[423]
L_s	L_3M_3	5296.50	[104]	1.26(-4)	[488]	5.369(-5)	10.72	[423]
		5298.8 ± 1.4	[145]					
$L_{\alpha 2}$	L_3M_4	5608.40 ± 0.70	[50]	4.37(-2)	[488]	1.193(-2)	4.50	[473]
		5608.40	[104]				4.53	[423]
		5609.053 ± 0.061	[366]					
$L_{\alpha 1}$	L_3M_5	5636.10 ± 0.50	[50]	0.386	[488]	1.056(-1)	4.13	[473]
		5636.00	[104]				4.21	[423]
		5635.970 ± 0.033	[366]					
$L_{\beta 6}$	L_3N_1	6369.72 ± 0.14	[50]	4.04(-3)	[488]	1.123(-3)	13.22	[423]
		6369.60	[104]					
	L_3N_2	6411 ± 15	[145]			8.338(-6)	15.37	[423]
	L_3N_3	6469.72 ± 0.68	[443]			8.053(-6)		
$L_{\beta 15,2}$	$L_3N_{4,5}$	6587.17 ± 0.14	[50]					
		6589.8 ± 3.2	[145]					
$L_{\beta 15}$	L_3N_4	1566.50	[423]	7.20(-3)	[488]	1.968(-3)	5.13	[423]
$L_{\beta 2}$	L_3N_5	6586.90	[104]	6.44(-2)	[488]	1.760(-2)	7.42	[473]
		6587.17 ± 0.14	[145]				4.83	[423]
$L_{u'}$	L_3N_6	6711.75 ± 0.73	[204]			1.501(-6)		
		6706.19 ± 0.71	[145]					
$L_{\beta 7}$	L_3O_1	6679.10 ± 0.11	[50]	8.56(-4)	[488]	2.019(-4)	3.88	[423]
		6679.00	[104]					
	$L_3O_{2,3}$	6694.30	[501]					
$L_{\beta 5}$	$L_3O_{4,5}$	6712.60 ± 0.33	[50]	8.10(-3)	[488]			
		6712.50	[104]					

M series

line	transition	E/eV		I/eV/$\hbar$		TPIV	Γ/eV	
$M_{\gamma_{2,1}}$	$M_3N_{4,5}$	1291.00 ± 1.00 1294.20	[50] [263]					
$M_{\gamma 2}$	M_3N_4			8.10(-3)	[335]	2.475(-4)	9.14	[423]
$M_{\gamma 1}$	M_3N_5	1291.50	[104]	4.07(-3)	[335]	2.011(-3)	8.84	[423]
M_{ξ_2}	M_4N_2					3.90(-3)	13.19	[423]
M_δ	M_4N_3					5.36(-4)	7.23	[423]
M_β	M_4N_6	1099.80 ± 1.00 1102.10	[50] [263]	5.01(-4)	[335]	1.346(-3)		
M_η	M_4O_2					6.81(-4)		
M_{ξ_1}	M_5N_3	831.00 ± 2.20 833.20	[50] [263]	4.07(-3)	[335]	4.276(-3)	6.92	[423]
$M_{\alpha_{2,1}}$	$M_5N_{6,7}$	1081.00 ± 3.00 1082.80	[50] [423]					
N series								
	$N_{4,5}N_{6,7}$	126.00 ± 1.30	[50]					
	$N_{4,5}O_{2,3}$	105.60	[50]					

level characteristics

level	E_B/eV		ω_{nlj}		Γ/eV		AE/eV	
K	46835.00	[503]	0.927	[301]	19.70	[301]	46834.2	[51]
	46834.20 ± 0.50	[493]	0.9290	[423]	19.23	[423]	46849 ± 13	[50]
			0.928	[38]	19.8	[94]	46839.02 ± 0.91	[145]
			0.9255	[39]				
			0.929	[310]				
			0.933	[33]				
			0.922	[242]				

level	E_B/eV		ω_{nlj}		Γ/eV		AE/eV	
L_1	7737.00	[503]	7.10(-2)	[301]	4.80	[301]	7736.8	[51]
	7736.80 ± 0.50	[493]	7.00(-2)	[423]	4.33	[423]	7747.93 ± 0.72	[50]
			7.50(-2)	[436]	3.3	[94]	7737.9 ± 1.2	[145]
			6.70(-2)	[525]				
L_2	7312.00	[503]	0.140	[301]	4.15	[301]	7311.8	[51]
	7311.80 ± 0.40	[493]	0.1543	[423]	3.53	[423]	7313.30 ± 0.64	[50]
			0.155	[436]	3.7	[94]	7314.76 ± 0.44	[145]
			0.146	[525]				
L_3	6717.00	[503]	0.139	[301]	3.86	[301]	6716.2	[51]
	6716.20 ± 0.50	[493]	0.1476	[423]	3.35	[423]	6717.36 ± 0.54	[50]
			0.15	[436]	3.53	[94]	6719.67 ± 0.37	[145]
M_1	1724.00	[503]	1.19(-3)	[512]	14.00	[74]	1722.8	[51]
	1722.80 ± 0.80	[493]	1.15(-3)	[423]	15.08	[423]		
					12.4	[94]		
M_2	1542.00	[503]	9.93(-3)	[512]	7.30	[74]	1540.7	[51]
	1540.70 ± 1.20	[493]	2.84(-3)	[423]	7.02	[423]		
					5.2	[94]		
M_3	1421.00	[503]	1.23(-3)	[512]	7.30	[74]	1419.8	[51]
	1419.80 ± 1.10	[493]	3.14(-3)	[423]	7.36	[423]		
					8.1	[94]		
M_4	1107.00	[503]	8.31(-3)	[512]	1.00	[74]	1106.0	[51]
	1106.00 ± 0.80	[493]	6.56(-3)	[423]	1.18	[423]		
					0.86	[94]		
M_5	1081.00	[503]	4.11(-3)	[512]	1.60	[74]	1080.2	[51]
	1080.20 ± 0.60	[493]	6.69(-3)	[423]	0.86	[423]		
					0.86	[94]		
N_1	347.00	[503]	1.6(-4)	[423]	11.00	[74]		
	345.70 ± 0.90	[493]			9.87	[423]		
					4.4	[94]		
N_2	267.00	[503]	3.2(-4)	[423]	7.20	[74]		
	265.60 ± 1.90	[493]			12.02	[423]		
					5.16	[94]		
N_3	249.00	[503]	1.1(-4)	[423]	7.20	[74]		
	247.40 ± 1.50	[493]			6.06	[423]		
					2.2	[94]		
N_4	130.00	[503]	1.0(-5)	[423]	1.30	[74]		
	129.00 ± 1.20	[51]			1.78	[423]		
	127.50 ± 1.20	[493]			1.78	[94]		

level	E_B/eV		ω_{nlj}		Γ/eV		AE/eV	
N_5	130.00	[503]	1.0(-5)	[423]	1.30	[74]		
	129.00 ± 1.20	[51]			1.48	[423]		
	123.30 ± 1.20	[493]			1.48	[94]		
N_6	7.00	[503]						
	5.50 ± 1.10	[493]						
O_1	39.00	[503]			0.52	[423]		
	37.40 ± 1.50	[493]						
O_2	22.00	[503]						
	21.30 ± 1.50	[51]						
	23.60 ± 1.30	[493]						
O_3	22.00	[503]						
	21.30 ± 1.50	[51]						
	18.90 ± 1.30	[493]						
IP	5.6436	[222]						

Eu   Z=63

[Xe] 4f^7 6s^2

Europium $A = 151.964(1)$ [222] $\varrho = 5.253$ g/cm^3 [547]
$\varrho = 5.30$ g/cm^3 [483]

X-ray transitions

line	transition	E/eV		I/eV/$\hbar$		TPIV	Γ/eV	
K series								
K$_{\alpha3}$	KL$_1$	40470.1 $\pm$ 1.6	[145]				24.95	[423]
K$_{\alpha2}$	KL$_2$	40902.33 $\pm$ 0.40	[50]	5.48	[488]	2.673(-1)	28.20	[301]
		40902.00	[104]	5.54	[487]		24.10	[423]
K$_{\alpha1}$	KL$_3$	41542.63 $\pm$ 0.41	[50]	9.89	[488]	4.821(-1)	27.90	[301]
		41542.00	[104]	10.00	[487]		26.00	[473]
							23.93	[423]
	KM$_1$	46718.4 $\pm$ 1.8	[145]				36.06	
K$_{\beta3}$	KM$_2$	46904.0 $\pm$ 1.3	[50]	0.998	[488]	4.866(-2)	27.86	[423]
		46904.00	[104]	1.050	[487]			
K$_{\beta1}$	KM$_3$	47038.4 $\pm$ 1.3	[50]	1.931	[488]	9.415(-2)	28.25	[423]
		47038.00	[104]	2.031	[487]			
K$_{\beta_5^{II}}$	KM$_4$	47358.3 $\pm$ 1.7	[145]			7.142(-4)	21.78	[423]
K$_{\beta_5^{I}}$	KM$_5$	47389.4 $\pm$ 1.7	[145]			9.450(-4)	21.40	[423]
	KN$_1$	48157.3 $\pm$ 4.3	[145]				30.86	[423]
K$_{\beta2}$	KN$_{2,3}$	48256.6 $\pm$ 2.2	[50]	0.2174	[488]	1.060(-2)	33.06	[423]
K$_{\beta_2^{II}}$	KN$_2$	48213.70	[104]	0.2344	[487]		33.06	[423]
		48230 $\pm$ 16	[145]					
K$_{\beta_2^{I}}$	KN$_3$	48256.50	[104]	0.422	[488]	2.057(-2)	27.07	[423]
		48261.1 $\pm$ 1.5	[145]	0.458	[487]			
K$_{\beta_4^{II}}$	KN$_4$	48382.0 $\pm$ 3.5	[145]			1.581(-4)	22.72	[423]
K$_{\beta_4^{I}}$	KN$_5$	48384.3 $\pm$ 1.2	[145]			2.086(-4)	22.38	[423]
	KO$_{2,3}$	48497.00 $\pm$ 1.30	[50]	8.65(-2)	[488]			
				9.69(-2)	[487]			
	KO$_2$					2.190(-3)		
	KO$_3$					4.471(-3)		

line	transition	E/eV		I/eV/$\hbar$		TPIV	Γ/eV	
L series								
	L_1M_2	6248.4 ± 2.1	[145]				6.03	[423]
$L_{\beta 4}$	L_1M_2	6437.81 ± 0.55	[50]	0.1128	[488]	2.277(-2)	11.74	[423]
		6439.10	[104]					
$L_{\beta 3}$	L_1M_3	6371.57 ± 0.58	[50]	0.1627	[488]	3.286(-2)	12.13	[423]
		6571.50	[104]					
$L_{\beta 10}$	L_1M_4	6889.81 ± 0.70	[50]	2.33(-3)	[488]	4.698(-4)	5.66	[423]
		6890.60	[104]					
$L_{\beta 9}$	L_1M_5	6919.15 ± 0.71	[50]	3.48(-3)	[488]	7.030(-4)	5.27	[423]
		6920.20	[104]					
	L_1N_1	7687.6 ± 2.4	[497]				14.83	[423]
	L_1N_1	7688.10	[501]				14.83	[423]
$L_{\gamma 2}$	L_1N_2	7768.10 ± 0.81	[50]	2.652(-2)	[488]	5.355(-3)	16.94	[423]
		7767.80	[104]			5.355(-3)		
$L_{\gamma 3}$	L_1N_3	7794.11 ± 0.81	[50]	3.98(-2)	[488]	8.035(-3)	10.95	[423]
		7796.20	[104]					
$L_{\gamma 11}$	$L_1N_{4,5}$	7912.00	[104]	4.83(-4)	[488]			
		7914.56 ± 0.36	[497]					
	L_1N_4	7911.9 ± 3.8	[145]			6.343(-5)	6.60	[423]
$L_{\gamma 11}$	L_1N_5	7914.2 ± 1.4	[145]			9.762(-5)	6.26	[423]
	L_1N_6	8040.1 ± 1.5	[145]					
	L_1N_7	8042.8 ± 1.5	[145]					
$L_{\gamma 4}$	$L_1O_{2,3}$	8030.40 ± 0.50	[50]	4.42(-3)	[488]			
		8030.45	[104]					
	L_1O_2					7.495(-4)		
	L_1O_3					1.094(-3)		
L_η	L_2M_1	5816.67 ± 0.81	[50]	1.384(-2)	[488]	5.043(-3)	19.18	[423]
		5816.50	[104]					
	L_2M_2	6003.0 ± 1.6	[145]				10.98	[423]
$L_{\beta 17}$	L_2M_3	6137.7 ± 1.5	[497]			9.775(-5)	11.37	[423]
		6137.3 ± 1.5	[145]					
$L_{\beta 1}$	L_2M_4	6455.72 ± 0.56	[50]	0.511	[488]	1.344(-1)	4.90	[423]
		6456.60	[104]					
	L_2M_5	6485.9 ± 1.2	[145]				4.52	[423]

line	transition	E/eV		I/eV/$\hbar$		TPIV	Γ/eV	
$L_{\gamma 5}$	L_2N_1	7255.47 ± 0.70 7256.70	[50] [104]	3.28(-3)	[488]	9.546(-4)	13.98	[423]
	L_2N_2	7331.60 ± 2.0	[501]				16.18	[423]
	L_2N_3	7357.6 ± 1.0	[145]			2.021(-5)	10.19	[423]
$L_{\gamma 1}$	L_2N_4	7479.12 ± 0.43 7480.10	[50] [104]	8.82(-2)	[488]	2.323(-2)	5.84	[423]
	L_2N_5	7480.82 ± 0.67	[145]				5.50	[423]
	$L_2N_{6,7}$	7612.87 ± 0.94	[207]					
L_ν	L_2N_6	7606.73 ± 0.76	[145]			1.387(-5)		
	L_2N_7	7609.40 ± 0.75	[145]					
$L_{\gamma 8}$	L_2O_1	7584.90 ± 1.00 7584.70	[50] [104]	5.54(-4)	[488]	1.592(-4)	4.64	[423]
	$L_2O_{2,3}$	7595.70	[501]					
$L_{\gamma 6}$	L_2O_4	7614.70 ± 1.00 7614.90	[50] [104]					
L_l	L_3M_1	5177.15 ± 0.64 5177.20	[50] [104]	1.903(-2)	[488]	9.315(-3)	19.02	[423]
L_t	L_3M_2	5363.2 ± 1.5	[145]			6.171(-5)	10.82	[423]
L_s	L_3M_3	5498.6 ± 1.2 5497.5 ± 1.4	[50] [145]			5.642(-5)	13.33	[423]
$L_{\alpha 2}$	L_3M_4	5816.61 ± 0.45 5816.50	[50] [104]	4.72(-2)	[488]	1.261(-2)	4.73	[423]
$L_{\alpha 1}$	L_3M_5	5846.46 ± 0.26 5845.70	[50] [104]	0.418	[488]	1.115(-1)	4.35	[423]
$L_{\beta 6}$	L_3N_1	6614.56 ± 0.59 6617.20	[50] [104]	4.42(-3)	[488]	1.198(-3)	13.81	[423]
	L_3N_2	6687 ± 15	[145]			9.159(-6)	16.02	[423]
	L_3N_3	6717.81 ± 0.94	[145]			8.855(-6)	10.03	[423]
$L_{\beta 15}$	L_3N_4	6840.80 6838.6 ± 2.9	[104] [145]	7.82(-3)	[488]	2.090(-3)	5.68	[423]
$L_{\beta 2}$	L_3N_5	6843.10 6841.01 ± 0.59	[104] [145]	7.00(-2)	[488]	1.869(-2)	5.34	[423]
$L_{\beta 15,2}$	$L_3N_{4,5}$	6841.83 ± 0.63	[50]					
	$L_4N_{6,7}$	6971.57 ± 0.79	[50]					
$L_{u'}$	L_3N_6	6966.92 ± 0.62	[145]			2.022(-6)		
L_u	L_3N_7	6969.59 ± 0.67	[145]			1.156(-5)		
$L_{\beta 7}$	L_3O_1	6945.30 ± 0.80 6945.30	[50] [104]	7.44(-4)	[488]	2.181(-4)	4.48	[423]
	$L_3O_{2,3}$	6953.40	[501]					

line	transition	E/eV		I/eV/$\hbar$		TPIV	Γ/eV	
$L_{\beta 5}$	$L_3O_{4,5}$	6976.30 ± 0.80	[50]					
		6976.30	[104]					

M series

$M_{\gamma 2}$	M_3N_4			1.03(-3)	[116]	2.633(-4)	9.99	[423]
				8.99(-3)	[335]			
$M_{\gamma 1}$	M_3N_5	1346.00	[104]	8.71(-3)	[116]	2.134(-3)	9.65	[423]
				8.99(-3)	[335]			
$M_{\gamma 2,1}$	$M_3N_{4,5}$	1346.00 ± 1.30	[50]					
		1349.10	[263]					
$M_{\xi 2}$	M_4N_2	877.15	[423]			3.961(-3)	13.86	[423]
M_{δ}	M_4N_3	906.90	[423]			5.320(-4)	7.87	[423]
M_{β}	M_4N_6	1153.30 ± 0.75	[50]	6.20(-3)	[116]	2.002(-3)		[423]
		1153.60	[104]	5.35(-3)	[335]			
		1155.50	[263]					
M_{η}	M_4O_2	1136.59			[423]	7.042(-4)		
$M_{\xi 1}$	M_5N_3	872.00 ± 1.20	[50]	7.05(-4)	[116]	4.258(-3)	7.49	[423]
		871.90	[104]	5.35(-3)	[335]			
		873.70	[263]					
$M_{\alpha 2.1}$	$M_5N_{6,7}$	1131.00 ± 3.00	[50]	5.89(-3)	[116]			
		1125.20	[104]					
		1133.30	[263]					
	M_5N_6					1.172(-4)		
	M_5N_7					2.273(-3)		
	M_5O_3	1107.50	[104]	4.16(-5)	[116]	8.822(-4)		

N series

	$N_4O_{2,3}$	110.70 ± 0.60	[50]					

level characteristics

level	E_B/eV		ω_{nlj}		Γ/eV		AE/eV	
K	48519.00	[503]	0.932	[301]	21.00	[301]	48519.0	[51]
	48519.00 ± 0.40	[493]	0.925	[364]	20.49	[423]	48519.7 ± 2.8	[50]
			0.931	[38]	21.1	[94]	48523.77 ± 0.96	[145]
			0.9289	[39]				
			0.9321	[423]				
			0.932	[310]				
			0.929	[112]				
			0.929	[242]				
L_1	8052.00	[503]	7.50(-2)	[301]	4.91	[301]	8052.0	[51]
	8052.00 ± 0.40	[493]	7.30(-2)	[423]	4.37	[423]	8060.75 ± 0.78	[50]
			7.80(-2)	[436]	3.6	[94]	8053.7 ± 1.2	[145]
			7.93(-2)	[244]				
L_2	7618.00	[503]	0.149	[301]	4.23	[301]	7617.1	[51]
	7617.10 ± 0.40	[493]	0.1640	[423]	3.61	[423]	7619.83 ± 0.69	[50]
			0.164	[436]	3.77	[94]	7620.28 ± 0.45	[145]
			0.167	[244]				
L_3	6977.00	[503]	0.147	[301]	3.91	[301]	6976.9	[51]
	6976.90 ± 0.40	[493]	0.1558	[423]	3.44	[423]	6980.59 ± 0.58	[50]
			0.158	[436]	3.62	[94]	6980.47 ± 0.37	[145]
			0.16	[244]				
M_1	1800.00	[503]	8.70(-4)	[38]	14.95	[350]	1800.0	[51]
	1800.00 ± 0.50	[493]	1.25(-3)	[512]	15.57	[423]		
			1.23(-3)	[423]	12.6	[94]		
M_2	1614.00	[503]	1.47(-3)	[38]	8.37	[350]	1613.9	[51]
	1613.90 ± 0.70	[493]	1.00(-2)	[512]	7.37	[423]		
			2.96(-3)	[423]	5.4	[94]		
M_3	1481.00	[503]	1.26(-3)	[38]	7.81	[350]	1480.6	[51]
	1480.60 ± 0.60	[493]	1.29(-3)	[512]	7.76	[423]		
			3.31(-3)	[423]	8.2	[94]		
M_4	1161.00	[503]	4.10(-3)	[38]	1.86	[350]	1160.6	[51]
	1160.60 ± 0.60	[493]	8.34(-3)	[512]	1.29	[423]		
			7.30(-3)	[423]	0.9	[94]		

level	E_B/eV		ω_{nlj}		Γ/eV		AE/eV	
M_5	1131.00	[503]	5.90(-3)	[38]	1.14	[350]	1130.9	[51]
	1130.90 ± 0.60	[493]	5.55(-3)	[512]	0.90	[423]		
			7.53(-3)	[423]	0.9	[94]		
N_1	360.00	[503]	1.7(-4)	[423]	4.6	[94]		
	360.20 ± 0.70	[493]			10.37	[423]		
N_2	284.00	[503]	3.5(-4)	[423]	5.2	[94]		
	283.90 ± 1.00	[493]			12.57	[423]		
N_3	257.00	[503]	1.2(-4)	[423]	2.35	[94]		
	256.60 ± 0.80	[493]			6.58	[423]		
N_4	134.00	[503]	2.0(-5)	[423]	2.2	[94]		
	133.20 ± 0.60	[51]			2.23	[423]		
N_5	134.00	[503]	2.0(-5)	[423]	1.9	[94]		
	133.20 ± 0.60	[51]			1.89	[423]		
N_6	1.50 ± 3.20	[493]						
N_7	0.00 ± 3.20	[51]						
O_1	32.00	[503]			1.03	[423]		
	31.80 ± 0.70	[493]						
O_2	22.00	[503]						
	22.00 ± 0.60	[51]						
	25.20 ± 0.60	[493]						
O_3	22.00	[503]						
	22.00 ± 0.60	[51]						
	20.40 ± 0.60	[493]						
IP	5.6704	[222]						

Gd   Z=64

[Xe] $4f^7 5d^1 6s^2$

Gadolinium

$A = 157.25(3)$ [222] $\varrho = 7.898$ g/cm^3 [547]

$\varrho = 7.80$ g/cm^3 [483]

X-ray transitions

line	transition	E/eV		I/eV/$\hbar$		TPIV	Γ/eV	
K series								
$K_{\alpha 3}$	KL_1	41865.8 ± 1.6	[145]				26.43	[423]
$K_{\alpha 2}$	KL_2	42309.30 ± 0.43	[50]	5.86	[488]	2.682(-1)	26.70	[301]
		42309.00	[104]	5.93	[487]		29.50	[473]
							25.54	[423]
$K_{\alpha 1}$	KL_3	42996.72 ± 0.44	[50]	10.54	[488]	4.822(-1)	26.40	[301]
		42997.00	[104]	10.66	[487]		28.00	[473]
							25.38	[423]
	KM_1	48357.9 ± 2.0	[145]				37.64	[423]
$K_{\beta 3}$	KM_2	48555.8 ± 5.6	[50]	1.071	[488]	4.901(-2)	28.00	[473]
		48556.00	[104]	1.127	[487]			
							29.51	[423]
$K_{\beta 1}$	KM_3	48696.9 ± 5.7	[50]	2.072	[488]	9.479(-2)	29.37	[473]
		48700.00	[104]	2.178	[487]		29.95	[423]
$K_{\beta 5}$	$KM_{4,5}$	49053.3 ± 8.6	[50]	2.153(-2)	[488]			
		49053.00	[104]					
$K_{\beta_5^{II}}$	KM_4	49020.8 ± 1.8	[145]			7.476(-4)	23.16	[423]
$K_{\beta_5^{I}}$	KM_5	49054.6 ± 1.7	[145]			9.850(-4)	22.78	[423]
	KN_1	49860.6 ± 4.3	[145]				32.47	[423]
$K_{\beta_2^{II}}$	KN_2	49935.3 ± 4.7	[145]	0.2536	[487]	1.077(-2)	30.63	[423]
$K_{\beta_2^{I}}$	KN_3	49969.7 ± 1.5	[145]	0.496	[487]	2.090(-2)	28.69	[423]
$K_{\beta 2}$	$KN_{2,3}$	49960.6 ± 8.9	[50]	0.457	[488]			
		49960.00	[104]					
$K_{\beta_4^{II}}$	KN_4	50090.7 ± 3.3	[145]			1.685(-4)	23.98	[423]
$K_{\beta_4^{I}}$	KN_5	50098.6 ± 1.2	[145]			2.214(-4)	23.92	[423]
	$KO_{2,3}$	50221.00 ± 0.60	[50]	0.1107	[487]			
		50221.00	[104]	9.87(-2)	[488]	0.94		
	KO_2					2.268(-3)		
	KO_3					4.674(-3)		

line	transition	E/eV		I/eV/$\hbar$		TPIV	Γ/eV	
	L series							
	L_1M_1	6492.2 ± 2.2	[145]				20.40	[423]
$L_{\beta4}$	L_1M_2	6687.3 ± 1.1	[50]	0.1225	[488]	2.394(-2)	9.08	[473]
		6687.10	[104]				12.27	[423]
$L_{\beta3}$	L_1M_3	6831.0 ± 1.1	[50]	0.1748	[488]	3.415(-2)	11.20	[473]
		6381.10	[104]				12.71	[423]
$L_{\beta10}$	L_1M_4	7160.4 ± 1.8	[50]	2.61(-3)	[488]	5.107(-4)	5.93	[423]
		7160.30	[104]					
$L_{\beta9}$	L_1M_5	7191.6 ± 1.9	[50]	3.91(-3)	[488]	7.641(-4)	5.55	[423]
		7191.70	[104]					
	L_1N_1	7994.8 ± 4.4	[145]				15.23	[423]
$L_{\gamma2}$	L_1N_2	8087.0 ± 1.6	[50]	2.906(-2)	[488]	5.679(-3)	13.39	[423]
		8087.20	[104]					
$L_{\gamma3}$	L_1N_3	8105.0 ± 1.6	[50]	4.32(-2)	[488]	8.445(-3)	11.46	[423]
		8105.20	[104]					
$L_{\gamma11}$	$L_1N_{4,5}$	8237.30	[104]	5.51(-4)	[488]			
		8237.2 ± 1.0	[380]					
	L_1N_4	8224.9 ± 3.4	[145]			6.978(-5)	6.74	[423]
	L_1N_5	8232.8 ± 1.4	[145]			1.076(-4)	6.60	[423]
	L_1N_6	8365.2 ± 1.5	[145]					
	L_1N_7	8368.5 ± 1.5	[145]					
$L_{\gamma4}$	$L_1O_{2,3}$	8355.00 ± 1.10	[50]	6.21(-3)	[488]			
		8355.00	[104]					
	L_1O_2					7.844(-4)		
	L_1O_3					1.130(-3)		
	$L_1O_{4,5}$	8373.00 ± 1.70	[50]	5.10(-6)	[488]			
		8373.00	[104]					
L_η	L_2M_1	6049.69 ± 0.44	[50]	1.494(-2)	[488]	5.180(-3)	19.52	[423]
		6049.30	[104]					
	L_2M_2	6240.9 ± 1.6	[145]				11.38	[423]
$L_{\beta17}$	L_2M_3	6386.0 ± 1.5	[145]			1.046(-4)	11.82	[423]
$L_{\beta1}$	L_2M_4	6713.4 ± 1.1	[50]	0.553	[488]	1.423(-1)	4.63	[473]
		6713.20	[104]				5.04	[423]
	L_2M_5	6743.40	[104]	1.05(-5)	[488]		4.66	[423]
		6743.32 ± 0.74	[145]					

line	transition	E/eV		I/eV/ℏ		TPIV	Γ/eV	
$L_{\gamma 5}$	L_2N_1	7554.4 ± 1.4	[50]	3.57(-3)	[488]	1.006(-3)	14.35	[423]
		7554.60	[104]					
	L_2N_2	7642.60	[104]	5.03(-8)	[488]		12.51	[423]
		7642.67 ± 0.47	[386]					
	L_2N_3	7655.20	[104]	8.70(-5)	[488]	2.241(-5)	10.57	[423]
		7657.7 ± 1.2	[137]					
$L_{\gamma 1}$	L_2N_4	7785.9 ± 1.4	[50]	9.70(-2)	[488]	2.499(-2)	7.83	[473]
		7785.70	[104]				5.86	[423]
	L_2N_5	7788.91 ± 0.63	[145]				5.80	[423]
L_ν	L_2N_6	7921.33 ± 0.71	[145]			1.856(-5)		
	L_2N_7	7924.64 ± 0.71	[145]					
$L_{\gamma 8}$	L_2O_1	7894.00 ± 1.00	[50]	6.25(-4)	[488]	1.705(-4)	5.25	[423]
		7893.40	[104]					
$L_{\gamma 6}$	L_2O_4	7925.00 ± 1.00	[50]	5.79(-4)	[488]			
		7924.90	[104]					
L_l	L_3M_1	5362.09 ± 0.69	[50]	2.075(-2)	[488]	9.558(-3)	19.36	[423]
		5362.20	[104]					
L_t	L_3M_2	5553.7 ± 1.5	[145]			6.483(-5)	11.21	[423]
L_s	L_3M_3	5698.20	[104]	1.58(-4)	[488]	5.928(-5)	11.66	[423]
		5698.3 ± 1.3	[385]					
$L_{\alpha 2}$	L_3M_4	6024.99 ± 0.87	[50]	5.10(-2)	[488]	1.329(-2)	4.90	[473]
		6025.00	[104]				4.88	[423]
$L_{\alpha 1}$	L_3M_5	6057.37 ± 0.88	[50]	0.451	[488]	1.175(-1)	4.46	[473]
		6057.30	[104]				4.50	[423]
$L_{\beta 6}$	L_3N_1	6867.3 ± 1.1	[50]	4.85(-3)	[488]	1.277(-3)	14.19	[423]
		6867.00	[104]					
	L_3N_2	6938.5 ± 4.0	[145]			1.001(-5)	12.34	[423]
	L_3N_3	6972.94 ± 0.89	[145]			9.755(-6)		
$L_{\beta 15}$	L_3N_4	7097.90	[104]	8.56(-3)	[488]	2.233(-3)	5.69	[423]
		7093.9 ± 2.6	[145]					
$L_{\beta 2}$	L_3N_5	7103.10	[104]	7.66(-2)	[488]	1.998(-2)	7.70	[473]
		7101.77 ± 0.55	[145]				5.55	[423]
$L_{\beta 2,15}$	$L_3N_{4,5}$	7103.0 ± 1.2	[50]					
$L_{u'}$	L_3N_6	7234.19 ± 0.63	[145]			2.663(-6)		
L_u	L_3N_7	7237.50 ± 0.63	[145]			1.530(-5)		

line	transition	E/eV		I/eV/$\hbar$		TPIV	Γ/eV	
$L_{\beta 7}$	L_3O_1	7207.10 ± 0.85 7207.20	[50] [104]	8.46(-4)	[488]	2.356(-4)	5.09	[423]
$L_{\beta 5}$	$L_3O_{4,5}$	7237.40 ± 0.85 7237.50	[50] [104]	4.90(-4)	[488]			

M series

line	transition	E/eV		I/eV/$\hbar$		TPIV	Γ/eV	
$M_{\gamma 2,1}$	$M_3N_{4,5}$	1402.00 ± 1.50 1404.10	[50] [263]					
$M_{\gamma 2}$	M_3N_4			1.007(-2)	[335]	2.811(-4)	10.26	[423]
$M_{\gamma 1}$	M_3N_5	1401.80	[104]	1.007(-2)	[335]	2.275(-3)	10.12	[423]
$M_{\xi 2}$	M_4N_2					3.939(-3)	10.13	[423]
M_δ	M_4N_3					5.117(-4)	8.19	[423]
M_β	M_4N_6	1209.10 ± 0.70 1209.90 1212.00	[50] [104] [104]	6.41(-3)	[335]	2.840(-3)		
M_η	M_4O_2					7.134(-4)		
$M_{\xi 1}$	M_5N_3	914.00 ± 1.40 913.70 915.70	[50] [104] [263]	5.93(-4)	[335]	4.125(-3)	7.81	[423]
$M_{\alpha 1.2}$	$M_5N_{6,7}$	1185.00 ± 3.40 1189.50 1187.60	[50] [104] [263]	6.41(-3)	[335]			
	M_5N_6					1.681(-4)		
	M_5N_7					3.255(-3)		
	M_5O_3	1165.00	[104]			8.874(-4)		

level characteristics

level	E_B/eV		ω_{nlj}		Γ/eV		AE/eV	
K	50239.00	[503]	0.935	[301]	22.30	[301]	50239.1	[51]
	50239.10 ± 0.50	[493]	0.934	[38]	21.83	[423]	50233.9 ± 3.0	[50]
			0.9320	[39]	22.4	[94]	50251.67 ± 0.97	[145]
			0.9350	[423]				
			0.935	[310]				
			0.922	[33]				
			0.935	[242]				
L_1	8376.00	[503]	7.90(-2)	[301]	5.05	[301]	8375.6	[51]
	8375.60 ± 0.50	[493]	7.65(-2)	[423]	4.60	[423]	8386.25 ± 0.84	[50]
			8.30(-2)	[436]	3.8	[94]	8385.9 ± 1.2	[145]
L_2	7931.00	[503]	0.158	[301]	4.32	[301]	7930.3	[51]
	7930.30 ± 0.40	[493]	0.1738	[423]	3.71	[423]	7931.32 ± 0.75	[50]
			0.175	[436]	3.87	[94]	7942.02 ± 0.41	[145]
L_3	7243.00	[503]	0.155	[301]	4.01	[301]	7242.8	[51]
	7242.80 ± 0.40	[493]	0.1643	[423]	3.55	[423]	7243.23 ± 0.63	[50]
			0.167	[436]	3.72	[94]	7254.88 ± 0.33	[145]
M_1	1881.00	[503]	1.31(-3)	[512]	16.00	[74]	1880.8	[51]
	1880.80 ± 0.50	[493]	1.30(-3)	[423]	15.81	[423]		
					12.8	[94]		
M_2	1689.00	[503]	0.010	[512]	8.30	[74]	1688.3	[51]
	1688.30 ± 0.70	[493]	3.10(-3)	[423]	7.68	[423]		
					5.6	[94]		
M_3	1544.00	[503]	1.35(-3)	[512]	8.30	[74]	1544.0	[51]
	1544.00 ± 0.80	[493]	3.51(-3)	[423]	8.11	[423]		
					8.3	[94]		
M_4	1218.00	[503]	8.51(-3)	[512]	1.20	[74]	1217.2	[51]
	1217.20 ± 0.60	[493]	8.10(-3)	[423]	1.33	[423]		
					0.95	[94]		
M_5	1186.00	[503]	6.99(-3)	[512]	1.90	[74]	1185.2	[51]
	1185.20 ± 0.60	[493]	8.43(-3)	[423]	0.95	[423]		
					0.95	[94]		

level	E_B/eV		ω_{nlj}		Γ/eV	AE/eV
N_1	376.00	[503]	1.9(-4)	[423]	12.00	[74]
	375.80 $\pm$ 0.70	[493]			10.64	[423]
					4.9	[94]
N_2	289.00	[503]	3.9(-4)	[423]	9.00	[74]
	288.50 $\pm$ 1.20	[493]			8.80	[423]
					5.23	[94]
N_3	271.00	[503]	1.3(-4)	[423]	9.00	[74]
	270.90 $\pm$ 0.90	[493]			6.86	[423]
					2.5	[94]
N_4	141.00	[503]	2.0(-5)	[423]	2.40	[74]
	149.50 $\pm$ 0.80	[493]			2.15	[423]
					2.45	[94]
N_5	141.00	[503]	2.0(-5)	[423]	2.40	[74]
	140.50 $\pm$ 0.80	[51]			2.00	[423]
	134.50 $\pm$ 0.80	[493]			2.2	[94]
N_6	0.10 $\pm$ 3.50	[51]				
	2.00 $\pm$ 3.50	[493]				
N_7	0.10 $\pm$ 3.50	[51]				
O_1	36.00	[503]			1.54	[423]
	36.10 $\pm$ 0.80	[493]				
O_2	21.00	[503]				
	20.30 $\pm$ 1.20	[51]				
	24.30 $\pm$ 1.20	[493]				
O_3	21.00	[503]				
	18.30 $\pm$ 1.20	[493]				
IP	6.1501	[222]				

Tb   Z=65

[Xe] $4f^9 6s^2$

Terbium $A = 158.92534(2)$ [222] $\varrho = 8.234$ g/cm^3 [547]
$\varrho = 8.19$ g/cm^3 [483]

X-ray transitions

line	transition	E/eV		I/eV/$\hbar$		TPIV	Γ/eV	
K series								
K_{α_3}	KL_1	43291.4 ± 1.7	[145]				27.94	[423]
$K_{\alpha 2}$	KL_2	43744.62 ± 0.46	[50]	6.26	[488]	2.692(-1)	28.20	[301]
		43745.00	[104]	6.33	[487]		26.99	[423]
$K_{\alpha 1}$	KL_3	44482.75 ± 0.47	[50]	11.23	[488]	4.825(-1)	27.90	[301]
		44482.00	[104]	11.35	[487]		26.85	[423]
	KM_1	50035.5 ± 2.2	[145]				39.75	[423]
K_{β_3}	KM_2	50229.8 ± 6.0	[50]	1.148	[488]	4.935(-2)	31.70	[423]
		50230.00	[104]	1.205	[487]			
K_{β_1}	KM_3	50382.9 ± 6.1	[50]	2.22	[488]	9.545(-2)	31.79	[423]
		50383.00	[104]	2.331	[487]			
$K_{\beta_5^{II}}$	KM_4	50722.7 ± 1.9	[145]			7.827(-4)	24.77	[423]
$K_{\beta_5^I}$	KM_5	50757.8 ± 1.8	[145]			1.027(-3)	24.23	[423]
	KN_1	51600.7 ± 4.3	[145]				34.56	[423]
$K_{\beta_2^{II}}$	KN_2	51678.3 ± 4.6	[145]	0.2712	[487]	1.083(-2)	37.64	[423]
$K_{\beta_2^I}$	KN_3	51714.0 ± 1.6	[145]	0.529	[487]	2.099(-2)	30.77	[423]
$K_{\beta 2}$	$KN_{2,3}$	51724 ± 64	[50]	0.488	[488]			
		51742.00	[104]					
$K_{\beta_4^{II}}$	KN_4	51837.5 ± 3.1	[145]			1.756(-4)	26.64	[423]
$K_{\beta beta_4^I}$	KN_5	51846.4 ± 1.3	[145]			2.297(-4)	26.06	[423]
	$KO_{2,3}$	51965.00 ± 6.50	[50]	9.78(-2)	[488]			
		51966.00	[104]	0.1091	[487]			
	KO_2					2.324(-3)		[423]
	KO_3					4.804(-3)		[423]
L series								
	$L_1 M_1$	6744.2 ± 2.3	[145]				21.24	[423]
$L_{\beta 4}$	$L_1 M_2$	6940.3 ± 1.1	[50]	0.1330	[488]	2.518(-2)	13.20	[423]
		6940.50	[104]					

line	transition	E/eV		I/eV/ℏ		TPIV	Γ/eV	
$L_{\beta3}$	L_1M_3	7096.1 ± 1.2 7095.80	[50] [104]	0.1876	[488]	3.552(-2)	13.29	[423]
$L_{\beta10}$	L_1M_4	7436.1 ± 2.0 7436.10	[50] [104]	2.93(-3)	[488]	5.551(-4)	6.27	[423]
$L_{\beta9}$	L_1M_5	7469.50 7466.5 ± 2.0	[104] [145]	4.39(-3)	[488]	8.307(-4)	5.72	[423]
	L_1N_1	8313.88 ± 0.56 8309.3 ± 4.4	[379] [145]				16.05	[423]
$L_{\gamma2}$	L_1N_2	8397.6 ± 1.7 8397.60	[50] [104]	3.15(-2)	[488]	5.965(-3)	19.13	[423]
$L_{\gamma3}$	L_1N_3	8423.9 ± 1.7 8423.70	[50] [104]	4.63(-2)	[488]	8.769(-3)	12.26	[423]
$L_{\gamma11}$	$L_1N_{4,5}$	8559.00 8558.88 ± 0.59	[104] [379]	6.13(-3)	[488]			
	L_1N_4	8546.2 ± 3.2	[145]			7.523(-5)	8.14	[423]
	L_1N_5	8555.0 ± 1.5	[145]			1.160(-4)	7.56	[423]
	L_1N_6	8694.9 ± 1.5	[145]					
	L_1N_7	8697.5 ± 1.5	[145]					
	L_1O_1	8669.60	[104]	4.63(-8)	[488]		7.03	[423]
$L_{\gamma4}$	$L_1O_{2,3}$	8685.00 ± 1.20 8684.90	[50] [104]	6.15(-3)	[488]			
	L_1O_2					8.176(-4)		
	L_1O_3					1.164(-3)		
	L_1O_4	8713.80	[104]					
	$L_1O_{4,5}$	8714.00 ± 1.80	[50]					
L_η	L_2M_1	6283.95 ± 0.94 6284.10	[50] [104]	1.613(-2)	[488]	5.310(-3)	20.30	[423]
	L_2M_2	6487.4 ± 1.6	[145]				12.25	[423]
$L_{\beta17}$	L_2M_3	6643.5 ± 1.6	[145]			1.118(-4)	12.34	[423]
$L_{\beta1}$	L_2M_4	6977.8 ± 1.7 6975.30	[50] [104]	0.598	[488]	1.514(-1)	5.33	[423]
	L_2M_5	7009.40	[104]	1.23(-5)	[488]		4.78	[423]
	L_2M_5	7009.37 ± 0.79 7012.3 ± 1.2	[136] [145]					

line	transition	E/eV		I/eV/$\hbar$		TPIV	Γ/eV	
$L_{\gamma 5}$	L_2N_1	7853.4 ± 1.5	[50]	3.85(-3)	[488]	1.057(-3)	15.11	[423]
		7853.50	[104]					
	L_2N_2	7936.00	[104]	6.09(-8)	[488]		18.19	[423]
		7935.81 ± 0.51	[382]					
	L_2N_3	7967.00	[104]	9.74(-5)	[488]	2.466(-5)	11.32	[423]
		7968.42 ± 0.49	[136]					
$L_{\gamma 1}$	L_2N_4	8101.8 ± 1.6	[50]	0.1044	[488]	2.644(-2)	7.20	[423]
		8102.50	[104]					
	L_2N_5	8100.81 ± 0.68	[145]				6.61	[423]
L_{v}	L_2N_6	8240.74 ± 0.75	[145]			2.405(-5)		
	L_2N_7	8243.33 ± 0.77	[145]					
$L_{\gamma 8}$	L_2O_1	8212.00 ± 1.10	[50]	6.41(-4)	[488]	1.821(-4)	6.09	[423]
		8212.40	[104]					
$L_{\gamma 6}$	L_2O_4	8246.00 ± 1.10	[50]					
		8246.30	[104]					
L_{l}	L_3M_1	5546.81 ± 0.73	[50]	2.26(-3)	[488]	9.828(-3)	20.16	[423]
L_{t}	L_3M_2	5750.1 ± 1.5	[145]			6.793(-5)	12.11	[423]
L_{s}	L_3M_3	5906.30	[104]	1.76(-4)	[488]	6.213(-5)	12.20	[423]
		5906.2 ± 1.5	[145]					
$L_{\alpha 2}$	L_3M_4	6238.10 ± 0.93	[50]	5.50(-2)	[488]	1.403(-2)	5.18	[423]
		6238.00	[104]					
$L_{\alpha 1}$	L_3M_5	6272.82 ± 0.94	[50]	0.486	[488]	1.241(-1)	4.64	[423]
		6272.80	[104]					
$L_{\beta 6}$	L_3N_1	7116.4 ± 1.2	[50]	5.28(-3)	[488]	1.358(-3)	14.97	[423]
		7116.30	[104]					
	L_3N_2	7198.40	[104]	4.30(-5)	[488]	1.098(-5)	18.05	[423]
		7195.4 ± 3.9	[145]					
	L_3N_3	7227.00	[104]	4.17(-5)	[488]	1.064(-5)	11.18	[423]
		7226.9 ± 1.3	[136]					
		7231.17 ± 0.92	[145]					
$L_{\beta 15}$	L_3N_4	7365.10	[104]	9.17(-3)	[488]	2.341(-3)	7.05	[423]
		7354.6 ± 2.4	[145]					
$L_{\beta 2}$	L_3N_5	7363.50 ± 0.60	[145]	8.20(-2)	[488]	2.094(-2)	6.47	[423]
$L_{\beta 15,2}$	$L_3N_{4,5}$	7366.7 ± 1.3	[50]					
$L_{u'}$	L_3N_6	7503.43 ± 0.68	[145]			3.433(-6)		
L_{u}	L_3N_7	7506.02 ± 0.69	[145]			1.965(-5)		
$L_{\beta 7}$	L_3O_1	7475.30 ± 0.90	[50]	8.76(-4)	[488]	2.536(-4)	5.94	[423]
		7475.30	[104]					
$L_{\beta 5}$	$L_3O_{4,5}$	7509.40 ± 0.90	[50]					
		7507.30	[104]					

line	transition	E/eV		I/eV/ℏ		TPIV	Γ/eV	
M series								
$M_{\gamma_{2,1}}$	$M_3N_{4,5}$	1461.00 ± 1.50	[50]	1.10(-2)	[335]	2.391(-3)	11.42	[423]
		1463.80	[263]					
M_{γ_1}	M_3N_5	1461.10	[104]			2.391(-3)	11.40	[423]
M_{ξ_2}	M_4N_2	963.00	[423]			3.769(-3)	16.06	[423]
M_δ	M_4N_3	997.96	[423]			4.779(-4)	9.10	[423]
M_β	M_4N_6	1266.10 ± 0.80	[50]	8.85(-3)	[335]	3.878(-3)		
		1266.10	[104]					
		1269.00	[263]					
M_η	M_4O_2	1246.76	[423]			6.980(-4)		
M_{ξ_1}	M_5N_3	955.00 ± 1.50	[50]	6.35(-4)	[335]	3.851(-3)	8.55	[423]
		955.50	[104]					
		957.40	[263]					
$M_{\alpha_{2,1}}$	$M_5N_{6,7}$	1240.00 ± 2.50	[50]	8.58(-3)	[335]			
		1242.30	[263]					
	M_5N_6					2.301(-4)		
	M_5N_7					4.452(-3)		
N series								
	$N_{4,5}N_{6,7}$	144.00	[50]					
	$N_{4,5}O_{2,3}$	121.30	[50]					

level characteristics

level	E_B/eV		ω_{nlj}		Γ/eV		AE/eV	
K	51996.00	[503]	0.938	[301]	23.80	[301]	51995.7	[51]
	51995.70 ± 0.50	[493]	0.937	[38]	23.22	[423]	52003.8 ± 3.2	[50]
			0.9349	[39]	23.8	[94]	51999.5 ± 1.1	[145]
			0.9376	[423]				
			0.938	[242]				

level	E_B/eV		ω_{nlj}		Γ/eV		AE/eV	
L_1	8708.00	[503]	8.30(-2)	[301]	5.19	[301]	8708.0	[51]
	8708.00 ± 0.50	[493]	0.18	[344]	4.72	[423]	8717.03 ± 0.91	[50]
			8.00(-2)	[423]	4	[94]	8708.1 ± 1.2	
			8.70(-2)	[436]				
			9.00(-2)	[476]				
			0.102	[477]				
L_2	8252.00	[503]	0.167	[301]	4.43	[301]	8251.6	[51]
	8251.60 ± 0.40	[493]	0.165	[344]	3.77	[423]	8252.83 ± 0.81	[50]
			0.166	[107]	3.93	[94]	8253.93 ± 0.46	[145]
			0.1846	[423]				
			0.186	[436]				
			0.194	[476]				
			0.209	[477]				
L_3	7515.00	[503]	0.164	[301]	4.12	[301]	7514.0	[51]
	7514.00 ± 0.40	[493]	0.188	[344]	3.63	[423]	7515.45 ± 0.67	[50]
			0.1730	[423]	3.8	[94]	7516.62 ± 0.39	[145]
			0.175	[436]				
M_1	1968.00	[503]	1.37(-3)	[512]	13	[94]	1967.5	[51]
	1967.50 ± 0.60	[493]	1.42(-3)	[423]	16.53	[423]		
M_2	1768.00	[503]	1.10(-2)	[512]	5.8	[94]	1767.7	[51]
	1767.70 ± 0.90	[493]	3.22(-3)	[423]	8.48	[423]		
M_3	1612.00	[503]	1.41(-3)	[512]	8.2	[94]	1611.3	[51]
	1611.30 ± 0.80	[493]	3.68(-3)	[423]	8.57	[423]		
M_4	1276.00	[503]	8.62(-3)	[512]	1.01	[94]	1275.0	[51]
	1275.00 ± 0.60	[493]	8.91(-3)	[423]	1.55	[423]		
M_5	1242.00	[503]	8.43(-3)	[512]	1.01	[94]	1241.2	[51]
	1241.20 ± 0.70	[493]	9.41(-3)	[423]	1.01	[423]		
N_1	398.00	[503]	2.0(-4)	[423]	5.1	[94]		
	397.90 ± 0.80	[493]			11.34	[423]		
N_2	311.00	[503]	4.2(-4)	[423]	5.26	[94]		
	310.20 ± 1.20	[493]			14.42	[423]		
N_3	286.00	[503]	1.5(-4)	[423]	2.65	[94]		
	285.00 ± 1.00	[493]			7.55	[423]		
N_4	148.00	[503]	3.0(-5)	[423]	2.7	[94]		
	147.00 ± 0.80	[51]			3.42	[423]		
	154.50 ± 0.80	[493]						

level	E_B/eV		ω_{nlj}		Γ/eV		AE/eV
N_5	148.00	[503]	3.0(-5)	[423]	2.4	[94]	
	147.00 ± 0.80	[51]			2.84	[423]	
	142.00 ± 0.80	[493]					
N_6	3.00	[503]					
	2.60 ± 1.50	[51]					
	4.00 ± 1.50	[493]					
N_7	3.00	[503]					
	2.60 ± 1.50	[51]					
	1.60 ± 1.50	[493]					
O_1	40.00	[503]			2.31	[423]	
	39.00 ± 0.80	[493]					
O_2	26.00	[503]					
	25.40 ± 0.80	[51]					
	26.30 ± 2.30	[493]					
O_3	26.00	[503]					
	25.40 ± 0.80	[51]					
	21.30 ± 2.30	[493]					
IP	5.8638	[222]					

Dy   Z=66

[Xe] $4f^{10}\,6s^2$

Dysprosium $A = 162.500(3)$ [222] $\varrho = 8.54$ g/cm^3 [547]
$\varrho = 8.35$ g/cm^3 [483]

X-ray transitions

line	transition	E/eV		I/eV/$\hbar$		TPIV	Γ/eV	
K series								
K_{α_3}	KL$_1$	44744.3 ± 1.7	[145]				29.54	[423]
$K_{\alpha 2}$	KL$_2$	45208.27 ± 0.49	[50]	6.68	[488]	2.700(-1)	29.80	[301]
		45208.00	[104]				32.20	[473]
							28.54	[423]
$K_{\alpha 1}$	KL$_3$	45998.94 ± 0.51	[50]	11.94	[488]	4.826(-1)	29.40	[301]
		45999.00	[104]				33.90	[473]
							28.41	[423]
	KM$_1$	51745.4 ± 2.3	[145]				41.67	[423]
$K_{\beta 3}$	KM$_2$	51958.1 ± 6.4	[50]	1.229	[488]	4.967(-2)	32.00	[473]
		51961.40	[104]				33.54	[423]
$K_{\beta 1}$	KM$_3$	52119.7 ± 6.5	[50]	2.376	[488]	9.602(-2)	32.73	[473]
		52121.00	[104]				33.66	[423]
$K_{\beta 5}$	KM$_{4,5}$	52494.8 ± 9.9	[50]	2.645(-2)	[488]			
		52495.00	[104]					
$K_{\beta_5^{II}}$	KM$_4$	52457.4 ± 1.9	[145]			8.178(-4)	26.38	[423]
$K_{\beta_5^{I}}$	KM$_5$	52494.8 ± 1.9	[145]			1.069(-3)	25.75	[423]
	KN$_1$	53372.0 ± 2.7	[145]				36.51	[423]
$K_{\beta 2}$	KN$_{2,3}$	53510 ± 68	[50]	0.524	[488]			
		53496.00	[104]					
$K_{\beta_2^{II}}$	KN$_2$	53451.7 ± 2.9	[145]			1.093(-2)	39.53	[423]
$K_{\beta_2^{I}}$	KN$_3$	53483.2 ± 1.2	[145]			2.118(-2)	32.69	[423]
$K_{\beta_4^{II}}$	KN$_4$	53619.4 ± 2.8	[145]			1.846(-4)	28.79	[423]
$K_{\beta_4^{I}}$	KN$_5$	53629.4 ± 1.3	[145]			2.404(-4)	28.07	[423]
	KO$_{2,3}$	53774.00 ± 7.00	[50]	0.1037	[488]			
		53775.00	[104]					
	KO$_2$					2.381(-3)		
	KO$_3$					4.938(-3)		

line	transition	E/eV		I/eV/ℏ		TPIV	Γ/eV	
	L series							
	L_1M_1	7001.1 ± 2.5	[145]				21.84	[423]
$L_{\beta 4}$	L_1M_2	7203.96 ± 0.43	[50]	0.1442	[488]	2.623(-2)	9.60	[473]
		7203.90	[104]				13.71	[423]
$L_{\beta 3}$	L_1M_3	7370.20 ± 0.90	[50]	0.2010	[488]	3.657(-2)	11.50	[473]
		7370.50	[104]				13.83	[423]
$L_{\beta 10}$	L_1M_4	7712.79 ± 0.58	[50]	2.28(-3)	[488]	5.972(-4)	6.54	[423]
		7711.40	[104]					
$L_{\beta 9}$	L_1M_5	7749.57 ± 0.59	[50]			8.940(-4)	5.92	[423]
		7746.40	[104]	4.91(-3)	[488]			
	L_1N_1	8604.00	[104]	3.05(-7)	[488]		16.68	[423]
		8627.7 ± 2.9	[145]					
$L_{\gamma 2}$	L_1N_2	8714.09 ± 0.63	[50]	3.43(-2)	[488]	6.232(-3)	19.69	[423]
		8714.00	[104]					
$L_{\gamma 3}$	L_1N_3	8753.34 ± 0.64	[50]	4.98(-2)	[488]	9.067(-3)	12.85	[423]
		8753.20	[104]					
	L_1N_4	8875.1 ± 2.9	[145]			8.106(-5)	8.96	[423]
	L_1N_5	8889.3 ± 3.2	[389]			1.252(-4)	8.24	[423]
		8885.1 ± 1.4	[145]					
	L_1N_6	9035.2 ± 1.5	[145]					
	L_1N_7	9037.6 ± 1.5	[145]					
$L_{\gamma 4}$	$L_1O_{2,3}$	9019.50 ± 0.50	[50]	6.53(-3)	[488]			
		9019.50	[104]					
	L_1O_2					8.457(-4)		
	L_1O_3					1.188(-3)		
L_{η}	L_2M_1	6534.22 ± 0.36	[50]	1.739(-2)	[488]	5.446(-3)	20.84	[423]
		6534.20	[104]					
	L_2M_2	6738.6 ± 1.6	[145]				12.71	[423]
$L_{\beta 17}$	L_2M_3	6905.70	[104]	4.41(-4)	[488]	1.195(-4)	12.83	[423]
		6905.61 ± 0.77	[138]					
$L_{\beta 1}$	L_2M_4	7247.80 ± 0.44	[50]	0.646	[488]	1.607(-1)	5.03	[473]
		7247.70	[104]				5.55	[423]
	L_2M_5	7277.30	[104]	1.44(-5)	[488]		4.92	[423]
		7277.2 ± 1.3	[138]					

line	transition	E/eV		$I/\text{eV}/\hbar$		TPIV	Γ/eV	
$L_{\gamma 5}$	L_2N_1	8166.19 ± 0.56 8166.20	[50] [104]	4.17(-3)	[488]	1.111(-3)	15.68	[423]
	L_2N_2	8242.6 ± 2.2	[145]				18.70	[423]
	L_2N_3	8286.50 8286.4 ± 1.1	[104] [138]	1.09(-4)	[488]	2.721(-5)	11.86	[423]
$L_{\gamma 1}$	L_2N_4	8418.94 ± 0.59 8418.00	[50] [104]	0.1132	[488]	2.820(-2)	8.30 7.95	[473] [423]
	L_2N_5	8420.27 ± 0.63	[145]				7.62	[423]
L_ν	L_2N_6	8570.38 ± 0.71	[145]			3.048(-5)		
	L_2N_7	8572.85 ± 0.72	[145]					
$L_{\gamma 8}$	L_2O_1	8575.30 ± 0.40 8521.70	[50] [104]	6.88(-4)	[488]	1.944(-4)	6.94	[423]
$L_{\gamma 6}$	L_2O_4	8575.30 ± 0.40 8579.90	[50] [104]					
L_l	L_3M_1	5743.05 ± 0.26 5741.90	[50] [104]	2.426(-2)	[488]	1.011(-2)	20.71	[423]
L_t	L_3M_2	5948.3 ± 1.5	[145]			7.118(-5)	12.57	[423]
L_s	L_3M_3	6114.70 6114.8 ± 1.5	[104] [384]	1.97(-4)	[488]	6.511(-5)	12.70	[423]
$L_{\alpha 2}$	L_3M_4	6457.72 ± 0.15 6457.70	[50] [104]	5.92(-2)	[488]	1.478(-2)	5.35 5.41	[473] [423]
$L_{\alpha 1}$	L_3M_5	6495.27 ± 0.15 6495.30	[50] [104]	0.523	[488]	1.306(-1)	4.81 4.79	[473] [423]
$L_{\beta 6}$	L_3N_1	7370.56 ± 0.45 7370.50	[50] [104]	5.76(-3)	[488]	1.445(-3)	15.55	[423]
	L_3N_2	7439.00	[104]	4.80(-5)	[488]	1.199(-5)	18.57	[423]
	L_3N_3	7494.80 7494.77 ± 0.91	[104] [138]	4.65(-5)	[488]	1.162(-5)	11.73	[423]
$L_{\beta 15,2}$	$L_3N_{4,5}$	7635.84 ± 0.49	[50]	9.90(-3)	[488]	2.472(-3)		
$L_{\beta 15}$	L_3N_4	7620.0 ± 2.1	[145]			2.472(-3)	7.82	[423]
$L_{\beta 2}$	L_3N_5	7629.92 ± 0.55	[145]	8.85(-2)	[488]	2.211(-2)	7.90 7.10	[473] [423]
$L_{u'}$	L_3N_6	7780.03 ± 0.62	[145]			4.329(-6)		
L_u	L_3N_7	7782.50 ± 0.65	[145]			2.456(-5)		

line	transition	E/eV		I/eV/$\hbar$		TPIV	Γ/eV	
$L_{\beta7}$	L_3O_1	7727.20 $\pm$ 0.30	[50]	9.48(-4)	[488]	2.730(-4)	6.81	[423]
		7726.40	[104]					
	$L_3O_{2,3}$	7764.60	[104]	6.44(-6)	[488]			
	L_3O_2					2.786(-6)		
	L_3O_3					2.365(-6)		
$L_{\beta5}$	$L_3O_{4,5}$	7805.50 $\pm$ 0.30	[50]					
		7785.90	[104]					

M series

$M_{\gamma2,1}$	$M_3N_{4,5}$	1522.00 $\pm$ 2.00	[50]	1.205(-2)	[335]			
		1522.00	[104]					
		1525.00	[263]					
	M_3N_4					3.160(-4)	13.07	[423]
	M_3N_5					2.540(-4)	12.35	[423]
$M_{\xi2}$	M_4N_2	1007.22	[423]			3.545(-3)	16.53	[423]
Mδ	M_4N_3	1045.04	[423]			4.385(-4)	9.69	[423]
M_β	M_4N_6	1325.00 $\pm$ 1.00	[50]	1.058(-2)	[335]	3.008(-4)		
		1324.70	[104]					
		1327.50	[263]					
M_η	M_4O_2	1303.54	[423]			6.710(-4)		
$M_{\xi1}$	M_5N_3	998.00 $\pm$ 1.60	[50]	6.83(-4)	[335]	3.501(-3)	9.06	[423]
		997.70	[104]					
		999.90	[263]					
$M_{\alpha1,2}$	$M_5N_{6,7}$	1293.00 $\pm$ 2.70	[50]	1.058(-2)	[335]			
		1295.60	[104]					
	M_5N_6					3.008(-4)		
	M_5N_7					5.817(-3)		
	M_5O_3	1272.10	[104]			8.290(-4)		

N series

	$N_{4,5}N_{6,7}$	149.00 $\pm$ 1.80	[50]	
	$N_{4,5}O_{2,3}$	128.00	[50]	

level characteristics

level	E_B/eV		ω_{nlj}		Γ/eV		AE/eV	
K	53788.00	[503]	0.941	[301]	25.20	[301]	53788.5	[51]
	53788.50 ± 0.50	[493]	0.943	[212]	24.69	[423]	53793.1 ± 3.5	[50]
			0.940	[38]	25.3	[94]	53792.3 ± 1.1	[145]
			0.9376	[39]				
			0.9401	[253]				
			0.941	[242]				
L_1	9047.00	[503]	8.90(-2)	[301]	5.25	[301]	9045.8	[51]
	9045.80 ± 0.50	[493]	8.29(-2)	[423]	4.85	[423]	9055.09 ± 0.98	[50]
			9.10(-2)	[436]	4.3	[94]	9048.0 ± 1.2	[145]
L_2	8581.00	[503]	0.178	[301]	4.55	[301]	8580.6	[51]
	8580.60 ± 0.40	[493]	0.1959	[423]	3.86	[423]	8583.06 ± 0.188	[50]
			0.197	[436]	4.01	[423]	8583.26 ± 0.42	[145]
L_3	7790.00	[503]	0.174	[301]	4.17	[301]	7790.1	[51]
	7790.10 ± 0.40	[493]	0.1820	[423]	3.73	[423]	7789.79 ± 0.72	[50]
			0.184	[436]	3.9	[94]	7792.90 ± 0.34	[145]
M_1	2047.00	[503]	1.43(-3)	[512]	17.00	[74]	2046.8	[51]
	2046.80 ± 0.40	[493]	1.52(-3)	[423]	16.99	[423]		
					13.2	[94]		
M_2	1842.00	[503]	1.10(-2)	[512]	10.00	[74]	1841.8	[51]
	1841.80 ± 0.50	[493]	3.35(-3)	[423]	8.86	[423]		
					6	[94]		
M_3	1676.00	[503]	1.47(-3)	[512]	10.00	[74]	1675.6	[51]
	1675.60 ± 0.90	[493]	3.89(-3)	[423]	8.97	[423]		
					8	[94]		
M_4	1332.00	[503]	9.27(-3)	[512]	1.40	[74]	1332.5	[51]
	1332.50 ± 0.40	[493]	9.81(-3)	[423]	1.69	[423]		
					1.065	[94]		
M_5	1295.00	[503]	9.87(-3)	[512]	2.30	[74]	1294.9	[51]
	1294.90 ± 0.40	[493]	1.05(-2)	[423]	1.06	[423]		
					1.065	[94]		

level	E_B/eV		ω_{nlj}		Γ/eV		AE/eV	
N_1	416.00	[503]	2.2(-4)	[423]	14.00	[74]		
	416.30 ± 0.50	[493]			11.82	[423]		
					5.4	[94]		
N_2	332.00	[503]	4.6(-4)	[423]	10.00	[74]		
	331.80 ± 0.60	[493]			14.84	[423]		
					5.3	[94]		
N_3	293.00	[503]	1.6(-4)	[423]	10.00	[74]		
	292.90 ± 0.60	[493]			8.00	[423]		
					2.8	[94]		
N_4	154.00	[503]	3.0(-5)	[423]	2.95	[94]		
	154.20 ± 0.50	[51]			4.10	[423]		
	161.40 ± 0.50	[493]			4.00	[74]		
N_5	154.00	[503]	3.0(-5)	[423]	2.6	[94]		
	154.20 ± 0.50	[51]			3.38	[423]		
	149.40 ± 0.50	[493]						
N_6	4.00	[503]						
	4.20 ± 1.60	[51]						
	5.50 ± 1.60	[493]						
N_7	4.00	[503]						
	4.20 ± 1.60	[51]						
	3.30 ± 1.60	[493]						
O_1	63.00	[503]	3.0(-5)	[423]	3.08	[423]		
	62.90 ± 0.50	[493]						
O_2	26.00	[503]						
	26.30 ± 0.60	[51]						
	28.20 ± 2.10	[493]						
O_3	26.00	[503]						
	22.90 ± 2.10	[493]						
IP	5.9389	[222]						

Ho   Z=67

[Xe] $4f^{11}$ $6s^2$

Holmium

$A = 164.93032(2)$ [222] $\varrho = 8.781$ g/cm³ [547]
$\varrho = 8.65$ g/cm³ [483]

X-ray transitions

line	transition	E/eV		I/eV/$\hbar$		TPIV	Γ/eV	
K series								
K_{α_3}	KL_1	46225.6 ± 1.8	[145]				31.20	[423]
K_{α_2}	KL_2	46699.70 ± 0.35	[50]	7.13	[488]	2.707(-1)	31.50	[301]
		46700.00	[104]				30.16	[423]
		46699.98 ± 0.15	[80]					
K_{α_1}	KL_3	47546.70 ± 0.40	[50]	12.69	[488]	4.821(-1)	31.50	[301]
		47547.00	[104]				30.04	[423]
		47547.10 ± 0.77	[80]					
	KM_1	53489.7 ± 2.5	[145]				43.66	[423]
K_{β_3}	KM_2	53711.00 ± 4.70	[50]	1.314	[488]	4.967(-2)	35.44	[423]
		53712.00	[104]					
		53711.3 ± 6.9	[80]					
K_{β_1}	KM_3	53877.00 ± 4.70	[50]	2.54	[488]	9.602(-2)	35.61	[423]
		53878.00	[104]					
		53877.1 ± 7.0	[80]					
K_{β_5}	$KM_{4,5}$	54247 ± 11	[50]	2.921(-2)	[488]			
		54246.00	[104]					
$K_{\beta_5^{II}}$	KM_4	54227.4 ± 2.0	[145]			8.528(-4)	28.06	[423]
$K_{\beta_5^{I}}$	KM_5	54267.1 ± 1.9	[145]			1.110(-3)	27.34	[423]
	KN_1	55188.2 ± 4.1	[145]				38.51	[423]
K_{β_2}	$KN_{2,3}$	55325 ± 73	[50]	0.562	[488]			
$K_{\beta_2^{II}}$	KN_2	55271.2 ± 4.4	[145]			1.102(-2)	42.38	[423]
$K_{\beta_2^{I}}$	KN_3	55312.1 ± 1.6	[145]			2.134(-2)	34.65	[423]
$K_{\beta_4^{II}}$	KN_4	55442.0 ± 2.6	[145]			1.936(-4)	30.93	[423]
$K_{\beta_4^{I}}$	KN_5	554543.7 ± 1.3	[145]			2.509(-4)	30.16	[423]
	$KO_{2,3}$	55584.00 ± 7.50	[50]	0.1098	[488]			
	KO_2					2.438(-3)		
	KO_3					5.071(-3)		

line	transition	E/eV		I/eV/$\hbar$		TPIV	Γ/eV	
	L series							
	L_1M_1	7264.1 ± 2.6	[145]				22.43	[423]
$L_{\beta 4}$	L_1M_2	7471.1 ± 1.3	[50]	0.1561	[488]	2.758(-2)	14.21	[423]
		7471.10	[104]					
$L_{\beta 3}$	L_1M_3	7651.8 ± 1.4	[50]	0.2150	[488]	3.798(-2)	14.38	[423]
		7652.20	[104]					
$L_{\beta 10}$	L_1M_4	8006.00 ± 1.60	[50]	3.67(-3)	[488]	6.481(-4)	6.83	[423]
		8006.10	[104]					
		8004.08 ± 0.15	[440]					
$L_{\beta 9}$	L_1M_5	8044.61 ± 0.14	[440]			9.701(-4)	6.10	[423]
	L_1N_1	8962.6 ± 4.2	[145]				17.28	[423]
$L_{\gamma 2}$	L_1N_2	9051.1 ± 2.0	[50]	3.72(-2)	[488]	6.571(-3)	21.15	[423]
		9050.90	[104]					
$L_{\gamma 3}$	L_1N_3	9087.6 ± 2.0	[50]	5.35(-2)	[488]	9.456(-3)	13.41	[423]
		9087.60	[104]					
$L_{\gamma 11}$	$L_1N_{4,5}$	9232.10	[104]	4.99(-4)	[488]			
	L_1N_4	9216.4 ± 2.7	[145]			8.809(-5)	9.70	[423]
	L_1N_5	9228.1 ± 1.5	[145]			1.362(-4)	8.93	[423]
	L_1N_6	9382.7 ± 1.5	[145]					
	L_1N_7	9385.0 ± 1.6	[145]					
	L_1O_1	9345.10	[104]	6.67(-8)	[488]		8.84	[423]
$L_{\gamma 4}$	$L_1O_{2,3}$	9374.00 ± 1.40	[50]	6.92(-3)	[488]			
		9374.50	[104]					
	L_1O_2					8.826(-4)		
	L_1O_3					1.223(-3)		
	$L_1O_{4,5}$	9387.00 ± 2.00	[50]					
		9386.90	[104]					
L_η	L_2M_1	6788.30 ± 0.80	[50]	1.873(-2)	[488]	5.582(-3)	21.39	[423]
		6788.40	[104]					
		9786.94 ± 0.27	[366]					
	L_2M_2	6995.9 ± 1.6	[145]				13.17	[423]
$L_{\beta 17}$	L_2M_3	7176.5 ± 1.7	[145]			1.276(-4)	13.34	[423]
$L_{\beta 1}$	L_2M_4	7525.30 ± 0.90	[50]	0.696	[488]	1.700(-1)	5.79	[423]
		7525.50	[104]					
		7525.67 ± 0.15	[366]					
	L_2M_5	7566.1 ± 1.2	[145]				5.06	[423]
$L_{\gamma 5}$	L_2N_1	8481.5 ± 1.7	[50]	4.50(-3)	[488]	1.167(-3)	16.24	[423]
		8481.60	[104]					
	L_1N_2	8570.2 ± 3.7	[145]				20.10	[423]

line	transition	E/eV		I/eV/$\hbar$		TPIV	Γ/eV	
	L_2N_3	8586.50	[104]	1.22(-4)	[488]	2.990(-5)	12.37	[423]
		8611.15 $\pm$ 0.93	[145]					
$L_{\gamma 1}$	L_2N_4	8747.2 $\pm$ 1.8	[50]	0.1226	[488]	2.995(-2)	8.66	[423]
	L_2N_5	8752.71 $\pm$ 0.63	[145]				7.89	[423]
L_ν	L_2N_6	8907.28 $\pm$ 0.70	[145]			3.808(-5)		
	L_2N_7	8909.60 $\pm$ 0.74	[145]					
$L_{\gamma 8}$	L_2O_1	8867.00 $\pm$ 1.30	[50]	7.38(-4)	[488]	2.074(-4)	7.80	[423]
		8866.90	[104]					
$L_{\gamma 6}$	L_2O_4	8905.00 $\pm$ 1.30	[50]					
		8905.00	[104]					
L_l	L_3M_1	5943.40 $\pm$ 0.60	[50]	2.676(-2)	[488]	1.039(-2)	21.27	[423]
		5943.50	[104]					
		5939.963 $\pm$ 0.71	[366]					
L_t	L_3M_2	6161.20	[104]	2.39(-4)	[488]	7.463(-5)	13.05	[423]
		6149.4 $\pm$ 1.6	[145]					
L_s	L_3M_3	6333.50	[104]	2.19(-4)	[488]	6.819(-5)	13.22	[423]
		6330.0 $\pm$ 1.6	[145]					
$L_{\alpha 2}$	L_3M_4	6679.50 $\pm$ 0.70	[50]	6.36(-2)	[488]	1.555(-2)	5.67	[423]
		6679.60	[104]					
		6678.484 $\pm$ 0.054	[366]					
$L_{\alpha 1}$	L_3M_5	6719.80 $\pm$ 0.70	[50]	0.562	[488]	1.374(-1)	4.95	[423]
		6719.90	[104]					
		6719.675 $\pm$ 0.062	[366]					
$L_{\beta 6}$	L_3N_1	7635.8 $\pm$ 1.4	[50]	6.27(-3)	[488]	1.537(-3)	16.12	[423]
		7636.00	[104]					
	L_3N_2	7727.00	[104]	5.35(-5)	[488]	1.306(-5)	19.99	[423]
		7727.16 $\pm$ 0.48	[387]					
	L_3N_3	7763.00	[104]	5.19(-5)	[488]	1.267(-5)	12.25	[423]
		7763.52 $\pm$ 0.49	[387]					
$L_{\beta 15}$	L_3N_4	7911.00 $\pm$ 1.00	[50]	1.067(-2)	[488]	2.606(-3)	8.54	[423]
		7902.86 $\pm$ 0.25	[387]					
$L_{\beta 2}$	L_3N_5	7911.20 $\pm$	[104]	9.54(-2)	[488]	2.331(-2)	7.77	[423]
		7911.35 $\pm$ 0.25	[387]					
$L_{\beta 15,2}$	$L_3N_{4,5}$	7903.60	[104]					
$L_{u'}$	L_3N_6	8060.83 $\pm$ 0.62	[145]			5.378(-6)		
L_u	L_3N_7	8063.15 $\pm$ 0.65	[145]			3.056(-5)		
$L_{\beta 5}$	$L_3O_{4,5}$	8062.00 $\pm$ 1.00	[50]					
		8062.30	[104]					

line	transition	E/eV		I/eV/$\hbar$		TPIV	Γ/eV	
M series								
$M_{\gamma_{2,1}}$	$M_3N_{4,5}$	1567.00 ± 2.00	[50]	1.413(-2)	[116]			
		1576.30	[104]	1.321(-2)	[335]			
		1579.40	[263]					
M_{γ_2}	M_3N_4	1554.90	[423]			3.363(-4)	14.11	[423]
M_{γ_1}	M_3N_5	1562.30	[423]			2.696(-3)	13.34	[423]
M_{ξ_2}	M_4N_2	1052.15	[423]			3.276(-3)	18.01	[423]
M_δ	M_4N_3	1093.00	[423]			3.934(-4)	10.27	[423]
M_β	M_4N_6	1383.00 ± 0.60	[50]	1.453(-2)	[116]	6.399(-3)		
		1382.90	[104]	1.287(-2)	[335]			
		1385.30	[263]					
M_η	M_4O_2	1361.30	[423]			6.399(-4)		
M_{ξ_1}	M_5N_3	1045.00 ± 0.90	[50]	9.93(-4)	[488]	3.158(-3)	9.55	[423]
		1045.10	[104]	7.35(-4)	[335]			
$M_{\alpha 1,2}$	$M_5N_{6,7}$	1348.00 ± 3.00	[50]	1.360(-2)	[116]			
		1350.60	[263]	1.287(-2)	[335]			
M_{α_2}	M_5N_6	1344.50	[423]			3.755(-4)		
M_{α_1}	M_5N_7	1345.60	[423]			7.259(-3)		

level characteristics

level	E_B/eV		ω_{nlj}		Γ/eV		AE/eV	
K	56618.00	[503]	0.944	[301]	26.80	[301]	55617.7	[51]
	55617.70 ± 0.50	[493]	0.943	[38]	26.22	[423]	55619.9 ± 3.7	[50]
			0.9401	[39]	26.9	[94]	55620.8 ± 1.1	[145]
			0.9415	[423]				
			0.944	[310]				
			0.939	[33]				
			0.94	[112]				

level	E_B/eV		ω_{nlj}		Γ/eV		AE/eV	
L_1	9395.00	[503]	9.40(-2)	[301]	5.33	[301]	9394.2	[51]
	9394.20 ± 0.40	[493]	0.112	[348]	4.98	[423]	9399.7 ± 1.1	[50]
			9.40(-29	[128]	4.5	[94]	9395.2 ± 1.2	
			8.67(-2)	[423]				
			9.50(-2)	[436]				
			9.64(-29	[244]				
L_2	8919.00	[503]	0.189	[301]	4.66	[301]	8917.8	[51]
			0.203	[348]	3.94	[423]	8916.38 ± 0.95	[50]
	8917.80 ± 0.40	[493]	0.170	[240]			8919.77 ± 0.42	[145]
			0.2071	[423]	4.09	[94]		
			0.208	[436]				
			0.21	[244]				
			0.226	[476]				
L_3	8071.00	[503]	0.182	[301]	4.26	[301]	8071.1	[51]
			0.201	[348]	3.82	[423]	8067.56 ± 0.78	[50]
	8071.10 ± 0.40	[493]	0.22	[269]	4	[94]	8073.31 ± 0.34	[145]
			0.169	[197]				
			0.1913	[423]				
			0.193	[436]				
			0.196	[244]				
			0.9415	[423]				
M_1	2128.00	[503]	1.08(-3)	[38]	18.30	[350]	2128.3	[51]
	2128.30 ± 0.60	[493]	1.49(-3)	[512]	17.45	[423]		
			1.60(-3)	[423]	13.4	[423]		
			0.9415	[423]				
M_2	1923.00	[503]	1.85(-3)	[51]	10.40	[350]	1922.8	[51]
	1922.80 ± 1.00	[493]	1.20(-2)	[512]	9.22	[423]		
			3.52(-3)	[423]	6.3	[94]		
			1.72(-3)	[245]				
M_3	1741.00	[503]	1.45(-3)	[38]	10.20	[350]	1741.2	[51]
	1741.20 ± 0.90	[493]	1.53(-3)	[512]	9.40	[423]		
			4.12(-3)	[423]	7.8	[94]		
			1.68(-3)	[245]				
M_4	1391.00	[503]	1.45(-3)	[38]	2.41	[350]	1391.5	[51]
	1391.50 ± 0.70	[493]	9.86(-3)	[512]	1.85	[423]		
			1.08(-2)	[423]	1.13	[94]		
M_5	1351.00	[503]	6.70(-3)	[38]	1.38	[350]	1351.4	[51]
	1351.40 ± 0.80	[493]	1.10(-2)	[512]	1.13	[423]		
			1.16(-2)	[423]	1.13	[94]		
N_1	436.00	[503]	2.4(-4)	[423]	5.6	[94]		
	435.70 ± 0.80	[493]			12.30	[423]		
N_2	343.00	[503]	5.0(-4)	[423]	5.33	[94]		
	343.50 ± 1.40	[493]			16.16	[423]		
N_3	306.00	[503]	1.8(-4)	[423]	2.95	[94]		
	306.60 ± 0.90	[493]			8.43	[423]		

level	E_B/eV		ω_{nlj}		Γ/eV		AE/eV	
N_4	161.00	[503]	4.0(-5)	[423]	3.15	[94]		
	161.00 ± 1.00	[51]			4.72	[423]		
	161.80 ± 1.00	[493]						
N_5	161.00	[503]	4.0(-5)	2.8		[94]		
	156.50 ± 1.00	[493]			3.95	[423]		
N_6	4.00	[503]						
	3.70 ± 3.00	[51]						
	4.80 ± 3.00	[493]						
N_7	4.00	[503]						
	3.70 ± 3.00	[51]						
	2.80 ± 3.00	[493]						
O_1	51.00	[503]	1.0(-6)	[423]	3.85	[423]		
	51.20 ± 1.30	[493]						
O_2	20.00	[503]						
	20.30 ± 1.50	[51]						
	24.90 ± 1.40	[493]						
O_3	20.00	[503]						
	19.50 ± 1.40	[493]						
IP	6.0215	[222]						

Er   Z=68

[Xe] $4f^{12} 6s^2$

Erbium

$A = 167.259(3)$ [222] $\varrho = 9.045$ g/cm^3 [547]

$\varrho = 9.01$ g/cm^3 [483]

X-ray transitions

line	transition	E/eV		I/eV/$\hbar$		TPIV	Γ/eV	
K series								
K_{α_3}	KL_1	47736.8 ± 1.9	[145]				32.95	[423]
K_{α_2}	KL_2	48221.10 ± 0.40	[50]	7.59	[488]	2.717(-1)	33.10	[301]
		48222.00	[104]	7.66	[487]		35.50	[473]
		48221.90 ± 0.20	[277]				31.86	[423]
		48221.61 ± 0.20	[144]					
K_{α_1}	KL_3	49127.70 ± 0.40	[50]	13.47	[488]	4.821(-1)	32.70	[301]
		49128.00	[104]	13.61	[487]		37.40	[473]
		49127.56 ± 0.12	[277]				31.76	[423]
		49127.24 ± 0.12	[144]					
	KM_1	55270.1 ± 2.7	[145]				45.71	[423]
K_{β_3}	KM_2	55494.00 ± 5.00	[50]	1.403	[488]	5.022(-2)	35.70	[473]
		55495.00	[104]	1.469	[487]		37.41	[423]
		55480.08 ± 0.35	[277]					
		55479.72 ± 0.35	[144]					
K_{β_1}	KM_3	55681.00 ± 5.00	[50]	2.712	[488]	9.706(-2)	36.20	[473]
		55682.00	[104]	2.840	[487]		37.48	[423]
		55673.88 ± 0.18	[277]					
		55673.52 ± 0.18	[144]					
K_{β_5}	$KM_{4,5}$	56040 ± 11	[50]	3.22(-2)	[488]			
		56038.00	[104]					
$K_{\beta_5^{II}}$	KM_4	56034.5 ± 2.0	[145]			8.893(-4)	29.84	[423]
$K_{\beta_5^{I}}$	KM_5	56076.6 ± 2.0	[145]			1.153(-3)	29.03	[423]
	KN_1	57037.5 ± 4.1	[145]				40.61	[423]
$K_{\beta_2^{II}}$	KN_2	57123.3 ± 4.3	[145]	0.334	[487]	1.112(-2)	44.51	[423]
$K_{\beta_2^{I}}$	KN_3	57167.0 ± 1.7	[145]	0.651	[487]	2.151(-2)	36.68	[423]
K_{β_2}	$KN_{2,3}$	57214 ± 78	[50]	0.601	[488]			
$K_{\beta_4^{II}}$	KN_4	57301.1 ± 2.3	[145]			2.029(-4)	33.45	[423]

line	transition	E/eV		I/eV/$\hbar$		TPIV	Γ/eV	
$K_{\beta_4^I}$	KN_5	57313.3 ± 1.4	[145]			2.618(-4)	32.38	[423]
	$KO_{2,3}$	57450.00 ± 8.00	[50]	0.1162	[488]			
		57448.00	[104]	0.1285	[487]			
	KO_2					2.477(-3)		
	KO_3					5.136(-3)		

L series

line	transition	E/eV		I/eV/$\hbar$		TPIV	Γ/eV	
	L_1M_1	7540.60	[104]	1.10(-6)	[488]		22.99	[423]
		7533.4 ± 2.8	[145]					
$L_{\beta 4}$	L_1M_2	7745.30 ± 0.50	[50]	0.1688	[488]	2.885(-2)	10.03	[473]
		7744.70	[104]				14.70	[423]
		7744.75 ± 0.14	[366]					
$L_{\beta 3}$	L_1M_3	7939.20 ± 0.50	[50]	0.2296	[488]	3.925(-2)	11.85	[473]
		7938.00	[104]				14.77	[423]
		7939.007 ± 0.086	[366]					
$L_{\beta 10}$	L_1M_4	8298.1 ± 2.5	[50]	4.09(-3)	[488]	6.998(-4)	7.12	[423]
		8301.70	[104]					
$L_{\beta 9}$	L_1M_5	8340.66 ± 0.56	[50]	6.13(-3)	[488]	1.048(-3)	6.31	[423]
		8347.00	[104]					
	L_1N_1	9300.7 ± 4.2	[145]				6.40	[423]
	L_1N_1	9292.00	[104]	4.43(-7)	[488]		17.89	[423]
$L_{\gamma 2}$	L_1N_2	9385.5 ± 2.1	[50]	4.03(-2)	[488]	6.894(-3)	21.79	[423]
		9385.90	[104]					
$L_{\gamma 3}$	L_1N_3	9431.2 ± 1.1	[50]	5.74(-2)	[488]	9.808(-3)	13.97	[423]
		9431.10	[104]					
	L_1N_4	9564.3 ± 2.4	[145]			9.525(-5)	10.74	[423]
	L_1N_5	9576.5 ± 1.5	[145]			1.475(-4)	9.67	[423]
	$L_1N_{4,5}$	9569.15 ± 0.74	[394]					
	L_1N_6	9738.6 ± 1.6	[145]					
	L_1N_7	9740.7 ± 1.6	[145]					
	L_1O_1	9690.20	[104]	7.98(-8)	[488]		9.39	[423]
$L_{\gamma 4}$	$L_1O_{2,3}$	9722.00 ± 1.50	[104]	5.36(-3)	[488]			
		9723.40	[104]					
	L_1O_2					9.169(-4)		
	L_1O_3					1.253(-3)		
	$L_1O_{4,5}$	9752.30	[104]					

line	transition	E/eV		I/eV/$\hbar$		TPIV	Γ/eV	
L_η	L_2M_1	7057.90 ± 0.40	[50]	2.015(-3)	[488]	5.720(-3)	21.90	[423]
		7057.20	[104]					
		7045.167 ± 0.090	[366]					
	L_2M_2	7259.3 ± 1.7	[145]				13.62	[423]
$L_{\beta 17}$	L_2M_3	7425.20	[104]	5.48(-4)	[488]	1.369(-4)	13.68	[423]
		7461.31 ± 0.45	[393]					
$L_{\beta 1}$	L_2M_4	7810.90 ± 0.50	[50]	0.750	[488]	1.796(-1)	6.04	[423]
		7810.90	[104]					
		7810.19 ± 0.42	[366]					
	L_2M_5	7853.6 ± 1.2	[145]				5.22	[423]
$L_{\gamma 5}$	L_2N_1	8813.7 ± 2.8	[50]	4.85(-3)	[488]	1.225(-3)	16.81	[423]
		8813.30	[104]					
	L_2N_2	8900.3 ± 3.5	[145]				20.71	[423]
	L_2N_3	8946.80	[104]	1.37(-4)	[488]	3.275(-5)	12.88	[423]
		8946.94 ± 0.65	[381]					
$L_{\gamma 1}$	L_2N_4	9088.9 ± 2.0	[50]	0.1326	[488]	3.176(-2)	9.65	[423]
		9084.00	[104]					
	L_2N_5	9090.26 ± 0.63	[145]				8.58	[423]
L_ν	L_2N_6	9252.38 ± 0.70	[145]			4.622(-5)		
	L_2N_7	9254.50 ± 0.74	[145]					
$L_{\gamma 8}$	L_2O_1	9255.00 ± 2.00	[50]	7.91(-4)	[488]	2.205(-4)	8.31	[423]
		9209.20	[104]					
	$L_2O_{2,3}$	9234.50	[104]	1.75(-5)	[488]			
	L_2O_3					7.272(-6)		
$L_{\gamma 6}$	L_2O_4	9255.00 ± 2.00	[104]					
		9254.40	[104]					
L_l	L_3M_1	6152.00 ± 3.00	[50]	2.907(-2)	[488]	1.068(-2)	21.80	[423]
		6153.20	[104]					
		6138.86 ± 0.14	[366]					
L_t	L_3M_2	6353.6 ± 1.6	[145]			7.874(-5)	13.51	[423]
L_s	L_3M_3	6547.6 ± 1.6	[145]			7.157(-5)	13.57	[423]
$L_{\alpha 2}$	L_3M_4	6905.00 ± 0.80	[50]	6.83(-2)	[488]	1.634(-2)	5.73	[473]
		6906.90	[104]				5.93	[423]
		6904.50 ± 0.17	[366]					
$L_{\alpha 1}$	L_3M_5	6948.70 ± 0.35	[50]	0.604	[488]	1.443(-1)	5.17	[473]
		6949.90	[104]				5.12	[423]
		6947.913 ± 0.077	[366]					
$L_{\beta 6}$	L_3N_1	7909.6 ± 1.5	[50]	6.83(-3)	[488]	1.636(-3)	16.70	[423]
		7908.40	[104]					
	L_3N_2	7994.6 ± 3.5	[145]			1.420(-5)	20.60	[423]
	L_3N_3	8038.29 ± 0.84	[145]			1.379(-5)	12.77	[423]

line	transition	E/eV		I/eV/$\hbar$		TPIV	Γ/eV	
$L_{\beta15,2}$	$L_3N_{4,5}$	8189.11 ± 0.72	[50]	1.147(-2)	[488]	2.744(-3)		
$L_{\beta15}$	L_3N_4	8189.70	[104]			2.744(-3)	9.54	[423]
		8172.4 ± 1.5	[145]					
$L_{\beta2}$	L_3N_5	8189.00 ± 0.50	[50]	0.1026	[488]	2.455(-2)	8.47	[423]
		8189.80	[104]					
$L_{u'}$	L_3N_6	8346.63 ± 0.63	[145]			6.647(-6)		
L_u	L_3N_7	8348.75 ± 0.65	[145]			3.729(-5)		
$L_{\beta7}$	L_3O_1	8298.00 ± 2.00	[50]	1.11(-3)	[488]	3.152(-4)	8.20	[423]
		8296.00	[104]					
$L_{\beta5}$	$L_3O_{4,5}$	8350.00 ± 2.00	[50]					
		8350.20	[104]					

M series

line	transition	E/eV		I/eV/$\hbar$		TPIV	Γ/eV	
	M_2N_4	1830	[565]			2.512(-3)	15.20	[423]
$M_{\gamma2}$	M_3N_4	1632.00 ± 2.20	[50]	1.44(-2)	[335]	2.512(-3)	15.20	[423]
		1635.70	[263]					
$M_{\gamma1}$	M_3N_5	1643.00 ± 2.00	[50]	1.44(-2)	[335]	2.857(-3)	14.20	[423]
		1646.50	[263]					
$M_{\xi2}$	M_4N_2	1097.96	[423]			2.992(-3)	18.68	[423]
M_{δ}	M_4N_3	1142.03	[423]			3.515(-4)	10.86	[423]
M_{β}	M_4N_6	1443.00 ± 0.50	[50]	1.55(-2)	[335]	7.822(-3)		
		1443.40	[104]					
		1446.70	[263]					
M_{η}	M_4O_2	1420.27	[423]			6.031(-4)		
$M_{\xi1}$	M_5N_3	1090.10 ± 1.00	[50]	7.89(-4)	[335]	2.814(-3)	10.05	[423]
		1090.30	[104]					
		1092.40	[263]					
$M_{\alpha1.2}$	$M_5N_{6,7}$	1406.00 ± 1.60	[50]	1.55(-2)	[335]			
		1408.90	[263]					
	M_5N_6	1401.06	[423]			4.532(-4)		
	M_5N_7	1402.24	[423]			8.762(-3)		

N series

line	transition	E/eV						
	N_4N_6	171.00 ± 2.00	[50]					
	$N_5N_{6,7}$	163.00 ± 1.50	[50]					

level characteristics

level	E_B/eV		ω_{nlj}		Γ/eV		AE/eV	
K	57486.00	[503]	0.947	[301]	28.40	[301]	57485.5	[51]
	57485.50 ± 0.50	[493]	0.945	[38]	27.83	[423]	57485.2 ± 2.0	[330]
			0.9425	[39]	28.5	[94]	57487.4 ± 1.2	[145]
			0.9438	[423]				
L_1	9752.00	[503]	0.100	[301]	5.43	[301]	9751.3	[51]
	9751.30 ± 0.40	[493]	9.03(-2)	[423]	5.12	[423]	9757.8 ± 1.1	[50]
			0.105	[436]	4.7	[94]	9750.6 ± 1.3	[145]
L_2	9265.00	[503]	0.200	[301]	4.73	[301]	9264.3	[51]
	9264.30 ± 0.40	[493]	0.185	[240]	4.03	[423]	9262.1 ± 1.0	[50]
			0.2189	[423]	4.18	[94]	9264.40 ± 0.42	[145]
			0.219	[436]				
L_3	8358.00	[503]	0.192	[301]	4.35	[301]	8357.9	[51]
	8357.90 ± 0.40	[493]	0.172	[240]	3.92	[423]	8357.42 ± 0.83	[50]
			0.2008	[423]	4.1	[94]	8358.66 ± 0.34	[145]
			0.203	[436]				
M_1	2207.00	[503]	1.55(-3)	[512]	19.00	[74]	2206.5	[51]
	2206.70 ± 1.50	[493]	1.70(-3)	[423]	17.87	[423]		
					13.6	[94]		
M_2	2006.00	[503]	1.20(-2)	[512]	11.00	[74]	2005.8	[51]
	2005.80 ± 0.60	[493]	3.68(-3)	[423]	9.58	[423]		
					6.6	[94]		
M_3	1812.00	[503]	1.59(-3)	[512]	11.00	[74]	1811.8	[51]
	1811.80 ± 0.60	[493]	4.36(-3)	[423]	9.65	[423]		
					7.5	[94]		
M_4	1453.00	[503]	1.10(-2)	[512]	1.40	[74]	1453.3	[51]
	1453.30 ± 0.30	[493]	1.18(-2)	[423]	2.01	[423]		
					1.2	[94]		
M_5	1409.00	[503]	1.308-2)	[512]	2.50	[74]	1409.3	[51]
	1409.30 ± 0.50	[493]	1.27(-2)	[423]	1.20	[423]		
					1.2	[94]		

level	E_B/eV		ω_{nlj}		Γ/eV		AE/eV
N_1	449.00	[503]	2.6(-4)	[423]	15.00	[74]	
	449.10 ± 1.00	[493]			12.78	[423]	
					5.8	[94]	
N_2	366.00	[503]	5.4(-4)	[423]	12.00	[74]	
	366.20 ± 1.50	[493]			16.68	[423]	
					5.36	[94]	
N_3	320.00	[503]	1.9(-4)	[423]	12.00	[74]	
	320.00 ± 0.70	[493]			8.85	[423]	
					3.15	[94]	
N_4	177.00	[503]	5.0(-5)	[423]	5.80	[74]	
	176.70 ± 1.20	[493]			5.62	[423]	
					3.35	[94]	
N_5	168.00	[503]	5.0(-5)	[423]	5.80	[74]	
	167.60 ± 1.50	[493]			4.55	[423]	
					2.95	[94]	
N_6	4.00	[503]					
	4.30 ± 1.40	[493]					
N_7	4.00	[503]					
	3.60 ± 1.40	[493]					
O_1	60.00	[503]	1.0(-6)	[423]	4.28	[423]	
	59.80 ± 1.70	[493]					
O_2	29.00	[503]					
	29.40 ± 1.60	[51]					
	27.90 ± 3.00	[493]					
O_3	29.00	[503]					
	22.30 ± 3.00	[493]					
IP	6.1077	[222]					

Tm   Z=69

[Xe] 4f^{13} 6s^2

Thulium

$A = 168.93421(2)$ [222] $\varrho = 9.314$ g/cm^3 [547]

$\varrho = 9.20$ g/cm^3 [483]

X-ray transitions

line	transition	E/eV		I/eV/$\hbar$		TPIV	Γ/eV	
K series								
K$_{\alpha_3}$	KL$_1$	49276.7 $\pm$ 1.9	[145]				35.67	[423]
		49273.3	[104]					
K$_{\alpha2}$	KL$_2$	49772.60 $\pm$ 0.40	[50]	8.08	[488]	2.727(-1)	34.90	[301]
		49773.00	[104]				33.63	[423]
		49773.00 $\pm$ 0.12	[277]					
		49772.67 $\pm$ 0.12	[78]					
K$_{\alpha1}$	KL$_3$	50741.60 $\pm$ 0.40	[50]	14.28	[488]	4.820(-1)	34.60	[301]
		50742.00	[104]				33.54	[423]
		50742.80 $\pm$ 0.09	[277]					
		50741.475 $\pm$ 0.092	[78]					
	KM$_1$	57085.6 $\pm$ 2.8	[145]				47.77	[423]
K$_{\beta3}$	KM$_2$	57304.00 $\pm$ 1.00	[50]	1.496	[488]	5.051(-2)	39.45	[423]
		57304.00	[104]					
		57303.0 $\pm$ 7.9	[78]					
K$_{\beta1}$	KM$_3$	57517.00 $\pm$ 1.00	[50]	2.892	[488]	9.762(-2)	39.57	[423]
		57517.00	[104]					
		57509.14 $\pm$ 0.15	[277]					
		57508.75 $\pm$ 0.15	[78]					
K$_{\beta5}$	KM$_{4,5}$	57924.8 $\pm$ 8.0	[50]	3.54(-2)	[488]	0.25		
		57942.00	[104]					
K$_{\beta_5^{II}}$	KM$_4$	57877.6 $\pm$ 2.1	[145]			9.264(-4)	31.68	[423]
K$_{\beta_5^I}$	KM$_5$	57922.3 $\pm$ 2.0	[145]			1.196(-3)	30.78	[423]
	KN$_1$	58923.4 $\pm$ 4.0	[145]				42.77	[423]
K$_{\beta2}$	KN$_{2,3}$	59095 $\pm$ 83	[50]	0.642	[488]			
K$_{\beta_2^{II}}$	KN$_2$	59012.3 $\pm$ 4.3	[145]			1.121(-2)	47.50	[423]
K$_{\beta_2^I}$	KN$_3$	59058.9 $\pm$ 1.7	[145]			2.168(-2)	38.76	[423]
K$_{\beta_4^{II}}$	KN$_4$	59197.7 $\pm$ 2.1	[145]			2.125(-4)	35.87	[423]

line	transition	E/eV		I/eV/ℏ		TPIV	Γ/eV	
$K_{\beta_4^I}$	KN_5	59209.8 ± 1.5	[145]			2.727(-4)	34.69	[423]
	$KO_{2,3}$	59346.00 ± 6.00	[50]	0.1228	[488]			
		59346.00	[104]					
	KO_2					2.508(-3)		
	KO_3					5.201(-3)		

L series

line	transition	E/eV		I/eV/ℏ		TPIV	Γ/eV	
	L_1M_1	7814.0 ± 2.5	[383]				24.42	[423]
		7808.9 ± 2.9	[145]					
$L_{\beta 4}$	L_1M_2	8025.8 ± 1.5	[50]	0.1823	[488]	2.639(-2)	16.10	[423]
		8025.90	[104]					
$L_{\beta 3}$	L_1M_3	8230.9 ± 1.6	[50]	0.2448	[488]	3.543(-2)	16.23	[423]
		8230.70	[104]					
$L_{\beta 10}$	L_1M_4	8600.8 ± 1.5	[50]	4.56(-3)	[488]	6.601(-4)	8.33	[423]
		8603.70	[104]					
$L_{\beta 9}$	L_1M_5	8648.6 ± 1.5	[50]	6.83(-3)	[488]	9.882(-4)	7.43	[423]
		8648.20	[104]					
	L_1N_1	9643.00	[104]	5.32(-7)	[488]		19.43	[423]
		9646.7 ± 4.1	[145]					
$L_{\gamma 2}$	L_1N_2	9730.2 ± 2.3	[50]	4.37(-2)	[488]	6.320(-3)	24.15	[423]
		9730.00	[104]					
$L_{\gamma 3}$	L_1N_3	9779.3 ± 2.3	[50]	6.14(-2)	[488]	8.887(-3)	15.41	[423]
		9778.80	[104]					
$L_{\gamma 11}$	$L_1N_{4,5}$	9927.00	[104]	9.63(-4)	[488]			
	L_1N_4	9921.0 ± 2.2	[145]			8.995(-5)	12.52	[423]
	L_1N_5	9933.1 ± 1.5	[145]			1.393(-4)	11.35	[423]
	L_1N_6	10102.9 ± 1.6	[145]					
	L_1N_7	10104.8 ± 1.7	[145]					
	L_1O_1	10054.20	[104]	9.52(-8)	[488]		10.86	[423]
$L_{\gamma 4}$	$L_1O_{2,3}$	10084.00 ± 1.60	[50]	7.75(-3)	[488]			
		10084.00	[104]					
	L_1O_2					8.324(-4)		
	L_1O_3					1.122(-3)		
	$L_1O_{4,5}$	10110.00 ± 2.50	[50]					
		10110.00	[104]					

line	transition	E/eV		I/eV/$\hbar$		TPIV	Γ/eV	
L_η	L_2M_1	7308.80 ± 0.90	[50]	2.165(-2)	[488]	5.858(-3)	22.38	[423]
		7309.00	[104]					
		7309.30 ± 0.56	[366]					
	L_2M_2	7529.0 ± 1.7	[145]				14.06	[423]
$L_{\beta17}$	L_2M_3	7728.00	[104]	6.09(-3)	[488]	1.475(-4)	14.18	[423]
		7728.6 ± 2.4	[383]					
$L_{\beta1}$	L_2M_4	8101.00 ± 1.10	[50]	0.806	[488]	1.894(-1)	6.29	[423]
		8101.20	[104]					
		8102.265 ± 0.037	[366]					
	L_2M_5	8146.00	[104]	2.24(-5)	[488]		5.39	[423]
		8146.0 ± 1.6	[135]					
$L_{\gamma5}$	L_2N_1	9144.6 ± 2.0	[50]	5.23(-3)	[488]	1.284(-3)	17.38	[423]
		9144.60	[104]					
	L_2N_2	9225.30	[104]	1.29(-7)	[488]		22.11	[423]
		9225.2 ± 2.1	[135]					
	L_2N_3	9292.00	[104]	1.52(-4)	[488]	3.585(-5)	13.37	[423]
		9285.08 ± 0.92	[145]					
$L_{\gamma1}$	L_2N_4	9426.2 ± 2.1	[50]	0.1431	[488]	3.364(-2)	10.48	[423]
		9425.80	[104]					
	L_2N_5	9436.02 ± 0.63	[145]				9.30	[423]
L_ν	L_2N_6	9605.83 ± 0.72	[145]			5.592(-5)		
	L_2N_7	9607.72 ± 0.76	[145]					
$L_{\gamma8}$	L_2O_1	10052.60	[104]	8.46(-4)	[488]	2.338(-4)	8.82	[423]
$L_{\gamma6}$	L_2O_4	9607.00 ± 1.50	[50]					
		9607.90	[104]					
L_l	L_3M_1	6341.96 ± 0.83	[50]	3.15(-2)	[488]	1.097(-2)	22.28	[423]
		6342.00	[104]					
L_t	L_3M_2	6557.40	[104]	2.95(-4)	[488]	8.300(-5)	13.97	[423]
		6557.5 ± 1.7	[383]					
L_s	L_3M_3	6768.9 ± 1.6	[145]				14.54	[423]
$L_{\alpha2}$	L_3M_4	7133.10 ± 0.80	[50]	7.33(-2)	[488]	1.715(-2)	6.20	[423]
		7133.30	[104]					
		7133.715 ± 0.078	[366]					
$L_{\alpha1}$	L_3M_5	7179.90 ± 0.80	[50]	0.647	[488]	1.515(-1)	5.30	[423]
		7177.80	[104]					
		7180.113 ± 0.029	[366]					

line	transition	E/eV		I/eV/ℏ		TPIV	Γ/eV	
$L_{\beta 6}$	L_3N_1	8177.2 ± 1.6	[50]	7.42(-3)	[488]	1.739(-3)	17.29	[423]
		8177.10	[104]					
	L_3N_2	8263.10	[104]	6.85(-5)	[488]	1.541(-5)	22.02	[423]
		8263.1 ± 1.7	[135]					
	L_3N_3	8316.72 ± 0.83	[145]			1.498(-5)		
$L_{\beta 15,2}$	$L_3N_{4,5}$	8468.7 ± 1.7	[50]	1.232(-2)	[488]	2.885(-3)		
$L_{\beta 15}$	L_3N_4	8455.6 ± 1.2	[145]			2.885(-3)	10.39	[423]
$L_{\beta 2}$	L_3N_5	8468.40	[104]	0.112	[488]	2.580(-2)	9.21	[423]
		8467.65 ± 0.55	[145]					
$L_{u'}$	L_3N_6	8637.46 ± 0.63	[145]			8.082(-6)		
L_u	L_3N_7	8639.35 ± 0.67	[145]			4.538(-5)		
$L_{\beta 7}$	L_3O_1	8584.90	[104]	1.19(-3)	[488]	3.381(-4)	8.72	[423]
$L_{\beta 5}$	$L_3O_{4,5}$	8641.00 ± 1.20	[50]					
		8640.90	[104]					

M series

line	transition	E/eV		I/eV/ℏ		TPIV	Γ/eV	
$M_{\gamma 1}$	M_3N_4	1678.15	[423]			3.800(-4)	16.43	[423]
$M_{\gamma 2}$	M_3N_5	1686.77	[423]			3.027(-3)	15.25	[423]
$M_{\xi 2}$	M_4N_2	1144.60	[423]			2.749(-3)	20.16	[423]
M_δ	M_4N_3	1192.10	[423]			3.162(-4)	11.42	[423]
M_β	M_4N_6	1503.00 ± 1.30	[50]	1.842(-2)	[335]	9.278(-3)		
M_η	M_4O_2	1480.40	[423]			5.684(-4)		
$M_{\xi 1}$	M_5N_3	1137.50	[263]			2.547(-3)	10.52	[423]
$M_{\alpha 1,2}$	$M_5N_{6,7}$	1462.00 ± 2.00	[50]	1.842(-2)	[335]			
		1465.50	[104]					
$M_{\alpha 2}$	M_5N_6	1458.80	[423]			5.296(-4)		
$M_{\alpha 1}$	M_5N_7	1460.10	[423]			1.025(-2)		

level characteristics

level	E_B/eV		ω_{nlj}		Γ/eV		AE/eV	
K	59390.00	[503]	0.949	[301]	30.10	[301]	59389.6	[51]
	59389.60 ± 0.50	[493]	0.948	[38]	29.51	[423]	59379 ± 21	[50]
			0.9447	[39]	30.2	[94]	59391.1 ± 1.3	[145]
			0.9460	[423]				
			0.949	[310]				

level	E_B/eV		ω_{nlj}		Γ/eV		AE/eV	
L_1	10116.00	[503]	0.106	[301]	5.47	[301]	10115.7	[51]
	10115.70 ± 0.40	[493]	8.21(-2)	[423]	6.16	[423]	10121.0 ± 1.2	[50]
			0.109	[436]	4.9	[94]	10114.4 ± 1.3	[145]
L_2	9618.00	[503]	0.211	[301]	4.79	[301]	9616.9	[51]
	9616.90 ± 0.40	[493]	0.2307	[423]	4.12	[423]	9617.0 ± 1.1	[50]
			0.231	[436]	4.26	[94]	9617.34 ± 0.43	[145]
L_3	8648.00	[503]	0.201	[301]	4.48	[301]	8648.0	[51]
	8648.00 ± 0.40	[493]	0.2106	[423]	4.03	[423]	8649.53 ± 0.89	[50]
			0.212	[436]	4.2	[94]	8648.97 ± 0.34	[145]
M_1	2307.00	[503]	1.61(-3)	[512]	13.8	[94]	2306.8	[51]
	2306.80 ± 0.70	[493]	1.82(-3)	[423]	18.26	[423]		
M_2	2090.00	[503]	1.30(-2)	[512]	6.85	[94]	2089.8	[51]
	2089.80 ± 1.10	[493]	3.86(-3)	[423]	9.94	[423]		
M_3	1885.00	[503]	1.95(-3)	[512]	7.1	[94]	1884.5	[51]
	1884.50 ± 1.10	[493]	4.60(-3)	[423]	10.06	[423]		
M_4	1515.00	[503]	1.10(-2)	[512]	1.27	[94]	1514.6	[51]
	1514.60 ± 0.70	[493]	1.30(-2)	[423]	2.17	[423]		
M_5	1468.00	[503]	1.40(-2)	[512]	1.27	[94]	1467.7	[51]
	1467.70 ± 0.90	[493]	1.40(-2)	[423]	1.27	[423]		
N_1	472.00	[503]	2.7(-4)	[423]	6.1	[94]		
	471.70 ± 0.90	[493]			13.26	[423]		
N_2	386.00	[503]	5.8(-4)	[423]	5.4	[94]		
	385.90 ± 1.60	[493]			17.99	[423]		
N_3	337.00	[503]	2.1(-4)	[423]	3.3	[94]		
	336.60 ± 1.60	[493]			9.25	[423]		
N_4	180.00	[503]	6.0(-5)	[423]	3.55	[94]		
	179.60 ± 1.20	[51]			6.36	[423]		
	185.50 ± 1.20	[493]						
N_5	180.00	[503]	6.0(-5)	[423]	3.1	[94]		
	175.70 ± 1.20	[493]			5.15	[423]		
N_6	5.00	[503]						
	5.30 ± 1.90	[51]						
	6.20 ± 1.90	[493]						

level	E_B/eV		ω_{nlj}		Γ/eV		AE/eV
N_7	5.00	[503]					
	5.30 ± 1.90	[51]					
	4.70 ± 1.90	[493]					
O_1	35.00	[503]	1.0(-6)	[423]	4.70	[423]	
	53.20 ± 3.00	[493]					
O_2	32.00	[503]					
	32.30 ± 1.60	[51]					
	36.20 ± 1.60	[493]					
O_3	32.00	[503]					
	32.30 ± 1.60	[51]					
	30.40 ± 1.60	[493]					
IP	6.18431	[222]					

Yb   Z=70   [Xe] $4f^{14}\,6s^2$

Ytterbium

$A = 173.04(3)$ [222] $\varrho = 6.972$ g/cm³ [547]
$\varrho = 7.02$ g/cm³ [483]

X-ray transitions

line	transition	E/eV		I/eV/$\hbar$		TPIV	Γ/eV	
K series								
K_{α_3}	KL_1	50987.00	[104]				39.31	[423]
		50846.5 ± 2.0	[145]					
K_{α_2}	KL_2	51354.60 ± 0.63	[50]	8.53	[488]	2.735(-1)	36.70	[301]
		51355.00	[104]	8.67	[487]		40.60	[473]
							35.48	[423]
K_{α_1}	KL_3	52389.48 ± 0.66	[50]	15.13	[488]	4.819(-1)	36.50	[301]
		52389.00	[104]	15.28	[487]		42.00	[473]
							35.54	[423]
	KM_1	58937.5 ± 3.0	[145]				49.96	[423]
K_{β_3}	KM_2	59152 ± 42	[50]	1.594	[488]	5.079(-2)	41.15	[473]
		59137.65	[104]	1.666	[487]		41.59	[423]
		59160.1 ± 2.6	[145]					
K_{β_1}	KM_3	59367.1 ± 8.4	[50]	3.08	[488]	9.815(-2)	41.43	[473]
		59383.3 ± 2.7	[145]	3.22	[116]		41.77	[423]
K_{β_5}	$KM_{4,5}$	59782.2 ± 8.5	[50]	3.89(-2)	[488]			
		59782.00	[104]					
$K_{\beta_5^{II}}$	KM_4	59758.2 ± 2.2	[145]			9.643(-4)	33.61	[423]
$K_{\beta_5^{I}}$	KM_5	59805.4 ± 2.1	[145]			1.239(-3)	32.61	[423]
	KN_1	60847.3 ± 3.9	[145]				45.01	[423]
$K_{\beta_2^{II}}$	KN_2	60939.1 ± 4.1	[145]	0.381	[488]	1.130(-2)	50.14	[423]
$K_{\beta_2^{I}}$	KN_3	60988.5 ± 1.8	[145]	0.742	[488]	2.185(-2)	40.91	[423]
K_{β_2}	$KN_{2,3}$	60985 ± 89	[50]	0.355	[488]			
		60882.40	[104]					
$K_{\beta_4^{II}}$	KN_4	61132.8 ± 1.9	[145]			2.221(-4)	38.52	[423]

line	transition	E/eV		I/eV/$\hbar$		TPIV	Γ/eV	
$K_{\beta_4^I}$	KN_5	61144.8 ± 1.5	[145]			2.838(-4)	37.11	[423]
	$KO_{2,3}$	61298.00 ± 60.00	[50]	0.1296	[488]	0.86		
		61299.00	[104]	0.1430	[487]			
	KO_2					2.540(-3)	31.31	[423]
	KO_3					5.265(-3)	31.27	[423]

L series

line	transition	E/eV		I/eV/$\hbar$		TPIV	Γ/eV	
	L_1M_1	8091.0 ± 3.1	[145]				26.74	[423]
$L_{\beta4}$	L_1M_2	8313.26 ± 0.25	[50]	0.1967	[488]	2.276(-2)	11.00	[473]
		8313.10	[104]				18.36	[423]
$L_{\beta3}$	L_1M_3	8536.79 ± 0.43	[50]	0.2607	[488]	3.016(-2)	12.20	[473]
		8536.20	[104]				18.55	[423]
$L_{\beta10}$	L_1M_4	8910.00 ± 0.70	[50]	5.07(-3)	[488]	5.870(-4)	10.39	[423]
		8910.00	[104]					
		8913.31 ± 0.14	[440]					
$L_{\beta9}$	L_1M_5	8959.70 ± 0.70	[50]	7.59(-3)	[488]	8.792(-4)	9.39	[423]
		8958.50	[104]					
		8960.64 ± 0.19	[440]					
	L_1N_1	10037.00	[104]	6.38(-7)	[488]		21.79	[423]
		10000.8 ± 4.0	[145]					
$L_{\gamma2}$	L_1N_2	10089.79 ± 0.85	[50]	4.72(-2)	[488]	5.464(-3)	26.92	[423]
		10105.00	[543]					
$L_{\gamma3}$	L_1N_3	10143.20 ± 0.61	[50]	6.56(-2)	[488]	7.593(-3)	17.69	[423]
		10143.40	[104]					
		10155.00	[543]					
$L_{\gamma11}$	$L_1N_{4,5}$	10289.00	[104]	1.07(-3)	[488]			
		10297.90 ± 0.86	[392]					
	L_1N_4	10286.3 ± 1.9	[145]			8.007(-5)	15.30	[423]
	L_1N_5	10298.3 ± 1.6	[145]			1.242(-4)	13.89	[423]
	L_1N_6	10475.9 ± 1.7	[145]					
	L_1N_7	10477.7 ± 1.7	[145]					
	L_1O_1	10431.20 ± 0.90	[50]	1.13(-7)	[488]		13.16	[423]
		10431.30	[104]					
$L_{\gamma4}$	$L_1O_{2,3}$	10460.30 ± 0.90	[50]	6.16(-3)	[488]			
		10462.00	[104]					
$L_{\gamma_4'}$	L_1O_2	10469.00	[543]			7.132(-4)	8.08	[423]

line	transition	E/eV		I/eV/$\hbar$		TPIV	Γ/eV	
L_{γ_4}	L_1O_3	10475.00	[543]			9.462(-4)	8.05	[423]
	$L_1O_{4,5}$	10483.30 ± 0.90	[50]	8.18(-3)	[488]			
		10483.00	[104]					
L_η	L_2M_1	7580.24 ± 0.34	[50]	2.325(-2)	[488]	6.004(-3)	22.91	[423]
		7580.20	[104]					
		7568.00	[543]					
	L_2M_2	7805.29 ± 0.65	[50]	4.36(-7)	[488]		14.54	[423]
		7805.90	[104]					
		7805.4 ± 1.8	[145]					
$L_{\beta_{17}}$	L_2M_3	8024.08 ± 0.52	[392]			1.586(-4)	14.71	[423]
		8028.6 ± 1.8	[145]					
L_{β_1}	L_2M_4	8401.88 ± 0.42	[50]	0.866	[488]	1.995(-1)	5.90	[473]
		8401.90	[104]				6.56	[423]
		8399.00	[543]					
	L_2M_5	8450.8 ± 1.2	[145]				5.56	[423]
L_{γ_5}	L_2N_1	9491.1 ± 1.1	[50]	5.63(-3)	[488]	1.346(-3)	17.96	[423]
		9491.00	[104]					
	L_2N_2	9584.4 ± 3.2	[145]				23.08	[423]
	L_2N_3	9632.70	[104]	1.70(-4)	[488]	3.912(-5)	13.86	[423]
		9633.83 ± 0.89	[145]					
L_{γ_1}	L_2N_4	9780.18 ± 0.57	[50]	0.1543	[488]	3.555(-2)	9.20	[473]
		9780.70	[104]				11.46	[423]
		9774.00	[543]					
	L_2N_5	9790.13 ± 0.65	[145]				10.05	[423]
L_ν	L_2N_6	9972.80	[104]	2.88(-4)	[488]	6.631(-5)		
	L_2N_7	9969.53 ± 0.78	[145]					
L_{γ_8}	L_2O_1	9924.60 ± 0.40	[50]	9.04(-4)	[488]	2.476(-4)	9.33	[423]
		9924.90	[104]					
	$L_2O_{2,3}$	9956.10 ± 0.80	[50]	2.12(-5)	[488]			
		9956.00	[104]					
	L_2O_2						4.25	[423]
	L_2O_3					8.595(-6)	4.22	[423]
L_{γ_6}	L_2O_4	9976.60 ± 0.25	[50]					
		9976.60	[104]					
L_l	L_3M_1	6545.54 ± 0.26	[50]	3.42(-2)	[488]	1.126(-2)	22.83	[423]
		6545.50	[104]					
		6535.00	[543]					
L_t	L_3M_2	6771.62 ± 0.49	[50]	3.27(-4)	[488]	8.742(-5)	14.46	[423]
		6771.60	[104]					
L_s	L_3M_3	6990.90	[104]	2.98(-4)	[488]	7.948(-5)	14.64	[423]
		6994.2 ± 1.7	[145]					

line	transition	E/eV		I/eV/ℏ		TPIV	Γ/eV	
$L_{\alpha 2}$	$L_3 M_4$	7367.40 ± 0.32	[50]	7.84(-2)	[488]	1.799(-2)	6.22	[473]
		7367.70	[104]				6.48	[423]
		7366.00	[543]					
$L_{\alpha 1}$	$L_3 M_5$	7415.70 ± 0.26	[50]	0.693	[488]	1.587(-1)	5.40	[473]
		7415.40	[104]				5.48	[423]
		7414.00	[543]					
$L_{\beta 6}$	$L_3 N_1$	8456.61 ± 0.85	[50]	8.05(-3)	[488]	1.847(-3)	17.88	[423]
		8456.00	[104]					
	$L_3 N_2$	8550.0 ± 3.1	[145]			1.669(-5)	23.00	[423]
	$L_3 N_3$	8599.50	[104]	7.09(-5)	[488]	1.625(-5)	13.78	[423]
		8599.39 ± 0.79	[145]					
$L_{\beta 15,2}$	$L_3 N_{4,5}$	8758.91 ± 0.46	[50]	1.321(-2)	[488]	3.028(-3)		
$L_{\beta 15}$	$L_3 N_4$	8743.73 ± 0.91	[145]			3.028(-3)	11.39	[423]
$L_{\beta 2}$	$L_3 N_5$	8759.60	[104]	0.1182	[488]	2.710(-2)	8.58	[473]
		8755.69 ± 0.55	[145]				9.98	[423]
L_u	$L_3 N_{6,7}$	8934.50	[50]	2.38(-4)	[488]			
$L_{u'}$	$L_3 N_6$	8933.35 ± 0.64	[145]			9.734(-6)		
L_u	$L_3 N_7$	8935.09 ± 0.68	[145]			5.453(-5)		
$L_{\beta 7}$	$L_3 O_1$	8888.90 ± 0.70	[50]	1.29(-3)	[488]	3.623(-4)	9.25	[423]
		8888.90	[104]					
	$L_3 O_{2,3}$	8920.90 ± 0.70	[50]	9.38(-6)	[488]			
		8921.60	[104]					
	$L_3 O_2$					3.702(-6)	4.17	[423]
	$L_3 O_3$					3.316(-6)	4.14	[423]
$L_{\beta 5}$	$L_3 O_{4,5}$	8939.00 ± 0.50	[50]					
		8940.30	[104]					

M series

	transition	E/eV		I/eV/ℏ		TPIV	Γ/eV	
	$M_3 N_1$	1464.00 ± 1.60	[50]	2.17(-3)	[335]	8.460(-4)	24.25	[423]
		1467.30	[263]					
$M_{\gamma 2}$	$M_3 N_4$	1741.47	[423]			4.028(-4)	17.76	[423]
$M_{\gamma 1}$	$M_3 N_5$	1765.00 ± 2.00	[50]	1.716(-2)	[335]	3.200(-3)	16.35	[423]
		1765.00	[104]					
		1768.70	[263]					
$M_{\xi 2}$	$M_4 N_2$	1192.06	[423]			2.552(-3)	21.22	[423]
M_δ	$M_4 N_3$	1243.18	[423]			2.876(-4)	12.00	[423]
M_β	$M_4 N_6$	1567.50 ± 0.40	[50]	2.17(-2)	[335]	1.076(-2)		
		1566.10	[104]					
		1571.40	[263]					

line	transition	E/eV		I/eV/$\hbar$		TPIV	Γ/eV	
M_η	M_4O_2	1541.66	[423]			5.411(-4)	2.39	[423]
$M_{\xi 1}$	M_5N_3	1183.00 ± 1.00	[50]	9.06(-4)	[335]	2.328(-3)	11.00	[423]
		1183.10	[104]					
		1185.30	[263]					
$M_{\alpha 2,1}$	$M_5N_{6,7}$	1521.40 ± 0.90	[50]	2.172(-2)	[335]			
		1525.00	[263]					
$M_{\alpha 2}$	M_5N_6	1517.41	[423]			6.069(-4)		
$M_{\alpha 1}$	M_5N_7	1518.83	[423]			1.175(-2)		

N series

	N_4N_6	190.00 ± 2.00	[50]			5.181(-5)		
	$N_5N_{6,7}$	179.00 ± 1.30	[50]					

level characteristics

level	E_B/eV		ω_{nlj}		Γ/eV		AE/eV	
K	61332.00	[503]	0.9467	[39]	31.26	[423]	61332.3	[51]
	61332.30 ± 0.50	[493]	0.963	[293]	32	[94]	61305 ± 22	[50]
			0.951	[301]	31.90	[301]	61333.3 ± 1.3	[145]
			0.950	[38]				
			0.9481	[423]				
			0.951	[310]				
			0.947	[112]				
			0.9661	[155]				
L_1	10488.00	[503]	0.112	[301]	5.53	[301]	10486.4	[51]
	10486.40 ± 0.40	[493]	0.112	[128]	8.04	[423]	10491.0 ± 1.3	[50]
			7.04(-2)	[423]	5.2	[94]	10486.8 ± 1.4	[145]
			0.114	[436]				
L_2	9978.00	[503]	0.222	[301]	4.82	[301]	9978.2	[51]
	9978.20 ± 0.40	[493]	0.2429	[423]	4.21	[423]	9976.0 ± 1.2	[50]
			0.246	[244]	4.36	[94]	9978.70 ± 0.44	[145]
L_3	8943.00	[503]	0.210	[301]	4.60	[301]	8943.6	[51]
	8943.60 ± 0.40	[493]	0.2206	[423]	4.13	[423]	8944.04 ± 0.94	[50]
			0.224	[244]	4.31	[94]	8944.26 ± 0.35	[145]
M_1	2397.00	[503]	1.15(-3)	[38]	20.80	[350]	2398.2	[51]
	2398.10 ± 0.40	[493]	1.67(-3)	[512]	18.70	[423]		
			1.94(-3)	[423]	13.9	[94]		
			1.18(-3)	[245]				
M_2	2172.00	[503]	1.97(-3)	[38]	11.80	[350]	2173.0	[51]
	2173.00 ± 0.40	[493]	1.30(-2)	[512]	10.32	[423]		
			4.08(-3)	[423]	7.1	[94]		
			2.01(-3)	[245]				

level	E_B/eV		ω_{nlj}		Γ/eV		AE/eV	
M_3	1949.00	[503]	1.66(-3)	[38]	11.60	[350]	1949.8	[51]
	1949.00 ± 0.50	[493]	1.71(-3)	[512]	10.50	[423]		
			4.86(-3)	[423]	6.7	[94]		
			2.15(-3)	[245]				
M_4	1576.00	[503]	8.60(-3)	[38]	3.10	[350]	1576.3	[51]
	1576.30 ± 0.40	[493]	1.20(-2)	[512]	2.35	[423]		
			1.42(-2)	[423]	1.35	[94]		
M_5	1527.00	[503]	1.49(-2)	[38]	1.56	[350]	1527.8	[51]
	1527.80 ± 0.40	[493]	1.60(-2)	[512]	1.35	[423]		
			1.53(-2)	[423]	1.35	[94]		
N_1	487.00	[503]	2.9(-4)	[423]	16.00	[74]		
	487.20 ± 0.60	[493]			13.75	[423]		
					6.3	[94]		
N_2	396.00	[503]	6.3(-4)	[423]	13.00	[74]		
	396.70 ± 0.70	[493]			18.88	[423]		
					5.5	[94]		
N_3	343.00	[503]	2.3(-4)	[423]	13.00	[74]		
	343.50 ± 0.50	[493]			9.65	[423]		
					3.5	[94]		
N_4	197.00	[503]	7.0(-5)	[739	8.00	[74]		
	198.10 ± 0.50	[493]			7.26	[423]		
					3.7	[94]		
N_5	184.00	[503]	7.0(-5)	[423]	8.00	[74]		
	184.90 ± 1.30	[493]			5.85	[423]		
					3.2	[94]		
N_6	6.00	[503]			0.0002	[74]		
	6.30 ± 1.00	[51]			0.03	[94]		
	7.00 ± 1.00	[493]						
N_7	6.00	[503]			0.03	[94]		
	6.30 ± 1.00	[51]			0.0002	[74]		
	5.80 ± 1.00	[493]						
O_1	53.00	[503]	1.0(-6)	[423]	5.12	[423]		
	54.10 ± 0.50	[493]						
O_2	23.00	[503]			0.04	[423]		
	23.40 ± 0.60	[51]						
	27.40 ± 0.60	[493]						
O_3	23.00	[503]			0.01	[423]		
	23.40 ± 0.60	[51]						
	21.40 ± 0.60	[493]						
IP	6.25416	[222]						

Lu   Z=71

[Xe] $4f^{14}$ $5d^1$ $6s^2$

Lutetium $A = 174.967(1)$ [222] $\varrho = 9.835$ g/cm^3 [547]
$\varrho = 9.79$ g/cm^3 [483]

X-ray transitions

line	transition	E/eV		I/eV/ℏ		TPIV	Γ/eV	
K series								
K_{α_3}	KL_1	52445.5 ± 2.1	[145]				41.75	[423]
$K_{\alpha 2}$	KL_2	52965.57 ± 0.67	[50]	9.12	[488]	2.743(-1)	38.70	[301]
		52966.00	[104]				37.43	[423]
$K_{\alpha 1}$	KL_3	54070.39 ± 0.70	[50]	16.09	[488]	4.817(-1)	38.40	[301]
		54070.00	[104]				37.35	[423]
	KM_1	60823.9 ± 3.1	[145]				51.68	[423]
$K_{\beta 3}$	KM_2	61048 ± 18	[50]	1.696	[488]	5.105(-2)	43.76	[423]
		61092.80	[104]					
		61051.8 ± 2.7	[145]					
$K_{\beta 1}$	KM_3	61283 ± 13	[50]	3.28	[488]	9.863(-2)	44.02	[423]
		61338.00	[104]					
		61291.2 ± 2.7	[145]					
$K_{\beta 5}$	$KM_{4,5}$	61731.9 ± 9.1	[50]	4.26(-2)	[488]			
		61733.00	[104]					
$K_{\beta_5^{II}}$	KM_4	61674.4 ± 2.2	[145]			1.002(-3)	35.55	[423]
$K_{\beta_5^{I}}$	KM_5	61725.3 ± 2.2	[145]			1.281(-3)	34.54	[423]
	KN_1	62809.2 ± 4.0	[145]				47.12	[423]
$K_{\beta 2}$	$KN_{2,3}$	62967 ± 95	[50]	0.734	[488]			
		62968.00	[104]					
$K_{\beta_2^{II}}$	KN_2	62903.9 ± 4.2	[145]			1.141(-2)	49.62	[423]
$K_{\beta_2^{I}}$	KN_3	62957.6 ± 1.8	[145]			2.207(-2)	42.94	[423]
$K_{\beta_4^{II}}$	KN_4	63106.2 ± 2.5	[145]			2.333(-4)	39.53	[423]
$K_{\beta_4^{I}}$	KN_5	63119.0 ± 1.6	[145]			2.967(-4)	39.30	[423]
	$KO_{2,3}$	63293.00 ± 7.00	[50]	0.1459	[488]			
		63293.00	[104]					
	KO_2					2.573(-3)	33.67	[423]
	KO_3					5.331(-3)	33.32	[423]

line	transition	E/eV		I/eV/$\hbar$		TPIV	Γ/eV	

L series

line	transition	E/eV		I/eV/$\hbar$		TPIV	Γ/eV	
	L_1M_1	8378.80	[104]	1.88(-6)	[488]		27.21	[423]
		8378.4 ± 3.1	[145]					
$L_{\beta 4}$	L_1M_2	8606.54 ± 0.44	[50]	0.2119	[488]	2.360(-2)	19.29	[423]
		8606.50	[104]					
$L_{\beta 3}$	L_1M_3	8847.03 ± 0.47	[50]	0.2771	[488]	3.086(-2)	19.54	[423]
		8846.20	[104]					
$L_{\beta 10}$	L_1M_4	9231.7 ± 2.0	[50]	5.63(-3)	[488]	6.272(-4)	11.08	[423]
		9230.90	[104]					
$L_{\beta 9}$	L_1M_5	9281.5 ± 1.0	[50]	8.43(-3)	[488]	9.392(-4)	10.07	[423]
		9282.10	[104]					
	L_1N_1	10363.00	[104]	7.64(-7)	[488]		22.65	[423]
		10363.7 ± 4.0	[145]					
$L_{\gamma 2}$	L_1N_2	10460.0 ± 2.6	[50]	5.11(-2)	[488]	5.695(-3)	25.15	[423]
		10462.40	[104]					
$L_{\gamma 3}$	L_1N_3	10511.16 ± 0.53	[50]	7.03(-2)	[488]	7.823(-3)	18.47	[423]
		10510.90	[104]					
	L_1N_4	10667.3 ± 1.2	[50]	7.75(-4)	[488]	8.629(-5)	15.06	[423]
		10665.70	[104]					
		10661.0 ± 2.5	[145]					
$L_{\gamma 11}$	L_1N_5	10678.3 ± 1.2	[50]	1.20(-3)	[488]	1.340(-4)	14.83	[423]
		10676.40	[104]					
	L_1N_6	10858.7 ± 1.7	[145]					
	L_1N_7	10860.6 ± 1.7	[145]					
	L_1O_1	10815.90	[104]	1.40(-7)	[488]		13.93	[423]
$L_{\gamma 4}$	$L_1O_{2,3}$	10842.50 ± 1.00	[50]	9.23(-3)	[488]			
		10844.00	[104]					
	L_1O_2					7.794(-4)	9.20	[423]
	L_1O_3					1.029(-3)	8.85	[423]
	$L_1O_{4,5}$	10863.00	[104]	8.01(-6)	[488]			
	L_1O_4					3.628(-7)		
	L_1O_5					5.289(-7)		
L_η	L_2M_1	7857.43 ± 0.74	[50]	2.493(-2)	[488]	6.190(-3)	22.90	[423]
		7858.00	[104]					
	L_2M_2	8085.82 ± 0.70	[50]	5.19(-7)	[488]		14.97	[423]
		8085.80	[104]					
$L_{\beta 17}$	L_3M_3	8323.17 ± 0.56	[391]			1.704(-4)	15.20	[423]
		8325.9 ± 1.8	[145]					
$L_{\beta 1}$	L_2M_4	8709.13 ± 0.27	[50]	0.929	[488]	2.099(-1)	6.77	[423]
		8709.20	[104]					

line	transition	E/eV		I/eV/$\hbar$		TPIV	Γ/eV	
	L_2M_5	8760.0 ± 1.2	[145]				5.75	[423]
$L_{\gamma 5}$	L_2N_1	9843.0 ± 1.2	[50]	6.06(-3)	[488]	1.409(-3)	18.33	[423]
		9842.00	[104]					
	L_2N_2	9920.00	[104]	1.86(-7)	[488]		20.83	[423]
		9938.7 ± 3.3	[145]					
	L_2N_3	9979.00	[104]	1.89(-4)	[488]	4.279(-5)	14.15	[423]
		9992.28 ± 0.91	[145]					
$L_{\gamma 1}$	L_2N_4	10143.53 ± 0.49	[50]	0.1671	[488]	3.777(-2)	10.74	[423]
		10143.00	[104]					
	L_2N_5	10153.73 ± 0.64	[145]				10.52	[423]
L_{ν}	L_2N_6	10338.91 ± 0.74	[145]			7.767(-5)		
	L_2N_7	10340.83 ± 0.78	[145]					
$L_{\gamma 8}$	L_2O_1	10291.50 ± 0.90	[50]	1.01(-3)	[488]	2.620(-4)	9.61	[423]
		10292.00	[104]					
	$L_2O_{2,3}$	10319.80 ± 0.90	[50]	2.49(-5)	[488]			
		10320.00	[104]					
	L_2O_2						4.88	[423]
	L_2O_3					9.243(-6)	4.53	[423]
$L_{\gamma 6}$	L_2O_4	10343.10 ± 0.90	[50]	7.33(-4)	[488]	1.656(-4)		
		10344.00	[104]					
L_l	L_3M_1	6752.85 ± 0.54	[50]	3.70(-2)	[488]	1.156(-2)	22.82	[423]
		6752.10	[104]					
L_t	L_3M_2	6980.99 ± 0.58	[50]	3.62(-4)	[488]	9.202(-5)	14.90	[423]
		6981.10	[104]					
L_s	L_3M_3	7313.00	[104]	3.29(-4)	[488]	8.366(-5)	15.16	[423]
		7221.7 ± 1.7	[145]					
$L_{\alpha 2}$	L_3M_4	7604.92 ± 0.35	[50]	8.39(-2)	[488]	1.879(-2)	6.69	[423]
		7605.00	[104]					
$L_{\alpha 1}$	L_3M_5	7655.55 ± 0.21	[50]	0.741	[488]	1.659(-1)	5.68	[423]
		7655.50	[104]					

line	transition	E/eV		I/eV/$\hbar$		TPIV	Γ/eV	
$L_{\beta6}$	L_3N_1	8737.92 ± 0.91 8739.00	[50] [104]	8.75(-3)	[488]	1.960(-3)	18.26	[423]
	L_3N_2	8834.5 ± 3.2	[145]			1.805(-5)	20.76	[423]
	L_3N_3	8881.60 8888.09 ± 0.81	[104] [145]	7.87(-5)	[488]	1.763(-5)	14.08	[423]
$L_{\beta15}$	L_3N_4	9039.91 ± 0.98 9039.40	[50] [104]	1.424(-2)	[488]	3.191(-3)	10.67	[423]
$L_{\beta2}$	L_3N_5	9049.01 ± 0.29 9049.10	[50] [104]	0.1275	[488]	2.856(-2)	10.44	[423]
$L_{u'}$	L_3N_6	9234.72 ± 0.65	[145]			1.138(-5)		
L_u	L_3N_7	9236.64 ± 0.68	[145]			6.386(-5)		
$L_{\beta7}$	L_3O_1	9187.30 ± 0.40 9187.00	[50] [104]	1.44(-3)	[488]	3.879(-4)	9.54	[423]
	$L_3O_{2,3}$	9216.30 ± 0.60 9216.40	[50] [104]	1.09(-5)	[488]			
	L_3O_2					3.964(-6)	4.81	[423]
	L_3O_3					3.597(-6)	4.46	[423]
$L_{\beta5}$	$L_3O_{4,5}$	9239.70 ± 0.50 9239.80	[50] [104]	5.80(-4)	[488]			
	L_3O_4					1.389(-5)		
	L_3O_5					1.162(-4)		

M series

line	transition	E/eV		I/eV/$\hbar$		TPIV	Γ/eV	
$M_{\gamma2}$	M_3N_4	1834.00	[104]	1.88(-2)	[335]	4.255(-4)	17.33	[423]
$M_{\gamma1}$	M_3N_5	1832.00 ± 2.00 1835.70	[50] [104]			3.371(-3)	17.10	[423]
$M_{\xi2}$	M_4N_2	1240.61	[423]			2.398(-3)	18.96	[423]
M_δ	M_4N_3	1295.75	[423]			2.646(-4)	12.28	[423]
M_β	M_4N_6	1631.20 ± 0.40 1631.00 1633.50	[50] [104] [263]	2.49(-2)	[335]	1.224(-2)		
M_η	M_4O_2	1609.28	[423]			5.209(-4)	3.01	[423]
$M_{\xi1}$	M_5N_3	1242.75	[423]			2.172(-3)	11.26	[423]
$M_{\alpha2,1}$	$M_5N_{6,7}$	1581.30 ± 0.40	[50]	2.49(-2)	[335]			
$M_{\alpha2}$	M_5N_6					6.833(-4)		
$M_{\alpha1}$	M_5N_7					1.325(-2)		

level characteristics

level	E_B/eV		ω_{nlj}		Γ/eV		AE/eV	
K	63314.00	[503]	0.953	[301]	33.11	[423]	63313.8	[51]
	63313.80 ± 0.50	[493]	0.952	[38]	33.9	[94]	63305 ± 24	[50]
			0.9487	[39]			63322.7 ± 1.4	[145]
			0.9500	[423]				
			0.953	[310]				
L_1	10870.00	[503]	0.120	[301]	5.59	[301]	10870.4	[51]
	10870.40 ± 0.40	[493]	7.28(-2)	[423]	8.64	[423]	10873.7 ± 1.4	[50]
			0.12	[436]	5.4	[94]	10877.2 ± 1.4	[145]
L_2	10349.00	[503]	0.234	[301]	4.92	[301]	10348.6	[51]
	10348.60 ± 0.40	[493]	0.2560	[423]	4.32	[423]	10344.8 ± 1.3	[50]
			0.256	[436]	4.46	[94]	10357.43 ± 0.44	[145]
L_3	9244.00	[503]	0.220	[301]	4.68	[301]	9244.1	[51]
	9244.10 ± 0.40	[493]	0.251	[435]	4.25	[423]	9249.0 ± 1.0	[50]
			0.2308	[423]	4.43	[94]	9253.24 ± 0.35	[145]
			0.231	[436]				
M_1	2491.00	[503]	1.73(-3)	[512]	14.1	[94]	2491.2	[51]
	2491.20 ± 0.50	[493]	2.14(-3)	[423]	18.58	[423]		
M_2	2264.00	[503]	1.40(-2)	[512]	7.3	[94]	2263.5	[51]
	2263.50 ± 0.40	[493]	4.27(-3)	[423]	10.65	[423]		
M_3	2024.00	[503]	1.77(-3)	[512]	6	[94]	2023.6	[51]
	2023.60 ± 0.50	[493]	5.13(-3)	[423]	10.91	[423]		
M_4	1640.00	[503]	1.40(-2)	[512]	1.43	[94]	1639.4	[51]
	1639.40 ± 0.40	[493]	1.55(-2)	[423]	2.45	[423]		
M_5	1589.00	[503]	1.70(-2)	[512]	1.43	[94]	1588.5	[51]
	1588.50 ± 0.40	[493]	1.67(-2)	[423]	1.43	[423]		
N_1	506.00	[503]	3.2(-4)	[739	6.6	[94]		
	506.20 ± 0.60	[493]			14.01	[423]		
N_2	410.00	[503]	6.8(-4)	[423]	5.5	[94]		
	410.10 ± 0.80	[493]			16.51	[423]		
N_3	359.00	[503]	2.5(-4)	[423]	3.65	[94]		
	359.30 ± 0.50	[493]			9.83	[423]		

level	E_B/eV		ω_{nlj}		Γ/eV	AE/eV	
N_4	205.00	[503]	8.0(-5)	[423]	3.8	[94]	
	204.80 ± 0.50	[493]			6.42	[423]	
N_5	195.00	[503]	8.0(-5)	[423]	3.3	[94]	
	195.00 ± 0.40	[493]			6.20	[423]	
N_6	7.00	[503]			0.03	[94]	
	6.90 ± 0.50	[51]					
	7.80 ± 0.50	[493]					
N_7	7.00	[503]			0.03	[94]	
	6.20 ± 0.50	[51]					
O_1	57.00	[503]	2.0(-6)	[423]	5.29	[423]	
	56.80 ± 0.50	[493]					
O_2	28.00	[503]			0.56	[423]	
	28.00 ± 0.60	[51]					
	33.00 ± 0.60	[493]					
O_3	28.00	[503]			0.21	[423]	
	25.50 ± 0.60	[493]					
O_4	5.00	[503]					
	4.60 ± 0.60	[493]					
IP	5.4259	[222]					

Hf   Z=72

[Xe] $4f^{14} 5d^2 6s^2$

Hafnium $A = 178.49(2)$ [222] $\varrho = 13.27\,\text{g/cm}^3$ [547]
$\varrho = 13.3\,\text{g/cm}^3$ [483]

X-ray transitions

line	transition	E/eV		I/eV/$\hbar$		TPIV	Γ/eV	
K series								
K_{α_3}	KL_1	54080.40	[104]				44.17	[423]
		54074.8 ± 2.1	[145]					
$K_{\alpha2}$	KL_2	54612.0 ± 1.1	[50]	9.68	[488]	2.752(-1)	40.70	[301]
		54613.00	[104]	9.77	[487]		44.30	[473]
							39.47	[423]
$K_{\alpha1}$	KL_3	55790.8 ± 1.1	[50]	16.93	[488]	4.815(-1)	40.50	[301]
		55792.00	[104]	17.09	[487]		45.30	[473]
							39.41	[423]
	KM_1	62747.2 ± 3.2	[145]				53.22	[423]
$K_{\beta3}$	KM_2	62980.8 ± 1.9	[50]	1.803	[488]	5.125(-2)	46.10	[473]
		63178.00	[104]	1.884	[487]		46.00	[423]
$K_{\beta1}$	KM_3	63234 ± 14	[50]	3.48	[488]	9.906(-2)	46.00	[473]
		63400.00	[104]	3.64	[487]		46.35	[423]
$K_{\beta_5^{II}}$	KM_4	63628.8 ± 2.3	[145]			1.040(-3)	37.56	[423]
$K_{\beta_5^{I}}$	KM_5	63683.2 ± 2.2	[145]			1.324(-3)	36.56	[423]
	KN_1	64810.0 ± 4.1	[145]				49.25	[423]
$K_{\beta_2^{II}}$	KN_2	64907.3 ± 4.3	[145]	0.435	[487]	1.153(-2)	49.32	[423]
$K_{\beta_2^{I}}$	KN_3	64965.3 ± 1.9	[145]	0.849	[487]	2.231(-2)	45.00	[423]
$K_{\beta2}$	$KN_{2,3}$	64980.00 ± 7.00	[50]	0.785	[488]			
		64975.00	[104]					
$K_{\beta_4^{II}}$	KN_4	65119.3 ± 2.6	[145]			2.450(-4)	41.41	[423]
$K_{\beta_4^{I}}$	KN_5	65132.6 ± 1.7	[145]			3.104(-4)	41.50	[423]
	$KO_{2,3}$			0.1798	[488]			
	KO_2					2.605(-3)	36.12	[423]
	KO_3					5.397(-3)	35.46	[423]
L series								
	L_1M_1	8668.58 ± 0.81	[50]	2.23(-6)	[488]		27.31	[423]
		8668.60	[104]					
$L_{\beta4}$	L_1M_2	8905.50 ± 0.47	[50]	0.2281	[488]	2.404(-2)	12.80	[473]
		8905.50	[104]				20.09	[423]
$L_{\beta3}$	L_1M_3	9163.51 ± 0.50	[50]	0.2941	[488]	3.099(-2)	12.40	[473]
		9163.50	[104]				20.44	[423]
$L_{\beta10}$	L_1M_4	9550.40 ± 0.98	[50]			6.581(-4)	11.65	[423]
		9550.40	[104]					

line	transition	E/eV		I/eV/$\hbar$		TPIV	Γ/eV	
L_{β_9}	L_1M_5	9609.17 ± 0.99	[50]			9.851(-4)	10.65	[423]
	L_1N_1	10735.2 ± 4.0	[145]				23.34	[423]
L_{γ_2}	L_1N_2	10833.64 ± 0.70 10833.60	[50] [104]	5.54(-2)	[488]	5.833(-3)	23.41	[423]
L_{γ_3}	L_1N_3	10890.83 ± 0.71 10895.00	[50] [104]	7.52(-2)	[488]	7.923(-3)	19.09	[423]
	L_1N_4	11045.2 ± 1.3 11045.20	[50] [104]	8.66(-4)	[488]	9.130(-5)	15.86	[423]
$L_{\gamma_{11}}$	L_1N_5	11055.4 ± 1.3	[50]	1.35(-3)	[488]	1.421(-4)	15.59	[423]
	L_1N_6	11250.5 ± 1.7	[145]					
	L_1N_7	11252.7 ± 1.7	[145]					
	L_1O_1	11203.40 ± 0.90 11203.50	[50] [104]	1.73(-7)	[488]		14.60	[423]
$L_{\gamma_{4'}}$	L_1O_2	11232.60 ± 0.50 11233.00	[50] [104]	7.92(-3)	[488]	8.346(-4)	10.21	[423]
L_{γ_4}	L_1O_3	11240.00	[104]	1.307(-2)	[488]	1.093(-3)	9.55	[423]
$L_{\gamma_{4',4}}$	$L_1O_{2,3}$	11240.10 ± 0.50	[50]					
	$L_1O_{4,5}$	11262.00 ± 0.90 11262.00	[50] [104]	2.20(-5)	[488]			
	L_1O_4					9.338(-7)		
	L_1O_5					1.386(-6)		
L_η	L_2M_1	8139.33 ± 0.40 8139.30	[50] [104]	2.671(-2)	[488]	6.378(-3)	22.62	[423]
	L_2M_2	8373.56 ± 0.75 8373.60	[50] [104]	6.17(-7)	[488]		15.40	[423]
$L_{\beta_{17}}$	L_2M_3	8631.28 ± 0.80 8631.30	[50] [104]	8.28(-5)	[488]	1.829(-4)	15.75	[423]
L_{β_1}	L_2M_4	9022.80 ± 0.49 9022.80	[50] [104]	0.995	[488]	2.197(-1)	6.36 6.95	[473] [423]
	L_2M_5	9076.9 ± 1.3	[145]				6.95	[423]
L_{γ_5}	L_2N_1	10201.20 ± 0.62 10201.00	[50] [104]	6.53(-3)	[488]	1.474(-3)	18.65	[423]
	L_2N_2	10301.0 ± 3.3	[145]	2.22(-7)	[488]		18.72	[423]
	L_2N_3	10358.90 ± 0.90	[145]	2.11(-4)	[488]	4.659(-5)	14.40	[423]
L_{γ_1}	L_2N_4	10515.89 ± 0.66 10516.00	[50] [104]	0.1810	[488]	3.996(-2)	9.63 11.17	[473] [423]
	L_2N_5	10525.9 ± 1.2 10526.00	[50] [104]	6.89(-6)	[488]		10.89	[423]
L_ν	L_2N_6	10703.8 ± 1.2 10704.00	[50] [104]	4.07(-4)	[488]	8.990(-5)		
	L_2N_7	10721.18 ± 0.78	[145]					
L_{γ_8}	L_2O_1	10675.40 ± 0.50 10676.00	[50] [104]	1.12(-3)	[488]	2.770(-4)	9.90	[423]
	L_2O_2	10700.00	[104]	3.41(-5)	[488]		5.52	[423]
	L_2O_3	10709.00	[104]	2.91(-5)	[488]	9.930(-6)	4.86	[423]

line	transition	E/eV		I/eV/$\hbar$		TPIV	Γ/eV	
$L_{\gamma 6}$	L_2O_4	10732.5 ± 0.50	[50]	1.93(-3)	[488]	4.270(-4)		
	$L_2O_{4,5}$	10733.00	[104]					
L_l	L_3M_1	6959.63 ± 0.29	[50]	4.01(-2)	[488]	1.186(-2)	22.55	[423]
		6960.00	[104]					
L_t	L_3M_2	7195.52 ± 0.56	[50]	3.99(-4)	[488]	9.678(-5)	15.33	[423]
		7195.50	[104]					
L_s	L_3M_3	7453.28 ± 0.60	[50]	6.63(-4)	[488]	8.800(-5)	15.68	[423]
		7453.30	[104]					
$L_{\alpha 2}$	L_3M_4	7844.70 ± 0.37	[50]	8.96(-2)	[488]	1.962(-2)	6.70	[473]
		7844.70	[104]				6.89	[423]
$L_{\alpha 1}$	L_3M_5	7899.08 ± 0.37	[50]	0.791	[488]	1.732(-1)	5.83	[473]
		7899.10	[104]				5.89	[423]
$L_{\beta 6}$	L_3N_1	9022.80 ± 0.49	[50]	9.50(-3)	[488]	2.081(-3)	18.58	[423]
		9022.80	[104]					
	L_3N_2	9123.93 ± 0.89	[50]	8.91(-5)	[488]	1.951(-5)	18.65	[423]
		9124.00	[104]					
	L_3N_3	9180.27 ± 0.91	[50]	8.72(-5)	[488]	1.909(-5)	14.33	[423]
		9180.30	[104]					
$L_{\beta 15}$	L_3N_4	9337.21 ± 0.52	[50]	1.534(-2)	[488]	3.359(-3)	11.10	[423]
		9337.20	[104]					
$L_{\beta 2}$	L_3N_5	9347.35 ± 0.52	[50]	0.1374	[488]	3.009(-2)	8.92	[473]
		9347.40	[104]				10.82	[423]
L_u	$L_3N_{6,7}$	9525.01 ± 0.97	[50]	3.38(-4)	[488]			
$L_{U'}$	L_3N_6	9541.24 ± 0.66	[145]			1.319(-5)		
L_u	L_3N_7	9543.46 ± 0.69	[145]			7.405(-5)		
$L_{\beta 7}$	L_3O_1	9495.80 ± 0.40	[50]	1.62(-3)	[488]	4.149(-4)	9.84	[423]
		9495.90	[104]					
	L_3O_2	9522.20	[104]	1.25(-5)	[488]	4.241(-6)	5.45	[423]
	L_3O_3	9530.80	[104]	1.19(-5)	[488]	3.898(-6)	4.79	[423]
$L_{\beta 5}$	$L_3O_{4,5}$	9554.60 ± 0.40	[50]	1.54(-3)	[488]			
		9554.70	[104]					
	L_3O_4					3.560(-5)		
	L_3O_5					3.028(-4)		

M series

line	transition	E/eV		I/eV/$\hbar$		TPIV	Γ/eV	
	M_3N_1	1572.00 ± 2.00	[50]	2.50(-2)	[335]	9.324(-4)	25.52	[423]
		1571.90	[104]					
		1575.40	[263]					
$M_{\gamma 2}$	M_3N_4	1873.41	[423]			4.482(-4)	18.05	[423]
$M_{\gamma 1}$	M_3N_5	1895.00 ± 1.00	[50]	2.058(-2)	[335]	3.543(-3)	17.77	[423]
		1895.00	[104]					
		1898.70	[263]					
$M_{\xi 2}$	M_4N_2	1280.00 ± 1.00	[50]	1.04(-3)	[335]	2.291(-3)	16.80	[423]
		1280.00	[104]					
		1282.20	[263]					
M_δ	M_4N_3	1282.20	[263]			2.478(-4)	12.48	[423]

line	transition	E/eV		I/eV/$\hbar$		TPIV	Γ/eV	
M_β	M_4N_6	1697.60 ± 0.20	[50]	2.8316(-3)	[335]	1.375(-2)		
		1697.00	[104]					
		1700.70	[263]					
M_η	M_4O_2	1679.16	[423]			5.099(-4)	3.60	[423]
$M_{\xi 1}$	M_5N_3	1280.00 ± 1.00	[50]	1.04(-4)	[335]	2.090(-3)	11.48	[423]
		1280.00	[104]					
		1282.20	[263]					
$M_{\alpha 2}$	M_5N_6	1642.90	[104]	2.831(-2)	[335]	7.579(-4)		
$M_{\alpha 1}$	M_5N_7	1644.40	[104]	2.831(-2)	[335]	1.472(-2)		
$M_{\alpha 2,1}$	$M_5N_{6,7}$	1644.60 ± 0.20	[50]					
		1648.70	[263]					

level characteristics

level	E_B/eV		ω_{nlj}		Γ/eV		AE/eV	
K	65351.00	[503]	0.9505	[30.F]	35.04	[423]	65350.8	[51]
	65350.80 ± 0.60	[493]	0.954	[38]	35.9	[94]	65316 ± 25	[50]
			0.955	[301]	35.70	[301]	65352.0 ± 1.4	[145]
			0.9517	[423]				
			0.955	[310]				
L_1	11272.00	[503]	0.128	[301]	5.63	[301]	11270.7	[51]
	11270.70 ± 0.40	[493]	7.39(-2)	[423]	9.13	[423]	11268.585 ± 0.050	[294]
			0.125	[436]	5.7	[94]	11277.2 ± 1.4	[145]
			0.125	[567]				
L_2	10739.00	[503]	0.246	[301]	5.02	[301]	10739.4	[51]
	10739.40 ± 0.40	[493]			4.44	[423]	10735.875 ± 0.020	[294]
			0.2685	[423]	4.57	[94]	10745.60 ± 0.45	
			0.268	[436]				
			0.243	[567]				
L_3	9561.00	[503]	0.231	[301]	4.80	[301]	9560.7	[51]
	9560.70 ± 0.40	[493]	0.228	[435]	4.37	[423]	9558.286 ± 0.050	[294]
			0.2413	[423]	4.55	[94]	9567.88 ± 0.36	
			0.241	[436]				
			0.222	[567]				
M_1	2601.00	[503]	2.24(-3)	[423]	20.00	[74]	2600.9	[51]
	2600.90 ± 0.40	[493]	1.79(-3)	[512]	18.18	[423]		
					14.2	[94]		
M_2	2365.00	[503]	4.46(-3)	[423]	11.00	[74]	2365.4	[51]
	2365.40 ± 0.40	[493]	1.40(-2)	[512]	10.96	[423]		
					7.5	[94]		

level	E_B/eV		ω_{nlj}		Γ/eV		AE/eV	
M_3	2108.00	[503]	5.40(-3)	[423]	12.00	[74]	2107.6	[51]
	2107.60 ± 0.40	[493]	1.83(-3)	[512]	11.31	[423]		
					5.6	[94]		
M_4	1716.00	[503]	1.69(-2)	[423]	1.60	[74]	1716.4	[51]
	1716.40 ± 0.40	[493]	1.50(-2)	[512]	2.52	[423]		
					1.52	[94]		
M_5	1662.00	[503]	1.81(-2)	[423]	3.20	[74]	1661.7	[51]
	1661.70 ± 0.40	[493]	1.90(-2)	[512]	1.52	[423]		
					1.52	[94]		
N_1	538.00	[503]	3.4(-4)	[423]	16.00	[74]		
	538.10 ± 0.40	[493]			14.21	[423]		
					6.8	[94]		
N_2	437.00	[503]	7.3(-4)	[423]	13.00	[74]		
	437.00 ± 0.50	[493]			14.28	[423]		
					5.6	[94]		
N_3	380.00	[503]	2.7(-4)	[423]	13.00	[74]		
	380.40 ± 0.50	[493]			9.96	[423]		
					3.85	[94]		
N_4	224.00	[503]	1.0(-4)	[423]	7.30	[74]		
	223.80 ± 0.90	[493]			6.73	[423]		
					3.9	[94]		
N_5	214.00	[503]	9.0(-5)	[423]	7.30	[74]		
	213.70 ± 0.50	[493]			6.46	[423]		
					3.5	[94]		
N_6	19.00	[503]			0.02	[74]		
	17.10 ± 0.50	[51]			0.07	[94]		
	18.20 ± 0.50	[493]						
N_7	18.00	[503]			0.03	[94]		
	17.10 ± 0.50	[51]			0.02	[74]		
	16.30 ± 0.50	[493]						
O_1	65.00	[503]	2.0(-6)	[423]	5.47	[423]		
	64.90 ± 0.40	[493]						
O_2	38.00	[503]			1.08	[423]		
	38.10 ± 0.60	[51]						
	38.20 ± 0.60	[493]						
O_3	31.00	[503]			0.42	[423]		
	30.60 ± 0.60	[51]						
	29.00 ± 1.50	[493]						
O_4	7.00	[503]						
	6.00 ± 0.40	[493]						
IP	6.82507	[222]						

Ta   Z=73   [Xe] $4f^{14}\,5d^3\,6s^2$

Tantalum $A = 180.9479(1)$ [222] $\varrho = 16.60\,\text{g/cm}^3$ [547]

X-ray transitions

line	transition	E/eV		I/eV/$\hbar$		TPIV	Γ/eV	
K series								
K_{α_3}	KL_1	55735.0 ± 2.1	[145]				47.28	[423]
$K_{\alpha 2}$	KL_2	56277.6 ± 1.5	[50]	10.26	[488]	2.760(-1)	42.90	[301]
		56278.00	[104]	10.36	[487]		41.61	[423]
$K_{\alpha 1}$	KL_3	57533.2 ± 1.6	[50]	17.89	[488]	4.811(-1)	42.60	[301]
		57533.00	[104]	18.06	[487]		41.55	[423]
	KM_1	64708.3 ± 3.2	[145]				54,59	[423]
$K_{\beta 3}$	KM_2	64949.6 ± 1.0	[50]	1.914	[488]	5.149(-2)	48.32	[423]
		64950.00	[104]	2.000	[487]			
$K_{\beta 1}$	KM_3	$65223.3, \pm 2.0$	[50]	3.70	[488]	9.948(-2)	47.24	[423]
		65224.00	[104]	3.86	[487]			
$K_{\beta_5^{II}}$	KM_4	65626.9 ± 3.1	[50]	4.01(-2)	[488]	1.078(-3)	39.61	[423]
		65627.80	[104]					
$K_{\beta_5^{I}}$	KM_5	65683.6 ± 3.1	[50]	5.08(-2)	[488]	1.367(-3)	38.66	[423]
		65684.60	[104]					
	KN_1	66866.7 ± 4.2	[145]				51.42	[423]
$K_{\beta_2^{II}}$	KN_2	66949.4 ± 4.8	[50]	0.433	[488]	1.164(-2)	50.60	[423]
		66950.20	[104]	0.464	[487]			
$K_{\beta_2^{I}}$	KN_3	67013.5 ± 4.3	[50]	0.839	[488]	2.256(-2)	47.07	[423]
		67014.40	[104]	0.907	[487]			
$K_{\beta 4}$	$KN_{4,5}$	67195.4 ± 5.4	[50]	2.162(-2)	[488]			
		67198.00	[104]					
$K_{\beta_4^{II}}$	KN_4	67242.1 ± 2.8	[145]			2.573(-4)	44.06	[423]
$K_{\beta_4^{I}}$	KN_5	67202.7 ± 1.7	[145]			3.244(-4)	43.73	[423]
	$KO_{2,3}$	67370.00 ± 2.60	[50]	0.1822	[488]			
		67370.90	[104]	0.2002	[487]			
	KO_2					2.637(-3)	38.65	[423]
	KO_3					5.443(-3)	37.68	[423]
L series								
	L_1M_1	8973.2 ± 3.1	[145]				27.75	[423]
$L_{\beta 4}$	L_1M_2	9212.47 ± 0.30	[50]	0.2453	[488]	2.344(-2)	14.10	[473]
		9213.20	[104]				21.49	[423]
$L_{\beta 3}$	L_1M_3	9487.62 ± 0.28	[104]	0.312	[488]	2.979(-2)	11.70	[473]
		9487.60	[104]				20.41	[423]
$L_{\beta 10}$	L_1M_4	9889.3 ± 2.3	[50]	6.92(-3)	[488]	6.605(-4)	12.78	[423]
		9888.70	[104]					

line	transition	E/eV		I/eV/$\hbar$		TPIV	Γ/eV	
$L_{\beta 9}$	L_1M_5	9945.6 ± 2.4	[50]	1.035(-2)	[488]	9.896(-4)	11.83	[423]
		9946.70	[104]					
	L_1N_1	11117.4 ± 1.3	[50]	1.09(-6)	[488]		24.59	[423]
		11117.40	[104]					
$L_{\gamma 2}$	L_1N_2	11217.1 ± 2.1	[50]	5.99(-2)	[488]	5.721(-3)	11.60	[473]
		11215.70	[104]				23.77	[423]
		11229.00	[543]					
$L_{\gamma 3}$	L_1N_3	11277.68 ± 0.61	[50]	8.04(-2)	[488]	7.681(-3)	10.00	[473]
		11277.00	[104]				20.24	[423]
		11292.00	[543]					
	L_1N_4	11439.9 ± 1.1	[50]	9.68(-4)	[488]	9.253(-5)	17.23	[423]
		11439.60	[104]					
$L_{\gamma 11}$	L_1N_5	11458.1 ± 1.1	[50]	1.51(-3)	[488]	1.443(-4)	16.90	[423]
		11458.00	[104]					
	$L_1N_{6,7}$	11657.00 ± 1.00	[50]	1.66(-5)	[488]			
		11657.00	[104]					
	L_1N_6	11667.8 ± 1.7	[145]					
	L_1N_7	11670.4 ± 1.8	[145]					
	L_1O_1	$11611.80, \pm 1.00$	[50]	2.13(-7)	[488]		15.87	[423]
		11612.00	[104]					
$L_{\gamma 4'}$	L_1O_2	$11636.60, \pm 0.30$	[50]	8.92(-3)	[488]	8.524(-4)	11.83	[423]
		11637.00	[104]					
		11655.00	[543]					
$L_{\gamma 4}$	L_1O_3	$11645.10, \pm 0.30$	[50]	1.158(-2)	[488]	1.107(-3)	10.85	[423]
		11645.00	[104]					
		11665.00	[543]					
	$L_1O_{4,5}$	11675.2 ± 1.00	[50]	4.28(-5)	[488]			
		11675.00	[104]					
	L_1O_4					1.636(-6)		
	L_1O_5					2.452(-6)		
L_η	L_2M_1	8428.09 ± 0.42	[50]	0.2859	[488]	6.569(-3)	22.09	[423]
		8428.00	[104]					
		8413.00	[543]					
	L_2M_2	8668.56 ± 0.49	[50]	7.31(-7)	[488]		15.83	[423]
		8667.20	[104]					
$L_{\beta 17}$	L_2M_3	8941.76 ± 0.54	[50]	9.14(-4)	[488]	1.972(-4)	14.75	[423]
		8941.40	[104]					
$L_{\beta 1}$	L_2M_4	9343.19 ± 0.31	[50]	1.065	[488]	2.295(-1)	6.60	[473]
		9342.90	[104]				7.11	[423]
		9338.00	[543]					
	L_2M_5	9399.94 ± 0.95	[50]	3.92(-5)	[488]		6.16	[423]
		9399.00	[104]					

line	transition	E/eV		I/eV/$\hbar$		TPIV	Γ/eV	
$L_{\gamma 5}$	L_2N_1	10570.6 ± 1.3	[50]	7.02(-3)	[488]	1.541(-3)	18.92	[423]
		10571.00	[104]					
	L_2N_2	10674.04 ± 0.87	[50]	2.66(-7)	[488]		18.10	[423]
	L_2N_3	10731.6 ± 1.4	[50]	2.35(-4)	[488]	5.064(-5)	14.58	[423]
		10732.00	[104]					
$L_{\gamma 1}$	L_2N_4	10895.33 ± 0.43	[50]	0.1959	[488]	4.224(-2)	9.50	[473]
		10895.00	[104]				11.56	[423]
		10887.00	[543]					
	L_2N_5	10906.84 ± 0.77	[50]	7.99(-6)	[488]		11.23	[423]
		10906.00	[104]					
L_{ν}	L_2N_6	11107.82 ± 0.83	[50]	4.87(-4)	[488]	1.032(-4)		
		11111.00	[104]					
	L_2N_7	11126.71 ± 0.80	[145]					
$L_{\gamma 8}$	L_2O_1	11064.60 ± 1.00	[50]	1.24(-3)	[488]	2.926(-4)	10.20	[423]
		11065.00	[104]					
	L_2O_2	11090.70 ± 0.90	[50]	4.26(-8)	[488]		6.16	[423]
		11091.00	[104]					
	L_2O_3	11100.10 ± 0.90	[50]	3.38(-5)	[488]	1.066(-5)	5.18	[423]
		11100.00	[104]					
$L_{\gamma 6}$	L_2O_4	11130.60 ± 0.30	[50]	3.62(-3)	[488]	7.815(-4)		
		11131.00	[104]					
L_l	L_3M_1	7173.20 ± 0.31	[50]	4.33(-2)	[488]	1.216(-2)	22.02	[423]
		7173.30	[104]					
		7160.00	[543]					
L_t	L_3M_2	7412.13 ± 0.35	[50]	4.39(-4)	[488]	1.017(-4)	15.76	[423]
		7412.30	[104]					
L_s	L_3M_3	7686.65 ± 0.38	[50]	3.99(-4)	[488]	9.249(-5)	14.68	[423]
		7688.20	[104]					
$L_{\alpha 2}$	L_3M_4	8087.93 ± 0.16	[50]	9.55(-2)	[488]	2.044(-2)	7.05	[423]
		8088.00	[104]					
		8085.00	[543]					
$L_{\alpha 1}$	L_3M_5	8146.17 ± 0.16	[50]	0.843	[488]	1.805(-1)	6.10	[423]
		8146.30	[104]					
$L_{\beta 6}$	L_3N_1	9315.40 ± 0.83	[50]			2.207(-3)	18.86	[423]
		9316.30	[104]					
	L_3N_2	9416.1 ± 1.1	[50]			2.103(-5)	18.04	[423]
		9415.60	[104]					
	L_3N_3	9474.4 ± 1.1	[50]			2.065(-5)	14.51	[423]
		9474.40	]					
$L_{\beta 15}$	L_3N_4	9639.50 ± 0.55	[50]	1.65(-2)	[488]	3.533(-3)	11.50	[423]
		9639.80	[104]					
$L_{\beta 2}$	L_3N_5	9651.89 ± 0.22	[50]	0.1480	[488]	3.167(-2)	9.80	[473]
		9651.90	[104]				11.17	[423]
$L_{u'}$	L_3N_6	9857.30	[104]			1.515(-5)		
		9868.99 ± 0.67	[145]					
L_u	L_3N_7	9871.52 ± 0.70	[145]			8.515(-5)		
	$L_3N_{6,7}$	9857.23 ± 0.46	[50]	7.09(-5)	[488]			
$L_{\beta 7}$	L_3O_1	9809.80 ± 0.40	[50]	1.81(-3)	[488]	4.433(-4)	10.14	[423]

line	transition	E/eV		I/eV/$\hbar$		TPIV	Γ/eV	
		9809.90	[104]					
	$L_3O_{2,3}$	9839.00 ± 2.50	[50]	1.44(-5)	[488]			
	L_3O_2	9837.70	[104]			4.532(-6)	6.10	[423]
	L_3O_3	9845.20	[104]	1.38(-5)	[488]	4.219(-6)	5.12	[423]
$L_{\beta5}$	$L_3O_{4,5}$	9875.00 ± 0.80	[50]	2.90(-3)	[488]			
		9875.80	[104]					
	L_3O_4					6.480(-5)		
	L_3O_5					5.557(-4)		

M series

line	transition	E/eV		I/eV/$\hbar$		TPIV	Γ/eV	
	M_1N_3	2295.00 ± 9.00	[50]			8.802(-4)	27.55	[423]
		2302.00	[263]					
	M_2N_4	2226.00 ± 1.60	[50]			3.136(-3)	18.27	[423]
		2226.00	[104]					
		2231.00	[263]					
	M_3N_1	1629.00 ± 2.00	[50]			9.761(-4)	24.55	[423]
		1631.40	[263]					
	M_3N_3	2295.00 ± 8.00	[50]	2.54(-3)	[335]		20.21	[423]
$M_{\gamma2}$	M_3N_4	1951.00 ± 1.50	[50]	2.251(-2)	[335]	4.711(-4)	17.20	[423]
		1952.00	[104]					
		1955.00	[263]					
$M_{\gamma1}$	M_3N_5	1964.00 ± 1.20	[50]	2.251(-2)	[335]	3.715(-3)	16.86	[423]
		1964.00	[104]					
		1968.30	[263]					
	M_3O_1	2126.00 ± 7.00	[50]			1.795(-4)	15.84	[423]
		2126.00	[104]					
		2130.00	[263]					
	$M_3O_{4,5}$	2190.00 ± 12.00	[50]					
		2190.00	[104]					
	M_3O_4					9.490(-6)		
	M_3O_5	2186.00	[104]			7.222(-5)		
$M_{\xi2}$	M_4N_2	1328.80 ± 0.70	[104]	1.12(-3)	[335]	2.240(-3)	16.10	[423]
		1328.80	[104]					
		1331.70	[263]					
M_δ	M_4N_3	1393.00 ± 3.00	[50]	1.12(-3)	[335]	2.375(-4)	12.58	[423]
		1394.10	[104]					
		1396.20	[263]					
M_β	M_4N_6	1765.50 ± 0.30	[50]	3.198(-2)	[335]	1.529(-2)		
		1765.50	[104]					
		1768.70	[263]					
M_η	M_4O_2	1748.00 ± 5.00	[50]			5.107(-4)	4.16	[423]
		1747.50	[104]					
	$M_4O_{2,3}$	1751.20	[263]					
	M_4O_3					5.417(-5)	3.18	[423]
$M_{\xi1}$	M_5N_3	1330.80 ± 0.60	[50]	3.96(-4)	[335]	2.007(-3)	2.23	[423]
		1330.80	[104]					
		1333.20	[263]					
	M_5N_4	1708.00	[104]				8.61	[423]

line	transition	E/eV		I/eV/$\hbar$		TPIV	Γ/eV	
	M_5N_5	1709.60	[104]				8.28	[423]
$M_{\alpha_{2,1}}$	$M_5N_{6,7}$	1709.60 ± 0.25	[50]					
		1712.50	[263]					
$M_{\alpha2}$	M_5N_6			3.198(-2)	[488]	8.356(-4)		
$M_{\alpha1}$	M_5N_7			3.198(-2)	[488]	1.625(-2)		
	M_5O_3	1700.00 ± 5.00	[50]			5.402(-4)		
		1700.00	[104]					
		1703.10	[263]					

N series

line	transition	E/eV		I/eV/$\hbar$		TPIV	Γ/eV	
	N_4N_6	213.00 ± 0.40	[50]			8.751(-5)		
	$N_5N_{6,7}$	202.80 ± 0.70	[50]					
	N_5N_6					4.012(-6)		
	N_5N_7					8.358(-5)		

level characteristics

level	E_B/eV		ω_{nlj}		Γ/eV		AE/eV	
K	67417.00	[503]	0.957	[301]	37.70	[301]	67416.4	[51]
	67416.40 ± 0.60	[493]	0.956	[38]	37.05	[423]	67403.7 ± 5.4	[50]
			0.9522	[39]	37.9	[94]	67431.9 ± 1.5	[145]
			0.9534	[423]				
			0.957	[310]				
			0.962	[33]				
L_1	11680.00	[503]	0.137	[301]	5.58	[301]	11681.5	[51]
	11681.50 ± 0.30	[493]	7.18(-2)	[423]	10.22	[423]	11682.1 ± 1.6	[50]
			0.131	[436]	6	[94]	11696.9 ± 1.4	[145]
			0.154	[578]				
L_2	11136.00	[503]	0.258	[301]	5.15	[301]	11136.1	[51]
	11136.10 ± 0.30	[493]	0.257	[361]	4.56	[423]	11132.5 ± 1.5	[50]
			0.25	[548]	4.69	[94]	11153.23 ± 0.56	[145]
			0.2813	[423]				
			0.26	[578]				
			0.262	[567]				
L_3	9881.00	[503]	0.243	[301]	4.88	[301]	9881.1	[51]
	9881.10 ± 0.30	[493]	0.25(3)	[212]	4.68	[94]	9876.7 ± 1.2	[50]
			0.27	[548]	4.49	[423]	9898.04 ± 0.36	[145]
			0.228	[361]				
			0.2519	[423]				
			0.251	[436]				
			0.241	[578]				
			0.233	[567]				
M_1	2708.00	[503]	1.45(-3)	[38]	19.30	[350]	2708.0	[51]
	2708.00 ± 0.40	[493]	1.85(-3)	[512]	17.53	[423]		
			2.43(-3)	[423]	14.3	[94]		
M_2	2469.00	[503]	2.64(-3)	[38]	12.00	[350]	2468.7	[51]
	2468.70 ± 0.30	[493]	1.50(-2)	[512]	11.27	[423]		
			2.64(-3)	[423]	7.8	[94]		

level	E_B/eV		ω_{nlj}		Γ/eV		AE/eV	
M_3	2194.00	[503]	2.14(-3)	[38]	10.80	[350]	2194.0	[51]
	2194.00 $\pm$ 0.30	[493]	2.22(-3)	[512]	10.19	[423]		
			5.69(-3)	[423]	5.7	[94]		
M_4	1793.00	[503]	1.30(-2)	[38]	3.25	[350]	1793.2	[51]
	1793.20 $\pm$ 0.30	[493]	1.60(-2)	[512]	2.56	[423]		
			1.83(-2)	[423]	1.61	[94]		
M_5	1735.00	[503]	2.05(-2)	[38]	1.80	[350]	1735.1	[51]
	1735.10 $\pm$ 0.30	[493]	2.00(-2)	[512]	1.61	[423]		
			1.96(-2)	[423]	1.61	[94]		
M_5	1735.00	[503]	2.05(-2)	[38]	1.80	[350]	1.7351	[51]
	1735.10 $\pm$ 0.30	[493]	2.00(-2)	[512]	1.61	[423]		
			1.96(-2)	[423]	1.61	[94]		
N_1	566.00	[503]	3.6(-4)	[423]	7	[94]		
	565.50 $\pm$ 0.50	[493]			14.36	[423]		
N_2	465.00	[503]	7.8(-4)	[423]	5.7	[94]		
	464.80 $\pm$ 0.50	[493]			13.55	[423]		
N_3	405.00	[503]	2.9(-4)	[423]	4	[94]		
	404.50 $\pm$ 0.40	[493]			10.02	[423]		
N_4	242.00	[503]	1.1(-4)	[423]	4	[94]		
	241.30 $\pm$ 0.40	[493]			7.01	[423]		
N_5	230.00	[503]	1.1(-4)	[423]	3.65	[94]		
	229.30 $\pm$ 0.30	[493]			6.67	[423]		
N_6	27.00	[503]			0.08	[94]		
	27.50 $\pm$ 0.20	[493]						
N_7	25.00	[503]			0.04	[94]		
	25.60 $\pm$ 0.20	[493]						
O_1	71.00	[503]	2.0(-6)	[423]	5.65	[423]		
	71.10 $\pm$ 0.50	[493]						
O_2	45.00	[503]			1.60	[423]		
	44.90 $\pm$ 0.40	[51]						
	43.70 $\pm$ 1.40	[493]						
O_3	37.00	[503]			0.63	[423]		
	36.40 $\pm$ 0.40	[51]						
	34.70 $\pm$ 2.50	[493]						
O_4	6.00	[503]						
	5.70 $\pm$ 0.40	[493]						
IP	7.5496	[222]						

W   Z=74   [Xe] 4f^{14} 5d^4 6s^2

Tungsten $A = 183.84(1)$ [222] $\varrho = 19.30 \, \text{g/cm}^3$ [547]

X-ray transitions

line	transition	E/eV		I/eV/$\hbar$		TPIV	Γ/eV	
K series								
K_{α_3}	KL$_1$	57420 ± 16	[50]	8.77(-3)	[488]		50.78	[423]
		57421.00	[104]					
$K_{\alpha 2}$	KL$_2$	57981.70 ± 0.55	[50]	10.87	[488]	2.765(-1)	45.20	[301]
		57982.00	[104]	10.97	[487]		43.84	[423]
		57982.60 ± 0.80	[143]					
		57982.60 ± 0.80	[277]					
		57981.77 ± 0.14	[144]					
$K_{\alpha 1}$	KL$_3$	59318.24 ± 0.06	[50]	4.801(-1)			44.90	[488]
		59319.00	[104]	19.06	[487]		47.75	[473]
		59319.23 ± 0.05	[143]				43.78	[423]
		59319.23 ± 0.05	[277]					
		59318.847 ± 0.050	[144]					
	KM$_1$	66706.9 ± 3.3	[145]				56.54	[423]
$K_{\beta 3}$	KM$_2$	66951.40 ± 0.70	[50]	2.120	[487]	5.162(-2)	50.70	[423]
		66956.30	[104]					
		66952.40 ± 1.07	[277]					
		66952.19 ± 0.25	[144]					
$K_{\beta 1}$	KM$_3$	67244.30 ± 0.70	[50]	3.92	[488]	9.976(-2)	51.53	[279]
		67243.00	[104]	4.10	[487]		49.27	[423]
		67245.45 ± 1.08	[277]					
$K_{\beta_5^{II}}$	KM$_4$	67652.3 ± 2.7	[50]	4.39(-2)	[488]	1.116(-3)	41.66	[423]
		67653.00	[104]					
		67245.0 ± 1.1	[144]					
$K_{\beta_5^{I}}$	KM$_5$	67715.9 ± 3.8	[50]	5.54(-2)	[488]	1.407(-3)	40.85	[423]
		67716.40	[104]					
	KN$_1$	68932.0 ± 4.3	[145]				53.64	[423]
$K_{\beta_2^{II}}$	KN$_2$	69032.5 ± 5.7	[50]	0.462	[488]	1.175(-2)	52.76	[423]
		69039.00	[104]	0.495	[487]			
$K_{\beta_2^{I}}$	KN$_3$	69101.3 ± 4.0	[50]	0.896	[488]	2.278(-2)	49.16	[423]
		69106.00	[104]	0.969	[487]			
$K_{\beta 4}$	KN$_{4,5}$	69295 ± 11	[50]	2.391(-2)	[488]			
		69295.00	[104]					
$K_{\beta_4^{II}}$	KN$_4$	69267.3 ± 3.2	[145]			2.695(-4)	46.41	[423]
$K_{\beta_4^{I}}$	KN$_5$	69281.8 ± 1.8	[145]			3.383(-4)	46.02	[423]
	KO$_{2,3}$	69479.00 ± 2.00	[50]	0.2023	[488]			
		69488.00	[104]	0.2218	[487]			

line	transition	E/eV		I/eV/$\hbar$		TPIV	Γ/eV	
	KO$_2$					2.670(-3)	41.28	[423]
	KO$_3$					5.470(-3)	39.99	[423]

L series

line	transition	E/eV		I/eV/$\hbar$		TPIV	Γ/eV	
	L$_1$M$_1$	9278.10 ± 0.68	[50]				29.01	[423]
L$_{\beta 4}$	L$_1$M$_2$	9525.23 ± 0.54	[50]	0.2636	[488]	2.268(-2)	15.10	[279]
		9523.60	[104]				14.60	[473]
							23.17	[423]
L$_{\beta 3}$	L$_1$M$_3$	9818.91 ± 0.46	[50]	0.330	[488]	2.839(-2)	12.30	[279]
		9817.10	[104]				13.10	[473]
							21.74	[423]
L$_{\beta 10}$	L$_1$M$_4$	10227.90 ± 0.30	[50]	7.65(-3)	[488]	6.581(-4)	14.13	[423]
		10227.00	[104]					
		10228.29 ± 0.24	[440]					
L$_{\beta 9}$	L$_1$M$_5$	10290.70 ± 0.60	[50]	0.1145	[488]	9.849(-4)	13.32	[423]
		10290.10	[104]					
		10291.13 ± 0.22	[440]					
	L$_1$N$_1$	11502.8 ± 1.1	[218]				26.11	[423]
L$_{\gamma 2}$	L$_1$N$_2$	11610.50 ± 0.44	[50]	6.47(-2)	[488]	5.567(-3)	12.10	[473]
		11607.70	[104]				25.23	[423]
		11622.00	[543]					
L$_{\gamma 3}$	L$_1$N$_3$	11680.49 ± 0.73	[50]	8.58(-2)	[488]	7.386(-3)	11.00	[473]
		11674.40	[104]				21.63	[423]
		11691.00	[543]					
	L$_1$N$_4$	11843.9 ± 3.3	[50]	1.08(-3)	[488]	9.306(-5)	18.87	[423]
		11847.00	[104]					
L$_{\gamma 11}$	L$_1$N$_5$	11861.9 ± 1.1	[50]	1.69(-3)	[488]	1.453(-4)	18.48	[423]
		11857.00	[104]					
	L$_1$N$_6$	12063.2 ± 1.8	[145]					
	L$_1$N$_7$	12065.8 ± 1.7	[145]					
	L$_1$O$_1$	12017.00 ± 3.50	[50]	2.68(-7)	[488]		17.44	[423]
		12017.00	[104]					
L$_{\gamma 4'}$	L$_1$O$_2$	12053.00 ± 0.40	[50]	1.002(-2)	[488]	8.620(-4)	13.74	[423]
		12053.00	[104]					
		12069.00	[543]					
L$_{\gamma 4}$	L$_1$O$_3$	12063.40 ± 0.40	[50]	1.288(-2)	[488]	1.108(-3)	12.46	[423]
		12063.00	[104]					
		12082.00	[543]					
	L$_1$O$_{4,5}$	12095.00 ± 2.40	[50]	7.15(-5)	[488]			
		12094.00	[104]					
	L$_1$O$_4$					2.465(-6)		
	L$_1$O$_5$					3.716(-6)		
L$_\eta$	L$_2$M$_1$	8724.42 ± 0.25	[50]	3.06(-2)	[488]	6.763(-3)	22.07	[423]
		8723.70	[104]					
		8708.00	[543]					
	L$_2$M$_2$	8952.90	[104]	8.65(-7)	[488]		16.23	[423]
		8953.03 ± 0.65	[388]					
L$_{\beta 17}$	L$_2$M$_3$	9268.72 ± 0.48	[50]	1.01(-3)	[488]	2.120(-4)	14.80	[423]

line	transition	E/eV		I/eV/$\hbar$		TPIV	Γ/eV	
		9262.00	[104]					
$L_{\beta 1}$	L_2M_4	9672.58 ± 0.10	[50]	1.138	[488]	2.394(-1)	6.50	[279]
		9672.50	[104]				6.90	[473]
		9667.00	[543]				7.19	[423]
	L_2M_5	9735.08 ± 0.73	[50]	4.49(-5)	[488]		6.38	[423]
		9741.00	[104]					
$L_{\gamma 5}$	L_2N_1	10948.91 ± 0.39	[50]	7.55(-3)	[488]	1.611(-3)	19.18	[423]
		10948.00	[104]					
	L_2N_2	11045.20 ± 0.96	[50]	3.17(-7)	[488]		18.29	[423]
		11052.00	[104]					
	L_2N_3	11120.5 ± 3.0	[50]	2.61(-4)	[488]	5.488(-5)	14.70	[423]
		11120.00	[104]					
$L_{\gamma 1}$	L_2N_4	11286.00 ± 0.46	[50]	0.2118	[488]	4.455(-2)	9.80	[279]
		11286.00	[104]				10.28	[473]
		11277.00	[543]				11.94	[423]
	L_2N_5	11299.89 ± 0.66	[145]				11.55	[423]
	$L_2N_{6,7}$	11510.00 ± 1.10	[50]					
	L_2N_6	11507.48 ± 0.79	[145]			1.175(-4)		
	L_2N_7	11510.08 ± 0.72	[145]					
L_ν	$L_2N_{6,7}$	11511.00	[104]	5.59(-3)	[488]			
$L_{\gamma 8}$	L_2O_1	11467.70 ± 0.40	[50]	1.37(-3)	[488]	3.089(-4)	10.51	[423]
		11467.00	[104]					
	L_2O_3	11488.00 ± 2.10	[50]	3.91(-5)	[488]	1.143(-5)	5.52	[423]
		11491.00	[104]					
$L_{\gamma 6}$	L_2O_4	11538.70 ± 0.55	[50]	5.86(-3)	[488]	1.221(-3)	7.00	[473]
		11539.00	[104]					
L_l	L_3M_1	7387.82 ± 0.65	[50]	4.68(-2)	[488]	1.253(-2)	22.01	[423]
		7387.70	[104]					
		7372.00	[543]					
L_t	L_3M_2	7631.41 ± 0.86	[50]	4.83(-4)	[488]	1.072(-4)	16.17	[423]
		7631.20	[104]					
L_s	L_3M_3	7926.44 ± 0.93	[50]	4.39(-4)	[488]	9.713(-5)	14.74	[423]
		7927.20	[104]					
$L_{\alpha 2}$	L_3M_4	8335.34 ± 0.17	[50]	0.1018	[488]	2.127(-2)	7.20	[473]
		8335.40	[104]				7.14	[423]
		8331.00	[543]					
$L_{\alpha 1}$	L_3M_5	8398.242 ± 0.054	[50]	0.898	[488]	1.877(-1)	6.50	[473]
		8397.40	[104]				6.32	[423]
		8393.00	[543]					
$L_{\beta 6}$	L_3N_1	9608.199 ± 0.074	[50]	1.119(-2)	[488]	2.338(-3)	19.12	[423]
		9611.10	[104]					
	L_3N_2	9712.7 ± 2.3	[50]	1.08(-4)	[488]	2.262(-5)	18.23	[423]
		9712.00	[104]					
	L_3N_3	9784.00 ± 1.50	[50]	1.07(-4)	[488]	2.231(-5)	14.64	[423]
		9784.00	[104]					
$L_{\beta 15}$	L_3N_4	9947.95 ± 0.35	[50]	1.775(-2)	[488]	3.711(-3)	11.88	[423]

line	transition	E/eV		I/eV/$\hbar$		TPIV	Γ/eV	
		9947.40	[104]					
$L_{\beta 2}$	$L_3 N_5$	9964.133 ± 0.078	[50]	0.1593	[488]	3.328(-2)	9.80	[279]
		9961.20	[104]				11.49	[423]
L_u	$L_3 N_{6,7}$	10173.49 ± 0.62	[50]	4.65(-4)	[488]			
		10173.00	[104]					
$L_{u'}$	$L_3 N_6$	10170.58 ± 0.69	[145]			1.731(-5)		
L_u	$L_3 N_7$	10173.18 ± 0.68	[145]			9.720(-5)		
$L_{\beta 7}$	$L_3 O_1$	10129.20 ± 0.30	[50]	2.01(-3)	[488]	4.731(-4)	10.45	[423]
		10129.00	[104]					
	$L_3 O_{2,3}$	10153.00 ± 2.00	[50]	1.65(-5)	[488]			
		10154.00	[104]					
	$L_3 O_2$					4.838(-6)	6.75	[423]
	$L_3 O_3$					4.560(-6)	5.46	[423]
$L_{\beta 5}$	$L_3 O_{4,5}$	10200.40 ± 0.25	[50]	4.68(-3)	[488]		6.50	[279]
		10200.00	[104]					
	$L_3 O_4$					1.017(-4)		
	$L_3 O_5$					8.690(-4)		

M series

line	transition	E/eV		I/eV/$\hbar$		TPIV	Γ/eV	
	$M_1 N_3$	2397.00 ± 4.00	[50]	1.200(-2)	[116]	9.076(-4)	27.40	[423]
		2396.00	[104]	2.412(-2)	[335]			
		2402.00	[263]					
	$M_1 O_2$					2.116(-4)	19.51	[423]
	$M_1 O_3$	2799.00	[263]			1.964(-4)	18.22	[423]
	$M_1 O_{2,3}$	2790.00 ± 13.00	[50]	4.17(-3)	[116]			
	$M_2 N_1$	1973.00 ± 6.00	[50]	4.35(-3)	[116]	6.166(-4)	26.04	[423]
		1971.20	[263]	2.88(-3)	[335]			
	$M_2 N_4$	2314.00 ± 2.00	[50]	2.237(-2)	[116]	3.245(-3)	18.70	[423]
		2316.10	[104]	2.458(-2)	[335]			
		2319.00	[263]					
	$M_3 N_1$	1684.00 ± 2.00	[50]	5.88(-3)	[116]	1.020(-3)	24.61	[423]
		1684.00	[104]	2.88(-3)	[116]			
		1686.90	[263]					
$M_{\gamma 2}$	$M_3 N_4$	2021.00 ± 1.30	[50]	2.85(-4)	[116]	4.942(-4)	17.37	[423]
		2021.00	[104]	2.457(-2)	[335]			
		2026.00	[263]					
$M_{\gamma 1}$	$M_3 N_5$	2035.00 ± 1.00	[50]	2.258(-2)	[116]	3.887(-3)	6.00	[279]
		2036.00	[104]	2.458(-2)	[335]		16.98	[423]
		2040.00	[263]					
	$M_3 O_1$	2203.00 ± 3.00	[50]	9.58(-4)	[116]	1.896(-4)	15.94	[423]
		2202.00	[104]					
		2208.00	[263]					
	$M_3 O_5$	2275.20	[104]	7.97(-4)	[116]	1.136(-4)		
$M_{\xi 2}$	$M_4 N_2$	1378.70 ± 0.80	[50]	2.04(-3)	[116]	2.188(-3)	16.11	[423]
		1378.20	[104]	1.20(-3)	[335]			
		1382.20	[263]					
M_{δ}	$M_4 N_3$	1446.00 ± 1.40	[50]	2.67(-4)	[116]	2.274(-4)	0.84	[473]

line	transition	E/eV		I/eV/ℏ		TPIV	Γ/eV	
		1445.60	[104]	1.20(-3)	[335]		12.52	[423]
		1448.40	[263]					
M_β	M_4N_6	1834.90 ± 0.30	[50]	5.09(-6)	[116]	1.694(-2)		
		1835.00	[104]	3.592(-2)	[335]		2.14	[473]
M_η	$M_4O_{2,3}$	1822.00 ± 2.40	[50]	1.37(-4)	[116]		3.01	[473]
	M_4O_2					5.106(-4)	4.63	[423]
	M_4O_3					5.383(-5)	3.34	[423]
		1821.00	[104]					
		1825.40	[263]					
$M_{\xi1}$	M_5N_3	1383.50 ± 0.60	[50]	1.73(-3)	[116]	1.925(-3)	2.30	[473]
		1383.50	[104]	1.20(-3)	[335]		11.71	[423]
		1386.80	[263]					
$M_{\alpha2}$	M_5N_6	1773.10 ± 0.50	[50]	1.84(-3)	[116]	9.166(-4)	0.31	[473]
		1773.50	[104]	3.592(-2)	[335]			
		1776.80	[263]					
$M_{\alpha1}$	M_5N_7	1775.40 ± 0.30	[50]	3.565(-2)	[116]	1.784(-2)	1.68	[473]
		1775.40	[104]	3.592(-2)	[335]			
		1779.10	[263]					
	M_5O_3	1770.00 ± 2.30	[50]	1.11(-4)	[116]	5.187(-4)	2.53	[423]
		1770.00	[104]					
		1773.70	[263]					
N series								
	N_2N_4	229.50 ± 0.90	[50]			6.797(-4)	20.86	[423]
	N_4N_6	222.10 ± 0.40	[50]			1.006(-4)		
	N_5N_6	208.00 ± 1.00	[50]			4.595(-6)		
	N_5N_7	212.20	[50]			9.608(-5)		

level characteristics

level	E_B/eV		ω_{nlj}		Γ/eV		AE/eV	
K	69525.00	[503]	0.958	[301]	39.90	[301]	69525.0	[51]
	69525.00 ± 0.30	[493]	0.957	[38]	39.15	[423]	69508.5 ± 5.8	[50]
			0.9538	[39]	40.1	[94]	69533.0 ± 1.6	[145]
			0.0.9539	[423]				
			0.956	[33]				
			0.954	[112]				
			0.961	[242]				
L_1	12099.00	[503]	0.147	[301]	5.61	[301]	12099.8	[51]
	12099.30 ± 0.30	[493]	0.138	[107]	11.62	[423]	12099.73 ± 0.87	[50]
			6.92(-2)	[423]	6.3	[94]	12106.9 ± 1.4	[145]
			0.136	[436]				
			0.137	[244]				
			0.13	[567]				

level	E_B/eV		ω_{nlj}		Γ/eV		AE/eV	
L_2	11542.00	[503]	0.270	[301]	5.33	[301]	11544.0	[51]
	11544.00 ± 0.30	[493]	0.271	[107]	4.69	[423]	11538.6 ± 1.6	[50]
			0.2942	[423]	4.82	[94]	11551.16 ± 0.47	[145]
			0.291	[436]				
			0.275	[578]				
			0.274	[567]				
L_3	10205.00	[503]	0.255	[301]	4.98	[301]	10206.8	[51]
	10206.80 ± 0.30	[493]	0.253	[107]	4.63	[423]	10200.1 ± 1.2	[50]
			0.268	[348]	4.81	[94]	10214.26 ± 0.36	[145]
			0.272	[435]				
			0.2627	[423]				
			0.261	[436]				
			0.264	[244]				
			0.246	[578]				
			0.245	[567]				
M_1	2820.00	[503]	1.91(-3)	[512]	21.00	[74]	2819.6	[51]
	2819.60 ± 0.40	[493]	2.54(-3)	[423]	17.39	[423]		
			1.63(-3)	[245]	14.5	[94]		
M_2	2575.00	[503]	1.60(-2)	[512]	10.00	[74]	2574.9	[51]
	2574.90 ± 0.30	[493]	4.85(-3)	[423]	11.54	[423]		
			2.52(-3)	[245]	8.1	[94]		
M_3	2281.00	[503]	2.56(-3)	[512]	13.00	[74]	2281.0	[51]
	2281.00 ± 0.30	[493]	5.99(-3)	[423]	10.12	[423]		
			2.94(-3)	[245]	6.4	[94]		
M_4	1872.00	[503]	1.80(-2)	[512]	1.90	[74]	1871.6	[51]
	1871.60 ± 0.30	[493]	1.99(-2)	[423]	2.51	[423]		
					1.7	[94]		
M_5	1810.00	[503]	2.10(-2)	[512]	3.50	[74]	1809.2	[51]
	1809.20 ± 0.30	[493]	2.12(-2)	[423]	1.70	[423]		
					1.7	[94]		
N_1	595.00	[503]	3.9(-4)	[423]	16.00	[74]		
	595.00 ± 0.40	[493]			14.49	[423]		
					7.3	[94]		
N_2	492.00	[503]	8.3(-4)	[423]	13.00	[74]		
	491.60 ± 0.40	[493]			13.60	[423]		
					5.8	[94]		
N_3	426.00	[503]	3.1(-4)	[423]	13.00	[74]		
	425.30 ± 0.50	[493]			10.01	[423]		
					4.2	[94]		
N_4	259.00	[503]	1.3(-4)	[423]	7.80	[74]		
	258.80 ± 0.40	[493]			7.25	[423]		
					4.1	[94]		
N_5	246.00	[503]	1.2(-4)	[423]	7.80	[74]		
	245.40 ± 0.40	[493]			6.86	[423]		
					3.8	[94]		

level	E_B/eV		ω_{nlj}		Γ/eV	AE/eV
N_6	37.00	[503]				
	36.50 ± 0.40	[51]			0.07	[74]
	37.40 ± 0.60	[493]			0.1	[94]
N_7	34.00	[503]				
	33.60 ± 0.40	[51]			0.07	[74]
	35.10 ± 1.00	[493]			0.06	[94]
O_1	77.00	[503]	3.0(-6)	[423]	5.82	[423]
	77.10 ± 0.40	[493]				
O_2	74.00	[503]			2.12	[423]
	46.80 ± 0.50	[51]				
	46.70 ± 0.50	[493]				
O_3	37.00	[503]			0.83	[423]
	35.60 ± 0.50	[51]				
	36.50 ± 2.30	[493]				
O_4	6.00	[503]				
	6.10 ± 0.40	[493]				
IP	7.8640	[222]				

Re	Z=75	[Xe] $4f^{14} 5d^5 6s^2$

Rhenium $A = 186.207(1)$ [222] $\varrho = 21.06\,\mathrm{g/cm^3}$ [547]
$\varrho = 20.53\,\mathrm{g/cm^3}$ [483]

X-ray transitions

line	transition	E/eV		I/eV/$\hbar$		TPIV	Γ/eV	
K series								
K_{α_3}	KL_1	59150.0 ± 2.2	[145]				54.00	[423]
$K_{\alpha 2}$	KL_2	59718.57 ± 0.43	[50]	11.51	[488]	2.773(-1)	47.60	[301]
		59719.00	[104]					
$K_{\alpha 1}$	KL_3	61141.00 ± 0.89	[50]	19.91	[488]	4.795(-1)	47.20	[301]
		61141.40	[104]					
	KM_1	68745.6 ± 3.4	[145]				59.02	[423]
$K_{\beta 3}$	KM_2	68995.2 ± 1.7	[50]	2.515	[488]	5.182(-2)	53.13	[423]
		68995.70	[104]					
$K_{\beta 1}$	KM_3	69310.3 ± 1.7	[50]	4.16	[488]	1.001(-1)	51.73	[423]
		69310.30	[104]					
$K_{\beta_5^{II}}$	KM_4	69719.6 ± 5.8	[50]	4.80(-2)	[488]	1.155(-3)	43.80	[423]
$K_{\beta_5^{I}}$	KM_5	69786.3 ± 5.8	[50]	6.02(-2)	[488]	1.450(-3)	43.12	[423]
	KN_1	71053.6 ± 4.4	[145]				55.97	[423]
$K_{\beta_2^{II}}$	KN_2	71152.0 ± 6.0	[50]	0.492	[488]	1.186(-2)	54.98	[423]
		71147.90	[104]					
$K_{\beta_2^{I}}$	KN_3	71232.1 ± 3.6	[50]	0.956	[488]	2.304(-2)	51.28	[423]
		71230.20	[104]					
$K_{\beta 4}$	$KN_{4,5}$	71410 ± 12	[50]	2.639(-3)	[488]		48.81	[423]
$K_{\beta_4^{II}}$	KN_4	71402.1 ± 3.2	[145]			2.825(-4)		
$K_{\beta_4^{I}}$	KN_5	71417.6 ± 1.9	[145]			3.531(-4)	48.35	[423]
	$KO_{2,3}$	71633.00 ± 4.00	[50]	0.2237	[488]			
	KO_2					2.702(-3)	44.79	[423]
	KO_3					5.496(-3)	43.49	[423]
L series								
	L_1M_1	9595.6 ± 3.2	[145]				30.36	[423]
$L_{\beta 4}$	L_1M_2	9846.35 ± 0.58	[50]	0.2830	[488]	2.250(-2)	16.40	[473]
		9846.00	[104]					
$L_{\beta 3}$	L_1M_3	10159.90 ± 0.62	[50]	0.349	[488]	2.775(-2)	13.60	[473]
		10160.00	[104]					
$L_{\beta 10}$	L_1M_4	10577.07 ± 0.67	[50]	8.45(-3)	[488]	6.718(-4)	15.14	[423]
		10577.00	[104]					
$L_{\beta 9}$	L_1M_5	10643.45 ± 0.54	[50]	1.265(-2)	[488]	1.006(-3)	14.46	[423]

line	transition	E/eV		I/eV/ℏ		TPIV	Γ/eV	
		10643.20	[104]					
	L_1N_1	11898.5 ± 1.7	[50]	1.54(-6)	[488]		27.31	[423]
		11899.00	[104]					
$L_{\gamma 2}$	L_1N_2	12009.95 ± 0.86	[50]	6.99(-2)	[488]	5.561(-3)	13.20	[473]
		12009.30	[104]				26.32	[423]
$L_{\gamma 3}$	L_1N_3	12082.5 ± 1.2	[50]	9.16(-2)	[488]	7.284(-3)	12.00	[473]
		12081.90	[104]				22.62	[423]
	L_1N_4	12252.4 ± 1.8	[50]	1.21(-3)	[488]	9.598(-5)	20.15	[423]
		12252.00	[104]					
$L_{\gamma 11}$	L_1N_5	12265.8 ± 1.8	[50]	1.89(-3)	[488]	1.501(-4)	19.69	[423]
		12266.00	[104]					
	L_1N_6	12483.0 ± 2.3	[145]					
	L_1N_7	12486.0 ± 2.3	[145]					
	L_1O_1	12442.00 ± 1.20	[50]	3.19(-7)	[488]		20.98	[423]
		12442.00	[104]					
$L_{\gamma 4'}$	L_1O_2	12481.30 ± 0.60	[50]			8.917(-4)	16.13	[423]
$L_{\gamma 4}$	L_1O_3	12492.00 ± 0.60	[50]	1.10(-4)	[488]	1.134(-3)	14.83	[423]
		12492.00	[104]					
	$L_1O_{4,5}$	12524.00 ± 1.30	[50]	1.426(-2)	[488]			
	L_1O_4					3.618(-6)		
	L_1O_5					5.482(-6)		
L_η	L_2M_1	9027.27 ± 0.49	[50]	2.37(-2)	[488]	6.959(-3)	22.51	[423]
		9027.10	[104]					
	L_2M_2	9275.9 ± 1.0	[50]	1.02(-6)	[488]		16.62	[423]
		9276.00	[104]					
$L_{\beta 17}$	L_2M_3	9591.0 ± 1.1	[50]	1.11(-4)	[488]	2.279(-4)	15.23	[423]
		9591.00	[104]					
$L_{\beta 1}$	L_2M_4	10010.04 ± 0.24	[50]	1.215	[488]	2.497(-1)	6.50	[473]
		10010.00	[104]				7.29	[423]
	L_2M_5	10075.8 ± 1.2	[50]	5.13(-5)	[488]		6.61	[423]
		10075.00	[104]					
$L_{\gamma 5}$	L_2N_1	11334.18 ± 0.77	[50]	8.11(-3)	[488]	1.682(-3)	19.46	[423]
		11334.00	[104]					
	L_2N_2	11438.5 ± 1.6	[50]	3.77(-7)	[488]		18.47	[423]
		11438.00	[104]					
	L_2N_3	11515.0 ± 1.6	[50]	2.90(-4)	[488]	5.954(-5)	14.77	[423]
		11515.00	[104]					
$L_{\gamma 1}$	L_2N_4	11685.53 ± 0.82	[50]	0.2289	[488]	4.704(-2)	9.50	[473]
		11685.00	[104]				12.30	[423]
	L_2N_5	11699.29 ± 0.67	[145]				16.66	[423]
L_ν	L_2N_6	11916.8 ± 1.7	[50]	6.48(-4)	[488]	1.332(-4)		
		11917.00	[104]					
	L_2N_7	11917.7 ± 1.4	[145]					
$L_{\gamma 8}$	L_2O_1	11875.80 ± 0.60	[50]	1.51(-3)	[488]	13.257(-4)	13.13	[423]
		11876.00	[104]					

line	transition	E/eV		I/eV/$\hbar$		TPIV	Γ/eV	
	L_2O_3	11925.00 ± 1.10	[50]	4.51(-5)	[488]	1.223(-5)	6.98	[423]
		11925.00	[104]					
$L_{\gamma6}$	L_2O_4	11956.00 ± 1.00	[50]	8.69(-3)	[488]	1.805(-3)		
		11955.00	[104]					
L_l	L_3M_1	7603.67 ± 0.35	[50]	5.05(-2)	[488]	1.293(-2)	22.45	[423]
		7603.50	[104]					
L_t	L_3M_2	7852.45 ± 0.74	[50]	5.31(-4)	[488]	1.136(-4)	16.56	[423]
		7853.50	[104]					
L_s	L_3M_3	8168.55 ± 0.80	[50]	4.81(-4)	[488]	1.028(-4)	15.17	[423]
		8168.00	[104]					
$L_{\alpha2}$	L_3M_4	8586.27 ± 0.44	[50]	0.1084	[488]	2.210(-2)	6.44	[423]
		8586.30	[104]					
$L_{\alpha1}$	L_3M_5	8652.55 ± 0.36	[50]	0.965	[488]	1.949(-1)	6.55	[423]
		8652.30	[104]					
$L_{\beta6}$	L_3N_1	9910.66 ± 0.59	[50]	1.212(-2)	[488]	2.473(-3)	19.40	[423]
		9910.20	[104]					
	L_3N_2	10018.4 ± 3.2	[145]			2.429(-5)	18.41	[423]
	L_3N_3	10093.8 ± 1.2	[50]	1.18(-4)	[488]	2.403(-5)	14.71	[423]
		10093.00	[104]					
$L_{\beta15}$	L_3N_4	10261.82 ± 0.63	[50]	1.608(-2)	[488]	3.889(-3)	9.44	[423]
		10262.00	[104]					
$L_{\beta2}$	L_3N_5	10275.35 ± 0.50	[50]	0.1713	[488]	3.493(-2)	16.60	[423]
		10275.00	[104]					
L_u	$L_3N_{6,7}$	10493.6 ± 1.3	[50]	5.40(-4)	[488]			
L'_u	L_3N_6	10495.00	[104]			1.963(-5)		
		10491.9 ± 1.2	[145]					
L_u	L_3N_7	10494.9 ± 1.3	[145]			1.102(-4)		
$L_{\beta7}$	L_3O_1	10452.90 ± 0.50	[50]			5.044(-4)	13.07	[423]
$L_{\beta5}$	$L_3O_{4,5}$	10531.80 ± 0.50	[50]	6.92(-3)	[488]			
	L_3O_4	10532.00	[104]			1.486(-4)		
	L_3O_5					1.272(-3)		

M series

line	transition	E/eV		I/eV/$\hbar$		TPIV	Γ/eV	
$M_{\gamma2}$	M_3N_4	2090.00 ± 1.80	[50]	2.68(-2)	[335]	5.171(-4)		
		2090.00	[104]					
		2095.00	[263]					
$M_{\gamma1}$	M_3N_5	2106.70 ± "0.70	[50]			4.057(-3)		
		2111.00	[263]					
$M_{\xi2}$	M_4N_2	1431.00 ± 0.80	[50]	1.29(-3)	[335]	2.133(-3)		
		1431.00	[104]					
		1433.30	[263]					
M_δ	M_4N_3	1505.00 ± 1.50	[50]	1.29(-3)	[335]	2.216(-4)		
		1504.80	[104]					
		1508.30	[263]					
M_β	M_4N_6	1906.10 ± 0.30	[50]			1.869(-2)		
		1910.10	[263]					
M_η	M_4O_2	1900.83	[423]			5.096(-4)		

line	transition	E/eV		I/eV/$\hbar$		TPIV	Γ/eV
$M_{\xi 1}$	M_5N_3	1436.80 ± 0.70	[50]	1.29(-3)	[335]	1.844(-3)	
		1436.80	[104]				
		1440.00	[263]				
$M_{\alpha 2}$	M_5N_6	1842.50 ± 0.30	[50]	4.01(-2)	[335]	1.000(-3)	
$M_{\alpha 1}$	M_5N_7	1899.00	[104]	4.01(-2)	[335]	1.949(-2)	
$M_{\alpha 1.2}$	$M_5N_{6,7}$	1842.50 ± 0.30	[50]				
		1846.40	[263]				

level characteristics

level	E_B/eV		ω_{nlj}		Γ/eV		AE/eV	
K	71677.00	[503]	0.959	[301]	42.10	[301]	71676.4	[51]
	71676.40 ± 0.40	[493]	0.959	[38]	41.33	[423]	71657.8 ± 6.1	[50]
			0.9553	[39]	42.2	[94]	71687.5 ± 1.7	[145]
			0.9551	[423]				
			0.959	[310]				
L_1	12527.00	[503]	0.144	[301]	6.18	[301]	12526.7	[51]
	12526.70 ± 0.40	[493]	6.85(-2)	[423]	12.67	[423]	12531.1 ± 1.9	[50]
			8.40(-2)	[436]	6.7	[94]	12537.5 ± 1.4	[145]
			0.116	[578]				
L_2	11957.00	[503]	0.283	[301]	5.48	[301]	11958.7	[51]
	11958.70 ± 0.30	[493]	0.3080	[423]	4.82	[423]	11954.7 ± 1.7	[50]
			0.304	[436]	4.95	[94]	11969.17 ± 0.47	[145]
			0.293	[578]				
L_3	10535.00	[503]	0.268	[301]	5.04	[301]	10535.3	[51]
	10535.30 ± 0.30	[493]	0.2736	[423]	4.76	[423]	10531.1 ± 1.3	[50]
			0.271	[436]	4.95	[94]	10546.35 ± 0.36	[145]
			0.254	[578]				
M_1	2932.00	[503]	1.97(-3)	[512]	17.69	[423]	2931.7	[51]
	2931.70 ± 0.40	[493]	2.69(-3)	[423]	14.6	[94]		
M_2	2682.00	[503]	2.40(-2)	[512]	11.80	[423]	2681.6	[51]
	2681.60 ± 0.40	[493]	1.60(-2)	[423]	8.4	[94]		
M_3	2367.00	[503]	2.9(-3)	[512]	10.40	[423]	2367.3	[51]
	2367.30 ± 0.30	[493]	6.30(-3)	[423]	6.9	[94]		
M_4	1949.00	[503]	2.00(-2)	[512]	2.47	[423]	1948.9	[51]
	1948.90 ± 0.30	[493]	2.16(-2)	[423]	1.79	[94]		
M_5	1883.00	[503]	2.30(-2)	[512]	1.79	[423]	1882.9	[51]
	1882.90 ± 0.30	[493]	2.28(-2)	[423]	1.79	[94]		

level	E_B/eV		ω_{nlj}		Γ/eV		AE/eV
N_1	625.00	[503]	4.1(-4)	[423]	14.64	[423]	
	625.00 ± 0.40	[493]			7.5	[94]	
N_2	518.00	[503]	8.8(-4)	[423]	13.65	[423]	
	517.90 ± 0.50	[493]			5.9	[94]	
N_3	445.00	[503]	3.3(-4)	[423]	9.95	[423]	
	444.40 ± 0.50	[493]			4.4	[94]	
N_4	274.00	[503]	1.5(-4)	[423]	7.48	[423]	
	273.70 ± 0.50	[493]			4.1	[94]	
N_5	260.00	[503]	1.4(-4)	[423]	7.02	[423]	
	260.20 ± 0.40	[493]			3.9	[94]	
N_6	47.00	[503]			0.15	[94]	
	48.10 ± 0.20	[493]					
N_7	45.00	[503]			0.11	[94]	
	45.70 ± 0.20	[493]					
O_1	83.00	[503]	3.0(-6)	[423]	8.31	[423]	
	82.80 ± 0.50	[493]					
O_2	46.00	[503]			3.46	[423]	
	45.60 ± 0.70	[51]					
	48.40 ± 2.40	[493]					
O_3	35.00	[503]			2.16	[423]	
	34.60 ± 0.60	[51]					
	36.80 ± 2.10	[493]					
O_4	4.00	[503]					
	3.50 ± 0.50	[51]					
	3.80 ± 0.50	[493]					
O_5	4.00	[493]					
	3.50 ± 0.50	[51]					
	2.50 ± 0.50	[493]					
P_1	7.88	[126]					
IP	7.8335	[222]					

Os   Z=76

[Xe] $4f^{14}\,5d^6\,6s^2$

Osmium

$A = 190.23(3)$ [222] $\varrho = 22.57\,\text{g/cm}^3$ [547]
$\varrho = 22.50\,\text{g/cm}^3$ [483]

X-ray transitions

line	transition	E/eV		I/eV/$\hbar$		TPIV	Γ/eV	
K series								
K_{α_3}	KL_1	60905.6 ± 2.3	[145]				56,73	[423]
K_{α_2}	KL_2	61487.27 ± 0.90	[50]	12.18	[488]	2.780(-1)	50.000	[301]
		61488.00	[104]				49.40	[473]
K_{α_1}	KL_3	63001.07 ± 0.95	[50]	20.98	[488]	4.788(-1)	49.60	[301]
		63001.00	[104]				53.00	[473]
	KM_1	70823.2 ± 3.5	[145]				61.59	[423]
K_{β_3}	KM_2	71078.1 ± 1.8	[50]	2.278	[488]	5.199(-2)	55.95	[473]
		71076.90	[104]					
K_{β_1}	KM_3	71413.9 ± 1.8	[50]	4.40	[488]	1.005(-1)	55.90	[473]
		71411.90	[104]					
$K_{\beta_5^{II}}$	KM_4	71823.8 ± 6.2	[50]	5.23(-2)	[488]	1.195(-3)	46.01	[423]
$K_{\beta_5^I}$	KM_5	71894.7 ± 6.2	[50]	6.54(-2)	[488]	1.492(-3)	45.49	[423]
	KN_1	73217.2 ± 4.5	[145]				58.46	[423]
$K_{\beta_2^{II}}$	KN_2	73318.9 ± 6.4	[50]	0.524	[488]	1.197(-2)	57.30	[423]
		73316.10	[104]					
$K_{\beta_2^I}$	KN_3	73403.2 ± 3.9	[50]	1.020	[488]	2.327(-2)	53.46	[423]
		73402.30	[104]					
K_{β_4}	$KN_{4,5}$	73615 ± 13	[50]	2.928(-2)	[488]			
$K_{\beta_4^{II}}$	KN_4	73579.9 ± 3.5	[145]			2.958(-4)	51.29	[423]
$K_{\beta_4^I}$	KN_5	73595.9 ± 2.0	[145]			3.680(-4)	50.74	[423]
	$KO_{2,3}$	73808.00 ± 4.40	[50]	0.2465	[488]			
	KO_2					2.734(-3)	48.41	[423]
	KO_3					5.521(-3)	47.09	[423]
L series								
	L_1M_1	9917.6 ± 3.2	[145]				31.10	[423]
L_{β_4}	L_1M_2	10175.50 ± 0.62	[50]	0.304	[488]	2.316(-2)	17.10	[279]
		10175.40	[104]				16.50	[473]
L_{β_3}	L_1M_3	10510.99 ± 0.92	[50]	0.368	[488]	2.810(-2)	14.70	[279]
		10511.00	[104]				14.60	[473]
$L_{\beta_{10}}$	L_1M_4	10937.72 ± 0.71	[50]	9.32(-3)	[488]	7.114(-4)	15.52	[423]
		10937.00	[104]					
L_{β_9}	L_1M_5	11007.25 ± 0.87	[50]	1.392(-2)	[488]	1.065(-3)	15.00	[423]
		11007.80	[104]					
	L_1N_1	12311.6 ± 4.2	[145]				27.97	[423]

line	transition	E/eV		I/eV/$\hbar$		TPIV	Γ/eV	
$L_{\gamma 2}$	$L_1 N_2$	12422.46 ± 0.92 12422.50	[50] [104]	7.54(-2)	[488]	5.757(-3)	14.00	[473]
$L_{\gamma 3}$	$L_1 N_3$	12499.98 ± 0.93 12499.90	[50] [104]	9.76(-2)	[488]	7.445(-3)	12.60	[473]
	$L_1 N_4$	12687.5 ± 5.8 12687.00	[50] [104]	1.34(-3)	[488]	1.026(-4)	20.80	[423]
$L_{\gamma 11}$	$L_1 N_5$	12696.6 ± 5.8 12696.00	[50] [104]	2.11(-3)	[488]	1.608(-4)	20.25	[423]
	$L_1 N_6$	12913.5 ± 2.3	[145]					
	$L_1 N_7$	12916.6 ± 2.4	[145]					
	$L_1 O_1$	12872.10 ± 0.90 12872.20	[50] [104]	3.88(-7)	[488]		23.91	[423]
$L_{\gamma 4'}$	$L_1 O_2$	12910.00 ± 1.10 12910.00	[50] [104]	1.25(-2)	[488]	9.535(-4)	17.92	[423]
$L_{\gamma 4}$	$L_1 O_3$	12923.00 ± 1.10 12923.00	[50] [104]	1.57(-2)	[488]	1.199(-3)	16.60	[423]
	$L_1 O_{4,5}$ $L_1 O_4$ $L_1 O_5$	12968.30 ± 0.70 12998.00	[50] [104]	1.59(-4)	[488]	0.04 5.127(-6) 7.804(-6)		
L_η	$L_2 M_1$	9337.07 ± 0.73 9337.10	[50] [104]	3.49(-2)	[488]	7.158(-3)	22.93	[423]
	$L_2 M_2$	9585.8 ± 2.2 9586.00	[50] [104]	1.20(-6)	[488]		17.02	[423]
$L_{\beta 17}$	$L_2 M_3$	9934.5 ± 2.4 9935.00	[50] [104]	1.22(-4)	[488]	2.445(-4)	15.67	[423]
$L_{\beta 1}$	$L_2 M_4$	10355.42 ± 0.65 10355.00	[50] [104]	1.296	[488]	2.600(-1)	6.60 7.42	[279] [473]
	$L_2 M_5$	10420.70 ± 0.91 10421.00	[50] [104]	5.85(-5)	[488]		6.83	[423]
$L_{\gamma 5}$	$L_2 N_1$	11730.42 ± 0.82 11730.00	[50] [104]	8.71(-4)	[488]	1.755(-3)	19.80	[423]
	$L_2 N_2$	11839.7 ± 3.3	[145]				18.64	[423]
	$L_2 N_3$	11924.47 ± 0.85 11924.00	[50] [104]	3.21(-4)	[488]	6.444(-5)	14.80	[423]
$L_{\gamma 1}$	$L_2 N_4$	12095.48 ± 0.87 12095.00	[50] [104]	0.2471	[488]	4.955(-2)	9.40 10.65	[279] [473]
	$L_2 N_5$	12109.35 ± 0.67	[145]				12.08	[423]
L_ν	$L_2 N_6$	12336.5 ± 3.6 12338.0	[50] [104]	7.48(-4)	[488]	1.500(-4)		
	$L_2 N_7$	12335.7 ± 1.4	[145]					
$L_{\gamma 8}$	$L_2 O_1$	12301.20 ± 0.60 12302.00	[50] [104]	1.66(-4)	[488]	3.433(-4)	15.74	[423]
	$L_2 O_3$	12340.00 ± 2.50 12340.00	[50] [104]	5.17(-5)	[488]	1.309(-5)	8.43	[423]
$L_{\gamma 6}$	$L_2 O_4$	12384.80 ± 0.60 12385.00	[50] [104]	1.22(-2)	[488]	2.519(-3)	7.00	[473]

line	transition	E/eV		I/eV/ℏ		TPIV	Γ/eV	
L_l	L_3M_1	7822.33 ± 0.51 7822.30	[50] [104]	5.44(-2)	[488]	1.333(-2)	22.19	[423]
L_t	L_3M_2	8078.6 ± 1.6 8079.00	[50] [104]	5.82(-4)	[488]	1.202(-4)	16.98	[423]
L_s	L_3M_3	8414.1 ± 1.7 8414.00	[50] [104]	5.27(-4)	[488]	1.088(-4)	15.63	[423]
$L_{\alpha2}$	L_3M_4	8841.10 ± 0.47 8841.00	[50] [104]	0.1152	[488]	2.293(-2)	7.20 6.56	[473] [423]
$L_{\alpha1}$	L_3M_5	8911.83 ± 0.47 8911.80	[50] [104]	1.0016	[488]	2.021(-1)	7.04 6.79	[473] [423]
$L_{\beta6}$	L_3N_1	10217.00 ± 0.62 10217.00	[50] [104]	1.31(-2)	[488]	2.612(-3)	19.76	[423]
	L_3N_2	10324.46 ± 0.89 10325.00	[50] [104]	1.31(-4)	[488]	2.602(-5)	18.60	[423]
	L_3N_3	10400.00 10404.12 ± 0.78	[104] [145]	1.30(-4)	[488]	2.586(-5)	14.76	[423]
$L_{\beta15}$	L_3N_4	10581.68 ± 0.67 10582.00	[50] [104]	2.0047(-2)	[488]	4.074(-3)	9.36	[423]
$L_{\beta2}$	L_3N_5	10598.7 ± 1.1 10598.00	[50] [104]	0.1840	[488]	3.661(-2)	9.90 9.60	[279] [473]
L_u $L_{u'}$	$L_3N_{6,7}$ L_3N_6	10824.65 ± 0.98 10825.00 10819.2 ± 1.3	[50] [104] [145]	1.11(-4)	[488]	2.213(-5)		
L_u	L_3N_7	10822.4 ± 1.3	[145]			1.241(-4)		
$L_{\beta7}$	L_3O_1	10787.20 ± 0.80 10787.00	[50] [104]	2.49(-3)	[488]	5.372(-4)	15.68	[423]
$L_{\beta5}$	$L_3O_{4,5}$ L_3O_4	10871.10 ± 1.00 10871.00	[50] [104]	9.68(-3)	[488]	2.063(-4)		
	L_3O_5					1.767(-3)		

M series

	transition	E/eV		I/eV/ℏ		TPIV		
	M_1N_3	2590.00 ± 11.00 2589.00 2594.00	[50] [104] [263]	2.816(-2)	[335]	9.760(-4)		
	M_2N_1	2133.00 ± 7.00 2132.00 2137.00	[50] [104] [263]	3.30(-3)	[335]	6.502(-4)		
	M_2N_4	2502.00 ± 2.00 2503.00 2507.00	[50] [104] [263]	3.36(-4)	[335]	3.451(-3)		
	M_3N_1	1798.00 ± 5.00 1798.00 1801.60	[50] [104] [263]	3.30(-3)	[335]	1.108(-3)		
$M_{\gamma2}$	M_3N_4	2166.00 ± 2.00 2166.00 2171.00	[50] [104] [263]	2.918(-2)	[335]	5.399(-4)		
$M_{\gamma1}$	M_3N_5	2182.00 ± 1.50 2182.00 2187.00	[50] [104] [263]	2.918(-2)	[335]	4.225(-3)		

line	transition	E/eV		I/eV/$\hbar$		TPIV	Γ/eV
$M_{\xi2}$	M_4N_2	1483.10 ± 0.90	[50]	1.38(-3)	[335]	2.102(-3)	
		1483.20	[104]				
		1486.60	[263]				
M_δ	M_4N_3	1578.27	[423]			2.154(-4)	
	M_4N_4	1978.30 ± 0.30	[50]	4.466(-2)	[335]		
M_β	M_4N_6	1978.30 ± 0.30	[50]			2.055(-2)	
		1978.30	[104]				
		1982.50	[263]				
M_η	M_4O_2	1978.49	[423]			5.072(-4)	
$M_{\xi1}$	M_5N_3	1491.90 ± 0.70	[50]	1.38(-3)	[335]	1.767(-3)	
		1495.60	[263]				
$M_{\alpha1.2}$	$M_5N_{6,7}$	1910.20 ± 0.30	[50]				
	M_5N_6	1910.00	[104]			1.086(-3)	
	M_5N_7	1913.30	[263]			2.118(-2)	
N series							
	N_4N_6	238.80 ± 0.50	[50]			1.305(-4)	
	$N_5N_{6,7}$	226.60 ± 0.80	[50]				
	N_5N_6					5.901(-6)	
	N_5N_7					1.244(-4)	

level characteristics

level	E_B/eV		ω_{nlj}		Γ/eV		AE/eV	
K	73871.00	[503]	0.961	[301]	44.40	[301]	73870.8	[51]
	73870.80 ± 0.50	[493]	0.961(18)	[38]	43.62	[423]	73856.2 ± 6.5	[50]
			0.9562	[423]	44.6	[94]	73884.0 ± 1.7	[145]
L_1	12968.00	[503]	0.130	[301]	7.25	[301]	12968.0	[51]
	12968.00 ± 0.40	[493]	8.80(-2)	[436]	13.11	[423]	12971.6 ± 2.0	[50]
			7.028-2)	[423]	7.2	[94]	12978.4 ± 1.5	[145]
L_2	12385.00	[503]	0.295	[301]	5.59	[301]	12385.0	[51]
	12385.00 ± 0.40	[493]	0.3218	[423]	4.95	[423]	12380.9 ± 1.8	[50]
			0.318	[436]	5.09	[94]	12397.42 ± 0.48	[145]
L_3	10871.00	[503]	0.281	[301]	5.16	[301]	10870.9	[51]
	10870.90 ± 0.30	[493]	0.290	[435]	4.91	[423]	10868.0 ± 1.ö4	[50]
			0.2846	[423]	5.09	[94]	10884.14 ± 0.37	[145]
			0.282	[436]				
M_1	3049.00	[503]	1.67(-3)	[38]	20.40	[350]	3048.5	[51]
	3048.50 ± 0.40	[493]	2.03(-3)	[512]	17.98	[423]		
			2.81(-3)	[423]	14.7	[94]		
M_2	2792.00	[503]	3.25(-3)	[38]	13.90	[350]	2792.2	[51]

level	E_B/eV		ω_{nlj}		Γ/eV		AE/eV	
	2792.20 ± 0.30	[493]	1.70(-2)	[512]	12.06	[423]		
			5.26(-3)	[423]	8.6	[94]		
M_3	2458.00	[503]	3.20(-3)	[38]	9.34	[350]	2457.2	[51]
	2457.20 ± 0.40	[493]	3.24(-3)	[512]	10.72	[423]		
			6.62(-3)	[423]	7.5	[94]		
M_4	2031.00	[503]	1.37(-2)	[38]	4.18	[350]	2038.0	[51]
	2030.80 ± 0.30	[493]	2.10(-2)	[512]	2.40	[423]		
			2.34(-2)	[423]	1.89	[94]		
M_5	1960.00	[503]	2.32(-2)	[38]	2.25	[350]	1960.1	[51]
	1960.10 ± 0.30	[493]	2.40(-2)	[512]	1.89	[423]		
			2.45(-2)	[423]	1.89	[94]		
N_1	655.00	[503]	4.4(-4)	[423]	16.00	[74]		
	654.30 ± 0.50	[493]			14.85	[423]		
					7.7	[94]		
N_2	547.00	[503]			13.00	[74]		
	546.50 ± 0.50	[493]	9.3(-4)	[423]	13.70	[423]		
					6	[94]		
N_3	469.00	[503]	3.5(-4)	[423]	13.00	[74]		
	468.20 ± 0.60	[493]			9.86	[423]		
					4.6	[94]		
N_4	290.00	[503]	1.6(-4)	[739	8.00	[74]		
	289.40 ± 0.50	[493]			7.68	[423]		
					4.1	[94]		
N_5	273.00	[503]	1.6(-4)	[423]	8.00	[74]		
	272.80 ± 0.60	[493]			7.13	[423]		
					3.9	[94]		
N_6	52.00	[503]			0.12	[74]		
	53.80 ± 0.20	[493]			0.22	[94]		
N_7	50.00	[503]			0.12	[74]		
	51.00 ± 0.20	[493]			0.18	[94]		
O_1	84.00	[503]	4.0(-6)	[423]	10.79	[423]		
	83.70 ± 0.60	[493]						
O_2	58.00	[503]	1.0(-6)	[423]	4.80	[423]		
	58.00 ± 1.10	[493]						
O_3	46.00	[503]			3.48	[423]		
	45.40 ± 1.00	[493]						
O_4	0.40 ± 0.70	[493]						
O_5	0.00 ± 0.70	[493]						
IP	8.4382	[222]						

Ir   Z=77   [Xe] 4f^{14} 5d^7 6s^2

Iridium A= 192.217(3) [222] ϱ= 22.42 g/cm^3 [547]

X-ray transitions

line	transition	E/eV		I/eV/$\hbar$		TPIV	Γ/eV	
K series								
K$_{\alpha_3}$	KL$_1$	62694.2 ± 2.3	[145]				59.12	[423]
K$_{\alpha2}$	KL$_2$	63287.29 ± 0.96	[50]	12.87	[488]	2.787(-1)	52.50	[301]
		63288.50	[104]				51.08	[423]
K$_{\alpha1}$	KL$_3$	64896.2 ± 1.0	[50]	22.09	[488]	4.781(-1)	52.10	[301]
		64897.20	[104]				51.04	[423]
	KM$_1$	72941.1 ± 3.6	[145]				64.20	[423]
K$_{\beta3}$	KM$_2$	73203.4 ± 1.3	[50]	2.403	[488]	5.214(-2)	58.24	[423]
		73202.90	[104]					
K$_{\beta1}$	KM$_3$	73561.7 ± 1.3	[50]	4.660	[488]	1.008(-1)	56.97	[423]
		73561.90	[50]					
K$_{\beta_5^{II}}$	KM$_4$	73980 ± 13	[50]	5.700(-2)	[488]	1.234(-3)	48.26	[423]
		73983.80	[104]					
		73996.1 ± 2.7	[145]					
K$_{\beta_5^I}$	KM$_5$	74075.5 ± 5.9	[50]	7.090(-2)	[488]	1.533(-3)	47.97	[423]
		74076.90	[104]					
	KN$_1$	75418.4 ± 4.7	[145]				61.03	[423]
K$_{\beta_2^{II}}$	KN$_2$	75529.9 ± 6.8	[50]	0.558	[488]	1.208(-2)	59.70	[423]
		75530.70	[104]					
K$_{\beta_2^I}$	KN$_3$	75619.3 ± 4.8	[50]	1.087	[488]	2.353(-2)	55.73	[423]
		75620.80	[104]					
K$_{\beta4}$	KN$_{4,5}$	75821 ± 14	[50]	3.200(-2)	[488]			
		75821.00	[104]					
K$_{\beta_4^{II}}$	KN$_4$	75795.7 ± 3.8	[145]			3.095(-4)	53.93	[423]
K$_{\beta_4^I}$	KN$_5$	75812.4 ± 2.2	[145]			3.832(-4)	53.20	[423]
	KO$_{2,3}$	76053.00 ± 2.30	[50]	0.2658	[488]			
		76055.40	[104]					
	KO$_2$					2.766(-3)	52.13	[423]
	KO$_3$					5.546(-3)	50.79	[423]
L series								
	L$_1$M$_1$	10244.8 ± 2.5	[50]	5.120(-6)	[488]		31.35	[423]
		10245.00	[104]					
L$_{\beta4}$	L$_1$M$_2$	10510.72 ± 0.40	[50]	0.325	[488]	2.513(-2)	17.70	[473]
		10510.20	[104]				25.40	[423]
L$_{\beta3}$	L$_1$M$_3$	10867.54 ± 0.42	[50]	0.388	[488]	3.000(-2)	14.90	[473]
		10867.00	[104]				24.12	[423]

line	transition	E/eV		I/eV/$\hbar$		TPIV	Γ/eV	
$L_{\beta 10}$	L_1M_4	11301.74 ± 0.61	[50]	1.027(-2)	[488]	7.933(-4)	15.42	[423]
		11301.00	[104]					
$L_{\beta 9}$	L_1M_5	11377.13 ± 0.77	[50]	1.538(-2)	[488]	1.187(-3)	15.13	[423]
		11376.90	[104]					
	L_1N_1	12695.3 ± 3.8	[50]	2.170(-2)	[104]		28.18	[423]
		12691.00	[104]					
$L_{\gamma 2}$	L_1N_2	12841.92 ± 0.59	[50]	8.130(-2)	[488]	6.280(-3)	14.60	[473]
		12840.60	[104]				26.85	[423]
$L_{\gamma 3}$	L_1N_3	12924.1 ± 1.0	[50]	0.1093	[488]	8.025(-3)	13.30	[473]
		12922.80	[104]				22.89	[423]
	L_1N_4	13107.3 ± 4.1	[50]	1.500(-3)	[488]	1.157(-4)	21.08	[423]
		13108.00	[104]					
$L_{\gamma 11}$	L_1N_5	13125.,4 ± 4.1	[50]	2.35(-3)	[488]	1.815(-4)	20.36	[423]
		13126.00	[104]					
	L_1N_6	13349.0 ± 2.5	[145]					
	L_1N_7	13352.4 ± 2.5	[145]					
	L_1O_1	13271.00	[104]	4.680(-7)	[488]		26.42	[423]
$L_{\gamma 4'}$	L_1O_2	13355.50 ± 0.40	[50]	1.370(-2)	[488]	1.059(-3)	19.28	[423]
		13355.00	[104]					
$L_{\gamma 4}$	L_1O_3	13368.10 ± 0.40	[50]	1.690(-2)	[488]	1.306(-3)	17.94	[423]
		13367.00	[104]					
	$L_1O_{4,5}$	13413.00 ± 4.40	[50]	2.500(-4)	[488]			
	L_1O_4					6.984(-6)		
	L_1O_5					1.067(-5)		
L_η	L_2M_1	9652.34 ± 0.33	[50]	3.720(-2)	[488]	7.367(-3)	23.30	[423]
		9651.60	[104]					
	L_2M_2	9917.0 ± 3.5	[50]	1.420(-6)	[488]		17.36	[423]
		9909.00	[104]					
$L_{\beta 17}$	L_2M_3	10272.8 ± 2.5	[50]	1.340(-3)	[488]	2.614(-4)	16.08	[423]
		10274.00	[104]					
$L_{\beta 1}$	L_2M_4	10708.35 ± 0.41	[50]	1.382	[488]	2.699(-1)	6.80	[473]
		10708.00	[104]				7.38	[423]
	L_2M_5	10791.4 ± 2.8	[50]	6.670(-5)	[488]		7.08	[423]
		10792.50	[104]					
$L_{\gamma 5}$	L_2N_1	12134.31 ± 0.88	[50]	9.350(-3)	[488]	1.832(-3)	20.14	[423]
		12133.00	[104]					
	L_2N_2	12251.2 ± 3.6	[50]	5.310(-7)	[488]		18.81	[423]
		12251.00	[104]					
	L_2N_3	12331.6 ± 5.4	[50]	3.560(-4)	[488]	6.949(-5)	14.84	[423]
		12332.00	[104]					
$L_{\gamma 1}$	L_2N_4	12512.72 ± 0.56	[50]	0.2665	[488]	5.205(-2)	9.60	[473]
		12511.30	[104]				13.04	[423]
	L_2N_5	12524.29 ± 0.81	[145]					
L_ν	L_2N_6	12760.5 ± 1.2	[50]	5.580(-4)	[488]	1.677(-4)		
		12759.00	[104]					

line	transition	E/eV		I/eV/$\hbar$		TPIV	Γ/eV	
	L_2N_7	12758.5 ± 1.6	[145]					
$L_{\gamma 8}$	L_2O_1	12727.90 ± 0.40	[50]	1.810(-3)	[488]	3.619(-4)	18.38	[423]
		12727.00	[104]					
	$L_2O_{2,3}$	12784.30 ± 0.70	[50]	5.780(-5)	[488]			
		12783.00	[104]					
	L_2O_2						11.24	[423]
	L_2O_3					1.401(-5)	9.90	[423]
$L_{\gamma 6}$	L_2O_4	12820.10 ± 0.50	[50]	1.873(-2)	[488]	3.366(-3)		
		12819.00	[104]					
L_l	L_3M_1	8045.89 ± 0.23	[50]	5.870(-2)	[488]	1.375(-2)	23.26	[423]
		8045.30	[104]					
L_t	L_3M_2	8304.2 ± 2.5	[50]	6.370(-4)	[488]	1.271(-4)	17.32	[423]
		8310.00	[104]					
L_s	L_3M_3	8659.2 ± 1.8	[50]	5.770(-4)	[488]	1.151(-4)	16.04	[423]
		8659.00	[104]					
$L_{\alpha 2}$	L_3M_4	9099.62 ± 0.49	[50]	0.1234	[488]	2.374(-2)	7.34	[423]
		9099.10	[104]					
$L_{\alpha 1}$	L_3M_5	9175.18 ± 0.30	[50]	1.079	[488]	2.093(-1)	8.10	[473]
		9174.50	[104]				7.05	[423]
$L_{\beta 6}$	L_3N_1	10525.17 ± 0.40	[50]	1.421(-2)	[488]	2.756(-3)	20.10	[423]
		10524.00	[104]					
	L_3N_2	10638.15 ± 0.68	[50]	1.430(-4)	[488]	2.782(-5)	18.77	[423]
		10640.00	[104]					
	L_3N_3	10725.1 ± 4.1	[50]	1.430(-4)	[488]	2.777(-5)	14.81	[423]
		10725.00	[104]					
$L_{\beta 15}$	L_3N_4	10903.67 ± 0.43	[50]	2.195(-2)	[488]	4.258(-3)	13.00	[423]
		10903.00	[104]					
$L_{\beta 2}$	L_3N_5	10920.47 ± 0.43	[50]	0.1975	[488]	3.832(-2)	9.70	[473]
		10920.00	[104]				12.28	[423]
L_u	$L_3N_{6,7}$	11155.01 ± 0.59	[50]	1.280(-4)	[488]			
		11153.00	[104]					
$L_{u'}$	L_3N_6	11146.8 ± 1.4	[145]			2.479(-5)		
L_u	L_3N_7	11150.1 ± 1.4	[145]			1.390(-4)		
$L_{\beta 7}$	L_3O_1	11120.50 ± 0.30	[50]	2.730(-3)	[488]	5.721(-4)	18.34	[423]
		11120.00	[104]					
	$L_3O_{2,3}$	11177.20 ± 0.60	[50]	2.370(-5)	[488]			
		11174.00	[104]					
	L_3O_2					5.855(-6)	11.20	[423]
	L_3O_3					5.723(-6)	9.77	[423]
$L_{\beta 5}$	$L_3O_{4,5}$	11211.40 ± 0.30	[50]	1.440(-2)	[488]			
		11211.00	[104]					
	L_3O_4					2.714(-4)		
	L_3O_5					2.333(-3)		

M series

| | M_1N_3 | 2677.00 ± 5.00 | [50] | 3.637(-2) | [335] | 1.035(-3) | 27.95 | [423] |

line	transition	E/eV		I/eV/$\hbar$		TPIV	Γ/eV	
		2683.00	[263]					
	M_2N_4	2594.00 ± 2.00	[50]	3.173(-2)	[335]	3.548(-3)	20.20	[423]
		2594.00	[104]					
		2599.00	[263]					
	M_3N_1	1859.00 ± 2.50	[50]	3.530(-3)	[335]	1.154(-3)	26.02	[423]
		1860.00	[104]					
		1863.00	[263]					
$M_{\gamma2}$	M_3N_4	2238.00 ± 2.00	[50]	3.173(-2)	[335]	5.628(-4)	18.92	[423]
		2238.00	[104]					
		2242.00	[263]					
$M_{\gamma1}$	M_3N_5	2254.00 ± 1.60	[50]	3.173(-2)	[335]		9.50	[279]
		2254.00	[104]			4.394(-3)	18.20	[423]
		2259.00	[263]					
	$M_3O_{4,5}$	2546.00 ± 5.00	[50]					
		2546.00						
		2552.00	[263]					
	M_3O_4					3.959(-5)		
	M_3O_5					3.018(-4)		
$M_{\xi2}$	M_5N_2	1537.30 ± 1.00	[50]	1.480(-3)	[335]		15.70	[423]
		1537.30	[104]					
		1540.20	[263]					
M_δ	M_4N_3	1622.00 ± 2.00	[50]	1.480(-3)	[335]	2.122(-4)	12.03	[423]
		1622.00	[104]					
		1625.00	[263]					
M_β	M_4N_6	2053.50 ± 0.30	[50]	4.949(-2)	[335]	2.253(-2)	2.20	[279]
		2054.00						
		2058.00	[263]					
M_η	M_4O_2	2057.50	[423]			5.043(-4)	8.42	[423]
$M_{\xi1}$	M_5N_3	1545.80 ± 0.80	[50]	1.480(-3)	[335]	1.715(-3)	11.73	[423]
		1546.20	[104]					
		1549.80	[263]					
$M_{\alpha2}$	M_5N_6	1975.80 ± 1.00	[50]	4.949(-2)	[335]	1.174(-3)		
		1976.20	[104]					
		1979.90	[263]					
$M_{\alpha1}$	M_5N_7	1979.90 ± 0.30	[50]	4.949(-2)	[335]	2.289(-2)	2.50	[473]
		1980.00	[104]					
		1984.10	[263]					

N series

line	transition	E/eV	
	N_4N_6	247.00	[50]
	$N_5N_{6,7}$	234.80	[50]

level characteristics

level	E_B/eV		ω_{nlj}		Γ/eV		AE/eV	
K	76111.00	[503]	0.962	[301]	46.80	[301]	76111.0	[51]
	76111.00 ± 0.50	[493]	0.962	[38]	45.98	[423]	76100.1 ± 6.9	[50]
			0.958	[39]	47	[94]	76117.5 ± 1.8	[145]
			0.9572	[423]				
L_1	13419.00	[503]	0.120	[301]	8.30	[301]	13418.5	[51]
	13418.50 ± 0.30	[493]	7.59(-2)	[423]	13.14	[423]	13423.8 ± 2.2	[50]
			9.30(-2)	[436]	7.9	[94]	13423.3 ± 1.5	[145]
			0.128	[578]				
			0.139	[567]				
L_2	12824.00	[503]	0.308	[301]	5.68	[301]	12824.1	[51]
	12824.10 ± 0.30	[493]	0.3354	[423]	5.10	[423]	12820.0 ± 2.0	[50]
			0.331	[436]	5.23	[94]	12754.38 ± 0.50	[145]
			0.323	[578]				
			0.323	[567]				
L_3	11215.00	[503]	0.294	[301]	5.25	[301]	11215.2	[51]
	11215.20 ± 0.30	[493]	0.262	[462]	5.06	[423]	11212.0 ± 1.5	[50]
			0.2958	[423]	5.24	[94]	11046.07 ± 0.78	[145]
			0.274	[436]				
			0.282	[567]				
M_1	3174.00	[503]	2.09(-3)	[512]	18.21	[423]	3173.7	[51]
	3173.70 ± 1.70	[493]	3.03(-3)	[423]	14.8	[94]		
M_2	2909.00	[503]	1.80(-2)	[512]	12.26	[423]	2908.7	[51]
	2908.70 ± 0.30	[493]	5.48(-2)	[423]	8.9	[94]		
M_3	2551.00	[503]	3.58(-3)	[512]	10.98	[423]	2550.7	[51]
	2550.70 ± 0.30	[493]	6.94(-3)	[423]	8	[94]		
M_4	2116.00	[503]	2.30(-2)	[512]	2.28	[423]	2116.1	[51]
	2116.10 ± 0.30	[493]	2.54(-2)	[423]	1.99	[94]		
M_5	2041.00	[503]	2.60(-2)	[512]	1.99	[423]	2040.4	[51]
	2040.40 ± 0.30	[493]	2.62(-2)	[423]	1.99	[94]		
N_1	690.00	[503]	4.7(-4)	[423]	15.04	[423]		
	690.10 ± 0.40	[493]			8	[94]		
N_2	577.00	[503]	9.8(-4)	[423]	13.71	[423]		
	577.10 ± 0.40	[493]			6.1	[94]		
N_3	495.00	[503]	3.8(-4)	[423]	9.75	[423]		
	494.30 ± 0.60	[493]			4.75	[94]		
N_4	312.00	[503]	1.8(-4)	[423]	7.94	[423]		
	311.40 ± 0.40	[493]			4.1	[94]		

level	E_B/eV		ω_{nlj}		Γ/eV		AE/eV
N_5	295.00	[503]	1.8(-4)	[423]	7.22	[423]	
	294.90 ± 0.40	[493]			4	[94]	
N_6	63.00	[503]			0.31	[94]	
	63.40 ± 0.40	[51]					
	63.80 ± 0.40	[493]					
N_7	60.00	[503]			0.27	[94]	
	60.50 ± 0.40	[51]					
	60.80 ± 0.30	[493]					
O_1	96.00	[503]	1.0(-5)	[423]	13.28	[423]	
	95.20 ± 0.40	[493]					
O_2	63.00	[503]	1.0(-6)	[423]	6.14	[423]	
	63.00 ± 0.60	[493]					
O_3	51.00	[503]			4.80	[423]	
	50.50 ± 0.60	[51]					
	49.60 ± 1.00	[493]					
O_4	4.00	[503]					
	4.20 ± 0.40	[493]					
O_5	3.20 ± 0.40	[493]					
IP	8.9670	[222]					

Pt   Z=78

[Xe] $4f^{14}\,5d^9\,6s^1$

Platinum $A = 195.0078(2)$ [222] $\varrho = 21.41\,\mathrm{g/cm^3}$ [547]
$\varrho = 21.40\,\mathrm{g/cm^3}$ [483]

X-ray transitions

line	transition	E/eV		I/eV/ℏ		TPIV	Γ/eV	
K series								
K_{α_3}	KL_1	64515.5 ± 2.4	[145]				61.83	[423]
$K_{\alpha 2}$	KL_2	65123.3 ± 2.0	[50]	13.60	[488]	2.794(-1)	53.69	[423]
		65123.00	[104]	13.71	[487]		54.30	[473]
$K_{\alpha 1}$	KL_3	66832.9 ± 2.1	[50]	23.24	[488]	4.774(-1)	53.66	[423]
		66832.00	[104]	23.45	[487]		60.30	[473]
	KM_1	75099.4 ± 3.6	[145]				66.91	[423]
$K_{\beta 3}$	KM_2	75368.7 ± 2.0	[50]	2.545	[488]	5.230(-2)	60.94	[423]
		75369.90	[104]	2.653	[487]			
$K_{\beta 1}$	KM_3	75749.1 ± 2.1	[50]	4.92	[488]	1.011(-1)	59.64	[423]
		75750.00	[104]	5.13	[487]			
$K_{\beta_5^{II}}$	KM_4	76198 ± 14	[50]	6.20(-2)	[488]	1.275(-3)	50.64	[423]
		76202.50	[104]					
$K_{\beta_5^{I}}$	KM_5	76273 ± 21	[50]	7.67(-2)	[488]	1.575(-3)	50.53	[423]
		76278.10	[104]					
	KN_1	77672.2 ± 4.7	[145]				63.81	[423]
$K_{\beta_2^{II}}$	KN_2	77785.5 ± 7.2	[50]	0.593	[488]	1.219(-2)	62.10	[423]
		77788.80	[104]	0.636	[487]			
$K_{\beta_2^{I}}$	KN_3	77878.3 ± 7.2	[50]	1.157	[488]	2.377(-2)	58.09	[423]
		77882.00	[104]	1.250	[487]			
$K_{\beta 4}$	$KN_{4,5}$	78070 ± 15	[50]					
		78070.00	[104]					
$K_{\beta_4^{II}}$	KN_4	78064.5 ± 3.8	[145]			3.235(-4)	56.49	[423]
$K_{\beta_4^{I}}$	KN_5	78081.9 ± 2.1	[145]			3.985(-4)	55.71	[423]
	$KO_{2,3}$	78341.00 ± 5.00	[50]	0.2932	[488]			
		78346.00	[104]					
	KO_2					2.799(-3)	55.93	[423]
	KO_3					5.572(-3)	54.57	[423]
L series								
	L_1M_1	10600.2 ± 1.2	[50]	6.002(-6)	[488]		31.84	[423]
		10530.00	[104]					
$L_{\beta 4}$	L_1M_2	10854.41 ± 0.70	[50]	0.348	[488]	2.640(-2)	25.87	[423]
		10854.40	[104]				18.000	[473]
$L_{\beta 3}$	L_1M_3	11230.89 ± 0.75	[50]	0.409	[488]	3.100(-2)	24.57	[423]
		11231.00	[2				16.10	[473]

line	transition	E/eV		I/eV/$\hbar$		TPIV	Γ/eV	
$L_{\beta 10}$	L_1M_4	11676.3 ± 1.1	[50]	1.13(-2)	[488]	8.566(-4)	15.57	[423]
		11675.00	[104]					
$L_{\beta 9}$	L_1M_5	11756.79 ± 0.22	[50]	1.692(-2)	[488]	1.283(-3)	15.46	[423]
		11756.90	[104]					
	L_1N_1	13112.9 ± 4.1	[50]	2.56(-6)	[488]		28.73	[423]
		13111.00	[104]					
$L_{\gamma 2}$	L_1N_2	13270.5 ± 1.1	[50]	8.76(-2)	[488]	6.640(-3)	27.03	[423]
		13270.60	[104]					
		13289.00	[543]					
$L_{\gamma 3}$	L_1N_3	13361.5 ± 1.1	[50]	0.1104	[488]	8.366(-3)	23.02	[423]
		13361.40	[104]					
		13380.00	[543]					
	L_1N_4	13549.0 ± 3.4	[145]			1.263(-4)	21.43	[423]
$L_{\gamma 11}$	L_1N_5	13560.4 ± 4.4	[50]	2.62(-3)	[488]	1.985(-4)	20.64	[423]
		13563.00	[104]					
	L_1N_6	13805.0 ± 2.4	[145]					
	L_1N_7	13808.5 ± 2.4	[145]					
	L_1O_1	13784.00 ± 3.00	[50]	5.68(-7)	[488]		29.15	[423]
		13789.00	[104]					
$L_{\gamma 4'}$	L_1O_2	13814.50 ± 0.60	[50]	1.528(-2)	[488]	1.158(-3)	20.86	[423]
		13814.00	[104]					
		13835.00	[543]					
$L_{\gamma 4}$	L_1O_3	13828.10 ± 0.60	[50]	1.867(-2)	[488]	1.415(-3)	19.50	[423]
		13827.00	[104]					
		13851.00	[543]					
	L_1O_4	13864.00 ± 1.60	[50]	3.17(-4)	[488]	9.253(-6)		
		13863.00	[104]					
	L_1O_5	13878.00 ± 1.60	[50]	3.17(-4)	[488]	1.420(-5)		
		13879.00	[104]					
L_η	L_2M_1	9975.2 ± 2.4	[50]	3.79(-2)	[488]	7.597(-3)	23.70	[423]
		9976.10	[104]					
		9959.00	[543]					
	L_2M_2	10221 ± 12	[50]	1.66(-6)	[488]		17.74	[423]
		10230.00	[104]					
$L_{\beta 17}$	L_2M_3	10626.8 ± 1.3	[50]	1.49(-3)	[488]	2.790(-4)	16.44	[423]
		10627.00	[104]					
$L_{\beta 1}$	L_2M_4	11070.84 ± 0.29	[50]	1.471	[488]	2.797(-1)	7.43	[423]
		11071.00	[104]				8.00	[473]
		11065.00	[543]					
	L_2M_5	11140.5 ± 3.0	[50]	7.59(-5)	[488]		7.33	[423]
$L_{\gamma 5}$	L_2N_1	12552.6 ± 3.8	[50]	1.0003(-2)	[488]	1.909(-3)	20.59	[423]
		12552.00	[104]					
	L_2N_2	12661.6 ± 3.8	[50]	6.29(-7)	[488]		18.90	[423]
		12619.00	[104]					
	L_2N_3	12758.93 ± 0.78	[50]	3.43(-4)	[488]	7.480(-5)	14.88	[423]
		12758.00	[104]					

line	transition	E/eV		I/eV/$\hbar$		TPIV	Γ/eV	
$L_{\gamma 1}$	L_2N_4	12942.19 ± 0.60	[50]	0.2871	[488]	5.459(-2)	13.28	[423]
		12942.00	[104]				11.20	[473]
		12933.00	[543]					
	L_2N_5	12959.33 ± 0.76	[145]				12.52	[423]
L_ν	L_2N_6	13199.3 ± 1.0	[50]	9.82(-4)	[488]	1.866(-4)	5.61	[423]
		13198.00	[104]					
	L_2N_7	13201.5 ± 1.5	[145]					
$L_{\gamma 8}$	L_2O_1	13173.00 ± 1.40	[50]	1.99(-3)	[488]	3.806(-4)	21.01	[423]
		13172.00	[104]					
$L_{\gamma 6}$	L_2O_4	13271.00 ± 3.00	[50]	2.293(-2)	[488]	4.260(-3)		
		13269.00	[104]					
L_l	L_3M_1	8268.2 ± 1.6	[50]	6.32(-2)	[488]	1.416(-2)	23.67	[423]
		8268.70	[104]					
		8252.00	[543]					
L_t	L_3M_2	8532.9 ± 1.7	[50]	6.96(-4)	[488]	1.343(-4)	17.70	[423]
L_s	L_3M_3	8922.8 ± 1.9	[50]	6.30(-4)	[488]	1.215(-4)	16.41	[423]
$L_{\alpha 2}$	L_3M_4	9361.92 ± 0.21	[50]	0.1299	[488]	2.456(-2)	7.40	[423]
		9362.00	[104]					
		9358.00	[543]					
$L_{\alpha 1}$	L_3M_5	9442.39 ± 0.32	[50]	1.145	[488]	2.165(-1)	7.30	[423]
		9442.30	[104]				7.60	[473]
		9438.00	[543]					
$L_{\beta 6}$	L_3N_1	10841.88 ± 0.70	[50]	1.507(-2)	[488]	2.906(-3)	20.56	[423]
		10841.00	[104]					
	L_3N_2	10962.2 ± 2.9	[50]	1.57(-4)	[488]	2.968(-5)	18.86	[423]
		10958.00	[104]					
	L_3N_3	11044.2 ± 2.9	[50]	1.57(-4)	[488]	2.975(-5)	14.85	[423]
		11046.00	[104]					
$L_{\beta 15}$	L_3N_4	11233.4 ± 2.4	[145]			4.446(-3)	13.26	[423]
$L_{\beta 2}$	L_3N_5	11250.66 ± 0.45	[50]	0.2117	[488]	4.004(-2)	12.47	[423]
		11250.00	[104]				9.95	[473]
L_u	$L_3N_{6,7}$	11490.91 ± 0.79	[50]	1.46(-4)	[488]			
		11491.00	[104]					
$L_{u'}$	L_3N_6	11489.5 ± 1.3	[145]			2.768(-5)	5.58	[423]
L_u	L_3N_7	11492.9 ± 1.3	[145]			1.551(-4)	5.52	[423]
$L_{\beta 7}$	L_3O_1	11461.90 ± 0.30	[50]	3.003(-3)	[488]	6.081(-4)	20.98	[423]
		11461.00	[104]					
	$L_3O_{2,3}$	11521.00 ± 3.20	[50]	2.68(-5)	[488]			
		11521.00						
	L_3O_2					6.226(-6)	12.69	[423]
	L_3O_3					6.157(-6)	11.34	[423]
$L_{\beta 5}$	$L_3O_{4,5}$	11561.00 ± 2.20	[50]	1.785(-2)	[488]			
		11560.00	[104]					
	L_3O_4					3.417(-4)		
	L_3O_5					2.951(-3)		

line	transition	E/eV		I/eV/$\hbar$		TPIV	Γ/eV	
	M series							
	M_1N_3	2780.00 ± 5.60	[50]	3.273(-2)	[335]	1.077(-3)	28.10	[423]
		2780.00	[104]					
		2786.00	[263]					
	M_2N_4	2695.00 ± 2.30	[50]	3.445(-2)	[335]	3.642(-3)	20.53	[423]
		2696.00	[104]					
		2701.00	[263]					
	M_3N_1	1921.00 ± 2.70	[50]	3.77(-3)	[335]	1.198(-3)	26.55	[423]
		1920.60	[104]					
		1924.60	[263]					
$M_{\gamma 2}$	M_3N_4	2314.00 ± 2.20	[50]	3.445(-2)	[335]	5.849(-4)	19.23	[423]
		2314.00	[104]					
		2319.00	[263]					
$M_{\gamma 1}$	M_3N_5	2331.00 ± 1.80	[50]	3.445(-2)	[335]	4.556(-3)	18.45	[423]
		2330.00	[104]					
		2336.00	[263]					
	M_3O_1	2543.00 ± 4.70	[50]			2.369(-4)	26.96	[423]
		2543.00	[104]					
		2548.00	[263]					
	$M_3O_{4,5}$	2641.00 ± 4.50	[50]					
		2647.00	[263]					
	M_3O_4					4.958(-5)		
	M_3O_5	2643.00	[104]			3.783(-4)		
$M_{\xi 2}$	M_4N_2	1592.00 ± 1.00	[50]	1.58(-3)	[335]	2.054(-3)	15.84	[423]
		1592.00	[104]					
		1595.70	[263]					
M_δ	M_4N_3	1682.00 ± 2.00	[50]	1.58(-3)	[335]	2.091(-4)	11.83	[423]
		1682.00	[104]					
		1684.60	[263]					
M_β	M_4N_6	2127.30 ± 0.40	[50]	5.462(-2)	[335]	2.462(-2)	2.55	[423]
		2127.00	[104]					
		2132.00	[263]					
M_η	M_4O_2	2139.06	[423]			5.006(-4)	9.67	[423]
$M_{\xi 1}$	M_5N_3	1602.20 ± 0.80	[50]	1.58(-3)	[335]	1.663(-3)	11.72	[423]
		1602.00	[104]					
		1606.00	[263]					
$M_{\alpha 2}$	M_5N_6	2047.00 ± 1.00	[50]	5.416(-2)	[335]	1.263(-3)	2.45	[423]
		2047.00	[104]					
		2051.00	[263]					
$M_{\alpha 1}$	M_5N_7	2050.50 ± 0.34	[50]	5.416(-2)	[335]	2.464(-2)	2.40	[423]
		2050.00	[104]					
		2055.00	[263]					
	M_5O_3	2071.00 ± 3.00	[50]			4.358(-4)	8.21	[423]
		2071.00	[104]					
		2075.00	[263]					

line	transition	E/eV		I/eV/$\hbar$	TPIV	Γ/eV	
N series							
N_4N_6	258.00 $\pm$ 1.10	[50]			1.637(-4)	8.40	[423]
$N_5N_{6,7}$	243.60 $\pm$ 0.50	[50]					
N_5N_6					7.350(-6)	7.62	[423]
N_5N_7					1.562(-4)	7.57	[423]

level characteristics

level	E_B/eV		ω_{nlj}		Γ/eV		AE/eV	
K	78395.00	[503]	0.963	[301]	49.30	[301]	78394.8	[51]
	78394.80 $\pm$ 0.70	[493]	0.967	[225]	49.5	[94]	78380.5 $\pm$ 7.3	[50]
			0.963(13)	[38]	48.45	[423]	78404.2 $\pm$ 1.9	[145]
			0.9592	[39]				
			0.9581	[423]				
L_1	13880.00	[503]	0.114	[301]	9.39	[301]	13879.9	[51]
	13880.10 $\pm$ 0.40	[493]	7.94(-2)	[423]	13.38	[423]	13880.69 $\pm$ 0.30	[294]
			7.40(-2)	[436]	8.8	[94]	13888.7 $\pm$ 1.4	[145]
			0.115	[578]				
			0.13	[567]				
L_2	13273.00	[503]	0.321	[301]	5.86	[301]	13272.6	[51]
	13272.60 $\pm$ 0.40	[493]	0.3491	[423]	5.24	[423]	13271.894	[294]
							$\pm$ 0.030	
			0.344	[436]	5.38	[94]	13281.67 $\pm$ 0.51	[145]
			0.341	[578]				
			0.349	[567]				
L_3	11564.00	[503]	0.306	[301]	5.31	[301]	11563.7	[51]
			0.31	[269]	5.21	[423]	11562.755 $\pm$ 0.020	[294]
	11563.70 $\pm$ 0.40	[493]	0.262	[435]	5.39	[94]	11573.11 $\pm$ 0.38	[145]
			0.3070	[423]				
			0.303	[436]				
			0.284	[578]				
			0.294	[567]				
M_1	3298.00	[503]	2.15(-3)	[512]	23.00	[74]	3296.0	[51]
	3297.20 $\pm$ 0.90	[493]	3.21(-3)	[423]	18.46	[423]		
			2.14(-3)	[245]	14.9	[94]		
M_2	3027.00	[503]	1.90(-2)	[512]	12.49	[423]	3026.5	[51]
	3026.50 $\pm$ 0.40	[51]	5.70(-3)	[423]	9.00	[74]		
	3026.70 $\pm$ 0.50	[493]	3.18(-3)	[245]	9.2	[94]		
M_3	2646.00	[503]	3.92(-3)	[512]	11.19	[423]	2645.4	[51]
	2645.40 $\pm$ 0.40	[51]	7.28(-3)	[423]	15.00	[74]		
	2645.70 $\pm$ 0.50	[493]	4.60(-3)	[245]	8.3	[94]		
M_4	2202.00	[503]	2.60(-2)	[512]	2.19	[423]	2201.9	[51]
	2201.90 $\pm$ 0.30	[51]	2.74(-2)	[423]	2.30	[74]		
	2201.70 $\pm$ 0.40	[493]			2.08	[94]		
M_5	2121.00	[503]	2.70(-2)	[512]	2.08	[423]	2121.6	[51]
	2121.60 $\pm$ 0.30	[51]	2.80(-2)	[423]	2.50	[74]		
	2121.40 $\pm$ 0.40	[493]			2.08	[94]		

level	E_B/eV		ω_{nlj}		Γ/eV		AE/eV
N_1	724.00	[503]	5.0(-4)	[423]	15.35	[423]	
	722.00 ± 0.60	[51]			17.00	[50]	
	722.80 ± 1.20	[493]			8.25	[94]	
N_2	608.00	[503]	1.03(-3)	[423]	13.65	[423]	
	609.00 ± 0.60	[51]			13.00	[74]	
	608.40 ± 1.00	[493]			6.25	[94]	
N_3	519.00	[503]	4.0(-4)	[423]	9.64	[423]	
	519.00 ± 0.60	[493]			13.00	[74]	
					4.9	[94]	
N_4	331.00	[503]	2.1(-4)	[423]	8.04	[423]	
	330.80 ± 0.50	[51]			8.30	[74]	
	330.70 ± 0.60	[493]			4.1	[94]	
N_5	314.00	[503]	2.0(-4)	[423]	7.26	[423]	
	313.30 ± 0.40	[51]			8.30	[74]	
	313.40 ± 0.50	[493]			3.95	[94]	
N_6	74.00	[503]	1.0(-6)	[423]	0.36	[423]	
	74.30 ± 0.40	[493]			0.22	[74]	
					0.35	[94]	
N_7	70.00	[503]	1.0(-6)	[423]	0.31	[423]	
	71.10 ± 0.50	[51]			0.22	[74]	
	70.90 ± 0.40	[493]			0.31	[94]	
O_1	102.00	[503]	1.0(-5)	[423]	15.77	[423]	
	102.70 ± 0.40	[493]					
O_2	66.00	[503]	1.0(-6)	[423]	7.48	[423]	
	65.30 ± 0.70	[493]					
O_3	51.00	[503]			6.12	[423]	
	51.70 ± 0.70	[51]					
	51.60 ± 0.70	[493]					
O_4	2.00	[503]					
	2.20 ± 1.30	[51]					
	2.80 ± 0.90	[493]					
O_5	2.00	[503]					
	2.20 ± 1.30	[51]					
	1.40 ± 0.90	[493]					
IP	8.9587	[222]					

Au   Z=79

[Xe] $4f^{14}\,5d^{10}\,6s^1$

Gold

$A = 196.96655(2)$ [222] $\varrho = 19.29\,\text{g/cm}^3$ [547]
$\varrho = 19.30\,\text{g/cm}^3$ [483]

X-ray transitions

line	transition	E/eV		I/eV/$\hbar$		TPIV	Γ/eV	
K series								
$K_{\alpha 3}$	KL_1	66400 ± 21	[50]	1.81(-2)	[488]		64.58	[423]
		66400.90	[104]					
		66372.5 ± 2.4	[145]					
$K_{\alpha 2}$	KL_2	66989.50 ± 0.70	[50]	14.36	[488]	2.801(-1)	58.00	[301]
		66991.90	[104]	14.48	[487]		56.42	[423]
		66991.16 ± 0.22	[277]					
		66990.73 ± 0.22	[144]					
$K_{\alpha 1}$	KL_3	68803.70 ± 0.80	[50]	24.43	[488]	4.766(-1)	57.40	[301]
		68806.60	[104]	24.64	[487]		56.40	[423]
		68804.94 ± 0.18	[277]					
		68804.50 ± 0.18	[144]					
	KM_1	77301.2 ± 3.7	[145]				69.68	[423]
$K_{\beta 3}$	KM_2	77580.00 ± 1.00	[50]	2.687	[488]	5.242(-2)	63.73	[423]
		77581.90	[104]	2.800	[487]			
		77575.51 ± 0.61	[277]					
		77575.01 ± 0.61	[144]					
$K_{\beta 1}$	KM_3	77984.00 ± 1.40	[50]	5.20	[488]	1.014(-1)	61.89	[423]
		77985.90	[104]	5.41	[487]			
		77980.30 ± 0.38	[277]					
		77979.80 ± 0.38	[144]					
$K_{\beta_5^{II}}$	KM_4	78439.0 ± 5.1	[50]	6.74(-2)	[488]	1.315(-3)	53.27	[423]
		78440.00	[104]					
$K_{\beta_5^{I}}$	KM_5	78529.5 ± 3.7	[50]	8.29(-2)	[488]	1.617(-3)	53.21	[423]
		78530.00	[104]					
	KN_1	79963.3 ± 4.8	[145]				66.63	[423]
$K_{\beta_2^{II}}$	KN_2	80076.5 ± 5.1	[50]	0.630	[488]	1.229(-2)	64.63	[423]
		80079.00	[104]	0.675	[487]			
$K_{\beta_2^{I}}$	KN_3	80186.2 ± 6.9	[50]	1.231	[488]	2.402(-2)	60.53	[423]
		80188.00	[104]	1.330	[487]			
$K_{\beta 4}$	$KN_{4,5}$	80391.1 ± 3.9	[50]	3.85(-2)	[488]			
		80390.00	[104]					
$K_{\beta_4^{II}}$	KN_4	80371.0 ± 4.2	[145]			3.378(-4)	59.14	[423]
$K_{\beta_4^{I}}$	KN_5	80389.0 ± 2.3	[145]			4.142(-4)	58.30	[423]
	$KO_{2,3}$	80667.00 ± 4.00	[50]	0.320	[488]			
		80669.00	[104]	0.345	[487]			
	KO_2					2.831(-3)	59.51	[423]
	KO_3					5.597(-3)	59.33	[423]

line	transition	E/eV		I/eV/$\hbar$		TPIV	Γ/eV	

L series

line	transition	E/eV		I/eV/$\hbar$		TPIV	Γ/eV	
	L_1M_1	10921.15 ± 0.71	[50]	7.05(-6)	[488]		32.33	[423]
		10920.40	[104]					
$L_{\beta4}$	L_1M_2	11204.81 ± 0.45	[50]	0.373	[488]	2.745(-2)	20.70	[473]
		11204.20	[104]				26.37	[423]
$L_{\beta3}$	L_1M_3	11610.5 ± 1.4	[50]	0.430	[488]	3.168(-2)	18.40	[473]
		11610.00	[104]				24.53	[423]
$L_{\beta10}$	L_1M_4	12061.8 ± 1.2	[50]	1.243(-2)	[488]	9.151(-4)	15.89	[423]
		12062.00	[104]					
$L_{\beta9}$	L_1M_5	12147.6 ± 1.2	[50]	1.861(-2)	[488]	1.370(-3)	15.85	[423]
		12147.30	[104]					
	L_1N_1	13578.2 ± 2.2	[50]	3.09(-6)	[488]		29.26	[423]
		13577.70	[104]					
$L_{\gamma2}$	L_1N_2	13709.70 ± 0.67	[50]	9.43(-2)	[488]	6.946(-3)	18.10	[473]
		13708.70	[104]				27.27	[423]
		13730.00	[543]					
$L_{\gamma3}$	L_1N_3	13809.1 ± 1.1	[50]	0.1173	[488]	8.633(-3)	16.40	[473]
		13807.70	[104]				23.17	[423]
		13828.00	[543]					
	L_1N_4	13999.3 ± 1.6	[50]	1.85(-3)	[488]	1.363(-4)	21.78	[423]
		13998.10	[104]					
$L_{\gamma11}$	L_1N_5	14019.9 ± 1.6	[50]	2.91(-3)	[488]	2.146(-4)	20.94	[423]
		14018.90	[104]					
	L_1N_6	14263.1 ± 2.6	[145]				13.99	[423]
	L_1N_7	14266.8 ± 2.5	[145]				13.93	[423]
	L_1O_1	14238.50 ± 0.80	[50]	6.85(-7)	[488]		31.90	[423]
		14238.00	[104]					
$L_{\gamma4'}$	L_1O_2	14280.90 ± 0.70	[50]	1.688(-2)	[488]	1.243(-3)	22.15	[423]
		14280.00	[104]					
		14304.00	[543]					
$L_{\gamma4}$	L_1O_3	14299.60 ± 0.70	[50]	2.036(-2)	[488]	1.499(-3)	21.97	[423]
		14299.00	[104]					
		14320.00	[543]					
	L_1O_4	14350.00 ± 0.80	[50]	4.15(-4)	[488]	1.184(-5)		
		14349.00	[104]					
	$L_1O_{4,5}$	14349.70 ± 0.80	[50]					
	L_1O_5	14358.00	[104]	4.15(-4)	[488]	1.823(-5)		
L_η	L_2M_1	10308.41 ± 0.38	[50]	4.23(-2)	[488]	7.850(-3)	24.07	[423]
		10308.00	[104]					
		10289.00	[543]					
	L_2M_2	10589.5 ± 1.3	[50]	1.95(-6)	[488]		18.11	[423]
		10587.00	[104]					
$L_{\beta17}$	L_2M_3	10991.54 ± 0.72	[50]	1.61(-3)	[488]	2.971(-4)	16.27	[423]
		10991.00	[104]					
$L_{\beta1}$	L_2M_4	11442.45 ± 0.47	[50]	1.565	[488]	2.895(-1)	8.50	[473]
		11441.00	[104]				7.63	[423]
		11436.00	[543]					

line	transition	E/eV		I/eV/ℏ		TPIV	Γ/eV	
	L_2M_5	11526.8 ± 3.2	[50]	8.62(-5)	[488]		7.65	[423]
		11525.00	[104]					
$L_{\gamma 5}$	L_2N_1	12974.43 ± 0.60	[50]	1.075(-2)	[488]	1.988(-3)	21.00	[423]
		12974.00	[104]					
	L_2N_2	13080.00	[104]	7.43(-7)	[488]		19.01	[423]
		13089.4 ± 3.5	[145]					
	L_2N_3	13186.8 ± 4.2	[50]	4.35(-4)	[488]	8.035(-5)	14.91	[423]
		13185.00	[104]					
$L_{\gamma 1}$	L_2N_4	13381.79 ± 0.64	[50]	0.309	[488]	5.714(-2)	11.50	[473]
		13381.00	[104]				13.52	[423]
		13372.00	[543]					
	L_2N_5	13396.01 ± 0.71	[145]				12.69	[423]
L_ν	L_2N_6	13648.9 ± 1.1	[50]	1.12(-3)	[488]	2.067(-4)	5.79	[423]
		13648.00	[104]					
	L_2N_7	13646.2 ± 1.5	[145]				5.73	[423]
$L_{\gamma 8}$	L_2O_1	13626.00 ± 0.75	[50]	2.18(-3)	[488]	4.019(-4)	23.83	[423]
		13625.00	[104]					
	L_2O_2	13662.00 ± 1.10	[50]	1.44(-7)	[488]		13.89	[423]
		13662.00	[104]					
	L_2O_3	13679.00 ± 1.10	[50]	9.54(-5)	[488]	1.597(-5)	13.71	[423]
		13679.00	[104]					
$L_{\gamma 6}$	L_2O_4	13730.40 ± 0.50	[50]	2.905(-2)	[488]	5.222(-3)		
		13729.00	[104]					
L_l	L_3M_1	8494.03 ± 0.78	[50]	6.80(-2)		1.459(-2)	24.09	[423]
		8493.60	[104]					
		8477.00	[543]					
L_t	L_3M_2	8770.31 ± 0.64	[50]	7.60(-4)	[488]	1.418(-4)	18.09	[423]
		8769.80	[104]					
L_s	L_3M_3	9174.97 ± 0.70	[50]	6.88(-4)	[488]	1.283(-4)	16.25	[423]
		9174.40	[104]					
$L_{\alpha 2}$	L_3M_4	9628.05 ± 0.33	[50]	0.1377	[488]	2.538(-2)	7.61	[423]
		9627.20	[104]					
		9624.00	[543]					
$L_{\alpha 1}$	L_3M_5	9713.44 ± 0.34	[50]	1.214	[488]	2.236(-1)	8.60	[473]
		9712.90	[104]				7.57	[423]
		9709.00	[543]					
$L_{\beta 6}$	L_3N_1	11160.33 ± 0.45	[50]	1.68(-2)	[488]	3.060(-3)	20.98	[423]
		11160.00	[104]					
	L_3N_2	11274.4 ± 1.1	[50]	1.71(-4)	[488]	3.162(-5)	18.99	[423]
		11274.00	[104]					
	L_3N_3	11371.8 ± 1.1	[50]	1.73(-4)	[488]	3.186(-5)	14.89	[423]
		11371.00	[104]					
$L_{\beta 15}$	L_3N_4	11566.81 ± 0.80	[50]	2.515(-2)	[488]	4.635(-3)	13.50	[423]
		11567.00	[104]					
$L_{\beta 2}$	L_3N_5	11584.75 ± 0.48	[50]	0.2267	[488]	4.178(-2)	11.20	[473]
		11584.00	[104]				12.66	[423]
L_u	$L_3N_{6,7}$	11835.80 ± 0.84	[50]	1.67(-4)	[488]			
		11835.00	[104]					
$L_{u'}$	L_3N_6	11828.7 ± 1.4	[145]			3.075(-5)	5.77	[423]

line	transition	E/eV		I/eV/$\hbar$		TPIV	Γ/eV	
L_u	L_3N_7	11832.3 ± 1.4	[145]			1.722(-4)	5.71	[423]
$L_{\beta 7}$	L_3O_1	11810.60 ± 0.90	[50]	3.34(-3)	[488]	6.462(-4)	23.81	[423]
		11810.00	[104]					
	$L_3O_{2,3}$	11865.00 ± 2.30	[50]	2.96(-5)	[488]	6.620(-6)		
		11865.00	[104]					
$L_{\beta 5}$	$L_3O_{4,5}$	11916.30 ± 0.35	[50]	2.25(-2)	[488]			
		11916.00	[104]					
	L_3O_4					4.156(-4)		
	L_3O_5					3.609(-3)		
	L_3P_1	11935.00	[104]	1.42(-4)	[488]	4.876(-5)		
	$L_3P_{2,3}$	11935.50 ± 0.80	[50]					

M series

line	transition	E/eV		I/eV/$\hbar$		TPIV	Γ/eV	
	M_1N_3	2883.00 ± 6.00	[50]	3.523(-2)	[335]	1.120(-3)	28.18	[423]
		2883.00	[104]					
		2889.00	[263]					
	M_2N_4	2797.00 ± 2.50	[50]	3.265(-2)	[116]	3.735(-3)	20.83	[423]
		2797.00	[104]	3.735(-2)	[335]			
		2803.00	[263]					
	M_3N_1	1981.00 ± 2.80	[50]	8.79(-3)	[116]	1.246(-3)	26.47	[423]
		1982.00	[104]	4.03(-3)	[335]		10.00	[473]
		1985.00	[263]					
$M_{\gamma 2}$	M_3N_4	2391.00 ± 2.30	[50]	4.30(-4)	[116]	6.083(-4)	18.99	[423]
		2391.00	[104]	3.7 [488]35(-2)	[335]			
		2396.00	[263]					
$M_{\gamma 1}$	M_3N_5	2410.00 ± 1.90	[50]	3.374(-2)	[116]	4.726(-3)	18.15	[423]
		2409.00	[104]	3.735(-2)	[335]			
		2415.00	[263]					
	M_3O_1	2636.00 ± 5.00	[50]	1.58(-3)	[116]	2.501(-4)	29.30	[423]
		2636.00	[104]					
		2642.00	[263]					
	M_3O_4	2742.00 ± 3.60	[50]	4.93(-4)	[116]	6.013(-5)		
	M_3O_5	2742.00 ± 4.00	[50]	3.78(-3)	[116]	4.595(-4)		
	$M_3O_{4,5}$	2747.00	[263]					
$M_{\zeta 2}$	M_4N_2	1648.00 ± 1.10	[50]	3.05(-3)	[116]	2.045(-3)	15.84	[423]
		1648.00	[104]	1.69(-3)	[335]			
		1650.90	[263]					
M_δ	M_4N_3	1746.00 ± 2.00	[50]	3.70(-4)	[116]	2.060(-4)	11.74	[423]
		1746.00	[104]	1.69(-3)	[335]			
		1748.70	[263]					
M_β	M_4N_6	2204.60 ± 0.40	[50]	6.34(-2)	[116]	2.685(-2)	8.50	[279]
		2205.00	[104]	6.01(-2)	[335]		2.62	[423]
		2209.00	[263]					
M_η	M_4O_2	2222.15	[423]			4.964(-4)	10.72	[423]
$M_{\zeta 1}$	M_5N_3	1660.50 ± 0.90	[50]	2.45(-3)	[116]	1.612(-3)	11.69	[423]
		1661.00	[104]	1.69(-3)	[335]			
		1664.20	[263]					

line	transition	E/eV		I/eV/$\hbar$		TPIV	Γ/eV	
$M_{\alpha 2}$	M_5N_6	2118.00 ± 1.10	[50]	2.97(-3)	[116]	1.354(-3)	2.57	[423]
		2118.00	[104]	6.01(-2)	[335]		2.50	[473]
		2122.00	[263]					
$M_{\alpha 1}$	M_5N_7	2122.90 ± 0.40	[50]	5.787(-2)	[116]	2.643(-2)	2.50	[473]
		2123.00	[104]	6.010(-2)	[335]		2.51	[423]
		2127.00	[263]					
	M_5O_3	2150.00 ± 3.40	[50]	3.37(-4)	[116]	4.166(-4)	10.49	[423]
		2149.00	[104]					
		2154.00	[263]					
N series								
N_4N_6		265.00 ± 1.10	[50]			1.825(-4)	8.51	[423]
$N_5N_{6,7}$		251.00 ± 0.50	[50]				7.63	[423]

level characteristics

level	E_B/eV		ω_{nlj}		Γ/eV		AE/eV	
K	80725.00	[503]	0.964	[301]	52.00	[301]	80724.9	[51]
	80724.90(0.5)	[493]	0.964(17)	[38]	51.02	[423]	80721.3 ± 3.9	[50]
			0.9604	[39]	52.1	[94]	80734.7 ± 2.1	[145]
			0.969	[423]				
			0.964	[39]				
			0.954	[242]				
L_1	14353.00	[503]	0.107	[301]	10.50	[301]	14352.8	[51]
	14352.80(0.4)	[493]	0.105	[348]	13.60	[423]	14355.29 ± 0.50	[50]
			8.23(-2)	[423]	9.8	[94]	14362.2 ± 1.6	[294]
			7.80(-2)	[436]				
			0.124	[578]				
			0.137	[567]				
L_2	13733.00	[503]	0.334	[301]	6.00	[301]	13733.6	[51]
	13733.60(0.3)	[493]	0.357	[348]	5.40	[423]	13734.194 ± 0.070	[294]
			0.3627	[423]	5.53	[94]	13741.67 ± 0.52	[145]
			0.358	[436]				
			0.342	[92]				
L_3	11918.00	[503]	0.320	[301]	5.41	[301]	11918.7	[51]
			0.31(4)	[269]	5.37	[423]	11919.694 ± 0.060	[294]
	11918.70(0.3)	[493]	0.317	[435]	5.54	[94]	11927.78 ± 0.39	[145]
			0.3183	[423]				
			0.313	[436]				
			0.296	[578]				
			0.286	[92]				
M_1	3425.00	[503]	2.13(-3)	[38]	20.90	[350]	3424.9	[51]
	3424.90(0.3)	[493]	2.21(-3)	[512]	18.67	[423]		
			3.40(-3)	[423]	15	[94]		
M_2	3150.00	[503]	4.32(-3)	[38]	10.70	[350]	3147.8	[51]
	3147.80(0.4)	[493]	5.92(-3)	[423]	12.71	[423]		
					9.5	[94]		
M_3	2743.00	[503]	4.20(-3)	[38]	12.10	[350]	2743.0	[51]
	2743.00(0.4)	[493]	4.26(-3)	[512]	10.86	[423]		
			7.64(-3)	[423]	8.5	[94]		

level	E_B/eV		ω_{nlj}		Γ/eV		AE/eV	
M_4	2291.00	[503]	2.64(-2)	[38]	12.80	[350]	2291.1	[51]
	2291.10(0.3)	[493]	2.80(-2)	[512]	2.24	[423]		
			2.96(-2)	[423]	2.18	[94]		
M_5	2206.00	[503]	2.56(-2)	[38]	12.66	[350]	2205.7	[51]
	2205.70(0.3)	[493]	2.90(-2)	[512]	2.18	[423]		
			2.98(-2)	[423]	2.18	[94]		
N_1	759.00	[503]	5.2(-4)	[423]	15.60	[423]		
	758.80(0.4)	[493]			8.5	[94]		
N_2	644.00	[503]	1.07(-3)	[423]	13.61	[423]		
	643.70(0.5)	[493]			6.4	[94]		
N_3	546.00	[503]	4.2(-4)	[423]	9.51	[423]		
	545.40(0.5)	[493]			5.05	[94]		
N_4	352.00	[503]	2.3(-4)	[423]	8.12	[423]		
	352.00(0.4)	[493]			4.1	[94]		
N_5	334.00	[503]	2.2(-4)	[423]	7.28	[423]		
	333.90(0.4)	[493]			3.9	[94]		
N_6	87.00	[503]	1.0(-6)	[423]	0.39	[423]		
	86.40 ± 0.40	[51]			0.37	[94]		
	87.30 ± 0.60	[493]						
N_7	83.00	[503]	1.0(-7)	[423]	0.33	[423]		
	82.80 ± 0.50	[51]			0.33	[94]		
	83.70 ± 0.60	[493]						
O_1	108.00	[503]	1.0(-5)	[423]	18.43	[423]		
	107.80 ± 0.70	[493]						
O_2	72.00	[503]	2.0(-6)	[423]	8.49	[423]		
	71.70 ± 0.70	[493]						
O_3	54.00	[503]			8.31	[423]		
	53.70 ± 0.70	[51]						
	56.90 ± 2.30	[493]						
O_4	3.00	[503]						
	2.50 ± 0.50	[51]						
	3.30 ± 0.50	[493]						
O_5	3.00	[503]						
	1.80 ± 0.50	[493]						
IP	9.2255	[222]						

Hg   Z=80

[Xe] $4f^{14}$ $5d^{10}$ $6s^2$

Mercury $A = 200.59(2)$ [222] $\varrho = 13.522\,\text{g/cm}^3$ [547]
$\varrho = 13.546\,\text{g/cm}^3$ [483]

X-ray transitions

line	transition	E/eV		I/eV/$\hbar$		TPIV	Γ/eV	
K series								
K_{α_3}	KL_1	68260.5 ± 2.5	[145]				67.45	[423]
K_{α_2}	KL_2	68895.1 ± 1.7	[50]	15.14	[488]	2.807(-1)	59.23	[423]
		68894.90	[104]	15.27	[487]		68.20	[473]
							60.80	[301]
K_{α_1}	KL_3	70819.5 ± 1.8	[50]	25.66	[488]	4.758(-1)	59.21	[423]
		70819.90	[104]	25.88	[487]		64.75	[473]
							60.10	[301]
	KM_1	79541.6 ± 3.8	[145]				72.54	[423]
K_{β_3}	KM_2	79823.3 ± 2.3	[50]	2.835	[488]	5.256(-2)	66.60	[423]
		79822.90	[104]	2.952	[487]		68.95	[473]
K_{β_1}	KM_3	80254.2 ± 2.3	[50]	5.46	[488]	1.017(-1)	63.79	[423]
		80253.90	[104]	5.71	[487]		65.75	[473]
K_{β_5}	$KM_{4,5}$	80754 ± 16	[50]	8.95(-2)	[488]			
		80760.00	[104]					
$K_{\beta_5^{II}}$	KM_4	80717.1 ± 3.0	[145]			1.356(-3)	55.98	[423]
$K_{\beta_5^{I}}$	KM_5	80807.6 ± 2.9				1.659(-3)	55.95	[423]
	KN_1	82302.4 ± 4.9	[145]				68.83	[423]
$K_{\beta_2^{II}}$	KN_2	82435 ± 16	[50]	0.668	[488]	1.239(-2)	67.34	[423]
		82434.00	[104]	0.716	[487]			
$K_{\beta_2^{I}}$	KN_3	82545 ± 16	[50]	1.309	[488]	2.427(-2)	63.03	[423]
		82540.00	[104]	1.413	[487]			
K_{β_4}	$KN_{4,5}$	82776 ± 16	[50]	4.22(-2)	[488]			
		82780.00	[104]					
$K_{\beta_4^{II}}$	KN_4	82726.0 ± 4.2	[145]			3.524(-4)	61.82	[423]
$K_{\beta_4^{I}}$	KN_5	82745.3 ± 2.3	[145]			4.301(-4)	60.94	[423]
	$KO_{2,3}$	83040.00 ± 11.00	[50]	0.351	[488]			
		83040.00	[104]	0.380	[487]			
	KO_2					2.863(-3)	62.60	[423]
	KO_3					5.622(-3)	63.50	[423]
L series								
	L_1M_1	11272.1 ± 3.0	[50]	8.26(-6)	[488]		32.64	[423]
		11272.00	[104]					

line	transition	E/eV		I/eV/ℏ		TPIV	Γ/eV	
$L_{\beta 4}$	$L_1 M_2$	11563.1 ± 1.1	[50]	0.399	[488]	2.894(-2)	26.70	[423]
		11563.00	[104]				19.70	[473]
$L_{\beta 3}$	$L_1 M_3$	11995.4 ± 1.2	[50]	0.452	[488]	3.280(-2)	23.88	[423]
		11995.00	[104]				17.40	[473]
$L_{\beta 10}$	$L_1 M_4$	12446.0 ± 2.5	[50]	1.366(-2)	[488]	9.909(-4)	16.07	[423]
		12446.00	[104]					
		12456.97 ± 0.52	[440]					
$L_{\beta 9}$	$L_1 M_5$	12560.0 ± 2.5	[50]	2.044(-2)	[488]	1.484(-3)	16.05	[423]
		12558.00	[104]					
		12547.52 ± 0.29	[440]					
	$L_1 N_1$	14045.8 ± 4.7	[50]	3.57(-6)	[488]		28.91	[423]
		14045.00	[104]					
$L_{\gamma 2}$	$L_1 N_2$	14162.3 ± 1.7	[50]	0.1015	[488]	7.363(-3)	27.43	[423]
		14162.10	[104]					
$L_{\gamma 3}$	$L_1 N_3$	14264.8 ± 1.7	[50]	0.1244	[488]	9.027(-3)	23.13	[423]
		14264.60	[104]					
	$L_1 N_4$	14465.5 ± 3.6	[145]			1.492(-4)	21.92	[423]
$L_{\gamma 11}$	$L_1 N_5$	14474.3 ± 1.8	[50]	3.24(-3)	[488]	2.352(-4)	21.04	[423]
		14473.20	[104]					
	$L_1 N_6$	14739.1 ± 2.6	[145]					
	$L_1 N_7$	14743.2 ± 2.6	[145]					
	$L_1 O_1$	14570.00 ± 3.50	[50]	8.28(-7)	[488]		34.53	[423]
		14670.00	[104]					
$L_{\gamma 4'}$	$L_1 O_2$	14757.00 ± 1.20	[50]	1.874(-2)	[488]	1.360(-3)	22.69	[423]
		14757.00	[104]					
$L_{\gamma 4}$	$L_1 O_3$	14778.00 ± 1.20	[50]	2.237(-2)	[488]	1.624(-3)	23.60	[423]
		14783.00	[104]					
	$L_1 O_{4,5}$	14847.00 ± 3.60	[50]	5.11(-4)	[488]			
		14847.00	[104]					
	$L_1 O_4$					1.482(-5)		
	$L_1 O_5$					2.293(-5)		
L_η	$L_2 M_1$	10651.4 ± 1.4	[50]	4.51(-2)	[488]	8.106(-3)	24.43	[423]
		10651.00	[104]					
	$L_2 M_2$	10888.1 ± 7.1	[50]	2.28(-6)	[488]		18.49	[423]
		10888.00	[104]					
$L_{\beta 17}$	$L_2 M_3$	11357.9 ± 7.7	[50]	1.75(-3)	[488]	3.160(-4)	15.67	[423]
		11370.00	[104]					
$L_{\beta 1}$	$L_2 M_4$	11822.70 ± 0.83	[50]	1.684	[488]	2.991(-1)	7.87	[423]
		11823.00	[104]				8.70	[473]
	$L_2 M_5$	11913.2 ± 1.3	[145]				7.84	[423]
$L_{\gamma 5}$	$L_2 N_1$	13410.3 ± 1.5	[50]	1.16(-2)	[488]	2.069(-3)	20.70	[423]
		13410.00	[104]					
	$L_2 N_2$	13530.8 ± 3.6	[145]				19.22	[423]
	$L_2 N_3$	13640.3 ± 1.6	[50]	4.79(-4)	[488]	8.621(-5)	14.92	[423]
		13641.00	[104]					

line	transition	E/eV		I/eV/ℏ		TPIV	Γ/eV	
$L_{\gamma 1}$	L_2N_4	13830.2 ± 1.1	[50]	0.322	[488]	5.975(-2)	13.71	[423]
		13830.00	[104]				11.80	[473]
	L_2N_5	13850.97 ± 0.72	[145]				12.83	[423]
L_{ν}	L_2N_6	14107.3 ± 1.7	[50]	1.27(-3)	[488]	2.280(-4)	5.89	[423]
		14107.00	[104]					
	L_2N_7	14109.3 ± 1.6	[145]				5.87	[423]
$L_{\gamma 8}$	L_2O_1	14090.00 ± 1.10	[50]	2.38(-3)	[488]	4.258(-4)	26.32	[423]
		14090.00	[104]					
	L_2O_2	14114.00 ± 1.60	[50]	1.76(-7)	[488]		14.48	[423]
		14114.00	[104]					
	L_2O_3	14156.00 ± 1.60	[50]	8.61(-5)	[488]	1.719(-5)	15.39	[423]
		14156.00	[104]					
$L_{\gamma 6}$	L_2O_4	14199.00 ± 1.10	[50]	3.45(-2)	[488]	6.200(-3)		
		14199.00	[104]					
	L_2P_1					4.116(-5)		
L_l	L_3M_1	8721.32 ± 0.91	[50]	7.91(-2)	[488]	1.502(-2)	24.41	[423]
		8721.00	[104]					
L_t	L_3M_2	9019.5 ± 1.9	[50]	8.29(-4)	[488]	1.496(-4)	18.47	[423]
		9019.00	[104]					
L_s	L_3M_3	9455.6 ± 2.1	[50]	7.50(-4)	[488]	1.353(-4)	15.65	[423]
		9456.00	[104]					
$L_{\alpha 2}$	L_3M_4	9897.68 ± 0.82	[50]	0.1459	[488]	2.620(-2)	7.84	[423]
		9897.60	[104]				8.80	[473]
$L_{\alpha 1}$	L_3M_5	9988.91 ± 0.60	[50]	1.285	[488]	2.307(-1)	7.82	[423]
		9888.80	[104]				8.10	[473]
$L_{\beta 6}$	L_3N_1	11482.5 ± 1.1	[50]	1.793(-2)	[488]	3.219(-3)	20.68	[423]
		11482.00	[104]					
	L_3N_2	11642.6 ± 3.2	[50]	1.87(-4)	[488]	3.361(-5)	19.20	[423]
		11643.00	[104]					
	L_3N_3	11713.0 ± 1.6	[50]	1.90(-4)	[488]	3.404(-5)	14.90	[423]
		11712.00	[104]					
$L_{\beta 15}$	L_3N_4	11904.1 ± 1.2	[50]	2.687(-2)	[488]	4.825(-3)	13.68	[423]
		11904.00	[104]					
$L_{\beta 2}$	L_3N_5	11924.2 ± 1.2	[50]	0.2425	[488]	4.355(-2)	12.81	[423]
		11925.00	[104]				10.40	[473]
$L_{u'}$	L_3N_6	12182.7 ± 1.2	[50]	1.90(-4)	[488]	3.404(-5)	5.87	[423]
		12183.00	[104]					
L_u	L_3N_7	12194.1 ± 1.2	[50]	1.06(-3)	[488]	1.904(-4)	5.85	[423]
		12194.00	[104]					
$L_{\beta 7}$	L_3O_1	12162.50 ± 0.80	[50]	3.69(-3)	[488]	6.859(-4)	26.30	[423]
		12163.00	]					
	L_3O_2	12207.90 ± 0.80	[50]	3.37(-5)	[488]	7.026(-6)	14.46	[423]
		12208.00	[104]					
	L_3O_3	12226.40 ± 0.80	[50]	3.36(-5)	[488]	7.111(-6)	15.37	[423]
		12227.00	[104]					
$L_{\beta 5}$	$L_3O_{4,5}$	12276.90 ± 0.90	[50]	2.675(-2)	[488]			

line	transition	E/eV		I/eV/$\hbar$		TPIV	Γ/eV	
		12277.00	[104]					
	L_3O_4					4.900(-4)		
	L_3O_5					4.285(-3)		

M series

line	transition	E/eV		I/eV/$\hbar$		TPIV	Γ/eV	
	M_3N_1	2036.00 ± 7.00	[50]	4.30(-3)	[335]	1.295(-3)	25.26	[423]
M_{γ_2}	M_3N_4	2458.69	[423]			6.327(-4)	18.26	[423]
M_{γ_1}	M_3N_5	2487.50 ± 1.00	[50]	4.045(-2)	[335]	4.904(-3)	17.38	[423]
		2496.00	[263]					
M_{ξ_2}	M_4N_2	1721.63	[423]			2.034(-3)	15.97	[423]
M_δ	M_4N_3	1805.00 ± 5.00	[50]	1.80(-3)	[335]	2.027(-4)	11.67	[423]
M_β	M_4N_6	2282.50 ± 0.40	[50]	6.595(-2)	[335]	2.915(-2)	2.64	[423]
		2283.00	[104]					
		2288.00	[263]					
M_η	M_4O_2	2307.46	[423]			4.979(-4)	11.23	[423]
	M_4O_3					5.111(-5)	12.13	[423]
M_{ξ_1}	M_5N_3	1722.50	[263]			1.588(-3)	11.64	[423]
M_{α_2}	M_5N_6	2187.30	[423]			1.445(-3)	2.61	[423]
M_{α_1}	M_5N_7	2195.30 ± 0.35	[50]	6.595(-2)	[335]	2.822(-2)	2.59	[423]
		2196.00	[104]					
		2201.00	[263]					
	M_5O_3					3.998(-4)	12.11	[423]

N series

line	transition	E/eV		I/eV/$\hbar$		TPIV	Γ/eV	
	N_4N_6	274.00 ± 1.80	[50]			2.026(-4)	8.48	[423]
	$N_5N_{6,7}$	259.00 ± 1.60	[50]					
	N_5N_6					8.963(-6)	7.60	[423]
	N_5N_7					1.919(-4)	7.58	[423]

level characteristics

level	E_B/eV		ω_{nlj}		Γ/eV		AE/eV	
K	83103.00	[503]	0.965	[301]	54.60	[301]	83102.3	[51]
	83102.30 ± 0.80	[493]	0.952	[372]	53.68	[423]	83109.2 ± 8.2	[50]
			0.966	[38]	54.8	[94]	83111.3 ± 2.1	[145]
			0.9615	[39]				
			0.970	[414]				
			0.9598	[423]				
			0.962	[112]				
			0.9615	[39]				
L_1	14839.00	[503]	0.107	[301]	11.30	[301]	14839.3	[51]
	14839.30 ± 1.00	[493]	8.63(-2)	[423]	13.78	[423]	14842.8 ± 2.6	[50]
			9.80(-2)	[107]	10.5	[94]	14850.8 ± 1.5	[145]
			8.20(-2)	[436]				
			8.20(-2)	[244]				

level	E_B/eV		ω_{nlj}		Γ/eV		AE/eV	
			0.098	[107]				
L_2	14209.00	[503]	0.347	[301]	6.17	[301]	14208.7	[51]
	14208.70 ± 0.70	[493]	0.39	[548]	5.56	[423]	14214.92 ± 0.24	[50]
			0.352	[107]	5.69	[94]	14216.92 ± 0.53	[145]
			0.319	[413]				
			0.3764	[423]				
			0.368	[244]				
			0.372	[578]				
L_3	12284.00	[503]	0.333	[301]	5.50	[301]	12283.9	[51]
	12289.90 ± 0.40	[493]	0.32	[269]	5.54	[423]	12286.4 ± 1.8	[50]
			0.40	[548]	5.71	[94]	12292.28 ± 0.39	[145]
			0.367	[435]				
			0.30	[413]				
			0.321	[107]				
			0.3297	[423]				
			0.31	[578]				
M_1	3562.00	[503]	2.75(-3)	[512]	23.00	[74]	3561.6	[51]
	3561.60 ± 1.10	[493]	3.59(-3)	[423]	18.88	[423]		
			2.47(-3)	[245]	15.1	[94]		
M_2	3279.00	[503]	2.108-2)	[512]	10.00	[74]	3278.5	[51]
	3278.50 ± 1.30	[493]	6.16(-3)	[423]	12.93	[423]		
			3.68(-3)	[245]	9.8	[94]		
M_3	2847.00	[503]	4.6(-3)	[512]	15.00	[74]	2847.1	[51]
	2847.10 ± 0.40	[493]	8.01(-3)	[423]	10.11	[423]		
			5.27(-3)	[245]	8.6	[94]		
M_4	2385.00	[503]	3.00(-2)	[512]	2.30	[74]	2384.9	[51]
	2384.90 ± 0.30	[493]	3.19(-2)	[423]	2.30	[423]		
					2.28	[94]		
M_5	2295.00	[503]	3.00(-2)	[512]	2.70	[74]	2294.9	[51]
	2294.90 ± 0.30	[493]	3.17(-2)	[423]	2.28	[423]		
					2.28	[94]		
N_1	800.00	[503]	5.5(-4)	[423]	16.00	[74]		
	800.30 ± 1.00	[493]			15.14	[423]		
					8.8	[94]		
N_2	677.00	[503]	1.12(-3)	[423]	13.00	[74]		
	676.90 ± 2.40	[493]			13.67	[423]		
					6.55	[94]		
N_3	571.00	[503]	4.5(-4)	[423]	13.00	[74]		
	571.00 ± 1.40	[493]			9.36	[423]		
					5.3	[94]		
N_4	379.00	[503]	2.6(-4)	[423]	8.00	[74]		
	378.30 ± 1.00	[493]			8.15	[423]		
					4	[94]		
N_5	360.00	[503]	2.4(-4)	[423]	8.00	[74]		

level	E_B/eV		ω_{nlj}		Γ/eV		AE/eV
	359.80 ± 1.20	[493]			7.27	[423]	
					3.85	[94]	
N_6	103.00	[503]	2.0(-6)	[423]			
	102.20 ± 0.50	[51]			0.20	[74]	
	103.30 ± 0.70	[493]			0.33	[423]	
					0.33	[94]	
N_7	99.00	[503]	2.0(-6)	[423]			
	98.30 ± 0.50	[51]			0.20	[74]	
	99.40 ± 0.60	[493]			0.31	[423]	
					0.31	[94]	
O_1	120.00	[503]	1.0(-5)	[423]	20.76	[423]	
	120.30 ± 1.30	[493]					
O_2	81.00	[503]	3.0(-6)	[423]	8.92	[423]	
	80.50 ± 1.30	[493]					
O_3	58.00	[503]	1.0(-6)	[423]	9.83	[423]	
	57.60 ± 1.30	[51]					
	61.80 ± 1.40	[493]					
O_4	7.00	[503]					
	6.40 ± 1.40	[51]					
	7.50 ± 1.40	[493]					
O_5	7.00	[503]					
	6.40 ± 1.40	[51]					
	5.70 ± 1.40	[493]					
IP	10.43750	[222]					

Tl   Z=81

[Xe] $4f^{14}$ $5d^{10}$ $6s^2$ $6p^1$

Thallium $A = 204.3833(2)$ [222] $\varrho = 11.83\,\text{g/cm}^3$ [547]
$\varrho = 11.86\,\text{g/cm}^3$ [483]

X-ray transitions

line	transition	E/eV		I/eV/$\hbar$		TPIV	Γ/eV	
K series								
K_{α_3}	KL_1	70185.1 ± 2.5	[145]				70.35	[423]
K_{α_2}	KL_2	70832.5 ± 1.2	[50]	15.96	[488]	2.815(-1)	63.80	[301]
		70831.90	[104]	16.09	[487]		62.18	[423]
K_{α_1}	KL_3	72872.5 ± 1.3	[50]	26.93	[488]	4.748(-1)	63.10	[301]
		72870.90	[104]	27.16	[487]		62.15	[423]
	KM_1	81827.0 ± 3.9	[145]				75.5	[423]
K_{β_3}	KM_2	82118.4 ± 4.8	[50]	2.988	[488]	5.268(-2)	69.58	[423]
		82118.90	[104]	3.11	[487]			
K_{β_1}	KM_3	82576.7 ± 4.1	[50]	5.79	[488]	1.020(-1)	66.78	[423]
		82576.90	[104]	6.01	[487]			
K_{β_5}	$KM_{4,5}$	83114.8 ± 8.2	[50]					
$K_{\beta_5^{II}}$	KM_4	83045.6 ± 3.1	[145]	7.92(-2)	[488]	1.398(-3)	58.86	[423]
$K_{\beta_5^{I}}$	KM_5	83141.7 ± 3.0	[145]				58.83	[423]
	KN_1	84685.5 ± 5.0	[145]					
$K_{\beta_2^{II}}$	KN_2	84838.0 ± 8.6	[50]	0.709	[488]	1.249(-2)	70.19	[423]
				0.759	[487]			
$K_{\beta_2^{I}}$	KN_3	84948.5 ± 8.6	[50]	1.390	[488]	2.450(-2)	65.68	[423]
				1.490	[487]			
K_{β_4}	$KN_{4,5}$	85194 ± 17	[50]	4.61(-3)	[488]			
$K_{\beta_4^{II}}$	KN_4	85125.4 ± 4.4	[145]			3.674(-4)	64.54	[423]
$K_{\beta_4^{I}}$	KN_5	85145.9 ± 4.3	[145]			4.462(-4)	63.69	[423]
	$KO_{2,3}$	85451.00 ± 6.00	[50]	5.41(-3)	[488]			
	KO_2					2.895(-3)	65.62	[423]
	KO_3					5.647(-3)	68.00	[423]
	$KP_{2,3}$			6.00(-3)	[488]			
	KP_2					2.457(-5)		
	KP_3					3.497(-5)		

line	transition	E/eV		I/eV/$\hbar$		TPIV	Γ/eV	
	L series							
	L_1M_1	11648.1 $\pm$ 3.2	[50]	9.66(-6)	[488]		32.95	[423]
		11648.00	[104]					
$L_{\beta 4}$	L_1M_2	11930.78 $\pm$ 0.51	[50]	0.426	[488]	3.064(-2)	21.60	[473]
		11929.80	[104]				27.04	[423]
$L_{\beta 3}$	L_1M_3	12390.55 $\pm$ 0.55	[50]	0.474	[488]	3.410(-2)	19.70	[473]
		12390.00	[104]				24.23	[423]
$L_{\beta 10}$	L_1M_4	12862.7 $\pm$ 1.4	[50]	1.499(-2)	[488]	1.078(-3)	16.31	[423]
		12861.00	[104]					
$L_{\beta 9}$	L_1M_5	12958.7 $\pm$ 1.4	[50]	2.243(-2)	[488]	1.613(-3)	16.28	[423]
		12957.40	[104]					
	L_1N_1	14502.6 $\pm$ 2.5	[50]	4.20(-6)	[488]		29.06	[423]
		14501.80	[104]					
$L_{\gamma 2}$	L_1N_2	14625.2 $\pm$ 1.3	[50]	0.1090	[488]	7.840(-3)	19.30	[473]
		14623.50	[104]				27.65	[423]
$L_{\gamma 3}$	L_1N_3	14737.0 $\pm$ 1.0	[50]	0.1318	[488]	9.478(-3)	17.80	[473]
		14735.40	[104]				23.13	[423]
	L_1N_4	14937.4 $\pm$ 1.9	[50]	2.28(-3)	[488]	1.639(-4)	22.00	[423]
		13937.30	[104]					
$L_{\gamma 11}$	L_1N_5	14959.4 $\pm$ 1.3	[50]	3.60(-3)	[488]	2.588(-4)	21.14	[423]
		14958.40	[104]					
	L_1N_6	15222.7 $\pm$ 2.6	[145]				14.20	[423]
	L_1N_7	15227.1 $\pm$ 2.5	[145]				14.18	[423]
	L_1O_1	15198.00 $\pm$ 2.00	[50]	9.98(-7)	[488]		37.31	[423]
		15196.70	[104]					
$L_{\gamma 4'}$	L_1O_2	15248.20 $\pm$ 0.90	[50]	2.075(-2)	[488]	1.492(-3)	23.07	[423]
		15247.00	[104]					
$L_{\gamma 4}$	L_1O_3	15271.60 $\pm$ 0.90	[50]	2.452(-2)	[488]	1.763(-3)	25.45	[423]
		15271.00	[104]					
	$L_1O_{4,5}$	15332.70 $\pm$ 1.00	[50]	6.26(-4)	[488]	0.13		
		15332.00	[104]					
	L_1O_4					1.792(-5)		
	L_1O_5					2.785(-5)		
L_{η}	L_2M_1	10994.36 $\pm$ 0.43	[50]	4.80(-2)	[488]	8.384(-3)	24.77	[423]
		10994.00	[104]					
	L_2M_2	11274.2 $\pm$ 1.5	[50]	2.67(-6)	[488]		18.86	[423]
		10274.00	[104]					
$L_{\beta 17}$	L_2M_3	11739.7 $\pm$ 1.2	[50]	1.92(-3)	[488]	3.355(-4)	16.06	[423]
		11738.40	[104]					
$L_{\beta 1}$	L_2M_4	12213.44 $\pm$ 0.71	[50]	1.767	[488]	3.088(-1)	9.00	[473]
		12212.00	[104]				8.14	[423]

line	transition	E/eV		I/eV/$\hbar$		TPIV	Γ/eV	
	L$_2$M$_5$	12309.36 ± 0.90	[50]	1.08(-4)	[488]		8.11	[423]
		12309.00	[104]					
L$_{\gamma 5}$	L$_2$N$_1$	13852.77 ± 0.92	[50]	1.231(-2)	[488]	2.152(-3)	20.88	[423]
		13852.00	[104]					
	L$_2$N$_2$	14057 ± 47	[50]	1.03(-6)	[488]		19.47	[423]
		13958.00	[104]					
		13979.1 ± 3.7	[145]					
	L$_2$N$_3$	14089.5 ± 1.2	[50]	5.28(-4)	[488]	9.232(-5)	14.96	[423]
		14089.00	[104]					
L$_{\gamma 1}$	L$_2$N$_4$	14291.58 ± 0.73	[50]	0.357	[488]	6.236(-2)	11.70	[473]
		14290.00	[104]				13.83	[423]
	L$_2$N$_5$	14313.2 ± 2.6	[145]				12.97	[423]
	L$_2$N$_{6,7}$	14577.9 ± 1.3	[50]	1.43(-3)	[488]			
		14577.00	[104]					
L$_\nu$	L$_2$N$_6$	14575.0 ± 1.6	[145]			2.505(-4)	6.02	[423]
	L$_2$N$_7$	14579.4 ± 1.6					6.00	[423]
L$_{\gamma 8}$	L$_2$O$_1$	14564.00 ± 3.40	[50]	2.61(-3)	[488]	4.508(-4)	29.14	[423]
		14561.00	[104]					
	L$_2$O$_2$	14604.00 ± 1.70	[50]	2.14(-7)	[488]		14.90	[423]
		14603.00	[104]					
L$_{\gamma 6}$	L$_2$O$_4$	14685.00 ± 3.50	[50]	4.07(-2)	[488]	7.115(-3)		
		14682.00	[104]					
L$_l$	L$_3$M$_1$	8953.28 ± 0.29	[50]	7.91(-2)	[488]	1.545(-2)	24.75	[423]
		8952.70	[104]					
L$_t$	L$_3$M$_2$	9241.79 ± 0.51	[50]	9.02(-4)	[488]	1.577(-4)	18.84	[423]
		9241.20	[104]					
L$_s$	L$_3$M$_3$	9700.75 ± 0.56	[50]	8.16(-4)	[488]	1.426(-4)	16.03	[423]
		9700.10	[104]					
L$_{\alpha 2}$	L$_3$M$_4$	10172.91 ± 0.37	[50]	0.1344	[488]	2.701(-2)	8.12	[423]
		10172.00	[104]					
L$_{\alpha 1}$	L$_3$M$_5$	10268.62 ± 0.50	[50]	1.360	[488]	2.378(-1)	9.40	[473]
		10268.00	[104]				8.09	[423]
L$_{\beta 6}$	L$_3$N$_1$	11812.00 ± 0.83	[50]	1.934(-2)	[488]	3.382(-3)	20.86	[423]
		11812.00	[104]					
	L$_3$N$_2$	11938.0 ± 3.5	[145]			3.556(-5)	19.45	[423]
	L$_3$N$_3$	12053.5 ± 1.7	[50]	2.08(-4)	[488]	3.633(-5)	14.93	[423]
		12052.00	[104]					
L$_{\beta 15}$	L$_3$N$_4$	12251.10 ± 0.54	[50]	2.869(-2)	[488]	5.017(-3)	13.80	[423]
		12250.00	[104]					
L$_{\beta 2}$	L$_3$N$_5$	12271.71 ± 0.54	[50]	0.2591	[488]	4.531(-2)	11.50	[473]
		12271.00	[104]				12.94	[423]

line	transition	E/eV		I/eV/$\hbar$		TPIV	Γ/eV	
L_u	$L_3N_{6,7}$	12538.7 ± 1.9	[50]	1.20(-3)	[488]			
		12537.00	[104]					
$L_{u'}$	L_3N_6	12533.9 ± 1.5	[145]			3.754(-5)	6.00	[423]
L_u	L_3N_7	12538.3 ± 1.4	[145]			2.097(-4)	5.97	[423]
$L_{\beta 7}$	L_3O_1	12521.20 ± 0.60	[50]	4.07(-3)	[488]	7.277(-4)	29.12	[423]
		12520.00	[104]					
	L_3O_2	12556.00 ± 0.60	[50]	3.78(-5)	[488]	7.455(-6)	14.88	[423]
		12556.00	[104]					
	L_3O_3	12582.00 ± 0.60	[50]	3.80(-5)	[488]	7.631(-6)	17.26	[423]
		12581.00	[104]					
$L_{\beta 5}$	$L_3O_{4,5}$	12643.60 ± 0.40	[50]	3.16(-2)	[488]			
		12643.00	[104]					
	L_3O_4					5.644(-4)		
	L_3O_5					4.969(-3)		
	L_3P_1	12660.00	[104]	4.66(-4)	[488]	8.150(-5)		
	$L_3P_{2,3}$	12660.70 ± 0.60	[50]	6.95(-7)	[488]			
	L_3P_2					6.919(-8)		
	L_3P_3					5.237(-8)		

M series

line	transition	E/eV		I/eV/$\hbar$		TPIV	Γ/eV	
	M_1N_3	3089.00 ± 7.00	[50]	4.069(-2)	[335]	1.178(-3)	28.27	[423]
		3089.00	[104]					
		3096.00	[263]					
	M_2N_4	3013.00 ± 3.00	[50]	4.374(-2)	[335]	3.934(-3)	21.23	[423]
		3010.00	[104]					
		3019.00	[263]					
	M_2O_4	2941.00 ± 3.00	[50]			5.225(-4)		
	M_3N_1	2107.00 ± 3.00	[50]	4.58(-3)	[335]	1.348(-3)	25.48	[423]
		2108.00	[104]					
		2111.00	[104]					
$M_{\gamma 2}$	M_3N4	2548.00 ± 2.60	[50]	4.374(-2)	[335]	6.582(-4)	18.42	[423]
		2548.00	[104]					
		2554.00	[263]					
$M_{\gamma 1}$	M_3N_5	2571.00 ± 2.10	[50]	4.374(-2)	[335]	5.087(-3)	17.57	[423]
		2570.00	[104]					
		2576.00	[263]					
	$M_3O_{4,5}$	2941.00 ± 4.00	[50]					
	M_3O_4					8.174(-5)		
	M_3O_5					6.266(-4)		
		2947.00	[104]					

line	transition	E/eV		I/eV/$\hbar$		TPIV	Γ/eV	
	M_3O_5	2941.00	[104]					
$M_{\xi 2}$	M_4N_2	1763.00 ± 1.30	[50]	1.11(-5)	[335]	2.019(-3)	16.16	[423]
		1763.00	[104]					
		1767.00	[263]					
$M\delta$	M_4N_3	1893.38	[423]			2.023(-4)	11.64	[423]
M_β	M_4N_6	2362.10 ± 0.50	[50]	7.282(-2)	[335]	3.099(-2)	2.70	[423]
		2362.00	[104]					
		2367.00	[263]					
M_η	M_4O_2	2386.00 ± 4.10	[50]			4.983(-4)	11.58	[423]
		2386.00	[104]					
		2391.00	[263]					
$M_{\xi 1}$	M_5N_3	1778.00 ± 1.00	[50]	1.11(-5)	[335]	1.562(-3)	11.61	[423]
		1778.00	[104]					
		1781.40	[263]					
$M_{\alpha 2}$	M_5N_6	2265.60 ± 0.80	[50]	7.219(-2)	[335]	1.538(-3)	2.67	[423]
		2266.00	[104]					
		2270.00	[263]					
$M_{\alpha 1}$	M_5N_7	2270.60 ± 0.40	[50]	7.219(-2)	[335]	3.004(-2)	2.65	[423]
		2270.00	[104]					
		2275.00	[263]					

N series

	transition	E/eV				TPIV	Γ/eV	
	$N_5N_{6,7}$	267.00 ± 1.20	[50]					
	N_5N_6					9.795(-6)	7.53	[423]
	N_5N_7					2.106(-4)	7.51	[423]
	N_6O_4	107.50 ± 0.20	[50]			3.202(-6)		
	N_6O_5	109.68 ± 0.10	[50]			2.129(-7)		
	N_7O_5	105.30 ± 0.10	[50]			3.084(-6)		

level characteristics

level	E_B/eV		ω_{nlj}		Γ/eV		AE/eV	
K	85531.00	[503]	0.9225	[39]	56.45	[423]	85530.4	[51]
	85530.40 ± 0.60	[493]	0.966	[301]	57.40	[301]	85534.5 ± 8.7	[50]
			0.9606	[423]	57.6	[94]	85538.2 ± 2.2	[145]
L_1	14347.00	[503]	0.107	[301]	12.00	[301]	15346.7	[51]
	14346.70 ± 0.40	[493]	7.00(-2)	[574]	13.90	[423]	15342.4 ± 2.8	[50]
			9.12(-2)	[423]	11.1	[94]	15353.1 ± 1.5	[145]
			8.80(-2)	[436]				
			0.127	[578]				
L_2	14698.00	[503]	0.360	[301]	6.32	[301]	14697.9	[51]
	14697.90 ± 0.30	[493]	0.319	[574]	5.73	[423]	14700.3 ± 2.6	[50]
			0.390	[423]	5.87	[94]	14705.47 ± 0.56	[145]
			0.384	[436]				
			0.389	[578]				
L_3	12657.00	[503]	0.347	[301]	5.65	[301]	12657.5	[51]
	12657.50 ± 0.30	[493]	0.386	[435]	5.71	[423]	12660.3 ± 1.9	[50]
			0.37	[269]	5.89	[94]	12664.38 ± 0.40	[145]
			0.306	[574]				
			0.3410	[423]				
			0.332	[436]				
			0.324	[578]				
M_1	3704.00	[503]	2.97(-3)	[512]	15.1	[94]	3704.1	[51]
	3704.10 ± 0.40	[493]	3.74(-3)	[423]	19.04	[423]		
M_2	3416.00	[503]	2.20(-2)	[512]	10.1	[94]	3415.7	[51]
	3415.70 ± 0.30	[493]	6.39(-3)	[423]	13.13	[423]		
M_3	2957.00	[503]	4.94(-3)	[512]	8.7	[94]	2956.6	[51]
	2956.60 ± 0.30	[493]	8.39(-3)	[423]	10.33	[423]		
M_4	2485.00	[503]	3.30(-2)	[512]	2.38	[94]	2485.1	[51]
	2485.10 ± 0.30	[493]	3.38(-2)	[423]	2.41	[423]		
M_5	2390.00	[503]	3.10(-2)	[512]	2.38	[94]	2389.3	[51]
	2389.30 ± 0.30	[493]	3.35(-2)	[423]	2.38	[423]		
N_1	846.00	[503]	5.8(-4)	[423]	9.1	[94]		

level	E_B/eV		ω_{nlj}		Γ/eV		AE/eV
	845.50 ± 0.50	[493]			15.15	[423]	
N_2	722.00	[503]	1.16(-3)	[423]	6.7	[94]	
	721.30 ± 0.80	[493]			13.75	[423]	
N_3	609.00	[503]	4.7(-4)	[423]	5.6	[94]	
	609.00 ± 0.50	[493]			9.23	[423]	
N_4	407.00	[503]	2.8(-4)	[423]	3.9	[94]	
	406.60 ± 0.40	[493]			8.10	[423]	
N_5	386.00	[503]	2.7(-4)	[423]	3.8	[94]	
	386.20 ± 0.50	[493]			7.24	[423]	
N_6	122.00	[503]	3.0(-6)	[423]	0.29	[94]	
]	122.80 ± 0.40	[51]			0.29	[423]	
	123.00 ± 0.30	[493]					
N_7	118.00	[503]	3.0(-6)	[423]	0.27	[94]	
	118.50 ± 0.40	[51]			0.27	[423]	
	118.70 ± 0.30	[493]					
O_1	137.00	[503]	1.0(-5)	[423]	23.41	[423]	
	136.30 ± 0.70	[493]					
O_2	100.00	[503]	4.0(-6)	[423]	9.17	[423]	
	99.60 ± 0.60	[493]					
O_3	76.00	[503]	1.0(-6)	[423]	11.55	[423]	
	75.40 ± 0.60	[51]					
	74.50 ± 1.30	[493]					
O_4	16.00	[503]					
	15.30 ± 0.40	[493]					
O_5	13.00	[503]					
	13.10 ± 0.40	[493]					
IP	6.1082	[222]					

Pb   Z=82

[Xe] $4f^{14} 5d^{10} 6s^2 6p^2$

Lead $A = 207.2(1)$ [222] $\varrho = 11.33\,\text{g/cm}^3$ [547]
$\varrho = 11.34\,\text{g/cm}^3$ [483]

X-ray transitions

line	transition	E/eV		I/eV/$\hbar$		TPIV	Γ/eV	
K series								
K_{α_3}	KL_1	72145.0 ± 2.6	[145]				73.18	[423]
K_{α_2}	KL_2	72804.20 ± 0.90	[50]	16.82	[488]	2.822(-1)	66.80	[301]
		72804.40	[104]	16.95	[487]		69.00	[473]
		72805.90 ± 0.24	[277]				65.23	[423]
		72805.42 ± 0.24	[144]					
K_{α_1}	KL_3	74969.40 ± 0.90	[50]	28.25	[488]	4.739(-1)	66.20	[301]
		74970.90	[104]	28.49	[487]		68.30	[473]
		74970.60 ± 0.17	[277]				65.21	[423]
		74970.11 ± 0.17	[144]					
	KM_1	84156.2 ± 4.0	[145]					
K_{β_3}	KM_2	84450.00 ± 2.30	[50]	3.147	[488]	5.279(-2)	74.90	[473]
		84450.60	[104]	3.27	[487]		72.65	[423]
		84451.00 ± 0.60	[277]					
		84450.45 ± 0.60	[145]					
K_{β_1}	KM_3	84936.00 ± 3.50	[50]	6.10	[488]	1.023(-1)	42.20	[473]
		84936.94	[104]	6.34	[487]		69.91	[423]
		84939.63 ± 0.34	[277]					
		84939.08 ± 0.34	[144]					
$K_{\beta_5^{II}}$	KM_4	85434 ± 17	[50]	8.58(-2)	[488]	1.439(-3)	61.84	[423]
		85430.00	[104]					
$K_{\beta_5^{I}}$	KM_5	85535 ± 26	[50]	0.1038	[488]	1.742(-3)	61.81	[423]
		85530.00	[104]					
	KN_1	87114.5 ± 5.1	[145]				74.64	[423]
$K_{\beta_2^{II}}$	KN_2	87238 ± 18	[50]	0.750	[488]	1.259(-2)	73.29	[423]
		87230.00	[104]	0.803	[487]			
$K_{\beta_2^{I}}$	KN_3	87366.9 ± 9.1	[50]	1.475	[488]	2.474(-2)	68.47	[423]
		87364.00	[104]	1.591	[487]			
K_{β_4}	$KN_{4,5}$	87589 ± 27	[50]	5.04(-2)	[488]			
		87590.00	[104]					

line	transition	E/eV		I/eV/$\hbar$		TPIV	Γ/eV	
$K_{\beta_4^{II}}$	KN_4	87571.0 ± 4.4	[145]			3.826(-4)	67.16	[423]
$K_{\beta_4^{I}}$	KN_5	87593.2 ± 4.4	[145]			4.623(-4)	66.47	[423]
	$KO_{2,3}$	87922.00 ± 5.00	[50]	0.421	[488]			
		87923.00	[104]	0.457	[487]			
	KO_2					2.927(-3)	67.33	[423]
	KO_3					5.672(-3)	72.70	[423]
	$KP_{2,3}$	88060.00 ± 60.00	[50]	9.93(-3)	[488]			
				1.51(-2)	[487]			
	KP_2					6.575(-5)		
	KP_3					1.008(-4)		

L series

line	transition	E/eV		I/eV/$\hbar$		TPIV	Γ/eV	
	L_1M_1	12010.3 ± 3.4	[50]	1.13(-5)	[488]		33.06	[423]
		12010.00	[104]					
$L_{\beta 4}$	L_1M_2	12305.9 ± 1.8	[50]	0.455	[488]	3.309(-2)	22.20	[279]
		12306.30	[104]				21.30	[473]
							27.18	[423]
$L_{\beta 3}$	L_1M_3	12793.4 ± 1.4	[50]	0.479	[488]	3.614(-2)	18.30	[279]
		12793.00	[104]				18.65	[473]
							24.44	[423]
$L_{\beta 10}$	L_1M_4	13275.8 ± 4.2	[50]	1.643(-2)	[488]	1.195(-3)	16.37	[423]
		13276.00	[104]					
$L_{\beta 9}$	L_1M_5	13377.5 ± 2.1	[50]	2.459(-2)	[488]	1.788(-3)	16.33	[423]
		13377.20	[104]					
	L_1N_1	14963.0 ± 1.9	[50]	4.93(-6)	[488]		29.16	[423]
		14962.00	[104]					
$L_{\gamma 2}$	L_1N_2	15101.4 ± 5.4	[50]	0.1171	[488]	8.516(-3)	19.90	[473]
		15095.60	[104]				27.81	[423]
		15118.00	[543]					
$L_{\gamma 3}$	L_1N_3	15218.2 ± 2.8	[50]	0.1394	[488]	1.014(-2)	19.40	[473]
		15214.60	[104]				22.99	[423]
		15238.00	[543]					
	L_1N_4	15427.6 ± 2.0	[50]	2.52(-3)	[488]	1.836(-4)	21.66	[423]
		15426.50	[104]					
$L_{\gamma 11}$	L_1N_5	15452.8 ± 2.6	[50]	3.99(-3)	[488]	2.905(-4)	21.00	[423]
		15452.30	[104]					
	$L_1N_{6,7}$	15725.8 ± 3.0	[50]	7.55(-5)	[488]			
		15725.00	[104]					
	L_1N_6	15717.7 ± 2.6	[145]				14.11	[423]
	L_1N_7	15722.6 ± 2.6	[145]				14.09	[423]

line	transition	E/eV		I/eV/$\hbar$		TPIV		Γ/eV	
	L_1O_1	15699.00 $\pm$ 2.00	[50]	1.20(-6)	[488]			40.05	[423]
		15698.80	[104]						
$L_{\gamma4'}$	L_1O_2	15752.00 $\pm$ 1.40	[50]	2.294(-2)	[488]	1.668(-3)		21.86	[423]
		15751.00	[104]						
		15779.00	[543]						
$L_{\gamma4}$	L_1O_3	15777.00 $\pm$ 2.00	[50]	2.683(-2)	[488]	1.951(-3)		27.22	[423]
		15775.00	[104]						
		15802.00	[543]						
	$L_1O_{4,5}$	15843.00 $\pm$ 1.40	[50]	7.54(-4)	[488]				
		15842.00	[104]						
	L_1O_4					2.141(-5)			
	L_1O_5					3.345(-5)			
L_η	L_2M_1	11349.4 $\pm$ 1.1	[50]	5.10(-2)	[488]	8.663(-3)		25.10	[423]
		11348.00	[104]						
		11329.00	[543]						
	L_2M_2	11648.1 $\pm$ 3.2	[50]	3.11(-6)	[488]			19.22	[423]
		11648.00	[104]						
$L_{\beta17}$	L_2M_3	12127.8 $\pm$ 1.8	[50]	2.09(-3)	[488]	3.555(-4)		16.49	[423]
		12127.00	[104]						
$L_{\beta1}$	L_2M_4	12613.80 $\pm$ 0.57	[50]	1.875	[488]	3.184(-1)		8.40	[279]
		12612.00	[104]					9.357	[473]
		12607.00	[543]					8.41	[423]
	L_2M_5	12720.0 $\pm$ 1.9	[50]	1.25(-4)	[488]			8.38	[423]
		12718.00	[104]						
$L_{\gamma5}$	L_2N_1	14307.6 $\pm$ 1.2	[50]	1.317(-2)	[488]	2.236(-3)		21.21	[423]
		14308.00	[104]						
	L_2N_2	14441.7 $\pm$ 7.5	[50]	1.22(-6)	[488]			19.86	[423]
		14409.00	[104]						
	L_2N_3	14553.3 $\pm$ 1.8	[50]	5.82(-4)	[488]	9.871(-5)		15.04	[423]
		14552.00	[104]						
$L_{\gamma1}$	L_2N_4	14764.55 $\pm$ 0.78	[50]	0.383	[488]	6.502(-2)		11.40	[279]
		14764.00	[104]					12.30	[473]
		14753.00	[543]					13.70	[423]
	L_2N_5	14791.5 $\pm$ 5.2	[50]	2.86(-5)	[488]			13.05	[423]
		14791.00	[104]						
L_ν	L_2N_6	15060 $\pm$ 19	[50]	1.61(-3)	[488]	2.741(-4)		6.16	[423]
		15059.00	[104]						
		15056.1 $\pm$ 1.6	[145]						
	L_2N_7	15061.0 $\pm$ 1.6	[145]					6.13	[423]

line	transition	E/eV		I/eV/ℏ		TPIV	Γ/eV	
$L_{\gamma 8}$	L_2O_1	15052.70 ± 0.90 15051.00	[50] [104]	2.58(-3)	[488]	4.770(-4)	32.10	[423]
	L_2O_3	15120.00 ± 1.80 15119.00	[50] [104]	1.12(-4)	[488]	1.989(-5)	19.27	[423]
$L_{\gamma 6}$	L_2O_4	15178.30 ± 0.90 15177.00	[50] [104]	4.74(-2)	[488]	8.046(-3)		
	L_2P_1	15196.90 ± 0.90 15196.00	[50] [104]	3.65(-4)	[488]	6.190(-5)		
L_l	L_3M_1	9184.56 ± 0.70 9184.50 9168.00	[50] [104] [543]	8.55(-2)	[488]	1.590(-2)	25.08	[423]
L_t	L_3M_2	9481.16 ± 0.75 9480.60	[50] [104]	9.81(-4)	[488]	1.671(-4)	19.20	[423]
L_s	L_3M_3	9967.63 ± 0.83 9967.00	[50] [104]	8.87(-4)	[488]	1.510(-4)	16.47	[423]
$L_{\alpha 2}$	L_3M_4	10449.59 ± 0.65 10449.00 10446.00	[50] [104] [543]	0.1633	[488]	2.782(-2)	9.35 8.39	[473] [423]
$L_{\alpha 1}$	L_3M_5	10551.60 ± 0.27 10551.00 10547.00	[50] [104] [543]	1.438	[488]	2.449(-1)	9.50 8.36	[279] [423]
$L_{\beta 6}$	L_3N_1	12143.2 ± 1.8 12143.00	[50] [104]	2.085(-2)	[488]	3.552(-3)	21.19	[423]
	L_3N_2	12270.6 ± 1.3 12269.00	[50] [104]	2.22(-4)	[488]	3.778(-5)	19.84	[423]
	L_3N_3	12392.0 ± 1.8 12391.00	[50] [104]	2.27(-4)	[488]	3.871(-5)	15.02	[423]
$L_{\beta 15}$	L_3N_4	12601.2 ± 1.3 12601.00	[50] [104]	3.06(-2)	[488]	5.210(-3)	13.68	[423]
$L_{\beta 2}$	L_3N_5	12622.8 ± 1.3 12622.00	[50] [104]	0.2767	[488]	4.713(-2)	11.80 10.75 13.03	[279] [473] [423]
L_u	$L_3N_{6,7}$	12897.0 ± 1.4 12896.00	[50] [104]	2.42(-4)	[488]			
$L_{u'}$	L_3N_6	12892.6 ± 1.5	[145]			4.127(-5)	6.14	[423]
L_u	L_3N_7	12897.5 ± 1.5	[145]			2.304(-4)	6.11	[423]
$L_{\beta 7}$	L_3O_1	12888.00 ± 1.30 12888.00	[50] [104]	1.35(-3)	[488]	7.715(-4)	32.08	[423]
	L_3O_2	12934.00 ± 1.30	[50]	4.23(-5)	[488]	7.905(-6)	13.89	[423]

line	transition	E/eV		I/eV/ℏ		TPIV	Γ/eV	
		12934.00	[104]					
	L_3O_3	12945.00 ± 1.40	[50]	4.33(-5)	[488]	8.184(-6)	19.25	[423]
		12934.00	[104]					
$L_{\beta5}$	$L_3O_{4,5}$	13015.00 ± 1.40	[50]	3.68(-2)	[488]			
		13016.00	[104]					
	L_3O_4					6.355(-4)		
	L_3O_5					5.631(-3)		
	L_3P_1	13034.00	[104]	5.72(-4)	[488]	9.743(-5)		
	$L_3P_{2,3}$	13034.40 ± 1.00	[50]	2.08(-6)	[488]			
	L_3P_2					1.943(-7)		
	L_3P_3					1.594(-7)		

M series

line	transition	E/eV		I/eV/ℏ		TPIV	Γ/eV	
	M_1N_3	3202.00 ± 7.40	[50]	1.858(-2)	[116]	1.213(-3)	28.34	[423]
		3202.00	[104]	4.366(-2)	[335]			
		3209.00	[263]					
	M_2N_1	2664.00 ± 4.60	[50]	7.67(-3)	[116]	7.584(-4)	28.63	[423]
		2663.00	[104]	4.88(-3)	[335]			
		2669.00	[263]					
	M_2N_4	3124.00 ± 3.90	[50]	4.000(-2)	[116]	4.034(-3)	21.13	[423]
		3121.00	[104]	4.723(-2)	[335]			
		3131.00	[263]					
	M_2O_4	3017.00 ± 4.00	[50]	6.12(-3)	[116]	5.876(-4)		
	M_3N_1	2174.00 ± 3.00	[50]	1.105(-2)	[116]	1.405(-3)	25.89	[423]
		2173.00	[104]	1.35(-4)	[335]			
		2178.00	[263]					
$M_{\gamma2}$	M_3N4	2630.00 ± 1.70	[50]	5.44(-3)	[116]	6.848(-4)	18.39	[423]
		2630.00	[104]	4.723(-2)	[335]			
		2635.00	[263]					
$M_{\gamma1}$	M_3N_5	2652.70 ± 0.60	[50]	4.218(-2)	[116]	5.279(-3)	17.73	[423]
		2652.00	[104]	4.723(-2)	[335]			
		2658.00	[263]					
	M_3O_1	2921.00 ± 6.20	[50]	2.15(-3)	[116]	2.984(-4)	36.78	[423]
		2921.00	[104]					
		2928.00	[263]					
	$M_3O_{4,5}$	3047.00 ± 4.50	[50]					
		3053.00	[263]					
	M_3O_4					9.270(-5)		
	M_3O_5	3045.00	[104]	6.04(-3)	[116]	7.100(-4)		

line	transition	E/eV		I/eV/$\hbar$		TPIV	Γ/eV	
$M_{\xi 2}$	M_4N_2	1823.00 ± 1.30	[50]	3.86(-3)	[116]	2.020(-3)	16.47	[423]
		1823.00	[104]	3.40(-5)	[335]			
		1826.50	[263]					
M_δ	M_4N_3	1942.00 ± 2.10	[50]	4.46(-4)	[116]	2.020(-4)	11.65	[423]
		1942.00	[104]	3.40(-5)	[335]			
		1946.10	[263]					
M_β	M_4N_6	2442.70 ± 0.50	[50]	8.245(-2)	[116]	3.290(-2)	2.77	[423]
		2443.00	[104]	7.881(-2)	[335]			
		2448.00	[263]					
M_η	M_4O_2	2477.00 ± 4.50	[50]	5.98(-4)	[116]	4.984(-4)	10.52	[423]
		2477.00	[104]					
		2483.00	[263]					
$M_{\xi 1}$	M_5N_3	1839.50 ± 0.80	[50]	2.99(-3)	[116]	1.537(-3)	11.62	[423]
		1840.00	[104]	3.40(-4)	[335]			
		1843.00	[263]					
$M_{\alpha 2}$	M_5N_6	2339.70 ± 0.90	[50]	3.84(-3)	[116]	1.630(-3)	2.74	[423]
		2340.00	[104]	7.88(-2)	[335]			
		2345.00	[263]					
$M_{\alpha 1}$	M_5N_7	2345.50 ± 0.40	[50]	7.51(-2)	[116]	3.187(-2)	2.71	[423]
		2346.00	[104]	7.88(-2)	[335]			
		2350.00	[263]					
	M_5O_3	2399.00 ± 4.20	[50]	4.73(-4)	[116]	3.758(-4)	15.85	[423]
		2399.00	[104]					
		2404.00	[263]					

N series

line	transition	E/eV		I/eV/$\hbar$		TPIV	Γ/eV	
	N_4N_6	293.00 ± 1.40	[50]			2.542(-4)	8.06	[423]
	$N_5N_{6,7}$	275.60 ± 0.60	[50]					
	N_5N_6					1.063(-5)	7.40	[423]
	N_5N_7					2.295(-4)	7.38	[423]
	N_6O_4	121.10 ± 0.10	[50]			4.880(-6)		
	N_6O_5	123.70 ± 0.25	[50]			3.242(-7)		
	N_7O_5	118.90 ± 0.10	[50]			4.700(-6)		

level characteristics

level	E_B/eV		ω_{nlj}		Γ/eV		AE/eV	
K	88005.00	[503]	0.967	[301]	60.40	[301]	88004.5	[51]
	88004.50 ± 0.70	[493]	0.972	[225]	59.33	[423]	88005.6 ± 4.6	[50]
			0.968	[38]	60.6	[94]	88012.8 ± 2.3	[145]
			0.9634	[39]				
			0.9613	[423]				
			0.967	[310]				
			0.9658	[309]				
			0.961	[33]				
			0.956	[242]				
L_1	15861.00	[503]	0.112	[301]	12.20	[301]	15860.8	[51]
	15860.80 ± 0.50	[493]	0.09	[550]	13.85	[423]	15857.99 ± 0.10	[50]
			7.00(-2)	[549]	11.8	[94]	15867.7 ± 1.5	[145]
			9.81(-2)	[423]				
			9.30(-2)	[436]				
			0.135	[579]				
L_2	15200.00	[503]	0.376	[301]	6.43	[301]	15200.0	[51]
	15200.00 ± 0.40	[493]	0.363	[549]	5.90	[423]	15198.993 ± 0.030	[294]
			0.4037	[423]	5.04	[94]	15206.12 ± 0.56	[145]
			0.405	[578]				
			0.408	[567]				
L_3	13035.00	[503]	0.360	[301]	5.81	[301]	13035.2	[51]
	13035.20 ± 0.30	[493]	0.32	[523]	5.88	[423]	13035.064 ± 0.030	[294]
			0.35(5)	[269]	6,03	[94]	13042.60 ± 0.40	[145]
			0.354	[435]				
			0.315	[549]				
			0.3523	[423]				
			0.326	[579]				
			0.346	[567]				
M_1	3851.00	[503]	3.19(-3)	[512]	23.00	[74]	3850.7	[51]
	3850.70 ± 0.50	[493]	3.95(-3)	[423]	19.20	[423]		
					15.2	[94]		
M_2	3554.00	[503]	2.308-29	[512]	10.00	[74]	3554.2	[51]
	3554.20 ± 0.30	[493]	6.64(-3)	[423]	13.32	[423]		
					10.4	[94]		

level	E_B/eV		ω_{nlj}		Γ/eV		AE/eV	
M_3	3067.00	[503]	5.28(-3)	[512]	15.00	[74]	3066.4	[51]
	3066.40 ± 0.40	[493]	8.79(-3)	[423]	10.58	[423]		
					8.7	[94]		
M_4	2586.00	[503]	3.60(-2)	[512]	2.50	[74]	2585.6	[51]
	2585.60 ± 0.30	[493]	3.57(-2)	[423]	2.51	[423]		
					2.48	[94]		
M_5	2484.00	[503]	3.30(-2)	[512]	2.80	[74]	2484.0	[51]
	2484.00 ± 0.30	[493]	3.54(-2)	[423]	2.48	[423]		
					2.48	[94]		
N_1	894.00	[503]	6.1(-4)	[423]	16.00	[74]		
	893.60 ± 0.70	[493]			15.31	[423]		
					9.35	[94]		
N_2	764.00	[503]	1.19(-3)	[423]	13.00	[74]		
	763.90 ± 0.80	[493]			13.96	[423]		
					6.9	[94]		
N_3	645.00	[503]	5.0(-4)	[423]	13.00	[74]		
	644.50 ± 0.60	[493]			9.14	[423]		
					5.8	[94]		
N_4	435.00	[503]	3.2(-4)	[423]	7.80	[74]		
	435.20 ± 0.50	[493]			7.80	[423]		
					3.8	[94]		
N_5	413.00	[503]	2.9(-4)	[423]	7.80	[74]		
	412.90 ± 0.60	[493]			7.14	[423]		
					3.8	[94]		
N_6	143.00	[503]	1.0(-5)	[423]	0.16	[74]		
	142.90 ± 0.40	[51]			0.26	[423]		
	141.80 ± 0.80	[493]			0.26	[94]		
N_7	138.00	[503]	1.0(-5)	[739	0.16	[74]		
	138.10 ± 0.40	[51]			0.23	[423]		
	136.90 ± 0.80	[493]			0.23	[94]		
O_1	148.00	[503]						
	147.30 ± 0.80	[493]	1.0(-5)	[423]	26.19	[423]		
O_2	105.00	[503]	1.0(-5)	[423]	8.00	[423]		
	104.80 ± 1.00	[493]						

level	E_B/eV			ω_{nlj}		Γ/eV		AE/eV
O_3	86.00	[503]		1.0(-6)	[423]	13.73	[423]	
	86.00 ± 1.00	[51]						
	84.50 ± 1.80	[493]						
O_4	22.00	[503]						
	21.80 ± 0.40	[493]						
O_5	20.00	[503]						
	19.20 ± 0.40	[493]						
P_1	3.00	[503]						
	3.10 ± 1.00	[493]						
P_2	1.00	[503]						
	0.70 ± 1.00	[51]						
	7.417	[126]						
IP	7.41666	[222]						

Bi   Z=83

[Xe] $4f^{14}\,5d^{10}\,6s^2\,6p^3$

Bismuth $A = 208.98038(2)$ [222] $\varrho = 9.78\,\text{g/cm}^3$ [547]

X-ray transitions

line	transition	E/eV		I/eV/$\hbar$		TPIV	Γ/eV	
K series								
K_{α_3}	KL_1	74141.0 ± 2.7	[145]				76.36	[423]
$K_{\alpha2}$	KL_2	77814.80 ± 0.90	[50]	17.91	[488]	2.828(-1)	70.10	[301]
		74817.20	[104]				68.41	[423]
		74816.21 ± 0.92	[80]					
$K_{\alpha1}$	KL_3	77107.90 ± 1.00	[50]	29.61	[488]	4.729(-1)	69.40	[301]
		77110.00	[104]				68.39	[423]
		77109.2 ± 2.2	[80]					
	KM_1	86529.6 ± 4.1	[145]				81.68	[423]
$K_{\beta3}$	KM_2	86834.00 ± 4.30	[50]	3.311	[488]	5.288(-2)	75.83	[423]
		86834.50	[104]					
		86835.7 ± 6.7	[80]					
$K_{\beta1}$	KM_3	87343.00 ± 1.80	[50]	6.42	[488]	1.025(-1)	73.15	[423]
		87342.94	[104]					
		87344.1 ± 3.3	[80]					
$K_{\beta5}$	$KM_{4,5}$	87862.2 ± 9.2	[50]	0.1116	[488]			
$K_{\beta_5^{II}}$	KM_4	87839.7 ± 3.3	[145]			1.480(-3)	64.94	[423]
$K_{\beta_5^{I}}$	KM_5	87947.6 ± 3.2	[145]			1.782(-3)	64.91	[423]
	KN_1	89590.3 ± 5.2	[145]				77.76	[423]
$K_{\beta_2^{II}}$	KN_2	89731.7 ± 9.6	[50]	0.794	[488]	1.268(-2)	76.47	[423]
		89727.00	[104]					
$K_{\beta_2^{I}}$	KN_3	89861.8 ± 9.6	[50]	1.564	[488]	2.498(-2)	71.41	[423]
		89858.00	[104]					
$K_{\beta4}$	$KN_{4,5}$	90110 ± 19	[50]	5.49(-2)	[488]			
$K_{\beta_4^{II}}$	KN_4	90064.5 ± 4.7	[145]			3.980(-4)	70.02	[423]
$K_{\beta_4^{I}}$	KN_5	90088.0 ± 4.5	[145]			4.786(-4)	69.33	[423]
	$KO_{2,3}$	90435.00 ± 7.00	[50]	0.460	[488]			
	KO_2					2.959(-3)	70.50	[423]
	KO_3					5.696(-3)	77.39	[423]

line	transition	E/eV		I/eV/$\hbar$		TPIV	Γ/eV	
	L series							
	L_1M_1	12392 ± 16	[50]	1.31(-5)	[488]		33.38	[423]
		12392.00	[104]					
$L_{\beta4}$	L_1M_2	12691.40 ± 0.77	[50]	0.486	[488]	3.461(-2)	27.54	[423]
		12691.00	[104]					
$L_{\beta3}$	L_1M_3	13209.99 ± 0.62	[50]	0.5204	[488]	3.706(-2)	24.86	[423]
		13210.00	[104]					
$L_{\beta10}$	L_1M_4	13700.0 ± 0.6	[50]	1.80(-2)	[488]	1.282(-3)	16.65	[423]
		13700.00	[104]					
		13698.87 ± 0.32	[440]					
$L_{\beta9}$	L_1M_5	13808.00 ± 0.50	[50]	2.693(-2)	[488]	1.918(-3)	16.62	[423]
		13807.70	[104]					
		13806.82 ± 0.20	[440]					
	L_1N_1	15455.3 ± 2.9	[50]	5.79(-6)	[488]		31.74	[423]
		15453.70	[104]					
$L_{\gamma2}$	L_1N_2	15582.52 ± 0.87	[50]	0.1257	[488]	8.957(-3)	28.19	[423]
		15581.60	[104]					
$L_{\gamma3}$	L_1N_3	15710.5 ± 1.5	[50]	0.1474	[488]	1.050(-2)	23.12	[423]
		15708.00	[104]					
	L_1N_4	15905 ± 15	[50]	2.79(-3)	[488]	1.991(-4)	21.73	[423]
		15924.30	[104]					
$L_{\gamma11}$	L_1N_5	15950.8 ± 1.5	[50]	4.43(-3)	[488]	3.155(-4)	21.05	[423]
		15949.30	[104]					
	$L_1N_{6,7}$	16226 ± 16	[50]	8.67(-5)	[488]			
	L_1N_6	16224.1 ± 2.7	[145]				14.26	[423]
	L_1N_7	16229.3 ± 2.6	[145]				14.24	[423]
		16224.99	[104]					
	L_1O_1	16180.00	[104]	1.44(-6)	[488]		42.19	[423]
$L_{\gamma4'}$	L_1O_2	16270.90 ± 0.60	[50]	2.532(-2)	[488]	1.804(-3)	22.21	[423]
		16270.00	[104]					
$L_{\gamma4}$	L_1O_3	16294.70 ± 0.60	[50]	2.928(-2)	[488]	2.086(-3)	29.11	[423]
		16294.00	[104]					
	$L_1O_{4,5}$	16358.00 ± 1.10	[50]	8.99(-4)	[488]			
		16358.00	[104]					
	L_1O_4					2.490(-5)		
	L_1O_5					3.912(-5)		
	$L_1P_{2,3}$	16380.20 ± 0.60	[50]	2.38(-3)	[488]			
		16379.00	[104]					
	L_1P_2					8.601(-5)		
	L_1P_3					8.343(-5)		

line	transition	E/eV		I/eV/ℏ		TPIV	Γ/eV	
L_η	L_2M_1	11712.36 ± 0.49	[50]	5.43(-2)	[488]	8.942(-3)	25.42	[423]
		11712.00	[104]					
	L_2M_2	11984 ± 15	[50]	3.63(-6)	[488]	3.762(-4)	19.58	[423]
		11987.00	[104]					
$L_{\beta17}$	L_2M_3	12534.48 ± 0.94	[50]	2.83(-3)	[488]		16.90	[423]
		12535.00	[104]					
$L_{\beta1}$	L_2M_4	13023.65 ± 0.18	[50]	1.989	[488]	3.278(-1)	9.60	[473]
		13024.00	[104]				8.69	[423]
	L_2M_5	13131.1 ± 1.0	[50]	1.42(-4)	[488]		8.66	[423]
		13131.00	[104]					
$L_{\gamma5}$	L_2N_1	14773.3 ± 1.3	[50]	1.408(-2)	[488]	2.320(-3)	23.78	[423]
		14773.00	[104]					
	L_2N_2	14859 ± 24	[50]	1.43(-6)	[488]		20.23	[423]
		14858.00	[104]					
	L_2N_3	15031.80 ± 2.7	[50]	6.93(-4)	[488]	1.054(-4)	15.17	[423]
		15029.00	[104]					
$L_{\gamma1}$	L_2N_4	15247.92 ± 0.56	[50]	0.411	[488]	6.769(-2)	12.50	[473]
		15247.00	[104]				13.78	[423]
	L_2N_5	15271.2 ± 2.7	[145]				13.09	[423]
L_v	L_2N_6	15552.0 ± 2.6	[50]	1.81(-3)	[488]	2.991(-4)	6.30	[423]
	L_2N_7	15553.5 ± 1.6	[145]				6.28	[423]
$L_{\gamma8}$	L_2O_1	15551.00 ± 2.00	[50]	3.11(-3)	[488]	5.044(-4)	34.24	[423]
		15551.00	[104]					
	L_2O_3	15617.80 ± 1.00	[50]	1.27(-4)	[488]	2.138(-5)	21.15	[423]
		15618.00	[104]					
$L_{\gamma6}$	L_2O_4	15685.30 ± 0.60	[50]	5.46(-2)	[488]	8.994(-3)		
		15684.00	[104]					
L_l	L_3M_1	9420.43 ± 0.74	[50]	9.005(-2)	[488]	1.635(-2)	25.40	[423]
		9419.90	[104]					
L_t	L_3M_2	9725.6 ± 1.1	[50]	1.007(-3)	[488]	1.767(-4)	19.57	[423]
		9725.10	[104]					
L_s	L_3M_3	10242.2 ± 1.3	[50]	9.46(-4)	[488]	1.597(-4)	16.88	[423]
		10241.00	[104]					
$L_{\alpha2}$	L_3M_4	10731.06 ± 0.14	[50]	0.1726	[488]	2.860(-2)	8.67	[423]
		10731.00	[104]					
$L_{\alpha1}$	L_3M_5	10838.94 ± 0.28	[50]	1.519	[488]	2.519(-1)	9.80	[473]
		10839.00	[104]				8.64	[423]
$L_{\beta6}$	L_3N_1	12481.74 ± 0.56	[50]	2.247(-2)	[488]	3.725(-3)	23.76	[423]
		12481.00	[104]					
	L_3N_2	12615.21 ± 0.95	[50]	2.41(-4)	[488]	3.993(-5)	20.21	[423]

line	transition	E/eV		I/eV/$\hbar$		TPIV	Γ/eV	
		12615.00	[104]					
	L_3N_3	12739.52 ± 0.97	[50]	2.48(-4)	[488]	4.118(-5)	15.15	[423]
		12738.00	[104]					
$L_{\beta 15}$	L_3N_4	12955.0 ± 1.0	[50]	3.26(-2)	[488]	5.403(-3)	13.76	[423]
		12955.00	[104]					
$L_{\beta 2}$	L_3N_5	12980.00 ± 0.80	[50]	0.2951	[488]	4.891(-2)	12.10	[473]
		12980.00	[104]				13.07	[423]
	$L_3N_{6,7}$	13259.4 ± 1.0	[50]	2.72(-4)	[488]			
		13259.00	[104]					
$L_{u'}$	L_3N_6	13256.2 ± 1.5	[145]			4.521(-5)	6.28	[423]
L_u	L_3N_7	13261.5 ± 1.5	[145]			2.522(-4)	6.26	[423]
$L_{\beta 7}$	L_3O_1	13259.30 ± 0.70	[50]	4.009(-3)	[488]	8.173(-4)	34.22	[423]
		13259.00	[104]					
	L_3O_2	13298.00 ± 3.00	[50]	4.72(-5)	[488]	8.377(-6)	14.24	[423]
		13304.00	[104]					
	L_3O_3	13328.00 ± 3.00	[50]	4.005(-5)	[488]	8.768(-6)	21.13	[423]
		13325.00	[104]					
$L_{\beta 5}$	$L_3O_{4,5}$	13395.30 ± 0.70	[50]	4.23(-2)	[488]			
	L_3O_4	13395.00	[104]			7.068(-4)		
	L_3O_5					6.298(-3)		
	$L_3P_{2,3}$	13415.90 ± 0.60	[50]	4.17(-7)	[488]			
	L_3P_2					3.711(-7)		
	L_3P_3					3.203(-7)		

M series

line	transition	E/eV		I/eV/$\hbar$		TPIV	Γ/eV	
	M_1N_2	3185.00 ± 7.40	[50]	4.681(-2)	[335]	1.672(-3)	33.49	[423]
		3186.00	[104]					
		3192.00	[263]					
	M_1N_3	3315.00 ± 8.00	[50]	4.681(-2)	[335]	1.244(-3)	28.43	[423]
		3315.00	[104]					
		3322.00	[263]					
	M_2N_4	3234.00 ± 3.40	[50]	5.0094(-2)	[335]	4.137(-3)	33.49	[423]
		3231.00	[104]					
		3241.00	[263]					
	M_3N_1	2239.00 ± 3.20	[50]	5.20(-3)	[335]	1.464(-3)	28.43	[423]
		2239.00	[104]					
		2244.00	[263]					
$M_{\gamma 2}$	M_3N4	2712.00 ± 3.00	[50]	5.0094(-2)	[335]	7.124(-4)	18.52	[423]
		2713.00	[104]					
		2718.00	[263]					
$M_{\gamma 1}$	M_3N_5	2735.00 ± 1.20	[50]	5.0094(-2)	[335]	5.477(-3)	17.83	[423]
		2736.00	[104]					
		2741.00	[263]					

line	transition	E/eV		I/eV/$\hbar$		TPIV	Γ/eV	
	M_3O_1	3021.00 ± 6.60	[50]			3.172(-4)	38.98	[423]
		3021.00	[104]					
		3027.00	[263]					
	$M_3O_{4,5}$	3153.00 ± 4.80	[50]					
	M_3O_4	3160.00	[104]			1.036(-4)		
	M_3O_5	3153.00 ± 5.00	[50]			7.950(-4)		
		3151.00	[104]					
$M_{\xi 2}$	M_4N_2	1883.00 ± 1.40	[50]	6.97(-5)	[335]	2.030(-3)	16.76	[423]
		1883.00	[104]					
		1886.00	[263]					
M_δ	M_4N_3	2012.00 ± 2.60	[50]	6.97(-5)	[335]	2.015(-4)	11.70	[423]
		2012.00	[104]					
		2016.00	[263]					
M_β	M_4N_6	2525.50 ± 0.50	[50]	3.584(-2)	[335]	3.487(-2)	2.83	[423]
		2526.00	[104]					
		2531.00	[263]					
M_η	M_4O_2	2571.00 ± 1.60	[50]			4.982(-4)	10.79	[423]
		2571.00	[104]					
		2676.00	[263]					
	$M_4P_{2,3}$	2700.00 ± 12.00	[50]					
	M_4P_2					1.396(-5)		
	M_4P_3					1.183(-6)		
$M_{\xi 1}$	M_5N_3	1901.00 ± 1.20	[50]	6.97(-5)	[335]	1.524(-3)	11.66	[423]
		1901.00	[104]					
		1905.40	[263]					
$M_{\alpha 2}$	M_5N_6	2417.00 ± 1.00	[50]	8.58(-2)	[488]	1.724(-3)	2.80	[423]
		2417.00	[104]					
		2422.00	[263]					
$M_{\alpha 1}$	M_5N_7	2422.60 ± 0.50	[50]	8.58(-2)	[488]	3.370(-2)	2.78	[423]
		2422.00	[104]					
		2428.00	[263]					

N series

line	transition	E/eV		I/eV/$\hbar$		TPIV	Γ/eV	
	$N_1P_{2,3}$	932.00 ± 4.20	[50]					
	N_1P_2						1.130(-5)	
	N_1P_3						3.976(-6)	
	N_6O_4	135.40 ± 0.15	[50]				7.140(-6)	
	N_7O_5	133.00 ± 0.15	[50]				6.882(-6)	

level characteristics

level	E_B/eV		ω_{nlj}		Γ/eV		AE/eV	
K	90526.00	[503]	0.9643	[39]	62.32	[423]	90525.9	[51]
	90525.90 ± 0.70	[493]	0.968	[301]	63.40	[301]	90537.7 ± 9.8	[50]
			0.968	[310]	63.6	[94]	90536.5 ± 2.4	[145]
			0.9620	[423]				
			0.966	[242]				
L_1	16388.00	[503]	0.117	[301]	12.40	[301]	16387.5	[51]
	16387.50 ± 0.40	[493]	0.120(1)	[461]	14.04	[423]	16376.0 ± 3.2	[50]
			0.120	[348]	12.3	[94]	16395.4 ± 1.6	[145]
			0.095(5)	[182]				
			0.1023	[423]				
			0.0.132	[578]				
			0.138	[579]				
			9.80(-2)	[436]				
L_2	15709.00	[503]	0.387	[301]	6.67	[301]	15711.1	[51]
	15711.10 ± 0.30	[493]	0.320	[461]	6.08	[423]	15719.8 ± 2.9	[50]
			0.417	[348]	6.22	[94]	15719.65 ± 0.57	[145]
			0.0.4172	[423]				
			0.418	[578]				
			0.428	[579]				
L_3	13418.00	[503]	0.373	[301]	5.98	[301]	13418.6	[51]
	13418.60 ± 0.30	[493]	0.362(29)	[435]	6.06	[423]	13426.7 ± 2.2	[50]
			0.389	[348]	6.27	[94]	13427.59 ± 0.41	[145]
			0.3636	[423]				
			0.354	[253]				
			0.354	[244]				
			0.370(5)	[269]				
			0.400(5)	[461]				
			0.340(18)	[182]				
M_1	3999.00	[503]	2.890(-3)	[38]	21.70	[350]	3999.1	[51]
	3999.10 ± 0.30	[493]	3.41(-3)	[512]	19.34	[423]		
			4.17(-3)	[423]	15.2	[94]		
M_2	3697.00	[503]	6.52(-3)	[38]	14.60	[350]	3696.3	[51]
	3696.30 ± 0.30	[493]	2.48-2)	[512]	13.50	[423]		
			6.89(-3)	[423]	10.7	[94]		
M_3	3177.00	[503]	5.33(-3)	[38]	10.70	[350]	3176.9	[51]
	3176.90 ± 0.30	[493]	5.62(-3)	[512]	10.82	[423]		
			9.2(-4)	[423]	8.6	[94]		
M_4	2688.00	[503]	3.30(-2)	[38]	2.88	[350]	2687.6	[51]
	2687.60 ± 0.30	[493]	3.80(-2)	[512]	2.61	[423]		
			3.67(-2)	[423]	2.58	[94]		

level	E_B/eV		ω_{nlj}		Γ/eV		AE/eV	
M_5	2580.00	[503]	3.25(-2)	[38]	2.74	[350]	2579.6	[51]
	2579.60 ± 0.30	[493]	3.40(-2)	[512]	2.58	[423]		
			3.73(-2)	[423]	2.58	[94]		
N_1	939.00	[503]	6.5(-4)	[423]	9.6	[94]		
	938.20 ± 0.30	[493]			15.43	[423]		
N_2	806.00	[503]	1.22(-3)	[423]	7.2	[94]		
	805.30 ± 0.30	[493]			14.15	[423]		
N_3	679.00	[503]	5.2(-4)	[423]	5.95	[94]		
	678.90 ± 0.30	[493]			9.08	[423]		
N_4	464.00	[503]	3.5(-4)	[423]	3.8	[94]		
	463.60 ± 0.30	[493]			7.69	[423]		
N_5	440.00	[503]	3.2(-5)	[423]	3.8	[94]		
	440.00 ± 0.30	[493]			7.01	[423]		
N_6	163.00	[503]	1.0(-5)	[423]	0.22	[94]		
	161.90 ± 0.50	[51]			0.22	[423]		
	162.30 ± 0.30	[493]						
N_7	158.00	[503]	1.0(-5)	[423]	0.20	[94]		
	157.40 ± 0.60	[51]			0.20	[423]		
	157.20 ± 0.30	[493]						
O_1	160.00	[503]	1.0(-5)	[423]	28.16	[423]		
	159.30 ± 0.70	[493]						
O_2	117.00	[503]	1.0(-5)	[423]	8.17	[423]		
	116.80 ± 0.70	[493]						
O_3	93.00	[503]	1.0(-6)	[423]	15.07	[423]		
	92.80 ± 0.60	[51]						
	92.90 ± 0.70	[493]						
O_4	27.00	[503]						
	26.50 ± 0.50	[493]						
O_5	25.00	[503]						
	24.40 ± 0.60	[493]						
P_1	8.00	[503]						
P_2	3.00	[503]						
	2.70 ± 0.70	[51]						
	7.50 ± 7.50	[493]						
IP	7.2856	[222]						

Po   Z=84

[Xe] 4f^{14} 5d^{10} 6s^2 6p^4

Polonium

A = [208.9824] [222] ϱ = 9.34 g/cm^3 [547]

ϱ = 9.32 g/cm^3 [483]

X-ray transitions

line	transition	E/eV		I/eV/$\hbar$		TPIV	Γ/eV	
K series								
K$_{\alpha_3}$	KL$_1$	76173.9 ± 2.7	[145]				79.62	[423]
K$_{\alpha_2}$	KL$_2$	76864.4 ± 7.1	[50]	18.63	[488]	2.834(-1)	73.40	[301]
		77087.00	[104]				76.30	[473]
							71.69	[423]
K$_{\alpha_1}$	KL$_3$	79292.9 ± 7.5	[50]	31.01	[488]	4.718(-1)	72.70	[301]
		79306.00	[104]				73.20	[473]
							71.67	[423]
	KM$_1$	88948.8 ± 4.1	[145]				84.88	[423]
K$_{\beta_3}$	KM$_2$	89247 ± 19	[50]	3.481	[488]	5.296(-2)	82.25	[473]
							79.12	[423]
K$_{\beta_1}$	KM$_3$	89797 ± 19	[50]	6.75	[488]	1.028(-1)	78.60	[473]
		89701.00	[104]				76.41	[423]
K$_{\beta_5^{II}}$	KM$_4$	90307.3 ± 3.4	[145]			1.522(-3)	68.14	[423]
K$_{\beta_5^I}$	KM$_5$	90421.5 ± 3.3	[145]			1.822(-3)	68.11	[423]
	KN$_1$	92115.1 ± 5.3	[145]				80.54	[423]
K$_{\beta_2^{II}}$	KN$_2$	92262 ± 20	[50]	0.839	[488]	1.277(-2)	79.76	[423]
K$_{\beta_2^I}$	KN$_3$	92400 ± 20	[50]	1.657	[488]	2.521(-2)	74.51	[423]
K$_{\beta_4^1{}^{II}}$	KN$_4$	92607.3 ± 4.8	[145]			4.139(-4)	73.00	[423]
K$_{\beta_4^I}$	KN$_5$	92632.5 ± 4.6	[145]			4.950(-4)	72.28	[423]
L series								
	L$_1$M$_1$	12774.9 ± 3.2	[145]				33.65	[423]
L$_{\beta_4}$	L$_1$M$_2$	13085.2 ± 6.1	[50]	0.518	[488]	3.651(-2)	22.70	[473]
		13085.20	[104]				27.89	[423]
L$_{\beta_3}$	L$_1$M$_3$	13637.9 ± 6.7	[50]	0.544	[488]	3.831(-2)	19.90	[473]
		13638.00	[104]				25.18	[423]
L$_{\beta_{10}}$	L$_1$M$_4$	14133.3 ± 2.5	[145]			1.388(-3)	16.91	[423]
L$_{\beta_9}$	L$_1$M$_5$	14247.6 ± 2.4	[145]			2.076(-3)	16.88	[423]

line	transition	E/eV		I/eV/$\hbar$		TPIV	Γ/eV	
	L_1N_1	15941.2 ± 4.3	[145]				29.31	[423]
$L_{\gamma 2}$	L_1N_2	16060 ± 31	[50]	0.1349	[488]	9.498(-3)	28.53	[423]
		16070.00	[104]					
$L_{\gamma 3}$	L_1N_3	16218.00	[104]	0.1556	[488]	1.096(-2)	23.28	[423]
		16212.9 ± 3.8	[145]					
	L_1N_4	16433.4 ± 3.8	[145]			2.178(-4)	21.77	[423]
	L_1N_5	16458.6 ± 3.7	[145]			3.456(-4)	21.05	[423]
	L_1N_6	16743.3 ± 2.7	[145]				14.39	[423]
	L_1N_7	16749.0 ± 2.6	[145]				14.38	[423]
L_η	L_2M_1	12084.7 ± 2.5	[145]			9.224(-3)	26.45	[423]
	L_2M_2	12393.8 ± 2.1	[145]				19.97	[423]
$L_{\beta 17}$	L_2M_3	12945.7 ± 2.3	[145]			3.977(-4)	17.25	[423]
$L_{\beta 1}$	L_2M_4	13447.1 ± 4.3	[50]	2.107	[488]	3.372(-1)	10.10	[473]
		13444.00	[104]				8.98	[423]
	L_2M_5	13557.4 ± 1.4	[145]				8.95	[423]
$L_{\gamma 5}$	L_2N_1	15251.0 ± 3.4	[145]			2.406(-3)	21.38	[423]
	L_2N_2	15388.3 ± 3.9	[145]				20.60	[423]
	L_2N_3	15522.7 ± 2.8	[145]			1.123(-4)	15.35	[423]
$L_{\gamma 1}$	L_2N_4	15744.2 ± 2.7	[50]	0.440	[488]	7.039(-2)	13.05	[473]
		15743.00	[104]				13.84	[423]
	L_2N_5	15768.4 ± 2.7	[145]				13.12	[423]
L_ν	L_2N_6	16053.1 ± 1.7	[145]			3.252(-4)	6.46	[423]
	L_2N_7	16058.8 ± 1.7	[145]				6.45	[423]
$L_{\gamma 6}$	L_2O_4	16218.00 ± 4.00	[50]	6.22(-2)	[488]	9.959(-3)		
		16217.98	[104]					
L_l	L_3M_1	9664.2 ± 5.6	[50]	9.71(-2)	[488]	1.681(-2)	25.70	[423]
		9665.00	[104]					
L_t	L_3M_2	9966.7 ± 2.0	[145]			1.865(-4)	19.95	[423]
L_s	L_3M_3	10518.6 ± 2.1	[145]			1.687(-4)	17.23	[423]
$L_{\alpha 2}$	L_3M_4	11015.95 ± 0.72	[50]	0.1822	[488]	2.940(-2)	9.95	[473]
		11015.00	[104]				8.97	[423]
$L_{\alpha 1}$	L_3M_5	11130.87 ± 0.59	[50]	1.604	[488]	2.587(-1)	9.50	[473]
		11131.00	[104]				8.93	[423]
$L_{\beta 6}$	L_3N_1	12818.7 ± 3.9	[50]	2.419(-2)	[488]	3.902(-3)	21.36	[423]
		12819.00	[104]					
	L_3N_2	12961.3 ± 3.7	[145]			4.215(-5)	20.58	[423]

line	transition	E/eV		I/eV/$\hbar$		TPIV	Γ/eV	
	L_3N_3	13095.6 ± 2.7	[145]			4.376(-5)	15.33	[423]
$L_{\beta15}$	L_3N_4	13314.3 ± 4.2	[50]	3.47(-3)	[488]	5.597(-3)	13.82	[423]
		13314.00	[104]					
$L_{\beta2}$	L_3N_5	13340.5 ± 1.1	[50]	0.314	[488]	5.073(-2)	13.10	[423]
		13341.00	[104]					
$L_{u'}$	L_3N_6	13626.0 ± 1.5	[145]			4.939(-5)	6.44	[423]
L_u	L_3N_7	13631.7 ± 1.5	[145]			2.752(-4)	6.43	[423]
$L_{\beta5}$	$L_3O_{4,5}$	13782.00 ± 3.00	[50]	4.80(-2)	[488]			
		13783.00	[104]					
	L_3O_4					7.791(-4)		
	L_3O_5					6.973(-3)		

M series

line	transition	E/eV		I/eV/$\hbar$		TPIV	Γ/eV	
$M_{\gamma2}$	M_3N_4	2785.94	[423]			7.409(-4)	18.56	[423]
$M_{\gamma1}$	M_3N_5	2824.00	[263]			5.681(-3)	17.83	[423]
$M_{\xi2}$	M_4N_2	1959.20	[423]			2.038(-3)	17.05	[423]
M_δ	M_4N_3	2098.54	[423]			2.010(-4)	11.80	[423]
M_β	M_4N_6	2605.00	[263]			3.683(-2)	2.91	[423]
$M\eta$	M_4O_2	2667.77	[423]			5.008(-4)	11.25	[423]
$M_{\xi1}$	M_5N_3	1968.00	[263]			1.522(-3)	11.77	[423]
$M_{\alpha1.2}$	$M_5N_{6,7}$	2500.00	[263]					
$M_{\alpha2}$	M_5N_6	2489.60	[423]			1.817(-3)	2.87	[423]
$M_{\alpha1}$	M_5N_7	2495.80	[423]			3.552(-2)	2.86	[423]

level characteristics

level	E_B/eV		ω_{nlj}		Γ/eV		AE/eV	
K	93105.00	[503]	0.968	[301]	68.60	[301]	93105.0	[51]
	93105.00 ± 3.80	[51]	0.9652	[39]	65.42	[423]	93109.9 ± 2.5	[145]
	93099.90 ± 1.20	[493]	0.9626	[423]	66.8	[94]		
L_1	16939.00	[503]	0.122	[301]	12.60	[301]	16939.3	[51]
	16939.30 ± 9.80	[51]	0.1076	[423]	14.19	[423]	16936.0 ± 1.6	[145]
	16927.90 ± 1.60	[493]	0.103	[436]	12.7	[94]		
L_2	16244.00	[503]	0.401	[301]	6.83	[301]	16244.3	[51]

level	E_B/eV		ω_{nlj}		Γ/eV		AE/eV	
	16244.30 ± 2.40	[51]	0.4306	[423]	6.27	[423]	16245.83 ± 0.59	[145]
	16238.00 ± 1.20	[493]	0.424	[436]	6.41	[94]		
L_3	13814.00	[503]	0.386	[301]	6.13	[301]	13813.8	[51]
	13813.80 ± 1.00	[51]	0.3747	[423]	6.25	[423]	13818.74 ± 0.42	[145]
	13810.60 ± 1.20	[493]	0.363	[436]	6.46	[94]		
M_1	4149.00	[503]	3.63(-3)	[512]	19.45	[423]	4149.4	[51]
	4149.40 ± 3.90	[51]	4.4(-3)	[423]	23.00	[74]		
	4153.50 ± 1.50	[493]			15.3	[94]		
M_2	3854.00	[503]	2.50(-2)	[512]	13.70	[423]	3854.1	[51]
	3854.00 ± 9.80	[51]	7.15(-3)	[423]	11.00	[74]		
	3844.30 ± 1.50	[493]			11.1	[94]		
M_3	3302.00	[503]	5.96(-3)	[512]	10.99	[423]	3301.9	[51]
	3301.90 ± 9.90	[51]	9.62(-3)	[423]	15.00	[74]		
	3293.40 ± 1.20	[493]			8.5	[94]		
M_4	2798.00	[503]	4.10(-2)	[512]	2.72	[423]	2798.0	[51]
	2798.00 ± 1.20	[51]	3.97(-2)	[423]	2.70	[74]		
	2793.60 ± 1.20	[493]			2.68	[94]		
M_5	2683.00	[503]	3.60(-2)	[512]	2.68	[423]	2683.0	[51]
	2683.00 ± 1.10	[51]	3.92(-2)	[423]	3.00	[74]		
	2679.20 ± 1.00	[493]			2.68	[94]		
N_1	995.00	[503]	6.8(-4)	[423]	15.11	[423]		
	995.30 ± 2.90	[51]			15.00	[74]		
	987.50 ± 1.20	[493]			9.9	[94]		
N_2	851.00	[503]	1.25(-3)	[423]	14.34	[423]		
	851.00 ± 1.20	[51]			13.00	[74]		
	850.90 ± 1.90	[493]			7.35	[94]		
N_3	705.00	[503]	5.5(-4)	[423]	9.09	[423]		
	705.00 ± 1.40	[51]			13.00	[74]		
	715.20 ± 1.00	[493]			6.2	[94]		
N_4	500.00	[503]	3.9(-4)	[423]	7.58	[423]		
	500.20 ± 2.40	[51]			7.30	[74]		
	495.70 ± 1.20	[493]			3.9	[94]		
N_5	473.00	[503]	3.5(-4)	[423]	6.85	[423]		
	473.40 ± 1.30	[51]			7.30	[74]		
	469.90 ± 1.00	[493]			3.8	[94]		

level	E_B/eV		ω_{nlj}		Γ/eV		AE/eV
N_6	184.00	[503]	1.0(-5)	[423]	0.19	[423]	
	184.60 ± 1.00	[493]			0.13	[74]	
					0.19	[94]	
N_7	178.90 ± 1.00	[493]	1.0(-5)	[423]	0.18	[423]	
					0.130	[74]	
					0.18	[94]	
O_1	177.00	[503]	1.0(-5)	[423]	29.51	[423]	
	177.50 ± 1.70	[493]					
O_2	132.00	[503]	1.0(-5)	[423]	8.53	[423]	
	131.80 ± 3.20	[493]					
O_3	104.00	[503]	2.0(-6)	[423]	16.74	[423]	
	103.70 ± 2.30	[493]					
O_4	31.00	[503]					
	31.40 ± 3.20	[51]					
	33.80 ± 1.10	[493]					
O_5	31.00	[503]					
	31.40 ± 3.20	[51]					
	30.60 ± 1.40	[493]					
P_1	12.00	[503]					
	11.00 ± 6.50	[493]					
P_2	5.00	[503]					
	3.20 ± 6.50	[493]					
P_3	5.00	[503]					
	1.40 ± 9.00	[493]					
	8.42	[126]					
IP	8.41671	[222]					

At   Z=85

[Xe] $4f^{14}\, 5d^{10}\, 6s^2\, 6p^5$

Astatine $A = [209.9871]\, [222]$ $\varrho = 8.75\ \mathrm{g/m^3}\ [246]$

X-ray transitions

line	transition	E/eV		I/eV/$\hbar$		TPIV	Γ/eV	
K series								
K_{α_3}	KL_1	78243.8 ± 2.8	[145]				82.99	[423]
K_{α_2}	KL_2	78944 ± 15	[50]	19.59	[488]	2.841(-1)	76.80	[301]
		78948.5 ± 2.8	[145]	19.74	[487]		75.10	[423]
K_{α_1}	KL_3	81514 ± 16	[50]	32.46	[488]	4.708(-1)	76.10	[301]
		81517.4 ± 2.7	[145]	32.72	[487]		75.08	[423]
	KM_1	91413.9 ± 4.3	[145]				88.20	[423]
K_{β_3}	KM_2	91723 ± 40	[50]	3.658	[488]	5.305(-2)	82.51	[423]
		91729.9 ± 4.2	[145]	3.80	[487]			
K_{β_1}	KM_3	92304 ± 41	[50]	7.10	[488]	1.029(-1)	79.76	[423]
		92315.8 ± 4.4	[145]	7.36	[487]			
$K_{\beta_5^{II}}$	KM_4	92822.6 ± 3.5	[145]			1.534(-3)	71.46	[423]
$K_{\beta_5^{I}}$	KM_5	92943.4 ± 3.5	[145]			1.862(-3)	71.42	[423]
	KN_1	94687.9 ± 5.4	[145]				83.61	[423]
$K_{\beta_2^{II}}$	KN_2	94846 ± 43	[50]	0.886	[488]	1.285(-2)	70.09	[423]
		94829.1 ± 6.0	[145]	0.946	[487]			
$K_{\beta_2^{I}}$	KN_3	94991 ± 43	[50]	1.754	[488]	2.544(-2)	77.84	[423]
		94972.8 ± 4.9	[145]	1.883	[487]			
$K_{\beta_4^{II}}$	KN_4	95198.7 ± 4.9	[145]			4.299(-4)	76.03	[423]
$K_{\beta_4^{I}}$	KN_5	95225.8 ± 4.8	[145]			5.116(-4)	75.31	[423]
	$KO_{2,3}$			0.590	[487]			
	KO_2					3.023(-3)	77.40	[423]
	KO_3					5.810(-3)	86.98	[423]
	$KP_{2,3}$			6.36(-2)	[487]			
	KP_2					2.632(-4)		
	KP_3					4.419(-4)		
L series								
	L_1M_1	13170.1 ± 3.2	[145]				33.91	[423]

line	transition	E/eV		I/eV/$\hbar$		TPIV	Γ/eV	
L_{β_4}	L_1M_2	13486.1 ± 3.2	[145]			3.821(-2)	28.21	[423]
L_{β_3}	L_1M_3	14067.3 ± 2.1	[50]	0.568	[488]	3.927(-2)	25.48	[423]
$L_{\beta_{10}}$	L_1M_4	13876.00 ± 1.50	[50]	2.231	[488]	3.463(-1)	9.28	[423]
L_{β_9}	L_1M_5	14699.7 ± 2.4	[145]			2.228(-3)	17.13	[423]
	L_1N_1	16444.1 ± 4.3	[145]				29.32	[423]
L_{γ_2}	L_1N_2	16585.4 ± 4.9	[145]			9.996(-3)	28.85	[423]
L_{γ_3}	L_1N_3	16729.1 ± 3.8	[145]			1.135(-2)	23.55	[423]
	L_1N_4	16955.0 ± 3.8	[145]			2.362(-4)	21.74	[423]
	L_1N_5	16982.0 ± 3.7	[145]			3.754(-4)	21.02	[423]
	L_1N_6	17274.3 ± 2.7	[145]				14.52	[423]
	L_1N_7	17280.6 ± 2.7	[145]				14.51	[423]
L_η	L_2M_1	12465.3 ± 2.5	[145]			9.501(-3)	26.02	[423]
	L_2M_2	12781.4 ± 2.2	[145]				20.32	[423]
$L_{\beta_{17}}$	L_2M_3	13367.3 ± 2.3	[145]			4.194(-4)	17.58	[423]
L_{β_1}	L_2M_4	13876.2 ± 2.1	[50]			3.463(-1)	9.28	[423]
		13874.1 ± 1.5	[145]					
	L_2M_5	13994.9 ± 1.4	[145]				9.24	[423]
L_{γ_5}	L_2N_1	15739.4 ± 3.3	[145]			2.491(-3)	21.43	[423]
	L_2N_2	15880.6 ± 4.0	[145]				20.96	[423]
	L_2N_3	16024.3 ± 2.8	[145]			1.195(-4)	15.66	[423]
L_{γ_1}	L_2N_4	16251.7 ± 2.8	[50]	0.471	[488]	7.308(-2)	13.86	[423]
		16250.2 ± 2.8	[145]					
	L_2N_5	16277.3 ± 2.7	[145]				13.13	[423]
L_ν	L_2N_6	16569.6 ± 1.7	[145]			3.523(-4)	6.63	[423]
	L_2N_7	16575.9 ± 1.7	[145]				6.62	[423]
L_l	L_3M_1	9896.4 ± 2.0	[145]			1.728(-2)	26.00	[423]
L_t	L_3M_2	10212.4 ± 2.0	[145]			1.965(-4)	20.30	[423]
L_s	L_3M_3	10798.4 ± 2.1	[145]			1.778(-4)	17.56	[423]
L_{α_2}	L_3M_4	11304.93 ± 0.76	[50]	0.1923	[488]	3.017(-2)	9.26	[423]
		11305.2 ± 1.3	[145]					
L_{α_1}	L_3M_5	11426.94 ± 0.78	[50]	1.692	[488]	2.655(-1)	9.22	[423]
		11426.0 ± 1.3	[145]					
L_{β_6}	L_3N_1	13170.4 ± 3.2	[145]			4.084(-3)	21.41	[423]
	L_3N_2	13311.7 ± 3.8	[145]			4.438(-5)	20.94	[423]
	L_3N_3	13455.4 ± 2.6	[145]			4.642(-5)	15.64	[423]

line	transition	E/eV		I/eV/ℏ	TPIV	Γ/eV	
$L_{\beta_{15}}$	L_3N_4	13681.3 ± 2.7	[145]		5.788(-3)	13.84	[423]
L_{β_2}	L_3N_5	13708.4 ± 2.6	[145]		5.252(-2)	13.12	[423]
$L_{u'}$	L_3N_6	14000.7 ± 1.6	[145]		5.377(-5)	6.61	[423]
L_u	L_3N_7	14007.0 ± 1.6	[145]		2.992(-4)	6.59	[423]
	M series						
M_{γ_2}	M_3N_4	2871.42	[423]		7.701(-4)	18.52	[423]
M_{γ_1}	M_3N_5	2912.00	[263]		5.890(-3)	17.80	[423]
M_{ξ_2}	M_4N_2	2020.75	[423]		2.057(-3)	4.28	[423]
M_β	M_4N_6	2695.00	[263]		3.876(-2)	2.99	[423]
M_{ξ_1}	M_5N_3	2032.00	[263]		1.520(-3)	11.98	[423]
$M_{\alpha_{1.2}}$	$M_5N_{6,7}$	2583.00	[263]				
	M_5N_6				1.909(-3)	2.95	[423]
	M_5N_7				3.734(-2)	2.94	[423]

level characteristics

level	E_B/eV		ω_{nlj}		Γ/eV		AE/eV	
K	95730.00	[503]	0.969	[301]	69.80	[301]	95729.9	[51]
	95729.90 ± 7.70	[51]	0.9659	[39]	68.64	[423]	95733.5 ± 2.7	[145]
	95724.00 ± 1.30	[493]	0.9632	[423]	70	[94]		
L_1	17493.00	[503]	0.128	[301]	12.80	[301]	17493.0	[51]
	17493.00 ± 2.90	[51]	0.129	[107]	14.35	[423]	17489.7 ± 1.6	[145]
	17481.50 ± 1.60	[493]	0.1123	[423]	13	[94]		
			0.109	[436]	6.6			
L_2	16785.00	[503]	0.415	[301]	7.01	[301]	16484.7	[51]
	16784.70 ± 2.50	[51]	0.422	[107]	6.46	[423]	16784.96 ± 0.60	[145]
	16777.30 ± 1.20	[493]	0.4439	[423]	6.6	[94]		
			0.438	[436]				
L_3	14214.00	[503]	0.399	[301]	6.29	[301]	14213.4	[51]
	14213.50 ± 2.00	[51]	0.380	[107]	6.44	[423]	14216.04 ± 0.45	[145]
	14208.00 ± 1.20	[493]	0.3858	[423]	6.66	[94]		
			0.374	[436]				
M_1	4317.00	[503]	3.85(-3)	[512]	19.56	[423]	4317.0	[51]

level	E_B/eV		ω_{nlj}		Γ/eV		AE/eV	
	4317.00 ± 2.00	[51]	4.64(-3)	[423]	15.3	[94]		
	4311.70 ± 1.50	[493]	3.46(-3)	[245]				
M_2	4008.00	[503]	2.60(-2)	[512]	13.87	[423]	4008.0	[51]
	4008.00 ± 2.80	[51]	7.41(-3)	[423]	11.4	[94]		
	3995.80 ± 1.50	[493]	5.34(-3)	[245]				
M_3	3426.00	[503]	6.3(-3)	[512]	11.12	[423]	3426.0	[51]
	3426.00 ± 2.90	[51]	1.01(-2)	[423]	8.4	[94]		
	3410.50 ± 1.20	[493]	7.39(-3)	[245]				
M_4	2909.00	[503]	4.40(-2)	[512]	2.82	[423]	2908.7	[51]
	2906.70 ± 2.10	[51]	4.16(-2)	[423]	2.78	[94]		
	2901.80 ± 1.10	[493]						
M_5	2787.00	[503]	3.70(-2)	[512]	2.78	[423]	2786.7	[51]
	2786.70 ± 2.10	[51]	4.16(-2)	[423]	2.78	[94]		
	2780.70 ± 1.10	[493]						
N_1	1042.00	[503]	7.2(-4)	[423]	14.97	[423]	1042.0	[51]
	1042.00 ± 2.00	[51]	10.1	[94]				
	1038.20 ± 1.20	[493]						
N_2	886.00	[503]	1.28(-3)	[423]	14.51	[423]		
	886.00 ± 3.00	[51]			7.6	[94]		
	897.70 ± 1.80	[493]						
N_3	740.00	[503]	5.8(-4)	[423]	9.20	[423]		
	740.00 ± 3.00	[51]			6.4	[94]		
	753.70 ± 1.00	[493]						
N_4	533.00	[503]	4.2(-4)	[739	7.40	[423]		
	533.20 ± 3.20	[51]			3.9	[94]		
	527.60 ± 1.20	[493]						
N_5	507.00	[503]	3.8(-4)		6.68	[423]		
	500.10 ± 1.00	[493]			3.85	[94]		
N_6	210.00	[503]	2.0(-5)	[423]	0.17	[423]		
	207.00 ± 1.00	[493]			0.17	[94]		
N_7	200.80 ± 1.50	[493]	1.0(-5)	[423]	0.16	[423]		
					0.16	[94]		
O_1	195.00	[503]	1.0(-5)	[739	26.23	[423]		
	193.40 ± 2.00	[493]						

level	E_B/eV		ω_{nlj}		Γ/eV		AE/eV
O_2	148.00	[503]	1.0(-5)	[423]	8.76	[423]	
	145.60 ± 3.00	[493]					
O_3	115.00	[503]	2.0(-6)	[423]	18.34	[423]	
	113.60 ± 2.40	[493]					
O_4	40.00	[503]					
	40.90 ± 1.00	[493]					
O_5	40.00	[503]					
	37.40 ± 1.20	[493]					
P_1	18.00	[503]					
	15.00 ± 5.30	[493]					
P_2	8.00	[503]					
	5.70 ± 6.00	[493]					
P_3	8.00	[503]					
	2.80 ± 5.80	[493]					
	9.00	[126]					
IP	9.65	[247]					

Rn   Z=86

[Xe] $4f^{14}\,5d^{10}\,6s^2\,6p^6$

Radon

$A = [222.0176]\,[222]$ $\varrho = 9.23(\text{-}4)\ \text{g/cm}^3\ [547]$

X-ray transitions

line	transition	E/eV		I/eV/$\hbar$		TPIV	Γ/eV	
K series								
K_{α_3}	KL_1	80351.3 ± 2.9	[145]				86.47	[423]
K_{α_2}	KL_2	81066 ± 24	[50]	20.59	[488]	2.848(-1)	80.50	[301]
		81070.7 ± 3.0	[145]				89.50	[473]
							78.64	[423]
K_{α_1}	KL_3	83783 ± 25	[50]	33.95	[488]	4.696(-1)	79.70	[301]
		83788.6 ± 2.8	[145]				80.00	[473]
							78.61	[423]
	KM_1	93925.8 ± 4.4	[145]				91.64	[423]
K_{β_3}	KM_2	94247 ± 53	[50]	3.480	[488]	5.310(-2)	91.20	[473]
		94248.8 ± 4.4	[145]				86.02	[423]
K_{β_1}	KM_3	94867 ± 54	[50]	7.46	[488]	1.032(-1)	85.50	[473]
		94870.2 ± 4.4	[145]				83.15	[423]
$K_{\beta_5^{II}}$	KM_4	95386.7 ± 3.6	[145]			1.606(-3)	74.89	[423]
$K_{\beta_5^{I}}$	KM_5	95514.5 ± 3.6	[145]			1.901(-3)	74.84	[423]
	KN_1	97310.5 ± 5.5	[145]				87.04	[423]
$K_{\beta_2^{II}}$	KN_2	97478 ± 57	[50]	0.935	[488]	1.293(-2)	86.65	[423]
		97455.1 ± 6.1	[145]					
$K_{\beta_2^{I}}$	KN_3	97639 ± 57	[50]	1.856	[488]	2.567(-2)	81.37	[423]
		97609.1 ± 5.0	[145]					
$K_{\beta_4^{II}}$	KN_4	97840.1 ± 4.9	[145]			4.462(-4)	79.17	[423]
$K_{\beta_4^{I}}$	KN_5	97868.9 ± 4.7	[145]			5.284(-4)	78.45	[423]
L series								
	L_1M_1	13574.5 ± 3.2	[145]				34.16	[423]
L_{β_4}	L_1M_2	13897.5 ± 3.2	[145]			4.004(-2)	28.84	[423]
L_{β_3}	L_1M_3	14512.7 ± 2.3	[50]	0.592	[488]	4.027(-2)	21.00	[473]
		14518.9 ± 3.3	[145]				25.68	[423]
$L_{\beta_{10}}$	L_1M_4	15035.4 ± 2.5	[145]			1.601(-3)	17.41	[423]
L_{β_9}	L_1M_5	15163.1 ± 2.4	[145]			2.393(-3)	17.36	[423]

line	transition	E/eV		I/eV/$\hbar$		TPIV	Γ/eV	
	L_1N_1	16959.2 ± 4.3	[145]				29.56	[423]
$L_{\gamma 2}$	L_1N_2	17103.8 ± 4.9	[423]			1.053(-2)	29.17	[423]
$L_{\gamma 3}$	L_1N_3	17257.8 ± 3.9	[145]			1.175(-2)	23.88	[423]
	L_1N_4	17488.8 ± 3.7	[145]			2.565(-4)	21.69	[423]
	L_1N_5	17517.5 ± 3.6	[145]			4.082(-4)	20.97	[423]
	L_1N_6	17817.9 ± 2.7	[145]				14.65	[423]
	L_1N_7	17824.8 ± 2.7	[145]				14.64	[423]
L_η	L_2M_1	12855.1 ± 2.6	[145]			9.779(-3)	26.34	[423]
	L_2M_2	13178.1 ± 2.2	[145]				20.72	[423]
$L_{\beta 17}$	L_2M_3	13799.5 ± 2.3	[145]			4.419(-4)	17.85	[423]
$L_{\beta 1}$	L_2M_4	14315.8 ± 2.2	[50]	2.361	[488]	3.553(-1)	10.65	[473]
		14316.0 ± 1.5	[145]				9.59	[423]
	L_1M_5	14443.7 ± 1.4	[145]				9.54	[423]
$L_{\gamma 5}$	L_2N_1	16239.8 ± 3.4	[145]			2.578(-3)	21.74	[423]
	L_2N_2	16384.4 ± 3.9	[145]				21.34	[423]
	L_2N_3	16538.4 ± 2.9	[145]			1.270(-4)	16.06	[423]
$L_{\gamma 1}$	L_2N_4	16770.7 ± 3.0	[50]	0.504	[488]	7.578(-2)	13.55	[473]
		16769.4 ± 2.7	[145]				13.87	[423]
	L_2N_5	16798.1 ± 2.6	[145]				13.15	[423]
L_ν	L_2N_6	17098.5 ± 1.7	[145]			3.807(-4)	6.83	[423]
	L_2N_7	17105.4 ± 1.7	[145]				6.82	[423]
L_l	L_3M_1	10137.2 ± 2.0	[145]			1.785(-2)	26.31	[423]
L_t	L_3M_2	10460.2 ± 2.0	[145]			2.068(-4)	20.69	[423]
L_s	L_3M_3	11081.6 ± 2.1	[145]			1.871(-4)	17.82	[423]
$L_{\alpha 2}$	L_3M_4	11598.08 ± 0.80	[50]	0.2027	[488]	3.092(-2)	10.50	[473]
		11598.1 ± 1.3	[145]				9.56	[423]
$L_{\alpha 1}$	L_3M_5	11727.09 ± 0.82	[50]	1.783	[488]	2.721(-1)	10.03	[473]
		11725.9 ± 1.2	[145]				9.52	[423]
$L_{\beta 6}$	L_3N_1	13521.9 ± 3.2	[145]			4.267(-3)	21.71	[423]
	L_3N_2	13666.6 ± 3.7	[145]			4.665(-5)	21.31	[423]
	L_3N_3	13820.6 ± 2.7	[145]			4.915(-5)	16.03	[423]
$L_{\beta 15}$	L_3N_4	14051.5 ± 2.6	[145]			5.977(-3)	13.84	[423]
$L_{\beta 2}$	L_3N_5	14080.3	[145]			5.430(-2)	13.12	[423]
$L_{u'}$	L_3N_6	14380.7 ± 1.6	[145]			5.836(-5)	6.80	[423]
L_u	L_3N_7	14387.6 ± 1.6	[145]			3.238(-4)	6.79	[423]

line	transition	E/eV		I/eV/$\hbar$	TPIV	Γ/eV	
	M series						
M_{γ_2}	M_3N_4	2958.40	[423]		7.997(-4)	18.38	[423]
M_{γ_1}	M_3N_5	3001.00	[263]		6.101(-3)	17.66	[423]
M_{ξ_2}	M_4N_2	2083.09	[423]		2.090(-3)		[423]
M_δ	M_4N_3	2242.21	[423]		2.008(-4)		[423]
M_β	M_4N_6	2780.00	[263]		4.070(-2)	3.08	[423]
M_η	M_4O_2	2859.31	[423]		5.088(-4)	11.79	[423]
M_{ξ_1}	M_5N_3	2097.00	[263]		1.518(-3)	12.27	[423]
$M_{\alpha_{1,2}}$	$M_5N_{6,7}$	2661.00	[263]				
	M_5N_6				2.002(-3)	3.04	[423]
	M_5N_7				3.915(-2)	3.03	[423]

level characteristics

level	E_B/eV		ω_{nlj}		Γ/eV		AE/eV	
K	98404.00	[503]	0.969	[301]	73.30	[301]	98404.0	[51]
	98404.00 ± 1.20	[51]	0.9667	[39]	71.97	[423]	98408.1 ± 2.8	[145]
	98397.20 ± 1.50	[493]	0.9638	[423]	73.4	[94]		
L_1	18049.00	[503]	0.134	[301]	13.10	[301]	18049.0	[51]
	18049.00 ± 3.80	[51]	0.1173	[423]	14.50	[423]	18056.8 ± 1.6	[145]
	18048.70 ± 1.50	[493]	0.114	[436]	13.2	[94]		
L_2	17337.00	[503]	0.429	[301]	7.20	[301]	17337.1	[51]
	17337.10 ± 3.40	[51]	0.4571	[423]	6.66	[423]	17337.38 ± 0.62	[145]
	17329.70 ± 1.10	[493]	0.451	[436]	6.81	[94]		
L_3	14619.00	[503]	0.411	[301]	6.41	[301]	14619.4	[51]
	14619.40 ± 3.00	[51]	0.3967	[423]	6.64	[423]	14619.53 ± 0.43	[145]
	14611.40 ± 1.20	[493]	0.384	[436]	6.87	[94]		
M_1	4482.00	[503]	3.95(-3)	[38]	20.20	[350]	4482.0	[51]
	4474.30 ± 1.50	[493]	4.07(-3)	[512]	19.67	[423]		
			4.89(-3)	[423]	15.3	[94]		
M_2	4159.00	[503]	9.75(-3)	[38]	14.05	[423]	4159.0	[51]
	4159.00 ± 3.80	[51]	2.7(-2)	[512]	13.90	[350]		

level	E_B/eV		ω_{nlj}		Γ/eV		AE/eV	
	4151.50 ± 1.50	[493]	7.68(-3)	[423]	11.7	[94]		
M_3	3538.00	[503]	6.30(-3)	[38]	11.18	[423]	3538.0	[51]
	3538.00 ± 3.80	[51]	6.64(-3)	[512]	11.90	[350]		
	3530.50 ± 1.50	[493]	1.05(-2)	[423]	8.3	[94]		
M_4	3022.00	[503]	3.55(-2)	[38]	2.92	[423]	3021.5	[51]
	3021.50 ± 3.10	[51]	4.80(-2)	[512]	3.04	[350]		
	3012.30 ± 1.10	[493]	4.36(-2)	[423]	2.88	[94]		
M_5	2892.00	[503]	3.62(-2)	[38]	2.88	[423]	2892.4	[51]
	2892.40 ± 3.10	[51]	3.90(-29	[512]	2.81	[350]		
	2884.20 ± 1.10	[493]	4.31(-2)	[423]	2.88	[94]		
N_1	1097.00	[503]	7.5(-4)	[423]	15.08	[423]	1097.0	[51]
	1090.50 ± 1.20	[493]			14.00	[74]		
					10.4	[94]		
N_2	929.00	[503]	1.30(-3)	[423]	14.68	[423]		
	929.00 ± 4.00	[51]			14.00	[74]		
	946.20 ± 1.70	[493]			7.8	[94]		
N_3	768.00	[503]	6.0(-4)	[423]	9.40	[423]		
	768.00 ± 4.00	[51]			14.00	[74]		
	791.20 ± 1.00	[493]			6.6	[94]		
N_4	567.00	[503]	4.6(-4)	[423]	7.21	[423]		
	566.60 ± 4.00	[51]			6.90	[74]		
	560.40 ± 1.20	[493]			4	[94]		
N_5	541.00	[503]	4.1(-4)	[423]	6.48	[423]		
	531.10 ± 1.00	[493]			6.90	[74]		
					3.9	[94]		
N_6	238.00	[503]	2.0(-5)	[423]	0.16	[423]		
	230.10 ± 1.10	[493]			0.10	[74]		
N_7	223.60 ± 1.10	[493]	2.0(-5)	[423]	0.15	[423]		
					0.10	[74]		
					0.16	[94]		
O_1	214.00	[503]	2.0(-5)	[423]	23.93	[423]		
	209.60 ± 2.20	[493]						
O_2	166.00	[503]	1.0(-5)	[423]	8.87	[423]		
	159.50 ± 2.80	[493]						

level	E_B/eV		ω_{nlj}		Γ/eV		AE/eV
O_3	127.00	[503]	3.0(-6)	[423]	19.89	[423]	
	123.90 ± 2.50	[493]					
O_4	48.00	[503]					
	48.00 ± 1.00	[493]					
O_5	48.00	[503]					
	44.20 ± 1.10	[493]					
P_1	26.00	[503]					
	18.70 ± 5.00	[493]					
P_2	11.00	[503]					
	7.60 ± 5.00	[493]					
P_3	11.00	[503]					
	4.10 ± 7.50	[493]					
	10.749	[126]					
IP	10.74850	[222]					

Fr   Z=87

[Rn] 7s^1

Francium

$A = [223.0197]\ [222]$ $\varrho = 2.50\ \text{g/cm}^{-3}\ [342]$

X-ray transitions

line	transition	E/eV		I/eV/$\hbar$		TPIV	Γ/eV	
K series								
K$_{\alpha_3}$	KL$_1$	82498.8 ± 2.9	[145]				90.06	[423]
K$_{\alpha2}$	KL$_2$	83232 ± 25	[50]	21.63	[488]	2.855(-1)	84.20	[301]
		82233.3 ± 3.1	[145]				82.29	[423]
K$_{\alpha1}$	KL$_3$	86105 ± 27	[50]	35.49	[488]	4.685(-1)	83.40	[301]
		86107.4 ± 3.0	[145]				82.26	[423]
	KM$_1$	96487.3 ± 4.5	[145]				95.15	[423]
K$_{\beta3}$	KM$_2$	96808 ± 56	[50]	4.028	[488]	5.317(-2)	89.64	[423]
		96817.3 ± 4.6	[145]					
K$_{\beta1}$	KM$_3$	97478 ± 57	[50]	7.83	[488]	1.033(-1)	86.82	[423]
		97476.7 ± 4.7	[145]					
K$_{\beta_5^{II}}$	KM$_4$	98002.4 ± 3.8	[145]			1.649(-3)	78.43	[423]
K$_{\beta_5^{I}}$	KM$_5$	98137.4 ± 3.8	[145]			1.941(-3)	78.39	[423]
	KN$_1$	99984.7 ± 5.4	[145]				90.65	[423]
K$_{\beta_2^{II}}$	KN$_2$	100155 ± 60	[50]	0.986	[488]	1.301(-2)	90.10	[423]
		100134.0 ± 6.4	[145]					
K$_{\beta_2^{I}}$	KN$_3$	100326 ± 60	[50]	1.062	[488]	2.590(-2)	85.04	[423]
		100298.1 ± 5.2	[145]					
K$_{\beta_4^{II}}$	KN$_4$	100535.0 ± 5.1	[145]			4.628(-4)	82.37	[423]
K$_{\beta_4^{I}}$	KN$_5$	100566.1 ± 5.1	[145]			5.448(-4)	81.71	[423]
L series								
	L$_1$M$_1$	13988.5 ± 3.3	[145]				34.38	[423]
L$_{\beta4}$	L$_1$M$_2$	14318.5 ± 3.3	[145]			4.221(-2)	28.87	[423]
L$_{\beta3}$	L$_1$M$_3$	14975.7 ± 2.4	[50]	0.617	[488]	4.151(-2)	26.04	[423]
		14977.9 ± 3.4	[145]					
L$_{\beta10}$	L$_1$M$_4$	15503.6 ± 2.6	[145]			1.730(-3)	17.66	[423]
L$_{\beta9}$	L$_1$M$_5$	15638.6 ± 2.5	[145]			2.584(-3)	17.62	[423]
	L$_1$N$_1$	17485.9 ± 4.2	[145]				29.88	[423]

line	transition	E/eV		I/eV/$\hbar$		TPIV	Γ/eV	
L_{γ_2}	L_1N_2	17635.2 ± 5.1	[145]			$1.115(-2)$	29.31	[423]
L_{γ_3}	L_1N_3	17799.3 ± 3.9	[145]			$1.223(-2)$	24.26	[423]
	L_1N_4	18036.2 ± 3.9	[145]			$2.800(-4)$	21.60	[423]
	L_1N_5	18067.3 ± 3.8	[145]			$4.465(-4)$	20.94	[423]
	L_1N_6	18375.3 ± 3.0	[145]				16.29	[423]
	L_1N_7	18382.3 ± 2.8	[145]				16.29	[423]
L_η	L_2M_1	13254.0 ± 2.6	[145]			$1.005(-2)$	26.61	[423]
	L_2M_2	13584.0 ± 2.2	[145]				21.10	[423]
$L_{\beta_{17}}$	L_2M_3	14243.4 ± 2.4	[145]			$4.646(-4)$	18.18	[423]
L_{β_1}	L_2M_4	14770.4 ± 2.3 14764.00 14769.1 ± 1.5	[50] [104] [145]	2.496	[488]	$3.640(-1)$	9.89	[423]
	L_2M_5	14904.1 ± 1.5	[145]				9.85	[423]
L_{γ_5}	L_2N_1	16751.5 ± 3.3	[145]			$2.663(-3)$	22.11	[423]
	L_2N_2	16900.7 ± 4.0	[145]				21.55	[423]
	L_2N_3	17064.8 ± 2.8	[145]			$1.347(-4)$	16.49	[423]
L_{γ_1}	L_2N_4	17303.4 ± 3.2 17304.00 $17301.7 \pm 2.l8$	[50] [104] [145]	0.538	[488]	$7.846(-2)$	13.84	[423]
	L_2N_5	17332.8 ± 2.8	[145]				13.17	[423]
L_ν	L_2N_6	17640.8 ± 1.9	[145]			$4.099(-4)$	7.02	[423]
	L_2N_7	17647.8 ± 1.8	[145]				7.02	[423]
L_l	L_3M_1	10379.8 ± 2.0	[145]			$1.843(-2)$	26.58	[423]
L_t	L_3M_2	10709.9 ± 2.0	[145]			$2.173(-4)$	21.07	[423]
L_s	L_3M_3	$11369.3 \pm 2.2.$	[145]			$1.967(-4)$	18.15	[423]
L_{α_2}	L_3M_4	11895.07 ± 0.84	[50]	0.2135	[488]	$3.166(-2)$	9.86	[423]
L_{α_1}	L_3M_5	12031.40 ± 0.86 12035.50	[50] [104]	1.878	[488]	$2.785(-1)$	9.82	[423]
L_{β_6}	L_3N_1	13877.3 ± 3.1	[145]			$4.454(-3)$	22.08	[423]
	L_3N_2	14026.6 ± 3.8	[145]			$4.894(-5)$	21.52	[423]
	L_3N_3	14190.6 ± 2.6	[145]			$5.196(-5)$	16.46	[423]
$L_{\beta_{15}}$	L_3N_4	14427.5 ± 2.6	[145]			$6.165(-3)$	13.81	[423]
L_{β_2}	L_3N_5	14450 ± 50 14454.00 14458.6 ± 2.6	[50] [104] [145]	0.378	[488]	$5.608(-2)$	13.14	[423]
$L_{u'}$	L_3N_6	14766.6 ± 1.7	[145]			$6.318(-5)$	6.99	[423]

line	transition	E/eV		I/eV/$\hbar$	TPIV	Γ/eV	
L_u	L_3N_7	14773.6 ± 1.6	[145]		3.508(-4)	6.99	[423]
M series							
M_{γ_2}	M_3N_4	3046.90	[423]		8.305(-4)	18.36	[423]
M_{γ_1}	M_3N_5	3092.00	[263]		6.315(-3)	17.70	[423]
M_{ξ_2}	M_4N_2	2146.21	[423]		2.122(-3)	17.70	[423]
M_δ	M_4N_3	2316.10	[423]		2.022(-4)	12.64	[423]
M_β	M_4N_6	2877.00	[263]		4.280(-2)	3.17	[423]
M_η	M_4O_2	2957.90	[423]		5.135(-4)	11.98	[423]
M_{ξ_1}	M_5N_3	2164.00	[263]		1.515(-3)	12.60	[423]
$M_{\alpha_{1.2}}$	$M_5N_{6,7}$	2743.00	[263]			3.13	[423]
	M_5N_6				2.092(-3)		
	M_5N_7				4.094(-2)		

level characteristics

level	E_B/eV		ω_{nlj}		Γ/eV		AE/eV	
K	101137.00	[503]	0.970	[301]	76.80	[301]	101137.0	[51]
	101137.00 ± 1.30	[51]	0.9674	[39]	75.42	[423]	101141.2 ± 3.0	[145]
	101129.90 ± 1.60	[493]	0.9644	[423]	76.9	[94]		
L_1	18639.00	[503]	0.139	[301]	13.30	[301]	18639.0	[51]
	18639.00 ± 4.00	[51]	0.1232	[423]	14.64	[423]	18642.4 ± 1.7	[145]
	18634.10 ± 1.40	[493]	0.12	[436]	13.5	[94]		
L_2	17906.00	[503]	0.443	[301]	7.47	[301]	17906.5	[51]
	17906.50 ± 3.50	[51]	0.4700	[423]	6.87	[423]	17907.94 ± 0.64	[145]
	17900.50 ± 1.10	[493]	0.464	[436]	7.02	[94]		
L_3	15031.00	[503]	0.424	[301]	8.65	[301]	15031.2	[51]
	15031.20 ± 3.00	[51]	0.4075	[423]	6.85	[423]	15033.78 ± 0.44	[145]
	15025.60 ± 1.20	[493]	0.394	[423]	7.08	[94]		
M_1	4652.00	[503]	4.29(-3)	[512]	19.74	[423]	4652.0	[51]
	4645.70 ± 1.50	[493]	5.15(-3)	[423]	15.4	[94]		
M_2	4327.00	[503]	2.8(-29)	[512]	14.22	[423]	4327.0	[51]
	4327.00 ± 4.00	[51]	7.96(-3)	[423]	12.1	[94]		

level	E_B/eV		ω_{nlj}		Γ/eV		AE/eV	
	4316.00 ± 1.50	[493]						
M_3	3663.00	[503]	6.98(-3)	[512]	11.40	[423]	3663.0	[51]
	3663.00 ± 4.00	[51]	1.10(-2)	[423]	8.2	[94]		
	3657.30 ± 1.50	[493]						
M_4	3136.00	[503]	5.1(-2)	[512]	3.02	[423]	3136.2	[51]
	3136.20 ± 3.10	[51]	4.58(-2)	[423]	2.98	[94]		
	3129.70 ± 1.00	[493]						
M_5	3000.00	[503]	4.00(-2)	[512]	2.98	[423]	2999.7	[51]
	2999.90 ± 3.10	[51]	4.49(-2)	[423]	2.98	[94]		
	2994.90 ± 1.20	[493]						
N_1	1153.00	[503]	7.9(-4)	[423]	15.24	[423]	1153.0	[51]
	1149.00 ± 1.20	[493]			10.7	[94]		
N_2	980.00	[503]	1.33(-3)	[423]	14.68	[423]		
	980.00 ± 4.20	[51]			8.1	[94]		
	1000.90 ± 1.60	[493]						
N_3	810.00	[503]	6.3(-4)	[423]	9.62	[423]		
	810.00 ± 4.30	[51]			6.8	[94]		
	835.10 ± 1.00	[493]						
N_4	603.00	[503]	4.9(-4)	[423]	6.96	[423]		
	603.30 ± 4.10	[51]			4.1	[94]		
	598.70 ± 1.20	[493]						
N_5	577.00	[503]	4.4(-4)	[423]	6.30	[423]		
	577.00 ± 3.40	[51]			3.95	[94]		
	567.50 ± 1.00	[493]						
N_6	268.00	[503]	3.0(-5)	[423]	0.15	[423]		
	258.60 ± 1.10	[493]			0.15	[94]		
N_7	268.00	[503]	2.0(-5)	[739	0.15	[423]		
	251.60 ± 1.20	[493]			0.15	[94]		
O_1	234.00	[503]	2.0(-5)	[423]	23.23	[423]		
	230.90 ± 2.40	[493]						
O_2	182.00	[503]	2.0(-5)	[423]	8.96	[423]		
	176.70 ± 2.60	[493]						
O_3	140.00	[503]	4.0(-6)	[423]	21.16	[423]		
	138.70 ± 2.60	[493]						

level	E_B/eV		ω_{nlj}	Γ/eV	AE/eV
O_4	58.00	[503]			
	60.00 ± 1.00	[493]			
O_5	58.00	[503]			
	55.60 ± 1.10	[493]			
P_1	34.00	[503]			
	26.30 ± 4.50	[493]			
P_2	15.00	[503]			
	13.20 ± 4.50	[493]			
P_3	15.00	[503]			
	8.80 ± 6.50	[493]			
Q_1	4.00	[126]			
IP	4.0727	[222]			

Ra   Z=88

[Rn] $7s^2$

Radium

$A = 226.0254$ [260] $\varrho = 4.00$ g/cm^3 [547]
$\varrho = 5.00$ g/cm^3 [483]

X-ray transitions

line	transition	E/eV		I/eV/$\hbar$		TPIV	Γ/eV	
K series								
K_{α_3}	KL_1	84685.4 ± 3.0	[145]				94.40	[423]
$K_{\alpha2}$	KL_2	85430.00 ± 12.00	[50]	22.70	[488]	2.861(-1)	88.10	[301]
		85436 ± 12	[43]				91.20	[473]
							86.47	[423]
$K_{\alpha1}$	KL_3	88470.00 ± 13.00	[50]	37.07	[488]	4.673(-1)	87.20	[301]
		88476 ± 12	[43]				87.00	[473]
							86.07	[423]
	KM_1	99097.5 ± 4.6	[145]				97.78	[423]
$K_{\beta3}$	KM_2	99430.00 ± 25.00	[50]	4.222	[488]	5.322(-2)	98.95	[473]
		99434 ± 12	[43]				93.38	[423]
$K_{\beta1}$	KM_3	100130.00 ± 12.00	[50]	8.21	[488]	1.036(-1)	94.20	[473]
		100136 ± 12	[43]				90.65	[423]
$K_{\beta_5^{II}}$	KM_4	100669.1 ± 3.9	[145]			1.692(-3)	82.11	[423]
$K_{\beta_5^{I}}$	KM_5	100811.5 ± 3.8	[145]			1.979(-3)	82.09	[423]
	KN_1	102710.7 ± 5.4	[145]				92.87	[423]
$K_{\beta_2^{II}}$	KN_2	100890.00 ± 25.00	[50]	1.038	[488]	1.309(-2)	93.85	[423]
		102861 ± 12	[43]					
$K_{\beta_2^{I}}$	KN_3	101070.00 ± 25.00	[50]	2.072	[488]	2.612(-2)	88.54	[423]
		103045 ± 12	[43]					
$K_{\beta_4^{II}}$	KN_4	103282.7 ± 5.4	[145]			4.797(-4)	85.95	[423]
$K_{\beta_4^{I}}$	KN_5	103316.0 ± 5.4	[145]			5.616(-4)	85.44	[423]
L series								
	L_1M_1	14412.1 ± 3.3	[145]				34.13	[423]
$L_{\beta4}$	L_1M_2	14747.3 ± 1.3	[50]	0.668	[488]	4.339(-2)	25.00	[473]
		14747.30	[104]				29.76	[423]
$L_{\beta3}$	L_1M_3	15445.1 ± 1.4	[50]	0.642	[488]	4.169(-2)	22.00	[473]
		15445.00	[104]				27.03	[423]

line	transition	E/eV		I/eV/$\hbar$		TPIV	Γ/eV	
$L_{\beta 10}$	L_1M_4	15988.2 ± 1.5	[50]	2.806(-2)	[488]	1.822(-3)	18.49	[423]
		15988.00	[104]					
$L_{\beta 9}$	L_1M_5	16131.6 ± 1.6	[50]	4.19(-2)	[488]	2.721(-3)	18.47	[423]
		16132.00	[104]					
	L_1N_1	18036.4 ± 3.9	[50]	1.26(-5)	[488]		29.25	[423]
		18040.00	[104]					
$L_{\gamma 2}$	L_1N_2	18179.5 ± 2.0	[50]	0.1773	[488]	1.152(-2)	30.23	[423]
		18179.00	[104]					
$L_{\gamma 3}$	L_1N_3	18357.4 ± 2.0	[50]	0.1910	[488]	1.241(-2)	24.92	[423]
		18357.00	[104]					
	L_1N_4	18599.2 ± 4.1	[50]	4.59(-3)	[488]	2.979(-4)	22.06	[423]
		18600.00	[104]					
$L_{\gamma 11}$	L_1N_5	18632.8 ± 4.1	[50]	7.32(-3)	[488]	4.754(-4)	21.61	[423]
		18630.00	[104]					
	L_1N_6	18945.9 ± 3.0	[145]				15.54	[423]
	L_1N_7	18953.1 ± 2.8	[145]				15.55	[423]
$L_{\gamma 4'}$	L_1O_2	19036.00 ± 1.50	[50]	4.02(-2)	[488]	2.610(-3)	24.23	[423]
		19036.00	[104]					
$L_{\gamma 4}$	L_1O_3	19084.00 ± 1.50	[50]	4.35(-3)	[488]	2.828(-3)	35.76	[423]
		19085.00	[104]					
	$L_1O_{4,5}$	19167.00 ± 3.00	[50]	1.93(-3)	[488]			
		19170.00	[104]					
	L_1O_4					4.805(-5)		
	L_1O_5					7.721(-5)		
	$L_1P_{2,3}$	19218.00 ± 1.50	[50]	1.298(-2)	[488]			
		19218.00	[104]					
	L_1P_2					4.457(-4)		
	L_1P_3					4.331(-4)		
L_{η}	L_2M_1	13663.2 ± 1.1	[50]	7.31(-2)	[488]	9.935(-3)	26.23	[423]
		13663.00	[104]					
	L_2M_2	13999.7 ± 2.3	[145]				21.84	[423]
$L_{\beta 17}$	L_2M_3	14693.3 ± 2.6	[50]	3.45(-3)	[488]	4.697(-4)	19.10	[423]
		14692.00	[104]					
$L_{\beta 1}$	L_2M_4	15235.9 ± 1.40	[50]	2.638	[488]	3.586(-1)	11.60	[473]
		15236.00	[104]				10.56	[423]
	L_2M_5	15376.4 ± 1.5	[145]				10.55	[423]
$L_{\gamma 5}$	L_2N_1	17274.0 ± 1.8	[50]	1.946(-2)	[488]	2.646(-3)	21.32	[423]
		17274.00	[104]					

line	transition	E/eV		I/eV/ℏ		TPIV	Γ/eV	
	L$_2$N$_2$	17429.7 ± 4.2	[145]					
	L$_2$N$_3$	17603.6 ± 3.7	[50]	1.01(-3)	[488]	1.373(-4)	16.99	[423]
		17600.00	[104]					
L$_{\gamma1}$	L$_2$N$_4$	17848.7 ± 1.9	[50]	0.574	[488]	7.810(-2)	14.30	[473]
		17849.00	[104]				14.13	[423]
	L$_2$N$_5$	17885.5 ± 3.8	[50]	6.25(-5)	[488]		13.68	[423]
		17880.00	[104]					
L$_\nu$	L$_2$N$_6$	18230.00 ± 3.00	[50]	3.12(-3)	[488]	4.238(-4)	7.61	[423]
		18196.2 ± 2.1	[145]					
	L$_2$N$_7$	18203.5 ± 1.8	[145]					
L$_{\gamma8}$	L$_2$O$_1$	18230.00	[104]	4.73(-3)	[488]	6.424(-4)	29.80	[423]
	L$_2$O$_2$	18286.00 ± 2.70	[50]	7.78(-7)	[488]		16.30	[423]
		18290.00	[104]					
	L$_2$O$_3$	18330.00 ± 2.70	[50]	2.29(-4)	[488]	3.119(-5)	27.83	[423]
		18330.00	[104]					
L$_{\gamma6}$	L$_2$O$_4$	18414.00 ± 1.60	[50]	9.91(-2)	[488]	1.348(-2)		
		18415.00	[104]					
	L$_2$P$_1$	18439.00 ± 2.70	[50]	9.00(-4)	[488]	1.224(-4)		
		18440.00	[104]					
	L$_2$P$_{2,3}$	18466.00 ± 2.80	[50]	3.46(-5)	[488]			
		18470.00	[104]					
	L$_2$P$_3$					4.686(-6)		
L$_l$	L$_3$M$_1$	10622.29 ± 0.67	[50]	0.1277	[488]	1.902(-2)	25.83	[423]
		10622.00	[104]					
L$_t$	L$_3$M$_2$	110948.00	[104]	1.58(-3)	[488]	2.278(-4)	21.43	[423]
		10961.4 ± 2.1	[145]					
L$_s$	L$_3$M$_3$	11660.4 ± 2.2	[145]			2.064(-4)	18.70	[423]
L$_{\alpha2}$	L$_3$M$_4$	12196.26 ± 0.89	[50]	0.2248	[488]	3.237(-2)	11.20	[473]
		12196.00	[104]				10.16	[423]
L$_{\alpha1}$	L$_3$M$_5$	12339.86 ± 0.91	[50]	1.977	[488]	2.847(-1)	11.00	[473]
		12340.00	[104]				10.14	[423]
L$_{\gamma6}$	L$_3$N$_1$	14236.4 ± 1.2	[50]	3.23(-2)	[488]	4.644(-3)	20.92	[423]
		14236.00	[104]					
	L$_3$N$_2$	14386.4 ± 2.5	[50]	2.56(-4)	[488]	5.124(-5)	21.90	[423]
		14387.00	[104]					
	L$_3$N$_3$	14565.6 ± 2.5	[50]	3.81(-4)	[488]	5.486(-5)	16.59	[423]
		14566.00	[104]					
L$_{\beta15}$	L$_3$N$_4$	14808.8 ± 1.3	[50]	4.41(-2)	[488]	6.349(-3)	13.73	[423]
		14809.00	[104]					

line	transition	E/eV		I/eV/$\hbar$		TPIV	Γ/eV	
$L_{\beta2}$	L_3N_5	14841.6 ± 1.3	[50]	0.402	[488]	5.782(-2)	12.20	[473]
		14842.00	[104]				13.27	[423]
	$L_3N_{6,7}$	15145.7 ± 2.7	[50]	2.62(-3)	[488]			
		15146.00	[104]					
$L_{u'}$	L_3N_6	15157.7 ± 2.7	[145]			6.820(-5)	7.21	[423]
L_u	L_3N_7	15165.2 ± 1.6	[145]			3.781(-4)	7.22	[423]
$L_{\beta7}$	L_3O_1	15190.00 ± 2.00	[50]	7.75(-3)	[488]	1.117(-3)	29.39	[423]
		15190.00	[104]					
$L_{\beta5}$	$L_3O_{4,5}$	15377.10 ± 1.00	[50]	7.52(-2)	[488]			
		15377.00	[104]					
	L_3O_4					1.074(-3)		
	L_3O_5					9.754(-3)		
	L_3P_1	15402.00 ± 2.00	[50]	1.47(-3)	[488]	2.122(-4)		
		15402.00	[104]					
	$L_3P_{2,3}$	15425.00 ± 2.00	[50]	2.51(-5)	[488]			
		15425.00	[104]					
	L_3P_2					1.780(-6)		
	L_3P_3					1.828(-6)		

M series

line	transition	E/eV		I/eV/$\hbar$		TPIV	Γ/eV	
$M_{\gamma2}$	M_3N_4	3136.93	[423]			8.611(-4)	18.31	[423]
$M_{\gamma1}$	M_3N_5	3185.00	[263]			6.534(-3)	17.85	[423]
$M_{\xi2}$	M_4N_2	2210.10	[423]			2.155(-3)	17.93	[423]
M_δ	M_4N_3	2391.39	[423]			2.036(-4)	12.62	[423]
M_β	M_4N_6	2966.00	[263]			4.503(-2)	3.24	[423]
M_η	M_4O_2	3058.38	[423]			5.222(-4)	11.93	[423]
$M_{\xi1}$	M_5N_3	2231.00	[263]			1.5524(-3)	12.60	[423]
$M_{\alpha2,1}$	$M_5N_{6,7}$	2824.00	[263]					
$M_{\alpha2}$	M_5N_6					2.182(-3)	3.22	[423]
$M_{\alpha1}$	M_5N_7					4.270(-2)	3.23	[423]

level characteristics

level	E_B/eV		ω_{nlj}		Γ/eV		AE/eV	
K	103922.00	[503]	0.970	[301]	80.40	[301]	103921.9	[51]
	103921.90 ± 7.20	[51]	0.9680	[39]	79.01	[423]	103927.7 ± 3.0	[145]
	103916.20 ± 1.70	[493]	0.9645	[423]	80.6	[94]		
			0.968	[112]				
L_1	19237.00	[503]	0.146	[301]	13.40	[301]	19236.7	[51]
	19236.70 ± 1.50	[493]	0.1262	[423]	15.39	[423]	19237.0 ± 4.4	[50]
			0.126	[436]	13.7	[94]	19242.3 ± 1.7	
L_2	18484.00	[503]	0.456	[301]	7.68	[301]	18484.3	[51]
	18484.30 ± 1.50	[493]	0.4647	[423]	7.46	[423]	18485.5 ± 4.1	[50]
			0.472	[244]	7.5	[94]	18492.61 ± 0.67	[145]
L_3	15444.00	[503]	0.437	[301]	6.82	[301]	15444.4	[51]
	15444.40 ± 1.50	[493]	0.4181	[423]	7.06	[423]	15443.7 ± 2.8	[50]
			0.404	[436]	7.29	[94]	15454.32 ± 0.45	
M_1	4822.00	[503]	4.51(-3)	[512]	18.77	[423]	4822.0	[51]
	4822.00 ± 1.50	[493]	5.42(-3)	[423]	22.00	[74]		
			4.43(-3)	[245]	15.4	[94]		
M_2	4490.00	[503]	2.9(-2)	[512]	14.37	[423]	4489.5	[51]
	4489.50 ± 1.80	[51]	8.24(-3)	[423]	12.00	[74]		
	4485.00 ± 1.50	[493]	6.38(-3)	[245]	12.5	[94]		
M_3	3792.00	[503]	7.32(-3)	[512]	11.64	[423]	3791.8	[51]
	3791.80 ± 1.70	[51]	1.14(-2)	[423]	14.00	[74]		
	3786.60 ± 1.50	[493]	9.56(-3)	[245]	8.2	[94]		
M_4	3248.00	[503]	5.5(-2)	[512]	3.10	[423]	3248.4	[51]
	3248.40 ± 1.60	[493]	4.808-2)	[423]	3.00	[74]		
					3.08	[94]		
M_5	3105.00	[503]	4.20(-2)	[512]	3.08	[423]	3104.9	[51]
	3104.90 ± 1.60	[493]	4.68(-2)	[423]	3.20	[74]		
					3.08	[94]		
N_1	1208.00	[503]	8.2(-4)	[423]	13.86	[423]	1208.4	[51]
	1208.40 ± 1.60	[493]			12.00	[74]		
					10.95	[94]		
N_2	1058.00	[503]	1.35(-3)	[423]	14.84	[423]	1057.6	[51]
	1057.60 ± 1.80	[493]			10.00	[74]		
					8.3	[94]		
N_3	879.00	[503]	6.6(-4)	[423]	9.53	[423]		
	879.10 ± 1.80	[493]			10.00	[74]		
					7	[94]		
N_4	636.00	[503]	5.3(-4)	[423]	6.67	[423]		
	635.90 ± 1.60	[493]			6.30	[74]		
					4.15	[94]		
N_5	603.00	[503]	4.7(-4)	[423]	6.21	[423]		

level	E_B/eV		ω_{nlj}		Γ/eV		AE/eV
	602.70 ± 1.70	[493]			6.30	[74]	
					4	[94]	
N_6	299.00	[503]	3.0(-5)	[423]	0.15	[423]	
	298.90 ± 2.40	[51]			0.13	[74]	
	287.90 ± 1.10	[493]			0.15	[94]	
N_7	299.00	[503]	3.0(-5)	[423]	0.16	[423]	
	298.90 ± 2.40	[51]			0.13	[74]	
	280.40 ± 1.20	[493]			0.16	[94]	
O_1	254.00	[503]	2.0(-5)	[423]	22.33	[423]	
	254.40 ± 2.10	[493]					
O_2	200.00	[503]	2.0(-5)	[423]	8.84	[423]	
	200.40 ± 2.00	[493]					
O_3	153.00	[503]	1.0(-5)	[423]	20.37	[423]	
	152.80 ± 2.00	[493]					
O_4	68.00	[503]					
	67.20 ± 1.70	[51]					
	69.40 ± 1.70	[493]					
O_5	68.00	[503]					
	67.20 ± 1.70	[51]					
	63.80 ± 1.70	[493]					
P_1	44.00	[503]					
	43.50 ± 2.20	[51]					
	35.50 ± 4.00	[493]					
P_2	44.00	[503]					
	43.50 ± 1.80	[51]					
	35.50 ± 4.00	[493]					
P_3	19.00	[503]					
	18.80 ± 1.80	[51]					
	13.70 ± 5.20	[493]					
Q_1	5.279	[126]					
IP	5.2784	[222]					

Ac   Z=89

[Rn] $6d^1 7s^2$

Actinium $A = 227.028$ [260] $\varrho = 10.09$ g/cm^3 [547]
$\varrho = 10.10$ g/cm^3 [483]

X-ray transitions

line	transition	E/eV		I/eV/$\hbar$		TPIV	Γ/eV	
K series								
K_{α_3}	KL$_1$	86913.0 ± 3.1	[145]				98.24	[423]
$K_{\alpha 2}$	KL$_2$	87676 ± 18	[50]	23.82	[488]	2.869(-1)	92.00	[301]
		87678.2 ± 3.3	[145]				90.39	[423]
$K_{\alpha 1}$	KL$_3$	90884.8 ± 7.9	[50]	38.70	[488]	4.662(-1)	91.10	[301]
		90895 ± 12	[43]				89.98	[423]
		90880.00 ± 11.00	[298]					
	KM$_1$	101758.9 ± 4.8	[145]				101.57	[423]
$K_{\beta 3}$	KM$_2$	102102 ± 25	[50]	4.422	[488]	5.326(-2)	97.23	[423]
		102103.3 ± 4.8	[145]					
$K_{\beta 1}$	KM$_3$	102847 ± 25	[50]	8.61	[488]	1.037(-1)	94.54	[423]
		102844.2 ± 4.9	[145]					
$K_{\beta_5^{II}}$	KM$_4$	103389.0 ± 4.0	[145]			1.736(-3)	85.9	[423]
$K_{\beta_5^I}$	KM$_5$	103539.3 ± 4.0	[145]			2.017(-3)	85.87	[423]
	KN$_1$	105491.2 ± 5.5	[145]				96.49	[423]
$K_{\beta_2^{II}}$	KN$_2$	105679 ± 27	[50]	1.092	[488]	1.316(-2)	97.71	[423]
		105648.6 ± 6.6	[145]					
$K_{\beta_2^I}$	KN$_3$	105868 ± 27	[50]	2.187	[488]	2.634(-2)	92.51	[423]
		105835.8 ± 5.3	[145]					
$K_{\beta_4^{II}}$	KN$_4$	106083.6 ± 5.3	[145]			4.968(-4)	89.09	[423]
$K_{\beta_4^I}$	KN$_5$	106119.2 ± 5.3	[145]			5.784(-4)	88.99	[423]
L series								
	L$_1$M$_1$	14845.8 ± 3.3	[145]				34.41	[423]
$L_{\beta 4}$	L$_1$M$_2$	15190.3 ± 3.3	[145]			4.568(-2)	30.07	[423]
$L_{\beta 3}$	L$_1$M$_3$	15931.5 ± 2.7	[50]	0.667	[488]	4.285(-2)	15.24	[423]
		15931.1 ± 3.4	[145]					
$L_{\beta 10}$	L$_1$M$_4$	16476.0 ± 2.6	[145]			1.966(-3)	18.74	[423]
$L_{\beta 9}$	L$_1$M$_5$	16626.3 ± 2.5	[145]			2.934(-3)	18.72	[423]
	L$_1$N$_1$	18578.2 ± 2.4	[145]				29.33	[423]

line	transition	E/eV		I/eV/$\hbar$		TPIV	Γ/eV	
L_{γ_2}	L_1N_2	18735.6 ± 5.1	[145]			1.218(-2)	30.55	[423]
L_{γ_3}	L_1N_3	18922.8 ± 3.8	[145]			1.288(-2)	25.34	[423]
	L_1N_4	19170.6 ± 3.8	[145]			3.247(-4)	21.93	[423]
	L_1N_5	19206.2 ± 3.8	[145]			5.190(-4)	21.83	[423]
	L_1N_6	19530.3 ± 3.2	[145]				15.69	[423]
	L_1N_7	19537.8 ± 2.9	[145]				15.71	[423]
L_η	L_2M_1	14080.7 ± 2.6	[145]			1.021(-2)	27.36	[423]
	L_2M_2	14425.2 ± 2.3	[145]				22.22	[423]
$L_{\beta_{17}}$	L_2M_3	15166.0 ± 2.5	[145]			4.933(-4)	19.53	[423]
L_{β_1}	L_2M_4	15713.3 ± 2.7	[50]	2.785	[488]	3.671(-1)	10.89	[423]
		15710.8 ± 1.6	[145]					
	L_2M_5	15861.1 ± 1.5	[145]				10.87	[423]
L_{γ_5}	L_2N_1	17813.0 ± 3.1	[145]			2.731(-3)	21.48	[423]
	L_2N_2	17970.5 ± 4.1	[145]				21.07	[423]
	L_2N_3	18157.6 ± 2.9	[145]			1.454(-4)	17.49	[423]
L_{γ_1}	L_2N_4	18408.4 ± 3.6	[50]	0.613	[488]	8.077(-2)	14.08	[423]
		18405.5 ± 2.8	[145]					
	L_2N_5	18441.0 ± 2.8	[145]				13.98	[423]
L_ν	L_2N_6	18765.1 ± 2.2	[145]			4.545(-4)	7.84	[423]
	L_2N_7	18772.7 ± 1.9	[145]				7.86	[423]
L_l	L_3M_1	10870.1 ± 2.1	[145]			1.963(-2)	26.95	[423]
L_t	L_3M_2	11214.6 ± 2.1	[145]			2.386(-4)	21.81	[423]
L_s	L_3M_3	11955.4 ± 2.2	[145]			2.163(-4)	19.11	[423]
L_{α_2}	L_3M_4	12500.99 ± 0.93	[50]	0.2365	[488]	3.307(-2)	10.48	[423]
		12500.2 ± 1.3	[145]					
L_{α_1}	L_3M_5	$12652.16 \pm 0.96 \pm 0.65$	[50]	2.079	[488]	2.907(-1)	10.46	[423]
		12650.5 ± 1.3	[145]					
L_{β_6}	L_3N_1	14602.4 ± 2.9	[145]			4.837(-3)	21.07	[423]
	L_3N_2	14759.9 ± 3.9	[145]			5.354(-5)	20.66	[423]
	L_3N_3	14947.0 ± 2.6	[145]			5.783(-5)	17.08	[423]
$L_{\beta_{15}}$	L_3N_4	15194.9 ± 2.6	[145]			6.530(-3)		
L_{β_2}	L_3N_5	15230.4 ± 2.6	[145]			5.956(-2)	13.57	[423]
$L_{u'}$	L_3N_6	15554.5 ± 2.0	[145]			7.343(-5)	7.43	[423]
L_u	L_3N_7	15562.1 ± 1.6	[145]			4.066(-4)	7.45	[423]

line	transition	E/eV		I/eV/$\hbar$	TPIV	Γ/eV	
M series							
M_{γ_2}	M_3N_4	3228.39	[423]		8.935(-4)	18.23	[423]
M_{γ_1}	M_3N_5	3279.00	[263]		6.757(-3)	18.13	[423]
M_{ξ_2}	M_4N_2	2274.80	[423]		2.187(-3)	18.22	[423]
M_δ	M_4N_3	2468.16	[423]		2.049(-4)	13.01	[423]
M_β	M_4N_6	3061.00	[263]		4.706(-2)	3.35	[423]
M_η	M_4O_2	3160.81	[423]		5.307(-4)	11.92	[423]
M_{ξ_1}	M_5N_3	2299.00	[263]		1.540(-3)	12.99	[423]
$M_{\alpha_{1,2}}$	$M_5N_{6,7}$	2910.00	[263]				
	M_5N_6				2.270(-3)	3.33	[423]
	M_5N_7				4.442(-2)	3.34	[423]

level characteristics

level	E_B/eV		ω_{nlj}		Γ/eV		AE/eV	
K	106755.00	[503]	0.971	[301]	84.10	[301]	106755.3	[51]
	106755.30 ± 5.50	[51]	0.9686	[39]	82.70	[423]	106768.2 ± 3.2	[145]
	106756.30 ± 1.40	[493]	0.9654	[423]	84.4	[94]		
L_1	19840.00	[503]	0.153	[301]	13.60	[301]	19840.0	[51]
	19840.00 ± 1.80	[51]	0.1324	[423]	15.54	[423]	19855.2 ± 1.7	[145]
	19845.90 ± 1.30	[493]	0.133	[436]	14	[94]		
L_2	19083.00	[503]	0.468	[301]	7.95	[301]	19083.2	[51]
	19083.20 ± 2.80	[51]	0.4773	[423]	7.69	[423]	19090.04 ± 0.69	[145]
	19083.00 ± 1.10	[493]	0.49	[436]	8	[94]		
L_3	15871.00	[503]	0.450	[301]	6.98	[301]	15871.0	[51]
	15871.00 ± 2.00	[51]	0.4284	[423]	7.28	[423]	15879.45 ± 0.46	[145]
	15871.20 ± 1.20	[493]	0.414	[436]	7.51	[94]		
5 M_1	5002.00	[503]	4.73(-3)	[512]	18.87	[423]	5002.0	[51]
	5000.60 ± 2.20	[493]	5.7(-3)	[423]	15.4	[94]		
M_2	4656.00	[503]	3.1(-2)	[512]	14.53	[423]	4656.0	[51]
	4656.00 ± 1.80	[51]	8.53(-3)	[423]	12.9	[94]		
	4656.80 ± 1.50	[493]						
M_3	3909.00	[503]	7.66(-3)	[512]	11.84	[423]	3909.0	[51]
	3909.00 ± 1.80	[51]	1.19(-2)	[423]	8	[94]		
	3916.70 ± 1.60	[493]						
M_4	3370.00	[503]	5.8(-2)	[512]	3.20	[423]	3370.2	[51]
	3370.20 ± 2.10	[51]	5.01(-2)	[423]	3.18	[94]		
	3370.10 ± 1.00	[493]						

level	E_B/eV		ω_{nlj}		Γ/eV		AE/eV	
M_5	3219.00	[503]	4.3(-2)	[512]	3.18	[423]	3219.0	[51]
	3219.20 ± 2.10	[51]	4.86(-2)	[423]	3.18	[94]		
	3219.70 ± 1.40	[493]						
N_1	1269.00	[503]	8.7(-4)	[423]	13.79	[423]	1269.0	[51]
	1269.40 ± 1.20	[493]			11.2	[94]		
N_2	1080.00	[503]	1.37(-3)	[423]	15.02	[423]	1080.0	[51]
	1080.00 ± 1.90	[51]			8.5	[94]		
N_3	890.00	[503]	6.8(-4)	[423]	9.81	[423]		
	890.00 ± 1.90	[51]			7.25	[94]		
N_4	675.00	[503]	5.7(-4)	[423]	6.39	[423]		
	674.90 ± 3.70	[51]			4.2	[94]		
	673.90 ± 1.50	[493]						
N_5	639.00	[503]	5.0(-4)	[423]	6.29	[423]		
	641.10 ± 1.20	[493]			4.05	[94]		
N_6	319.00	[503]	4.0(-5)	[423]	0.15	[423]		
	316.40 ± 1.50	[493]			0.15	[94]		
N_7	319.00	[503]	4.0(-5)	[423]	0.17	[423]		
	308.40 ± 1.30	[493]			0.17	[94]		
O_1	272.00	[503]	2.0(-5)	[423]	21.18	[423]		
	273.50 ± 2.50	[493]						
O_2	215.00	[503]	3.0(-5)	[739	8.72	[423]		
	216.90 ± 2.30	[493]						
O_3	167.00	[503]	1.0(-5)	[423]	12.63	[423]		
	167.80 ± 2.50	[493]						
O_4	80.00	[503]						
	83.30 ± 0.70	[493]						
O_5	80.00	[503]						
	77.70 ± 1.00	[493]						
P_1	39.80 ± 3.50	[493]						
P_2	24.10 ± 3.40	[493]						
P_3	17.00 ± 4.50	[493]						
Q_1	5.17	[126]						
IP	5.17	[222]						

Th   Z=90

[Rn] 6d^2 7s^2

Thorium

$A = 232.0381(1)$ [222] $\varrho = 11.00 \text{ g/cm}^3$ [547]

$\varrho = 11.70 \text{ g/cm}^3$ [483]

X-ray transitions

line	transition	E/eV		I/eV/ℏ		TPIV	Γ/eV	
K series								
K_{α_3}	KL$_1$	89176.6 ± 3.2	[145]				102.19	[423]
K_{α_2}	KL$_2$	89953.0 ± 1.0	[50]	24.98	[488]	2.877(-1)	96.20	[301]
		89954.94	[104]	25.15	[487]		97.00	[473]
		89957.47 ± 0.50	[143]				94.42	[423]
		89957.62 ± 0.20	[277]					
		89957.00 ± 2.00	[43]					
		89957.04 ± 0.20	[144]					
K_{α_1}	KL$_3$	93350.00 ± 1.40	[50]	40.38	[488]	4.650(-1)	95.20	[301]
		93350.94	[104]	40.68	[487]		94.70	[473]
		93348.26 ± 0.42	[143]				94.01	[423]
		93347.98 ± 0.25	[277]					
		93348.00 ± 2.00	[43]					
		93347.38 ± 0.25	[144]					
	KM$_1$	104465.6 ± 4.9	[145]				105.31	[423]
K_{β_3}	KM$_2$	104831.00 ± 3.00	[50]	4.628	[488]	5.330(-2)	105.00	[473]
		104829.94	[104]	4.79	[487]		101.19	[423]
		104817.21 ± 0.69	[277]					
		104822.00 ± 2.00	[43]					
		104816.53 ± 0.69	[144]					
K_{β_1}	KM$_3$	105609.00 ± 1.00	[50]	9.03	[488]	1.040(-1)	99.70	[473]
		105609.94	[104]	9.34	[487]		98.49	[423]
		105602.19 ± 0.53	[277]					
		105606.00 ± 2.00	[43]					
		105601.51 ± 0.53	[144]					
K_{β_5}	KM$_{4,5}$	106270 ± 12	[50]	0.1545	[488]			
$K_{\beta_5^{II}}$	KM$_4$	106161.00 ± 10.00	[43]			1.779(-3)	89.81	[423]
		106275.00	[104]					
		106156.4 ± 4.2	[145]					
$K_{\beta_5^{I}}$	KM$_5$	106318.00 ± 9.00	[43]			2.055(-3)	89.79	[423]
		106314.9 ± 4.1	[145]					
	KN$_1$	108320.0 ± 5.7	[145]				100.30	[423]
$K_{\beta_2^{II}}$	KN$_2$	108509 ± 14	[50]	1.148	[488]	1.322(-2)	101.71	[423]

line	transition	E/eV		I/eV/$\hbar$		TPIV	Γ/eV	
		108483.00	[104]	1.221	[487]			
		108471.00 ± 4.00	[43]					
		108481.2 ± 6.7	[145]					
$K_{\beta_2^I}$	KN_3	108718 ± 13	[50]	2.307	[488]	2.657(-2)	95.69	[423]
		108712.00	[104]	2.469	[487]			
		108692.00 ± 3.00	[43]					
		108680.9 ± 5.5	[145]					
$K_{\beta4}$	$KN_{4,5}$	109082 ± 28	[50]	9.63(-2)	[488]			
$K_{\beta_4^{II}}$	KN_4	108934.2 ± 5.4	[145]			5.142(-4)	92.71	[423]
$K_{\beta_4^I}$	KN_5	108972.2 ± 5.4	[145]			5.953(-4)	92.89	[423]
	$KO_{2,3}$	109500.00 ± 10.00	[50]	0.799	[488]			
		109510.00	[104]	0.852	[487]			
	KO_2					3.196(-3)	95.07	[423]
	KO_3					6.260(-3)	90.70	[423]
	$KP_{2,3}$			0.1652	[488]			
	KP_2					5.588(-4)		
	KP_3					1.029(-3)		

L series

line	transition	E/eV		I/eV/$\hbar$		TPIV	Γ/eV	
	L_1M_1	15275.70	[104]	3.72(-5)	[488]		34.49	[423]
		15289.0 ± 3.3	[145]					
$L_{\beta4}$	L_1M_2	15643.90 ± 0.80	[50]	0.765	[488]	4.783(-2)	26.35	[473]
		15642.00	[104]				30.37	[423]
		15639.54 ± 0.35	[439]					
$L_{\beta3}$	L_1M_3	16425.80 ± 0.70	[50]	0.692	[488]	4.383(-2)	22.85	[473]
		16424.98	[104]				27.66	[423]
		16423.855 ± 0.070	[439]					
$L_{\beta10}$	L_1M_4	16981.00 ± 2.30	[50]	3.33(-2)	[488]	2.109(-3)	18.98	[423]
		16976.00	[104]					
		16980.26 ± 0.21	[439]					
$L_{\beta9}$	L_1M_5	17139.00 ± 2.00	[50]	4.97(-2)	[488]	3.147(-3)	18.96	[423]
		17133.00	[104]					
		17138.89 ± 0.12	[439]					
	L_1N_1	19146.4 ± 2.2	[50]	1.71(-5)	[488]		29.49	[423]
$L_{\gamma2}$	L_1N_2	19305.00 ± 1.20	[50]	0.2026	[488]	1.282(-2)	30.88	[423]
		19304.00	[104]					
		19331.00	[543]					
		19302.987 ± 0.050	[439]					
$L_{\gamma3}$	L_1N_3	19507.00 ± 1.20	[50]	0.2101	[488]	1.330(-2)	24.86	[423]
		19505.00	[104]					
		19535.00	[543]					
		19503.445 ± 0.060	[439]					

line	transition	E/eV		I/eV/$\hbar$		TPIV	Γ/eV	
	L_1N_4	19755.00 ± 3.10	[50]	5.56(-3)	[488]	3.521(-4)	21.88	[423]
		19750.00	[104]					
		19756.75 ± 0.83	[439]					
$L_{\gamma 11}$	L_1N_5	19794.00 ± 2.80	[50]	8.90(-3)	[488]	5.633(-4)	22.07	[423]
		19791.00	[104]					
		19793.91 ± 0.50	[439]					
	$L_1N_{6,7}$	20127.0 ± 4.8	[50]	2.29(-4)	[488]			
		20129.99	[104]					
	L_1N_6	20138.2 ± 3.3	[145]				15.83	[423]
	L_1N_7	20146.2 ± 2.9	[145]				15.86	[423]
	L_1O_1	20174.00 ± 3.30	[50]	4.86(-6)	[488]		34.44	[423]
		20170.00	[104]					
$L_{\gamma 4'}$	L_1O_2	20242.00 ± 1.30	[50]	4.78(-2)	[488]	3.023(-3)	24.24	[423]
		20240.00	[104]					
		20273.00	[543]					
$L_{\gamma 4}$	L_1O_3	20292.00 ± 1.30	[50]	5.02(-2)	[488]	3.176(-3)	19.88	[423]
		20290.98	[104]					
		20319.00	[543]					
	$L_1O_{4,5}$	20383.00 ± 3.40	[50]	2.53(-3)	[488]	0.37		
		20379.99	[104]					
	L_1O_4					6.125(-5)		
	L_1O_5					9.907(-5)		
	$L_1P_{2,3}$	20424.00 ± 2.70	[50]	1.74(-2)	[488]	2.51		
		20422.00	[104]					
	L_1P_2					5.516(-4)		
	L_1P_3					5.500(-4)		
L_η	L_2M_1	14509.90 ± 0.70	[50]	8.21(-2)	[488]	1.048(-2)	26.73	[423]
		14509.00	[104]					
		14486.00	[543]					
		14510.327 ± 0.080	[439]					
	L_2M_2	14869.6 ± 2.6	[50]	1.03(-5)	[488]		22.60	[423]
		14867.00	[104]					
$L_{\beta 17}$	L_2M_3	15643.1 ± 1.2	[50]	4.05(-3)	[488]	5.177(-4)	19.90	[423]
$L_{\beta 1}$	L_2M_4	16202.20 ± 0.20	[50]	2.939	[488]	3.756(-1)	12.40	[473]
		16201.99	[104]				11.22	[423]
		16194.00	[543]					
		16201.556 ± 0.030	[439]					
	L_2M_5	16359.00 ± 2.20	[50]	3.22(-4)	[488]		11.20	[423]
		16359.98	[104]					
		16358.1 ± 1.0	[439]					
$L_{\gamma 5}$	L_2N_1	18370.00 ± 1.10	[50]	2.21(-2)	[488]	2.819(-3)	21.72	[423]
		18368.98	[104]					

line	transition	E/eV		I/eV/$\hbar$		TPIV	Γ/eV	
		18364.35 ± 0.11	[439]					
	L_2N_2	18524.5 ± 4.1	[145]				23.11	[423]
	L_2N_3	18728.4 ± 4.2	[50]	4.24(-6)	[488]	1.539(-4)	17.10	[423]
		18729.99	[104]					
$L_{\gamma1}$	L_2N_4	18982.50 ± 0.90	[50]	0.659	[488]	8.348(-2)	15.00	[473]
		18980.98	[104]				14.12	[423]
		18966.00	[543]					
		18978.259 ± 0.020	[439]					
	L_2N_5	19012.8 ± 4.3	[50]	8.03(-5)	[488]		14.30	[423]
		19010.0	[104]					
	$L_2N_{6,7}$	19014.0 ± 3.00	[50]	3.81(-3)	[488]			
L_ν	L_2N_6	19348.00 ± 0.15	[439]			4.864(-4)	8.07	[423]
		19358.2 ± 2.3	[145]					
	L_2N_7	19366.2 ± 1.9					8.10	[423]
$L_{\gamma8}$	L_2O_1	19403.00 ± 1.50	[50]	5.33(-3)	[488]	7.070(-4)	26.68	[423]
		19400.98	[104]					
	L_2O_2	19466.00 ± 3.10	[50]	1.10(-6)	[488]		16.48	[423]
		19469.98	[104]					
	L_2O_3	19506.00 ± 3.10	[50]	2.86(-4)	[488]	3.661(-5)	12.11	[423]
$L_{\gamma6}$	L_2O_4	19599.00 ± 1.20	[50]	0.1218	[488]	1.557(-2)		
		19596.98	[104]					
	L_2P_1	19629.00 ± 3.10	[50]	1.16(-3)	[488]	1.482(-4)		
		19630.00	[104]					
	$L_2P_{2,3}$	19642.00 ± 3.10	[50]	4.98(-5)	[488]			
		19640.00	[104]					
	L_2P_3					5.783(-6)		
	L_2P_4	19682.00 ± 2.80	[50]	2.25(-3)	[488]			
L_l	L_3M_1	11118.60 ± 0.40	[50]	0.1460	[488]	2.025(-2)	26.32	[423]
		11118.00	[104]					
		11096.00	[543]					
		11118.06 ± 0.18	[439]					
L_t	L_3M_2	11478.80 ± 1.00	[50]	1.84(-3)	[488]	2.494(-4)	22.19	[423]
		11479.00	[104]					
		11469.5 ± 1.1	[398]					
L_s	L_3M_3	12261.00 ± 1.20	[50]	1.67(-3)	[488]	2.263(-4)	19.49	[423]
		12259.00	[104]					
		12255.98 ± 0.96	[398]					
$L_{\alpha2}$	L_3M_4	12809.60 ± 0.30	[50]	0.2487	[488]	3.373(-2)	11.80	[473]
		12809.00	[104]				10.81	[423]
		12804.00	[543]					
		12809.498 ± 0.030	[439]					

line	transition	E/eV		I/eV/ℏ		TPIV	Γ/eV	
$L_{\alpha1}$	L_3M_5	12968.70 ± 0.40	[50]	2.186	[488]	2.965(-1)	11.90	[473]
		12968.00	[104]				10.79	[423]
		12961.00	[543]					
		12967.937 ± 0.020	[439]					
$L_{\beta6}$	L_3N_1	14975.00 ± 1.40	[50]	3.71(-2)	[488]	5.031(-3)	21.31	[423]
		14976.00	[104]					
		14973.424 ± 0.050	[439]					
	L_3N_2	15138.00 ± 3.70	[50]	4.12(-4)	[488]	5.583(-5)	22.71	[423]
		15140.00	[104]					
		15132.2 ± 1.3	[398]					
	L_3N_3	15341.00 ± 1.90	[50]	4.49(-4)	[488]	6.086(-5)	18.36	[423]
		15339.00	[104]					
		15334.0 ± 1.2	[398]					
$L_{\beta15}$	L_3N_4	15587.50 ± 1.00	[50]	4.94(-2)	[488]	6.708(-3)	13.71	[423]
		15587.00	[104]					
		15586.910 ± 0.070	[439]					
$L_{\beta2}$	L_3N_5	15623.70 ± 0.60	[50]	0.451	[488]	6.125(-2)	12.80	[473]
		15624.00	[104]				13.89	[423]
		15623.960 ± 0.030	[439]					
L_u	$L_3N_{6,7}$	15964.00 ± 1.00	[50]	5.81(-4)	[488]			
		15964.00	[104]					
		15955.85 ± 0.58	[439]					
$L_{u'}$	L_3N_6	15966.9 ± 2.1	[145]			7.886(-5)	7.66	[423]
L_u	L_3N_7	15974.8 ± 1.7	[145]			4.362(-4)	7.69	[423]
$L_{\beta7}$	L_3O_1	16010.50 ± 0.80	[50]	9.20(-3)	[488]	1.248(-3)	26.26	[423]
		16009.99	[104]					
	L_3O_2	16074.00 ± 2.10	[50]	9.32(-5)	[488]	1.264(-5)	16.07	[423]
		16073.99	[104]					
	L_3O_3	16123.00 ± 2.10	[50]	1.24(-4)	[488]	1.417(-5)	11.70	[423]
		16120.00	[104]					
$L_{\beta5}$	$L_3O_{4,5}$	16213.00 ± 1.10	[50]	9.15(-2)	[488]			
		16212.00	[104]					
	L_3O_4					1.224(-3)		
	L_3O_5					1.119(-2)		
	L_3P_1	16241.00 ± 1.10	[50]	1.92(-3)	[488]	2.539(-4)		
		16239.99	[104]					
	$L_3P_{2,3}$	16260.00 ± 4.30	[50]	3.48(-5)	[488]			
		16260.00	[104]					
	L_3P_2					2.249(-6)		
	L_3P_3					2.375(-6)		
	$L_3P_{4,5}$	16295.00 ± 1.90	[50]	1.60(-3)	[488]			

line	transition	E/eV		I/eV/$\hbar$		TPIV	Γ/eV	
	M series							
	M_1N_2	3957.00	[104]	7.43(-2)	[335]	2.400(-3)	34.01	[423]
	M_1N_3	4230.00 ± 12.00	[50]	7.43(-2)	[335]	1.391(-3)	27.99	[423]
		4211.00	[104]					
	M_1O_3	5080.00 ± 20.00	[50]			4.078(-4)	23.01	[423]
		5077.00	[104]					
	M_2N_1	3505.00 ± 9.00	[50]	7.89(-3)	[335]	9.542(-4)	28.49	[423]
		3505.00	[104]					
	M_2N_4	4117.00 ± 2.70	[50]	8.37(-2)	[335]	4.829(-3)	20.89	[423]
		4116.00	[104]					
	M_2O_4	4735.00 ± 9.00	[50]			1.097(-3)		
		4735.00	[104]					
	M_3N_1	2714.00 ± 3.00	[50]	7.89(-3)	[335]	1.955(-3)	25.79	[423]
		2717.00	[104]					
		2720.00	[263]					
$M_{\gamma 2}$	M_3N_4	3335.00 ± 2.70	[50]	8.37(-2)	[335]	9.272(-4)	18.18	[423]
		3335.00	[104]					
		3342.00	[263]					
$M_{\gamma 1}$	M_3N_5	3370.00 ± 1.80	[50]	8.37(-2)	[335]	6.988(-3)	18.37	[423]
		3369.00	[104]					
		3377.00	[104]					
	M_3O_1	3780.00 ± 10.00	[50]			4.745(-4)	30.74	[423]
		3777.70	[104]					
		3785.00	[263]					
	$M_3O_{4,5}$	3959.00 ± 4.00	[50]					
		3967.00	[263]					
	M_3O_4					1.890(-4)		
	M_3O_5	3959.00 ± 4.00	[50]			1.456(-3)		
		3960.00	[104]					
$M_{\xi 2}$	M_4N_2	2322.00 ± 2.20	[50]	6.35(-4)	[335]	2.229(-3)	18.50	[423]
		2322.00	[104]					
		2327.00	[263]					
M_δ	M_4N_3	2524.00 ± 2.60	[50]	6.35(-4)	[335]	2.062(-4)	12.48	[423]
		2524.00	[104]					
		2530.00	[263]					
M_β	M_4N_6	3145.80 ± 0.80	[50]	0.1479	[335]	4.904(-2)	3.45	[423]
		[3145.00	[104]					
		3152.00	[263]					

line	transition	E/eV		I/eV/$\hbar$		TPIV	Γ/eV	
M_η	M_4O_2	3256.00 ± 3.40	[50]			5.391(-4)	11.87	[423]
		3253.00	[104]					
		3263.00	[263]					
	M_5N_2	2322.0 ± 2.00	[50]	6.35(-4)	[335]		18.48	[423]
M_{ξ_1}	M_5N_3	2364.00 ± 2.30	[50]	6.35(-4)	[335]	1.555(-3)	12.46	[423]
		2366.00	[104]					
		2369.00	[263]					
$M_{\alpha2}$	M_5N_6	2987.00 ± 1.40	[50]	0.1479	[335]	2.355(-3)	3.43	[423]
		2986.00	[104]					
		2993.00	[263]					
		2996.00	[104]					
		3003.00	[263]					
	M_5P_3	3298.00 ± 8.00	[50]			5.433(-5)		

N series

	transition	E/eV				TPIV	Γ/eV	
	N_1P_2	1313.00 ± 10.00	[50]			6.129(-5)		
	N_1P_3	1131.90 ± 7.20	[50]			1.845(-5)		
	N_2O_4	1072.00 ± 4.60	[50]			6.896(-5)		
	N_2P_1	1120.00 ± 7.10	[50]			4.742(-5)		
	N_3O_5	897.00 ± 6.50	[50]			1.591(-4)		
	N_4N_6	369.30 ± 1.00	[50]			4.377(-4)	6.35	[423]
	$N_5N_{6,7}$	341.40 ± 0.80	[50]					
	N_5N_6					1.707(-5)	6.54	[423]
	N_5N_7					3.826(-4)	6.57	[423]
	N_6O_4	250.50 ± 0.50	[50]			5.053(-5)		
	N_6O_5	257.20 ± 0.50	[50]			3.269(-6)		
	N_7O_5	247.90 ± 0.50	[50]			4.934(-5)		

O series

	transition	E/eV						
	$O_3P_{4,5}$	181.70 ± 0.80	[50]					
	$O_{4,5}Q_{2,3}$	68.00 ± 2.00	[50]					

level characteristics

level	E_B/eV		ω_{nlj}		Γ/eV		AE/eV	
K	109651.00	[503]	0.971	[301]	88.00	[301]	109650.9	[51]
	109650.90 ± 0.90	[51]	0.9691	[39]	86.51	[423]	109649.0 ± 1.0	[50]
	109649.10 ± 1.50	[493]	0.969	[112]	88.2	[94]	109658.2 ± 3.3	[145]
			0.974	[33]				
			0.9660	[423]				
			0.971	[310]				
L_1	20472.00	[503]	0.161	[301]	13.70	[301]	20472.1	[51]
	20472.10 ± 0.50	[493]	0.1382	[423]	15.68	[423]	20462.5 ± 5.0	[50]
			0.139	[436]	14.3	[94]	20481.6 ± 1.7	[145]
			0.153	[51]				
L_2	19693.00	[503]	0.479	[301]	8.18	[301]	19693.2	[51]
	19693.20 ± 0.40	[493]	0.4900	[423]	7.92	[423]	19682.9 ± 4.6	[50]
			0.503	[436]	8.5	[94]	19701.59 ± 0.72	[145]
			0.498	[244]				
			0.50	[51]				
L_3	16300.00	[503]	0.463	[301]	7.13	[301]	16300.3	[51]
	16300.30 ± 0.30	[493]	0.4386	[423]	7.51	[423]	16298.5 ± 3.2	[50]
			0.424	[436]	7.74	[94]	16310.27 ± 0.47	[145]
			0.423	[244]				
			0.43	[304]				
M_1	5182.00	[503]	5.99(-3)	[423]	18.81	[423]	5182.3	[51]
	5182.30 ± 0.30	[493]	4.97(-3)	[245]	22.70	[487]		
					15.5	[94]		
M_2	4831.00	[503]	4.53(-3)	[38]	14.68	[423]	4830.4	[51]
	4830.40 ± 0.40	[493]	8.83(-3)	[423]	15.50	[487]		
					13.2	[94]		
M_3	4046.00	[503]	1.40(-2)	[38]	11.98	[423]	4046.1	[51]
	4046.10 ± 0.40	[493]	1.24(-2)	[423]	12.90	[487]		
			1.16(-2)	[245]	8	[94]		
M_4	3491.00	[503]	5.22(-2)	[423]	3.30	[423]	3490.8	[51]
	3490.80 ± 0.30	[493]	5.82(-2)	[38]	3.22	[487]		
					3.28	[94]		
M_5	3332.00	[503]	4.97(-2)	[38]	3.28	[423]	3332.0	[51]
	3332.00 ± 0.30	[493]	0.044	[512]	2.92	[487]		
			5.04(-2)	[423]	3.28	[94]		
N_1	1330.00	[503]	9.1(-4)	[423]	13.81	[423]	1329.5	[51]
	1329.50 ± 0.40	[493]			11.00	[74]		
					11.5	[94]		
N_2	1168.00	[503]	1.39(-3)	[423]	15.20	[423]	1168.2	[51]
	1168.20 ± 0.40	[493]			8.00	[74]		
					8.75	[94]		

level	E_B/eV		ω_{nlj}		Γ/eV	AE/eV
N_3	968.00	[503]	7.1(-4)	[423]	9.18	[423]
	967.30 ± 0.40	[51]			8.00	[74]
	967.20 ± 0.50	[493]			7.5	
N_4	714.00	[503]	6.1(-4)	[423]	6.20	[423]
	714.10 ± 0.40	[51]			5.30	[74]
	713.70 ± 0.70	[493]			4.3	[94]
N_5	677.00	[503]	5.3(-4)	[423]	6.39	[423]
	676.40 ± 0.40	[51]			5.30	[74]
	676.60 ± 0.90	[493]			4.1	[94]
N_6	344.00	[503]	5.0(-5)	[423]	0.15	[423]
	344.40 ± 0.40	[493]			0.14	[74]
					0.15	[94]
N_7	335.00	[503]	5.0(-5)	[423]	0.18	[423]
	335.00 ± 0.70	[493]			0.14	[74]
					0.18	[94]
O_1	290.00	[503]	2.0(-5)	[423]	18.76	[423]
	290.20 ± 0.80	[493]				
O_2	229.00	[503]	3.0(-5)	[423]	8.56	[423]
	229.40 ± 1.10	[51]				
	232.00 ± 2.70	[493]				
O_3	182.00	[503]	1.0(-5)	[423]	4.20	[423]
	181.80 ± 0.40	[51]				
	180.80 ± 1.30	[493]				
O_4	95.00	[503]				
	94.30 ± 0.40	[51]				
	94.10 ± 0.30	[493]				
O_5	88.00	[503]				
	87.90 ± 0.30	[51]				
	87.30 ± 0.50	[493]				
P_1	60.00	[503]				
	59.50 ± 1.10	[51]				
	41.40 ± 0.50	[493]				
P_2	49.00	[503]				
	49.00 ± 2.50	[51]				
	25.80 ± 0.40	[493]				
P_3	43.00	[503]				
	43.00 ± 2.50	[51]				
	17.30 ± 0.40	[493]				
P_4	2.00	[503]				
Q_1	6.08	[126]				
IP	6.3067	[222]				

Pa   Z=91

[Rn] $5f^2 6d^1 7s^2$

Protactinium $A = 231.03588(2)$ [222] $\varrho = 15.40$ g/cm³ [547]

$\varrho = 15.37$ g/cm³ [483]

X-ray transitions

line	transition	E/eV		I/eV/ℏ		TPIV	Γ/eV	
K series								
K_{α_3}	KL_1	91488.0 ± 3.3	[145]				106.24	[423]
$K_{\alpha 2}$	KL_2	92287.00 ± 6.00	[50]	26.18	[488]	2.884(-1)	100.70	[301]
		92286.00 ± 3.00	[277]				99.67	[423]
		92283.4 ± 2.0	[43]					
$K_{\alpha 1}$	KL_3	95868.00 ± 2.20	[50]	42.10	[488]	4.639(-1)	99.30	[301]
		95863.00 ± 3.00	[277]				98.15	[423]
		95866.4 ± 2.0	[43]					
		95865.00 ± 3.00	[298]					
	KM_1	107232.3 ± 5.0	[145]				109.16	[423]
$K_{\beta 3}$	KM_2	107600.00 ± 20.00	[50]	4.841	[488]	5.334(-2)	105.28	[423]
		107585.3 ± 2.0	[43]					
$K_{\beta 1}$	KM_3	108427.00 ± 8.00	[50]	9.45	[488]	1.042(-1)	101.99	[423]
		108417.3 ± 2.0	[43]					
$K_{\beta_5^{II}}$	KM_4	108986.2 ± 4.3	[145]			1.823(-3)	93.84	[423]
$K_{\beta_5^{I}}$	KM_5	109153.5 ± 4.3	[145]			2.093(-3)	93.81	[423]
	KN_1	111223.1 ± 5.8	[145]				104.31	[423]
$K_{\beta_2^{II}}$	KN_2	111405 ± 30	[50]	1.206	[488]	1.329(-2)	105.67	[423]
		111387.6 ± 6.9	[145]					
$K_{\beta_2^{I}}$	KN_3	111625 ± 30	[50]	2.431	[488]	2.678(-2)	99.73	[423]
		111600.5 ± 5.6	[145]					
$K_{\beta_4^{II}}$	KN_4	111859.4 ± 5.5	[145]			5.318(-4)	96.63	[423]
$K_{\beta_4^{I}}$	KN_5	111899.8 ± 5.6	[145]			6.121(-4)	96.81	[423]
L series								
	$L_1 M_1$	15744.4 ± 3.3	[145]				34.56	[423]
$L_{\beta 4}$	$L_1 M_2$	16104.1	[104]	0.804	[488]	5.040(-2)	30.69	[423]
		16103.7 ± 3.1	[50]					
$L_{\beta 3}$	$L_1 M_3$	16930.5 ± 1.7	[50]	0.718	[488]	4.499(-2)	27.39	[423]

line	transition	E/eV		I/eV/ℏ		TPIV	Γ/eV	
		16929.99	[104]					
$L_{\beta 10}$	L_1M_4	17491.9 ± 7.3	[50]	3.63(-2)	[488]	2.276(-3)	19.25	[423]
		17489.99	[104]					
$L_{\beta 9}$	L_1M_5	17666.3 ± 3.7	[50]	5.41(-2)	[488]	3.394(-3)	19.21	[423]
		17665.00	[104]					
	L_1N_1	19735.1 ± 4.1	[145]				29.71	[423]
$L_{\gamma 2}$	L_1N_2	19872.1 ± 4.7	[50]	0.2164	[488]	1.357(-2)	31.08	[423]
		19870.00	[104]					
$L_{\gamma 3}$	L_1N_3	20097.6 ± 4.8	[50]	0.2199	[488]	1.379(-2)	25.14	[423]
		20100.00	[104]					
	L_1N_4	20371.5 ± 3.8	[145]			3.836(-4)	22.03	[423]
	L_1N_5	20411.9 ± 3.8	[145]			6.146(-4)	22.21	[423]
	L_1N_6	20751.5 ± 3.4	[145]				16.11	[423]
	L_1N_7	20760.4 ± 2.9	[145]				16.07	[423]
$L_{\gamma 4}$	$L_1O_{2,3}$	20882.00 ± 3.50	[50]	5.32(-2)	[488]	7.41		
		20879.99	[104]					
	L_1O_2					3.241(-3)	25.06	[423]
	L_1O_3					3.336(-3)	22.50	[423]
L_{η}	L_2M_1	14946.6 ± 2.7	[50]	8.69(-2)	[488]	9.843(-3)	28.00	[423]
		14946.00	[104]					
	L_2M_2	15307.6 ± 2.4	[145]				24.12	[423]
$L_{\beta 17}$	L_2M_3	16138.8 ± 2.5	[145]			4.962(-4)	20.83	
$L_{\beta 1}$	L_2M_4	16702.0 ± 1.7	[50]	3.10	[488]	3.511(-1)	12.68	[423]
		16700.00	[104]					
	L_2M_5	16869.9 ± 1.5	[145]				12.65	[423]
$L_{\gamma 5}$	L_2N_1	18928.6 ± 4.3	[50]	2.35(-2)	[488]	2.658(-3)	23.16	[423]
		18929.99	[104]					
	L_2N_2	19104.1 ± 4.2	[145]				24.52	[423]
	L_2N_3	19317.0 ± 2.9	[145]			1.487(-4)	18.57	[423]
$L_{\gamma 1}$	L_2N_4	19568.5 ± 4.1	[50]	0.696	[488]	7.882(-2)	15.47	[423]
		19561.00	[104]					
	L_2N_5	19616.3 ± 2.9	[145]				15.65	[423]
L_{ν}	L_2N_6	19955.9 ± 2.4	[145]			4.745(-4)	9.55	[423]
	L_2N_7	19964.8 ± 2.0	[145]				9.51	[423]
$L_{\gamma 6}$	L_2O_4	20216.00 ± 3.30	[50]	0.1318	[488]	1.492(-2)		
		20219.98	[104]					
L_l	L_3M_1	11366.2 ± 1.5	[50]	0.1560	[488]	2.087(-2)	26.48	[423]

line	transition	E/eV		I/eV/$\hbar$		TPIV	Γ/eV	
		11367.00	[104]					
L_t	L_3M_2	11726.4 ± 2.1	[145]			2.606(-4)	22.29	[423]
L_s	L_3M_3	12557.6 ± 2.3	[145]			2.367(-4)	19.00	[423]
$L_{\alpha2}$	L_3M_4	13122.3 ± 1.0	[50]	0.2613	[488]	3.441(-2)	11.16	[423]
		13125.00	[104]					
$L_{\alpha1}$	L_3M_5	13290.8 ± 1.1	[50]	2.296	[488]	3.024(-1)	11.12	[423]
		13291.00	[104]					
$L_{\beta6}$	L_3N_1	15346.2 ± 2.8	[50]	3.97(-2)	[488]	5.229(-3)	21.63	[423]
		15347.00	[104]					
	L_3N_2	15522.8 ± 3.9	[145]			5.815(-5)	22.69	[423]
	L_3N_3	15735.8 ± 2.6	[145]			6.400(-5)	16.74	[423]
$L_{\beta15}$	L_3N_4	15994.7 ± 2.5	[145]			6.883(-3)	13.64	[423]
$L_{\beta2}$	L_3N_5	16024.6 ± 3.1	[50]	0.478	[488]	6.294(-2)	14.13	[423]
		16024.99	[104]					
$L_{u'}$	L_3N_6	16374.7 ± 2.2	[145]			8.447(-5)	7.72	[423]
L_u	L_3N_7	16383.6 ± 1.7	[145]			4.665(-4)	7.68	[423]
$L_{\beta7}$	L_3O_1	16431.00 ± 4.40	[50]	9.96(-3)	[488]	1.312(-3)	29.52	[423]
		16429.99	[104]					
$L_{\beta5}$	$L_3O_{4,5}$	16636.00 ± 4.50	[50]	9.82(-2)	[488]			
		16636.0	[104]					
	L_3O_4					1.274(-3)		
	L_3O_5					1.166(-2)		

M series

line	transition	E/eV		I/eV/$\hbar$		TPIV	Γ/eV	
	M_2N_1	3603.00 ± 5.00	[50]	8.34(-3)	[335]	9.853(-4)	28.76	[423]
		3603.00	[104]					
		3610.00	[263]					
	M_2N_4	4260.00 ± 3.00	[50]	8.94(-2)	[335]	4.934(-3)	21.07	[423]
		4261.00	[104]					
		4269.00	[263]					
	M_2O_4	4906.00 ± 7.80	[50]			1.143(-3)		
		4906.00	[104]					
		4916.00	[263]					
	M_3N_1	2786.00 ± 2.50	[50]	8.34(-3)	[335]	2.044(-3)	25.47	[423]
		2786.00	[104]					
		2792.00	[263]					
$M_{\gamma2}$	M_3N_4	3430.00 ± 2.00	[50]	8.94(-2)	[335]	9.639(-4)	17.78	[423]
		3430.00	[104]					
		3437.00	[263]					
$M_{\gamma1}$	M_3N_5	3465.70 ± 1.00	[50]	8.94(-2)	[335]	7.251(-3)	17.96	[423]

line	transition	E/eV		I/eV/$\hbar$		TPIV	Γ/eV	
		3466.00	[104]					
		3473.00	[263]					
	M_3O_1	3820.00 ± 11.00	[50]			5.010(-4)	33.35	[423]
		3821.00	[104]					
		3829.00	[263]					
	$M_3O_{4,5}$	4081.00 ± 2.70	[50]					
		4081.00	[104]					
		4089.00	[263]					
	M_3O_4					1.995(-4)		
	M_3O_5					1.531(-3)		
$M_{\xi2}$	M_4N_2	2387.60 ± 0.90	[50]	6.51(-4)	[335]	2.286(-3)	18.68	[423]
		2388.00	[104]					
		2393.00	[263]					
M_δ	M_4N_3	2625.53	[423]			2.071(-4)	12.74	[423]
M_β	M_4N_6	3239.70 ± 0.80	[50]	0.1586	[335]	5.018(-2)	3.72	[423]
		3240.00	[104]					
		3247.00	[263]					
M_η	M_4O_2	3559.00 ± 1.80	[50]			5.468(-4)	12.67	[423]
		3359.00	[104]					
		3366.00	[263]					
$M_{\xi1}$	M_5N_3	2435.00 ± 1.00	[50]	6.51(-4)	[335]	1.569(-3)	12.71	[423]
		2435.00	[104]					
		2440.00	[263]					
$M_{\alpha2}$	M_5N_6	3072.00 ± 2.30	[50]	0.1586	[335]	2.411(-3)	3.68	[423]
		3072.00	[104]					
		3079.00	[263]					
$M_{\alpha1}$	M_5N_7	3082.30 ± 0.80	[50]	0.1586	[335]	4.723(-2)	3.64	[423]
		3082.00	[104]					
		3089.00	[263]					

level characteristics

level	E_B/eV		ω_{nlj}		Γ/eV		AE/eV	
K	112601.00	[503]	0.972	[301]	91.90	[301]	112601.4	[51]
	112601.40 ± 2.40	[51]	0.9696	[39]	90.42	[423]	112601.7 ± 3.5	[145]
	112596.10 ± 2.10	[493]	0.9665	[423]	92.1	[94]		
L_1	21105.00	[503]	0.162	[301]	14.30	[301]	21104.6	[51]
	21104.60 ± 1.80	[51]	0.147	[436]	15.82	[423]	21113.7 ± 1.7	[145]
	21111.40 ± 1.20	[493]	0.1449	[423]	14.7	[94]		
L_2	20314.00	[503]	0.472	[301]	8.75	[301]	20313.7	[51]
	20313.70 ± 1.50	[493]	0.495	[436]	9.26	[423]	20318.12 ± 0.73	[145]
			0.494	[244]	9.1	[94]		
			0.4594	[423]				
L_3	16733.00	[503]	0.476	[301]	7.33	[301]	16733.1	[51]
	16733.10 ± 1.40	[51]	0.46	[81]	7.73	[423]	16736.90 ± 0.48	[145]
	16729.10 ± 1.20	[493]	0.434	[436]	7.97	[94]		
			0.4485	[423]				
M_1	5357.00	[503]	6.3(-3)	[423]	18.74	[423]	5366.9	[51]
	5366.90 ± 1.60	[493]			15.5	[94]		
M_2	5001.00	[503]	9.13(-3)	[423]	14.86	[423]	5000.9	[51]
	5000.90 ± 2.30	[51]			13.6	[94]		
	5002.70 ± 1.70	[493]						
M_3	4174.00	[503]	1.29(-2)	[423]	11.57	[423]	4173.8	[51]
	4173.80 ± 1.80	[493]			7.9	[94]		
M_4	3611.00	[503]	5.39(-2)	[423]	3.43	[423]	3611.2	[51]
	3611.20 ± 1.40	[51]			3.39	[94]		
	3606.40 ± 1.10	[493]						
M_5	3442.00	[503]	5.22(-2)	[423]	3.39	[423]	3441.8	[51]
	3441.80 ± 1.40	[51]			3.39	[94]		
	3439.40 ± 1.00	[493]						
N_1	1387.00	[503]	9.4(-4)	[423]	13.90	[423]	1387.1	[51]
	1387.10 ± 1.90	[493]			11.6	[94]		
N_2	1224.00	[503]	1.4(-3)	[423]	15.26	[423]	1224.3	[51]

level	E_B/eV		ω_{nlj}		Γ/eV		AE/eV	
	1224.30 ± 1.60	[493]			9.2	[94]		
N_3	1007.00	[503]	7.3(-4)	[423]	9.32	[423]	1006.7	[51]
	1006.70 ± 1.70	[493]			7.75	[94]		
N_4	743.00	[503]	6.5(-4)	[423]	6.21	[423]		
	743.40 ± 2.10	[493]			4,4	[94]		
N_5	708.00	[503]	5.6(-4)	[423]	6.39	[423]		
	708.20 ± 1.80	[493]			4.2	[94]		
N_6	371.00	[503]	7.0(-5)	[423]	0.29	[423]		
	371.20 ± 1.60	[493]			0.29	[94]		
N_7	360.00	[503]	6.0(-5)	[423]	0.25	[423]		
	359.50 ± 1.60	[493]			0.25	[94]		
O_1	310.00	[503]	2.0(-5)	[423]	21.79	[423]		
	309.60 ± 4.30	[493]						
O_2	223.00	[503]			9.24	[423]		
	222.90 ± 3.90	[51]						
	244.60 ± 2.00	[493]						
O_3	226.00	[503]			6.68	[423]		
	223.90 ± 3.90	[51]						
	186.30 ± 2.50	[493]						
O_4	94.00	[503]						
	94.10 ± 2.80	[51]						
	97.30 ± 0.60	[493]						
O_5	94.00	[503]						
	94.10 ± 2.80	[51]						
	89.20 ± 1.10	[493]						
P_1	46.70 ± 3.20	[493]						
P_2	28.10 ± 3.40	[493]						
P_3	18.90 ± 3.50	[493]						
Q_1	5.89	[126]						
IP	5.89	[222]						

U   Z=92

[Rn] 5f³ 6d¹ 7s²

Uranium $A = 238.0289(1)$ [260] $\varrho = 19.00$ g/cm³ [547]
$\varrho = 17.3$ g/cm³ [222]

X-ray transitions

line	transition	E/eV		I/eV/$\hbar$		TPIV	Γ/eV	
K series								
K_{α_3}	KL$_1$	93842.0 ± 3.4	[145]				110.35	[423]
K_{α_2}	KL$_2$	94665.00 ± 3.00	[50]	27.42	[488]	2.893(-1)	105.40	[301]
		94664.94	[43]	27.61	[104]		106.00	[473]
		94653.00 ± 0.53	[143]				104.30	[222]
		94651.45 ± 0.56	[277]				103.97	[423]
		94656.00 ± 2.00	[43]					
		94650.84 ± 0.56	[144]					
K_{α_1}	KL$_3$	98439.0 ± 2.30	[50]	43.87	[488]	4.628(-1)	103.50	[301]
		98435.00 ± 2.00	[43]	44.19	[487]		103.00	[473]
		98439.94	[104]				102.44	[423]
		98.434.14 ± 0.48	[143]					
		98432.21 ± 0.28	[277]					
		98431.58 ± 0.28	[144]					
	KM$_1$	110052.1 ± 5.1	[145]				113.39	[423]
K_{β_3}	KM$_2$	110406.0 ± 4.0	[50]	5.059	[488]	5.337(-2)	120.00	[473]
		110416.00 ± 3.00	[43]	5.23	[487]		107.30	[222]
		110409.94	[104]				109.49	[423]
		110416.38 ± 0.65	[277]					
		110415.67 ± 0.65	[144]					
K_{β_1}	KM$_3$	111300.0 ± 5.0	[50]	9.89	[488]	1.044(-1)	115.00	[473]
		111299.94	[104]	10.22	[487]	105.00	[222]	
		111300 ± 2.00	[43]				106.22	[423]
		111295.80 ± 0.65	[277]					
		111295.08 ± 0.65	[144]					
$K_{\beta_5^{II}}$	KM$_4$	111868.00 ± 5.00	[43]			1.867(-3)	98.01	[423]
		111871.8 ± 4.5	[145]					
$K_{\beta_5^{I}}$	KM$_5$	112043.00 ± 5.00	[43]			2.130(-3)	97.97	[423]
		112048.1 ± 4.4	[145]					
K_{β_5}	KM$_{4,5}$	120009 ± 15	[50]	0.177	[488]	0.46		
	KN$_1$	114160.4 ± 6.0	[145]				108.54	[423]

line	transition	E/eV		I/eV/ℏ		TPIV	Γ/eV	
$K_{\beta_2^{II}}$	KN_2	114407 ± 16	[50]	1.265	[488]	1.335(-2)	109.80	[423]
		114329.2 ± 7.1	[145]					
K_{β_2}	$KN_{2,3}$	114117.00	[104]	1.343	[487]			
$K_{\beta_2^I}$	KN_3	114607 ± 16	[50]	2.559	[488]	2.700(-2)	103.97	[423]
		114120.00	[104]	2.736	[487]			
		114556.2 ± 5.8	[145]					
K_{β_4}	$KN_{4,5}$	115011 ± 32	[50]	0.117	[488]			
$K_{\beta_4^{II}}$	KN_4	114820.7 ± 5.7	[145]			5.495(-4)	100.68	[423]
$K_{\beta_5^I}$	KN_5	114863.6 ± 5.7	[145]			6.288(-4)	101.02	[423]
	$KO_{2,3}$	115390.0 ± 11.0	[50]	0.909	[488]			
	KO_2					3.266(-3)	104.06	[423]
	KO_3					6.478(-3)	101.47	[423]
	$KP_{2,3}$			0.1779	[488]			
	KP_2					5.823(-4)		
	KP_3					1.097(-3)		

L series

line	transition	E/eV		I/eV/ℏ		TPIV	Γ/eV	
	L_1M_1	16575.30 ± 0.20	[50]	4.97(-5)	[488]		34.80	[423]
		16210.1 ± 2.2	[145]					
L_{β_4}	L_1M_2	16574.8	[104]	0.855	[488]	5.376(-2)	32.30	[279]
		16575.51 ± 0.30	[50]				27.50	[473]
		16576.2 ± 3.4	[145]				30.91	[423]
L_{β_3}	L_1M_3	17455.17 ± 0.73	[50]	0.743	[488]	4.676(-2)	18.80	[279]
		17454.98	[104]				23.70	[473]
		17456.5 ± 3.6					27.64	[423]
$L_{\beta_{10}}$	L_1M_4	18031.2 ± 1.9	[50]	3.95(-2)	[488]	2.485(-3)	19.43	[423]
		18032.00	[104]					
$L_{\beta-9}$	L_1M_5	18205.55 ± 0.32	[50]	5.89(-2)	[488]	3.704(-3)	19.38	[423]
		18205.50	[104]					
	L_1N_1	20318.3 ± 4.2	[145]				29.96	[423]
L_{γ_2}	L_1N_2	20484.92 ± 0.45	[50]	0.2310	[488]	1.453(-2)	39.40	[473]
		20484.80	[104]				31.22	[423]
L_{γ_3}	L_1N_3	20712.95 ± 0.46	[50]	0.2299	[488]	1.447(-2)	32.40	[473]
		20712.80	[104]				25.39	[423]
	L_1N_4	20979.8 ± 2.6	[1.2]	6.73(-3)	[488]	4.233(-4)	22.10	[423]
$L_{\gamma_{11}}$	L_1N_5	21018.9 ± 2.6	[50]	1.08(-2)	[488]	6.784(-4)	22.43	[423]
		21018.00	[104]					
	L_1N_6	21368.9 ± 3.5	[145]				16.26	[423]
	L_1N_7	21378.7 ± 3.0	[145]				16.20	[423]

line	transition	E/eV		I/eV/$\hbar$		TPIV	Γ/eV	
$L_{\gamma_{4'}}$	L_1O_2	21498.40 ± 0.30 21497.98	[50] [104]	5.60(-2)	[488]	3.521(-3)	25.48	[423]
L_{γ_4}	L_1O_3	21562.00 ± 3.40 21562.99	[50] [104]	5.65(-2)	[488]	3.555(-3)	22.89	[423]
	$L_1O_{4,5}$	21657.00 ± 3.80 21659.99	[50] [104]	3.18(-3)	[488]			
	L_1O_4					7.628(-5)		
	L_1O_5					1.237(-4)		
$L_{\gamma_{13}}$	$L_1P_{2,3}$	21729.00 ± 3.80 21728.99	[50] [104]	1.92(-2)	[488]			
	L_1P_2					6.366(-4)		
	L_1P_3					6.173(-4)		
L_η	L_2M_1	15399.81 ± 0.57 15400.00	[50] [104]	9.20(-2)	[488]	1.010(-2)	28.42	[423]
	L_2M_2	15765.1 ± 2.4	[145]				24.53	[423]
$L_{\beta_{17}}$	L_2M_3	16641.00 ± 1.10 16642.00	[50] [104]	4.73(-3)	[488]	5.200(-4)	21.26	[423]
		16645.3 ± 2.6	[145]					
L_{β_1}	L_2M_4	17220.15 ± 0.28 17219.98	[50] [104]	3.267	[488]	3.589(-1)	14.30 13.50 13.05	[279] [473] [423]
	L_2M_5	17395.0 ± 1.6	[145]				13.01	[423]
L_{γ_5}	L_2N_1	19507.27 ± 0.91 19506.99	[50] [104]	2.50(-2)	[488]	2.742(-3)	23.57	[423]
	L_2N_2	19676.1 ± 4.2	[145]				24.85	[423]
	L_2N_3	19907.2 ± 4.7 19905.99	[50] [104]	1.43(-3)	[488]	1.571(-4)	19.01	[423]
L_{γ_1}	L_2N_4	20167.27 ± 0.44 20168.00	[50] [104]	0.740	[488]	8.129(-2)	15.90 15.70 15.72	[279] [473] [423]
	L_2N_5	20210.5 ± 2.9	[145]				16.06	[423]
L_ν	L_2N_6	20557.5 ± 5.0 20554.99	[50] [104]	4.60(-3)	[488]	5.055(-4)	9.87	[423]
	L_2N_7	20567.5 ± 2.0	[145]				9.81	[423]
L_{γ_8}	L_2O_1	20621.00 ± 1.70 20620.98	[50] [104]	4.41(-3)	[488]	7.041(-4)	32.02	[423]
	L_2O_3	20758.00 ± 1.70 20757.99	[50] [104]	3.50(-4)	[488]	3.843(-5)	16.51	[423]

line	transition	E/eV		I/eV/$\hbar$		TPIV	Γ/eV	
L_{γ_6}	L_2O_4	20842.60 ± 0.30	[50]	0.1433	[488]	1.574(-2)	16.40	[473]
		20842.98	[104]					
	$L_2P_{2,3}$	20906.00 ± 7.10	[50]	5.77(-5)	[488]			
		20899.99	[104]					
	L_2P_3					6.591(-6)		
	L_2P_4	20942.00 ± 1.80	[50]					
	$L_2P_{4,5}$	20941.98	[104]	1.43(-4)	[488]			
L_l	L_3M_1	11618.41 ± 0.32	[50]	0.1665	[488]	2.151(-2)	26.89	[423]
		11618.00	[104]					
L_t	L_3M_2	11983.5 ± 1.5	[50]	2.13(-3)	[488]	2.718(-4)	22.99	[423]
		11983.00	[104]					
L_s	L_3M_3	12865.45 ± 0.81	[50]	1.94(-3)	[488]	2.472(-4)	19.73	[423]
		12864.0	[104]					
L_{α_2}	L_3M_4	13438.97 ± 0.19	[50]	0.2744	[488]	3.505(-2)	14.40	[279]
		13439.00	[104]				12.40	[473]
							11.52	[423]
L_{α_1}	L_3M_5	13614.87 ± 0.20	[50]	2.411	[488]	3.079(-1)	13.10	[279]
		13615.00	[104]				12.40	[473]
							11.47	[423]
L_{β_6}	L_3N_1	15726.21 ± 0.59	[1.2]	4.25(-2)	[488]	5.429(-3)	19.40	[473]
	L_3N_2	15895.63 ± 0.99	[50]	4.73(-4)	[488]	6.043(-5)	22.03	[423]
		15893.00	[104]					
	L_3N_3	16123.56 ± 0.90	[50]	5.26(-4)	[488]	6.720(-5)	17.48	[423]
		16122.98	[104]					
$L_{\beta_{15}}$	L_3N_4	16385.86 ± 0.29	[50]	5.52(-2)	[488]	7.053(-3)	14.19	[423]
		16385.98	[104]					
L_{β_2}	L_3N_5	16428.44 ± 0.29	[50]	0.506	[488]	6.458(-2)	16.10	[279]
		16427.99	[104]				13.30	[473]
							14.52	[423]
L_u	$L_3N_{6,7}$	16786.06 ± 0.30	[50]	3.90(-3)	[488]			
		16427.99	[104]					
$L_{u'}$	L_3N_6	16777.3 ± 3.2	[145]			9.023(-5)	8.34	[423]
L_u	L_3N_7	16787.1 ± 1.7	[145]			4.979(-4)	8.28	[423]
L_{β_7}	L_3O_1	16845.00 ± 1.40	[50]	1.08(-2)	[488]	1.377(-3)	30.48	[423]
		16844.98	[104]					
	L_3O_2	16907.00 ± 2.30	[50]	1.10(-4)	[488]	1.401(-5)	17.57	[423]
		16907.99	[104]					
	L_3O_3	16962.00 ± 2.30	[50]	1.26(-4)	[488]	1.606(-5)	14.98	[423]
		16963.99	[104]					

line	transition	E/eV		I/eV/$\hbar$		TPIV	Γ/eV	
L_{β_5}	L_3O_4	17070.10 ± 0.20	[50]	0.1061	[488]	1.334(-3)	11.90	[279]
		17069.99	[104]					
	L_3P_1	17096.00 ± 1.20	[50]	2.21(-3)	[488]	2.872(-4)		
		17095.99	[104]					
	$L_3P_{2,3}$	17118.00 ± 1.20	[50]	3.98(-5)	[488]			
		17118.98	[104]					
	L_3P_2					2.568(-6)		
	L_3P_3					2.772(-6)		
	$L_3P_{4,5}$	17162.00 ± 1.20	[50]	7.49(-4)	[488]			
		17162.99	[104]					

M series

line	transition	E/eV		I/eV/$\hbar$		TPIV	Γ/eV	
	M_1N_2	4250.00 ± 30.00	[50]	4.49(-2)	[116]	2.663(-3)	34.25	[423]
		4274.40	[104]	8.40(-2)	[335]			
		4262.00	[263]					
	M_1N_3	4500.00 ± 13.0	[50]	2.40(-2)	[279]	1.418(-3)	28.42	[423]
		4506.20	[104]	8.40(-2)	[335]			
		4513.00	[263]					
	M_1O_3	5380.00 ± 16.0	[50]	7.02(-3)	[116]	4.329(-4)	25.92	[423]
		5382.00	[104]					
		5393.00	[263]					
	M_1P_3	5500.00 ± 15.00	[50]			7.733(-5)		
		5515.00	[263]					
	M_2N_1	3724.00 ± 4.50	[1.2]	1.34(-2)	[279]	1.017(-3)	29.09	[423]
		3732.00	[263]	8.82(-3)	[335]			
	M_2N_4	4401.00 ± 3.10	[50]	6.48(-2)	[116]	5.032(-3)	18.1 ± 0.2	[438]
		4400.50	[104]	9.55(-2)	[335]		21.24	[423]
		4411.00	[263]					
		4400.84 ± 0.04	[438]					
	M_2O_1	4844.00	[104]	3.29(-3)	[116]	2.644(-4)	37.54	[423]
	M_2O_4	5075.00 ± 8.30	[1.2]	1.47(-2)	[116]	1.197(-3)	18.7 ± 0.3	[438]
		5085.00	[263]					
		5075.95 ± 0.07	[438]					
	M_2P_1	5119.00	[104]	9.99(-4)	[116]	5.526(-5)		
	M_3N_1	2863.00 ± 1.30	[50]	2.18(-2)	[116]	2.136(-3)	20.2 ± 0.8	[438]
		2859.00	[104]	8.82(-3)	[335]		25.82	[423]
		2869.00	[263]					
		2863.01 ± 0.14	[438]					
M_{γ_2}	M_3N_4	3521.00 ± 2.00	[50]	1.02(-2)	[116]	1.001(-3)	13.4 ± 0.3	[438]
		3522.40	[104]	9.55(-2)	[335]		17.97	[423]
		3528.00	[263]					

line	transition	E/eV		I/eV/$\hbar$		TPIV	Γ/eV	
		3521.03 ± 0.07	[438]					
M_{γ_1}	M_3N_5	3563.00 ± 1.00	[50]	7.67(-2)	[116]	7.507(-3)	18.31	[423]
		3564.10	[104]	9.55(-2)	[335]			
		3571.00	[263]					
		3565.10 ± 0.30	[275]					
	M_3N_7	3920.00	[104]	1.11(-3)	[116]		12.06	[423]
	M_3O_1	3980.00 ± 9.00	[50]	5.12(-3)	[116]	5.288(-4)	38.4 ± 3.2	[438]
		3978.00	[104]				34.27	[423]
		3988.00	[263]					
		3979.94 ± 0.33	[438]					
	M_3O_4	4200.00	[104]	2.05(-3)	[116]	2.118(-4)	8.1 ± 1.0	[438]
		4195.84 ± 0.37	[438]					
	$M_3O_{4,5}$	4205.00 ± 2.90	[104]					
		4214.00	[263]					
	M_3O_5	4205.60	[104]	1.57(-2)	[116]	1.622(-3)	8.2 ± 0.1	
		4204.16 ± 0.03	[438]					
	M_3P_1	4231.10	[104]	9.99(-4)	[116]	1.025(-4)		
M_{ξ_2}	M_4N_2	2454.80 ± 1.00	[50]	9.99(-3)	[116]	2.345(-3)	15.3 ± 1.0	[438]
		2455.00	[104]	7.04(-4)	[335]		13 ± 2	[275]
		2460.00	[263]				18.88	[423]
		2455.70 ± 0.30	[275]					
		2455.64 ± 0.19	[438]					
M_δ	M_4N_3	2681.00 ± 2.90	[1.2]	6.83(-4)	[116]	2.081(-4)	13.05	[423]
		2687.00	[263]	7.08(-4)	[335]			
M_β	M_4N_6	3336.70 ± 0.90	[50]	0.1657	[116]	5.159(-2)	3.6 ± 0.1	[438]
		3336.20	[104]	0.1700	[335]		4.3 ± 0.1	[275]
		3344.00	[263]				3.91	[423]
		3336.70 ± 0.10	[275]					
		3336.70 ± 0.20	[401]					
		3336.73 ± 0.01	[438]					
M_η	M_4O_2	3466.60 ± 1.00	[50]	1.53(-3)	[116]	5.547(-4)	13.14	[423]
		3468.20	[104]					
		3474.00	[263]					
	M_4O_3	3531.30	[104]	1.40(-4)	[116]	4.887(-5)	10.55	[423]
	$M_4P_{2,3}$	3698.10	[104]	2.80(-4)	[116]			
	M_4P_2					9.488(-5)		
	M_4P_3					7.909(-6)		
M_{ξ_1}	M_5N_3	2507.00 ± 1.00	[1.2]	4.89(-3)	[116]	1.583(-3)	12.8 ± 0.3	[438]
		2512.00	[263]	7.08(-4)	[335]		15 ± 2	[275]
		2506.80 ± 0.20	[275]				13.01	[423]

line	transition	E/eV		I/eV/$\hbar$		TPIV	Γ/eV	
		2506.98 ± 0.06	[438]					
M_{α_2}	M_5N_6	3159.50 ± 0.80	[50]	7.62(-3)	[116]	2.476(-3)	3.6 ± 0.1	[438]
		3162.00	[104]	0.1699	[335]		4.1 ± 0.6	[275]
		3166.00	[263]				3.87	[423]
		3160.00 ± 0.50	[275]					
		3161.00 ± 0.30	[401]					
		$3160.91 \pm "s,0.04$	[438]					
M_{α_1}	M_5N_7	3170.80 ± 0.80	[50]	0.1494	[116]	4.848(-2)	3.5 ± 0.1	[438]
		3172.00	[104]	0.1699	[335]		4.1 ± 0.6	[275]
		3177.00	[263]				3.81	[423]
		3171.40 ± 0.10	[275]					
		3171.60 ± 0.20	[401]					
		3171.76 ± 0.01	[438]					

N series

line	transition	E/eV		I/eV/$\hbar$		TPIV	Γ/eV	
	N_1O_3	1229.00 ± 8.50	[50]			7.383(-5)	2.11	[423]
	N_1P_2	1410.00 ± 11.00	[50]			6.873(-5)		
	N_1P_3	1420.00 ± 11.40	[50]			1.839(-5)		
	$N_1P_{4,5}$	1440.00 ± 12.00	[50]					
	N_2P_1	1192.00 ± 8.00	[50]			4.942(-5)		
	N_3O_5	961.00 ± 6.70	[50]			1.791(-4)		
	N_4N_6	390.00 ± 1.20	[50]			4.651(-4)	6.59	[423]
	N_4O_4	286.00 ± 1.30	[50]					
	$N_5N_{6,7}$	357.00 ± 1.00	[50]					
	N_5N_6					1.755(-5)	6.92	[423]
	N_5N_7					3.977(-4)	6.86	[423]
	N_6O_5	295.00 ± 1.40	[50]			4.730(-6)		

level characteristics

level	E_B/eV		ω_{nlj}		Γ/eV		AE/eV	
K	115600.60	[503]	0.972	[301]	96.10	[301]	115606.1	[51]
	115600.00 ± 2.40	[493]	0.9701	[39]	94.46	[423]	115601.1 ± 1.0	[433]
			0.970	[38]	96.3	[94]		
			0.9670	[423]				
			0.970	[225]				
L_1	21758.00	[503]	0.176	[301]	14.00	[301]	21757.4	[51]
	21757.40 ± 0.30	[493]	0.149	[436]	15.88	[423]	21770.4 ± 5.7	[50]
			0.1538	[423]	16	[94]	21766.1 ± 1.8	
			0.173	[304]				
L_2	20948.00	[503]	0.467	[301]	9.32	[301]	20947.6	[51]
	20947.60 ± 0.30	[493]	0.506	[436]	9.50	[423]	20946.5 ± 5.2	[50]
			0.4710	[423]	10	[94]	20954.95 ± 0.77	[145]
			0.49	[304]				
L_3	17168.00	[503]	0.489	[301]	7.43	[301]	17166.3	[51]
			0.500	[435]	7.97	[423]	17171.37 ± 0.50	[50]
	17166.30 ± 0.30	[493]	0.44	[240]	8.2	[94]	17174.51 ± 0.50	[145]
			0.4581	[423]				
			0.44	[304]				
M_1	5548.00	[503]	5.73(-3)	[245]	23.00	[74]	5548.0	[51]
	5548.00 ± 0.40	[493]	6.61(-3)	[423]	18.92	[423]		
					0.15.5	[94]		
M_2	5181.00	[503]	8.38(-3)	[245]	12.00	[74]	5182.2	[51]
	5182.20 ± 0.40	[493]	9.44(-3)	[423]	15.02	[423]		
					14.1	[94]		
M_3	4304.00	[503]	1.3(-2)	[245]	17.00	[74]	4303.4	[51]
	4303.40 ± 0.30	[493]	1.35(-2)	[423]	11.76	[423]		
					7.9	[94]		
M_4	3728.00	[503]	5.58(-2)	[423]	3.60	[74]	3727.6	[51]
	3727.60 ± 0.30	[493]			3.55	[423]		
					3.5	[94]		
M_5	3552.00	[503]	5.39(-2)	[423]	4.10	[74]	3551.7	[51]
	3551.70 ± 0.30	[493]			3.50	[423]		
					3.5	[94]		

level	E_B/eV		ω_{nlj}		Γ/eV		AE/eV	
N_1	1442.00	[503]	9.8(-4)	[423]	12.00	[74]	1440.8	[51]
	1440.80 ± 0.40	[493]			14.06	[423]		
					12.2	[94]		
N_2	1273.00	[503]	1.42(-3)	[423]	9.6	[94]	1272.6	[51]
	1272.60 ± 0.30	[51]			7.80	[74]		
	1271.80 ± 1.10	[493]			9.6	[94]		
N_3	1045.00	[503]	7.6(-4)	[423]	7.80	[74]	1944.9	[51]
	1044.90 ± 1.20	[493]			9.51	[423]		
					8	[94]		
N_4	780.00	[503]	6.9(-4)	[423]	6.22	[423]		
	780.40 ± 0.30	[51]			4.40	[74]		
	780.20 ± 1.10	[493]			4.5	[94]		
N_5	738.00	[503]	5.9(-4)	[423]	4.40	[74]		
	737.70 ± 1.10	[493]			6.55	[423]		
					4.25	[94]		
N_6	392.00	[503]	8.0(-5)	[423]	0.37	[423]		
	391.30 ± 0.60	[51]			0.19	[74]		
	390.70 ± 1.30	[493]			0.37	[94]		
N_7	381.00	[503]	7.0(-5)	[423]	0.31	[423]		
	380.90 ± 0.90	[51]			0.19	[74]		
	379.90 ± 1.20	[493]			0.31	[94]		
O_1	324.00	[503]	3.0(-5)	[423]	22.51	[423]		
	323.70 ± 1.10	[51]						
	323.30 ± 1.40	[493]						
O_2	260.00	[503]	5.08-5)	[423]	9.60	[423]		
	259.20 ± 0.50	[51]						
	259.30 ± 1.60	[493]						
O_3	195.00	[503]	1.0(-5)	[423]	7.01	[423]		
	159.10 ± 1.30	[51]						
	195.90 ± 3.10	[493]						
O_4	105.00	[503]						
	105.00 ± 0.50	[51]						
	104.40 ± 0.90	[493]						
O_5	96.00	[503]						
	96.30 ± 1.40	[51]						

level	E_B/eV	ω_{nlj}	Γ/eV	AE/eV	
	95.20 ± 2.10	[493]			
P_1	71.00	[503]			
	70.70 ± 1.20	[51]			
	49.50 ± 2.00	[493]			
P_2	43.00	[503]			
	42.30 ± 0.90	[51]			
	30.80 ± 0.80	[493]			
P_3	33.00	[503]			
	32.30 ± 0.90	[51]			
P_4	4.00	[503]			
Q_1	6.194	[126]			
IP	6.19405	[222]			

: Note: Transition energies from Refs. [50,144,145] are tabulated for ^{238}U.

Np   Z=93

[Rn] $5f^4 6d^1 7s^2$

Neptunium

$A = 237.0482$ [260] $\varrho = 19.50$ g/cm^3 [547]

X-ray transitions

line	transition	E/eV		I/eV/$\hbar$		TPIV	Γ/eV	
K series								
K_{α_3}	KL$_1$	96240.6 ± 3.5	[145]				1114.66	[423]
		96232 ± 34	[580]					
$K_{\alpha 2}$	KL$_2$	97077.50 ± 33.40		28.71	[488]	2.902(-1)	103.90	[74]
		97088.10	[104]				108.38	[423]
		97095.00 ± 4.00	[374]					
		97068.4 ± 3.0	[43]					
$K_{\alpha 1}$	KL$_3$	101068.00±33.40	[51]	45.69	[488]	4.617(-1)	98.50	[74]
		101077.70	[104]				106.85	[423]
		101085.00 ± 4.00	[374]					
		101056.00 ± 3.00	[298]					
		101056.3 ± 3.0	[43]					
	KM$_1$	112928.1 ± 5.3	[145]				117.70	[423]
$K_{\beta 3}$	KM$_2$	113311.80±33.70	[51]	5.284	[488]	5.340(-2)	113.80	[423]
		113295.20	[104]					
		113308.00 ± 4.00	[43]					
$K_{\beta 1}$	KM$_3$	114243.30±33.50	[51]	10.35	[488]	1.045(-1)	110.53	[423]
		114248.30	[104]					
		114243.3 ± 3.0	[43]					
$K_{\beta_5^{II}}$	KM$_4$	114827.70±33.40	[51]	0.1892	[488]	1.912(-3)	102.30	[423]
		114814.5 ± 4.6	[145]					
$K_{\beta_5^{I}}$	KM$_5$	115012.20±33.40	[51]	0.2144	[488]	2.167(-3)	102.24	[423]
		115000.2 ± 4.6	[145]					
		114989 ± 39	[580]					
	KN$_1$	117167.4 ± 6.2	[145]				112.90	[423]
$K_{\beta_2^{II}}$	KN$_2$	117350.30±33.80	[51]	1.326	[488]	1.341(-2)	114.04	[423]
		117340.3 ± 7.3	[145]					
		117332 ± 35	[580]					
$K_{\beta_2^{I}}$	KN$_3$	117591.20±33.70	[51]	2.693	[488]	2.722(-2)	108.15	[423]
		117581.9 ± 5.9	[145]					
		117569 ± 35	[580]					

line	transition	E/eV		I/eV/$\hbar$		TPIV	Γ/eV	
$K_{\beta_4^{II}}$	KN_4	117862.10±33.50	[51]	0.1200	[488]	5.676(-4)	104.81	[423]
		117852.1 ± 5.8	[145]					
$K_{\beta_4^{I}}$	KN_5	117907.70±33.40	[51]	0.1200	[488]	6.456(-4)	105.32	[423]
		117897.6 ± 5.9	[145]					

L series

line	transition	E/eV		I/eV/$\hbar$		TPIV	Γ/eV	
	L_1L_2	827.00	[122]			1.293(-5)	25.77	[423]
		4818.00	[122]					
	L_1M_1	16703.60±4.50	[51]	5.73(-5)	[488]		35.10	[423]
		16683 ± 20	[580]					
$L_{\beta4}$	L_1M_2	17060.77 ± 0.69	[50]	0.908	[488]	5.666(-2)	31.20	[423]
		17060.70	[104]					
		17061.00	[122]					
		17066.00 ± 6.00	[336]					
$L_{\beta3}$	L_1M_3	17989.3 ± 3.5	[50]	0.768	[488]	4.791(-2)	27.93	[423]
		18001.40	[104]					
		179995.00	[122]					
		17995.00 ± 6.00	[336]					
$L_{\beta10}$	L_1M_4	18576.50±1.30	[51]	4.30(-2)	[488]	2.680(-3)	19.70	[423]
		18578.00	[122]					
		18578.00 ± 8.00	[336]					
		18573.9 ± 2.7	[145]					
$L_{\beta9}$	L_1M_5	18761.00 ± 1.30	[51]	6.40(-2)	[488]	3.991(-3)	19.64	[423]
		18763.00	[122]					
		18761.00 ± 8.00	[336]					
	L_1N_1	20926.10 ± 1.70	[51]	2.67(-5)	[488]		30.29	[423]
		20922 ± 21	[580]					
		20926.8 ± 4.2	[145]					
$L_{\gamma2}$	L_1N_2	21104± 16	[50]	0.2464	[488]	32.08	31.44	[423]
		21108.50	[104]					
		21100.00	[122]					
		21109.00 ± 8.00	[336]					
$L_{\gamma3}$	L_1N_3	21338 ± 16	[50]	0.2401	[488]	1.497(-2)	25.55	[423]
		21340.40	[104]					
		21342.00	[122]					
		21344.00 ± 8.00	[336]					
	L_1N_4	21610.90 ± 1.40	[51]	7.39(-3)	[488]	4.607(-4)	22.21	[423]
		21611.4 ± 3.9	[145]					
$L_{\gamma11}$	L_1N_5	21656.5±1.30	[51]	1.19(-2)	[488]	7.394(-4)	22.72	[423]
		21657.00	[122]					
		21690.00 ± 20.00	[336]					
		21657.0 ± 3.9	[145]					

line	transition	E/eV		I/eV/$\hbar$		TPIV	Γ/eV	
	L_1N_6	22011.80 ± 1.70	[51]	3.33(-4)	[488]		16.47	[423]
		22012.0 ± 3.6	[145]					
	L_1N_7	22022.40 ± 1.40	[51]	3.33(-4)	[488]		16.39	[423]
		22022.7 ± 3.1	[145]					
$L_{\gamma_{4'}}$	L_1O_2	22143.80 ± 1.70	[51]	6.05(-2)	[488]	3.773(-3)	25.98	[423]
		22145.00	[122]					
$L_{\gamma_{4',4}}$	$L_1O_{2,3}$	22200.00 ± 20.00	[50]					
		22170.00 ± 8.00	[336]					
$L_{\gamma 4}$	L_1O_3	22200.0 ± 19.90	[50]	5.98(-2)	[488]	3.732(-3)	23.59	[423]
	L_1O_4	22318.00 ± 1.60	[51]	3.56(-3)	[488]	8.470(-5)		
		22222.00	[122]					
	L_1O_5	22325.5 ± 1.40	[51]	3.56(-3)	[488]	1.374(-4)		
$L_{\gamma_{13}}$	L_1P_2	22399.00	[122]			6.742(-4)		
	$L_1P_{2,3}$	22375.00 ± 10.00	[336]					
	L_2L_3	3981.00	[122]			6.348(-4)	17.96	[423]
L_η	L_2M_1	15876.00 ± 4.10	[50]	9.73(-2)	[488]	1.037(-2)	28.81	[423]
		15876.40	[104]					
		15862.00	[122]					
		15870.00 ± 8.00	[336]					
		15864 ± 21	[580]					
		15860.5 ± 2.6	[145]					
	L_2M_2	16234.30 ± 1.10	[51]	1.58(-5)	[488]		24.91	[423]
		16242 ± 21	[580]					
		16232.6 ± 2.4	[145]					
$L_{\beta 17}$	L_2M_3	17165.8 ± 0.90	[51]	5.11(-3)	[488]	5.448(-4)	21.64	[423]
		17168.00	[122]					
		17168 ± 21	[580]					
		17163.8 ± 2.6	[145]					
L_{β_1}	L_2M_4	17750.36 ± 0.34	[50]	3.441	[488]	3.670(-1)	13.41	[423]
		17750.30	[104]					
		17752.00	[122]					
		17751.00 ± 2.00	[336]					
	L_2M_5	17938 ± 28	[580]				13.37	[423]
		17932.7 ± 1.6	[145]					
$L_{\gamma 5}$	L_2N_1	20107 ± 19	[50]	2.65(-2)	[488]	2.827(-3)	24.01	[423]
		20124.50	[41]					
		20101.00	[122]					
		20099.00 ± 6.00	[336]					
		20099.9 ± 3.2	[145]					
	L_2N_2	20272.80 ± 1.90	[51]	6.64(-6)	[488]		25.16	[423]

line	transition	E/eV		I/eV/$\hbar$		TPIV	Γ/eV	
		20281 ± 21	[580]					
		20272.7 ± 4.3	[145]					
	L_2N_3	20513.70 ± 1.10	[51]	1.56(-3)	[488]	1.659(-4)	19.26	[423]
		20518 ± 21	[580]					
		20514.3 ± 2.9	[145]					
$L_{\gamma 1}$	L_2N_4	20785.04 ± 0.46	[50]	0.787	[488]	8.388(-2)	15.93	[423]
		20784.5 ± 2.8	[145]					
		20783.00 ± 3.00	[336]					
	L_2N_5	20830.0 ± 2.9	[145]				16.43	[423]
L_ν	L_2N_6	21187.00	[122]			5.379(-4)	10.19	[423]
		21185.0 ± 2.6	[145]					
	L_2N_7	21195.7 ± 2.0	[145]				9.98	[423]
$L_{\gamma 8}$	L_2O_1	21262.00	[122]			7.342(-4)	33.16	[423]
	L_2O_2	21326.40 ± 1.20	[51]	1.80(-6)	[488]		19.70	[423]
	L_2O_3	21394.00 ± 1.10	[51]	3.86(-4)	[488]	4.115(-5)	17.31	[423]
$L_{\gamma 6}$	L_2O_4	21488.00 ± 2.00	[50]	0.1554	[488]	1.657(-2)		
		21487.90	[104]					
		21492.00	[122]					
		21487.00 ± 6.00	[336]					
	L_2P_1	21551.00	[122]			1.518(-4)		
	L_2P_4	21590.00	[336]					
L_l	L_3M_1	11888.6 ± 9.1	[50]	0.1777	[488]	2.216(-2)	27.28	[423]
		11889.80	[104]					
		11871.00	[122]					
		11888.00	[336]					
L_t	L_3M_2	12243.80 ± 1.10	[51]	2.29(-3)	[488]	2.830(-4)	23.39	[423]
		12244.00	[122]					
		12263.00 ± 2.00	[336]					
		12247 ± 22	[580]					
L_s	L_3M_3	13175.30 ± 0.90	[51]	2.08(-3)	[488]	2.579(-4)	20.11	[423]
		13177.00	[122]					
		13161.00 ± 20.00	[336]					
		13173 ± 21	[580]					
$L_{\alpha 2}$	L_3M_4	13759.84 ± 0.20	[50]	0.288	[488]	3.566(-2)	11.88	[423]
		13759.80	[104]					
		13761.00	[122]					
		13760.00 ± 6.00	[336]					
$L_{\alpha 1}$	L_3M_5	13944.26 ± 0.21	[50]	2.529	[488]	3.132(-1)	11.83	[423]
		13940.40	[104]					
		13945.00	[122]					

line	transition	E/eV		I/eV/$\hbar$		TPIV	Γ/eV	
		13952.00	[336]					
$L_{\beta 6}$	L_3N_1	16113 ± 18	[50]	4.55(-2)	[488]	5.629(-3)	22.48	[423]
		16111.00	[122]					
		16128.00 ± 8.00	[336]					
	L_3N_2	16282.30 ± 1.20	[51]	5.06(-4)	[488]	6.269(-5)	23.63	[423]
		16286 ± 22	[580]					
	L_3N_3	16523.20 ± 1.10	[51]	5.69(-4)	[488]	7.044(-5)	17.73	[423]
$L_{\beta 15}$	L_3N_4	16794.10 ± 0.90	[51]	5.83(-2)	[488]	7.218(-3)	14.40	[423]
		16794.8 ± 2.6	[145]					
		16795.00	[122]					
$L_{\beta 15,2}$	$L_3N_{4,5}$	16862.00 ± 6.00	[336]					
$L_{\beta 2}$	L_3N_5	16840.16 ± 0.30	[50]	0.534	[488]	6.618(-2)	14.90	[423]
		16840.10	[104]					
		16840.00	[122]					
$L_{u'}$	L_3N_6	17195.00 ± 1.20	[51]	7.77(-4)	[488]	9.621(-5)	8.66	[423]
		17196.00	[122]					
		17195.3 ± 2.3	[145]					
L_u	L_3N_7	17205.60 ± 0.90	[51]	4.28(-3)	[488]	5.301(-4)	8.58	[423]
		17207.00	[122]					
		17206.1 ± 1.7	[145]					
$L_{\beta 7}$	L_3O_1	17271.00	[122]			1.444(-3)	31.61	[423]
	L_3O_2	17326.60 ± 1.20	[51]	1.19(-4)	[488]	1.470(-5)		[423]
	L_3O_3	17403.90 ± 1.10	[51]	1.38(-4)	[488]	1.706(-5)		[423]
$L_{\beta 5}$	L_3O_4	17501.00	[122]			1.391(-3)		
	$L_3O_{4,5}$	17508.10 ± 0.50	[50]					
		17454.00 ± 24.00	[336]					
	L_3O_5	17508.20	[104]			1.276(-2)		
		17509.00	[122]					

M series

	M_1M_3	1306.00	[122]			1.285(-3)	30.97	[423]
	M_1N_2	4412.00	[122]			2.813(-3)	34.48	[423]
	M_1N_3	4654.00	[122]			1.430(-3)	28.59	[423]
	M_1O_2	5456.00	[122]			7.552(-4)	29.02	[423]
	M_1O_3	5533.00	[122]			4.455(-4)	26.63	[423]
	M_1P_2	5710.00	[122]			1.384(-4)		
	M_1P_3	5722.00	[122]			7.750(-5)		
	M_2N_1	3867.00	[122]			1.050(-3)	29.43	[423]
		3849.30 ± 3.60	[358]					

line	transition	E/eV		I/eV/$\hbar$	TPIV	Γ/eV	
	M_2N_4	4551.00	[122]		5.127(-3)	21.35	[423]
		4548.20 ± 5.00	[358]				
	M_2O_1	5027.00	[122]		2.760(-4)	38.59	[423]
	M_2O_4	5257.00	[122]		1.249(-3)		
	M_2P_1	5317.00	[122]		5.732(-5)		
	M_2Q_1	5360.00	[122]		5.838(-6)		
	M_3M_4	583.00	[122]		1.368(-5)	15.57	[423]
	M_3M_5	768.00	[121]		3.191(-4)	15.51	[423]
	M_3N_1	2933.00	[122]		2.231(-3)	26.16	[423]
M_{γ_2}	M_3N_4	3618.80 ± 1.00	[51]		1.040(-3)	18.08	[423]
		3617.00	[122]				
M_{γ_1}	M_3N_5	3664.40 ± 0.90	[51]		7.773(-3)	18.59	[423]
		3663.00	[122]				
		3662.80 ± 3.25	[358]				
	M_3O_1	4094.00	[122]		5.576(-4)	35.32	[423]
		4096.00 ± 4.10	[358]				
	M_3O_4	4324.00	[122]		2.242(-4)		
	M_3O_5	4332.00	[122]		1.714(-3)		
		4330.00 ± 4.50	[358]				
	M_3P_1	4383.00	[122]		1.096(-4)		
	M_3Q_1	4430.00	[122]		1.261(-5)		
	M_4N_1	2350.00	[122]			17.93	[423]
M_{ξ_2}	M_4N_2	2522.60 ± 1.20	[51]		2.403(-3)	19.08	[423]
		2522.00	[122]				
M_δ	M_4N_3	2764.00	[122]		2.090(-4)	13.19	[423]
M_β	M_4N_6	3435.30 ± 1.20	[51]		5.293(-2)	4.11	[423]
		3436.00	[122]				
		3434.20 ± 0.22	[358]				
M_η	M_4O_2	3566.00	[122]		5.666(-4)	13.62	[423]
		3595.90 ±	[358]				
	M_4O_3	3644.00	[122]		4.904(-5)	11.23	[423]
	M_4O_6	3750.00	[122]		1.406(-3)		
	M_4P_2	3821.00	[122]		9.704(-5)		
	M_4P_3	3832.00	[122]		7.931(-6)		
M_{ξ_1}	M_5N_3	2579.00 ± 1.10	[51]		1.596(-3)	13.13	[423]
		2579.00	[122]				
$M_{xi_{1,2}}$	$M_{5,4}N_{3,2}$	2551.00	[122]				

line	transition	E/eV		I/eV/$\hbar$	TPIV	Γ/eV	
	M_5N_2					19.03	[423]
M_{α_2}	M_5N_6	3250.80 ± 1.20	[51]		2.531(-3)	4.06	[423]
		3251.00	[122]				
M_{α_1}	M_5N_7	3261.40 ± 0.90	[51]		4.958(-2)	3.98	[423]
		3262.00	[122]				
		3260.50 ± 0.20	[358]				
	M_5O_3	3459.00	[122]		3.740(-4)	11.18	[423]
	M_5O_6	3570.00	[122]		6.604(-5)		
	M_5P_3	3648.00	[122]		6.014(-5)		
	M_5Q_7	3570.00	[122]				

level characteristics

level	E_B/eV		ω_{nlj}		Γ/eV		AE/eV	
K	118678.00 ± 33.00	[51]	0.9706	[39]	100.00	[301]	118678	[51]
	118668.60 ± 3.10	[493]	0.973	[301]	98.63	[423]	118688.7 ± 6.8	[580]
			0.969	[253]			118674.2 ± 3.8	[145]
L_1	22426.80 ± 0.90	[51]	0.187	[301]	14.00	[301]	22426.8	[51]
	22428.10 ± 1.60	[493]	0.157	[436]	16.03	[423]	22437.5 ± 9.5	[580]
							22433.6 ± 1.8	[145]
L_2	21600.50 ± 0.40	[51]	0.466	[301]	9.91	[301]	21600.5	[51]
	21601.30 ± 0.70	[493]	0.519	[436]	9.75	[423]	21615.0 ± 9.9	[580]
							21606.62 ± 0.79	[145]
L_3	17610.00 ± 0.40	[51]	0.502	[301]	7.59	[301]	17610.0	[51]
	17610.60 ± 0.70	[493]	0.454	[436]	8.22	[423]	17608.04 ± 0.50	[580]
							17616.94 ± 0.50	[145]
M_1	5723.20 ± 3.60	[51]	6.96(-3)	[423]	19.07	[423]		
	5739.60 ± 3.10	[493]						
M_2	5366.20 ± 0.70	[51]	9.76(-3)	[423]	15.17	[423]		
	5366.70 ± 0.90	[493]						
M_3	4434.70 ± 0.50	[51]	1.40(-2)	[423]	11.90	[423]		
	4433.40 ± 0.90	[493]						
M_4	3850.30 ± 0.40	[51]	5.77(-2)	[423]	3.67	[423]		
	3849.80 ± 0.50	[493]						
M_5	3665.80 ± 0.40	[51]	6.0(-2)	[38]	3.61	[423]		
	3665.20 ± 0.70	[493]	5.55(-2)	[423]				
N_1	1500.70 ± 0.80	[51]	1.02(-3)	[423]	14.26	[423]		
	1500.10 ± 1.50	[493]						
N_2	1327.70 ± 0.80	[51]	1.43(-3)	[423]	15.41	[423]		

level	E_B/eV		ω_{nlj}		Γ/eV		AE/eV
	1327.70 ± 1.40	[493]					
N_3	1086.80 ± 0.70	[51]	7.8(-4)	[423]	9.52	[423]	
	1086.00 ± 1.80	[493]					
N_4	815.90 ± 0.50	[51]	7.3(-4)	[423]	6.18	[423]	
	816.10 ± 0.70	[493]					
N_5	770.30 ± 0.40	[51]	6.2(-4)	[423]	6.69	[423]	
	770.80 ± 0.70	[493]					
N_6	415.00 ± 0.80	[51]	9.0(-5)	[423]	0.44	[423]	
	414.30 ± 0.90	[493]					
N_7	404.40 ± 0.50	[51]	9.0(-5)	[423]	0.36	[423]	
	403.40 ± 0.90	[493]					
O_1	339.80 ± 3.50	[493]	3.0(-5)	[423]	23.42	[423]	
O_2	283.40 ± 0.80	[51]	5.0(-5)	[423]	9.96	[423]	
O_3	206.10 ± 0.70	[51]	1.0(-5)	[423]	7.57	[423]	
	206.20 ± 0.90	[493]					
O_4	109.30 ± 0.70	[51]					
O_5	101.30 ± 0.50	[51]					
	101.20 ± 0.60	[493]					
P_1	50.00 ± 3.50	[493]					
P_2	29.30 ± 1.50	[493]					
P_3	15.50 ± 1.00	[493]					
Q_1	6.266	[126]					
IP	6.2657	[222]					

: Note: Transition energies and edge energies from Refs. [43,50,580] are tabulated for ^{237}Np.

Pu   Z=94

[Rn] $5f^6\,7s^2$

Plutonium

$A = 244.0642$ [222] $\varrho = 19.78$ g/cm^3 [547]
$\varrho = 16.63$ g/cm^3 [222]

X-ray transitions

line	transition	E/eV		I/eV/$\hbar$		TPIV	Γ/eV	
K series								
K_{α_3}	KL$_1$	98682.4 ± 3.6	[145]				119.07	[423]
K_{α_2}	KL$_2$	99529.60	[104]	30.04	[488]	2.910(-1)	103.90	[74]
		99523.80 ± 1.20	[143]				112.90	[423]
		99523.80 ± 1.20	[277]					
		99529.4 ± 2.0	[43]					
K_{α_1}	KL$_3$	103745.00	[104]	47.55	[488]	4.606(-1)	98.50	[74]
		103734.70 ± 0.60	[143]				111.38	[423]
		103740.3 ± 2.0	[43]					
		103735.00 ± 3.00	[298]					
	KM$_1$	115858.7 ± 5.4	[145]				123.25	[423]
K_{β_3}	KM$_2$	116261.20	[104]	5.514	[488]	5.342(-2)	118.24	[423]
		116241.3 ± 2.0	[43]					
K_{β_1}	KM$_3$	117252.80	[104]	10.81	[488]	1.048(-1)	114.88	[423]
		117232.2 ± 2.0	[43]					
$K_{\beta_5^{II}}$	KM$_4$	117845.40 ± 44.60	[51]	0.2020	[488]	1.957(-3)	106.72	[423]
		117817.5 ± 4.8	[145]					
$K_{\beta_5^{I}}$	KM$_5$	118039.90 ± 44.60	[51]	0.2274	[488]	2.203(-3)	106.65	[423]
		118013.2 ± 4.8	[145]					
	KN$_1$	120232.8 ± 6.4	[145]				117.35	[423]
$K_{\beta_2^{II}}$	KN$_2$	120405.2 ± 3.0	[43]	2.095	[488]	1.346(-2)	118.35	[423]
		120409.5 ± 7.5	[145]					
$K_{\beta_2^{I}}$	KN$_3$	120674.2 ± 3.0	[43]	4.503	[488]	2.743(-2)	112.57	[423]
		120666.5 ± 6.1	[145]					
$K_{\beta_4^{II}}$	KN$_4$	120969.10 ± 44.60	[51]	0.2442	[488]	5.858(-4)	109.19	[423]
		120942.4 ± 6.0	[145]					
$K_{\beta_4^{I}}$	KN$_5$	121016.60 ± 44.60	[51]	0.2442	[488]	6.620(-4)	109.70	[423]
		120990.6 ± 6.0	[145]					
L series								
	L$_1$M$_1$	17164.30 ± 3.00	[51]	6.61(-5)	[488]		36.48	[423]
		17176.4 ± 3.3	[145]					

line	transition	E/eV		I/eV/$\hbar$		TPIV	Γ/eV	
L$_{\beta4}$	L$_1$M$_2$	17556.00 ± 0.50	[50]	0.965	[488]	5.972(-2)	31.48	[423]
		17556.20	[104]					
		17557.1 ± 3.5	[145]					
L$_{\beta3}$	L$_1$M$_3$	18540.50 ± 0.60	[50]	0.793	[488]	4.908(-2)	28.11	[423]
		18540.70	[104]					
		18542.5 ± 3.7	[145]					
L$_{\beta10}$	L$_1$M$_4$	19126.00 ± 3.00	[50]	4.67(-2)	[488]	2.890(-3)	19.95	[423]
		19133.30	[104]					
		19135.1 ± 2.7	[145]					
L$_{\beta9}$	L$_1$M$_5$	19323.00 ± 3.00	[50]	6.95(-2)	[488]	4.302(-3)	19.88	[423]
		19328.40	[104]					
		19330.8 ± 2.6	[145]					
	L$_1$N$_1$	20926.10 ± 1.70	[51]	2.67(-5)	[488]		30.58	[423]
		21550.4 ± 4.3	[145]					
L$_{\gamma2}$	L$_1$N$_2$	21725.10 ± 0.80	[50]	0.2627	[488]	1.625(-2)	31.59	[423]
		21725.30	[104]					
		21727.1 ± 5.4	[145]					
L$_{\gamma3}$	L$_1$N$_3$	21982.40 ± 0.35	[50]	0.2503	[488]	1.549(-2)	25.81	[423]
		21982.60	[104]					
		21984.2 ± 3.9	[145]					
	L$_1$N$_4$	21610.90 ± 1.40	[51]	7.39(-3)	[488]	5.019(-4)	22.43	[423]
		22260.0 ± 3.9	[145]					
L$_{\gamma11}$	L$_1$N$_5$	22295.80 ± 2.20	[51]	1.30(-2)	[488]	8.061(-4)	22.93	[423]
		22308.2 ± 3.9	[145]					
	L$_1$N$_6$	22670.7 ± 3.7	[145]				16.79	[423]
	L$_1$N$_7$	22682.8 ± 3.1	[145]				16.59	[423]
L$_{\gamma4'}$	L$_1$O$_2$	22823.00 ± 4.20	[50]	6.52(-2)	[488]	4.035(-3)	26.72	[423]
		22821.20	[104]					
L$_{\gamma4}$	L$_1$O$_3$	22891.00 ± 4.20	[50]	6.31(-2)	[488]	3.901(-3)	26.90	[423]
		22890.90	[104]					
L$_\eta$	L$_2$M$_1$	16333.00 ± 2.20	[50]	0.1029	[488]	1.063(-2)	30.32	[423]
		16334.1 ± 2.5	[145]					
	L$_2$M$_2$	16725.00 ± 2.40	[51]	1.82(-5)	[488]		25.31	[423]
		16714.8 ± 2.4	[145]					
L$_{\beta17}$	L$_2$M$_3$	17709.60 ± 2.20	[51]	5.51(-3)	[488]	5.690(-4)	21.95	[423]
		17700.2 ± 2.6	[145]					
L$_{\beta1}$	L$_2$M$_4$	18293.70 ± 0.54	[50]	3.623	[488]	3.740(-1)	13.79	[423]
		18294.30	[104]					
		18292.8 ± 1.6	[145]					
	L$_2$M$_5$	18488.10 ± 1.30	[51]	5.05(-4)	[51]		13.72	[423]

line	transition	E/eV		I/eV/$\hbar$		TPIV	Γ/eV	
		18488.5 ± 1.6	[145]					
$L_{\gamma 5}$	L_2N_1	20704.00 ± 3.50	[50]	2.82(-2)	[488]	2.908(-3)	24.42	[423]
		20705.40	[104]					
		20708.1 ± 3.3	[145]					
	L_2N_2	20894.10 ± 2.50	[51]	7.69(-6)	[488]		25.42	[423]
		20884.8 ± 4.3	[145]					
	L_2N_3	21151.40 ± 2.30	[51]	1.69(-3)	[488]	1.747(-4)	19.64	[423]
		21141.9 ± 2.9	[145]					
$L_{\gamma 1}$	L_2N_4	21417.30 ± 0.33	[50]	0.836	[488]	8.631(-2)	16.26	[423]
		21417.50	[104]					
		21417.7 ± 2.8	[145]					
	L_2N_5	21646.80 ± 1.30	[51]	1.31(-4)	[488]		16.77	[423]
		21465.9 ± 2.8	[145]					
L_ν	L_2N_6	21820.40 ± 2.40	[51]	5.52(-3)	[488]	5.697(-4)	10.63	[423]
		21828.4 ± 2.6	[145]					
	L_2N_7	21840.5 ± 2.0	[145]				10.43	[423]
$L_{\gamma 8}$	L_2O_1	21914.30 ± 3.10	[51]	7.39(-3)	[488]	7.626(-4)	35.78	[423]
		21916.90	[104]					
	L_2O_2	21992.10 ± 5.40	[51]	2.11(-6)	[488]		20.55	[423]
	L_2O_3	22059.70 ± 5.40	[51]	4.24(-4)	[488]	4.376(-5)	20.74	[423]
$L_{\gamma 6}$	L_2O_4	22150.50 ± 0.80	[50]	0.1671	[488]	1.725(-2)		
		22150.50	[104]					
L_l	L_3M_1	12124.00 ± 1.20	[50]	0.1895	[488]	2.281(-2)	28.79	[423]
		12124.50	[104]					
		12124.7 ± 2.1	[145]					
L_t	L_3M_2	12515.60 ± 2.30	[51]	2.45(-3)	[488]	2.943(-4)	23.79	[423]
		12505.4 ± 2.1	[145]					
L_s	L_3M_3	13500.20 ± 2.10	[51]	2.24(-3)	[488]	2.687(-4)	20.43	[423]
		13490.8 ± 2.3	[145]					
$L_{\alpha 2}$	L_3M_4	14084.20 ± 0.30	[50]	0.302	[488]	3.626(-2)	12.27	[423]
		14084.40	[104]					
		14083.4 ± 1.4	[145]					
$L_{\alpha 1}$	L_3M_5	14278.60 ± 0.30	[50]	2.652	[488]	3.184(-1)	12.20	[423]
		14279.20	[104]					
		14279.1 ± 1.3	[145]					
$L_{\beta 6}$	L_3N_1	16498.30 ± 0.44	[50]	4,55(-2)	[488]	5.830(-3)	22.89	[423]
		16498.50	[104]					
		16498.7 ± 3.0	[145]					
	L_3N_2	16675.5 ± 4.1	[145]			6.492(-5)	23.91	[423]
	L_3N_3	16932.5 ± 2.6	[145]			7.376(-5)	18.13	[423]

line	transition	E/eV		I/eV/$\hbar$		TPIV	Γ/eV	
$L_{\beta15}$	L_3N_4	17208.00 ± 2.40	[50]	6.15(-2)	[488]	7.378(-3)	14.74	[423]
		17207.50	[104]					
		17208.3 ± 2.5	[145]					
$L_{\beta2}$	L_3N_5	17255.30 ± 0.50	[50]	0.564	[488]	6.774(-2)	15.25	[423]
		17255.60	[104]					
		17256.6 ± 2.5	[145]					
$L_{u'}$	L_3N_6	17611.00 ± 2.30	[51]	8.53(-4)	[488]	1.023(-4)	9.11	[423]
		17619.1 ± 2.4	[145]					
L_u	L_3N_7	17624.40 ± 2.70	[51]	4.69(-3)	[488]	5.631(-4)	8.90	[423]
		17631.1 ± 1.8	[145]					
	$L_3N_{6,7}$	17635.00 ± 2.50	[50]					
		17633.50	[104]					
$L_{\beta7}$	L_3O_1	17705.00 ± 2.50	[50]	1.26(-2)	[488]	1.510(-3)	34.26	[423]
		17707.70	[104]					
$L_{\beta5}$	$L_3O_{4,5}$	17950.60 ± 0.50	[50]	0.1218	[488]			
		17950.80	[104]					
	L_3O_4					1.437(-3)		
	L_3O_5					1.319(-2)		

M series

line	transition	E/eV		I/eV/$\hbar$		TPIV	Γ/eV	
	M_2N_1	3970.10 ± 3.80	[358]			1.085(-3)	29.75	[423]
$M_{\gamma2}$	M_3N_4	3707.70 ± 2.10	[51]			1.081(-3)	18.23	[423]
		4697.80 ± 5.30	[358]					
$M_{\gamma1}$	M_3N_5	3755.20 ± 2.10	[51]			8.053(-3)	18.74	[423]
		3765.10 ± 3.40	[358]					
	M_3O_1	4212.90 ± 4.30	[358]			5.877(-4)	37.75	[423]
	M_3O_5	4457.00 ± 4.80	[358]			1.795(-3)		
$M_{\xi2}$	M_4N_2	2600.50 ± 2.40	[51]			2.468(-3)	19.24	[423]
M_δ	M_4N_3	2871.80	[423]			2.107(-4)	13.46	[423]
M_β	M_4N_6	3526.80 ± 2.30	[51]			5.397(-2)	4.45	[423]
		3533.25 ± 0.23	[358]					
M_η	M_4O_2	3707.70 ± 3.30	[358]			5.815(-4)	14.37	[423]
$M_{\xi1}$	M_5N_3	2663.30 ± 2.20	[51]			1.613(-3)	13.39	[423]
$M_{\alpha2}$	M_5N_6	3332.30 ± 2.30	[51]			2.581(-3)	4.37	[423]
$M_{\alpha1}$	M_5N_7	3345.70 ± 2.70	[51]					
		3350.80 ± 0.20	[358]					

level characteristics

level	E_B/eV		ω_{nlj}		Γ/eV		AE/eV	
K	121818.00 ± 40.00	[51]	0.9710	[39]	96.00	[74]	121818	[51]
	121791.10 ± 2.10	[493]	0.973	[301]	105.00	[301]	121794.1 ± 4.0	[145]
			0.969	[253]	102.92	[423]		
L_1	23097.20 ± 1.60	[51]	0.205	[301]	17.50	[74]	23097.2	[51]
	23097.80 ± 1.80	[493]	0.165	[436]	13.50	[301]	23109.5 ± 6.4	[50]
			0.1691	[423]	16.15	[423]	23111.8 ± 1.8	
L_2	22266.20 ± 0.70	[51]	0.464	[301]	10.00	[74]	22266.2	[51]
	22266.80 ± 0.90	[493]	0.473	[436]	10.50	[301]	22251.0 ± 5.9	[50]
			0.473	[244]	9.99	[423]	22269.47 ± 0.59	[145]
L_3	18056.80 ± 0.60	[51]	0.514	[301]	8.40	[74]	18056.8	[51]
	18057.00 ± 0.70	[493]	0.463	[436]	7.82	[301]	18055.99 ± 0.35	[50]
			0.4764	[423]	8.47	[423]	18060.10 ± 0.51	[145]
M_1	5932.90 ± 1.40	[51]	7.1(-3)	[423]	22.00	[74]		
					20.33	[423]		
M_2	5541.20 ± 1.70	[51]	1.01(-2)	[423]	12.00	[74]		
					15.33	[423]		
M_3	4556.60 ± 1.50	[51]	1.46(-2)	[423]	16.00	[74]		
					11.96	[423]		
M_4	3972.60 ± 0.60	[51]	5.94(-2)	[423]	4.00	[74]		
					11.96	[423]		
M_5	3778.10 ± 0.60	[51]	5.708-2)	[423]	4.30	[74]		
					3.73	[423]		
N_1	1558.60 ± 0.80	[51]	1.07(-3)	[423]	12.00	[74]		
	1559.30 ± 1.50	[493]			14.43	[423]		
N_2	1372.10 ± 1.80	[51]	1.44(-3)	[423]	8.00	[74]		
	1377.40 ± 4.10	[493]			15.44	[423]		
N_3	1114.80 ± 1.60	[51]	8.1(-4)	[423]	8.00	[74]		
	1120.90 ± 2.80	[493]			9.66	[423]		
N_4	848.90 ± 0.60	[51]	7.7(-4)	[423]	5.00	[74]		

level	E_B/eV		ω_{nlj}		Γ/eV		AE/eV
	848.90 ± 1.00	[493]			6.27	[423]	
N_5	801.40 ± 0.60	[51]	6.5(-4)	[423]	5.00	[74]	
	801.50 ± 2.10	[493]			6.78	[423]	
N_6	445.80 ± 1.70	[51]	1.1(-4)	[423]	0.40	[74]	
	437.40 ± 2.10	[493]			0.64	[423]	
N_7	432.40 ± 2.10	[51]	1.1(-4)	[423]	0.40	[74]	
	425.20 ± 2.20	[493]			0.44	[423]	
O_1	351.90 ± 2.40	[51]	3.0(-5)	[423]	25.79	[423]	
O_2	274.10 ± 4.70	[51]	6.0(-5)	[423]	10.57	[423]	
	282.50 ± 4.60	[493]					
O_3	206.50 ± 4.70	[51]	1.0(-5)	[423]	10.75	[423]	
	215.30 ± 4.90	[493]					
O_4	116.00 ± 1.20	[51]					
O_5	105.40 ± 1.00	[51]					
	105.20 ± 1.00	[493]					
P_1	48.60 ± 4.00	[493]					
P_2	30.60 ± 1.00	[493]					
P_3	18.40 ± 1.00	[493]					
Q_1	6.06	[126]					
IP	6.0262	[222]					

: Note: Transition energies and edge energies from Refs. [43, 50, 145] are tabulated for ^{244}Pu (isotope with the longest life-time).

Am   Z=95

Americium $A = 243.0614$ [222] $\varrho = 11.70$ g/cm^3 [547]   [Rn] 5f^7 7s^2
$\varrho = 13.67$ g/cm^3 [248]

X-ray transitions

line	transition	E/eV		I/eV/$\hbar$		TPIV	Γ/eV	
K series								
K_{α_3}	KL$_1$	101172.2 ± 3.7	[145]				123.57	[423]
		101174.9 ± 2.7	[432]					
$K_{\alpha 2}$	KL$_2$	102044.00	[104]	31.42	[488]	2.919(-1)	129.50	[74]
		102032.00 ± 3.00	[43]				117.59	[423]
		102030.3 ± 3.8	[257]					
$K_{\alpha 1}$	KL$_3$	106481.60	[104]	49.46	[488]	4.595(-1)	115.00	[74]
		106474.00 ± 3.00	[43]				116.07	[423]
		106472.00 ± 3.00	[298]					
		106471.3 ± 4.2	[257]					
	KM$_1$	118849.4 ± 5.5	[145]				127.81	[423]
$K_{\beta 3}$	KM$_2$	119254.40	[104]	5.751	[488]	5.343(-2)	122.81	[423]
		119240.00 ± 2.00	[43]					
		119237.7 ± 2.4	[432]					
$K_{\beta 1}$	KM$_3$	120315.50	[104]	11.30	[488]	1.050(-1)	118.83	[423]
		120280.00 ± 2.00	[43]					
		120280.5 ± 3.5	[432]					
$K_{\beta_5^{II}}$	KM$_4$	120834.90 ± 56.00	[51]	0.2155	[488]	2.003(-3)	111.27	[423]
		120882.3 ± 4.9	[145]					
$K_{\beta_5^{I}}$	KM$_5$	121140.10 ± 56.00	[51]	0.2410	[488]	2.239(-3)	111.19	[423]
		121087.9 ± 4.9	[145]					
	KN$_1$	123361.6 ± 6.6	[145]				121.82	[423]
$K_{\beta_2^{II}}$	KN$_2$	123548.00 ± 3.00	[43]	1.453	[488]	1.351(-2)	122.91	[423]
		123542.7 ± 7.8	[145]					
		123541.5 ± 2.8	[432]					
$K_{\beta_2^{I}}$	KN$_3$	123817.00 ± 3.00	[43]	2.975	[488]	2.764(-2)	117.16	[423]
		123815.9 ± 6.3	[145]					
		123817.5 ± 2.9	[432]					
$K_{\beta_4^{II}}$	KN$_4$	124148.30 ± 56.00	[51]	0.1381	[488]	6.043(-4)	113.72	[423]
		124097.7 ± 6.3	[145]					

line	transition	E/eV		I/eV/$\hbar$		TPIV	Γ/eV	
$K_{\beta4}$	KN_5	124199.40 ± 56.00	[51]	0.1381	[488]	6.790(-4)	114.21	[423]
		124148.6 ± 6.2	[145]					

L series

line	transition	E/eV		I/eV/$\hbar$		TPIV	Γ/eV	
	L_1M_1	17652.40 ± 3.00	[51]	7.61(-5)	[488]		36.70	[423]
		17674.5 ± 1.2	[432]					
		17677.2 ± 3.3	[145]					
$L_{\beta4}$	L_1M_2	18062.96 ± 0.78	[50]	1.025	[488]	6.309(-2)	31.69	[423]
		18062.90	[104]					
		18065.7 ± 3.5	[145]					
$L_{\beta3}$	L_1M_3	19106.24 ± 0.87	[50]	0.818	[488]	5.036(-2)	27.72	[423]
		19106.00	[104]					
$L_{\beta10}$	L_1M_4	19680.80 ± 2.20	[51]	5.07(-2)	[488]	3.125(-3)	20.16	[423]
		19710.1 ± 2.7	[145]					
$L_{\beta9}$	L_1M_5	19886.00 ± 2.60	[51]	7.55(-2)	[488]	4.647(-3)	20.08	[423]
		19915.7 ± 2.7	[145]					
	L_1N_1	22155.80 ± 2.70	[51]	3.58(-5)	[488]		30.70	[423]
		22187.5 ± 2.0	[432]					
		22189.4 ± 4.4	[145]					
$L_{\gamma2}$	L_1N_2	22365.3 ± 2.9	[50]	0.2799	[488]	1.724(-2)	31.80	[423]
		22359.70	[104]					
		22370.4 ± 5.6	[145]					
$L_{\gamma3}$	L_1N_3	22637.20	[51]	0.2606	[488]	1.605(-2)	26.05	[423]
		22642.2 ± 3.1	[432]					
		22643.6 ± 4.0	[145]					
	L_1N_4	22925.5 ± 3.9	[145]	8.90(-3)	[488]	5.478(-4)	22.61	[423]
$L_{\gamma11}$	L_1N_5	22945.30 ± 2.60	[51]	1.43(-2)	[488]	8.811(-4)	23.10	[423]
		22976.4 ± 3.9	[145]					
	L_1N_6	23346.5 ± 3.8	[145]				16.94	[423]
	L_1N_7	23359.4 ± 3.1	[145]				16.73	[423]
	L_1O_4	23657.10 ± 2.90	[51]	4.41(-3)	[488]	1.035(-4)		
	L_1O_5	23669.60 ± 2.70	[51]	4.41(-3)	[488]	1.681(-4)		
L_η	L_2M_1	16823.00 ± 8.50	[51]	0.1088	[488]	1.092(-2)	30.71	[423]
		16819.2 ± 1.3	[432]					
		16818.6 ± 2.6	[145]					
	L_2M_2	17206.5 ± 2.1	[432]	2.10(-5)	[488]		25.71	[423]
		17207.2 ± 2.4	[145]					
$L_{\beta17}$	L_2M_3	18250.0 ± 4.1	[432]	5.94(-3)	[488]	5.959(-4)	21.73	[423]
		18248.9 ± 2.6	[145]					
$L_{\beta1}$	L_2M_4	18852.18 ± 0.38	[50]	3.812	[488]	3.824(-1)	14.17	[423]
		18851.6 ± 1.7	[145]					

line	transition	E/eV		I/eV/$\hbar$		TPIV	Γ/eV	
	L_2M_5	18852.10	[104]	5.64(-4)	[488]		14.09	[423]
		19057.2 $\pm$ 1.6	[145]					
$L_{\gamma5}$	L_2N_1	21332.0 $\pm$ 2.0	[432]	2.99(-2)	[488]	3.000(-3)	24.72	[423]
		21330.9 $\pm$ 3.3	[145]					
	L_2N_2	21510.1 $\pm$ 3.0	[432]	8.91(-6)	[488]		25.81	[423]
		21511.9 $\pm$ 4.5	[145]					
	L_2N_3	21787.0 $\pm$ 3.1	[432]	1.84(-3)	[488]	1.843(-4)	20.06	[423]
		21785.1 $\pm$ 2.9	[145]					
$L_{\gamma1}$	L_2N_4	22065.39 $\pm$ 0.52	[50]	0.887	[488]	8.902(-2)	16.62	[423]
		22065.30	[104]					
	L_2N_5	22117.9 $\pm$ 2.9	[145]				17.11	[423]
L_ν	L_2N_6	22488.0 $\pm$ 2.7	[145]			6.045(-4)	10.96	[423]
	L_2N_7	22500.9 $\pm$ 2.1	[145]				10.74	[423]
$L_{\gamma6}$	L_2O_4	22828.20 $\pm$ 0.80	[50]	0.1803	[488]	1.809(-2)		
		22828.20	[104]					
L_l	L_3M_1	12378.2 $\pm$ 1.4	[50]	0.2020	[488]	2.349(-2)	29.19	[423]
		12383.30	[104]					
L_t	L_3M_2	12765.3 $\pm$ 2.2	[432]	2.63(-3)	[488]	3.056(-4)	24.19	[423]
		12767.5 $\pm$ 2.1	[145]					
L_s	L_3M_3	13809.2 $\pm$ 4.1	[432]	2.40(-3)	[488]	2.795(-4)	20.21	[423]
		13809.2 $\pm$ 2.3	[145]					
$L_{\alpha2}$	L_3M_4	14412.09 $\pm$ 0.22	[50]	0.317	[488]	3.682(-2)	12.65	[423]
		14412.10	[104]					
$L_{\alpha1}$	L_3M_5	14617.33 $\pm$ 0.23	[50]	2.780	[488]	3.232(-1)	12.57	[423]
		14617.30	[104]					
$L_{\beta6}$	L_3N_1	16887.52 $\pm$ 0.65	[50]	5.19(-2)	[488]	6.031(-3)	23.20	[423]
		16887.20	[104]					
	L_3N_2	17068.7 $\pm$ 3.1	[51]	5.77(-4)	[488]	6.707(-5)	24.29	[423]
		17072.3 $\pm$ 4.2	[145]					
	L_3N_3	17346.2 $\pm$ 3.2	[432]	6.63(-4)	[488]	7.710(-5)	18.54	[423]
		17345.5 $\pm$ 2.6	[145]					
$L_{\beta15}$	L_3N_4	17625.90 $\pm$ 0.74	[50]	6.48(-2)	[488]	7.531(-3)	15.10	[423]
		16625.90	[104]					
$L_{\beta2}$	L_3N_5	17676.66 $\pm$ 0.34	[50]	0.595	[488]	6.925(-2)	15.59	[423]
		17676.60	[104]					
$L_{u'}$	L_3N_6	18048.4 $\pm$ 2.4	[145]			1.086(-4)	9.45	[423]
L_u	L_3N_7	18061.3 $\pm$ 1.8	[145]			5.967(-4)	9.23	[423]
$L_{\beta5}$	$L_3O_{4,5}$	18399.60 $\pm$ 0.50	[50]	0.1306	[488]	64.70		

line	transition	E/eV		I/eV/ℏ	TPIV	Γ/eV	
		18399.70	[104]				
	L_3O_4				1.490(-3)		
	L_3O_5				1.370(-2)		

M series

line	transition	E/eV		I/eV/ℏ	TPIV	Γ/eV	
	M_2N_1	4090.60 ± 4.00	[358]		1.120(-3)	29.94	[423]
	M_2N_4	4847.00 ± 5.70	[358]		5.309(-3)	21.84	[423]
$M_{\gamma2}$	M_3N_4	3788.30 ± 3.10	[51]		1.122(-3)	17.87	[423]
$M_{\gamma1}$	M_3N_5	3867.30 ± 3.60	[358]		8.334(-3)	18.36	[423]
	M_3O_1	4332.10 ± 4.50	[358]		6.190(-4)	38.45	[423]
	M_3O_5	4586.90 ± 5.10	[358]		1.889(-3)		
$M_{\xi2}$	M_4N_2	2680.30 ± 9.30	[51]		2.526(-3)	19.50	[423]
M_{δ}	M_4N_3	2956.60	[423]		2.126(-4)	13.75	[423]
M_{β}	M_4N_6	3633.80 ± 0.25	[358]		5.507(-2)	4.65	[423]
M_{η}	M_4O_2	3812.60 ± 3.50	[358]		5.949(-4)	14.86	[423]
$M_{\xi1}$	M_5N_3	2751.20	[51]		1.633(-3)	13.67	[423]
$M_{\alpha2}$	M_5N_6	3437.90 ± 1.70	[51]		2.625(-3)	4.57	[423]
$M_{\alpha1}$	M_5N_7	3442.70 ± 0.22	[358]		5.142(-2)	4.35	[423]

level characteristics

level	E_B/eV		ω_{nlj}		Γ/eV		AE/eV	
K	125027.00 ± 55.00	[51]	0.9713	[39]	109.00	[301]	125027	[51]
	124980.90 ± 4.70	[493]	0.974	[301]			124986.1 ± 4.5	[432]
			0.9687	[423]	107.34	[423]	124984.8 ± 4.2	[145]
L_1	23772.90 ± 2.00	[51]	0.218	[301]	13.30	[301]	23772.9	[51]
	23805.10	[493]	0.173	[436]	16.23	[423]	23808.0 ± 3.0	[432]
							23812.5 ± 1.9	[145]
L_2	22944.00 ± 1.00	[51]	0.471	[301]	10.90	[301]	22944	[51]
	22949.40 ± 2.50	[493]	0.487	[436]	10.24	[423]	22952.0 ± 3.0	[432]
							22954.03 ± 0.80	[145]
L_3	18504.10 ± 0.90	[51]	0.526	[301]	8.04	[301]	18504.1	[51]
	18506.20 ± 3.00	[493]	0.473	[436]	8.73	[423]	18510.0 ± 3.0	[432]
							18514.38 ± 0.50	[145]
M_1	6120.50 ± 7.50	[51]	6.94(-3)	[245]	20.47	[423]		
	6132.60 ± 5.30	[493]	7.40(-3)	[423]				
M_2	5710.20 ± 2.10	[51]	9.82(-3)	[245]	15.47	[423]		
	5747.00 ± 2.00	[493]	1.04(-2)	[423]				
M_3	4667.00 ± 2.10	[51]	1.54(-2)	[245]	11.49	[423]		

level	E_B/eV		ω_{nlj}		Γ/eV		AE/eV
	4706.00 ± 3.00	[493]	1.51(-2)	[423]			
M_4	4092.10 ± 1.00	[51]	6.11(-2)	[423]	3.93	[423]	
M_5	3886.90 ± 1.00	[51]	5.85(-2)	[423]	3.85	[423]	
N_1	1617.10 ± 1.10	[51]	1.11(-3)	[423]	14.48	[423]	
	1619.20 ± 2.90	[493]					
N_2	1411.80 ± 8.30	[51]	1.45(-3)	[423]	15.57	[423]	
	1435.10 ± 5.00	[493]					
N_3	1135.70	[51]	8.3(-4)	[423]	9.82	[423]	
	1168.00 ± 3.00	[493]					
N_4	878.70 ± 1.00	[51]	8.1(-4)	[423]	6.38	[423]	
	880.40 ± 1.90	[493]					
N_5	827.60 ± 1.00	[51]	6.8(-4)	[423]	6.87	[423]	
	830.00 ± 2.50	[493]					
N_6	463.30 ± 0.70	[493]	1.2(-4)	[423]	0.72	[423]	
N_7	449.00 ± 0.70	[493]	1.3(-4)	[423]	0.50	[423]	
O_1	373.00 ± 3.00	[493]	3.0(-5)	[423]	26.96	[423]	
O_2	303.00 ± 4.00	[493]	7.0(-5)	[423]	10.93	[423]	
O_3	216.40 ± 1.20	[493]	2.0(-5)	[423]	12.09	[423]	
O_4	115.80 ± 1.30	[51]					
	118.00 ± 1.30	[493]					
O_5	103.30 ± 1.10	[51]					
	107.90 ± 2.10	[493]					
P_1	50.40 ± 0.80	[493]					
P_2	31.10 ± 1.00	[493]					
P_3	18.10 ± 1.00	[493]					
Q_1	5.99	[126]					
IP	5.9738	[222]					

: Note: Transition energies and edge energies from Refs. [50,145,257,432] are tabulated for ^{241}Am.

Cm   Z=96

[Rn] 5f^7 6d^1 7s^2

Curium

$A = 247.0704$ [222] $\varrho = 13.67$ g/cm^{-3} [342]
$\varrho = 13.51$ g/cm^{-3} [248]

X-ray transitions

line	transition	E/eV		I/eV/$\hbar$	TPIV	Γ/eV	
K series							
K$_{\alpha_3}$	KL$_1$	103707.4 ± 3.8	[145]			128.19	[423]
K$_{\alpha2}$	KL$_2$	104591.0 ± 2.0	[43]	2.929(-1)	122.33	[423]	
		104589.0 ± 5.0	[117]				
		104589.0 ± 5.0	[150]				
		104591.0 ± 2.0	[277]				
		104589.0 ± 5.0	[257]				
		104590.0 ± 2.0	[43]				
K$_{\alpha1}$	KL$_3$	109273.0 ± 2.0	[43]	4.584(-1)	120.74	[423]	
		109273.0 ± 2.0	[277]				
		109268.0 ± 3.0	[298]				
		109272.9 ± 5.0	[257]				
		109272.3 ± 2.0	[43]				
	KM$_1$	121897.0 ± 5.7	[145]			132.27	[423]
K$_{\beta3}$	KM$_2$	122303.0 ± 2.0	[43]	5.343(-2)	127.28	[423]	
		122303.0 ± 2.0	[277]				
		122288.9 ± 5.0	[117]				
		122302.2 ± 2.0	[43]				
K$_{\beta1}$	KM$_3$	123404.0 ± 2.0	[43]	1.052(-1)	123.30	[423]	
		123404.0 ± 2.0	[277]				
		123406.9 ± 5.0	[117]				
		123393.2 ± 6.1	[43]				
K$_{\beta_5^{II}}$	KM$_4$	124001.00 ± 5.00	[43]	2.048(-3)	115.75	[423]	
		124007.0 ± 5.1	[145]				
K$_{\beta_5^{I}}$	KM$_5$	124215.0 ± 2.0	[43]	2.274(-3)	115.67	[423]	
		124223.0 ± 5.1	[145]				
	KN$_1$	126550.9 ± 6.8	[145]			126.28	[423]
K$_{\beta_2^{II}}$	KN$_2$	126727.2 ± 3.0	[43]	1.355(-2)	127.37	[423]	
		126736.1 ± 8.1	[145]				
K$_{\beta_2^{I}}$	KN$_3$	1227039.2 ± 2.0	[43]	2.785(-2)	121.62	[423]	
		127026.4 ± 6.5	[43]				

line	transition	E/eV		I/eV/$\hbar$	TPIV	Γ/eV	
	$KN_{2,3}$	126982 ± 15	[117]				
$K_{\beta_4^{II}}$	KN_4	127314.2 ± 6.4	[145]		6.228(-4)	118.18	[423]
$K_{\beta4}$	KN_5	127367.9 ± 6.3	[145]		6.956(-4)	118.67	[423]
	KO,KP	112804	[117]				

L series

line	transition	E/eV		I/eV/$\hbar$	TPIV	Γ/eV	
	L_1M_1	18189.6 ± 3.3	[145]			36.87	[423]
$L_{\beta4}$	L_1M_2	18568.0 ± 25.0	[117]		6.635(-2)	31.87	[423]
		18585.7 ± 3.3	[145]				
$L_{\beta3}$	L_1M_3	19680.0 ± 25.0	[117]		5.133(-2)	27.89	[423]
		19686.9 ± 3.8	[145]				
$L_{\beta10}$	L_1M_4	20287.0 ± 22.0	[117]		3.361(-3)	20.34	[423]
		20299.7 ± 2.8	[145]				
$L_{\beta9}$	L_1M_5	20500.0 ± 24.0	[117]		4.996(-3)	20.25	[423]
		20515.7 ± 2.8	[145]				
	L_1N_1	22843.5 ± 4.5	[145]			30.87	[423]
$L_{\gamma2}$	L_1N_2	23018.0 ± 25.0	[117]		1.819(-2)	31.96	[423]
		23028.8 ± 5.7	[145]				
$L_{\gamma3}$	L_1N_3	23318.0 ± 25.0	[117]		1.653(-2)	26.21	[423]
		23319.10 ± 4.0	[145]				
	L_1N_4	23606.8 ± 4.1	[145]		5.952(-4)	22.76	[423]
$L_{\gamma11}$	L_1N_5	23606.8 ± 4.1	[145]		9.575(-4)	23.26	[423]
	L_1N_6	24038.4 ± 3.8	[145]			17.11	[423]
	L_1N_7	24052.4 ± 3.2	[145]			16.89	[423]
$L_{\gamma4'}$	L_1O_2	24219.00 ± 26.0	[117]		4.623(-3)	27.32	[423]
$L_{\gamma4}$	L_1O_3	24286.00 ± 24.00	[117]		4.277(-3)	28.48	[423]
L_η	L_2M_1	17331.0 ± 16.0	[117]		1.118(-2)	31.00	[423]
		17315.1 ± 2.5	[145]				
	L_2M_2	17711.3 ± 2.4	[145]			26.00	[423]
$L_{\beta17}$	L_2M_3	18812.5 ± 2.6	[145]		6.216(-4)	22.02	[423]
$L_{\beta1}$	L_2M_4	19426.0 ± 2.0	[117]		3.894(-1)	14.47	[423]
		19425.2 ± 1.7	[145]				
	L_2M_5	19641.2 ± 1.7	[145]			14.38	[423]
$L_{\gamma5}$	L_2N_1	21980.0 ± 8.0	[117]		3.084(-3)	25.00	[423]
		21969.1 ± 3.4	[145]				
	L_2N_2	22154.3 ± 4.6	[145]			26.09	[423]
	L_2N_3	22444.6 ± 2.9	[145]		1.938(-4)	20.34	[423]
$L_{\gamma1}$	L_2N_4	22730.0 ± 2.0	[117]		9.147(-2)	16.90	[423]

line	transition	E/eV		I/eV/$\hbar$	TPIV	Γ/eV	
		22732.4 ± 3.0	[145]				
	L_2N_5	22786.1 ± 2.9	[145]			17.39	[423]
L_v	L_2N_6	23163.9 ± 2.7	[145]		6.383(-4)	11.24	[423]
	L_2N_7	23177.9 ± 2.1	[145]			11.02	[423]
$L_{\gamma6}$	L_2O_4	23519.00 ± 2.00	[117]		1.899(-2)		
L_l	L_3M_1	12650.0 ± 2.0	[117]		2.423(-2)	29.42	[423]
		12633.8 ± 2.1	[145]				
L_t	L_3M_2	13030.0 ± 2.1	[145]		3.167(-4)	24.41	[423]
L_s	L_3M_3	14131.2 ± 2.4	[145]		2.903(-4)	20.44	[423]
$L_{\alpha2}$	L_3M_4	14745.0 ± 14.0	[117]		3.733(-2)	12.88	[423]
		14743.9 ± 1.4	[145]				
$L_{\alpha1}$	L_3M_5	14959.0 ± 2.0	[117]		3.277(-1)	12.80	[423]
		14959.5 ± 1.4	[145]				
$L_{\beta6}$	L_3N_1	17299.0 ± 16.0	[117]		6.229(-3)	23.41	[423]
		17287.8 ± 3.1	[145]				
	L_3N_2	17473.0 ± 4.3	[145]		6.915(-5)	24.51	[423]
	L_3N_3	17763.3 ± 2.6	[145]		8.048(-5)	18.76	[423]
$L_{\beta15}$	L_3N_4	18049.0 ± 16.0	[117]		7.677(-3)	15.31	[423]
		18051.1 ± 2.7	[145]				
$L_{\beta2}$	L_3N_5	18113.0 ± 2.0	[117]		7.069(-2)	15.81	[423]
		18104.8 ± 2.6	[145]				
$L_{u'}$	L_3N_6	18482.6 ± 2.4	[145]		1.151(-4)	9.66	[423]
L_u	L_3N_7	18496.6 ± 1.8	[145]		6.310(-4)	9.42	[423]

level characteristics

level	E_B/eV		ω_{nlj}		Γ/eV		AE/eV	
K	128220	[51]	0.9693	[423]	111.80	[423]	128242.8 ± 4.4	[145]
							128241.3 ± 2.5	[433]
			0.974	[301]			128200	[463]
			0.971	[253]			128200	[51]
L_1	24460	[51]	0.181	[436]	16.39	[423]	24515 ± 21	[145]
			0.1860	[423]			24460	[463]
							24460	[51]

level	E_B/eV		ω_{nlj}		Γ/eV		AE/eV	
L_2	23779	[51]	0.501	[436]	10.52	[423]	23651 ± 11	[433]
							23660.98 ± 0.80	[145]
			0.497	[244]			23779	[463]
			0.5168	[423]			23779	[51]
L_3	18930	[51]	0.477	[244]	8.94	[423]	1897.0 ± 11	[433]
			0.482	[436]			18979.68 ± 0.50	[145]
							18930	[463]
			0.0.4933	[423]			18930	[51]
M_1	6288	[51]	7.77(-3)	[423]	20.48	[423]		
M_2	5895	[51]	1.08(-2)	[423]	15.48	[423]		
M_3	4797	[51]	1.57(-2)	[423]	11.50	[423]		
M_4	4227	[51]	6.28(-2)	[423]	3.94	[423]		
M_5	3971	[51]	5.99(-2)	[423]	3.86	[423]		
N_1	1643	[51]	1.16(-3)	[423]	14.48	[423]		
N_2	1440	[139	1.47(-3)	[423]	15.56	[423]		
N_3	1154	[51]	8.5(-4)	[423]	9.82	[423]		
N_4			8.5(-4)	[423]	6.38	[423]		
N_5			7.1(-4)	[423]	6.87	[423]		
N_6			1.4(-4)	[423]	0.72	[423]		
N_7			1.6(-4)	[423]	0.50	[423]		
O_1	385	[51]	3(-5)	[423]	26.96	[423]		
O_2			8(-5)	[423]	10.93	[423]		
O_3			2(-5)	[423]	12.09	[423]		
O_4								
O_5								
IP	6.02	[222]						

Note: Transition energies and edge energies from [117,145,257] are given for ^{245}Cm.

Bk   Z=97   [Rn] 5f⁹ 7s²

$[Rn] 5f^9 7s^2$

Berkelium

$A = 247.0703$ [239] $\varrho = 14.79$ g/cm^{-3} [547]
$\varrho = 13.25$ g/cm^{-3} [248]

X-ray transitions

line	transition	E/eV		I/eV/$\hbar$	TPIV	Γ/eV	
K series							
K$_{\alpha_3}$	KL$_1$	106318 $\pm$ 65	[239]			133.21	[423]
		106290.5 $\pm$ 4.0	[145]				
K$_{\alpha2}$	KL$_2$	107194.3 $\pm$ 5.0	[43]		2.941(-1)	127.72	[423]
		107165.0 $\pm$ 6.0	[149]				
		107181.0 $\pm$ 6.0	[277]				
		107164.4 $\pm$ 6.0	[257]				
		107194.3 $\pm$ 5.0	[43]				
K$_{\alpha1}$	KL$_3$	112127.3 $\pm$ 5.0	[43]		4.577(-1)	126.03	[423]
		112112.0 $\pm$ 6.0	[149]				
		112130.0 $\pm$ 3.0	[277]				
		112126.0 $\pm$ 5.0	[257]				
		112126.0 $\pm$ 5.0	[298]				
		112111.4 $\pm$ 6.0	[257]				
		112127.3 $\pm$ 5.0	[43]				
	KM$_1$	125005.5 $\pm$ 5.8	[145]			137.12	[423]
K$_{\beta3}$	KM$_2$	125414.2 $\pm$ 7.0	[43]		5.347(-2)	132.55	[423]
		125478.0 $\pm$ 10.0	[150]				
		125409.4 $\pm$ 7.0	[43]				
K$_{\beta1}$	KM$_3$	126577.2 $\pm$ 7.0	[43]		1.054(-1)	128.70	[423]
		126582.0 $\pm$ 10.0	[150]				
		126577.2 $\pm$ 7.0	[43]				
K$_{\beta_5^{II}}$	KM$_4$	127220.0	[98]		2.094(-3)	121.01	[423]
		127196.0 $\pm$ 5.3	[145]				
K$_{\beta_5^{I}}$	KM$_5$	127453.0	[98]		2.310(-3)	120.89	[423]
		127422.6 $\pm$ 5.3	[145]				
	KN$_1$	129810.4 $\pm$ 7.1	[145]			131.69	[423]
K$_{\beta_2^{II}}$	KN$_2$	130032.0	[98]		1.360(-2)	132.94	[423]
		130000.0 $\pm$ 8.3	[145]				
K$_{\beta_2^{I}}$	KN$_3$	130350.0	[98]		2.808(-2)	127.11	[423]
		130308.10 $\pm$ 6.7	[145]				

line	transition	E/eV		I/eV/$\hbar$	TPIV	Γ/eV	
$K_{\beta_4^{II}}$	KN_4	130601.5 ± 6.7	[145]		6.42(-4)	123.413	[423]
$K_{\beta_4^I}$	KN_5	130658.4 ± 6.5	[145]		7.124(-4)	123.07	[423]
L series							
	L_1M_1	18719.00 ± 40.00	[239]			36.93	[423]
		18715.10 ± 3.40	[145]				
$L_{\beta 4}$	L_1M_2	19128.00 ± 53.00	[239]		6.992(-2)	32.36	[423]
		19118.90 ± 3.60	[145]				
$L_{\beta 3}$	L_1M_3	20298 ± 53	[239]		5.241(-2)	28.51	[423]
		20282.6 ± 3.8	[145]				
$L_{\beta 10}$	L_1M_4	20905.5 ± 2.8	[145]		3.622(-3)	20.82	[423]
$L_{\beta 9}$	L_1M_5	21132.1 ± 2.7	[145]		5.377(-3)	20.70	[423]
	L_1N_1	23521 ± 42	[239]			31.502	[423]
		23520.0 ± 4.6	[145]				
$L_{\gamma 2}$	L_1N_2	23709.6 ± 5.9	[145]		1.924(-2)	32.75	[423]
$L_{\gamma 3}$	L_1N_3	24017.7 ± 4.1	[145]		1.705(-2)	26.92	[423]
	L_1N_4	24311.6 ± 4.2	[145]			23.23	[423]
$L_{\gamma 11}$	L_1N_5	24368.0 ± 4.1	[145]		1.043(-3)	22.88	[423]
	L_1N_6	24753.9 ± 3.8	[145]			17.56	[423]
	L_1N_7	24768.5 ± 3.3	[145]			17.22	[423]
L_η	L_2M_1	17829 ± 40	[239]		1.123(-2)	31.45	[423]
		17824.8 ± 2.5	[145]				
	L_2M_2	18128 ± 53	[239]			26.873	[423]
		18228.6 ± 2.6	[145]				
$L_{\beta 17}$	L_2M_3	19408 ± 53	[239]		6.361(-4)	23.03	[423]
		19392.3 ± 3.7	[145]				
$L_{\beta 1}$	L_2M_4	20015.2 ± 1.7	[145]		3.891(-1)	15.33	[423]
	L_2M_5	20241.9 ± 1.6	[145]			15.22	[423]
$L_{\gamma 5}$	L_2N_1	22631 ± 42	[239]		3.109(-3)	26.02	[423]
		22629.7 ± 3.5	[145]				
	L_2N_2	22819.3 ± 4.8	[145]			27.27	[423]
	L_2N_3	23127.4 ± 3.0	[145]		1.997(-4)	21.44	[423]
$L_{\gamma 1}$	L_2N_4	23421.3 ± 3.0	[145]		9.218(-2)	17.74	[423]
	L_2N_5	23477.7 ± 2.9	[145]			17.39	[423]
L_ν	L_2N_6	23863.7 ± 2.7	[145]		6.603(-4)	12.08	[423]
	L_2N_7	23878.2 ± 2.1	[145]			11.74	[423]
L_l	L_3M_1	12896 ± 43	[239]		2.496(-2)	29.75	[423]
		12889.9 ± 2.1	[145]				

line	transition	E/eV		I/eV/ℏ	TPIV	Γ/eV	
L_t	L_3M_2	13305 ± 55	[239]		3.278(-4)	25.18	[423]
		13293.7 ± 2.1	[145]				
L_s	L_3M_3	14475 ± 55	[239]		3.012(-4)	21.34	[423]
		14457.4 ± 2.4	[145]				
$L_{\alpha2}$	L_3M_4	15080.3 ± 1.4	[145]		3.782(-2)	13.64	[423]
$L_{\alpha1}$	L_3M_5	15306.9 ± 1.3	[145]		3.320(-1)	13.52	[423]
$L_{\beta6}$	L_3N_1	17697 ± 45	[239]		6.427(-3)	24.32	[423]
		17694.8 ± 3.2	[145]				
	L_3N_2	17884.4 ± 4.4	[145]		7.116(-5)	25.57	[423]
	L_3N_3	18192.5 ± 2.7	[145]		8.391(-5)	19.74	[423]
$L_{\beta15}$	L_3N_4	18486.4 ± 2.7	[145]		7.817(-3)	16.04	[423]
$L_{\beta2}$	L_3N_5	18542.8 ± 2.6	[145]		7.207(-2)	15.70	[423]
$L_{u'}$	L_3N_6	18928.7 ± 2.4	[145]		1.215(-4)	10.38	[423]
L_u	L_3N_7	18943.3 ± 1.8	[145]		6.668(-4)	10.04	[423]

level characteristics

level	E_B/eV		ω_{nlj}		Γ/eV		AE/eV	
K	131590 ± 40	[239]	0.975	[301]	117.06	[423]	131561.4 ± 4.5	[145]
			0.9705	[423]			131590	[51]
							131555.6 ± 4.7	[433]
L_1	25275 ± 17	[239]	0.236	[301]	16.51	[423]	25.272 ± 25	[239]
			0.1950	[423]			25271.0 ± 2.2	[145]
L_2	24385 ± 17	[239]	0.485	[301]	11.02	[423]	24382 ± 25	[239]
			0.5177	[423]			24380.68 ± 0.90	[145]
L_3	19452 ± 20	[239]	0.55	[301]	9.33	[423]	19449 ± 30	[239]
			0.5012	[423]			19445.76 ± 0.50	[145]
M_1	6556 ± 21	[239]	8.15(-3)	[423]	20.42	[423]		
M_2	6147 ± 31	[239]	1.11(-2)	[423]	15.85	[423]		
M_3	4977 ± 31	[239]	1.63(-2)	[423]	12.01	[423]		
M_4	4366	[51]	6.43(-2)	[423]	4.31	[423]		

level	E_B/eV		ω_{nlj}		Γ/eV		AE/eV
M_5	4132	[51]	6.12(-2)	[423]	4.31	[423]	
N_1	1755 ± 22	[239]	1.2(-3)	[423]	15.22	[423]	
N_2	1554	[51]	1.48(-3)	[423]	16.24	[423]	
N_3	1235	[51]	8.7(-4)	[423]	10.41	[423]	
N_4			8.9(-4)	[423]	6.71	[423]	
N_5			7.3(-4)	[423]	6.37	[423]	
N_6			1.6(-4)	[423]	1.05	[423]	
N_7			1.8(-4)	[423]	0.71	[423]	
O_1	398 ± 22	[239]	3.0(-5)	[423]	30.28	[423]	
O_2			1.0(-4)	[423]	11.98	[423]	
O_3			2.0(-5)	[423]	17.88	[423]	
IP	6.23	[222]					

Note: Transition energies and edge energies from [43,145,239,433] are given for ^{249}Bk.

Cf   Z=98

[Rn] $5f^{10}\,7s^2$

Californium

$A = 251.0796$ [239] $\varrho = 15.10\ \text{g/cm}^3$ [342]

X-ray transitions

line	transition	E/eV		I/eV/$\hbar$	TPIV	Γ/eV	
K series							
K_{α_3}	KL_1	108947.0 ± 17.0	[180]			138.08	[423]
		108922.8 ± 4.1	[145]				
K_{α_2}	KL_2	109837.3 ± 8.0	[43]		2.949(-1)	132.71	[423]
		109818.0 ± 5.0	[149]				
		109829.0 ± 4.7	[145]				
K_{α_1}	KL_3	115035.3 ± 8.0	[43]		4.563(-1)	130.94	[423]
		115031.0 ± 5.0	[149]				
		115045.0 ± 6.0	[298]				
		115030.0 ± 4.6	[145]				
	KM_1	128184.1 ± 6.0	[145]			141.82	[423]
K_{β_3}	KM_2	128599.0 ± 7.0	[150]		5.342(-2)	137.25	[423]
		128595.6 ± 6.3	[145]				
K_{β_1}	KM_3	129816.0 ± 7.0	[150]		1.056(-1)	133.41	[423]
		129845.0 ± 12.0	[180]				
		129825.1 ± 6.5	[145]				
$K_{\beta_5^{II}}$	KM_4	130475.0	[98]		2.139(-3)	125.719	[423]
		130458.3 ± 5.5	[145]				
$K_{\beta_5^{I}}$	KM_5	130720.0	[98]		2.343(-3)	125.60	[423]
		130695.8 ± 5.5	[145]				
	KN_1	133140.1 ± 7.4	[145]			136.38	[423]
$K_{\beta_2^{II}}$	KN_2	133375.0	[98]		1.362(-2)	137.63	[423]
		133334.1 ± 8.7	[145]				
$K_{\beta_2^{I}}$	KN_3	133694.0	[98]		2.827(-2)	131.80	[423]
		133661.0 ± 6.9	[145]				
$K_{\beta_4^{II}}$	KN_4	133961.0 ± 6.9	[145]		6.607(-4)	128.10	[423]
$K_{\beta_4^{I}}$	KN_5	134020.4 ± 6.7	[145]		7.279(-4)	127.76	[423]
L series							
	L_1M_1	19259 ± 14	[5]			37.12	[423]
		19257.6 ± 8.6	[180]				

line	transition	E/eV		I/eV/$\hbar$	TPIV	Γ/eV	
$L_{\beta 4}$	L_1M_2	19678 ± 51	[5]		7.340(-2)	32.55	[423]
		19676.20 ± 6.30	[180]				
$L_{\beta 3}$	L_1M_3	20902 ± 58	[5]		5.321(-2)	28.70	[423]
		20892.0 ± 9.4	[180]				
$L_{\beta 10}$	L_1M_4	21556 ± 70	[5]		3.886(-3)	21.02	[423]
		21535.5 ± 2.9	[145]				
$L_{\beta 9}$	L_1M_5	21776 ± 70	[5]		5.763(-3)	20.89	[423]
		21773.1 ± 2.8	[145]				
	L_1N_1	24201.9 ± 6.3	[180]			31.68	[423]
		24237 ± 61	[5]				
		24217.3 ± 4.7	[145]				
$L_{\gamma 2}$	L_1N_2	24411 ± 72	[5]		2.025(-2)	32.93	[423]
		24404.8 ± 8.2	[180]				
		24411.3 ± 6.0	[145]				
$L_{\gamma 3}$	L_1N_3	24745 ± 72	[5]		1.748(-2)	27.10	[423]
		24721.2 ± 9.2	[180]				
		24738.3 ± 4.2	[145]				
	L_1N_4	25038.3 ± 4.2	[145]		7.013(-4)	23.41	[423]
$L_{\gamma 11}$	L_1N_5	25097.6 ± 4.1	[145]		1.130(-3)	23.06	[423]
	L_1N_6	25491.5 ± 3.8	[145]			17.74	[423]
	L_1N_7	25507.0 ± 3.3	[145]			17.35	[423]
L_η	L_2M_1	18366 ± 12	[5]		1.151(-2)	31.75	[423]
		118347.1 ± 8.7	[180]				
		18355.1 ± 2.4	[145]				
	L_2M_2	18772 ± 19	[5]			27.18	[423]
		18767.3 ± 6.1	[180]				
		18766.6 ± 2.5	[145]				
$L_{\beta 17}$	L_2M_3	20001 ± 16	[5]		6.636(-4)	23.33	[423]
		19990.0 ± 8.5	[180]				
		19996.1 ± 1.7	[145]				
$L_{\beta 1}$	L_2M_4	20656 ± 70	[5]		3.967(-1)	15.65	[423]
		20629.3 ± 1.7	[145]				
	L_2M_5	20876 ± 70	[5]			15.524	[423]
		20866.8 ± 1.6	[145]				
$L_{\gamma 5}$	L_2N_1	23345 ± 62	[5]		3.200(-3)	26.31	[423]
		23311.1 ± 3.5	[145]				
	L_2N_2	23485 ± 23	[5]			27.56	[423]
		23501.1 ± 7.6	[180]				
		23505.1 ± 4.9	[145]				

line	transition	E/eV		I/eV/$\hbar$	TPIV	Γ/eV	
	L_2N_3	23822 ± 24	[5]		2.101(-4)	21.73	[423]
		23819.1 ± 8.4	[180]				
		23832.1 ± 3.0	[145]				
$L_{\gamma 1}$	L_2N_4	24132.0 ± 3.2	[145]		9.481(-2)	18.03	[423]
	L_2N_5	24191.4 ± 3.0	[145]			17.68	[423]
L_ν	L_2N_6	24585.2 ± 2.7	[145]		6.960(-4)	12.36	[423]
	L_2N_7	24600.7 ± 2.2	[145]			12.02	[423]
L_l	L_3M_1	13145.0 ± 14.0	[180]		2.570(-2)	29.99	[423]
		13141 ± 14	[5]				
		13154.1 ± 2.1	[145]				
L_t	L_3M_2	13568 ± 15	[5]		3.389(-4)	25.42	[423]
		13557.0 ± 8.2	[180]				
		13565.6 ± 2.1	[145]				
L_s	L_3M_3	14797 ± 12	[5]		3.122(-4)	21.57	[423]
		14785.0 ± 8.5	[180]				
		14795.1 ± 2.4	[145]				
$L_{\alpha 2}$	L_3M_4	15434 ± 69	[5]		3.829(-2)	13.89	[423]
		15428.3 ± 1.4	[145]				
$L_{\alpha 1}$	L_3M_5	15654 ± 69	[5]		3.360(-1)	13.77	[423]
		15665.8 ± 1.3	[145]				
$L_{\beta 6}$	L_3N_1	18128 ± 37	[5]		6.625(-3)	24.55	[423]
		18090.8 ± 8.5	[180]				
		18110.1 ± 3.2	[145]				
	L_3N_2	18278 ± 18	[5]		7.313(-5)	25.80	[423]
		18304.1 ± 4.5	[145]				
	L_3N_3	18610 ± 18	[5]		8.736(-5)	19.97	[423]
		18613.9 ± 8.4	[180]				
		18631.1 ± 2.7	[145]				
$L_{\beta 15}$	L_3N_4	18931.0 ± 2.8	[145]		7.95(-3)	16.269	[423]
$L_{\beta 2}$	L_3N_5	18990.4 ± 2.6	[145]		7.341(-2)	15.92	[423]
$L_{u'}$	L_3N_6	19384.2 ± 2.3	[145]		1.282(-4)	10.60	[423]
L_u	L_3N_7	19399.7 ± 1.9	[145]		7.008(-4)	10.26	[423]

level characteristics

level	E_B/eV		ω_{nlj}		Γ/eV		AE/eV	
K	135960	[51]	0.975	[301]	121.39	[423]	134956.9 ± 4.7	[145]
			0.9705	[423]			135960.0	[51]
			0.974	[253]				
L_1	26110	[51]	0.244	[301]	16.69	[423]	26002.4 ± 0.9	[451]
			0.2037	[423]			26016 ±	[180]
			0.197	[244]			26110.0	[51]
							26032.9 ± 2.0	[145]
L_2	25250	[51]	0.49	[301]	11.32	[423]	25097.8 ± 4.5	[451]
			0.5292	[423]			25108 ±	[180]
			0.524	[244]			25250.0	[51]
							25126.63 ± 0.90	[145]
L_3	19930	[51]	056	[301]	9.55	[423]	19901 ±	[180]
			0.5087	[423]			19901.5 ± 0.5	[451]
			0.494	[244]			19930.0	[51]
							19925.62 ± 0.50	[145]
M_1	6754	[51]	8.5(-3)	[423]	20.43	[423]		
M_2	6359	[51]	1.15(-2)	[423]	15.86	[423]		
M_3	5109	[51]	1.17(-2)	[423]	12.02	[423]		
M_4	4497	[51]	6.57(-2)	[423]	4.33	[423]		
M_5	4253	[51]	6.24(-2)	[423]	4.21	[423]		
N_1	1799	[51]	1.25(-3)	[423]	15.00	[423]		
N_2	1616	[51]	1.50(-3)	[423]	16.24	[423]		
N_3	1279	[51]	8.8(-4)	[423]	10.41	[423]		
N_4			9.3(-4)	[423]	6.72	[423]		
N_5			7.6(-4)	[423]	6.37	[423]		
N_6			1.7(-4)	[423]	1.05	[423]		
N_7			2.1(-4)	[423]	0.71	[423]		
O_1	419	[51]	3.0(-5)	[423]	30.28	[423]		
O_2			1.1(-4)	[423]	11.98	[423]		
O_3			2.0(-5)	[423]	17.88	[423]		
IP	6.30	[222]						

Note: Transition energies and edge energies from [5, 51, 145, 180, 301] are given for [251]Cf.

Es   Z=99

[Rn] 5f^{11} 7s^2

Einsteinium

$A = 252.0830$ [239] $\varrho = 13.5$ g/cm^{-3} [342]

X-ray transitions

line	transition	E/eV		I/eV/ℏ	TPIV	Γ/eV	
K series							
K$_{\alpha_3}$	KL$_1$	111607.8 ± 4.3	[145]			143.21	[423]
K$_{\alpha_2}$	KL$_2$	112581.0	[98]		2.961(-1)	137.97	[423]
		112501.0 ± 10.0	[149]				
		112500.9 ± 10.0	[257]				
		112530.0 ± 4.8	[145]				
K$_{\alpha_1}$	KL$_3$	118057.0	[98]		4.555(-1)	136.21	[423]
		118018.0 ± 10.0	[149]				
		118026.0 ± 6.0	[298]				
	KM$_1$	131422.2 ± 6.2	[145]			146.89	[423]
K$_{\beta_3}$	KM$_2$	131874.0	[98]		5.344(-2)	142.44	[423]
		131848.0 ± 20.0	[150]				
		131841.6 ± 6.4	[145]				
K$_{\beta_1}$	KM$_3$	133193.0	[98]		1.058(-1)	138.52	[423]
		133188.0 ± 20.0	[150]				
		133140.1 ± 6.7	[145]				
K$_{\beta_5^{II}}$	KM$_4$	133815.0	[98]		2.186(-3)	130.84	[423]
		133783.7 ± 5.7	[145]				
K$_{\beta_5^{I}}$	KM$_5$	134071.0	[98]		2.377(-3)	130.69	[423]
		134032.5 ± 5.7	[145]				
	KN$_1$	136523.0 ± 7.7	[145]			141.66	[423]
K$_{\beta_2^{II}}$	KN$_2$	136766.0	[98]		1.366(-2)	142.79	[423]
		136721.4 ± 9.0	[145]				
K$_{\beta_2^{I}}$	KN$_3$	137124.0	[98]		2.849(-2)	136.97	[423]
		137068.3 ± 7.1	[145]				
K$_{\beta_4^{II}}$	KN$_4$	137374.4 ± 7.1	[145]		6.803(-4)	133.19	[423]
K$_{\beta_4^{I}}$	KN$_5$	137436.8 ± 6.9	[145]		7.444(-4)	132.86	[423]
L series							
	L$_1$M$_1$	19814.4 ± 3.3	[145]			37.38	[423]
L$_{\beta_4}$	L$_1$M$_2$	20233.8 ± 3.7	[145]		7.711(-2)	32.93	[423]
L$_{\beta_3}$	L$_1$M$_3$	21532.3 ± 3.9	[145]		5.403(-2)	29.01	[423]
L$_{\beta_{10}}$	L$_1$M$_4$	22175.9 ± 3.0	[145]		4.172(-3)	21.33	[423]

line	transition	E/eV		I/eV/$\hbar$	TPIV	Γ/eV	
$L_{\beta 9}$	$L_1 M_5$	22424.7 ± 2.8	[145]		6.184(-3)	21.18	[423]
	$L_1 N_1$	24915.2 ± 4.8	[145]			32.15	[423]
$L_{\gamma 2}$	$L_1 N_2$	25113.6 ± 6.2	[145]		2.133(-2)	33.28	[423]
$L_{\gamma 3}$	$L_1 N_3$	25460.5 ± 4.2	[145]		1.792(-2)	27.46	[423]
	$L_1 N_4$	25766.6 ± 4.4	[145]		7.602(-4)	23.69	[423]
$L_{\gamma 11}$	$L_1 N_5$	25829.0 ± 4.1	[145]		1.225(-3)	23.35	[423]
	$L_1 N_6$	26230.8 ± 3.8	[145]			18.01	[423]
	$L_1 N_7$	26247.3 ± 3.4	[145]			17.63	[423]
L_η	$L_2 M_1$	18892.2 ± 2.4	[145]		1.175(-2)	32.14	[423]
	$L_2 M_2$	19311.6 ± 2.5	[145]			27.694	[423]
$L_{\beta 17}$	$L_2 M_3$	20610.1 ± 2.8	[145]		6.899(-4)	23.78	[423]
$L_{\beta 1}$	$L_2 M_4$	21253.7 ± 1.8	[145]		4.028(-1)	16.10	[423]
	$L_2 M_5$	21502.5 ± 1.7	[145]			15.94	[423]
$L_{\gamma 5}$	$L_2 N_1$	23993.0 ± 3.6	[145]		3.280(-3)	26.91	[423]
	$L_2 N_2$	24191.3 ± 5.0	[145]			28.04	[423]
	$L_2 N_3$	24538.3 ± 3.0	[145]		2.200(-4)	22.22	[423]
$L_{\gamma 1}$	$L_2 N_4$	24844.3 ± 3.2	[145]		9.710(-2)	18.45	[423]
	$L_2 N_5$	24906.8 ± 2.9	[145]			18.11	[423]
L_ν	$L_2 N_6$	25308.5 ± 2.7	[145]		7.299(-4)	12.77	[423]
	$L_2 N_7$	25325.1 ± 2.3	[145]			12.39	[423]
L_l	$L_3 M_1$	13411.7 ± 2.1	[145]		2.643(-2)	30.38	[423]
L_t	$L_3 M_2$	13831.1 ± 2.1	[145]		3.497(-4)	25.93	[423]
L_s	$L_3 M_3$	15129.6 ± 2.4	[145]		3.232(-4)	22.01	[423]
$L_{\alpha 2}$	$L_3 M_4$	15773.2 ± 1.4	[145]		3.872(-2)	14.33	[423]
$L_{\alpha 1}$	$L_3 M_5$	16022.0 ± 1.3	[145]		3.397(-1)	14.18	[423]
$L_{\beta 6}$	$L_3 N_1$	18512.5 ± 3.3	[145]		6.816(-3)	25.15	[423]
		18710.8 ± 4.6	[145]				
	$L_3 N_2$	18710.8 ± 4.6	[145]		7.494(-5)	26.28	[423]
	$L_3 N_3$	19057.8 ± 2.7	[145]		9.084(-5)	20.46	[423]
$L_{\beta 15}$	$L_3 N_4$	19363.8 ± 2.8	[145]		8.071(-3)	16.68	[423]
$L_{\beta 2}$	$L_3 N_5$	19426.3 ± 2.5	[145]		7.467(-2)	16.35	[423]
$L_{u'}$	$L_3 N_6$	19828.1 ± 2.4	[145]		1.350(-4)	11.01	[423]
L_u	$L_3 N_7$	19844.6 ± 1.9	[145]		7.366(-4)	10.63	[423]

level characteristics

level	E_B/eV		ω_{nlj}		Γ/eV		AE/eV	
K	139490	[51]	0.975 0.976	[301] [253]	126.36	[423]	138399.9 ± 5.0 138.391.5 ± 6.3 [180] 138490.0	[145] [51]
L_1	26900	[51]	0.253	[301]	16.85	[423]	26792.1 ± 2.1 [145] 26900.0	 [51]
L_2	26020	[51]	0.497	[301]	11.61	[423]	25869.90 ± 0.90 26.0200	[145] [51]
L_3	20410	[51]	0.570	[301]	9.85	[423]	20389.42 ± 0.60 [145] 20410.0	 [51]
M_1	6977	[51]	8.57(-3)	[423]	20.53	[423]		
M_2	6574	[51]	1.16(-2)	[423]	16.08	[423]		
M_3	5252	[51]	1.74(-2)	[423]	12.16	[423]		
M_4	4630	[51]	6.58(-2)	[423]	4.48	[423]		
M_5	4374	[51]	6.19(-2)	[423]	4.33	[423]		
N_1	1868	[51]	1.3(-3)	[423]	15.30	[423]		
N_2	1680	[51]	1.5(-3)	[423]	16.46	[423]		
N_3	1321	[51]	9.0(-4)	[423]	10.61	[423]		
N_4			9.6(-4)	[423]	6.83	[423]		
N_5			7.8(-4)	[423]	6.50	[423]		
N_6			1.9(-4)	[423]	1.16	[423]		
N_7			2.5(-4)	[423]	0.78	[423]		
O_1	435	[51]	3.0(-5)	[423]	31.25	[423]		
O_2			1.3(-4)	[423]	12.27	[423]		
O_3			3.0(-5)	[423]	20.15	[423]		
IP	6.42	[222]						

Note: Transition energies and edge energies from [51,98,145,149,180,257,298] are given for ^{251}Es.

Fm   Z=100   [Rn] 5f^{12} 7s^2

Fermium $A = 257.0951$ [239] $\varrho = 13.60$ g/cm^{-3} [342]

X-ray transitions

line	transition	E/eV		I/eV/$\hbar$	TPIV	Γ/eV	
K series							
K$_{\alpha_3}$	KL$_1$	114390 ± 15	[431]			148.45	[423]
		114343.0 ± 4.4	[145]				
K$_{\alpha 2}$	KL$_2$	115320.0	[98]		2.971(-1)	143.35	[423]
		115280.0 ± 90.0	[149]				
		115319 ± 15	[257]				
K$_{\alpha 1}$	KL$_3$	121090.0	[98]		4.542(-1)	141.59	[423]
		121070.0 ± 90.0	[149]				
		121070.0 ± 13.0	[298]				
		121095 ± 15	[257]				
	KM$_1$	134723.3 ± 6.4	[145]			152.08	[423]
K$_{\beta 3}$	KM$_2$	135171.0	[98]		5.338(-2)	147.64	[423]
		135184 ± 15	[431]				
K$_{\beta 1}$	KM$_3$	136563.0	[98]		1.060(-1)	143.74	[423]
		116555 ± 15	[431]				
K$_{\beta_5^{II}}$	KM$_4$	137195.0	[98]		2.231(-3)	136.07	[423]
		137217 ± 17	[431]				
K$_{\beta_5^{I}}$	KM$_5$	137464.0	[98]		2.408(-3)	135.89	[423]
		137479 ± 17	[431]				
	KN$_1$	139980.7 ± 7.9	[145]			146.88	[423]
K$_{\beta_2^{II}}$	KN$_2$	140216.0	[98]		1.368(-2)	148.22	[423]
		140220 ± 16	[431]				
K$_{\beta_2^{I}}$	KN$_3$	140596.0	[98]		2.868(-2)	142.29	[423]
		140592 ± 16	[431]				
K$_{\beta_4^{II}}$	KN$_4$	140863.5 ± 7.4	[145]		6.991(-4)	138.34	[423]
K$_{\beta_4^{I}}$	KN$_5$	140929.3 ± 7.2	[145]		7.602(-4)	138.02	[423]
L series							
	L$_1$M$_1$	20373 ± 12	[431]			37.65	[423]
		20380.3 ± 3.4	[145]				
L$_{\beta 4}$	L$_1$M$_2$	20794 ± 11	[431]		8.099(-2)	33.23	[423]

line	transition	E/eV		I/eV/ℏ	TPIV	Γ/eV	
		20807.5 ± 3.6	[145]				
L$_{\beta 3}$	L$_1$M$_3$	22165 ± 11	[431]		5.469(-2)	29.32	[423]
		22178.7 ± 4.0	[145]				
L$_{\beta 10}$	L$_1$M$_4$	22827 ± 14	[431]		4.475(-3)	21.65	[423]
		22832.7 ± 2.9	[145]				
L$_{\beta 9}$	L$_1$M$_5$	23089 ± 14	[431]		6.626(-3)	21.48	[423]
		23093.3 ± 2.9	[145]				
	L$_1$N$_1$	25633 ± 14	[431]			32.46	[423]
		25637.7 ± 5.0	[145]				
L$_{\gamma 2}$	L$_1$N$_2$	25830 ± 12	[431]		2.246(-2)	33.80	[423]
		25840.3 ± 6.2	[145]				
L$_{\gamma 3}$	L$_1$N$_3$	26202 ± 12	[431]		1.834(-2)	27.87	[423]
		26208.5 ± 4.3	[145]				
	L$_1$N$_4$	26520.5 ± 4.5	[145]		8.239(-4)	23.93	[423]
L$_{\gamma 11}$	L$_1$N$_5$	26584 ± 14	[431]		1.328(-3)	23.60	[423]
		26586.4 ± 4.1	[145]				
	L$_1$N$_6$	26996.0 ± 3.9	[145]			18.30	[423]
	L$_1$N$_7$	27013.6 ± 3.6	[145]			17.77	[423]
	L$_1$N$_{6,7}$	26934 ± 17	[431]				
L$_\eta$	L$_2$M$_1$	19444 ± 11	[431]		1.200(-2)	32.38	[423]
		19442.4 ± 2.4	[145]				
	L$_2$M$_2$	19865.0 ± 9.9	[431]			28.13	[423]
		19869.6 ± 2.5	[145]				
L$_{\beta 17}$	L$_2$M$_3$	21236.0 ± 9.9	[431]		7.163(-4)	24.22	[423]
		21240.8 ± 2.8	[145]				
L$_{\beta 1}$	L$_2$M$_4$	21898 ± 13	[431]		4.087(-1)	16.55	[423]
		21894.8 ± 1.8	[145]				
	L$_2$M$_5$	22160 ± 13	[431]			16.38	[423]
		22155.4 ± 1.7	[145]				
L$_{\gamma 5}$	L$_2$N$_1$	24704 ± 13	[431]		3.360(-3)	27.36	[423]
		24699.8 ± 3.7	[145]				
	L$_2$N$_2$	24901 ± 11	[431]			28.71	[423]
		24902.4 ± 5.1	[145]				
	L$_2$N$_3$	25273 ± 11	[431]		2.301(-4)	22.77	[423]
		25270.6 ± 3.1	[145]				
L$_{\gamma 1}$	L$_2$N$_4$	25585 ± 13	[431]		9.931(-2)	18.82	[423]
		25582.6 ± 3.2	[145]				
	L$_2$N$_5$	25655 ± 13	[431]			18.51	[423]

line	transition	E/eV		I/eV/$\hbar$	TPIV	Γ/eV	
		25648.5 ± 2.9	[145]				
L_v	L_2N_6	26058.1 ± 2.7	[145]		7.644(-4)	13.20	[423]
	L_2N_7	26075.7 ± 2.3	[145]			12.67	[423]
	$L_2N_{6,7}$	26005 ± 17	[431]				
L_l	L_3M_1	13668 ± 11	[431]		2.715(-2)	30.79	[423]
		13668.8 ± 2.1	[145]				
L_t	L_3M_2	14089.0 ± 9.9	[431]		3.604(-4)	26.36	[423]
		14095.9 ± 2.1	[145]				
L_s	L_3M_3	15460.0 ± 9.9	[431]		3.342(-4)	22.45	[423]
		15467.1 ± 2.4	[145]				
$L_{\alpha2}$	L_3M_4	16122 ± 13	[431]		3.912(-2)	14.78	[423]
		16121.2 ± 1.4	[145]				
$L_{\alpha1}$	L_3M_5	16384 ± 13	[431]		3.431(-1)	14.61	[423]
		16381.8 ± 1.3	[145]				
$L_{\beta6}$	L_3N_1	18928 ± 13	[431]		7.009(-3)	25.60	[423]
		18926.2 ± 3.4	[145]				
	L_3N_2	19125 ± 11	[431]		7.671(-5)	27.08	[423]
		19128.8 ± 4.8	[145]				
	L_3N_3	19125 ± 11	[431]		9.428(-5)	21.00	[423]
		19128.8 ± 4.8	[145]				
$L_{\beta15}$	L_3N_4	19809 ± 13	[431]		8.189(-3)	17.06	[423]
		19809.0 ± 2.9	[145]				
$L_{\beta2}$	L_3N_5	19879 ± 13	[431]		7.586(-2)	16.74	[423]
		19874.8 ± 2.6	[145]				
$L_{u'}$	L_3N_6	20284.4 ± 2.3	[145]		1.418(-4)	11.44	[423]
L_u	L_3N_7	20302.1 ± 2.0	[145]		7.729(-4)	10.91	[423]
	$L_3N_{6,7}$	20229 ± 17	[431]				

level characteristics

level	E_B/eV		ω_{nlj}		Γ/eV		AE/eV	
K	143090	[51]	0.976	[301]	131.44	[423]	141930.4 ± 7.1	[431]
			0.972	[423]			141927.3 ± 5.2	[145]
							143090.0	[51]
L_1	27700	[51]	0.263	[301]	17.02	[423]	27573 ± 8	[431]
			0.216	[244]			27584.4 ± 2.2	[145]

level	E_B/eV		ω_{nlj}		Γ/eV		AE/eV	
			0.2225	[423]			27700.0	[51]
L_2	26810	[51]	0.506	[301]	11.92	[423]	26644 ± 7	[431]
			0.539	[244]			26646.5 ± 1.0	[145]
			0.5481	[423]			26810.0	[51]
L_3	20900	[51]	0.579	[301]	10.15	[423]	20868 ± 7	[431]
			0.512	[244]			20872.81 ± 0.60	[145]
			0.5224	[423]			20900.0	[51]
M_1	7205	[51]	9.13(-3)	[423]	20.63	[423]		
M_2	6793	[51]	1.22(-2)	[423]	16.21	[423]		
M_3	5397	[51]	1.83(-2)	[423]	12.30	[423]		
M_4	4766	[51]	6.83(-2)	[423]	4.63	[423]		
M_5	4498	[51]	6.35(-2)	[423]	4.46	[423]		
N_1	1937	[51]	1.35(-3)	[423]	15.45	[423]		
N_2	1747	[51]	1.54(-3)	[423]	16.79	[423]		
N_3	1366	[51]	9.1(-4)	[423]	10.85	[423]		
N_4			1.0(-3)	[423]	6.91	[423]		
N_5			8.0(-4)	[423]	6.59	[423]		
N_6			2.2(-4)	[423]	1.28	[423]		
N_7			2.9(-4)	[423]	0.85	[423]		
O_1	454	[51]	3.0(-5)	[423]	32.13	[423]		
O_2			1.4(-4)	[423]	12.55	[423]		
O_3			3.0(-5)	[423]	22.60	[423]		
IP	6.50	[222]						

Note: Transition energies and edge energies from [51,98,145,149,180,257,298] are given for ^{254}Fm.

6 X-Ray Transition Energies: Ordered by Energy/Wavelength

In the following we give ordered by the wavelength and transition energy emission lines and absorption edges for all elements up to americium (Z=95). To restrict the volume of the table we give for each quantity only one selected value what in a first step is sufficient to indentify lines or edges in a spectrum of interest.

Tabulated are

λ – wavelength of the tabulated transition or absorption edge;
E – energy of the tabulated transition or absorption edge;
Z – atomic number.

λ [pm]	E [keV]	Z	element	transition	notation	reference
10.0299	123.6152	95	Am	KN_2	$K_{\beta_{21}}$	[51]
10.2719	120.7032	94	Pu	KN_3	$K_{\beta_{22}}$	[51]
10.2939	120.4459	94	Pu	KN_2	$K_{\beta_{21}}$	[51]
10.3012	120.3600	95	Am	KM_3	K_{β_1}	[51]
10.3913	119.3168	95	Am	KM_2	K_{β_3}	[51]
10.5438	117.5912	93	Np	KN_3	$K_{\beta_{21}}$	[51]
10.5654	117.3503	93	Np	KN_2	$K_{\beta_{22}}$	[51]
10.5734	117.2614	94	Pu	KM_3	K_{β_1}	[51]
10.6629	116.2768	94	Pu	KM_2	K_{β_3}	[51]
10.7235	115.606	92	U	abs. edge	K	[50]
10.8190	114.6	92	U	KN_3	$K_{\beta_{21}}$	[50]
10.8379	114.4	92	U	KN_2	$K_{\beta_{22}}$	[50]
10.8527	114.2433	93	Np	KM_3	K_{β_1}	[51]
10.9420	113.3118	93	Np	KM_2	K_{β_3}	[51]
11.0110	112.601	91	Pa	abs. edge	K	[51]
11.1078	111.62	91	Pa	KN_3	$K_{\beta_{21}}$	[50]
11.1297	111.4	91	Pa	KN_2	$K_{\beta_{22}}$	[50]
11.1397	111.3	92	U	KM_3	K_{β_1}	[50]
11.2295	110.41	92	U	KM_2	K_{β_3}	[50]
11.3078	109.646	90	Th	abs. edge	K	[50]
11.3228	109.5	90	Th	$KO_{2,3}$		[50]
11.4044	108.717	90	Th	KN_3	$K_{\beta_{21}}$	[50]
11.4260	108.511	90	Th	KN_2	$K_{\beta_{22}}$	[50]

λ [pm]	E [keV]	Z	element	transition	notation	reference
11.4349	108.427	91	Pa	KM$_3$	K$_{\beta_1}$	[50]
11.5228	107.6	91	Pa	KM$_2$	K$_{\beta_3}$	[50]
11.6140	106.755	89	Ac	abs. edge	K	[51]
11.6393	106.5229	95	Am	KL$_3$	K$_{\alpha_1}$	[51]
11.7122	105.86	89	Ac	KN$_3$	K$_{\beta_{21}}$	[50]
11.7310	105.69	90	Th	KM$_3$	K$_{\beta_1}$	[50]
11.7332	105.67	89	Ac	KN$_2$	K$_{\beta_{22}}$	[50]
11.8272	104.831	90	Th	KM$_2$	K$_{\beta_3}$	[50]
11.9306	103.922	88	Ra	abs. edge	K	[51]
11.9491	103.7612	94	Pu	KL$_3$	K$_{\alpha_1}$	[51]
12.0292	103.07	88	Ra	KN$_3$	K$_{\beta_{21}}$	[50]
12.0503	102.89	88	Ra	KN$_2$	K$_{\beta_{22}}$	[50]
12.0550	102.85	89	Ac	KM$_3$	K$_{\beta_1}$	[50]
12.1435	102.1	89	Ac	KM$_2$	K$_{\beta_3}$	[50]
12.1455	102.083	95	Am	KL$_2$	K$_{\alpha_2}$	[51]
12.2591	101.137	87	Fr	abs. edge	K	[51]
12.2675	101.068	93	Np	KL$_2$	K$_{\alpha_2}$	[51]
12.2891	100.89	88	Ra	KN$_2$	K$_{\beta_{22}}$	[50]
12.3577	100.33	87	Fr	KN$_3$	K$_{\beta_{21}}$	[50]
12.3787	100.16	87	Fr	KN$_2$	K$_{\beta_{22}}$	[50]
12.3824	100.13	88	Ra	KM$_3$	K$_{\beta_1}$	[50]
12.4543	99.5518	94	Pu	KL$_2$	K$_{\alpha_2}$	[51]
12.4696	99.43	88	Ra	KM$_2$	K$_{\beta_3}$	[50]
12.5951	98.439	92	U	KL$_3$	K$_{\alpha_1}$	[50]
12.5996	98.404	86	Ra	abs. edge	K	[51]
12.6982	97.646	86	Ra	KN$_3$	K$_{\beta_{21}}$	[50]
12.7203	97.47	86	Ra	KN$_2$	K$_{\beta_{22}}$	[50]
12.7203	97.47	87	Fr	KM$_3$	K$_{\beta_1}$	[50]
12.7718	97.0775	93	Np	KL$_2$	K$_{\alpha_2}$	[51]
12.8071	96.81	87	Fr	KM$_2$	K$_{\beta_3}$	[50]
12.9329	95.868	91	Pa	KL$_3$	K$_{\alpha_1}$	[50]
12.9516	95.7299	85	At	abs. edge	K	[51]
13.0524	94.99	85	At	KN$_3$	K$_{\beta_{21}}$	[50]
13.0690	94.97	86	Ra	KM$_3$	K$_{\beta_1}$	[50]
13.0731	94.84	85	At	KN$_2$	K$_{\beta_{22}}$	[50]
13.0973	94.665	92	U	KL$_2$	K$_{\alpha_2}$	[50]
13.1563	94.24	86	Ra	KM$_2$	K$_{\beta_3}$	[50]
13.2818	93.35	90	Th	KL$_3$	K$_{\alpha_1}$	[50]
13.3167	93.105	84	Po	abs. edge	K	[51]
13.4183	92.4	84	Po	KN$_3$	K$_{\beta_{21}}$	[50]
13.4328	92.3	85	At	KM$_3$	K$_{\beta_1}$	[50]
13.4347	92.287	91	Pa	KL$_2$	K$_{\alpha_2}$	[50]

λ [pm]	E [keV]	Z	element	transition	notation	reference
13.4387	92.26	84	Po	KN_2	$K_{\beta_{22}}$	[50]
13.5178	91.72	85	At	KM_2	K_{β_3}	[50]
13.6421	90.884	89	Ac	KL_3	K_{α_1}	[50]
13.6949	90.534	83	Bi	abs. edge	K	[51]
13.7099	90.435	83	Bi	$KO_{2,3}$		[50]
13.7593	90.11	83	Bi	$KN_{4,5}$	K_{β_4}	[50]
13.7833	89.953	90	Th	KL_2	K_{α_2}	[50]
13.7970	89.864	83	Bi	KN_3	$K_{\beta_{21}}$	[50]
13.8068	89.80	84	Po	KM_3	K_{β_1}	[50]
13.8171	89.733	83	Bi	KN_2	$K_{\beta_{22}}$	[50]
13.8919	89.25	84	Po	KM_2	K_{β_3}	[50]
14.0144	88.47	88	Ra	KL_3	K_{α_1}	[50]
14.0884	88.005	82	Pb	abs. edge	K	[51]
14.0796	88.06	82	Pb	$KP_{2,3}$		[50]
14.1017	87.922	82	Pb	$KO_{2,3}$		[50]
14.1117	87.860	83	Bi	$KM_{4,5}$	K_{β_5}	[50]
14.1423	87.67	89	Ac	KL_2	K_{α_2}	[50]
14.1552	87.59	82	Pb	$KN_{4,5}$	K_{β_4}	[50]
14.1947	87.346	82	Pb	KN_3	$K_{\beta_{21}}$	[50]
14.1952	87.343	83	Bi	KM_3	K_{β_1}	[50]
14.2136	87.23	82	Pb	KN_2	$K_{\beta_{22}}$	[50]
14.2784	86.834	83	Bi	KM_2	K_{β_3}	[50]
14.4001	86.10	87	Fr	KL_3	K_{α_1}	[50]
14.4956	85.533	81	Tl	abs. edge	K	[51]
14.4961	85.53	82	Pb	KM_5	$K_{\beta_{51}}$	[50]
14.5095	85.451	81	Tl	$KO_{2,3}$		[50]
14.5131	85.43	88	Ra	KL_2	K_{α_2}	[50]
14.5131	85.43	82	Pb	KM_4	$K_{\beta_{52}}$	[50]
14.5540	85.19	81	Tl	$KN_{4,5}$	K_{β_4}	[50]
14.5958	84.946	81	Tl	KN_3	$K_{\beta_{21}}$	[50]
14.5975	84.936	82	Pb	KM_3	K_{β_1}	[50]
14.6147	84.836	81	Tl	KN_2	$K_{\beta_{22}}$	[50]
14.6815	84.450	82	Pb	KM_2	K_{β_3}	[50]
14.7989	83.78	88	Ra	KL_3	K_{α_1}	[50]
14.8967	83.23	87	Fr	KL_2	K_{α_2}	[50]
14.9175	83.114	81	Tl	KM_4	$K_{\beta_{52}}$	[50]
14.9184	83.109	80	Hg	abs. edge	K	[51]
14.9308	83.04	80	Hg	$KO_{2,3}$		[50]
14.9777	82.78	80	Hg	$KN_{4,5}$	K_{β_4}	[50]
15.0147	82.576	81	Tl	KM_3	K_{β_1}	[50]
15.0212	82.54	80	Hg	KN_3	$K_{\beta_{21}}$	[50]
15.0413	82.43	80	Hg	KN_2	$K_{\beta_{22}}$	[50]

λ [pm]	E [keV]	Z	element	transition	notation	reference
15.0984	82.118	81	Tl	KM_2	K_{β_3}	[50]
15.2092	81.52	85	At	KL_3	K_{α_1}	[50]
15.2936	81.07	86	Rn	KL_2	K_{α_2}	[50]
15.3542	80.75	80	Hg	$KM_{4,5}$	K_{β_5}	[50]
15.3599	80.720	79	Au	abs. edge	K	[51]
15.3700	80.667	79	Au	$KO_{2,3}$		[50]
15.4228	80.391	79	Au	$KN_{4,5}$	K_{β_4}	[50]
15.4493	80.253	80	Hg	KM_3	K_{β_1}	[50]
15.4624	80.185	79	Au	KN_3	$K_{\beta_{21}}$	[50]
15.4827	80.08	79	Au	KN_2	$K_{\beta_{22}}$	[50]
15.5327	79.822	80	Hg	KM_2	K_{β_3}	[50]
15.6369	79.290	84	Po	KL_3	K_{α_1}	[50]
15.7043	78.95	85	At	KL_2	K_{α_2}	[50]
15.7885	78.529	79	Au	KM_5	$K_{\beta_{51}}$	[50]
15.8068	78.438	79	Au	KM_4	$K_{\beta_{52}}$	[50]
15.8183	78.381	78	Pt	abs. edge	K	[51]
15.8263	78.341	78	Pt	$KO_{2,3}$		[50]
15.8815	78.069	78	Pt	$KN_{4,5}$	K_{β_4}	[50]
15.8988	77.984	79	Au	KM_3	K_{β_1}	[50]
15.9204	77.878	78	Pt	KN_3	$K_{\beta_{21}}$	[50]
15.9395	77.785	78	Pt	KN_2	$K_{\beta_{22}}$	[50]
15.9816	77.580	79	Au	KM_2	K_{β_3}	[50]
16.0794	77.1079	83	Bi	KL_3	K_{α_1}	[50]
16.1309	76.862	84	Po	KL_2	K_{α_2}	[50]
16.2561	76.27	78	Pt	KM_5	$K_{\beta_{51}}$	[50]
16.2712	76.199	78	Pt	KM_4	$K_{\beta_{52}}$	[50]
16.2922	76.101	77	Ir	abs. edge	K	[51]
16.3025	76.053	77	Ir	$KO_{2,3}$		[50]
16.3524	75.821	77	Ir	$KN_{4,5}$	K_{β_4}	[50]
16.3681	75.748	78	Pt	KM_3	K_{β_1}	[50]
16.3960	75.619	77	Ir	KN_3	$K_{\beta_{21}}$	[50]
16.4156	75.529	77	Ir	KN_2	$K_{\beta_{22}}$	[50]
16.4506	75.368	78	Pt	KM_3	K_{α_1}	[50]
16.5723	74.8148	83	Bi	KL_2	K_{α_2}	[50]
16.7378	74.075	77	Ir	KM_5	$K_{\beta_{51}}$	[50]
16.7593	73.980	77	Ir	KM_4	$K_{\beta_{52}}$	[50]
16.7874	73.856	76	Os	abs. edge	K	[51]
16.7983	73.808	76	Os	$KO_{2,3}$		[50]
16.8424	73.615	76	Os	$KN_{4,5}$	K_{β_4}	[50]
16.8548	73.5608	77	Ir	KM_3	K_{β_1}	[50]
16.8913	73.402	76	Os	KN_3	$K_{\beta_{21}}$	[50]
16.9106	73.318	76	Os	KN_2	$K_{\beta_{22}}$	[50]

λ [pm]	E [keV]	Z	element	transition	notation	reference
16.9372	73.2027	77	Ir	KM_2	K_{β_3}	[50]
17.0142	72.8715	81	Tl	KL_3	K_{α_1}	[50]
17.0300	72.8042	82	Pb	KL_2	K_{α_2}	[50]
17.2453	71.895	76	Os	KM_5	$K_{\beta_{51}}$	[50]
17.2624	71.824	76	Os	KM_4	$K_{\beta_{52}}$	[50]
17.3024	71.658	75	Re	abs. edge	K	[51]
17.3084	71.633	75	Re	$KO_{2,3}$		[50]
17.3617	71.413	76	Os	KM_3	K_{β_1}	[50]
17.3624	71.410	75	Re	$KN_{4,5}$	K_{β_4}	[50]
17.4058	71.232	75	Re	KN_3	$K_{\beta_{21}}$	[50]
17.4256	71.151	75	Re	KN_2	$K_{\beta_{22}}$	[50]
17.4438	71.077	76	Os	KM_2	K_{β_3}	[50]
17.5042	70.8319	81	Tl	KL_2	K_{α_2}	[50]
17.5073	70.819	80	Hg	KL_3	K_{α_1}	[50]
17.7665	69.786	75	Re	KM_5	$K_{\beta_{51}}$	[50]
17.7836	69.719	75	Re	KM_4	$K_{\beta_{52}}$	[50]
17.8375	69.508	74	W	abs. edge	K	[51]
17.8450	69.479	74	W	$KO_{2,3}$		[50]
17.8885	69.310	75	Re	KM_3	K_{β_1}	[50]
17.8926	69.294	74	W	$KN_{4,5}$	K_{β_4}	[50]
17.9426	69.101	74	W	KN_3	$K_{\beta_{21}}$	[50]
17.9608	69.031	74	W	KN_2	$K_{\beta_{22}}$	[50]
17.9704	68.994	75	Re	KM_2	K_{β_3}	[50]
17.9963	68.895	80	Hg	KL_2	K_{α_2}	[50]
18.0201	68.8037	79	Au	KL_3	K_{α_1}	[50]
18.3098	67.7153	74	W	KM_5	$K_{\beta_{51}}$	[50]
18.3268	67.6523	74	W	KM_4	$K_{\beta_{52}}$	[50]
18.3946	67.403	73	Ta	abs. edge	K	[51]
18.4036	67.370	73	Ta	$KO_{2,3}$		[50]
18.4380	67.2443	74	W	KM_3	K_{β_1}	[50]
18.4518	67.194	73	Ta	$KN_{4,5}$	K_{β_4}	[50]
18.5017	67.013	73	Ta	KN_3	$K_{\beta_{21}}$	[50]
18.5082	66.9895	79	Au	KL_2	K_{α_2}	[50]
18.5187	66.9514	74	W	KM_2	$K_{\beta_{22}}$	[50]
18.5518	66.832	78	Pt	KL_3	K_{α_1}	[50]
18.6725	66.40	79	Au	KL_1	K_{α_3}	[50]
18.8763	65.683	73	Ta	KM_5	$K_{\beta_{51}}$	[50]
18.8927	65.626	73	Ta	KM_4	$K_{\beta_{52}}$	[50]
18.9841	65.31	72	Hf	abs. edge	K	[51]
19.0094	65.223	73	Ta	KM_3	K_{β_1}	[50]
19.0389	65.122	78	Pt	KL_2	K_{α_2}	[50]
19.0805	64.98	72	Hf	$KN_{2,3}$	K_{β_2}	[50]

λ [pm]	E [keV]	Z	element	transition	notation	reference
19.0897	64.9488	73	Ta	KM$_2$	K$_{\beta_3}$	[50]
19.1053	64.8956	77	Ir	KL$_3$	K$_{\alpha_1}$	[50]
19.5838	63.31	71	Lu	abs. edge	K	[51]
19.5891	63.293	71	Lu	KO$_{2,3}$		[50]
19.5910	63.2867	77	Ir	KL$_2$	K$_{\alpha_2}$	[50]
19.6074	63.234	72	Hf	KM$_3$	K$_{\beta_1}$	[50]
19.6800	63.0005	76	Os	KL$_3$	K$_{\alpha_1}$	[50]
19.6864	62.98	72	Hf	KM$_2$	K$_{\beta_3}$	[50]
19.6896	62.97	71	Lu	KN$_{2,3}$	K$_{\beta_2}$	[50]
20.0844	61.732	71	Lu	KM$_{4,5}$	K$_{\beta_5}$	[50]
20.1646	61.4867	76	Os	KL$_2$	K$_{\alpha_2}$	[50]
20.2260	61.30	70	Yb	abs. edge	K	[51]
20.2266	61.298	70	Yb	KO$_{2,3}$		[50]
20.2316	61.283	71	Lu	KM$_3$	K$_{\beta_1}$	[50]
20.2788	61.1403	75	Re	KL$_3$	K$_{\alpha_1}$	[50]
20.3088	61.05	71	Lu	KM$_3$	K$_{\beta_3}$	[50]
20.3648	60.882	70	Yb	KN$_{2,3}$	K$_{\beta_2}$	[50]
20.7396	59.782	70	Yb	KM$_{4,5}$	K$_{\beta_5}$	[50]
20.7618	59.7179	75	Re	KL$_2$	K$_{\alpha_2}$	[50]
20.8800	59.38	69	Tm	abs. edge	K	[51]
20.8835	59.37	70	Yb	KM$_3$	K$_{\beta_1}$	[50]
20.8919	59.346	69	Tm	KO$_{2,3}$		[50]
20.9017	59.31824	74	W	KL$_3$	K$_{\alpha_1}$	[50]
20.9641	59.14	70	Yb	KM$_2$	K$_{\beta_3}$	[50]
20.9824	59.09	69	Tm	KN$_{2,3}$	K$_{\beta_2}$	[50]
21.3835	57.9817	74	W	KL$_2$	K$_{\alpha_2}$	[50]
21.4052	57.923	69	Tm	KM$_{4,5}$	K$_{\beta_5}$	[50]
21.5507	57.532	73	Ta	KL$_3$	K$_{\alpha_1}$	[50]
21.5563	57.517	69	Tm	KM$_3$	K$_{\beta_1}$	[50]
21.5681	57.4855	68	Er	abs. edge	K	[51]
21.5814	57.450	68	Er	KO$_{2,3}$		[50]
21.5923	57.421	74	W	KL$_1$	K$_{\alpha_3}$	[50]
21.6364	57.304	69	Tm	KM$_2$	K$_{\beta_3}$	[50]
21.6719	57.21	68	Er	KN$_{2,3}$	K$_{\beta_2}$	[50]
22.0312	56.277	73	Ta	KL$_2$	K$_{\alpha_2}$	[50]
22.1244	56.04	68	Er	KM$_{4,5}$	K$_{\beta_5}$	[50]
22.2235	55.7902	72	Hf	KL$_3$	K$_{\alpha_1}$	[50]
22.2671	55.681	68	Er	KM$_3$	K$_{\beta_1}$	[50]
22.2919	55.619	67	Ho	abs. edge	K	[51]
22.3059	55.584	67	Ho	KO$_{2,3}$		[50]
22.3421	55.494	68	Er	KM$_2$	K$_{\beta_3}$	[50]
22.4124	55.32	67	Ho	KN$_{2,3}$	K$_{\beta_2}$	[50]

λ [pm]	E [keV]	Z	element	transition	notation	reference
22.7032	54.6114	72	Hf	KL_2	K_{α_2}	[50]
22.8561	54.246	67	Ho	$KM_{4,5}$	K_{β_5}	[50]
22.9263	54.08	72	Hf	KL_1	K_{α_3}	[104]
22.9306	54.0698	71	Lu	KL_3	K_{α_1}	[50]
23.0126	53.877	67	Ho	KM_3	K_{β_1}	[50]
23.0486	53.793	66	Dy	abs. edge	K	[51]
23.0567	53.774	66	Dy	$KO_{2,3}$		[50]
23.0838	53.711	67	Ho	KM_2	K_{β_3}	[50]
23.1765	53.496	66	Dy	$KN_{2,3}$	K_{β_2}	[50]
23.4089	52.965	71	Lu	KL_2	K_{α_2}	[50]
23.6189	52.494	66	Dy	$KM_{4,5}$	K_{β_5}	[50]
23.6663	52.3889	70	Yb	KL_3	K_{α_1}	[50]
23.7889	52.119	66	Dy	KM_3	K_{β_1}	[50]
23.8424	52.002	65	Tb	abs. edge	K	[51]
23.8594	51.965	65	Tb	$KO_{2,3}$		[50]
23.8630	51.957	66	Dy	KM_2	K_{β_3}	[50]
23.9622	51.742	65	Tb	$KN_{2,3}$	K_{β_2}	[104]
24.1432	51.3540	70	Yb	KL_2	K_{α_2}	[50]
24.4346	50.7416	69	Tm	KL_3	K_{α_1}	[50]
24.6090	50.382	65	Tb	KM_3	K_{β_1}	[50]
24.6820	50.233	64	Gd	abs. edge	K	[51]
24.6840	50.229	65	Tb	KM_2	K_{β_3}	[50]
24.6879	50.221	64	Gd	$KO_{2,3}$		[50]
24.8174	49.959	64	Gd	$KM_{2,3}$	K_{β_2}	[50]
24.9103	49.7726	69	Tm	KL_2	K_{α_2}	[50]
25.1629	49.273	69	Tm	KL_1	K_{α_3}	[104]
25.2373	49.1217	68	Er	KL_3	K_{α_1}	[50]
25.2763	49.052	64	Gd	$KM_{4,5}$	K_{β_5}	[50]
25.4605	48.697	64	Gd	KM_3	K_{β_1}	[50]
25.5350	48.555	64	Gd	KM_2	K_{β_3}	[50]
25.5539	48.519	63	Eu	abs. edge	K	[51]
25.5655	48.497	63	Eu	$KO_{2,3}$		[50]
25.6932	48.256	63	Eu	$KN_{2,3}$	K_{β_2}	[50]
25.7118	48.2211	68	Er	KL_2	K_{α_2}	[50]
26.0765	47.5467	67	Ho	KL_3	K_{α_1}	[50]
26.3586	47.0379	63	Eu	KM_3	K_{β_1}	[50]
26.4340	46.9036	63	Eu	KM_2	K_{β_3}	[50]
26.4649	46.849	62	Sm	abs. edge	K	[51]
26.4929	46.801	62	Sm	$KO_{2,3}$		[50]
26.5495	46.6997	67	Ho	KL_2	K_{α_2}	[50]
26.6177	46.580	62	Sm	$KN_{2,3}$	K_{β_2}	[50]
26.9542	45.9984	66	Dy	KL_3	K_{α_1}	[50]

λ [pm]	E [keV]	Z	element	transition	notation	reference
27.1118	45.731	62	Sm	$KM_{4,5}$	K_{β_5}	[50]
27.3017	45.413	62	Sm	KM_3	K_{β_1}	[50]
27.3764	45.289	62	Sm	KM_2	K_{β_3}	[50]
27.4256	45.2078	66	Dy	KL_2	K_{α_2}	[50]
27.4316	45.198	61	Pm	abs. edge	K	[51]
27.5891	44.940	61	Pm	$KN_{2,3}$	K_{β_2}	[50]
27.8734	44.4816	65	Tb	KL_3	K_{α_1}	[50]
28.2903	43.826	61	Pm	KM_3	K_{β_1}	[50]
28.3433	43.7441	65	Tb	KL_2	K_{α_2}	[50]
28.3641	43.713	61	Pm	KM_2	K_{β_3}	[50]
28.4539	43.574	60	Nd	abs. edge	K	[51]
28.6142	43.33	60	Nd	$KN_{2,3}$	K_{β_2}	[50]
28.8363	42.9962	64	Gd	KL_3	K_{α_1}	[50]
29.3048	42.3089	64	Gd	KL_2	K_{α_2}	[50]
29.3308	42.2713	60	Nd	KM_3	K_{β_1}	[50]
29.4037	42.1665	60	Nd	KM_2	K_{β_3}	[50]
29.5189	42.002	59	Pr	abs. edge	K	[51]
29.6807	41.773	59	Pr	$KN_{2,3}$	K_{β_2}	[50]
29.8456	41.5422	63	Eu	KL_3	K_{α_1}	[50]
30.3128	40.9019	63	Eu	KL_2	K_{α_2}	[50]
30.4772	40.7482	59	Pr	KM_3	K_{β_1}	[50]
30.4985	40.6529	59	Pr	KM_2	K_{β_3}	[50]
30.6492	40.453	58	Ce	abs. edge	K	[51]
30.6689	40.427	58	Ce	$KO_{2,3}$		[50]
30.7373	40.337	58	Ce	$KN_{4,5}$	K_{β_4}	[50]
30.8168	40.233	58	Ce	$KN_{2,3}$	K_{β_2}	[50]
30.9051	40.1181	62	Sm	KL_3	K_{α_1}	[50]
31.3426	39.558	58	Ce	KM_5	$K_{\beta_{51}}$	[50]
31.3577	39.539	58	Ce	KM_4	$K_{\beta_{52}}$	[50]
31.3709	39.5224	62	Sm	KL_2	K_{α_2}	[50]
31.5827	39.2573	58	Ce	KM_3	K_{β_1}	[50]
31.6530	39.1701	58	Ce	KM_2	K_{β_3}	[50]
31.8450	38.934	57	La	abs. edge	K	[51]
31.8654	38.909	57	La	$KO_{2,3}$		[50]
31.9319	38.828	57	La	$KN_{4,5}$	K_{β_4}	[50]
32.0128	38.7299	57	La	$KN_{2,3}$	K_{β_2}	[50]
32.0171	38.7247	61	Pm	KL_3	K_{α_1}	[50]
32.4813	38.1712	61	Pm	KL_2	K_{α_2}	[50]
32.5472	38.094	57	La	KM_5	$K_{\beta_{51}}$	[50]
32.5643	38.074	57	La	KM_4	$K_{\beta_{52}}$	[50]
32.7994	37.8010	57	La	KM_3	K_{β_1}	[50]
32.8697	37.7202	57	La	KM_2	K_{β_3}	[50]

λ [pm]	E [keV]	Z	element	transition	notation	reference
33.1051	37.452	56	Ba	abs. edge	K	[51]
33.1281	37.426	56	Ba	$KO_{2,3}$		[50]
33.1857	37.3610	60	Nd	KL_3	K_{α_1}	[50]
33.2302	37.311	56	Ba	$KN_{4,5}$	K_{β_4}	[50]
33.2784	37.257	56	Ba	$KN_{2,3}$	K_{β_2}	[50]
33.6483	36.8474	60	Nd	KL_2	K_{α_2}	[50]
33.8148	36.666	56	Ba	KM_5	$K_{\beta_{51}}$	[50]
33.8360	36.643	56	Ba	KM_4	$K_{\beta_{52}}$	[50]
34.0823	36.3782	56	Ba	KM_3	K_{β_1}	[50]
34.1519	36.3040	56	Ba	KM_2	K_{β_3}	[50]
34.4152	36.0263	59	Pr	KL_3	K_{α_1}	[50]
34.4528	35.987	55	Cs	abs. edge	K	[51]
34.6115	35.822	55	Cs	$KN_{2,3}$	K_{β_2}	[50]
34.8761	35.5502	59	Pr	KL_3	K_{α_2}	[50]
35.4376	34.9869	55	Cs	KM_3	K_{β_1}	[50]
35.5061	34.9194	55	Cs	KM_2	K_{β_3}	[50]
35.7103	34.7197	55	Ce	KL_3	K_{α_1}	[50]
35.844	34.59	54	Xe	abs. edge	K	[51]
36.0265	34.415	54	Xe	$KN_{2,3}$	K_{β_2}	[50]
36.1695	34.2789	55	Ce	KL_2	K_{α_2}	[50]
36.8740	33.624	54	Xe	KM_3	K_{β_1}	[50]
36.9421	33.562	54	Xe	KM_2	K_{β_3}	[50]
37.0749	33.4418	57	La	KL_3	K_{α_1}	[50]
37.3827	33.1665	53	I	abs. edge	K	[51]
37.5235	33.042	53	I	$KN_{2,3}$	K_{β_2}	[50]
37.5325	33.0341	57	La	KL_2	K_{α_2}	[50]
38.3918	32.2947	53	I	KM_3	K_{β_1}	[50]
38.4577	32.2394	53	I	KM_2	K_{β_3}	[50]
38.5124	32.1936	56	Ba	KL_3	K_{α_1}	[50]
38.9681	31.8171	56	Ba	KL_2	K_{α_2}	[50]
38.9754	31.8114	52	Te	abs. edge	K	[51]
38.9751	31.8114	52	Te	$KO_{2,3}$		[50]
39.1116	31.7004	52	Te	$KN_{2,3}$	K_{β_2}	[50]
40.0008	30.9957	52	Te	KM_3	K_{β_1}	[50]
40.0303	30.9728	55	Cs	KL_3	K_{α_1}	[50]
40.0672	30.9443	52	Te	KM_2	K_{β_3}	[50]
40.4848	30.6251	55	Cs	KL_2	K_{α_2}	[50]
40.6696	30.4860	51	Sb	abs. edge	K	[51]
40.6676	30.4875	51	Sb	$KO_{2,3}$		[50]
40.7029	30.461	51	Sb	KN_5	K_{β_4}	[50]
40.7471	30.428	51	Sb	KN_4	$K_{\beta_{4'}}$	[50]
40.7987	30.3895	51	Sb	$KN_{2,3}$	K_{β_2}	[50]

λ [pm]	E [keV]	Z	element	transition	notation	reference
41.3793	29.9632	51	Sb	KM_5	$K_{\beta_{51}}$	[50]
41.3891	29.956	51	Sb	KM_4	$K_{\beta_{52}}$	[50]
41.6351	29.779	54	Xe	KL_3	K_{α_1}	[50]
41.7099	29.7256	51	Sb	KM_3	K_{β_1}	[50]
41.7751	29.6792	51	Sb	KM_2	K_{β_3}	[50]
42.0888	29.458	54	Xe	KL_2	K_{α_1}	[50]
42.4684	29.1947	50	Sn	abs. edge	K	[51]
42.4680	29.195	50	Sn	$KO_{2,3}$		[50]
42.4971	29.175	50	Sn	$KN_{4,5}$	K_{β_4}	[50]
42.5930	29.1093	50	Sn	$KN_{2,3}$	K_{β_2}	[50]
43.1763	28.716	50	Sn	KM_5	$K_{\beta_{51}}$	[50]
43.1854	28.710	50	Sn	KM_4	$K_{\beta_{52}}$	[50]
43.3333	28.6120	53	I	KL_3	K_{α_1}	[50]
43.5240	28.4860	50	Sn	KM_3	K_{β_1}	[50]
43.5892	28.4440	50	Sn	KM_2	K_{β_3}	[50]
43.7844	28.3172	53	I	KL_2	K_{α_2}	[50]
44.3723	27.9420	49	In	abs. edge	K	[51]
44.3755	27.940	49	In	$KO_{2,3}$		[50]
44.3946	27.928	49	In	$KN_{4,5}$	K_{β_4}	[50]
44.5017	27.8608	49	In	$KN_{2,3}$	K_{β_2}	[50]
45.0872	27.499	49	In	KM_5	$K_{\beta_{51}}$	[50]
45.1003	27.491	49	In	KM_4	$K_{\beta_{52}}$	[50]
45.1310	27.4723	52	Te	KL_3	K_{α_1}	[50]
45.4560	27.2759	49	In	KM_3	K_{β_1}	[50]
45.5264	27.2337	49	In	KM_2	K_{β_3}	[50]
45.5799	27.2017	52	Te	KL_2	K_{α_2}	[50]
46.4088	26.7159	48	Cd	abs. edge	K	[51]
46.5344	26.6438	48	Cd	$KN_{2,3}$	K_{β_2}	[50]
47.0370	26.3591	51	Sb	KL_3	K_{α_1}	[50]
47.4843	26.1108	51	Sb	KL_2	K_{α_2}	[50]
47.5121	26.0955	48	Cd	KM_3	K_{β_1}	[50]
47.5746	26.0612	48	Cd	KM_2	K_{β_3}	[50]
48.5902	25.5165	47	Ag	abs. edge	K	[51]
48.5988	25.5l2	47	Ag	$KN_{4,5}$	K_{β_4}	[50]
48.7049	25.4564	47	Ag	$KN_{2,3}$	K_{β_2}	[50]
49.0617	25.2713	50	Sn	KL_3	K_{α_1}	[50]
49.3081	25.145	47	Ag	$KM_{4,5}$	K_{β_5}	[50]
49.5069	25.0440	50	Sn	KL_2	K_{α_2}	[50]
49.7086	24.9424	47	Ag	KM_3	K_{β_1}	[50]
49.7703	24.9115	47	Ag	KM_2	K_{β_3}	[50]
50.9221	24.348	46	Pd	abs. edge	K	[51]
50.9263	24.346	46	Pd	$KN_{4,5}$	K_{β_4}	[50]

λ [pm]	E [keV]	Z	element	transition	notation	reference
51.0246	24.2991	46	Pd	$KN_{2,3}$	K_{β_2}	[50]
51.2130	24.2097	49	In	KL_3	K_{α_1}	[50]
51.6562	24.0020	49	In	KL_2	K_{α_2}	[50]
51.6713	23.995	46	Pd	$KM_{4,5}$	K_{β_5}	[50]
52.0537	23.8187	46	Pd	KM_3	K_{β_1}	[50]
52.1141	23.7911	46	Pd	KM_2	K_{β_3}	[50]
53.3963	23.2198	45	Rh	abs. edge	K	[51]
53.3961	23.217	45	Rh	$KN_{4,5}$	K_{β_4}	[50]
53.5028	23.1736	48	Cd	KL_3	K_{α_1}	[50]
53.5046	23.1728	45	Rh	KN_3	$K_{\beta_{21}}$	[50]
53.5157	23.168	45	Rh	KN_2	$K_{\beta_{21}}$	[50]
53.6523	23.109	94	Pu	abs. edge	L_1	[51]
53.9439	22.9841	48	Cd	KL_2	K_{α_2}	[50]
54.0351	22.9453	95	Am	L_1N_5	$L_{\gamma_{11}}$	[51]
54.1018	22.917	45	Rh	KM_5	$K_{\beta_{51}}$	[50]
54.1207	22.909	45	Rh	KM_4	$K_{\beta_{52}}$	[50]
54.1633	22.891	94	Pu	L_1O_3	L_{γ_4}	[50]
54.3123	22.8282	95	Am	L_2O_4	L_{γ_6}	[50]
54.3247	22.823	94	Pu	L_1O_2	$L_{\gamma_{4'}}$	[50]
54.5623	22.7236	45	Rh	KM_3	K_{β_1}	[50]
54.6217	22.6989	45	Rh	KM_2	K_{β_3}	[50]
54.7706	22.6372	95	Am	L_1N_3	L_{γ_3}	[51]
55.4468	22.3611	95	Am	L_1N_2	L_{γ_2}	[51]
55.6092	22.2958	94	Pu	L_1N_5	$L_{\gamma_{11}}$	[51]
55.7162	22.253	94	Pu	abs. edge	L_2	[51]
55.8492	22.20	93	Np	$L_1O_{2,3}$	L_{γ_4}	[50]
55.9426	22.16292	47	Ag	KL_3	K_{α_1}	[50]
55.9748	22.1502	94	Pu	L_2O_4	L_{γ_6}	[50]
56.0529	22.1193	44	Ru	abs. edge	K	[51]
56.0603	22.1164	95	Am	L_2N_5		[51]
56.0917	22.104	44	Ru	$KN_{4,5}$	K_{β_4}	[50]
56.1680	22.074	44	Ru	$KN_{2,3}$	K_{β_2}	[50]
56.1904	22.0652	95	Am	L_2N_4	L_{γ_1}	[50]
56.3818	21.9903	47	Ag	KL_2	K_{α_2}	[50]
56.4020	21.9824	94	Pu	L_1N_3	L_{γ_3}	[50]
56.5781	21.914	94	Pu	L_2O_1	L_{γ_8}	[50]
56.7854	21.834	44	Ru	KM_5	$K_{\beta_{51}}$	[50]
56.7984	21.829	44	Ru	KM_4	$K_{\beta_{52}}$	[50]
56.8208	21.8204	94	Pu	L_2N_4	L_ν	[51]
56.8523	21.8033	95	Am	L_2N_3		[51]
56.9497	21.771	92	U	abs. edge	L_1	[51]
57.0598	21.729	92	U	$L_1P_{2,3}$	$L_{\gamma_{13}}$	[50]

λ [pm]	E [keV]	Z	element	transition	notation	reference
57.0700	21.7231	94	Pu	L_1N_2	L_{γ_2}	[50]
57.2495	21.657	92	U	$L_1O_{4,5}$		[50]
57.2500	21.6568	44	Ru	KM_3	K_{β_1}	[50]
57.2508	21.6565	93	Np	L_1N_5	$L_{\gamma_{11}}$	[51]
57.3088	21.6346	44	Ru	KM_2	K_{β_3}	[50]
57.5017	21.562	92	U	L_1O_3	L_{γ_4}	[50]
57.5813	21.5322	95	Am	L_2N_2		[51]
57.6718	21.4984	92	U	L_1O_2	$L_{\gamma_{4'}}$	[50]
57.6997	21.488	93	Np	L_2O_4	L_{γ_6}	[50]
57.7621	21.4648	94	Pu	L_2N_5		[51]
57.8902	21.4173	94	Pu	L_2N_4	L_{γ_1}	[50]
58.0999	21.34	93	Np	L_1N_3	L_{γ_3}	[50]
58.5468	21.1771	46	Pd	KL_3	K_{α_1}	[50]
58.6088	21.1547	93	Np	L_2N_6	L_ν	[51]
58.7329	21.11	93	Np	L_1N_2	L_{γ_2}	[50]
58.7480	21.1046	91	Pa	abs. edge	L_1	[51]
58.9079	21.0473	43	Tc	abs. edge	K	[51]
58.9841	21.0201	46	Pd	KL_2	K_{α_2}	[50]
58.9872	21.019	92	U	L_1N_5	$L_{\gamma_{11}}$	[50]
59.0265	21.005	43	Tc	KN_2	$K_{\beta_{22}}$	[50]
59.0996	20.979	92	U	L_1N_4		[50]
59.1956	20.945	92	U	abs. edge	L_2	[51]
59.2040	20.942	92	U	$L_2P_{4,5}$		[50]
59.3060	20.906	92	U	$L_2P_{2,3}$		[50]
59.3741	20.882	91	Pa	$L_1O_{2,3}$	L_{γ_4}	[50]
59.4864	20.8426	92	U	L_2O_4	L_{γ_6}	[50]
59.6519	20.7848	93	Np	L_2N_4	L_{γ_1}	[50]
59.7288	20.758	92	U	L_2O_3		[50]
59.8595	20.7127	92	U	L_1N_3	L_{γ_3}	[50]
59.8847	20.704	94	Pu	L_2N_1	L_{γ_5}	[50]
60.1256	20.621	92	U	L_2O_1	L_{γ_8}	[50]
60.1315	20.619	43	Tc	$KM_{2,3}$	$K_{\beta_{1,3}}$	[50]
60.0190	20.599	43	Tc	KM_2	K_{β_3}	[50]
60.3158	20.556	92	U	L_2N_6	L_ν	[50]
60.5257	20.4847	92	U	L_1N_2	L_{γ_2}	[50]
60.5870	20.464	90	Th	abs. edge	L_1	[51]
60.7056	20.424	90	Th	$L_1P_{2,3}$	$L_{\gamma_{13}}$	[50]
60.8277	20.383	90	Th	$L_1O_{4,5}$		[50]
61.0352	20.3137	91	Pa	abs. edge	L_2	[51]
61.1005	20.292	90	Th	L_1O_3	L_{γ_4}	[50]
61.2514	20.242	90	Th	L_1O_2	$L_{\gamma_{4'}}$	[50]
61.3299	20.2161	45	Rh	KL_3	K_{α_1}	[50]

λ [pm]	E [keV]	Z	element	transition	notation	reference
61.3302	20.216	91	Pa	L_2O_4	L_{γ_6}	[50]
61.4579	20.174	90	Th	L_1O_1		[50]
61.4789	20.1671	92	U	L_2N_4	L_{γ_1}	[50]
61.5983	20.128	90	Th	$L_1N_{6,7}$		[50]
61.6229	20.12	93	Np	L_2N_1	L_{γ_5}	[50]
61.6903	20.098	91	Pa	L_1N_3	L_{γ_3}	[50]
61.7650	20.0737	45	Rh	KL_2	K_{α_2}	[50]
61.9941	19.9995	42	Mo	abs. edge	K	[51]
62.0050	19.996	42	Mo	$KN_{4,5}$	K_{β_4}	[50]
62.1006	19.9652	42	Mo	KN_3	$K_{\beta_{21}}$	[50]
62.1074	19.963	42	Mo	KN_2	$K_{\beta_{22}}$	[50]
62.2822	19.907	92	U	L_2N_3		[50]
62.3480	19.886	95	Am	L_1M_5	L_{β_9}	[51]
62.3919	19.872	91	Pa	L_1N_2	L_{γ_2}	[50]
62.4925	19.84	89	Ac	abs. edge	L_1	[51]
62.6377	19.794	90	Th	L_1N_5		[50]
62.6947	19.776	42	Mo	KM_5	$K_{\beta_{51}}$	[50]
62.7106	19.771	42	Mo	KM_4	$K_{\beta_{52}}$	[50]
62.7614	19.755	90	Th	L_1N_4		[50]
62.9910	19.683	90	Th	abs. edge	L_2	[51]
62.9942	19.682	90	Th	L_2P_4		[50]
62.9981	19.6808	95	Am	L_1M_4	$L_{\beta_{10}}$	[51]
63.1224	19.642	90	Th	$L_2P_{2,3}$		[50]
63.1642	19.629	90	Th	L_2P_1		[50]
63.2309	19.6083	42	Mo	KM_3	K_{β_1}	[50]
63.2609	19.599	90	Th	L_2O_4	L_{γ_6}	[50]
63.2890	19.5903	42	Mo	KM_2	K_{β_3}	[50]
63.3612	19.568	91	Pa	L_2N_4	L_{γ_1}	[50]
63.5586	19.5072	92	U	L_2N_1	L_{γ_5}	[50]
63.5593	19.507	90	Th	L_1N_3	L_{γ_3}	[50]
63.5626	19.506	90	Th	L_2O_3		[50]
63.6932	19.466	90	Th	L_2O_2		[50]
63.9000	19.403	90	Th	L_2O_1	L_{γ_8}	[50]
64.0651	19.353	90	Th	L_2N_6	L_ν	[50]
64.1646	19.323	94	Pu	L_1M_5	L_{β_9}	[50]
64.2243	19.305	90	Th	L_1N_2	L_{γ_2}	[50]
64.3103	19.2792	44	Ru	KL_3	K_{α_1}	[50]
64.4548	19.236	88	Ra	abs. edge	L_1	[51]
64.5151	19.218	88	Ra	$L_1P_{2,3}$	$L_{\gamma_{13}}$	[50]
64.6868	19.167	88	Ra	$L_1O_{4,5}$		[50]
64.7428	19.1504	44	Ru	KL_2	K_{α_2}	[50]
64.7577	19.146	90	Th	L_1N_1		[50]

λ [pm]	E [keV]	Z	element	transition	notation	reference
64.8255	19.126	94	Pu	L_1M_4	$L_{\beta_{10}}$	[50]
64.8937	19.1059	95	Am	L_1M_3	L_{β_3}	[50]
64.9681	19.084	88	Ra	L_1O_3	L_{γ_4}	[50]
64.9708	19.0832	89	Ac	abs. edge	L_2	[51]
65.1319	19.036	88	Ra	L_1O_2	$L_{\gamma_{4'}}$	[50]
65.2073	19.014	90	Th	$L_2N_{6,7}$	L_ν	[50]
65.3004	18.9869	41	Nb	abs. edge	K	[51]
65.3155	18.9825	90	Th	L_2N_4	L_{γ_1}	[50]
65.3172	18.981	41	Nb	$KN_{4,5}$	K_{β_4}	[50]
65.4171	18.953	41	Nb	$KN_{2,3}$	K_{β_2}	[50]
65.4966	18.930	91	Pa	L_2N_1	L_{γ_5}	[50]
65.7677	18.8520	95	Am	L_2M_4	L_{β_1}	[50]
66.0057	18.784	41	Nb	$KM_{4,5}$	K_{β_5}	[50]
66.0761	18.764	93	Np	L_1M_5	L_{β_9}	[51]
66.1995	18.729	90	Th	$L_{2,3}$		[50]
66.5192	18.639	87	Fr	abs. edge	L_1	[51]
66.5406	18.633	88	Ra	L_1N_5		[50]
66.5781	18.6225	41	Nb	KM_3	K_{β_1}	[50]
66.6361	18.6063	41	Nb	KM_2	K_{β_3}	[50]
66.6587	18.600	88	Ra	L_1N_4		[50]
66.7430	18.5765	93	Np	L_1M_4	$L_{\beta_{10}}$	[51]
66.8726	18.5405	94	Pu	L_1M_3	L_{β_3}	[50]
67.0698	18.486	88	Ra	abs. edge	L_2	[51]
67.0886	18.4808	89	Ac	L_2N_4	L_{γ_1}	[50]
67.1424	18.466	88	Ra	$L_2P_{2,3}$		[50]
67.2407	18.439	88	Ra	L_2P_1		[50]
67.3320	18.414	88	Ra	L_2O_4	L_{γ_6}	[50]
67.3540	18.408	89	Ac	L_2N_4	L_{γ_1}	[50]
67.3847	18.3996	95	Am	$L_3O_{4,5}$	L_{β_3}	[50]
67.4933	18.370	90	Th	L_2N_1	L_{γ_5}	[50]
67.5039	18.3671	43	Tc	KL_3	K_{α_1}	[50]
67.5411	18.357	88	Ra	L_1N_3	L_{γ_3}	[50]
67.6405	18.330	88	Ra	L_2O_3		[50]
67.7748	18.2937	94	Pu	L_2M_4	L_{β_1}	[50]
67.8033	18.286	88	Ra	L_2O_2		[50]
67.8367	18.277	95	Am	L_2M_3	$L_{\beta_{17}}$	[51]
67.9341	18.2508	43	Tc	KL_2	K_{α_2}	[50]
68.0116	18.230	88	Ra	L_2N_6	L_ν	[50]
68.1035	18.2054	92	U	L_1M_5	L_{β_9}	[50]
68.2025	18.179	88	Ra	L_1N_2	L_{γ_2}	[50]
68.6416	18.0627	95	Am	L_1M_2	L_{β_4}	[50]
68.6936	18.049	86	Rn	abs. edge	L_1	[51]

λ [pm]	E [keV]	Z	element	transition	notation	reference
68.7431	18.036	88	Ra	L_1N_1		[50]
68.7622	18.031	92	U	L_1M_4	$L_{\beta_{10}}$	[50]
68.8849	17.9989	40	Zr	abs. edge	K	[51]
68.9036	17.994	40	Zr	$KN_{4,5}$	K_{β_4}	[50]
68.9109	17.9921	93	Np	L_1M_3	L_{β_3}	[51]
68.9956	17.970	40	Zr	$KN_{2,3}$	K_{β_2}	[50]
69.2403	17.9065	87	Fr	abs. edge	L_2	[51]
69.3274	17.884	88	Ra	L_2N_5		[50]
69.4633	17.849	88	Ra	L_2N_4	L_{γ_1}	[50]
69.5959	17.815	40	Zr	$KM_{4,5}$	K_{β_5}	[50]
69.8500	17.7502	93	Np	L_2M_4	L_{β_1}	[50]
70.0102	17.7096	94	Pu	L_2M_3	$L_{\beta_{17}}$	[51]
70.0284	17.705	94	Pu	L_3O_1	L_{β_7}	[50]
70.1413	17.6765	95	Am	L_3N_5	L_{β_2}	[50]
70.1758	17.6678	40	Zr	KM_3	K_{β_1}	[50]
70.1789	17.667	91	Pa	L_1M_5	L_{β_9}	[50]
70.2306	17.654	40	Zr	KM_2	K_{β_3}	[50]
70.3063	17.635	94	Pu	$L_3N_{4,5}$	L_u	[50]
70.3430	17.6258	95	Am	L_3N_4	$L_{\beta_{15}}$	[50]
70.4301	17.604	88	Ra	L_2N_3		[50]
70.6227	17.5560	94	Pu	L_1M_2	L_{β_4}	[50]
70.8159	17.5081	93	Np	$L_3O_{4,5}$	L_{β_5}	[50]
70.8770	17.493	85	At	abs. edge	L_1	[51]
70.8810	17.492	91	Pa	L_1M_4	$L_{\beta_{10}}$	[50]
70.9324	17.47934	42	Mo	KL_3	K_{α_1}	[50]
71.0313	17.4550	92	U	L_1M_3	L_{β_3}	[50]
71.3612	17.3743	42	Mo	KL_2	K_{α_2}	[50]
71.5143	17.3371	86	Rn	abs. edge	L_2	[51]
71.6553	17.303	87	Fr	L_2N_4	L_{γ_1}	[50]
71.7756	17.274	88	Ra	L_2N_1	L_{γ_5}	[50]
71.8534	17.2553	94	Pu	L_3N_5	L_{β_2}	[50]
72.0006	17.2200	92	U	L_2M_4	L_{β_1}	[50]
72.0509	17.208	94	Pu	L_3N_4	$L_{\beta_{15}}$	[50]
72.0610	17.2056	93	Np	L_3N_7	L_u	[51]
72.2314	17.165	92	U	abs. edge	L_3	[51]
72.2280	17.1658	93	Np	L_2M_3	$L_{\beta_{17}}$	[51]
72.2440	17.162	92	U	$L_3P_{4,5}$		[50]
72.3409	17.139	90	Th	L_1M_5	L_{β_9}	[50]
72.4297	17.118	92	U	$L_3P_{2,3}$		[50]
72.5229	17.096	92	U	L_3P_1		[50]
72.6329	17.0701	92	U	$L_3O_{4,5}$	L_{β_5}	[50]
72.6734	17.0606	93	Np	L_1M_2	L_{β_4}	[51]

λ [pm]	E [keV]	Z	element	transition	notation	reference
72.7698	17.038	39	Y	abs. edge	K	[51]
72.7783	17.036	39	Y	$KN_{4,5}$	K_{β_4}	[50]
72.8664	17.0154	39	Y	$KN_{2,3}$	K_{β_2}	[50]
73.0140	16.981	90	Th	L_1M_4	$L_{\beta_{10}}$	[50]
73.0958	16.962	92	U	L_3O_3		[50]
73.1938	16.9393	84	Po	abs. edge	L_1	[51]
73.2340	16.930	91	Pa	L_1M_3	L_{β_3}	[50]
73.3336	16.907	92	U	L_3O_2		[50]
73.4205	16.8870	95	Am	L_3N_1	L_{β_6}	[50]
73.4552	16.879	39	Y	$KM_{4,5}$	K_{β_5}	[50]
73.6035	16.845	92	U	L_3O_1	L_{β_7}	[50]
73.6259	16.8400	93	Np	L_3N_5	L_{β_2}	[50]
73.8266	16.7941	93	Np	L_3N_4	$L_{\beta_{15}}$	[51]
73.8627	16.7859	92	U	$L_3O_{6,7}$	L_u	[50]
73.8710	16.784	85	At	abs. edge	L_2	[51]
73.9327	16.770	86	Rn	L_2N_4	L_{γ_1}	[50]
74.0749	16.7378	39	Y	KM_3	K_{β_1}	[50]
74.0957	16.7331	91	Pa	abs. edge	L_3	[51]
74.1281	16.7258	39	Y	KM_2	K_{β_3}	[50]
74.2337	16.702	91	Pa	L_2M_4	L_{β_1}	[50]
74.5058	16.641	92	U	L_2M_3	$L_{\beta_{17}}$	[50]
74.5282	16.636	91	Pa	$L_3O_{4,5}$	L_{β_5}	[50]
74.6220	16.6151	41	Nb	KL_3	K_{α_1}	[50]
74.8012	16.5753	92	U	L_1M_2	L_{β_4}	[50]
75.0470	16.5210	41	Nb	KL_2	K_{α_2}	[50]
75.1503	16.4983	94	Pu	L_3N_1	L_{β_6}	[50]
75.4580	16.431	91	Pa	L_3O_1	L_{β_7}	[50]
75.4704	16.4283	92	U	L_3N_5	L_{β_2}	[50]
75.4819	16.4258	90	Th	L_1M_3	L_{β_3}	[50]
75.7115	16.376	83	Bi	abs. edge	L_1	[51]
75.6667	16.3857	92	U	L_3N_4	$L_{\beta_{15}}$	[50]
75.6921	16.3802	83	Bi	$L_1P_{2,3}$	$L_{\gamma_{13}}$	[50]
75.7902	16.359	90	Th	L_2M_5		[50]
75.7948	16.358	83	Bi	$L_1O_{4,5}$		[50]
75.9109	16.333	94	Pu	L_2M_1	L_η	[50]
76.0692	16.299	90	Th	abs. edge	L_3	[51]
76.0878	16.295	90	Th	L_3P_4		[50]
76.0893	16.2947	83	Bi	L_1O_3	L_{γ_4}	[50]
76.2005	16.2709	83	Bi	L_1O_2	$L_{\gamma_{4'}}$	[50]
76.2516	16.260	90	Th	$L_3P_{2,3}$		[50]
76.2938	16.251	85	At	L_2N_4	L_{γ_1}	[50]
76.3267	16.244	84	Po	abs. edge	L_2	[51]

λ [pm]	E [keV]	Z	element	transition	notation	reference
76.3408	16.241	90	Th	L_2P_1		[50]
76.3926	16.23	83	Bi	$L_1N_{6,7}$		[50]
76.4491	16.218	84	Po	L_1N_3	L_{γ_3}	[50]
76.4727	16.213	90	Th	$L_3O_{4,5}$	L_{β_5}	[50]
76.5237	16.2022	90	Th	L_2M_4	L_{β_1}	[50]
76.6286	16.18	83	Bi	L_1O_1		[104]
76.8614	16.131	88	Ra	L_1M_5	L_{β_9}	[50]
76.8662	16.13	93	Np	L_3N_1	L_{β_6}	[50]
76.9138	16.123	90	Th	L_3O_3		[50]
76.9139	16.120	92	U	L_3N_3		[50]
76.9760	16.107	38	Sr	abs. edge	K	[51]
76.9903	16.104	91	Pa	L_1M_2	L_{β_4}	[50]
76.9903	16.104	38	Sr	$KN_{4,5}$	K_{β_4}	[50]
77.0831	16.0846	38	Sr	$KN_{2,3}$	K_{β_2}	[50]
77.1340	16.074	90	Th	L_3O_2		[50]
77.1532	16.07	84	Po	L_1N_2	L_{γ_2}	[50]
77.3746	16.024	91	Pa	L_3N_5	L_{β_2}	[50]
77.4399	16.0105	90	Th	L_3O_1	L_{β_7}	[50]
77.5489	15.988	88	Ra	L_1M_4	$L_{\beta_{10}}$	[50]
77.6411	15.969	38	Sr	$KM_{4,5}$	K_{β_5}	[50]
77.6654	15.964	90	Th	$L_3N_{6,7}$	L_u	[50]
77.6800	15.961	83	Bi	L_1N_5		[50]
77.8263	15.931	89	Ac	L_1M_3	L_{β_3}	[50]
77.9585	15.904	83	Bi	L_1N_4		[50]
78.0173	15.892	92	U	L_3N_2		[50]
78.0960	15.876	93	Np	L_2M_1	L_η	[50]
78.1205	15.871	89	Ac	abs. edge	L_3	[51]
78.1353	15.855	82	Pb	abs. edge	L_1	[51]
78.3586	15.843	82	Pb	$L_1O_{4,5}$		[50]
78.2947	15.8357	38	Sr	KM_3	K_{β_1}	[50]
78.3481	15.8249	38	Sr	KM_2	K_{β_3}	[50]
78.5860	15.777	82	Pb	L_1O_3	L_{γ_4}	[50]
78.5955	15.7751	40	Zr	KL_3	K_{α_1}	[50]
78.7107	15.752	82	Pb	L_1O_2	$L_{\gamma_{4'}}$	[50]
78.7507	15.744	84	Po	L_2N_4	L_{γ_1}	[50]
78.8409	15.7260	92	U	L_3N_1	L_{β_6}	[50]
78.8459	15.725	82	Pb	$L_1N_{6,7}$		[50]
78.9061	15.713	89	Ac	L_2M_4	L_{β_1}	[50]
78.8760	15.719	83	Bi	abs. edge	L_2	[51]
78.9201	15.7102	83	Bi	L_1N_3	L_{γ_3}	[50]
78.9764	15.699	82	Pb	L_1O_1		[50]
79.0172	15.6909	40	Zr	KL_2	K_{α_2}	[50]

λ [pm]	E [keV]	Z	element	transition	notation	reference
79.0454	15.6853	83	Bi	L_2O_4	L_{γ_6}	[50]
79.2597	15.6429	90	Th	L_1M_2	L_{β_4}	[50]
79.2597	15.6429	90	Th	L_2M_3	$L_{\beta_{17}}$	[50]
79.3571	15.6237	90	Th	L_3N_5	L_{β_2}	[50]
79.3871	15.6178	83	Bi	L_2O_3		[50]
79.5414	15.5875	90	Th	L_3N_4	$L_{\beta_{15}}$	[50]
79.5674	15.5824	83	Bi	L_1N_2	L_{γ_2}	[50]
79.7229	15.552	83	Bi	L_2N_6	L_ν	[50]
79.7281	15.551	83	Bi	L_2O_1	L_{γ_8}	[50]
80.2026	15.456	83	Bi	L_1N_1		[50]
80.2357	15.453	82	Pb	L_1N_5		[50]
80.2758	15.4449	88	Ra	L_1M_3	L_{β_3}	[50]
80.2805	15.444	88	Ra	abs. edge	L_3	[51]
80.3689	15.427	82	Pb	L_1N_4		[50]
80.3793	15.425	88	Ra	$L_3P_{2,3}$		[50]
80.4994	15.402	88	Ra	L_3P_1		[50]
80.5144	15.3997	92	U	L_3M_1	L_η	[50]
80.6297	15.3771	88	Ra	$L_3O_{4,5}$	L_{β_5}	[50]
80.7879	15.347	91	Pa	L_3N_1	L_{β_6}	[50]
80.8195	15.341	90	Th	L_3N_3		[50]
80.8632	15.3327	81	Tl	$L_1O_{4,5}$		[50]
81.1649	15.2757	90	Th	L_1M_1		[50]
81.1867	15.2716	81	Tl	L_1O_3	L_{γ_4}	[50]
81.3113	15.2482	81	Tl	L_1O_2	$L_{\gamma_{4'}}$	[50]
81.3140	15.2477	83	Bi	L_2N_4	L_{γ_1}	[50]
81.3775	15.2358	88	Ra	L_2M_4	L_{β_1}	[50]
81.4727	15.218	82	Pb	L_1N_3	L_{γ_3}	[50]
81.5423	15.205	37	Rb	$KN_{4,5}$	L_{β_4}	[50]
81.5408	15.2053	82	Pb	abs. edge	L_2	
81.5569	15.2023	37	Rb	abs. edge	K	
81.5799	15.198	81	Tl	L_1O_1		[50]
81.5858	15.1969	82	Pb	L_2P_1		[50]
81.6229	15.190	88	Ra	L_3O_1	L_{β_7}	[50]
81.6476	15.1854	37	Rb	$KN_{2,3}$	L_{β_2}	[50]
81.6858	15.1783	82	Rb	L_2O_4	L_{γ_6}	[50]
81.8600	15.146	88	Ra	$L_3N_{6,7}$	L_u	[50]
81.9032	15.138	90	Th	L_3N_2		[50]
82.0007	15.120	82	Pb	L_2O_3		[50]
82.1039	15.101	82	Pb	L_1N_2	L_{γ_2}	[50]
82.2191	15.085	37	Rb	$KM_{4,5}$	L_{β_5}	[50]
82.3274	15.060	82	Pb	L_2N_6	L_ν	[50]
82.3674	15.0527	82	Pb	L_2O_1	L_{γ_8}	[50]

λ [pm]	E [keV]	Z	element	transition	notation	reference
82.4852	15.0312	87	Fr	abs. edge	L_3	[51]
82.4863	15.031	83	Bi	L_2N_3		[50]
82.7892	14.976	87	Fr	L_1M_3	L_{β_3}	[50]
82.7947	14.975	90	Th	L_3N_1	L_{β_6}	[50]
82.8611	14.963	82	Pb	L_1N_1		[50]
82.8706	14.9613	37	Rb	KM_3	L_{β_1}	[50]
82.8816	14.9593	81	Th	L_1N_5		[50]
82.8866	14.9584	39	Y	KL_3	L_{α_1}	[50]
82.9238	14.9517	37	Rb	KM_2	L_{β_3}	[50]
82.9554	14.946	91	Pa	L_2M_1	L_η	[50]
83.0054	14.937	81	Tl	L_1N_4		[50]
83.3071	14.8829	39	Y	KL_2	L_{α_2}	[50]
83.3800	14.8699	90	Th	L_2M_2		[50]
83.4355	14.86	83	Bi	L_2N_2		[50]
83.5367	14.842	80	Hg	abs. edge	L_1	[51]
83.5085	14.847	80	Hg	$L_1O_{4,5}$		[50]
83.5400	14.8414	88	Ra	L_3N_5	L_{β_3}	[50]
83.7251	14.8086	88	Ra	L_3N_4	$L_{\beta_{15}}$	[50]
83.8247	14.791	82	Pb	L_2N_5		[50]
83.8984	14.778	80	Hg	L_1O_3	L_{γ_4}	[50]
83.9257	14.7732	83	Bi	L_2N_1	L_{γ_5}	[50]
83.9439	14.7704	87	Fr	L_2M_4	L_{β_1}	[50]
83.9757	14.7644	82	Pb	L_2N_4	L_{γ_1}	[50]
84.0178	14.757	80	Hg	L_1O_2	$L_{\gamma_{4'}}$	[50]
84.0737	14.7472	88	Ra	L_1M_2	L_{β_4}	[50]
84.1130	14.7368	81	Tl	L_1N_3	L_{γ_3}	[50]
84.3494	14.699	81	Tl	abs. edge	L_2	[50]
84.3896	14.692	88	Ra	L_2M_3	$L_{\beta_{17}}$	[50]
84.4298	14.685	81	Tl	L_2O_4	L_{γ_6}	[50]
84.5161	14.670	80	Hg	L_1O_1		[50]
84.7756	14.6251	81	Tl	L_1N_2	L_{γ_2}	[50]
84.8086	14.6194	86	Rn	abs. edge	L_3	[51]
84.8614	14.6172	95	Am	L_3M_5	L_{α_1}	[50]
84.8981	14.604	81	Tl	L_2O_2		[50]
85.0512	14.5777	81	Tl	$L_2N_{6,7}$	L_ν	[50]
85.1195	14.566	88	Ra	L_3N_3		[50]
85.1312	14.564	81	Tl	L_2O_1	L_{γ_8}	[50]
85.1956	14.553	82	Po	L_2N_3		[50]
85.4363	14.512	86	Rn	L_1M_3	L_{β_3}	[50]
85.4486	14.5099	90	Th	L_2M_1	L_η	[50]
85.4893	14.503	81	Tl	L_1N_1		[50]
85.6606	14.474	80	Hg	L_1N_1		[50]

λ [pm]	E [keV]	Z	element	transition	notation	reference
85.8029	14.450	87	Fr	L_3N_5	L_{β_2}	[50]
85.8504	14.442	82	Pb	L_2N_2		[50]
86.0297	14.4119	95	Am	L_3M_4	L_{α_2}	[50]
86.1786	14.387	88	Ra	L_3N_2		[50]
86.3526	14.358	79	Au	L_1O_5		[50]
86.3786	14.3537	79	Au	abs. edge	L_1	[51]
86.4026	14.3497	79	Au	L_1O_4		[50]
86.5335	14.328	36	Kr	$KN_{4,5}$	K_{β_4}	[50]
86.5552	14.3244	36	Kr	abs. edge	K	[51]
86.6060	14.316	86	Rn	L_2M_4	L_{β_1}	[50]
86.6120	14.315	36	Kr	$KN_{2,3}$	K_{β_2}	[50]
86.6574	14.3075	82	Pb	L_2N_1	L_{γ_5}	[50]
86.7054	14.2956	79	Au	L_1O_3	L_{γ_4}	[50]
86.7545	14.2915	81	Tl	L_2N_4	L_{γ_1}	[50]
86.8188	14.2809	79	Au	L_1O_2	$L_{\gamma_{4'}}$	[50]
86.8329	14.2786	94	Pu	L_3M_5	L_{α_1}	[50]
86.9156	14.265	80	Hg	L_1N_3	L_{γ_3}	[50]
87.0774	14.2385	79	Au	L_1O_1		[50]
87.0804	14.238	36	Kr	$KM_{4,5}$	K_{β_5}	[50]
87.0914	14.2362	88	Ra	L_3N_1	L_{β_6}	[50]
87.2306	14.2135	85	At	abs. edge	L_3	[51]
87.2214	14.215	80	Hg	abs. edge	L_2	[51]
87.3196	14.199	80	Hg	L_2O_4	L_{γ_6}	[50]
87.5292	14.1650	38	Sr	KL_3	K_{α_1}	[50]
87.5478	14.162	80	Hg	L_1N_2	L_{γ_2}	[50]
87.5849	14.156	80	Hg	L_2O_3		[50]
87.8455	14.114	80	Hg	L_2O_2		[50]
87.8579	14.1123	36	Kr	KM_3	K_{β_1}	[50]
87.8891	14.107	80	Hg	L_2N_6	L_ν	[50]
87.9078	14.104	36	Kr	KM_2	K_{β_3}	[50]
87.9458	14.0979	38	Sr	KL_2	K_{α_2}	[50]
87.9951	14.090	80	Hg	L_2O_1	L_{γ_8}	[50]
87.9995	14.0893	81	Tl	L_2N_3		[50]
88.0314	14.0842	94	Pu	L_3M_4	L_{α_2}	[50]
88.0314	14.0842	94	Pu	L_2N_3		[50]
88.1390	14.067	85	At	L_1M_3	L_{β_3}	[50]
88.2771	14.045	80	Hg	L_1N_1		[50]
88.4345	14.020	79	Au	L_1N_5	$L_{\gamma_{11}}$	[50]
88.5672	13.999	79	Au	L_1N_4		[50]
88.8110	13.959	81	Tl	L_2N_2		[50]
88.9159	13.9441	93	Np	L_3M_5	L_{α_1}	[50]
89.3072	13.883	78	Pt	abs. edge	L_1	[51]

λ [pm]	E [keV]	Z	element	transition	notation	reference
89.3393	13.878	78	Pt	L_1O_5		[50]
89.3522	13.876	85	At	L_2M_4	L_{β_1}	[50]
89.4296	13.864	78	Pt	L_1O_4		[50]
89.5032	13.8526	81	Tl	L_2N_1	L_{γ_5}	[50]
89.6035	13.8371	95	Am	L_3M_3	L_s	[51]
89.6488	13.8301	80	Hg	L_2N_4	L_{γ_1}	[50]
89.6617	13.8281	78	Pt	L_1O_3	L_{γ_4}	[50]
89.7500	13.8145	78	Pt	L_1O_2	$L_{\gamma_{4'}}$	[50]
89.7597	13.813	84	Po	abs. edge	L_3	[51]
89.7857	13.8090	79	Au	L_1N_3	L_{γ_3}	[50]
89.7942	13.8077	83	Bi	L_1M_2	L_{β_9}	[50]
89.9486	13.784	78	Pt	L_1O_1		[50]
89.9617	13.782	84	Po	$L_3O_{4,5}$	L_{β_5}	[50]
90.1075	13.7597	92	Np	L_3M_4	L_{α_2}	[51]
90.2623	13.7361	79	Au	abs. edge	L_2	[51]
90.2997	13.7304	79	Au	L_2O_4	L_{γ_6}	[50]
90.4374	13.7095	79	Au	L_1O_2	L_{γ_2}	[50]
90.4988	13.7002	83	Bi	L_1M_4	$L_{\beta_{10}}$	[50]
90.6384	13.6791	79	Au	L_2O_3		[50]
90.7452	13.6630	88	Ra	L_2M_1	L_η	[50]
90.7518	13.662	79	Au	L_2O_2		[50]
90.8403	13.6487	79	Au	L_2O_6	L_ν	[50]
90.8982	13.640	80	Hg	L_2N_3		[50]
90.9115	13.638	84	Po	L_1M_3	L_{β_3}	[50]
90.9916	13.6260	79	Au	L_2O_1	L_{γ_8}	[50]
91.0671	13.6147	92	U	L_3M_5	L_{α_1}	[50]
91.3133	13.578	79	Au	L_1N_1		[50]
91.4345	13.560	78	Pt	L_1N_1	$L_{\gamma_{11}}$	[50]
91.8395	13.5002	94	Pu	L_3M_3	L_S	[51]
92.0454	13.470	35	Br	abs. edge	K	[51]
92.0488	13.4695	35	Br	$KN_{2,3}$	K_{β_2}	[50]
92.2028	13.447	84	Po	L_2M_4	L_{β_1}	[50]
92.2591	13.4388	92	U	L_3M_4	L_{α_2}	[50]
92.3471	13.426	83	Bi	abs. edge	L_3	[51]
92.3677	13.423	77	Ir	abs. edge	L_1	[51]
92.4166	13.4159	83	Bi	$L_3P_{2,3}$		[50]
92.4365	13.413	77	Ir	$L_1O_{4,5}$		[50]
92.4572	13.410	80	Hg	L_2N_1	L_{γ_5}	[50]
92.5538	13.396	35	Br	$KM_{4,5}$	K_{β_5}	[50]
92.5587	13.3953	37	Rb	KL_3	K_{α_1}	[50]
92.5587	13.3953	83	Bi	$L_3O_{4,5}$	L_{β_5}	[50]
92.6528	13.3817	79	Au	L_2N_4	L_{γ_1}	[50]

λ [pm]	E [keV]	Z	element	transition	notation	reference
92.6853	13.377	82	Pb	L_1M_5	L_{β_3}	[50]
92.7470	13.3681	77	Ir	L_1O_3	L_{γ_4}	[50]
92.7942	13.3613	78	Pt	L_1N_3	L_{γ_3}	[50]
92.8345	13.3555	77	Ir	L_1O_2	$L_{\gamma_{4'}}$	[50]
92.9396	13.3404	84	Po	L_3N_5	L_{β_2}	[50]
92.9717	13.3358	37	Rb	KL_2	K_{α_2}	[50]
93.0261	13.328	83	Bi	L_3O_3		[50]
93.1239	13.314	84	Po	L_3N_4	$L_{\beta_{15}}$	[50]
93.2359	13.298	83	Bi	L_3O_2		[50]
93.2822	13.2914	35	Br	KM_3	K_{β_1}	[50]
93.2871	13.2907	91	Pa	L_3M_5	L_{α_1}	[50]
93.3307	13.2845	35	Br	KM_2	K_{β_3}	[50]
93.3975	13.275	82	Pb	L_1M_4	$L_{\beta_{10}}$	[50]
93.4165	13.2723	78	Pt	abs. edge	L_2	[51]
93.4257	13.271	77	Ir	L_1O_1		[104]
93.4257	13.271	78	Pt	L_2O_4	L_{γ_6}	[50]
93.4299	13.2704	78	Pt	L_1N_2	L_{γ_2}	[50]
93.5081	13.2593	83	Bi	L_3O_1	L_{β_7}	[50]
93.8584	13.2098	83	Bi	L_1M_3	L_{β_3}	[50]
93.9338	13.1992	78	Pt	L_2N_6	L_ν	[50]
94.0279	13.186	79	Au	L_2N_3		[50]
94.1043	13.1753	93	Np	L_3M_3	L_s	[50]
94.1207	13.173	78	Pt	L_2O_1	L_{γ_8}	[50]
94.4217	13.1310	83	Bi	L_2M_5		[50]
94.4577	13.126	77	Ir	L_1N_5	$L_{\gamma_{11}}$	[50]
94.4850	13.1222	91	Pa	L_3M_4	L_{α_2}	[50]
94.5531	13.113	78	Pt	L_1N_1		[50]
94.5874	13.108	77	Ir	L_1N_4		[50]
94.7964	13.086	84	Po	L_1M_2	L_{β_4}	[50]
94.7899	13.08	79	Au	L_2N_2		[50]
95.0763	13.0406	82	Pb	abs. edge	L_3	[51]
95.1215	13.0344	82	Pb	$L_3P_{2,3}$		[50]
95.1266	13.0337	82	Pb	L_3P_1		[50]
95.2011	13.0235	83	Bi	L_2M_4	L_{β_1}	[50]
95.2633	13.015	82	Pb	$L_3O_{4,5}$	L_{β_5}	[50]
95.5209	12.9799	83	Bi	L_3N_5	L_{β_2}	[50]
95.5621	12.9743	79	Au	L_2N_1	L_{γ_5}	[50]
95.6034	12.9687	90	Th	L_3M_5	L_{α_1}	[50]
95.6063	12.9683	76	Os	$L_1O_{4,5}$		[50]
95.5791	12.972	76	Os	abs. edge	L_1	[50]
95.6786	12.9585	81	Tl	L_1M_5	L_{β_9}	[50]
95.7052	12.9549	83	Bi	L_2N_4	$L_{\beta_{15}}$	[50]

λ [pm]	E [keV]	Z	element	transition	notation	reference
95.7784	12.945	82	Pb	L_3O_3		[50]
95.8006	12.9420	78	Pt	L_2N_4	L_{γ_1}	[50]
95.8599	12.934	82	Pb	L_3O_2		[50]
95.9340	12.9240	77	Ir	L_1N_3	L_{γ_3}	[50]
95.9415	12.923	76	Os	L_1O_3	L_{γ_4}	[50]
96.0381	12.910	76	Os	L_1O_2	$L_{\gamma_{4'}}$	[50]
96.1364	12.8968	82	Pb	$L_3N_{6,7}$	L_u	[50]
96.2021	12.888	82	Pb	L_3O_1	L_{β_7}	[50]
96.3208	12.8721	76	Os	L_1O_1		[50]
96.3665	12.866	92	U	L_3M_3	L_s	[50]
96.3920	12.8626	81	Tl	L_1M_4	$L_{\beta_{10}}$	[50]
96.5481	12.8418	77	Ir	L_1N_2	L_{γ_2}	[50]
96.7123	12.820	77	Ir	abs. edge	L_2	[51]
96.7115	12.8201	77	Ir	L_2O_4	L_{γ_6}	[50]
96.7198	12.819	84	Po	L_3N_1	L_{β_6}	[50]
96.7908	12.8096	90	Th	L_3M_4	L_{α_2}	[50]
96.9096	12.7939	95	Am	L_3M_2	L_t	[51]
96.9141	12.7933	82	Pb	L_1M_3	L_{β_3}	[50]
96.9823	12.7843	77	Ir	$L_2O_{2,3}$		[50]
96.7167	12.7603	77	Ir	L_2N_6	L_ν	[50]
97.1762	12.7588	78	Pt	L_2N_3		[50]
97.3242	12.7394	83	Bi	L_3N_3		[50]
97.4121	12.7279	77	Ir	L_2O_1	L_{γ_8}	[50]
97.4726	12.720	82	Pb	L_2M_5		[50]
97.6569	12.696	76	Os	L_1N_5	$L_{\gamma_{11}}$	[50]
97.6645	12.695	77	Ir	L_1N_1		[50]
97.6938	12.6912	83	Bi	L_1M_2	L_{β_4}	[50]
97.7261	12.689	76	Os	L_1N_4		[50]
97.9268	12.661	78	Pt	L_2N_2		[50]
97.9291	12.6607	81	Tl	$L_3P_{2,3}$		[50]
97.9346	12.660	81	Tl	L_3P_1		[50]
97.9346	12.660	81	Tl	abs. edge	L_3	[51]
97.9772	12.6545	34	Se	abs. edge	K	[51]
97.9949	12.6522	34	Se	$KN_{2,3}$	K_{β_2}	[50]
97.9965	12.6520	89	Ac	L_3M_5	L_{α_1}	[50]
98.0197	12.649	36	Kr	KL_3	K_{α_1}	[50]
98.0616	12.6436	81	Tl	$L_3O_{4,5}$	L_{β_5}	[50]
98.2247	12.6226	82	Pb	L_3N_5	L_{β_2}	[50]
98.2831	12.6151	83	Bi	L_3N_2		[50]
98.2940	12.6137	82	Pb	L_2M_4	L_{β_1}	[50]
98.4001	12.6011	82	Pb	L_3N_4	$L_{\beta_{15}}$	[50]
98.4165	12.598	36	Kr	KL_2	K_{α_2}	[50]

λ [pm]	E [keV]	Z	element	transition	notation	reference
98.4400	12.595	34	Se	$KM_{4,5}$	K_{β_5}	[50]
98.5417	12.5820	81	Tl	L_3O_3		[50]
98.7143	12.560	80	Hg	L_1M_5	L_{β_9}	[50]
98.7457	12.556	81	Tl	L_3O_2		[50]
98.7772	12.552	78	Pt	L_2N_1	L_{γ_5}	[50]
98.8875	12.538	81	Tl	$L_3N_{6,7}$	L_u	[50]
98.9159	12.5344	83	Bi	L_2M_3	$L_{\beta_{17}}$	[50]
98.9507	12.530	75	Re	abs. edge	L_1	[50]
98.9980	12.524	75	Re	$L_1O_{4,5}$		[50]
99.0202	12.5212	81	Tl	L_3O_1	L_{β_7}	[50]
99.0645	12.5156	94	Pu	L_3M_2	L_t	[51]
99.0882	12.5126	77	Ir	L_2N_4	L_{γ_1}	[50]
99.1818	12.5008	89	Ac	L_3M_4	L_{α_2}	[50]
99.1897	12.4998	76	Os	L_1N_3	L_{γ_3}	[50]
99.2207	12.4959	34	Se	KM_3	K_{β_1}	[50]
99.2516	12.492	75	Re	L_1O_3	L_{γ_4}	[50]
99.2707	12.4896	34	Se	KM_2	K_{β_3}	[50]
99.3343	12.4816	83	Bi	L_3N_1	L_{β_6}	[50]
99.3383	12.4811	75	Re	L_1O_2	$L_{\gamma_{4'}}$	[104]
99.6185	12.446	80	Hg	L_1M_4	$L_{\beta_{10}}$	[50]
99.6505	12.442	75	Re	L_1O_1		[50]
99.8077	12.4224	76	Os	L_1N_2	L_{γ_2}	[50]
100.0526	12.392	82	Pb	L_3N_3		[50]
100.0655	12.3904	81	Tl	L_1M_3	L_{β_3}	[50]
100.0687	12.39	83	Bi	L_1M_1		[50]
100.1415	12.381	76	Os	abs. edge	L_2	[51]
100.1107	12.3848	76	Os	L_2O_4	L_{γ_6}	[50]
100.1205	12.3836	95	Am	L_3M_1	L_1	[51]
100.4742	12.340	76	Os	L_2O_3		[50]
100.4766	12.3397	88	Ra	L_3M_5	L_{α_1}	[50]
100.4986	12.337	76	Os	L_2N_6	L_v	[50]
100.5394	12.332	77	Ir	L_2N_3		[50]
100.7248	12.3093	81	Tl	L_2M_5		[50]
100.7518	12.306	82	Pb	L_1M_2	L_{β_4}	[50]
100.7911	12.3012	76	Os	L_2O_1	L_{γ_8}	[50]
100.9158	12.286	80	Hg	abs. edge	L_3	[51]
100.9906	12.2769	80	Hg	$L_3O_{4,5}$	L_{β_5}	[50]
101.0350	12.2715	81	Tl	L_3N_5	L_{β_2}	[50]
101.0433	12.2705	82	Pb	L_3N_2		[50]
101.0803	12.266	75	Re	L_1N_5	$L_{\gamma_{11}}$	[50]
101.1216	12.261	90	Th	L_3M_3	L_s	[50]
101.1958	12.252	75	Re	L_1N_4		[50]

λ [pm]	E [keV]	Z	element	transition	notation	reference
101.2041	12.251	77	Ir	L_2N_2		[50]
101.2041	12.2510	81	Tl	L_3N_4	$L_{\beta_{15}}$	[50]
101.2637	12.2438	93	Np	L_3M_2	L_t	[51]
101.4677	12.2264	80	Hg	L_3O_3		[50]
101.5165	l2.2133	81	Tl	L_2M_4	L_{β_1}	[50]
101.5614	12.2079	80	Hg	L_3O_2		[50]
101.6588	12.1962	88	Ra	L_3M_4	L_{α_2}	[50]
101.6772	12.1940	80	Hg	L_3N_7	L_u	[50]
101.7723	12.1826	80	Hg	L_3N_6	L_u	[50]
101.9405	12.1625	80	Hg	L_3O_1	L_{β_7}	[50]
102.0672	12.1474	79	Au	L_1M_5	L_{β_9}	[50]
102.1042	12.143	81	Pb	L_3N_1	L_{β_6}	[50]
102.1783	12.1342	77	Ir	L_2N_1	L_{γ_5}	[50]
102.2389	12.127	81	Pb	L_2M_3	$L_{\beta_{17}}$	[50]
102.2643	12.124	94	Pu	L_3M_1	L_l	[50]
102.4705	12.0996	74	W	abs. edge	L_1	[51]
102.5069	12.0953	79	0s	L_2N_4	L_{γ_1}	[50]
102.5094	12.095	74	W	$L_1O_{4,5}$		[50]
102.6163	12.0824	75	Re	L_1N_3	L_{γ_3}	[50]
102.7779	12.0634	74	W	L_1O_3	L_{γ_4}	[50]
102.7924	12.0617	79	Au	L_1M_4	$L_{\beta_{10}}$	[50]
102.8666	12.0530	74	W	L_1O_2	$L_{\gamma_{4'}}$	[50]
102.8666	12.053	81	Tl	L_3N_3		[50]
103.0522	12.0313	87	Fr	L_3M_5	L_{α_1}	[50]
103.1748	12.017	74	W	L_1O_1		[50]
103.2349	12.010	82	Pb	L_1M_1		[50]
103.2366	12.0098	75	Re	L_1N_2	L_{γ_2}	[50]
103.3614	11.9953	80	Hg	L_1M_3	L_{β_3}	[50]
103.4762	11.982	92	U	L_3M_2	L_t	[50]
103.4934	11.98	83	Bi	L_2M_2		[50]
103.7186	11.954	75	Re	abs. edge	L_2	[51]
103.7012	11.956	75	Re	L_2O_4	L_{γ_6}	[50]
103.8793	11.9355	79	Au	$L_3P_{2,3}$		[50]
103.8854	11.9348	79	Au	L_3P_1		[50]
103.9220	11.9306	81	Tl	L_1M_2	L_{β_4}	[50]
103.9708	11.925	75	Re	L_2O_3		[50]
103.9769	11.9243	76	0s	L_2N_3		[50]
103.9777	11.9242	35	Br	KL_3	K_{α_1}	[50]
103.9786	11.9241	80	Hg	L_3N_5	L_{β_2}	[50]
104.0040	11.9212	79	Au	abs. edge	L_3	[50]
104.0406	11.917	75	Re	L_2N_6	L_ν	[50]
104.0467	11.9163	79	Au	$L_3O_{4,5}$	L_{β_5}	[50]

λ [pm]	E [keV]	Z	element	transition	notation	reference
104.1542	11.9040	80	Hg	L_3N_4	$L_{\beta_{15}}$	[50]
104.1980	11.899	75	Re	L_1N_1		[50]
104.2330	11.8950	87	Fr	L_3M_4	L_{α_2}	[50]
104.2J69	11.890	93	Np	L_3M_1	L_l	[50]
104.3857	11.8776	35	Br	KL_2	K_{α_2}	[50]
104.4015	11.8758	75	Re	L_2O_1	L_{γ_8}	[50]
104.4966	11.865	33	As	abs. edge	K	[51]
104.4965	11.865	79	Au	$L_3O_{2,3}$		[50]
104.5036	11.8642	33	As	$KN_{2,3}$	K	[50]
104.5759	11.856	74	W	L_1N_5	$L_{\gamma_{11}}$	[50]
104.6818	11.844	74	W	L_1N_4		[50]
104.7552	11.8357	79	Au	$L_3N_{6,7}$	L_u	[50]
104.8713	11.8226	80	Hg	L_2M_4	L_{β_1}	[50]
104.8766	11.822	33	As	$KM_{4,5}$	K_{β_5}	[50]
104.9762	11.8118	81	Tl	L_3N_1	L_{β_6}	[50]
104.9779	11.8106	79	Au	L_3O_1	L_{β_7}	[50]
105.4502	11.7577	78	Pt	L_1M_5	L_{β_9}	[50]
105.6119	11.7397	81	Tl	L_2M_3	$L_{\beta_{17}}$	[50]
105.6965	11.7303	76	0s	L_2N_1	L_{γ_5}	[50]
105.7262	11.7270	86	Rn	L_3M_5	L_{α_1}	[50]
105.7334	11.7262	33	As	KM_3	K_{β_1}	[50]
105.7867	11.7203	33	As	KM_2	K_{β_3}	[50]
105.8526	11.713	80	Hg	L_3N_3		[50]
105.8598	11.7122	83	Bi	L_2M_1	L_η	[50]
106.1026	11.6854	75	Re	L_2N_4	L_{γ_1}	[50]
106.1335	11.682	73	Ta	abs. edge	L_1	[51]
106.1862	11.6762	78	Pt	L_1M_4	$L_{\beta_{10}}$	[50]
106.1953	11.6752	73	Ta	$L_1O_{4,5}$		[50]
106.2035	11.6743	74	W	L_1N_3	L_{γ_3}	[50]
106.3611	11.6570	73	Ta	$L_1N_6, 7$		[50]
106.4433	11.648	81	Tl	L_1M_1		[50]
106.4433	11.648	82	Pb	L_2M_2		[50]
106.4698	11.6451	73	Ta	L_1O_3	L_{γ_4}	[50]
106.4982	11.642	80	Hg	L_3N_2		[50]
106.5476	11.6366	73	Ta	L_1O_2	$L_{\gamma_{4'}}$	[50]
106.7154	11.6183	92	U	L_3M_1	L_l	[50]
106.7751	11.6118	73	Ta	L_1O_1		[50]
106.7889	11.6103	79	Au	L_1M_3	L_{β_3}	[50]
106.8101	11.6080	74	W	L_1N_2	L_{γ_2}	[50]
106.9031	11.5979	86	Rn	L_3M_4	L_{α_2}	[50]
107.0249	11.5847	79	Au	L_3N_5	L_{β_2}	[50]
107.1915	11.5667	79	Au	L_3N_4	$L_{\beta_{15}}$	[50]

λ [pm]	E [keV]	Z	element	transition	notation	reference
107.2351	11.562	78	Pt	abs. edge	L_3	[51]
107.2258	11.5630	80	Hg	L_1M_2	L_{β_4}	[50]
107.2443	11.561	78	Pt	$L_3O_{4,5}$	L_{β_5}	[50]
107.4581	11.538	74	W	abs. edge	L_2	[51]
107.4516	11.5387	74	W	L_2O_4	L_{γ_6}	[50]
107.5700	11.526	79	Au	L_2M_5		[50]
107.6167	11.521	78	Pt	$L_3O_{2,3}$		[50]
107.6727	11.515	75	Re	L_2N_3		[50]
107.7195	11.510	74	W	$L_2N_{6,7}$	L_v	[50]
107.8995	11.4908	78	Pt	$L_3N_{6,7}$	L_u	[50]
107.9258	11.488	74	W	L_2O_3		[50]
107.9784	11.4824	80	Hg	L_3N_1	L_{β_6}	[50]
108.0123	11.4788	90	Th	L_3M_2	L_t	[50]
108.1168	11.4677	74	W	L_2O_1	L_{γ_8}	[50]
108.1715	11.4619	78	Pt	K_3O_1	L_{β_7}	[50]
108.2084	11.4580	73	Ta	L_1N_5	$L_{\gamma_{11}}$	[50]
108.3568	11.4423	79	Au	L_2M_4	L_{β_1}	[50]
108.3805	11.4398	73	Ta	L_1N_4	L_{γ_1}	[50]
108.3976	11.438	75	Re	L_2N_2		[50]
108.5038	11.4268	85	At	L_3M_5	L_{α_1}	[50]
108.9022	11.385	80	Hg	L_2M_3	$L_{\beta_{17}}$	[50]
108.9788	11.3770	77	Ir	L_1M_5	L_{β_9}	[50]
109.0296	11.3717	79	Au	L_3N_3		[50]
109.0842	11.366	91	Pa	L_3N_1	L_l	[50]
109.1611	11.358	80	Hg	L_2M_3	$L_{\beta_{17}}$	[50]
109.2448	11.3493	82	Pb	L_2M_1		[50]
109.3913	11.3341	75	Re	L_2N_1	L_{γ_5}	[50]
109.6748	11.3048	85	At	L_3M_4	L_{α_2}	[50]
109.7058	11.3016	77	Ir	L_1M_4	$L_{\beta_{10}}$	[50]
109.9885	11.2859	74	W	L_2N_4	L_{γ_1}	[50]
109.9393	11.2776	73	Ta	L_1N_3	L_{γ_3}	[50]
109.9715	11.2743	79	Au	L_3N_2		[50]
109.9744	11.274	81	Tl	L_2M_2		[50]
109.9939	11.272	80	Hg	L_1M_1		[50]
109.9745	11.274	72	Hf	abs. edge	L_1	[51]
110.0896	11.2622	72	Hf	$L_1O_{4,5}$		[50]
110.2042	11.2505	78	Pt	L_3N_5	L_{β_2}	[50]
110.3061	11.2401	72	Hf	L_1O_3	L_{γ_4}	[50]
110.3797	11.2326	72	Hf	L_1O_2	$L_{\gamma_{4'}}$	[50]
110.3974	11.2308	78	Pt	L_1M_3	L_{β_3}	[50]
110.4801	11.2224	34	Se	KL_3	K_{α_1}	[50]
110.5333	11.217	73	Ta	L_1N_2	L_{γ_2}	[50]

λ [pm]	E [keV]	Z	element	transition	notation	reference
110.5826	11.212	77	Ir	abs. edge	L_3	[51]
110.5885	11.2114	77	Ir	$L_3O_{4,5}$	L_{β_5}	[50]
110.6546	11.2047	79	Au	L_1M_2	L_{β_4}	[50]
110.6674	11.2034	72	Hf	L_1O_1		[50]
110.8852	11.1814	34	Se	KL_2	K_{α_2}	[50]
110.9268	11.1772	77	Ir	$L_3O_{2,3}$		[50]
111.0958	11.1602	79	Au	L_3N_1	L_{β_6}	[50]
111.1486	11.1549	77	Ir	$L_3N_{6,7}$	L_u	[50]
111.2973	11.140	78	Pt	L_2M_5		[50]
111.3773	11.132	73	Ta	abs. edge	L_2	[51]
111.3893	11.1308	84	Po	L_3M_5	L_{α_1}	[50]
111.3913	11.1306	73	Ta	L_2O_4	L_{γ_6}	[50]
111.4942	11.1205	77	Ir	L_3O_1	L_{β_7}	[50]
111.4974	11.120	74	W	L_2N_3		[50]
111.5115	11.1186	90	Th	L_3M_1	L_l	[50]
111.5245	11.1173	73	Ta	L_1N_1		[50]
111.5847	11.1113	73	Ta	L_2N_6	L_v	[50]
111.6622	11.1036	32	Ge	abs. edge	K	[51]
111.6903	11.1008	32	Ge	$KN_{2,3}$	K_{β_2}	[50]
111.6973	11.1001	73	Ta	L_2O_3		[50]
111.7920	11.0907	73	Ta	L_2O_2		[50]
111.9555	11.0745	32	Ge	$KM_{4,5}$	K_{β_5}	[50]
111.9940	11.0707	78	Pt	L_2M_4	L_{β_1}	[50]
112.0557	11.0646	73	Ta	L_2O_1	L_{γ_8}	[50]
112.1500	11.0553	72	Hf	L_1N_5	$L_{\gamma_{11}}$	[50]
112.1853	11.052	74	W	L_2N_2		[50]
112.2535	11.0451	72	Hf	L_1N_4		[50]
112.2647	11.044	78	Pt	L_3N_3		[50]
112.5521	11.0158	84	Po	L_3M_4	L_{α_2}	[50]
112.6411	11.0071	76	0s	L_1M_5	L_{β_9}	[50]
112.6800	11.0033	72	Hf	L_1N_1		[104]
112.7722	10.9943	81	Tl	L_2M_1	L_η	[50]
112.8009	10.9915	79	Au	L_2M_3	$L_{\beta_{17}}$	[50]
112.8975	10.9821	32	Ge	KM_3	K_{β_1}	[50]
112.9397	10.9780	32	Ge	KM_2	K_{β_3}	[50]
113.1045	10.962	78	Pt	L_3N_2		[50]
113.2388	10.9490	74	W	L_2N_1	L_{γ_5}	[50]
113.2460	10.9483	88	Ra	L_3M_2	L_t	[50]
113.3568	10.9376	76	0s	L_1M_4	$L_{\beta_{10}}$	[50]
113.5291	10.9210	79	Au	L_1M_1		[50]
113.5364	10.9203	77	Ir	L_3N_5	L_{β_2}	[50]
113.6905	10.9055	73	Ta	L_2N_5		[50]

λ [pm]	E [keV]	Z	element	transition	notation	reference
113.7103	10.9036	77	Ir	L_3N_4	$L_{\beta_{15}}$	[50]
113.7980	10.8952	73	Ta	L_2N_4	L_{γ_1}	[50]
113.8450	10.8907	72	Hf	L_1N_3	L_{γ_3}	[50]
113.8732	10.888	80	Hg	L_2M_2		[50]
114.0199	10.8740	71	Lu	abs. edge	L_1	[51]
114.0502	10.8711	76	Os	$L_3O_{4,5}$	L_{β_5}	[50]
114.0797	10.8683	76	Os	abs. edge	L_3	[51]
114.0891	10.8674	77	Ir	L_1M_3	L_{β_3}	[50]
114.1342	10.8631	71	Lu	$L_1O_{4,5}$		[104]
114.2268	10.8543	78	Pt	L_1M_2	L_{β_4}	[50]
114.3511	10.8423	71	Lu	$L_1O_{2,3}$	$L_{\gamma_4\gamma_{4'}}$	[50]
114.3585	10.8418	78	Pt	L_3N_1	L_{β_6}	[50]
114.3901	10.8388	83	Bi	L_3M_5	L_{α_1}	[50]
114.4461	10.8335	72	Hf	L_1N_2	L_{γ_2}	[50]
114.5412	10.8245	76	Os	$L_3N_{6,7}$	L_u	[50]
114.6376	10.8154	71	Lu	L_1O_1		[104]
114.8968	10.191	77	Ir	L_2M_5		[50]
114.9313	10.1812	76	Os	L_3O_1	L_{β_7}	[50]
115.4833	10.7362	72	Hf	abs. edge	L_2	[51]
115.5231	10.1325	72	Hf	$L_2O_{4,5}$	L_{γ_6}	[50]
115.5328	10.1316	73	Ta	L_2N_3		[50]
115.5403	10.13091	83	Bi	L_3M_4	L_{α_2}	[50]
115.6039	10.725	77	Ir	L_3N_3		[50]
115.1166	10.109	72	Hf	L_2O_3		[50]
115.1842	10.1083	77	Ir	L_2M_4	L_{β_1}	[50]
115.8339	10.1031	72	Hf	L_2N_6	L_{γ_8}	[50]
115.8151	10.6999	72	Hf	L_2O_2		[50]
116.0041	10.688	73	Ta	L_2N_2		[50]
116.1105	10.6182	71	Lu	L_1N_5	$L_{\gamma_{11}}$	[50]
116.1410	10.6154	72	Hf	L_2O_1	L_{γ_8}	[50]
116.2303	10.6612	71	Lu	L_1N_4		[50]
116.4016	10.6512	80	Hg	L_2M_1	L_η	[50]
116.4913	10.6433	75	Re	L_1M_5	L_{β_9}	[50]
116.5493	10.6380	77	Ir	L_3N_2		[50]
116.6154	10.626S	78	Pt	L_2M_3	$L_{\beta_{17}}$	[50]
116.7221	10.6222	88	Ra	L_3M_1	L_l	[50]
116.9660	10.6001	78	Pt	L_1M_1		[50]
116.9831	10.5985	76	Os	L_3N_5	L_{β_2}	[50]
117.0864	10.5892	79	Au	L_2M_2		[50]
117.1705	10.5816	76	Os	L_3N_4	$L_{\beta_{15}}$	[50]
117.2215	10.5110	75	Re	L_1M_4	$L_{\beta_{10}}$	[50]
117.2969	10.5102	73	Ta	L_2N_1	L_{γ_5}	[50]

λ [pm]	E [keV]	Z	element	transition	notation	reference
117.5048	10.5515	82	Pb	$L_3 M_5$	L_{α_1}	[50]
117.5915	10.54372	33	As	KL_3	K_{α_1}	[50]
117.7380	10.5306	75	Re	abs. edge	L_3	[51]
117.7246	10.5318	75	Re	$L_3 O_{4,5}$	L_{β_5}	[50]
117.7911	10.5258	72	Hf	$L_2 N_5$		[50]
117.1995	10.5251	77	Ir	$L_3 N_1$	L_{β_6}	[50]
117.9031	10.5158	72	Hf	$L_2 N_4$	L_{γ_1}	[50]
117.9515	10.5110	71	Lu	$L_1 N_3$	L_{γ_3}	[50]
117.9598	10.5108	76	0s	$L_1 M_3$	L_{β_3}	[50]
117.9620	10.5106	77	Ir	$L_1 M_2$	L_{β_4}	[50]
117.9914	10.50199	33	As	KL_2	K_{α_2}	[50]
118.1587	10.4931	75	Re	$L_3 N_{6,7}$	L_u	[50]
118.1892	10.4904	70	Yb	abs. edge	L_1	[51]
118.2693	10.4833	70	Yb	$L_1 O_{4,5}$		[50]
118.5293	10.4603	70	Yb	$L_1 O_{2,3}$	$L_{\gamma_4 \gamma_{4'}}$	[50]
118.5321	10.460	71	Lu	$L_1 N_2$	L_{γ_2}	[50]
118.6132	10.4529	75	Re	$L_3 O_1$	L_{β_7}	[104]
118.6518	10.4495	82	Pb	$L_3 M_4$	L_{α_2}	[50]
118.8599	10.4312	70	Yb	$L_1 O_1$		[50]
118.9820	10.4205	76	0s	$L_2 M_5$		[50]
119.2119	10.4004	76	0s	$L_3 N_3$		[104]
119.5822	10.3682	31	Ga	abs. edge	K	[51]
119.6041	10.3663	31	Ga	$KN_{2,3}$	K_{β_2}	[50]
119.6456	10.3621	71	Lu	$L_1 N_1$		[104]
119.7311	10.3553	76	0s	$L_2 M_4$	L_{β_1}	[50]
119.8527	10.3448	71	Lu	abs. edge	L_2	[51]
119.8156	10.348	31	Ga	$KM_{4,5}$	K_{β_5}	[50]
119.8723	10.3431	71	Lu	$L_2 O_4$	L_{γ_6}	[50]
120.0895	10.3244	76	0s	$L_3 N_2$		[50]
120.1430	10.3198	71	Lu	$L_2 O_{2,3}$		[50]
120.2770	10.3083	79	Au	$L_2 M_1$	L_η	[50]
120.4734	10.2915	71	Lu	$L_2 O_1$	L_{γ_8}	[50]
120.4828	10.2907	74	W	$L_1 M_5$	L_{β_9}	[50]
120.5026	10.289	70	Yb	$L_1 M_{4,5}$		[104]
120.6645	10.2752	75	Re	$L_3 N_5$	L_{β_2}	[50]
120.6903	10.273	77	Ir	$L_2 M_3$	$L_{\beta_{17}}$	[50]
120.7432	10.2685	81	Tl	$L_3 M_5$	L_{α_1}	[50]
120.7938	10.2642	31	Ga	KM_3	K_{β_1}	[50]
120.8232	10.2617	75	Re	$L_3 N_4$	$L_{\beta_{15}}$	[50]
120.8397	10.2603	31	Ga	KM_2	K_{β_3}	[50]
121.0202	10.245	77	Ir	$L_1 M_1$		[50]
121.0544	10.2421	83	Bi	$L_3 M_3$	L_s	[50]

λ [pm]	E [keV]	Z	element	transition	notation	reference
121.2225	10.2279	74	W	L_1M_4	$L_{\beta_{10}}$	[50]
121.2569	10.225	78	Pt	L_2M_2		[50]
121.3530	10.2169	76	Os	L_3N_1	L_{β_6}	[50]
121.5553	10.1999	74	W	abs. edge	L_3	[51]
121.5410	10.2011	72	Hf	L_2N_1	L_{γ_5}	[50]
121.5493	10.2004	74	W	$L_3O_{4,5}$	L_{β_5}	[50]
121.8479	10.1754	76	Os	L_1M_2	L_{β_4}	[50]
121.8731	10.1733	74	W	$L_3N_{6,7}$	L_u	[50]
121.8791	10.1728	81	Tl	L_3M_4	L_{α_2}	[50]
122.0350	10.1598	75	Re	L_1M_3	L_{β_3}	[50]
122.1144	10.1532	74	W	$L_3O_{2,3}$		[50]
122.2323	10.1434	71	Lu	L_2N_4	L_{γ_1}	[50]
122.2360	10.1431	70	Yb	L_1N_3	L_{γ_3}	[50]
122.4037	10.1292	74	W	L_3O_1	L_{β_7}	[50]
122.5078	10.1206	69	Tm	abs. edge	L_1	[51]
122.6362	10.110	69	Tm	$L_1O_{4,5}$		[50]
122.8391	10.0933	75	Re	L_3N_3		[50]
122.8829	10.0897	70	Yb	L_1N_2	L_{γ_2}	[50]
122.9524	10.084	69	Tm	$L_1O_{2,3}$	$L_{\gamma_4\gamma_{4'}}$	[50]
123.0585	10.0753	75	Re	L_2M_5		[50]
123.3168	10.0542	69	Tm	L_1O_1		[104]
123.3364	10.0526	69	Tm	L_2O_1		[104]
123.8613	10.0100	75	Re	L_2M_4	L_{β_1}	[50]
123.9393	10.0037	70	Yb	L_1N_1		[104]
124.1248	9.9888	80	Hg	L_3M_5	L_{α_1}	[50]
124.2461	9.979	71	Lu	L_2N_3		[104]
124.2822	9.9761	70	Yb	abs. edge	L_2	[51]
124.2760	9.9766	70	Yb	L_2O_4	L_{γ_6}	[50]
124.2959	9.975	78	Pt	L_2M_1	L_η	[50]
124.3233	9.9728	70	Yb	L_2N_6	L_ν	[50]
124.3894	9.9675	82	Pb	L_3M_3	L_s	[50]
124.4643	9.9615	74	W	L_3N_5	L_{β_2}	[50]
124.5319	9.9561	70	Yb	$L_2O_{2,3}$		[50]
124.6358	9.9478	74	W	L_3N_4	$L_{\beta_{15}}$	[50]
124.6583	9.946	73	Ta	L_1M_5	L_{β_9}	[50]
124.8089	9.934	76	Os	L_2M_3	$L_{\beta_{17}}$	[50]
124.8969	9.927	69	Tm	$L_1N_{4,5}$		[104]
124.9271	9.9246	70	Yb	L_2O_1	L_{γ_8}	[50]
124.9850	9.92	71	Lu	L_2N_2		[104]
125.0228	9.917	77	Ir	L_2M_2		[50]
125.1048	9.9105	75	Re	L_3N_1	L_{β_6}	[50]
125.2679	9.8976	80	Hg	L_3M_4	L_{α_2}	[50]

λ [pm]	E [keV]	Z	element	transition	notation	reference
125.3768	9.889	73	Ta	L_1M_4	$L_{\beta_{10}}$	[50]
125.4096	9.88642	32	Ge	KL_3	K_{α_1}	[50]
125.5343	9.8766	73	Ta	abs. edge	L_3	[51]
125.5546	9.8750	73	Ta	$L_3O_{4,5}$	L_{β_5}	[50]
125.7813	9.8572	73	Ta	$L_3N_{6,7}$	L_u	[50]
125.8054	9.85532	32	Ge	KL_2	K_{α_2}	[50]
125.9206	9.8463	75	Re	L_1M_2	L_{β_4}	[50]
125.9653	9.8428	71	Lu	L_2N_1	L_{γ_5}	[50]
126.0140	9.839	73	Ta	$L_3O_{2,3}$		[50]
126.2732	9.8188	74	W	L_1M_3	L_{β_3}	[50]
126.3891	9.8098	73	Ta	L_3O_1	L_{β_7}	[50]
126.7224	9.784	74	W	L_3N_3		[50]
126.7729	9.7801	70	Yb	L_2N_4	L_{γ_1}	[50]
126.7872	9.779	69	Tm	L_1N_3	L_{γ_3}	[50]
127.1343	9.7523	68	Er	$L_1O_{4,5}$		[104]
127.0679	9.7574	68	Er	abs. edge	L_1	[51]
127.2818	9.741	74	W	L_2M_5	L_{γ_2}	[50]
127.4256	9.730	69	Tm	L_1N_2		[50]
127.4885	9.7252	83	Bi	L_3M_2	L_t	[50]
127.5305	9.722	68	Er	$L_1O_{2,3}$	$L_{\gamma_4\gamma_{4'}}$	[50]
127.6447	9.7133	79	Au	L_3M_5	L_{α_1}	[50]
127.6618	9.712	74	W	L_3N_2		[50]
127.8105	9.7007	81	Tl	L_3M_3	L_s	[50]
127.9490	9.6902	88	Er	L_1O_1		[104]
128.1852	9.67235	74	W	L_2M_4	L_{β_1}	[50]
128.2959	9.664	84	Po	L_3M_1	L_l	[50]
128.3398	9.6607	30	Zn	abs. edge	K	[51]
128.3756	9.6580	30	Zn	$KN_{2,3}$	K_{β_2}	[50]
128.4527	9.6522	77	Ir	L_2M_1	L_η	[50]
128.4581	9.6518	73	Ta	L_3N_5	L_{β_2}	[50]
128.4807	9.6501	30	Zn	$KM_{4,5}$	K_{β_5}	[50]
128.5753	9.643	69	Tm	L_1N_1		[104]
128.6233	9.6394	73	Ta	L_3N_4	$L_{\beta_{15}}$	[50]
128.7128	9.6327	70	Yb	L_2N_3		[104]
128.7756	9.6280	79	Au	L_3M_4	L_{α_2}	[50]
128.9216	9.6171	69	Tm	abs. edge	L_2	[51]
128.9940	9.6117	74	W	L_3N_1	L_{β_6}	[50]
129.0289	9.6091	72	Hf	L_1M_5	L_{β_9}	[104]
129.0571	9.607	69	Tm	L_2O_4	L_{γ_6}	[50]
129.2724	9.5910	75	Re	L_2M_3	$L_{\beta_{17}}$	[50]
129.3398	9.586	76	Os	L_2M_2		[50]
129.5290	9.5720	30	Zn	$KM_{2,3}$	$K_{\beta_{1,3}}$	[50]

λ [pm]	E [keV]	Z	element	transition	notation	reference
129.7228	9.5577	72	Hf	abs. edge	L_3	[51]
129.7649	9.5546	72	Hf	$L_3O_{4,5}$	L_{β_5}	[50]
129.8220	9.5504	72	Hf	L_1M_4	$L_{\beta_{10}}$	[104]
130.0889	9.5308	72	Hf	L_3O_3		[104]
130.1654	9.5252	74	W	L_1M_2	L_{β_4}	[50]
130.1695	9.5249	72	Hf	$L_3N_{6,7}$	L_u	[50]
130.2064	9.5222	72	Hf	L_3O_2		[50]
130.5684	9.4958	72	Hf	L_3O_1	L_{β_7}	[50]
130.6345	9.4910	70	Yb	L_2N_1	L_{γ_5}	[50]
130.6826	9.4875	73	Ta	L_1M_3	L_{β_3}	[50]
130.7709	9.4811	82	Pb	L_3M_2	L_t	[50]
130.8633	9.4744	73	Ta	L_3N_3		[104]
131.1318	9.455	80	Hg	L_3M_3	L_s	[50]
131.3082	9.4423	78	Pt	L_3M_5	L_{α_1}	[50]
131.4669	9.4309	68	Er	L_1N_3	L_{γ_3}	[50]
131.5353	9.426	69	Tm	L_2N_4	L_{γ_1}	[50]
131.6135	9.4204	83	Bi	L_3M_1	L_l	[50]
131.6806	9.4156	73	Ta	L_3N_2		[104]
131.9019	9.3998	73	Ta	L_2M_5		[50]
131.9076	9.3994	67	Ho	abs. edge	L_1	[51]
132.0818	9.387	67	Ho	$L_1O_{4,5}$		[50]
132.1099	9.385	68	Er	L_1N_2	L_{γ_2}	[50]
132.2649	9.374	67	Ho	$L_1O_{2,3}$	$L_{\gamma_4\gamma_{4'}}$	[50]
132.4373	9.3618	78	Pt	L_3M_4	L_{α_2}	[50]
132.6428	9.3473	72	Hf	L_3N_5	L_{β_2}	[50]
132.6740	9.3451	67	Ho	L_1O_1		[104]
132.7024	9.3431	73	Ta	L_2M_4	L_{β_1}	[50]
132.7877	9.3371	72	Hf	L_3N_4	$L_{\beta_{15}}$	[50]
132.7891	9.3370	76	Os	L_2O_1	L_η	[50]
133.0984	9.3153	73	Ta	L_3N_1	L_{β_6}	[50]
133.4322	9.292	68	Er	L_1N_1		[104]
133.4322	9.292	69	Tm	L_2N_3		[104]
133.5817	9.2816	71	Lu	L_1M_5	L_{β_9}	[50]
133.6479	9.277	74	W	L_1M_1		[50]
133.6609	9.2761	75	Re	L_1M_2		[50]
133.8615	9.2622	68	Er	abs. edge	L_2	[51]
133.8788	9.261	74	W	L_2M_3	$L_{\beta_{17}}$	[50]
133.9743	9.2544	68	Er	L_2O_4	L_{γ_6}	[50]
134.0134	9.25174	31	Ga	KL_3	K_{α_1}	[50]
134.0525	9.2490	71	Lu	abs. edge	L_3	[51]
134.1584	2.2417	81	Tl	L_3M_2	L_t	[50]
134.1874	9.2397	71	Lu	$L_3O_{4,5}$	L_{β_5}	[50]

λ [pm]	E [keV]	Z	element	transition	notation	reference
134.2630	9.2345	68	Er	$L_2O_{2,3}$		[104]
134.2979	9.2321	67	Ho	$L_1N_{4,5}$		[104]
134.2993	9.232	71	Lu	L_1M_4	$L_{\beta_{10}}$	[50]
134.3969	9.2253	69	Tm	L_2N_2		[104]
134.4042	9.22482	31	Ga	KL_2	K_{α_2}	[50]
134.5281	9.2163	71	Lu	$L_3O_{2,3}$		[50]
134.5851	9.2124	73	Ta	L_1M_2	L_{β_4}	[50]
134.6933	9.205	68	Er	L_2O_1	L_{γ_8}	[50]
134.9528	9.1873	71	Lu	L_3O_1	L_{β_7}	[50]
134.9939	9.1845	82	Pb	L_3M_1	L_l	[50]
135.0571	9.1802	72	Hf	L_3N_3		[50]
135.1322	9.1751	77	Ir	L_2M_5	L_{α_1}	[50]
135.1352	9.1749	79	Au	L_3M_3	L_s	[50]
135.3048	9.1634	72	Hf	L_1M_3	L_{β_3}	[50]
135.5918	9.144	69	Tm	L_2N_1	L_{γ_5}	[50]
135.8905	9.1239	72	Hf	L_3N_2		[50]
136.2549	9.0995	77	Ir	L_3M_4	L_{α_2}	[50]
136.4123	9.089	68	Er	L_2N_4	L_{γ_1}	[50]
136.4423	9.087	67	Ho	L_1N_3	L_{γ_3}	[50]
136.6985	9.051	67	Ho	L_1N_2	L_{γ_2}	[50]
136.9276	9.0548	66	Dy	abs. edge	L_1	[51]
137.0168	9.0489	71	Lu	L_3N_5	L_{β_2}	[50]
137.1608	9.0394	71	Lu	L_3N_4	$L_{\beta_{15}}$	[104]
137.3462	9.0272	75	Re	L_2M_1	L_η	[50]
137.4147	9.0227	72	Hf	L_2M_4	L_{β_1}	[50]
137.4147	9.0227	72	Hf	L_3N_1	L_{β_6}	[50]
137.4635	9.0195	66	Dy	$L_1O_{2,3}$	$L_{\gamma_4\gamma_{4'}}$	[50]
137.4711	9.019	80	Hg	L_3M_2	L_t	[50]
138.0635	8.9803	29	Cu	abs. edge	K	[51]
138.1127	8.9771	29	Cu	$KN_{2,3}$	K_{β_2}	[50]
138.1142	8.997	29	Cu	$KM_{4,5}$	K_{β_5}	[50]
138.3809	8.9597	70	Yb	L_1M_5	L_{β_9}	[50]
138.4814	8.9532	81	Tl	L_3M_1	L_l	[50]
138.4860	8.9529	74	W	L_2M_2		[104]
138.5805	8.9468	68	Er	L_2N_3		[104]
138.6223	8.9441	70	Yb	abs. edge	L_3	[51]
138.6424	8.9428	73	Ta	L_2M_3	$L_{\beta_{17}}$	[50]
138.7014	8.939	70	Yb	$L_3O_{5,6}$		[50]
138.7712	8.9345	70	Yb	$L_3N_{6,7}$	L_u	[104]
138.9501	8.923	78	Pt	L_3M_3	L_s	[50]
138.9828	8.9209	70	Yb	$L_3O_{2,3}$		[50]
139.0530	8.9164	67	Ho	abs. edge	L_2	[51]

λ [pm]	E [keV]	Z	element	transition	notation	reference
139.1236	8.9117	76	Os	L_3M_5	L_{α_1}	[50]
139.1528	8.91	70	Yb	L_1M_4	$L_{\beta_{10}}$	[50]
139.2247	8.90454	72	Hf	L_1M_2	L_{β_4}	[50]
139.2263	8.90529	29	Cu	KM_3	K_{β_1}	[50]
139.2309	8.905	67	Ho	L_2O_4	L_{γ_6}	[50]
139.2638	8.9029	29	Cu	KM_2	K_{β_3}	[50]
139.4800	8.8891	66	Dy	L_1N_5	$L_{\gamma_{11}}$	[104]
139.4831	8.8889	70	Yb	L_3O_1	L_{β_7}	[50]
139.5978	8.8816	71	Lu	L_3N_3		[104]
139.8276	8.867	67	Ho	L_2O_1	L_{γ_8}	[50]
140.1453	8.8469	71	Lu	L_1M_3	L_{β_3}	[50]
140.2388	8.841	76	Os	L_3M_4	L_{α_2}	[50]
140.6684	8.814	68	Er	L_2N_1	L_{γ_5}	[50]
141.3710	8.7702	79	Au	L_3M_2	L_t	[50]
141.5550	8.7588	70	Yb	L_2N_4	$L_{\beta_{15}}$	[50]
141.5550	8.7588	70	Yb	L_2N_5	L_{β_2}	[50]
141.6455	8.7532	66	Dy	L_1N_3	L_{γ_3}	[50]
141.7459	8.747	67	Ho	L_2N_4	L_{γ_1}	[50]
141.8984	8.7376	71	Lu	L_3N_1	L_{β_6}	[50]
142.1147	8.7243	74	W	L_2M_1	L_γ	[50]
142.1687	8.721	80	Hg	L_3M_1	L_1	[50]
142.2827	8.714	66	Dy	L_1N_2	L_{γ_2}	[50]
142.2827	8.714	65	Tb	$L_1N_{4,5}$		[50]
142.3497	8.7099	71	Lu	L_2M_4	L_{β_1}	[50]
142.2387	8.7167	65	Tb	abs. edge	L_1	
142.7578	8.685	65	Tb	$L_1O_{2,3}$	$L_{\gamma_4\gamma_{4'}}$	[50]
143.0114	8.6696	65	Tb	L_1O_1		
143.0295	8.6685	72	Hf	L_1M_1		[50]
143.0526	8.6671	75	Ta	L_2M_2		[50]
143.1865	8.659	77	Ir	L_3M_3	L_s	[50]
143.2940	8.6525	75	Re	L_3M_5	L_{α_1}	[50]
143.3421	8.6496	69	Tm	abs. edge	L_3	[51]
143.3686	8.648	69	Tm	L_3M_5	L_{β_9}	[50]
143.4847	8.641	69	Tm	$L_3O_{4,5}$	L_{β_5}	[50]
143.5196	8.63886	30	Zn	KL_3	K_{α_1}	[50]
143.6477	8.6312	72	Hf	L_2M_3	$L_{\beta_{17}}$	[50]
143.9044	8.61578	30	Zn	KL_2	K_{α_2}	[50]
144.0616	8.6064	71	Lu	L_1M_2	L_{β_4}	[50]
144.1018	8.604	66	Dy	L_1N_5		[104]
144.1018	8.604	69	Tm	L_1M_4	$L_{\beta_{10}}$	[50]
144.1772	8.5995	70	Yb	L_3N_3		[104]
144.4005	8.5862	75	Re	L_3M_4	L_{α_2}	[50]

λ [pm]	E [keV]	Z	element	transition	notation	reference
144.4224	8.5849	69	Tm	L_3O_1	L_{β_7}	[50]
144.4544	8.5830	66	Dy	abs. edge	L_2	[51]
144.5841	8.5753	66	Dy	L_2O_5	L_{γ_6}	[50]
144.5841	8.5753	66	Dy	L_2O_1	L_{γ_8}	[50]
144.8594	8.559	65	Tb	$L_1N_{4,5}$		[104]
145.2378	8.5367	70	Yb	L_1M_3	L_{β_3}	[50]
145.3008	8.533	78	Pt	L_3M_2	L_t	[50]
145.9697	8.4939	79	Au	L_3M_1	L_l	[50]
146.1917	8.481	67	Ho	L_2N_1	L_{γ_5}	[50]
146.4161	8.468	69	Tm	L_3N_5	L_{β_2}	[50]
146.4161	8.468	69	Tm	L_3N_4	$L_{\beta_{15}}$	[50]
146.6187	8.4563	70	Yb	L_3N_1	L_{β_6}	[50]
147.1110	8.428	73	Ta	L_2M_1	L_η	[50]
147.1983	8.423	65	Tb	L_1N_3	L_{γ_3}	[50]
147.2718	8.4188	66	Dy	L_2N_4	L_{γ_1}	[50]
147.3558	8.414	76	Os	L_3M_3	L_s	[50]
147.5698	8.4018	70	Yb	L_2M_4	L_{β_1}	[50]
147.6365	8.398	65	Tb	L_1N_2	L_{γ_2}	[50]
147.6436	8.3976	74	W	L_3M_4	L_{α_1}	[50]
147.9748	8.3788	71	Lu	L_1M_1		[50]
147.8408	8.3864	64	Gd	abs. edge	L_1	[51]
148.0685	8.3735	72	Hf	L_2M_2		[50]
148.0773	8.373	64	Gd	$L_1O_{4,5}$	L_3	[50]
148.3520	8.3575	68	Er	abs. edge	L_3	[51]
148.3964	8.355	64	Gd	$L_1O_{2,3}$	$L_{\gamma_4\gamma_{4'}}$	[50]
148.4852	8.35	68	Er	$L_2O_{4,5}$	L_{β_5}	[50]
148.5564	8.346	68	Er	L_1M_5	L_{β_9}	[50]
148.7489	8.3352	74	W	L_3M_4	L_{α_2}	[50]
148.8123	8.33165	28	Ni	abs. edge	K	[51]
148.8668	8.3286	28	Ni	$KM_{4,5}$	K_{β_5}	[50]
149.1282	8.314	65	Tb	L_1N_1		[104]
149.1425	8.3132	70	Yb	L_1M_2	L_{β_4}	[50]
149.2538	8.307	69	Tm	L_3N_3		[104]
149.3078	8.304	77	Ir	L_3M_2	L_t	[50]
149.4157	8.298	68	Er	L_1M_4	$L_{\beta_{10}}$	[50]
149.4157	8.298	68	Er	L_3O_1	L_{β_7}	[50]
149.6231	8.2865	66	Dy	L_2N_3		[104]
149.9579	8.268	78	Pt	L_3M_1	L_l	[50]
150.0177	8.26466	28	Ni	$KM_{2,3}$	$K_{\beta_{1,3}}$	[50]
150.0468	8.2631	69	Tm	L_2N_2		[104]
150.2359	8.2527	65	Tb	abs. edge	L_2	[51]
150.3579	8.246	65	Tb	L_2O_5	L_{γ_6}	[50]

λ [pm]	E [keV]	Z	element	transition	notation	reference
150.5168	8.2373	64	Gd	$L_1N_{4,5}$		[104]
150.6320	8.231	69	Tm	L_1M_3	L_{β_3}	[50]
150.9805	8.212	65	Tb	L_2O_1	L_{γ_8}	[50]
151.4045	8.189	68	Er	L_3N_5	L_{β_2}	[50]
151.4045	8.189	68	Er	L_3N_4	$L_{\beta_{15}}$	[50]
151.6267	8.177	69	Tm	L_3N_1	L_{β_6}	[50]
151.7901	8.1682	75	Re	L_3M_3	L_s	[50]
151.8291	8.1661	66	Dy	L_2N_1	L_{γ_5}	[50]
152.2019	8.1461	73	Ta	L_3M_5	L_{α_1}	[50]
152.2037	8.146	69	Tm	L_2M_5		[104]
152.3290	8.1393	72	Hf	L_2M_1	L_η	[50]
152.9737	8.105	64	Gd	L_1N_3	L_{γ_3}	[50]
153.0303	8.102	65	Tb	L_2N_4	L_{γ_1}	[50]
153.0492	8.101	69	Tm	L_2M_4	L_{β_1}	[50]
153.2971	8.0879	73	Ta	L_3M_4	L_{α_2}	[50]
153.3142	8.087	64	Gd	L_1N_2	L_{γ_2}	[50]
153.3369	8.0858	71	Lu	L_2M_2		[50]
153.4660	8.079	76	Os	L_3M_2	L_t	[50]
153.6829	8.0676	67	Ho	abs. edge	L_3	[51]
153.7896	8.062	67	Ho	$L_3O_{4,5}$	L_{β_5}	[50]
153.8144	8.0607	63	Eu	abs. edge	L_1	[51]
154.0609	8.04778	29	Cu	KL_3	K_{α_1}	[50]
154.0992	8.0458	77	Ir	L_3M_1	L_l	[50]
154.3948	8.0304	63	Eu	$L_1O_{2,3}$	$L_{\gamma_4\gamma_{4'}}$	[50]
154.4448	8.02783	29	Cu	KL_2	K_{α_2}	[50]
154.4794	8.026	69	Tm	L_1M_2	L_{β_4}	[50]
154.8653	8.006	67	Ho	L_1M_4	$L_{\beta_{10}}$	[50]
155.6136	7.9675	65	Tb	L_2N_3		[104]
156.1683	7.9392	68	Er	L_1M_3	L_{β_3}	[50]
156.2313	7.936	65	Tb	L_2N_2		[104]
156.3298	7.9310	64	Gd	abs. edge	L_2	[51]
156.4284	7.926	74	W	L_2M_3	L_s	[50]
156.4482	7.925	64	Gd	L_2O_4	L_{γ_6}	[50]
156.7052	7.912	63	Eu	$L_1N_{4,5}$		[50]
156.7250	7.911	67	Ho	L_3N_5	L_{β_2}	[50]
156.7250	7.911	67	Ho	L_3N_4	$L_{\beta_{15}}$	[50]
156.7647	7.909	68	Er	L_3N_1	L_{β_6}	[50]
156.9631	7.899	72	Hf	L_3M_5	L_{α_1}	[50]
157.0625	7.894	64	Gd	L_2O_1	L_{γ_8}	[50]
157.7921	7.8575	71	Lu	L_2M_1	L_η	[50]
157.8725	7.8535	65	Tb	L_2N_1	L_{γ_5}	[50]
157.9027	7.852	75	Re	L_3M_2	L_t	[50]

λ [pm]	E [keV]	Z	element	transition	notation	reference
158.0516	7.8446	72	Hf	L_2M_4	L_{α_2}	[50]
158.5042	7.8222	76	Os	L_3M_1	L_l	[50]
158.6705	7.814	69	Tm	L_1M_1		[104]
158.7335	7.8109	68	Er	L_2M_4	L_{β_1}	[50]
158.8433	7.8055	66	Dy	$L_3O_{4,5}$	L_{β_5}	[50]
158.8494	7.8052	70	Yb	L_2M_2		[50]
159.0349	7.7961	63	Eu	L_1N_3	L_{γ_3}	[50]
159.1656	7.7897	66	Dy	abs. edge	L_3	[51]
159.2452	7.7858	64	Gd	L_2N_4	L_{γ_1}	[50]
159.6163	7.7677	63	Eu	L_1N_2	L_{γ_2}	[50]
159.6800	7.7646	66	Dy	$L_3O_{2,3}$		[104]
159.7130	7.763	67	Ho	L_3N_3		[104]
159.9788	7.7501	66	Dy	L_1M_5	L_{β_9}	[50]
160.0779	7.7453	68	Er	L_1M_2	L_{β_4}	[50]
160.0263	7.7478	62	Sm	abs. edge	L_1	[104]
160.3326	7.733	62	Sm	L_1N_6		[104]
160.4363	7.728	69	Tm	L_2M_3	$L_{\beta_{17}}$	[104]
160.4529	7.7272	66	Dy	L_3O_1	L_{β_7}	[50]
160.4571	7.727	67	Ho	L_2N_2		[104]
160.7337	7.7137	62	Sm	$L_1O_{2,3}$	$L_{\gamma_4\gamma_{4'}}$	[50]
160.7483	7.713	66	Dy	L_1M_4	$L_{\beta_{10}}$	[50]
160.8205	7.70954	27	Co	abs. edge	K	[51]
160.8967	7.7059	27	Co	$KM_{4,5}$	K_{β_5}	[50]
161.2689	7.6881	73	Ta	L_3M_3	L_s	[50]
161.9557	7.6555	71	Lu	L_3M_5	L_{α_1}	[50]
161.9620	7.6552	64	Gd	L_2N_3		[104]
162.0319	7.6519	67	Ho	L_1M_3	L_{β_3}	[50]
162.0848	7.64943	27	Co	$KM_{2,3}$	$K_{\beta_{1,3}}$	[50]
162.2290	7.6426	64	Gd	L_2M_2		[104]
162.3714	7.6359	67	Ho	L_3N_1	L_{β_6}	[50]
162.3756	7.6357	66	Dy	L_3N_5	L_{β_2}	[50]
162.3156	7.6357	66	Dy	L_3N_4	$L_{\beta_{15}}$	[50]
162.4544	7.632	74	W	L_3M_2	L_t	[50]
162.7124	7.6199	63	Eu	abs. edge	L_2	[51]
162.8234	7.6147	63	Eu	L_2O_2	L_{γ_6}	[50]
163.0118	7.6059	62	Sm	$L_1N_{4,5}$		[50]
163.0333	7.6049	71	Lu	L_2M_4	L_{α_2}	[50]
163.0611	7.6036	75	Re	L_3M_1	L_l	[50]
163.4632	7.5849	63	Eu	L_2O_1	L_{γ_8}	[50]
163.5645	7.5802	70	Yb	L_2M_1	L_η	[50]
164.1253	7.5543	64	Gd	L_2N_1	L_{γ_5}	[50]
164.3058	7.546	68	Er	L_1M_1		[104]

λ [pm]	E [keV]	Z	element	transition	notation	reference
164.7578	7.5253	67	Ho	L_2M_4	L_{β_1}	[50]
164.9770	7.5153	65	Tb	abs. edge	L_3	[51]
165.1066	7.5094	65	Tb	$L_3O_{4,5}$	L_{β_5}	[50]
165.4283	7.4948	66	Dy	L_3N_3		[104]
165.6072	7.4867	62	Sm	L_1N_3	L_{γ_3}	[50]
165.7489	7.4803	63	Eu	L_2N_4	L_{γ_1}	[50]
165.7955	7.47825	21	Ni	KL_3	K_{α_1}	[50]
165.8598	7.4753	65	Tb	L_3O_1	L_{β_7}	[50]
165.9591	7.4708	67	Ho	L_1M_2	L_{β_4}	[50]
165.9886	7.4695	65	Tb	L_1M_5	L_{β_9}	[104]
166.0486	7.4668	62	Sm	L_1N_2	L_{γ_2}	[50]
166.1799	7.46089	28	Ni	KL_2	K_{α_2}	[50]
166.3516	7.4532	72	Hf	L_3M_3	L_s	[50]
166.6691	7.439	66	Dy	L_3N_2		[104]
166.7364	7.436	65	Tb	L_1M_4	$L_{\beta_{10}}$	[50]
166.7364	7.436	61	Pm	abs. edge	L_1	[51]
166.9782	7.4252	68	Er	L_2M_3	$L_{\beta_{17}}$	[104]
167.1951	7.4156	70	Yb	L_3M_5	L_{α_1}	[50]
167.2695	7.4123	73	Ta	L_3M_2	L_t	[50]
167.8242	7.3878	74	W	L_3M_1	L_l	[50]
168.2181	7.3705	66	Dy	L_3N_1	L_{β_6}	[50]
168.2250	7.3702	66	Dy	L_1M_3	L_{β_3}	[50]
168.2912	7.3673	70	Yb	L_3M_4	L_{α_2}	[50]
168.3049	7.3667	65	Tb	L_3N_5	L_{β_2}	[50]
168.3049	7.3667	65	Tb	L_3N_4	$L_{\beta_{15}}$	[50]
169.5362	7.3132	62	Sm	abs. edge	L_2	[51]
169.6382	7.3088	69	Tm	L_2M_1	L_{η}	[50]
169.6661	7.3076	62	Sm	L_2O_4	L_{γ_6}	[50]
170.1223	7.288	62	Sm	L_2O_2		[104]
170.3725	7.2773	66	Dy	L_2M_5		[104]
170.6421	7.2658	62	Sm	L_2O_1	L_{γ_8}	[104]
170.8585	7.2566	63	Eu	L_2N_1	L_{γ_5}	[50]
171.0683	7.2477	66	Dy	L_2M_4	L_{β_1}	[50]
171.1793	7.2430	64	Gd	abs. edge	L_3	[51]
171.3118	7.2374	64	Gd	$L_3O_{4,5}$	L_{β_5}	[50]
171.5583	7.227	65	Tb	L_3N_3		[104]
171.8913	7.213	71	Lu	L_3M_3	L_s	[104]
172.0320	7.2071	64	Gd	L_3O_1	L_{β_7}	[50]
172.1084	7.2039	66	Dy	L_1M_2	L_{β_4}	[50]
172.2399	7.1984	65	Tb	L_3N_2		[50]
172.3117	7.1954	72	Hf	L_3M_2	L_t	[50]
172.3932	7.192	64	Gd	L_1M_5	L_{β_9}	[50]

λ [pm]	E [keV]	Z	element	transition	notation	reference
172.6834	7.1799	69	Tm	L_3M_5	L_{α_1}	[50]
172.7294	7.178	62	Sm	L_2N_4	L_{γ_1}	[50]
172.8474	7.1731	73	Ta	L_3M_1	L_l	[50]
173.1636	7.16	64	Gd	L_1M_4	$L_{\beta_{10}}$	[50]
173.8164	7.1331	69	Tm	L_3M_4	L_{α_2}	[50]
173.9069	7.1294	60	Nd	abs. edge	L_1	[51]
174.2270	7.1163	65	Tb	L_3N_1	L_{β_6}	[50]
174.3520	7.1112	26	Fe	abs. edge	K	[51]
174.3740	7.1103	64	Gd	L_3N_5	L_{β_2}	[50]
174.3740	7.1103	64	Gd	L_3N_4	$L_{\beta_{15}}$	[50]
174.4280	7.1081	26	Fe	$KM_{4,5}$	K_{β_5}	[50]
174.4550	7.107	60	Nd	L_1O_2	L_{γ_4}	[50]
174.7279	7.0959	65	Tb	L_1M_3	L_{β_3}	[50]
174.9177	7.0882	60	Nd	L_1O_1		[104]
175.5005	7.0643	62	Sm	L_2N_3		[104]
175.6661	7.05798	26	Fe	$KM_{2,3}$	$K_{\beta_{1,3}}$	[50]
175.6686	7.0579	68	Er	L_2M_1	L_η	[50]
176.7682	7.014	61	Pm	abs. edge	L_2	[51]
176.8841	7.0094	65	Tb	L_2M_5		[104]
176.9296	7.0076	60	Nd	$L_1N_{4,5}$		[104]
177.3522	6.9909	70	Yb	L_3M_3	L_s	[104]
177.6037	6.981	71	Lu	L_3M_2	L_t	[50]
177.6801	6.978	65	Tb	L_2M_4	L_{β_1}	[50]
177.6140	6.9806	63	Eu	abs. edge	L_3	[51]
177.7234	6.9763	63	Eu	$L_3O_{4,5}$	L_{β_5}	[50]
177.9402	6.9678	62	Sm	L_2N_1	L_{γ_5}	[50]
178.1498	6.9596	72	Hf	L_3M_1	L_l	[50]
178.4293	6.9487	68	Er	L_3M_5	L_{α_1}	[50]
178.5166	6.9453	63	Eu	L_3O_1	L_{β_7}	[50]
178.6453	6.9403	65	Tb	L_1M_2	L_{β_4}	[50]
178.9030	6.9303	27	Co	KL_3	K_{α_1}	[50]
179.1694	6.920	63	Eu	L_1M_5	L_{β_9}	[50]
179.2911	6.91530	27	Co	KL_2	K_{α_2}	[50]
179.5403	6.9057	66	Dy	L_2M_3	$L_{\beta_{17}}$	[104]
179.5582	6.9050	68	Er	L_3M_4	L_{α_2}	[50]
179.6366	6.902	60	Nd	L_1N_3	L_{γ_3}	[50]
179.8972	6.892	61	Pm	L_2N_4	L_{γ_1}	[50]
179.9494	6.89	63	Eu	L_1M_4	$L_{\beta_{10}}$	[50]
180.1325	6.883	60	Nd	L_1N_2	L_{γ_2}	[50]
180.5495	6.8671	64	Gd	L_3N_1	L_{β_6}	[50]
181.1801	6.8432	63	Eu	L_3N_5	L_{β_2}	[50]
181.1801	6.8432	63	Eu	L_3N_4	$L_{\beta_{15}}$	[50]

λ [pm]	E [keV]	Z	element	transition	notation	reference
181.4241	6.834	59	Pr	abs. edge	L_1	[51]
181.5010	6.8311	64	Gd	$L_1 M_3$	L_{β_3}	[50]
181.9298	6.815	59	Pr	$L_1 O_{2,3}$	$L_{\gamma_4 \gamma_{4'}}$	[50]
182.6454	6.7883	67	Ho	$L_2 M_1$	L_η	[50]
183.0904	6.7718	70	Yb	$L_3 M_2$	L_t	[50]
183.6056	6.7528	71	Lu	$L_3 M_1$	L_l	[50]
183.8615	6.7434	64	Gd	$L_2 M_5$		[104]
184.4085	6.7234	60	Nd	abs. edge	L_2	[51]
184.5072	6.7198	67	Ho	$L_3 M_5$	L_{α_1}	[50]
184.5072	6.7198	60	Nd	$L_2 N_{6,7}$	L_5	[104]
184.5374	6.7187	59	Pr	$L_1 N_{4,5}$		[104]
184.5787	6.7172	62	Sm	abs. edge	L_3	[51]
184.6886	6.7132	64	Gd	$L_2 M_4$	L_{β_1}	[50]
184.7051	6.7126	62	Sm	$L_3 O_{4,5}$	L_{β_5}	[50]
185.4095	6.6871	64	Gd	$L_1 M_2$	L_{β_4}	[50]
185.5232	6.683	60	Nd	$L_2 O_1$	L_{γ_8}	[50]
185.6204	6.6795	67	Ho	$L_3 M_4$	L_{α_2}	[50]
185.6316	6.6791	62	Sm	$L_3 O_1$	L_{β_7}	[50]
186.1751	6.6596	62	Sm	$L_1 M_5$	L_{β_9}	[104]
186.9978	6.6303	62	Sm	$L_1 M_4$	$L_{\beta_{10}}$	[104]
187.3737	6.617	63	Eu	$L_3 N_1$	L_{β_6}	[50]
187.4020	6.616	59	Pr	$L_1 N_3$	L_{γ_3}	[50]
187.7966	6.6021	60	Nd	$L_2 N_4$	L_{γ_1}	[50]
187.9133	6.598	59	Pr	$L_1 N_2$	L_{γ_2}	[50]
188.2271	6.587	62	Sm	$L_3 N_5$	L_{β_2}	[50]
188.2271	6.587	62	Sm	$L_3 N_4$	$L_{\beta_{15}}$	[50]
188.6768	6.5713	63	Eu	$L_1 M_3$	L_{β_3}	[50]
189.0767	6.5574	69	Tm	$L_3 M_2$	L_t	[50]
189.3482	6.548	58	Ce	abs. edge	L_1	[51]
189.4205	6.545	70	Yb	$L_3 M_1$	L_l	[50]
189.6494	6.5376	25	Mn	abs. edge	K	[51]
189.7190	6.5352	25	Mn	$K M_{4,5}$	K_{β_5}	[50]
189.7480	6.5342	66	Dy	$L_2 M_1$	L_η	[50]
189.9283	6.528	58	Ce	$L_1 O_{2,3}$	$L_{\gamma_4 \gamma_{4'}}$	[50]
190.3686	6.5129	58	Ce	$L_1 O_1$		[104]
190.8874	6.4952	66	Dy	$L_3 M_5$	L_{α_1}	[50]
191.0256	6.49045	25	Mn	$K M_{2,3}$	$K_{\beta_{1,3}}$	[50]
191.6368	6.4698	62	Sm	$L_3 N_3$	L_3	[104]
191.9127	6.4605	61	Pm	abs. edge	L_3	[51]
191.9959	6.4577	66	Dy	$L_3 M_4$	L_{α_2}	[50]
192.0345	6.4564	63	Eu	$L_2 M_4$	L_{β_1}	[50]
192.5535	6.439	59	Pr	abs. edge	L_2	[51]

λ [pm]	E [keV]	Z	element	transition	notation	reference
192.5236	6.44	58	Ce	$L_1N_{4,5}$		[104]
192.5564	6.4389	63	Eu	L_1M_2	L_{β_4}	[50]
192.6223	6.4367	59	Pr	$L_2N_{6,7}$	L_5	[104]
193.5454	6.406	60	Nd	L_2N_1	L_{γ_5}	[50]
193.6119	6.40384	26	Fe	KL_3	K_{α_1}	[50]
193.6361	6.403	59	Pr	L_2O_1	L_{γ_8}	[50]
194.0057	6.39084	26	Fe	KL_2	K_{α_2}	[50]
194.6484	6.3697	62	Sm	L_2N_1	L_{β_6}	[50]
195.5016	6.3419	69	Tm	L_3M_1	L_l	[50]
195.5325	6.3409	58	Ce	L_1N_3	L_{γ_3}	[50]
195.5911	6.339	61	Pm	L_3N_5	L_{β_2}	[50]
195.5911	6.339	61	Pm	L_3N_4	$L_{\beta_{15}}$	[50]
195.7609	6.3335	67	Ho	L_3M_3	L_s	[104]
196.0240	6.325	58	Ce	L_1N_2	L_{γ_2}	[50]
196.1139	6.3321	59	Pr	L_2N_4	L_{γ_1}	[50]
196.2412	6.318	62	Sm	L_1M_3	L_{β_3}	[50]
197.3061	6.2839	65	Tb	L_2M_1	L_η	[50]
197.6552	6.2728	65	Tb	L_3M_5	L_{α_1}	[50]
197.8066	6.268	57	La	abs. edge	L_1	[51]
198.3128	6.252	57	La	$L_1O_{2,3}$	$L_{\gamma_4\gamma_{4'}}$	[50]
199.0770	6.228	65	Tb	L_3M_4	L_{α_2}	[50]
199.6798	6.2092	60	Nd	abs. edge	L_3	[51]
199.8117	6.2051	62	Sm	L_2M_4	L_{β_1}	[50]
199.8182	6.2049	60	Nd	$L_3N_{6,7}$	L_u	[104]
200.0955	6.1963	62	Sm	L_1M_2	L_{β_4}	[50]
200.9224	6.1708	60	Nd	L_3O_1	L_{β_7}	[50]
201.2420	6.161	58	Ce	abs. edge	L_2	[51]
201.2224	6.1616	58	Ce	$L_2N_{6,7}$	L_5	[50]
201.5364	6.152	68	Er	L_3M_1	L_l	[50]
201.5626	6.1512	67	Ho	L_3M_2	L_t	[104]
201.6544	6.1484	60	Nd	L_1M_5	L_{β_9}	[50]
202.0619	6.136	59	Pr	L_2N_1	L_{γ_5}	[50]
202.3752	6.1265	60	Nd	L_1M_4	$L_{\beta_{10}}$	[50]
202.3917	6.126	58	Ce	L_2O_1	L_{γ_8}	[50]
202.7657	6.1147	66	Dy	L_3M_3	L_s	[104]
203.6082	6.0894	60	Nd	L_3N_4	L_{β_2}	[50]
203.6082	6.0894	60	Nd	L_3N_5	$L_{\beta_{15}}$	[50]
204.1244	6.074	57	La	L_1N_3	L_{γ_3}	[50]
204.2253	6.071	61	Pm	L_1M_3	L_{β_3}	[50]
204.5960	6.06	57	La	L_1N_2	L_{γ_2}	[50]
204.6906	6.052	64	Gd	L_3M_5	L_{α_1}	[50]
204.8664	6.052	58	Ce	L_2N_4	L_{γ_1}	[50]

λ [pm]	E [keV]	Z	element	transition	notation	reference
204.9511	6.0495	64	Gd	L_2M_1	L_η	[50]
205.7854	6.025	64	Gd	L_3M_4	L_{α_2}	[50]
206.7799	5.996	56	Ba	abs. edge	L_1	[51]
207.0285	5.9888	24	Cr	abs. edge	K	[51]
207.0941	5.9869	24	Cr	$KM_{4,5}$	K_{β_5}	[50]
207.5656	5.9733	56	Ba	$L_1O_{2,3}$	$L_{\gamma_4\gamma_4'}$	[50]
207.9242	5.963	59	Pr	abs. edge	L_3	[51]
207.9939	5.961	61	Pm	L_2M_4	L_{β_1}	[50]
208.0463	5.9595	59	Pr	$L_3N_{6,7}$	L_u	[104]
208.4941	5.94671	24	Cr	$KM_{2,3}$	$K_{\beta_{1,3}}$	[50]
208.6098	5.9434	67	Ho	L_3M_1	L_l	[50]
209.1553	5.9279	59	Pr	L_3O_1	L_{β_7}	[50]
209.9202	5.9063	65	Tb	L_3M_3	L_s	[104]
210.0376	5.903	59	Pr	L_1M_5	L_{β_9}	[50]
210.1871	5.89875	25	Mn	KL_3	K_{α_1}	[50]
210.3940	5.8930	60	Nd	L_3N_1	L_{β_6}	[50]
210.4404	5.8917	62	Sm	L_2M_3	$L_{\beta_{17}}$	[50]
210.5369	5.889	57	La	abs. edge	L_2	[51]
210.5726	5.888	57	La	L_2O_2		[50]
210.5834	5.88705	25	Mn	KL_2	K_{α_2}	[50]
210.7158	5.884	59	Pr	L_1M_4	$L_{\beta_{10}}$	[50]
211.0350	5.8751	58	Ce	L_2N_1	L_{γ_5}	[50]
211.9405	5.85	59	Pr	L_3N_4	$L_{\beta_{15}}$	[50]
212.0964	5.8457	63	Eu	L_3M_5	L_{α_1}	[50]
212.6894	5.8294	60	Nd	L_1M_3	L_{β_3}	[50]
213.1575	5.8166	63	Eu	L_2M_1	L_η	[50]
213.1575	5.8166	63	Eu	L_3M_4	L_{α_2}	[50]
213.4290	5.8092	56	Ba	L_1N_3	L_{γ_3}	[50]
213.8819	5.7969	56	Ba	L_1N_2	L_{γ_2}	[50]
214.1922	5.7885	57	La	L_2N_4	L_{γ_1}	[50]
215.8855	5.7431	66	Dy	L_3M_1	L_l	[50]
216.6437	5.723	58	Ce	abs. edge	L_3	[51]
216.6967	5.7216	60	Nd	L_1M_2	L_{β_4}	[50]
216.6967	5.7216	60	Nd	L_2M_4	L_{β_1}	[50]
216.7156	5.7211	58	Ce	$L_3N_{6,7}$	L_u	[104]
216.7195	5.721	55	Cs	abs. edge	L_1	[51]
217.0153	5.7132	58	Ce	L_3O_1	L_{β_7}	[50]
217.4187	5.7026	55	Cs	$L_1O_{2,3}$	$L_{\gamma_4\gamma_4'}$	[50]
217.5866	5.6982	64	Gd	L_3M_3	L_s	[104]
218.8611	5.665	58	Ce	L_1M_5	L_{β_9}	[50]
219.0551	5.66	59	Pr	L_3N_1	L_{β_6}	[50]
219.5983	5.646	58	Ce	L_1M_4	$L_{\beta_{10}}$	[50]

λ [pm]	E [keV]	Z	element	transition	notation	reference
219.9840	5.6361	62	Sm	L_3M_5	L_{α_1}	[50]
220.4848	5.6233	56	Ba	abs. edge	L_2	[51]
220.5749	5.621	57	La	L_2N_1	L_{γ_5}	[50]
220.8736	5.6134	58	Ce	L_3N_4	L_{β_2}	[50]
220.8736	5.6134	58	Ce	L_3N_5	$L_{\beta_{15}}$	[50]
221.0705	5.6084	62	Sm	L_3M_4	L_{α_2}	[50]
221.7268	5.5918	59	Pr	L_1M_3	L_{β_3}	[50]
221.8299	5.5892	62	Sm	L_2M_1	L_{η}	[50]
222.2753	5.578	56	Ba	L_3O_1	L_{γ_8}	[104]
223.2881	5.5527	55	Cs	L_1N_3	L_{γ_3}	[50]
223.4772	5.548	92	U	abs. edge	M_1	[51]
223.5296	5.5467	65	Tb	L_3M_1	L_l	[50]
223.7192	5.542	55	Cs	L_1N_2	L_{γ_2}	[50]
224.2601	5.5311	56	Ba	L_2N_4	L_{γ_1}	[50]
225.5055	5.4981	59	Pr	L_1M_2	L_{β_9}	[50]
225.8835	5.4889	59	Pr	L_2M_4	L_{β_1}	[50]
226.0853	5.484	57	La	abs. edge	L_3	[51]
226.3329	5.478	57	La	$L_3O_{4,5}$	L_{β_5}	[51]
226.9170	5.4639	23	V	abs. edge	K	[51]
226.9585	5.4629	23	V	$KM_{4,5}$	K_{β_5}	[50]
227.3790	5.4528	54	Xe	abs. edge	L_1	[51]
227.4957	5.45	57	La	L_3O_1	L_{β_7}	[50]
228.1656	5.434	57	La	L_1M_5	L_{β_9}	[50]
228.1908	5.4334	58	Ce	L_3N_1	L_{β_6}	[50]
228.2286	5.4325	61	Pm	L_3M_5	L_{α_1}	[50]
228.4472	5.42729	23	V	$KM_{2,3}$	$K_{\beta_{1,3}}$	[50]
228.9662	5.415	57	La	L_1M_4	$L_{\beta_{10}}$	[50]
228.9788	5.41472	24	Cr	KL_3	K_{α_1}	[50]
229.2710	5.4078	61	Pm	L_3M_4	L_{α_2}	[50]
229.3686	5.405509	24	Cr	KL_2	K_{α_2}	[50]
230.3059	5.3835	57	La	L_3N_4	L_{β_2}	[50]
230.3059	5.3835	57	La	L_3N_5	$L_{\beta_{15}}$	[50]
230.4557	5.38	92	U	M_1O_3		[50]
230.8548	5.3707	56	Ba	L_2N_1	L_{γ_5}	[50]
231.0182	5.3669	91	Pa	abs. edge	M_1	[51]
231.0957	5.3651	58	Ce	L_1M_3	L_{β_3}	[50]
231.1225	5.3621	64	Gd	L_3M_1	L_l	[50]
231.3977	5.3581	55	Cs	abs. edge	L_2	[51]
234.0889	5.2965	62	Sm	L_3M_3	L_s	[104]
234.8026	5.2804	55	Cs	L_2N_4	L_{γ_1}	[50]
234.9762	5.2765	58	Ce	L_1M_2	L_{β_4}	[50]
235.6147	5.2622	58	Ce	L_2M_4	L_{β_1}	[50]

λ [pm]	E [keV]	Z	element	transition	notation	reference
236.2973	5.2470	56	Ba	abs. edge	L_3	[51]
237.0472	5.2304	60	Nd	L_3M_5	L_{α_1}	[50]
237.6515	5.2171	56	Ba	L_1M_5	L_{β_9}	[50]
237.9115	5.1114	57	La	L_3N_1	L_{β_6}	[50]
238.0713	5.2079	56	Ba	L_3O_1	L_{β_7}	[50]
238.0805	5.2077	60	Nd	L_3M_4	L_{α_2}	[50]
238.7037	5.1941	56	Ba	L_1M_4	$L_{\beta_{10}}$	[50]
238.8005	5.192	53	I	abs. edge	L_1	[51]
239.1320	5.1848	53	I	$L_1O_{2,3}$	$L_{\gamma_4\gamma_{4'}}$	[50]
239.2474	5.1823	90	Th	abs. edge	M_1	[51]
239.2520	5.1822	92	U	abs. edge	M_2	[51]
239.4831	5.1772	63	Eu	L_3M_1	L_l	[50]
240.4444	5.1565	56	Ba	L_2N_4	L_{β_2}	[50]
240.4444	5.1565	56	Ba	L_2N_5	$L_{\beta_{15}}$	[50]
240.9491	5.1457	60	Nd	L_2M_1	L_η	[50]
241.0568	5.1434	57	La	L_1M_3	L_{β_3}	[50]
241.7478	5.1287	55	Cs	L_2N_1	L_{γ_5}	[50]
242.2059	5.119	92	U	M_2P_1		[50]
242.9320	5.1037	54	Xe	abs. edge	L_2	[51]
244.0653	5.08	90	Th	M_1O_3		[50]
244.3058	5.075	92	U	M_2O_4		[50]
244.7543	5.0657	53	I	$L_1N_{2,3}$	$L_{\gamma_{2,3}}$	[50]
244.9332	5.062	57	La	L_1M_2	$L_{\beta_{4'}}$	[50]
245.8999	5.0421	57	La	L_2M_4	L_{β_1}	[50]
246.3102	5.0337	59	Pr	L_3M_5	L_{α_1}	[50]
247.3026	5.0135	59	Pr	L_3M_4	L_{α_2}	[50]
247.4113	5.0113	55	Cs	abs. edge	L_3	[51]
247.8415	5.0026	55	Cs	L_1M_5	L_{β_9}	[50]
247.8712	5.002	89	Ac	abs. edge	M_1	[51]
247.9257	5.0009	91	Pa	abs. edge	M_2	[51]
248.2434	4.9945	62	Sm	L_3M_1	L_l	[50]
248.2732	4.9939	56	Ba	L_3N_1	L_{β_6}	[50]
248.5021	4.9893	55	Cs	L_3O_1	L_{β_7}	[50]
249.2064	4.9752	55	Cs	L_1M_4	$L_{\beta_{10}}$	[50]
249.7426	4.96452	22	Ti	abs. edge	K	[51]
249.8543	4.9623	22	Ti	$KM_{4,5}$	K_{β_5}	[50]
250.3638	4.95220	23	V	K_3	K_{α_1}	[50]
250.7484	4.94464	23	V	K_2	K_{α_2}	[50]
250.9974	4.9397	52	Te	abs. edge	L_1	[51]
251.1397	4.9369	52	Te	$L_1O_{2,3}$	$L_{\gamma_4\gamma_{4'}}$	[50]
251.1906	4.9359	55	Cs	L_3N_5	L_{β_2}	[50]
251.1906	4.9359	55	Cs	L_3N_4	$L_{\beta_{15}}$	[50]

λ [pm]	E [keV]	Z	element	transition	notation	reference
251.2364	4.935	59	Pr	L_2M_1	L_η	[50]
251.3994	4.93181	22	Ti	$KM_{2,3}$	$K_{\beta_{1,3}}$	[50]
251.6495	4.9269	56	Ba	L_1M_3	L_{β_3}	[50]
252.7215	4.906	91	Pa	M_2O_4		[50]
255.4290	4.8540	53	I	abs. edge	L_2	[51]
255.5394	4.8519	56	Ba	L_1M_2	L_{β_4}	[50]
255.9562	4.844	92	U	M_2O_1		[104]
256.1571	4.8402	58	Ce	L_2M_5	L_{α_1}	[50]
256.6768	4.8304	90	Th	abs. edge	M_2	[51]
256.7512	4.829	52	Te	$L_1N_{2,3}$	$L_{\gamma_{2,3}}$	[50]
256.6829	4.82753	56	Ba	L_2M_4	L_{β_1}	[50]
257.0707	4.823	58	Ce	L_3M_4	L_{α_2}	[50]
257.1240	4.822	88	Ra	abs. edge	M_1	[51]
258.2540	4.8009	53	I	L_2N_4	L_{γ_1}	[50]
259.2639	4.7822	54	Xe	abs. edge	L_3	[51]
259.3235	4.7811	55	Cs	L_3N_1	L_{β_6}	[50]
261.8483	4.735	90	Th	M_2O_4		[50]
262.0420	4.7315	58	Ce	L_2M_1	L_η	[50]
262.8642	4.7167	55	Cs	L_1M_3	L_{β_3}	[50]
263.8881	4.6984	51	Sb	abs. edge	L_1	[51]
263.9836	4.6967	51	Sb	$L_1O_{2,3}$	$L_{\gamma_4\gamma_{4'}}$	[50]
265.7205	4.666	53	I	L_2N_1	L_{γ_5}	[50]
266.2912	4.656	89	Ac	abs. edge	M_2	[51]
266.5202	4.652	87	Fr	abs. edge	M_1	[51]
266.5775	4.65097	57	La	L_3M_5	L_{α_1}	[50]
266.6692	4.6494	55	Cs	L_1M_2	L_{β_4}	[50]
267.5439	4.63423	57	La	L_3M_4	L_{α_2}	[50]
267.6132	4.633	60	Nd	L_3M_1	L_l	[50]
268.3778	4.6198	55	Cs	L_2M_4	L_{β_1}	[50]
268.7968	4.6126	52	Te	abs. edge	L_2	[51]
269.5389	4.5999	51	Sb	$L_1N_{2,3}$	$L_{\gamma_{2,3}}$	[50]
271.2489	4.5709	52	Te	L_2N_4	L_{γ_1}	[50]
271.2617	4.569	53	I	L_1M_5	L_{β_9}	[50]
271.9749	4.5587	53	I	abs. edge	L_3	[51]
272.1221	4.5564	53	I	L_1M_4	$L_{\beta_{10}}$	[50]
272.8847	4.5435	53	I	L_3O_1	L_{β_7}	[50]
274.0004	4.525	57	La	L_2M_1	L_η	[50]
274.8630	4.51084	22	Ti	KL_3	K_{α_1}	[50]
275.0642	4.5075	53	I	L_3N_5	L_{β_2}	[50]
275.0642	4.5075	53	I	L_3N_4	$L_{\beta_{15}}$	[50]
275.2229	4.50486	22	Ti	KL_3	K_{α_2}	[50]
275.5226	4.50	92	U	M_1N_3		[50]

λ [pm]	E [keV]	Z	element	transition	notation	reference
276.1978	4.489	21	Sc	abs. edge	K	[51]
276.1670	4.4865	88	Ra	abs. edge	M_2	[51]
276.3517	4.4865	21	Sc	$KM_{4,5}$	K_{β_5}	[50]
276.6291	4.482	86	Rn	abs. edge	M_1	[51]
277.6015	4.46626	56	Ba	L_3M_5	L_{α_1}	[50]
277.6949	4.4648	50	Sn	abs. edge	L_1	[51]
277.7570	4.4638	50	Sn	$L_1O_{2,3}$	$L_{\gamma_4\gamma_{4'}}$	[50]
277.9625	4.4605	21	Sc	KM_3	K_{β_1}	[50]
278.4182	4.4532	59	Pr	L_3M_1	L_l	[50]
278.5620	4.45090	56	Ba	L_3M_4	L_{α_2}	[50]
279.0134	4.4437	52	Te	L_2N_1	L_{γ_5}	[50]
281.7205	4.401	92	U	M_2N_4		[50]
282.9485	4.3819	51	Sb	abs. edge	L_2	[51]
283.2781	4.3768	50	Sn	$L_1N_{2,3}$	$L_{\gamma_{2,3}}$	[50]
283.6800	4.3706	53	I	L_3N_1	L_{β_6}	[50]
283.9074	4.3671	52	Te	L_1M_5	L_{β_9}	[50]
284.6896	4.3551	52	Te	L_1M_4	$L_{\beta_{10}}$	[50]
285.1676	4.34779	51	Sb	L_2N_4	L_{γ_1}	[50]
285.5617	4.3418	52	Te	abs. edge	L_3	[51]
286.2804	4.3309	56	Ba	L_2M_1	L_η	[50]
286.3531	4.3298	52	Te	L_3O_1	L_{β_7}	[50]
286.5384	4.327	87	Fr	abs. edge	M_2	[51]
287.2022	4.317	85	At	abs. edge	M_1	[51]
287.4419	4.3134	53	I	L_1M_3	L_{β_3}	[50]
288.4047	4.299	92	U	abs. edge	M_3	[51]
288.2237	4.3017	52	Te	L_3N_5	L_{β_2}	[50]
288.2237	4.3017	52	Te	L_3N_4	$L_{\beta_{15}}$	[50]
289.1783	4.2875	58	Ce	L_3M_1	L_l	[50]
289.2457	4.2865	55	Cs	L_3M_5	L_{α_1}	[50]
290.2139	4.2722	55	Cs	L_3M_4	L_{α_2}	[50]
291.0450	4.26	91	Pa	M_2N_4		[50]
291.2159	4.2575	53	I	L_1M_2	L_{β_4}	[50]
291.7298	4.25	92	U	M_1N_2		[50]
292.6043	4.2373	49	In	abs. edge	L_1	[51]
292.6456	4.2367	49	In	$L_1O_{2,3}$	$L_{\gamma_4\gamma_{4'}}$	[50]
293.0330	4.2311	92	U	M_3P_1		[104]
293.1092	4.23	90	Th	M_1N_3		[50]
293.1993	4.2287	51	Sb	L_2N_1	L_{γ_5}	[50]
293.7550	4.22072	53	J	L_2M_4	L_{β_1}	[50]
294.8518	4.205	92	U	M_3O_5		[50]
294.8518	4.205	92	U	M_3O_4		[50]
297.0559	4.1738	91	Pa	abs. edge	M_3	[51]

λ [pm]	E [keV]	Z	element	transition	notation	reference
297.0986	4.1732	52	Te	L_3N_1	L_{β_6}	[50]
297.2695	4.1708	51	Sb	L_1M_5	L_{β_9}	[50]
297.9267	4.1616	51	Sb	L_1M_4	$L_{\beta_{10}}$	[50]
298.0055	4.1605	49	In	$L_1N_{2,3}$	$L_{\gamma_{2,3}}$	[50]
298.1130	4.159	87	Fr	abs. edge	M_2	[51]
298.2349	4.1573	50	Sn	abs. edge	L_2	[51]
298.8315	4.149	84	Po	abs. edge	M_1	[51]
299.3293	4.1421	55	Cs	L_2M_1	L_η	[50]
300.0392	4.1323	51	Sb	abs. edge	L_3	[51]
300.1263	4.13112	50	Sn	L_2N_4	L_{γ_1}	[50]
300.5337	4.1255	51	Sb	L_3O_1	L_{β_7}	[50]
300.6430	4.124	57	La	L_3M_1	L_l	[50]
300.9057	4.1204	52	Te	L_1M_3	L_{β_3}	[50]
301.1542	4.117	90	Th	M_2N_4		[50]
301.6744	4.1099	54	Xe	L_3M_5	L_{α_2}	[50]
302.3439	4.1008	51	Sb	L_3N_4	$L_{\beta_{15}}$	[50]
302.3454	4.10078	51	Sb	L_3N_5	L_{β_2}	[50]
303.0978	4.0906	21	Sc	KL_3	K_{α_1}	[50]
303.4316	4.0861	21	Sc	KL_2	K_{α_1}	[50]
303.8108	4.081	91	Pa	$M_2O_{4,5}$		[50]
304.6693	4.0695	52	Te	L_1M_3	L_{β_4}	[50]
306.8181	4.041	90	Th	abs. edge	M_3	[51]
307.0385	4.0381	20	Ca	abs. edge	K	[51]
307.4648	4.0325	20	Ca	$KM_{4,5}$	K_{β_5}	[50]
307.6877	4.02958	52	Te	L_2M_4	L_{β_1}	[50]
308.4822	4.0192	50	Sn	L_2N_1	L_{γ_5}	[50]
308.4976	4.0190	48	Cd	abs. edge	L_1	[51]
308.9819	4.0127	20	Ca	$KM_{2,3}$	$K_{\beta_{1,3}}$	[50]
309.3443	4.008	85	At	abs. edge	M_2	[51]
309.4215	4.007	83	Bi	abs. edge	M_1	[51]
311.5206	3.9800	51	Sb	L_3N_1	L_{β_6}	[50]
311.5206	3.980	92	U	M_3O_1		[50]
312.1794	3.9716	50	Sn	L_1M_4	$L_{\beta_{10}}$	[50]
313.1730	3.959	90	Th	M_3O_5		[50]
313.1730	3.959	90	Th	M_3O_4		[50]
313.3313	3.957	90	Th	M_1N_2		[104]
313.5611	3.9541	56	Ba	L_2M_1	L_l	[50]
313.7833	3.9513	48	Cd	$L_1N_{2,3}$	$L_{\gamma_{2,3}}$	[50]
314.7392	3.9393	49	In	abs. edge	L_2	[51]
314.8711	3.93765	53	I	L_2M_5	L_{α_1}	[50]
315.2673	3.9327	51	Sb	L_1M_3	L_{β_3}	[50]
315.5803	3.9288	50	Sn	abs. edge	L_2	[51]

λ [pm]	E [keV]	Z	element	transition	notation	reference
315.6526	3.9279	50	Sn	L_3O_1	L_{β_7}	[50]
315.8022	3.92604	53	I	L_3M_4	L_{α_2}	[50]
316.2234	3.92081	49	In	L_2N_4	L_{γ_1}	[50]
316.2888	3.92	92	U	M_3N_7		[104]
317.1788	3.909	89	Ac	abs. edge	M_3	[51]
317.5151	3.90486	50	Sn	L_2N_5	L_{β_2}	[50]
317.5151	3.90486	50	Sn	L_2N_4	$L_{\beta_{15}}$	[50]
319.0232	3.8864	51	Sb	L_1M_2	L_{β_4}	[50]
321.7052	3.854	84	Po	abs. edge	M_2	[51]
321.7052	3.854	82	Pb	abs. edge	M_1	[51]
322.5782	3.8457	51	Sb	L_2M_4	L_{β_1}	[50]
322.8286	3.8394	95	Am	M_3N_5	M_{γ_1}	[51]
324.5685	3.82	91	Pa	M_3O_1		[50]
324.9173	3.8159	49	In	L_2N_1	L_{γ_5}	[50]
325.6598	3.8072	47	Ag	abs. edge	L_1	[51]
326.7076	3.795	55	Cs	L_3M_1	L_l	[50]
326.7756	3.7942	49	In	L_1M_5	L_{β_9}	[50]
326.9134	3.7926	50	Sn	L_3N_1	L_{β_6}	[50]
326.9824	3.7918	88	Ra	abs. edge	M_3	[51]
327.2845	3.7883	95	Am	M_3N_4	M_{γ_2}	[51]
327.4141	3.7868	49	In	L_1M_4	$L_{\beta_{10}}$	[50]
327.9945	3.7801	53	I	L_2M_1	L_{η}	[50]
328.0031	3.78	90	Th	M_3O_1		[50]
328.9342	3.16933	52	Te	L_3M_5	L_{α_1}	[50]
329.8531	3.7588	52	Te	L_3M_4	L_{α_2}	[50]
330.1694	3.7552	94	Pu	M_3N_5	M_{γ_1}	[51]
330.6272	3.75	50	Sn	L_1M_2	L_{β_3}	[50]
330.6448	3.7498	47	Ag	L_1N_3	L_{γ_3}	[50]
331.2278	3.7432	47	Ag	L_1N_2	L_{γ_2}	[50]
332.3822	3.7302	49	In	abs. edge	L_3	[51]
332.4000	3.73	49	In	L_3O_1	L_{β_7}	[50]
332.5783	3.7280	48	Cd	abs. edge	L_2	[51]
332.9355	3.724	92	U	M_2N_1		[50]
333.2935	3.720	92	U	abs. edge	M_4	[51]
333.5751	3.71686	48	Cd	L_2N_4	L_{γ_1}	[50]
333.8491	3.7181	49	In	L_3N_5	L_{β_2}	[50]
333.8491	3.7181	49	In	L_3N_4	$L_{\beta_{15}}$	[50]
334.3451	3.7083	50	Sn	L_1M_2	L_{β_4}	[50]
334.3992	3.7077	94	Pu	M_3N_4	M_{γ_2}	[51]
334.6429	3.705	81	Tl	abs. edge	M_1	[51]
335.2673	3.6981	92	U	$M_4P_{2,3}$		[50]
335.8503	3.69168	20	Ca	KL_3	K_{α_1}	[50]

λ [pm]	E [keV]	Z	element	transition	notation	reference
335.9122	3.691	83	Bi	abs. edge	M_2	[51]
336.1773	3.68809	20	Ca	KL_2	K_{α_2}	[50]
338.3506	3.6644	93	Np	M_2N_5	M_{γ_1}	[51]
338.4799	3.663	87	Fr	abs. edge	M_3	[51]
338.4984	3.66280	50	Sn	L_2M_4	L_{β_1}	[50]
340.0302	3.6463	95	Am	M_4N_6	M_β	[51]
342.5621	3.6135	48	Cd	L_2N_1	L_{γ_5}	[50]
342.6141	3.6188	93	Np	M_3N_4	M_{γ_2}	[51]
343.0264	3.61445	48	Cd	L_1M_5	L_{β_9}	[50]
343.3351	3.6112	91	Pa	abs. edge	M_4	[51]
343.6206	3.60823	49	In	L_3N_1	L_{β_6}	[50]
343.6873	3.6075	48	Cd	L_1M_4	$L_{\beta_{10}}$	[50]
343.6587	3.6078	19	K	abs. edge	K	[51]
343.7350	3.607	46	Pd	abs. edge	L_1	[51]
343.8436	3.60586	52	Te	L_2M_1	L_η	[50]
343.9524	3.60472	51	Sb	L_2M_5	L_{α_1}	[50]
344.1165	3.603	91	Pa	M_2N_1		[50]
344.1452	3.6027	19	K	KM_4	K_{β_5}	[50]
344.8516	3.59532	51	Sb	L_3M_4	L_{α_2}	[50]
345.4011	3.5896	19	K	KM_3	K_{β_1}	[50]
346.9952	3.57311	49	In	L_1M_3	L_{β_3}	[50]
347.7846	3.565	80	Hg	abs. edge	M_1	[51]
347.9797	3.563	92	U	M_3N_5	M_γ	[50]
348.9297	3.5533	46	Pd	$L_1N_{2,3}$	$L_{\gamma_{2,3}}$	[50]
349.2541	3.550	82	Pb	abs. edge	M_2	[51]
349.7467	3.545	92	U	abs. edge	M_5	[51]
350.4386	3.538	86	Rn	abs. edge	M_3	[51]
350.4783	3.5376	48	Cd	abs. edge	L_3	[51]
350.7083	3.53528	49	In	L_1M_2	L_{β_4}	[50]
351.1035	3.5313	92	U	M_4O_3	M_η	[104]
351.4200	3.52812	48	Cd	L_3N_5	L_{β_2}	[50]
351.4200	3.52812	48	Cd	L_3N_4	$L_{\beta_{15}}$	[50]
351.5516	3.5268	94	Pu	M_4N_6	M_β	[51]
351.6513	3.5258	47	Ag	abs. edge	L_2	[51]
352.1306	3.521	92	U	M_3N_4		[50]
352.2717	3.51959	47	Ag	L_2N_4	L_{γ_1}	[50]
353.7380	3.505	90	Th	M_2N_1		[50]
355.5427	3.48721	49	In	L_2M_4	L_{β_1}	[50]
355.7611	3.48502	53	I	L_3M_1	L_l	[50]
355.7681	3.485	90	Th	abs. edge	M_4	[51]
357.6564	3.4666	92	U	M_4O_2		[50]
357.7493	3.4657	91	Pa	M_3N_5	M_{γ_1}	[50]

λ [pm]	E [keV]	Z	element	transition	notation	reference
360.0056	3.44398	50	Sn	L_3M_5	L_{α_1}	[50]
360.2336	3.4418	91	Pa	abs. edge	M_5	[51]
360.5091	3.43917	47	Ag	L_1M_5	L_{β_9}	[50]
360.7776	3.43661	51	Sb	L_2M_1	L_η	[50]
360.9026	3.43542	50	Sn	L_3M_4	L_{α_2}	[50]
360.9152	3.4353	93	Np	M_4N_6	M_β	[51]
361.1707	3.43287	47	Ag	L_1M_4	$L_{\beta_{10}}$	[50]
361.4728	3.430	91	Pa	M_3N_4	M_{γ_2}	[50]
361.4792	3.42994	48	Cd	L_3N_1	L_{β_6}	[50]
361.6500	3.42832	47	Ag	L_2N_1	L_{γ_5}	[50]
361.6838	3.428	79	Au	abs. edge	M_1	[51]
361.8949	3.426	85	At	abs. edge	M_3	[51]
362.8481	3.417	45	Rh	abs. edge	L_1	[51]
363.3798	3.412	81	Tl	abs. edge	M_2	[51]
364.5069	3.40145	48	Cd	L_1M_3	L_{β_3}	[50]
367.8867	3.3702	89	Ac	abs. edge	M_4	[51]
367.9086	3.37	90	Th	M_3N_5	M_{γ_1}	[50]
368.2156	3.36719	48	Cd	L_1M_2	L_{β_4}	[50]
368.5648	3.364	45	Rh	$L_1N_{2,3}$	$L_{\gamma_{2,3}}$	[50]
369.1134	3.359	91	Pa	M_4O_2		[50]
369.4433	3.356	92	U	M_4N_4	M_β	[50]
369.9990	3.35096	47	Ag	abs. edge	L_3	[51]
370.3472	3.34781	47	Ag	L_3N_5	L_{β_2}	[50]
370.3472	3.34781	47	Ag	L_3N_4	$L_{\beta_{15}}$	[50]
370.5808	3.3457	94	Pu	M_5N_7	M_{α_1}	[51]
371.7084	3.33555	52	Te	L_3M_1	L_l	[50]
371.7697	3.335	90	Th	M_3N_4	M_{γ_2}	[50]
371.9816	3.33031	46	Pd	abs. edge	L_2	[51]
372.0709	3.3323	94	Pu	M_5N_6	M_{α_2}	[51]
372.4733	3.3287	46	Pd	L_2N_4	L_{γ_1}	[50]
372.8878	3.325	90	Th	abs. edge	M_5	[51]
373.8356	3.31657	48	Cd	L_2M_4	L_{β_1}	[50]
374.0126	3.315	83	Bi	M_1N_3		[50]
374.1481	3.3138	19	K	KL_3	K_{α_1}	[50]
374.4532	3.3111	19	K	KL_2	K_{α_2}	[50]
375.4965	3.3019	84	Po	abs. edge	M_3	[51]
376.1687	3.296	78	Pt	abs. edge	M_1	[51]
377.2101	3.2869	49	In	L_3M_5	L_{α_1}	[50]
378.0843	3.2793	49	In	L_3M_4	L_{α_2}	[50]
378.3497	3.277	80	Hg	abs. edge	M_2	[51]
378.8931	3.2723	50	Sn	L_2M_1	L_η	[50]
379.2060	3.2696	46	Pd	L_1M_5	L_{β_9}	[50]

λ [pm]	E [keV]	Z	element	transition	notation	reference
379.8915	3.2637	46	Pd	L_1M_4	$L_{\beta_{10}}$	[50]
380.1594	3.2614	93	Np	M_5N_7	M_{α_1}	[51]
380.7864	3.25603	47	Ag	L_3N_1	L_{β_6}	[50]
380.7899	3.256	90	Th	M_4O_2	M_η	[50]
381.3990	3.2508	93	Np	M_5N_6	M_{α_2}	[51]
381.7278	3.248	88	Ra	abs. edge	M_4	[51]
382.2338	3.2437	46	Pd	L_2N_1	L_{γ_5}	[50]
382.7058	3.2397	91	Pa	M_4N_6	M_β	[50]
383.3210	3.2345	47	Ag	L_1M_3	L_{β_3}	[50]
383.3803	3.234	83	Bi	M_2N_4		[50]
383.4989	3.233	44	Ru	abs. edge	L_1	[51]
385.1668	3.219	89	Ac	abs. edge	M_5	[51]
387.0304	3.2035	47	Ag	L_1M_2	L_{β_4}	[50]
387.1029	3.20290	18	Ar	abs. edge	K	[51]
387.2117	3.202	82	Pb	M_1N_3		[50]
388.6074	3.1905	18	Ar	$KM_{2,3}$	$K_{\beta_{1,3}}$	[50]
388.8389	3.1886	51	Sb	L_3M_1	L_l	[50]
389.2784	3.185	83	Bi	M_1N_2		[50]
389.7802	3.1809	44	Ru	$L_1N_{2,3}$	$L_{\gamma_{2,3}}$	[50]
390.3816	3.176	83	Bi	abs. edge	M_3	[51]
390.7137	3.1733	46	Pd	abs. edge	L_3	[51]
390.8985	3.1718	46	Pd	L_3N_5	L_{β_2}	[50]
390.8985	3.1718	46	Pd	L_3N_4	$L_{\beta_{15}}$	[50]
391.0218	3.1708	92	U	M_5N_7	M_{α_1}	[50]
391.4910	3.167	77	Ir	abs. edge	M_1	[51]
392.4203	3.1595	92	U	M_5N_6	M_{α_2}	[50]
393.2293	3.153	83	Bi	M_3O_5		[50]
393.2293	3.153	83	Bi	M_3N_4		[50]
393.4913	3.1509	47	Ag	L_2M_4	L_{β_1}	[50]
393.6038	3.150	79	Au	abs. edge	M_2	[51]
394.1293	3.1458	90	Th	M_4N_6	M_β	[50]
394.2546	3.1448	45	Re	abs. edge	L_2	[51]
394.3800	3.1438	45	Rh	L_2N_4	L_{γ_1}	[50]
395.3357	3.1362	87	Fr	abs. edge	M_4	[51]
395.6511	3.1337	48	Cd	L_3M_5	L_{α_1}	[50]
396.5115	3.1269	48	Cd	L_3M_4	L_{α_2}	[50]
396.8796	3.124	82	Pb	M_2N_4		[50]
398.3460	3.1125	49	In	L_2M_1	L_η	[50]
399.3210	3.1049	88	Ra	abs. edge	M_5	[51]
401.3765	3.089	81	Tl	M_1N_3		[50]
401.6365	3.087	46	Pd	L_3N_1	L_{β_6}	[50]
402.2489	3.0823	91	Pa	M_5N_7	M_{α_1}	[50]

λ [pm]	E [keV]	Z	element	transition	notation	reference
403.4663	3.073	46	Pd	L_1M_3	L_{β_3}	[50]
403.5976	3.072	91	Pa	M_5N_6	M_{α_2}	[50]
404.5194	3.065	45	Rh	L_2N_1	L_{γ_5}	[50]
404.7571	3.0632	82	Pb	abs. edge	M_3	[51]
405.8435	3.055	43	Tc	abs. edge	L_1	[51]
406.9091	3.047	82	Pb	M_3O_5		[50]
407.1228	3.0454	46	Pd	L_1M_2	L_{β_4}	[50]
407.1763	3.045	76	0s	abs. edge	M_1	[51]
407.1897	3.0449	50	Sn	L_3M_1	L_l	[50]
409.3272	3.029	78	Pt	abs. edge	M_2	[51]
410.3432	3.0215	86	Rn	abs. edge	M_4	[51]
410.4111	3.021	83	Bi	M_3O_1		[50]
410.9552	3.017	82	Pb	M_2O_4		[50]
411.5008	3.013	81	Tl	M_2N_4		[50]
412.9949	3.0021	45	Rh	abs. edge	L_3	[51]
413.1049	3.0013	45	Rh	L_3N_5	L_{β_2}	[50]
413.1049	3.0013	45	Rh	L_3N_4	$L_{\beta_{15}}$	[50]
413.2977	2.9999	87	Fr	abs. edge	M_5	[51]
413.8219	2.9961	90	Th	M_5N_7	M_{α_1}	[50]
414.6371	2.99021	46	Pd	L_2M_4	L_{β_1}	[50]
415.0826	2.987	90	Th	M_5N_6	M_{α_2}	[50]
415.4568	2.98431	47	Ag	L_3M_5	L_{α_1}	[50]
416.3078	2.97821	47	Ag	L_3M_4	L_{α_2}	[50]
417.9793	2.9663	44	Ru	abs. edge	L_2	[51]
418.2330	2.9645	44	Ru	L_2N_4	L_{γ_1}	[50]
419.1946	2.95770	18	Ar	KL_3	K_{α_1}	[50]
419.3293	2.95675	48	Cd	L_2M_1	L_η	[50]
419.4822	2.95563	18	Ar	KL_2	K_{α_2}	[50]
419.7908	2.9535	81	Tl	abs. edge	M_3	[51]
421.5749	2.941	81	Tl	M_3O_5		[50]
423.5914	2.927	75	Re	abs. edge	M_1	[51]
424.1855	2.9229	45	Rh	L_3N_1	L_{β_6}	[50]
424.4614	2.921	82	Pb	M_3O_1		[50]
425.2330	2.9157	45	Rh	L_1M_3	L_{β_9}	[50]
426.0660	2.910	77	Ir	abs. edge	M_2	[51]
426.2564	2.9087	85	At	abs. edge	M_4	[51]
426.8874	2.90440	49	In	L_3M_1	L_l	[50]
428.6585	2.8924	86	Rn	abs. edge	M_5	[51]
428.8216	2.8913	44	Ru	L_2N_1	L_{γ_5}	[50]
428.8958	2.8908	45	Rh	L_1M_2	L_{β_4}	[50]
430.3547	2.881	42	Mo	abs. edge	L_1	[51]
430.0561	2.883	79	Au	M_1N_3		[50]

λ [pm]	E [keV]	Z	element	transition	notation	reference
433.0604	2.863	92	U	M_3N_1		[50]
435.5095	2.8469	80	Hg	abs. edge	M_3	[51]
436.7814	2.83861	46	Pd	L_3M_5	L_{α_1}	[50]
436.9214	2.8377	44	Ru	abs. edge	L_3	[51]
437.1833	2.836	44	Ru	L_3N_4	$L_{\beta_{15}}$	[50]
437.4286	2.83441	45	Rh	L_2M_4	L_{β_1}	[50]
437.6015	2.83329	46	Pd	L_3M_4	L_{α_2}	[50]
438.0173	2.8306	42	Mo	$L_1N_{2,3}$	$L_{\gamma_{2,3}}$	[50]
439.7262	2.81960	17	Cl	abs. edge	K	[51]
440.3509	2.8156	17	Cl	$KM_{2,3}$	$K_{\beta_{1,3}}$	[50]
440.7579	2.813	74	W	abs. edge	M_1	[51]
441.8417	2.8061	47	Ag	L_2M_1	L_η	[50]
443.1208	2.798	84	Po	abs. edge	M_4	[51]
443.2792	2.797	79	Au	M_2N_4		[50]
443.2792	2.797	76	Os	abs. edge	M_2	[51]
443.6282	2.7948	43	Tc	abs. edge	L_2	[51]
444.3914	2.79	74	W	$M_1O_{2,3}$		[50]
444.9176	2.7867	85	At	abs. edge	M_5	[51]
445.0294	2.786	91	Pa	M_3N_1		[50]
445.9899	2.78	78	Pt	M_1N_3		[50]
448.0286	2.76735	48	Cd	L_3N_1	L_l	[50]
448.6690	2.7634	44	Ru	L_1M_3	L_{β_3}	[50]
448.6690	2.7634	44	Ru	L_3N_1	L_{β_6}	[50]
450.6586	2.7512	95	Am	M_5N_3	M_{ξ_1}	[51]
451.8576	2.7439	79	Au	abs. edge	M_3	[51]
452.1706	2.742	79	Au	M_3O_5		[50]
452.1706	2.742	79	Au	M_3O_4		[50]
452.3191	2.7411	44	Ru	L_1M_2	L_{β_4}	[50]
453.3279	2.735	83	Bi	M_3N_5	M_{γ_1}	[50]
456.8356	2.714	90	Th	M_3N_1		[50]
457.1725	2.712	83	Bi	M_3N_4	M_{γ_2}	[50]
457.3412	2.711	83	Bi	abs. edge	M_4	[51]
457.5100	2.710	41	Nb	abs. edge	L_1	[51]
458.5251	2.704	73	Ta	abs. edge	M_1	[51]
459.2044	2.7	83	Bi	$M_4P_{2,3}$		[50]
459.7596	2.69674	45	Rh	L_3M_5	L_{α_1}	[50]
460.0564	2.695	78	Pt	M_2N_4		[50]
460.5519	2.6921	45	Rh	L_3M_4	L_{α_2}	[50]
461.9419	2.684	75	Re	abs. edge	M_2	[51]
462.0744	2.68323	44	Ru	L_2M_4	L_{β_1}	[50]
462.1140	2.683	84	Po	abs. edge	M_5	[51]
462.4587	2.681	92	U	M_4N_3	M_δ	[50]

λ [pm]	E [keV]	Z	element	transition	notation	reference
462.5796	2.6803	95	Am	M_4N_2	M_{ξ_2}	[51]
462.9768	2.6780	43	Tc	abs. edge	L_3	[51]
463.1497	2.677	77	Ir	M_1N_3		[50]
465.4099	2.664	82	Pb	M_2N_1		[50]
465.4448	2.6638	41	Nb	$L_1N_{2,3}$	$L_{\gamma_{2,3}}$	[50]
465.5322	2.6633	94	Pu	M_5N_3	M_{ξ_1}	[51]
466.0572	2.6603	46	Pd	L_2N_1	L_η	[50]
467.3924	2.6527	82	Pb	M_3N_5	M_{γ_1}	[50]
468.5937	2.6459	78	Pt	abs. edge	M_3	[51]
469.4630	2.641	78	Pt	M_3O_5		[50]
469.4630	2.641	78	Pt	M_3O_4		[50]
470.3535	2.636	79	Au	M_3O_1		[50]
470.7643	2.6337	47	Ag	L_3M_1	L_l	[50]
471.4266	2.63	82	Pb	M_3N_4	M_{γ_1}	[50]
471.8931	2.6274	42	Mo	abs. edge	L_2	[51]
472.5946	2.6235	42	Mo	L_2N_4	L_{γ_1}	[50]
472.7947	2.62239	17	Cl	KL_3	K_{α_1}	[50]
473.0851	2.62078	17	Cl	KL_2	K_{α_2}	[50]
475.7682	2.606	82	Pb	abs. edge	M_4	[51]
476.3166	2.603	83	Bi	abs. edge	M_5	[51]
476.7011	2.6009	72	Hf	abs. edge	M_1	[51]
476.7745	2.6005	94	Pu	M_4N_2	M_{ξ_2}	[51]
477.9691	2.594	77	Ir	M_2N_4		[50]
478.7073	2.59	76	0s	M_1N_3		[50]
480.7491	2.579	93	Np	M_5N_3	M_{ξ_1}	[51]
481.4959	2.575	74	W	abs. edge	M_2	[51]
482.2450	2.571	83	Bi	M_4O_2		[50]
482.2450	2.571	81	Tl	M_3N_5	M_{γ_1}	[50]
483.7125	2.5632	42	Mo	L_2N_1	L_{γ_5}	[50]
484.5917	2.55855	44	Ru	L_3N_5	L_{α_1}	[50]
485.3961	2.55431	44	Ru	L_3N_4	L_{α_2}	[50]
486.1211	2.5505	77	Ir	abs. edge	M_3	[51]
486.5981	2.548	81	Tl	M_3N_4	M_{γ_2}	[50]
486.9803	2.546	77	Ir	$M_3O_{4,5}$		[50]
487.5548	2.543	78	Pt	M_3O_1		[50]
488.7464	2.5368	43	Tc	L_2M_4	L_{β_1}	[50]
487.9386	2.541	40	Zr	abs. edge	L_1	[51]
491.0305	2.525	83	Bi	M_4N_6	M_β	[50]
491.2250	2.524	90	Th	M_4N_3	M_δ	[50]
491.4977	2.5226	93	Np	M_4N_2	M_{ξ_2}	[51]
491.3418	2.5234	42	Mo	abs. edge	L_3	[51]
492.1805	2.5191	45	Rh	L_2M_1	L_η	[50]

λ [pm]	E [keV]	Z	element	transition	notation	reference
492.3368	2.5183	42	Mo	L_3N_5	L_{β_2}	[50]
492.3368	2.5183	42	Mo	L_3N_4	$L_{\beta_{15}}$	[50]
494.5560	2.507	92	U	M_5N_3	M_{ξ_1}	[50]
495.2672	2.5034	46	Pd	L_3M_1	L_l	[50]
495.3661	2.5029	40	Zr	$L_1N_{2,3}$	$L_{\gamma_{2,3}}$	[50]
495.5443	2.502	76	0s	M_2N_4		[50]
495.5444	2.502	82	Pb	abs. edge	M_5	[51]
497.5329	2.492	41	Nb	L_1M_5	L_{β_9}	[50]
497.6926	2.4912	71	Lu	abs. edge	M_1	[51]
498.4329	2.4875	80	Hg	M_3N_5	M_{γ_1}	[50]
498.5331	2.487	41	Nb	L_1M_4	$L_{\beta_{10}}$	[50]
498.9143	2.4851	81	Tl	abs. edge	M_4	[51]
500.5458	2.477	82	Pb	M_4O_2		[50]
501.3554	2.473	42	Mo	L_1M_3	L_{β_3}	[50]
501.8668	2.47048	16	S	abs. edge	K	[51]
501.9644	2.470	73	Ta	abs. edge	M_2	[51]
503.1663	2.4641	41	Nb	abs. edge	L_2	[51]
503.1785	2.46404	16	S	KM_3	K_{β_1}	[50]
503.6363	2.4618	41	Nb	L_2N_4	L_{γ_1}	[50]
504.4150	2.458	76	0s	abs. edge	M_3	[51]
504.8874	2.4557	42	Mo	L_1M_2	L_{β_4}	[50]
504.8874	2.4557	42	Mo	L_3N_1	L_{β_6}	[50]
505.0725	2.4548	92	U	M_4N_2		[50]
507.7199	2.442	82	Pb	M_4N_6	M_β	[50]
509.1794	2.435	91	Pa	M_5N_3	M_{ξ_1}	[50]
510.5212	2.4286	83	Bi	M_3N_6	M_{α_1}	[50]
511.4901	2.424	43	Tc	L_3M_5	L_{α_1}	[50]
512.9714	2.4170	83	Bi	M_5N_6	M_{α_2}	[50]
514.4614	2.410	79	Au	M_3N_5	M_{γ_1}	[50]
515.3167	2.406	41	Nb	L_2N_1	L_{γ_5}	[50]
515.3167	2.406	81	Tl	abs. edge	M_5	[51]
515.7454	2.404	80	Hg	abs. edge	M_4	[51]
516.8203	2.399	82	Pb	M_5O_3		[50]
517.0143	2.3981	70	Yb	abs. edge	M_1	[51]
517.2515	2.397	74	W	M_1N_3		[50]
517.7246	2.39481	42	Mo	L_2M_4	L_{β_1}	[50]
518.5495	2.391	79	Au	M_3N_4	M_{γ_2}	[50]
519.2879	2.3876	91	Pa	M_4N_2		[50]
519.6362	2.386	81	Tl	M_4O_2	M_η	[50]
520.5154	2.38197	44	Ru	L_2M_1	L_η	[50]
521.6037	2.377	39	Y	abs. edge	L_1	[51]
521.7134	2.3765	45	Rh	L_3M_1	L_l	[50]

λ [pm]	E [keV]	Z	element	transition	notation	reference
523.0119	2.3706	41	Nb	abs. edge	L_3	[51]
523.3651	2.369	75	Re	abs. edge	M_3	[51]
523.8073	2.3670	41	Nb	L_3N_5	L_{β_2}	[50]
523.8073	2.3670	41	Nb	L_3N_4	$L_{\beta_{15}}$	[50]
524.1616	2.3654	72	Hf	abs. edge	M_2	[51]
524.4720	2.364	90	Th	M_5N_3	M_{ξ_1}	[50]
524.8939	2.3621	81	Tl	M_4N_6	M_β	[50]
528.3160	2.3468	39	Y	$L_1N_{2,3}$	$L_{\gamma_{2,3}}$	[50]
528.5637	2.3457	82	Pb	M_5N_7	M_{α_1}	[50]
529.9192	2.3397	82	Pb	M_5N_6	M_{α_2}	[50]
531.0313	2.3348	41	Nb	L_1M_3	L_{β_3}	[50]
531.8970	2.331	78	Pt	M_3N_5	M_{γ_1}	[50]
533.9586	2.322	90	Th	M_4N_2		[104]
533.9586	2.322	90	Th	M_5N_2	M_{ξ_1}	[50]
534.5572	2.3194	41	Nb	L_1M_2	L_{β_4}	[50]
535.8046	2.314	74	W	M_2N_4		[50]
535.8046	2.314	78	Pt	M_3N_4	M_{γ_2}	[50]
536.1522	2.3125	41	Nb	L_3N_1	L_{β_6}	[50]
537.2348	2.30784	16	S	KL_3	K_{α_1}	[50]
537.4304	2.307	79	Au	abs. edge	M_4	[51]
537.4770	2.3068	69	Tm	abs. edge	M_1	[51]
537.8267	2.3053	40	Zr	abs. edge	L_2	[51]
537.5143	2.30664	16	S	KL_2	K_{α_2}	[50]
538.4340	2.3027	40	Zr	L_2N_4	L_{γ_1}	[50]
540.2405	2.295	73	Ta	M_1N_3		[50]
536.0363	2.313	80	Hg	abs. edge	M_5	[51]
540.6740	2.29316	42	Mo	L_3M_5	L_{α_1}	[50]
541.4556	2.28985	42	Mo	L_3M_4	L_{α_2}	[50]
543.1991	2.2825	80	Hg	M_4N_6	M_β	[50]
543.5325	2.2811	74	W	abs. edge	M_3	[51]
544.9419	2.2752	74	W	M_3O_5		[50]
546.0459	2.2706	81	Tl	M_5N_7	M_{α_1}	[50]
547.2510	2.2656	81	Tl	M_5N_7	M_{α_2}	[50]
547.3235	2.2653	71	Lu	abs. edge	M_2	[51]
549.2389	2.2574	41	Nb	L_2M_4	L_{β_1}	[50]
549.7991	2.2551	40	Zr	L_2N_1	L_{γ_5}	[50]
550.0674	2.254	77	Ir	M_3N_5	M_{γ_1}	[50]
550.3604	2.2528	44	Ru	L_3M_1	L_l	[50]
553.7525	2.239	83	Bi	M_3N_1		[50]
554.0000	2.238	77	Ir	M_3N_4	M_{γ_2}	[50]
556.9865	2.226	71	Ta	M_2N_4		[50]
557.8637	2.2225	40	Zr	abs. edge	L_3	[51]

λ [pm]	E [keV]	Z	element	transition	notation	reference
558.4919	2.220	79	Au	abs. edge	M_5	[51]
558.6428	2.2194	40	Zr	L_3N_5	L_{β_2}	[50]
558.6428	2.2194	40	Zr	L_3N_4	$L_{\beta_{15}}$	[50]
559.2476	2.217	78	Pt	abs. edge	M_4	[51]
559.2476	2.217	38	Sr	abs. edge	L_1	[51]
561.9089	2.2065	68	Er	abs. edge	M_1	[51]
562.3931	2.2046	79	Au	M_4N_6	M_β	[50]
562.8016	2.203	74	W	M_3O_1		[50]
563.3130	2.2010	40	Zr	L_1M_3	L_{β_3}	[50]
564.4671	2.1965	38	Sr	$L_1N_{2,3}$	$L_{\gamma_{2,3}}$	[50]
564.7756	2.1953	50	Hg	M_5N_7	M_{α_1}	[50]
565.1103	2.194	73	Ta	abs. edge	M_3	[51]
566.1424	2.19	71	Ta	M_3O_5		[50]
566.1424	2.19	71	Ta	M_3O_4		[50]
566.8413	2.1873	40	Zr	L_1M_2	L_{β_4}	[50]
568.2181	2.182	76	0s	M_3N_5	M_{γ_1}	[50]
570.3091	2.174	82	Pb	M_3N_1		[50]
570.5715	2.173	70	Yb	abs. edge	M_2	[51]
571.0445	2.1712	40	Zr	L_3N_1	L_{β_6}	[50]
572.4155	2.166	76	0s	M_3N_4	M_{γ_2}	[50]
572.4446	2.16589	41	Nb	L_3M_5	L_{α_1}	[50]
573.2094	2.1630	41	Nb	L_3M_4	L_{α_2}	[50]
575.6045	2.1540	39	Y	abs. edge	L_2	[51]
578.4241	2.1435	15	P	abs. edge	K	[51]
579.6409	2.139	15	P	$KM_{2,3}$	$K_{\beta_{1,3}}$	[50]
581.2714	2.133	76	0s	M_2N_1		[50]
581.2714	2.133	78	Pt	abs. edge	M_5	[51]
582.5550	2.1283	67	Ho	abs. edge	M_1	[51]
582.8289	2.1273	78	Pt	M_4N_6	M_β	[50]
583.1853	2.126	71	Ta	M_3O_1		[50]
583.1853	2.126	77	Ir	abs. edge	M_4	[51]
583.6245	2.1244	40	Zr	L_2M_4	L_{β_1}	[50]
584.0369	2.1229	79	Au	M_5N_7	M_{α_1}	[50]
584.7806	2.1202	42	Mo	L_2M_1	L_η	[50]
585.3881	2.118	79	Au	M_5N_6	M_{α_2}	[50]
586.2184	2.115	79	Au	M_5O_3		[50]
587.4683	2.1102	39	Y	L_2N_1	L_{γ_5}	[50]
588.2767	2.1076	72	Hf	abs. edge M_3		[51]
588.4442	2.107	81	Tl	M_3N_1		[50]
588.7236	2.106	75	Re	M_3N_5	M_{γ_1}	[104]
593.2306	2.09	75	Re	M_3N_4	M_{γ_2}	[50]
593.2874	2.0898	69	Tm	abs. edge	M_2	[51]

λ [pm]	E [keV]	Z	element	transition	notation	reference
596.2547	2.0794	39	Y	abs. edge	L_3	[51]
598.3264	2.0722	39	Y	L_1M_3	L_{β_3}	[50]
598.6731	2.071	78	Pt	M_5O_3		[50]
600.9947	2.063	37	Rb	abs. edge	L_1	[51]
601.8699	2.0600	39	Y	L_1M_2	L_{β_4}	[50]
603.7750	2.0535	77	Ir	M_4N_6	M_β	[50]
604.5994	2.0507	37	Rb	$L_1N_{2,3}$	$L_{\gamma_{2,3}}$	[50]
604.6583	2.0505	78	Pt	M_5N_7	M_{α_1}	[50]
605.3965	2.048	77	Ir	abs. edge	M_5	[51]
605.6922	2.047	78	Pt	M_5N_6	M_{α_2}	[50]
605.7514	2.0468	66	Dy	abs. edge	M_1	[51]
607.0683	2.04236	40	Zr	L_3M_5	L_{α_1}	[50]
607.1753	2.042	76	0s	abs. edge	M_4	[51]
607.8003	2.0399	40	Zr	L_3M_4	L_{α_2}	[50]
608.9646	2.036	80	Hg	M_3N_1		[50]
609.2638	2.035	74	W	M_3N_5	M_{γ_1}	[50]
609.4435	2.0344	39	Y	L_3N_1	L_{β_6}	[50]
612.6961	2.0236	71	Lu	abs. edge	M_3	[51]
613.4844	2.021	74	W	M_3N_4	M_{γ_2}	[50]
615.1036	2.01568	42	Mo	L_3M_1	L_l	[50]
615.7084	2.0137	15	P	KL_3	K_{α_1}	[50]
616.0143	2.0127	15	P	KL_2	K_{α_2}	[50]
616.2286	2.012	83	Bi	M_4N_3	M_δ	[50]
617.3025	2.0085	38	Sr	abs. edge	L_2	[51]
618.1334	2.0058	68	Er	abs. edge	M_2	[51]
621.1061	1.99620	41	Nb	L_2N_1	L_η	[50]
621.2181	1.99584	39	Y	L_2M_4	L_{β_1}	[50]
625.8717	1.981	79	Au	M_3N_1		[50]
626.2195	1.9799	77	Ir	M_5N_7	M_{α_1}	[50]
626.7259	1.9783	76	0s	M_4N_6	M_β	[50]
627.5189	1.9758	77	Ir	M_5N_6	M_{α_2}	[50]
628.4095	1.973	74	W	M_2N_1		[50]
629.6350	1.96916	38	Sr	L_2N_1	L_{γ_5}	[50]
630.1662	1.9675	65	Tb	abs. edge	M_1	[51]
630.3264	1.967	76	0s	abs. edge	M_5	[51]
631.2892	1.964	73	Ta	M_3N_5	M_{γ_1}	[50]
633.2237	1.958	75	Re	abs. edge	M_4	[51]
635.4956	1.951	73	Ta	M_3N_4	M_{γ_2}	[50]
635.8867	1.9498	70	Yb	abs. edge	M_3	[51]
636.7391	1.94719	38	Sr	L_1M_3	L_{β_3}	[50]
638.4407	1.942	82	Pb	M_4N_3	M_δ	[50]
638.7368	1.9411	38	Sr	abs. edge	L_3	[51]

λ [pm]	E [keV]	Z	element	transition	notation	reference
640.2871	1.93643	38	Sr	L_1M_2	L_{β_4}	[50]
644.8158	1.9228	67	Ho	abs. edge	M_2	[51]
644.8964	1.92256	39	Y	L_3M_5	L_{α_1}	[50]
645.4200	1.921	78	Pt	M_3N_1		[50]
645.5881	1.92047	39	Y	L_3M_4	L_{α_2}	[50]
647.4423	1.915	36	Kr	abs. edge	L_1	[51]
649.0692	1.9102	76	Os	$M_5N_{6,7}$	$M_{\alpha_{1,2}}$	[50]
650.4653	1.9061	75	Re	M_4N_6	M_β	[104]
651.7818	1.90225	41	Nb	L_3M_1	L_l	[50]
651.9326	1.90181	38	Sr	L_3N_1	L_{β_6}	[50]
652.2104	1.901	83	Bi	M_5N_3	M_{ξ_1}	[50]
652.8973	1.899	75	Re	M_5N_7	M_{α_1}	[104]
654.2754	1.895	72	Hf	M_3N_5	M_{γ_1}	[50]
657.9209	1.8845	69	Tm	abs. edge	M_3	[51]
658.4450	1.883	83	Bi	M_4N_2		[50]
658.7949	1.882	75	Re	abs. edge	M_5	[51]
659.2152	1.8808	64	Gd	abs. edge	M_1	[51]
659.4957	1.880	74	W	abs. edge	M_4	[51]
660.7117	1.87654	40	Zr	L_2M_1	L_η	[50]
662.4132	1.87172	38	Sr	L_2M_4	L_{β_1}	[50]
664.4081	1.8661	37	Rb	abs. edge	L_2	[51]
666.9456	1.859	77	Ir	M_3N_1		[50]
672.9183	1.8425	75	Re	M_5N_6	M_{α_2}	[50]
673.1740	1.8418	66	Dy	abs. edge	M_2	[51]
673.8326	1.8400	14	Si	abs. edge	K	[51]
674.0157	1.8395	82	Pb	M_5N_3	M_{ξ_1}	[50]
675.3227	1.83594	14	Si	KM	K_β	[50]
675.5509	1.83532	37	Rb	L_2N_1	L_{γ_5}	[50]
675.7054	1.8349	74	W	M_4N_6	M_β	[50]
676.7751	1.832	71	Lu	M_3N_5	M_{γ_1}	[50]
678.7796	1.82659	37	Rb	L_1M_3	L_{β_3}	[50]
680.1162	1.823	82	Pb	M_4N_2		[50]
680.4895	1.822	74	W	$M_4O_{2,3}$	M_η	[50]
682.0956	1.81771	37	Rb	L_1M_2	L_{β_4}	[50]
683.4906	1.814	74	W	abs. edge	M_5	[51]
684.3205	1.8118	68	Er	abs. edge	M_3	[51]
686.2523	1.8067	37	Rb	abs. edge	L_3	[51]
686.3055	1.80656	38	Sr	L_3M_5	L_{α_1}	[50]
686.9976	1.80474	80	Hg	M_4N_3	M_δ	[50]
687.0128	1.8047	38	Sr	L_3M_4	L_{α_2}	[50]
687.2794	1.804	73	Ta	abs. edge	M_4	[51]
688.8066	1.80	63	Eu	abs. edge	M_1	[51]

λ [pm]	E [keV]	Z	element	transition	notation	reference
689.5728	1.798	76	0s	M_3N_1		[50]
691.8778	1.79201	40	Zr	L_3M_1	L_l	[50]
696.1550	1.781	35	Br	abs. edge	L_1	[51]
697.3295	1.778	81	Ti	M_5N_3	M_{ξ_1}	[50]
698.3507	1.7754	74	W	M_5N_7	M_{α_1}	[50]
698.4413	1.77517	37	Rb	L_3N_1	L_{β_6}	[50]
699.2566	1.7731	74	W	M_5N_6	M_{α_2}	[50]
700.4813	1.77	74	W	M_5O_3		[50]
701.3927	1.7677	65	Tb	abs. edge	M_2	[51]
702.2667	1.7655	73	Ta	M_4N_6	M_β	[50]
702.4657	1.765	70	Yb	M_3N_5	M_{γ_1}	[50]
703.2626	1.763	81	Tl	M_4N_2		[50]
704.0813	1.76095	39	Y	L_2M_1	L_η	[50]
707.6094	1.75217	37	Rb	L_2M_4	L_{β_1}	[50]
709.0946	1.7485	73	Ta	M_4O_2	M_η	[50]
710.1099	1.746	79	Au	M_4N_3	M_δ	[50]
711.3322	1.743	73	Ta	abs. edge	M_5	[51]
712.0675	1.7412	67	Ho	abs. edge	M_3	[51]
712.5668	1.73998	14	Si	KL_3	K_{α_1}	[50]
712.8126	1.73938	14	Si	KL_2	K_{α_2}	[50]
716.8018	1.7297	36	Kr	abs. edge	L_2	[51]
719.6726	1.7228	62	Sm	abs. edge	M_1	[51]
722.5244	1.716	72	Hf	abs. edge	M_4	[51]
725.0596	1.71	36	Kr	L_2N_3		[50]
725.2292	1.7096	73	Ta	M_5N_7	M_{α_1}	[50]
725.2292	1.7096	73	Ta	M_5N_6	M_{α_2}	[50]
726.3339	1.707	36	Kr	L_1M_3	L_{β_3}	[50]
728.0399	1.703	36	Kr	L_2N_1	L_{γ_5}	[50]
729.3247	1.7	73	Ta	M_5O_3		[50]
730.6140	1.697	36	Kr	L_1M_2	L_{β_4}	[50]
730.6140	1.697	72	Hf	M_4N_6	M_β	[50]
731.8518	1.69413	37	Rb	L_3M_5	L_{α_1}	[50]
732.5306	1.69256	37	Rb	L_3M_4	L_{α_2}	[50]
734.3789	1.6883	64	Gd	abs. edge	M_2	[51]
735.6600	1.68536	39	Y	L_3M_1	L_l	[50]
736.2541	1.684	74	W	M_3N_1		[50]
737.1296	1.682	78	Pt	M_4N_3	M_δ	[50]
739.2392	1.6772	36	Kr	abs. edge	L_3	[51]
739.9450	1.6756	66	Dy	abs. edge	M_3	[51]
746.1346	1.6617	72	Hf	abs. edge	M_5	[51]
746.6738	1.6605	79	Au	M_5N_3	M_{ξ_1}	[50]
750.2886	1.6525	34	Se	abs. edge	L_1	[51]

λ [pm]	E [keV]	Z	element	transition	notation	reference
750.9703	1.651	36	Kr	L_3N_1	L_{β_6}	[50]
751.7307	1.64933	38	Sr	L_2M_1	L_η	[50]
752.0636	1.6486	61	Pm	abs. edge	M_1	[51]
752.3373	1.648	79	Au	M_4N_2		[50]
753.8927	1.6446	72	Hf	M_5N_7	M_{α_1}	[50]
753.8927	1.6446	72	Hf	M_5N_6	M_{α_2}	[50]
754.6269	1.643	68	Er	M_3N_5	M_{γ_1}	[50]
756.2840	1.6394	71	La	abs. edge	M_4	[51]
757.5779	1.6366	36	Kr	L_2M_4	L_{β_1}	[50]
759.7132	1.632	68	Er	M_2N_4		[50]
760.0858	1.6312	71	Lu	M_4N_6	M_β	[50]
761.1123	1.629	73	Ta	M_3N_1		[104]
764.3970	1.622	77	Ir	M_4N_3	M_δ	[50]
768.2334	1.6139	63	Eu	abs. edge	M_3	[51]
769.4731	1.6113	65	Tb	abs. edge	M_2	[51]
773.8434	1.6022	78	Pt	M_5N_3	M_{ξ_1}	[50]
775.3921	1.599	35	Br	abs. edge	L_2	[51]
776.8496	1.596	35	Br	L_1M_2	L_{β_4}	[50]
778.8015	1.592	78	Pt	M_4N_2		[50]
780.5174	1.5885	71	Lu	abs. edge	M_5	[51]
781.7477	1.586	36	Kr	$L_3M_{4,5}$	$L_{\alpha_{1,2}}$	[50]
783.6501	1.58215	38	Sr	L_3M_1	L_l	[50]
784.0713	1.5813	71	Lu	$M_5N_{6,7}$	$M_{\alpha_{1,2}}$	[50]
786.5584	1.5763	70	Yb	abs. edge	M_4	[51]
787.0577	1.5753	60	Nd	abs. edge	M_1	[51]
788.7099	1.572	72	Hf	M_3N_1		[50]
790.9741	1.5675	70	Yb	M_4N_6	M_β	[50]
791.2265	1.567	67	Ho	$M_3N_{6,7}$	M_γ	[50]
794.8381	1.55988	13	Al	abs. edge	K	[51]
796.0782	1.55745	13	Al	KM	K_β	[50]
798.3593	1.5530	35	Br	abs. edge	L_3	[51]
802.0778	1.5458	77	Ir	M_5N_3	M_{ξ_1}	[50]
803.0129	1.544	64	Gd	abs. edge	M_3	[51]
804.1774	1.54177	37	Rb	L_2M_1	L_η	[50]
804.4718	1.5412	64	Gd	$M_3O_{4,5}$		[104]
804.7329	1.5407	62	Sm	abs. edge	M_2	[51]
806.5127	1.5373	77	Ir	M_4N_2		[50]
811.5276	1.5278	70	Yb	abs. edge	M_5	[51]
810.7317	1.5293	33	As	abs. edge	L_1	[51]
812.5381	1.52590	35	Br	L_2M_4	L_{β_1}	[50]
814.6202	1.522	66	Dy	$M_3N_{4,5}$	M_γ	[50]
814.9415	1.5214	70	Yb	$M_5N_{6,7}$	$M_{\alpha_{1,2}}$	[50]

λ [pm]	E [keV]	Z	element	transition	notation	reference
818.6002	1.5146	69	Tm	abs. edge	M_4	[51]
820.5506	1.511	59	Pr	abs. edge	M_1	[51]
823.8219	1.505	75	Re	M_4N_3		[50]
824.9181	1.503	69	Tm	M_4N_6	M_β	[50]
831.0557	1.4919	76	Os	M_5N_3	M_{ξ_1}	[50]
832.1154	1.49	34	Se	$L_1M_{2,3}$	$L_{\beta_{3,4}}$	[50]
833.9624	1.48670	13	Al	KL_3	K_{α_1}	[50]
834.2037	1.48627	13	Al	KL_2	K_{α_2}	[50]
835.9867	1.4831	76	Os	M_4N_2		[50]
836.3928	1.48238	37	Rb	L_3M_1	L_l	[50]
837.3983	1.4806	63	Eu	abs. edge	M_3	[51]
837.4945	1.48043	35	Br	$L_3M_{4,5}$	L_{α_2}	[50]
840.7486	1.4747	34	Se	abs. edge	L_2	[51]
842.1763	1.4722	63	Eu	$L_3O_{4,5}$		[104]
842.6342	1.4714	61	Pm	abs. edge	M_2	[51]
848.6905	1.4609	69	Tm	abs. edge	M_5	[51]
846.8934	1.464	70	Yb	M_3N_1		[50]
848.0519	1.462	69	Tm	M_5N_6	M_γ	[50]
848.6324	1.461	65	Tb	M_3N_5	M_{γ_1}	[50]
857.4356	1.446	74	W	M_4N_3	M_δ	[50]
859.2183	1.443	68	Er	M_3N_6	M_β	[50]
859.6949	1.4422	92	U	$N_1P_{4,5}$		[104]
860.1124	1.4415	68	Er	abs. edge	M_4	[51]
862.9259	1.4368	75	Re	M_5N_3	M_{ξ_1}	[50]
864.2492	1.4346	58	Ce	abs. edge	M_1	[51]
864.6109	1.4340	34	Se	abs. edge	L_3	[51]
866.4234	1.431	75	Re	M_4N_2		[50]
873.2582	1.4198	62	Sm	abs. edge	M_3	[51]
873.6089	1.41923	34	Se	L_2M_4	L_{β_1}	[50]
876.2204	1.415	92	U	N_1P_3		[104]
877.3365	1.4132	32	Ge	abs. edge	L_1	[51]
880.8269	1.4076	92	U	N_1P_2		[50]
881.8293	1.406	68	Er	$M_5N_{6,7}$	$M_{\alpha_{1,2}}$	[50]
883.8408	1.4028	60	Nd	abs. edge	M_2	[51]
884.3452	1.402	64	Gd	M_3N_5	M_{γ_1}	[50]
884.3452	1.402	64	Gd	M_3N_4	M_{γ_2}	[50]
884.7870	1.4013	68	Er	abs. edge	M_5	[51]
890.0588	1.393	73	Ta	M_4N_3	M_δ	[50]
891.0183	1.3915	67	Ho	abs. edge	M_4	[51]
893.0077	1.3884	33	As	$L_2M_{2,3}$	$L_{\beta_{3,4}}$	[50]
896.1705	1.3835	74	W	M_5N_3	M_{ξ_1}	[50]
896.4945	1.383	67	Ho	M_4N_6	M_β	[50]

λ [pm]	E [keV]	Z	element	transition	notation	reference
899.0298	1.37910	34	Se	$L_3M_{4,5}$	$L_{\alpha_{1,2}}$	[50]
899.2906	1.3787	74	W	M_4N_2		[50]
910.7852	1.3613	57	La	abs. edge	M_1	[51]
912.5281	1.3587	33	As	abs. edge	L_2	[51]
913.7386	1.3569	61	Pm	abs. edge	M_3	[51]
917.4574	1.3514	67	Ho	abs. edge	M_5	[51]
919.7715	1.3480	67	Ho	$M_5N_{6,7}$	$M_{\alpha_{1,2}}$	[50]
921.1381	1.346	63	Eu	$M_5N_{6,7}$	M_{γ_1}	[50]
921.1381	1.346	63	Eu	M_3N_5	M_{γ_2}	[50]
925.5389	1.3396	35	Br	L_2M_1	L_η	[50]
927.0614	1.3374	59	Pr	abs. edge	M_2	[51]
930.4705	1.3325	66	Dy	abs. edge	M_4	[51]
931.6591	1.3308	13	Ta	M_5N_3	M_{ξ_1}	[50]
933.0614	1.3288	13	Ta	M_4N_2		[50]
935.7373	1.325	66	Dy	M_4N_6	M_β	[50]
936.7979	1.3235	33	As	abs. edge	L_3	[51]
939.9939	1.319	90	Th	N_1P_3		[50]
941.4214	1.317	33	As	L_2M_4	L_{β_1}	[50]
944.2894	1.313	90	Th	N_1P_2		[50]
951.2517	1.30339	12	Mg	abs. edge	K	[51]
951.6825	1.3028	31	Ga	abs. edge	L_1	[51]
952.1212	1.3022	12	Mg	KM	K_β	[50]
955.6435	1.2974	60	Nd	abs. edge	M_3	[51]
957.4886	1.2949	66	Dy	abs. edge	M_5	[51]
958.0805	1.2941	32	Ge	L_1M_3	L_{β_3}	[50]
958.5249	1.2935	35	Br	L_3M_1	L_l	[50]
958.8955	1.2930	66	Dy	$M_5N_{6,7}$	$M_{\alpha_{1,2}}$	[50]
959.0439	1.2928	56	Ba	abs. edge	M_1	[51]
960.3811	1.291	62	Sm	M_3N_5	M_{γ_1}	[50]
960.3811	1.291	62	Sm	M_3N_4	M_{γ_2}	[50]
964.0401	1.2861	32	Ge	L_1M_2	L_{β_4}	[50]
967.1232	1.2820	33	As	$L_3M_{4,5}$	$L_{\alpha_{1,2}}$	[50]
968.6343	1.28	12	Hf	M_4N_2		[50]
968.6343	1.28	12	Hf	M_5N_3		[50]
972.4329	1.275	65	Tb	abs. edge	M_4	[51]
974.1137	1.2728	58	Ce	abs. edge	M_2	[51]
974.6497	1.2721	66	Dy	M_5O_3		[50]
979.2686	1.2661	65	Tb	M_4N_6	M_β	[50]
989.0331	1.2536	12	Mg	$KL_{2,3}$	$K_{\alpha_{1,2}}$	[50]
992.3579	1.2494	32	Ge	abs. edge	L_2	[51]
996.1851	1.2446	34	Se	L_2M_1	L_η	[50]
998.1098	1.2422	59	Pr	abs. edge	M_3	[51]

λ [pm]	E [keV]	Z	element	transition	notation	reference
998.9139	1.2412	65	Tb	abs. edge	M_5	[51]
999.8806	1.24	65	Tb	$M_5N_{6,7}$	$M_{\alpha_{1,2}}$	[50]
1006.7820	1.2315	11	Lu	M_5N_3	M_{ξ_1}	[50]
1012.4547	1.2246	32	Ge	L_2N_1	L_{γ_5}	[104]
1017.5232	1.2185	32	Ge	L_2M_4	L_{β_1}	[50]
1018.6099	1.2172	64	Gd	abs. edge	M_4	[51]
1018.6936	1.2171	55	Cs	abs. edge	M_1	[51]
1018.7773	1.2170	32	Ge	abs. edge	L_3	[51]
1025.4338	1.2091	64	Gd	M_4N_6	M_β	[50]
1029.4354	1.2044	57	La	abs. edge	M_2	[51]
1029.4354	1.2044	34	Se	L_3M_1	L_l	[50]
1035.7995	1.197	31	Ga	$L_1M_{2,3}$	$L_{\beta_{3,4}}$	[50]
1040.1443	1.192	92	U	N_2P_1		[50]
1043.6465	1.188	32	Ge	$L_3M_{4,5}$	$L_{\alpha_{1,2}}$	[50]
1045.9345	1.1854	58	Ce	abs. edge	M_3	[51]
1046.1120	1.1852	64	Gd	abs. edge	M_5	[51]
1046.2886	1.185	64	Gd	$M_5N_{6,7}$	$M_{\alpha_{1,2}}$	[50]
1048.0575	1.183	10	Yb	M_5N_3	M_{ξ_1}	[50]
1050.7220	1.180	60	Nd	M_3N_5	M_{γ_1}	[50]
1050.7220	1.180	60	Nd	M_3N_4	M_{γ_2}	[50]
1064.2506	1.165	64	Gd	M_5O_3		[104]
1071.1464	1.1575	63	Eu	abs. edge	M_4	[51]
1073.4649	1.1550	33	As	L_2M_1	L_η	[50]
1075.0473	1.1533	63	Eu	M_4N_6	M_β	[50]
1079.6343	1.1484	54	Xe	abs. edge	M_1	[493]
1082.8402	1.1450	31	Ga	abs. edge	L_2	[51]
1090.7469	1.1367	56	Ba	abs. edge	M_2	[51]
1096.2440	1.131	63	Eu	$M_5N_{6,7}$	$M_{\alpha_{1,2}}$	[50]
1099.8421	1.1273	59	Pr	$M_3N_{4,5}$	M_γ	[50]
1101.3075	1.1258	63	Eu	abs. edge	M_5	[51]
1102.2866	1.1248	31	Ga	L_2M_4	L_{β_1}	[50]
1103.6603	1.1234	57	La	abs. edge	M_3	[51]
1107.2084	1.1198	33	As	L_3M_1	L_l	[50]
1110.0833	1.1169	31	Ga	abs. edge	L_3	[51]
1119.5052	1.1075	63	Eu	M_5O_3		[50]
1120.0108	1.107	30	Zn	$L_1M_{2,3}$	$L_{\beta_{3,4}}$	[50]
1127.3431	1.0998	62	Sm	M_4N_6	M_β	[50]
1128.8828	1.0983	62	Sm	abs. edge	M_4	[51]
1129.2941	1.09792	31	Ga	$L_3M_{4,5}$	$L_{\alpha_{1,2}}$	[50]
1130.6329	1.0966	29	Cu	abs. edge	L_1	[51]
1137.3745	1.0901	68	Er	M_5N_3	M_{ξ_1}	[50]
1146.9491	1.081	62	Sm	$M_5N_{6,7}$	$M_{\alpha_{1,2}}$	[50]

λ [pm]	E [keV]	Z	element	transition	notation	reference
1153.4580	1.0749	58	Ce	$M_3N_{4,5}$	M_γ	[50]
1155.2851	1.0732	62	Sm	abs. edge	M_5	[51]
1156.9345	1.07167	11	Na	abs. edge	K	[51]
1156.4705	1.0721	53	I	abs. edge	M_1	[51]
1157.5502	1.0711	11	Na	KM	K_β	[50]
1160.9101	1.068	32	Ge	L_2M_1	L_η	[50]
1164.1803	1.065	55	Cs	abs. edge	M_2	[51]
1167.2491	1.0622	56	Ba	abs. edge	M_3	[51]
1179.4635	1.0512	61	Pm	abs. edge	M_4	[51]
1186.4612	1.045	67	Ho	M_5N_3	M_{ξ_1}	[50]
1186.2002	1.04523	30	Zn	abs. edge	L_2	[51]
1191.0202	1.041	11	Na	$KL_{2,3}$	$K_{\alpha_{1,2}}$	[50]
1196.5373	1.0362	32	Ge	L_3M_1	L_l	[50]
1198.2720	1.0347	30	Zn	L_2M_4	L_{β_1}	[50]
1207.2561	1.027	57	La	$M_3N_{4,5}$	M_γ	[50]
1207.3736	1.0269	61	Pm	abs. edge	M_5	[51]
1212.2135	1.0228	29	Cu	$L_1M_{2,3}$	$L_{\beta_{3,4}}$	[50]
1213.1506	1.02201	30	Zn	abs. edge	L_3	[51]
1225.5135	1.0117	30	Zn	$L_3M_{4,5}$	$L_{\alpha_{1,2}}$	[50]
1229.8899	1.0081	28	Ni	abs. edge	L_1	[51]
1241.0931	0.999	54	Xe	abs. edge	M_2	[51]
1242.3367	0.998	66	Dy	M_5N_3	M_{ξ_1}	[50]
1242.8348	0.9976	55	Cs	abs. edge	M_3	[51]
1243.5827	0.997	60	Nd	M_4N_6	M_β	[50]
1245.9572	0.9951	60	Nd	abs. edge	M_4	[51]
1259.7561	0.9842	31	Ga	L_2M_1	L_η	[50]
1267.7423	0.978	60	Nd	$M_5N_{6,7}$	$M_{\alpha_{1,2}}$	[50]
1273.7333	0.9734	60	Nd	abs. edge	M_5	[51]
1274.2569	0.973	56	Ba	M_3N_5	M_{γ_1}	[50]
1274.2569	0.973	56	Ba	M_3N_4	M_{γ_2}	[50]
1295.2904	0.9572	31	Ga	L_3M_1	L_l	[50]
1297.5950	0.9555	65	Tb	M_5N_3	M_{ξ_1}	[50]
1301.4359	0.95268	29	Cu	abs. edge	L_2	[51]
1305.1074	0.950	59	Pr	M_4N_6	M_β	[50]
1305.3822	0.9489	29	Cu	L_2M_4	L_{β_1}	[50]
1312.2904	0.9448	59	Pr	abs. edge	M_4	[51]
1313.8201	0.9437	51	Sb	abs. edge	M_1	[51]
1317.5898	0.941	28	Ni	$L_1M_{2,3}$	$L_{\beta_{3,4}}$	[50]
1323.2145	0.937	54	Xe	abs. edge	M_3	[51]
1328.1757	0.9335	83	Bi	$N_1P_{2,3}$		[104]
1328.8020	0.93306	29	Cu	abs. edge	L_3	[51]
1339.3670	0.9257	59	Pr	abs. edge	M_5	[51]

λ [pm]	E [keV]	Z	element	transition	notation	reference
1332.4578	0.9305	53	I	abs. edge	M_2	[51]
1334.6093	0.929	59	Pr	$M_5N_{6,7}$	$M_{\alpha_{1,2}}$	[50]
1339.5117	0.9256	27	Co	abs. edge	L_1	[51]
1343.7217	0.9227	29	Cu	$L_3M_{4,5}$	$L_{\alpha_{1,2}}$	[50]
1356.5120	0.914	64	Gd	M_5N_3		[50]
1368.4901	0.906	30	Zn	L_2M_1	L_η	[50]
1374.5588	0.902	58	Ce	M_4N_6	M_β	[50]
1375.6263	0.9013	58	Ce	abs. edge	M_4	[51]
1402.5475	0.884	30	Zn	L_3M_1	L_l	[50]
1402.8649	0.8838	50	Sn	abs. edge	M_1	[51]
1403.659	0.8833	58	Ce	abs. edge	M_5	[51]
1404.1359	0.883	58	Ce	$M_5N_{6,7}$	$M_{\alpha_{1,2}}$	[50]
1417.6218	0.8746	53	I	abs. edge	M_3	[51]
1421.8486	0.872	63	Eu	M_5N_3	L_{ξ_1}	[50]
1424.1351	0.8706	28	Ni	abs. edge	L_2	[51]
1425.1172	0.870	27	Co	$L_1M_{2,3}$	$L_{\beta_{3,4}}$	[50]
1425.6088	0.8697	52	Te	abs. edge	M_2	[51]
1427.0856	0.8688	28	Ni	L_2M_4	L_{β_1}	[50]
1430.2315	0.866889	10	Ne	abs. edge	K	[51]
1438.3434	0.862	58	Ce	$M_5O_{2,3}$		[50]
1446.7351	0.857	10	Ne	KM	K_β	[50]
1451.8173	0.854	57	La	M_4N_6	M_β	[50]
1452.4977	0.8536	28	Ni	abs. edge	L_3	[51]
1456.0799	0.8515	28	Ni	$L_3M_{4,5}$	$L_{\alpha_{1,2}}$	[50]
1461.0559	0.8486	10	Ne	$KL_{2,3}$	$K_{\alpha_{1,2}}$	[50]
1461.2280	0.8485	57	La	abs. edge	M_4	[51]
1465.3729	0.8461	26	Fe	abs. edge	L_1	[51]
1488.4178	0.833	57	La	$M_5N_{6,7}$	$M_{\alpha_{1,2}}$	[50]
1490.2067	0.832	29	Cu	L_2M_1	L_η	[50]
1490.7443	0.8317	57	La	abs. edge	M_5	[51]
1492.0000	0.831	62	Sm	M_5N_3	M_{ξ_1}	[50]
1501.7587	0.8256	49	In	abs. edge	M_1	[51]
1514.4155	0.8187	52	Te	abs. edge	M_3	[51]
1527.0094	0.8119	51	Sb	abs. edge	M_2	[51]
1528.6056	0.8111	29	Cu	L_3M_1	L_1	[50]
1556.2345	0.7967	56	Ba	abs. edge	M_4	[51]
1561.9199	0.7938	27	Co	abs. edge	L_2	[51]
1565.4697	0.792	26	Fe	$L_1M_{2,3}$	$L_{\beta_{3,4}}$	[50]
1566.6566	0.7914	27	Co	L_2M_4	L_{β_1}	[50]
1571.4220	0.789	56	Ba	M_4O_3	M_η	[50]
1589.3501	0.7801	56	Ba	abs. edge	M_5	[51]
1591.5943	0.779	56	Ba	M_4O_3	M_η	[50]

λ [pm]	E [keV]	Z	element	transition	notation	reference
1591.5944	0.7790	27	Co	abs. edge	L_3	[51]
1593.6401	0.778	52	Te	M_3N_5	M_{γ_1}	[50]
1593.6401	0.778	52	Te	M_3N_4	M_{γ_2}	[50]
1597.3357	0.7762	27	Co	$L_3M_{4,5}$	$L_{\alpha_{1,2}}$	[50]
1597.7474	0.776	51	Sb	M_2N_4		[50]
1610.1974	0.770	48	Cd	abs. edge	M_1	[51]
16l2.2913	0.769	25	Mn	abs. edge	L_1	[51]
1619.4514	0.7656	51	Sb	abs. edge	M_3	[51]
1620.7216	0.765	56	Ba	M_5O_3		[50]
1627.1024	0.762	28	Ni	L_2M_1	L_η	[50]
1639.1486	0.7564	50	Sn	abs. edge	M_2	[51]
1646.5498	0.753	60	Nd	M_5N_3	M_{ξ_1}	[50]
1669.3847	0.7427	28	Ni	L_3M_1	L_l	[50]
1676.6085	0.7395	55	Cs	abs. edge	M_4	[51]
1691.4761	0.733	51	Sb	M_3N_5	M_{γ_1}	[50]
1691.4761	0.733	51	Sb	M_3N_4	M_{γ_2}	[50]
1691.4761	0.733	50	Sn	M_2N_4		[50]
1708.9621	0.7255	55	Cs	abs. edge	M_5	[51]
1719.6283	0.721	25	Mn	$L_1M_{2,3}$	$L_{\beta_{3,4}}$	[50]
1720.1054	0.7208	26	Fe	abs. edge	L_2	[51]
1725.6117	0.7185	26	Fe	L_2M_4	L_{β_1}	[50]
1728.0167	0.7175	47	Ag	abs. edge	M_1	[51]
1735.5151	0.7144	50	Sn	abs. edge	M_3	[51]
1736.4874	0.714	59	Pr	M_5N_3	M_{ξ_1}	[50]
1752.6887	0.7074	26	Fe	abs. edge	L_3	[51]
1758.6553	0.705	26	Fe	$L_3M_{4,5}$	$L_{\alpha_{1,2}}$	[50]
1763.6586	0.703	52	Te	M_2N_1		[50]
1765.6679	0.7022	49	In	abs. edge	M_2	[51]
1786.5303	0.694	27	Co	L_2M_1	L_η	[50]
1794.2865	0.691	24	Cr	abs. edge	L_2	[51]
1794.2865	0.691	50	Sn	M_3N_5	M_{γ_1}	[50]
1794.2865	0.691	50	Sn	M_3N_4	M_{γ_2}	[50]
1795.3258	0.6906	54	Xe	abs. edge	M_4	[493]
1808.9466	0.6854	9	F	abs. edge	K	[51]
1829.2299	0.6778	27	Co	L_3M_1	L_l	[50]
1831.9326	0.6768	9	F	$KL_{2,3}$	$K_{\alpha_{1,2}}$	[50]
1834.1006	0.676	58	Ce	M_5N_3		[50]
1844.1945	0.6723	54	Xe	abs. edge	M_5	[51]
1850.8016	0.6699	46	Pd	abs. edge	M_1	[51]
1866.4037	0.6643	49	In	abs. edge	M_3	[51]
1884.2735	0.658	51	Sb	M_2N_1		[50]
1884.2735	0.658	47	Ag	$M_1N_{2,3}$		[50]

λ [pm]	E [keV]	Z	element	transition	notation	reference
1895.7982	0.654	24	Cr	$L_1M_{2,3}$	$L_{\beta_{3,4}}$	[50]
1903.3651	0.6514	25	Mn	abs. edge	L_2	[51]
1905.4126	0.6507	48	Cd	abs. edge	M_2	[51]
1910.9926	0.6488	25	Mn	L_2M_4	L_{β_1}	[50]
1913.3518	0.648	52	Te	M_3N_1		[50]
1936.3611	0.6403	25	Mn	abs. edge	L_3	[51]
1940.3005	0.639	48	Cd	M_2N_5		[50]
1943.3417	0.638	57	La	M_5N_3	M_{ξ_1}	[50]
1945.1710	0.6374	25	Mn	$L_3M_{4,5}$	$L_{\alpha_{1,2}}$	[50]
1964.9002	0.631	53	I	abs. edge	$M_{4,5}$	[51]
1973.6581	0.6282	23	V	abs. edge	L_1	[51]
1974.2866	0.628	26	Fe	L_2M_1	L_η	[50]
1977.1201	0.6271	45	Rh	abs. edge	M_1	[51]
2002.9919	0.619	50	Sn	M_2N_1		[50]
2011.1143	0.6165	48	Cd	abs. edge	M_3	[51]
2012.7467	0.616	46	Pd	$M_1N_{2,3}$		[50]
2015.3641	0.6152	26	Fe	L_3M_1	L_l	[50]
2025.9020	0.612	51	Sb	M_3N_1		[50]
2045.9604	0.606	48	Cd	M_3N_5	M_{γ_1}	[50]
2045.9604	0.606	48	Cd	M_3N_4	M_{γ_2}	[50]
2058.1872	0.6024	47	Ag	abs. edge	M_2	[51]
2062.9817	0.601	56	Ba	M_5N_3		[50]
2066.4200	0.6	47	Ag	M_2N_4		[50]
2073.3311	0.598	24	Cr	abs. edge	L_3	[51]
2119.4051	0.585	44	Ru	abs. edge	M_1	[51]
2119.4051	0.585	23	V	$L_1M_{2,3}$	$L_{\beta_{3,4}}$	[50]
2124.1254	0.5837	24	Cr	abs. edge	L_2	[51]
2127.4056	0.5828	24	Cr	L_2M_4	L_{β_1}	[50]
2128.5013	0.5825	52	Te	abs. edge	M_4	[51]
2133.9966	0.581	52	Te	M_4O_2	M_η	[50]
2156.2643	0.575	50	Sn	M_3N_1		[50]
2164.5461	0.5728	24	Cr	$L_3M_{4,5}$	$L_{\alpha_{1,2}}$	[50]
2167.1945	0.5721	52	Te	abs. edge	M_5	[51]
2169.8495	0.5714	47	Ag	abs. edge	M_3	[51]
2179.0017	0.569	52	Te	M_5O_3		[50]
2182.8380	0.568	47	Ag	M_3N_5	M_{γ_1}	[50]
2182.8380	0.568	47	Ag	M_3N_4	M_{γ_2}	[50]
2191.7129	0.5675	25	Mn	L_2M_1	L_η	[50]
2199.4891	0.5637	22	Ti	abs. edge	L_1	[51]
2214.0214	0.56	46	Pd	M_2N_4		[50]
2217.5854	0.5591	46	Pd	abs. edge	M_2	[51]
2228.7471	0.5563	25	Mn	L_3M_1	L_l	[50]

λ [pm]	E [keV]	Z	element	transition	notation	reference
2279.1397	0.544	43	Tc	abs. edge	M_1	[493]
2296.0222	0.54	48	Cd	M_2M_1		[50]
2309.2792	0.5369	51	Sb	abs. edge	M_4	[51]
2331.8638	0.5317	8	O	abs. edge	K	[51]
2332.7413	0.5315	46	Pd	abs. edge	M_3	[51]
2334.9378	0.531	46	Pd	$M_3N_{4,5}$	M_γ	[50]
2350.4303	0.5275	51	Sb	abs. edge	M_5	[51]
2362.0728	0.5249	8	O	KL	K_α	[50]
2379.7543	0.521	45	Rh	abs. edge	M_2	[51]
2382.0403	0.5205	23	V	abs. edge	L_2	[51]
2388.0046	0.5192	23	V	L_2M_4	L_{β_1}	[50]
2417.3367	0.5l29	23	V	abs. edge	L_3	[51]
2424.9012	0.5113	23	V	$L_3M_{4,5}$	$L_{\alpha_{1,2}}$	[50]
2426.3249	0.511	50	Sn	abs. edge	$M_{4,5}$	[51]
2430.1294	0.5102	24	Cr	L_2M_1	L_η	[50]
2445.4674	0.507	48	Cd	M_3N_1		[50]
2457.0987	0.5046	42	Mo	abs. edge	M_1	[51]
2477.7218	0.5004	21	Sc	abs. edge	L_1	[51]
2478.2171	0.5003	24	Cr	L_3M_1	L_l	[50]
2498.6941	0.4962	45	Rh	abs. edge	M_3	[51]
2499.7016	0.496	45	Rh	$M_3N_{4,5}$	M_γ	[50]
2525.1568	0.491	50	Sn	M_4O_2	M_η	[50]
2551.1358	0.486	44	Ru	M_4N_4		[50]
2562.2071	0.4839	50	Sn	$M_{4,5}O_3$		[104]
2568.0447	0.4828	44	Ru	abs. edge	M_2	[51]
2593.8326	0.478	47	Ag	M_3N_1		[50]
2615.7215	0.474	46	Pd	M_2N_1		[50]
2637.9830	0.47	42	Mo	$M_1N_{2,3}$		[104]
2646.9940	0.4684	41	Nb	abs. edge	M_1	[51]
2650.3890	0.4678	21	Sc	$L_1M_{2,3}$	$L_{\beta_{3,4}}$	[104]
2669.7933	0.4644	52	Te	$M_{4,5}N_{2,3}$	M_ξ	[104]
2683.6623	0.462	44	Ru	M_3N_4	M_{γ_2}	[50]
2691.8194	0.4606	44	Ru	abs. edge	M_3	[51]
2704.7382	0.4584	22	Ti	$L_2M_{4,5}$	L_{β_1}	[50]
2728.5475	0.4544	22	Ti	abs. edge	$L_{2,3}$	[51]
2733.9625	0.4535	23	V	L_2M_1	L_η	[50]
2741.8222	0.4522	22	Ti	$L_3M_{4,5}$	L_α	[50]
2750.3372	0.4508	49	In	abs. edge	M_4	[51]
2776.8241	0.4465	23	V	L_3M_1	L_l	[50]
2786.1843	0.445	46	Pd	M_3N_1		[50]
2786.8105	0.4449	43	Tc	abs. edge	M_2	[51]
2798.1313	0.4431	49	In	abs. edge	M_5	[51]

λ [pm]	E [keV]	Z	element	transition	notation	reference
2805.0950	0.442	45	Rh	M_2N_1		[50]
2812.7314	0.4408	48	Cd	abs. edge	$M_{4,5}$	[51]
2832.0055	0.4378	20	Ca	abs. edge	L_1	[51]
2837.1899	0.437	41	Nb	$M_1N_{2,3}$		[104]
2881.3665	0.4303	40	Zr	abs. edge	M_1	[51]
2890.7717	0.4289	51	Sb	$M_{4,5}N_{2,3}$	M_ξ	[104]
2917.2988	0.425	43	Tc	abs. edge	M_3	[51]
2973.2662	0.417	45	Rh	M_3N_1		[50]
3026.2436	0.4097	42	Mo	abs. edge	M_2	[51]
3038.8529	0.408	48	Cd	M_4O_2	M_η	[50]
3048.5665	0.4067	21	Sc	abs. edge	L_2	[51]
3076.5558	0.403	48	Cd	M_5O_3		[50]
3082.6753	0.4022	21	Sc	abs. edge	L_3	[51]
3082.6753	0.4022	47	Ag	abs. edge	M_4	[51]
3089.5888	0.4013	22	Ti	L_2M_1	L_η	[50]
3102.7327	0.3996	21	Sc	L_2M_4	L_{β_1}	[50]
3114.4235	0.3981	47	Ag	abs. edge	M_5	[51]
3116.7722	0.3978	50	Sn	$M_{4,5}N_{2,3}$	M_ξ	[104]
3132.5215	0.3958	22	Ti	L_3M_1	L_l	[50]
3135.6904	0.3954	21	Sc	$L_3M_{4,5}$	L_α	[50]
3150.0305	0.3936	39	Y	abs. edge	M_1	[51]
3159.6636	0.3924	7	N	KL	K_α	[50]
3160.4690	0.3923	42	Mo	abs. edge	M_3	[51]
3228.7812	0.384	44	Ru	M_2N_1		[50]
3276.5645	0.3784	41	Nb	abs. edge	M_2	[51]
3287.8600	0.3771	19	K	abs. edge	L_1	[51]
3306.2720	0.375	41	Nb	M_2N_4		[50]
3357.3030	0.3693	90	Th	N_4N_6		[104]
3415.5702	0.363	41	Nb	abs. edge	M_3	[51]
3468.1175	0.3575	38	Sr	abs. edge	M_1	[51]
3482.7303	0.356	41	Nb	$M_3N_{4,5}$	M_γ	[50]
3513.3239	0.3529	21	Sc	L_2M_1	L_η	[50]
3513.3239	0.3529	20	Ca	abs. edge	L_3	[51]
3532.3419	0.351	42	Mo	M_2N_1		[50]
3542.4343	0.35	20	Ca	abs. edge	L_2	[51]
3559.7244	0.3483	21	Sc	L_3M_1	L_l	[50]
3594.8159	0.3449	20	Ca	$L_2M_{4,5}$	L_β	[50]
3602.1267	0.3442	40	Zr	abs. edge	M_2	[51]
3631.6696	0.3414	90	Th	$N_5N_{6,7}$		[104]
3632.7337	0.3413	20	Ca	$L_3M_{4,5}$	$L_{\alpha_{1,2}}$	[50]
3646.6235	0.34	46	Pd	abs. edge	M_4	[51]
3666.0319	0.3382	48	Cd	$M_{4,5}N_{2,3}$	M_ξ	[104]

λ [pm]	E [keV]	Z	element	transition	notation	reference
3704.3681	0.3347	46	Pd	abs. edge	M_5	[51]
3745.7764	0.331	42	Mo	M_3N_1		[50]
3751.4433	0.3305	40	Zr	abs. edge	M_3	[51]
3838.5511	0.323	41	Nb	M_2N_1		[50]
3849.2766	0.3221	37	Rb	abs. edge	M_1	[51]
3874.5375	0.32	18	Ar	abs. edge	L_1	[51]
3968.7964	0.3124	39	Y	abs. edge	M_2	[51]
3977.7093	0.3117	45	Rh	abs. edge	M_4	[51]
3982.8204	0.3113	47	Ag	$M_{4,5}N_{2,3}$	M_ξ	[104]
4038.6058	0.307	45	Rh	abs. edge	M_5	[51]
4046.5143	0.3064	20	Ca	L_2M_1	L_η	[50]
4065.0885	0.305	41	Nb	M_3N_1		[50]
4095.9762	0.3027	20	Ca	L_3M_1	L_l	[50]
4128.7113	0.3003	39	Y	abs. edge	M_3	[51]
4160.5772	0.298	19	K	$L_3N_{2,3}$		[104]
4208.5947	0.2946	19	K	abs. edge	$L_{2,3}$	[51]
4224.3680	0.2935	82	Pb	N_4N_6		[104]
4244.6148	0.2921	36	Kr	abs. edge	M_1	[493]
4327.5811	0.2865	92	U	$N_{6,7}O_{4,5}$		[104]
4336.6632	0.2859	46	Pd	$M_{4,5}N_2$		[50]
4354.9420	0.2847	19	K	L_1N_3		[104]
4365.6760	0.284	80	Hg	N_4N_6		[104]
4368.1370	0.28384	6	C	abs. edge	K	[51]
4371.8336	0.2836	44	Ru	abs. edge	M_4	[51]
4378.0085	0.2832	46	Pd	$M_{4,5}N_3$		[104]
4431.2080	0.2798	38	Sr	abs. edge	M_2	[51]
4437.5519	0.2794	44	Ru	abs. edge	M_5	[51]
4476.0000	0.277	6	C	KL	K_α	[50]
4495.4750	0.2758	82	Pb	$N_5N_{6,7}$		[104]
4518.4111	0.2744	51	Sb	M_2N_4		[104]
4588.6454	0.2702	17	Cl	abs. edge	L_1	[51]
4607.4024	0.2691	38	Sr	abs. edge	M_3	[51]
4634.9607	0.2675	81	Tl	$N_5N_{6,7}$		[104]
4643.6404	0.267	80	Hg	$N_5N_{6,7}$		[104]
4678.6868	0.265	79	Au	N_4N_6		[104]
4723.2457	0.2625	19	K	L_2M_1	L_η	[50]
4730.4540	0.2621	50	Sn	M_2M_4		[104]
4761.3364	0.2604	45	Rh	$M_{4,5}N_{2,3}$	M_ξ	[104]
4774.1702	0.25971	19	K	L_3M_1	L_l	[50]
4805.6279	0.258	78	Pt	N_4N_6		[104]
4818.7019	0.2573	90	Th	N_4O_5		[104]
4833.7310	0.2565	35	Br	abs. edge	M_1	[51]

λ [pm]	E [keV]	Z	element	transition	notation	reference
4835.6162	0.2564	52	Te	abs. edge	M_4	[51]
4843.1719	0.256	39	Y	M_2N_1		[50]
4843.1719	0.256	39	Y	M_3N_1		[50]
4902.5385	0.2529	52	Te	abs. edge	M_5	[51]
4939.6494	0.251	79	Au	$N_5N_{6,7}$		[104]
4953.4638	0.2503	90	Th	N_6O_4		[104]
4999.4032	0.248	90	Th	N_7O_5		[104]
5011.5279	0.2474	37	Rb	abs. edge	M_2	[51]
5013.5544	0.2473	18	Ar	abs. edge	L_2	[51]
5019.6437	0.247	77	Ir	N_4N_6		[104]
5029.8255	0.2465	52	Te	M_3M_5		[104]
5056.4926	0.2452	18	Ar	abs. edge	L_3	[51]
5081.3606	0.244	78	Pt	$N_5N_{6,7}$		[104]
5131.8377	0.2416	38	Sr	M_2N_1		[104]
5174.6744	0.2396	48	Cd	M_2M_4		[104]
5187.6653	0.239	76	Os	N_4N_6		[104]
5198.5409	0.2385	37	Rb	abs. edge	M_3	[51]
5222.6285	0.2374	51	Sb	M_3M_5		[104]
5231.4430	0.237	44	Ru	$M_{4,5}N_{2,3}$	M_ζ	[104]
5275.9659	0.235	77	Ir	$N_5N_{6,7}$		[104]
5355.7322	0.2315	34	Se	abs. edge	M_1	[51]
5360.3631	0.2313	38	Sr	M_3N_1		[50]
5383.6387	0.2303	42	Mo	abs. edge	M_4	[51]
5400.0522	0.2296	47	Ag	M_2M_4		[104]
5407.1173	0.2293	74	W	N_2N_4		[104]
5409.4764	0.2292	16	S	abs. edge	L_1	[51]
5414.2009	0.229	50	Sn	M_3M_5		[104]
5461.9031	0.227	76	Os	$N_5N_{6,7}$		[104]
5461.9031	0.227	42	Mo	abs. edge	M_5	[51]
5567.3641	0.2227	36	Kr	abs. edge	M_2	[51]
5584.9189	0.222	74	W	N_4N_6		[104]
5592.4763	0.2217	18	Ar	L_2M_1	L_η	[50]
5633.1304	0.2201	18	Ar	L_3M_1	L_1	[50]
5661.4246	0.219	46	Pd	M_2M_4		[104]
5703.0911	0.2174	37	Rb	M_2N_1		[50]
5780.1958	0.2145	74	W	$N_5N_{6,7}$		[104]
5799.1207	0.2138	36	Kr	abs. edge	M_3	[51]
5809.9906	0.2134	73	Ta	N_4N_6		[104]
5845.6011	0.2121	48	Cd	M_3M_5		[104]
5890.0332	0.2105	74	W	N_5O_3		[104]
5952.2419	0.2083	37	Rb	M_3N_1		[50]
5955.1008	0.2082	45	Rh	M_2M_4		[104]

λ [pm]	E [keV]	Z	element	transition	notation	reference
5978.0713	0.2074	41	Nb	abs. edge	M_4	[51]
6039.2206	0.2053	73	Ta	$N_5N_{6,7}$		[104]
6059.8827	0.2046	41	Nb	abs. edge	M_5	[51]
6062.8459	0.2045	47	Ag	M_3M_5		[104]
6092.6388	0.2035	33	As	abs. edge	M_1	[51]
6150.0595	0.2016	17	Cl	abs. edge	L_2	[51]
6199.2600	0.200	17	Cl	abs. edge	L_3	[51]
6230.4120	0.199	44	Ru	M_2M_4		[104]
6284.0953	0.1973	46	Pd	M_3M_5		[104]
6300.0610	0.1968	72	Hf	$N_5N_{6,7}$		[104]
6302.2638	0.1967	71	Lu	N_4N_6		[104]
6358.2154	0.195	73	Ta	N_5O_3		[104]
6358.2154	0.195	36	Kr	M_2N_1		[104]
6437.4455	0.1926	42	Mo	$M_{4,5}N_{2,3}$	M_ξ	[104]
6450.8428	0.1922	90	Th	O_2P_1		[104]
6546.2091	0.1894	45	Rh	M_3M_5		[104]
6549.6672	0.1893	35	Br	abs. edge	M_2	[51]
6549.6672	0.1893	15	P	abs. edge	L_1	[51]
6584.4503	0.1883	71	Lu	N_5N_7		[50]
6594.9574	0.188	5	B	abs. edge	K	[51]
6630.2246	0.187	36	Kr	M_3N_1		[104]
6734.6659	0.1841	17	Cl	L_2M_1	L_η	[50]
6764.0589	0.1833	5	B	KL	K_α	[50]
6789.9890	0.1826	17	Cl	L_3M_1	L_l	[50]
6797.4342	0.1824	40	Zr	abs. edge	M_4	[51]
6812.3736	0.182	72	Hf	N_5O_3		[104]
6831.1405	0.1815	35	Br	abs. edge	M_3	[51]
6831.1405	0.1815	90	Th	O_3P_1		[104]
6834.9063	0.1814	44	Ru	$M_3M_{4,5}$		[104]
6888.0666	0.18	40	Zr	abs. edge	M_5	[51]
6888.0666	0.18	32	Ge	abs. edge	M_1	[51]
6891.8955	0.1799	42	Mo	M_2M_4		[104]
7221.0367	0.1717	41	Nb	$M_{4,5}N_{2,3}$	M_ξ	[104]
7306.1402	0.1697	71	Lu	$N_{4,5}O_{2,3}$		[104]
7371.2961	0.1682	34	Se	abs. edge	M_2	[51]
7491.5528	0.1655	42	Mo	$M_3M_{4,5}$		[104]
7523.3738	0.1648	16	S	abs. edge	L_2	[51]
7550.8648	0.1642	16	S	abs. edge	L_3	[51]
7658.1346	0.1619	34	Se	abs. edge	M_3	[51]
7658.1346	0.1619	40	Zr	M_2M_4		[104]
7686.6212	0.1613	35	Br	M_2N_1		[50]
7768.4962	0.1596	39	Y	abs. edge	M_4	[51]

λ [pm]	E [keV]	Z	element	transition	notation	reference
7797.8113	0.159	70	Yb	$M_{4,5}O_{2,3}$		[50]
7820.9298	0.15853	41	Nb	$M_3N_{4,5}$		[104]
7842.2011	0.1581	31	Ga	abs. edge	M_1	[51]
7877.0775	0.1574	39	Y	abs. edge	M_5	[51]
7978.4556	0.1554	35	Br	M_3N_1		[50]
8175.2077	0.15166	40	Zr	$M_{4,5}N_{2,3}$	M_ξ	[104]
8249.1816	0.1503	69	Tm	$N_{4,5}O_{2,3}$		[104]
8337.9421	0.1487	16	S	$L_{2,3}M_1$	L_l	[50]
8349.1717	0.1485	39	Y	M_2M_4		[104]
8468.9344	0.1464	33	As	abs. edge	M_2	[51]
8640.0836	0.1435	38	Sr	M_2M_4		[104]
8670.2937	0.143	68	Er	$N_{4,5}O_{2,3}$		[104]
8731.3521	0.142	39	Y	$M_3M_{4,5}$		[104]
8824.5694	0.1405	33	As	abs. edge	M_3	[51]
9123.2671	0.1359	30	Zn	abs. edge	M_1	[51]
9160.3398	0.13535	83	Bi	$N_6O_{4,5}$		[104]
9173.8956	0.13515	38	Sr	$M_3M_{4,5}$		[104]
9177.2909	0.1351	37	Rb	M_2M_4		[104]
9184.0889	0.135	38	Sr	abs. edge	M_4	[51]
9211.3818	0.1346	67	Ho	$N_{4,5}O_{2,3}$		[104]
9315.1916	0.1331	38	Sr	abs. edge	M_5	[51]
9315.8915	0.13309	83	Bi	N_7O_5		[104]
9358.7862	0.13248	39	Y	$M_{4,5}N_{2,3}$	M_ξ	[104]
9392.8182	0.132	15	P	abs. edge	$L_{2,3}$	[493]
9686.3437	0.128	66	Dy	$N_{4,5}O_{2,3}$		[104]
9693.9171	0.1279	32	Ge	abs. edge	M_2	[51]
9701.5023	0.1278	37	Rb	$M_3M_{4,5}$		[104]
10020.626	0.12373	82	Pb	N_6O_5		[104]
10121.241	0.1225	65	Tb	$N_{4,5}O_{2,3}$		[104]
10229.802	0.1212	82	Pb	O_1P_2		[104]
10238.249	0.1211	82	Pb	N_6O_4		[104]
10263.675	0.1208	32	Ge	abs. edge	M_3	[51]
11060.2319	0.1121	29	Cu	abs. edge	M_1	[51]
10384.020	0.1194	15	P	$L_{2,3}M$		[50]
10430.319	0.11887	82	Pb	N_7O_5		[104]
10480.575	0.1183	82	Pb	O_1P_3		[104]
10679.173	0.1161	64	Gd	$N_{4,5}O_{2,3}$		[104]
10800.104	0.1148	38	Sr	$M_4N_{2,3}$	M_{ξ_2}	[50]
10875.895	0.114	38	Sr	M_5N_3	M_{ξ_1}	[50]
10941.158	0.11332	35	Br	M_2M_4		[104]
11089.911	0.1118	37	Rb	abs. edge	M_4	[51]
11089.911	0.1118	28	Ni	abs. edge	M_1	[51]

λ [pm]	E [keV]	Z	element	transition	notation	reference
11109.785	0.1116	63	Eu	$N_{4,5}O_{2,3}$		[104]
11138.750	0.11131	34	Se	$M_2M_{4,5}$		[104]
11169.8378	0.111	4	Be	abs. edge	K	[51]
11240.725	0.1103	37	Rb	abs. edge	M_5	[51]
11303.236	0.10969	81	Tl	N_6O_5		[104]
11380.009	0.10895	35	Br	$M_3M_{4,5}$		[104]
11427.207	0.1085	4	Be	KL	K_α	[50]
11533.507	0.1075	81	Tl	N_6O_4		[104]
11609.101	0.1068	31	Ga	abs. edge	M_2	[51]
11638.524	0.10653	34	Se	$M_3M_{4,5}$		[104]
11729.915	0.1057	62	Sm	$N_{4,5}O_{2,3}$		[104]
11774.473	0.1053	81	Tl	N_7O_5		[104]
11979.246	0.1035	33	As	$M_2M_{4,5}$		[104]
12049.096	0.1029	31	Ga	abs. edge	M_3	[51]
12312.334	0.1007	27	Co	abs. edge	M_1	[51]
12485.921	0.0993	33	As	$M_3M_{4,5}$		[104]
12324.5726	0.1006	14	Si	abs. edge	$L_{2,3}$	[51]
12677.423	0.0978	37	Rb	M_4N_3	M_δ	[50]
12781.979	0.097	37	Rb	M_4N_2		[50]
12861.535	0.0964	37	Rb	M_5N_3	M_{ξ_1}	[50]
12901.686	0.0961	60	Nd	$N_{4,5}O_{2,3}$		[104]
12928.592	0.0959	32	Ge	$M_2M_{4,5}$		[104]
13346.093	0.0929	26	Fe	abs. edge	M_1	[51]
13447.419	0.0922	32	Ge	$M_3M_{4,5}$		[104]
13550.295	0.0915	14	Si	$L_{2,3}M$		[50]
13609.791	0.0911	59	Pr	$N_{4,5}O_{2,3}$		[104]
13700.0221	0.0905	30	Zn	abs. edge	M_2	[51]
13868.591	0.0894	12	Mg	abs. edge	L_1	[51]
13992.236	0.08861	31	Ga	$M_2M_{4,5}$		[104]
14089.227	0.088	36	Kr	abs. edge	M_4	[51]
14089.227	0.088	36	Kr	abs. edge	M_5	[51]
14249.5345	0.08701	13	Al	abs. edge	L_1	[51]
14383.4339	0.0862	30	Zn	abs. edge	M_3	[51]
14433.667	0.0859	58	Ce	$N_{4,5}O_{2,3}$		[104]
14440.391	0.08586	35	Br	M_1M_3		[104]
14509.678	0.08545	31	Ga	$M_3M_{4,5}$		[104]
14638.158	0.0847	79	Au	N_6O_5		[104]
14777.735	0.0839	25	Mn	abs. edge	M_1	[51]
15231.597	0.0814	79	Au	N_7O_5		[104]
15269.113	0.0812	57	La	$N_{4,5}O_{2,3}$		[104]
15615.264	0.0794	35	Br	M_1N_2		[104]
15694.329	0.079	30	Zn	$M_2M_{4,5}$		[50]

λ [pm]	E [keV]	Z	element	transition	notation	reference
15905.734	0.07795	56	Ba	N_4O_3		[104]
15956.9112	0.0777	29	Cu	abs. edge	M_2	[51]
16324.582	0.07595	56	Ba	N_4O_2		[104]
16458.941	0.07533	56	Ba	N_5O_3		[104]
16732.146	0.0741	24	Cr	abs. edge	M_1	[51]
16822.958	0.0737	78	Pt	N_6O_5		[104]
17035.6142	0.07278	13	Al	abs. edge	$L_{2,3}$	[51]
17125.028	0.0724	13	Al	$L_{2,3}M$		[50]
17220.167	0.072	29	Cu	$M_{2,3}M_{4,5}$		[50]
17340.587	0.0715	79	Au	O_2O_4		[50]
17611.535	0.0704	78	Pt	N_7O_5		[50]
17686.904	0.0701	35	Br	abs. edge	M_4	[51]
17968.870	0.069	35	Br	abs. edge	M_5	[51]
18206.344	0.0681	28	Ni	abs. edge	M_2	[51]
18368.178	0.0675	55	Cs	N_4O_3		[50]
18644.391	0.0665	23	V	abs. edge	M_1	[51]
18839.8724	0.06581	28	Ni	abs. edge	M_3	[493]
18859.933	0.06574	55	Cs	N_4O_2		[104]
18900.183	0.0656	78	Pt	O_2O_4		[104]
18957.982	0.0654	35	Br	M_4N_3	M_δ	[50]
19016.135	0.0652	55	Cs	N_5O_3		[104]
19045.346	0.0651	28	Ni	$M_{2,3}M_{4,5}$		[50]
19104.037	0.0649	35	Br	M_4N_2		[50]
19252.360	0.0644	35	Br	M_5N_3	M_{ξ_1}	[50]
19586.919	0.0633	11	Na	abs. edge	L_1	[51]
19965.411	0.0621	77	Ir	N_6O_5		[104]
20325.4426	0.061	27	Co	abs. edge	$M_{2,3}$	[493]
20561.393	0.0603	22	Ti	abs. edge	M_1	[51]
20802.886	0.0596	77	Ir	N_7O_5		[104]
20837.849	0.0595	27	Co	abs. edge	M_2	[51]
21376.759	0.058	27	Co	$M_{2,3}M_{4,5}$		[50]
21866.878	0.0567	34	Se	abs. edge	M_4	[51]
22061.423	0.0562	76	0s	O_2O_4		[104]
22645.6986	0.05475	3	Li	abs. edge	K	[51]
22778.8352	0.05443	34	Se	abs. edge	M_5	[51]
23045.576	0.0538	34	Se	M_5N_3	M_{ξ_1}	[50]
23045.576	0.0538	21	Sc	abs. edge	M_1	[51]
23174.804	0.0535	79	Au	$O_3O_{4,5}$		[104]
23526.603	0.0527	76	0s	N_6O_5		[104]
24168.655	0.0513	3	Li	$KL_{2,3}$	$K_{\alpha_{1,2}}$	[50]
24931.6710	0.04973	12	Mg	abs. edge	L_2	[51]
24310.823	0.051	26	Fe	$M_{2,3}M_{4,5}$		[50]

λ [pm]	E [keV]	Z	element	transition	notation	reference
24310.8235	0.051	26	Fe	abs. edge	$M_{2,3}$	[51]
24747.545	0.0501	76	Os	N_7O_5		[104]
25072.8413	0.04945	12	Mg	abs. edge	L_3	[51]
25098.219	0.0494	75	Re	O_2O_4		[104]
25149.128	0.0493	12	Mg	$L_{2,3}M$		[50]
25303.102	0.049	78	Pt	$O_3O_{4,5}$		[104]
25511.358	0.0486	25	Mn	abs. edge	M_2	[51]
25511.358	0.0486	25	Mn	abs. edge	M_3	[51]
27130.241	0.0457	77	Ir	$O_3O_{4,5}$		[104]
27369.801	0.0453	74	W	O_2O_4		[104]
27552.267	0.045	25	Mn	$M_{2,3}M_{4,5}$		[50]
27552.267	0.045	10	Ne	abs. edge	L_1	[51]
28371.899	0.0437	20	Ca	abs. edge	M_1	[51]
28968.508	0.0428	13	Al	$L_1L_{2,3}$		[50]
29172.988	0.0425	24	Cr	abs. edge	M_2	[51]
29172.988	0.0425	24	Cr	abs. edge	M_3	[51]
29380.379	0.0422	75	Re	N_4O_5		[50]
29450.166	0.0421	76	Os	$O_3O_{4,5}$		[50]
30093.495	0.0412	33	As	abs. edge	M_4	[51]
30240.293	0.041	73	Ta	O_2O_4		[50]
30314.423	0.0409	33	As	abs. edge	M_5	[493]
30996.300	0.04	24	Cr	$M_{2,3}M_{4,5}$		[50]
32627.684	0.038	79	Re	$O_3O_{4,5}$		[104]
32800.317	0.0378	23	V	abs. edge	M_2	[51]
33509.513	0.037	72	Hf	O_2O_4		[104]
33509.513	0.037	23	V	$M_{2,3}M_{4,5}$		[50]
35424.343	0.035	23	V	abs. edge	M_3	[493]
35833.873	0.0346	22	Ti	abs. edge	M_2	[51]
36573.805	0.0339	19	K	abs. edge	M_1	[51]
37121.317	0.0334	74	W	$O_3O_{4,5}$		[104]
37685.471	0.0329	71	Lu	O_2O_4		[104]
37915.963	0.0327	11	Na	L_1L_3		[104]
38388.511	0.0323	21	Sc	abs. edge	M_2	[51]
38388.511	0.0323	21	Sc	abs. edge	M_3	[51]
38504.720	0.0322	22	Ti	abs. edge	M_3	[493]
39611.885	0.0313	73	Ta	$O_3O_{4,5}$		[104]
39995.226	0.031	9	F	abs. edge	L_1	[51]
40518.0392	0.0306	11	Na	abs. edge	$L_{2,3}$	[51]
40650.885	0.0305	11	Na	$L_{2,3}M$		[104]
40784.605	0.0304	11	Na	$L_1L_{2,3}$		[50]
43200.418	0.0287	32	Ge	abs. edge	M_4	[51]
43200.418	0.0287	32	Ge	abs. edge	M_5	[51]

λ [pm]	E [keV]	Z	element	transition	notation	reference
44280.428	0.028	72	Hf	$O_3O_{4,5}$		[50]
48813.071	0.0254	20	Ca	abs. edge	M_2	[51]
48813.071	0.0254	20	Ca	abs. edge	M_3	[51]
49006.008	0.0253	18	Ar	abs. edge	M_1	[51]
49396.494	0.0251	71	Lu	$O_3O_{4,5}$		[50]
50400.488	0.0246	2	He	abs. edge	K	[51]
52314.430	0.0237	8	O	abs. edge	L_1	[51]
53212.532	0.0233	73	Ta	N_6O_5		[50]
53906.609	0.023	20	Ca	$M_{2,3}N_1$		[50]
59040.571	0.021	73	Ta	N_7O_5		[50]
67751.475	0.0183	10	Ne	abs. edge	L_2	[51]
67751.475	0.0183	10	Ne	abs. edge	L_3	[51]
69265.475	0.0179	19	K	$M_{2,3}N_1$		[50]
69654.607	0.0178	19	K	abs. edge	M_2	[51]
69654.607	0.0178	19	K	abs. edge	M_3	[51]
71255.862	0.0174	31	Ga	abs. edge	M_4	[51]
71255.862	0.0174	31	Ga	abs. edge	M_5	[51]
76534.074	0.0162	15	P	abs. edge	M_1	[51]
78471.645	0.0158	16	S	abs. edge	M_1	[51]
91165.588	0.0136	1	H	abs. edge	K	[51]
99988.064	0.0124	18	Ar	abs. edge	M_2	[51]
99988.064	0.0124	18	Ar	abs. edge	M_3	[51]
129151.25	0.0096	15	P	abs. edge	M_2	[51]
134766.52	0.0092	7	N	abs. edge	L_2	[51]
134766.52	0.0092	7	N	abs. edge	L_3	[51]
144168.84	0.0086	9	F	abs. edge	L_2	[51]
144168.84	0.0086	9	F	abs. edge	L_3	[51]
151201.46	0.0082	16	S	abs. edge	M_3	[51]
153068.15	0.0081	30	Zn	abs. edge	M_4	[51]
153068.15	0.0081	30	Zn	abs. edge	M_5	[51]
158955.38	0.0078	16	S	abs. edge	M_2	[51]
163138.42	0.0076	14	Si	abs. edge	M_1	[51]
174627.04	0.0071	8	O	abs. edge	L_2	[51]
174627.04	0.0071	8	O	abs. edge	L_3	[51]
182331.18	0.0068	17	Cl	abs. edge	M_2	[51]
182331.18	0.0068	17	Cl	abs. edge	M_3	[51]
187856.36	0.0066	21	Sc	abs. edge	M_4	[51]
187856.36	0.0066	21	Sc	abs. edge	M_5	[51]
193726.87	0.0064	6	C	abs. edge	L_2	[51]
193726.87	0.0064	6	C	abs. edge	L_3	[51]
263798.30	0.0047	5	B	abs. edge	$L_{2,3}$	[51]
335095.13	0.0037	22	Ti	abs. edge	$M_{4,5}$	[51]

λ [pm]	E [keV]	Z	element	transition	notation	reference
344403.33	0.0036	28	Ni	abs. edge	M_4	[51]
344403.33	0.0036	26	Fe	abs. edge	$M_{4,5}$	[51]
375712.73	0.0033	27	Co	abs. edge	M_5	[493]
375712.73	0.0033	25	Mn	abs. edge	$M_{4,5}$	[51]
375712.73	0.0033	28	Ni	abs. edge	M_5	[493]
413284.00	0.003	14	Si	abs. edge	M_2	[51]
427535.17	0.0029	27	Co	abs. edge	M_4	[51]
539066.09	0.0023	24	Cr	abs. edge	M_4	[51]
563569.09	0.0022	24	Cr	abs. edge	M_5	[493]
563569.09	0.0022	23	V	abs. edge	M_4	[51]
590405.71	0.0021	12	Mg	abs. edge	M_1	[493]
774907.50	0.0016	29	Cu	abs. edge	$M_{4,5}$	[51]
1771217.1	0.0007	13	Al	abs. edge	M_1	[493]
1771217.1	0.0007	11	Na	abs. edge	M_1	[493]

7 K-Shell Intensity Ratios and K-Vacancy Decay Rates

In the following total K-vacancy decay rates and intensity ratios for the X-ray K-series as calculated by [489] are tabulated. The total rates *total* are given in eV/$\hbar$.

Z	total	K_β/K_α	$K_{\alpha_2}/K_{\alpha_1}$	K_{β_3}/K_{β_1}	$K_{\beta_1'}/K_{\alpha_1}$	$K_{\beta_2'}/K_{\alpha_1}$
14	0.0238	0.0294	0.5037	0.5052	0.0443	
16	0.0461	0.0659	0.5053	0.5047	0.0992	
18	0.0799	0.1088	0.5049	0.5041	0.1638	
20	0.1312	0.1315	0.5061	0.5043	0.1982	
22	0.2020	0.1355	0.5076	0.5054	0.2043	
24	0.2970	0.1337	0.5091	0.5070	0.2018	
26	0.424	0.1391	0.5107	0.5079	0.2102	
28	0.586	0.1401	0.5124	0.5093	0.2119	
30	0.790	0.1410	0.5142	0.5108	0.2135	
32	1.051	0.1504	0.5149	0.5105	0.2229	0.0049
34	1.375	0.1624	0.5158	0.5116	0.2331	0.0131
36	1.766	0.1727	0.5186	0.5111	0.2381	0.0240
38	2.239	0.1831	0.5205	0.5115	0.2463	0.0320
40	2.800	0.1913	0.5225	0.5120	0.2543	0.0370
42	3.46	0.1981	0.5247	0.5125	0.2617	0.0403
44	4.23	0.2045	0.5269	0.5130	0.2684	0.0439
46	5.12	0.2100	0.5293	0.5135	0.2745	0.0466
48	6.14	0.2161	0.5317	0.5142	0.2802	0.0507
50	7.32	0.2230	0.5343	0.5148	0.2857	0.0564
52	8.67	0.2299	0.5370	0.5153	0.2906	0.0628
54	10.19	0.2368	0.5398	0.5157	0.2951	0.0695

Z	total	K_β/K_α	$K_{\alpha_2}/K_{\alpha_1}$	K_{β_3}/K_{β_1}	$K_{\beta_1'}K_{\alpha_1}$	$K_{\beta_2'}/K_{\alpha_1}$
56	11.91	0.2433	0.5428	0.5160	0.2997	0.0756
58	13.81	0.2470	0.5459	0.5164	0.3042	0.0777
60	15.93	0.2504	0.5491	0.5167	0.3086	0.0792
62	18.29	0.2534	0.5524	0.5169	0.3127	0.0807
64	20.89	0.2570	0.5559	0.5171	0.3166	0.0832
66	23.75	0.2588	0.5596	0.5173	0.3203	0.0832
68	26.88	0.2612	0.5634	0.5174	0.3240	0.0843
70	30.32	0.2634	0.5673	0.5175	0.3274	0.0853
72	34.09	0.2666	0.5714	0.5176	0.3307	0.0883
74	38.20	0.2698	0.5757	0.5176	0.3338	0.0913
76	42.66	0.2728	0.5802	0.5175	0.3369	0.0943
78	47.50	0.2758	0.5850	0.5173	0.3399	0.0972
80	52.72	0.2788	0.5899	0.5170	0.3430	0.1004
82	58.37	0.2821	0.5950	0.5165	0.3459	0.1043
84	64.46	0.2855	0.6004	0.5161	0.3489	0.1084
86	71.00	0.2890	0.6061	0.5154	0.3518	0.1126
88	78.00	0.2923	0.6120	0.5146	0.3547	0.1167
90	85.47	0.2952	0.6182	0.5134	0.3577	0.1205
92	93.41	0.2975	0.6247	0.5122	0.3606	0.1233
94	101.89	0.2996	0.6315	0.5107	0.3636	0.1259
96	110.81	0.3019	0.6387	0.5090	0.3665	0.1290
98	120.29	0.3037	0.6462	0.5070	0.3695	0.1315

Z	$K_{\alpha_3}/K_{\alpha_1}$	K_{β_1}/K_{α_1}	K_{β_2}/K_{β_1}	K_{β_5}/K_{β_1}	K_{β_4}/K_{β_1}	$K\text{-}O_{2,3}/K_{\alpha_1}$
14		0.0279				
16		0,0619				
18		0.1025				
20		0.1262				
22		0.1304		0.0001		
24		0.1289		0.0003		
26		0.1348		0.0005		
28		0.1361		0.0009		
30		0.1372		0.0015		
32		0.1436	0.0330	0.0020		
34		0.1504	0.0834	0.0026		
36		0.1537	0.1493	0.0033		
38	0.00001	0.1592	0.1938	0.0040		
40	0.00001	0.1646	0.2164	0.0047	0.0001	
42	0.00001	0.1695	0.2281	0.0055	0.0002	
44	0.00001	0.1740	0.2423	0.0064	0.0004	
46	0.00002	0.1781	0.2510	0.0072	0.0007	
48	0.00002	0.1819	0.2668	0.0082	0.0010	
50	0.00003	0.1854	0.2844	0.0092	0.0013	0.0015
52	0.00004	0.1885	0.2991	0.0102	0.0017	0.0041
54	0.00005	0.1914	0.3104	0.0113	0.0021	0.0075
56	0.00007	0.1943	0.3239	0.0124	0.0025	0.0100
58	0.00009	0.1969	0.3305	0.0136	0.0029	0.0100
60	0.00011	0.1994	0.3355	0.0148	0.0032	0.0098
62	0.00014	0.2019	0.3393	0.0161	0.0035	0.0097
64	0.00017	0.2044	0.3440	0.0174	0.0039	0.0104
66	0.00021	0.2066	0.3446	0.0187	0.0042	0.0096
68	0.00026	0.2087	0.3467	0.0201	0.0046	0.0094
70	0.00031	0.2108	0.3485	0.0215	0.0049	0.0094
72	0.00038	0.2130	0.3526	0.0228	0.0054	0.0105
74	0.00046	0.2150	0.3574	0.0242	0.0058	0.0116

Z	$K_{\alpha_3}/K_{\alpha_1}$	K_{β_1}/K_{α_1}	K_{β_2}/K_{β_1}	K_{β_5}/K_{β_1}	K_{β_4}/K_{β_1}	$K\text{-}O_{2,3}/K_{\alpha_1}$
76	0.00056	0.2169	0.3624	0.0256	0.0063	0.0127
78	0.00067	0.2188	0.3677	0.0270	0.0069	0.0135
80	0.00081	0.2206	0.3728	0.0285	0.0074	0.0147
82	0.00097	0.2224	0.3779	0.0299	0.0079	0.0160
84	0.00116	0.2241	0.3822	0.0314	0.0085	0.0174
86	0.00139	0.2258	0.3866	0.0328	0.0091	0.0186
88	0.00165	0.2277	0.3908	0.0342	0.0097	0.0198
90	0.00197	0.2295	0.3951	0.0356	0.0103	0.0209
92	0.00234	0.2312	0.3992	0.0370	0.0109	0.0219
94	0.00278	0.2331	0.4029	0.0384	0.0115	0.0229
96	0.00330	0.2349	0.4063	0.0398	0.0121	0.0238
98	0.00391	0.2368	0.4095	0.0411	0.0127	0.0246

8 Atomic Scattering Factors

Given are mean atomic scattering factors (in electrons) for free atoms and chemical significant ions (after [129])[1].

The intervals of $\sin \vartheta / \lambda$ were were selected in a way that a linear interpolation between the tabulated values is allowed. For carbon and silicon valence state scattering facors are given additionally (indicated by C_{val} and Si_{val}).

---

[1] With IUCR copyright permission

$\sin\vartheta/\lambda$	$_1$H	$_1$H^{1-}	$_2$He	$_3$Li	$_3$Li^{1+}	$_4$Be	$_4$Be^{2+}	$_5$B
0.00	1.000	2.000	2.000	3.000	2.000	4.000	2.000	5.000
0.01	0.998	1.986	1.998	2.986	1.999	3.987	2.000	4.988
0.02	0.991	1.946	1.993	2.947	1.997	3.950	1.999	4.954
0.03	0.980	1.883	1.984	2.884	1.994	3.889	1.997	4.897
0.04	0.966	1.802	1.972	2.802	1.990	3.807	1.995	4.820
0.05	0.947	1.708	1.957	2.708	1.984	3.707	1.992	4.724
0.06	0.925	1.606	1.939	2.606	1.977	3.592	1.988	4.613
0.07	0.900	1.501	1.917	2.502	1.968	3.468	1.983	4.488
0.08	0.872	1.396	1.893	2.400	1.959	3.336	1.978	4.352
0.09	0.842	1.293	1.866	2.304	1.948	3.201	1.973	4.209
0.10	0.811	1.195	1.837	2.215	1.936	3.065	1.966	4.060
0.11	0.778	1.102	1.806	2.135	1.923	2.932	1.959	3.908
0.12	0.744	1.014	1.772	2.065	1.909	2.804	1.952	3.756
0.13	0.710	0.933	1.737	2.004	1.894	2.683	1.944	3.606
0.14	0.676	0.858	1.701	1.950	1.878	2.569	1.935	3.459
0.15	0.641	0.789	1.663	1.904	1.861	2.463	1.925	3.316
0.16	0.608	0.725	1.624	1.863	1.843	2.365	1.915	3.179
0.17	0.574	0.667	1.584	1.828	1.824	2.277	1.905	3.048
0.18	0.542	0.613	1.543	1.796	1.804	2.197	1.894	2.924
0.19	0.511	0.565	1.502	1.768	1.784	2.125	1.882	2.808
0.20	0.481	0.520	1.460	1.742	1.762	2.060	1.870	2.699
0.22	0.424	0.442	1.377	1.693	1.718	1.951	1.845	2.503
0.24	0.373	0.377	1.295	1.648	1.672	1.864	1.817	2.336
0.25	0.350	0.348	1.254	1.626	1.648	1.828	1.803	2.263
0.26	0.328	0.322	1.214	1.604	1.624	1.795	1.788	2.195
0.28	0.287	0.277	1.136	1.559	1.574	1.739	1.758	2.077
0.30	0.251	0.238	1.060	1.513	1.523	1.692	1.726	1.979
0.32	0.220	0.206	0.988	1.465	1.472	1.652	1.693	1.897
0.34	0.193	0.178	0.920	1.417	1.420	1.616	1.659	1.829
0.35	0.180	0.166	0.887	1.393	1.394	1.600	1.641	1.799
0.36	0.169	0.155	0.856	1.369	1.368	1.583	1.623	1.771
0.38	0.148	0.136	0.795	1.320	1.317	1.551	1.588	1.723
0.40	0.130	0.119	0.738	1.270	1.266	1.520	1.551	1.681
0.42	0.115	0.104	0.686	1.221	1.215	1.489	1.514	1.644
0.44	0.101	0.092	0.636	1.173	1.166	1.458	1.477	1.611
0.45	0.095	0.086	0.613	1.149	1.141	1.443	1.458	1.596
0.46	0.090	0.081	0.591	1.125	1.117	1.427	1.439	1.581
0.48	0.079	0.072	0.548	1.078	1.070	1.395	1.402	1.553
0.50	0.071	0.064	0.509	1.033	1.024	1.362	1.364	1.526
0.55	0.053	0.048	0.423	0.924	0.915	1.279	1.271	1.463
0.60	0.040	0.037	0.353	0.823	0.815	1.195	1.179	1.402
0.65	0.031	0.028	0.295	0.732	0.724	1.112	1.091	1.339
0.70	0.024	0.022	0.248	0.650	0.643	1.030	1.007	1.276
0.80	0.015	0.014	0.177	0.512	0.507	0.876	0.853	1.147
0.90	0.010	0.009	0.129	0.404	0.400	0.740	0.718	1.020
1.00	0.007	0.006	0.095	0.320	0.317	0.622	0.603	0.900
1.10	0.005	0.005	0.072	0.255	0.252	0.522	0.506	0.790
1.20	0.003	0.003	0.055	0.205	0.202	0.439	0.425	0.690
1.30	0.003	0.002	0.042	0.165	0.163	0.369	0.357	0.602
1.40	0.002	0.002	0.033	0.134	0.133	0.311	0.301	0.524
1.50	0.001	0.001	0.026	0.110	0.109	0.263	0.255	0.457
1.60			0.021	0.091	0.090	0.223	0.216	0.398
1.70			0.017	0.075	0.075	0.190	0.184	0.347
1.80			0.014	0.063	0.062	0.163	0.157	0.304
1.90			0.011	0.053	0.052	0.139	0.135	0.266
2.00			0.010	0.044	0.044	0.120	0.116	0.233

$\sin\vartheta/\lambda$	$_6$C	$_6$C$_{val}$	$_7$N	$_8$O	$_8$O^{1-}	$_9$F	$_9$F^{1-}	$_{10}$Ne
0.00	6.000	6.000	7.000	8.000	9.000	9.000	10.000	10.000
0.01	5.990	5.989	6.991	7.992	8.986	8.993	9.988	9.993
0.02	5.958	5.956	6.963	7.967	8.945	8.970	9.953	9.973
0.03	5.907	5.903	6.918	7.926	8.878	8.933	9.895	9.938
0.04	5.837	5.829	6.855	7.869	8.785	8.881	9.816	9.891
0.05	5.749	5.738	6.776	7.798	8.670	8.815	9.716	9.830
0.06	5.645	5.629	6.682	7.712	8.534	8.736	9.597	9.757
0.07	5.526	5.507	6.574	7.612	8.381	8.645	9.461	9.672
0.08	5.396	5.372	6.453	7.501	8.211	8.541	9.309	9.576
0.09	5.255	5.227	6.321	7.378	8.029	8.427	9.144	9.469
0.10	5.107	5.074	6.180	7.245	7.836	8.302	8.967	9.351
0.11	4.952	4.916	6.030	7.103	7.635	8.168	8.781	9.225
0.12	4.794	4.754	5.875	6.954	7.429	8.026	8.586	9.090
0.13	4.633	4.591	5.714	6.798	7.218	7.876	8.386	8.948
0.14	4.472	4.428	5.551	6.637	7.005	7.721	8.181	8.799
0.15	4.311	4.267	5.385	6.472	6.792	7.560	7.973	8.643
0.16	4.153	4.109	5.218	6.304	6.579	7.395	7.762	8.483
0.17	3.998	3.954	5.051	6.134	6.368	7.226	7.551	8.318
0.18	3.847	3.805	4.886	5.964	6.160	7.055	7.341	8.150
0.19	3.701	3.661	4.723	5.793	5.956	6.883	7.131	7.978
0.20	3.560	3.523	4.563	5.623	5.756	6.709	6.924	7.805
0.22	3.297	3.266	4.254	5.289	5.371	6.362	6.517	7.454
0.24	3.058	3.035	3.963	4.965	5.008	6.020	6.126	7.102
0.25	2.949	2.930	3.825	4.808	4.836	5.851	5.937	6.928
0.26	2.846	2.831	3.693	4.655	4.670	5.685	5.753	6.754
0.28	2.658	2.651	3.445	4.363	4.357	5.363	5.399	6.412
0.30	2.494	2.495	3.219	4.089	4.068	5.054	5.067	6.079
0.32	2.351	2.358	3.014	3.834	3.804	4.761	4.756	5.758
0.34	2.227	2.241	2.831	3.599	3.564	4.484	4.467	5.451
0.35	2.171	2.188	2.747	3.489	3.452	4.353	4.330	5.302
0.36	2.120	2.139	2.667	3.383	3.345	4.225	4.199	5.158
0.38	2.028	2.050	2.522	3.186	3.147	3.983	3.951	4.880
0.40	1.948	1.974	2.393	3.006	2.969	3.759	3.724	4.617
0.42	1.880	1.907	2.278	2.844	2.808	3.551	3.514	4.370
0.44	1.821	1.849	2.178	2.697	2.663	3.360	3.322	4.139
0.45	1.794	1.822	2.132	2.629	2.597	3.270	3.233	4.029
0.46	1.770	1.798	2.089	2.564	2.533	3.183	3.147	3.923
0.48	1.725	1.752	2.011	2.445	2.417	3.022	2.987	3.722
0.50	1.685	1.711	1.942	2.338	2.313	2.874	2.841	3.535
0.55	1.603	1.624	1.802	2.115	2.097	2.559	2.531	3.126
0.60	1.537	1.552	1.697	1.946	1.934	2.309	2.288	2.790
0.65	1.479	1.488	1.616	1.816	1.808	2.112	2.096	2.517
0.70	1.426	1.428	1.551	1.714	1.710	1.956	1.945	2.296
0.80	1.322	1.315	1.445	1.568	1.567	1.735	1.729	1.971
0.90	1.219	1.204	1.353	1.463	1.463	1.588	1.585	1.757
1.00	1.114	1.096	1.265	1.377	1.376	1.482	1.481	1.609
1.10	1.012	0.992	1.177	1.298	1.296	1.398	1.397	1.502
1.20	0.914	0.894	1.090	1.221	1.219	1.324	1.322	1.418
1.30	0.822	0.802	1.004	1.145	1.143	1.254	1.252	1.346
1.40	0.736	0.718	0.921	1.070	1.067	1.186	1.184	1.280
1.50	0.659	0.642	0.843	0.997	0.994	1.120	1.117	1.218
1.60	0.588		0.769	0.926		1.055		1.158
1.70	0.525		0.700	0.857		0.990		1.099
1.80	0.468		0.636	0.792		0.928		1.041
1.90	0.418		0.578	0.731		0.868		0.984
2.00	0.373		0.525	0.674		0.810		0.929

$\sin\vartheta/\lambda$	$_{11}$Na	$_{11}$Na^{1+}	$_{12}$Mg	$_{12}$Mg^{2+}	$_{13}$Al	$_{13}$Al^{3+}	$_{14}$Si	$_{14}$Si$_{val}$
0.00	11.000	10.000	12.000	10.000	13.000	10.000	14.000	14.000
0.01	10.980	9.995	11.978	9.997	12.976	9.997	13.976	13.973
0.02	10.922	9.981	11.914	9.986	12.903	9.989	13.904	13.894
0.03	10.830	9.958	11.811	9.969	12.786	9.976	13.787	13.766
0.04	10.709	9.925	11.674	9.945	12.629	9.957	13.628	13.593
0.05	10.568	9.883	11.507	9.914	12.439	9.933	13.434	13.381
0.06	10.412	9.833	11.319	9.876	12.222	9.904	13.209	13.138
0.07	10.249	9.773	11.116	9.832	11.987	9.870	12.961	12.870
0.08	10.084	9.705	10.903	9.782	11.739	9.831	12.695	12.586
0.09	9.920	9.630	10.687	9.725	11.485	9.787	12.417	12.293
0.10	9.760	9.546	10.472	9.662	11.230	9.738	12.134	11.995
0.11	9.605	9.455	10.262	9.594	10.978	9.684	11.849	11.700
0.12	9.455	9.357	10.059	9.519	10.733	9.625	11.567	11.410
0.13	9.309	9.253	9.864	9.440	10.498	9.563	11.292	11.130
0.14	9.166	9.142	9.678	9.355	10.273	9.495	11.025	10.862
0.15	9.027	9.026	9.502	9.265	10.059	9.424	10.769	10.608
0.16	8.888	8.904	9.334	9.171	9.857	9.349	10.525	10.368
0.17	8.751	8.777	9.175	9.072	9.667	9.270	10.293	10.143
0.18	8.613	8.647	9.023	8.969	9.487	9.187	10.074	9.933
0.19	8.475	8.512	8.876	8.862	9.318	9.101	9.868	9.737
0.20	8.335	8.374	8.735	8.751	9.158	9.011	9.673	9.553
0.22	8.052	8.089	8.465	8.521	8.862	8.823	9.319	9.222
0.24	7.764	7.795	8.205	8.280	8.592	8.623	9.004	8.931
0.25	7.618	7.646	8.078	8.156	8.465	8.520	8.859	8.798
0.26	7.471	7.496	7.951	8.030	8.341	8.414	8.722	8.671
0.28	7.176	7.195	7.698	7.774	8.103	8.198	8.467	8.435
0.30	6.881	6.894	7.446	7.513	7.873	7.975	8.231	8.214
0.32	6.588	6.597	7.194	7.251	7.648	7.747	8.011	8.005
0.34	6.298	6.304	6.943	6.987	7.426	7.515	7.800	7.803
0.35	6.156	6.160	6.817	6.856	7.316	7.399	7.698	7.704
0.36	6.015	6.018	6.691	6.725	7.205	7.282	7.597	7.606
0.38	5.739	5.739	6.442	6.465	6.985	7.047	7.398	7.410
0.40	5.471	5.471	6.194	6.210	6.766	6.813	7.202	7.215
0.42	5.214	5.212	5.951	5.959	6.548	6.581	7.008	7.021
0.44	4.967	4.964	5.712	5.715	6.330	6.350	6.815	6.826
0.45	4.848	4.845	5.595	5.595	6.222	6.237	6.719	6.729
0.46	4.731	4.728	5.480	5.477	6.115	6.124	6.622	6.632
0.48	4.506	4.503	5.253	5.247	5.902	5.901	6.431	6.437
0.50	4.293	4.290	5.034	5.025	5.692	5.683	6.240	6.244
0.55	3.811	3.808	4.520	4.508	5.186	5.162	5.769	5.766
0.60	3.398	3.395	4.059	4.046	4.713	4.681	5.312	5.303
0.65	3.048	3.046	3.652	3.641	4.277	4.243	4.878	4.865
0.70	2.754	2.753	3.297	3.288	3.883	3.851	4.470	4.455
0.80	2.305	2.305	2.729	2.724	3.221	3.195	3.750	3.734
0.90	1.997	1.997	2.311	2.315	2.712	2.693	3.164	3.150
1.00	1.784	1.785	2.022	2.023	2.330	2.319	2.702	2.691
1.10	1.634	1.635	1.812	1.813	2.049	2.041	2.346	2.338
1.20	1.524	1.524	1.660	1.662	1.841	1.837	2.076	2.069
1.30	1.438	1.438	1.546	1.548	1.687	1.685	1.872	1.867
1.40	1.367	1.367	1.459	1.460	1.571	1.570	1.717	1.713
1.50	1.304	1.304	1.387	1.388	1.481	1.479	1.598	1.595
1.60	1.247	1.246	1.326	1.326	1.408		1.505	
1.70	1.191	1.191	1.270	1.270	1.346		1.430	
1.80	1.137	1.137	1.219	1.218	1.292		1.367	
1.90	1.084	1.084	1.169	1.168	1.243		1.313	
2.00	1.032	1.032	1.120	1.119	1.195		1.264	

$\sin\vartheta/\lambda$	$_{14}Si^{4+}$	$_{15}P$	$_{16}S$	$_{17}Cl$	$_{17}Cl^{1-}$	$_{18}Ar$	$_{19}K$	$_{19}K^{1+}$
0.00	10.000	15.000	16.000	17.000	18.000	18.000	19.000	18.000
0.01	9.998	14.977	15.979	16.980	17.972	17.981	18.963	17.986
0.02	9.991	14.909	15.915	16.919	17.888	17.924	18.854	17.943
0.03	9.981	14.798	15.809	16.820	17.751	17.830	18.683	17.872
0.04	9.966	14.646	15.665	16.683	17.563	17.700	18.462	17.774
0.05	9.947	14.458	15.484	16.511	17.330	17.536	18.204	17.649
0.06	9.924	14.237	15.271	16.306	17.057	17.340	17.924	17.499
0.07	9.896	13.990	15.030	16.073	16.750	17.116	17.630	17.325
0.08	9.865	13.721	14.764	15.814	16.415	16.865	17.332	17.129
0.09	9.829	13.435	14.478	15.533	16.058	16.591	17.032	16.912
0.10	9.790	13.138	14.177	15.234	15.685	16.298	16.733	16.677
0.11	9.747	12.834	13.865	14.921	15.301	15.988	16.436	16.426
0.12	9.700	12.527	13.546	14.597	14.911	15.665	16.138	16.160
0.13	9.649	12.223	13.224	14.266	14.519	15.331	15.841	15.882
0.14	9.595	11.922	12.902	13.932	14.130	14.991	15.543	15.594
0.15	9.537	11.629	12.583	13.597	13.747	14.647	15.243	15.297
0.16	9.476	11.345	12.270	13.263	13.371	14.301	14.941	14.994
0.17	9.411	11.072	11.964	12.934	13.006	13.957	14.638	14.688
0.18	9.343	10.811	11.668	12.611	12.653	13.615	14.334	14.378
0.19	9.272	10.563	11.382	12.297	12.313	13.279	14.031	14.069
0.20	9.199	10.327	11.109	11.991	11.987	12.949	13.728	13.760
0.22	9.043	9.894	10.598	11.413	11.379	12.315	13.130	13.150
0.24	8.877	9.510	10.138	10.881	10.832	11.721	12.550	12.560
0.25	8.790	9.335	9.927	10.633	10.580	11.441	12.268	12.275
0.26	8.701	9.170	9.727	10.398	10.343	11.172	11.994	11.997
0.28	8.518	8.869	9.363	9.964	9.908	10.671	11.468	11.467
0.30	8.327	8.600	9.039	9.576	9.524	10.216	10.977	10.972
0.32	8.131	8.357	8.752	9.231	9.184	9.807	10.521	10.515
0.34	7.929	8.134	8.494	8.923	8.884	9.441	10.103	10.097
0.35	7.821	8.029	8.376	8.782	8.746	9.272	9.908	9.901
0.36	7.724	7.928	8.262	8.649	8.616	9.113	9.722	9.715
0.38	7.516	7.133	8.051	8.403	8.311	8.820	9.375	9.369
0.40	7.306	7.547	7.856	8.181	8.162	8.558	9.061	9.056
0.42	7.095	7.367	7.673	7.979	7.965	8.322	8.778	8.773
0.44	6.884	7.190	7.501	7.794	7.785	8.110	8.522	8.518
0.45	6.779	7.103	7.417	7.706	7.699	8.011	8.403	8.399
0.46	6.674	7.017	7.335	7.621	7.616	7.917	8.290	8.287
0.48	6.465	6.845	7.174	7.459	7.451	7.739	8.080	8.077
0.50	6.259	6.674	7.017	7.305	7.305	7.575	7.889	7.886
0.55	5.755	6.250	6.633	6.941	6.945	7.207	7.474	7.474
0.60	5.277	5.829	6.254	6.595	6.600	6.815	7.125	7.125
0.65	4.830	5.418	5.877	6.254	6.259	6.560	6.814	6.814
0.70	4.418	5.020	5.505	5.915	5.920	6.252	6.523	6.523
0.80	3.701	4.284	4.790	5.245	5.248	5.639	5.961	5.962
0.90	3.124	3.649	4.138	4.607	4.608	5.036	5.406	5.406
1.00	2.673	3.122	3.570	4.023	4.024	4.460	4.859	4.859
1.10	2.326	2.698	3.092	3.509	3.509	3.931	4.337	4.336
1.20	2.063	2.364	2.699	3.070	3.070	3.462	3.855	3.854
1.30	1.864	2.104	2.384	2.704	2.705	3.056	3.423	3.423
1.40	1.712	1.903	2.133	2.405	2.405	2.713	3.045	3.045
1.50	1.595	1.747	1.935	2.162	2.162	2.421	2.722	2.122
1.60		1.626	1.779	1.967	1.968	2.192	2.450	2.449
1.10		1.530	1.655	1.811	1.811	2.000	2.221	2.221
1.80		1.453	1.557	1.686	1.686	1.844	2.033	2.033
1.90		1.389	1.477	1.585	1.585	1.717	1.876	1.877
2.00		1.333	1.411	1.502	1.502	1.614	1.748	1.749

$\sin\vartheta/\lambda$	$_{20}$Ca	$_{20}$Ca^{2+}	$_{21}$Sc	$_{21}$Sc^{3+}	$_{22}$Ti	$_{22}$Ti^{2+}	$_{22}$Ti^{3+}	$_{22}$Ti^{4+}
0.00	20.000	18.000	21.000	18.000	22.000	20.000	19.000	18.000
0.01	19.959	17.989	20.962	17.991	21.964	19.988	18.990	17.992
0.02	19.838	17.955	20.848	17.963	21.856	19.951	18.962	17.969
0.03	19.645	17.899	20.665	17.917	21.682	19.891	18.914	17.930
0.04	19.392	17.821	20.422	17.853	21.451	19.807	18.848	17.877
0.05	19.091	17.121	20.131	17.771	21.171	19.701	18.764	17.808
0.06	18.758	17.601	19.805	17.672	20.854	19.572	18.662	17.725
0.07	18.405	17.462	19.455	17.556	20.511	19.423	18.543	17.628
0.08	18.045	17.303	19.091	17.424	20.150	19.253	18.407	17.516
0.09	17.685	17.127	18.723	17.278	19.781	19.065	18.255	17.392
0.10	17.331	16.935	18.356	17.116	19.410	18.860	18.089	17.255
0.11	16.981	16.727	17.995	16.941	19.041	18.639	17.909	17.106
0.12	16.655	16.506	17.643	16.754	18.678	18.404	17.716	16.946
0.13	16.334	16.272	17.301	16.555	18.322	18.156	17.510	16.775
0.14	16.024	16.028	16.968	16.345	17.974	17.896	17.294	16.593
0.15	15.723	15.774	16.645	16.126	17.635	17.626	17.067	16.403
0.16	15.430	15.512	16.330	15.898	17.304	17.348	16.832	16.205
0.17	15.142	15.244	16.023	15.662	16.980	17.062	16.589	15.998
0.18	14.859	14.970	15.722	15.421	16.663	16.771	16.339	15.785
0.19	14.580	14.692	15.426	15.173	16.351	16.475	16.083	15.566
0.20	14.304	14.412	15.135	14.922	16.044	16.176	15.822	15.342
0.22	13.760	13.850	14.564	14.410	15.444	15.574	15.291	14.881
0.24	13.225	13.292	14.006	13.893	14.859	14.972	14.752	14.408
0.25	12.961	13.017	13.732	13.634	14.572	14.673	14.482	14.170
0.26	12.701	12.745	13.462	13.377	14.289	14.377	14.213	13.930
0.28	12.194	12.217	12.933	12.869	13.735	13.797	13.680	13.452
0.30	11.705	11.713	12.423	12.374	13.198	13.236'	13.157	12.979
0.32	11.240	11.235	11.934	11.896	12.682	12.697	12.650	12.515
0.34	10.800	10.787	11.467	11.438	12.187	12.184	12.162	12.064
0.35	10.590	10.575	11.244	11.218	11.949	11.938	11.926	11.844
0.36	10.388	10.370	11.027	11.004	11.717	11.698	11.696	11.628
0.38	10.004	9.984	10.613	10.595	11.271	11.242	11.254	11.211
0.40	9.650	9.629	10.226	10.212	10.852	10.815	10.837	10.815
0.42	9.324	9.303	9.866	9.855	10.459	10.417	10.446	10.439
0.44	9.025	9.006	9.534	9.524	10.093	10.047	10.080	10.086
0.45	8.885	8.867	9.377	9.368	9.920	9.873	9.907	9.917
0.46	8.752	8.734	9.227	9.218	9.753	9.706	9.740	9.754
0.48	8.502	8.487	8.946	8.937	9.438	9.391	9.426	9.445
0.50	8.275	8.262	8.687	8.678	9.148	9.102	9.135	9.158
0.55	7.788	7.781	8.132	8.121	8.518	8.477	8.503	8.529
0.60	7.392	7.389	7.682	7.670	8.007	7.972	7.990	8.012
0.65	7.057	7.058	7.312	7.298	7.588	7.560	7.571	7.588
0.70	6.762	6.764	6.996	6.982	7.240	7.216	7.222	7.234
0.80	6.228	6.231	6.460	6.445	6.676	6.656	6.658	6.664
0.90	5.717	5.719	5.975	5.961	6.200	6.179	6.182	6.189
1.00	5.209	5.209	5.501	5.488	5.752	5.728	5.734	5.745
1.10	4.710	4.710	5.030	5.017	5.310	5.282	5.291	5.306
1.20	4.233	4.232	4.570	4.556	4.872	4.840	4.852	4.870
1.30	3.791	3.790	4.131	4.115	4.445	4.411	4.425	4.443
1.40	3.391	3.390	3.722	3.706	4.038	4.004	4.017	4.035
1.50	3.039	3.038	3.352	3.335	3.660	3.626	3.638	3.655
1.60	2.733	2.732	3.023		3.316			
1.70	2.470	2.470	2.733		3.006			
1.80	2.250	2.250	2.485		2.734			
1.90	2.063	2.064	2.271		2.496			
2.00	1.908	1.909	2.090		2.290			

$\sin\vartheta/\lambda$	$_{23}$V	$_{23}$V^{2+}	$_{23}$V^{3+}	$_{23}$V^{5+}	$_{24}$Cr	$_{24}$Cr^{2+}	$_{24}$Cr^{3+}	$_{25}$Mn
0.00	23.000	21.000	20.000	18.000	24.000	22.000	21.000	25.000
0.01	22.966	20.988	19.990	17.993	23.971	21.988	20.990	24.969
0.02	22.864	20.952	19.961	17.974	23.885	21.952	20.961	24.876
0.03	22.698	20.892	19.913	17.941	23.746	21.892	20.913	24.726
0.04	22.477	20.808	19.846	17.895	23.558	21.808	20.845	24.523
0.05	22.208	20.702	19.760	17.837	23.329	21.702	20.759	24.274
0.06	21.902	20.573	19.657	17.766	23.065	21.574	20.655	23.988
0.07	21.567	20.424	19.536	17.682	22.772	21.425	20.534	23.671
0.08	21.212	20.255	19.398	17.587	22.459	21.256	20.395	23.331
0.09	20.846	20.066	19.244	17.480	22.129	21.067	20.240	22.976
0.10	20.474	19.861	19.075	17.362	21.789	20.861	20.069	22.611
0.11	20.102	19.639	18.892	17.234	21.441	20.638	19.884	22.240
0.12	19.733	19.402	18.695	17.095	21.089	20.400	19.685	21.868
0.13	19.369	19.152	18.485	16.946	20.734	20.148	19.474	21.497
0.14	19.011	18.890	18.265	16.789	20.378	19.884	19.250	21.128
0.15	18.661	18.618	18.033	16.622	20.022	19.609	19.016	20.764
0.16	18.317	18.336	17.793	16.448	19.667	19.324	18.772	20.404
0.17	17.980	18.047	17.544	16.266	19.312	19.030	18.519	20.049
0.18	17.649	17.751	17.287	16.078	18.960	18.729	18.258	19.699
0.19	17.323	17.450	17.025	15.883	18.609	18.423	17.991	19.354
0.20	17.003	17.146	16.757	15.683	18.260	18.112	17.718	19.012
0.22	16.376	16.529	16.210	15.268	17.570	17.481	17.157	18.342
0.24	15.765	15.910	15.653	14.839	16.893	16.845	16.585	17.686
0.25	15.465	15.602	15.373	14.620	16.561	16.527	16.297	17.364
0.26	15.169	15.296	15.093	14.399	16.232	16.210	16.008	17.045
0.28	14.589	14.694	14.537	13.955	15.588	15.584	15.431	16.417
0.30	14.026	14.107	13.989	13.509	14.965	14.972	14.862	15.806
0.32	13.482	13.541	13.455	13.067	14.365	14.378	14.303	15.211
0.34	12.959	12.998	12.938	12.631	13.790	13.805	13.759	14.634
0.35	12.705	12.736	12.687	12.417	13.513	13.528	13.494	14.353
0.36	12.458	12.481	12.441	12.205	13.242	13.257	13.234	14.078
0.38	11.982	11.991	11.967	11.792	12.720	12.734	12.730	13.543
0.40	11.530	11.530	11.517	11.392	12.227	12.238	12.248	13.031
0.42	11.105	11.096	11.092	11.010	11.762	11.770	11.790	12.543
0.44	10.705	10.692	10.692	10.644	11.326	11.330	11.357	12.080
0.45	10.515	10.500	10.502	10.469	11.118	11.121	11.150	11.858
0.46	10.332	10.315	10.318	10.298	10.917	10.918	10.950	11.642
0.48	9.984	9.965	9.969	9.970	10.536	10.533	10.567	11.228
0.50	9.660	9.641	9.645	9.662	10.180	10.174	10.210	10.840
0.55	8.952	8.935	8.936	8.973	9.400	9.386	9.419	9.973
0.60	8.373	8.359	8.354	8.396	8.756	8.737	8.764	9.245
0.65	7.898	7.889	7.878	7.915	8.227	8.205	8.224	8.639
0.70	7.506	7.501	7.485	7.515	7.791	7.766	7.779	8.137
0.80	6.892	6.892	6.870	6.888	7.118	7.091	7.095	7.368
0.90	6.406	6.407	6.384	6.399	6.606	6.578	6.580	6.808
1.00	5.972	5.973	5.950	5.968	6.172	6.143	6.145	6.359
1.10	5.553	5.553	5.531	5.556	5.768	5.738	5.742	5.962
1.20	5.139	5.137	5.116	5.147	5.372	5.341	5.348	5.586
1.30	4.730	4.727	4.705	4.741	4.982	4.949	4.958	5.215
1.40	4.333	4.330	4.307	4.344	4.597	4.564	4.573	4.849
1.50	3.956	3.952	3.929	3.965	4.226	4.191	4.202	4.490
1.60	3.604	3.600			3.874			4.144
1.70	3.281	3.278			3.545			3.814
1.80	2.992	2.989			3.244			3.506
1.90	2.733	2.731			2.971			3.221
2.00	2.506	2.505			2.727			2.963

$\sin \vartheta / \lambda$	$_{25}Mn^{2+}$	$_{25}Mn^{3+}$	$_{25}Mn^{4+}$	$_{26}Fe$	$_{26}Fe^{2+}$	$_{26}Fe^{3+}$	$_{27}Co$	$_{27}Co^{2+}$
0.00	23.000	22.000	21.000	26.000	24.000	23.000	27.000	25.000
0.01	22.988	21.990	20.992	25.970	23.989	22.991	26.972	24.989
0.02	22.953	21.961	20.968	25.882	23.954	22.962	26.887	24.954
0.03	22.894	21.913	20.927	25.738	23.895	22.914	26.749	24.897
0.04	22.812	21.846	20.871	25.543	23.814	22.848	26.562	24.818
0.05	22.707	21.760	20.799	25.304	23.711	22.763	26.331	24.716
0.06	22.581	21.656	20.712	25.026	23.587	22.660	26.063	24.593
0.07	22.433	21.534	20.610	24.719	23.441	22.539	25.764	24.450
0.08	22.266	21.395	20.493	24.387	23.276	22.401	25.440	24.287
0.09	22.080	21.240	20.363	24.038	23.091	22.247	25.098	24.104
0.10	21.875	21.070	20.218	23.678	22.889	22.078	24.744	23.904
0.11	21.654	20.884	20.061	23.310	22.669	21.893	24.380	23.687
0.12	21.418	20.684	19.891	22.939	22.435	21.695	24.011	23.455
0.13	21.167	20.472	19.710	22.568	22.185	21.483	23.641	23.207
0.14	20.904	20.247	19.517	22.197	21.923	21.258	23.270	22.946
0.15	20.629	20.011	19.315	21.829	21.648	21.023	22.900	22.673
0.16	20.344	19.765	19.102	21.465	21.363	20.776	22.533	22.389
0.17	20.050	19.509	18.881	21.104	21.068	20.521	22.168	22.095
0.18	19.748	19.246	18.652	20.748	20.765	20.256	21.806	21.791
0.19	19.440	18.975	18.415	20.395	20.455	19.984	21.448	21.481
0.20	19.126	18.697	18.172	20.046	20.140	19.705	21.093	21.164
0.22	18.488	18.127	17.669	19.359	19.494	19.130	20.393	20.514
0.24	17.841	17.543	17.149	18.685	18.838	18.538	19.704	19.850
0.25	17.517	17.247	16.884	18.354.	18.508	18.238	19.364	19.516
0.26	17.193	16.951	16.617	18.025	18.178	17.937	19.027	19.180
0.28	16.551	16.357	16.079	17.378	17.520	17.331	18.361	18.510
0.30	15.920	15.768	15.540	16.744	16.871	16.727	17.709	17.845
0.32	15.304	15.187	15.005	16.127	16.234	16.130	17.072	17.191
0.34	14.707	14.619	14.477	15.527	15.614	15.543	16.450	16.550
0.35	14.417	14.341	14.217	15.233	15.312	15.254	16.145	16.236
0.36	14.132	14.068	13.961	14.945	15.014	14.970	15.845	15.927
0.38	13.581	13.536	13.458	14.384	14.436	14.414	15.260	15.324
0.40	13.055	13.024	12.972	13.845	13.881	13.877	14.695	14.743
0.42	12.556	12.536	12.504	13.328	13.352	13.361	14.151	14.186
0.44	12.083	12.072	12.057	12.835	12.848	12.868	13.630	13.653
0.45	11.857	11.848	11.841	12.598	12.606	12.630	13.379	13.396
0.46	11.638	11.632	11.630	12.370	12.367	12.398	13.133	13.146
0.48	11.219	11.216	11.225	11.922	11.919	11.953	12.659	12.664
0.50	10.827	10.826	10.843	11.502	11.494	11.531	12.209	12.207
0.55	9.956	9.954	9.982	10.557	10.542	10.581	11.188	11.176
0.60	9.229	9.223	9.252	9.753	9.737	9.772	10.309	10.293
0.65	8.626	8.615	8.641	9.077	9.063	9.092	9.561	9.546
0.70	8.128	8.111	8.132	8.512	8.501	8.523	8.930	8.917
0.80	7.365	7.341	7.352	7.645	7.640	7.651	7.955	7.948
0.90	6.808	6.779	6.785	7.023	7.023	7.026	7.259	7.257
1.00	6.360	6.330	6.334	6.545	6.546	6.548	6.738	6.739
1.10	5.963	5.933	5.938	6.143	6.144	6.145	6.318	6.320
1.20	5.585	5.555	5.562	5.775	5.775	5.778	5.950	5.951
1.30	5.213	5.183	5.193	5.420	5.419	5.423	5.605	5.605
1.40	4.846	4.815	4.826	5.070	5.068	5.074	5.270	5.268
1.50	4.487	4.454	4.467	4.725	4.722	4.729	4.939	4.936
1.60	4.140			4.388	4.384	4.392	4.613	4.609
1.70	3.810			4.062	4.058	4.066	4.295	4.291
1.80	3.502			3.753	3.749	3.757	3.989	3.985
1.90	3.218			3.463	3.459	3.467	3.697	3.694
2.00	2.960			3.195	3.192	3.199	3.424	3.421

$\sin\vartheta/\lambda$	$_{27}Co^{3+}$	$_{28}Ni$	$_{28}Ni^{2+}$	$_{28}Ni^{3+}$	$_{29}Cu$	$_{29}Cu^{1+}$	$_{29}Cu^{2+}$	$_{30}Zn$
0.00	24.000	28.000	26.000	25.000	29.000	28.000	27.000	30.000
0.01	23.990	27.973	25.989	24.991	28.977	27.987	26.989	29.975
0.02	23.962	27.892	25.955	24.962	28.908	27.946	26.956	29.900
0.03	23.914	27.759	25.899	24.915	28.794	27.878	26.901	29.777
0.04	23.848	27.579	25.821	24.850	28.640	27.783	26.824	29.609
0.05	23.764	27.356	25.721	24.766	28.448	27.663	26.726	29.401
0.06	23.661	27.096	25.600	24.665	28.223	27.518	26.608	29.157
0.07	23.541	26.806	25.459	24.546	27.971	27.349	26.469	28.883
0.08	23.404	26.490	25.299	24.410	27.694	27.157	26.311	28.583
0.09	23.250	26.156	25.119	24.258	27.397	26.944	26.134	28.263
0.10	23.081	25.807	24.921	24.090	27.084	26.711	25.939	27.927
0.11	22.896	25.448	24.707	23.907	26.758	26.459	25.728	27.579
0.12	22.698	25.083	24.477	23.709	26.422	26.190	25.500	27.222
0.13	22.486	24.714	24.232	23.498	26.077	25.905	25.258	26.859
0.14	22.261	24.344	23.973	23.275	25.726	25.606	25.001	26.492
0.15	22.024	23.973	23.702	23.039	25.370	25.294	24.732	26.124
0.16	21.777	23.604	23.419	22.792	25.009	24.972	24.451	25.754
0.17	21.520	23.237	23.126	22.535	24.645	24.639	24.159	25.385
0.18	21.253	22.872	22.824	22.268	24.278	24.297	23.857	25.017
0.19	20.978	22.510	22.513	21.993	23.910	23.949	23.547	24.649
0.20	20.696	22.150	22.195	21.710	23.540	23.594	23.229	24.283
0.22	20.114	21.438	21.543	21.125	22.798	22.872	22.574	23.556
0.24	19.513	20.737	20.875	20.518	22.057	22.139	21.900	22.836
0.25	19.207	20.390	20.536	20.209	21.687	21.770	21.558	22.478
0.26	18.899	20.046	20.197	19.897	21.319	21.401	21.214	22.122
0.28	18.280	19.365	19.516	19.268	20.589	20.666	20.523	21.417
0.30	17.659	18.696	18.839	18.636	19.869	19.939	19.832	20.720
0.32	17.043	18.040	18.169	18.005	19.162	19.224	19.146	20.034
0.34	16.435	17.398	17.510	17.380	18.472	18.524	18.469	19.359
0.35	16.135	17.084	17.187	17.071	18.133	18.180	18.135	19.027
0.36	15.838	16.773	16.867	16.765	17.799	17.842	17.805	18.698
0.38	15.258	16.165	16.242	16.164	17.145	17.180	17.157	18.051
0.40	14.694	15.576	15.637	15.578	16.514	16.541	16.528	17.421
0.42	14.151	15.008	15.054	15.010	15.904	15.925	15.919	16.809
0.44	13.629	14.461	14.495	14.463	15.318	15.333	15.332	16.216
0.45	13.376	14.196	14.224	14.197	15.034	15.047	15.046	15.926
0.46	13.129	13.937	13.959	13.936	14.757	14.767	14.767	15.642
0.48	12.652	13.435	13.448	13.432	14.219	14.225	14.227	15.090
0.50	12.200	12.956	12.962	12.950	13.707	13.710	13.711	14.559
0.55	11.171	11.862	11.854	11.847	12.533	12.530	12.526	13.328
0.60	10.286	10.909	10.895	10.887	11.507	11.502	11.491	12.235
0.65	9.534	10.090	10.075	10.062	10.621	10.614	10.597	11.276
0.70	8.900	9.392	9.378	9.360	9.861	9.855	9.831	10.442
0.80	7.921	8.301	8.292	8.265	8.663	8.659	8.625	9.108
0.90	7.224	7.519	7.516	7.482	7.799	7.797	7.757	8.132
1.00	6.703	6.944	6.944	6.906	7.166	7.165	7.123	7.417
1.10	6.283	6.495	6.497	6.457	6.681	6.681	6.637	6.879
1.20	5.913	6.118	6.119	6.078	6.285	6.285	6.240	6.453
1.30	5.566	5.776	5.776	5.734	5.939	5.939	5.892	6.096
1.40	5.228	5.451	5.450	5.407	5.617	5.617	5.568	5.775
1.50	4.895	5.133	5.131	5.086	5.308	5.307	5.256	5.473
1.60		4.819	4.816		5.005	5.003		5.180
1.70		4.511	4.507		4.705	4.704		4.892
1.80		4.211	4.207		4.413	4.411		4.610
1.90		3.922	3.918		4.128	4.127		4.332
2.00		3.647	3.643		3.855	3.853		4.063

sin ϑ/λ	$_{30}Zn^{2+}$	$_{31}Ga$	$_{31}Ga^{3+}$	$_{32}Ge$	$_{32}Ge^{4+}$	$_{33}As$	$_{34}Se$	$_{35}Br$
0.00	28.000	31.000	28.000	32.000	28.000	33.000	34.000	35.000
0.01	27.989	30.971	27.991	31.970	27.992	32.970	33.970	34.971
0.02	27.957	30.883	27.964	31.878	27.969	32.879	33.881	34.883
0.03	27.903	30.740	27.919	31.729	27.931	32.730	33.734	34.739
0.04	27.828	30.546	27.856	31.526	27.877	32.527	33.532	34.540
0.05	27.732	30.308	27.776	31.276	27.808	32.274	33.280	34.291
0.06	27.615	30.031	27.678	30.984	27.724	31.977	32.982	33.995
0.07	27.479	29.724	27.564	30.657	27.625	31.642	32.645	33.658
0.08	27.323	29.391	27.433	30.302	27.512	31.276	32.273	33.284
0.09	27.149	29.040	27.286	29.926	27.386	30.884	31.872	32.880
0.10	26.958	28.675	27.123	29.534	27.245	30.473	31.449	32.450
0.11	26.749	28.302	26.946	29.133	27.091	30.049	31.009	32.000
0.12	26.525	27.924	26.754	28.725	26.924	29.616	30.557	31.535
0.13	26.286	27.543	26.548	28.316	26.745	29.179	30.099	31.060
0.14	26.032	27.162	26.330	27.908	26.554	28.742	29.637	30.578
0.15	25.766	26.783	26.099	27.504	26.351	28.307	29.175	30.095
0.16	25.488	26.406	25.856	27.104	26.137	27.877	28.718	29.613
0.17	25.198	26.033	25.603	26.709	25.913	27.454	28.266	29.136
0.18	24.899	25.663	25.339	26.322	25.680	27.039	27.822	28.664
0.19	24.591	25.297	25.066	25.941	25.437	26.633	27.387	28.202
0.20	24.275	24.935	24.784	25.567	25.185	26.235	26.962	27.749
0.22	23.622	24.221	24.197	24.839	24.658	25.469	26.145	26.876
0.24	22.949	23.520	23.585	24.135	24.104	24.739	25.372	26.052
0.25	22.606	23.174	23.270	23.791	23.818	24.386	25.001	25.658
0.26	22.261	22.830	22.952	23.452	23.526	24.041	24.641	25.276
0.28	21.566	22.151	22.305	22.787	22.931	23.370	23.947	24.545
0.30	20.869	21.481	21.649	22.136	22.321	22.724	23.288	23.857
0.32	20.175	20.820	20.988	21.498	21.702	22.097	22.656	23.206
0.34	19.488	20.169	20.327	20.870	21.077	21.486	22.048	22.587
0.35	19.149	19.847	19.997	20.560	20.764	21.185	21.751	22.288
0.36	18.812	19.527	19.669	20.253	20.451	20.888	21.459	21.995
0.38	18.150	18.897	19.019	19.645	19.826	20.301	20.887	21.425
0.40	17.504	18.278	18.379	19.047	19.205	19.725	20.328	20.874
0.42	16.876	17.673	17.751	18.459	18.593	19.159	19.780	20.338
0.44	16.269	17.083	17.139	17.882	17.989	18.602	19.242	19.816
0.45	15.974	16.794	16.839	17.598	17.692	18.326	18.977	19.558
0.46	15.683	16.508	16.544	17.317	17.398	18.054	18.713	19.304
0.48	15.120	15.950	15.967	16.765	16.821	17.516	18.193	18.801
0.50	14.580	15.410	15.409	16.227	16.259	16.989	17.682	18.307
0.55	13.331	14.142	14.106	14.947	14.929	15.721	16.444	17.107
0.60	12.227	12.996	12.937	13.770	13.716	14.535	15.269	15.958
0.65	11.263	11.974	11.902	12.702	12.625	13.440	14.166	14.865
0.70	10.429	11.073	10.995	11.745	11.656	12.442	13.145	13.837
0.80	9.097	9.604	9.526	10.151	10.058	10.741	11.362	12.001
0.90	8.126	8.510	8.441	8.937	8.853	9.411	9.928	10.480
1.00	7.414	7.702	7.642	8.028	7.956	8.396	8.809	9.262
1.10	6.879	7.099	7.045	7.348	7.286	7.631	7.952	8.312
1.20	6.455	6.633	6.582	6.830	6.774	7.050	7.299	7.580
1.30	6.096	6.254	6.203	6.419	6.365	6.597	6.795	7.016
1.40	5.775	5.926	5.872	6.076	6.021	6.231	6.395	6.574
1.50	5.472	5.627	5.569	5.774	5.715	5.917	6.063	6.216
1.60	5.178	5.342		5.493		5.636	5.775	5.913
1.70	4.890	5.065		5.224		5.372	5.511	5.645
1.80	4.606	4.792		4.961		5.117	5.262	5.398
1.90	4.329	4.523		4.702		4.867	5.020	5.162
2.00	4.059	4.260		4.447		4.621	4.782	4.932

$\sin\vartheta/\lambda$	$_{35}Br^{1-}$	$_{36}Kr$	$_{37}Rb$	$_{37}Rb^{1+}$	$_{38}Sr$	$_{38}Sr^{2+}$	$_{39}Y$	$_{39}Y^{3+}$
0.00	36.000	36.000	37.000	36.000	38.000	36.000	39.000	36.000
0.01	35.961	35.972	36.952	35.977	37.946	35.981	38.947	35.983
0.02	35.845	35.886	36.809	35.908	37.786	35.923	38.792	35.933
0.03	35.656	35.744	36.583	35.794	37.532	35.827	38.543	35.850
0.04	35.398	35.549	36.291	35.635	37.197	35.694	38.212	35.735
0.05	35.077	35.304	35.948	35.435	36.802	35.524	37.816	35.588
0.06	34.703	35.011	35.571	35.195	36.363	35.320	37.369	35.411
0.07	34.282	34.677	35.171	34.917	35.897	35.084	36.889	35.204
0.08	33.824	34.305	34.758	34.605	35.418	34.816	36.387	34.970
0.09	33.336	33.899	34.336	34.262	34.937	34.520	35.876	34.710
0.10	32.827	33.467	33.907	33.891	34.458	34.198	35.364	34.425
0.11	32.303	33.011	33.473	33.496	33.986	33.851	34.855	34.118
0.12	31.771	32.537	33.034	33.079	33.522	33.484	34.354	33.791
0.13	31.236	32.051	32.588	32.646	33.066	33.098	33.861	33.445
0.14	30.703	31.555	32.137	32.199	32.616	32.696	33.378	33.082
0.15	30.175	31.055	31.681	31.740	32.171	32.281	32.904	32.705
0.16	29.657	30.553	31.220	31.275	31.730	31.854	32.437	32.316
0.17	29.149	30.053	30.757	30.805	31.292	31.420	31.977	31.916
0.18	28.654	29.558	30.293	30.333	30.856	30.979	31.523	31.509
0.19	28.172	29.070	29.830	29.862	30.421	30.535	31.075	31.094
0.20	27.706	28.590	29.368	29.393	29.988	30.089	30.631	30.675
0.22	26.817	27.663	28.459	28.471	29.128	29.198	29.758	29.830
0.24	25.988	26.784	27.576	27.579	28.280	28.322	28.904	28.986
0.25	25.595	26.364	27.148	27.147	27.863	27.892	28.485	28.567
0.26	25.215	25.957	26.729	26.726	27.452	27.469	28.071	28.152
0.28	24.491	25.181	25.922	25.916	26.648	26.647	27.263	27.337
0.30	23.812	24.453	25.158	25.150	25.875	25.861	26.483	26.548
0.32	23.170	23.771	24.437	24.428	25.135	25.113	25.734	25.789
0.34	22.559	23.128	23.758	23.749	24.430	24.404	25.018	25.063
0.35	22.264	22.820	23.432	23.424	24.090	24.064	24.673	24.712
0.36	21.975	22.520	23.116	23.109	23.760	23.734	14.336	24.370
0.38	21.412	21.941	22.510	22.503	23.125	23.100	23.687	23.712
0.40	20.867	21.388	21.934	21.929	22.522	22.500	23.071	23.086
0.42	20.335	20.855	21.386	21.381	21.950	21.931	22.485	22.492
0.44	19.816	20.339	20.860	20.857	21.404	21.389	21.928	21.927
0.45	19.560	20.087	20.605	20.603	21.141	21.128	21.660	21.654
0.46	19.306	19.838	20.354	20.353	20.883	20.872	21.398	21.388
0.48	18.806	19.349	19.866	19.865	20.383	20.376	20.890	20.874
0.50	18.313	18.870	19.391	19.391	19.902	19.898	20.404	20.382
0.55	17.114	17.709	18.252	18.253	18.764	18.765	19.263	19.231
0.60	15.964	16.594	17.167	17.169	17.696	17.700	18.204	18.166
0.65	14.870	15.524	16.125	16.127	16.678	16.684	17.203	17.163
0.70	13.840	14.504	15.126	15.128	15.702	15.707	16.246	16.208
0.80	12.002	12.645	13.272	13.273	13.872	13.875	14.443	14.415
0.90	10.479	11.057	11.645	11.645	12.230	12.231	12.798	12.784
1.00	9.261	9.752	10.270	10.270	10.806	10.805	11.339	11.340
1.10	8.311	8.711	9.147	9.147	9.612	9.611	10.088	10.100
1.20	7.580	7.898	8.252	8.251	8.640	8.638	9.046	9.067
1.30	7.016	7.266	7.548	7.548	7.863	7.862	8.200	8.225
1.40	6.573	6.773	6.996	6.997	7.249	7.249	7.523	7.548
1.50	6.216	6.380	6.562	6.561	6.764	6.764	6.985	7.008
1.60	5.913	6.056	6.210	6.209	6.376	6.375	6.554	6.575
1.70	5.645	5.778	5.913	5.913	6.055	6.056	6.205	6.222
1.80	5.398	5.528	5.656	5.656	5.785	5.785	5.914	5.927
1.90	5.162	5.295	5.420	5.421	5.544	5.545	5.662	5.672
2.00	4.932	5.071	5.200	5.201	5.323	5.324	5.440	5.443

$\sin\vartheta/\lambda$	$_{40}$Zr	$_{40}$Zr^{4+}	$_{41}$Nb	$_{41}$Nb^{3+}	$_{41}$Nb^{5+}	$_{42}$Mo	$_{42}$Mo^{3+}	$_{42}$Mo^{5+}
0.00	40.000	36.000	41.000	38.000	36.000	42.000	39.000	37.000
0.01	39.949	35.985	40.956	37.981	35.987	41.958	38.981	36.986
0.02	39.800	35.942	40.824	37.925	35.948	41.831	38.923	36.946
0.03	39.559	35.869	40.610	37.832	35.884	41.625	38.827	36.878
0.04	39.237	35.768	40.323	37.702	35.795	41.346	38.695	36.783
0.05	38.847	35.640	39.970	37.537	35.681	41.003	38.526	36.663
0.06	38.403	35.484	39.565	37.339	35.543	40.606	38.323	36.517
0.07	37.921	35.302	39.116	37.109	35.381	40.164	38.087	36.347
0.08	37.412	35.096	38.634	36.849	35.197	39.686	37.820	36.152
0.09	36.887	34.865	38.128	36.560	34.991	39.181	37.523	35.936
0.10	36.356	34.612	37.606	36.246	34.765	38.656	37.200	35.697
0.11	35.824	34.338	37.073	35.908	34.519	38.117	36.853	35.438
0.12	35.296	34.045	36.535	35.548	34.254	37.569	36.483	35.160
0.13	34.775	33.734	35.994	35.170	33.973	37.016	36.094	34.865
0.14	34.262	33.406	35.454	34.775	33.675	36.461	35.688	34.553
0.15	33.758	33.064	34.916	34.366	33.363	35.907	35.266	34.226
0.16	33.263	32.708	34.382	33.945	33.038	35.335	34.832	33.886
0.17	32.776	32.341	33.854	33.514	32.701	34.806	34.388	33.533
0.18	32.298	31.964	33.331	33.076	32.353	34.263	33.936	33.170
0.19	31.827	31.580	32.814	32.632	31.997	33.725	33.478	32.798
0.20	31.363	31.188	32.305	32.184	31.632	33.195	33.016	32.418
0.22	30.454	30.392	31.310	31.284	30.885	32.157	32.086	31.640
0.24	29.572	29.586	30.348	30.388	30.121	31.153	31.159	30.847
0.25	29.141	29.182	29.881	29.945	29.736	30.665	30.701	30.448
0.26	28.716	28.781	29.424	29.506	29.351	30.188	30.246	30.048
0.28	27.889	27.985	28.538	28.646	28.582	29.263	29.356	29.252
0.30	27.092	27.205	27.692	27.814	27.821	28.382	28.494	28.465
0.32	26.327	26.447	26.888	27.013	27.074	27.543	27.664	27.695
0.34	25.596	25.716	26.126	26.248	26.346	26.749	26.871	26.944
0.35	25.243	25.360	25.760	25.878	25.990	26.368	26.488	26.578
0.36	24.899	25.012	25.404	25.518	25.640	25.998	26.115	26.218
0.38	24.236	24.339	24.721	24.824	24.960	25.289	25.397	25.518
0.40	23.606	23.696	24.077	24.167	24.306	24.620	24.717	24.847
0.42	23.008	23.083	23.468	23.543	23.680	23.989	24.073	24.205
0.44	22.439	22.500	22.892	22.953	23.081	23.394	23.464	23.592
0.45	22.166	22.218	22.615	22.669	22.792	23.109	23.172	23.296
0.46	21.899	21.944	22.346	22.393	22.509	22.832	22.888	23.007
0.48	21.384	21.414	21.829	21.861	21.963	22.300	22.342	22.450
0.50	20.892	20.907	21.336	21.355	21.442	21.796	21.825	21.920
0.55	19.745	19.731	20.195	20.187	20.235	20.638	20.638	20.697
0.60	18.693	18.658	19.156	19.128	19.142	19.595	19.573	19.599
0.65	17.706	17.659	18.187	18.148	18.137	18.635	18.597	18.598
0.70	16.767	16.716	17.268	17.224	17.198	17.732	17.685	17.668
0.80	14.996	14.952	15.533	15.492	15.458	16.036	15.985	15.955
0.90	13.361	13.333	13.915	13.886	13.858	14.448	14.405	14.377
1.00	11.883	11.873	12.427	12.414	12.395	12.968	12.939	12.918
1.10	10.588	10.592	11.098	11.099	11.088	11.621	11.606	11.593
1.20	9.486	9.501	9.945	9.958	9.951	10.430	10.427	10.418
1.30	8.574	8.595	8.972	8.992	8.988	9.404	9.410	9.405
1.40	7.833	7.856	8.169	8.193	8.190	8.542	8.554	8.551
1.50	7.238	7.261	7.516	7.541	7.539	7.831	7.846	7.843
1.60	6.760	6.782	6.989	7.013	7.011	7.251	7.267	7.265
1.70	6.375	6.394	6.564	6.584	6.583	6.780	6.795	6.793
1.80	6.059	6.074	6.216	6.234	6.233	6.397	6.409	6.408
1.90	5.790	5.802	5.927	5.941	5.940	6.080	6.090	6.089
2.00	5.558	5.565	5.680	5.689	5.690	5.813	5.820	5.820

$\sin \vartheta/\lambda$	$_{42}Mo^{6+}$	$_{43}Tc$	$_{44}Ru$	$_{44}Ru^{3+}$	$_{44}Ru^{4+}$	$_{45}Rh$	$_{45}Rh^{3+}$	$_{45}Rh^{4+}$
0.00	36.000	43.000	44.000	41.000	40.000	45.000	42.000	41.000
0.01	35.988	42.955	43.960	40.980	39.983	44.961	41.980	40.983
0.02	35.954	42.821	43.842	40.922	39.933	44.847	41.922	40.932
0.03	35.897	42.603	43.649	40.824	39.849	44.660	41.824	40.848
0.04	35.817	42.308	43.386	40.689	39.733	44.405	41.689	40.730
0.05	35.715	41.945	43.061	40.517	39.585	44.088	41.516	40.581
0.06	35.591	41.526	42.681	40.309	39.406	43.717	41.308	40.400
0.07	35.446	41.059	42.254	40.067	39.197	43.299	41.066	40.188
0.08	35.280	40.557	41.789	39.793	38.959	42.842	40.791	39.948
0.09	35.095	40.028	41.292	39.489	38.695	42.351	40.485	39.680
0.10	34.890	39.480	40.770	39.156	38.404	41.834	40.150	39.385
0.11	34.667	38.921	40.229	38.798	38.090	41.296	39.789	39.067
0.12	34.428	38.355	39.674	38.416	37.754	40.741	39.404	38.725
0.13	34.172	37.787	39.108	38.012	37.397	40.173	38.997	38.363
0.14	33.900	37.221	38.536	37.590	37.022	39.597	38.569	37.981
0.15	33.615	36.658	37.959	37.151	36.630	39.015	38.125	37.582
0.16	33.317	36.100	37.381	36.698	36.223	38.429	37.665	37.168
0.17	33.006	35.548	36.803	36.233	35.804	37.841	37.193	36.740
0.18	32.685	35.003	36.228	35.758	35.374	37.254	36.710	36.301
0.19	32.354	34.466	35.655	35.276	34.934	36.668	36.218	35.852
0.20	32.015	33.936	35.088	34.789	34.488	36.086	35.720	35.395
0.22	31.316	32.900	33.971	33.804	33.579	34.937	34.713	34.463
0.24	30.595	31.897	32.886	32.819	32.659	33.815	33.701	33.518
0.25	30.229	31.409	32.356	32.329	32.199	33.267	33.198	33.045
0.26	29.862	30.930	31.837	31.844	31.741	32.728	32.697	32.572
0.28	29.123	29.998	30.829	30.889	30.833	31.680	31.711	31.635
0.30	28.387	29.104	29.866	29.962	29.943	30.675	30.751	30.715
0.32	27.658	28.250	28.949	29.067	29.078	29.717	29.823	29.819
0.34	26.941	27.435	28.079	28.210	28.242	28.807	28.932	28.951
0.35	26.589	27.042	27.662	27.796	27.836	28.370	28.500	28.530
0.36	26.241	26.660	27.257	27.392	27.439	27.944	28.079	28.117
0.38	25.562	25.925	26.480	26.614	26.671	27.130	27.268	27.318
0.40	24.904	25.229	25.749	25.878	25.940	26.363	26.499	26.557
0.42	24.270	24.571	25.062	25.181	25.245	25.642	25.772	25.833
0.44	23.662	23.949	24.415	24.524	24.586	24.964	25.086	25.148
0.45	23.366	23.651	24.106	24.209	24.271	24.640	24.757	24.819
0.46	23.078	23.361	23.807	23.904	23.963	24.327	24.438	24.499
0.48	22.518	22.806	23.235	23.319	23.374	23.729	23.829	23.886
0.50	21.983	22.280	22.696	22.767	22.817	23.167	23.254	23.307
0.55	20.744	21.080	21.476	21.516	21.549	21.900	21.957	21.995
0.60	19.627	20.012	20.403	20.416	20.434	20.798	20.826	20.850
0.65	18.608	19.042	19.438	19.430	19.436	19.820	19.824	19.835
0.70	17.664	18.142	18.551	18.528	18.525	18.932	18.918	18.919
0.80	15.937	16.477	16.922	16.884	16.872	17.326	17.292	17.282
0.90	14.357	14.925	15.405	15.367	15.354	15.845	15.807	15.795
1.00	12.902	13.466	13.968	13.939	13.929	14.440	14.407	14.396
1.10	11.581	12.116	12.620	12.605	12.597	13.107	13.086	13.078
1.20	10.411	10.900	11.385	11.382	11.378	11.866	11.859	11.853
1.30	9.400	9.833	10.282	10.291	10.288	10.740	10.744	10.740
1.40	8.547	8.919	9.323	9.339	9.338	9.743	9.756	9.754
1.50	7.841	8.154	8.506	8.528	8.527	8.880	8.899	8.898
1.60	7.263	7.521	7.823	7.847	7.846	8.148	8.171	8.170
1.70	6.792	7.004	7.258	7.282	7.282	7.535	7.559	7.559
1.80	6.407	6.582	6.794	6.817	6.817	7.028	7.051	7.051
1.90	6.089	6.234	6.412	6.433	6.433	6.608	6.631	6.630
2.00	5.820	5.946	6.097	6.114	6.114	6.262	6.281	6.281

$\sin\vartheta/\lambda$	$_{46}$Pd	$_{46}$Pd^{2+}	$_{46}$Pd^{4+}	$_{47}$Ag	$_{47}$Ag^{1+}	$_{47}$Ag^{2+}	$_{48}$Cd	$_{48}$Cd^{2+}
0.00	46.000	44.000	42.000	47.000	46.000	45.000	48.000	46.000
0.01	45.968	43.977	41.983	46.964	45.974	44.978	44.962	45.978
0.02	45.874	43.909	41.932	46.857	45.894	44.911	47.848	45.912
0.03	45.718	43.796	41.847	46.681	45.764	44.799	47.660	45.802
0.04	45.503	43.640	41.729	46.440	45.582	44.645	47.404	45.650
0.05	45.232	43.441	41.579	46.139	45.353	44.448	47.085	45.456
0.06	44.908	43.201	41.396	45.786	45.076	44.211	46.710	45.222
0.07	44.535	42.923	41.184	45.385	44.757	43.936	46.287	44.950
0.08	44.119	42.608	40.942	44.944	44.397	43.624	45.822	44.641
0.09	43.663	42.258	40.671	44;469	43.999	43.277	45.324	44.298
0.10	43.172	41.877	40.375	43.964	43.567	42.898	44.797	43.923
0.11	42.651	41.467	40.053	43.435	43.105	42.490	44.248	43.517
0.12	42.105	41.031	39.708	42.886	42.616	42.056	43.683	43.085
0.13	41.538	40.572	39.341	42.322	42.103	41.597	43.104	42.628
0.14	40.954	40.092	38.955	41.744	41.570	41.117	42.517	42.148
0.15	40.357	39.595	38.551	41.157	41.020	40.618	41.923	41.649
0.16	39.750	39.083	38.131	40.563	40.457	40.103	41.325	41.134
0.17	39.137	38.558	37.696	39.964	39.883	39.575	40.726	40.603
0.18	38.520	38.024	37.249	39.361	39.301	39.036	40.126	40.061
0.19	37.902	37.483	36.792	38.758	38.713	38.489	39.527	39.509
0.20	37.286	36.936	36.326	38.154	38.122	37.935	38.930	38.949
0.22	36.064	35.836	35.374	36.955	36.940	36.817	37.746	37.816
0.24	34.868	34.739	34.406	35.774	35.768	35.697	36.581	36.677
0.25	34.283	34.195	33.921	35.192	35.191	35.141	36.007	36.110
0.26	33.708	33.657	33.435	34.619	34.620	34.589	35.440	35.545
0.28	32.592	32.601	32.471	33.498	33.503	33.502	34.329	34.431
0.30	31.523	31.577	31.521	32.416	32.424	32.445	33.251	33.344
0.32	30.505	30.592	30.594	31.378	31.389	31.425	32.210	32.291
0.34	29.540	29.649	29.695	30.387	30.400	30.446	31.210	31.276
0.35	29.077	29.194	29.257	29.910	29.924	29.973	30.725	30.785
0.36	28.628	28.751	28.828	29.444	29.460	29.511	30.252	30.305
0.38	27.769	27.899	27.998	28.551	28.569	28.622	29.338	29.379
0.40	26.961	27.093	27.204	27.707	27.727	27.780	28.468	28.500
0.42	26.202	26.333	26.450	26.911	26.933	26.984	27.644	27.667
0.44	25.491	25.617	25.735	26.163	26.186	26.233	26.865	26.881
0.45	25.153	25.275	25.391	25.805	25.829	25.874	26.492	26.505
0.46	24.825	24.944	25.057	25.459	25.484	25.527	26.129	26.140
0.48	24.201	24.311	24.418	24.800	24.825	24.863	25.436	25.443
0.50	23.617	23.716	23.814	24.181	24.206	24.239	24.784	24.788
0.55	22.307	22.378	22.450	22.795	22.817	22.839	23.320	23.319
0.60	21.177	21.221	21.267	21.607	21.623	21.635	22.063	22.061
0.65	20.186	20.205	20.228	20.575	20.583	20.588	20.978	20.974
0.70	19.296	19.295	19.300	19.661	19.660	19.660	20.027	20.021
0.80	17.711	17.683	17.668	18.069	18.051	18.046	18.405	18.392
0.90	16.266	16.229	16.208	16.651	16.622	16.616	17.000	16.979
1.00	14.893	14.859	14.840	15.316	15.284	15.278	15.698	15.673
1.10	13.580	13.557	13.542	14.035	14.006	14.002	14.451	14.425
1.20	12.342	12.331	12.321	12.813	12.790	12.788	13.253	13.230
1.30	11.200	11.201	11.194	11.669	11.654	11.653	12.116	12.099
1.40	10.173	10.183	10.180	10.623	10.616	10.616	11.060	11.050
1.50	9.270	9.288	9.286	9.687	9.688	9.688	10.101	10.098
1.60	8.492	8.514	8.513	8.869	8.875	8.876	9.249	9.251
1.70	7.833	7.858	7.857	8.165	8.176	8.176	8.505	8.513
1.80	7.282	7.307	7.306	7.569	7.582	7.582	7.867	7.878
1.90	6.824	6.847	6.847	7.069	7.083	7.083	7.326	7.339
2.00	6.443	6.464	6.464	6.651	6.665	6.665	6.871	6.884

$\sin\vartheta/\lambda$	$_{49}$In	$_{49}$In^{3+}	$_{50}$Sn	$_{50}$Sn^{2+}	$_{50}$Sn^{4+}	$_{51}$Sb	$_{51}$Sb^{2+}	$_{51}$Sb^{5+}
0.00	49.000	46.000	50.000	48.000	46.000	51.000	48.000	46.000
0.01	48.957	45.981	49.955	47.975	45.984	50.955	47.978	45.985
0.02	48.828	45.924	49.821	47.898	45.934	50.819	47.911	45.940
0.03	48.618	45.829	49.601	47.771	45.852	50.596	47.801	45.865
0.04	48.332	45.697	49.303	47.596	45.737	50.293	47.647	45.760
0.05	47.980	45.529	48.934	47.373	45.590	49.915	47.452	45.627
0.06	47.570	45.325	48.504	47.106	45.411	49.474	47.218	45.464
0.07	47.112	45.087	48.022	46.797	45.203	48.977	46.945	45.274
0.08	46.614	44.816	47.498	46.449	44.964	48.434	46.636	45.057
0.09	46.086	44.513	46.942	46.066	44.698	47.856	46.293	44.813
0.10	45.534	44.181	46.361	45.650	44.404	47.250	45.920	44.544
0.11	44.964	43.821	45.764	45.206	44.084	46.625	45.517	44.251
0.12	44.383	43.435	45.155	44.736	43.739	45.988	45.089	43.935
0.13	43.793	43.024	44.541	44.244	43.371	45.344	44.638	43.596
0.14	43.199	42.591	43.924	43.733	42.981	44.699	44.167	43.237
0.15	42.603	42.138	43.309	43.206	42.572	44.056	43.677	42.859
0.16	42.006	41.667	42.696	42.667	42.143	43.419	43.172	42.462
0.17	41.410	41.180	42.088	42.117	41.698	42.789	42.655	42.049
0.18	40.817	40.678	41.486	41.560	41.237	42.168	42.127	41.621
0.19	40.226	40.165	40.891	40.998	40.763	41.556	41.590	41.178
0.20	39.639	39.641	40.302	40.431	40.276	40.955	41.047	40.723
0.22	38.478	38.570	39.145	39.296	39.274	39.783	39.950	39.781
0.24	37.337	37.481	38.016	38.164	38.242	38.652	38.847	38.806
0.25	36.774	36.933	37.462	37.604	37.718	38.100	38.298	38.309
0.26	36.218	36.385	36.915	37.047	37.192	37.556	37.750	37.807
0.28	35.125	35.295	35.841	35.950	36.135	36.495	36.668	36.796
0.30	34.059	34.220	34.794	34.878	35.082	35.465	35.605	35.780
0.32	33.025	33.167	33.775	33.836	34.041	34.464	34.569	34.770
0.34	32.025	32.145	32.786	32.826	33.019	33.491	33.561	33.771
0.35	31.538	31.647	32.303	32.333	32.517	33.016	33.069	33.278
0.36	31.060	31.158	31.828	31.850	32.023	32.547	32.585	32.790
0.38	30.134	30.209	30.902	30.910	31.057	31.631	31.643	31.834
0.40	29.247	29.302	30.011	30.008	30.127	30.745	30.737	30.905
0.42	28.401	28.438	29.154	29.144	29.235	29.888	29.866	30.009
0.44	27.596	27.618	28.334	28.318	28.383	29.063	29.031	29.146
0.45	27.209	27.224	27.938	27.920	27.972	28.663	28.628	28.729
0.46	26.832	26.842	27.551	27.532	27.571	28.270	28.234	28.321
0.48	26.108	26.109	26.805	26.785	26.802	27.511	27.472	27.532
0.50	25.425	25.418	26.096	26.075	26.074	26.784	26.747	26.782
0.55	23.881	23.865	24.482	24.464	24.430	25.113	25.088	25.073
0.60	22.552	22.533	23.081	23.067	23.019	23.646	23.634	23.590
0.65	21.405	21.389	21.868	21.859	21.810	22.366	22.367	22.310
0.70	20.408	20.394	20.815	20.810	20.767	21.253	21.261	21.205
0.80	18.736	18.724	19.073	19.074	19.052	19.424	19.433	19.397
0.90	17.329	17.315	17.646	17.649	17.646	17.958	17.957	17.947
1.00	16.053	16.034	16.384	16.386	16.395	16.696	16.684	16.690
1.10	14.840	14.818	15.201	15.203	15.215	15.537	15.516	15.529
1.20	13.670	13.649	14.062	14.063	14.074	14.429	14.403	14.416
1.30	12.548	12.530	12.962	12.962	12.970	13.355	13.329	13.339
1.40	11.492	11.479	11.913	11.913	11.917	12.321	12.300	12.305
1.50	10.518	10.510	10.933	10.932	10.933	11.341	11.326	11.328
1.60	9.639	9.637	10.034	10.033	10.033	10.431	10.422	10.422
1.70	8.860	8.864	9.227	9.227	9.225	9.602	9.599	9.597
1.80	8.184	8.191	8.516	8.515	8.513	8.861	8.863	8.860
1.90	7.603	7.613	7.897	7.897	7.896	8.208	8.215	8.213
2.00	7.110	7.122	7.367	7.367	7.366	7.642	7.652	7.650

$\sin\vartheta/\lambda$	$_{52}$Te	$_{53}$I	$_{53}$I^{1-}	$_{54}$Xe	$_{55}$Cs	$_{55}$Cs^{1+}	$_{56}$Ba	$_{56}$Ba^{2+}
0.00	52.000	53.000	54.000	54.000	55.000	54.000	56.000	54.000
0.01	51.954	52.955	53.943	53.956	54.932	53.963	55.925	53.967
0.02	51.818	52.820	53.772	53.821	54.732	53.850	55.703	53.869
0.03	51.594	52.597	53.493	53.601	54.417	53.665	55.350	53.708
0.04	51.288	52.292	52.114	53.297	54.008	53.408	54.888	53.484
0.05	50.906	51.911	52.646	52.917	53.527	53.084	54.345	53.200
0.06	50.458	51.460	52.101	52.467	52.996	52.698	53.743	52.861
0.07	49.951	50.950	51.492	51.954	52.430	52.254	53.106	52.468
0.08	49.395	50.387	50.834	51.388	51.839	51.758	52.450	52.027
0.09	48.800	49.781	50.136	50.775	51.229	51.217	51.786	51.543
0.10	48.174	49.142	49.413	50.125	50.603	50.635	51.122	51.018
0.11	47.526	48.476	48.672	49.447	49.963	50.020	50.460	50.460
0.12	46.863	47.793	47.924	48.747	49.309	49.377	49.802	49.872
0.13	46.193	47.099	47.175	48.033	48.645	48.714	49.146	49.259
0.14	45.519	46.400	46.432	47.311	47.971	48.035	48.492	48.627
0.15	44.848	45.702	45.698	46.588	47.291	47.345	47.839	47.980
0.16	44.182	45.008	44.978	45.868	46.606	46.651	47.186	47.322
0.17	43.526	44.323	44.273	45.155	45.921	45.955	46.533	46.657
0.18	42.879	43.648	43.585	44.453	45.237	45.262	45.882	45.989
0.19	42.245	42.987	42.916	43.763	44.559	44.575	45.232	45.321
0.20	41.623	42.340	42.265	43.088	43.888	43.897	44.586	44.657
0.22	40.419	41.091	41.019	41.788	42.578	42.577	43.309	43.348
0.24	39.267	39.904	39.841	40.557	41.320	41.312	42.064	42.077
0.25	38.709	39.333	39.276	39.967	40.713	40.703	41.456	41.459
0.26	38.163	38.776	38.726	39.393	40.121	40.110	40.859	40.855
0.28	37.102	37.702	37.665	38.294	38.982	38.971	39.702	39.688
0.30	36.079	36.675	36.650	37.251	37.904	37.893	38.598	38.579
0.32	35.090	35.690	35.676	36.259	36.881	36.872	37.546	37.525
0.34	34.131	34.741	34.735	35.310	35.909	35.902	36.545	36.525
0.35	33.663	34.279	34.276	34.850	35.440	35.434	36.063	36.043
0.36	33.202	33.824	33.824	34.399	34.981	34.977	35.593	35.574
0.38	32.299	32.936	32.941	33.520	34.094	34.091	34.685	34.668
0.40	31.424	32.075	32.082	32.671	33.241	33.240	33.818	33.802
0.42	30.575	31.238	31.248	31.847	32.419	32.419	32.986	32.972
0.44	29.753	30.427	30.437	31.047	31.624	31.625	32.187	32.173
0.45	29.352	30.030	30.040	30.656	31.236	31.238	31.798	31.785
0.46	28.959	29.640	29.650	30.271	30.854	30.856	31.415	31.403
0.48	28.194	28.877	28.887	29.517	30.107	30.110	30.670	30.659
0.50	27.458	28.141	28.149	28.785	29.382	29.385	29.948	29.939
0.55	25.748	26.412	26.418	27.054	27.661	27.664	28.238	28.231
0.60	24.226	24.851	24.855	25.470	26.072	26.074	26.652	26.649
0.65	22.885	23.459	23.460	24.038	24.619	24.620	25.189	25.189
0.70	21.711	22.228	22.227	22.758	23.303	23.303	23.851	23.854
0.80	19.783	20.193	20.191	20.618	21.072	21.071	21.547	21.555
0.90	18.262	18.599	18.598	18.943	19.310	19.309	19.701	19.709
1.00	16.986	17.293	17.292	17.591	17.900	17.900	18.224	18.227
1.10	15.841	16.150	16.150	16.438	16.722	16.721	17.008	17.003
1.20	14.759	15.090	15.091	15.390	15.676	15.676	15.953	15.941
1.30	13.712	14.072	14.072	14.396	14.700	14.701	14.988	14.970
1.40	12.698	13.082	13.082	13.432	13.759	13.760	14.067	14.048
1.50	11.726	12.125	12.126	12.494	12.845	12.844	13.175	13.154
1.60	10.811	11.214	11.214	11.592	11.956	11.956	12.305	12.285
1.70	9.966	10.360	10.360	10.736	11.104	11.104	11.461	11.447
1.80	9.201	9.576	9.577	9.940	10.303	10.302	10.661	10.649
1.90	8.518	8.868	8.868	9.212	9.558	9.559	9.907	9.902
2.00	7.921	8.239	8.239	8.556	8.881	8.882	9.213	9.213

$\sin \vartheta / \lambda$	$_{57}\text{La}$	$_{57}\text{La}^{3+}$	$_{58}\text{Ce}$	$_{58}\text{Ce}^{3+}$	$_{58}\text{Ce}^{4+}$	$_{59}\text{Pr}$	$_{59}\text{Pr}^{3+}$	$_{59}\text{Pr}^{4+}$
0.00	57.000	54.000	58.000	55.000	54.000	59.000	56.000	55.000
0.01	56.926	53.971	57.928	54.972	53.974	58.929	55.972	54.975
0.02	56.708	53.885	57.715	54.886	53.897	58.722	55.888	54.898
0.03	56.360	53.742	57.375	54.745	53.769	58.392	55.748	54.772
0.04	55.900	53.544	56.924	54.549	53.592	57.956	55.555	54.597
0.05	55.351	53.293	56.385	54.300	53.366	57.439	55.309	54.373
0.06	54.736	52.991	55.779	54.001	53.094	56.861	55.013	54.104
0.07	54.076	52.640	55.127	53.654	52.778	56.242	54.669	53.791
0.08	53.388	52.245	54.446	53.261	52.420	55.599	54.280	53.436
0.09	52.687	51.808	53.750	52.827	52.022	54.943	53.850	53.042
0.10	51.982	51.332	53.047	52.355	51.589	54.281	53.381	52.612
0.11	51.278	50.823	52.345	51.848	51.122	53.617	52.878	52.148
0.12	50.580	50.284	51.646	51.310	50.625	52.952	52.343	51.654
0.13	49.888	49.718	50.952	50.745	50.102	52.288	51.781	51.133
0.14	49.202	49.130	50.263	50.158	49.555	51.623	51.195	50.588
0.15	48.523	48.524	49.579	49.551	48.988	50.957	50.589	50.022
0.16	47.849	47.903	48.901	48.928	48.404	50.289	49.966	49.439
0.17	47.182	47.272	48.227	48.294	47.807	49.620	49.331	48.841
0.18	46.519	46.633	47.557	47.651	47.199	48.950	48.686	48.231
0.19	45.862	45.989	46.892	47.002	46.583	48.280	48.034	47.613
0.20	45.212	45.344	46.233	46.351	45.963	47.610	47.378	46.989
0.22	43.932	44.061	44.933	45.052	44.718	46.278	46.066	45.733
0.24	42.686	42.801	43.663	43.771	43.481	44.967	44.767	44.481
0.25	42.078	42.183	43.042	43.142	42.871	44.323	44.128	43.861
0.26	41.481	41.576	42.432	42.522	42.268	43.688	43.497	43.248
0.28	40.321	40.396	41.244	41.315	41.088	42.448	42.266	42.046
0.30	39.212	39.267	40.104	40.157	39.950	41.256	41.080	40.882
0.32	38.153	38.190	39.014	39.050	38.859	40.113	39.945	39.763
0.34	37.145	37.166	37.975	37.996	37.817	39.022	38.860	38.692
0.35	36.659	36.673	37.474	37.488	37.314	38.496	38.337	38.174
0.36	36.185	36.192	36.985	36.992	36.823	37.982	37.826	37.669
0.38	35.270	35.266	36.040	36.037	35.877	36.989	36.842	36.693
0.40	34.397	34.384	35.139	35.127	34.977	36.042	35.903	35.763
0.42	33.562	33.541	34.277	34.258	34.118	35.137	35.007	34.876
0.44	32.760	32.734	33.451	33.427	33.298	34.269	34.150	34.028
0.45	32.370	32.342	33.051	33.025	32.901	33.849	33.736	33.619
0.46	31.988	31.959	32.658	32.630	32.513	33.437	33.329	33.218
0.48	31.243	31.212	31.893	31.863	31.759	32.635	32.541	32.440
0.50	30.523	30.492	31.154	31.124	31.034	31.862	31.782	31.693
0.55	28.817	28.789	29.409	29.382	29.329	30.040	29.996	29.939
0.60	27.231	27.211	27.791	27.771	27.753	28.358	28.348	28.323
0.65	25.759	25.748	26.289	26.278	26.290	26.803	26.822	26.826
0.70	24.401	24.398	24.901	24.899	24.933	25.370	25.411	25.437
0.80	22.031	22.039	22.469	22.479	22.532	22.867	22.927	22.976
0.90	20.106	20.117	20.481	20.495	20.543	20.824	20.881	20.927
1.00	18.561	18.568	18.881	18.892	18.926	19.182	19.222	19.256
1.10	17.300	17.299	17.583	17.585	17.605	17.854	17.874	17.895
1.20	16.227	16.218	16.491	16.485	16.495	16.745	16.749	16.760
1.30	15.265	15.249	15.526	15.513	15.519	15.776	15.769	15.775
1.40	14.362	14.341	14.633	14.614	14.620	14.888	14.875	14.880
1.50	13.489	13.467	13.776	13.754	13.763	14.042	14.027	14.034
1.60	12.636	12.616	12.939	12.919	12.931	13.218	13.207	13.217
1.70	11.807	11.791	12.123	12.105	12.120	12.414	12.407	12.419
1.80	11.009	10.997	11.319	11.319	11.335	11.631	11.629	11.644
1.90	10.253	10.246	10.568	10.568	10.585	10.878	10.881	10.897
2.00	9.550	9.545	9.860	9.860	9.877	10.166	10.171	10.187

$\sin\vartheta/\lambda$	$_{60}$Nd	$_{60}$Nd^{3+}	$_{61}$Pm	$_{61}$Pm^{3+}	$_{62}$Sm	$_{62}$Sm^{3+}	$_{63}$Eu	$_{63}$Eu^{2+}
0.00	60.000	57.000	61.000	58.000	62.000	59.000	63.000	61.000
0.01	59.931	56.972	60.932	57.973	61.934	58.973	62.036	60.970
0.02	59.728	56.889	60.734	57.891	61.740	58.892	62.746	60.881
0.03	59.404	56.752	60.417	57.755	61.428	58.759	62.441	60.732
0.04	58.977	56.561	59.998	57.567	61.017	58.573	62.036	60.527
0.05	58.468	56.318	59.497	57.328	60.525	58.337	61.552	60.266
0.06	57.899	56.026	58.936	57.039	59.972	58.052	61.007	59.952
0.07	57.288	55.686	58.333	56.704	59.377	57.721	60.419	59.587
0.08	56.651	55.302	57.703	56.324	58.753	57.345	59.801	59.175
0.09	56.000	54.876	57.057	55.902	58.113	56.929	59.166	58.718
0.10	55.342	54.411	56.403	55.442	57.463	56.473	58.521	58.222
0.11	54.680	53.911	55.744	54.947	56.809	55.982	57.869	57.688
0.12	54.017	53.380	55.084	54.420	56.151	55.460	57.214	57.122
0.13	53.354	52.821	54.422	53.864	55.491	54.908	56.555	56.527
0.14	52.689	52.237	53.758	53.284	54.828	54.330	55.893	55.906
0.15	52.022	51.632	53.091	52.681	54.163	53.731	55.228	55.264
0.16	51.353	51.010	52.422	52.061	53.493	53.112	54.559	54.604
0.17	50.682	50.374	51.749	51.425	52.821	52.478	53.886	53.930
0.18	50.009	49.727	51.074	50.178	52.145	51.831	53.210	53.245
0.19	49.334	49.013	50.398	50.122	51.467	51.175	52.530	52.552
0.20	48.660	48.414	49.720	49.461	50.786	50.512	51.847	51.854
0.22	47.317	47.091	48.361	48.130	49.426	49.115	50.480	50.454
0.24	45.989	45.778	47.026	46.804	48.074	41.839	49.119	49.062
0.25	45.336	45.129	46.364	46.148	47.406	47.176	48.444	48.374
0.26	44.690	44.489	45.710	45.499	46.743	46.519	47.775	47.694
0.28	43.428	43.235	44.427	44.226	45.443	45.228	46.458	46.361
0.30	42.210	42.025	43.186	42.993	44.180	43.975	45.176	45.069
0.32	41.040	40.863	41.991	41.805	42.961	42.764	43.935	43.825
0.34	39.920	39.750	40.844	40.666	41.789	41.600	42.740	42.629
0.35	39.379	39.213	40.289	40.115	41.221	41.036	42.160	42.050
0.36	38.851	38.688	39.747	39.576	40.666	40.484	41.591	41.484
0.38	37.830	37.675	38.697	38.534	39.589	39.416	40.489	40.387
0.40	36.854	36.708	37.694	37.540	38.559	38.395	39.433	39.338
0.42	35.922	35.785	36.135	36.590	37.573	37.418	38.421	38.333
0.44	35.029	34.903	35.815	35.681	36.627	36.483	37.451	37.371
0.45	34.596	34.476	35.370	35.241	36.169	36.031	36.980	36.904
0.46	34.171	34.057	34.933	34.810	35.720	35.587	36.519	36.447
0.48	33.347	33.246	34.085	33.975	34.848	34.728	35.623	35.560
0.50	32.553	32.465	33.269	33.172	34.008	33.902	34.761	34.707
0.55	30.683	30.631	31.349	31.287	32.036	31.965	32.137	32.702
0.60	28.960	28.943	29.581	29.555	30.222	30.188	30.871	30.861
0.65	27.367	27.380	27.948	27.955	28.547	28.547	29.161	29.160
0.70	25.899	25.936	26.442	26.475	27.002	27.029	27.576	27.589
0.80	23.325	23.387	23.796	23.858	24.281	24.342	24.781	24.811
0.90	21.214	21.275	21.616	21.681	22.030	22.098	22.459	22.494
1.00	19.513	19.559	19.853	19.905	20.202	20.260	20.565	20.599
1.10	18.139	18.166	18.430	18.464	18.728	18.768	19.035	19.061
1.20	17.003	17.012	11.262	17.277	17.523	17.544	17.789	17.805
1.30	16.024	16.020	16.266	16.267	16.507	16.512	16.747	16.753
1.40	15.138	15.126	15.378	15.370	15.613	15.607	15.841	15.839
1.50	14.303	14.288	14.551	14.538	14.790	14.778	15.020	15.010
1.60	13.493	13.481	13.755	13.743	14.005	13.993	14.245	14.231
1.70	12.704	12.695	12.980	12.970	13.243	13.232	13.494	13.480
1.80	11.932	11.928	12.220	12.215	12.497	12.490	12.763	12.748
1.90	11.185	11.186	11.481	11.481	11.767	11.765	12.044	12.032
2.00	10.473	10.476	10.773	10.774	11.064	11.063	11.345	11.336

$\sin \vartheta/\lambda$	$_{63}Eu^{3+}$	$_{64}Gd$	$_{64}Gd^{3+}$	$_{65}Tb$	$_{65}Tb^{3+}$	$_{66}Dy$	$_{66}Dy^{3+}$	$_{67}Ho$
0.00	60.000	64.000	61.000	65.000	62.000	66.000	63.000	67.000
0.01	59.973	63.936	60.974	64.938	61.974	65.939	62.975	66.940
0.02	59.894	63.749	60.895	64.755	61.896	65.760	62.898	66.763
0.03	59.762	63.447	60.765	64.461	61.767	65.471	62.772	66.476
0.04	59.579	63.044	60.585	64.071	61.588	65.088	62.596	66.093
0.05	59.347	62.557	60.355	63.603	61.360	64.627	62.373	65.627
0.06	59.066	62.004	60.077	63.073	61.086	64.105	62.102	65.096
0.07	58.739	61.400	59.754	62.499	60.766	63.538	61.787	64.513
0.08	58.368	60.762	59.387	61.894	60.403	62.940	61.429	63.895
0.09	57.956	60.102	58.980	61.270	60.000	62.321	61.031	63.251
0.10	57.505	59.427	58.534	60.634	59.559	61.689	60.595	62.591
0.11	57.019	58.746	58.053	59.989	59.082	61.049	60.124	61.921
0.12	56.501	58.061	57.539	59.340	58.574	60.403	59.620	61.247
0.13	55.954	57.375	56.996	58.686	58.036	59.752	59.086	60.569
0.14	55.380	56.690	56.427	58.029	57.471	59.097	58.525	59.891
0.15	54.784	56.005	55.834	57.366	56.883	58.437	57.940	59.212
0.16	54.168	55.321	55.222	56.699	56.274	57.771	57.334	58.532
0.17	54.536	54.637	54.592	56.028	55.647	57.101	56.710	57.851
0.18	52.890	53.953	53.948	55.351	55.006	56.425	56.070	57.169
0.19	52.234	53.270	53.292	54.670	54.353	55.744	55.417	56.486
0.20	51.570	52.588	52.628	53.985	53.689	55.059	54.754	55.803
0.22	50.228	51.227	51.283	52.610	52.344	53.681	53.407	54.435
0.24	48.884	49.878	49.933	51.234	50.989	52.300	52.046	53.070
0.25	48.216	49.209	49.260	50.549	50.312	51.611	51.366	52.390
0.26	47.553	48.546	48.591	49.868	49.639	50.926	50.688	51.714
0.28	46.246	47.240	47.270	48.523	48.306	49.570	49.344	50.375
0.30	44.973	45.965	45.980	47.208	47.001	48.240	48.025	49.059
0.32	43.741	44.729	44.728	45.929	45.731	46.944	46.738	47.772
0.34	42.553	43.533	43.519	44.690	44.501	45.686	45.489	46.520
0.35	41.977	42.951	42.931	44.087	43.902	45.073	44.880	45.908
0.36	41.412	42.880	42.354	43.496	43.314	44.471	44.282	45.305
0.38	40.319	41.272	41.236	42.346	42.172	43.299	43.118	44.131
0.40	39.271	40.207	40.163	41.241	41.075	42.171	41.998	42.996
0.42	38.268	39.184	39.135	40.179	40.021	41.086	40.921	41.903
0.44	37.307	38.203	38.149	39.160	39.010	40.042	39.887	40.849
0.45	36.842	37.726	37.671	38.665	38.520	39.536	39.385	40.337
0.46	36.386	37.259	37.203	38.180	38.040	39.039	38.893	39.834
0.48	35.503	36.352	36.295	37.237	37.108	38.073	37.938	38.856
0.50	34.653	35.479	35.423	36.329	36.212	37.143	37.019	37.914
0.55	32.663	33.428	33.379	34.199	34.113	34.958	34.866	35.699
0.60	30.838	31.543	31.506	32.243	32.191	32.953	32.894	33.664
0.65	29.155	29.802	29.779	30.438	30.420	31.103	31.078	31.786
0.70	27.599	28.192	28.183	28.772	28.784	29.394	29.400	30.049
0.80	24.840	35.335	25.351	25.822	25.876	26.366	26.416	26.958
0.90	22.528	22.940	22.969	23.353	23.424	23.821	23.892	24.343
1.00	20.626	20.970	21.003	21.323	21.392	21.721	21.793	22.167
1.10	19.080	19.372	19.400	19.675	19.730	20.011	20.072	20.385
1.20	17.815	18.072	18.092	18.338	18.373	18.623	18.666	18.934
1.30	16.758	16.995	17.004	17.234	17.252	17.483	17.508	17.746
1.40	15.840	16.072	16.071	16.296	16.298	16.522	16.531	16.753
1.50	15.011	15.247	15.237	15.465	15.457	15.680	15.678	15.895
1.60	14.233	14.477	14.463	14.697	14.685	14.913	14.904	15.123
1.70	13.483	13.741	13.724	13.968	13.953	14.190	14.178	14.406
1.80	12.753	13.022	13.005	13.259	13.245	13.491	13.479	13.718
1.90	12.039	12.317	12.302	12.564	12.554	12.808	12.798	13.047
2.00	11.344	11.631	11.616	11.886	11.878	12.141	12.132	12.392

$\sin\vartheta/\lambda$	$_{67}\text{Ho}^{3+}$	$_{68}\text{Er}$	$_{68}\text{Er}^{3+}$	$_{69}\text{Tm}$	$_{69}\text{Tm}^{3+}$	$_{70}\text{Yb}$	$_{70}\text{Yb}^{2+}$	$_{70}\text{Yb}^{3+}$
0.00	64.000	68.000	65.000	69.000	66.000	70.000	68.000	67.000
0.01	63.975	67.941	64.975	68.943	65.976	69.944	67.973	66.976
0.02	63.900	67.769	64.901	68.773	65.903	69.777	67.892	66.904
0.03	63.775	67.491	64.779	68.500	65.782	69.509	67.759	66.785
0.04	63.602	67.120	64.608	68.136	65.613	69.151	67.573	66.619
0.05	63.382	66.673	64.391	67.696	65.399	68.717	67.337	66.407
0.06	63.115	66.166	64.128	67.195	65.139	68.223	67.051	66.151
0.07	62.804	65.613	63.821	66.649	64.836	67.684	66.719	65.851
0.08	62.451	65.028	63.472	66.070	64.491	67.112	66.342	65.511
0.09	62.058	64.420	63.083	65.468	64.107	66.516	65.922	65.131
0.10	61.626	63.798	62.657	64.852	63.685	65.904	65.464	64.714
0.11	61.160	63.167	62.195	64.224	63.228	65.281	64.968	64.262
0.12	60.660	62.528	61.701	63.589	62.739	65.650	64.439	63.777
0.13	60.131	61.884	61.176	62.948	62.219	64.012	63.879	63.262
0.14	59.574	61.234	60.624	62.301	61.671	63.368	63.292	62.719
0.15	58.993	60.578	60.047	61.648	61.099	62.718	62.679	62.151
0.16	58.391	59.917	59.448	60.989	60.504	62.062	62.046	61.561
0.17	57.769	59.249	58.830	60.324	59.889	61.399	61.393	60.950
0.18	57.132	58.576	58.196	59.653	59.258	60.729	60.724	60.321
0.19	56.481	57.897	57.547	58.975	58.611	60.053	60.043	59.678
0.20	55.819	57.213	56.886	58.292	57.953	59.371	59.350	59.022
0.22	54.471	55.833	55.540	56.912	56.608	57.992	57.943	57.679
0.24	53.107	54.445	54.174	55.521	55.241	56.601	56.521	56.312
0.25	52.424	53.750	53.489	54.825	54.554	55.903	55.809	55.624
0.26	51.742	53.058	52.804	54.130	53.866	55.206	55.098	54.935
0.28	50.389	51.683	51.442	52.748	52.498	53.817	53.687	53.560
0.30	49.057	50.329	50.099	51.384	51.145	52.444	52.297	52.198
0.32	47.755	49.004	48.783	50.046	49.817	51.095	50.937	50.858
0.34	46.489	47.712	47.501	48.739	48.520	49.774	49.611	49.548
0.35	45.871	47.081	46.874	48.099	47.884	49.127	48.962	48.904
0.36	45.263	46.459	46.256	47.469	47.258	48.488	48.323	48.270
0.38	44.078	45.246	45.052	46.237	46.035	47.239	47.076	47.030
0.40	42.936	44.075	43.889	45.046	44.853	46.029	45.871	45.828
0.42	41.837	42.945	42.768	43.896	43.711	44.859	44.707	44.667
0.44	40.780	41.857	41.689	42.786	42.611	43.728	43.585	43.545
0.45	40.267	41.327	41.164	42.246	42.075	43.178	43.038	42.999
0.46	39.764	40.808	40.649	41.715	41.550	42.637	42.502	42.463
0.48	38.786	39.797	39.649	40.682	40.527	41.583	41.458	41.419
0.50	37.844	38.822	38.685	39.686	39.542	40.565	40.450	40.412
0.55	35.637	36.531	36.424	37.342	37.228	38.169	38.080	38.046
0.60	33.615	34.425	34.352	35.187	35.106	35.964	35.901	35.874
0.65	31.753	32.483	32.444	33.198	33.151	33.929	33.890	33.873
0.70	30.032	30.688	30.680	31.359	31.344	32.045	32.029	32.022
0.80	26.970	27.497	27.540	28.086	28.123	.28.690	28.709	28.722
0.90	24.374	24.800	24.870	25.311	25.380	25.837	25.880	25.904
1.00	22.207	22.556	22.634	22.995	23.074	23.447	23.501	23.527
1.10	20.424	20.718	20.787	21.089	21.163	21.474	21.528	21.551
1.20	18.966	19.221	19.276	19.535	19.595	19.860	19.908	19.926
1.30	17.768	17.998	18.035	18.266	18.309	18.542	18.580	18.591
1.40	16.764	16.980	17.000	17.215	17.241	17.454	17.480	17.486
1.50	15.896	16.107	16.114	16.321	16.332	16.536	16.550	16.553
1.60	15.118	15.329	15.327	15.533	15.534	15.735	15.740	15.741
1.70	14.394	14.612	14.604	14.815	14.809	15.013	15.009	15.010
1.80	13.703	13.929	13.919	14.137	14.127	14.338	14.330	14.330
1.90	13.032	13.267	13.257	13.483	13.473	13.691	13.681	13.682
2.00	12.376	12.621	12.610	12.847	12.836	13.064	13.051	13.053

$\sin\vartheta/\lambda$	$_{71}$Lu	$_{71}$Lu^{3+}	$_{72}$Hf	$_{72}$Hf^{4+}	$_{73}$Ta	$_{73}$Ta^{5+}	$_{74}$W	$_{74}$W^{6+}
0.00	71.000	68.000	72.000	68.000	73.000	68.000	74.000	68.000
0.01	70.944	67.976	71.945	67.979	72.946	67.981	73.948	67.982
0.02	70.778	67.905	71.783	67.915	72.788	67.922	73.793	67.929
0.03	70.509	67.788	71.518	67.809	72.529	67.826	73.539	67.840
0.04	70.148	67.624	71.161	67.661	72.177	67.691	73.194	67.716
0.05	69.707	67.415	70.723	67.472	71.745	67.519	72.767	67.557
0.06	69.202	67.161	70.217	67.243	71.242	67.309	72.269	67.365
0.07	68.646	66.866	69.656	66.976	70.680	67.065	71.711	67.139
0.08	68.051	66.529	69.052	66.670	70.072	66.785	71.103	66.881
0.09	67.429	66.154	68.416	66.329	69.428	66.471	70.455	66.592
0.10	66.789	65.741	67.757	65.953	68.758	66.126	69.778	66.272
0.11	66.137	65.294	67.083	65.544	68.069	65.749	69.078	65.923
0.12	65.477	64.814	66.400	65.103	67.367	65.343	68.363	65.546
0.13	64.813	64.303	65.711	64.634	66.658	64.908	67.637	65.142
0.14	64.146	63.765	65.019	64.138	65.944	64.448	66.906	64.713
0.15	63.478	63.201	64.326	63.616	65.229	63.963	66.172	64.260
0.16	62.807	62.615	63.634	63.071	64.515	63.455	65.437	63.785
0.17	62.134	62.008	62.942	62.505	63.802	62.926	64.703	63.290
0.18	61.460	61.383	62.251	61.921	63.090	62.378	63.972	62.775
0.19	60.783	60.742	61.560	61.319	62.382	61.812	63.243	62.242
0.20	60.103	60.088	60.870	60.703	61.675	61.231	62.519	61.693
0.22	58.739	58.749	59.492	59.433	60.271	60.028	61.082	60.553
0.24	57.369	57.382	58.119	58.127	58.880	58.783	59.663	59.367
0.25	56.683	56.693	57.434	57.465	58.189	58.149	58.961	58.760
0.26	55.998	56.002	56.752	56.799	57.502	57.509	58.265	58.146
0.28	54.634	54.622	55.396	55.460	56.141	56.216	56.888	56.901
0.30	53.282	53.253	54.054	54.123	54.799	54.917	55.536	55.643
0.32	51.950	51.903	52.733	52.796	53.479	53.620	54.210	54.381
0.34	50.642	50.580	51.435	51.487	52.185	52.334	52.912	53.122
0.35	49.998	49.930	50.796	50.842	51.548	51.696	52.274	52.496
0.36	49.363	49.288	50.164	50.203	50.918	51.064	51.644	51.873
0.38	48.117	48.032	48.924	48.947	49.683	49.817	50.408	50.641
0.40	46.906	46.813	47.717	47.723	48.479	48.597	49.205	49.430
0.42	45.731	45.633	46.543	46.534	47.308	47.406	48.036	48.244
0.44	44.593	44.491	45.405	45.381	46.171	46.247	46.900	47.086
0.45	44.038	43.935	44.849	44.818	45.615	45.681	46.344	46.518
0.46	43.492	43.389	44.301	44.265	45.068	45.122	45.797	45.957
0.48	42.427	42.325	43.232	43.184	43.998	44.031	44.728	44.860
0.50	41.398	41.297	42.197	42.140	42.962	42.974	43.691	43.795
0.55	38.970	38.880	39.752	39.680	40.508	40.479	41.236	41.271
0.60	36.733	36.658	37.494	37.419	38.238	38.181	38.960	38.942
0.65	34.666	34.610	35.404	35.335	36.132	36.064	36.846	36.793
0.70	32.752	32.716	33.465	33.409	34.175	34.106	34.878	34.806
0.80	29.334	29.335	29.992	29.970	30.658	30.612	31.327	31.258
0.90	26.413	26.442	27.008	27.018	27.618	27.605	28.238	28.199
1.00	23.950	23.994	24.473	24.506	25.016	25.033	25.576	25.572
1.10	21.902	21.952	22.352	22.396	22.823	22.858	23.313	23.336
1.20	20.219	20.267	20.598	20.645	20.998	21.042	21.418	21.456
1.30	18.842	18.883	19.159	19.202	19.494	19.538	19.847	19.890
1.40	17.709	17.738	17.975	18.009	18.256	18.293	18.552	18.592
1.50	16.759	16.777	16.988	17.011	17.228	17.255	17.478	17.510
1.60	15.939	15.947	16.145	16.158	16.356	16.374	16.575	16.597
1.70	15.208	15.208	15.403	15.406	15.598	15.605	15.796	15.808
1.80	14.534	14.528	14.727	14.722	14.916	14.914	15.104	15.106
1.90	13.894	13.884	14.091	14.081	14.282	14.274	14.469	14.463
2.00	13.277	13.263	13.481	13.467	13.679	13.666	13.871	13.858

$\sin\vartheta/\lambda$	$_{75}$Re	$_{76}$Os	$_{76}$Os^{4+}	$_{77}$Ir	$_{77}$Ir^{3+}	$_{77}$Ir^{4+}	$_{78}$Pt	$_{78}$Pt^{2+}
0.00	75.000	76.000	72.000	77.000	74.000	73.000	78.000	76.000
0.01	74.949	75.950	71.976	76.951	73.972	72.975	77.955	75.968
0.02	74.797	75.801	71.904	76.806	73.889	72.902	77.820	75.874
0.03	74.548	75.558	71.784	76.567	73.752	72.780	77.599	75.717
0.04	74.209	75.225	71.617	76.240	73.561	72.611	77.295	75.499
0.05	73.788	74.810	71.404	75.832	73.318	72.395	76.914	75.222
0.06	73.295	74.323	71.147	75.352	73.024	72.133	76.462	74.889
0.07	72.740	73.772	70.847	74.806	72.682	71.828	75.946	74.502
0.08	72.132	73.167	70.506	74.206	72.294	71.481	75.373	74.065
0.09	71.482	72.518	70.125	73.558	71.863	71.094	74.751	73.580
0.10	70.799	71.832	69.707	72.872	71.392	70.669	74.086	73.052
0.11	70.091	71.119	69.254	72.156	70.883	70.208	73.386	72.485
0.12	69.365	70.384	68.769	71.416	70.339	69.715	72.656	71.881
0.13	68.625	69.634	68.253	70.658	69.764	69.190	71.902	71.245
0.14	67.878	68.874	67.711	69.887	69.162	68.638	71.130	70.582
0.15	67.126	68.107	67.143	69.108	68.534	68.060	70.343	69.894
0.16	66.372	67.337	66.552	68.324	67.884	67.460	69.546	69.185
0.17	65.619	66.566	65.942	67.538	67.215	66.839	68.742	68.459
0.18	64.868	65.797	65.313	66.752	66.530	66.200	67.934	67.719
0.19	64.121	65.031	64.670	65.969	65.832	65.546	67.125	66.968
0.20	63.378	64.269	64.014	65.189	65.123	64.879	66.317	66.210
0.22	61.906	62.761	62.671	63.645	63.684	63.515	64.709	64.679
0.24	60.457	61.278	61.302	62.127	62.228	62.124	63.125	63.144
0.25	59.742	60.548	60.612	61.380	61.500	61.423	62.344	62.381
0.26	59.034	59.825	59.920	60.641	60.773	60.721	61.571	61.621
0.28	57.637	58.403	58.537	59.189	59.328	59.319	60.056	60.121
0.30	56.270	57.013	57.164	57.773	57.905	57.927	58.582	58.652
0.32	54.932	55.658	55.809	56.395	56.510	56.555	57.152	57.220
0.34	53.627	54.339	54.478	55.056	55.148	55.209	55.769	55.830
0.35	52.986	53.692	53.823	54.401	54.481	54.548	55.094	55.151
0.36	52.354	53.055	53.175	53.756	53.823	53.894	54.432	54.483
0.38	51.114	51.807	51.906	52.496	52.538	52.613	53.141	53.182
0.40	49.910	50.596	50.671	51.274	51.293	51.369	51.897	51.925
0.42	48.739	49.422	49.473	50.091	50.089	50.163	50.697	50.714
0.44	47.603	48.283	48.312	48.946	48.926	48.995	49.540	49.545
0.45	47.048	47.726	47.745	48.387	48.359	48.425	48.977	48.977
0.46	46.501	47.179	47.187	47.837	47.802	47.865	48.424	48.419
0.48	45.432	46.109	46.099	46.765	46.717	46.773	47.347	47.333
0.50	44.396	45.072	45.046	45.726	45.668	45.717	46.308	46.286
0.55	41.940	42.617	42.562	43.269	43.197	43.227	43.860	43.820
0.60	39.662	40.340	40.270	40.994	40.918	40.932	41.601	41.549
0.65	37.544	38.222	38.147	38.878	38.805	38.807	39.502	39.443
0.70	35.569	36.244	36.172	36.901	36.835	36.830	37.539	37.479
0.80	31.993	32.654	32.602	33.305	33.261	33.249	33.958	33.905
0.90	28.865	29.495	29.471	30.125	30.104	30.094	30.766	30.732
1.00	26.148	26.732	26.734	27.323	27.324	27.317	27.930	27.918
1.10	23.821	24.345	24.368	24.882	24.902	24.898	25.437	25.447
1.20	21.856	22.314	22.350	22.789	22.821	22.820	23.281	23.307
1.30	20.219	20.610	20.652	21.019	21.057	21.058	21.445	21.481
1.40	18.864	19.194	19.234	19.541	19.579	19.581	19.902	19.942
1.50	17.742	18.019	18.054	18.312	18.347	18.348	18.616	18.654
1.60	16.801	17.038	17.066	17.287	17.315	17.317	17.545	17.578
1.70	15.998	16.206	16.225	16.422	16.443	16.444	16.644	16.670
1.80	15.293	15.483	15.493	15.678	15.691	15.691	15.875	15.893
1.90	14.653	14.835	14.838	15.018	15.024	15.024	15.202	15.211
2.00	14.057	14.239	14.234	14.418	14.417	14.416	14.595	14.597

sin ϑ/λ	$_{78}Pt^{4+}$	$_{79}Au$	$_{79}Au^{1+}$	$_{79}Au^{3+}$	$_{80}Hg$	$_{80}Hg^{1+}$	$_{80}Hg^{2+}$	$_{81}Tl$
0.00	74.000	79.000	78.000	76.000	80.000	79.000	78.000	81.000
0.01	73.975	78.957	77.964	75.972	79.956	78.962	77.968	80.950
0.02	73.901	78.826	77.855	75.888	79.819	78.850	77.875	80.799
0.03	73.778	78.609	77.676	75.750	79.595	78.664	77.719	80.553
0.04	73.606	78.311	77.428	75.557	79.286	78.406	77.503	80.217
0.05	73.387	77.936	77.113	75.311	78.899	78.080	77.229	79.798
0.06	73.123	77.491	76.736	75.015	78.439	77.689	76.897	79.305
0.07	72.814	76.981	76.299	74.669	77.913	77.238	76.512	78.748
0.08	72.462	76.414	75.807	74.276	77.330	76.731	76.076	78.134
0.09	72.070	75.797	75.264	73.839	76.696	76.173	75.591	77.473
0.10	71.639	75.135	74.676	73.361	76.018	75.570	75.062	76.773
0.11	71.173	74.437	74.046	72.843	75.303	74.925	74.492	76.042
0.12	70.673	73.706	73.380	72.290	74.559	74.245	73.884	75.284
0.13	70.141	72.950	72.683	71.705	73.790	73.535	73.243	74.507
0.14	69.581	72.173	71.958	71.089	73.001	72.798	72.571	73.715
0.15	68.995	71.380	71.211	70.448	72.198	72.041	71.874	72.912
0.16	68.386	70.575	70.446	69.783	71.385	71.266	71.153	72.101
0.17	67.756	69.761	69.665	69.097	70.564	70.477	70.413	71.285
0.18	67.107	68.941	68.874	68.395	69.740	69.679	69.658	70.467
0.19	66.443	68.119	68.075	67.678	68.914	68.874	68.889	69.648
0.20	65.766	67.296	67.271	66.949	68.088	68.065	68.111	68.830
0.22	64.380	65.657	65.658	65.466	66.447	66.445	66.536	67.205
0.24	62.968	64.039	64.054	63.965	64.828	64.836	64.952	65.600
0.25	62.256	63.241	63.260	63.213	64.029	64.040	64.162	64.807
0.26	61.543	62.452	62.472	62.462	63.239	63.251	63.376	64.022
0.28	60.119	60.902	60.923	60.969	61.687	61.698	61.821	62.478
0.30	58.707	59.395	59.413	59.499	60.177	60.184	60.296	60.970
0.32	57.315	57.935	57.947	58.058	58.711	58.714	58.810	59.503
0.34	55.951	56.523	56.529	56.653	57.292	57.290	57.367	58.079
0.35	55.281	55.835	55.839	55.965	56.600	56.596	56.663	57.383
0.36	54.619	55.160	55.161	55.288	55.920	55.914	55.972	56.698
0.38	53.323	53.846	53.841	53.967	54.595	54.586	54.625	55.362
0.40	52.065	52.581	52.571	52.689	53.318	53.306	53.327	54.072
0.42	50.847	51.363	51.348	51.457	52.088	52.073	52.079	52.826
0.44	49.669	50.191	50.172	50.269	50.902	50.885	50.879	51.625
0.45	49.095	49.622	49.600	49.691	50.326	50.308	50.296	51.041
0.46	48.531	49.063	49.040	49.124	49.761	49.742	49.725	50.467
0.48	47.431	47.976	47.950	48.021	48.661	48.640	48.615	49.352
0.50	46.370	46.929	46.899	46.958	47.601	47.578	47.548	48.276
0.55	43.871	44.469	44.432	44.464	45.113	45.085	45.050	45.753
0.60	41.573	42.207	42.163	42.175	42.829	42.795	42.764	43.442
0.65	39.447	40.110	40.062	40.060	40.718	40.679	40.656	41.313
0.70	37.470	38.153	38.103	38.093	38.753	38.709	38.696	39.337
0.80	33.885	34.581	34.534	34.518	35.176	35.131	35.133	35.755
0.90	30.713	31.387	31.352	31.338	31.980	31.944	31.952	32.561
1.00	27.905	28.530	28.513	28.504	29.112	29.090	29.100	29.687
1.10	25.440	25.998	26.000	25.996	26.554	26.549	26.557	27.109
1.20	23.305	23.789	23.807	23.806	24.303	24.315	24.319	24.824
1.30	21.482	21.892	21.921	21.923	22.354	22.378	22.379	22.827
1.40	19.945	20.287	20.322	20.325	20.692	20.723	20.722	21.110
1.50	18.658	18.943	18.978	18.981	19.290	19.324	19.322	19.652
1.60	17.580	17.821	17.853	17.856	18.116	18.148	18.146	18.424
1.70	16.672	16.880	16.907	16.909	17.131	17.160	17.157	17.394
1.80	15.894	16.081	16.101	16.102	16.298	16.320	16.319	16.524
1.90	15.211	15.388	15.401	15.401	15.581	15.597	15.596	15.780
2.00	14.596	14.770	14.777	14.777	14.949	14.958	14.958	15.131

$\sin \vartheta/\lambda$	$_{81}\text{Tl}^{1+}$	$_{81}\text{Tl}^{3+}$	$_{82}\text{Pb}$	$_{82}\text{Pb}^{2+}$	$_{82}\text{Pb}^{4+}$	$_{83}\text{Bi}$	$_{83}\text{Bi}^{3+}$	$_{83}\text{Bi}^{5+}$
0.00	80.000	78.000	82.000	80.000	78.000	83.000	80.000	78.000
0.01	79.961	77.975	81.949	79.966	77.975	82.947	79.969	77.977
0.02	79.845	77.891	81.792	79.864	77.899	82.784	79.878	77.908
0.03	79.653	77.753	81.536	79.695	77.774	82.518	79.727	77.793
0.04	79.388	77.560	81.186	79.461	77.599	82.154	79.516	77.633
0.05	79.052	77.314	80.750	79.164	77.376	81.700	79.249	77.428
0.06	78.650	77.017	80.237	78.807	77.106	81.167	78.926	77.180
0.07	78.186	76.670	79.656	78.392	76.790	80.563	78.550	76.889
0.08	77.665	76.276	79.018	77.924	76.430	79.901	78.124	76.558
0.09	77.093	75.836	78.332	77.406	76.028	79.189	77.651	76.187
0.10	76.474	75.355	77.607	76.843	75.586	78.438	77.134	75.778
0.11	75.814	74.833	76.851	76.238	75.106	77.657	76.577	75.333
0.12	75.119	74.275	76.071	75.597	74.590	76.852	75.983	74.854
0.13	74.394	73.683	75.274	74.922	74.041	76.032	75.355	74.342
0.14	73.644	73.060	74.464	74.220	73.461	75.202	74.698	73.800
0.15	72.873	72.409	73.645	73.493	72.853	74.365	74.014	73.231
0.16	72.085	71.733	72.822	72.745	72.220	73.527	73.308	72.635
0.17	71.286	71.035	71.997	71.981	71.563	72.689	72.581	72.016
0.18	70.477	70.319	71.172	71.204	70.885	71.855	71.839	71.376
0.19	69.663	69.586	70.349	70.417	70.190	71.026	71.083	70.716
0.20	68.847	68.841	69.530	69.623	69.479	70.203	70.317	70.039
0.22	67.214	67.320	67.907	68.023	68.020	68.578	68.764	68.643
0.24	65.597	65.776	66.310	66.425	66.527	66.987	67.199	67.204
0.25	64.797	65.001	65.423	65.631	65.772	66.204	66.416	66.474
0.26	64.005	64.226	64.743	64.841	65.015	65.430	65.636	65.739
0.28	62.448	62.685	63.210	63.284	63.501	63.909	64.090	64.260
0.30	60.931	61.163	61.712	61.759	61.995	62.425	62.569	62.781
0.32	59.458	59.670	60.253	60.275	60.509	60.977	61.081	61.312
0.34	58.031	58.214	58.833	58.833	59.052	59.566	59.631	59.863
0.35	57.335	57.502	58.138	58.128	58.336	58.875	58.922	59.149
0.36	56.651	56.800	57.453	57.436	57.629	58.193	58.223	58.442
0.38	55.318	55.432	56.116	56.085	56.247	56.859	56.858	57.055
0.40	54.032	54.110	54.820	54.781	54.908	55.563	55.538	55.706
0.42	52.792	52.837	53.567	53.523	53.614	54.306	54.263	54.398
0.44	51.596	51.613	52.356	52.309	52.367	53.089	53.033	53.134
0.45	51.015	51.018	51.766	51.719	51.762	52.495	52.435	52.518
0.46	50.444	50.435	51.187	51.140	51.167	51.910	51.847	51.914
0.48	49.332	49.304	50.058	50.013	50.014	50.771	50.704	50.740
0.50	48.261	48.217	48.969	48.927	48.905	49.669	49.602	49.611
0.55	45.742	45.677	46.411	46.377	46.318	47.077	47.018	46.974
0.60	43.429	43.364	44.069	44.040	43.969	44.700	44.653	44.584
0.65	41.294	41.241	41.914	41.888	41.822	42.517	42.481	42.407
0.70	39.311	39.275	39.921	39.895	39.844	40.501	40.473	40.409
0.80	35.718	35.714	36.322	36.293	36.278	36.879	36.857	36.830
0.90	32.523	32.538	33.127	33.096	33.108	33.680	33.657	33.663
1.00	29.659	29.679	30.252	30.227	30.249	30.805	30.784	30.807
1.10	27.097	27.114	27.662	27.648	27.669	28.208	28.194	28.219
1.20	24.828	24.839	25.350	25.349	25.364	25.875	25.871	25.890
1.30	22.846	22.850	23.313	23.325	23.332	23.804	23.811	23.821
1.40	21.139	21.138	21.546	21.568	21.568	21.992	22.009	22.011
1.50	19.686	19.682	20.034	20.062	20.058	20.429	20.453	20.449
1.60	18.460	18.454	18.754	18.784	18.778	19.097	19.124	19.117
1.70	17.426	17.420	17.674	17.705	17.697	17.969	17.997	17.989
1.80	16.550	16.546	16.764	16.790	16.784	17.017	17.043	17.035
1.90	15.800	15.797	15.989	16.010	16.005	16.207	16.229	16.223
2.00	15.143	15.141	15.317	15.332	15.329	15.510	15.527	15.523

$\sin\vartheta/\lambda$	$_{84}$Po	$_{85}$At	$_{86}$Rn	$_{87}$Fr	$_{88}$Ra	$_{88}$Ra^{2+}	$_{89}$Ac	$_{89}$Ac^{3+}
0.00	84.000	85.000	86.000	87.000	88.000	86.000	89.000	86.000
0.01	83.944	84.944	85.945	86.922	87.915	85.957	88.915	85.961
0.02	83.778	84.776	85.777	86.694	87.664	85.829	88.664	85.846
0.03	83.506	84.502	85.502	86.332	87.263	85.616	88.260	85.655
0.04	83.134	84.125	85.123	85.854	86.734	85.323	87.723	85.390
0.05	82.669	83.654	84.649	85.286	86.104	84.951	87.077	85.054
0.06	82.121	83.098	84.087	84.647	85.397	84.506	86.346	84.651
0.07	81.501	82.466	83.448	83.955	84.638	83.993	85.553	84.183
0.08	80.819	81.770	82.742	83.222	83.845	83.417	84.719	83.656
0.09	80.086	81.020	81.979	82.457	83.030	82.783	83.859	83.074
0.10	79.312	80.226	81.169	81.666	82.202	82.099	82.985	82.441
0.11	78.500	79.398	80.322	80.852	81.368	81.371	82.105	81.765
0.12	77.677	78.545	79.448	80.018	80.528	80.605	81.225	81.048
0.13	76.831	77.674	78.554	79.167	79.685	79.808	80.348	80.298
0.14	75.976	76.794	77.648	78.303	78.839	78.985	79.474	79.519
0.15	75.117	75.908	76.737	77.430	77.990	78.142	78.605	78.716
0.16	74.257	75.023	75.826	76.550	77.138	77.285	77.739	77.895
0.17	73.400	74.143	74.920	75.667	76.285	76.418	76.879	77.059
0.18	72.549	73.269	74.021	74.785	75.431	75.546	76.023	76.213
0.19	71.706	72.405	73.133	73.907	74.578	74.673	75.172	75.362
0.20	70.871	71.553	72.258	73.035	73.728	73.803	74.326	74.508
0.22	69.232	69.885	70.552	71.320	72.043	72.080	72.654	72.805
0.24	67.634	68.269	68.907	69.653	70.389	70.396	71.014	71.125
0.25	66.852	67.481	68.109	68.841	69.576	69.572	70.208	70.300
0.26	66.080	66.706	67.325	68.043	68.775	68.762	69.412	69.485
0.28	64.567	65.193	65.802	66.491	67.210	67.184	67.855	67.893
0.30	63.093	63.725	64.332	64.996	65.696	65.663	66.345	66.355
0.32	61.658	62.301	62.912	63.556	64.235	64.200	64.884	64.873
0.34	60.260	60.915	61.535	62.167	62.826	62.791	63.473	63.446
0.35	59.575	60.236	60.862	61.489	62.140	62.105	62.785	62.753
0.36	58.899	59.566	60.198	60.823	61.466	61.432	62.110	62.072
0.38	57.573	58.253	58.898	59.520	60.151	60.120	60.792	60.748
0.40	56.283	56.974	57.631	58.256	58.879	58.850	59.517	59.470
0.42	55.029	55.728	56.397	57.026	57.646	57.619	58.282	58.235
0.44	53.811	54.515	55.194	55.829	56.448	56.424	57.084	57.037
0.45	53.215	53.921	54.604	55.242	55.862	55.839	56.497	56.452
0.46	52.629	53.335	54.021	54.663	55.284	55.262	55.919	55.875
0.48	51.483	52.189	52.879	53.527	54.151	54.130	54.787	54.746
0.50	50.373	51.075	51.767	52.420	53.048	53.028	53.684	53.647
0.55	47.752	48.435	49.119	49.777	50.413	50.396	51.050	51.023
0.60	45.343	45.997	46.659	47.310	47.948	47.932	48.580	48.561
0.65	43.127	43.750	44.384	45.017	45.646	45.632	46.268	46.255
0.70	41.085	41.678	42.281	42.891	43.504	43.490	44.110	44.100
0.80	37.430	37.980	38.533	39.095	39.664	39.649	40.229	40.220
0.90	34.220	34.751	35.277	35.804	36.335	36.318	36.863	36.851
1.00	31.344	31.872	32.389	32.900	33.408	33.389	33.912	33.896
1.10	28.744	29.271	29.787	30.292	30.790	30.771	31.283	31.264
1.20	26.397	26.915	27.426	27.926	28.418	28.404	28.906	28.890
1.30	24.298	24.794	25.291	25.779	26.263	26.256	26.744	26.734
1.40	22.446	22.909	23.379	23.845	24.312	24.314	24.779	24.777
1.50	20.836	21.256	21.689	22.123	22.564	22.574	23.008	23.015
1.60	19.453	19.826	20.215	20.608	21.014	21.033	21.427	21.443
1.70	18.277	18.602	18.944	19.295	19.660	19.685	20.036	20.058
1.80	17.281	17.562	17.859	18.165	18.488	18.516	18.823	18.849
1.90	16.435	16.677	16.934	17.199	17.481	17.510	17.776	17.804
2.00	15.711	15.922	16.143	16.377	16.623	16.646	16.880	16.904

$\sin\vartheta/\lambda$	$_{90}$Th	$_{90}$Th^{4+}	$_{91}$Pa	$_{92}$U	$_{92}$U^{3+}	$_{92}$U^{4+}	$_{92}$U^{6+}	$_{93}$Np
0.00	90.000	86.000	91.000	92.000	89.000	88.000	86.000	93.000
0.01	89.916	85.965	90.919	91.922	88.961	87.965	85.970	92.922
0.02	89.669	85.860	90.678	91.687	88.846	87.860	85.881	92.691
0.03	89.269	85.686	90.290	91.307	88.654	87.686	85.733	92.318
0.04	88.735	85.444	89.772	90.798	88.389	87.444	85.527	91.817
0.05	88.085	85.137	89.144	90.180	88.051	87.137	85.264	91.208
0.06	87.344	84.767	88.427	89.474	87.646	86.766	84.947	90.510
0.07	86.533	84.337	87.644	88.699	87.175	86.335	84.577	89.742
0.08	85.672	83.851	86.813	87.874	86.643	85.847	84.157	88.923
0.09	84.779	83.313	85.950	87.014	86.054	85.305	83.689	88.067
0.10	83.867	82.725	85.066	86.130	85.414	84.714	83.176	87.186
0.11	82.946	82.094	84.170	85.232	84.727	84.077	82.622	86.288
0.12	82.025	81.423	83.269	84.326	83.998	83.399	82.029	85.380
0.13	81.107	80.717	82.366	83.417	83.233	82.685	81.401	84.467
0.14	80.196	79.981	81.463	82.505	82.436	81.938	80.741	83.550
0.15	79.294	79.218	80.563	81.595	81.612	81.163	80.052	82.632
0.16	78.400	78.433	79.665	80.685	80.766	80.364	79.339	81.715
0.17	77.516	77.631	78.771	79.779	79.903	79.546	78.605	80.799
0.18	76.642	76.815	77.881	78.875	79.027	78.712	77.852	79.885
0.19	75.777	75.990	76.995	77.975	78.142	77.866	77.084	78.973
0.20	74.922	75.158	76.115	77.080	77.253	77.013	76.305	78.066
0.22	73.242	73.488	74.375	75.308	75.471	75.294	74.722	76.267
0.24	71;602	71.827	72.668	73.568	73.705	73.578	73.125	74.496
0.25	70.798	71.006	71.829	72.712	72.834	72.727	72.327	73.624
0.26	70.005	70.193	71.001	71.866	71.972	71.884	71.532	72.763
0.28	68.454	68.598	69.380	70.211	70.286	70.227	69.957	71.074
0.30	66.951	67.050	67.810	68.607	68.654	68.616	68.413	69.436
0.32	65.497	65.554	66.294	67.058	67.081	67.057	66.908	67.853
0.34	64.091	64.112	64.832	65.564	65.569	65.555	65.448	66.326
0.35	63.405	63.410	64.121	64.838	64.835	64.825	64.736	65.584
0.36	62.731	62.722	63.423	64.126	64.117	64.109	64.036	64.857
0.38	61.416	61.384	62.066	62.742	62.723	62.720	62.672	63.443
0.40	60.143	60.094	60.758	61.409	61.383	61.384	61.357	62.083
0.42	58.910	58.850	59.495	60.125	60.095	60.099	60.090	60.775
0.44	57.713	57.646	58.274	58.886	58.854	58.861	58.867	59.514
0.45	57.127	57.059	57.679	58.283	58.251	58.259	58.271	58.901
0.46	56.550	56.481	57.093	57.689	57.657	57.667	57.686	58.298
0.48	55.419	55.350	55.948	56.531	56.501	56.513	56.544	57.124
0.50	54.317	54.252	54.836	55.410	55.381	55.397	55.439	55.989
0.55	51.684	51.633	52.191	52.748	52.725	52.749	52.815	53.303
0.60	49.211	49.176	49.719	50.268	50.251	50.282	50.364	50.808
0.65	46.889	46.869	47.405	47.950	47.938	47.972	48.062	48.483
0.70	44.716	44.706	45.241	45.784	45.774	45.807	45.895	46.312
0.80	40.795	40.794	41.333	41.869	41.860	41.882	41.942	42.390
0.90	37.391	37.387	37.930	38.454	38.443	38.449	38.468	38.966
1.00	34.413	34.402	34.946	35.458	35.443	35.435	35.419	35.961
1.10	31.770	31.753	32.292	32.794	32.776	32.762	32.724	33.289
1.20	29.387	29.370	29.897	30.391	30.373	30.357	30.314	30.879
1.30	27.219	27.206	27.714	28.199	28.184	28.170	28.132	28.680
1.40	25.244	25.238	25.720	26.192	26.183	26.173	26.146	26.662
1.50	23.454	23.456	23.905	24.360	24.357	24.352	24.335	24.813
1.60	21.846	21.858	22.266	22.699	22.703	22.701	22.695	23.128
1.70	20.421	20.439	20.807	21.207	21.219	21.220	21.221	21.609
1.80	19.170	19.194	19.518	19.886	19.902	19.904	19.910	20.253
1.90	18.083	18.111	18.394	18.723	18.745	18.748	18.756	19.055
2.00	17.149	17.174	17.423	17.713	'17.736	17.740	17.748	18.012

$\sin\vartheta/\lambda$	$_{93}Np^{3+}$	$_{93}Np^{4+}$	$_{93}Np^{6+}$	$_{94}Pu$	$_{94}Pu^{3+}$	$_{94}Pu^{4+}$	$_{94}Pu^{6+}$	$_{95}Am$
0.00	90.000	89.000	87.000	94.000	91.000	90.000	88.000	95.000
0.01	89.962	88.965	86.970	93.924	90.962	89.965	87.970	94.926
0.02	89.847	88.860	86.881	93.701	90.848	89.861	87.881	94.706
0.03	89.657	88.687	86.733	93.340	90.660	89.689	81.734	94.352
0.04	89.393	88.446	86.521	92.851	90.398	89.450	81.528	93.817
0.05	89.058	88.140	86.265	92.271	90.066	89.145	87.267	93.299
0.06	88.654	87.770	85.947	91.601	89.665	88.777	86.950	92.638
0.06	88.185	87.340	85.577	90.866	89.199	88.349	86.580	91.910
0.08	87.656	86.853	85.157	90.082	88.673	81.863	86.160	91.131
0.09	87.069	86.312	84.688	89.261	88.089	81.324	85.692	90.315
0.10	86.430	85.721	84.174	88.413	87.453	86.734	85.178	89.470
0.11	85.744	85.084	83.618	81.547	86.169	86.098	84.621	88.605
0.12	85.015	84.405	83.023	86.665	86.041	85.419	84.025	87.723
0.13	84.249	83.688	82.392	85.772	85.275	84.703	83.393	86.829
0.14	83.449	82.939	81.729	84.870	84.475	83.952	82.727	85.924
0.15	82.623	82.160	81.036	83.961	83.646	83.171	82.032	85.011
0.16	81.773	81.357	80.318	83.044	82.794	82.365	81.310	84.090
0.11	80.904	80.533	79.578	82.123	81.921	81.537	80.565	83.163
0.18	80.021	79.693	78.818	81.198	81.033	80.691	79.800	82.231
0.19	79.129	78.840	78.043	80.271	80.134	79.832	79.019	81.296
0.20	78.230	77.977	71.255	79.343	79.227	78.962	78.224	80.360
0.22	76.426	76.237	75.652	77.493	71.403	77.204	76.604	78.490
0.24	74.634	74.497	74.032	75.663	75.587	75.441	74.963	76.636
0.25	73.748	73.634	73.221	74.159	74.688	74.565	74.140	75.119
0.26	72.872	72.776	72.413	73.865	73.797	73.695	73.320	74.811
0.28	71.154	71.089	70.810	72.110	72.048	71.979	71.690	73.027
0.30	69.489	69.447	69.236	70.408	70.351	70.305	70.088	71.293
0.32	61.882	67.856	67.701	68.763	68.711	68.683	68.522	69.615
0.34	66.337	66.322	66.209	61.178	67.133	61.116	66.999	61.997
0.35	65.587	65.576	65.482	66.409	66.367	66.354	66.256	67.212
0.36	64.853	64.845	64.767	65.655	65.616	65.607	65.525	66.441
0.38	63.429	63.426	63.374	64.193	64.161	64.157	64.102	64.947
0.40	62.062	62.063	62.032	62.789	62.765	62.765	62.731	63.513
0.42	60.749	60.753	60.739	61.442	61.425	61.428	61.411	62.137
0.44	59.487	59.492	59.493	60.147	60.138	60.143	60.140	60.816
0.45	58.873	58.880	58.887	59.518	59.513	59.519	59.522	60.175
0.46	58.271	58.279	58.292	58.901	58.900	58.907	58.916	59.546
0.48	57.097	57.108	57.133	57.702	57.702	57.717	57.736	58.325
0.50	55.964	55.978	56.013	56.544	56.551	56.569	56.598	51.148
0.55	53.283	53.305	53.363	53.819	53.845	53.864	53.915	54.385
0.60	50.795	50.823	50.898	51.302	51.337	51.363	51.430	51.842
0.65	48.474	48.507	48.591	48.967	49.004	49.034	49.113	49.490
0.70	46.306	46.338	46.423	46.794	46.828	46.860	46.942	41.307
0.80	42.384	42.408	42.472	42.879	42.898	42.922	42.989	43.380
0.90	38.958	38.966	38.992	39.465	39.463	39.474	39.506	39.958
1.00	35.948	35.943	35.933	36.465	36.445	36.443	36.440	36.952
1.10	33.272	33.259	33.227	33.793	33.763	33.752	33.724	34.276
1.20	30.861	30.846	30.805	31.379	31.346	31.331	31.293	31.858
1.30	28.665	28.651	28.612	29.172	29.142	29.128	29.090	29.648
1.40	26.652	26.642	26.612	27.142	27.121	27.109	27.018	27.611
1.50	24.810	24.803	24.784	25.275	25.264	25.257	25.235	25.133
1.60	23.133	23.130	23.121	23.566	23.567	23.564	23.552	24.006
1.10	21.621	21.621	21.620	22.019	22.030	22.029	22.025	22.435
1.80	20.272	20.274	20.278	20.630	20.650	20.651	20.653	21.018
1.90	19.080	19.083	19.090	19.398	19.424	19.427	19.433	19.754
2.00	18.036	18.039	18.047	18.319	18.346	18.349	18.357	18.640

$\sin\vartheta/\lambda$	$_{96}$Cm	$_{97}$Bk	$_{98}$Cf
0.00	96.000	97.000	98.000
0.01	95.926	96.928	97.929
0.02	95.708	96.713	97.718
0.03	95.354	96.365	97.375
0.04	94.877	95.895	96.912
0.05	94.294	95.320	96.344
0.06	93.623	94.656	95.688
0.07	92.879	93.920	94.961
0.08	92.081	93.129	94.176
0.09	91.241	92.294	93.347
0.10	90.371	91.429	92.486
0.11	89.479	90.540	91.601
0.12	88.573	89.635	90.699
0.13	87.656	88.718	89.783
0.14	86.731	87.793	88.858
0.15	85.802	86.862	87.926
0.16	84.869	85.926	86.989
0.17	83.934	84.988	86.048
0.18	82.998	84.047	85.103
0.19	82.062	83.105	84.157
0.20	81.126	82.163	83.210
0.22	79.263	80.285	81.318
0.24	77.419	78.421	79.437
0.25	76.507	77.498	78.504
0.26	75.603	76.582	77.577
0.28	73.824	74.777	75.749
0.30	72.091	73.016	73.960
0.32	70.409	71.303	72.219
0.34	68.783	69.645	70.531
0.35	67.991	68.838	69.707
0.36	67.214	68.045	68.898
0.38	65.705	66.503	67.325
0.40	64.254	65.020	65.810
0.42	62.859	63.595	64.354
0.44	61.519	62.226	62.954
0.45	60.869	61.562	62.276
0.46	60.231	60.910	61.610
0.48	58.992	59.646	60.319
0.50	57.798	58.430	59.078
0.55	54.998	55.581	56.176
0.60	52.427	52.974	53.528
0.65	50.052	50.574	51.098
0.70	47.850	48.354	48.858
0.80	43.894	44.380	44.859
0.90	40.449	40.926	41.395
1.00	37.426	37.898	38.361
1.10	34.740	35.209	35.671
1.20	32.318	32.786	33.247
1.30	30.106	30.572	31.033
1.40	28.068	28.530	28.989
1.50	26.184	26.639	27.093
1.60	24.446	24.889	25.332
1.70	22.857	23.281	23.708
1.80	21.415	21.815	22.221
1.90	20.121	20.493	20.872
2.00	18.975	19.315	19.665

9 Analytical approximation of Atomic Scattering Factors

In this caption an analytical approximation after [129] of the atomic scattering factors given in chapter 8 is presented [1].

The approximation uses the following expression

$$f\left(\frac{\sin\vartheta}{\lambda}\right) = \sum_{i=1}^{4} a_i \exp\left(-b_i \frac{\sin^2\vartheta}{\lambda}\right) + c$$

This expression allows a precise approximation of atomic scattering factors in the region $\sin\vartheta/\lambda \leq 2.0$.

To characterize the precision of the analytically determined values we give

- f_{max} as mean deviation from the exact value at $\sin\vartheta/\lambda$ and
- f_m as mean value of the absolute deviation.

As a rule the maximum deviations were observed at very small or at very big values of $\sin\vartheta/\lambda$, which often are localized outside the range important for experiments.

For the determination of the coefficients a_i, b_i and c ($i = 1, ..., 4$) wave functions from Hartree-Fock, Dirac-Fock- or Dirac-Fock-Slater calculations are used.

Beside hydrogen for neutral atoms the calculations are exclusively performed with relativistic Dirac-Fock wavefunctions. For hydrogen and for ions in the range to rubidium (Z=37) nonrelativistic wavefunctions and for ions with atomic numbers above rubidium relativistic Dirac-Fock-Slater wavefunctions were applied.

---

[1] With IUCR copyright permission

Element	a_1	b_1	a_2	b_2	a_3	b_3
H	0.489918	20.6593	0.262003	7.74039	0.196767	49.5519
H^{1-}	0.897661	53.1368	0.565616	15.1870	0.415815	186.576
He	0.873400	9.10370	0.630900	3.35680	0.311200	22.9276
Li	1.12820	3.95460	0.750800	1.05240	0.617500	85.3905
Li^{1+}	0.696800	4.62370	0.788800	1.95570	0.341400	0.631600
Be	1.59190	43.6427	1.12780	1.86230	0.539100	103.483
Be^{2+}	6.26030	0.002700	0.884900	0.831300	0.799300	2.27580
B	2.05450	23.2185	1.33260	1.02100	1.09790	60.3498
C	2.31000	20.8439	1.02000	10.2075	1.58860	0.568700
C_{val}	2.26069	22.6907	1.56165	0.656665	1.05075	9.75618
N	12.2126	0.005700	3.13220	9.89330	2.01250	28.9975
O	3.04850	13.2771	2.28680	5.70110	1.54630	0.323900
O^{1-}	4.19160	12.8573	1.63969	4.17236	1.52673	47.0179
F	3.53920	10.2825	2.64120	4.29440	1.51700	0.261500
F^{1-}	3.63220	5.27756	3.51057	14.7353	1.26064	0.442258
Ne	3.95530	8.40420	3.11250	3.42620	1.45460	0.230600
Na	4.76260	3.28500	3.17360	8.84220	1.26740	0.313600
Na^{1+}	3.25650	2.66710	3.93620	6.11530	1.39980	0.200100
Mg	5.42040	2.82750	2.17350	79.2611	1.22690	0.380800
Mg^{2+}	3.49880	2.16760	3.83780	4.75420	1.32840	0.185000
Al	6.42020	3.03870	1.90020	0.742600	1.59360	31.5472
Al^{3+}	4.17448	1.93816	3.38760	4.14553	1.20296	0.228753
Si_v	6.29150	2.43860	3.03530	32.3337	1.98910	0.678500
Si_{val}	5.66269	2.66520	3.07164	38.6634	2.62446	0.916946
Si^{4+}	4.43918	1.64167	3.20345	3.43757	1.19453	0.214900
P	6.43450	1.90670	4.17910	27.1570	1.78000	0.526000
S	6.90530	1.46790	5.20340	22.2151	1.43790	0.253600
Cl	11.4604	0.010400	7.19640	1.16620	6.25560	18.5194
Cl^{1-}	18.2915	0.006600	7.20840	1.17170	6.53370	19.5424
Ar	7.48450	0.907200	6.77230	14.8407	0.653900	43.8983
K	8.21860	12.7949	7.43980	0.774800	1.05190	213.187
K^{1+}	7.95780	12.6331	7.49170	0.767400	6.35900	-0.00200
Ca	8.62660	10.442.1	7.38730	0.659900	1.58990	85.7484
Ca^{2+}	15.6348	-0.00740	7.95180	0.608900	8.43720	10.3116
Sc	9.18900	9.02130	7.36790	0.572900	1.64090	136.108
Sc^{3+}	13.4008	0.298540	8.02730	7.96290	1.65943	-0.28604
Ti	9.75950	7.85080	7.35580	0.500000	1.69910	35.6338
Ti^{2+}	9.11423	7.52430	7.62174	0.457585	2.27930	19.5361
Ti^{3+}	17.7344	0.220610	8.73816	7.04716	5.25691	-0.15762
Ti^{4+}	19.5114	0.178847	8.23473	6.67018	2.01341	-0.29263
V	10.2971	6.86570	7.35110	0.438500	2.07030	26.8938
V^{2+}	10.1060	6.88180	7.35410	0.440900	2.28840	20.3004
V^{3+}	9.43141	6.39535	7.74190	0.383349	2.15343	15.1908

a$_4$	b$_4$	c	f$_{max}$	sin ϑ/λ	f$_m$
0.049879	2.20159	0.001305	0.000	0.17	0.000
0.116973	3.56709	0.002389	0.002	0.09	0.001
0.178000	0.982100	0.006400	0.001	1.01	0.000
0.465300	168.261	0.037700	0.005	2.00	0.001
0.156300	10.0953	0.016700	0.001	1.78	0.000
0.702900	0.542000	0.038500	0.003	0.56	0.001
0.164700	5.11460	-6.1092	0.001	1.97	0.000
0.706800	0.140300	-0.19320	0.002	0.75	0.001
0.865000	51.6512	0.215600	0.006	2.00	0.001
0.839259	55.5949	0.286977	0.001	0.16	0.000
1.16630	0.582600	-11.529	0.007	0.11	0.002
0.867000	32.9089	0.250800	0.001	0.22	0.000
-20.307	-0.01404	21.9412	0.011	1.50	0.004
1.02430	26.1476	0.277600	0.001	0.01	0.000
0.940706	47.3437	0.653369	0.003	0.09	0.001
1.12510	21.7184	0.351500	0.002	0.25	0.001
1.11280	129.424	0.676000	0.009	0.13	0.002
1.00320	14.0390	0.404000	0.001	0.70	0.000
2.30730	7.19370	0.858400	0.015	0.08	0.003
0.846700	10.1411	0.485300	0.001	1.34	0.000
1.96460	85.0886	1.11510	0.018	2.00	0.005
0.528137	8.28524	0.706786	0.000	1.50	0.000
1.54100	81.6937	1.14070	0.009	2.00	0.002
1.39320	93.5458	1.24707	0.001	0.53	0.001
0.416530	6.65365	0.746297	0.000	1.50	0.000
1.49080	68.1645	1.11490	0.003	0.65	0.001
1.58630	56.1720	0.866900	0.005	0.67	0.002
1.64550	47.7784	-9.5574	0.007	0.78	0.003
2.33860	60.4486	-16.378	0.007	0.76	0.003
1.64420	33.3929	1.44450	0.029	2.00	0.006
0.865900	41.6841	1.42280	0.011	0.90	0.005
1.19150	31.9128	-4.9978	0.011	0.91	0.005
1.02110	178.437	1.37510	0.016	0.99	0.006
0.853700	25.9905	-14.875	0.017	2.00	0.004
1.46800	51.3531	1.33290	0.014	1.07	0.006
1.57936	16.0662	-6.6667	0.002	1.50	0.000
1.90210	116.105	1.28070	0.014	2.00	0.006
0.087899	61.6558	0.897155	0.006	1.50	0.001
1.92134	15.9768	-14.652	0.001	0.00	0.000
1.52080	12.9464	-13.280	0.002	1.50	0.000
2.05710	102.478	1.21990	0.014	2.00	0.005
0.022300	115.122	1.22980	0.015	2.00	0.004
0.016865	63.9690	0.656565	0.004	1.50	0.001

Element	a_1	b_1	a_2	b_2	a_3	b_3
V^{5+}	15.6887	0.679003	8.14208	5.40135	2.03081	9.97278
Cr	10.6406	6.10380	7.35370	0.392000	3.32400	20.2626
Cr^{2+}	9.54034	5.66078	7.75090	0.344261	3.58274	13.3075
Cr^{3+}	9.68090	5.59463	7.81136	0.334393	2.87603	12.8288
Mn	11.2819	5.34090	7.35730	0.343200	3.01930	17.8674
Mn^{2+}	10.8061	5.27960	7.36200	0.343500	3.52680	14.3430
Mn^{3+}	9.84521	4.91797	7.87194	0.294393	3.56531	10.8171
Mn^{4+}	9.96253	4.84850	7.97057	0.283303	2.76067	10.4852
Fe	11.7695	4.76110	7.35730	0.307200	3.52220	15.3535
Fe^{2+}	11.0424	4.65380	7.37400	0.305300	4.13460	12.0546
Fe^{3+}	11.1764	4.61470	7.38630	0.300500	3.39480	11.6729
Co	12.2841	4.27910	7.34090	0.278400	4.00340	13.5359
Co^{2+}	11.2296	4.12310	7.38830	0.272600	4.73930	10.2443
Co^{3+}	10.3380	3.90969	7.88173	0.238668	4.76795	8.35583
Ni	12.8376	3.87850	7.29200	0.256500	4.44380	12.1763
Ni^{2+}	11.4166	3.6!660	7.40050	0.244900	5.34420	8.87300
Ni^{3+}	10.7806	3.54770	7.75868	0.223140	5.22746	7.64468
Cu	13.3380	3.58280	7.16760	0.247000	5.61580	11.3966
Cu^{1+}	11.9475	3.36690	-7.35730	0.227400	6.24550	8.66250
Cu^{2+}	11.8168	3.37484	7.11181	0.244078	5.78135	7.98760
Zn	14.0743	3.26550	7.03180	0.233300	5.16520	10.3163
Zn^{2+}	11.9719	2.99460	7.38620	0.203100	6.46680	7.08260
Ga	15.2354	3.06690	6.70060	0.241200	4.35910	10.7805
Ga^{3+}	12.6920	2.81262	6.69883	0.227890	6.06692	6.36441
Ge	16.0816	2.85090	6.37470	0.251600	3.70680	11.4468
Ge^{4+}	12.9172	2.53718	6.70003	0.205855	6.06791	5.47913
As	16.6723	2.63450	6.07010	0.264700	3.43130	12.9479
Se	17.0006	2.40980	5.81960	0.272600	3.97310	15.2372
Br	17.1789	2.17230	5.23580	16.5796	5.63770	0.260900
Br^{1-}	17.1718	2.20590	6.33380	19.3345	5.57540	0.287100
Kr	17.3555	1.93840	6.72860	16.5623	5.54930	0.226100
Rb	17.1784	1.78880	9.64350	17.3151	5.13990	0.274800
Rb^{1+}	17.5816	1.71390	7.65980	14.7957	5.89810	0.160300
Sr	17.5663	1.55640	9.81840	14.0988	5.42200	0.166400
Sr^{2+}	18.0874	1.49070	8.13730	12.6963	2.56540	24.5651
Y	17.7760	1.40290	10.2946	12.8006	5.72629	0.125599
Y^{3+}	17.9268	1.35417	9.15310	11.2145	1.76795	22.6599
Zr	17.8765	1.27618	10.9480	11.9160	5.41732	0.117622
Zr^{4+}	18.1668	1.21480	10.0562	10.1483	1.01118	21.6054
Nb	17.6142	1.18865	12.0144	11.7660	4.04183	0.204785
Nb^{3+}	19.8812	0.019175	18.0653	1.13305	11.0177	10.1621
Nb^{5+}	17.9163	1.12446	13.3417	0.028781	10.7990	9.28206
Mo	3.70250	0.277200	17.2356	1.09580	12.8876	11.0040

a$_4$	b$_4$	c	f$_{max}$	sin ϑ/λ	f$_m$
-9.5760	0.940464	1.71430	0.000	0.34	0.000
1.49220	98.7399	1.18320	0.011	2.00	0.004
0.509107	32.4224	0.616898	0.002	1.50	0.000
0.113575	32.8761	0.518275	0.002	1.50	0.000
2.24410	83.7543	1.08960	0.009	2.00	0.004
0.218400	41.3235	1.08740	0.009	2.00	0.002
0.323613	24.1281	0.393974	0.001	1.50	0.000
0.054447	27.5730	0.251877	0.001	1.50	0.000
2.30450	76.8805	1.03690	0.011	0.08	0.004
0.439900	31.2809	1.00970	0.008	2.00	0.002
0.072400	38.5566	0.970700	0.008	2.00	0.002
2.34880	71.1692	1.01180	0.013	0.08	0.004
0.710800	25.6466	0.932400	0.006	2.00	0.001
0.725591	18.3491	0.286667	0.000	1.50	0.000
2.38000	66.3421	1.03410	0.014	0.08	0.004
0.977300	22.1626	0.861400	0.003	2.00	0.001
0.847114	16.9673	0.386044	0.000	0.57	0.000
1.67350	64.8126	1.19100	0.015	0.08	0.005
1.55780	25.8487	0.89000	0.003	0.24	0.001
1.14523	19.8970	1.14431	0.001	0.26	0.000
2.41000	58.7097	1.30410	0.016	0.08	0.005
1.39400	18.0995	0.780700	0.001	0.62	0.000
2.96230	61.4135	1.71890	0.025	0.08	0.008
1.00660	14.4122	1.53545	0.008	1.45	0.000
3.68300	54.7625	2.13130	0.024	0.08	0.008
0.859041	11.6030	1.45572	0.000	0.32	0.000
4.27790	47.7972	2.53100	0.019	0.09	0.008
4.35430	43.8163	2.84090	0.016	2.00	0.006
3.98510	41.4328	2.95570	0.012	2.00	0.004
3.72720	58.1535	3.17760	0.016	2.00	0.006
3.53750	39.3972	2.82500	0.008	2.00	0.002
1.52920	164.934	3.48730	0.028	0.12	0.008
2.78170	31.2087	2.07820	0.002	1.99	0.001
2.66940	132.376	2.50640	0.021	0.13	0.005
-34.193	-0.01380	41.4025	0.008	2.00	0.002
3.26588	104.354	1.91213	0.028	0.07	0.006
-33.108	-0.01319	40.2602	0.005	2.00	0.001
3.65721	87.6627	2.06929	0.035	0.07	0.008
-2.6479	-0.10276	9.41454	0.004	2.00	0.001
3.53346	69.7957	3.75591	0.042	0.08	0.011
1.94715	28.3389	-12.912	0.006	2.00	0.002
0.337905	25.7228	-6.3934	0.007	2.00	0.003
3.74290	61.6584	4.38750	0.046	0.08	0.012

Element	a_1	b_1	a_2	b_2	a_3	b_3
Mo^{3+}	21.1664	0.014734	18.2017	1.03031	11.7423	9.53659
Mo^{5+}	21.0149	0.014345	18.0992	1.02238	11.4632	8.78809
Mo^{6+}	17.8871	1.03649	11.1750	8.48061	6.57891	0.058881
Tc	19.1301	0.864132	11.0948	8.14487	4.64901	21.5707
Ru	19.2674	0.808520	12.9182	8.43467	4.86337	24.7997
Ru^{3+}	18.5638	0.847329	13.2885	8.37164	9.32602	0.017662
Ru^{4+}	18.5003	0.844582	13.1787	8.12534	4.71304	0.36495
Rh	19.2957	0.751536	14.3501	8.21758	4.73425	25.8749
Rh^{3+}	18.8785	0.764252	14.1259	7.84438	3.32515	21.2487
Rh^{4+}	18.8545	0.760825	13.9806	7.62436	2.53464	19.3317
Pd	19.3319	0.698655	15.5017	7.98929	5.29537	25.2052
Pd^{2+}	19.1701	0.696219	15.2096	7.55573	4.32234	22.5057
Pd^{4+}	19.2493	0.683839	14.7900	-7.14833	2.89289	17.9144
Ag	19.2808	0.644600	16.6885	7.47260	4.80450	24.6605
Ag^{1+}	19.1812	0.646179	15.9719	7.19123	5.27475	21.7326
Ag^{2+}	19.1643	0.645643	16.2456	7.18544	4.37090	21.4072
Cd	19.2214	0.594600	17.6444	6.90890	4.46100	24.7008
Cd^{2+}	19.1514	0.597922	17.2535	6.80639	4.47128	20.2521
In	19.1624	0.547600	18.5596	6.37760	4.29480	25.8499
In^{3+}	19.1045	0.551522	18.1108	6.32470	3.78897	17.3595
Sn	19.1889	5.83030	19.1005	0.503100	4.45850	26.8909
Sn^{2+}	19.1094.	0.503600	19.0548	5.83780	4.56480	23.3752
Sn^{4+}	19.9333	5.76400	19.7131	0.465500	3.41820	14.0049
Sb	19.6418	5.30340	19.0455	0.460700	5.03710	27.9074
Sb^{3+}	18.9755	0.467196	18.9330	5.22126	5;10789	19.5902
Sb^{5+}	19.8685	5.44853	19.0302	0.467973	2.41253	14.1259
Te	19.9644	4.81742	19.0138	0.420885	6.14487	28.5284
I	20.1472	4.34700	18.9949	0.381400	7.51380	27.7660
I^{1-}	20.2332	4.35790	18.9970	0.381500	7.80690	29.5259
Xe	20.2933	3.92820	19.0298	0.344000	8.97670	26.4659
Cs	20.3892	3.56900	19.1062	0.310700	10.6620	24.3879
Cs^{1+}	20.3524	3.55200	19.1278	0.308600	10.2821	23.7128
Ba	20.3361	3.21600	19.2970	0.275600	10.8880	20.2073
Ba^{2+}	20.1807	3.21367	19.1136	0.283310	10.9054	20.0558
La	20.5780	2.94817	19.5990	0.244475	11.3727	18.7726
La^{3+}	20.2489	2.92070	19.3763	0.250698	11.6323	17.8211
Ce	21.1671	2.81219	19.7695	0.226836	11.8513	17.6083
Ce^{3+}	20.8036	2.77691	19.5590	0.231540	11.9369	16.5408
Ce^{4+}	20.3235	2.65941	19.8186	0.218850	12.1233	15.7992
Pr	22.0440	2.77393	19.6697	0.222087	12.3856	16.7669
Pr^{3+}	21.3727	2.64520	19.7491	0.214299	12.1329	15.3230
Pr^{4+}	20.9413	2.54467	20.0539	0.202481	12.4668	14.8137
Nd	22.6845	2.66248	19.6847	0.210628	12.7740	15.8850

a$_4$	b$_4$	c	f$_{max}$	sin ϑ/λ	f$_m$
2.30951	26.6307	-14.421	0.009	2.00	0.003
0.740625	23.3452	-14.316	0.010	2.00	0.003
0.000000	0.000000	0.344941	0.014	0.00	0.006
2.71263	86.8472	5.40428	0.061	2.00	0.011
1.56756	94.2928	5.37874	0.041	2.00	0.006
3.00964	22.8870	-3.1892	0.013	2.00	0.004
2.18535	20.8504	1.42357	0.014	2.00	0.004
1.28918	98.6062	5.32800	0.021	2.00	0.004
-6.1989	-0.01036	11.8678	0.014	2.00	0.004
-5.6526	-0.01020	11.2835	0.014	2.00	0.003
0.605844	76.8986	5.26593	0.012	1.08	0.005
0.000000	0.000000	5.29160	0.011	2.00	0.004
-7.9492	0.005127	13.0174	0.014	2.00	0.003
1.04630	99.8156	5.17900	0.016	1.14	0.007
0.357534	66.1147	5.21572	0.012	1.13	0.005
0.000000	0.000000	5.21404	0.011	1.14	0.005
1.60290	87.4825	5.06940	0.020	2.00	0.008
0.000000	0.000000	5.11937	0.014	1.17	0.007
2.03960	92.8029	4.93910	0.027	2.00	0.009
0.000000	0.000000	4.99635	0.022	2.00	0.007
2.46630	83.9571	4.78210	0.032	2.00	0.009
0.487000	62.2061	4.78610	0.032	2.00	0.009
0.019300	-0.75830	3.91820	0.016	2.00	0.004
2.68270	75.2825	4.59090	0.035	2.00	0.009
0.288753	55.5113	4.69626	0.028	2.00	0.007
0.000000	0.000000	4.69263	0.030	2.00	0.008
2.52390	70.8403	4.35200	0.038	2.00	0.009
2.27350	66.8776	4.07120	0.037	2.00	0.009
2.88680	84.9304	4.07140	0.038	2.00	0.009
1.99000	64.2658	3.71180	0.038	2.00	0.009
1.49530	213.904	3.33520	0.032	2.00	0.010
0.961500	59.4565	3.27910	0.037	2.00	0.009
2.69590	167.202	2.77310	0.032	2.00	0.009
0.77634	51.7460	3.02902	0.029	2.00	0.007
3.28719	133.124	2.14678	0.032	2.00	0.009
0.336048	54.9453	2.40860	0.028	2.00	0.007
3.33049	127.113	1.86264	0.026	2.00	0.008
0.612376	43.1692	2.09013	0.023	2.00	0.005
0.144583	62.2355	1.59180	0.026	2.00	0.007
2.82428	143.644	2.05830	0.021	0.12	0.007
0.975180	36.4065	1.77132	0.019	2.00	0.004
0.296689	45.4643	1.24285	0.021	2.00	0.005
2.85137	137.903	1.98486	0.024	0.13	0.007

Element	a_1	b_1	a_2	b_2	a_3	b_3
Nd^{3+}	21.9610	2.52722	19.9339	0.199237	12.1200	14.1783
Pm	23.3405	2.56270	19.6095	0.202088	13.1235	15.1009
Pm^{3+}	22.5527	2.41740	20.1108	0.185769	12.0671	13.1275
Sm	24.0042	2.47274	19.4258	0.196451	13.4396	14.3996
Sm^{3+}	23.1504	2.31641	20.2599	0.174081	11.9202	12.1571
Eu	24.6274	2.38790	19.0886	0.194200	13.7603	13.7546
Eu^{2+}	24.0063	2.27783	19.9504	0.173530	11.8034	11.6096
Eu^{3+}	23.7497	2.22258	20.3745	0.163940	11.8509	11.3110
Gd	25.0709	2.25341	19.0798	0.181951	13.8518	12.9331
Gd^{3+}	24.3466	2.13553	20.4208	0.155525	11.8708	10.5782
Tb	25.8976	2.24256	18.2185	0.196143	14.3167	12.6648
Tb^{3+}	24.9559	2.05601	20.3271	0.149525	12.2471	10.0499
Dy	26.5070	2.18020	17.6383	0.202172	14.5596	12.1899
Dy^{3+}	25.5395	1.98040	20.2861	0.143384	11.9812	9.34972
Ho	26.9049	2.07051	17.2940	0.197940	14.5583	11.4407
Ho^{3+}	26.1296	1.91072	20.0994	0.139358	11.9788	8.80018
Er	27.6563	2.07356	16.4285	0.223545	14.9779	11.3604
Er^{3+}	26.7220	1.84659	19.7748	0.137290	12.1506	8.36225
Tm	28.1819	2.02859	15.8851	0.238849	15.1542	10.9975
Tm^{3+}	27.3083	1.78711	19.3320	0.136974	12.3339	7.96778
Yb	28.6641	1.98890	15.4345	0.257119	15.3087	10.6647
Yb^{2+}	28.1209	1.78503	17.6817	0.159970	13.3335	8.18304
Yb^{3+}	27.8917	1.73272	18.7614	0.138790	12.6072	7.64412
Lu	28.9476	1.90182	15.2208	9.98519	15.1000	0.261033
Lu^{3+}	28.4628	1.68216	18.1210	0.142292	12.8429	7.33727
Hf	29.1440	1.83262	15.1726	9.59990	14.7586	0.275116
Hf^{4+}	28.8131	1.59136	18.4601	0.128903	12.7285	6.76232
Ta	29.2024	1.77333	15.2293	9.37046	14.5135	0.295977
Ta^{5+}	29.1587	1.50711	18.8407	0.116741	12.8268	6.31524
W	29.0818	1.72029	15.4300	9.22590	14.4327	0.321703
W^{6+}	29.4936	1.42755	19.3763	0.104621	13.0544	5.93667
Re	28.7621	1.67191	15.7189	9.09227	14.5564	0.350500
Os	28.1894	1.62903	16.1550	8.97948	14.9305	0.382661
Os^{4+}	30.4190	1.37113	15.2637	6.84706	14.7458	0.165191
Ir	27.3049	1.59279	16.7296	8.86553	15.6115	0.417916
Ir^{3+}	30.4156	1.34323	15.8620	7.10909	13.6145	0.204633
Ir^{4+}	30.7058	1.30923	15.5512	6.71983	14.2326	0.167252
Pt	27.0059	1.51293	17.7639	8.81174	15.7131	0.424593
Pt^{2+}	29.8429	1.32927	16.7224	7.38979	13.2153	0.263297
Pt^{4+}	30.9612	1.24813	15.9829	6.60834	13.7348	0.168640
Au	16.8819	0.461100	18.5913	8.62160	25.5582	1.48260
Au^{1+}	28.0109	1.35321	17.8204	7.73950	14.3359	0.356752
Au^{3+}	30.6886	1.21990	16.9029	6.82872	12.7801	0.212867

a_4	b_4	c	f_{max}	$\sin\vartheta/\lambda$	f_m
1.51031	30.8717	1.47588	0.015	2.00	0.003
2.87516	132.721	2.02876	0.026	0.13	0.008
2.07492	27.4491	1.19499	0.012	2.00	0.002
2.89604	128.007	2.20963	0.029	0.13	0.009
2.71488	24.8242	0.954586	0.009	2.00	0.002
2.92270	123.174	2.57450	0.031	0.14	0.010
3.87243	26.5156	1.36389	0.004	2.00	0.002
3.26503	22.9966	0.759344	0.006	2.00	0.001
3.54545	101.398	2.41960	0.036	0.15	0.011
3.71490	21.7029	0.645089	0.004	2.00	0.001
2.95354	115.362	3.58024	0.035	0.14	0.012
3.77300	21.2773	0.691967	0.005	0.00	0.001
2.96577	111.874	4.29728	0.037	0.15	0.013
4.50073	19.5810	0.689690	0.003	0.00	0.001
3.63837	92.6566	4.56796	0.040	0.15	0.013
4.93676	18.5908	0.852795	0.003	0.00	0.001
2.98233	105.703	5.92046	0.040	0.15	0.015
5.17379	17.8974	1.17613	0.003	0.00	0.001
2.98706	102.961	6.75621	0.041	0.15	0.016
5.38348	17.2922	1.63929	0.003	0.00	0.001
2.98963	100.417	7.56672	0.042	0.15	0.016
5.14657	20.3900	3.70983	0.008	0.00	0.003
5.47647	16.8153	2.26001	0.003	0.00	0.002
3.71601	84.3298	7.97628	0.043	0.16	0.016
5.59415	16.3535	2.97573	0.004	0.14	0.002
4.30013	72.0290	8.58154	0.047	0.08	0.016
5.59927	14.0366	2.39699	0.002	0.00	0.001
4.76492	63.3644	9.24354	0.049	0.08	0.017
5.38695	12.4244	1.78555	0.002	2.00	0.001
5.11982	57.0560	9.88750	0.051	0.09	0.017
5.06412	11.1972	1.01074	0.001	0.00	0.000
5.44174	52.0861	10.4720	0.052	0.09	0.017
5.67589	48.1647	11.0005	0.051	0.09	0.017
5.06795	18.0030	6.49804	0.006	0.29	0.003
5.83377	45.0011	11.4722	0.050	0.09	0.017
5.82008	20.3254	8.27903	0.009	0.28	0.004
5.53672	17.4911	6.96824	0.006	0.29	0.003
5.78370	38.6103	11.6883	0.046	0.10	0.016
6.35234	22.9426	9.85329	0.014	0.00	0.006
5.92034	16.9392	7.39534	0.006	0.14	0.003
5.86000	36.3956	12.0658	0.045	0.10	0.015
6.58077	26.4043	11.2299	0.023	0.12	0.009
6.52354	18.6590	9.09680	0.009	0.14	0.004

Element	a_1	b_1	a_2	b_2	a_3	b_3
Hg	20.6809	0.545000	19.0417	8.44840	21.6575	1.57290
Hg^{1+}	25.0853	1.39507	18.4973	7.65105	16.8883	0.443378
Hg^{2+}	29.5641	1.21152	18.0600	7.05639	12.8374	0.284738
Tl	27.5446	0.655150	19.1584	8.70751	15.5380	1.96347
Tl^{1+}	21.3985	1.47110	20.4723	0.517394	18.7478	7.43463
Tl^{3+}	30.8695	1.10080	18.3841	6.53852	11.9328	0.219074
Pb	31.0617	0.690200	13.0637	2.35760	18.4420	8.61800
Pb^{2+}	21.7886	1.33660	19.5682	0.488383	19.1406	6.77270
Pb^{4+}	32.1244	1.00566	18.8003	6.10926	12.0175	0.147041
Bi	33.3689	0.704000	12.9510	2.92380	16.5877	8.79370
Bi^{3+}	21.8053	1.23560	19.5026	6.24149	19.1053	0.469999
Bi^{5+}	33.5364	0.916540	25.0946	0.39042	18.2497	5.71414
Po	34.6726	0.700999	15.4733	3.55078	13.1138	9.55642
At	35.3163	0.685870	19.0211	3.97458	9.49887	11.3824
Rn	35.5631	0.663100	21.2816	4.06910	8.00370	14.0422
Fr	35.9299	0.646453	23.0547	4.17619	12.1439	23.1052
Ra	35.7630	0.616341	22.9064	3.87135	12.4739	19.9887
Ra^{2+}	35.2150	0.604909	21.6700	3.57670	7.91342	12.6010
Ac	35.6597	0.589092	23.1032	3.65155	12.5977	18.5990
Ac^{2+}	35.1736	0.579689	22.1112	3.41437	8.19216	12.9187
Th	35.5645	0.563359	23.4219	3.46204	12.7473	17.8309
Th^{4+}	35.1007	0.555054	22.4418	3.24498	9.78554	13.4661
Pa	35.8847	0.547751	23.2948	3.41519	14.1891	16.9235
U	36.0228	0.529300	23.4128	3.32530	14.9491	16.0927
U^{3+}	35.5747	0.520480	22.5259	3.12293	12.2165	12.7148
U^{4+}	35.3715	0.516598	22.5326	3.05053	12.0291	12.5723
U^{6+}	34.8509	0.507079	22.7584	2.89030	14.0099	13.1767
Np	36.1874	0.511929	23.5964	3.25396	15.6402	15.3622
Np^{3+}	35.7074	0.502322	22.6130	3.03807	12.9898	12.1449
Np^{4+}	35.5103	0.498626	22.5787	2.96627	12.7766	11.9484
Np^{6+}	35.0136	0.489810	22.7286	2.81099	14.3884	12.3300
Pu	36.5254	0.499384	23.8083	3.26371	16.7707	14.9455
Pu^{3+}	35.8400	0.484938	22.7169	2.96118	13.5807	11.5331
Pu^{4+}	35.6493	0.481422	22.6460	2.89020	13.3595	11.3160
Pu^{6+}	35.1736	0.473204	22.7181	2.73848	14.7635	11.5530
Am	36.6706	0.483629	24.0992	3.20647	17.3415	14.3136
Cm	36.6488	0.465154	24.4096	3.08997	17.3990	13.4346
Bk	36.7881	0.451018	24.7736	3.04619	17.8919	12.8946
Cf	36.9185	0.437533	25.1995	3.00775	18.3317	12.4044

a_4	b_4	c	f_{max}	$\sin \vartheta/\lambda$	f_m
5.96760	38.3246	12.6089	0.046	0.10	0.017
6.48216	28.2262	12.0205	0.027	0.12	0.011
6.89912	20.7482	10.6268	0.013	0.00	0.006
5.52593	45.8149	13.1746	0.059	0.09	0.021
6.82847	28.8482	12.5258	0.028	0.12	0.011
7.00574	17.2114	9.80270	0.008	0.01	0.004
5.96960	47.2579	13.4118	0.060	2.00	0.021
7.01107	23.8132	12.4734	0.020	2.00	0.008
6.96886	14.7140	8.08428	0.005	0.31	0.002
6.46220	48.0093	13.5782	0.065	2.00	0.020
7.10295	20.3185	12.4711	0.015	2.00	0.006
6.91555	12.8285	-6.7994	0.003	0.00	0.001
7.02588	47.0045	13.6770	0.066	2.00	0.018
7.42518	45.4715	13.7108	0.062	2.00	0.015
7.44330	44.2473	13.6905	0.054	2.00	0.012
2.11253	150.645	13.7247	0.055	2.00	0.017
3.21097	142.325	13.6211	0.037	2.00	0.012
7.65078	29.8436	13.5431	0.029	2.00	0.006
4.08655	117.020	13.5266	0.030	0.06	0.009
7.05545	25.9443	13.4637	0.021	2.00	0.004
4.80703	99.1722	13.4314	0.031	0.07	0.008
5.29444	23.9533	13.3760	0.014	2.00	0.002
4.17287	105.251	13.4287	0.033	0.06	0.010
4.18800	100.613	13.3966	0.035	0.07	0.010
5.37073	26.3394	13.3092	0.009	2.00	0.002
4.79840	23.4582	13.2671	0.007	2.00	0.001
1.21457	25.2017	13.1665	0.003	2.00	0.001
4.18550	97.4908	13.3573	0.037	0.07	0.011
5.43127	25.4928	13.2544	0.006	2.00	0.002
4.92159	22.7502	13.2116	0.005	2.00	0.001
1.75669	22.6581	13.1130	0.002	2.00	0.001
3.47947	105.980	13.3812	0.038	0.14	0.013
5.66016	24.3992	13.1991	0.005	2.00	0.001
5.18831	21.8301	13.1555	0.003	2.00	0.001
2.28678	20.9303	13.0582	0.001	1.36	0.000
3.49331	102.273	13.3592	0.040	0.07	0.013
4.21665	88.4834	13.2887	0.041	0.07	0.013
4.23284	86.0030	13.2754	0.042	0.07	0.014
4.24391	83.7881	13.2674	0.043	0.07	0.014

10 Mass Attenuation Coefficients for Selected X-Ray Transitions of Elements up to Plutonium

For X-ray energies below 1 keV photoabsorption cross-sections as given by [234] and for energies above 1 keV experimental photon attenuation coefficients given by [547] were used. All quantitites are given in cm^2/g.

Underlined values indicate the position of absorption edges of the element.

Z	transition	E [keV]	λ [pm]	$_1$H	$_2$He	$_3$Li	$_4$Be	$_5$B
4	Be K$_\alpha$	0.1085	11427.207	8.95 03	4.22 04	1.12 05	6.50 03	1.12 04
38	Sr M$_\xi$	0.1140	10875.895	7.68 03	3.37 04	1.02 05	1.69 05	1.02 04
39	Y M$_\xi$	0.1328	9339.235	4.78 03	2.53 04	7.07 04	1.23 05	7.23 03
16	S L$_l$	0.1487	8337.942	3.35 03	1.88 04	5.30 04	9.63 04	5.58 03
40	Zr M$_\xi$	0.1511	8205.506	3.19 03	1.80 04	5.12 04	9.35 04	5.42 03
41	Nb M$_\xi$	0.1717	7221.037	2.13 03	1.28 04	3.71 04	6.99 04	3.87 03
5	B K$_\alpha$	0.1833	6764.059	1.73 03	1.07 04	3.16 04	6.06 04	3.35 03
42	Mo M$_\xi$	0.1926	6437.445	1.47 03	9.37 03	2.75 04	5.32 04	8.36 04
6	C K$_\alpha$	0.2770	4476.000	4.62 02	3.35 03	1.03 04	2.20 04	3.70 04
47	Ag M$_\xi$	0.3117	3977.709	3.16 02	2.32 03	7.36 03	1.62 04	2.77 04
7	N K$_\alpha$	0.3924	3159.664	1.49 02	1.14 03	3.81 03	8.88 03	1.58 04
22	Ti L$_l$	0.3953	3136.484	1.45 02	1.11 03	3.73 03	8.69 03	1.55 04
22	Ti L$_\alpha$	0.4522	2741.822	9.35 01	7.28 02	2.51 03	5.97 03	1.10 04
23	V L$_\alpha$	0.5113	2424.901	6.25 01	4.94 02	1.73 03	4.22 03	7.95 03
8	O K$_\alpha$	0.5249	2362.072	5.73 01	4.55 02	1.60 03	3.92 03	7.42 03
25	Mn L$_l$	0.5563	2228.747	4.74 01	3.80 02	1.35 03	3.34 03	6.33 03
24	Cr L$_\alpha$	0.5728	2164.549	4.31 01	3.45 02	1.23 03	3.06 03	5.87 03
25	Mn L$_\alpha$	0.6374	1945.171	3.03 01	2.46 02	9.83 02	2.26 03	4.39 03
9	F K$_\alpha$	0.6768	1831.932	2.48 01	2.01 02	7.35 02	1.88 03	3.68 03
26	Fe L$_\alpha$	0.7050	1758.655	2.17 01	1.78 02	6.55 02	1.68 03	3.31 03
27	Co L$_\alpha$	0.7762	1597.335	1.57 01	1.31 02	4.88 02	1.27 03	2.35 03
28	Ni L$_\alpha$	0.8515	1456.080	1.16 01	9.73 01	3.66 02	9.63 02	1.95 03

Z	transition	E [keV]	λ [pm]	$_1$H	$_2$He	$_3$Li	$_4$Be	$_5$B
29	Cu L$_\alpha$	0.9297	1336.044	8.64 00	7.39 01	2.80 02	7.45 02	1.51 03
30	Zn L$_\alpha$	1.0117	1225.513	6.98 00	6.02 01	2.27 02	5.99 02	1.22 03
11	Na K$_\alpha$	1.0410	1191.020	6.41 00	5.49 01	2.07 02	5.49 02	1.12 03
11	Na K$_\beta$	1.0711	1157.550	5.88 00	5.00 01	1.89 02	5.04 02	1.03 03
12	Mg K$_\alpha$	1.2536	989.033	3.67 00	3.01 01	1.15 02	3.12 02	6.45 02
33	As L$_\alpha$	1.2820	967.123	3.43 00	2.80 01	1.07 02	2.92 02	6.03 02
12	Mg K$_\beta$	1.3022	952.121	3.28 00	2.66 01	1.02 02	2.78 02	5.76 02
33	As L$_{\beta_1}$	1.3170	941.421	3.17 00	2.57 01	2.85 02	2.69 02	5.57 02
66	Dy M$_\beta$	1.3250	935.737	3.11 00	2.52 01	9.66 01	2.64 02	5.47 02
67	Ho M$_\alpha$	1.3480	919.771	2.95 00	2.38 01	9.15 01	2.51 02	5.19 02
34	Se L$_\alpha$	1.3791	899.029	2.76 00	2.21 01	8.51 01	2.34 02	4.85 02
67	Ho M$_\beta$	1.3830	896.494	2.74 00	2.19 01	8.44 01	2.32 02	4.81 02
68	Er M$_\alpha$	1.4060	881.829	2.60 00	2.08 01	8.01 01	2.20 02	4.58 02
34	Se L$_{\beta_1}$	1.4192	873.627	2.53 00	2.02 01	7.78 01	2.14 02	4.45 02
68	Er M$_\beta$	1.4430	859.218	2.41 00	1.91 01	7.38 01	2.04 02	4.24 02
69	Tm M$_\alpha$	1.4620	848.051	2.32 00	1.83 01	7.08 01	1.96 02	4.08 02
35	Br L$_\alpha$	1.4804	837.511	2.23 00	1.76 01	6.80 01	1.88 02	3.93 02
13	Al K$_{\alpha_1}$	1.4867	833.962	2.20 00	1.73 01	6.71 01	1.86 02	3.88 02
69	Tm M$_\beta$	1.5030	824.918	2.13 00	1.67 01	6.49 01	1.80 02	3.75 02
70	Yb M$_\alpha$	1.5214	814.941	2.07 00	1.61 01	6.24 01	1.73 02	3.62 02
35	Br L$_{\beta_1}$	1.5259	812.538	2.06 00	1.60 01	6.18 01	1.72 02	3.59 02
13	Al K$_\beta$	1.5574	796.103	1.96 00	1.50 01	5.79 01	1.61 02	3.37 02
70	Yb M$_\beta$	1.5675	790.974	1.93 00	1.47 01	5.67 01	1.58 02	3.31 02
71	Lu M$_\alpha$	1.5813	784.071	1.88 00	1.43 01	5.52 01	1.54 02	3.22 02
36	Kr L$_\alpha$	1.5860	781.747	1.87 00	1.41 01	5.47 01	1.52 02	3.19 02
71	Lu M$_\beta$	1.6312	760.085	1.75 00	1.29 01	5.00 01	1.40 02	2.93 02
36	Kr L$_{\beta_1}$	1.6366	757.577	1.73 00	1.28 01	4.95 01	1.38 02	2.90 02
72	Hf M$_{\alpha_1}$	1.6446	753.892	1.71 00	1.26 01	4.87 01	1.36 02	2.86 02
37	Rb L$_{\alpha_1}$	1.6941	731.864	1.59 00	1.15 01	4.43 01	1.24 02	2.61 02
72	Hf M$_\beta$	1.6976	730.355	1.58 00	1.14 01	4.40 01	1.23 02	2.60 02
73	Ta M$_{\alpha_1}$	1.7096	725.229	1.56 00	1.11 01	4.30 01	1.21 02	2.54 02
14	Si K$_{\alpha_1}$	1.7400	712.558	1.49 00	1.05 01	4.07 01	1.14 02	2.41 02
37	Rb L$_{\beta_1}$	1.7522	707.597	1.46 00	1.03 01	3.98 01	1.12 02	2.36 02
73	Ta M$_\beta$	1.7655	702.266	1.44 00	1.00 01	3.89 01	1.09 02	2.31 02
74	W M$_{\alpha_1}$	1.7754	698.350	1.42 00	9.87 00	3.82 01	1.07 02	2.27 02
38	Sr L$_{\alpha_1}$	1.8066	686.290	1.36 00	9.34 00	3.61 01	1.02 02	2.15 02
74	W M$_\beta$	1.8349	675.705	1.31 00	8.89 00	3.44 01	9.69 01	2.05 02
14	Si K$_\beta$	1.8359	675.337	1.31 00	8.87 00	3.43 01	9.68 01	2.05 02
75	Re M$_{\alpha_1}$	1.8420	673.100	1.29 03	8.78 03	3.39 01	9.58 01	2.03 02
38	Sr L$_{\beta_1}$	1.8717	662.420	1.24 00	8.34 00	3.23 01	9.11 01	1.93 02
75	Re M$_\beta$	1.9061	650.465	1.19 00	7.87 00	3.04 01	8.61 01	1.83 02
76	Os M$_{\alpha_1}$	1.9102	649.069	1.18 00	7.82 00	3.02 01	8.55 01	1.82 02
39	Y L$_{\alpha_1}$	1.9220	644.882	1.17 00	7.66 00	2.96 01	8.38 01	1.78 02
76	Os M$_\beta$	1.9783	626.725	1.09 00	7.00 00	2.70 01	7.67 01	1.64 02

Z	transition	E [keV]	λ [pm]	$_1$H	$_2$He	$_3$Li	$_4$Be	$_5$B
77	Ir M$_{\alpha_1}$	1.9799	626.219	1.08 00	6.98 00	2.70 01	7.65 01	1.63 02
39	Y L$_{\beta_1}$	1.9958	621.230	1.06 00	6.80 00	2.63 01	7.47 01	1.59 02
15	P K$_{\alpha_1}$	2.0137	615.708	1.05 00	6.62 00	2.56 01	7.26 01	1.55 02
40	Zr L$_{\alpha_1}$	2.0424	607.056	1.02 00	6.35 00	2.44 01	6.95 01	1.48 02
78	Pt M$_{\alpha_1}$	2.0505	604.658	1.02 00	6.27 00	2.41 01	6.87 01	1.47 02
77	Ir M$_\beta$	2.0535	603.775	1.01 00	6.24 00	2.40 01	6.83 01	1.46 02
79	Au M$_{\alpha_1}$	2.1229	584.036	9.63 -1	5.65 00	2.16 01	6.16 01	1.32 02
40	Zr L$_{\beta_1}$	2.1244	583.624	9.62 -1	5.64 00	2.16 01	6.15 01	1.31 02
78	Pt M$_\beta$	2.1273	582.828	9.60 -1	5.62 00	2.15 01	6.12 01	1.31 02
15	P K$_{\beta_{1,3}}$	2.1390	579.640	9.52 -1	5.52 00	2.11 01	6.02 01	1.29 02
41	Nb L$_{\alpha_1}$	2.1659	572.441	9.33 -1	5.32 00	2.03 01	5.79 01	1.24 02
80	Hg M$_{\alpha_1}$	2.1953	564.775	9.14 -1	5.11 00	1.95 01	5.55 01	1.19 02
79	Au M$_\beta$	2.2046	562.393	9.08 -1	5.04 00	1.92 01	5.48 01	1.17 02
41	Nb L$_{\beta_1}$	2.2574	549.238	8.75 -1	4.70 00	1.78 01	5.09 01	1.09 02
81	Tl M$_{\alpha_1}$	2.2706	546.045	8.67 -1	4.62 00	1.75 01	5.00 01	1.07 02
80	Hg M$_\beta$	2.2825	543.199	8.60 -1	4.55 00	1.72 01	4.92 01	1.05 02
42	Mo L$_{\alpha_1}$	2.2932	540.664	8.54 -1	4.48 00	1.69 01	4.85 01	1.04 02
16	S K$_{\alpha_1}$	2.3080	537.197	8.45 -1	4.40 00	1.66 01	4.75 01	1.02 02
82	Pb M$_{\alpha_1}$	2.3457	528.563	8.24 -1	4.19 00	1.58 01	4.52 01	9.69 01
81	Tl M$_\beta$	2.3621	524.893	8.15 -1	4.10 00	1.54 01	4.42 01	9.49 01
42	Mo L$_{\beta_1}$	2.3948	517.726	7.98 -1	3.94 00	1.48 01	4.24 01	9.10 01
83	Bi M$_{\alpha_1}$	2.4226	511.785	7.83 -1	3.80 00	1.42 01	4.09 01	8.76 01
43	Tc L$_{\alpha_1}$	2.4240	511.490	7.83 -1	3.80 00	1.42 01	4.08 01	8.76 01
82	Pb M$_\beta$	2.3327	507.574	7.73 -1	3.71 00	1.39 01	3.99 01	8.56 01
16	S K$_\beta$	2.4640	503.186	7.63 -1	3.61 00	1.35 01	3.88 01	8.33 01
83	Bi M$_{\beta_1}$	2.5255	490.933	7.34 -1	3.36 00	1.25 01	3.59 01	7.73 01
43	Tc L$_{\beta_1}$	2.5368	488.746	7.29 -1	3.31 00	1.23 01	3.54 01	7.62 01
44	Ru L$_{\alpha_1}$	2.5586	484.582	7.19 -1	3.23 00	1.20 01	3.45 01	7.42 01
17	Cl K$_{\alpha_1}$	2.6224	472.792	6.92 -1	3.00 00	1.11 01	3.20 01	6.86 01
44	Ru L$_{\beta_1}$	2.6832	462.079	6.67 -1	2.80 00	1.03 01	2.98 01	6.41 01
45	Rh L$_{\alpha_1}$	2.6967	459.766	6.62 -1	2.76 00	1.01 01	2.93 01	6.32 01
17	Cl K$_\beta$	2.8156	440.350	6.19 -1	2.42 00	8.85 00	2.56 01	5.53 01
45	Rh L$_{\beta_1}$	2.8344	437.430	6.13 -1	2.37 00	8.66 00	2.51 01	5.42 01
46	Pd L$_{\alpha_1}$	2.8386	436.782	6.11 -1	2.36 00	8.62 00	2.50 01	5.39 01
18	Ar K$_{\alpha_1}$	2.9577	419.194	5.73 -1	2.09 00	7.57 00	2.20 01	4.75 01
47	Ag L$_{\alpha_1}$	2.9843	415.458	5.56 -1	2.03 00	7.36 00	2.14 01	4.63 01
46	Pd L$_{\beta_1}$	2.9902	414.638	5.63 -1	2.02 00	7.31 00	2.13 01	4.60 01
90	Th M$_{\alpha_1}$	2.9961	413.821	5.62 -1	2.01 00	7.27 00	2.11 01	4.57 01
91	Pa M$_{\alpha_1}$	3.0823	402.248	5.49 -1	1.86 00	6.66 00	1.93 01	4.19 01
48	Cd L$_{\alpha_1}$	3.1337	395.651	5.43 -1	1.78 00	6.33 00	1.84 01	3.98 01
90	Th M$_\beta$	3.1458	394.129	5.41 -1	1.77 00	6.25 00	1.81 01	3.93 01
47	Ag L$_{\beta_1}$	3.1509	393.491	5.41 -1	1.76 00	6.22 00	1.81 01	3.01 01
92	U M$_{\alpha_1}$	3.1708	391.021	5.38 -1	1.73 00	6.10 00	1.88 01	3.84 01
18	Ar K$_\beta$	3.1905	388.607	5.36 -1	1.70 00	5.99 00	1.74 01	3.76 01

Z	transition	E [keV]	λ [pm]	$_1$H	$_2$He	$_3$Li	$_4$Be	$_5$B
91	Pa M_β	3.2397	382.705	5.30 -1	1.63 00	5.71 00	1.65 01	3.59 01
49	In L_{α_1}	3.2869	377.210	5.24 -1	1.57 00	5.46 00	1.58 01	3.43 01
19	K K_{α_1}	3.3138	374.148	5.21 -1	1.54 00	5.33 00	1.54 01	3.35 01
48	Cd L_{β_1}	3.3160	373.832	5.21 -1	1.54 00	5.32 00	1.54 01	3.34 01
92	U M_β	3.3360	371.658	5.19 -1	1.51 00	5.22 00	1.51 01	3.28 01
50	Sn L_{α_1}	3.4440	360.003	5.07 -1	1.39 00	4.73 00	1.37 01	2.97 01
49	In L_{β_1}	3.4872	355.543	5.02 -1	1.35 00	4.55 00	1.31 01	2.80 01
19	K K_β	3.5896	345.401	4.91 -1	1.25 00	4.17 00	1.20 01	2.62 01
51	Sb L_{α_1}	3.6047	343.954	4.90 -1	1.23 00	4.11 00	1.18 01	2.58 01
50	Sn L_{β_1}	3.6628	338.498	4.84 -1	1.18 00	3.92 00	1.13 01	2.40 01
20	Ca K_{α_1}	3.6917	335.848	4.82 -1	1.16 00	3.82 00	1.10 01	2.40 01
52	Te L_{α_1}	3.7693	328.934	4.74 -1	1.10 00	3.58 00	1.03 01	2.25 01
51	Sb L_{β_1}	3.8436	322.575	4.68 -1	1.04 00	3.38 00	9.69 00	2.12 01
53	I L_{α_1}	3.9377	314.867	4.59 -1	9.77 -1	3.13 00	8.99 00	1.96 01
20	Ca K_β	4.0127	308.981	4.54 -1	9.31 -1	2.96 00	8.47 00	1.85 01
52	Te L_{β_1}	4.0296	307.686	4.53 -1	9.23 -1	2.92 00	8.36 00	1.83 01
21	Sc K_{α_1}	4.0906	303.097	4.50 -1	8.93 -1	2.80 00	7.99 00	1.75 01
54	Xe L_{α_1}	4.1099	301.674	4.50 -1	8.84 -1	2.76 00	7.87 00	1.72 01
53	I L_{β_1}	4.2207	293.755	4.45 -1	8.35 -1	2.55 00	7.26 00	1.59 01
55	Cs L_{α_1}	4.2865	289.245	4.43 -1	8.08 -1	2.44 00	6.92 00	1.51 01
21	Sc K_β	4.4605	277.962	4.37 -1	7.41 -1	2.18 00	6.13 00	1.34 01
56	Ba L_{α_1}	4.4663	277.601	4.36 -1	7.39 -1	2.17 00	6.11 00	1.33 01
22	Ti K_{α_1}	4.5108	274.863	4.35 -1	7.24 -1	2.11 00	5.92 00	1.29 01
55	Cs L_{β_1}	4.6198	268.377	4.31 -1	6.87 -1	1.97 00	5.51 00	1.20 01
57	La L_{α_1}	4.6510	266.577	4.30 -1	6.77 -1	1.93 00	5.39 00	1.18 01
56	Ba L_{β_1}	4.8275	256.841	4.24 -1	6.25 -1	1.73 00	4.81 00	1.05 01
58	Ce L_{α_1}	4.8402	256.157	4.24 -1	6.21 -1	1.72 00	4.78 00	1.04 01
22	Ti $K_{\beta_{1,3}}$	4.9318	251.399	4.21 -1	5.97 -1	1.63 00	4.51 00	9.82 00
23	V K_{α_1}	4.9522	250.363	4.20 -1	5.92 -1	1.61 00	4.45 00	9.70 00
59	Pr L_{α_1}	5.0337	246.310	4.18 -1	5.73 -1	1.53 00	4.24 00	9.22 00
57	La L_{β_1}	5.0421	245.899	4.18 -1	5.71 -1	1.53 00	4.22 00	9.18 00
60	Nd L_{α_1}	5.2304	237.047	4.15 -1	5.36 -1	1.38 00	3.78 00	8.20 00
58	Ce L_{β_1}	5.2622	235.614	4.15 -1	5.30 -1	1.36 00	3.71 00	8.05 00
24	Cr K_{α_1}	5.4147	228.978	4.12 -1	5.05 -1	1.26 00	3.41 00	7.37 00
23	V $K_{\beta_{1,3}}$	5.4273	228.447	4.12 -1	5.03 -1	1.25 00	3.38 00	7.32 00
61	Pm L_{α_1}	5.4325	228.228	4.12 -1	5.02 -1	1.25 00	3.37 00	7.30 00
59	Pr L_{β_1}	5.4889	225.883	4.11 -1	4.93 -1	1.21 00	3.27 00	7.07 00
62	Sm L_{α_1}	5.6361	219.984	4.09 -1	4.71 -1	1.13 00	3.02 00	6.52 00
60	Nd L_{β_1}	5.7216	216.696	4.08 -1	4.59 -1	1.09 00	2.89 00	6.23 00
63	Eu L_{α_1}	5.8457	212.096	4.06 -1	4.42 -1	1.02 00	2.71 00	5.83 00
25	Mn K_{α_1}	5.8988	210.187	4.05 -1	4.36 -1	1.00 00	2.64 00	5.67 00
24	Cr $K_{\beta_{1,3}}$	5.9467	208.494	4.05 -1	4.29 -1	9.78 -1	2.57 00	5.53 00
61	Pm L_{β_1}	5.9610	207.993	4.04 -1	4.28 -1	9.72 -1	2.56 00	5.49 00
64	Gd L_{α_1}	6.0572	204.690	4.03 -1	4.18 -1	9.34 -1	2.44 00	5.23 00

Z	transition	E [keV]	λ [pm]	$_1$H	$_2$He	$_3$Li	$_4$Be	$_5$B
62	Sm L$_{\beta_1}$	6.2051	199.811	4.02 -1	4.06 -1	8.84 -1	2.28 00	4.87 00
65	Tb L$_{\alpha_1}$	6.2726	197.655	4.02 -1	4.01 -1	8.63 -1	2.21 00	4.72 00
26	Fe K$_{\alpha_1}$	6.4038	193.611	4.01 -1	3.91 -1	8.23 -1	2.09 00	4.43 00
63	Eu L$_{\beta_1}$	6.4564	192.034	4.01 -1	3.87 -1	8.08 -1	2.04 00	4.33 00
25	Mn K$_{\beta_{1,3}}$	6.4905	191.025	4.00 -1	3.84 -1	7.99 -1	2.01 00	4.26 00
66	Dy L$_{\alpha_1}$	6.4952	190.887	4.00 -1	3.84 -1	7.97 -1	2.01 00	4.25 00
64	Gd L$_{\beta_1}$	6.7132	184.688	3.99 -1	3.69 -1	7.40 -1	1.83 00	3.85 00
67	Ho L$_{\alpha_1}$	6.7198	184.507	3.99 -1	3.68 -1	7.38 -1	1.82 00	3.84 00
27	Co K$_{\alpha_1}$	6.9303	178.903	3.98 -1	3.55 -1	6.88 -1	1.67 00	3.51 00
68	Er L$_{\alpha_1}$	6.9487	178.429	3.97 -1	3.54 -1	6.84 -1	1.66 00	3.48 00
65	Tb L$_{\beta_1}$	6.9780	177.680	3.97 -1	3.52 -1	6.77 -1	1.64 00	3.44 00
26	Fe K$_{\beta_{1,3}}$	7.0580	175.666	3.97 -1	3.47 -1	6.60 -1	1.59 00	3.32 00
69	Tm L$_{\alpha_1}$	7.1799	172.583	3.96 -1	3.40 -1	6.35 -1	1.51 00	3.16 00
66	Dy L$_{\beta_1}$	7.2477	171.068	3.96 -1	3.36 -1	6.21 -1	1.47 00	3.07 00
70	Yb L$_{\alpha_1}$	7.4156	167.195	3.95 -1	3.27 -1	5.90 -1	1.38 00	2.87 00
28	Ni K$_{\alpha_1}$	7.4782	165.795	3.94 -1	3.23 -1	5.79 -1	1.35 00	2.80 00
67	Ho L$_{\beta_1}$	7.5253	164.757	3.94 -1	3.21 -1	5.70 -1	1.33 00	2.75 00
27	Co K$_{\beta_{1,3}}$	7.6494	162.084	3.93 -1	3.15 -1	5.50 -1	1.27 00	2.62 00
71	Lu L$_{\alpha_1}$	7.6555	161.955	3.93 -1	3.14 -1	5.49 -1	1.26 00	2.61 00
68	Er L$_{\beta_1}$	7.8109	158.733	3.92 -1	3.07 -1	5.24 -1	1.19 00	2.46 00
72	Hf L$_{\alpha_1}$	7.8990	156.963	3.92 -1	3.03 -1	5.11 -1	1.16 00	2.38 00
29	Cu K$_{\alpha_1}$	8.0478	154.060	3.91 -1	2.97 -1	4.91 -1	1.10 00	2.25 00
69	Tm L$_{\beta_1}$	8.1010	153.049	3.91 -1	2.95 -1	4.86 -1	1.08 00	2.21 00
73	Ta L$_{\alpha_1}$	8.1461	152.201	3.91 -1	2.94 -1	4.81 -1	1.07 00	2.18 00
28	Ni K$_{\beta_{1,3}}$	8.2647	150.017	3.90 -1	2.91 -1	4.96 -1	1.03 00	2.09 00
74	W L$_{\alpha_1}$	8.3976	147.643	3.90 -1	2.87 -1	4.56 -1	9.91 -1	2.00 00
70	Yb L$_{\beta_1}$	8.4018	147.569	3.90 -1	2.87 -1	4.56 -1	9.90 -1	2.00 00
30	Zn K$_{\alpha_1}$	8.6389	143.519	3.89 -1	2.81 -1	4.34 -1	9.92 -1	1.85 00
75	Re L$_{\alpha_1}$	8.6525	143.294	3.89 -1	2.81 -1	4.33 -1	9.21 -1	1.84 00
71	Lu L$_{\beta_1}$	8.7090	142.364	3.89 -1	2.79 -1	4.28 -1	9.07 -1	1.81 00
29	Cu K$_{\beta_1}$	8.9053	139.226	3.88 -1	2.75 -1	4.12 -1	8.58 -1	1.70 00
76	Os L$_{\alpha_1}$	8.9117	139.126	3.88 -1	2.74 -1	4.11 -1	8.57 -1	1.70 00
72	Hf L$_{\beta_1}$	9.0227	137.414	3.88 -1	2.72 -1	4.03 -1	8.31 -1	1.64 00
77	Ir L$_{\alpha_1}$	9.1751	135.132	3.88 -1	2.68 -1	3.91 -1	7.98 -1	1.56 00
31	Ga K$_{\alpha_1}$	9.2517	134.013	3.87 -1	2.67 -1	3.86 -1	7.82 -1	1.53 00
73	Ta L$_{\beta_1}$	9.3431	132.702	3.87 -1	2.65 -1	3.79 -1	7.63 -1	1.49 00
78	Pt L$_{\alpha_1}$	9.4423	131.308	3.87 -1	2.63 -1	3.72 -1	7.44 -1	1.44 00
30	Zn K$_{\beta_{1,3}}$	9.5720	129.529	3.87 -1	2.60 -1	3.63 -1	7.19 -1	1.39 00
74	W L$_{\beta_1}$	9.6724	128.184	3.86 -1	2.58 -1	3.57 -1	7.01 -1	1.35 00
79	Au L$_{\alpha_1}$	9.7133	127.644	3.86 -1	2.57 -1	3.54 -1	6.94 -1	1.33 00
31	Ge K$_{\alpha_1}$	9.8864	125.409	3.86 -1	2.54 -1	3.44 -1	6.65 -1	1.27 00
80	Hg L$_{\alpha_1}$	9.9888	124.124	3.85 -1	2.52 -1	3.37 -1	6.48 -1	1.24 00
75	Re L$_{\beta_1}$	10.010	123.124	3.85 -1	2.51 -1	3.36 -1	6.45 -1	1.23 00
31	Ga K$_{\beta_1}$	10.264	120.796	3.85 -1	2.49 -1	3.28 -1	6.16 -1	1.16 00

Z	transition	E [keV]	λ [pm]	$_1$H	$_2$He	$_3$Li	$_4$Be	$_5$B
81	Tl L$_{\alpha_1}$	10.269	120.737	3.85 -1	2.49 -1	3.27 -1	6.16 -1	1.16 00
76	Os L$_{\beta_1}$	10.355	119.734	3.85 -1	2.48 -1	3.24 -1	6.06 -1	1.14 00
33	As K$_{\alpha_1}$	10.544	117.588	3.84 -1	2.46 -1	3.18 -1	5.87 -1	1.09 00
82	Pb L$_{\alpha_1}$	10.552	117.499	3.84 -1	2.46 -1	3.18 -1	5.86 -1	1.09 00
77	Ir L$_{\beta_1}$	10.708	115.787	3.84 -1	2.44 -1	3.13 -1	5.71 -1	1.05 00
83	Bi L$_{\alpha_1}$	10.839	114.388	3.84 -1	2.43 -1	3.09 -1	5.58 -1	1.02 00
32	Ge K$_{\beta_1}$	10.982	112.898	3.83 -1	2.42 -1	3.05 -1	5.45 -1	9.90 -1
78	Pt L$_{\beta_1}$	11.071	111.990	3.83 -1	2.41 -1	3.02 -1	5.37 -1	9.72 -1
84	Po L$_{\alpha_1}$	11.131	111.387	3.83 -1	2.40 -1	3.00 -1	5.32 -1	9.60 -1
34	Se K$_{\alpha_1}$	11.222	110.484	3.83 -1	2.40 -1	2.98 -1	5.24 -1	9.42 -1
85	At L$_{\alpha_1}$	11.427	108.501	3.82 -1	2.38 -1	2.92 -1	5.07 -1	9.03 -1
79	Au L$_{\beta_1}$	11.442	108.359	3.82 -1	2.38 -1	2.92 -1	5.06 -1	9.00 -1
33	As K$_{\beta_1}$	11.726	105.735	3.82 -1	2.34 -1	2.84 -1	4.84 -1	8.50 -1
86	Rn L$_{\alpha_1}$	11.727	105.726	3.82 -1	2.34 -1	2.84 -1	4.83 -1	8.50 -1
80	Hg L$_{\beta_1}$	11.823	104.867	3.82 -1	2.34 -1	2.82 -1	4.76 -1	8.34 -1
35	Br K$_{\alpha_1}$	11.924	103.979	3.81 -1	2.34 -1	2.79 -1	4.69 -1	8.18 -1
87	Fr L$_{\alpha_1}$	12.031	103.054	3.81 -1	2.33 -1	2.77 -1	4.61 -1	8.01 -1
81	Tl L$_{\beta_1}$	12.213	101.519	3.81 -1	2.31 -1	2.72 -1	4.49 -1	7.74 -1
88	Ra L$_{\alpha_1}$	12.340	100.474	3.81 -1	2.30 -1	2.69 -1	4.41 -1	7.55 -1
34	Se K$_{\beta_1}$	12.496	99.219	3.80 -1	2.29 -1	2.66 -1	4.31 -1	7.33 -1
82	Pb L$_{\beta_1}$	12.614	98.291	3.80 -1	2.28 -1	2.63 -1	4.23 -1	7.18 -1
36	Kr K$_{\alpha_1}$	12.649	98.019	3.80 -1	2.28 -1	2.62 -1	4.21 -1	7.13 -1
89	Ac L$_{\alpha_1}$	12.652	97.996	3.80 -1	2.28 -1	2.62 -1	4.21 -1	7.13 -1
90	Th L$_{\alpha_1}$	12.969	95.601	3.80 -1	2.25 -1	2.55 -1	4.02 -1	6.73 -1
83	Bi L$_{\beta_1}$	13.024	95.197	3.80 -1	2.25 -1	2.54 -1	3.99 -1	6.66 -1
91	Pa L$_{\alpha_1}$	13.291	93.285	3.79 -1	2.23 -1	2.49 -1	3.85 -1	6.35 -1
35	Br K$_{\beta_1}$	13.291	93.285	3.79 -1	2.23 -1	2.49 -1	3.85 -1	6.35 -1
37	Rb K$_{\alpha_1}$	13.395	92.560	3.79 -1	2.22 -1	2.47 -1	3.79 -1	6.24 -1
84	Po L$_{\beta_1}$	13.447	92.202	3.79 -1	2.22 -1	2.46 -1	3.77 -1	6.18 -1
92	U L$_{\alpha_1}$	13.615	91.065	3.79 -1	2.21 -1	2.42 -1	3.68 -1	6.01 -1
85	At L$_{\beta_1}$	13.876	89.352	3.78 -1	2.19 -1	2.38 -1	3.56 -1	5.75 -1
36	Kr K$_{\beta_1}$	14.112	87.857	3.78 -1	2.18 -1	2.33 -1	3.45 -1	5.53 -1
38	Sr K$_{\alpha_1}$	14.165	87.529	3.78 -1	2.17 -1	.2.32 -1	3.43 -1	5.48 -1
86	Rn L$_{\beta_1}$	14.316	86.606	3.77 -1	2.16 -1	2.30 -1	3.36 -1	5.35 -1
87	Fr L$_{\beta_1}$	14.770	83.943	3.77 -1	2.14 -1	2.22 -1	3.18 -1	4.97 -1
39	Y K$_{\alpha_1}$	14.958	82.888	3.76 -1	2.12 -1	2.19 -1	3.10 -1	4.83 -1
37	Rb K$_{\beta_1}$	14.961	82.872	3.76 -1	2.12 -1	2.19 -1	3.10 -1	4.83 -1
88	Ra L$_{\beta_1}$	15.236	81.376	3.76 -1	2.11 -1	2.17 -1	3.04 -1	4.68 -1
89	Ac L$_{\beta_1}$	15.713	78.906	3.75 -1	2.10 -1	2.13 -1	2.94 -1	4.45 -1
40	Zr K$_{\alpha_1}$	15.775	78.596	3.75 -1	2.09 -1	2.13 -1	2.92 -1	4.42 -1
38	Sr K$_{\beta_1}$	15.836	78.293	3.75 -1	2.09 -1	2.12 -1	2.91 -1	4.39 -1
90	Th L$_{\beta_1}$	16.202	76.524	3.74 -1	2.08 -1	2.10 -1	2.84 -1	4.24 -1
41	Nb K$_{\alpha_1}$	16.615	74.622	3.74 -1	2.07 -1	2.07 -1	2.77 -1	4.07 -1
91	Pa L$_{\beta_1}$	16.702	74.233	3.74 -1	2.06 -1	2.06 -1	2.75 -1	4.03 -1

Z	transition	E [keV]	λ [pm]	$_1$H	$_2$He	$_3$Li	$_4$Be	$_5$B
39	Y K$_{\beta_1}$	16.738	74.074	3.74 -1	2.06 -1	2.06 -1	2.74 -1	4.02 -1
92	U L$_{\beta_1}$	17.220	72.000	3.73 -1	2.05 -1	2.03 -1	2.66 -1	3.84 -1
42	Mo K$_{\alpha_1}$	17.479	70.933	3.73 -1	2.04 -1	2.01 -1	2.62 -1	3.75 -1
40	Zr K$_{\beta_1}$	17.668	70.175	3.72 -1	2.04 -1	2.00 -1	2.59 -1	3.68 -1
43	Tc K$_{\alpha_1}$	18.367	67.504	3.71 -1	2.01 -1	1.96 -1	2.48 -1	3.46 -1
41	Nb K$_{\beta_1}$	18.623	66.576	3.71 -1	2.01 -1	1.94 -1	2.45 -1	3.38 -1
44	Ru K$_{\alpha_1}$	19.279	64.311	3.70 -1	1.99 -1	1.90 -1	2.36 -1	3.20 -1
42	Mo K$_{\beta_1}$	19.608	63.231	3.70 -1	1.98 -1	1.89 -1	2.31 -1	3.11 -1
45	Rh K$_{\alpha_1}$	20.216	61.330	3.69 -1	1.97 -1	1.86 -1	2.25 -1	2.98 -1
43	Tc K$_{\beta_1}$	20.619	60.131	3.68 -1	1.96 -1	1.85 -1	2.23 -1	2.93 -1
46	Pd K$_{\alpha_1}$	21.177	58.547	3.67 -1	1.95 -1	1.83 -1	2.19 -1	2.86 -1
44	Ru K$_{\beta_1}$	21.657	57.249	3.67 -1	1.95 -1	1.82 -1	2.16 -1	2.80 -1
47	Ag K$_{\alpha_1}$	22.163	55.942	3.66 -1	1.94 -1	1.81 -1	2.14 -1	2.74 -1
45	Rh K$_{\beta_1}$	22.724	54.561	3.65 -1	1.93 -1	1.79 -1	2.11 -1	2.68 -1
48	Cd K$_{\alpha_1}$	23.174	53.501	3.65 -1	1.93 -1	1.78 -1	2.08 -1	2.63 -1
46	Pd K$_{\beta_1}$	23.819	52.053	3.64 -1	1.92 -1	1.77 -1	2.05 -1	2.56 -1
49	In K$_{\alpha_1}$	24.210	51.212	3.63 -1	1.91 -1	1.76 -1	2.03 -1	2.52 -1
49	In K$_{\beta_1}$	27.276	45.455	3.60 -1	1.88 -1	1.70 -1	1.90 -1	2.26 -1
52	Te K$_{\alpha_1}$	27.472	45.131	3.59 -1	1.88 -1	1.69 -1	1.89 -1	2.24 -1
50	Sn K$_{\beta_1}$	28.486	43.524	3.58 -1	1.87 -1	1.68 -1	1.85 -1	2.17 -1
53	I K$_{\alpha_1}$	28.612	43.333	3.58 -1	1.86 -1	1.67 -1	1.85 -1	2.16 -1
51	Sb K$_{\beta_1}$	29.726	41.709	3.57 -1	1.85 -1	1.65 -1	1.81 -1	2.08 -1
54	Xe K$_{\alpha_1}$	29.779	41.635	3.57 -1	1.85 -1	1.65 -1	1.81 -1	2.08 -1
55	Cs K$_{\alpha_1}$	30.973	40.030	3.55 -1	1.84 -1	1.64 -1	1.78 -1	2.04 -1
52	Te K$_{\beta_1}$	30.996	40.000	3.55 -1	1.84 -1	1.64 -1	1.78 -1	2.03 -1
56	Ba K$_{\alpha_1}$	32.194	38.511	3.54 -1	1.83 -1	1.63 -1	1.76 -1	2.00 -1
53	I K$_{\beta_1}$	32.295	38.391	3.54 -1	1.83 -1	1.62 -1	1.76 -1	1.99 -1
57	La K$_{\alpha_1}$	33.442	37.074	3.53 -1	1.82 -1	1.61 -1	1.74 -1	1.96 -1
54	Xe K$_{\beta_1}$	33.624	36.874	3.52 -1	1.82 -1	1.61 -1	1.74 -1	1.96 -1
58	Ce K$_{\alpha_1}$	34.279	36.169	3.52 -1	1.82 -1	1.60 -1	1.72 -1	1.94 -1
55	Cs K$_{\beta_1}$	34.987	35.437	3.51 -1	1.81 -1	1.60 -1	1.71 -1	1.92 -1
59	Pr K$_{\alpha_1}$	36.026	34.415	3.50 -1	1.80 -1	1.59 -1	1.70 -1	1.89 -1
56	Ba K$_{\beta_1}$	36.378	34.082	3.49 -1	1.80 -1	1.58 -1	1.69 -1	1.88 -1
60	Nd K$_{\alpha_1}$	36.847	33.648	3.49 -1	1.80 -1	1.58 -1	1.69 -1	1.87 -1
57	La K$_{\beta_1}$	37.801	32.799	3.48 -1	1.79 -1	1.57 -1	1.67 -1	1.85 -1
61	Pm K$_{\alpha_1}$	38.725	32.016	3.47 -1	1.78 -1	1.56 -1	1.66 -1	1.83 -1
58	Ce K$_{\beta_1}$	39.257	31.582	3.47 -1	1.78 -1	1.56 -1	1.65 -1	1.82 -1
62	Sm K$_{\alpha_1}$	40.118	30.905	3.46 -1	1.77 -1	1.55 -1	1.64 -1	1.80 -1
59	Pr K$_{\beta_1}$	40.748	30.427	3.45 -1	1.77 -1	1.55 -1	1.64 -1	1.79 -1
63	Eu K$_{\alpha_1}$	41.542	29.845	3.44 -1	1.76 -1	1.54 -1	1.63 -1	1.78 -1
60	Nd K$_{\beta_1}$	42.271	29.331	3.43 -1	1.76 -1	1.54 -1	1.62 -1	1.77 -1
64	Gd K$_{\alpha_1}$	42.996	28.836	3.42 -1	1.75 -1	1.53 -1	1.62 -1	1.76 -1
61	Pm K$_{\beta_1}$	43.826	28.290	3.42 -1	1.74 -1	1.53 -1	1.61 -1	1.74 -1
65	Tb K$_{\alpha_1}$	44.482	27.873	3.41 -1	1.74 -1	1.52 -1	1.60 -1	1.73 -1

Z	transition	E [keV]	λ [pm]	$_1$H	$_2$He	$_3$Li	$_4$Be	$_5$B
62	Sm K$_{\beta_1}$	45.413	27.301	3.40 -1	1.73 -1	1.52 -1	1.59 -1	1.72 -1
66	Dy K$_{\alpha_1}$	45.998	26.954	3.39 -1	1.73 -1	1.52 -1	1.59 -1	1.71 -1
63	Eu K$_{\beta_1}$	47.038	26.358	3.38 -1	1.72 -1	1.51 -1	1.58 -1	1.70 -1
67	Ho K$_{\alpha_1}$	47.547	26.076	3.38 -1	1.72 -1	1.51 -1	1.58 -1	1.69 -1
64	Gd K$_{\beta_1}$	48.697	25.460	3.36 -1	1.71 -1	1.50 -1	1.57 -1	1.68 -1
68	Er K$_{\alpha_1}$	49.128	25.237	3.36 -1	1.71 -1	1.50 -1	1.56 -1	1.68 -1
65	Tb K$_{\beta_1}$	50.382	24.609	3.35 -1	1.70 -1	1.49 -1	1.55 -1	1.66 -1
69	Tm K$_{\alpha_1}$	50.742	24.434	3.34 -1	1.70 -1	1.49 -1	1.55 -1	1.65 -1
66	Dy K$_{\beta_1}$	52.119	23.788	3.33 -1	1.69 -1	1.48 -1	1.54 -1	1.65 -1
70	Yb K$_{\alpha_1}$	52.389	23.666	3.33 -1	1.69 -1	1.48 -1	1.54 -1	1.65 -1
67	Ho K$_{\beta_1}$	53.877	23.012	3.31 -1	1.68 -1	1.47 -1	1.53 -1	1.63 -1
68	Er K$_{\beta_1}$	55.681	22.167	3.30 -1	1.67 -1	1.46 -1	1.52 -1	1.62 -1
72	Hf K$_{\alpha_1}$	55.790	22.223	3.30 -1	1.67 -1	1.46 -1	1.52 -1	1.62 -1
69	Tm K$_{\beta_1}$	57.517	21.556	3.28 -1	1.67 -1	1.45 -1	1.51 -1	1.61 -1
73	Ta K$_{\alpha_1}$	57.532	21.550	3.28 -1	1.67 -1	1.45 -1	1.51 -1	1.61 -1
74	W K$_{\alpha_1}$	59.318	20.901	3.27 -1	1.66 -1	1.44 -1	1.50 -1	1.59 -1
70	Yb K$_{\beta_1}$	59.370	20.883	3.27 -1	1.66 -1	1.44 -1	1.50 -1	1.59 -1
75	Re K$_{\alpha_1}$	61.140	20.278	3.25 -1	1.65 -1	1.43 -1	1.50 -1	1.58 -1
71	Lu K$_{\beta_1}$	61.283	20.231	3.25 -1	1.65 -1	1.43 -1	1.50 -1	1.58 -1
76	Os K$_{\alpha_1}$	63.001	19.679	3.23 -1	1.64 -1	1.43 -1	2.49 -1	1.57 -1
72	Hf K$_{\beta_1}$	63.234	19.607	3.23 -1	1.64 -1	1.42 -1	1.49 -1	1.57 -1
77	Ir K$_{\alpha_1}$	64.896	19.105	3.21 -1	1.63 -1	1.42 -1	1.48 -1	1.55 -1
73	Ta K$_{\beta_1}$	65.223	19.009	3.21 -1	1.63 -1	1.41 -1	1.48 -1	1.55 -1
78	Pt K$_{\alpha_1}$	66.832	18.551	3.20 -1	1.62 -1	1.41 -1	1.47 -1	1.54 -1
74	W K$_{\beta_1}$	67.244	18.438	3.19 -1	1.62 -1	1.41 -1	1.47 -1	1.54 -1
79	Au K$_{\alpha_1}$	68.804	18.020	3.18 -1	1.61 -1	1.40 -1	1.46 -1	1.53 -1
75	Re K$_{\beta_1}$	69.310	17.888	3.17 -1	1.61 -1	1.40 -1	1.46 -1	1.53 -1
80	Hg K$_{\alpha_1}$	70.819	17.507	3.16 -1	1.60 -1	1.39 -1	1.45 -1	1.52 -1
76	Os K$_{\beta_1}$	71.413	17.361	3.16 -1	1.60 -1	1.39 -1	1.45 -1	1.52 -1
91	Tl K$_{\alpha_1}$	72.872	17.014	3.14 -1	1.59 -1	1.38 -1	1.44 -1	1.51 -1
77	Ir K$_{\beta_1}$	73.651	16.854	3.14 -1	1.59 -1	1.38 -1	1.44 -1	1.59 -1
82	Pb K$_{\alpha_1}$	74.969	16.538	3.13 -1	1.59 -1	1.37 -1	1.43 -1	1.50 -1
78	Pt K$_{\beta_1}$	75.748	16.368	3.12 -1	1.58 -1	1.37 -1	1.43 -1	1.49 -1
83	Bi K$_{\alpha_1}$	77.108	16.079	3.11 -1	1.58 -1	1.36 -1	1.42 -1	1.49 -1
79	Au K$_{\beta_1}$	77.948	15.906	3.10 -1	1.57 -1	1.36 -1	1.42 -1	1.48 -1
84	Po K$_{\alpha_1}$	79.290	15.636	3.09 -1	1.57 -1	1.36 -1	1.41 -1	1.47 -1
80	Hg K$_{\beta_1}$	80.253	15.449	3.09 -1	1.56 -1	1.35 -1	1.41 -1	1.47 -1
85	At K$_{\alpha_1}$	81.520	15.209	3.08 -1	1.56 -1	1.35 -1	1.40 -1	1.46 -1
81	Tl K$_{\beta_1}$	82.576	15.014	3.07 -1	1.55 -1	1.35 -1	1.40 -1	1.46 -1
86	Rn K$_{\alpha_1}$	83.780	14.798	3.06 -1	1.55 -1	1.34 -1	1.39 -1	1.45 -1
82	Pb K$_{\beta_1}$	84.936	14.597	3.05 -1	1.54 -1	1.34 -1	1.39 -1	1.45 -1
87	Fr K$_{\alpha_1}$	86.100	14.400	3.04 -1	1.54 -1	1.33 -1	1.38 -1	1.44 -1
83	Bi K$_{\beta_1}$	87.343	14.195	3.03 -1	1.53 -1	1.33 -1	1.38 -1	1.44 -1
88	Ra K$_{\alpha_1}$	88.470	14.014	3.02 -1	1.53 -1	1.33 -1	1.37 -1	1.44 -1

Z	transition	E [keV]	λ [pm]	$_1$H	$_2$He	$_3$Li	$_4$Be	$_5$B
84	Po K$_{\beta_1}$	89.800	13.806	3.01 -1	1.52 -1	1.32 -1	1.37 -1	1.43 -1
89	Ac K$_{\alpha_1}$	90.884	13.642	3.01 -1	1.52 -1	1.32 -1	1.36 -1	1.43 -1
85	At K$_{\beta_1}$	92.300	13.432	3.00 -1	1.51 -1	1.31 -1	1.36 -1	1.42 -1
90	Tm K$_{\alpha_1}$	93.350	13.281	2.99 -1	1.51 -1	1.31 -1	1.35 -1	1.42 -1
86	Rn K$_{\beta_1}$	94.870	13.068	2.98 -1	1.51 -1	1.31 -1	1.35 -1	1.41 -1
91	Pa K$_{\alpha_1}$	95.868	12.932	2.97 -1	1.50 -1	1.30 -1	1.34 -1	1.41 -1
92	U K$_{\alpha_1}$	98.439	12.595	2.96 -1	1.49 -1	1.30 -1	1.34 -1	1.40 -1
88	Ra K$_{\beta_1}$	100.130	12.382	2.94 -1	1.49 -1	1.29 -1	1.33 -1	1.39 -1
89	Ac K$_{\beta_1}$	102.850	12.054	2.92 -1	1.48 -1	1.28 -1	1.32 -1	1.38 -1
90	Tm K$_{\beta_1}$	105.610	11.739	2.90 -1	1.47 -1	1.27 -1	1.31 -1	1.37 -1
91	Pa K$_{\beta_1}$	108.430	11.434	2.88 -1	1.46 -1	1.26 -1	1.30 -1	1.36 -1
92	U K$_{\beta_1}$	111.300	11.139	2.87 -1	1.45 -1	1.25 -1	1.29 -1	1.35 -1

Z	transition	E [keV]	λ [pm]	$_6$C	$_7$N	$_8$O	$_9$F	$_{10}$Ne
4	Be K$_\alpha$	0.1085	11427.207	1.92 04	3.64 04	5.40 04	7.37 04	1.00 05
38	Sr M$_\xi$	0.1140	10875.895	1.74 04	3.27 04	4.89 04	6.64 04	9.17 04
39	Y M$_\xi$	0.1228	9339.235	1.31 04	2.34 04	3.47 04	4.82 04	6.88 04
16	S L	0.1487	8337.942	1.02 04	1.81 04	2.68 04	3.75 04	5.49 04
40	Zr M$_\xi$	0.1511	8205.506	9.84 03	1.75 04	2.60 04	3.62 04	5.33 04
41	Nb M$_\xi$	0.1717	7221.037	7.37 03	1.30 04	1.82 04	2.71 04	4.08 04
5	B K$_\alpha$	0.1833	6764.058	6.35 03	1.12 04	1.65 04	2.33 04	3.54 04
42	Mo M$_\xi$	0.1926	6473.445	5.67 03	1.00 04	1.46 04	2.08 04	3.18 04
6	C K$_\alpha$	0.2770	4476.000	<u>2.35 03</u>	4.22 03	6.04 03	8.75 03	1.36 04
47	Ag M$_\xi$	0.3117	3977.709	4.50 04	3.18 03	4.49 03	6.55 03	1.01 04
7	N K$_\alpha$	0.3924	3159.664	2.55 04	1.81 03	2.53 03	3.70 03	5.62 03
22	Ti L	0.3953	3136.484	2.50 04	<u>1.78 03</u>	2.48 03	3.63 03	5.50 03
22	Ti L$_\alpha$	0.4522	2741.822	1.80 04	2.48 04	1.76 03	2.59 03	3.85 03
23	V L$_\alpha$	0.5113	2424.901	1.32 04	1.84 04	1.28 03	1.90 03	2.77 03
8	O K$_\alpha$	0.5249	2362.072	1.24 04	1.73 04	<u>1.20 03</u>	1.77 03	2.58 03
25	Mn L$_l$	0.5563	2228.747	1.07 04	1.51 04	2.01 04	1.53 03	2.21 03
24	Cr L$_\alpha$	0.5728	2164.546	9.90 03	1.40 04	1.87 04	1.42 03	2.04 03
25	Mn L$_\alpha$	0.6374	1945.171	7.50 03	1.07 04	1.45 04	1.08 03	1.54 03
9	F K$_\alpha$	0.6768	1831.932	6.37 03	9.16 03	1.24 04	<u>9.22 02</u>	1.30 03
26	Fe L$_\alpha$	0.7050	1756.655	5.79 03	8.34 03	1.13 04	1.29 04	1.18 03
27	Co L$_\alpha$	0.7762	1597.335	4.45 03	6.50 03	8.88 03	1.06 04	9.18 02
28	Ni L$_\alpha$	0.8515	1456.080	3.46 03	5.07 03	6.96 03	8.39 03	<u>7.15 02</u>
29	Cu L$_\alpha$	0.9297	1336.044	2.71 03	4.02 03	5.60 03	6.78 03	9.19 03
30	Zn L$_\alpha$	1.0117	1225.513	2.16 03	3.35 03	4.59 03	5.90 03	7.70 03
11	Na K$_\alpha$	1.0410	1191.020	1.99 03	3.09 03	4.25 03	5.46 03	7.14 03
11	Na K$_\beta$	1.0711	1157.550	1.83 03	2.85 03	3.93 03	5.06 03	6.63 03
12	Mg K$_\alpha$	1.2536	989.033	1.16 03	1.83 03	2.56 03	3.30 03	4.39 03
33	As L$_\alpha$	1.2820	967.123	1.09 03	1.72 03	2.41 03	3.11 03	4.14 03
12	Mg K$_\beta$	1.3022	952.121	1.04 03	1.64 03	2.31 03	2.98 03	3.97 03

Z	transition	E [keV]	λ [pm]	$_6$C	$_7$N	$_8$O	$_9$F	$_{10}$Ne
33	As L$_\beta$	1.3170	941.421	1.01 03	1.59 03	2.24 03	2.89 03	3.85 03
66	Dy M$_\beta$	1.3250	935.737	9.90 02	1.56 03	2.21 03	2.85 03	3.79 03
67	Ho M$_\alpha$	1.3480	919.771	9.42 02	1.49 03	2.10 03	2.72 03	3.63 03
34	Se L$_\alpha$	1.3791	899.029	8.82 02	1.40 03	1.98 03	2.55 03	3.42 03
67	Ho M$_\beta$	1.3830	896.494	8.75 02	1.38 03	1.96 03	2.53 03	3.39 03
68	Er M$_\alpha$	1.4060	881.829	8.34 02	1.32 03	1.88 03	2.42 03	3.25 03
34	Se L$_{\beta_1}$	1.4292	873.627	8.12 02	1.29 03	1.83 03	2.36 03	3.17 03
68	Er M$_\beta$	1.4430	859.218	7.74 02	1.23 03	1.75 03	2.26 03	3.03 03
69	Tm M$_\alpha$	1.4620	848.051	7.45 02	1.18 03	1.69 03	2.18 03	2.93 03
35	Br L$_\alpha$	1.4804	837.511	7.19 02	1.14 03	1.63 03	2.11 03	2.84 03
13	Al K$_{\alpha_1}$	1.4867	833.962	7.10 02	1.13 03	1.61 03	2.08 03	2.80 03
69	Tm M$_\beta$	1.5030	824.918	6.88 02	1.09 03	1.56 03	2.02 03	2.72 03
70	Yb M$_\alpha$	1.5214	814.941	6.64 02	1.06 03	1.51 03	1.96 03	2.64 03
35	Br L$_{\beta_1}$	1.5259	812.538	6.58 02	1.05 03	1.50 03	1.94 03	2.61 03
13	Al K$_\beta$	1.5574	796.103	6.20 02	9.87 02	1.42 03	1.88 03	2.47 03
70	Yb M$_\beta$	1.5675	790.974	6.08 02	9.69 02	1.39 03	1.80 03	2.43 03
71	Lu M$_\alpha$	1.5813	784.071	5.93 02	9.44 02	1.36 03	1.76 03	2.37 03
36	Kr L$_\alpha$	1.5860	781.747	5.87 02	9.36 02	1.34 03	1.74 03	2.35 03
71	Lu M$_\beta$	1.6312	760.085	5.41 02	8.63 02	1.24 03	1.61 03	2.18 03
36	Kr L$_{\beta_1}$	1.6366	757.577	5.36 02	8.55 02	1.23 03	1.60 03	2.16 03
72	Hf M$_{\alpha_1}$	1.6446	753.892	5.28 02	8.43 02	1.21 03	1.57 03	2.13 03
37	Rb L$_{\alpha_1}$	1.6941	731.864	4.84 02	7.73 02	1.12 03	1.45 03	1.97 03
72	Hf M$_\beta$	1.6976	730.355	4.81 02	7.68 02	1.11 03	1.44 03	1.96 03
73	Ta M$_{\alpha_1}$	1.7096	725.229	4.71 02	7.53 02	1.09 03	1.41 03	1.92 03
14	Si K$_{\alpha_1}$	1.7400	712.558	4.48 02	7.15 02	1.04 03	1.35 03	1.83 03
37	Rb L$_{\beta_1}$	1.7522	707.597	4.38 02	7.01 02	1.02 03	1.32 03	1.80 03
73	Ta M$_\beta$	1.7655	702.266	4.29 02	6.86 02	9.95 02	1.29 03	1.76 03
74	W M$_{\alpha_1}$	1.7754	698.350	4.22 02	6.75 02	9.79 02	1.27 03	1.73 03
38	Sr L$_{\alpha_1}$	1.8066	686.290	4.01 02	6.41 03	9.32 03	1.21 03	1.65 03
74	W M$_\beta$	1.8349	675.705	3.83 02	6.13 02	8.92 02	1.16 03	1.58 03
14	Si K$_\beta$	1.8354	675.337	3.82 02	6.12 02	8.91 02	1.16 03	1.58 03
75	Re M$_{\alpha_1}$	1.8420	673.100	3.79 02	6.06 02	8.83 02	1.15 03	1.57 03
38	Sr L$_{\beta_1}$	1.8717	662.420	3.61 02	5.79 02	8.44 02	1.10 03	1.50 03
75	Re M$_\beta$	1.9061	650.465	3.42 02	5.49 02	8.02 02	1.04 03	1.43 03
76	Os M$_{\alpha_1}$	1.9102	649.069	3.40 02	3.45 02	7.97 02	1.04 03	1.42 03
39	Y L$_{\alpha_1}$	1.9226	644.882	3.34 02	5.35 02	7.82 02	1.02 03	1.40 03
76	Os M$_\beta$	1.9783	626.725	3.07 02	4.93 02	7.22 02	9.41 02	1.29 03
77	Ir M$_{\alpha_1}$	1.9799	626.219	3.06 02	4.92 02	7.20 02	9.39 02	1.29 03
39	Y L$_{\beta_1}$	1.9958	621.230	2.99 02	4.80 02	7.04 02	9.18 02	1.26 03
15	P K$_{\alpha_1}$	2.0137	615.708	2.91 02	4.68 02	6.86 02	8.95 02	1.23 03
40	Zr L$_{\alpha_1}$	2.0424	607.056	2.79 02	4.49 02	6.59 02	8.60 02	1.18 03
78	Pt M$_{\alpha_1}$	2.0505	604.658	2.76 02	4.44 02	6.51 02	8.50 02	1.17 03
77	Ir M$_\beta$	2.0535	603.775	2.75 02	4.42 02	6.49 02	8.47 02	1.16 03
79	Au M$_{\alpha_1}$	2.1229	584.036	2.49 02	4.01 02	5.89 02	7.70 02	1.06 03

Z	transition	E [keV]	λ [pm]	$_6$C	$_7$N	$_8$O	$_9$F	$_{10}$Ne
40	Zr L$_{\beta_1}$	2.1244	583.624	2.48 02	4.00 02	5.88 02	7.69 02	1.06 03
78	Pt M$_\beta$	2.1273	582.828	2.47 02	3.98 02	5.86 02	7.66 02	1.05 03
15	P K$_{\beta_{1,3}}$	2.1390	579.640	2.43 02	3.92 02	5.77 02	7.54 02	1.04 03
41	Nb L$_{\alpha_1}$	2.1659	572.441	2.34 02	3.78 02	5.56 02	7.27 02	1.00 03
80	Hg M$_{\alpha_1}$	2.1953	564.775	2.25 02	3.63 02	5.35 02	7.00 02	9.66 02
79	Au M$_\beta$	2.2046	562.393	2.22 02	3.58 02	5.29 02	6.92 02	9.54 02
41	Nb L$_{\beta_1}$	2.2574	549.238	2.07 02	3.34 02	4.94 02	6.47 02	8.93 02
81	Tl M$_{\alpha_1}$	2.2706	546.045	2.03 02	3.29 02	4.85 02	6.36 02	8.79 02
80	Hg M$_\beta$	2.2825	543.199	2.00 02	3.24 02	4.78 02	6.27 02	8.66 02
42	Mo L$_{\alpha_1}$	2.2932	540.664	1.97 02	3.19 02	4.72 02	6.18 02	8.55 02
16	S K$_{\alpha_1}$	2.3080	537.197	1.94 02	3.13 02	4.63 02	6.07 02	8.39 02
82	Pb M$_{\alpha_1}$	2.3457	528.563	1.84 02	2.99 02	4.42 02	5.80 02	8.02 02
81	Tl M$_\beta$	2.3621	524.893	1.81 02	2.93 02	4.33 02	5.68 02	7.87 02
42	Mo L$_{\beta_1}$	2.3948	517.726	1.73 02	2.81 02	4.16 02	5.46 02	7.57 02
83	Bi M$_{\alpha_1}$	2.4226	511.785	1.67 02	2.72 02	4.03 02	5.29 02	7.33 02
43	Tc L$_{\alpha_1}$	2.4240	511.490	1.67 02	2.71 02	4.02 02	5.28 02	7.32 02
82	Pb M$_\beta$	2.4427	507.574	1.63 02	2.65 02	3.93 02	5.16 02	7.16 02
16	S K$_\beta$	2.4640	503.186	1.59 02	2.58 02	3.83 02	5.04 02	6.99 02
83	Bi M$_{\beta_1}$	2.5255	490.933	1.48 02	2.40 02	3.57 02	4.70 02	6.52 02
43	Tc L$_{\beta_1}$	2.5368	488.746	1.46 02	2.37 02	3.52 02	4.64 02	6.44 02
44	Ru L$_{\alpha_1}$	2.5586	484.582	1.42 02	2.31 02	3.44 02	4.53 02	6.29 02
17	Cl K$_{\alpha_1}$	2.6224	472.792	1.32 02	2.15 02	3.20 02	4.22 02	5.87 02
44	Ru L$_{\beta_1}$	2.6832	462.079	1.23 02	2.01 02	3.00 02	3.95 02	5.51 02
45	Rh L$_{\alpha_1}$	2.6967	459.766	1.21 02	1.98 02	2.95 02	3.90 02	5.43 02
17	Cl K$_\beta$	2.8156	440.350	1.07 02	1.75 02	2.61 02	3.44 02	4.81 02
45	Rh L$_{\beta_1}$	2.8344	437.430	1.05 02	1.71 02	2.56 02	3.38 02	4.72 02
46	Pd L$_{\alpha_1}$	2.8386	436.782	1.04 02	1.70 02	2.55 02	3.37 02	4.70 02
18	Ar K$_{\alpha_1}$	2.9577	419.194	9.21 01	1.51 02	2.26 02	2.99 02	4.19 02
47	Ag L$_{\beta_1}$	2.9843	415.458	8.96 01	1.47 02	2.20 02	2.92 02	4.09 02
46	Pd L$_{\beta_1}$	2.9902	414.638	8.91 01	1.46 02	2.19 02	2.90 02	4.07 02
90	Th M$_\beta$	2.9961	413.821	8.86 01	1.45 62	2.18 02	2.89 02	4.04 02
91	Pa M$_{\alpha_1}$	3.0823	402.248	8.13 01	1.34 02	2.01 02	2.66 02	3.73 02
48	Cd L$_{\alpha_1}$	3.1337	395.651	7.73 01	1.27 02	1.91 02	2.53 02	3.56 02
90	Th M$_\beta$	3.1458	394.129	7.64 01	1.26 02	1.89 02	2.50 02	3.52 02
47	Ag L$_{\beta_1}$	3.1509	393.491	7.60 01	1.23 02	1.88 02	2.49 02	3.50 02
92	U M$_{\alpha_1}$	3.1708	391.021	7.46 01	1.23 02	1.84 02	2.45 02	3.44 02
18	Ar K$_\beta$	3.1905	388.607	7.32 01	1.20 02	1.81 02	2.40 02	3.38 02
91	Pa M$_\beta$	3.2397	382.705	6.98 01	1.15 02	1.73 02	2.30 02	3.23 02
49	In L$_{\alpha_1}$	3.2869	377.210	6.68 01	1.10 02	1.66 02	2.20 02	3.10 02
19	K K$_{\alpha_1}$	3.3138	374.148	6.52 01	1.07 02	1.62 02	2.15 02	3.03 02
48	Cd L$_{\beta_1}$	3.3166	373.832	6.50 01	1.07 02	1.62 02	2.15 02	3.03 02
92	U M$_\beta$	3.3360	371.659	6.39 01	1.05 02	1.59 02	2.11 02	2.98 02
50	Sn L$_{\alpha_1}$	3.4440	360.003	5.80 01	9.57 01	1.45 02	1.92 02	2.72 02
49	In L$_{\beta_1}$	3.4872	355.543	5.58 01	9.22 01	1.39 02	1.85 02	2.62 02

Z	transition	E [keV]	λ [pm]	$_6$C	$_7$N	$_8$O	$_9$F	$_{10}$Ne
19	K K$_\beta$	3.5896	345.401	5.11 01	8.45 01	1.28 02	1.70 02	2.41 02
51	Sb L$_{\alpha_1}$	3.6047	343.954	5.05 01	8.35 01	1.26 02	1.68 02	2.39 02
50	Sn L$_{\beta_1}$	3.6628	338.498	4.81 01	7.96 01	1.21 02	1.61 02	2.28 02
20	Ca K$_{\alpha_1}$	3.6917	335.848	4.69 01	7.77 01	1.18 02	1.57 02	2.23 02
52	Te L$_{\alpha_1}$	3.7693	328.934	4.40 01	7.30 01	1.11 02	1.48 02	2.10 02
51	Sb L$_{\beta_1}$	3.8436	322.575	4.15 01	6.88 01	1.05 02	1.40 02	1.99 02
53	I L$_{\alpha_1}$	3.9377	314.867	3.86 01	6.40 01	9.74 01	1.30 02	1.85 02
20	Ca K$_\beta$	4.0127	308.981	3.64 01	6.05 01	9.21 01	1.23 02	1.76 02
52	Te L$_{\beta_1}$	4.0296	307.686	3.59 01	5.97 01	9.09 01	1.22 02	1.73 02
21	Sc K$_{\alpha_1}$	4.0906	303.097	3.43 01	5.71 01	8.70 01	1.16 02	1.66 02
54	Xe L$_{\alpha_1}$	4.1099	301.674	3.38 01	5.63 01	8.58 01	1.15 02	1.64 02
53	I L$_{\beta_1}$	4.2207	293.755	3.12 01	5.19 01	7.92 01	1.06 02	1.52 02
55	Cs L$_{\alpha_1}$	4.2865	289.245	2.97 01	4.95 01	7.56 01	1.01 02	1.45 02
21	Sc K$_\beta$	4.4605	277.962	2.63 01	4.39 01	6.72 01	9.01 01	1.29 02
56	Ba L$_{\alpha_1}$	4.4663	277.601	2.62 01	4.37 01	6.69 01	8.98 01	1.29 02
22	Ti K$_{\alpha_1}$	4.5108	274.863	2.54 01	4.24 01	6.50 01	8.72 01	1.25 02
55	Cs L$_{\beta_1}$	4.6198	268.377	2.36 01	3.95 01	6.05 01	8.13 01	1.16 02
57	La L$_{\alpha_1}$	4.6510	266.577	2.32 01	3.87 01	5.93 01	7.97 01	1.14 02
56	Ba L$_{\beta_1}$	4.8273	256.841	2.07 01	3.46 01	5.31 01	7.14 01	1.02 02
58	Ce L$_{\alpha_1}$	4.8402	256.157	2.05 01	3.43 01	5.27 01	7.08 01	1.02 02
22	Ti K$_{\beta_{1,3}}$	4.9318	251.399	1.93 01	3.24 01	4.98 01	6.70 01	9.63 01
23	V K$_{\alpha_1}$	4.9522	250.363	1.91 01	3.20 01	4.92 01	6.62 01	9.51 01
59	Pr L$_{\alpha_1}$	5.0337	246.310	1.82 01	3.05 01	4.68 01	6.31 01	9.07 01
57	La L$_{\beta_1}$	5.0421	245.899	1.81 01	3.03 01	4.66 01	6.28 01	9.02 01
60	Nd L$_{\alpha_1}$	5.2304	237.047	1.62 01	2.71 01	4.17 01	5.63 01	8.10 01
58	Ce L$_{\beta_1}$	5.2622	235.614	1.59 01	2.66 01	4.10 01	5.53 01	7.96 01
24	Cr K$_{\alpha_1}$	5.4147	228.978	1.45 01	2.44 01	3.76 01	5.08 01	7.31 01
23	V K$_{\beta_{1,3}}$	5.4273	228.447	1.44 01	2.42 01	3.74 01	5.04 01	7.26 01
61	Pm L$_{\alpha_1}$	5.4325	228.228	1.44 01	2.42 01	3.72 01	5.03 01	7.24 01
59	Pr L$_{\beta_1}$	5.4889	225.883	1.40 01	2.34 01	3.61 01	4.88 01	7.02 01
62	Sm L$_{\alpha_1}$	5.6361	219.984	1.29 01	2.16 01	3.33 01	4.51 01	6.50 01
60	Nd L$_{\beta_1}$	5.7210	216.696	1.23 01	2.07 01	3.19 01	4.31 01	6.21 01
63	Eu L$_{\alpha_1}$	5.8457	212.096	1.15 01	1.94 01	2.99 01	4.04 01	5.83 01
25	Mn K$_{\alpha_1}$	5.8988	210.187	1.12 01	1.88 01	2.91 01	3.93 01	5.68 01
24	Cr K$_{\beta_{1,3}}$	5.9467	208.494	1.09 01	1.84 01	2.84 01	3.84 01	5.55 01
61	Pm L$_{\beta_1}$	5.9610	207.993	1.08 01	1.82 01	2.82 01	3.81 01	5.51 01
64	Gd L$_{\alpha_1}$	6.0572	204.690	1.03 01	1.74 01	2.68 01	3.64 01	5.25 01
62	Sm L$_{\beta_1}$	6.2051	199.811	9.60 00	1.62 01	2.50 01	3.38 01	4.89 01
65	Tb L$_{\alpha_1}$	6.2728	197.655	9.30 00	1.56 01	2.42 01	3.27 01	4.74 01
26	Fe K$_{\alpha_1}$	6.4038	193.611	8.73 00	1.47 01	2.27 01	3.08 01	4.43 01
63	Eu L$_{\beta_1}$	6.4564	192.034	8.52 00	1.43 01	2.21 01	3.00 01	4.35 01
15	Mn K$_{\beta_{1,3}}$	6.4905	191.025	8.39 00	1.41 01	2.18 01	2.96 01	4.28 01
66	Dy L$_{\alpha_1}$	6.4952	190.887	8.37 00	1.41 01	2.17 01	2.95 01	4.27 01
64	Gd L$_{\beta_1}$	6.7132	184.688	7.57 00	1.27 01	1.97 01	2.67 01	3.87 01

Z	transition	E [keV]	λ [pm]	$_6$C	$_7$N	$_8$O	$_9$F	$_{10}$Ne
67	Ho L_{α_1}	6.7198	184.507	7.55 00	1.27 01	1.96 01	2.67 01	3.86 01
27	Co K_{α_1}	6.9303	178.903	6.88 00	1.16 01	1.79 01	2.43 01	3.53 01
68	Er L_{α_1}	6.9487	178.429	6.83 00	1.15 01	1.77 01	2.41 01	3.50 01
65	Tb L_{β_1}	6.9780	177.680	6.74 00	1.13 01	1.75 01	2.38 01	3.46 01
26	Fe $K_{\beta_{1,3}}$	7.0580	175.666	6.51 00	1.09 01	1.69 01	2.30 01	3.34 01
69	Tm L_{α_1}	7.1799	172.683	6.18 00	1.04 01	1.61 01	2.19 01	3.18 01
66	Dy L_{β_1}	7.2477	171.068	6.01 00	1.01 01	1.56 01	2.13 01	2.09 01
70	Yb L_{α_1}	7.4150	167.195	5.61 00	9.42 00	1.46 01	1.98 01	2.89 01
28	Ni K_{α_1}	7.4782	165.795	5.47 00	9.18 00	1.42 01	1.94 01	2.82 01
67	Ho L_{β_1}	7.5253	164.757	5.37 00	9.01 00	1.39 01	1.90 01	2.76 01
27	Co $K_{\beta_{1,3}}$	7.6494	162.084	5.11 00	8.57 00	1.33 01	1.81 01	2.63 01
71	Lu L_{α_1}	7.6555	161.955	5.10 00	8.55 00	1.32 01	1.80 01	2.63 01
68	Er L_{β_1}	7.8109	158.733	4.80 01	8.04 00	1.25 01	1.70 01	2.48 01
72	Hf L_{α_1}	7.8990	156.963	4.64 00	7.78 00	1.20 01	1.64 01	2.39 01
29	Cu K_{α_1}	8.0478	154.060	4.38 00	7.35 00	1.14 01	1.55 01	2.27 01
69	Tm L_{β_1}	8.1010	153.049	4.29 00	7.21 00	1.12 01	1.52 01	2.22 01
73	Ta L_{α_1}	8.1461	152.201	4.21 00	7.09 00	1.10 01	1.50 01	2.19 01
28	Ni $K_{\beta_{1,3}}$	8.2647	150.017	4.02 00	6.79 00	1.05 01	1.44 01	2.09 01
74	W L_{α_1}	8.3970	147.643	3.82 00	6.47 00	1.00 01	1.37 01	2.00 01
70	Yb L_{β_1}	8.4018	147.560	3.82 00	6.46 00	1.00 01	1.37 01	1.99 01
30	Zn K_{α_1}	8.6389	143.519	3.49 00	5.95 00	9.21 00	1.26 01	1.84 01
75	Re L_{α_1}	8.6525	143.294	3.47 00	5.92 00	9.17 00	1.25 01	1.83 01
71	Lu L_{β_1}	8.7090	142.364	3.40 00	5.81 00	8.99 00	1.23 01	1.79 01
29	Cu K_{β_1}	8.9053	139.226	3.17 00	5.43 00	8.41 00	1.15 01	1.68 01
77	Ir L_{α_1}	9.1751	135.132	2.88 00	4.97 00	7.69 00	1.05 01	1.53 01
31	Ga K_{α_1}	9.2517	134.013	2.80 00	4.85 00	7.50 00	1.02 01	1.50 01
73	Ta L_{β_1}	9.3431	132.702	2.72 00	4.71 00	7.29 00	9.94 00	1.45 01
78	Pt L_{α_1}	9.4423	131.308	2.63 00	4.36 00	7.06 00	9.63 00	1.41 01
30	Zn $K_{\beta_{1,3}}$	9.5720	129.529	2.51 00	4.38 00	6.78 00	9.25 00	1.35 01
74	W L_{β_1}	9.6724	128.184	2.43 00	4.25 00	6.57 00	8.97 00	1.31 01
79	Au L_{α_1}	9.7133	127.644	2.40 00	4.19 00	6.49 00	8.85 00	1.29 01
31	Ge K_{β_1}	9.8864	125.409	2.27 00	3.98 00	6.15 00	8.40 00	1.23 01
80	Hg L_{α_1}	9.9888	124.124	2.19 00	3.86 00	5.97 00	8.14 00	1.19 01
75	Re L_{β_1}	10.010	123.861	2.18 00	3.83 00	5.93 00	8.09 00	1.18 01
31	Ga K_{β_1}	10.264	120.796	2.05 00	3.57 00	5.51 00	7.52 00	1.10 01
81	Tl L_{α_1}	10.269	120.737	2.05 00	3.57 00	3.51 00	7.51 00	1.10 01
76	Os L_{β_1}	10.355	119.734	2.00 00	3.49 00	5.38 00	7.33 00	1.07 01
33	As K_{α_1}	10.544	117.588	1.92 00	3.31 00	5.10 00	6.95 00	1.02 01
82	Pb L_{α_1}	10.552	117.499	1.91 00	3.31 00	5.09 00	6.93 00	1.01 01
77	Ir L_{β_1}	10.708	115.787	1.84 00	3.17 00	4.88 00	6.64 00	9.71 00
83	Bi L_{α_1}	10.839	114.388	1.79 00	3.07 00	4.71 00	6.41 00	9.37 00
32	Ge K_{β_1}	10.982	112.898	1.73 00	2.96 00	4.53 00	6.17 00	9.02 00
78	Pt L_{β_1}	11.071	111.990	1.70 00	2.89 00	4.43 00	6.02 00	8.81 00
84	Po L_{α_1}	11.131	111.387	1.67 00	2.85 00	4.36 00	5.93 00	8.67 00

Z	transition	E [keV]	λ [pm]	$_6$C	$_7$N	$_8$O	$_9$F	$_{10}$Ne
34	Se K$_{\alpha_1}$	11.222	110.484	1.64 00	2.78 00	4.26 00	5.79 00	8.47 00
85	At L$_{\alpha_1}$	11.427	108.501	1.57 00	2.65 00	4.04 00	5.49 00	8.03 00
79	Au L$_{\beta_1}$	11.442	108.359	1.56 00	2.64 00	4.03 00	5.47 00	8.00 00
33	As K$_{\beta_1}$	11.726	105.735	1.47 00	2.46 00	3.75 00	5.09 00	7.44 00
86	Rn L$_{\alpha_1}$	11.727	105.726	1.47 00	2.46 00	3.75 00	5.09 00	7.44 00
80	Hg L$_{\beta_1}$	11.823	104.867	1.44 00	2.40 00	3.66 00	4.97 00	7.26 00
35	Br K$_{\alpha_1}$	11.924	103.979	1.41 00	2.35 00	3.57 00	4.85 00	7.08 00
87	Fr L$_{\alpha_1}$	12.031	103.054	1.38 00	2.29 00	3.48 00	4.72 00	6.90 00
81	Tl L$_{\beta_1}$	12.213	101.519	1.33 00	2.20 00	3.33 00	4.52 00	6.60 00
88	Ra L$_{\alpha_1}$	12.340	100.474	1.30 00	2.13 00	3.23 00	4.38 00	6.41 00
34	Se K$_{\beta_1}$	12.496	99.219	1.26 00	2.06 00	3.12 00	4.23 00	6.17 00
82	Pb L$_{\beta_1}$	12.614	98.291	1.23 00	2.01 00	3.03 00	4.11 00	6.01 00
36	Kr K$_{\alpha_1}$	12.649	98.019	1.22 00	1.99 00	3.01 00	4.08 00	5.96 00
89	Ac L$_{\alpha_1}$	12.652	97.996	1.22 00	1.99 00	3.01 00	4.08 00	5.95 00
90	Th L$_{\alpha_1}$	12.969	95.601	1.14 00	1.86 00	2.80 00	3.79 00	5.54 00
83	Bi L$_{\beta_1}$	13.024	95.197	1.13 00	1.83 00	2.77 00	3.74 00	5.47 00
91	Pa L$_{\alpha_1}$	13.291	93.285	1.08 00	1.73 00	2.61 00	3.53 00	5.15 00
35	Br K$_{\beta_1}$	13.291	93.285	1.08 00	1.73 00	2.61 00	3.53 00	5.15 00
37	Rb K$_{\alpha_1}$	13.395	92.560	1.06 00	1.69 00	2.55 00	3.43 00	5.03 00
84	Po L$_{\beta_1}$	13.447	92.202	1.05 00	1.68 00	2.52 00	3.41 00	4.98 00
92	U L$_{\alpha_1}$	13.615	91.065	1.01 00	1.62 00	2.43 00	3.29 00	4.80 00
85	At L$_{\beta_1}$	13.876	89.352	9.68 -1	1.34 00	2.30 00	3.11 00	4.54 00
36	Kr K$_{\beta_1}$	14.112	87.857	9.28 -1	1.46 00	2.19 00	2.96 00	4.32 00
38	Sr K$_{\alpha_1}$	14.165	87.529	9.19 -1	1.45 00	2.17 00	2.93 00	4.27 00
86	Rn L$_{\beta_1}$	14.316	86.606	8.95 -1	1.41 00	2.10 00	2.84 00	4.14 00
87	Fr L$_{\beta_1}$	14.770	83.943	8.28 -1	1.29 00	1.92 00	2.59 00	3.78 00
39	Y K$_{\alpha_1}$	14.958	82.888	8.03 -1	1.24 00	1.85 00	2.50 00	3.64 00
37	Rb K$_{\beta_1}$	14.961	82.872	8.02 -1	1.24 00	1.85 00	2.49 00	3.64 00
88	Ra L$_{\beta_1}$	15.236	81.376	7.72 -1	1.19 00	1.76 00	2.37 00	3.46 00
89	Ac L$_{\beta_1}$	15.713	78.906	7.24 -1	1.10 00	1.63 00	2.18 00	3.17 00
40	Zr K$_{\alpha_1}$	15.775	78.596	7.19 -1	1.09 00	1.61 00	2.16 00	3.13 00
38	Sr K$_{\beta_1}$	15.836	78.293	7.13 -1	1.08 00	1.59 00	2.14 00	3.10 00
90	Th L$_{\beta_1}$	16.202	76.524	6.80 -1	1.03 00	1.50 00	2.01 00	2.91 00
41	Nb K$_{\alpha_1}$	16.615	74.622	6.46 -1	9.66 -1	1.41 00	1.87 00	2.71 00
91	Pa K$_{\beta_1}$	16.702	74.233	6.39 -1	9.54 -1	1.39 00	1.85 00	2.67 00
39	Y K$_{\beta_1}$	16.738	74.074	6.36 -1	9.49 -1	1.38 00	1.84 00	2.65 00
92	U L$_{\beta_1}$	17.220	72.000	6.00 -1	8.86 -1	1.28 00	1.70 00	2.45 00
42	Mo K$_{\alpha_1}$	17.479	70.933	5.82 -1	8.55 -1	1.23 00	1.63 00	2.33 00
40	Zr K$_{\beta_1}$	17.668	70.175	5.69 -1	8.34 -1	1.20 00	1.59 00	2.28 00
43	Tc K$_{\alpha_1}$	18.367	67.504	5.25 -1	7.59 -1	1.08 00	1.43 00	2.04 00
41	Nb K$_{\beta_1}$	18.623	66.576	5.11 -1	7.35 -1	1.04 00	1.37 00	1.96 00
44	Ru K$_{\alpha_1}$	19.279	64.311	4.75 -1	6.76 -1	9.53 -1	1.25 00	1.78 00
42	Mo K$_{\beta_1}$	19.608	63.231	4.59 -1	6.49 -1	9.12 -1	1.19 00	1.70 00
45	Rh K$_{\alpha_1}$	20.216	61.330	4.34 -1	6.08 -1	8.47 -1	1.10 00	1.56 00

Z	transition	E [keV]	λ [pm]	$_6$C	$_7$N	$_8$O	$_9$F	$_{10}$Ne
43	Tc K$_{\beta_1}$	20.619	60.131	4.23 -1	5.88 -1	8.14 -1	1.06 00	1.49 00
46	Pd K$_{\alpha_1}$	21.177	58.547	4.08 -1	5.61 -1	7.71 -1	9.94 -1	1.40 00
44	Ru K$_{\beta_1}$	21.657	57.240	3.96 -1	5.40 -1	7.37 -1	9.44 -1	1.32 00
47	Ag K$_{\alpha_1}$	22.163	55.942	3.84 -1	5.19 -1	7.03 -1	8.96 -1	1.25 00
45	Rh K$_{\beta_1}$	22.724	54.561	3.72 -1	4.97 -1	6.68 -1	8.47 -1	1.18 00
48	Cd K$_{\alpha_1}$	23.174	53.501	3.62 -1	4.81 -1	6.42 -1	8.10 -1	1.12 00
46	Pd K$_{\beta_1}$	23.819	52.053	3.49 -1	4.58 -1	6.07 -1	7.61 -1	1.05 00
49	In K$_{\alpha_1}$	24.210	51.212	3.42 -1	4.46 -1	5.87 -1	7.33 -1	1.01 00
47	Ag K$_{\beta_1}$	24.942	49.709	3.28 -1	4.24 -1	5.53 -1	6.85 -1	9.37 -1
50	Sn K$_{\alpha_1}$	25.271	49.062	3.23 -1	4.14 -1	5.38 -1	6.65 -1	9.07 -1
48	Cd K$_{\beta_1}$	26.096	47.511	3.09 -1	3.92 -1	5.05 -1	6.18 -1	8.39 -1
51	Sb K$_{\alpha_1}$	26.359	47.037	3.05 -1	3.85 -1	4.94 -1	6.04 -1	8.18 -1
49	In K$_{\beta_1}$	27.276	45.455	2.91 -1	3.63 -1	4.61 -1	5.59 -1	7.53 -1
52	Te K$_{\alpha_1}$	27.472	45.131	2.89 -1	3.59 -1	4.55 -1	5.50 -1	7.40 -1
50	Sn K$_{\beta_1}$	28.486	43.524	2.75 -1	3.37 -1	4.22 -1	5.06 -1	6.77 -1
53	I K$_{\alpha_1}$	28.612	43.333	2.73 -1	3.34 -1	4.19 -1	5.01 -1	6.70 -1
51	Sb K$_{\beta_1}$	29.726	41.709	2.60 -1	3.13 -1	3.87 -1	4.60 -1	6.10 -1
54	Xe K$_{\alpha_1}$	29.779	41.635	2.59 -1	3.12 -1	3.86 -1	4.58 -1	6.08 -1
55	Cs K$_{\alpha_1}$	30.973	40.030	2.51 -1	2.98 -1	3.64 -1	4.28 -1	5.62 -1
52	Te K$_{\beta_1}$	30.996	40.000	2.51 -1	2.98 -1	3.64 -1	4.27 -1	5.62 -1
56	Ba K$_{\alpha_1}$	32.194	38.511	2.44 -1	2.87 -1	3.46 -1	4.02 -1	5.23 -1
53	I K$_{\beta_1}$	32.295	38.391	2.43 -1	2.86 -1	3.45 -1	4.00 -1	5.20 -1
57	La K$_{\alpha_1}$	33.442	37.074	2.37 -1	2.76 -1	3.29 -1	3.79 -1	4.87 -1
54	Xe K$_{\beta_1}$	33.624	36.874	2.36 -1	2.75 -1	3.27 -1	3.76 -1	4.83 -1
58	Ce K$_{\alpha_1}$	34.279	36.169	2.33 -1	2.69 -1	3.19 -1	3.64 -1	4.65 -1
55	Cs K$_{\beta_1}$	34.987	35.437	2.30 -1	2.64 -1	3.10 -1	3.53 -1	4.48 -1
59	Pr K$_{\alpha_1}$	36.026	34.415	2.25 -1	2.56 -1	2.99 -1	3.37 -1	4.24 -1
56	Ba K$_{\beta_1}$	36.378	34.082	2.23 -1	2.53 -1	2.95 -1	3.31 -1	4.17 -1
60	Nd K$_{\alpha_1}$	36.847	33.648	2.21 -1	2.50 -1	2.90 -1	3.25 -1	4.07 -1
57	La K$_{\beta_1}$	37.801	32.799	2.17 -1	2.44 -1	2.80 -1	3.12 -1	3.88 -1
61	Pm K$_{\alpha_1}$	38.725	32.016	2.14 -1	2.38 -1	2.71 -1	3.00 -1	3.71 -1
58	Ce K$_{\beta_1}$	39.257	31.582	2.11 -1	2.34 -1	2.67 -1	2.94 -1	3.62 -1
62	Sm K$_{\alpha_1}$	40.118	30.905	2.08 -1	2.30 -1	2.59 -1	2.84 -1	3.48 -1
63	Eu K$_{\alpha_1}$	41.542	29.845	2.05 -1	2.24 -1	2.52 -1	2.73 -1	3.32 -1
60	Nd K$_{\beta_1}$	42.271	29.331	2.03 -1	2.22 -1	2.48 -1	2.68 -1	3.25 -1
64	Gd K$_{\alpha_1}$	42.996	28.836	2.02 -1	2.19 -1	2.44 -1	2.63 -1	3.17 -1
61	Pm K$_{\beta_1}$	43.826	28.290	2.00 -1	2.17 -1	2.40 -1	2.58 -1	3.09 -1
65	Tb K$_{\alpha_1}$	44.482	27.873	1.98 -1	2.15 -1	2.37 -1	2.54 -1	3.03 -1
62	Sm K$_{\beta_1}$	45.413	27.301	1.96 -1	2.12 -1	2.33 -1	2.48 -1	2.95 -1
66	Dy K$_{\alpha_1}$	45.998	26.954	1.95 -1	2.10 -1	2.30 -1	2.44 -1	2.90 -1
63	Eu K$_{\beta_1}$	47.038	26.358	1.93 -1	2.07 -1	2.26 -1	2.38 -1	2.82 -1
67	Ho K$_{\alpha_1}$	47.547	26.076	1.92 -1	2.05 -1	2.24 -1	2.36 -1	2.78 -1
64	Gd K$_{\beta_1}$	48.697	25.460	1.90 -1	2.02 -1	2.19 -1	2.29 -1	2.69 -1
68	Er K$_{\alpha_1}$	49.128	25.237	1.89 -1	2.01 -1	2.17 -1	2.27 -1	2.66 -1

Z	transition	E [keV]	λ [pm]	$_6$C	$_7$N	$_8$O	$_9$F	$_{10}$Ne
65	Tb K_{β_1}	50.382	24.609	1.87 -1	1.98 -1	2.13 -1	2.22 -1	2.58 -1
69	Tm K_{α_1}	50.742	24.434	1.87 -1	1.97 -1	2.12 -1	2.20 -1	2.56 -1
66	Dy K_{β_1}	52.119	23.788	1.85 -1	1.95 -1	2.09 -1	2.16 -1	2.50 -1
70	Yb K_{α_1}	52.389	23.666	1.84 -1	1.94 -1	2.08 -1	2.15 -1	2.48 -1
67	Ho K_{β_1}	53.877	23.012	1.82 -1	1.92 -1	2.04 -1	2.10 -1	2.42 -1
71	Lu K_{α_1}	54.070	22.930	1.82 -1	1.91 -1	2.04 -1	2.10 -1	2.41 -1
68	Er K_{β_1}	55.681	22.267	1.80 -1	1.89 -1	2.00 -1	2.05 -1	2.34 -1
72	Hf K_{α_1}	55.790	22.223	1.80 -1	1.89 -1	2.00 -1	2.04 -1	2.34 -1
69	Tm K_{β_1}	57.517	21.556	1.78 -1	1.86 -1	1.96 -1	2.00 -1	2.27 -1
73	Ta K_{α_1}	57.532	21.550	1.78 -1	1.86 -1	1.96 -1	2.00 -1	2.27 -1
74	W K_{α_1}	59.318	20.901	1.76 -1	1.83 -1	1.93 -1	1.95 -1	2.20 -1
70	Yb K_{β_1}	59.370	20.883	1.76 -1	1.83 -1	1.92 -1	1.95 -1	2.20 -1
75	Re K_{α_1}	61.140	20.278	1.74 -1	1.81 -1	1.90 -1	1.91 -1	2.15 -1
71	Lu K_{β_1}	61.283	20.231	1.74 -1	1.81 -1	1.89 -1	1.91 -1	2.14 -1
76	Os K_{α_1}	63.001	19.679	1.73 -1	1.79 -1	1.87 -1	1.88 -1	2.10 -1
72	Hf K_{β_1}	63.234	19.607	1.73 -1	1.79 -1	1.87 -1	1.88 -1	2.10 -1
77	Ir K_{α_1}	64.896	19.105	1.71 -1	1.77 -1	1.85 -1	1.85 -1	2.06 -1
73	Ta K_{β_1}	65.223	19.009	1.71 -1	1.77 -1	1.84 -1	1.84 -1	2.05 -1
78	Pt K_{α_1}	66.832	18.551	1.70 -1	1.75 -1	1.82 -1	1.82 -1	2.02 -1
74	W K_{β_1}	67.244	18.438	1.70 -1	1.75 -1	1.82 -1	1.81 -1	2.01 -1
79	Au K_{α_1}	68.804	18.020	1.68 -1	1.73 -1	1.80 -1	1.79 -1	1.98 -1
75	Re K_{β_1}	69.310	17.888	1.68 -1	1.73 -1	1.79 -1	1.78 -1	1.97 -1
80	Hg K_{α_1}	70.819	17.507	1.67 -1	1.72 -1	1.78 -1	1.76 -1	1.94 -1
76	Os K_{β_1}	71.413	17.361	1.67 -1	1.71 -1	1.77 -1	1.75 -1	1.93 -1
81	Tl K_{α_1}	72.872	17.014	1.66 -1	1.70 -1	1.75 -1	1.73 -1	1.91 -1
77	Ir K_{β_1}	73.651	16.854	1.65 -1	1.69 -1	1.75 -1	1.72 -1	1.89 -1
82	Pb K_{α_1}	74.969	16.538	1.64 -1	1.68 -1	1.73 -1	1.71 -1	1.87 -1
78	Pt K_{β_1}	75.748	16.368	1.64 -1	1.68 -1	1.72 -1	1.70 -1	1.86 -1
83	Bi K_{α_1}	77.108	16.079	1.63 -1	1.66 -1	1.71 -1	1.68 -1	1.83 -1
79	Au K_{β_1}	77.948	15.906	1.62 -1	1.66 -1	1.70 -1	1.67 -1	1.82 -1
84	Po K_{α_1}	79.290	15.636	1.61 -1	1.65 -1	1.69 -1	1.65 -1	1.80 -1
80	Hg K_{β_1}	80.253	15.449	1.61 -1	1.64 -1	1.68 -1	1.64 -1	1.78 -1
85	At K_{α_1}	81.520	15.209	1.60 -1	1.63 -1	1.67 -1	1.63 -1	1.77 -1
81	Tl K_{β_1}	82.576	15.014	1.66 -1	1.63 -1	1.66 -1	1.62 -1	1.76 -1
86	Rn K_{α_1}	83.780	14.798	1.59 -1	1.62 -1	1.66 -1	1.61 -1	1.75 -1
82	Pb K_{β_1}	84.936	14.597	1.58 -1	1.61 -1	1.65 -1	1.60 -1	1.74 -1
87	Fr K_{α_1}	86.100	14.400	1.58 -1	1.60 -1	1.64 -1	1.60 -1	1.73 -1
83	Bi K_{β_1}	87.343	14.195	1.57 -1	1.60 -1	1.63 -1	1.59 -1	1.71 -1
88	Ra K_{α_1}	88.470	14.014	1.57 -1	1.59 -1	1.62 -1	1.58 -1	1.70 -1
84	Po K_{β_1}	89.800	13.806	1.56 -1	1.58 -1	1.61 -1	1.57 -1	1.69 -1
89	Ac K_{α_1}	90.884	13.642	1.55 -1	1.58 -1	1.61 -1	1.56 -1	1.68 -1
85	At K_{β_1}	92.300	13.432	1.55 -1	1.57 -1	1.60 -1	1.55 -1	1.67 -1
90	Th K_{α_1}	93.350	13.281	1.54 -1	1.56 -1	1.59 -1	1.54 -1	1.66 -1
86	Rn K_{β_1}	94.870	13.068	1.54 -1	1.56 -1	1.58 -1	1.53 -1	1.65 -1

Z	transition	E [keV]	λ [pm]	$_6$C	$_7$N	$_8$O	$_9$F	$_{10}$Ne
91	Pa K$_{\alpha_1}$	95.868	12.932	1.53 -1	1.55 -1	1.58 -1	1.53 -1	1.64 -1
87	Fr K$_{\beta_1}$	97.470	12.720	1.52 -1	1.54 -1	1.57 -1	1.52 -1	1.63 -1
92	U K$_{\alpha_1}$	98.439	12.595	1.52 -1	1.54 -1	1.56 -1	1.51 -1	1.62 -1
88	Ra K$_{\beta_1}$	100.130	12.382	1.51 -1	1.53 -1	1.55 -1	1.50 -1	1.60 -1
89	Ac K$_{\beta_1}$	102.850	12.054	1.50 -1	1.52 -1	1.54 -1	1.48 -1	1.59 -1
90	Th K$_{\beta_1}$	105.610	11.739	1.49 -1	1.51 -1	1.53 -1	1.47 -1	1.57 -1
91	Pa K$_{\beta_1}$	108.430	11.434	1.48 -1	1.49 -1	1.51 -1	1.46 -1	1.56 -1
92	U K$_{\beta_1}$	111.300	11.139	1.47 -1	1.48 -1	1.50 -1	1.44 -1	1.54 -1

Z	transition	E [keV]	λ [pm]	$_{11}$Na	$_{12}$Mg	$_{13}$Al	$_{14}$Si	$_{15}$P
4	Be K$_\alpha$	0.1085	11421.207	1.22 05	1.21 05	1.01 05	1.15 05	9.05 04
38	Sr M$_\xi$	0.1140	10815.895	1.11 05	1.25 05	1.14 05	1.13 05	8.91 04
39	Y M$_\xi$	0.1328	9339.235	8.42 04	9.54 04	9.58 04	1.09 05	8.69 04
16	S L$_l$	0.1487	8331.942	6.13 04	8.32 04	8.51 04	1.06 05	8.46 04
40	Zr M$_\xi$	0.1511	8205.506	6.52 04	8.14 04	8.49 04	1.06 05	8.34 04
41	Nb M$_\xi$	0.1717	7221.037	5.03 04	6.12 04	1.43 04	9.32 04	1.18 04
5	B K$_\alpha$	0.1833	6764.059	4.39 04	5.95 04	6.40 04	8.40 04	6.62 04
42	Mo M$_\xi$	0.1926	6437.445	3.95 04	5.39 04	6.02 04	7.74 04	7.33 04
6	C K$_\alpha$	0.2770	4476.000	1.10 04	2.39 04	3.10 04	3.68 04	4.13 04
47	Ag M$_\xi$	0.3117	3977.709	1.29 04	1.88 04	2.30 04	2.83 04	3.26 04
7	N K$_\alpha$	0.3924	3159.664	1.33 03	1.10 04	1.38 04	1.65 04	2.06 04
22	Ti L$_l$	0.3953	3136.484	1.20 03	1.01 04	1.36 04	1.62 04	2.02 04
22	Ti L$_\alpha$	0.4522	2741.822	5.14 03	1.64 03	9.15 03	1.18 04	1.50 04
23	V L$_\alpha$	0.5113	2424.901	3.11 03	5.54 03	1.18 03	9.38 03	1.12 04
8	O K$_\alpha$	0.5249	2362.072	3.52 03	5.11 03	6.12 03	8.19 03	1.05 04
25	Mn L$_l$	0.5563	2228.747	3.04 03	4.46 03	5.80 03	1.63 03	9.15 03
24	Cr L$_\alpha$	0.5728	2164.546	2.82 03	4.12 03	5.31 03	1.03 03	8.41 03
25	Mn L$_\alpha$	0.6374	1945.171	2.11 03	3.11 03	4.04 03	5.38 03	6.53 03
9	F K$_\alpha$	0.6768	1831.932	1.82 03	2.62 03	3.41 03	4.54 03	5.53 03
26	Fe L$_\alpha$	0.7050	1758.655	1.65 03	2.36 03	3.08 03	4.11 03	5.01 03
21	Co L$_\alpha$	0.7762	1597.335	1.21 03	1.82 03	2.38 03	3.19 03	3.90 03
28	Ni L$_\alpha$	0.8515	1456.080	9.48 02	1.41 03	1.85 03	2.48 03	3.04 03
29	Cu L$_\alpha$	0.9297	1336.044	1.15 02	1.11 03	1.45 03	1.96 03	2.41 03
30	Zn L$_\alpha$	1.0011	1225.513	6.09 02	9.25 02	1.15 03	1.54 03	1.81 03
11	Na K$_\alpha$	1.0410	1191.020	5.91 02	8.52 02	1.01 03	1.43 03	1.14 03
11	Na K$_\beta$	1.0711	1157.550	5.86 02	7.85 02	9.86 02	1.32 03	1.61 03
12	Mg K$_\alpha$	1.2536	989.033	5.46 03	5.00 02	6.41 02	8.63 02	1.06 03
33	As L$_\alpha$	1.2820	967.123	5.14 03	4.69 02	6.03 02	8.12 02	9.96 02
12	Mg K$_\beta$	1.3022	952.121	4.94 03	4.49 02	5.78 02	7.79 02	9.55 02
33	As L$_{\beta_1}$	1.3170	941.421	4.79 03	5.95 03	5.55 02	7.57 02	9.27 02
66	Dy M$_\beta$	1.3250	935.737	4.72 03	5.86 03	5.51 02	7.43 02	9.12 02

Z	transition	E [keV]	λ [pm]	$_{11}$Na	$_{12}$Mg	$_{13}$Al	$_{14}$Si	$_{15}$P
67	Ho M$_\alpha$	1.3480	919.771	4.51 03	5.61 03	5.26 02	7.09 02	8.71 02
34	Se L$_\alpha$	1.3791	899.029	4.25 03	5.29 03	4.94 02	6.67 02	8.20 02
67	Ho M$_\beta$	1.3830	896.494	4.22 03	5.25 03	4.90 02	6.62 02	8.14 02
68	Er M$_\alpha$	1.4060	881.829	4.04 03	5.03 03	4.69 02	6.33 02	7.79 02
34	Se L$_{\beta_1}$	1.4192	873.627	3.94 03	4.91 03	4.57 02	6.17 02	7.59 02
68	Er M$_\beta$	1.4430	859.218	3.77 03	4.71 03	4.37 02	5.90 02	7.26 02
69	Tm M$_\alpha$	1.4620	848.051	3.64 03	4.55 03	4.21 02	5.70 02	7.02 02
35	Br L$_\alpha$	1.4804	837.511	3.53 03	4.41 03	4.07 02	5.51 02	6.79 02
13	Al K$_{\alpha_1}$	1.4867	833.962	3.49 03	4.36 03	4.03 02	5.45 02	6.71 02
69	Tm M$_\beta$	1.5030	824.918	3.39 03	4.24 03	3.91 02	5.29 02	6.52 02
70	Yb M$_\alpha$	1.5214	814.941	3.28 03	4.10 03	3.78 02	5.12 02	6.31 02
35	Br L$_{\beta_1}$	1.5259	812.538	3.25 03	4.07 03	3.75 02	5.08 02	6.26 02
13	Al K$_\beta$	1.5574	796.103	3.08 03	3.86 03	<u>3.54 02</u>	4.81 02	5.93 02
70	Yb M$_\beta$	1.5675	790.974	3.02 03	3.79 03	4.43 03	4.73 02	5.83 02
71	Lu M$_\alpha$	1.5813	784.071	2.95 03	3.71 03	4.33 03	4.62 02	5.69 02
36	Kr L$_\alpha$	1.5860	781.747	2.93 03	3.68 03	4.30 03	4.58 02	5.65 02
71	Lu M$_\beta$	1.6312	760.085	2.72 03	3.42 03	4.00 03	4.25 02	5.24 02
36	Kr L$_{\beta_1}$	1.6366	757.577	2.69 03	3.39 03	3.97 03	4.21 02	5.20 02
72	Hf M$_{\alpha_1}$	1.6446	753.892	2.66 03	3.34 03	3.92 03	4.16 02	5.13 02
37	Rb L$_{\alpha_1}$	1.6941	731.864	2.45 03	3.09 03	3.63 03	3.84 02	4.74 02
72	Hf M$_\beta$	1.6970	730.355	2.44 03	3.08 03	3.61 03	3.82 02	4.72 02
73	Ta M$_{\alpha_1}$	1.7096	725.229	2.39 03	3.02 03	3.54 03	3.75 02	4.63 02
14	Si K$_{\alpha_1}$	1.7400	712.558	2.28 03	2.88 03	3.39 03	3.58 02	4.42 02
57	Rb L$_{\beta_1}$	1.7522	707.597	2.24 03	2.83 03	3.33 03	3.51 02	4.34 02
73	Ta M$_\beta$	1.7655	702.266	2.19 03	2.78 03	3.26 03	3.44 02	4.25 02
74	W M$_{\alpha_1}$	1.7754	698.350	2.16 03	2.73 03	3.22 03	3.39 02	4.19 02
58	Sr L$_{\alpha_1}$	1.8066	686.290	2.06 03	2.61 03	3.07 03	3.24 02	4.00 02
74	W M$_\beta$	1.8349	675.705	1.98 03	2.51 03	2.95 03	3.11 02	3.84 02
14	Si K$_\beta$	1.8359	675.337	1.97 03	2.50 03	2.95 03	<u>3.10 02</u>	3.83 02
75	Re M$_\beta$	1.8420	673.100	1.96 03	2.48 03	2.92 03	3.71 03	3.80 02
38	Sr L$_{\beta_1}$	1.8717	662.420	1.87 03	2.38 03	2.81 03	3.56 03	3.64 02
75	Re M$_\beta$	1.9061	650.465	1.78 03	2.27 03	2.68 03	3.40 03	3.47 02
76	Os M$_{\alpha_1}$	1.9102	649.069	1.77 03	2.26 03	2.66 03	3.38 03	3.45 02
39	Y L$_{\alpha_1}$	1.9226	644.882	1.74 03	2.22 03	2.62 03	3.32 03	3.39 02
76	Os M$_\beta$	1.9783	626.725	1.61 03	2.06 03	2.43 03	3.09 03	3.14 02
77	Ir M$_{\alpha_1}$	1.9799	626.219	1.61 03	2.05 03	2.43 03	3.08 03	3.14 02
39	Y L$_{\beta_1}$	1.9958	621.230	1.58 03	2.01 03	2.38 01	3.02 03	3.07 02
15	P K$_{\alpha_1}$	2.0137	615.708	1.54 03	1.96 03	2.32 03	2.95 03	3.00 02
40	Zr L$_{\alpha_1}$	2.0424	607.056	1.48 03	1.69 03	2.24 03	2.84 03	2.89 02
78	Pt M$_{\alpha_1}$	2.0505	604.658	1.46 03	1.87 03	2.21 03	2.81 03	2.85 02
77	Ir M$_\beta$	2.9535	603.775	1.46 03	1.86 03	2.21 03	2.80 03	2.84 02
79	Au M$_{\alpha_1}$	2.1229	584.036	1.33 03	1.70 03	2.02 03	2.56 03	2.84 02
40	Zr L$_{\beta_1}$	2.1244	583.624	1.33 03	1.70 03	2.02 03	2.55 03	2.60 02
78	Pt M$_\beta$	2.1273	582.828	1.32 03	1.69 03	2.01 03	2.54 03	2.59 02

Z	transition	E [keV]	λ [pm]	$_{11}$Na	$_{12}$Mg	$_{13}$Al	$_{14}$Si	$_{15}$P
15	P K$_{\beta_{1,3}}$	2.1390	579.640	1.30 03	1.66 03	1.98 03	2.51 03	<u>2.55 02</u>
41	Nb L$_{\alpha_1}$	2.1659	572.441	1.26 03	1.61 03	1.92 03	2.42 03	2.77 03
80	Hg M$_{\alpha_1}$	2.1953	564.775	1.21 03	1.55 03	1.85 03	2.34 03	2.67 03
79	Au M$_\beta$	2.2046	562.393	1.20 03	1.53 03	1.83 03	2.31 03	2.64 03
41	Nb L$_{\beta_1}$	2.2574	549.238	1.12 03	1.44 03	1.72 03	2.17 03	2.48 03
81	Tl M$_{\alpha_1}$	2.2706	546.045	1.10 03	1.41 03	1.69 03	2.14 03	2.44 03
80	Hg M$_\beta$	2.2825	543.199	1.09 03	1.39 03	1.67 03	2.11 03	2.41 03
42	Mo L$_{\alpha_1}$	2.2932	540.664	1.07 03	1.38 03	1.65 03	2.08 03	2.38 03
16	S K$_{\alpha_1}$	2.3080	537.197	1.05 03	1.35 03	1.62 03	2.04 03	2.34 03
82	Pb M$_{\alpha_1}$	2.3457	528.563	1.01 03	1.29 03	1.55 03	1.96 03	2.24 03
81	Tl M$_\beta$	2.3621	524.893	9.87 02	1.27 03	1.52 03	1.92 03	2.20 03
42	Mo L$_{\beta_1}$	2.3948	517.726	9.50 02	1.22 03	1.47 03	1.85 03	2.12 03
83	Bi M$_{\alpha_1}$	2.4226	511.785	9.20 02	1.18 03	1.42 03	1.80 03	2.06 03
43	Tc L$_{\alpha_1}$	2.4240	511.490	9.19 02	1.18 03	1.42 03	1.79 03	2.02 03
82	Pb M$_\beta$	2.4427	507.574	8.99 02	1.16 03	1.39 03	1.76 03	2.02 03
16	S K$_\beta$	2.4640	503.186	8.78 02	1.13 03	1.36 03	1.72 03	1.97 03
83	Bi M$_{\beta_1}$	2.5255	490.933	8.20 02	1.06 03	1.28 03	1.61 03	1.83 03
43	Tc L$_{\beta_1}$	2.5368	488.746	8.10 02	1.05 03	1.26 03	1.59 03	1.82 03
44	Ru L$_{\alpha_1}$	2.5586	484.582	7.91 02	1.02 03	1.23 03	1.55 03	1.78 03
17	Cl K$_{\alpha_1}$	2.6224	472.792	7.38 02	9.55 02	1.15 03	1.45 03	1.67 03
44	Ru L$_{\beta_1}$	2.6832	462.079	6.93 02	8.97 02	1.08 03	1.37 03	1.57 03
45	Rh L$_{\alpha_1}$	2.6967	459.766	6.83 02	8.85 02	1.07 03	1.35 03	1.55 03
17	Cl K$_{\beta_1}$	2.8156	440.350	6.06 02	7.86 02	9.56 02	1.20 03	1.38 03
45	Rh L$_{\beta_1}$	2.8344	437.430	5.95 02	7.72 02	9.39 02	1.18 03	1.36 03
46	Pd L$_{\alpha_1}$	2.8386	436.782	5.93 02	7.69 02	9.35 02	1.11 03	1.35 03
18	Ar K$_{\alpha_1}$	2.9577	419.194	5.29 02	6.87 02	8.39 02	1.05 03	1.22 03
47	Ag L$_{\alpha_1}$	2.9843	415.458	5.16 02	6.71 02	8.19 02	1.03 03	1.19 03
46	Pd L$_{\beta_1}$	2.9902	414.638	5.13 02	6.67 02	8.15 02	1.02 03	1.18 03
90	Th M$_{\alpha_1}$	2.9961	413.821	5.10 02	6.64 02	8.11 02	1.02 03	1.17 03
91	Pa M$_{\alpha_1}$	3.0823	402.248	4.71 02	6.13 02	7.50 02	9.39 02	1.09 03
48	Cd L$_{\alpha_1}$	3.1337	395.651	4.49 02	5.85 02	7.17 02	8.97 02	1.04 03
90	Th M$_\beta$	3.1458	394.129	4.44 02	5.79 02	7.09 02	8.87 02	1.03 03
47	Ag L$_{\beta_1}$	3.1509	393.491	4.42 02	5.76 02	7.06 02	8.84 02	1.02 03
92	U M$_{\alpha_1}$	3.1708	391.021	4.34 02	5.66 02	6.94 02	8.68 02	1.01 03
18	Ar K$_\beta$	3.1905	388.607	4.27 02	5.56 02	6.82 02	8.54 02	9.90 02
91	Pa M$_\beta$	3.2397	382.705	4.08 02	5.33 02	6.54 02	8.18 02	9.49 02
49	In L$_{\alpha_1}$	3.2869	377.210	3.92 02	5.12 02	6.28 02	7.86 02	9.13 02
19	K K$_{\alpha_1}$	3.3138	374.148	3.83 02	5.00 02	6.14 02	7.69 02	8.93 02
48	Cd L$_{\beta_1}$	3.3166	373.832	3.82 02	4.99 02	6.13 02	7.67 02	8.90 02
92	U M$_\beta$	3.3360	371.658	3.76 02	4.91 02	6.03 02	7.54 02	8.76 02
50	Sn L$_{\alpha_1}$	3.4440	360.003	3.43 02	4.49 02	5.53 02	6.91 02	8.04 02
49	In L$_{\beta_1}$	3.4872	355.543	3.31 02	4.33 02	5.34 02	6.67 02	7.77 02
19	K K$_\beta$	3.5896	345.401	3.05 02	4.00 02	4.93 02	6.16 02	7.18 02
51	Sb L$_{\alpha_1}$	3.6047	343.954	3.02 02	3.95 02	4.87 02	6.09 02	7.10 02

Z	transition	E [keV]	λ [pm]	$_{11}$Na	$_{12}$Mg	$_{13}$Al	$_{14}$Si	$_{15}$P
50	Sn L_{β_1}	3.6628	338.498	2.88 02	3.77 02	4.66 02	5.83 02	6.80 02
20	Ca K_{α_1}	3.6917	335.848	2.82 02	3.69 02	4.56 02	5.70 02	6.65 02
52	Te L_{α_1}	3.7693	328.934	2.66 02	3.48 02	4.31 02	5.38 02	6.28 02
51	Sb L_{β_1}	3.8436	322.575	2.51 02	3.30 02	4.09 02	5.10 02	5.96 02
53	I L_{α_1}	3.9377	314.867	2.35 02	3.08 02	3.82 02	4.77 02	5.58 02
20	Ca K_{β}	4.0127	308.981	2.22 02	2.92 02	3.63 02	4.53 02	5.30 02
52	Te L_{β_1}	4.0296	307.686	2.20 02	2.89 02	3.59 02	4.47 02	5.24 02
21	Sc K_{α_1}	4.0906	303.097	2.10 02	2.77 02	3.44 02	4.29 02	5.02 02
54	Xe L_{α_1}	4.1099	301.674	2.07 02	2.73 02	3.39 02	4.23 02	4.96 02
53	I L_{β_1}	4.2207	293.755	1.92 02	2.53 02	3.15 02	3.93 02	4.61 02
55	Cs L_{α_1}	4.2865	289.245	1.84 02	2.42 02	3.01 02	3.76 02	4.41 02
21	Sc K_{β}	4.4605	277.962	1.64 02	2.16 02	2.69 02	3.37 02	3.95 02
56	Ba L_{α_1}	4.4663	277.601	1.63 02	2.15 02	2.68 02	3.35 02	3.94 02
22	Ti K_{α_1}	4.5108	274.863	1.59 02	2.09 02	2.61 02	3.26 02	3.83 02
55	Cs L_{β_1}	4.6198	268.377	1.48 02	1.96 02	2.44 02	3.05 02	3.58 02
57	La L_{α_1}	4.6510	266.577	1.45 02	1.92 02	2.39 02	2.99 02	3.52 02
56	Ba L_{β_1}	4.8273	256.841	1.30 02	1.73 02	2.16 02	2.70 02	3.17 02
58	Ce L_{α_1}	4.8402	256.157	1.29 02	1.71 02	2.14 02	2.68 02	3.15 02
22	Ti $K_{\beta_{1,3}}$	4.9318	251.399	1.23 02	1.63 02	2.03 02	2.54 02	2.99 02
23	V K_{α_1}	4.9522	250.363	1.21 02	1.61 02	2.01 02	2.51 02	2.95 02
59	Pr L_{α_1}	5.0337	246.310	1.16 02	1.53 02	1.92 02	2.40 02	2.82 02
51	La L_{β_1}	5.0421	245.899	1.15 02	1.53 02	1.91 02	2.39 02	2.81 02
60	Nd L_{α_1}	5.2304	237.047	1.03 02	1.31 02	1.12 02	2.15 02	2.53 02
58	Ce L_{β_1}	5.2622	235.614	1.02 02	1.35 02	1.69 02	2.11 02	2.49 02
24	Cr K_{α_1}	5.4147	228.987	9.34 01	1.24 02	1.56 02	1.94 02	2.30 02
23	V $K_{\beta_{1,3}}$	5.4273	228.447	9.28 01	1.23 02	1.55 02	1.93 02	2.28 02
61	Pm L_{α_1}	5.4325	228.228	9.25 01	1.23 02	1.54 02	1.93 02	2.28 02
59	Pr L_{β_1}	5.4889	225.883	8.91 01	1.19 02	1.50 02	1.81 02	2.21 02
62	Sm L_{α_1}	5.6361	219.984	8.31 01	1.11 02	1.39 02	1.13 02	2.05 02
60	Nd L_{β_1}	5.7210	216.696	1.95 01	1.06 02	1.33 02	1.66 02	1.91 02
63	Eu L_{α_1}	5.8457	212.096	1.46 01	9.96 01	1.25 02	1.56 02	1.85 02
25	Mn K_{α_1}	5.8988	210.187	1.21 01	9.10 01	1.22 02	1.52 02	1.80 02
24	Cr $K_{\beta_{1,3}}$	5.9467	208.494	1.10 01	9.48 01	1.19 02	1.48 02	1.16 02
61	Pm L_{β_1}	5.9610	207.993	1.05 01	9.41 01	1.18 02	1.41 02	1.15 02
64	Gd L_{α_1}	6.0572	204.690	6.13 01	8.99 01	1.13 02	1.41 02	1.61 02
62	Sm L_{β_1}	6.2051	199.811	6.26 01	8.38 01	1.05 02	1.31 02	1.56 02
65	Tb L_{α_1}	6.2728	197.655	6.01 01	8.11 01	1.02 02	1.21 02	1.51 02
26	Fe K_{α_1}	6.4038	193.611	5.11 01	1.64 01	9.61 01	1.20 02	1.43 02
63	Eu L_{β_1}	6.4564	192.034	5.51 01	1.46 01	9.39 01	1.11 02	1.39 02
25	Mn $K_{\beta_{1,3}}$	6.4905	191.025	5.49 01	1.34 01	9.24 01	1.15 02	1.31 02
66	Dy L_{α_1}	6.4952	190.887	5.41 01	1.33 01	9.22 01	1.15 02	1.31 02
64	Gd L_{β_1}	6.7132	184.688	4.91 01	6.66 01	8.38 01	1.05 02	1.25 02
61	Ho L_{α_1}	6.7198	184.507	4.95 01	6.64 01	8.36 01	1.04 02	1.24 02
21	Co K_{α_1}	6.9303	178.903	4.52 01	6.06 01	1.64 01	9.54 01	1.14 02

Z	transition	E [keV]	λ [pm]	$_{11}$Na	$_{12}$Mg	$_{13}$Al	$_{14}$Si	$_{15}$P
68	Er L$_{\alpha_1}$	6.9487	178.429	4.49 01	6.02 01	1.59 01	9.41 01	1.13 02
65	Tb L$_{\beta_1}$	6.9780	177.680	4.43 01	5.94 01	1.49 01	9.36 01	1.12 02
26	Fe K$_{\beta_{1,3}}$	7.0580	175.666	4.28 01	5.15 01	1.25 01	9.05 01	1.08 02
69	Tm L$_{\alpha_1}$	7.1799	172.683	4.01 01	5.41 01	6.90 01	8.62 01	1.03 02
66	Dy L$_{\beta_1}$	7.2477	171.068	3.96 01	5.32 01	6.11 01	8.38 01	1.00 02
70	Yb L$_{\alpha_1}$	7.4156	167.195	3.10 01	4.98 01	6.28 01	1.85 01	9.31 01
28	Ni K$_{\alpha_1}$	7.4182	165.795	3.61 01	4.86 01	6.13 01	1.66 01	9.15 01
61	Ho L$_{\beta_1}$	7.5253	164.757	3.55 01	4.11 01	6.02 01	1.52 01	8.99 01
21	Co K$_{\beta_{1,3}}$	7.6494	162.084	3.38 01	4.55 01	5.14 01	1.11 01	8.58 01
71	Lu L$_{\alpha_1}$	7.6555	161.955	3.31 01	4.54 01	5.13 01	1.16 01	8.56 01
68	Er L$_{\beta_1}$	7.8109	158.773	3.18 01	4.28 01	5.41 01	6.15 01	8.08 01
72	Hf L$_{\alpha_1}$	7.8990	156.963	3.01 01	4.14 01	5.23 01	6.54 01	1.82 01
29	Cu K$_{\alpha_1}$	8.0478	154.060	2.91 01	3.92 01	4.96 01	6.19 01	1.41 01
69	Tm L$_{\beta_1}$	8.1010	153.049	2.85 01	3.84 01	4.86 01	6.08 01	7.21 01
73	Ta L$_{\alpha_1}$	8.1461	152.201	2.81 01	3.18 01	4.18 01	5.98 01	1.16 01
28	Ni K$_{\beta_{1,3}}$	8.2641	150.017	2.69 01	3.62 01	4.58 01	5.13 01	6.86 01
74	W L$_{\alpha_1}$	8.3916	141.643	2.56 01	3.45 01	4.31 01	5.41 01	6.55 01
70	Yb L$_{\beta_1}$	8.4018	141.569	2.56 01	3.45 01	4.31 01	5.46 01	6.54 01
30	Zn K$_{\alpha_1}$	8.6389	143.519	2.36 01	3.18 01	4.02 01	5.03 01	6.03 01
75	Re L$_{\alpha_1}$	8.6525	143.294	2.35 01	3.16 01	4.00 01	5.01 01	6.01 01
71	Lu L$_{\beta_1}$	8.7090	142.364	2.30 01	3.20 01	3.93 01	4.92 01	5.89 01
29	Cu K$_{\beta_1}$	8.9053	139.226	2.16 01	2.90 01	3.68 01	4.61 01	5.53 01
76	Os L$_{\alpha_1}$	8.9117	138.126	2.15 01	2.90 01	3.61 01	4.60 01	5.51 01
72	Hf L$_{\beta_1}$	9.0227	137.414	2.01 01	2.79 01	3.54 01	4.43 01	5.32 01
77	Ir L$_{\alpha_1}$	9.1751	135.132	1.91 01	2.66 01	3.31 01	4.22 01	5.01 01
31	Ga K$_{\alpha_1}$	9.2517	134.013	1.93 01	2.59 01	3.29 01	4.12 01	4.95 01
13	Ta L$_{\beta_1}$	9.3431	132.702	1.81 01	2.52 01	3.19 01	4.00 01	4.81 01
18	Pt L$_{\alpha_1}$	9.4423	131.308	1.81 01	2.44 01	3.09 01	3.88 01	4.66 01
30	Zn K$_{\beta_{1,3}}$	9.5720	129.529	1.74 01	2.34 01	2.97 01	3.73 01	4.48 01
74	W L$_{\beta_1}$	9.6724	128.184	1.69 01	2.27 01	2.88 01	3.62 01	4.35 01
79	Au L$_{\alpha_1}$	9.7133	127.644	1.67 01	2.24 01	2.85 01	3.57 01	4.29 01
32	Ge K$_{\alpha_1}$	9.8864	125.409	1.58 01	2.13 01	2.70 01	3.39 01	4.08 01
80	Hg L$_{\alpha_1}$	9.9888	124.124	1.54 01	2.07 01	2.62 01	3.29 01	3.96 01
75	Re L$_{\beta_1}$	10.010	123.861	1.53 01	2.05 01	2.60 01	3.27 01	3.93 01
31	Ga K$_{\beta_1}$	10.264	120.796	1.42 01	1.91 01	2.42 01	3.04 01	3.66 01
81	Tl L$_{\alpha_1}$	10.269	120.737	1.42 01	1.90 01	2.42 01	3.03 01	3.65 01
76	Os L$_{\beta_1}$	10.355	119.734	1.38 01	1.86 01	2.36 01	2.96 01	3.56 01
33	As K$_{\alpha_1}$	10.544	117.588	1.31 01	1.76 01	2.23 01	2.81 01	3.38 01
82	Pb L$_{\alpha_1}$	10.552	117.499	1.31 01	1.76 01	2.23 01	2.80 01	3.37 01
77	Ir L$_{\beta_1}$	10.708	115.787	1.25 01	1.68 01	2.14 01	2.68 01	3.23 01
83	Bi L$_{\alpha_1}$	10.839	114.388	1.21 01	1.62 01	2.06 01	2.59 01	3.12 01
32	Ge K$_{\beta_1}$	10.982	112.898	1.16 01	1.56 01	1.98 01	2.49 01	3.00 01
78	Pt L$_{\beta_1}$	11.071	111.990	1.13 01	1.52 01	1.94 01	2.43 01	2.93 01
84	Po L$_{\alpha_1}$	11.131	111.387	1.12 01	1.50 01	1.91 01	2.39 01	2.89 01

Z	transition	E [keV]	λ [pm]	$_{11}$Na	$_{12}$Mg	$_{13}$Al	$_{14}$Si	$_{15}$P
34	Se K_{α_1}	11.222	110.484	1.09 01	1.46 01	1.86 01	2.34 01	2.82 01
85	At L_{α_1}	11.427	108.501	1.03 01	1.39 01	1.76 01	2.21 01	2.67 01
79	Au L_{β_1}	11.442	108.359	1.03 01	1.38 01	1.76 01	2.21 01	2.66 01
33	As K_{β_1}	11.726	105.735	9.57 00	1.29 01	1.63 01	2.05 01	2.48 01
86	Rn L_{α_1}	11.727	105.726	9.57 00	1.29 01	1.63 01	2.05 01	2.48 01
80	Hg L_{β_1}	11.823	104.867	9.35 00	1.26 01	1.60 01	2.00 01	2.42 01
35	Br K_{α_1}	11.924	103.979	9.11 00	1.22 01	1.56 01	1.95 01	2.36 01
87	Fr L_{α_1}	12.031	103.054	8.88 00	1.19 01	1.52 01	1.90 01	2.30 01
81	Tl L_{β_1}	12.213	101.519	8.49 00	1.14 01	1.45 01	1.82 01	2.20 01
88	Ra L_{α_1}	12.340	100.474	8.24 00	1.11 01	1.41 01	1.76 01	2.14 01
34	Se K_{β_1}	12.496	99.219	7.94 00	1.07 01	1.36 01	1.70 01	2.06 01
82	Pb L_{β_1}	12.614	98.291	7.72 00	1.04 01	1.32 01	1.65 01	2.00 01
36	Kr K_{α_1}	12.649	98.019	7.66 00	1.03 01	1.31 01	1.64 01	1.99 01
89	Ac L_{α_1}	12.652	97.996	7.65 00	1.03 01	1.31 01	1.64 01	1.99 01
90	Th L_{α_1}	12.969	95.601	7.12 00	9.55 00	1.22 01	1.52 01	1.85 01
83	Bi L_{β_1}	13.024	95.197	7.03 00	9.43 00	1.20 01	1.51 01	1.83 01
91	Pa L_{α_1}	13.291	93.285	6.62 00	8.89 00	1.13 01	1.42 01	1.72 01
35	Br K_{β_1}	13.291	93.185	6.62 00	8.89 00	1.13 01	1.42 01	1.72 01
37	Rb K_{α_1}	13.395	92.560	6.47 00	8.68 00	1.11 01	1.39 01	1.68 01
84	Po L_{β_1}	13.447	92.202	6.40 00	8.59 00	1.09 01	1.37 01	1.66 01
92	U L_{α_1}	13.615	91.065	6.17 00	8.28 00	1.05 01	1.32 01	1.60 01
85	At L_{β_1}	13.876	89.352	5.83 00	7.82 00	9.96 00	1.25 01	1.52 01
36	Kr K_{β_1}	14.112	87.857	5.55 00	7.44 00	9.48 00	1.19 01	1.44 01
38	Sr K_{β_1}	14.165	87.529	5.49 00	7.36 00	9.38 00	1.18 01	1.43 01
86	Rn L_{α_1}	14.316	86.606	5.32 00	7.14 00	9.09 00	1.14 01	1.39 01
87	Fr L_{β_1}	14.770	83.943	4.85 00	6.51 00	8.29 00	1.04 01	1.26 01
39	Y K_{α_1}	14.958	82.888	4.68 00	6.27 00	7.99 00	1.00 01	1.22 01
37	Rb K_{β_1}	14.961	82.872	4.67 00	6.27 00	7.98 00	1.00 01	1.22 01
88	Ra L_{β_1}	15.236	81.376	4.44 00	5.94 00	7.57 00	9.48 00	1.16 01
89	Ac L_{β_1}	15.713	78.906	4.06 00	5.43 00	6.92 00	8.67 00	1.06 01
40	Zr K_{α_1}	15.775	78.596	4.02 00	5.37 00	6.84 00	8.57 00	1.04 01
38	Sr K_{β_1}	15.836	78.293	3.97 00	5.31 00	6.76 00	8.47 00	1.03 01
90	Th L_{β_1}	16.202	76.524	3.72 00	4.97 00	6.32 00	7.93 00	9.66 00
41	Nb K_{α_1}	16.615	74.622	3.47 00	4.62 00	5.87 00	7.37 00	8.98 00
91	Pa L_{β_1}	16.702	74.233	3.41 00	4.55 00	5.79 00	7.25 00	8.84 00
39	Y K_{β_1}	16.738	74.074	3.39 00	4.52 00	5.72 00	7.21 00	8.79 00
92	U L_{β_1}	17.220	72.000	3.13 00	4.17 00	5.29 00	6.64 00	8.09 00
42	Mo K_{α_1}	17.479	70.933	3.00 00	3.99 00	5.07 00	6.35 00	7.74 00
40	Zr K_{β_1}	17.668	70.175	2.91 00	3.87 00	4.91 00	6.16 00	7.50 00
43	Tc K_{α_1}	18.367	67.504	2.60 00	3.46 00	4.38 00	5.50 00	6.70 00
41	Nb K_{β_1}	18.623	66.576	2.50 00	3.32 00	4.21 00	5.28 00	6.44 00
44	Ru K_{α_1}	19.279	64.311	2.27 00	3.00 00	3.80 00	4.77 00	5.82 00
42	Mo K_{β_1}	19.608	63.231	2.16 00	2.86 00	3.62 00	4.54 00	5.54 00
45	Rh K_{α_1}	20.216	61.330	1.99 00	2.62 00	3.32 00	4.16 00	5.07 00

Z	transition	E [keV]	λ [pm]	$_{11}$Na	$_{12}$Mg	$_{13}$Al	$_{14}$Si	$_{15}$P
43	Tc K$_{\beta_1}$	20.619	60.131	1.89 00	2.49 00	3.14 00	3.94 00	4.80 00
46	Pd K$_{\alpha_1}$	21.177	58.547	1.76 00	2.32 00	2.92 00	3.66 00	4.45 00
44	Ru K$_{\beta_1}$	21.657	57.240	1.67 00	2.19 00	2.75 00	3.44 00	4.18 00
47	Ag K$_{\alpha_1}$	22.163	55.942	1.57 00	2.05 00	2.58 00	3.22 00	3.92 00
45	Rh K$_{\beta_1}$	22.724	54.561	1.47 00	1.92 00	2.41 00	3.01 00	3.65 00
48	Cd K$_{\alpha_1}$	23.174	53.501	1.40 00	1.83 00	2.28 00	2.85 00	3.46 00
46	Pd K$_{\beta_1}$	23.819	52.053	1.30 00	1.70 00	2.11 00	2.64 00	3.20 00
49	In K$_{\alpha_1}$	24.210	51.212	1.25 00	1.62 00	2.02 00	2.52 00	3.06 00
47	Ag K$_{\beta_1}$	24.942	49.709	1.16 00	1.50 00	1.86 00	2.32 00	2.81 00
50	Sn K$_{\alpha_1}$	25.271	49.062	1.12 00	1.45 00	1.80 00	2.24 00	2.71 00
48	Cd K$_{\beta_1}$	26.096	47.511	1.03 00	1.33 00	1.65 00	2.05 00	2.47 00
51	Sb K$_{\alpha_1}$	26.359	47.037	1.00 00	1.30 00	1.60 00	1.99 00	2.41 00
49	In K$_{\beta_1}$	27.276	45.455	9.20 -1	1.18 00	1.46 00	1.81 00	2.19 00
52	Te K$_{\alpha_1}$	27.472	45.131	9.04 -1	1.16 00	1.43 00	1.77 00	2.14 00
50	Sn K$_{\beta_1}$	28.486	43.524	8.23 -1	1.05 00	1.29 00	1.60 00	1.93 00
53	I K$_{\alpha_1}$	28.612	43.333	8.14 -1	1.04 00	1.28 00	1.58 00	1.91 00
51	Sb K$_{\beta_1}$	29.726	41.709	7.38 -1	9.42 -1	1.15 00	1.43 00	1.72 00
54	Xe K$_{\alpha_1}$	29.779	41.635	7.34 -1	9.37 -1	1.15 00	1.42 00	1.71 00
55	Cs K$_{\alpha_1}$	30.973	40.030	6.75 -1	8.56 -1	1.04 00	1.28 00	1.54 00
52	Te K$_{\beta_1}$	30.996	40.000	6.74 -1	8.55 -1	1.04 00	1.28 00	1.54 00
56	Ba K$_{\alpha_1}$	32.194	38.511	6.24 -1	7.86 -1	9.49 -1	1.17 00	1.40 00
53	I K$_{\beta_1}$	32.295	38.391	6.20 -1	7.80 -1	9.42 -1	1.16 00	1.39 00
57	La K$_{\alpha_1}$	33.442	37.074	5.77 -1	7.22 -1	8.67 -1	1.06 00	1.27 00
54	Xe K$_{\beta_1}$	33.624	36.874	5.71 -1	7.14 -1	8.56 -1	1.05 00	1.25 00
58	Ce K$_{\alpha_1}$	34.279	36.169	5.49 -1	6.84 -1	8.18 -1	1.00 00	1.19 00
55	Cs K$_{\beta_1}$	34.987	35.437	5.27 -1	6.53 -1	7.79 -1	9.51 -1	1.13 00
59	Pr K$_{\alpha_1}$	36.026	34.415	4.96 -1	6.12 -1	7.27 -1	8.85 -1	1.05 00
56	Ba K$_{\beta_1}$	36.378	34.082	4.86 -1	5.99 -1	7.10 -1	8.64 -1	1.02 00
60	Nd K$_{\alpha_1}$	36.847	33.648	4.74 -1	5.82 -1	6.89 -1	8.37 -1	9.90 -1
57	La K$_{\beta_1}$	37.801	32.799	4.50 -1	5.50 -1	6.48 -1	7.86 -1	9.27 -1
61	Pm K$_{\alpha_1}$	38.725	32.016	4.28 -1	5.22 -1	6.12 -1	7.41 -1	8.72 -1
58	Ce K$_{\beta_1}$	39.257	31.582	4.16 -1	5.06 -1	5.93 -1	7.16 -1	8.42 -1
62	Sm K$_{\alpha_1}$	40.118	30.905	3.99 -1	4.83 -1	5.64 -1	6.80 -1	7.98 -1
59	Pr K$_{\beta_1}$	40.748	30.427	3.89 -1	4.70 -1	5.47 -1	6.58 -1	7.71 -1
63	Eu K$_{\alpha_1}$	41.542	29.845	3.78 -1	4.55 -1	5.27 -1	6.33 -1	7.38 -1
60	Nd K$_{\beta_1}$	42.271	29.331	3.68 -1	4.41 -1	5.10 -1	6.11 -1	7.10 -1
64	Gd K$_{\alpha_1}$	42.996	28.836	3.58 -1	4.28 -1	4.93 -1	5.90 -1	6.84 -1
61	Pm K$_{\beta_1}$	43.826	28.290	3.48 -1	4.14 -1	4.75 -1	5.67 -1	6.56 -1
65	Tb K$_{\alpha_1}$	44.482	27.873	3.40 -1	4.04 -1	4.62 -1	5.50 -1	6.35 -1
62	Sm K$_{\beta_1}$	45.413	27.301	3.29 -1	3.89 -1	4.44 -1	5.27 -1	6.06 -1
66	Dy K$_{\alpha_1}$	45.998	26.954	3.22 -1	3.81 -1	4.33 -1	5.14 -1	5.89 -1
63	Eu K$_{\beta_1}$	47.038	26.358	3.11 -1	3.66 -1	4.14 -1	4.91 -1	5.61 -1
67	Ho K$_{\alpha_1}$	47.547	26.076	3.06 -1	3.60 -1	4.06 -1	4.80 -1	5.48 -1
64	Gd K$_{\beta_1}$	48.697	25.460	2.95 -1	3.45 -1	3.88 -1	4.57 -1	5.09 -1

Z	transition	E [keV]	λ [pm]	$_{11}$Na	$_{12}$Mg	$_{13}$Al	$_{14}$Si	$_{15}$P
68	Er K$_{\alpha_1}$	49.128	25.237	2.91 -1	3.40 -1	3.81 -1	4.49 -1	5.09 -1
65	Tb K$_{\beta_1}$	50.382	24.609	2.80 -1	3.26 -1	3.64 -1	4.28 -1	4.83 -1
69	Tm K$_{\alpha_1}$	50.742	24.434	2.78 -1	3.23 -1	3.60 -1	4.22 -1	4.77 -1
66	Dy K$_{\beta_1}$	52.119	23.788	2.70 -1	3.12 -1	3.46 -1	4.03 -1	4.54 -1
70	Yb K$_{\alpha_1}$	52.389	23.666	2.68 -1	3.09 -1	3.43 -1	4.00 -1	4.49 -1
67	Ho K$_{\beta_1}$	53.877	23.012	2.59 -1	2.98 -1	3.29 -1	3.81 -1	4.27 -1
71	Lu Ka$_{\alpha_1}$	54.070	22.930	2.58 -1	2.96 -1	3.27 -1	3.79 -1	4.24 -1
68	Er K$_{\beta_1}$	55.681	22.267	2.50 -1	2.85 -1	3.13 -1	3.60 -1	4.02 -1
72	Hf Ka$_{\alpha_1}$	55.790	22.223	2.49 -1	2.84 -1	3.12 -1	3.59 -1	4.00 -1
69	Tm K$_{\beta_1}$	57.517	21.556	2.40 -1	2.73 -1	2.98 -1	3.41 -1	3.78 -1
73	Ta K$_{\alpha_1}$	57.532	21.550	2.40 -1	2.73 -1	2.97 -1	3.41 -1	3.78 -1
74	W K$_{\alpha_1}$	59.318	20.901	2.32 -1	2.62 -1	2.84 -1	3.24 -1	3.57 -1
70	Yb K$_{\beta_1}$	59.370	20.883	2.32 -1	2.61 -1	2.84 -1	3.23 -1	3.57 -1
75	Re K$_{\alpha_1}$	61.140	20.278	2.25 -1	2.53 -1	2.73 -1	3.10 -1	3.41 -1
71	Lu K$_{\beta_1}$	61.283	20.231	2.25 -1	2.52 -1	2.72 -1	3.09 -1	3.40 -1
76	Os K$_{\alpha_1}$	63.001	19.679	2.20 -1	2.46 -1	2.64 -1	2.99 -1	3.27 -1
72	Hf K$_{\beta_1}$	63.234	19.607	2.19 -1	2.45 -1	2.63 -1	2.98 -1	3.25 -1
77	Ir K$_{\alpha_1}$	64.896	19.105	2.15 -1	2.39 -1	2.56 -1	2.88 -1	3.13 -1
73	Ta K$_{\beta_1}$	65.223	19.009	2.14 -1	2.38 -1	2.54 -1	2.86 -1	3.11 -1
78	Pt K$_{\alpha_1}$	66.832	18.551	2.09 -1	2.32 -1	2.47 -1	2.78 -1	3.01 -1
74	W K$_{\beta_1}$	67.244	18.438	2.08 -1	2.31 -1	2.46 -1	2.76 -1	2.98 -1
79	Au K$_{\alpha_1}$	68.804	18.020	2.04 -1	2.26 -1	2.39 -1	2.68 -1	2.89 -1
75	Re K$_{\beta_1}$	69.310	17.888	2.03 -1	2.24 -1	2.38 -1	2.66 -1	2.86 -1
80	Hg K$_{\alpha_1}$	70.819	17.507	2.00 -1	2.20 -1	2.32 -1	2.59 -1	2.77 -1
76	Os K$_{\beta_1}$	71.413	17.361	1.98 -1	2.18 -1	2.30 -1	2.56 -1	2.74 -1
81	Tl K$_{\alpha_1}$	72.872	17.014	1.95 -1	2.14 -1	2.25 -1	2.50 -1	2.66 -1
77	Ir K$_{\beta_1}$	73.561	16.854	1.94 -1	2.l2 -1	2.22 -1	2.47 -1	2.63 -1
82	Pb K$_{\alpha_1}$	74.969	16.538	1.91 -1	2.08 -1	2.18 -1	2.41 -1	2.56 -1
78	Pt K$_{\beta_1}$	75.748	16.368	1.89 -1	2.06 -1	2.15 -1	2.38 -1	2.52 -1
83	Bi K$_{\alpha_1}$	77.108	16.079	1.86 -1	2.03 -1	2.11 -1	2.33 -1	2.46 -1
79	Au K$_{\beta_1}$	77.948	15.906	1.85 -1	2.01 -1	2.08 -1	2.30 -1	2.42 -1
84	Po K$_{\alpha_1}$	79.290	15.636	1.82 -1	1.97 -1	2.04 -1	2.25 -1	2.36 -1
80	Hg K$_{\beta_1}$	80.253	15.449	1.80 -1	1.95 -1	2.02 -1	2.22 -1	2.33 -1
85	At K$_{\alpha_1}$	81.520	15.209	1.79 -1	1.93 -1	2.00 -1	2.19 -1	2.29 -1
81	Tl K$_{\beta_1}$	82.576	15.014	1.78 -1	1.92 -1	1.98 -1	2.17 -1	2.26 -1
86	Rn K$_{\alpha_1}$	83.780	14.798	1.76 -1	1.90 -1	1.95 -1	2.14 -1	2.23 -1
82	Pb K$_{\beta_1}$	84.936	14.597	1.75 -1	1.88 -1	1.93 -1	2.12 -1	2.20 -1
87	Fr K$_{\alpha_1}$	86.100	14.400	1.73 -1	1.86 -1	1.91 -1	2.09 -1	2.17 -1
83	Bi K$_{\beta_1}$	87.343	14.195	1.72 -1	1.85 -1	1.89 -1	2.07 -1	2.14 -1
88	Ra K$_{\alpha_1}$	88.470	14.014	1.71 -1	1.83 -1	1.87 -1	2.04 -1	2.11 -1
84	Po K$_{\beta_1}$	89.800	13.806	1.69 -1	1.81 -1	1.85 -1	2.02 -1	2.08 -1
89	Ac K$_{\alpha_1}$	90.884	13.642	1.68 -1	1.80 -1	1.84 -1	1.99 -1	2.06 -1
85	At K$_{\beta_1}$	92.300	13.432	1.67 -1	1.78 -1	1.82 -1	1.97 -1	2.03 -1
90	Th K$_{\alpha_1}$	93.350	13.281	1.65 -1	1.77 -1	1.80 -1	1.95 -1	2.00 -1

Z	transition	E [keV]	λ [pm]	$_{11}$Na	$_{12}$Mg	$_{13}$Al	$_{14}$Si	$_{15}$P
86	Rn K$_{\beta_1}$	94.870	13.068	1.64 -1	1.75 -1	1.78 -1	1.92 -1	1.97 -1
91	Pa K$_{\alpha_1}$	95.868	12.932	1.63 -1	1.74 -1	1.76 -1	1.90 -1	1.95 -1
87	Fr K$_{\beta_1}$	97.470	12.720	1.61 -1	1.72 -1	1.74 -1	1.88 -1	1.92 -1
92	U K$_{\alpha_1}$	98.439	12.595	1.60 -1	1.71 -1	1.73 -1	1.86 -1	1.90 -1
88	Ra K$_{\beta_1}$	100.130	12.382	1.59 -1	1.69 -1	1.71 -1	1.83 -1	1.87 -1
89	Ac K$_{\beta_1}$	102.850	12.054	1.57 -1	1.67 -1	1.68 -1	1.81 -1	1.84 -1
90	Th K$_{\beta_1}$	105.610	11.739	1.55 -1	1.65 -1	1.66 -1	1.78 -1	1.81 -1
91	Pa K$_{\beta_1}$	108.430	11.434	1.54 -1	1.63 -1	1.64 -1	1.75 -1	1.78 -1
92	U K$_{\beta_1}$	111.300	11.139	1.52 -1	1.61 -1	1.61 -1	1.72 -1	1.75 -1

Z	transition	E [keV]	λ [pm]	$_{16}$S	$_{17}$Cl	$_{18}$Ar	$_{19}$K	$_{20}$Ca
4	Be K$_\alpha$	0.1095	11427.207	1.28 04	1.61 04	1.80 04	2.10 04	2.22 04
38	Sr M$_\xi$	0.1140	10875.895	1.20 04	1.52 04	1.75 04	2.01 04	2.14 04
39	Y M$_\xi$	0.1328	9339.235	9.64 03	1.25 04	1.49 04	1.74 04	1.87 04
16	S L$_l$	0.1487	8337.942	8.09 03	1.06 04	1.29 04	1.52 04	1.70 04
40	Zr M$_\xi$	0.1511	8205.506	<u>7.92 03</u>	1.03 04	1.26 04	1.49 04	1.67 04
41	Nb M$_\xi$	0.1717	7221.037	7.58 04	8.42 03	1.03 04	1.25 04	1.41 04
5	B K$_\alpha$	0.1833	6764.059	7.42 04	7.55 03	9.37 03	1.13 04	1.30 04
42	Mo M$_\xi$	0.1926	6437.445	<u>6.98 04</u>	<u>6.94 03</u>	<u>8.63 03</u>	1.05 04	1.21 04
6	C K$_\alpha$	0.2770	4476.000	4.79 04	5.08 04	4.56 04	<u>5.68 03</u>	6.84 03
47	Ag M$_\xi$	0.3117	3977.709	3.84 04	4.14 04	<u>3.93 04</u>	<u>7.38 04</u>	<u>5.64 03</u>
7	N K$_\alpha$	0.3924	3159.664	2.49 04	2.75 04	2.95 04	3.59 04	3.56 04
22	Ti L$_l$	0.3953	3136.484	2.45 04	2.71 04	2.91 04	3.55 04	<u>3.51 04</u>
22	Ti L$_\alpha$	0.4522	2741.822	1.83 04	2.01 04	2.23 04	2.69 04	2.93 04
23	V L$_\alpha$	0.5113	2424.901	1.38 04	1.50 04	1.71 04	2.05 04	2.33 04
8	O K$_\alpha$	0.5249	2362.072	1.30 04	1.41 04	1.62 04	1.94 04	2.20 04
25	Mn L$_l$	0.5563	2228.747	1.14 04	1.23 04	1.42 04	1.70 04	1.94 04
24	Cr L$_\alpha$	0.5728	2164.546	1.06 04	1.14 04	1.33 04	1.58 04	1.82 04
25	Mn L$_\alpha$	0.6374	1945.171	8.20 03	8.81 03	1.04 04	1.24 04	1.43 04
9	F K$_\alpha$	0.6768	1831.932	7.01 03	7.59 03	8.92 03	1.06 04	1.24 04
26	Fe L$_\alpha$	0.7050	1758.655	6.38 03	6.86 03	8.16 03	9.71 03	1.14 04
27	Co L$_\alpha$	0.7762	1597.335	4.95 03	5.39 03	6.45 03	7.68 03	9.01 03
28	Ni L$_\alpha$	0.8515	1456.080	3.89 03	4.26 03	5.07 03	6.09 03	7.16 03
29	Cu L$_\alpha$	0.9297	1336.044	3.08 03	3.40 03	4.05 03	4.89 03	5.77 03
30	Zn L$_\alpha$	1.0117	1225.513	2.38 03	2.78 03	3.19 03	4.03 03	4.90 03
11	Na K$_\alpha$	1.0410	1191.020	2.21 03	2.58 03	2.95 03	3.74 03	4.54 03
11	Na K$_\beta$	1.0711	1157.550	2.05 03	2.39 03	2.74 03	3.47 03	4.21 03
12	Mg K$_\alpha$	1.2536	989.033	1.35 03	1.58 03	1.79 03	2.28 03	2.76 03
33	As L$_\alpha$	1.2820	967.123	1.27 03	1.49 03	1.69 03	2.15 03	2.60 03
12	Mg K$_\beta$	1.3022	952.121	1.22 03	1.43 03	1.62 03	2.06 03	2.50 03
33	As L$_{\beta_1}$	1.3170	941.421	1.18 03	1.38 03	1.57 03	2.00 03	2.42 03
66	Dy M$_\beta$	1.3250	935.737	1.16 03	1.36 03	1.55 03	1.97 03	2.36 03

Z	transition	E [keV]	λ [pm]	$_{16}$S	$_{17}$Cl	$_{18}$Ar	$_{19}$K	$_{20}$Ca
67	Mo M_α	1.3480	919.771	1.11 03	1.30 03	1.48 03	1.88 03	2.28 03
34	Se L_α	1.3791	899.029	1.04 03	1.22 03	1.39 03	1.77 03	2.14 03
67	Ho M_β	1.3830	896.494	1.04 03	1.22 03	1.38 03	1.76 03	2.13 03
68	Er M_α	1.4060	881.829	9.91 02	1.16 03	1.32 03	1.68 03	2.04 03
34	Se L_{α_1}	1.4192	873.627	9.67 02	1.14 03	1.29 03	1.64 03	1.99 03
68	Er M_β	1.4430	859.218	9.25 02	1.09 03	1.23 03	1.57 03	1.90 03
69	Tm M_α	1.4620	848.051	8.93 02	1.05 03	1.19 03	1.52 03	1.83 03
35	Br L_α	1.4804	837.511	8.64 02	1.02 03	1.15 03	1.47 03	1.77 03
13	Al K_{α_1}	1.4867	833.962	8.54 02	1.00 03	1.14 03	1.45 03	1.75 03
69	Tm M_β	1.5030	824.918	8.29 02	9.75 02	1.10 03	1.41 03	1.70 03
70	Yb M_α	1.5214	814.941	8.03 02	9.44 02	1.07 03	1.37 03	1.65 03
35	Br L_{β_1}	1.5259	812.538	7.97 02	9.37 02	1.06 03	1.36 03	1.64 03
13	Al K_β	1.5574	796.103	7.54 02	8.87 02	1.00 03	1.28 03	1.55 03
70	Yb M_β	1.5675	790.974	7.41 02	8.72 02	9.87 02	1.26 03	1.52 03
71	Lu M_α	1.5813	784.071	7.24 02	8.52 02	9.64 02	1.23 03	1.49 03
36	Kr L_α	1.5860	781.747	7.19 02	8.46 02	9.56 02	1.22 03	1.48 03
71	Lu M_β	1.6312	760.085	6.67 02	7.85 02	8.87 02	1.14 03	1.37 03
36	Kr L_{β_1}	1.6366	757.577	6.61 02	7.78 02	8.79 02	1.13 03	1.36 03
72	Hf M_{α_1}	1.6446	753.892	6.52 02	7.68 02	8.68 02	1.11 03	1.34 03
37	Rb L_{α_1}	1.6941	731.864	6.03 02	7.10 02	8.02 02	1.03 03	1.24 03
72	Hf M_β	1.6976	730.355	5.99 02	7.06 02	7.97 02	1.02 03	1.23 03
73	Ta M_{α_1}	1.7096	725.229	5.88 02	6.93 02	7.82 02	1.00 03	1.21 03
14	Si K_{α_1}	1.7400	712.558	5.61 02	6.61 02	7.46 02	9.56 02	1.15 03
37	Rb L_{β_1}	1.7522	707.597	5.51 02	6.49 02	7.32 02	9.39 02	1.13 03
73	Ta M_β	1.7655	702.266	5.40 02	6.36 02	7.18 02	9.20 02	1.11 03
74	W M_{α_1}	1.7754	698.350	5.32 02	6.27 02	7.07 02	9.07 02	1.09 03
38	Sr L_{α_1}	1.8066	686.290	5.08 02	5.99 02	6.75 02	8.66 02	1.04 03
74	W M_β	1.8349	675.705	4.87 02	5.74 02	6.47 02	8.31 02	1.00 03
14	Si K_β	1.8359	675.337	4.86 02	5.74 02	6.47 02	8.29 02	1.00 03
75	Re M_{α_1}	1.8420	673.100	4.82 02	5.69 02	6.41 02	8.22 02	9.92 02
38	Sr L_{β_1}	1.8717	662.420	4.62 02	5.45 02	6.14 02	7.88 02	9.50 02
75	Re M_β	1.9061	650.465	4.40 02	5.19 02	5.85 02	7.51 02	9.05 02
76	Os M_{α_1}	1.9192	649.69	4.37 02	5.16 02	5.81 02	7.46 02	9.00 02
39	Y L_{β_1}	1.9226	644.882	4.30 02	5.07 02	5.71 02	7.34 02	8.85 02
76	Os M_β	1.9783	626.725	3.98 02	4.70 02	5.29 02	6.80 02	8.20 02
77	Ir M_{α_1}	1.9799	626.219	3.97 02	4.69 02	5.28 02	6.79 02	8.18 02
39	Y L_{α_1}	1.9958	621.230	3.89 02	4.60 02	5.17 02	6.64 02	8.01 02
15	P K_{α_1}	2.0137	615.708	3.80 02	4.49 02	5.05 02	6.49 02	7.82 02
40	Zr L_{α_1}	2.0424	607.056	3.66 02	4.33 02	4.86 02	6.25 02	7.54 02
78	Pt M_{α_1}	2.0505	604.658	3.62 02	4.28 02	4.81 02	6.19 02	7.46 02
77	Ir M_β	2.0535	603.775	3.61 02	4.26 02	4.79 02	6.16 02	7.43 02
79	Au M_{α_1}	2.1229	584.036	3.31 02	3.91 02	4.39 02	5.64 02	6.80 02
40	Zr L_{β_1}	2.1244	583.624	3.30 02	3.90 02	4.38 02	5.63 02	6.79 02
78	Pt M_β	2.1273	582.828	3.29 02	3.89 02	4.36 02	5.61 02	6.76 02

Z	transition	E [keV]	λ [pm]	$_{16}$S	$_{17}$Cl	$_{18}$Ar	$_{19}$K	$_{20}$Ca
15	P K$_{\beta_{1,3}}$	2.1390	579.640	3.24 02	3.83 02	4.30 02	5.53 02	6.67 02
41	Nb L$_{\alpha_1}$	2.1659	572.441	3.14 02	3.71 02	4.16 02	5.35 02	6.45 02
80	Hg M$_{\alpha_1}$	2.1953	564.775	3.03 02	3.58 02	4.01 02	5.17 02	6.22 02
79	Au M$_\beta$	2.2046	562.393	2.99 02	3.54 02	3.97 02	5.11 02	6.15 02
41	Nb L$_{\beta_1}$	2.2574	549.238	2.81 02	3.33 02	3.73 02	4.80 02	5.78 02
81	Tl M$_{\alpha_1}$	2.2706	546.045	2.77 02	3.28 02	3.67 02	4.73 02	5.69 02
80	Hg M$_\beta$	2.2825	543.199	2.73 02	3.23 02	3.62 02	4.66 02	5.61 02
42	Mo L$_{\alpha_1}$	2.2932	540.664	2.70 02	3.19 02	3.57 02	4.60 02	5.54 02
16	S K$_{\alpha_1}$	2.3080	537.197	2.65 02	3.14 02	3.51 02	4.53 02	5.45 02
82	Pb M$_{\alpha_1}$	2.3457	528.563	2.54 02	3.01 02	3.37 02	4.34 02	5.22 02
81	Tl M$_\beta$	2.3621	524.893	2.50 02	2.95 02	3.30 02	4.26 02	5.13 02
42	Mo L$_{\beta_1}$	2.3948	517.726	2.41 02	2.85 02	3.18 02	4.10 02	4.94 02
83	Bi M$_{\alpha_1}$	2.4226	511.785	2.34 02	2.76 02	3.09 02	3.98 02	4.79 02
43	Tc L$_{\alpha_1}$	2.4240	511.490	2.33 02	2.76 02	3.08 02	3.98 02	4.79 02
82	Pb M$_\beta$	2.4427	507.574	2.29 02	2.70 02	3.02 02	3.90 02	4.69 02
16	S K$_\beta$	2.4640	503.186	<u>2.24 02</u>	2.64 02	2.95 02	3.81 02	4.58 02
83	Bi K$_\beta$	2.5255	490.933	2.24 03	2.48 02	2.77 02	3.57 02	4.29 02
43	Tc L$_{\beta_1}$	2.5368	488.746	2.21 03	2.45 02	2.73 02	3.53 02	4.24 02
44	Ru L$_{\alpha_1}$	2.5586	484.582	2.16 03	2.39 02	2.67 02	3.45 02	4.15 02
17	Cl K$_{\alpha_1}$	2.6224	472.792	2.02 03	2.24 02	2.50 02	3.23 02	3.89 02
44	Ru L$_{\beta_1}$	2.6832	462.079	1.90 03	2.11 02	2.35 02	3.04 02	3.66 02
45	Rh L$_{\alpha_1}$	2.6967	459.766	1.88 03	2.09 02	2.32 02	3.00 02	3.61 02
17	Cl K$_\beta$	2.8156	440.350	1.68 03	<u>1.86 02</u>	2.07 02	2.68 02	3.22 02
45	Rh L$_{\beta_1}$	2.8344	437.430	1.65 03	1.83 03	2.03 02	2.63 02	3.16 02
46	Pd L$_{\alpha_1}$	2.8386	436.782	1.64 03	1.82 03	2.03 02	2.62 02	3.15 02
18	Ar K$_{\alpha_1}$	2.9577	419.194	1.47 03	1.64 03	1.82 02	2.35 02	2.82 02
47	Ag L$_{\alpha_1}$	2.9843	415.458	1.44 03	1.60 03	1.77 02	2.30 02	2.76 02
46	Pd L$_{\beta_1}$	2.9902	414.638	1.43 03	1.59 03	1.76 02	2.28 02	2.74 02
90	Th M$_{\alpha_1}$	2.9961	413.821	1.42 03	1.59 03	1.76 02	2.27 02	2.73 02
91	Pa M$_{\alpha_1}$	3.0823	402.248	1.32 03	1.47 03	1.63 02	2.11 02	2.53 02
48	Cd L$_{\alpha_1}$	3.1337	395.651	1.26 03	1.41 03	1.55 02	2.02 02	2.42 02
90	Th M$_\beta$	3.1458	394.129	1.25 03	1.39 03	1.54 02	2.00 02	2.40 02
47	Ag L$_{\beta_1}$	3.1509	393.491	1.24 03	1.39 03	1.53 02	1.99 02	2.39 02
92	U M$_{\alpha_1}$	3.1708	391.021	1.22 03	1.36 03	1.51 02	1.96 02	2.35 02
18	Ar K$_\beta$	3.1905	388.607	1.20 03	1.34 03	<u>1.48 02</u>	1.92 02	2.31 02
91	Pa M$_\beta$	3.2397	382.705	1.15 03	1.29 03	1.44 03	1.85 02	2.22 02
49	In L$_{\alpha_1}$	3.2869	377.210	1.11 03	1.24 03	1.39 03	1.78 02	2.14 02
19	K K$_{\alpha_1}$	3.3138	374.148	1.08 03	1.21 03	1.36 03	1.74 02	2.09 02
48	Cd L$_{\alpha_1}$	3.3166	373.832	1.08 03	1.21 03	1.35 03	1.74 02	2.08 02
92	U M$_\beta$	3.3360	371.658	1.06 03	1.19 03	1.33 03	1.71 02	2.05 02
50	Sn L$_{\alpha_1}$	3.4440	360.003	9.76 02	1.09 03	1.22 03	1.57 02	1.89 02
49	In L$_{\beta_1}$	3.4872	355.543	9.44 02	1.06 03	1.18 03	1.52 02	1.83 02
19	K K$_\beta$	3.5896	345.401	8.73 02	9.78 02	1.09 03	1.41 02	1.69 02
51	Sb L$_{\alpha_1}$	3.6047	343.954	8.63 02	9.67 02	1.08 03	1.39 02	1.67 02

Z	transition	E [keV]	λ [pm]	$_{16}$S	$_{17}$Cl	$_{18}$Ar	$_{19}$K	$_{20}$Ca
50	Sn L_{β_1}	3.6628	338.498	8.26 02	9.27 02	1.04 03	1.24 02	1.60 02
20	Ca K_{α_1}	3.6917	335.848	8.09 02	9.07 02	1.02 03	<u>1.22 02</u>	1.57 02
52	Te L_{α_1}	3.7693	328.934	7.65 02	8.58 02	9.60 02	1.15 03	1.48 02
51	Sb L_{β_1}	3.8436	322.575	7.26 02	8.15 02	9.11 02	1.09 03	1.41 02
53	I L_{α_1}	3.9377	314.867	6.80 02	7.64 02	8.54 02	1.02 03	1.32 02
20	Ca K_β	4.0127	308.981	6.46 02	7.26 02	8.12 02	9.74 02	1.26 02
52	Te L_{β_1}	4.0296	307.686	6.38 02	7.18 02	8.03 02	9.63 02	<u>1.24 02</u>
21	Sc K_{α_1}	4.0906	303.097	6.12 02	6.89 02	7.70 02	9.25 02	1.07 03
54	Xe L_{α_1}	4.1099	301.674	6.05 02	6.80 02	7.60 02	9.13 02	1.06 03
53	I L_{β_1}	4.2207	293.755	5.62 02	6.32 02	7.07 02	8.50 02	9.87 02
55	Cs L_{α_1}	4.2865	289.245	5.38 02	6.06 02	6.78 02	8.15 02	9.47 02
21	Sc K_β	4.4605	277.962	4.82 02	5.43 02	6.08 02	7.32 02	8.52 02
56	Ba L_{α_1}	4.4663	277.601	4.80 02	5.41 02	6.06 02	7.29 02	8.49 02
22	Ti K_{α_1}	4.5108	274.863	4.67 02	5.27 02	5.90 02	7.10 02	8.27 02
55	Cs L_{β_1}	4.6198	268.377	4.37 02	4.94 02	5.52 02	6.66 02	7.76 02
57	La L_{α_1}	4.6510	266.577	4.29 02	4.85 02	5.42 02	6.54 02	7.62 02
56	Ba L_{β_1}	4.8273	256.841	3.87 02	4.38 02	4.90 02	5.91 02	6.90 02
58	Ce L_{α_1}	4.8402	256.157	3.84 02	4.35 02	4.86 02	5.87 02	6.85 02
22	Ti $K_{\beta_{1,3}}$	4.9318	251.390	3.65 02	4.13 02	4.62 02	5.58 02	6.52 02
23	V K_{α_1}	4.9522	250.363	3.61 02	4.08 02	4.57 02	5.52 02	6.45 02
59	Pr L_{α_1}	5.0337	246.310	3.45 02	3.90 02	4.37 02	5.28 02	6.17 02
57	La L_{β_1}	5.0421	245.899	3.43 02	3.88 02	4.35 02	5.25 02	6.14 02
60	Nd L_{α_1}	5.2304	237.047	3.09 02	3.51 02	3.93 02	4.75 02	5.56 02
58	Ce L_{β_1}	5.2622	235.614	3.04 02	3.45 02	3.86 02	4.67 02	5.47 02
24	Cr K_{α_1}	5.4147	228.978	2.81 02	3.19 02	3.57 02	4.32 02	5.06 02
23	V $K_{\beta_{1,3}}$	5.4273	228.447	2.79 02	3.17 02	3.54 02	4.30 02	5.03 02
61	Pm L_{α_1}	5.4325	228.228	2.78 02	3.16 02	3.53 02	4.28 02	5.02 02
59	Pr L_{β_1}	5.4889	225.883	2.70 02	3.07 02	3.44 02	4.16 02	4.88 02
62	Sm L_{α_1}	5.6361	219.984	2.51 02	2.85 02	3.19 02	3.87 02	4.54 02
60	Nd L_{β_1}	5.7216	216.696	2.40 02	2.73 02	3.06 02	3.72 02	4.36 02
63	Eu L_{α_1}	5.8457	212.096	2.26 02	2.58 02	2.89 02	3.51 02	4.11 02
25	Mn K_{α_1}	5.8988	210.187	2.21 02	2.51 02	2.81 02	3.42 02	4.01 02
24	Cr $K_{\beta_{1,3}}$	5.9467	208.494	2.16 02	2.46 02	2.75 02	3.35 02	3.93 02
61	Pm L_{β_1}	5.9610	207.993	2.14 02	2.44 02	2.73 02	3.32 02	3.90 02
64	Gd L_{β_1}	6.0572	204.690	2.05 02	2.33 02	2.61 02	3.18 02	3.73 02
62	Sm L_{β_1}	6.2051	199.811	1.91 02	2.18 02	2.44 02	2.97 02	3.49 02
65	Tb L_{α_1}	6.2726	197.655	1.85 02	2.11 02	2.37 02	2.88 02	3.39 02
26	Fe K_{α_1}	6.4038	193.611	1.75 02	1.99 02	2.23 02	2.72 02	3.20 02
63	Eu L_{β_1}	6.4564	192.034	1.71 02	1.95 02	2.18 02	2.66 02	3.13 02
25	Mn $K_{\beta_{1,3}}$	6.4905	191.025	1.68 02	1.92 02	2.15 02	2.62 02	3.08 02
66	Dy L_{α_1}	6.4952	190.887	1.68 02	1.92 02	2.15 02	2.62 02	3.08 02
64	Gd L_{β_1}	6.7132	184.688	1.53 02	1.74 02	1.95 02	2.39 02	2.81 02
67	Ho L_{α_1}	6.7198	184.507	1.53 02	1.74 02	1.95 02	2.38 02	2.80 02
27	Co K_{α_1}	6.9303	178.903	1.40 02	1.59 02	1.79 02	2.19 02	2.57 02

Z	transition	E [keV]	λ [pm]	$_{16}$S	$_{17}$Cl	$_{18}$Ar	$_{19}$K	$_{20}$Ca
68	Er L$_{\alpha_1}$	6.9487	178.429	1.39 02	1.58 02	1.77 02	2.17 02	2.55 02
65	Tb L$_{\beta_1}$	6.9780	177.680	1.37 02	1.56 02	1.75 02	2.14 02	2.52 02
26	Fe K$_{\beta_{1,3}}$	7.0580	175.666	1.33 02	1.51 02	1.70 02	2.08 02	2.44 02
69	Tm L$_{\alpha_1}$	7.1799	172.683	1.27 02	1.44 02	1.62 02	1.98 02	2.33 02
66	Dy L$_{\beta_1}$	7.2477	171.068	1.23 02	1.41 02	1.57 02	1.93 02	2.27 02
70	Yb L$_{\alpha_1}$	7.4156	167.195	1.15 02	1.32 02	1.48 02	1.81 02	2.13 02
28	Ni K$_{\alpha_1}$	7.4782	165.795	1.13 02	1.29 02	1.44 02	1.77 02	2.08 02
67	Ho L$_{\beta_1}$	7.5253	164.757	1.11 02	1.26 02	1.42 02	1.74 02	2.05 02
27	Co K$_{\beta_{1,3}}$	7.6494	162.084	1.06 02	1.21 02	1.35 02	1.66 02	1.96 02
71	Lu L$_{\alpha_1}$	7.6555	161.955	1.06 02	1.20 02	1.35 02	1.66 02	1.95 02
68	Er L$_{\beta_1}$	7.8109	158.733	9.97 01	1.14 02	1.27 02	1.57 02	1.85 02
72	Hf L$_{\alpha_1}$	7.8990	156.963	9.66 01	1.10 02	1.23 02	1.52 02	1.79 02
29	Cu K$_{\alpha_1}$	8.0478	154.060	9.16 01	1.05 02	1.17 02	1.44 02	1.70 02
69	Tm L$_{\beta_1}$	8.1010	153.049	8.98 01	1.03 02	1.15 02	1.41 02	1.67 02
73	Ta L$_{\alpha_1}$	8.1461	152.201	8.84 01	1.01 02	1.13 02	1.39 02	1.64 02
28	Ni K$_{\beta_{1,3}}$	8.2647	150.017	8.48 01	9.69 01	1.09 02	1.34 02	1.58 02
74	W L$_{\alpha_1}$	8.3976	147.643	8.10 01	9.26 01	1.04 02	1.28 02	1.51 02
70	Yb L$_{\beta_1}$	8.4018	147.569	8.09 01	9.25 01	1.04 02	1.28 02	1.51 02
30	Zn K$_{\alpha_1}$	8.6389	143.519	7.47 01	8.54 01	9.56 01	1.18 02	1.39 02
75	Re L$_{\alpha_1}$	8.6525	143.294	7.43 01	8.50 01	9.52 01	1.17 02	1.39 02
71	Lu L$_{\beta_1}$	8.7090	142.364	1.30 01	8.35 01	9.34 01	1.15 02	1.36 02
29	Cu K$_{\beta_1}$	8.9053	139.226	6.84 01	7.83 01	8.76 01	1.08 02	1.28 02
76	Os L$_{\alpha_1}$	8.9117	139.126	6.83 01	7.81 01	8.75 01	1.08 02	1.28 02
72	Hf L$_{\beta_1}$	9.0227	137.414	6.59 01	7.54 01	8.44 01	1.04 02	1.23 02
77	Ir L$_{\alpha_1}$	9.1751	135.132	6.28 01	7.19 01	8.05 01	9.94 01	1.18 02
31	Ga K$_{\alpha_1}$	9.2517	134.013	6.13 01	7.02 01	7.86 01	9.71 01	1.15 02
73	Ta L$_{\beta_1}$	9.3431	132.702	5.96 01	6.83 01	7.64 01	9.45 01	1.12 02
78	Pt L$_{\alpha_1}$	9.4423	131.308	5.78 01	6.63 01	7.41 01	9.17 01	1.09 02
30	Zn K$_{\beta_{1,3}}$	9.5720	129.529	5.56 01	6.37 01	7.13 01	8.82 01	1.05 02
74	W L$_{\beta_1}$	9.6724	128.184	5.39 01	6.18 01	6.92 01	8.56 01	1.02 02
79	Au L$_{\alpha_1}$	9.7133	127.644	5.53 01	6.11 01	6.83 01	8.46 01	1.00 02
32	Ge K$_{\alpha_1}$	9.8864	125.409	5.07 01	5.81 01	6.50 01	8.05 01	9.56 01
80	Hg L$_{\alpha_1}$	9.9888	124.124	4.92 01	5.64 01	6.31 01	7.82 01	9.29 01
75	Re L$_{\beta_1}$	10.010	123.861	4.89 01	5.61 01	6.27 01	7.77 01	9.23 01
31	Ga K$_{\beta_1}$	10.264	120.796	4.54 01	5.21 01	5.83 01	7.23 01	8.60 01
81	Tl L$_{\alpha_1}$	10.269	120.737	4.54 01	5.21 01	5.82 01	7.22 01	8.58 01
76	Os L$_{\beta_1}$	10.355	119.734	4.43 01	5.08 01	5.68 01	7.05 01	8.38 01
33	As K$_{\alpha_1}$	10.544	117.588	4.20 01	4.82 01	5.39 01	6.69 01	7.96 01
82	Pb L$_{\alpha_1}$	10.552	117.499	4.19 01	4.81 01	5.38 01	6.68 01	7.94 01
77	Ir L$_{\beta_1}$	10.708	115.787	4.02 01	4.61 01	5.16 01	6.40 01	7.62 01
83	Bi L$_{\alpha_1}$	10.839	114.388	3.88 01	4.45 01	4.98 01	6.18 01	7.36 01
32	Ge K$_{\beta_1}$	10.982	112.898	3.73 01	4.28 01	4.80 01	5.96 01	7.09 01
78	Pt L$_{\beta_1}$	11.071	111.990	3.65 01	4.19 01	4.68 01	5.82 01	6.93 01
84	Po L$_{\alpha_1}$	11.131	111.387	3.59 01	4.12 01	4.61 01	5.73 01	6.82 01

Z	transition	E [keV]	λ [pm]	$_{16}$S	$_{17}$Cl	$_{18}$Ar	$_{19}$K	$_{20}$Ca
34	Se K$_{\alpha_1}$	11.222	110.484	3.50 01	4.02 01	4.50 01	5.60 01	6.67 01
85	At L$_{\alpha_1}$	11.427	108.501	3.32 01	3.82 01	4.27 01	5.31 01	6.33 01
79	Au L$_{\beta_1}$	11.442	108.359	3.31 01	3.80 01	4.26 01	5.29 01	6.31 01
33	As K$_{\beta_1}$	11.726	105.735	3.08 01	3.54 01	3.97 01	4.93 01	5.88 01
86	Rn L$_{\alpha_1}$	11.727	105.726	3.08 01	3.54 01	3.97 01	4.93 01	5.88 01
80	Hg L$_{\beta_1}$	11.823	104.867	3.01 01	3.46 01	3.87 01	4.82 01	5.75 01
35	Br K$_{\alpha_1}$	11.924	103.979	2.94 01	3.37 01	3.78 01	4.70 01	5.61 01
87	Fr L$_{\alpha_1}$	12.031	103.054	2.86 01	3.29 01	3.68 01	4.58 01	5.47 01
81	Tl L$_{\beta_1}$	12.213	101.519	2.74 01	3.15 01	3.53 01	4.39 01	5.24 01
88	Ra L$_{\alpha_1}$	12.340	100.474	2.66 01	3.05 01	3.42 01	4.26 01	5.09 01
34	Se K$_{\beta_1}$	12.496	99.219	2.56 01	2.94 01	3.30 01	4.11 01	4.91 01
82	Pb L$_{\beta_1}$	12.614	98.291	2.49 01	2.87 01	3.21 01	4.00 01	4.78 01
36	Kr K$_{\alpha_1}$	12.649	98.019	2.47 01	2.84 01	3.19 01	3.97 01	4.74 01
89	Ac L$_{\alpha_1}$	12.652	97.996	2.47 01	2.84 01	3.18 01	3.97 01	4.74 01
90	Th L$_{\alpha_1}$	12.969	95.601	2.30 01	2.64 01	2.96 01	3.69 01	4.41 01
83	Bi L$_{\beta_1}$	13.024	95.197	2.27 01	2.61 01	2.93 01	3.65 01	4.36 01
91	Pa L$_{\alpha_1}$	13.291	93.285	2.14 01	2.46 01	2.76 01	3.44 01	4.12 01
35	Br K$_{\beta_1}$	13.291	93.285	2.14 01	2.46 01	2.76 01	3.44 01	4.12 01
37	Rb K$_{\alpha_1}$	13.395	92.560	2.09 01	2.41 01	2.70 01	3.37 01	4.03 01
84	Po L$_{\beta_1}$	13.447	92.202	2.07 01	2.38 01	2.67 01	3.33 01	3.98 01
92	U L$_{\alpha_1}$	13.615	91.065	2.00 01	2.30 01	2.57 01	3.21 01	3.84 01
85	At L$_{\beta_1}$	13.876	89.352	1.89 01	2.17 01	2.44 01	3.04 01	3.64 01
36	Kr K$_{\beta_1}$	14.112	87.857	1.80 01	2.07 01	2.32 01	2.90 01	3.47 01
38	Sr K$_{\alpha_1}$	14.165	87.529	1.78 01	2.05 01	2.30 01	2.87 01	3.43 01
86	Rn L$_{\beta_1}$	14.316	86.606	1.73 01	1.98 01	2.23 01	2.78 01	3.33 01
87	Fr L$_{\beta_1}$	14.770	63.943	1.58 01	1.81 01	2.03 01	2.54 01	3.05 01
39	Y K$_{\alpha_1}$	14.958	82.888	1.52 01	1.75 01	1.96 01	2.45 01	2.94 01
37	Rb K$_{\beta_1}$	14.961	82.872	1.52 01	1.75 01	1.96 01	2.45 01	2.94 01
88	Ra L$_{\beta_1}$	15.236	81.376	1.44 01	1.66 01	1.86 01	2.33 01	2.79 01
89	Ac L$_{\beta_1}$	15.713	78.906	1.32 01	1.51 01	1.70 01	2.13 01	2.55 01
40	Zr K$_{\alpha_1}$	15.775	78.596	1.30 01	1.50 01	1.68 01	2.10 01	2.52 01
38	Sr K$_{\beta_1}$	15.836	78.293	1.29 01	1.48 01	1.66 01	2.08 01	2.49 01
90	Th L$_{\beta_1}$	16.202	76.524	1.20 01	1.39 01	1.55 01	1.95 01	2.33 01
41	Nb K$_{\alpha_1}$	16.615	74.622	1.12 01	1.29 01	1.44 01	1.81 01	2.17 01
91	Pa L$_{\beta_1}$	16.702	74.233	1.10 01	1.27 01	1.42 01	1.78 01	2.14 01
39	Y K$_{\beta_1}$	16.738	74.074	1.09 01	1.26 01	1.41 01	1.77 01	2.12 01
92	U L$_{\beta_1}$	17.220	72.000	1.01 01	1.16 01	1.30 01	1.63 01	1.95 01
42	Mo K$_{\alpha_1}$	17.479	70.933	9.65 00	1.11 01	1.25 01	1.57 01	1.87 01
40	Zr K$_{\beta_1}$	17.668	70.175	9.35 00	1.08 01	1.21 01	1.52 01	1.81 01
43	Tc K$_{\alpha_1}$	18.367	67.504	8.35 00	9.62 00	1.08 01	1.36 01	1.62 01
41	Nb K$_{\beta_1}$	18.623	66.576	8.02 00	9.24 00	1.04 01	1.30 01	1.56 01
44	Ru K$_{\alpha_1}$	19.279	64.311	7.25 00	8.36 00	9.36 00	1.18 01	1.41 01
42	Mo K$_{\beta_1}$	19.608	63.231	6.90 00	7.95 00	8.91 00	1.12 01	1.34 01
45	Rh K$_{\alpha_1}$	20.216	61.330	6.32 00	7.28 00	8.16 00	1.03 01	1.23 01

Z	transition	E [keV]	λ [pm]	$_{16}$S	$_{17}$Cl	$_{18}$Ar	$_{19}$K	$_{20}$Ca
43	Tc K$_{\beta_1}$	20.619	60.131	5.98 00	6.88 00	7.71 00	9.72 00	1.16 01
46	Pd K$_{\alpha_1}$	21.177	58.547	5.54 00	6.38 00	7.14 00	9.01 00	1.06 01
44	Ru K$_{\beta_1}$	21.657	57.249	5.20 00	5.98 00	6.69 00	8.45 00	1.01 01
47	Ag K$_{\alpha_1}$	22.163	55.942	4.87 00	5.60 00	6.26 00	7.91 00	9.43 00
45	Rh K$_{\beta_1}$	22.724	54.561	4.54 00	5.22 00	5.83 00	7.36 00	8.78 00
48	Cd K$_{\alpha_1}$	23.174	53.501	4.29 00	4.93 00	5.51 00	6.96 00	8.30 00
46	Pd K$_{\beta_1}$	23.819	52.053	3.97 00	4.56 00	5.09 00	6.44 00	7.67 00
49	In K$_{\alpha_1}$	24.210	51.212	3.79 00	4.35 00	4.86 00	6.14 00	7.32 00
47	Ag K$_{\beta_1}$	24.942	49.709	3.49 00	4.00 00	4.46 00	5.64 00	6.72 00
50	Sn K$_{\alpha_1}$	25.271	49.062	3.36 00	3.85 00	4.30 00	5.43 00	6.47 00
48	Cd K$_{\beta_1}$	26.096	47.511	3.07 00	3.52 00	3.92 00	4.96 00	5.90 00
51	Sb K$_{\alpha_1}$	26.359	47.037	2.98 00	3.42 00	3.81 00	4.82 00	5.73 09
49	In K$_{\beta_1}$	27.276	45.455	2.71 00	3.10 00	3.45 00	4.37 00	5.20 00
52	Te K$_{\alpha_1}$	27.472	45.131	2.65 00	3.04 00	3.38 00	4.28 00	5.09 00
50	Sn K$_{\beta_1}$	28.486	43.524	2.39 00	2.74 00	3.04 00	3.86 00	4.59 00
53	I K$_{\alpha_1}$	28.612	43.333	2.36 00	2.70 00	3.01 00	3.81 00	4.53 00
51	Sb K$_{\beta_1}$	29.726	41.709	2.12 00	2.42 00	2.69 00	3.41 00	4.06 00
54	Xe K$_{\alpha_1}$	29.779	41.635	2.11 00	2.41 00	2.68 00	3.40 00	4.04 00
55	Cs K$_{\alpha_1}$	30.973	40.030	1.90 00	2.17 00	2.41 00	3.05 00	3.62 00
52	Te K$_{\beta_1}$	30.996	40.000	1.90 00	2.16 00	2.40 00	3.04 00	3.61 00
56	Ba K$_{\alpha_1}$	32.194	38.511	1.72 00	1.96 00	2.17 00	2.74 00	3.25 00
53	I K$_{\beta_1}$	32.295	38.391	1.70 00	1.94 00	2.15 00	2.72 00	3.23 00
57	La K$_{\alpha_1}$	33.442	37.074	1.55 00	1.77 00	1.96 00	2.47 00	2.93 00
54	Xe K$_{\beta_1}$	33.624	36.874	1.53 00	1.74 00	1.93 00	2.43 00	2.89 00
58	Ce K$_{\alpha_1}$	34.279	36.169	1.46 00	1.66 00	1.83 00	2.31 00	2.74 00
55	Cs K$_{\beta_1}$	34.987	35.437	1.38 00	1.57 00	1.73 00	2.18 00	2.59 00
59	Pr K$_{\alpha_1}$	36.026	34.415	1.28 00	1.45 00	1.60 00	2.01 00	2.39 00
56	Ba K$_{\beta_1}$	36.378	34.082	1.25 00	1.41 00	1.56 00	1.96 00	2.32 00
60	Nd K$_{\alpha_1}$	36.847	33.648	1.21 00	1.37 00	1.50 00	1.89 00	2.24 00
57	La K$_{\beta_1}$	37.801	32.799	1.13 00	1.28 00	1.40 00	1.77 00	2.09 00
61	Pm K$_{\alpha_1}$	38.725	32.016	1.06 00	1.20 00	1.32 00	1.65 00	1.96 00
58	Ce K$_{\beta_1}$	39.257	31.582	1.02 00	1.15 00	1.27 00	1.59 00	1.88 00
62	Sm K$_{\alpha_1}$	40.118	30.905	9.66 -1	1.09 00	1.20 00	1.50 00	1.77 00
59	Pr K$_{\beta_1}$	40.748	30.427	9.32 -1	1.05 00	1.15 00	1.44 00	1.70 00
63	Eu K$_{\alpha_1}$	41.542	29.845	8.91 -1	1.00 00	1.10 00	1.37 00	1.62 00
60	Nd K$_{\beta_1}$	42.271	29.331	8.56 -1	9.61 -1	1.05 00	1.31 00	1.55 00
64	Gd K$_{\alpha_1}$	42.996	28.836	8.23 -1	9.23 -1	1.01 00	1.26 00	1.48 00
61	Pm K$_{\beta_1}$	43.826	28.290	7.87 -1	8.81 -1	9.60 -1	1.20 00	1.41 00
65	Tb K$_{\alpha_1}$	44.482	27.873	7.61 -1	8.50 -1	9.25 -1	1.15 00	1.36 00
62	Sm K$_{\beta_1}$	45.413	27.301	7.25 -1	8.09 -1	8.79 -1	1.10 00	1.28 00
66	Dy K$_{\alpha_1}$	45.998	26.954	7.04 -1	7.85 -1	8.52 -1	1.06 00	1.24 00
63	Eu K$_{\beta_1}$	47.038	26.358	6.68 -1	7.44 -1	8.06 -1	1.00 00	1.17 00
67	Ho K$_{\alpha_1}$	47.547	26.076	6.52 -1	7.25 -1	7.84 -1	9.75 -1	1.07 00
64	Gd K$_{\beta_1}$	48.697	25.460	6.17 -1	6.84 -1	7.39 -1	9.17 -1	1.07 00

Z	transition	E [keV]	λ [pm]	$_{16}$S	$_{17}$Cl	$_{18}$Ar	$_{19}$K	$_{20}$Ca
68	Er K$_{\alpha_1}$	49.128	25.237	6.04 -1	6.70 -1	7.23 -1	8.97 -1	1.05 00
65	Tb K$_{\beta_1}$	50.382	24.609	5.72 -1	6.32 -1	6.81 -1	8.43 -1	9.82 -1
69	Tm K$_{\alpha_1}$	50.742	24.434	5.64 -1	6.23 -1	6.70 -1	8.29 -1	9.65 -1
66	Dy K$_{\beta_1}$	52.119	23.788	5.34 -1	5.89 -1	6.32 -1	7.80 -1	9.06 -1
70	Yb K$_{\alpha_1}$	52.389	23.666	5.29 -1	5.82 -1	6.24 -1	7.71 -1	8.95 -1
67	Ho K$_{\beta_1}$	53.877	23.012	5.00 -1	5.49 -1	5.87 -1	7.23 -1	8.36 -1
71	Lu K$_{\alpha_1}$	54.070	22.930	4.97 -1	5.45 -1	5.82 -1	7.17 -1	8.31 -1
68	Er K$_{\beta_1}$	55.681	22.267	4.68 -1	5.13 -1	5.46 -1	6.71 -1	7.75 -1
72	Hf K$_{\alpha_1}$	55.790	22.223	4.67 -1	5.11 -1	5.44 -1	6.68 -1	7.71 -1
69	Tm K$_{\beta_1}$	57.517	21.556	4.39 -1	4.79 -1	5.08 -1	6.23 -1	7.18 -1
73	Ta K$_{\alpha_1}$	57.532	21.550	4.39 -1	4.79 -1	5.08 -1	6.22 -1	7.17 -1
74	W K$_{\alpha_1}$	59.318	20.901	4.13 -1	4.49 -1	4.75 -1	5.80 -1	6.67 -1
70	Yb K$_{\beta_1}$	59.370	20.883	4.12 -1	4.48 -1	4.74 -1	5.79 -1	6.66 -1
75	Re K$_{\alpha_1}$	61.140	20.278	3.92 -1	4.25 -1	4.48 -1	5.45 -1	6.25 -1
71	Lu K$_{\beta_1}$	61.283	20.231	3.91 -1	4.23 -1	4.46 -1	5.43 -1	6.22 -1
76	Os K$_{\alpha_1}$	63.001	19.679	3.75 -1	4.04 -1	4.24 -1	5.15 -1	5.89 -1
72	Hf K$_{\beta_1}$	63.234	19.607	3.72 -1	4.01 -1	4.21 -1	5.11 -1	5.84 -1
77	Ir K$_{\alpha_1}$	64.896	19.195	3.58 -1	3.84 -1	4.02 -1	4.87 -1	5.54 -1
73	Ta K$_{\beta_1}$	65.223	19.009	3.55 -1	3.81 -1	3.98 -1	4.82 -1	5.49 -1
78	Pt K$_{\alpha_1}$	66.832	18.551	3.42 -1	3.66 -1	3.81 -1	4.60 -1	5.23 -1
74	W K$_{\beta_1}$	67.244	18.438	3.39 -1	3.62 -1	3.77 -1	4.55 -1	5.16 -1
79	Au K$_{\alpha_1}$	68.804	18.020	3.27 -1	3.48 -1	3.62 -1	4.35 -1	4.93 -1
75	Re K$_{\beta_1}$	69.310	17.888	3.23 -1	3.44 -1	3.57 -1	4.29 -1	4.86 -1
80	Hg K$_{\alpha_1}$	70.819	17.507	3.13 -1	3.32 -1	3.44 -1	4.12 -1	4.65 -1
76	Os K$_{\beta_1}$	71.413	17.361	3.09 -1	3.27 -1	3.38 -1	4.05 -1	4.57 -1
81	Tl K$_{\alpha_1}$	72.872	17.014	2.99 -1	3.16 -1	3.26 -1	3.90 -1	4.39 -1
77	Ir K$_{\beta_1}$	73.561	16.854	2.95 -1	3.11 -1	3.21 -1	3.83 -1	4.31 -1
82	Pb K$_{\alpha_1}$	74.969	16.538	2.86 -1	3.01 -1	3.10 -1	3.70 -1	4.15 -1
78	Pt K$_{\beta_1}$	75.748	16.368	2.82 -1	2.96 -1	3.04 -1	3.62 -1	4.06 -1
83	Bi K$_{\alpha_1}$	77.108	16.079	2.74 -1	2.87 -1	2.95 -1	3.50 -1	3.92 -1
79	Au K$_{\beta_1}$	77.948	15.906	2.70 -1	2.82 -1	2.89 -1	3.43 -1	3.83 -1
84	Po K$_{\alpha_1}$	79.290	15.636	2.63 -1	2.74 -1	2.80 -1	3.32 -1	3.70 -1
80	Hg K$_{\beta_1}$	80.253	15.449	2.58 -1	2.69 -1	2.75 -1	3.25 -1	3.62 -1
85	At K$_{\alpha_1}$	81.520	15.209	2.54 -1	2.64 -1	2.69 -1	3.18 -1	3.53 -1
81	Tl K$_{\beta_1}$	82.576	15.014	2.50 -1	2.60 -1	2.65 -1	3.12 -1	3.46 -1
86	Rn K$_{\alpha_1}$	83.780	14.798	2.46 -1	2.55 -1	2.60 -1	3.05 -1	3.39 -1
82	Pb K$_{\beta_1}$	84.936	14.597	2.43 -1	2.51 -1	2.55 -1	2.99 -1	3.31 -1
87	Fr K$_{\alpha_1}$	86.100	14.400	2.39 -1	2.47 -1	2.50 -1	2.93 -1	3.24 -1
83	Bi K$_{\beta_1}$	87.343	14.195	2.35 -1	2.43 -1	2.46 -1	2.87 -1	3.17 -1
88	Ra K$_{\alpha_1}$	88.470	14.014	2.32 -1	2.39 -1	2.41 -1	2.82 -1	3.11 -1
84	Po K$_{\beta_1}$	89.800	13.806	2.28 -1	2.35 -1	2.37 -1	2.76 -1	3.04 -1
89	Ac K$_{\alpha_1}$	90.884	13.642	2.25 -1	2.31 -1	2.33 -1	2.71 -1	2.98 -1
85	At K$_{\beta_1}$	92.300	13.432	2.21 -1	2.27 -1	2.28 -1	2.65 -1	2.91 -1
90	Th K$_{\alpha_1}$	93.350	13.281	2.19 -1	2.24 -1	2.25 -1	2.61 -1	2.86 -1

Z	transition	E [keV]	λ [pm]	$_{16}$S	$_{17}$Cl	$_{18}$Ar	$_{19}$K	$_{20}$Ca
86	Rn K$_{\beta_1}$	94.870	13.068	2.15 -1	2.19 -1	2.20 -1	2.54 -1	2.79 -1
91	Pa K$_{\alpha_1}$	95.868	12.932	2.12 -1	2.16 -1	2.17 -1	2.51 -1	2.74 -1
87	Fr K$_{\beta_1}$	97.470	12.720	2.09 -1	2.12 -1	2.12 -1	2.45 -1	2.67 -1
92	U K$_{\alpha_1}$	98.439	12.595	2.06 -1	2.10 -1	2.09 -1	2.41 -1	2.63 -1
88	Ra K$_{\beta_1}$	100.130	12.382	2.03 -1	2.05 -1	2.05 -1	2.35 -1	2.57 -1
89	Ac K$_{\beta_1}$	102.850	12.054	1.99 -1	2.01 -1	2.00 -1	2.29 -1	2.50 -1
90	Th K$_{\beta_1}$	105.610	11.739	1.95 -1	1.97 -1	1.95 -1	2.23 -1	2.41 -1
91	Pa K$_{\beta_1}$	108.430	11.434	1.91 -1	1.93 -1	1.91 -1	2.18 -1	2.36 -1
92	U K$_{\beta_1}$	111.300	11.139	1.88 -1	1.89 -1	1.86 -1	2.12 -1	2.30 -1

Z	transition	E [keV]	λ [pm]	$_{21}$Sc	$_{22}$Ti	$_{23}$V	$_{24}$Cr	$_{25}$Mn
4	Be K$_\alpha$	0.1085	11427.207	2.50 04	2.75 04	3.07 04	4.02 04	4.18 04
38	Sr M$_\xi$	0.1140	10875.895	2.39 04	2.63 04	2.93 04	3.81 04	3.97 04
39	Y M$_\xi$	0.1328	9339.235	2.07 04	2.25 04	2.48 04	3.17 04	3.35 04
16	S L$_l$	0.1487	8337.342	1.86 04	2.00 04	2.20 04	2.77 04	2.95 04
40	Zr M$_\xi$	0.1511	8205.506	1.83 04	1.97 04	2.16 04	2.73 04	2.90 04
41	Nb M$_\xi$	0.1717	7221.037	1.54 04	1.66 04	1.82 04	2.26 04	2.42 04
5	B K$_\alpha$	0.1833	6764.059	1.42 04	1.53 04	1.67 04	2.07 04	2.22 04
42	Mo M$_\xi$	0.1926	6437.447	1.32 04	1.43 04	1.56 04	1.92 04	2.06 04
6	C K$_\alpha$	0.2770	4476.000	7.46 03	8.09 03	8.84 03	1.06 04	1.15 04
47	Ag M$_\xi$	0.3117	3977.709	6.03 03	6.57 03	7.18 03	8.54 03	9.29 03
7	N K$_\alpha$	0.3924	3159.664	3.99 03	4.36 03	4.79 03	5.63 03	6.16 03
22	Ti L$_l$	0.3953	3136.484	3.93 03	4.30 03	4.72 03	5.55 03	6.08 03
22	Ti L$_\alpha$	0.4522	2741.822	2.84 04	3.32 03	3.65 03	4.26 03	4.67 03
23	V L$_\alpha$	0.5113	2424.901	2.44 04	2.34 04	2.88 03	3.32 03	3.66 03
8	O K$_\alpha$	0.5249	2362.072	2.36 04	2.21 04	2.43 04	3.14 03	3.47 03
25	Mn L$_l$	0.5563	2228.747	2.08 04	1.95 04	2.13 04	2.78 03	3.08 03
24	Cr L$_\alpha$	0.5728	2164.546	1.95 04	2.12 04	1.99 04	2.61 03	2.90 03
25	Mn L$_\alpha$	0.6374	1945.171	1.54 04	1.67 04	1.82 04	1.87 04	2.32 03
9	F K$_\alpha$	0.6768	1831.932	1.33 04	1.45 04	1.58 04	1.59 04	1.61 04
26	Fe L$_\alpha$	0.7050	1758.655	1.22 04	1.34 04	1.45 04	1.67 04	1.61 04
27	Co L$_\alpha$	0.7762	1597.335	9.72 03	1.07 04	1.16 04	1.34 04	1.43 04
28	Ni L$_\alpha$	0.8515	1456.080	7.75 03	8.56 03	9.37 03	1.08 04	1.16 04
29	Cu L$_\alpha$	0.9297	1336.044	6.25 03	6.93 03	7.62 03	8.84 03	9.47 03
30	Zn L$_\alpha$	1.0117	1225.513	5.38 03	6.23 03	6.95 03	8.00 03	9.04 03
11	Na K$_\alpha$	1.0410	1191.020	4.98 03	5.77 03	6.44 03	7.41 03	8.37 03
11	Na K$_\beta$	1.0711	1157.550	4.62 03	5.34 03	5.96 03	6.87 03	7.75 03
12	Mg K$_\alpha$	1.2536	989.033	3.03 03	3.50 03	3.90 03	4.50 03	5.07 03
33	As L$_\alpha$	1.2820	967.123	2.86 03	3.30 03	3.67 03	4.24 03	4.77 03
12	Mg K$_\beta$	1.3022	952.121	2.7403	3.16 03	3.52 03	4.06 03	4.57 03
33	As L$_{\beta_1}$	1.3170	941.421	2.66 03	3.07 03	3.41 03	3.94 03	4.44 03
66	Dy M$_\beta$	1.3250	935.737	2.61 03	3.02 03	3.35 03	3.88 03	4.37 03

Z	transition	E [keV]	λ [pm]	$_{21}$Sc	$_{22}$Ti	$_{23}$V	$_{24}$Cr	$_{25}$Mn
67	Ho M_α	1.3480	919.771	2.50 03	2.88 03	3.20 03	3.70 03	4.17 03
34	Se L_α	1.3791	899.029	2.35 03	2.71 03	3.01 03	3.48 03	3.92 03
67	Ho M_β	1.3830	896.494	2.33 03	2.69 03	2.99 03	3.46 03	3.89 03
68	Er M_α	1.4060	881.829	2.23 03	2.57 03	2.86 03	3.31 03	3.72 03
34	Se L_{β_1}	1.4192	873.627	2.18 03	2.51 03	2.79 03	3.22 03	3.63 03
68	Er M_β	1.4430	859.218	2.08 03	2.40 03	2.66 03	3.08 03	3.47 03
69	Tm M_α	1.4620	848.051	2.01 03	2.32 03	2.57 03	2.98 03	3.35 03
35	Br L_α	1.4804	837.511	1.94 03	2.24 03	2.49 03	2.86 03	3.24 03
13	Al K_{α_1}	1.4867	833.962	1.92 03	2.21 03	2.46 03	2.85 03	3.20 03
69	Tm M_β	1.5030	824.918	1.87 03	2.15 03	2.39 03	2.76 03	3.11 03
70	Yb M_α	1.5214	814.941	1.81 03	2.08 03	2.31 03	2.68 03	3.01 03
35	Br L_{β_1}	1.5259	812.538	1.79 03	2.06 03	2.29 03	2.65 03	2.98 03
13	Al K_β	1.5574	796.103	1.70 03	1.95 03	2.17 03	2.51 03	2.82 03
70	Yb M_β	1.5675	790.974	1.67 03	1.92 03	2.13 03	2.47 03	2.77 03
71	Lu M_α	1.5813	784.071	1.63 03	1.88 03	2.08 03	2.41 03	2.71 03
36	Kr L_{β_1}	1.5860	781.747	1.62 03	1.86 03	2.06 03	2.39 03	2.69 03
71	Lu M_β	1.6312	760.085	1.50 03	1.73 03	1.91 03	2.22 03	2.49 03
36	Kr L_{β_1}	1.6366	757.577	1.49 03	1.71 03	1.90 03	2.20 03	2.47 03
72	Hf M_{α_1}	1.6446	753.892	1.47 03	1.69 03	1.87 03	2.17 03	2.44 03
37	Rb L_{α_1}	1.6941	731.864	1.36 03	1.56 03	1.73 03	2.01 03	2.25 03
72	Hf M_β	1.6976	730.355	1.35 03	1.55 03	1.72 03	1.99 03	2.24 03
73	Ta M_{α_1}	1.7096	725.229	1.32 03	1.52 03	1.69 03	1.96 03	2.19 03
14	Si K_{α_1}	1.7400	712.558	1.26 03	1.45 03	1.61 03	1.87 03	2.09 03
37	Rb L_{β_1}	1.7522	707.597	1.24 03	1.42 03	1.58 03	1.83 03	2.05 03
73	Ta M_β	1.7655	702.266	1.21 03	1.39 03	1.54 03	1.80 03	2.01 03
74	W M_{α_1}	1.7754	698.350	1.20 03	1.37 03	1.52 03	1.77 03	1.98 03
38	Sr L_{α_1}	1.8066	686.290	1.14 03	1.31 03	1.45 03	1.69 03	1.81 03
74	W M_β	1.8349	675.705	1.10 03	1.26 03	1.39 03	1.62 03	1.81 03
14	Si K_β	1.8359	675.337	1.09 03	1.26 03	1.39 03	1.62 03	1.81 03
75	Re M_{α_1}	1.8420	673.100	1.08 03	1.24 03	1.38 03	1.60 03	1.79 03
38	Sr L_{β_1}	1.8717	662.420	1.04 03	1.19 03	1.32 03	1.54 03	1.72 03
75	Re M_β	1.9061	650.465	9.90 02	1.13 03	1.26 03	1.46 03	1.63 03
76	Os M_{α_1}	1.9102	649.069	9.85 02	1.13 03	1.25 03	1.45 03	1.63 03
39	Y L_{α_1}	1.9226	644.882	9.68 02	1.11 03	1.23 03	1.43 03	1.60 03
76	Os M_β	1.9783	626.725	8.97 02	1.03 03	1.14 03	1.32 03	1.48 03
77	Ir M_{α_1}	1.9799	626.219	8.95 02	1.02 03	1.13 03	1.32 03	1.48 03
39	Y L_{β_1}	1.9958	621.230	8.76 02	1.00 03	1.11 03	1.29 03	1.44 03
15	P K_{α_1}	2.0137	615.708	8.55 02	9.79 02	1.08 03	1.26 03	1.41 03
40	Zr L_{α_1}	2.0424	607.056	8.24 02	9.42 02	1.04 03	1.22 03	1.36 03
78	Pt M_{α_1}	2.0505	604.658	8.15 02	9.32 02	1.03 03	1.20 03	1.34 03
77	Ir M_β	2.0535	603.775	8.12 02	9.29 02	1.03 03	1.20 03	1.34 03
79	Au M_{α_1}	2.1229	584.036	7.43 02	8.50 02	9.39 02	1.10 03	1.22 03
40	Zr L_{β_1}	2.1244	583.624	7.42 02	8.48 02	9.37 02	1.09 03	1.22 03
78	Pt M_β	2.1273	582.828	7.39 02	8.45 02	9.34 02	1.09 03	1.22 03

Z	transition	E [keV]	λ [pm]	$_{21}$Sc	$_{22}$Ti	$_{23}$V	$_{24}$Cr	$_{25}$Mn
15	P $K_{\beta_{1,3}}$	2.1390	579.640	7.28 02	8.33 02	9.20 02	1.07 03	1.20 03
41	Nb L_{α_1}	2.1659	572.441	7.05 02	8.05 02	8.90 02	1.04 03	1.16 03
80	Hg M_{α_1}	2.1953	564.775	6.80 02	7.77 02	8.58 02	1.00 03	1.12 03
79	Au M_β	2.2046	562.393	6.72 02	7.68 02	8.48 02	9.90 02	1.11 03
41	Nb L_{β_1}	2.2574	549.238	6.31 02	7.21 02	7.96 02	9.29 02	1.04 03
81	Tl M_{α_1}	2.2706	546.045	6.21 02	7.10 02	7.84 02	9.14 02	1.02 03
80	Hg M_β	2.2825	543.199	6.13 02	7.00 02	7.73 02	9.02 02	1.01 03
42	Mo L_{α_1}	2.2932	540.664	6.05 02	6.91 02	7.63 02	8.90 02	9.94 02
16	S K_{α_1}	2.3080	537.197	6.95 02	6.79 02	7.50 02	8.75 02	9.77 02
82	Pb M_{α_1}	2.3457	528.563	5.70 02	6.51 02	7.18 02	8.38 02	9.36 02
81	Tl M_β	2.3621	524.893	5.59 02	6.39 02	7.04 02	8.23 02	9.18 02
42	Mo L_{β_1}	2.3948	517.726	5.39 02	6.16 02	6.79 02	7.93 02	8.85 02
83	Bi M_{α_1}	2.4226	511.785	5.23 02	5.97 02	6.58 02	7.69 02	8.58 02
43	Tc L_{α_1}	2.4240	511.490	5.22 02	5.96 02	6.57 02	7.67 02	8.57 02
82	Pb M_β	2.4427	507.574	5.12 02	5.84 02	6.44 02	7.52 02	8.39 02
16	S K_β	2.4640	503.186	5.00 02	5.70 02	6.29 02	7.34 02	8.20 02
83	Bi M_{β_1}	2.5255	490.933	4.68 02	5.34 02	5.88 02	6.87 02	7.67 02
43	Tc L_{β_1}	2.5368	488.746	4.63 02	5.28 02	5.81 02	6.79 02	7.58 02
44	Ru L_{α_1}	2.5586	484.582	4.52 02	5.16 02	5.68 02	6.64 02	7.41 02
17	Cl K_{α_1}	2.6224	472.792	4.24 02	4.83 02	5.32 02	6.21 02	6.94 02
44	Ru L_{β_1}	2.6832	462.079	3.99 02	4.54 02	5.00 02	5.84 02	6.52 02
45	Rh L_{α_1}	2.6967	459.766	3.93 02	4.48 02	4.93 02	5.76 02	6.43 02
17	Cl K_β	2.8156	440.350	3.51 02	3.99 02	4.39 02	5.13 02	5.73 02
45	Rh L_{β_1}	2.8344	437.430	3.44 02	3.92 02	4.31 02	5.04 02	5.63 02
46	Pd L_{α_1}	2.8386	436.782	3.43 02	3.91 02	4.29 02	5.02 02	5.61 02
18	Ar K_{α_1}	2.9577	419.194	3.08 02	3.50 02	3.84 02	4.50 02	5.02 02
47	Ag L_{α_1}	2.9843	415.458	3.00 02	3.42 02	3.75 02	4.39 02	4.90 02
46	Pd L_{β_1}	2.9902	414.638	2.99 02	3.40 02	3.73 02	4.37 02	4.88 02
90	Th M_{α_1}	2.9961	413.821	2.97 02	3.38 02	3.71 02	4.35 02	4.85 02
91	Pa M_{α_1}	3.0823	402.248	2.76 02	3.13 02	3.44 02	4.03 02	4.50 02
48	Cd L_{α_1}	3.1337	395.651	2.64 02	3.00 02	3.29 02	3.86 02	4.30 02
90	Th M_β	3.1458	394.129	2.61 02	2.97 02	3.26 02	3.82 02	4.26 02
47	Ag L_{β_1}	3.1509	393.491	2.60 02	2.95 02	3.24 02	3.80 02	4.24 02
92	U M_{α_1}	3.1708	391.021	2.56 02	2.90 02	3.19 02	3.74 02	4.17 02
18	Ar K_β	3.1905	388.607	2.52 02	2.86 02	3.14 02	3.68 02	4.10 02
91	Pa M_β	3.2397	382.705	2.41 02	2.74 02	3.01 02	3.53 02	3.93 02
49	In L_{α_1}	3.2869	377.210	2.32 02	2.64 02	2.90 02	3.40 02	3.78 02
19	K K_{α_1}	3.3138	374.148	2.27 02	2.58 02	2.84 02	3.32 02	3.70 02
48	Cd L_{β_1}	3.3160	373.832	2.27 02	2.57 02	2.83 02	3.32 02	3.69 02
92	U M_β	3.3360	371.658	2.23 02	2.53 02	2.79 02	3.27 02	3.64 02
50	Sn L_{α_1}	3.4440	360.003	2.05 02	2.33 02	2.56 02	3.00 02	3.34 02
49	In L_{β_1}	3.4872	355.543	1.99 02	2.25 02	2.47 02	2.90 02	3.23 02
19	K K_β	3.5896	345.401	1.84 02	2.08 02	2.29 02	2.69 02	2.99 02
51	Sb L_{α_1}	3.6047	343.954	1.82 02	2.06 02	2.26 02	2.66 02	2.95 02

Z	transition	E [keV]	λ [pm]	$_{21}$Sc	$_{22}$Ti	$_{23}$V	$_{24}$Cr	$_{25}$Mn
50	Sn L_{β_1}	3.6628	338.498	1.74 02	1.97 02	2.17 02	2.55 02	2.83 02
20	Ca K_{α_1}	3.6917	335.848	1.71 02	1.93 02	2.12 02	2.49 02	2.77 02
52	Te L_{α_1}	3.7693	328.934	1.62 02	1.83 02	2.01 02	2.36 02	2.62 02
51	Sb L_{β_1}	3.8436	322.575	1.53 02	1.73 02	1.91 02	2.24 02	2.49 02
53	I L_{α_1}	3.9377	314.867	1.44 02	1.63 02	1.79 02	2.10 02	2.33 02
20	Ca K_β	4.0127	308.981	1.37 02	1.55 02	1.70 02	2.00 02	2.22 02
52	Te L_{β_1}	4.0296	307.686	1.35 02	1.53 02	1.68 02	1.98 02	2.19 02
21	Sc K_{α_1}	4.0906	303.097	1.30 02	1.47 02	1.61 02	1.90 02	2.11 02
54	Xe L_{α_1}	4.1099	301.674	1.29 02	1.45 02	1.59 02	1.87 02	2.08 02
53	I L_{β_1}	4.2207	293.755	1.20 02	1.35 02	1.48 02	1.75 02	1.94 02
55	Cs L_{α_1}	4.2865	289.245	1.15 02	1.30 02	1.42 02	1.68 02	1.86 02
21	Sc K_β	4.4605	277.962	1.04 02	1.17 02	1.28 02	1.51 02	1.68 02
56	Ba L_{α_1}	4.4663	277.601	<u>1.03 02</u>	1.16 02	1.28 02	1.50 02	1.67 02
22	Ti K_{α_1}	4.5108	274.863	8.71 02	1.13 02	1.24 02	1.46 02	1.63 02
55	Cs L_{β_1}	4.6198	268.377	8.18 02	1.06 02	1.17 02	1.37 02	1.53 02
57	La L_{α_1}	4.6510	266.577	8.03 02	1.04 02	1.14 02	1.35 02	1.50 02
56	Ba L_{β_1}	4.8273	256.841	7.28 02	9.46 01	1.04 02	1.22 02	1.36 02
58	Ce L_{α_1}	4.8402	256.157	7.23 02	9.39 01	1.03 02	1.21 02	1.35 02
22	Ti $K_{\beta_{1,3}}$	4.9318	251.399	6.88 02	8.94 01	9.79 01	1.15 02	1.28 02
23	V K_{α_1}	4.9522	250.363	6.80 02	<u>8.84 01</u>	9.68 01	1.14 02	1.27 02
59	Pr L_{α_1}	5.0337	246.310	6.51 02	7.05 02	9.27 01	1.09 02	1.22 02
57	La L_{β_1}	5.0421	245.899	6.49 02	7.02 02	9.22 01	1.09 02	1.21 02
60	Nd L_{α_1}	5.2304	237.047	5.88 02	6.38 02	8.37 01	9.86 01	1.10 02
58	Ce L_{β_1}	5.2622	235.614	5.79 02	6.28 02	8.23 01	9.70 01	1.08 02
24	Cr K_{α_1}	5.4147	228.978	5.36 02	5.83 02	7.63 01	9.00 01	1.00 02
23	V $K_{\beta_{1,3}}$	5.4273	228.447	5.33 02	5.80 02	7.58 01	8.94 01	9.94 01
61	Pm L_{α_1}	5.4325	228.228	5.31 02	5.78 02	<u>7.57 01</u>	8.92 01	9.91 01
59	Pr L_{β_1}	5.4889	225.883	5.17 02	5.63 02	6.24 02	8.68 01	9.64 01
62	Sm L_{α_1}	5.6361	219.984	4.82 02	5.25 02	5.82 02	8.09 01	8.98 01
60	Nd L_{β_1}	5.7216	216.696	4.63 02	5.05 02	5.59 02	7.77 01	8.63 01
63	Eu L_{α_1}	5.8457	212.096	4.37 02	4.78 02	5.28 02	7.35 01	8.14 01
25	Mn K_{α_1}	5.8988	210.187	4.26 02	4.66 02	5.16 02	7.17 01	7.95 01
24	Cr $K_{\beta_{1,3}}$	5.9467	208.494	4.17 02	4.57 02	5.05 02	7.02 01	7.78 01
61	Pm L_{β_1}	5.9610	207.993	4.15 02	4.54 02	5.01 02	<u>6.98 01</u>	7.73 01
64	Gd L_{α_1}	6.0572	204.690	3.97 02	4.35 02	4.80 02	5.55 02	7.41 01
62	Sm L_{β_1}	6.2051	199.811	3.71 02	4.08 02	4.50 02	5.21 02	6.95 01
65	Tb L_{α_1}	6.2128	197.655	3.61 02	3.96 02	4.37 02	5.06 02	6.75 01
26	Fe K_{α_1}	6.4038	193.611	3.41 02	3.75 02	4.14 02	4.79 02	6.39 01
63	Eu L_{β_1}	6.4564	192.034	3.33 02	3.67 02	4.04 02	4.68 02	6.26 01
25	Mn $K_{\beta_{1,3}}$	6.4905	191.025	3.28 02	3.61 02	3.99 02	4.62 02	6.17 01
66	Gd L_{α_1}	6.4952	190.887	3.28 02	3.61 02	3.98 02	4.61 02	<u>6.16 01</u>
64	Gd L_{β_1}	6.7132	184.688	2.99 02	3.30 02	3.64 02	4.22 02	4.56 02
67	Ho L_{α_1}	6.7198	184.507	2.98 02	3.29 02	3.63 02	4.21 02	4.54 02
27	Co K_{α_1}	6.9303	178.903	2.74 02	3.03 02	3.34 02	3.88 02	4.16 02

Z	transition	E [keV]	λ [pm]	$_{21}$Sc	$_{22}$Ti	$_{23}$V	$_{24}$Cr	$_{25}$Mn
68	Er L$_{\alpha_1}$	6.9487	178.429	2.72 02	3.01 02	3.32 02	3.85 02	4.15 02
65	Tb L$_{\beta_1}$	6.9780	177.680	2.69 02	2.98 02	3.28 02	3.80 02	4.11 02
26	Fe K$_{\beta_{1,3}}$	7.0580	175.666	2.61 02	2.89 02	3.18 02	3.69 02	3.98 02
69	Tm L$_{\alpha_1}$	7.1799	172.683	2.49 02	2.76 02	3.04 02	3.53 02	3.81 02
66	Dy L$_{\beta_1}$	7.2477	171.068	2.43 02	2.69 02	2.96 02	3.44 02	3.71 02
70	Yb L$_{\alpha_1}$	7.4156	167.195	2.28 02	2.53 02	2.78 02	3.23 02	3.49 02
28	Ni K$_{\alpha_1}$	7.4782	165.795	2.23 02	2.47 02	2.72 02	3.16 02	3.41 02
67	Ho L$_{\beta_1}$	7.5253	164.757	2.10 02	2.43 02	2.66 02	3.11 02	3.36 02
27	Co K$_{\beta_{1,3}}$	7.6494	162.084	2.09 02	2.33 02	2.56 02	2.98 02	3.21 02
71	Lu L$_{\alpha_1}$	7.6555	161.955	2.09 02	2.32 02	2.56 02	2.97 02	3.21 02
68	Er L$_{\beta_1}$	7.8109	158.733	1.97 02	2.20 02	2.42 02	2.81 02	3.04 02
72	Hf L$_{\alpha_1}$	7.8990	156.963	1.91 02	2.13 02	2.35 02	2.73 02	2.95 02
29	Cu K$_{\alpha_1}$	8.0478	154.060	1.82 02	2.03 02	2.23 02	2.60 02	2.80 02
69	Tm L$_{\beta_1}$	8.1010	153.049	1.79 02	1.99 02	2.19 02	2.55 02	2.75 02
73	Ta L$_{\alpha_1}$	8.1461	152.201	1.76 02	1.96 02	2.16 02	2.51 02	2.71 02
28	Ni K$_{\beta_{1,3}}$	8.2647	150.017	1.69 02	1.89 02	2.08 02	2.42 02	2.61 02
74	W L$_{\alpha_1}$	8.3976	147.643	1.62 02	1.81 02	1.99 02	2.31 02	2.49 02
70	Yb L$_{\beta_1}$	8.4018	147.569	1.61 02	1.80 02	1.98 02	2.31 02	2.49 02
30	Zn K$_{\alpha_1}$	8.6338	143.519	1.49 02	1.67 02	1.84 02	2.14 02	2.31 02
75	Re L$_{\alpha_1}$	8.6525	143.294	1.49 02	1.66 02	1.83 02	2.13 02	2.30 02
71	Lu L$_{\beta_1}$	8.7090	142.364	1.46 02	1.63 02	1.80 02	2.09 02	2.26 02
29	Cu K$_{\beta_1}$	8.9053	139.226	1.37 02	1.54 02	1.69 02	1.97 02	2.12 02
76	Os L$_{\alpha_1}$	8.9117	139.126	1.37 02	1.53 02	1.69 02	1.97 02	2.12 02
72	Hf L$_{\beta_1}$	9.0227	137.414	1.32 02	1.48 02	1.63 02	1.90 02	2.05 02
77	Ir L$_{\alpha_1}$	9.1751	135.132	1.26 02	1.42 02	1.56 02	1.82 02	1.96 02
31	Ga K$_{\alpha_1}$	9.2517	134.013	1.24 02	1.39 02	1.52 02	1.78 02	1.91 02
73	Ta L$_{\beta_1}$	9.3431	132.702	1.20 02	1.35 02	1.48 02	1.73 02	1.86 02
78	Pt L$_{\alpha_1}$	9.4423	131.308	1.17 02	1.31 02	1.44 02	1.68 02	1.81 02
30	Zn K$_{\beta_{1,3}}$	9.5720	129.529	1.12 02	1.26 02	1.39 02	1.62 02	1.74 02
74	W L$_{\beta_1}$	9.6724	128.184	1.09 02	1.23 02	1.35 02	1.57 02	1.69 02
79	Au L$_{\alpha_1}$	9.7133	127.644	1.08 02	1.21 02	1.33 02	1.55 02	1.67 02
32	Ge K$_{\alpha_1}$	9.8864	125.409	1.03 02	1.16 02	1.27 02	1.48 02	1.59 02
80	Hg L$_{\alpha_1}$	9.9888	124.124	9.99 01	1.12 02	1.23 02	1.44 02	1.55 02
75	Re L$_{\beta_1}$	10.010	123.861	9.93 01	1.12 02	1.23 02	1.43 02	1.54 02
31	Ga K$_{\beta_1}$	10.264	120.796	9.25 01	1.04 02	1.14 02	1.33 02	1.44 02
81	Tl L$_{\alpha_1}$	10.269	120.737	9.23 01	1.04 02	1.14 02	1.33 02	1.44 02
76	Os L$_{\beta_1}$	10.355	119.734	9.02 01	1.02 02	1.11 02	1.30 02	1.40 02
33	As K$_{\alpha_1}$	10.544	117.588	8.57 01	9.65 01	1.06 02	1.24 02	1.33 02
82	Pb L$_{\alpha_1}$	10.552	117.499	8.55 01	9.63 01	1.06 02	1.23 02	1.33 02
77	Ir L$_{\beta_1}$	10.708	115.787	8.20 01	9.24 01	1.01 02	1.18 02	1.28 02
83	Bi L$_{\alpha_1}$	10.839	114.388	7.92 01	8.93 01	9.81 01	1.15 02	1.23 02
32	Ge K$_{\beta_1}$	10.982	112.898	7.63 01	8.61 01	9.46 01	1.10 02	1.19 02
78	Pt L$_{\beta_1}$	11.071	111.990	7.46 01	8.41 01	9.24 01	1.08 02	1.16 02
84	Po L$_{\alpha_1}$	11.131	111.387	7.34 01	8.29 01	9.11 01	1.06 02	1.15 02

Z	transition	E [keV]	λ [pm]	$_{21}$Sc	$_{22}$Ti	$_{23}$V	$_{24}$Cr	$_{25}$Mn
34	Se K$_{\alpha_1}$	11.222	110.484	7.18 01	8.10 01	8.90 01	1.04 02	1.12 02
85	At L$_{\alpha_1}$	11.427	108.501	6.82 01	7.70 01	8.46 01	9.87 01	1.07 02
79	Au L$_{\beta_1}$	11.442	108.359	6.79 01	7.67 01	8.43 01	9.84 01	1.06 02
33	As K$_{\beta_1}$	11.726	105.735	6.34 01	7.16 01	7.87 01	9.18 01	9.92 01
86	Rn L$_{\alpha_1}$	11.727	105.726	6.33 01	7.16 01	7.87 01	9.18 01	9.92 01
80	Hg L$_{\beta_1}$	11.823	104.867	6.19 01	6.99 01	7.69 01	8.97 01	9.70 01
35	Br K$_{\alpha_1}$	11.924	103.979	6.04 01	6.83 01	7.51 01	8.76 01	9.47 01
87	Fr L$_{\alpha_1}$	12.031	103.054	5.89 01	6.66 01	7.32 01	8.54 01	9.24 01
81	Tl L$_{\beta_1}$	12.213	101.519	5.64 01	6.38 01	7.02 01	8.19 01	8.86 01
88	Ra L$_{\alpha_1}$	12.340	100.474	5.48 01	6.20 01	6.82 01	7.95 01	8.61 01
34	Se K$_{\beta_1}$	12.496	99.219	5.29 01	5.99 01	6.59 01	7.68 01	8.31 01
82	Pb L$_{\beta_1}$	12.614	98.291	5.15 01	5.83 01	6.41 01	7.48 01	8.10 01
36	Kr K$_{\alpha_1}$	12.649	98.019	5.11 01	5.79 01	6.36 01	7.42 01	8.03 01
89	Ac L$_{\alpha_1}$	12.652	97.996	5.11 01	5.78 01	6.36 01	7.42 01	8.03 01
90	Th L$_{\alpha_1}$	12.969	95.601	4.76 01	5.39 01	5.93 01	6.92 01	7.49 01
83	Bi L$_{\beta_1}$	13.024	95.197	4.70 01	5.33 01	5.86 01	6.84 01	7.41 01
91	Pa L$_{\alpha_1}$	13.291	93.285	4.44 01	5.03 01	5.54 01	6.46 01	7.00 01
35	Br K$_{\beta_1}$	13.291	93.285	4.44 01	5.03 01	5.54 01	6.46 01	7.00 01
37	Rb K$_{\alpha_1}$	13.395	92.560	4.34 01	4.92 01	5.42 01	6.32 01	6.85 01
84	Po L$_{\beta_1}$	13.447	92.202	4.29 01	4.87 01	5.36 01	6.25 01	6.78 01
92	U L$_{\alpha_1}$	13.615	91.065	4.15 01	4.70 01	5.18 01	6.04 01	6.55 01
85	At L$_{\beta_1}$	13.876	89.352	3.93 01	4.46 01	4.91 01	5.72 01	6.21 01
36	Kr K$_{\beta_1}$	14.112	87.857	3.74 01	4.25 01	4.68 01	5.46 01	5.92 01
38	Sr K$_{\alpha_1}$	14.165	87.529	3.71 01	4.21 01	4.63 01	5.40 01	5.86 01
86	Rn L$_{\beta_1}$	14.316	86.606	3.60 01	4.08 01	4.50 01	5.24 01	5.69 01
87	Fr L$_{\beta_1}$	14.770	83.943	3.29 01	3.74 01	4.12 01	4.80 01	5.22 01
39	Y K$_{\alpha_1}$	14.958	82.888	3.17 01	3.61 01	3.98 01	4.63 01	5.04 01
37	Rb K$_{\beta_1}$	14.961	82.872	3.17 01	3.61 01	3.98 01	4.63 01	5.03 01
88	Ra L$_{\beta_1}$	15.236	81.376	3.01 01	3.43 01	3.78 01	4.40 01	4.78 01
89	Ac L$_{\beta_1}$	15.713	78.906	2.76 01	3.14 01	3.46 01	4.03 01	4.38 01
40	Zr K$_{\alpha_1}$	15.775	78.596	2.73 01	3.10 01	3.42 01	3.98 01	4.33 01
38	Sr K$_{\beta_1}$	15.836	78.293	2.70 01	3.07 01	3.38 01	3.94 01	4.29 01
90	Th L$_{\beta_1}$	16.202	76.524	2.52 01	2.87 01	3.17 01	3.69 01	4.02 01
41	Nb K$_{\alpha_1}$	16.615	74.622	2.35 01	2.67 01	2.95 01	3.44 01	3.74 01
91	Pa L$_{\beta_1}$	16.702	74.233	2.31 01	2.63 01	2.91 01	3.39 01	3.69 01
39	Y K$_{\beta_1}$	16.738	74.074	2.30 01	2.62 01	2.89 01	3.36 01	3.66 01
92	U L$_{\beta_1}$	17.220	72.000	2.12 01	2.41 01	2.66 01	3.10 01	3.38 01
42	Mo K$_{\alpha_1}$	17.479	70.933	2.03 01	2.31 01	2.55 01	2.97 01	3.24 01
40	Zr K$_{\beta_1}$	17.668	70.175	1.97 01	2.24 01	2.47 01	2.88 01	3.14 01
43	Tc K$_{\alpha_1}$	18.367	67.504	1.76 01	2.01 01	2.22 01	2.58 01	2.82 01
41	Nb K$_{\beta_1}$	18.623	66.576	1.69 01	1.93 01	2.13 01	2.48 01	2.71 01
44	Ru K$_{\alpha_1}$	19.279	64.311	1.53 01	1.75 01	1.93 01	2.25 01	2.46 01
42	Mo K$_{\beta_1}$	19.608	63.231	1.46 01	1.66 01	1.84 01	2.14 01	2.34 01
45	Rh K$_{\alpha_1}$	20.216	61.330	1.34 01	1.52 01	1.68 01	1.96 01	2.15 01

Z	transition	E [keV]	λ [pm]	$_{21}$Sc	$_{22}$Ti	$_{23}$V	$_{24}$Cr	$_{25}$Mn
43	Tc K$_{\beta_1}$	20.619	60.131	1.26 01	1.44 01	1.59 01	1.86 01	2.03 01
46	Pd K$_{\alpha_1}$	21.177	58.547	1.17 01	1.33 01	1.47 01	1.72 01	1.88 01
44	Ru K$_{\beta_1}$	21.657	57.249	1.10 01	1.25 01	1.38 01	1.61 01	1.76 01
47	Ag K$_{\alpha_1}$	22.163	55.942	1.03 01	1.17 01	1.29 01	1.51 01	1.65 01
45	Rh K$_{\beta_1}$	22.724	54.561	9.55 00	1.09 01	1.20 01	1.40 01	1.54 01
48	Cd K$_{\alpha_1}$	23.174	53.501	9.03 00	1.03 01	1.14 01	1.33 01	1.45 01
46	Pd K$_{\beta_1}$	23.819	52.053	8.34 00	9.50 00	1.05 01	1.23 01	1.34 01
49	In K$_{\alpha_1}$	24.210	51.212	7.96 00	9.07 00	1.00 01	1.17 01	1.28 01
47	Ag K$_{\beta_1}$	24.942	49.709	7.30 00	8.32 00	9.21 00	1.07 01	1.18 01
50	Sn K$_{\alpha_1}$	25.271	49.062	7.03 00	8.01 00	8.87 00	1.03 01	1.13 01
48	Cd K$_{\beta_1}$	26.096	47.511	6.41 00	7.30 00	8.08 00	9.43 00	1.03 01
51	Sb K$_{\alpha_1}$	26.359	47.037	6.23 00	7.10 00	7.85 00	9.16 00	1.00 01
49	In K$_{\beta_1}$	27.276	45.455	5.64 00	6.43 00	7.12 00	8.30 00	9.11 00
52	Te K$_{\alpha_1}$	27.472	45.131	5.53 00	6.30 00	6.97 00	8.13 00	8.93 00
50	Sn K$_{\beta_1}$	28.486	43.524	4.98 00	5.68 00	6.28 00	7.33 00	8.05 00
53	I K$_{\alpha_1}$	28.612	43.333	4.92 00	5.60 00	6.20 00	7.23 00	7.94 00
51	Sb K$_{\beta_1}$	29.726	41.709	4.40 00	5.02 00	5.56 00	6.48 00	7.12 00
54	Xe K$_{\alpha_1}$	29.779	41.635	4.38 00	4.99 00	5.53 00	6.45 00	7.09 00
55	Cs K$_{\alpha_1}$	30.973	40.030	3.92 00	4.47 00	4.95 00	5.77 00	6.34 00
52	Te K$_{\beta_1}$	30.996	40.000	3.91 00	4.46 00	4.94 00	5.76 00	6.33 00
56	Ba K$_{\alpha_1}$	32.194	38.511	3.52 00	4.01 00	4.44 00	5.17 00	5.68 00
53	I K$_{\beta_1}$	32.295	38.391	3.49 00	3.97 00	4.40 00	5.13 00	5.63 00
57	La K$_{\alpha_1}$	33.442	37.074	3.16 00	3.60 00	3.99 00	4.65 00	5.11 00
54	Xe K$_{\beta_1}$	33.624	36.874	3.12 00	3.55 00	3.93 00	4.57 00	5.03 00
58	Ce K$_{\alpha_1}$	34.279	36.169	2.95 00	3.36 00	3.72 00	4.33 00	4.76 00
55	Cs K$_{\beta_1}$	34.987	35.437	2.79 00	3.17 00	3.51 00	4.09 00	4.49 00
59	Pr K$_{\alpha_1}$	36.026	34.415	2.57 00	2.92 00	3.24 00	3.77 00	4.14 00
56	Ba K$_{\beta_1}$	36.378	34.082	2.50 00	2.84 00	3.15 00	3.66 00	4.03 00
60	Nd K$_{\alpha_1}$	36.847	33.648	2.41 00	2.74 00	3.04 00	3.53 00	3.88 00
57	La K$_{\beta_1}$	37.801	32.799	2.24 00	2.55 00	2.83 00	3.29 00	3.61 00
61	Pm K$_{\alpha_1}$	38.725	32.016	2.10 00	2.38 00	2.64 00	3.07 00	3.37 00
58	Ce K$_{\beta_1}$	39.257	31.582	2.02 00	2.29 00	2.54 00	2.95 00	3.25 00
62	Sm K$_{\alpha_1}$	40.118	30.905	1.90 00	2.16 00	2.39 00	2.78 00	3.05 00
59	Pr K$_{\beta_1}$	40.748	30.427	1.83 00	2.07 00	2.29 00	2.66 00	2.93 00
63	Eu K$_{\alpha_1}$	41.542	29.845	1.73 00	1.96 00	2.17 00	2.53 00	2.78 00
60	Nd K$_{\beta_1}$	42.271	29.331	1.66 00	1.87 00	2.07 00	2.41 00	2.65 00
64	Gd K$_{\alpha_1}$	42.996	28.836	1.58 00	1.79 00	1.98 00	2.30 00	2.52 00
61	Pm K$_{\beta_1}$	43.826	28.290	1.51 00	1.70 00	1.88 00	2.18 00	2.39 00
65	Tb K$_{\alpha_1}$	44.482	27.873	1.45 00	1.63 00	1.80 00	2.09 00	2.30 00
62	Sm K$_{\beta_1}$	45.413	27.301	1.37 00	1.55 00	1.71 00	1.98 00	2.17 00
66	Dy K$_{\alpha_1}$	45.998	26.954	1.33 00	1.49 00	1.65 00	1.91 00	2.10 00
63	Eu K$_{\beta_1}$	47.038	26.358	1.25 00	1.41 00	1.55 00	1.79 00	1.97 00
67	Ho K$_{\alpha_1}$	47.547	26.076	1.22 00	1.37 00	1.50 00	1.74 00	1.91 00
64	Gd K$_{\beta_1}$	48.697	25.460	1.14 00	1.28 00	1.41 00	1.63 00	1.79 00

Z	transition	E [keV]	λ [pm]	$_{21}$Sc	$_{22}$Ti	$_{23}$V	$_{24}$Cr	$_{25}$Mn
68	Er K$_{\alpha_1}$	49.128	25.237	1.12 00	1.25 00	1.38 00	1.59 00	1.75 00
65	Tb K$_{\beta_1}$	50.382	24.609	1.05 00	1.17 00	1.29 00	1.49 00	1.63 00
69	Tm K$_{\alpha_1}$	50.742	24.434	9.63 -1	1.08 00	1.18 00	1.36 00	1.50 00
66	Dy K$_{\beta_1}$	52.119	23.788	9.63 -1	1.08 00	1.18 00	1.36 00	1.50 00
70	Yb K$_{\alpha_1}$	52.389	23.666	9.51 -1	1.06 00	1.17 00	1.35 00	1.48 00
67	Ho K$_{\beta_1}$	53.877	23.012	8.88 -1	9.90 -1	1.09 00	1.25 00	1.37 00
71	Lu K$_{\alpha_1}$	54.070	22.930	8.80 -1	9.81 -1	1.08 00	1.24 00	1.36 00
68	Er K$_{\beta_1}$	55.681	22.267	8.20 -1	9.11 -1	9.99 -1	1.15 00	1.26 00
72	Hf K$_{\alpha_1}$	55.790	22.223	8.16 -1	9.07 -1	9.95 -1	1.14 00	1.25 00
69	Tm K$_{\beta_1}$	57.517	21.556	7.57 -1	8.40 -1	9.21 -1	1.06 00	1.16 00
73	Ta K$_{\alpha_1}$	57.532	21.550	7.57 -1	8.39 -1	9.20 -1	1.06 00	1.16 00
74	W K$_{\alpha_1}$	59.318	20.901	7.03 -1	7.77 -1	8.52 -1	9.77 -1	1.07 00
70	Yb K$_{\beta_1}$	59.370	20.883	7.01 -1	7.76 -1	8.50 -1	9.75 -1	1.06 00
75	Re K$_{\alpha_1}$	61.140	20.278	6.57 -1	7.25 -1	7.93 -1	9.08 -1	9.90 -1
71	Lu K$_{\beta_1}$	61.283	20.231	6.54 -1	7.21 -1	7.89 -1	9.03 -1	9.85 -1
76	0s K$_{\alpha_1}$	63.001	19.679	6.17 -1	6.79 -1	7.41 -1	8.47 -1	9.22 -1
72	Hf K$_{\beta_1}$	63.234	19.607	6.12 -1	6.73 -1	7.35 -1	8.40 -1	9.14 -1
77	Ir K$_{\alpha_1}$	64.896	19.105	5.79 -1	6.36 -1	6.93 -1	7.91 -1	8.60 -1
73	Ta K$_{\beta_1}$	65.223	19.009	5.73 -1	6.29 -1	6.86 -1	7.81 -1	8.49 -1
78	Pt K$_{\alpha_1}$	66.832	18.551	5.44 -1	5.96 -1	6.49 -1	7.38 -1	8.02 -1
74	W K$_{\beta_1}$	67.244	18.438	5.37 -1	5.89 -1	6.40 -1	7.28 -1	7.90 -1
79	Au K$_{\alpha_1}$	68.804	18.020	5.l2 -1	5.60 -1	6.08 -1	6.90 -1	7.48 -1
75	Re K$_{\beta_1}$	69.310	17.888	5.04 -1	5.51 -1	5.98 -1	6.79 -1	7.35 -1
80	Hg K$_{\alpha_1}$	70.819	17.507	4.82 -1	5.25 -1	5.69 -1	6.45 -1	6.98 -1
76	0s K$_{\beta_1}$	71.413	17.361	4.73 -1	5.16 -1	5.59 -1	6.33 -1	6.85 -1
81	Tl K$_{\alpha_1}$	72.872	17.014	4.54 -1	4.93 -1	5.34 -1	5.04 -1	6.53 -1
77	Ir K$_{\beta_1}$	73.561	16.854	4.45 -1	4.83 -1	5.23 -1	5.91 -1	6.38 -1
82	Pb K$_{\alpha_1}$	74.969	16.538	4.27 -1	4.64 -1	5.01 -1	5.66 -1	6.10 -1
78	Pt K$_{\beta_1}$	75.748	16.368	4.18 -1	4.53 -1	4.89 -1	5.52 -1	5.95 -1
83	Bi K$_{\alpha_1}$	77.108	16.079	4.03 -1	4.36 -1	4.70 -1	5.30 -1	5.71 -1
79	Au K$_{\beta_1}$	77.948	15.906	3.94 -1	4.26 -1	4.59 -1	5.17 -1	5.56 -1
84	Po K$_{\alpha_1}$	79.290	15.636	3.80 -1	4.10 -1	4.41 -1	4.97 -1	5.34 -1
80	Hg K$_{\beta_1}$	80.253	15.449	3.71 -1	4.00 -1	4.30 -1	4.83 -1	5.19 -1
85	At K$_{\alpha_1}$	81.520	15.209	3.61 -1	3.89 -1	4.18 -1	4.69 -1	5.03 -1
81	Tl K$_{\beta_1}$	82.576	15.014	3.53 -1	3.80 -1	4.08 -1	4.57 -1	4.91 -1
86	Rn K$_{\alpha_1}$	83.780	14.798	3.45 -1	3.71 -1	3.97 -1	4.45 -1	4.77 -1
82	Pb K$_{\beta_1}$	84.936	14.597	3.37 -1	3.62 -1	3.87 -1	4.33 -1	4.64 -1
87	Fr K$_{\alpha_1}$	86.100	14.400	3.30 -1	3.53 -1	3.78 -1	4.22 -1	4.51 -1
83	Bi K$_{\beta_1}$	87.343	14.195	3.22 -1	3.45 -1	3.68 -1	4.10 -1	4.38 -1
88	Ra K$_{\alpha_1}$	88.470	14.014	3.15 -1	3.37 -1	3.59 -1	4.00 -1	4.27 -1
84	Po K$_{\beta_1}$	89.800	13.806	3.08 -1	3.28 -1	3.49 -1	3.88 -1	4.15 -1
89	Ac K$_{\alpha_1}$	90.884	13.642	3.01 -1	3.21 -1	3.42 -1	3.79 -1	4.05 -1
85	At K$_{\beta_1}$	92.300	13.432	2.94 -1	3.13 -1	3.32 -1	3.68 -1	3.93 -1
90	Th K$_{\alpha_1}$	93.350	13.281	2.88 -1	3.06 -1	3.25 -1	3.60 -1	3.84 -1

Z	transition	E [keV]	λ [pm]	$_{21}$Sc	$_{22}$Ti	$_{23}$V	$_{24}$Cr	$_{25}$Mn
86	Rn K$_{\beta_1}$	94.870	13.068	2.81 -1	2.98 -1	3.15 -1	3.49 -1	3.71 -1
91	Pa K$_{\alpha_1}$	95.868	12.932	2.76 -1	2.92 -1	3.09 -1	3.42 -1	3.64 -1
87	Fr K$_{\beta_1}$	97.470	12.720	2.68 -1	2.84 -1	3.00 -1	3.31 -1	3.52 -1
92	U K$_{\alpha_1}$	98.439	12.595	2.64 -1	2.79 -1	2.95 -1	3.25 -1	3.45 -1
88	Ra K$_{\beta_1}$	100.130	12.382	2.57 -1	2.71 -1	2.86 -1	3.14 -1	3.34 -1
89	Ac K$_{\beta_1}$	102.850	12.054	2.49 -1	2.62 -1	2.76 -1	3.03 -1	3.21 -1
90	Th K$_{\beta_1}$	105.610	11.739	2.42 -1	2.54 -1	2.66 -1	2.92 -1	3.08 -1
91	Pa K$_{\beta_1}$	108.430	11.434	2.35 -1	2.46 -1	2.57 -1	2.82 -1	2.97 -1
92	U K$_{\beta_1}$	111.500	11.139	2.28 -1	2.38 -1	2.48 -1	2.72 -1	2.86 -1

Z	transition	E [keV]	λ [pm]	$_{26}$Fe	$_{27}$Co	$_{28}$Ni	$_{29}$Cu	$_{30}$Zn
4	Be K$_\alpha$	0.1085	11427.207	5.43 04	5.71 04	6.44 04	7.44 04	7.32 04
38	Sr M$_\xi$	0.1140	10875.895	5.13 04	5.47 04	6.18 04	7.40 04	7.01 04
39	Y M$_\xi$	0.1328	9339.235	4.30 04	4.76 04	5.41 04	6.30 04	6.13 04
16	S L$_l$	0.1487	8331.342	3.11 04	4.13 04	4.72 04	5.33 04	5.57 04
40	Zr M$_\xi$	0.1511	8205.506	3.64 04	4.04 04	4.63 04	5.21 04	5.50 04
41	Nb M$_\xi$	0.1717	7221.037	3.04 04	3.40 04	3.91 04	4.40 04	4.80 04
5	B K$_\alpha$	0.1833	6764.059	2.76 04	3.09 04	3.57 04	4.04 04	4.42 04
42	Mo M$_\xi$	0.1926	6437.447	2.56 04	2.87 04	3.33 04	3.79 04	4.15 04
6	C K$_\alpha$	0.2170	4476.000	1.39 04	1.56 04	1.82 04	1.87 04	2.35 04
41	Ag M$_\xi$	0.3117	3977.709	1.12 04	1.25 04	1.46 04	1.51 04	1.90 04
7	N K$_\alpha$	0.3924	3159.664	7.19 03	8.02 03	9.34 03	9.98 03	1.22 04
22	Ti L$_l$	0.3953	3136.484	7.09 03	7.90 03	9.20 03	9.84 03	1.20 04
22	Ti L$_\alpha$	0.4522	2741.822	5.41 03	6.02 03	7.00 03	1.82 03	9.10 03
23	V L$_\alpha$	0.5113	2424.901	4.42 03	4.66 03	5.41 03	6.22 03	6.91 03
8	O K$_\alpha$	0.5249	2362.012	4.00 03	4.41 03	5.12 03	5.92 03	6.55 03
25	Mn L$_l$	0.5563	2228.747	3.55 03	3.91 03	4.52 03	5.21 03	5.77 03
24	Cr L$_\alpha$	0.5728	2164.546	3.33 03	3.67 03	4.25 03	4.88 03	5.40 03
25	Mn L$_\alpha$	0.6314	1945.171	2.68 03	2.94 03	3.36 03	3.75 03	4.28 03
9	F K$_\alpha$	0.6768	1831.932	2.33 03	2.57 03	2.94 03	3.19 03	3.72 03
26	Fe L$_\alpha$	0.7050	1758.655	2.15 03	2.37 03	2.69 03	2.69 03	3.42 03
21	Co L$_\alpha$	0.7762	1597.335	1.45 04	1.92 03	2.17 03	2.54 03	2.19 03
28	Ni L$_\alpha$	0.8515	1456.080	1.32 04	1.27 04	1.80 03	1.97 03	2.24 03
29	Cu L$_\alpha$	0.9297	1336.044	1.07 04	1.17 04	1.17 04	1.59 03	1.82 03
30	Zn L$_\alpha$	1.0117	1225.513	1.06 04	1.15 04	1.34 04	1.19 04	1.42 03
11	Na K$_\alpha$	1.0410	1191.020	9.77 03	1.06 04	1.24 04	1.10 04	8.40 03
11	Na K$_\beta$	1.0711	1157.550	9.04 03	9.85 03	1.15 04	1.02 04	1.09 04
12	Mg K$_\alpha$	1.2536	989.033	5.90 03	6.45 03	7.51 03	8.01 03	8.57 03
33	As L$_\alpha$	1.2820	967.123	5.56 03	6.07 03	7.07 03	7.54 03	8.08 03
12	Mg K$_\beta$	1.3022	952.121	5.33 03	5.82 03	6.78 03	7.23 03	7.74 03
33	As L$_{\beta_1}$	1.3170	941.421	5.17 03	5.65 03	6.58 03	7.01 03	7.51 03
66	Dy M$_\beta$	1.3250	935.737	5.08 03	5.55 03	6.47 03	6.90 03	7.39 03
67	Ho M$_\alpha$	1.3490	919.771	4.85 03	5.30 03	6.18 03	6.58 03	7.06 03

Z	transition	E [keV]	λ [pm]	$_{26}$Fe	$_{27}$Co	$_{28}$Ni	$_{29}$Cu	$_{30}$Zn
34	Se L$_\alpha$	1.3791	899.029	4.56 03	4.99 03	5.81 03	6.19 03	6.64 03
67	Ho M$_\beta$	1.3830	896.494	4.53 03	4.95 03	5.16 03	6.15 03	6.59 03
68	Er M$_\alpha$	1.4060	881.829	4.33 03	4.73 03	5.51 03	5.88 03	6.31 03
34	Se L$_{\beta_1}$	1.4192	813.627	4.22 03	4.62 03	5.37 03	5.73 03	6.15 03
68	Er M$_\beta$	1.4430	859.218	4.03 03	4.41 03	5.14 03	5.48 03	5.89 03
69	Tm M$_\alpha$	1.4620	848.051	3.89 03	4.26 03	4.96 03	5.29 03	5.68 03
35	Br L$_\alpha$	1.4804	837.511	3.76 03	4.12 03	4.79 03	5.12 03	5.50 03
13	Al K$_{\alpha_1}$	1.4867	833.962	3.72 03	4.01 03	4.74 03	5.06 03	5.43 03
69	Tm M$_\beta$	1.5030	824.918	3.61 03	3.95 03	4.60 03	4.91 03	5.28 03
70	Yb M$_\alpha$	1.5214	814.941	3.50 03	3.83 03	4.45 03	4.75 03	5.11 03
35	Br L$_{\beta_1}$	1.5259	812.538	3.47 03	3.80 03	4.42 03	4.71 03	5.07 03
13	Al K$_\beta$	1.5574	796.103	3.28 03	3.59 03	4.18 03	4.46 03	4.80 03
70	Yb M$_\beta$	1.5615	790.974	3.23 03	3.53 03	4.11 03	4.38 03	4.72 03
71	Lu M$_\alpha$	1.5813	784.071	3.15 03	3.45 03	4.01 03	4.28 03	4.61 03
36	Kr L$_\alpha$	1.5860	781.747	3.13 03	3.42 03	3.98 03	4.25 03	4.57 03
71	Lu M$_\beta$	1.6312	760.085	2.90 03	3.17 03	3.69 03	3.93 03	4.24 03
36	Kr L$_{\beta_1}$	1.6369	757.577	2.87 03	3.14 03	3.65 03	3.90 03	4.20 03
72	Hf M$_{\alpha_1}$	1.6446	753.892	2.83 03	3.10 03	3.61 03	3.85 03	4.15 03
37	Rb L$_{\alpha_1}$	1.6941	731.864	2.62 03	2.86 03	3.33 03	3.55 03	3.83 03
72	Hf M$_\beta$	1.6916	730.355	2.60 03	2.85 03	3.31 03	3.53 03	3.81 03
73	Ta M$_{\alpha_1}$	1.7096	725.229	2.55 03	2.79 03	3.25 03	3.46 03	3.74 03
14	Si K$_{\alpha_1}$	1.7400	712.558	2.43 03	2.66 03	3.10 03	3.30 03	3.57 03
37	Rb L$_{\beta_1}$	1.7522	707.597	2.39 03	2.61 03	3.04 03	3.24 03	3.50 03
73	Ta M$_\beta$	1.7655	702.266	2.34 03	2.56 03	2.98 03	3.18 03	3.43 03
74	W M$_{\alpha_1}$	1.7754	698.350	2.30 03	2.52 03	2.93 03	3.13 03	3.38 03
38	Sr L$_{\alpha_1}$	1.8069	686.290	2.20 03	2.41 03	2.80 03	2.98 03	3.23 03
74	W M$_\beta$	1.8349	675.705	2.11 03	2.31 03	2.68 03	2.86 03	3.10 03
14	Si K$_\beta$	1.8359	675.337	2.11 03	2.30 03	2.68 03	2.86 03	3.09 03
75	Re M$_{\alpha_1}$	1.8420	673.100	2.09 03	2.28 03	2.65 03	2.83 03	3.07 03
38	Sr L$_{\beta_1}$	1.8717	662.420	2.00 03	2.19 03	2.54 03	2.71 03	2.94 03
75	Re M$_\beta$	1.9061	650.465	1.90 03	2.08 03	2.42 03	2.58 03	2.80 03
76	Os M$_{\alpha_1}$	1.9102	649.069	1.89 03	2.07 03	2.41 03	2.57 03	2.78 03
39	Y L$_{\alpha_1}$	1.9226	644.882	1.86 03	2.03 03	2.36 03	2.52 03	2.73 03
76	Os M$_\beta$	1.9783	626.725	1.72 03	1.88 03	2.19 03	2.33 03	2.53 03
77	Ir M$_{\alpha_1}$	1.9799	626.219	1.72 03	1.88 03	2.18 03	2.33 03	2.53 03
39	Y L$_{\beta_1}$	1.9958	621.230	1.68 03	1.84 03	2.14 03	2.28 03	2.47 03
15	P K$_{\alpha_1}$	2.0137	615.708	1.64 03	1.80 03	2.08 03	2.22 03	2.42 03
40	Zr L$_{\alpha_1}$	2.0424	607.056	1.58 03	1.73 03	2.01 03	2.14 03	2.33 03
78	Pt M$_{\alpha_1}$	2.0505	604.658	1.56 03	1.71 03	1.99 03	2.12 03	2.30 03
77	Ir M$_\beta$	2.0535	603.775	1.56 03	1.70 03	1.98 03	2.11 03	2.29 03
79	Au M$_{\alpha_1}$	2.1229	584.036	1.42 03	1.56 03	1.81 03	1.93 03	2.10 03
40	Zr L$_{\beta_1}$	2.1244	583.624	1.42 03	1.56 03	1.80 03	1.93 03	2.09 03
78	Pt M$_\beta$	2.1273	582.828	1.41 03	1.55 03	1.80 03	1.92 03	2.09 03
15	P K$_{\beta_{1,3}}$	2.1390	579.640	1.39 03	1.53 03	1.77 03	1.89 03	2.06 03

Z	transition	E [keV]	λ [pm]	$_{26}$Fe	$_{27}$Co	$_{28}$Ni	$_{29}$Cu	$_{30}$Zn
41	Nb L$_{\alpha_1}$	2.1659	572.441	1.35 03	1.48 03	1.71 03	1.83 03	1.99 03
80	Hg M$_{\alpha_1}$	2.1953	564.775	1.30 03	1.42 03	1.65 03	1.76 03	1.92 03
79	Au M$_\beta$	2.2046	562.393	1.28 03	1.41 03	1.63 03	1.74 03	1.90 03
41	Nb L$_{\beta_1}$	2.2574	549.238	1.20 03	1.32 03	1.53 03	1.63 03	1.78 03
81	Tl M$_{\alpha_1}$	2.2706	546.045	1.19 03	1.30 03	1.51 03	1.61 03	1.75 03
80	Hg M$_\beta$	2.2825	543.199	1.17 03	1.28 03	1.49 03	1.59 03	1.73 03
42	Mo L$_{\alpha_1}$	2.2932	540.664	1.15 03	1.27 03	1.47 03	1.57 03	1.71 03
16	S K$_{\alpha_1}$	2.3080	537.197	1.13 03	1.24 03	1.44 03	1.54 03	1.68 03
82	Pb M$_{\alpha_1}$	2.3457	528.563	1.09 03	1.19 03	1.38 03	1.47 03	1.61 03
81	Tl M$_\beta$	2.3621	524.893	1.07 03	1.17 03	1.35 03	1.45 03	1.58 03
42	Mo L$_{\beta_1}$	2.3948	517.726	1.03 03	1.13 03	1.30 03	1.39 03	1.52 03
83	Bi M$_{\alpha_1}$	2.4226	511.785	9.95 02	1.09 03	1.26 03	1.35 03	1.47 03
43	Tc L$_{\alpha_1}$	2.4240	511.490	9.94 02	1.09 03	1.26 03	1.35 03	1.47 03
82	Pb M$_\beta$	2.4427	507.574	9.73 02	1.07 03	1.24 03	1.32 03	1.44 03
16	S K$_\beta$	2.4640	503.186	9.51 02	1.04 03	1.21 03	1.29 03	1.41 03
83	Bi M$_{\beta_1}$	2.5255	490.933	8.89 02	9.76 02	1.13 03	1.21 03	1.32 03
43	Tc L$_{\beta_1}$	2.5368	488.746	8.79 02	9.65 02	1.12 03	1.19 03	1.30 03
44	Ru L$_{\alpha_1}$	2.5586	484.582	8.59 02	9.43 02	1.09 03	1.17 03	1.27 03
17	Cl K$_{\alpha_1}$	2.6224	472.792	8.03 02	8.82 02	1.02 03	1.09 03	1.19 03
44	Ru L$_{\beta_1}$	2.6832	462.079	7.55 02	8.29 02	9.59 02	1.03 03	1.12 03
45	Rh L$_{\alpha_1}$	2.6967	459.766	7.45 02	8.18 02	9.46 02	1.01 03	1.11 03
17	Cl K$_\beta$	2.8156	440.350	6.63 02	7.29 02	8.42 02	9.02 02	9.86 02
45	Rh L$_{\beta_1}$	2.8344	437.430	6.51 02	7.16 02	8.27 02	8.86 02	9.68 02
46	Pd L$_{\alpha_1}$	2.8386	436.782	6.48 02	7.13 02	8.23 02	8.82 02	9.65 02
18	Ar K$_{\alpha_1}$	2.9577	419.194	5.80 02	6.38 02	7.37 02	7.90 02	8.64 02
47	Ag L$_{\alpha_1}$	2.9843	415.458	5.66 02	6.23 02	7.19 02	7.71 02	8.44 02
46	Pd L$_{\beta_1}$	2.9902	414.638	5.63 02	6.20 02	7.15 02	7.67 02	8.39 02
90	Th M$_{\alpha_1}$	2.9961	413.821	5.60 02	6.16 02	7.11 02	7.63 02	8.35 02
91	Pa M$_{\alpha_1}$	3.0823	402.248	5.19 02	5.71 02	6.59 02	7.07 02	7.74 02
48	Cd L$_{\alpha_1}$	3.1337	395.651	4.97 02	5.46 02	6.30 02	6.76 02	7.41 02
90	Th M$_\beta$	3.1458	394.129	4.92 02	5.41 02	6.24 02	6.69 02	7.33 02
47	Ag L$_{\beta_1}$	3.1509	393.491	4.89 02	5.39 02	6.21 02	6.66 02	7.30 02
92	U M$_{\alpha_1}$	3.1708	391.021	4.81 02	5.30 02	6.11 02	6.55 02	7.18 02
18	Ar K$_\beta$	3.1905	388.607	4.73 02	5.21 02	6.00 02	6.44 02	7.06 02
91	Pa M$_\beta$	3.2397	382.705	4.54 02	5.00 02	5.76 02	6.18 02	6.78 02
49	In L$_{\alpha_1}$	3.2869	377.210	4.37 02	4.81 02	5.54 02	5.95 02	6.52 02
19	K K$_{\alpha_1}$	3.3138	374.148	4.27 02	4.70 02	5.42 02	5.82 02	6.38 02
48	Cd L$_{\beta_1}$	3.3166	373.832	4.26 02	4.69 02	5.41 02	5.80 02	6.37 02
92	U M$_\beta$	3.3360	371.658	4.20 02	4.62 02	5.32 02	5.71 02	6.27 02
50	Sn L$_{\alpha_1}$	3.4440	360.003	3.85 02	4.24 02	4.89 02	5.24 02	5.76 02
49	In L$_{\beta_1}$	3.4872	355.543	3.73 02	4.10 02	4.72 02	5.07 02	5.57 02
19	K K$_\beta$	3.5896	345.401	3.45 02	3.80 02	4.37 02	4.69 02	5.16 02
51	Sb L$_{\alpha_1}$	3.6047	343.954	3.41 02	3.75 02	4.32 02	4.64 02	5.10 02
50	Sn L$_{\beta_1}$	3.6628	338.498	3.26 02	3.60 02	4.14 02	4.44 02	4.89 02

Z	transition	E [keV]	λ [pm]	$_{26}$Fe	$_{27}$Co	$_{28}$Ni	$_{29}$Cu	$_{30}$Zn
20	Ca K$_{\alpha_1}$	3.6917	335.848	3.20 02	3.52 02	4.05 02	4.35 02	4.79 02
52	Te L$_{\alpha_1}$	3.7693	328.934	3.02 02	3.33 02	3.83 02	4.11 02	4.53 02
51	Sb L$_{\beta_1}$	3.8436	322.575	2.87 02	3.16 02	3.63 02	5.90 02	4.30 02
53	I L$_{\alpha_1}$	3.9377	314.867	2.69 02	2.96 02	3.40 02	3.66 02	4.03 02
20	Ca K	4.0127	308.981	2.55 02	2.82 02	3.24 02	3.48 02	3.84 02
52	Te L$_{\beta_1}$	4.0296	307.686	2.53 02	2.79 02	3.20 02	3.44 02	3.79 02
21	Sc K$_{\alpha_1}$	4.0906	303.097	2.43 02	2.68 02	3.07 02	3.30 02	3.65 02
54	Xe L$_{\alpha_1}$	4.1099	301.674	2.40 02	2.64 02	3.03 02	3.26 02	3.60 02
53	I L$_{\beta_1}$	4.2207	293.755	2.23 02	2.46 02	2.83 02	3.04 02	3.35 02
55	Cs L$_{\alpha_1}$	4.2865	289.245	2.14 02	2.36 02	2.71 02	2.91 02	3.22 02
21	Sc K$_\beta$	4.4605	277.962	1.93 02	2.12 02	2.44 02	2.62 02	2.90 02
56	Ba L$_{\alpha_1}$	4.4663	277.601	1.92 02	2.12 02	2.43 02	2.61 02	2.89 02
22	Ti K$_{\alpha_1}$	4.5108	274.863	1.87 02	2.06 02	2.37 02	2.54 02	2.81 02
55	Cs L$_{\beta_1}$	4.6198	268.377	1.76 02	1.93 02	2.22 02	2.38 02	2.64 02
57	La L$_{\alpha_1}$	4.6510	266.577	1.73 02	1.90 02	2.18 02	2.34 02	2.59 02
56	Ba L$_{\beta_1}$	4.8273	256.841	1.56 02	1.72 02	1.97 02	2.11 02	2.35 02
58	Ce L$_{\alpha_1}$	4.8402	256.157	1.55 02	1.71 02	1.96 02	2.10 02	2.33 02
22	Ti K$_{\beta_{1,3}}$	4.9318	251.399	1.48 02	1.62 02	1.86 02	2.00 02	2.22 02
23	V K$_{\alpha_1}$	4.9522	250.363	1.46 02	1.60 02	1.84 02	1.97 02	2.19 02
59	Pr L$_{\alpha_1}$	5.0337	246.310	1.40 02	1.54 02	1.76 02	1.89 02	2.10 02
57	La L$_{\beta_1}$	5.0421	245.899	1.39 02	1.53 02	1.76 02	1.88 02	2.09 02
60	Nd L$_{\alpha_1}$	5.2304	237.047	1.26 02	1.39 02	1.59 02	1.71 02	1.90 02
58	Ce L$_{\beta_1}$	5.2622	235.614	1.24 02	1.36 02	1.56 02	1.68 02	1.87 02
24	Cr K$_{\alpha_1}$	5.4147	228.978	1.15 02	1.26 02	1.45 02	1.56 02	1.73 02
23	V K$_{\beta_{1,3}}$	5.4273	228.447	1.14 02	1.26 02	1.44 02	1.55 02	1.72 02
61	Pm L$_{\alpha_1}$	5.4325	228.228	1.13 02	1.25 02	1.44 02	1.54 02	1.71 02
59	Pr L$_{\beta_1}$	5.4889	225.883	1.10 02	1.22 02	1.40 02	1.50 02	1.67 02
62	Sm L$_{\alpha_1}$	5.6361	219.984	1.03 02	1.14 02	1.30 02	1.40 02	1.56 02
60	Nd L$_{\beta_1}$	5.7216	216.696	9.85 01	1.09 02	1.25 02	1.34 02	1.49 02
63	Eu L$_{\alpha_1}$	5.8457	212.096	9.29 01	1.03 02	1.18 02	1.27 02	1.41 02
25	Mn K$_{\alpha_1}$	5.8988	210.187	9.07 01	1.01 02	1.15 02	1.24 02	1.38 02
24	Cr K$_{\beta_{1,3}}$	5.9467	208.494	8.87 01	9.85 01	1.12 02	1.21 02	1.35 02
61	Pm L$_{\beta_1}$	5.9610	207.993	8.81 01	9.79 01	1.12 02	1.21 02	1.34 02
64	Gd L$_{\alpha_1}$	6.0572	204.690	8.44 01	9.38 01	1.07 02	1.16 02	1.29 02
62	Sm L$_{\beta_1}$	6.2051	199.811	7.93 01	8.80 01	1.00 02	1.08 02	1.21 02
65	Tb L$_{\alpha_1}$	6.2728	197.655	7.71 01	8.55 01	9.76 01	1.05 02	1.17 02
26	Fe K$_{\alpha_1}$	6.4038	193.611	7.30 01	8.09 01	9.24 01	9.96 01	1.11 02
63	Eu L$_{\beta_1}$	6.4564	192.034	7.15 01	7.92 01	9.04 01	9.75 01	1.09 02
25	Mn K$_{\beta_{1,3}}$	6.4905	191.025	7.05 01	7.81 01	8.92 01	9.61 01	1.07 02
66	Gd L$_{\alpha_1}$	6.4952	190.887	7.04 01	7.79 01	8.90 01	9.60 01	1.07 02
64	Gd L$_{\beta_1}$	6.7132	184.688	6.46 01	7.14 01	8.16 01	8.79 01	9.81 01
67	Ho L$_{\alpha_1}$	6.7198	184.507	6.44 01	7.12 01	8.14 01	8.77 01	9.79 01
27	Co K$_{\alpha_1}$	6.9303	178.903	5.94 01	6.56 01	7.50 01	8.08 01	9.02 01
68	Er L$_{\alpha_1}$	6.9487	178.429	5.90 01	6.52 01	7.45 01	8.02 01	8.96 01

Z	transition	E [keV]	λ [pm]	$_{26}$Fe	$_{27}$Co	$_{28}$Ni	$_{29}$Cu	$_{30}$Zn
65	Tb L$_{\beta_1}$	6.9780	177.680	5.83 01	6.44 01	7.36 01	7.93 01	8.86 01
26	Fe K$_{\beta_{1,3}}$	7.0580	175.666	<u>5.66 01</u>	6.25 01	7.15 01	7.70 01	8.60 01
69	Tm L$_{\alpha_1}$	7.1799	172.683	4.26 01	5.97 01	6.83 01	7.35 01	8.22 01
66	Dy L$_{\beta_1}$	7.2477	171.068	4.15 02	5.83 01	6.66 01	7.17 01	8.02 01
70	Yb L$_{\alpha_1}$	7.4156	167.195	3.91 02	5.48 01	6.27 01	6.75 01	7.56 01
28	Ni K$_{\alpha_1}$	7.4782	165.795	3.82 02	5.36 01	6.13 01	6.60 01	7.39 01
67	Ho L$_{\beta_1}$	7.5253	164.757	3.76 02	5.27 01	6.03 01	6.49 01	7.27 01
27	Co K$_{\beta_{1,3}}$	7.6494	162.084	3.60 02	5.05 01	5.78 01	6.21 01	6.96 01
71	Lu L$_{\alpha_1}$	7.6555	161.955	3.60 02	<u>5.04 01</u>	5.76 01	6.20 01	6.95 01
68	Er L$_{\beta_1}$	7.8109	158.733	3.41 02	3.81 02	5.47 01	5.88 01	6.59 01
72	Hf L$_{\alpha_1}$	7.8990	156.963	3.31 02	3.70 02	5.31 01	5.71 01	6.40 01
29	Cu K$_{\alpha_1}$	8.0478	154.060	3.15 02	3.52 02	5.05 01	5.43 01	6.09 01
69	Tm L$_{\beta_1}$	8.1010	153.049	3.10 02	3.46 02	4.96 01	5.34 01	5.99 01
73	Ta L$_{\alpha_1}$	8.1461	152.201	3.05 02	3.41 02	4.89 01	5.26 01	5.90 01
28	Ni K$_{\beta_{1,3}}$	8.2647	150.017	2.93 02	3.27 02	<u>4.70 01</u>	5.06 01	5.68 01
74	W L$_{\alpha_1}$	8.3976	147.643	2.81 02	3.14 02	3.36 02	4.85 01	5.45 01
70	Yb L$_{\beta_1}$	8.4018	147.569	2.81 02	3.13 02	3.35 02	4.85 01	5.44 01
30	Zn K$_{\alpha_1}$	8.6389	143.519	2.60 02	2.90 02	3.12 02	4.50 01	5.06 01
75	Re L$_{\alpha_1}$	8.6525	143.294	2.59 02	2.89 02	3.10 02	4.48 01	5.04 01
71	Lu L$_{\beta_1}$	8.7090	142.364	2.55 02	2.84 02	3.05 02	4.41 01	4.95 01
29	Cu K$_{\beta_1}$	8.9053	139.226	2.40 02	2.67 02	2.88 02	4.16 01	4.67 01
76	Os L$_{\alpha_1}$	8.9117	139.126	2.40 02	2.67 02	2.87 02	<u>4.15 01</u>	4.66 01
72	Hf L$_{\beta_1}$	9.0227	137.414	2.32 02	2.58 02	2.78 02	2.94 02	4.51 01
77	Ir L$_{\alpha_1}$	9.1751	135.132	2.21 02	2.47 02	2.66 02	2.81 02	4.32 01
31	Ga K$_{\alpha_1}$	9.2517	134.013	2.17 02	2.41 02	2.60 02	2.75 02	4.23 01
73	Ta L$_{\beta_1}$	9.3431	132.702	2.11 02	2.35 02	2.54 02	2.68 02	4.12 01
78	Pt L$_{\alpha_1}$	9.4423	131.308	2.05 02	2.28 02	2.47 02	2.60 02	4.01 01
30	Zn K$_{\beta_{1,3}}$	9.5720	129.529	1.98 02	2.20 02	2.38 02	2.51 02	<u>3.87 01</u>
74	W L$_{\beta_1}$	9.6724	128.184	1.92 02	2.14 02	2.32 02	2.44 02	2.74 02
79	Au L$_{\alpha_1}$	9.7133	127.644	1.90 02	2.11 02	2.29 02	2.42 02	2.71 02
32	Ge K$_{\alpha_1}$	9.8864	125.409	1.81 02	2.01 02	2.19 02	2.31 02	2.58 02
80	Hg L$_{\alpha_1}$	9.9888	124.124	1.76 02	1.96 02	2.13 02	2.24 02	1.51 02
75	Re L$_{\beta_1}$	10.010	123.861	1.75 02	1.95 02	2.12 02	2.23 02	2.50 02
31	Ga K$_{\beta_1}$	10.264	120.796	1.64 02	1.82 02	1.98 02	2.09 02	2.34 02
81	Tl L$_{\alpha_1}$	10.269	120.737	1.63 02	1.81 02	1.98 02	2.08 02	2.33 02
76	Os L$_{\beta_1}$	10.355	119.734	1.60 02	1.77 02	1.93 02	2.04 02	2.28 02
33	As K$_{\alpha_1}$	10.544	117.588	1.52 02	1.68 02	1.84 02	1.94 02	2.17 02
82	Pb L$_{\alpha_1}$	10.552	117.499	1.51 02	1.68 02	1.84 02	1.94 02	2.17 02
77	Ir L$_{\beta_1}$	10.708	115.787	1.45 02	1.61 02	1.77 02	1.86 02	2.08 02
83	Bi L$_{\alpha_1}$	10.839	114.388	1.41 02	1.56 02	1.71 02	1.80 02	2.02 02
32	Ge K$_{\beta_1}$	10.982	112.898	1.36 02	1.50 02	1.65 02	1.74 02	1.95 02
78	Pt L$_{\beta_1}$	11.071	111.990	1.33 02	1.47 02	1.61 02	1.70 02	1.90 02
84	Po L$_{\alpha_1}$	11.131	111.387	1.31 02	1.45 02	1.59 02	1.68 02	1.88 02
34	Se K$_{\alpha_1}$	11.222	110.484	1.28 02	1.42 02	1.56 02	1.64 02	1.84 02

Z	transition	E [keV]	λ [pm]	$_{26}$Fe	$_{27}$Co	$_{28}$Ni	$_{29}$Cu	$_{30}$Zn
85	At L$_{\alpha_1}$	11.427	108.501	1.22 02	1.35 02	1.48 02	1.56 02	1.75 02
79	Au L$_{\beta_1}$	11.442	108.359	1.21 02	1.34 02	1.48 02	1.56 02	1.74 02
33	As K$_{\beta_1}$	11.726	105.735	1.13 02	1.25 02	1.38 02	1.46 02	1.63 02
86	Rn L$_{\alpha_1}$	11.727	105.726	1.13 02	1.25 02	1.38 02	1.46 02	1.63 02
80	Hg L$_{\beta_1}$	11.823	104.867	1.11 02	1.22 02	1.35 02	1.42 02	1.59 02
35	Br K$_{\alpha_1}$	11.924	103.979	1.08 02	1.20 02	1.32 02	1.39 02	1.56 02
87	Fr L$_{\alpha_1}$	12.031	103.054	1.05 02	1.17 02	1.29 02	1.36 02	1.52 02
81	Tl L$_{\beta_1}$	12.213	101.519	1.01 02	1.12 02	1.24 02	1.31 02	1.46 02
88	Ra L$_{\alpha_1}$	12.340	100.474	9.83 01	1.09 02	1.20 02	1.27 02	1.42 02
34	Se K$_{\beta_1}$	12.496	99.219	9.50 01	1.05 02	1.16 02	1.23 02	1.37 02
82	Pb L$_{\beta_1}$	12.614	98.291	9.25 01	1.02 02	1.13 02	1.20 02	1.34 02
36	Kr K$_{\alpha_1}$	12.649	98.019	9.18 01	1.02 02	1.13 02	1.19 02	1.33 02
89	Ac L$_{\alpha_1}$	12.652	97.996	9.18 01	1.01 02	1.13 02	1.19 02	1.33 02
90	Th L$_{\alpha_1}$	12.969	95.601	8.57 01	9.47 01	1.05 02	1.11 02	1.24 02
83	Bi L$_{\beta_1}$	13.024	95.197	8.47 01	9.36 01	1.04 02	1.10 02	1.23 02
91	Pa L$_{\alpha_1}$	13.291	93.285	8.01 01	8.84 01	9.86 01	1.04 02	1.16 02
35	Br K$_{\beta_1}$	13.291	93.285	8.01 01	8.84 01	9.86 01	1.04 02	1.16 02
37	Rb K$_{\alpha_1}$	13.395	92.560	7.84 01	8.65 01	9.65 01	1.02 02	1.14 02
84	Po L$_{\beta_1}$	13.447	92.202	7.76 01	8.56 01	9.55 01	1.01 02	1.13 02
92	U L$_{\alpha_1}$	13.615	91.065	7.49 01	8.27 01	9.23 01	9.74 01	1.09 02
85	At L$_{\beta_1}$	13.876	89.352	7.11 01	7.84 01	8.77 01	9.26 01	1.03 02
36	Kr K$_{\beta_1}$	14.112	87.857	6.79 01	7.49 01	8.38 01	8.85 01	9.89 01
38	Sr K$_{\alpha_1}$	14.165	87.529	6.72 01	7.41 01	8.30 01	8.76 01	9.79 01
86	Rn L$_{\beta_1}$	14.316	86.606	6.52 01	7.19 01	8.06 01	8.51 01	9.51 01
87	Fr L$_{\beta_1}$	14.770	83.943	5.98 01	6.59 01	7.41 01	7.83 01	8.74 01
39	Y K$_{\alpha_1}$	14.958	82.888	5.78 01	6.37 01	7.16 01	7.56 01	8.45 01
37	Rb K$_{\beta_1}$	14.961	82.872	5.78 01	6.36 01	7.16 01	7.56 01	8.44 01
88	Ra L$_{\beta_1}$	15.236	81.376	5.49 01	6.04 01	6.81 01	7.19 01	8.03 01
89	Ac L$_{\beta_1}$	15.713	78.906	5.03 01	5.54 01	6.25 01	6.60 01	7.37 01
40	Zr K$_{\alpha_1}$	15.775	78.596	4.97 01	5.47 01	6.18 01	6.53 01	7.29 01
38	Sr K$_{\beta_1}$	15.836	78.293	4.92 01	5.41 01	6.12 01	6.46 01	7.21 01
90	Th L$_{\beta_1}$	16.202	76.524	4.61 01	5.07 01	5.74 01	6.06 01	6.77 01
41	Nb K$_{\alpha_1}$	16.615	74.622	4.30 01	4.72 01	5.35 01	5.66 01	6.31 01
91	Pa L$_{\beta_1}$	16.702	74.233	4.23 01	4.65 01	5.28 01	5.57 01	6.22 01
39	Y K$_{\beta_1}$	16.738	74.074	4.21 01	4.63 01	5.25 01	5.54 01	6.18 01
92	U L$_{\beta_1}$	17.220	72.000	3.88 01	4.27 01	4.85 01	5.12 01	5.71 01
42	Mo K$_{\alpha_1}$	17.479	70.933	3.72 01	4.09 01	4.65 01	4.91 01	5.48 01
40	Zr K$_{\beta_1}$	17.668	70.175	3.61 01	3.97 01	4.52 01	4.77 01	5.32 01
43	Tc K$_{\alpha_1}$	18.367	67.504	3.24 01	3.55 01	4.05 01	4.28 01	4.77 01
41	Nb K$_{\beta_1}$	18.623	66.576	3.11 01	3.42 01	3.90 01	4.12 01	4.59 01
44	Ru K$_{\alpha_1}$	19.279	64.311	2.82 01	3.10 01	3.54 01	3.75 01	4.17 01
42	Mo K$_{\beta_1}$	19.608	63.231	2.69 01	2.95 01	3.38 01	3.57 01	3.98 01
45	Rh K$_{\alpha_1}$	20.216	61.330	2.47 01	2.71 01	3.11 01	3.28 01	3.66 01
43	Tc K$_{\beta_1}$	20.619	60.131	2.33 01	2.56 01	2.94 01	3.11 01	3.46 01

Z	transition	E [keV]	λ [pm]	$_{26}$Fe	$_{27}$Co	$_{28}$Ni	$_{29}$Cu	$_{30}$Zn
46	Pd K$_{\alpha_1}$	21.177	58.547	2.16 01	2.37 01	2.72 01	2.88 01	3.21 01
44	Ru K$_{\beta_1}$	21.657	57.249	2.03 01	2.22 01	2.56 01	2.70 01	3.01 01
47	Ag K$_{\alpha_1}$	22.163	55.942	1.90 01	2.08 01	2.39 01	2.53 01	2.82 01
45	Rh K$_{\beta_1}$	22.724	54.561	1.77 01	1.94 01	2.23 01	2.36 01	2.63 01
48	Cd K$_{\alpha_1}$	23.174	53.501	1.67 01	1.83 01	2.11 01	2.24 01	2.48 01
46	Pd K$_{\beta_1}$	23.819	52.053	1.55 01	1.69 01	1.95 01	2.07 01	2.30 01
49	In K$_{\alpha_1}$	24.210	51.212	1.48 01	1.62 01	1.87 01	1.98 01	2.20 01
47	Ag K$_{\beta_1}$	24.942	49.709	1.36 01	1.48 01	1.72 01	1.82 01	2.02 01
50	Sn K$_{\alpha_1}$	25.271	49.062	1.31 01	1.43 01	1.65 01	1.75 01	1.95 01
48	Cd K$_{\beta_1}$	26.096	47.511	1.19 01	1.30 01	1.51 01	1.60 01	1.78 01
51	Sb K$_{\alpha_1}$	26.359	47.037	1.16 01	1.27 01	1.47 01	1.56 01	1.73 01
49	In K$_{\beta_1}$	27.276	45.455	1.05 01	1.15 01	1.33 01	1.41 01	1.57 01
52	Te K$_{\alpha_1}$	27.472	45.131	1.03 01	1.13 01	1.31 01	1.38 01	1.54 01
50	Sn K$_{\beta_1}$	28.486	43.524	9.31 00	1.02 01	1.18 01	1.25 01	1.39 01
53	I K$_{\alpha_1}$	28.612	43.333	9.19 00	1.00 01	1.16 01	1.23 01	1.37 01
51	Sb K$_{\beta_1}$	29.726	41.709	8.24 00	8.99 00	1.04 01	1.11 01	1.23 01
54	Xe K$_{\alpha_1}$	29.779	41.635	8.20 00	8.95 00	1.04 01	1.10 01	1.22 01
55	Cs K$_{\alpha_1}$	30.973	40.030	7.34 00	8.00 00	9.30 00	9.87 00	1.09 01
52	Te K$_{\beta_1}$	30.996	40.000	7.32 00	7.98 00	9.28 00	9.85 00	1.09 01
56	Ba K$_{\alpha_1}$	32.194	38.511	6.58 00	7.16 00	8.34 00	8.84 00	9.81 00
53	I K$_{\beta_1}$	32.295	38.391	6.52 00	7.10 00	8.26 00	8.77 00	9.72 00
57	La L$_{\alpha_1}$	33.442	37.074	5.91 00	6.43 00	7.49 00	7.94 00	8.81 00
54	Xe K$_{\beta_1}$	33.624	36.874	5.82 00	6.33 00	7.37 00	7.82 00	8.67 00
58	Ce K$_{\alpha_1}$	34.279	36.169	5.51 00	5.99 00	6.98 00	7.40 00	8.21 00
55	Cs K$_{\beta_1}$	34.987	35.437	5.20 00	5.65 00	6.59 00	6.98 00	7.75 00
59	Pr K$_{\alpha_1}$	36.026	34.415	4.78 00	5.20 00	6.07 00	6.43 00	7.14 00
56	Ba K$_{\beta_1}$	36.378	34.082	4.65 00	5.06 00	5.90 00	6.25 00	6.94 00
60	Nd K$_{\alpha_1}$	36.847	33.648	4.49 00	4.88 00	5.69 00	6.03 00	6.69 00
57	La K$_{\beta_1}$	37.801	32.799	4.17 00	4.54 00	5.30 00	5.61 00	6.23 00
61	Pm K$_{\alpha_1}$	38.725	32.016	3.90 00	4.24 00	4.95 00	5.24 00	5.82 00
58	Ce K$_{\beta_1}$	39.257	31.582	3.75 00	4.07 00	4.76 00	5.04 00	5.60 00
62	Sm K$_{\alpha_1}$	40.118	30.905	3.53 00	3.83 00	4.48 00	4.74 00	5.26 00
59	Pr K$_{\beta_1}$	40.748	30.427	3.37 00	3.67 00	4.29 00	4.54 00	5.04 00
63	Eu K$_{\alpha_1}$	41.542	29.845	3.19 00	3.48 00	4.06 00	4.30 00	4.78 00
60	Nd K$_{\beta_1}$	42.271	29.331	3.03 00	3.32 00	3.87 00	4.09 00	4.55 00
64	Gd K$_{\alpha_1}$	42.996	28.836	2.89 00	3.16 00	3.69 00	3.90 00	4.34 00
61	Pm K$_{\beta_1}$	43.826	28.290	2.73 00	3.00 00	3.50 00	3.70 00	4.11 00
65	Tb K$_{\alpha_1}$	44.482	27.873	2.62 00	2.88 00	3.36 00	3.55 00	3.95 00
62	Sm K$_{\beta_1}$	45.413	27.301	2.46 00	2.72 00	3.17 00	3.35 00	3.72 00
66	Dy K$_{\alpha_1}$	45.998	26.954	2.37 00	2.63 00	3.06 00	3.23 00	3.59 00
63	Eu K$_{\beta_1}$	47.038	26.358	2.23 00	2.47 00	2.87 00	3.04 00	3.38 00
67	Ho K$_{\alpha_1}$	47.547	26.076	2.16 00	2.40 00	2.79 00	2.95 00	3.28 00
64	Gd K$_{\beta_1}$	48.697	25.460	2.01 00	2.24 00	2.61 00	2.76 00	3.06 00
68	Er K$_{\alpha_1}$	49.128	25.237	1.96 00	2.19 00	2.54 00	2.69 00	2.99 00

Z	transition	E [keV]	λ [pm]	$_{26}$Fe	$_{27}$Co	$_{28}$Ni	$_{29}$Cu	$_{30}$Zn
65	Tb K$_{\beta_1}$	50.382	24.609	1.83 00	2.04 00	2.37 00	2.51 00	2.79 00
69	Tm K$_{\alpha_1}$	50.742	24.434	1.80 00	2.00 00	2.33 00	2.46 00	2.73 00
66	Dy K$_{\beta_1}$	52.119	23.788	1.68 00	1.86 00	2.17 00	2.29 00	2.54 00
70	Yb K$_{\alpha_1}$	52.389	23.666	1.66 00	1.84 00	2.14 00	2.26 00	2.51 00
67	Ho K$_{\beta_1}$	53.877	23.012	1.55 00	1.71 00	1.98 00	2.09 00	2.32 00
71	Lu K$_{\alpha_1}$	54.070	22.930	1.54 00	1.69 00	1.96 00	2.07 00	2.30 00
68	Er K$_{\beta_1}$	55.681	22.267	1.43 00	1.56 00	1.82 00	1.91 00	2.12 00
72	Hf K$_{\alpha_1}$	55.790	22.223	1.42 00	1.55 00	1.81 00	1.90 00	2.11 00
69	Tm K$_{\beta_1}$	57.517	21.556	1.32 00	1.43 00	1.67 00	1.75 00	1.94 00
73	Ta K$_{\alpha_1}$	57.532	21.550	1.32 00	1.43 00	1.66 00	1.75 00	1.94 00
74	W K$_{\alpha_1}$	59.318	20.901	1.22 00	1.32 00	1.53 00	1.61 00	1.79 00
70	Yb K$_{\beta_1}$	59.370	20.883	1.22 00	1.31 00	1.53 00	1.61 00	1.78 00
75	Re K$_{\alpha_1}$	61.140	20.278	1.13 00	1.22 00	1.42 00	1.49 00	1.65 00
71	Lu K$_{\beta_1}$	61.283	20.231	1.13 00	1.21 00	1.41 00	1.48 00	1.64 00
76	Os K$_{\alpha_1}$	63.001	19.679	1.05 00	1.13 00	1.32 00	1.38 00	1.53 00
72	Hf K$_{\beta_1}$	63.234	19.607	1.04 00	1.12 00	1.30 00	1.37 00	1.51 00
77	Ir K$_{\alpha_1}$	64.896	19.105	9.80 -1	1.05 00	1.22 00	1.28 00	1.41 00
73	Ta K$_{\beta_1}$	65.223	19.009	9.68 -1	1.04 00	1.21 00	1.26 00	1.40 00
78	Pt K$_{\alpha_1}$	66.832	18.551	9.12 -1	9.79 -1	1.13 00	1.19 00	1.31 00
74	W K$_{\beta_1}$	67.244	18.438	8.98 -1	9.65 -1	1.12 00	1.17 00	1.29 00
79	Au K$_{\alpha_1}$	68.804	18.020	8.50 -1	9.12 -1	1.05 00	1.10 00	1.22 00
75	Re K$_{\beta_1}$	69.310	17.888	8.35 -1	8.95 -1	1.03 00	1.08 00	1.19 00
80	Hg K$_{\alpha_1}$	70.819	17.507	7.92 -1	8.49 -1	9.79 -1	1.02 00	1.13 00
76	Os K$_{\beta_1}$	71.413	17.361	7.76 -1	8.32 -1	9.59 -1	1.00 00	1.11 00
81	Tl K$_{\alpha_1}$	72.872	17.014	7.39 -1	7.91 -1	9.11 -1	9.52 -1	1.05 00
77	Ir K$_{\beta_1}$	73.561	16.854	7.22 -1	7.73 -1	8.90 -1	9.30 -1	1.02 00
82	Pb K$_{\alpha_1}$	74.969	16.538	6.89 -1	7.38 -1	8.48 -1	8.86 -1	9.75 -1
78	Pt K$_{\beta_1}$	75.748	16.368	6.72 -1	7.19 -1	8.27 -1	8.63 -1	9.49 -1
83	Bi K$_{\alpha_1}$	77.108	16.079	6.44 -1	6.88 -1	7.90 -1	8.25 -1	9.07 -1
79	Au K$_{\beta_1}$	77.948	15.906	6.27 -1	6.70 -1	7.69 -1	8.02 -1	8.82 -1
84	Po K$_{\alpha_1}$	79.290	15.636	6.02 -1	6.42 -1	7.37 -1	7.68 -1	8.44 -1
80	Hg K$_{\beta_1}$	80.253	15.449	5.85 -1	6.24 -1	7.15 -1	7.45 -1	8.19 -1
85	At K$_{\alpha_1}$	81.520	15.209	5.66 -1	6.04 -1	6.91 -1	7.19 -1	7.89 -1
81	Tl K$_{\beta_1}$	82.576	15.014	5.51 -1	5.87 -1	6.72 -1	6.99 -1	7.66 -1
86	Rn K$_{\alpha_1}$	83.780	14.798	5.35 -1	5.70 -1	6.50 -1	6.76 -1	7.41 -1
82	Pb K$_{\beta_1}$	84.936	14.597	5.20 -1	5.33 -1	6.31 -1	6.56 -1	7.18 -1
87	Fr K$_{\alpha_1}$	86.100	14.400	5.05 -1	5.37 -1	6.12 -1	6.36 -1	6.96 -1
83	Bi K$_{\beta_1}$	87.343	14.195	4.90 -1	5.21 -1	5.93 -1	6.15 -1	6.73 -1
88	Ra K$_{\alpha_1}$	88.470	14.014	4.77 -1	5.07 -1	5.77 -1	5.98 -1	6.53 -1
84	Po K$_{\beta_1}$	89.800	13.806	4.63 -1	4.91 -1	5.58 -1	5.78 -1	6.31 -1
89	Ac K$_{\alpha_1}$	90.884	13.642	4.51 -1	4.79 -1	5.44 -1	5.62 -1	6.14 -1
85	At K$_{\beta_1}$	92.300	13.432	4.37 -1	4.63 -1	5.25 -1	5.43 -1	5.92 -1
90	Th K$_{\alpha_1}$	93.350	13.281	4.27 -1	4.52 -1	5.12 -1	5.29 -1	5.77 -1
86	Rn K$_{\beta_1}$	94.870	13.068	4.13 -1	4.37 -1	4.94 -1	5.10 -1	5.56 -1

Z	transition	E [keV]	λ [pm]	$_{26}$Fe	$_{27}$Co	$_{28}$Ni	$_{29}$Cu	$_{30}$Zn
91	Pa K$_{\alpha_1}$	95.868	12.932	4.04 -1	4.27 -1	4.83 -1	4.98 -1	5.42 -1
87	Fr K$_{\beta_1}$	97.470	12.720	3.90 -1	4.12 -1	4.66 -1	4.80 -1	5.22 -1
92	U K$_{\alpha_1}$	98.439	12.595	3.82 -1	4.04 -1	4.56 -1	4.69 -1	5.10 -1
88	Ra K$_{\beta_1}$	100.130	12.382	3.69 -1	3.90 -1	4.39 -1	4.52 -1	4.91 -1
89	Ac K$_{\beta_1}$	102.850	12.054	3.54 -1	3.73 -1	4.20 -1	4.31 -1	4.67 -1
90	Th K$_{\beta_1}$	105.610	11.739	3.40 -1	3.57 -1	4.01 -1	4.11 -1	4.45 -1
91	Pa K$_{\beta_1}$	108.430	11.434	3.26 -1	3.42 -1	3.83 -1	3.93 -1	4.24 -1
92	U K$_{\beta_1}$	111.500	11.139	3.13 -1	3.28 -1	3.67 -1	3.75 -1	4.04 -1

Z	transition	E [keV]	λ [pm]	$_{31}$Ga	$_{32}$Ge	$_{33}$As	$_{34}$Se	$_{35}$Br
4	Be K$_\alpha$	0.1085	11427.207	6.37 04	5.79 04	4.91 04	4.12 04	3.07 04
38	Sr M$_\xi$	0.1140	10875.895	6.12 04	5.69 04	5.22 04	4.35 04	3.25 04
39	Y M$_\xi$	0.1328	9339.235	5.41 04	5.57 04	5.12 04	4.63 04	3.96 04
16	S L$_l$	0.1487	8337.942	4.94 04	5.15 04	5.27 04	4.85 04	4.57 04
40	Zr M$_\xi$	0.1511	8205.506	4.89 04	5.10 04	5.23 04	4.84 04	4.60 04
41	Nb M$_\xi$	0.1717	7221.037	4.51 04	4.53 04	4.68 04	4.75 04	4.37 04
5	B K$_\alpha$	0.1833	6764.059	4.20 04	4.51 04	4.45 04	4.56 04	4.27 04
42	Mo M$_\xi$	0.1926	6437.445	3.96 04	4.24 04	4.25 04	4.40 04	4.49 04
6	C K$_\alpha$	0.2770	4476.000	2.33 04	2.55 04	2.79 04	3.02 04	3.26 04
47	Ag M$_\xi$	0.3117	3977.709	1.90 04	2.08 04	2.29 04	2.49 04	2.70 04
7	N K$_\alpha$	0.3924	3159.664	1.27 04	1.40 04	1.56 04	1.70 04	1.88 04
22	Ti L$_l$	0.3953	3136.484	1.25 04	1.39 04	1.54 04	1.68 04	1.85 04
22	Ti L$_\alpha$	0.4522	2741.822	9.62 03	1.06 04	1.19 04	1.30 04	1.44 04
23	V L$_\alpha$	0.5113	2424.901	7.48 03	8.30 03	9.28 03	1.01 04	1.13 04
8	O K$_\alpha$	0.5249	2362.072	7.09 03	7.87 03	8.81 03	9.63 03	1.07 04
25	Mn L$_l$	0.5563	2228.747	6.26 03	6.97 03	7.80 03	8.55 03	9.53 03
24	Cr L$_\alpha$	0.5728	2164.546	5.86 03	6.55 03	7.33 03	8.04 03	8.96 03
25	Mn L$_\alpha$	0.6374	1945.171	4.67 03	5.22 03	5.84 03	6.42 03	7.17 03
9	F K$_\alpha$	0.6768	1831.932	4.07 03	4.54 03	5.08 03	5.59 03	6.26 03
26	Fe L$_\alpha$	0.7050	1758.655	3.76 03	4.18 03	4.68 03	5.15 03	5.77 03
27	Co L$_\alpha$	0.7762	1597.335	3.04 03	3.39 03	3.79 03	4.16 03	4.67 03
28	Ni L$_\alpha$	0.8515	1456.080	2.45 03	2.74 03	3.07 03	3.37 03	3.78 03
29	Cu L$_\alpha$	0.9297	1336.044	1.99 03	2.23 03	2.51 03	2.76 03	3.10 03
30	Zn L$_\alpha$	1.0117	1225.513	1.69 03	1.97 03	2.06 03	2.59 03	2.56 03
11	Na K$_\alpha$	1.0410	1191.020	1.57 03	1.82 03	1.93 03	2.17 03	2.39 03
11	Na K$_\beta$	1.0711	1157.550	1.45 03	1.69 03	1.75 03	2.08 03	2.17 03
12	Mg K$_\alpha$	1.2536	989.033	7.93 03	8.77 03	1.51 03	1.60 03	1.77 03
33	As L$_\alpha$	1.2820	967.123	7.47 03	8.25 03	1.41 03	1.51 03	1.67 03
12	Mg K$_\beta$	1.3022	952.121	8.56 03	7.91 03	1.35 03	1.45 03	1.60 03
33	As L$_{\beta_1}$	1.3170	941.421	8.31 03	7.67 03	1.31 03	1.40 03	1.55 03
66	Dy M$_\beta$	1.3250	935.737	8.17 03	7.54 03	5.87 03	1.38 03	1.53 03

Z	transition	E [keV]	λ [pm]	$_{31}$Ga	$_{32}$Ge	$_{33}$As	$_{34}$Se	$_{35}$Br
67	Ho M$_\alpha$	1.3480	919.771	7.80 03	7.19 03	5.60 03	1.32 03	1.46 03
34	Se L$_\alpha$	1.3791	899.029	7.34 03	6.76 03	7.37 03	1.24 03	1.37 03
67	Ho M$_\beta$	1.3830	896.494	7.28 03	6.71 03	7.31 03	1.23 03	1.36 03
68	Er M$_\alpha$	1.4060	881.829	6.79 03	6.41 03	6.99 03	1.17 03	1.30 03
34	Se L$_{\beta_1}$	1.4192	873.627	6.80 03	7.50 03	6.82 03	1.14 03	1.27 03
68	Er M$_\beta$	1.4430	859.218	6.50 03	7.17 03	6.52 03	4.98 03	1.21 03
69	Tm M$_\alpha$	1.4620	848.051	6.27 03	6.91 03	6.29 03	4.81 03	1.17 03
35	Br L$_\alpha$	1.4804	837.511	6.07 03	6.68 03	6.08 03	6.51 03	1.13 03
13	Al K$_{\alpha_1}$	1.4867	833.962	6.00 03	6.61 03	6.01 03	6.43 03	1.12 03
69	Tm M$_\beta$	1.5030	824.918	5.83 03	6.41 03	5.84 03	6.25 03	1.09 03
70	Yb M$_\alpha$	1.5214	814.941	5.64 03	6.20 03	5.65 03	6.05 03	1.05 03
35	Br L$_{\beta_1}$	1.5259	812.538	5.59 03	6.15 03	5.61 03	6.00 03	1.04 03
13	Al K$_\beta$	1.5574	796.103	5.29 03	5.82 03	6.36 03	5.68 03	4.53 03
70	Yb M$_\beta$	1.5675	790.974	5.20 03	5.72 03	6.25 03	5.58 03	4.45 03
71	Lu M$_\alpha$	1.5813	784.071	5.08 03	5.59 03	6.11 03	5.45 03	4.35 03
36	Kr L$_\alpha$	1.5860	781.747	5.04 03	5.54 03	6.06 03	5.40 03	4.32 03
71	Lu M$_\beta$	1.6312	760.085	4.67 03	5.14 03	5.61 03	5.01 03	5.59 03
36	Kr L$_{\beta_1}$	1.6366	757.577	4.63 03	5.09 03	5.56 03	4.96 03	5.54 03
72	Hf M$_{\alpha_1}$	1.6446	753.982	4.57 03	5.02 03	5.49 03	4.90 03	5.47 03
37	Rb L$_{\alpha_1}$	1.6941	731.864	4.22 03	4.63 03	5.07 03	5.43 03	5.05 03
72	Hf M$_\beta$	1.6976	730.355	4.19 03	4.61 03	5.04 03	5.40 03	5.02 03
73	Ta M$_{\alpha_1}$	1.7096	725.229	4.12 03	4.52 03	4.94 03	5.29 03	4.92 03
14	Si K$_{\alpha_1}$	1.7400	712.558	3.92 03	4.31 03	4.71 03	5.05 03	4.70 03
37	Rb L$_{\beta_1}$	1.7522	707.597	3.85 03	4.23 03	4.62 03	4.95 03	4.61 03
73	Ta M$_\beta$	1.7655	702.266	3.77 03	4.14 03	4.53 03	4.85 03	4.51 03
74	W M$_{\alpha_1}$	1.7754	698.350	3.72 03	4.08 03	4.46 03	4.78 03	4.45 03
38	Sr L$_{\alpha_1}$	1.8066	686.290	3.55 03	3.89 03	4.26 03	4.56 03	5.09 03
74	W M$_\beta$	1.8349	675.705	3.40 03	3.73 03	4.08 03	4.37 03	4.88 03
14	Si K$_\beta$	1.8359	675.337	3.40 03	3.73 03	4.07 03	4.37 03	4.87 03
75	Re M$_{\alpha_1}$	1.8420	673.100	3.37 03	3.69 03	4.04 03	4.33 03	4.83 03
38	Sr L$_{\beta_1}$	1.8717	662.420	3.22 03	3.54 03	3.87 03	4.14 03	4.62 03
75	Re M$_\beta$	1.9061	650.465	3.07 03	3.37 03	3.68 03	3.94 03	4.40 03
76	Os M$_{\alpha_1}$	1.9102	649.069	3.05 03	3.35 03	3.66 03	3.92 03	4.37 03
39	Y L$_{\alpha_1}$	1.9226	644.882	3.00 03	3.29 03	3.60 03	3.85 03	4.30 03
76	Os M$_\beta$	1.9783	626.725	2.78 03	3.04 03	3.33 03	3.57 03	3.98 03
77	Ir M$_{\alpha_1}$	1.9799	626.219	2.77 03	3.04 03	3.32 03	3.56 03	3.97 03
39	Y L$_{\beta_1}$	1.9958	621.230	2.71 03	2.97 03	3.25 03	3.48 03	3.88 03
15	P K$_{\alpha_1}$	2.0137	615.708	2.65 03	2.90 03	3.17 03	3.40 03	3.79 03
40	Zr L$_{\alpha_1}$	2.0424	607.056	2.55 03	2.79 03	3.05 03	3.27 03	3.65 03
78	Pt M$_{\alpha_1}$	2.0505	604.658	2.52 03	2.76 03	3.02 03	3.24 03	3.61 03
77	Ir M$_\beta$	2.0535	603.775	2.51 03	2.75 03	3.01 03	3.22 03	3.59 03
79	Au M$_{\alpha_1}$	2.1229	584.036	2.30 03	2.51 03	2.75 03	2.95 03	3.28 03
40	Zr L$_{\beta_1}$	2.1244	583.624	2.29 03	2.51 03	2.75 03	2.94 03	3.28 03
78	Pt M$_\beta$	2.1273	582.828	2.28 03	2.50 03	2.74 03	2.93 03	3.27 03

Z	transition	E [keV]	λ [pm]	$_{31}$Ga	$_{32}$Ge	$_{33}$As	$_{34}$Se	$_{35}$Br
15	P K$_{\beta_{1,3}}$	2.1390	579.640	2.25 03	2.46 03	2.70 03	2.89 03	3.22 03
41	Nb L$_{\alpha_1}$	2.1659	572.441	2.18 03	2.38 03	2.61 03	2.79 03	3.11 03
80	Hg M$_{\alpha_1}$	2.1953	564.775	2.10 03	2.30 03	2.51 03	2.69 03	3.00 03
79	Au M$_\beta$	2.2046	562.393	2.07 03	2.27 03	2.49 03	2.66 03	2.97 03
41	Nb L$_{\beta_1}$	2.2574	549.238	1.95 03	2.13 03	2.33 03	2.50 03	2.78 03
81	Tl M$_{\alpha_1}$	2.2706	546.045	1.92 03	2.10 03	2.30 03	2.46 03	2.74 03
80	Hg M$_\beta$	2.2825	543.199	1.89 03	2.07 03	2.26 03	2.42 03	2.70 03
42	Mo L$_{\alpha_1}$	2.2932	540.664	1.87 03	2.04 03	2.24 03	2.39 03	2.67 03
16	S K$_{\alpha_1}$	2.3080	537.197	1.83 03	2.01 03	2.20 03	2.35 03	2.62 03
82	Pb M$_{\alpha_1}$	2.3457	528.563	1.76 03	1.92 03	2.10 03	2.25 03	2.51 03
81	Tl M$_\beta$	2.3621	524.893	1.72 03	1.88 03	2.06 03	2.21 03	2.46 03
42	Mo L$_{\beta_1}$	2.3948	517.726	1.66 03	1.82 03	1.99 03	2.13 03	2.37 03
83	Bi M$_{\alpha_1}$	2.4226	511.785	1.61 03	1.76 03	1.93 03	2.06 03	2.30 03
43	Tc L$_{\alpha_1}$	2.4240	511.490	1.61 03	1.76 03	1.93 03	2.06 03	2.29 03
82	Pb M$_\beta$	2.4427	507.574	1.58 03	1.72 03	1.89 03	2.02 03	2.25 03
16	S K$_\beta$	2.4640	503.186	1.54 03	1.68 03	1.84 03	1.97 03	2.19 03
83	Bi M$_{\beta_1}$	2.5255	490.933	1.44 03	1.57 03	1.72 03	1.84 03	2.05 03
43	Tc L$_{\beta_1}$	2.5368	488.746	1.42 03	1.55 03	1.70 03	1.82 03	2.03 03
44	Ru L$_{\alpha_1}$	2.5586	484.582	1.39 03	1.52 03	1.67 03	1.78 03	1.98 03
17	Cl K$_{\alpha_1}$	2.6224	472.792	1.30 03	1.42 03	1.56 03	1.67 03	1.83 03
44	Ru L$_{\beta_1}$	2.6832	462.079	1.22 03	1.33 03	1.47 03	1.57 03	1.74 03
45	Rh L$_{\alpha_1}$	2.6967	459.766	1.21 03	1.32 03	1.45 03	1.55 03	1.72 03
17	Cl K$_\beta$	2.8156	440.350	1.08 03	1.17 03	1.29 03	1.38 03	1.53 03
45	Rh L$_{\beta_1}$	2.8344	437.430	1.06 03	1.15 03	1.26 03	1.35 03	1.50 03
46	Pd L$_{\alpha_1}$	2.8386	436.782	1.05 03	1.15 03	1.26 03	1.35 03	1.50 03
18	Ar K$_{\alpha_1}$	2.9577	419.194	9.42 02	1.03 03	1.13 03	1.20 03	1.34 03
47	Ag L$_{\alpha_1}$	2.9843	415.458	9.20 02	1.00 03	1.10 03	1.18 03	1.31 03
46	Pd L$_{\beta_1}$	2.9902	414.638	9.15 02	9.96 02	1.09 03	1.17 03	1.30 03
90	Th M$_{\alpha_1}$	2.9961	413.821	9.10 02	9.91 02	1.09 03	1.16 03	1.29 03
91	Pa M$_{\alpha_1}$	3.0823	402.248	8.43 02	9.17 02	1.01 03	1.08 03	1.20 03
48	Cd L$_{\alpha_1}$	3.1337	395.651	8.06 02	8.77 02	9.64 02	1.03 03	1.15 03
90	Th M$_\beta$	3.1458	394.129	7.98 02	8.68 02	9.54 02	1.02 03	1.13 03
47	Ag L$_{\beta_1}$	3.1509	393.491	7.94 02	8.64 02	9.50 02	1.02 03	1.13 03
92	U M$_{\alpha_1}$	3.1708	391.021	7.81 02	8.49 02	9.34 02	9.98 02	1.11 03
18	Ar K$_\beta$	3.1905	388.607	7.68 02	8.35 02	9.18 02	9.82 02	1.09 03
91	Pa M$_\beta$	3.2397	382.705	7.37 02	8.01 02	8.81 02	9.42 02	1.05 03
49	In L$_{\alpha_1}$	3.2869	377.210	7.08 02	7.70 02	8.47 02	9.06 02	1.01 03
19	K K$_{\alpha_1}$	3.3138	374.148	6.93 02	7.53 02	8.29 02	8.86 02	9.85 02
48	Cd L$_{\beta_1}$	3.3166	373.832	6.91 02	7.51 02	8.27 02	8.84 02	9.82 02
92	U M$_\beta$	3.3360	371.658	6.80 02	7.40 02	8.14 02	8.70 02	9.67 02
50	Sn L$_{\alpha_1}$	3.4440	360.003	6.24 02	6.78 02	7.46 02	7.99 02	8.87 02
49	In L$_{\beta_1}$	3.4872	355.543	6.04 02	6.56 02	7.22 02	7.72 02	8.58 02
19	K K$_\beta$	3.5896	345.401	5.58 02	6.06 02	6.67 02	7.14 02	7.93 02
51	Sb L$_{\alpha_1}$	3.6047	343.954	5.52 02	5.99 02	6.60 02	7.06 02	7.84 02

Z	transition	E [keV]	λ [pm]	$_{31}$Ga	$_{32}$Ge	$_{33}$As	$_{34}$Se	$_{35}$Br
50	Sn L$_{\beta_1}$	3.6628	338.498	5.28 02	5.74 02	6.32 02	6.76 02	7.51 02
20	Ca K$_{\alpha_1}$	3.6917	335.848	5.17 02	5.61 02	6.18 02	6.62 02	7.35 02
52	Te L$_{\alpha_1}$	3.7693	328.934	4.89 02	5.31 02	5.84 02	6.26 02	6.95 02
51	Sb L$_{\beta_1}$	3.8436	322.575	4.64 02	5.03 02	5.54 02	5.94 02	6.59 02
53	I L$_{\alpha_1}$	3.9377	314.867	4.34 02	4.71 02	5.19 02	5.56 02	6.17 02
20	Ca K$_\beta$	4.0127	308.981	4.13 02	4.48 02	4.93 02	5.29 02	5.87 02
52	Te L$_{\beta_1}$	4.0296	307.686	4.08 02	4.42 02	4.88 02	5.23 02	5.80 02
21	Sc K$_{\alpha_1}$	4.0906	303.097	3.92 02	4.25 02	4.69 02	5.02 02	5.57 02
54	Xe L$_{\alpha_1}$	4.1099	301.674	3.87 02	4.20 02	4.63 02	4.96 02	5.50 02
53	I L$_{\beta_1}$	4.2207	293.755	3.61 02	3.90 02	4.31 02	4.62 02	5.12 02
55	Cs L$_{\alpha_1}$	4.2865	289.245	3.46 02	3.75 02	4.13 02	4.43 02	4.91 02
21	Sc K$_\beta$	4.4605	277.962	3.11 02	3.36 02	3.71 02	3.98 02	4.41 02
56	Ba L$_{\alpha_1}$	4.4663	277.601	3.10 02	3.35 02	3.70 02	3.97 02	4.39 02
22	Ti K$_{\alpha_1}$	4.5108	274.863	3.02 02	3.26 02	3.60 02	3.86 02	4.28 02
55	Cs L$_{\beta_1}$	4.6198	268.377	2.83 02	3.06 02	3.38 02	3.62 02	4.01 02
57	La L$_{\alpha_1}$	4.6510	266.577	2.78 02	3.01 02	3.32 02	3.56 02	3.94 02
56	Ba L$_{\beta_1}$	4.8273	256.841	2.52 02	2.72 02	3.00 02	3.22 02	3.56 02
58	Ce L$_{\alpha_1}$	4.8402	256.157	2.50 02	2.70 02	2.98 02	3.20 02	3.54 02
22	Ti K$_{\beta_{1,3}}$	4.9318	251.399	2.37 02	2.57 02	2.84 02	3.04 02	3.36 02
23	V K$_{\alpha_1}$	4.9522	250.363	2.35 02	2.54 02	2.80 02	3.01 02	3.33 02
59	Pr L$_{\alpha_1}$	5.0337	246.310	2.25 02	2.43 02	2.68 02	2.88 02	3.18 02
57	La L$_{\beta_1}$	5.0421	245.899	2.24 02	2.42 02	2.67 02	2.86 02	3.17 02
60	Nd L$_{\alpha_1}$	5.2304	237.047	2.03 02	2.19 02	2.42 02	2.60 02	2.87 02
58	Ce L$_{\beta_1}$	5.2622	235.614	2.00 02	2.16 02	2.38 02	2.56 02	2.82 02
24	Cr K$_{\alpha_1}$	5.4147	228.978	1.85 02	2.00 02	2.21 02	2.37 02	2.61 02
23	V K$_{\beta_{1,3}}$	5.4273	228.447	1.84 02	1.99 02	2.20 02	2.35 02	2.60 02
61	Pm L$_{\alpha_1}$	5.4325	228.228	1.83 02	1.98 02	2.19 02	2.35 02	2.59 02
59	Pr L$_{\beta_1}$	5.4889	225.883	1.79 02	1.93 02	2.13 02	2.28 02	2.52 02
62	Sm L$_{\alpha_1}$	5.6361	219.984	1.66 02	1.79 02	1.99 02	2.13 02	2.35 02
60	Nd L$_{\beta_1}$	5.7216	216.696	1.60 02	1.72 02	1.91 02	2.04 02	2.25 02
63	Eu L$_{\alpha_1}$	5.8457	212.096	1.51 02	1.63 02	1.80 02	1.93 02	2.13 02
25	Mn K$_{\alpha_1}$	5.8988	210.187	1.47 02	1.59 02	1.76 02	1.88 02	2.08 02
24	Cr K$_{\beta_{1,3}}$	5.9467	208.494	1.44 02	1.56 02	1.72 02	1.84 02	2.03 02
61	Pm L$_{\beta_1}$	5.9610	207.993	1.43 02	1.55 02	1.71 02	1.83 02	2.02 02
64	Gd L$_{\alpha_1}$	6.0572	204.690	1.37 02	1.48 02	1.64 02	1.75 02	1.93 02
62	Sm L$_{\beta_1}$	6.2051	199.811	1.29 02	1.39 02	1.54 02	1.64 02	1.81 02
65	Tb L$_{\alpha_1}$	6.2728	197.655	1.25 02	1.35 02	1.49 02	1.60 02	1.76 02
26	Fe K$_{\alpha_1}$	6.4038	193.611	1.18 02	1.27 02	1.41 02	1.51 02	1.67 02
63	Eu L$_{\beta_1}$	6.4564	192.034	1.16 02	1.25 02	1.38 02	1.48 02	1.63 02
25	Mn K$_{\beta_1}$	6.4905	191.025	1.14 02	1.23 02	1.36 02	1.46 02	1.61 02
66	Dy L$_{\alpha_1}$	6.4952	190.887	1.14 02	1.23 02	1.36 02	1.45 02	1.60 02
64	Gd L$_{\beta_1}$	6.7132	184.688	1.04 02	1.12 02	1.24 02	1.33 02	1.47 02
67	Ho L$_{\alpha_1}$	6.7198	184.507	1.04 02	1.12 02	1.24 02	1.33 02	1.46 02
27	Co K$_{\alpha_1}$	6.9303	178.903	9.59 01	1.03 02	1.14 02	1.22 02	1.35 02

Z	transition	E [keV]	λ [pm]	$_{31}$Ga	$_{32}$Ge	$_{33}$As	$_{34}$Se	$_{35}$Br
68	Er L_{α_1}	6.9487	178.429	9.53 01	1.02 02	1.13 02	1.21 02	1.34 02
65	Tb L_{β_1}	6.9780	177.680	9.42 01	1.01 02	1.12 02	1.20 02	1.32 02
26	Fe K_{β_1}	7.0580	175.666	9.14 01	9.82 01	1.09 02	1.16 02	1.28 02
69	Tm L_{α_1}	7.1799	172.683	8.73 01	9.38 01	1.04 02	1.11 02	1.23 02
66	Dy L_{β_1}	7.2477	171.068	8.52 01	9.15 01	1.01 02	1.08 02	1.20 02
70	Yb L_{α_1}	7.4156	167.195	8.01 01	8.60 01	9.54 01	1.02 02	1.13 02
28	Ni K_{α_1}	7.4782	165.795	7.84 01	8.41 01	9.33 01	9.96 01	1.10 02
67	Ho L_{β_1}	7.5253	164.757	7.71 01	8.27 01	9.17 01	9.80 01	1.08 02
27	Co $K_{\beta_{1,3}}$	7.6494	162.084	7.38 01	7.92 01	8.87 01	9.38 01	1.04 02
71	Lu L_{α_1}	7.6555	161.955	7.36 01	7.90 01	8.76 01	9.36 01	1.03 02
68	Er L_{β_1}	7.8109	158.733	6.98 01	7.48 01	8.30 01	8.86 01	9.80 01
72	Hf L_{α_1}	7.8990	156.963	6.77 01	7.26 01	8.06 01	8.60 01	9.51 01
29	Cu K_{α_1}	8.0478	154.060	6.45 01	6.91 01	7.67 01	8.18 01	9.05 01
69	Tm L_{β_1}	8.1010	153.049	6.33 01	6.79 01	7.53 01	8.04 01	8.89 01
73	Ta L_{α_1}	8.1461	152.201	6.24 01	6.69 01	7.42 01	7.93 01	8.76 01
28	Ni K_{β_1}	8.2647	150.017	6.01 01	6.44 01	7.14 01	7.63 01	8.43 01
74	W L_{α_1}	8.3976	147.643	5.76 01	6.17 01	6.85 01	7.31 01	8.08 01
70	Yb L_{β_1}	8.4018	147.569	5.75 01	6.16 01	6.84 01	7.31 01	8.07 01
30	Zn K_{α_1}	8.6389	143.519	5.35 01	5.73 01	6.35 01	6.79 01	7.50 01
75	Re L_{α_1}	8.6525	143.294	5.32 01	5.70 01	6.33 01	6.76 01	7.46 01
71	Lu L_{β_1}	8.7090	142.364	5.23 01	5.60 01	6.22 01	6.64 01	7.34 01
29	Cu K_{β_1}	8.9053	139.226	4.93 01	5.28 01	5.86 01	6.27 01	6.91 01
76	Os L_{α_1}	8.9117	139.126	4.92 01	5.27 01	5.85 01	6.25 01	6.90 01
72	Hf L_{β_1}	9.0227	137.414	4.77 01	5.10 01	5.66 01	6.05 01	6.68 01
77	Ir L_{α_1}	9.1751	135.132	4.56 01	4.88 01	5.41 01	5.79 01	6.39 01
31	Ga K_{α_1}	9.2517	134.013	4.46 01	4.77 01	5.30 01	5.67 01	6.25 01
73	Ta L_{β_1}	9.3431	132.702	4.35 01	4.65 01	5.16 01	5.52 01	6.09 01
78	Pt L_{α_1}	9.4423	131.308	4.23 01	4.52 01	5.02 01	5.37 01	5.92 01
30	Zn $K_{\beta_{1,3}}$	9.5720	129.529	4.08 01	4.36 01	4.84 01	5.18 01	5.71 01
74	W L_{β_1}	9.6724	128.184	3.97 01	4.24 01	4.71 01	5.04 01	5.55 01
79	Au L_{α_1}	9.7133	127.644	3.93 01	4.20 01	4.65 01	4.98 01	5.49 01
32	Ge K_{α_1}	9.8864	125.409	3.75 01	4.00 01	4.44 01	4.76 01	5.24 01
80	Hg L_{α_1}	9.9888	124.124	3.65 01	3.89 01	4.32 01	4.63 01	5.10 01
75	Re L_{β_1}	10.010	123.861	3.63 01	3.87 01	4.30 01	4.60 01	5.07 01
31	Ga K_{β_1}	10.264	120.796	3.39 01	3.62 01	4.02 01	4.31 01	4.74 01
81	Tl L_{α_1}	10.269	120.737	3.39 01	3.62 01	4.02 01	4.30 01	4.74 01
76	Os L_{β_1}	10.355	119.734	<u>3.31 01</u>	3.54 01	3.93 01	4.21 01	4.63 01
33	As K_{α_1}	10.544	117.588	2.25 02	3.37 01	3.75 01	4.01 01	4.42 01
82	Pb L_{α_1}	10.552	117.499	2.24 02	3.37 01	3.74 01	4.00 01	4.41 01
77	Ir L_{β_1}	10.708	115.787	2.16 02	3.24 01	3.60 01	3.85 01	4.24 01
83	Bi L_{α_1}	10.839	114.388	2.09 02	3.14 01	3.49 01	3.73 01	4.11 01
32	Ge K_{β_1}	10.982	112.898	2.01 02	3.03 01	3.37 01	3.60 01	3.97 01
78	Pt L_{β_1}	11.071	111.990	1.97 02	<u>2.97 01</u>	3.30 01	3.53 01	3.88 01
84	Po L_{α_1}	11.131	111.387	1.94 02	2.03 02	3.25 01	3.48 01	3.83 01

Z	transition	E [keV]	λ [pm]	$_{31}$Ga	$_{32}$Ge	$_{33}$As	$_{34}$Se	$_{35}$Br
34	Se K$_{\alpha_1}$	11.222	110.484	1.90 02	1.99 02	3.18 01	3.40 01	3.75 01
85	At L$_{\alpha_1}$	11.427	108.501	1.81 02	1.90 02	3.04 01	3.25 01	3.57 01
79	Au L$_{\beta_1}$	11.442	108.359	1.80 02	1.89 02	3.03 01	3.23 01	3.56 01
33	As K$_{\beta_1}$	11.726	105.735	1.69 02	1.77 02	2.84 01	3.03 01	3.34 01
86	Rn L$_{\alpha_1}$	11.727	105.726	1.69 02	1.77 02	2.84 01	3.03 01	3.34 01
80	Hg L$_{\beta_1}$	11.823	104.867	1.65 02	1.73 02	2.78 01	2.97 01	3.27 01
35	Br K$_{\alpha_1}$	11.924	103.979	1.62 02	1.69 02	1.88 02	2.90 01	3.19 01
87	Fr L$_{\alpha_1}$	12.031	103.054	1.58 02	1.65 02	1.84 02	2.83 01	3.12 01
81	Tl L$_{\beta_1}$	12.213	101.519	1.51 02	1.59 02	1.76 02	2.72 01	3.00 01
88	Ra L$_{\alpha_1}$	12.340	100.474	1.47 02	1.54 02	1.72 02	2.65 01	2.92 01
34	Se K$_{\beta_1}$	12.496	99.219	1.42 02	1.49 02	1.66 02	2.57 01	2.82 01
82	Pb L$_{\beta_1}$	12.614	98.291	1.39 02	1.45 02	1.62 02	2.50 01	2.75 01
36	Kr K$_{\alpha_1}$	12.649	98.019	1.38 02	1.44 02	1.60 02	<u>2.48 01</u>	2.73 01
89	Ac L$_{\alpha_1}$	12.652	97.996	1.38 02	1.44 02	1.60 02	2.48 01	2.73 01
90	Th L$_{\alpha_1}$	12.969	95.601	1.29 02	1.35 02	1.50 02	1.60 02	2.56 01
83	Bi L$_{\beta_1}$	13.024	95.197	1.27 02	1.33 02	1.48 02	1.58 02	2.53 01
91	Pa L$_{\alpha_1}$	13.291	93.285	1.21 02	1.26 02	1.40 02	1.49 02	2.40 01
35	Br K$_{\beta_1}$	13.291	93.285	1.21 01	1.26 02	1.40 02	1.49 02	2.40 01
37	Rb K$_{\alpha_1}$	13.395	92.560	1.18 02	1.24 02	1.37 02	1.46 02	2.35 01
84	Po L$_{\beta_1}$	13.447	92.202	1.17 02	1.22 02	1.36 02	1.45 02	<u>2.32 01</u>
92	U L$_{\alpha_1}$	13.615	91.065	1.13 02	1.18 02	1.32 02	1.40 02	1.52 02
85	At L$_{\beta_1}$	13.876	89.352	1.08 02	1.13 02	1.25 02	1.33 02	1.45 02
36	Kr K$_{\beta_1}$	14.112	87.857	1.03 02	1.08 02	1.19 02	1.27 02	1.38 02
38	Sr K$_{\alpha_1}$	14.165	87.529	1.02 02	1.06 02	1.18 02	1.26 02	1.37 02
86	Rn L$_{\beta_1}$	14.316	86.606	9.89 01	1.03 02	1.15 02	1.23 02	1.33 02
87	Fr L$_{\beta_1}$	14.770	83.943	9.09 01	9.51 01	1.06 02	1.13 02	1.23 02
39	Y K$_{\alpha_1}$	14.958	82.888	8.79 01	9.20 01	1.02 02	1.09 02	1.18 02
37	Rb K$_{\beta_1}$	14.961	82.872	8.79 01	9.19 01	1.02 02	1.09 02	1.18 02
88	Ra L$_{\beta_1}$	15.236	81.376	8.35 01	8.75 01	9.70 01	1.04 02	1.13 02
89	Ac L$_{\beta_1}$	15.713	78.906	7.67 01	8.04 01	8.92 01	9.54 01	1.04 02
40	Zr K$_{\alpha_1}$	15.775	78.596	7.58 01	7.96 01	8.83 01	9.44 01	1.03 02
38	Sr K$_{\beta_1}$	15.836	78.293	7.50 01	7.87 01	8.73 01	9.34 01	1.01 02
90	Th L$_{\beta_1}$	16.202	76.524	7.04 01	7.40 01	8.20 01	8.77 01	9.54 01
41	Nb K$_{\alpha_1}$	16.615	74.622	6.57 01	6.91 01	7.66 01	8.19 01	8.91 01
91	Pa L$_{\beta_1}$	16.702	74.233	6.47 01	6.81 01	7.55 01	8.07 01	8.78 01
39	Y K$_{\beta_1}$	16.738	74.074	6.43 01	6.77 01	7.51 01	8.02 01	8.73 01
92	U L$_{\beta_1}$	17.220	72.000	5.94 01	6.26 01	6.95 01	7.42 01	8.08 01
42	Mo K$_{\alpha_1}$	17.479	70.933	5.70 01	6.01 01	6.67 01	7.12 01	7.76 01
40	Zr K$_{\beta_1}$	17.668	70.175	5.54 01	5.84 01	6.47 01	6.92 01	7.54 01
43	Tc K$_{\alpha_1}$	18.367	67.504	4.97 01	5.25 01	5.82 01	6.22 01	6.78 01
41	Nb K$_{\beta_1}$	18.623	66.576	4.78 01	5.06 01	5.61 01	5.99 01	6.53 01
44	Ru K$_{\alpha_1}$	19.279	64.311	4.34 01	4.60 01	5.10 01	5.45 01	5.95 01
42	Mo K$_{\beta_1}$	19.608	63.231	4.14 01	4.40 01	4.87 01	5.20 01	5.68 01
45	Rh K$_{\alpha_1}$	20.216	61.330	3.80 01	4.04 01	4.48 01	4.78 01	5.22 01

Z	transition	E [keV]	λ [pm]	$_{31}$Ga	$_{32}$Ge	$_{33}$As	$_{34}$Se	$_{35}$Br
43	Tc K$_{\beta_1}$	20.619	60.131	3.60 01	3.83 01	4.24 01	4.52 01	4.95 01
46	Pd K$_{\alpha_1}$	21.177	58.547	3.34 01	3.55 01	3.93 01	4.20 01	4.59 01
44	Ru K$_{\beta_1}$	21.657	57.249	3.13 01	3.34 01	3.69 01	3.94 01	4.32 01
47	Ag K$_{\alpha_1}$	22.163	55.942	2.94 01	3.13 01	3.46 01	3.70 01	4.05 01
45	Rh K$_{\beta_1}$	22.724	54.561	2.74 01	2.92 01	3.23 01	3.45 01	3.78 01
48	Cd K$_{\alpha_1}$	23.174	53.501	2.59 01	2.76 01	3.06 01	3.26 01	3.58 01
46	Pd K$_{\beta_1}$	23.819	52.053	2.40 01	2.56 01	2.83 01	3.02 01	3.31 01
49	In K$_{\alpha_1}$	24.210	51.212	2.29 01	2.44 01	2.71 01	2.89 01	3.17 01
47	Ag K$_{\beta_1}$	24.942	49.709	2.11 01	2.25 01	2.49 01	2.66 01	2.92 01
50	Sn K$_{\alpha_1}$	25.271	39.062	2.03 01	2.17 01	2.40 01	2.56 01	2.81 01
48	Cd K$_{\beta_1}$	26.096	47.511	1.85 01	1.98 01	2.19 01	2.34 01	2.57 01
51	Sb K$_{\alpha_1}$	26.359	47.037	1.80 01	1.93 01	2.13 01	2.28 01	2.50 01
49	In K$_{\beta_1}$	27.276	45.455	1.64 01	1.75 01	1.94 01	2.07 01	2.28 01
52	Te K$_{\alpha_1}$	27.472	45.131	1.60 01	1.72 01	1.90 01	2.03 01	2.23 01
50	Sn K$_{\beta_1}$	28.486	43.524	1.45 01	1.55 01	1.72 01	1.83 01	2.02 01
53	I K$_{\alpha_1}$	28.612	43.333	1.43 01	1.53 01	1.70 01	1.81 01	1.99 01
51	Sb K$_{\beta_1}$	29.726	41.709	1.29 01	1.38 01	1.53 01	1.63 01	1.79 01
54	Xe K$_{\alpha_1}$	29.779	41.635	1.28 01	1.37 01	1.52 01	1.62 01	1.78 01
55	Cs K$_{\alpha_1}$	30.973	40.030	1.15 01	1.23 01	1.36 01	1.45 01	1.60 01
52	Te K$_{\beta_1}$	30.996	40.000	1.14 01	1.23 01	1.36 01	1.45 01	1.59 01
56	Ba K$_{\alpha_1}$	32.194	38.511	1.03 01	1.10 01	1.22 01	1.30 01	1.43 01
53	I K$_{\beta_1}$	32.295	38.391	1.02 01	1.09 01	1.21 01	1.29 01	1.42 01
57	La K$_{\alpha_1}$	33.442	37.074	9.23 00	9.91 00	1.09 01	1.17 01	1.29 01
54	Xe K$_{\beta_1}$	33.624	36.874	9.09 00	9.76 00	1.08 01	1.15 01	1.27 01
58	Ce K$_{\alpha_1}$	34.279	36.169	8.61 00	9.24 00	1.02 01	1.09 01	1.20 01
55	Cs K$_{\beta_1}$	34.987	35.437	8.13 00	8.73 00	9.64 00	1.03 01	1.13 01
59	Pr K$_{\alpha_1}$	36.026	34.415	7.49 00	8.04 00	8.87 00	9.49 00	1.04 01
56	Ba K$_{\beta_1}$	36.378	34.082	7.29 00	7.82 00	8.63 00	9.24 00	1.01 01
60	Nd K$_{\alpha_1}$	36.847	33.648	7.03 00	7.54 00	8.33 00	8.91 00	9.79 00
57	La K$_{\beta_1}$	37.801	32.799	6.54 00	7.02 00	7.75 00	8.29 00	9.11 00
61	Pm K$_{\alpha_1}$	38.725	32.016	6.11 00	6.56 00	7.24 00	7.75 00	8.51 00
58	Ce K$_{\beta_1}$	39.257	31.582	5.88 00	6.31 00	6.97 00	7.46 00	8.19 00
62	Sm K$_{\alpha_1}$	40.118	30.905	5.53 00	5.94 00	6.55 00	7.02 00	7.70 00
59	Pr K$_{\beta_1}$	40.748	30.427	5.30 00	5.69 00	6.27 00	6.72 00	7.37 00
63	Eu K$_{\alpha_1}$	41.542	29.845	5.02 00	5.39 00	5.94 00	6.36 00	6.99 00
60	Nd K$_{\beta_1}$	42.271	29.331	4.78 00	5.13 00	5.66 00	6.06 00	6.66 00
64	Gd K$_{\alpha_1}$	42.996	28.836	4.56 00	4.90 00	5.40 00	5.78 00	6.35 00
61	Pm K$_{\beta_1}$	43.826	28.290	4.32 00	4.64 00	5.12 00	5.48 00	6.02 00
65	Tb K$_{\alpha_1}$	44.482	27.873	4.15 00	4.45 00	4.91 00	5.25 00	5.77 00
62	Sm K$_{\beta_1}$	45.413	27.301	3.91 00	4.20 00	4.63 00	4.96 00	5.45 00
66	Dy K$_{\alpha_1}$	45.998	26.954	3.77 00	4.05 00	4.47 00	4.78 00	5.26 00
63	Eu K$_{\beta_1}$	47.038	26.358	3.55 00	3.81 00	4.20 00	4.49 00	4.94 00
67	Ho K$_{\alpha_1}$	47.547	26.076	3.44 00	3.70 00	4.07 00	4.36 00	4.80 00
64	Gd K$_{\beta_1}$	48.697	25.460	4.22 00	3.46 00	3.81 00	4.07 00	4.49 00

Z	transition	E [keV]	λ [pm]	$_{31}$Ga	$_{32}$Ge	$_{33}$As	$_{34}$Se	$_{35}$Br
68	Er K_{α_1}	49.128	25.237	3.14 00	3.37 00	3.72 00	3.97 00	4.38 00
65	Tb K_{β_1}	50.382	24.609	2.93 00	3.14 00	3.46 00	3.70 00	4.08 00
69	Tm K_{α_1}	50.742	24.434	2.87 00	3.08 00	3.40 00	3.63 00	4.00 00
66	Dy K_{β_1}	52.119	23.788	2.67 00	2.87 00	3.16 00	3.37 00	3.72 00
70	Yb K_{α_1}	52.389	23.666	2.63 00	2.83 00	3.11 00	3.33 00	3.66 00
67	Ho K_{β_1}	53.877	23.012	2.44 00	2.62 00	2.88 00	3.08 00	3.39 00
71	Lu K_{α_1}	54.070	22.930	2.41 00	2.59 00	2.85 00	3.05 00	3.36 00
68	Er K_{β_1}	55.681	22.267	2.23 00	2.39 00	2.63 00	2.81 00	3.10 00
72	Hf K_{α_1}	55.790	22.223	2.21 00	2.38 00	2.62 00	2.80 00	3.08 00
69	Tm K_{β_1}	57.517	21.556	2.04 00	2.19 00	2.41 00	2.57 00	2.83 00
73	Ta K_{α_1}	57.532	21.550	2.04 00	2.19 00	2.41 00	2.57 00	2.83 00
74	W K_{α_1}	59.318	20.901	1.87 00	2.01 00	2.21 00	2.36 00	2.60 00
70	Yb K_{β_1}	59.370	20.883	1.87 00	2.01 00	2.21 00	2.36 00	2.59 00
75	Re K_{α_1}	61.140	20.278	1.73 00	1.86 00	2.04 00	2.18 00	2.39 00
71	Lu K_{β_1}	61.283	20.231	1.72 00	1.85 00	2.03 00	2.16 00	2.38 00
76	Os K_{α_1}	63.001	19.679	1.60 00	1.72 00	1.89 00	2.01 00	2.21 00
72	Hf K_{β_1}	63.234	19.607	1.58 00	1.70 00	1.87 00	1.99 00	2.19 00
77	Ir K_{α_1}	64.896	19.105	1.48 00	1.59 00	1.74 00	1.86 00	2.04 00
73	Ta K_{β_1}	65.223	19.009	1.46 00	1.57 00	1.72 00	1.83 00	2.01 00
78	Pt K_{α_1}	66.832	18.551	1.37 00	1.47 00	1.61 00	1.72 00	1.89 00
74	W K_{β_1}	67.244	18.438	1.35 00	1.45 00	1.59 00	1.69 00	1.86 00
79	Au K_{α_1}	68.804	18.020	1.27 00	1.36 00	1.50 00	1.59 00	1.75 00
75	Re K_{β_1}	69.310	17.888	1.25 00	1.34 00	1.47 00	1.56 00	1.71 00
80	Hg K_{α_1}	70.819	17.507	1.18 00	1.27 00	1.39 00	1.47 00	1.62 00
76	Os K_{β_1}	71.413	17.361	1.16 00	1.24 00	1.36 00	1.44 00	1.58 00
81	Tl K_{α_1}	72.872	17.014	1.10 00	1.17 00	1.29 00	1.37 00	1.50 00
77	Ir K_{β_1}	73.561	16.854	1.07 00	1.15 00	1.25 00	1.33 00	1.46 00
82	Pb K_{α_1}	74.696	16.538	1.02 00	1.09 00	1.19 00	1.27 00	1.39 00
78	Pt K_{β_1}	75.748	16.368	9.93 -1	1.06 00	1.16 00	1.23 00	1.35 00
83	Bi K_{α_1}	77.108	16.079	9.48 -1	1.01 00	1.11 00	1.18 00	1.29 00
79	Au K_{β_1}	77.948	15.906	9.22 -1	9.86 -1	1.08 00	1.14 00	1.25 00
84	Po K_{α_1}	79.290	15.636	8.82 -1	9.43 -1	1.03 00	1.09 00	1.20 00
80	Hg K_{β_1}	80.253	15.449	8.56 -1	9.14 -1	9.97 -1	1.06 00	1.16 00
85	At K_{α_1}	81.520	15.209	8.25 -1	8.81 -1	9.60 -1	1.02 00	1.12 00
81	Tl K_{β_1}	82.576	15.014	8.00 -1	8.54 -1	9.31 -1	9.88 -1	1.08 00
86	Rn K_{α_1}	83.780	14.798	7.74 -1	8.26 -1	8.99 -1	9.53 -1	1.04 00
82	Pb K_{β_1}	84.936	14.597	7.49 -1	7.99 -1	8.69 -1	9.21 -1	1.01 00
87	Fr K_{α_1}	86.100	14.400	7.26 -1	7.74 -1	8.41 -1	8.91 -1	9.74 -1
83	Bi K_{β_1}	87.343	14.195	7.02 -1	7.48 -1	8.13 -1	8.60 -1	9.40 -1
88	Ra K_{α_1}	88.470	14.014	6.81 -1	7.26 -1	7.88 -1	8.33 -1	9.11 -1
84	Po K_{β_1}	89.800	13.806	6.58 -1	7.01 -1	7.60 -1	8.03 -1	8.78 -1
89	Ac K_{α_1}	90.884	13.642	6.40 -1	6.81 -1	7.38 -1	7.80 -1	8.52 -1
85	At K_{β_1}	92.300	13.432	6.17 -1	6.57 -1	7.11 -1	7.51 -1	8.20 -1
90	Th K_{α_1}	93.350	13.281	6.01 -1	6.39 -1	6.92 -1	7.30 -1	7.97 -1

Z	transition	E [keV]	λ [pm]	$_{31}$Ga	$_{32}$Ge	$_{33}$As	$_{34}$Se	$_{35}$Br
86	Rn K$_{\beta_1}$	94.780	13.068	5.78 -1	6.15 -1	6.65 -1	7.02 -1	7.66 -1
91	Pa K$_{\alpha_1}$	95.868	12.932	5.64 -1	6.00 -1	6.49 -1	6.84 -1	7.46 -1
87	Fr K$_{\beta_1}$	97.470	12.720	5.43 -1	5.77 -1	6.23 -1	6.56 -1	7.16 -1
92	U K$_{\alpha_1}$	98.439	12.595	5.31 -1	5.64 -1	6.09 -1	6.41 -1	6.99 -1
88	Ra K$_{\beta_1}$	100.130	12.382	5.10 -1	5.42 -1	5.84 -1	6.15 -1	6.70 -1
89	Ac K$_{\beta_1}$	102.850	12.054	4.85 -1	5.15 -1	5.54 -1	5.82 -1	6.34 -1
90	Th K$_{\beta_1}$	105.610	11.739	4.61 -1	4.89 -1	5.25 -1	5.51 -1	5.99 -1
91	Pa K$_{\beta_1}$	108.430	11.434	4.39 -1	4.64 -1	4.98 -1	5.22 -1	5.67 -1
92	U K$_{\beta_1}$	111.300	11.139	4.18 -1	4.41 -1	4.73 -1	4.94 -1	5.37 -1

Z	transition	E [keV]	λ [pm]	$_{36}$Kr	$_{37}$Rb	$_{38}$Sr	$_{39}$Y	$_{40}$Zr
4	Be K$_\alpha$	0.1085	11427.207	1.21 04	<u>5.02 03</u>	5.89 03	1.19 04	1.09 04
38	Sr M$_\xi$	0.1140	10875.895	1.45 04	7.05 03	6.22 03	1.12 04	1.02 04
39	Y M$_\xi$	0.1328	9339.235	2.48 04	2.64 04	<u>7.58 03</u>	1.06 04	9.27 03
16	S L$_l$	0.1487	8337.942	3.82 04	2.94 04	1.40 04	1.22 04	8.61 03
40	Zr M$_\xi$	0.1511	8205.506	3.92 04	2.98 04	1.51 04	<u>1.25 04</u>	8.52 03
41	Nb M$_\xi$	0.1717	7221.037	3.99 04	3.38 04	2.26 04	1.78 04	<u>7.96 03</u>
5	B K$_\alpha$	0.1833	6764.059	4.02 04	3.59 04	2.73 04	1.92 04	8.27 04
42	Mo M$_\xi$	0.1926	6337.445	<u>4.05 04</u>	<u>3.78 04</u>	3.22 04	2.04 04	8.53 04
6	C K$_\alpha$	0.2770	4476.000	3.33 04	3.55 04	<u>3.53 04</u>	<u>3.14 04</u>	3.16 04
47	Ag M$_\xi$	0.3117	3977.709	2.83 04	<u>3.02 04</u>	<u>3.07 04</u>	<u>2.86 04</u>	<u>2.25 04</u>
7	N K$_\alpha$	0.3924	3159.664	2.07 04	2.20 04	2.33 04	<u>2.38 04</u>	2.04 04
22	Ti L$_l$	0.3935	3136.484	2.04 04	2.17 04	2.30 04	2.51 04	<u>2.03 04</u>
22	Ti L$_\alpha$	0.4522	2741.822	1.59 04	1.70 04	1.81 04	1.98 04	1.71 04
23	V L$_\alpha$	0.5113	2424.901	1.25 04	1.35 04	1.44 04	1.59 04	1.52 04
8	O K$_\alpha$	0.5249	2362.072	1.19 04	1.28 04	1.38 04	1.51 04	1.48 04
25	Mn L$_l$	0.5563	2228.747	1.06 04	1.14 04	1.23 04	1.35 04	1.33 04
24	Cr L$_\alpha$	0.5728	2164.546	9.97 03	1.07 04	1.16 04	1.28 04	1.27 04
25	Mn L$_\alpha$	0.6374	1945.171	7.99 03	8.63 03	9.36 03	1.03 04	1.05 04
9	F K$_\alpha$	0.6768	1831.932	6.96 03	7.55 03	8.21 03	9.05 03	9.42 03
26	Fe L$_\alpha$	0.7050	1758.655	6.42 03	6.98 03	7.59 03	8.37 03	8.76 03
27	Co L$_\alpha$	0.7762	1597.335	5.19 03	5.67 03	6.17 03	6.82 03	7.39 03
28	Ni L$_\alpha$	0.8515	1456.080	4.21 03	4.61 03	5.03 03	5.57 03	6.04 03
29	Cu L$_\alpha$	0.9297	1336.044	3.45 03	3.78 03	4.13 03	4.58 03	4.97 03
30	Zn L$_\alpha$	1.0117	1225.513	3.26 03	3.66 03	3.95 03	4.30 03	4.65 03
11	Na K$_\alpha$	1.0410	1191.020	3.00 03	3.38 03	3.66 03	3.99 03	4.32 03
11	Na K$_\beta$	1.0711	1157.550	2.77 03	3.13 03	3.39 03	3.71 03	4.02 03
12	Mg K$_\alpha$	1.2530	989.033	1.78 03	2.03 03	2.22 03	2.46 03	2.69 03
33	As L$_\alpha$	1.2820	967.123	1.68 03	1.91 03	2.09 03	2.32 03	2.54 03
12	Mg K$_\beta$	1.3022	952.121	1.60 03	1.83 03	2.01 03	2.23 03	2.44 03
33	As L$_{\beta_1}$	1.3170	941.421	1.55 03	1.77 03	1.95 03	2.16 03	2.37 03
66	Dy M$_\beta$	1.3250	935.737	1.53 03	1.74 03	1.92 03	2.13 03	2.33 03

Z	transition	E [keV]	λ [pm]	$_{36}$Kr	$_{37}$Rb	$_{38}$Sr	$_{39}$Y	$_{40}$Zr
67	Ho M_α	1.3480	919.771	1.45 03	1.66 03	1.83 03	2.03 03	2.23 03
34	Se L_α	1.3791	899.029	1.36 03	1.56 03	1.72 03	1.92 03	2.10 03
67	Ho M_β	1.3830	896.494	1.35 03	1.55 03	1.71 03	1.90 03	2.09 03
68	Er M_α	1.4060	881.829	1.29 03	1.48 03	1.63 03	1.82 03	2.00 03
34	Se L_{β_1}	1.4192	873.627	1.26 03	1.44 03	1.59 03	1.78 03	1.96 03
68	Er M_β	1.4430	859.218	1.20 03	1.38 03	1.53 03	1.70 03	1.87 03
69	Tm M_α	1.4620	848.051	1.16 03	1.33 03	1.47 03	1.64 03	1.81 03
35	Br L_α	1.4804	837.511	1.12 03	1.29 03	1.42 03	1.59 03	1.75 03
13	Al K_{α_1}	1.4867	833.962	1.11 03	1.27 03	1.41 03	1.57 03	1.74 03
69	Tm M_β	1.5030	824.918	1.07 03	1.23 03	1.37 03	1.53 03	1.69 03
70	Yb M_α	1.5214	814.941	1.04 03	1.19 03	1.32 03	1.48 03	1.64 03
35	Br L_{β_1}	1.5259	812.538	1.03 03	1.18 03	1.31 03	1.47 03	1.62 03
13	Al K_β	1.5574	796.103	9.70 02	1.12 03	1.24 03	1.40 03	1.54 03
70	Yb M_β	1.5675	790.974	9.53 02	1.10 03	1.22 03	1.37 03	1.52 03
71	Lu M_α	1.5813	784.071	9.29 02	1.07 03	1.19 03	1.34 03	1.48 03
36	Kr L_α	1.5860	781.747	9.22 02	1.06 03	1.19 03	1.33 03	1.47 03
71	Lu M_β	1.6312	760.085	8.52 02	9.85 02	1.10 03	1.24 03	1.37 03
36	Kr L_{β_1}	1.6366	757.577	8.44 02	9.76 02	1.09 03	1.23 03	1.36 03
72	Hf M_{α_1}	1.6446	753.892	<u>8.32 02</u>	9.63 02	1.08 03	1.21 03	1.34 03
37	Rb L_{α_1}	1.6941	731.864	3.81 03	8.88 02	9.95 02	1.12 03	1.24 03
72	Hf M_β	1.6976	730.355	3.79 03	8.83 03	9.90 02	1.12 03	1.24 03
73	Ta M_{α_1}	1.7096	725.229	<u>3.72 03</u>	8.66 02	9.71 02	1.10 03	1.21 03
14	Si K_{α_1}	1.7400	712.558	4.96 03	8.25 02	9.27 02	1.05 03	1.16 03
37	Rb L_{β_1}	1.7522	707.597	4.87 03	8.09 02	9.10 02	1.03 03	1.14 03
73	Ta M_β	1.7655	702.266	4.77 03	7.92 02	8.92 02	1.01 03	1.12 03
74	W M_{α_1}	1.7754	698.350	4.70 03	<u>7.80 02</u>	8.78 02	9.95 02	1.10 03
38	Sr L_{α_1}	1.8066	686.290	4.48 03	3.61 03	8.39 02	9.51 02	1.05 03
74	W M_β	1.8349	675.705	4.30 03	3.46 03	8.05 02	9.14 02	1.01 03
14	Si K_β	1.8359	675.337	4.29 03	3.46 03	8.04 02	9.13 02	1.02 03
75	Re M_{α_1}	1.8420	673.100	4.25 03	<u>3.43 03</u>	7.97 02	9.05 02	1.00 03
38	Sr L_{β_1}	1.8717	662.420	4.07 03	4.60 03	7.64 02	8.68 02	9.62 02
75	Re M_β	1.9061	650.465	3.88 03	4.37 03	7.27 02	8.28 02	9.18 02
76	Os M_{α_1}	1.9102	649.069	<u>3.85 03</u>	4.35 03	7.23 02	8.24 02	9.13 02
39	Y L_{α_1}	1.9226	644.882	4.55 03	4.27 03	<u>7.11 02</u>	8.10 02	8.98 02
76	Os M_β	1.9783	626.725	4.21 03	3.95 03	3.07 03	7.52 02	8.35 02
77	Ir M_{α_1}	1.9799	626.219	4.20 03	3.94 03	3.06 03	7.51 02	8.33 02
39	Y L_{β_1}	1.9958	621.230	4.11 03	3.86 03	<u>2.99 03</u>	7.36 02	8.16 02
15	P K_{α_1}	2.0137	615.708	4.01 03	3.76 03	4.09 03	7.18 02	7.98 02
40	Zr L_{α_1}	2.0424	607.056	3.86 03	3.62 03	3.94 03	6.92 02	7.70 02
78	Pt M_{α_1}	2.0505	604.658	3.82 03	3.58 03	3.89 03	6.84 02	7.63 02
77	Ir M_β	2.0535	603.775	3.80 03	<u>3.57 03</u>	3.88 03	<u>6.82 02</u>	7.60 02
79	Au M_{α_1}	2.1229	584.036	3.47 03	3.91 03	3.54 03	2.77 03	6.99 02
40	Zr L_{β_1}	2.1244	583.624	3.47 03	3.90 03	3.53 03	2.77 03	6.98 02
78	Pt M_β	2.1273	582.828	3.46 03	3.89 03	3.52 03	2.76 03	6.96 02

Z	transition	E [keV]	λ [pm]	$_{36}$Kr	$_{37}$Rb	$_{38}$Sr	$_{39}$Y	$_{40}$Zr
15	P K$_{\beta_{1,3}}$	2.1390	579.640	3.40 03	3.83 03	3.47 03	<u>2.71 03</u>	6.86 02
41	Nb L$_{\alpha_1}$	2.1659	572.441	3.29 03	3.70 03	3.35 03	3.67 03	6.65 02
80	Hg M$_{\alpha_1}$	2.1953	564.775	3.17 03	3.57 03	3.23 03	3.54 03	6.43 02
79	Au M$_\beta$	2.2046	562.393	3.14 03	3.53 03	<u>3.19 03</u>	3.50 03	<u>6.36 02</u>
41	Nb L$_{\beta_1}$	2.2574	549.238	2.94 03	3.30 03	3.59 03	3.28 03	2.57 03
81	Tl M$_{\alpha_1}$	2.2706	546.045	2.90 03	3.25 03	3.54 03	3.23 03	2.52 03
80	Hg M$_\beta$	2.2825	543.199	2.86 03	3.21 03	3.49 03	3.18 03	2.49 03
42	Mo L$_{\alpha_1}$	2.2932	540.664	2.82 03	3.17 03	3.44 03	3.14 03	<u>2.46 03</u>
16	S K$_{\alpha_1}$	2.3080	537.197	2.77 03	3.11 03	3.38 03	3.08 03	3.38 03
82	Pb M$_{\alpha_1}$	2.3457	528.563	2.65 03	2.98 03	3.24 03	2.95 03	3.24 03
81	Tl M$_\beta$	2.3621	524.893	2.60 03	2.92 03	3.17 03	<u>2.89 03</u>	3.17 03
42	Mo L$_{\beta_1}$	2.3948	517.726	2.51 03	2.81 03	3.06 03	3.35 03	3.06 03
83	Bi M$_{\alpha_1}$	2.4226	511.785	2.43 03	2.73 03	2.96 03	3.25 03	2.96 03
43	Tc L$_{\alpha_1}$	2.4240	511.490	2.43 03	2.72 03	2.96 03	3.24 03	2.96 03
82	Pb M$_\beta$	2.4427	507.574	2.38 03	2.67 03	2.90 03	3.17 03	2.89 03
16	S K$_\beta$	2.4640	503.186	2.32 03	2.60 03	2.83 03	3.10 03	2.83 03
83	Bi M$_{\beta_1}$	2.5255	490.933	2.17 03	2.43 03	2.64 03	2.90 03	<u>2.64 03</u>
43	Tc L$_{\beta_1}$	2.5368	488.746	2.15 03	2.40 03	2.61 03	2.86 03	3.13 03
44	Ru L$_{\alpha_1}$	2.5586	484.582	2.10 03	2.35 03	2.55 03	2.80 03	3.05 03
17	Cl K$_{\alpha_1}$	2.6224	472.792	1.96 03	2.20 03	2.38 03	2.61 03	2.85 03
44	Ru L$_{\beta_1}$	2.6832	462.079	1.85 03	2.06 03	2.24 03	2.45 03	2.66 03
45	Rh L$_{\alpha_1}$	2.6967	459.766	1.82 03	2.04 03	2.21 03	2.42 03	2.64 03
17	Cl K$_\beta$	2.8156	440.350	1.62 03	1.81 03	1.96 03	2.15 03	2.35 03
45	Rh L$_{\beta_1}$	2.8344	437.430	1.59 03	1.78 03	1.93 03	2.11 03	2.31 03
46	Pd L$_{\alpha_1}$	2.8386	436.782	1.59 03	1.77 03	1.92 03	2.10 03	2.30 03
18	Ar K$_{\alpha_1}$	2.9577	419.194	1.42 03	1.58 03	1.71 03	1.88 03	2.05 03
47	Ag L$_{\alpha_1}$	2.9843	415.458	1.39 03	1.54 03	1.67 03	1.83 03	2.00 03
46	Pd L$_{\beta_1}$	2.9902	414.638	1.38 03	1.54 03	1.66 03	1.82 03	1.99 03
90	Th M$_{\alpha_1}$	2.9961	413.821	1.37 03	1.53 03	1.66 03	1.82 03	1.98 03
91	Pa M$_{\alpha_1}$	3.0823	402.248	1.27 03	1.41 03	1.53 03	1.68 03	1.83 03
48	Cd L$_{\alpha_1}$	3.1337	395.651	1.21 03	1.35 03	1.47 03	1.61 03	1.75 03
90	Th M$_\beta$	3.1458	394.129	1.20 03	1.34 03	1.45 03	1.59 03	1.73 03
47	Ag L$_{\beta_1}$	3.1509	393.491	1.20 03	1.33 03	1.44 03	1.58 03	1.73 03
92	U M$_{\alpha_1}$	3.1708	391.021	1.18 03	1.31 03	1.42 03	1.56 03	1.70 03
18	Ar K$_\beta$	3.1905	388.607	1.16 03	1.29 03	1.40 03	1.53 03	1.67 03
91	Pa M$_\beta$	3.2397	382.705	1.11 03	1.23 03	1.34 03	1.47 03	1.60 03
49	In L$_{\alpha_1}$	3.2869	377.210	1.07 03	1.19 03	1.29 03	1.41 03	1.54 03
19	K K$_{\alpha_1}$	3.3138	374.148	1.05 03	1.16 03	1.26 03	1.38 03	1.50 03
48	Cd L$_{\beta_1}$	3.3166	373.832	1.04 03	1.16 03	1.26 03	1.38 03	1.50 03
92	U M$_\beta$	3.3360	371.658	1.03 03	1.14 03	1.24 03	1.35 03	1.48 03
50	Sn L$_{\alpha_1}$	3.4440	360.003	9.42 02	1.04 03	1.13 03	1.24 03	1.35 03
49	In L$_{\beta_1}$	3.4872	355.543	9.11 02	1.01 03	1.10 03	1.20 03	1.31 03
19	K K$_\beta$	3.5896	345.401	8.43 02	9.33 02	1.01 03	1.11 03	1.21 03
51	Sb L$_{\alpha_1}$	3.6047	343.954	8.34 02	9.23 02	1.00 03	1.10 03	1.19 03

Z	transition	E [keV]	λ [pm]	$_{36}$Kr	$_{37}$Rb	$_{38}$Sr	$_{39}$Y	$_{40}$Zr
50	Sn L$_{\beta_1}$	3.6628	338.498	7.99 02	8.83 02	9.60 02	1.05 03	1.14 03
20	Ca K$_{\alpha_1}$	3.6917	335.848	7.82 02	8.65 02	9.40 02	1.03 03	1.12 03
52	Te L$_{\alpha_1}$	3.7693	328.934	7.39 02	8.17 02	8.88 02	9.72 02	1.06 03
51	Sb L$_{\beta_1}$	3.8436	322.575	7.02 02	7.75 02	8.42 02	9.21 02	1.00 03
53	I L$_{\alpha_1}$	3.9377	314.867	6.57 02	7.25 02	7.89 02	8.63 02	9.37 02
20	Ca K$_\beta$	4.0127	308.981	6.25 02	6.89 02	7.50 02	8.19 02	8.90 02
52	Te L$_{\beta_1}$	4.0296	307.686	6.18 02	6.81 02	7.41 02	8.10 02	8.80 02
21	Sc K$_{\alpha_1}$	4.0906	303.097	5.93 02	6.54 02	7.11 02	7.77 02	8.44 02
54	Xe L$_{\alpha_1}$	4.1099	301.674	5.86 02	6.45 02	7.02 02	7.68 02	8.33 02
53	I L$_{\beta_1}$	4.2207	293.755	5.45 02	6.00 02	6.53 02	7.14 02	7.75 02
55	Cs L$_{\alpha_1}$	4.2865	289.245	5.23 02	5.75 02	6.26 02	6.84 02	7.43 02
21	Sc K$_\beta$	4.4605	277.962	4.70 02	5.16 02	5.61 02	6.14 02	6.66 02
56	Ba L$_{\alpha_1}$	4.4663	277.601	4.68 02	5.15 02	5.59 02	6.12 02	6.63 02
22	Ti K$_{\alpha_1}$	4.5108	274.863	4.56 02	5.01 02	5.44 02	5.96 02	6.46 02
55	Cs L$_{\beta_1}$	4.6198	268.377	4.28 02	4.69 02	5.10 02	5.58 02	6.05 02
57	La L$_{\alpha_1}$	4.6510	266.577	4.20 02	4.61 02	5.01 02	5.48 02	5.93 02
56	Ba L$_{\beta_1}$	4.8273	256.841	3.80 02	4.16 02	4.52 02	4.95 02	5.36 02
58	Ce L$_{\alpha_1}$	4.8402	256.157	3.77 02	4.13 02	4.49 02	4.91 02	5.32 02
22	Ti K$_{\beta_{1,3}}$	4.9318	251.399	3.59 02	3.93 02	4.27 02	4.67 02	5.05 02
23	V K$_{\alpha_1}$	4.9522	250.363	3.55 02	3.88 02	4.22 02	4.62 02	5.00 02
59	Pr L$_{\alpha_1}$	5.0337	246.310	3.40 02	3.72 02	4.04 02	4.42 02	4.78 02
57	La L$_{\beta_1}$	5.0421	245.899	3.38 02	3.70 02	4.02 02	4.40 02	4.76 02
60	Nd L$_{\alpha_1}$	5.2304	237.047	3.06 02	3.35 02	3.64 02	3.98 02	4.30 02
58	Ce L$_{\beta_1}$	5.2622	235.614	3.01 02	3.30 02	3.58 02	3.92 02	4.23 02
24	Cr K$_{\alpha_1}$	5.4147	228.978	2.79 02	3.05 02	3.31 02	3.62 02	3.91 02
23	V K$_{\beta_{1,3}}$	5.4273	228.447	2.77 02	3.03 02	3.29 02	3.60 02	3.89 02
61	Pm L$_{\alpha_1}$	5.4325	228.228	2.77 02	3.02 02	3.28 02	3.59 02	3.88 02
59	Pr L$_{\beta_1}$	5.4889	225.883	2.69 02	2.94 02	3.19 02	3.49 02	3.77 02
62	Sm L$_{\alpha_1}$	5.6361	219.984	2.51 02	2.74 02	2.97 02	3.25 02	3.51 02
60	Nd L$_{\beta_1}$	5.7210	216.696	2.41 02	2.63 02	2.85 02	3.12 02	3.37 02
63	Eu L$_{\alpha_1}$	5.8457	212.096	2.27 02	2.48 02	2.69 02	2.94 02	3.18 02
25	Mn K$_{\alpha_1}$	5.8988	210.187	2.22 02	2.52 02	2.63 02	2.87 02	3.10 02
24	Cr K$_{\beta_{1,3}}$	5.9467	208.494	2.19 02	2.37 02	2.57 02	2.81 02	3.03 02
61	Pm L$_{\beta_1}$	5.9610	207.993	2.16 02	2.35 02	2.55 02	2.79 02	3.01 02
64	Gd L$_{\alpha_1}$	6.0572	204.690	2.07 02	2.25 02	2.45 02	2.67 02	2.88 02
62	Sm L$_{\beta_1}$	6.2051	199.811	1.94 02	2.11 02	2.29 02	2.51 02	2.70 02
65	Tb L$_{\alpha_1}$	6.2728	197.655	1.88 02	2.05 02	2.23 02	2.43 02	2.62 02
26	Fe K$_{\alpha_1}$	6.4038	193.611	1.78 02	1.94 02	2.10 02	2.30 02	2.48 02
63	Eu L$_{\beta_1}$	6.4564	192.034	1.74 02	1.90 02	2.06 02	2.25 02	2.42 02
25	Mn K$_{\beta_{1,3}}$	6.4905	191.025	1.72 02	1.87 02	2.03 02	2.22 02	2.39 02
66	Dy L$_{\alpha_1}$	6.4952	190.887	1.72 02	1.87 02	2.03 02	2.21 02	2.38 02
64	Gd L$_{\beta_1}$	6.7132	184.688	1.57 02	1.71 02	1.85 02	2.02 02	2.18 02
67	Ho L$_{\alpha_1}$	6.7198	184.507	1.57 02	1.70 02	1.85 02	2.02 02	2.17 02
27	Co K$_{\alpha_1}$	6.9303	178.903	1.44 02	1.57 02	1.70 02	1.86 02	2.00 02

Z	transition	E [keV]	λ [pm]	$_{36}$Kr	$_{37}$Rb	$_{38}$Sr	$_{39}$Y	$_{40}$Zr
68	Er L$_{\alpha_1}$	6.9487	178.429	1.43 02	1.56 02	1.69 02	1.84 02	1.98 02
65	Tb L$_{\beta_1}$	6.9780	177.680	1.42 02	1.54 02	1.67 02	1.82 02	1.96 02
26	Fe K$_{\beta_{1,3}}$	7.0580	175.666	1.38 02	1.49 02	1.62 02	1.77 02	1.90 02
69	Tm L$_{\alpha_1}$	7.1799	172.683	1.31 02	1.42 02	1.55 02	1.69 02	1.82 02
66	Dy L$_{\beta_1}$	7.2477	171.068	1.28 02	1.39 02	1.51 02	1.65 02	1.77 02
70	Yb L$_{\alpha_1}$	7.4156	167.195	1.21 02	1.31 02	1.42 02	1.55 02	1.66 02
28	Ni K$_{\alpha_1}$	7.4782	165.795	1.18 02	1.28 02	1.39 02	1.51 02	1.63 02
67	Ho L$_{\beta_1}$	7.5253	164.757	1.16 02	1.25 02	1.36 02	1.49 02	1.60 02
27	Co K$_{\beta_{1,3}}$	7.6494	162.084	1.11 02	1.20 02	1.30 02	1.42 02	1.53 02
71	Lu L$_{\alpha_1}$	7.6555	161.955	1.11 02	1.20 02	1.30 02	1.42 02	1.53 02
68	Er L$_{\beta_1}$	7.8109	158.733	1.05 02	1.13 02	1.23 02	1.34 02	1.44 02
72	Hf L$_{\alpha_1}$	7.8990	156.963	1.02 02	1.10 02	1.19 02	1.30 02	1.40 02
29	Cu K$_{\alpha_1}$	8.0478	154.060	9.69 01	1.05 02	1.14 02	1.24 02	1.33 02
69	Tm L$_{\beta_1}$	8.1010	153.049	9.52 01	1.03 02	1.12 02	1.22 02	1.31 02
73	Ta L$_{\alpha_1}$	8.1461	152.201	9.38 01	1.01 02	1.10 02	1.20 02	1.29 02
28	Ni K$_{\beta_{1,3}}$	8.2647	150.017	9.03 01	9.75 01	1.06 02	1.15 02	1.24 02
74	W L$_{\alpha_1}$	8.3976	147.643	8.66 01	9.34 01	1.01 02	1.11 02	1.19 02
70	Yb L$_{\beta_1}$	8.4018	147.569	8.65 01	9.33 01	1.01 02	1.10 02	1.19 02
30	Zn K$_{\alpha_1}$	8.6389	143.519	8.03 01	8.65 01	9.39 01	1.02 02	1.10 02
75	Re L$_{\alpha_1}$	8.6525	143.294	8.00 01	8.62 01	9.35 01	1.02 02	1.09 02
71	Lu L$_{\beta_1}$	8.7090	142.364	7.86 01	8.47 01	9.19 01	1.00 02	1.08 02
29	Cu K$_{\beta_1}$	8.9053	139.226	7.41 01	7.98 01	8.65 01	9.44 01	1.01 02
76	Os L$_{\alpha_1}$	8.9117	139.126	7.40 01	7.96 01	8.63 01	9.42 01	1.01 02
72	Hf L$_{\beta_1}$	9.0227	137.414	7.16 01	7.70 01	8.35 01	9.11 01	9.77 01
77	Ir L$_{\alpha_1}$	9.1751	135.132	6.85 01	7.36 01	7.98 01	8.71 01	9.34 01
31	Ga K$_{\alpha_1}$	9.2517	134.013	6.70 01	7.20 01	7.81 01	8.52 01	9.13 01
73	Ta L$_{\beta_1}$	9.3431	132.702	6.52 01	7.01 01	7.60 01	8.30 01	8.89 01
78	Pt L$_{\alpha_1}$	9.4423	131.308	6.34 01	6.82 01	7.39 01	8.06 01	8.64 01
30	Zn K$_{\beta_{1,3}}$	9.5720	129.529	6.12 01	6.57 01	7.12 01	7.77 01	8.32 01
74	W L$_{\beta_1}$	9.6724	128.184	5.95 01	6.39 01	6.92 01	7.56 01	8.09 01
79	Au L$_{\alpha_1}$	9.7135	127.644	5.88 01	6.32 01	6.85 01	7.47 01	8.00 01
32	Ge K$_{\alpha_1}$	9.8864	125.409	5.62 01	6.02 01	6.53 01	7.13 01	7.63 01
80	Hg L$_{\alpha_1}$	9.9888	124.124	5.46 01	5.86 01	6.35 01	6.93 01	7.42 01
75	Re L$_{\beta_1}$	10.010	123.861	5.43 01	5.83 01	6.31 01	6.89 01	7.37 01
31	Ga K$_{\beta_1}$	10.264	120.796	5.09 01	5.45 01	5.91 01	6.45 01	6.90 01
81	Tl L$_{\alpha_1}$	10.269	120.737	5.08 01	5.44 01	5.90 01	6.44 01	6.89 01
76	Os L$_{\beta_1}$	10.355	119.734	4.97 01	5.32 01	5.77 01	6.30 01	6.73 01
33	As K$_{\alpha_1}$	10.544	117.588	4.74 01	5.07 01	5.50 01	6.00 01	6.42 01
82	Pb L$_{\alpha_1}$	10.552	117.499	4.73 01	5.06 01	5.49 01	5.99 01	6.40 01
77	Ir L$_{\beta_1}$	10.708	115.787	4.55 01	4.87 01	5.28 01	5.76 01	6.16 01
83	Bi L$_{\alpha_1}$	10.839	114.388	4.41 01	4.71 01	5.11 01	5.58 01	5.96 01
32	Ge K$_{\beta_1}$	10.982	112.898	4.26 01	4.55 01	4.93 01	5.38 01	5.75 01
78	Pt L$_{\beta_1}$	11.071	111.990	4.71 01	4.46 01	4.83 01	5.27 01	5.63 01
84	Po L$_{\alpha_1}$	11.131	111.387	4.11 01	4.39 01	4.76 01	5.19 01	5.55 01

Z	transition	E [keV]	λ [pm]	$_{36}$Kr	$_{37}$Rb	$_{38}$Sr	$_{39}$Y	$_{40}$Zr
34	Se K$_{\alpha_1}$	11.222	110.484	4.02 01	4.30 01	4.66 01	5.08 01	5.43 01
85	At L$_{\alpha_1}$	11.427	108.501	3.84 01	4.10 01	4.44 01	4.84 01	5.17 01
79	Au L$_{\beta_1}$	11.442	108.359	3.82 01	4.08 01	4.42 01	4.83 01	5.15 01
33	As K$_{\beta_1}$	11.726	105.735	3.58 01	3.82 01	4.14 01	4.52 01	4.83 01
86	Rn L$_{\alpha_1}$	11.727	105.726	3.58 01	3.82 01	4.14 01	4.52 01	4.83 01
80	Hg L$_{\beta_1}$	11.823	104.867	3.51 01	3.74 01	4.05 01	4.42 01	4.72 01
35	Br K$_{\alpha_1}$	11.924	103.979	3.43 01	3.66 01	3.96 01	4.32 01	4.61 01
87	Fr L$_{\alpha_1}$	12.031	103.054	3.35 01	3.57 01	3.87 01	4.22 01	4.51 01
81	Tl L$_{\beta_1}$	12.213	101.519	3.22 01	3.43 01	3.72 01	4.06 01	4.33 01
88	Ra L$_{\alpha_1}$	12.340	100.474	3.13 01	3.34 01	3.62 01	3.95 01	4.21 01
34	Se K$_{\beta_1}$	12.496	99.219	3.03 01	3.23 01	3.50 01	3.82 01	4.07 01
82	Pb L$_{\beta_1}$	12.614	98.291	2.96 01	3.15 01	3.41 01	3.72 01	3.97 01
36	Kr K$_{\alpha_1}$	12.649	98.019	2.94 01	3.13 01	3.39 01	3.70 01	3.94 01
89	Ac L$_{\alpha_1}$	12.652	97.996	2.93 01	3.13 01	3.39 01	3.69 01	3.94 01
90	Th L$_{\alpha_1}$	12.969	95.601	2.75 01	2.93 01	3.17 01	3.46 01	3.68 01
83	Bi L$_{\beta_1}$	13.024	95.197	2.72 01	2.89 01	3.13 01	3.42 01	3.64 01
91	Pa L$_{\alpha_1}$	13.291	93.285	2.58 01	2.74 01	2.97 01	3.24 01	3.45 01
35	Br K$_{\beta_1}$	13.291	93.285	2.58 01	2.74 01	2.97 01	3.24 01	3.45 01
37	Rb K$_{\alpha_1}$	13.395	92.560	2.53 01	2.69 01	2.91 01	3.17 01	3.38 01
84	Po L$_{\beta_1}$	13.447	92.202	2.50 01	2.66 01	2.88 01	3.14 01	3.34 01
92	U L$_{\alpha_1}$	13.615	91.065	2.42 01	2.57 01	2.78 01	3.04 01	3.24 01
85	At L$_{\beta_1}$	13.876	89.352	2.31 01	2.44 01	2.65 01	2.89 01	3.07 01
36	Kr K$_{\beta_1}$	14.112	87.857	2.21 01	2.34 01	2.53 01	2.76 01	2.94 01
38	Sr K$_{\alpha_1}$	14.165	87.529	2.19 01	2.31 01	2.51 01	2.73 01	2.91 01
86	Rn L$_{\beta_1}$	14.316	86.606	<u>2.13 01</u>	2.25 01	2.44 01	2.66 01	2.83 01
87	Fr L$_{\beta_1}$	14.770	83.943	1.28 02	2.07 01	2.24 01	2.45 01	2.60 01
39	Y K$_{\alpha_1}$	14.958	82.888	1.24 02	2.00 01	2.17 01	2.36 01	2.51 01
37	Rb K$_{\beta_1}$	14.961	82.872	1.24 02	<u>2.00 01</u>	2.17 01	2.36 01	2.51 01
88	Ra L$_{\beta_1}$	15.236	81.376	1.18 02	1.27 02	2.06 01	2.26 01	2.40 01
89	Ac L$_{\beta_1}$	15.713	78.906	1.08 02	1.17 02	1.90 01	2.08 01	2.21 01
40	Zr K$_{\alpha_1}$	15.775	78.596	1.07 02	1.15 02	1.89 01	2.06 01	2.19 01
38	Sr K$_{\beta_1}$	15.836	78.293	1.06 02	1.14 02	<u>1.87 01</u>	2.04 01	2 16 01
90	Th L$_{\beta_1}$	16.202	76.524	9.99 01	1.07 02	1.16 02	1.92 01	2.04 01
41	Nb K$_{\alpha_1}$	16.615	74.622	9.33 01	1.00 02	1.08 02	1.80 01	1.90 01
91	Pa L$_{\beta_1}$	16.702	74.233	9.20 01	9.91 01	1.06 02	1.77 01	1.88 01
39	Y K$_{\beta_1}$	16.738	74.074	9.15 01	9.85 01	1.06 02	<u>1.76 01</u>	1.87 01
92	U L$_{\beta_1}$	17.220	72.000	8.47 01	9.13 01	9.81 01	1.06 02	1.73 01
42	Mo K$_{\alpha_1}$	17.479	70.933	8.14 01	8.77 01	9.43 01	1.02 02	1.66 01
40	Zr K$_{\beta_1}$	17.668	70.175	7.91 01	8.52 01	9.16 01	9.90 01	<u>1.62 01</u>
43	Tc K$_{\alpha_1}$	18.367	67.504	7.12 01	1.68 01	8.25 01	8.93 01	9.15 01
41	Nb K$_{\beta_1}$	18.623	66.576	6.86 01	7.40 01	7.95 01	8.60 01	8.82 01
44	Ru K$_{\alpha_1}$	19.279	64.311	6.25 01	6.74 01	7.24 01	7.84 01	8.06 01
42	Mo K$_{\beta_1}$	19.608	63.231	5.97 01	6.44 01	6.92 01	7.50 01	7.72 01
45	Rh K$_{\alpha_1}$	20.216	61.330	5.50 01	5.93 01	6.37 01	6.91 01	7.12 01

Z	transition	E [keV]	λ [pm]	$_{36}$Kr	$_{37}$Rb	$_{38}$Sr	$_{39}$Y	$_{40}$Zr
43	Tc K$_{\beta_1}$	20.619	60.131	5.21 01	5.62 01	6.04 01	6.54 01	6.76 01
46	Pd K$_{\alpha_1}$	21.l77	58.547	4.84 01	5.22 01	5.61 01	6.08 01	6.29 01
44	Ru K$_{\beta_1}$	21.657	57.249	4.55 01	4.91 01	5.28 01	5.72 01	5.93 01
47	Ag K$_{\alpha_1}$	22.163	55.942	4.26 01	4.60 01	4.95 01	5.37 01	5.57 01
45	Rh K$_{\beta_1}$	22.724	54.561	3.98 01	4.30 01	4.62 01	5.02 01	5.22 01
48	Cd K$_{\alpha_1}$	23.174	53.501	3.77 01	4.07 01	4.38 01	4.76 01	4.95 01
46	Pd K$_{\beta_1}$	23.819	52.053	3.49 01	3.78 01	4.06 01	4.41 01	4.60 01
49	In K$_{\alpha_1}$	24.210	51.212	3.34 01	3.61 01	3.89 01	4.22 01	4.41 01
47	Ag K$_{\beta_1}$	24.942	49.709	3.08 01	3.33 01	3.58 01	3.89 01	4.07 01
50	Sn K$_{\alpha_1}$	25.271	49.062	2.97 01	3.21 01	3.46 01	3.76 01	3.93 01
48	Cd K$_{\beta_1}$	26.096	47.511	2.72 01	2.94 01	3.16 01	3.44 01	3.61 01
51	Sb K$_{\alpha_1}$	26.359	47.037	2.64 01	2.86 01	3.08 01	3.35 01	3.51 01
49	In K$_{\beta_1}$	27.276	45.455	2.40 01	2.60 01	2.80 01	3.05 01	3.21 01
52	Te K$_{\alpha_1}$	27.472	45.131	2.36 01	2.55 01	2.75 01	2.99 01	3.15 01
50	Sn K$_{\beta_1}$	28.486	43.524	2.13 01	2.31 01	2.49 01	2.71 01	2.86 01
53	I K$_{\alpha_1}$	28.612	43.333	2.11 01	2.28 01	2.46 01	2.68 01	2.82 01
51	Sb K$_{\beta_1}$	29.726	41.709	1.89 01	2.05 01	2.21 01	2.41 01	2.55 01
54	Xe K$_{\alpha_1}$	29.779	41.635	1.89 01	2.04 01	2.20 01	2.40 01	2.54 01
55	Cs K$_{\alpha_1}$	30.973	40.030	1.69 01	1.83 01	1.97 01	2.15 01	2.28 01
52	Te K$_{\beta_1}$	30.996	40.000	1.69 01	1.83 01	1.97 01	2.15 01	2.28 01
56	Ba K$_{\alpha_1}$	32.194	38.511	1.52 01	1.64 01	1.77 01	1.93 01	2.05 01
53	I K$_{\beta_1}$	32.295	38.391	1.50 01	1.63 01	1.76 01	1.92 01	2.04 01
57	La K$_{\alpha_1}$	33.442	37.074	1.36 01	1.48 01	1.60 01	1.74 01	1.85 01
54	Xe K$_{\beta_1}$	33.624	36.874	1.34 01	1.45 01	1.57 01	1.71 01	1.82 01
58	Ce K$_{\alpha_1}$	34.279	36.169	1.27 01	1.38 01	1.49 01	1.62 01	1.73 01
55	Cs K$_{\beta_1}$	34.987	35.437	1.20 01	1.30 01	1.41 01	1.53 01	1.64 01
59	Pr K$_{\alpha_1}$	36.026	34.415	1.11 01	1.20 01	1.30 01	1.41 01	1.51 01
56	Ba K$_{\beta_1}$	36.378	34.082	1.08 01	1.17 01	1.26 01	1.38 01	1.47 01
60	Nd K$_{\alpha_1}$	36.847	33.648	1.04 01	1.13 01	1.22 01	1.33 01	1.42 01
57	La K$_{\beta_1}$	37.801	32.799	9.68 00	1.05 01	1.13 01	1.24 01	1.32 01
61	Pm K$_{\alpha_1}$	38.725	32.016	9.05 00	9.80 00	1.06 01	1.16 01	1.24 01
58	Ce K$_{\beta_1}$	39.257	31.582	8.71 00	9.43 00	1.02 01	1.11 01	1.20 01
62	Sm K$_{\alpha_1}$	40.118	30.905	8.20 00	8.88 00	9.61 00	1.05 01	1.13 01
59	Pr K$_{\beta_1}$	40.748	30.427	7.85 00	8.50 00	9.20 00	1.00 01	1.08 01
63	Eu K$_{\alpha_1}$	41.542	29.845	7.44 00	8.06 00	8.72 00	9.51 00	1.02 01
60	Nd K$_{\beta_1}$	42.271	29.331	7.08 00	7.68 00	8.31 00	9.06 00	9.74 00
64	Gd K$_{\alpha_1}$	42.996	28.836	6.75 00	7.32 00	7.92 00	8.64 00	9.29 00
61	Pm K$_{\beta_1}$	43.826	28.290	6.40 00	6.94 00	7.51 00	8.19 00	8.81 00
65	Tb K$_{\alpha_1}$	44.482	27.873	6.14 00	6.66 00	7.21 00	7.85 00	8.46 00
62	Sm K$_{\beta_1}$	45.413	27.301	5.79 00	6.29 00	6.80 00	7.41 00	7.98 00
66	Dy K$_{\alpha_1}$	45.998	26.854	5.59 00	6.07 00	6.56 00	7.15 00	7.71 00
63	Eu K$_{\beta_1}$	47.038	26.358	5.25 00	5.70 00	6.17 00	6.72 00	7.24 00
67	Ho K$_{\alpha_1}$	47.547	26.076	5.10 00	5.53 00	5.99 00	6.52 00	7.03 00
64	Gd K$_{\beta_1}$	48.697	25.460	4.77 00	5.18 00	5.60 00	6.10 00	6.58 00

Z	transition	E [keV]	λ [pm]	$_{36}$Kr	$_{37}$Rb	$_{38}$Sr	$_{39}$Y	$_{40}$Zr
68	Er K$_{\alpha_1}$	49.128	25.237	4.65 00	5.05 00	5.46 00	5.95 00	6.42 00
65	Tb K$_{\beta_1}$	50.382	24.609	4.33 00	4.71 00	5.09 00	5.54 00	5.99 00
69	Tm K$_{\alpha_1}$	50.742	24.434	4.25 00	4.62 00	4.99 00	5.44 00	5.87 00
66	Dy K$_{\beta_1}$	52.119	23.788	3.95 00	4.29 00	4.64 00	5.05 00	5.45 00
70	Yb K$_{\alpha_1}$	52.389	23.666	3.89 00	4.23 00	4.57 00	4.97 00	5.37 00
67	Ho K$_{\beta_1}$	53.877	23.012	3.60 00	3.92 00	4.23 00	4.60 00	4.98 00
71	Lu K$_{\alpha_1}$	54.070	22.930	3.57 00	3.88 00	4.19 00	4.56 00	4.93 00
68	Er K$_{\beta_1}$	55.681	22.267	3.29 00	3.58 00	3.86 00	4.20 00	4.54 00
72	Hf K$_{\alpha_1}$	55.790	22.223	3.27 00	3.56 00	3.84 00	4.18 00	4.52 00
69	Tm K$_{\beta_1}$	57.517	21.556	3.01 00	3.27 00	3.53 00	3.84 00	4.15 00
73	Ta K$_{\alpha_1}$	57.532	21.550	3.00 00	3.27 00	3.53 00	3.84 00	4.15 00
74	W K$_{\alpha_1}$	59.318	20.901	2.76 00	3.00 00	3.24 00	3.52 00	3.82 00
70	Yb K$_{\beta_1}$	59.370	20.883	2.75 00	3.00 00	3.23 00	3.52 00	3.81 00
75	Re K$_{\alpha_1}$	61.140	20.278	2.54 00	2.77 00	2.99 00	3.24 00	3.51 00
71	Lu K$_{\beta_1}$	61.283	20.231	2.53 00	2.75 00	2.97 00	3.22 00	3.49 00
76	Os K$_{\alpha_1}$	63.001	19.679	2.35 00	2.55 00	2.75 00	2.99 00	3.24 00
72	Hf K$_{\beta_1}$	63.234	19.607	2.32 00	2.53 00	2.73 00	2.96 00	3.21 00
77	Ir K$_{\alpha_1}$	64.896	19.105	2.17 00	2.35 00	2.54 00	2.76 00	2.99 00
73	Ta K$_{\beta_1}$	65.223	19.009	2.14 00	2.32 00	2.51 00	2.72 00	2.95 00
78	Pt K$_{\alpha_1}$	66.832	18.551	2.00 00	2.18 00	2.35 00	2.55 00	2.76 00
74	W K$_{\beta_1}$	67.244	18.438	1.97 00	2.14 00	2.31 00	2.51 00	2.72 00
79	Au K$_{\alpha_1}$	68.804	18.020	1.85 00	2.01 00	2.17 00	2.36 00	2.55 00
75	Re K$_{\beta_1}$	69.310	17.888	1.82 00	1.97 00	2.13 00	2.31 00	2.50 00
80	Hg K$_{\alpha_1}$	70.819	17.507	1.71 00	1.86 00	2.01 00	2.18 00	2.36 00
76	Os K$_{\beta_1}$	71.413	17.361	1.68 00	1.82 00	1.96 00	2.13 00	2.31 00
81	Tl K$_{\alpha_1}$	72.872	17.014	1.59 00	1.72 00	1.86 00	2.02 00	2.18 00
77	Ir K$_{\beta_1}$	73.561	16.854	1.55 00	1.68 00	1.81 00	1.96 00	2.13 00
82	Pb K$_{\alpha_1}$	74.969	16.538	1.47 00	1.60 00	1.72 00	1.87 00	2.02 00
78	Pt K$_{\beta_1}$	75.748	16.368	1.43 00	1.55 00	1.67 00	1.81 00	1.97 00
83	Bi K$_{\alpha_1}$	77.108	16.079	1.36 00	1.48 00	1.59 00	1.73 00	1.87 00
79	Au K$_{\beta_1}$	77.948	15.906	1.33 00	1.44 00	1.55 00	1.68 00	1.82 00
84	Po K$_{\alpha_1}$	79.290	15.636	1.27 00	1.37 00	1.48 00	1.60 00	1.74 00
80	Hg K$_{\beta_1}$	80.253	15.449	1.23 00	1.33 00	1.43 00	1.55 00	1.68 00
85	At K$_{\alpha_1}$	81.520	15.209	1.18 00	1.28 00	1.38 00	1.49 00	1.62 00
81	Tl K$_{\beta_1}$	82.576	15.014	1.14 00	1.24 00	1.33 00	1.44 00	1.56 00
86	Rn K$_{\alpha_1}$	83.780	14.798	1.10 00	1.19 00	1.28 00	1.39 00	1.51 00
82	Pb K$_{\beta_1}$	84.936	14.597	1.06 00	1.15 00	1.24 00	1.34 00	1.45 00
87	Fr K$_{\alpha_1}$	86.100	14.400	1.03 00	1.11 00	1.20 00	1.30 00	1.40 00
83	Bi K$_{\beta_1}$	87.343	14.195	9.93 -1	1.07 00	1.15 00	1.25 00	1.35 00
88	Ra K$_{\alpha_1}$	88.470	14.014	9.61 -1	1.04 00	1.12 00	1.21 00	1.31 00
84	Po K$_{\beta_1}$	89.800	13.806	9.26 -1	9.99 -1	1.08 00	1.17 00	1.26 00
89	Ac K$_{\alpha_1}$	90.884	13.642	8.99 -1	9.69 -1	1.04 00	1.13 00	1.22 00
85	At K$_{\beta_1}$	92.300	13.432	8.65 -1	9.32 -1	1.00 00	1.09 00	1.17 00
90	Th K$_{\alpha_1}$	93.350	13.281	8.41 -1	9.06 -1	9.74 -1	1.06 00	1.14 00

Z	transition	E [keV]	λ [pm]	$_{36}$Kr	$_{37}$Rb	$_{38}$Sr	$_{39}$Y	$_{40}$Zr
86	Rn K$_{\beta_1}$	94.870	13.068	8.07 -1	8.69 -1	9.35 -1	1.01 00	1.09 00
91	Pa K$_{\alpha_1}$	95.868	12.932	7.87 -1	8.47 -1	9.11 -1	9.88 -1	1.06 00
87	Fr K$_{\beta_1}$	97.470	12.720	7.55 -1	8.12 -1	8.73 -1	9.47 -1	1.02 00
92	U K$_{\alpha_1}$	98.439	12.595	7.36 -1	7.92 -1	8.51 -1	9.24 -1	9.90 -1
88	Ra K$_{\beta_1}$	100.130	12.382	7.06 -1	7.59 -1	8.16 -1	8.85 -1	9.48 -1
89	Ac K$_{\beta_1}$	102.850	12.054	6.67 -1	7.16 -1	7.68 -1	8.33 -1	8.91 -1
90	Th K$_{\beta_1}$	105.610	11.739	6.30 -1	6.76 -1	7.24 -1	7.84 -1	8.38 -1
91	Pa K$_{\beta_1}$	108.430	11.434	5.95 -1	6.38 -1	6.83 -1	7.39 -1	7.89 -1
92	U K$_{\beta_1}$	111.300	11.139	5.63 -1	6.03 -1	6.45 -1	6.96 -1	7.43 -1

Z	transition	E [keV]	λ [pm]	$_{41}$Nb	$_{42}$Mo	$_{43}$Tc	$_{44}$Ru	$_{45}$Rh
4	Be K$_\alpha$	0.1085	11427.207	1.06 04	7.97 03	7.30 03	4.43 03	3.91 03
38	Sr M$_\xi$	0.1140	10875.895	9.69 03	7.50 03	7.08 03	5.54 03	4.11 03
39	Y L$_\xi$	0.1328	9339.235	8.51 03	6.85 03	6.46 03	4.89 03	5.34 03
16	S L$_l$	0.1487	8337.942	7.75 03	6.41 03	6.04 03	5.17 03	6.15 03
40	Zr M$_\xi$	0.1511	8205.506	7.65 03	6.34 03	5.98 03	5.21 03	6.13 03
41	Nb M$_\xi$	0.1717	7221.037	7.14 03	5.77 03	5.55 03	5.34 03	5.98 03
5	B K$_\alpha$	0.1833	6764.059	6.56 03	5.61 03	5.33 03	5.36 03	6.00 03
42	Mo M$_\xi$	0.1926	6473.445	<u>6.12 03</u>	<u>5.48 03</u>	<u>5.18 03</u>	5.38 03	6.62 03
6	C K$_\alpha$	0.2770	4476.000	1.94 04	1.64 04	7.95 03	<u>5.52 03</u>	<u>4.65 03</u>
47	Ag M$_\xi$	0.3117	3977.709	<u>2.28 04</u>	2.04 04	1.16 04	6.43 03	<u>5.00 03</u>
7	N K$_\alpha$	0.3924	3159.664	2.14 04	<u>2.02 04</u>	2.29 04	1.41 04	1.83 04
22	Ti L$_l$	0.3953	3136.484	2.12 04	<u>2.00 04</u>	<u>2.32 04</u>	1.44 04	1.91 04
22	Ti L$_\alpha$	0.4522	2741.822	<u>1.78 04</u>	<u>1.73 04</u>	2.20 04	<u>2.06 04</u>	<u>2.16 04</u>
23	V L$_\alpha$	0.5113	2424.901	1.57 04	1.68 04	1.88 04	1.98 04	<u>2.00 04</u>
8	O K$_\alpha$	0.5249	2362.072	1.53 04	1.67 04	<u>1.81 04</u>	1.97 04	1.97 04
25	Mn L$_l$	0.5563	2228.747	1.39 04	1.52 04	1.75 04	1.78 04	1.83 04
24	Cr L$_\alpha$	0.5728	2164.549	1.33 04	1.44 04	1.65 04	<u>1.69 04</u>	<u>1.76 04</u>
25	Mn L$_\alpha$	0.6374	1945.171	1.12 04	1.21 04	1.34 04	1.47 04	1.57 04
9	F K$_\alpha$	0.6768	1831.932	1.02 04	1.09 04	1.19 04	1.30 04	1.39 04
26	Fe L$_\alpha$	0.7050	1758.655	9.52 03	1.02 04	1.10 04	1.20 04	1.29 04
27	Co L$_\alpha$	0.7762	1597.335	8.17 03	8.66 03	9.05 03	9.90 03	1.07 04
28	Ni L$_\alpha$	0.8515	1456.080	6.68 03	7.09 03	7.45 03	8.13 03	8.78 03
29	Cu L$_\alpha$	0.9297	1336.044	5.50 03	5.85 03	6.17 03	6.74 03	7.28 03
30	Zn L$_\alpha$	1.0117	1225.513	5.10 03	5.44 03	5.86 03	6.37 03	6.94 03
11	Na K$_\alpha$	1.0410	1191.020	4.74 03	5.06 03	5.46 03	5.93 03	6.46 03
11	Na K$_\beta$	1.0711	1157.550	4.41 03	4.71 03	5.08 03	5.53 03	6.02 03
12	Mg K$_\alpha$	1.2536	989.033	2.95 03	3.18 03	3.43 03	3.74 03	4.07 03
33	As L$_\alpha$	1.2820	967.123	2.78 03	3.00 03	3.25 03	3.54 03	3.85 03
12	Mg K$_\beta$	1.3022	952.121	2.62 03	2.89 03	3.12 03	3.40 03	3.71 03
33	As L$_{\beta_1}$	1.3117	941.421	2.60 03	2.81 03	3.04 03	3.31 03	3.60 03
66	Dy M$_{\beta_1}$	1.3250	935.737	2.56 03	2.77 03	2.99 03	3.26 03	3.55 03
67	Ho M$_\alpha$	1.3480	919.771	2.45 03	2.65 03	2.87 03	3.12 03	3.40 03

Z	transition	E [keV]	λ [pm]	$_{41}$Nb	$_{42}$Mo	$_{43}$Tc	$_{44}$Ru	$_{45}$Rh
34	Se L_α	1.3791	899.029	2.31 03	2.50 03	2.71 03	2.95 03	3.22 03
67	Ho M_β	1.3830	896.494	2.29 03	2.48 03	2.69 03	2.93 03	3.19 03
68	Er M_α	1.4060	881.829	2.20 03	2.38 03	2.58 03	2.81 03	3.06 03
34	Se L_{β_1}	1.4192	873.627	2.14 03	2.33 03	2.52 03	2.75 03	2.99 03
68	Er M_β	1.4430	859.218	2.05 03	2.23 03	2.42 03	2.64 03	2.87 03
69	Tm M_α	1.4620	848.051	1.99 03	2.16 03	2.34 03	2.55 03	2.78 03
35	Br L_α	1.4804	837.511	1.92 03	2.10 03	2.27 03	2.48 03	2.70 03
13	Al K_{α_1}	1.4867	833.962	1.90 03	2.07 03	2.25 03	2.45 03	2.67 03
69	Tm M_β	1.5030	824.918	1.85 03	2.02 03	2.19 03	2.38 03	2.60 03
70	Yb M_α	1.5214	814.941	1.79 03	1.96 03	2.12 03	2.31 03	2.52 03
35	Br L_{β_1}	1.5259	812.538	1.78 03	1.94 03	2.10 03	2.30 03	2.50 03
13	Al K_β	1.5574	796.103	1.69 03	1.85 03	2.00 03	2.18 03	2.38 03
70	Yb M_β	1.5675	790.974	1.66 03	1.82 03	1.97 03	2.15 03	2.34 03
71	Lu M_α	1.5813	784.071	1.62 03	1.78 03	1.92 03	2.10 03	2.29 03
36	Kr L_α	1.5860	781.747	1.61 03	1.76 03	1.91 03	2.08 03	2.27 03
71	Lu M_β	1.6312	760.085	1.50 03	1.64 03	1.78 03	1.94 03	2.12 03
36	Kr L_{β_1}	1.6366	757.577	1.49 03	1.63 03	1.77 03	1.93 03	2.10 03
72	Hf M_β	1.6446	753.892	1.47 03	1.61 03	1.75 03	1.90 03	2.08 03
37	Rb L_{β_1}	1.6941	731.864	1.36 03	1.49 03	1.62 03	1.77 03	1.93 03
72	Hf M_{α_1}	1.6976	730.355	1.35 03	1.49 03	1.61 03	1.76 03	1.92 03
73	Ta M_β	1.7096	725.229	1.33 03	1.46 03	1.58 03	1.73 03	1.88 03
14	Si K_{α_1}	1.7400	712.558	1.27 03	1.40 03	1.52 03	1.65 03	1.80 03
37	Rb L_{β_1}	1.7522	707.597	1.25 03	1.37 03	1.49 03	1.63 03	1.77 03
73	Ta M_β	1.7655	702.266	1.22 03	1.35 03	1.46 03	1.59 03	1.74 03
74	W M_{α_1}	1.7754	698.350	1.21 03	1.33 03	1.44 03	1.57 03	1.72 03
38	Sr L_{α_1}	1.8066	686.290	1.15 03	1.27 03	1.38 03	1.51 03	1.64 03
74	W M_β	1.8349	675.705	1.11 03	1.22 03	1.33 03	1.45 03	1.58 03
14	Si K_β	1.8359	675.337	1.11 03	1.22 03	1.33 03	1.45 03	1.58 03
75	Re M_{α_1}	1.8420	673.100	1.10 03	1.21 03	1.31 03	1.43 03	1.57 03
38	Sr L_{β_1}	1.8717	662.420	1.05 03	1.16 03	1.26 03	1.38 03	1.50 03
75	Re M_β	1.9061	650.465	1.00 03	1.11 03	1.21 03	1.32 03	1.44 03
76	Os M_{α_1}	1.9102	649.069	9.99 02	1.11 03	1.20 03	1.31 03	1.43 03
39	Y L_{β_1}	1.9226	644.882	9.83 02	1.09 03	1.18 03	1.29 03	1.41 03
76	Os M_β	1.9783	626.725	9.13 02	1.01 03	1.10 03	1.20 03	1.31 03
77	Ir M_{α_1}	1.9799	626.219	9.11 02	1.01 03	1.10 03	1.20 03	1.31 03
39	Y L_{β_1}	1.9958	621.230	8.93 02	9.91 02	1.08 03	1.17 03	1.28 03
15	P K_{α_1}	2.0137	615.708	8.73 02	9.69 02	1.05 03	1.15 03	1.25 03
40	Zr L_{α_1}	2.0424	607.056	8.42 02	9.36 02	1.02 03	1.11 03	1.21 03
78	Pt M_{α_1}	2.0505	604.658	8.33 02	9.26 02	1.01 03	1.10 03	1.20 03
77	Ir M_β	2.0535	603.775	8.30 02	9.23 02	1.00 03	1.09 03	1.20 03
79	Au M_{α_1}	2.1229	584.036	7.63 02	8.50 02	9.23 02	1.01 03	1.10 03
40	Zr L_{β_1}	2.1244	583.624	7.61 02	8.48 02	9.22 02	1.01 03	1.10 03
78	Pt M_β	2.1273	582.828	7.59 02	8.46 02	9.19 02	1.00 03	1.10 03
15	P $K_{\beta_{1,3}}$	2.1390	579.640	7.48 02	8.34 02	9.06 02	9.89 02	1.08 03

Z	transition	E [keV]	λ [pm]	$_{41}$Nb	$_{42}$Mo	$_{43}$Tc	$_{44}$Ru	$_{45}$Rh
41	Nb L$_{\alpha_1}$	2.1659	572.441	7.25 02	8.09 02	8.79 02	9.59 02	1.05 03
80	Hg M$_{\alpha_1}$	2.1953	564.775	7.00 02	7.82 02	8.50 02	9.28 02	1.01 03
79	Au M$_{\beta}$	2.2046	562.393	6.93 02	7.74 02	8.41 02	9.18 02	1.00 03
41	Nb L$_{\beta_1}$	2.2574	549.238	6.52 02	7.30 02	7.93 02	8.66 02	9.46 02
81	Tl M$_{\alpha_1}$	2.2706	546.045	6.43 02	7.19 02	7.82 02	8.54 02	9.32 02
80	Hg M$_{\beta}$	2.2825	543.199	6.34 02	7.10 02	7.71 02	8.43 02	9.20 02
42	Mo L$_{\alpha_1}$	2.2932	540.664	6.27 02	7.02 02	7.63 02	8.33 02	9.10 02
16	S K$_{\alpha_1}$	2.3080	537.197	6.16 02	6.91 02	7.50 02	8.20 02	8.95 02
82	Pb M$_{\alpha_1}$	2.3457	528.563	5.92 02	6.64 02	7.21 02	7.88 02	8.60 02
81	Tl M$_{\beta}$	2.3621	524.893	<u>5.81 02</u>	6.52 02	7.09 02	7.74 02	8.45 02
42	Mo L$_{\beta_1}$	2.3948	517.726	2.39 03	6.30 02	6.85 02	7.49 02	8.17 02
83	Bi M$_{\alpha_1}$	2.4226	511.785	2.31 03	6.13 02	6.65 02	7.28 02	7.94 02
43	Tc L$_{\alpha_1}$	2.4240	511.490	2.31 03	6.12 02	6.65 02	7.27 02	7.93 02
82	Pb M$_{\beta}$	2.4427	507.574	2.26 03	6.00 02	6.52 02	7.13 02	7.78 02
16	S K$_{\beta}$	2.4640	503.186	<u>2.21 03</u>	<u>5.87 02</u>	6.38 02	6.98 02	7.62 02
83	Bi M$_{\beta_1}$	2.5255	490.933	2.89 03	2.21 03	6.00 02	6.57 02	7.17 02
43	Tc L$_{\beta_1}$	2.5368	488.746	2.85 03	2.19 03	5.94 02	6.49 02	7.09 02
44	Ru L$_{\alpha_1}$	2.5586	484.582	2.78 03	2.14 03	5.81 02	6.36 02	6.94 02
17	Cl K$_{\alpha_1}$	2.6224	472.792	2.60 03	<u>2.00 03</u>	<u>5.47 02</u>	5.98 02	6.53 02
44	Ru L$_{\beta_1}$	2.6832	462.079	2.44 03	2.62 03	1.96 03	5.65 02	6.17 02
45	Rh L$_{\alpha_1}$	2.6967	459.766	<u>2.57 03</u>	2.59 03	<u>1.94 03</u>	5.58 02	5.09 02
17	Cl K$_{\beta}$	2.8156	440.350	2.57 03	2.29 03	2.40 03	5.02 02	5.48 02
45	Rh L$_{\beta_1}$	2.8344	437.430	2.52 03	2.25 03	2.36 03	<u>4.94 02</u>	5.39 02
46	Pd L$_{\alpha_1}$	2.8386	436.782	2.51 03	<u>2.24 03</u>	2.35 03	1.80 03	5.37 02
18	Ar K$_{\alpha_1}$	2.9577	419.194	2.24 03	2.40 03	2.10 03	<u>1.62 02</u>	4.85 02
47	Ag L$_{\alpha_1}$	2.9843	415.458	2.19 03	2.34 03	2.05 03	2.20 03	4.74 02
46	Pd L$_{\alpha_1}$	2.9902	414.638	2.18 03	2.33 03	2.04 03	2.19 03	4.72 02
90	Th M$_{\alpha_1}$	2.9961	413.821	2.17 03	2.32 03	<u>2.03 03</u>	2.18 03	<u>4.70 02</u>
91	Pa M$_{\alpha_1}$	3.0823	402.248	2.00 03	2.14 03	2.25 03	2.01 03	1.52 03
48	Cd L$_{\alpha_1}$	3.1337	395.651	1.91 03	2.05 03	2.15 03	1.93 03	1.46 03
90	Th M$_{\beta}$	3.1458	394.129	1.89 03	2.03 03	2.13 03	1.91 03	<u>1.44 03</u>
47	Ag L$_{\beta_1}$	3.1509	393.491	1.89 03	2.02 03	2.12 03	1.90 03	2.00 03
92	U M$_{\alpha_1}$	3.1708	391.021	1.85 03	1.98 03	2.08 03	1.86 03	1.97 03
18	Ar K$_{\beta}$	3.1905	388.607	1.82 03	1.95 03	2.05 03	1.83 03	1.94 03
91	Pa M$_{\beta}$	3.2397	382.705	1.75 03	1.87 03	1.96 03	2.11 03	1.86 03
49	In L$_{\alpha_1}$	3.2869	377.210	1.68 03	1.80 03	1.89 03	2.03 03	1.79 03
19	K K$_{\alpha_1}$	3.3138	374.148	1.64 03	1.76 03	1.84 03	1.98 03	1.75 03
48	Cd L$_{\beta_1}$	3.3166	373.832	1.64 03	1.75 03	1.84 03	<u>1.98 03</u>	1.74 03
92	U M$_{\beta}$	3.3360	371.658	1.61 03	1.73 03	1.81 03	1.95 03	<u>1.72 03</u>
50	Sn L$_{\alpha_1}$	3.4440	360.003	1.48 03	1.58 03	1.66 03	1.79 03	1.89 03
49	In L$_{\beta_1}$	3.4872	355.543	1.43 03	1.53 03	1.60 03	1.73 03	1.83 03
19	K K$_{\beta}$	3.5896	345.401	1.32 03	1.41 03	1.48 03	1.59 03	1.69 03
51	Sb L$_{\alpha_1}$	3.6047	343.954	1.30 03	1.40 03	1.46 03	1.58 03	1.67 03
50	Sn L$_{\beta_1}$	3.6628	338.498	1.25 03	1.34 03	1.40 03	1.51 03	1.60 03

Z	transition	E [keV]	λ [pm]	$_{41}$Nb	$_{42}$Mo	$_{43}$Tc	$_{44}$Ru	$_{45}$Rh
20	Ca K$_{\alpha_1}$	3.6917	335.848	1.22 03	1.31 03	1.37 03	1.48 03	1.56 03
52	Te L$_{\alpha_1}$	3.7693	328.934	1.15 03	1.23 03	1.30 03	1.40 03	1.48 03
51	Sb L$_{\beta_1}$	3.8436	322.575	1.09 03	1.17 03	1.23 03	1.32 03	1.40 03
53	I L$_{\alpha_1}$	3.9377	314.867	1.02 03	1.09 03	1.15 03	1.24 03	1.31 03
20	Ca K$_\beta$	4.0127	308.981	9.70 02	1.04 03	1.09 03	1.18 03	1.25 03
52	Te L$_{\beta_1}$	4.0296	307.686	9.59 02	1.03 03	1.08 03	1.16 03	1.23 03
21	Sc K$_{\alpha_1}$	4.0906	303.097	9.21 02	9.85 02	1.04 03	1.12 03	1.18 03
54	Xe L$_{\alpha_1}$	4.1099	301.674	9.09 02	9.73 02	1.02 03	1.10 03	1.17 03
53	I L$_{\beta_1}$	4.2207	293.755	8.45 02	9.04 02	9.51 02	1.02 03	1.09 03
55	Cs L$_{\alpha_1}$	4.2865	289.245	8.10 02	8.66 02	9.12 02	9.81 02	1.04 03
21	Sc K$_\beta$	4.4605	277.962	7.27 02	7.76 02	8.18 02	8.80 02	9.35 02
56	Ba L$_{\alpha_1}$	4.4663	277.601	7.24 02	7.73 02	8.15 02	8.77 02	9.31 02
22	Ti K$_{\alpha_1}$	4.5108	274.863	7.05 02	7.52 02	7.93 02	8.53 02	9.07 02
55	Cs L$_{\beta_1}$	4.6198	268.377	6.60 02	7.04 02	7.43 02	7.99 02	8.49 02
57	La L$_{\alpha_1}$	4.6510	266.577	6.48 02	6.91 02	7.30 02	7.85 02	8.34 02
56	Ba L$_{\beta_1}$	4.8273	256.841	5.86 02	6.24 02	6.59 02	7.09 02	7.54 02
58	Ce L$_{\beta_1}$	4.8402	256.157	5.82 02	6.19 02	6.55 02	7.03 02	7.48 02
22	Ti K$_{\beta_{1,3}}$	4.9318	251.399	5.53 02	5.88 02	6.22 02	6.68 02	7.11 02
23	V K$_{\alpha_1}$	4.9522	250.363	5.46 02	5.81 02	6.15 02	6.61 02	7.03 02
59	Pr L$_{\alpha_1}$	5.0337	246.310	5.22 02	5.56 02	5.88 02	6.32 02	6.72 02
57	La L$_{\beta_1}$	5.0421	245.899	5.20 02	5.53 02	5.86 02	6.29 02	6.69 02
60	Nd L$_{\alpha_1}$	5.2304	237.047	4.71 02	5.00 02	5.30 02	5.69 02	6.06 02
58	Ce L$_{\beta_1}$	5.2622	235.614	4.63 02	4.92 02	5.21 02	5.60 02	5.96 02
24	Cr K$_{\alpha_1}$	5.4147	228.978	4.28 02	4.55 02	4.82 02	5.18 02	5.52 02
23	V K$_{\beta_{1,3}}$	5.4273	228.447	4.25 02	4.52 02	4.79 02	5.15 02	5.48 02
61	Pm L$_{\alpha_1}$	5.4325	228.228	4.24 02	4.51 02	4.78 02	5.13 02	5.47 02
59	Pr L$_{\beta_1}$	5.4889	225.883	4.12 02	4.38 02	4.65 02	4.99 02	5.32 02
62	Sm L$_{\alpha_1}$	5.6361	219.984	3.84 02	4.08 02	4.32 02	4.64 02	4.95 02
60	Nd L$_{\beta_1}$	5.7216	216.696	3.68 02	3.91 02	4.15 02	4.46 02	4.75 02
63	Eu L$_{\alpha_1}$	5.8457	212.096	3.47 02	3.69 02	3.91 02	4.20 02	4.48 02
25	Mn K$_{\alpha_1}$	5.8988	210.187	3.39 02	3.60 02	3.82 02	4.10 02	4.38 02
24	Cr K$_{\beta_1}$	5.9467	208.494	3.31 02	3.52 02	3.73 02	4.01 02	4.28 02
61	Pm L$_{\beta_1}$	5.9610	207.993	3.29 02	3.50 02	3.71 02	3.99 02	4.25 02
64	Gd L$_{\alpha_1}$	6.0572	204.690	3.15 02	3.35 02	3.55 02	3.82 02	4.07 02
62	Sm L$_{\beta_1}$	6.2051	199.811	2.95 02	3.13 02	3.33 02	3.57 02	3.82 02
65	Tb L$_{\alpha_1}$	6.2728	197.655	2.87 02	3.04 02	3.23 02	3.47 02	3.71 02
26	Fe K$_{\alpha_1}$	6.4038	193.611	2.71 02	2.87 02	3.05 02	3.28 02	3.50 02
63	Eu L$_{\beta_1}$	6.4564	192.034	2.65 02	2.81 02	2.99 02	3.21 02	3.43 02
25	Mn K$_{\beta_{1,3}}$	6.4905	191.025	2.61 02	2.77 02	2.94 02	3.16 02	3.38 02
66	Dy L$_{\alpha_1}$	6.4952	190.887	2.61 02	2.76 02	2.94 02	3.16 02	3.37 02
64	Gd L$_{\beta_1}$	6.7132	184.688	2.38 02	2.53 02	2.68 02	2.89 02	3.08 02
67	Ho L$_{\alpha_1}$	6.7198	184.507	2.38 02	2.52 02	2.68 02	2.88 02	3.08 02
27	Co K$_{\alpha_1}$	6.9303	178.903	2.18 02	2.31 02	2.46 02	2.65 02	2.83 02
68	Er L$_{\alpha_1}$	6.9487	178.429	2.17 02	2.30 02	2.44 02	2.63 02	2.81 02

Z	transition	E [keV]	λ [pm]	$_{41}$Nb	$_{42}$Mo	$_{43}$Tc	$_{44}$Ru	$_{45}$Rh
65	Tb L_{β_1}	6.9780	177.680	2.14 02	2.27 02	2.42 02	2.60 02	2.78 02
26	Fe $K_{\beta_{1,3}}$	7.0580	175.666	2.08 02	2.20 02	2.34 02	2.52 02	2.69 02
69	Tm L_{α_1}	7.1799	172.683	1.98 02	2.10 02	2.24 02	2.41 02	2.57 02
66	Dy L_{β_1}	7.2477	171.068	1.93 02	2.05 02	2.18 02	2.34 02	2.51 02
70	Yb L_{α_1}	7.4156	167.196	1.82 02	1.92 02	2.05 02	2.20 02	2.36 02
28	Ni K_{α_1}	7.4782	165.795	1.78 02	1.88 02	2.00 02	2.15 02	2.30 02
67	Ho L_{β_1}	7.5253	164.757	1.75 02	1.85 02	1.97 02	2.12 02	2.26 02
27	Co $K_{\beta_{1,3}}$	7.6494	162.084	1.67 02	1.77 02	1.88 02	2.03 02	2.17 02
71	Lu L_{α_1}	7.6555	161.955	1.67 02	1.76 02	1.88 02	2.02 02	2.16 02
68	Er L_{β_1}	7.8109	158.733	1.58 02	1.67 02	1.78 02	1.91 02	2.05 02
72	Hf L_{α_1}	7.8990	156.963	1.53 02	1.62 02	1.73 02	1.86 02	1.99 02
29	Cu K_{α_1}	8.0478	154.060	1.45 02	1.54 02	1.64 02	1.76 02	1.89 02
69	Tm L_{β_1}	8.1010	153.049	1.43 02	1.51 02	1.61 02	1.73 02	1.86 02
73	Ta L_{α_1}	8.1461	152.201	1.41 02	1.49 02	1.59 02	1.71 02	1.83 02
28	Ni $K_{\beta_{1,3}}$	8.2647	150.017	1.35 02	1.43 02	1.53 02	1.64 02	1.76 02
74	W L_{α_1}	8.3976	147.643	1.30 02	1.37 02	1.46 02	1.57 02	1.68 02
70	Yb L_{β_1}	8.4018	147.569	1.30 02	1.37 02	1.46 02	1.57 02	1.68 02
30	Zn K_{α_1}	8.6389	143.519	1.20 02	1.27 02	1.35 02	1.46 02	1.56 02
75	Re L_{α_1}	8.6525	143.294	1.20 02	1.26 02	1.35 02	1.45 02	1.55 02
71	Lu L_{β_1}	8.7090	142.364	1.18 02	1.24 02	1.33 02	1.42 02	1.53 02
29	Cu K_{β_1}	8.9053	139.226	1.11 02	1.17 02	1.25 02	1.34 02	1.44 02
76	Os L_{α_1}	8.9117	139.126	1.11 02	1.17 02	1.25 02	1.34 02	1.44 02
72	Hf L_{β_1}	9.0227	137.414	1.07 02	1.13 02	1.20 02	1.29 02	1.39 02
77	Ir L_{α_1}	9.1751	135.132	1.02 02	1.08 02	1.15 02	1.24 02	1.33 02
31	Ga K_{α_1}	9.2517	134.013	9.99 01	1.05 02	1.13 02	1.21 02	1.30 02
73	Ta L_{β_1}	9.3431	132.702	9.73 01	1.03 02	1.10 02	1.18 02	1.26 02
78	Pt L_{α_1}	9.4423	131.308	9.46 01	9.97 01	1.07 02	1.14 02	1.25 02
30	Zn $K_{\beta_{1,3}}$	9.5720	129.529	9.12 01	9.61 01	1.03 02	1.10 02	1.18 02
74	W L_{β_1}	9.6724	128.184	8.86 01	9.34 01	9.98 01	1.07 02	1.15 02
79	Au L_{α_1}	9.7133	127.644	8.76 01	9.24 01	9.87 01	1.06 02	1.14 02
32	Ge K_{α_1}	9.8864	125.409	8.36 01	8.81 01	9.41 01	1.01 02	1.09 02
80	Hg L_{α_1}	9.9888	124.124	8.13 01	8.56 01	9.15 01	9.80 01	1.06 02
75	Re L_{β_1}	10.010	123.861	8.08 01	8.51 01	9.10 01	9.75 01	1.05 02
31	Ga K_{β_1}	10.264	120.796	7.55 01	7.96 01	8.51 01	9.11 01	9.82 01
81	Tl L_{α_1}	10.269	120.737	7.54 01	7.95 01	8.50 01	9.10 01	9.81 01
76	Os L_{β_1}	10.355	119.743	7.38 01	7.77 01	8.31 01	8.90 01	9.59 01
33	As K_{α_1}	10.544	117.588	7.03 01	7.40 01	7.91 01	8.48 01	9.14 01
82	Pb L_{α_1}	10.552	117.499	7.01 01	7.38 01	7.90 01	8.46 01	9.12 01
77	Ir L_{β_1}	10.708	115.787	6.74 01	7.09 01	7.59 01	8.14 01	8.77 01
83	Bi L_{α_1}	10.839	114.388	6.52 01	6.86 01	7.35 01	7.88 01	8.49 01
32	Ge K_{β_1}	10.982	112.898	6.30 01	6.63 01	7.10 01	7.61 01	8.20 01
78	Pt L_{β_1}	11.071	111.990	6.16 01	6.48 01	6.94 01	7.44 01	8.02 01
84	Po L_{α_1}	11.131	111.387	6.07 01	6.39 01	6.84 01	7.34 01	7.91 01
34	Se K_{α_1}	11.222	110.484	5.94 01	6.25 01	6.70 01	7.18 01	7.74 01

Z	transition	E [keV]	λ [pm]	$_{41}$Nb	$_{42}$Mo	$_{43}$Tc	$_{44}$Ru	$_{45}$Rh
85	At L$_{\alpha_1}$	11.427	108.501	5.66 01	5.95 01	6.38 01	6.84 01	7.37 01
79	Au L$_{\beta_1}$	11.442	108.359	5.64 01	5.93 01	6.36 01	6.81 01	7.34 01
33	As K$_{\beta_1}$	11.726	105.735	5.28 01	5.55 01	5.95 01	6.38 01	6.88 01
86	Rn L$_{\alpha_1}$	11.727	105.726	5.28 01	5.55 01	5.95 01	6.38 01	6.88 01
80	Hg L$_{\beta_1}$	11.823	104.867	5.16 01	5.43 01	5.82 01	6.24 01	6.73 01
35	Br K$_{\alpha_1}$	11.924	103.979	5.05 01	5.30 01	5.69 01	6.10 01	6.58 01
87	Fr L$_{\alpha_1}$	12.031	103.054	4.93 01	5.18 01	5.55 01	5.96 01	6.42 01
81	Tl L$_{\beta_1}$	12.213	101.519	4.73 01	4.97 01	5.33 01	5.72 01	6.17 01
88	Ra L$_{\alpha_1}$	12.340	100.474	4.60 01	4.83 01	5.19 01	5.57 01	6.00 01
34	Se K$_{\beta_1}$	12.496	99.219	4.45 01	4.67 01	5.02 01	5.38 01	5.80 01
82	Pb L$_{\beta_1}$	12.614	98.291	4.34 01	4.55 01	4.89 01	5.25 01	5.66 01
36	Kr K$_{\alpha_1}$	12.649	98.019	4.31 01	4.52 01	4.86 01	5.21 01	5.62 01
89	Ac L$_{\alpha_1}$	12.652	97.996	4.30 01	4.52 01	4.85 01	5.21 01	5.61 01
90	Th L$_{\alpha_1}$	12.969	95.601	4.03 01	4.22 01	4.54 01	4.87 01	5.25 01
83	Bi L$_{\beta_1}$	13.024	95.197	3.98 01	4.18 01	4.49 01	4.82 01	5.19 01
91	Pa L$_{\alpha_1}$	13.291	93.285	3.77 01	3.95 01	4.25 01	4.56 01	4.92 01
35	Br K$_{\beta_1}$	13.291	93.285	3.77 01	3.95 01	4.25 01	4.56 01	4.92 01
37	Rb K$_{\alpha_1}$	13.395	92.560	3.69 01	3.87 01	4.16 01	4.47 01	4.82 01
84	Po L$_{\beta_1}$	13.447	92.202	3.65 01	3.83 01	4.12 01	4.42 01	4.77 01
92	U L$_{\alpha_1}$	13.615	91.065	3.53 01	3.70 01	3.98 01	4.28 01	4.61 01
85	At L$_{\beta_1}$	13.876	89.352	3.36 01	3.52 01	3.79 01	4.07 01	4.38 01
36	Kr K$_{\beta_1}$	14.112	87.857	3.21 01	3.36 01	3.62 01	3.89 01	4.19 01
38	Sr K$_{\alpha_1}$	14.165	87.529	3.18 01	3.33 01	3.58 01	3.85 01	4.15 01
86	Rn L$_{\beta_1}$	14.316	86.606	3.09 01	3.23 01	3.48 01	3.74 01	4.15 01
87	Fr L$_{\beta_1}$	14.770	83.943	2.84 01	2.97 01	3.20 01	3.44 01	4.03 01
39	Y K$_{\alpha_1}$	14.958	82.888	2.74 01	2.87 01	3.10 01	3.33 01	3.58 01
37	Rb K$_{\beta_1}$	14.961	82.872	2.74 01	2.87 02	3.09 01	3.32 01	3.58 01
88	Ra L$_{\beta_1}$	15.236	81.376	2.61 01	2.73 01	2.95 01	3.17 01	3.41 01
89	Ac L$_{\beta_1}$	15.713	78.906	2.41 01	2.52 01	2.72 01	2.92 01	3.15 01
40	Zr K$_{\alpha_1}$	15.775	78.596	2.36 01	2.49 01	2.69 01	2.89 01	3.11 01
38	Sr K$_{\beta_1}$	15.836	78.293	2.36 01	2.47 01	2.66 01	2.86 01	3.08 01
90	Th L$_{\beta_1}$	16.202	76.524	2.22 01	2.32 01	2.50 01	2.69 01	2.90 01
41	Nb K$_{\alpha_1}$	16.615	74.622	2.08 01	2.17 01	2.34 01	2.51 01	2.71 01
91	Pa L$_{\beta_1}$	16.702	74.233	2.05 01	2.14 01	2.31 01	2.48 01	2.68 01
39	Y K$_{\beta_1}$	16.738	74.074	2.04 01	2.13 01	2.30 01	2.47 01	2.66 01
92	U L$_{\beta_1}$	17.220	72.000	1.89 01	1.97 01	2.13 01	2.29 01	2.47 01
42	Mo K$_{\alpha_1}$	17.479	70.933	1.82 01	1.90 01	2.05 01	2.20 01	2.37 01
40	Zr K$_{\beta_1}$	17.668	70.175	1.76 01	1.84 01	1.99 01	2.14 01	2.31 01
43	Tc K$_{\alpha_1}$	18.367	67.504	1.59 01	1.66 01	1.79 01	1.93 01	2.08 01
41	Nb K$_{\beta_1}$	18.623	66.576	<u>1.54 01</u>	1.60 01	1.73 01	1.86 01	2.01 01
44	Ru K$_{\alpha_1}$	19.279	64.311	8.50 01	1.46 01	1.58 01	1.69 01	1.83 01
42	Mo K$_{\beta_1}$	19.608	63.231	8.16 01	<u>1.40 01</u>	1.51 01	1.62 01	1.75 01
45	Rh K$_{\alpha_1}$	20.216	61.330	7.56 01	8.05 01	1.39 01	1.49 01	1.62 01
43	Tc K$_{\beta_1}$	20.619	60.131	7.18 01	7.64 01	<u>1.32 01</u>	1.42 01	1.53 01

Z	transition	E [keV]	λ [pm]	$_{41}$Nb	$_{42}$Mo	$_{43}$Tc	$_{44}$Ru	$_{45}$Rh
46	Pd K$_{\alpha_1}$	21.177	58.547	6.69 01	7.12 01	1.77 01	1.32 01	1.43 01
44	Ru K$_{\beta_1}$	21.657	57.249	6.31 01	6.71 01	7.31 01	<u>1.25 01</u>	1.35 01
47	Ag K$_{\alpha_1}$	22.163	55.942	5.93 01	6.32 01	6.87 01	7.35 01	1.27 01
45	Rh K$_{\beta_1}$	22.724	54.561	5.55 01	5.91 01	6.42 01	6.87 01	1.19 01
48	Cd K$_{\alpha_1}$	23.174	53.501	5.28 01	5.61 01	6.09 01	6.51 01	<u>1.13 01</u>
46	Pd K$_{\beta_1}$	23.819	52.053	4.91 01	5.22 01	5.65 01	6.04 01	6.29 01
49	In K$_{\alpha_1}$	24.210	51.212	4.70 01	5.00 01	5.41 01	5.78 01	6.02 01
47	Ag K$_{\beta_1}$	24.942	49.709	4.35 01	4.62 01	4.99 01	5.33 01	5.56 01
50	Sn K$_{\alpha_1}$	25.271	49.062	4.20 01	4.46 01	4.82 01	5.15 01	5.37 01
48	Cd K$_{\beta_1}$	26.096	47.511	3.86 01	4.10 01	4.41 01	4.72 01	4.93 01
51	Sb K$_{\alpha_1}$	26.359	47.037	3.76 01	3.99 01	4.30 01	4.59 01	4.80 01
49	In K$_{\beta_1}$	27.276	45.455	3.43 01	3.65 01	3.92 01	4.18 01	4.38 01
52	Te K$_{\alpha_1}$	27.472	45.131	3.37 01	3.58 01	3.84 01	4.10 01	4.30 01
50	Sn K$_{\beta_1}$	28.486	43.524	3.06 01	3.25 01	3.48 01	3.72 01	3.90 01
53	I K$_{\alpha_1}$	28.612	43.333	3.03 01	3.22 01	3.44 01	3.67 01	3.86 01
51	Sb K$_{\beta_1}$	29.726	41.709	2.74 01	2.91 01	3.11 01	3.31 01	3.48 01
54	Xe K$_{\alpha_1}$	29.779	41.635	2.72 01	2.89 01	3.09 01	3.30 01	3.47 01
55	Cs K$_{\alpha_1}$	30.973	40.030	2.45 01	2.60 01	2.77 01	2.96 01	3.12 01
52	Te K$_{\beta_1}$	30.996	40.000	2.44 01	2.60 01	2.77 01	2.95 01	3.11 01
56	Ba K$_{\alpha_1}$	32.194	38.511	2.21 01	2.34 01	2.49 01	2.66 01	2.81 01
53	I K$_{\beta_1}$	32.295	38.391	2.19 01	2.32 01	2.47 01	2.64 01	2.78 01
57	La K$_{\alpha_1}$	33.442	37.074	1.99 01	2.11 01	2.24 01	2.40 01	2.53 01
54	Xe K$_{\beta_1}$	33.624	36.874	1.96 01	2.08 01	2.21 01	2.36 01	2.49 01
58	Ce K$_{\alpha_1}$	34.279	36.169	1.86 01	1.98 01	2.10 01	2.24 01	2.37 01
55	Cs K$_{\beta_1}$	34.987	35.437	1.76 01	1.87 01	1.98 01	2.12 01	2.24 01
59	Pr K$_{\alpha_1}$	36.026	34.415	1.63 01	1.73 01	1.83 01	1.95 01	2.07 01
56	Ba K$_{\beta_1}$	36.378	34.082	1.58 01	1.68 01	1.78 01	1.90 01	2.01 01
60	Nd K$_{\alpha_1}$	36.847	33.648	1.53 01	1.62 01	1.72 01	1.84 01	1.95 01
57	La K$_{\beta_1}$	37.801	32.799	1.43 01	1.51 01	1.60 01	1.71 01	1.81 01
61	Pm K$_{\alpha_1}$	38.725	32.016	1.34 01	1.42 01	1.50 01	1.60 01	1.70 01
58	Ce K$_{\beta_1}$	39.257	31.582	1.29 01	1.37 01	1.44 01	1.54 01	1.64 01
62	Sm K$_{\alpha_1}$	40.118	30.905	1.22 01	1.29 01	1.36 01	1.45 01	1.54 01
59	Pr K$_{\beta_1}$	40.748	30.427	1.16 01	1.23 01	1.30 01	1.39 01	1.48 01
63	Eu K$_{\alpha_1}$	41.542	29.845	1.11 01	1.17 01	1.23 01	1.32 01	1.34 01
60	Nd K$_{\beta_1}$	42.271	29.331	1.05 01	1.12 01	1.17 01	1.26 01	1.34 01
64	Gd K$_{\alpha_1}$	42.996	28.836	1.01 01	1.07 01	1.12 01	1.20 01	1.28 01
61	Pm K$_{\beta_1}$	43.826	28.290	9.56 00	1.01 01	1.06 01	1.14 01	1.21 01
65	Tb K$_{\alpha_1}$	44.482	27.873	9.18 00	9.70 00	1.02 01	1.09 01	1.16 01
62	Sm K$_{\beta_1}$	45.413	27.301	8.67 00	9.16 00	9.62 00	1.03 01	1.10 01
66	Dy K$_{\alpha_1}$	45.998	26.954	8.38 00	8.85 00	9.28 00	9.93 00	1.06 01
63	Eu K$_{\beta_1}$	47.038	26.358	7.88 00	8.32 00	8.72 00	9.33 00	9.97 00
67	Ho K$_{\alpha_1}$	47.547	26.076	7.66 00	8.07 00	8.47 00	9.05 00	9.68 00
64	Gd K$_{\beta_1}$	48.697	25.460	7.17 00	7.56 00	7.92 00	8.47 00	9.06 00
68	Er K$_{\alpha_1}$	49.128	25.237	7.00 00	7.38 00	7.73 00	8.20 00	8.84 00

Z	transition	E [keV]	λ [pm]	$_{41}$Nb	$_{42}$Mo	$_{43}$Tc	$_{44}$Ru	$_{45}$Rh
65	Tb K_{β_1}	50.382	24.609	6.54 00	6.89 00	7.21 00	7.70 00	8.25 00
69	Tm K_{α_1}	50.742	24.343	6.41 00	6.75 00	7.07 00	7.55 00	8.09 00
66	Dy K_{β_1}	52.119	23.788	5.95 00	6.28 00	6.50 00	7.02 00	7.51 00
70	Yb K_{α_1}	52.389	23.666	5.87 00	6.19 00	6.47 00	6.92 00	7.41 00
67	Ho K_{β_1}	53.877	23.012	5.43 00	5.73 00	5.99 00	6.40 00	6.86 00
71	Lu K_{α_1}	54.070	22.930	5.38 00	5.68 00	5.93 00	6.34 00	6.79 00
68	Er K_{β_1}	55.681	22.267	4.96 00	5.24 00	5.47 00	5.85 00	6.26 00
72	Hf K_{α_1}	55.790	22.223	4.94 00	5.21 00	5.44 00	5.82 00	6.23 00
69	Tm K_{β_1}	57.517	21.556	4.54 00	4.80 00	5.00 00	5.35 00	5.72 00
73	Ta K_{α_1}	57.532	21.550	4.53 00	4.79 00	5.00 00	5.35 00	5.72 00
74	W K_{α_1}	59.318	20.901	4.17 00	4.41 00	4.60 00	4.91 00	5.25 00
70	Yb K_{β_1}	59.370	20.883	4.16 00	4.40 00	4.58 00	4.90 00	5.24 00
75	Re K_{α_1}	61.140	20.278	3.84 00	4.06 00	4.23 00	4.52 00	4.84 00
71	Lu K_{β_1}	61.283	20.231	3.81 00	4.04 00	4.20 00	4.40 00	4.81 00
76	Os K_{α_1}	63.001	19.679	3.54 00	3.75 00	3.90 00	4.17 00	4.40 00
72	Hf K_{β_1}	63.234	19.607	3.50 00	3.71 00	3.80 00	4.12 00	4.41 00
77	Ir K_{α_1}	64.896	19.105	3.27 00	3.46 00	3.60 00	3.84 00	4.11 00
73	Ta K_{β_1}	65.223	19.009	3.22 00	3.41 00	3.55 00	3.79 00	4.05 00
78	Pt K_{α_1}	66.832	18.551	3.02 00	3.19 00	3.32 00	3.55 00	3.79 00
74	W K_{β_1}	67.244	18.438	2.97 00	3.14 00	3.26 00	3.49 00	3.73 00
79	Au K_{α_1}	68.804	18.020	2.79 00	2.95 00	3.07 00	3.28 00	3.50 00
75	Re K_{β_1}	69.310	17.888	2.73 00	2.89 00	3.00 00	3.31 00	3.44 00
80	Hg K_{α_1}	70.819	17.507	2.58 00	2.73 00	2.83 00	3.03 00	3.24 00
76	Os K_{β_1}	71.413	17.361	2.52 00	2.67 00	2.77 00	2.96 00	3.17 00
81	Tl K_{α_1}	72.872	17.014	2.39 00	2.53 00	2.62 00	2.80 00	3.00 00
77	Ir K_{β_1}	73.561	16.854	2.33 00	2.46 00	2.55 00	2.73 00	2.92 00
82	Pb K_{α_1}	74.969	16.538	2.21 00	2.34 00	2.43 00	2.59 00	2.77 00
78	Pt K_{β_1}	75.748	16.368	2.15 00	2.28 00	2.36 00	2.52 00	2.70 00
83	Bi K_{α_1}	77.108	16.079	2.05 00	2.17 00	2.25 00	2.40 00	2.57 00
79	Au K_{β_1}	77.948	15.906	1.99 00	2.11 00	2.18 00	2.33 00	2.49 00
84	Po K_{α_1}	79.290	15.636	1.90 00	2.01 00	2.08 00	2.22 00	2.38 00
80	Hg K_{β_1}	80.253	15.449	1.84 00	1.95 00	2.01 00	2.15 00	2.30 00
85	At K_{α_1}	81.520	15.209	1.77 00	1.87 00	1.93 00	2.06 00	2.21 00
81	Tl K_{β_1}	82.576	15.014	1.71 00	1.81 00	1.87 00	2.00 00	2.14 00
86	Rn K_{α_1}	83.780	14.798	1.64 00	1.74 00	1.80 00	1.92 00	2.06 00
82	Pb K_{β_1}	84.936	14.597	1.59 00	1.68 00	1.74 00	1.85 00	1.96 00
87	Fr K_{α_1}	86.100	14.400	1.53 00	1.62 00	1.68 00	1.79 00	1.91 00
83	Bi K_{β_1}	87.343	14.195	1.48 00	1.56 00	1.61 00	1.72 00	1.84 00
88	Ra K_{α_1}	88.470	14.014	1.43 00	1.51 00	1.56 00	1.66 00	1.78 00
84	Po K_{β_1}	89.800	13.806	1.37 00	1.45 00	1.50 00	1.60 00	1.71 00
89	Ac K_{α_1}	90.884	13.642	1.33 00	1.41 00	1.45 00	1.55 00	1.66 00
85	At K_{β_1}	92.300	13.432	1.28 00	1.35 00	1.40 00	1.49 00	1.59 00
90	Th K_{α_1}	93.350	13.281	1.24 00	1.31 00	1.36 00	1.44 00	1.54 00
86	Rn K_{β_1}	94.870	13.068	1.19 00	1.26 00	1.30 00	1.38 00	1.48 00

Z	transition	E [keV]	λ [pm]	$_{41}$Nb	$_{42}$Mo	$_{43}$Tc	$_{44}$Ru	$_{45}$Rh
91	Pa K$_{\alpha_1}$	95.868	12.932	1.16 00	1.23 00	1.26 00	1.35 00	1.44 00
87	Fr K$_{\beta_1}$	97.470	12.720	1.11 00	1.17 00	1.21 00	1.29 00	1.38 00
92	U K$_{\alpha_1}$	98.439	12.595	1.08 00	1.14 00	1.18 00	1.25 00	1.34 00
88	Ra K$_{\beta_1}$	100.130	12.382	1.03 00	1.10 00	1.13 00	1.20 00	1.28 00
89	Ac K$_{\beta_1}$	102.850	12.054	9.71 -1	1.03 00	1.06 00	1.13 00	1.20 00
90	Th K$_{\beta_1}$	105.610	11.739	8.13 -1	9.66 -1	9.94 -1	1.06 00	1.13 00
91	Pa K$_{\beta_1}$	108.430	11.434	8.59 -1	9.07 -1	9.34 -1	9.92 -1	1.06 00
92	U K$_{\beta_1}$	111.300	11.139	8.08 -1	8.53 -1	8.77 -1	9.31 -1	9.94 -1

Z	transition	E [keV]	λ [pm]	$_{46}$Pd	$_{47}$Ag	$_{48}$Cd	$_{49}$In	$_{50}$Sn
4	Be K$_\alpha$	0.1085	11427.207	1.47 04	2.04 04	3.19 04	7.99 04	7.47 04
38	Sr M$_\xi$	0.1140	10875.895	1.14 04	1.25 04	2.58 04	6.41 04	6.19 04
39	Y M$_\xi$	0.1328	9339.235	7.90 03	4.65 03	1.11 04	2.29 04	3.99 04
16	S L$_l$	0.1487	8337.942	8.28 03	5.74 03	6.12 03	7.29 04	1.23 04
40	Zr M$_\xi$	0.1511	8205.506	8.34 03	5.99 03	6.14 03	6.33 03	1.02 04
41	Nb M$_\xi$	0.1717	7221.037	9.22 03	6.90 03	6.43 03	4.34 03	6.56 03
5	B K$_\alpha$	0.1833	6764.059	8.20 03	7.31 03	6.57 03	5.13 03	6.77 03
42	Mo M$_\xi$	0.1926	6473.445	7.46 03	7.28 03	6.69 03	5.80 03	6.96 03
6	C K$_\alpha$	0.2770	4476.000	7.46 03	8.19 03	5.79 03	6.14 03	6.33 03
47	Ag M$_\xi$	0.3117	3977.709	9.07 03	8.05 03	5.17 03	5.51 03	5.75 03
7	N K$_\alpha$	0.3924	3159.664	1.67 04	7.17 03	4.15 03	4.47 03	4.76 03
22	Ti L$_l$	0.3953	3136.484	1.68 04	7.55 03	4.12 03	4.44 03	4.73 03
22	Ti L$_\alpha$	0.4522	2741.822	1.93 04	1.92 04	2.66 04	3.06 04	3.94 03
23	V L$_\alpha$	0.5113	2424.901	1.79 04	1.86 04	2.12 04	2.35 04	2.43 04
8	O K$_\alpha$	0.5249	2362.072	1.76 04	1.85 04	2.02 04	2.22 04	2.31 04
25	Mn L$_l$	0.5563	2228.747	1.78 04	1.80 04	1.82 04	1.96 04	2.06 04
24	Cr L$_\alpha$	0.5728	2164.549	1.80 04	1.78 04	1.72 04	1.84 04	1.94 04
25	Mn L$_\alpha$	0.6374	1945.171	1.56 04	1.63 04	1.57 04	1.49 04	1.56 04
9	F K$_\alpha$	0.6768	1831.932	1.44 04	1.55 04	1.49 04	1.59 04	1.38 04
26	Fe L$_\alpha$	0.7050	1758.655	1.36 04	1.49 04	1.43 04	1.49 04	1.28 04
27	Co L$_\alpha$	0.7762	1597.335	1.14 04	1.21 04	1.27 04	1.25 04	1.27 04
28	Ni L$_\alpha$	0.8515	1456.080	9.40 03	1.00 04	1.04 04	1.11 04	1.09 04
29	Cu L$_\alpha$	0.9297	1336.044	7.81 03	8.33 03	8.72 03	9.20 03	9.62 03
30	Zn L$_\alpha$	1.0117	1225.513	7.53 03	8.13 03	8.69 03	7.70 03	8.06 03
11	Na K$_\alpha$	1.0410	1191.020	7.01 03	7.57 03	8.09 03	7.22 03	7.55 03
11	Na K$_\beta$	1.0711	1157.550	6.52 03	7.04 03	7.53 03	6.61 03	6.90 03
12	Mg K$_\alpha$	1.2530	989.033	4.39 03	4.74 03	5.08 03	5.48 03	5.86 03
33	As L$_\alpha$	1.2820	967.123	4.15 03	4.48 03	4.80 03	5.18 03	5.54 03
12	Mg K$_\beta$	1.3022	952.121	3.99 03	4.31 03	4.62 03	4.98 03	5.32 03
33	As L$_{\beta_1}$	1.3170	941.421	3.87 03	4.18 03	4.49 03	4.84 03	5.17 03
66	Dy M$_\beta$	1.3250	955.737	3.81 03	4.12 03	4.42 03	4.77 03	5.10 03
67	Ho M$_\alpha$	1.3480	919.771	3.65 03	3.95 03	4.24 03	4.57 03	4.88 03
34	Se L$_\alpha$	1.3791	899.029	3.45 03	3.73 03	4.00 03	4.31 03	4.61 03

Z	transition	E [keV]	λ [pm]	$_{46}$Pd	$_{47}$Ag	$_{48}$Cd	$_{49}$In	$_{50}$Sn
67	Ho M$_\beta$	1.3830	896.494	3.42 03	3.70 03	3.97 03	4.28 03	4.58 03
68	Er M$_\alpha$	1.4060	881.829	3.28 03	3.55 03	3.81 03	4.11 03	4.39 03
34	Se L$_{\beta_1}$	1.4192	873.627	3.21 03	3.47 03	3.72 03	4.01 03	4.29 03
68	Er M$_\beta$	1.4430	859.218	3.08 03	3.32 03	3.57 03	3.85 03	4.11 03
69	Tm M$_\alpha$	1.4620	848.051	2.98 03	3.22 03	3.46 03	3.73 03	3.98 03
35	Br L$_\alpha$	1.4804	837.511	2.88 03	3.12 03	3.35 03	3.61 03	3.85 03
13	Al K$_{\alpha_1}$	1.4867	833.962	2.85 03	3.08 03	3.31 03	3.57 03	3.81 03
69	Tm M$_\beta$	1.5030	824.918	2.78 03	3.00 03	3.23 03	3.48 03	3.71 03
70	Yb M$_\alpha$	1.5214	814.941	2.69 03	2.91 03	3.13 03	3.37 03	3.60 03
35	Br L$_{\beta_1}$	1.5259	812.538	2.67 03	2.89 03	3.11 03	3.35 03	3.57 03
13	Al K$_\beta$	1.5574	796.103	2.54 03	2.74 03	2.95 03	3.18 03	3.40 03
70	Yb M$_\beta$	1.5675	790.974	2.50 03	2.64 03	2.84 03	3.06 03	3.27 03
71	Lu M$_\alpha$	1.5813	784.071	2.44 03	2.64 03	2.84 03	3.06 03	3.27 03
36	Kr L$_\alpha$	1.5860	781.747	2.42 03	2.62 03	2.82 03	3.04 03	3.24 03
71	Lu M$_\beta$	1.6312	760.085	2.26 03	2.44 03	2.63 03	2.83 03	3.02 03
36	Kr L$_{\beta_1}$	1.6366	757.577	2.24 03	2.42 03	2.61 03	2.81 03	3.00 03
72	Hf M$_{\alpha_1}$	1.6446	753.892	2.21 03	2.39 03	2.58 03	2.78 03	2.96 03
37	Rb L$_{\alpha_1}$	1.6941	731.864	2.05 03	2.22 03	2.39 03	2.58 03	2.75 03
72	Hf M$_\beta$	1.6976	730.355	2.04 03	2.21 03	2.38 03	2.57 03	2.74 03
73	Ta M$_{\alpha_1}$	1.7096	725.229	2.00 03	2.17 03	2.34 03	2.52 03	2.69 03
14	Si K$_{\alpha_1}$	1.7400	712.558	1.92 03	2.07 03	2.24 03	2.41 03	2.57 03
37	Rb L$_{\beta_1}$	1.7522	707.597	1.88 03	2.04 03	2.20 03	2.37 03	2.53 03
73	Ta M$_\beta$	1.7655	702.266	1.85 03	2.00 03	2.16 03	2.33 03	2.48 03
74	W M$_{\alpha_1}$	1.7754	698.350	1.82 03	1.97 03	2.13 03	2.29 03	2.45 03
38	Sr L$_{\alpha_1}$	1.8066	686.290	1.74 03	1.88 03	2.04 03	2.20 03	2.34 03
74	W M$_\beta$	1.8349	675.705	1.68 03	1.81 03	1.96 03	2.11 03	2.25 03
14	Si K$_\beta$	1.8359	675.337	1.67 03	1.81 03	1.96 03	2.11 03	2.25 03
75	Re M$_{\alpha_1}$	1.8420	673.100	1.66 03	1.79 03	1.94 03	2.09 03	2.23 03
38	Sr L$_{\beta_1}$	1.8717	662.420	1.59 03	1.72 03	1.87 03	2.01 03	2.14 03
75	Re M$_\beta$	1.9061	650.465	1.52 03	1.65 03	1.78 03	1.92 03	2.05 03
76	Os M$_{\alpha_1}$	1.9102	649.069	1.51 03	1.64 03	1.77 03	1.91 03	2.04 03
39	Y L$_{\alpha_1}$	1.9226	644.882	1.49 03	1.61 03	1.74 03	1.88 03	2.00 03
76	Os M$_{\beta_1}$	1.9783	626.725	1.38 03	1.50 03	1.62 03	1.75 03	1.87 03
77	Ir M$_{\alpha_1}$	1.9799	626.219	1.38 03	1.49 03	1.62 03	1.75 03	1.86 03
39	Y L$_{\beta_1}$	1.9958	621.230	1.35 03	1.46 03	1.59 03	1.71 03	1.83 03
15	P K$_{\alpha_1}$	2.0137	615.708	1.32 03	1.43 03	1.55 03	1.68 03	1.79 03
40	Zr L$_{\alpha_1}$	2.0424	607.056	1.28 03	1.38 03	1.50 03	1.62 03	1.72 03
78	Pt M$_{\alpha_1}$	2.0505	604.658	1.27 03	1.37 03	1.49 03	1.60 03	1.71 03
77	Ir M$_\beta$	2.0535	603.775	1.26 03	1.36 03	1.48 03	1.60 03	1.70 03
79	Au M$_{\alpha_1}$	2.1229	584.036	1.16 03	1.25 03	1.36 03	1.47 03	1.57 03
40	Zr L$_{\beta_1}$	2.1244	583.624	1.16 03	1.25 03	1.36 03	1.47 03	1.56 03
78	Pt M$_\beta$	2.1273	582.828	1.15 03	1.25 03	1.36 03	1.46 03	1.56 03
15	P K$_{\beta_{1,3}}$	2.1390	579.640	1.14 03	1.23 03	1.34 03	1.44 03	1.54 03
41	Nb L$_{\alpha_1}$	2.1659	572.441	1.10 03	1.19 03	1.30 03	1.40 03	1.49 03

Z	transition	E [keV]	λ [pm]	$_{46}$Pd	$_{47}$Ag	$_{48}$Cd	$_{49}$In	$_{50}$Sn
80	Hg M$_{\alpha_1}$	2.1953	564.775	1.07 03	1.15 03	1.25 03	1.35 03	1.44 03
79	Au M$_\beta$	2.2046	562.393	1.05 03	1.14 03	1.24 03	1.34 03	1.42 03
41	Nb L$_{\beta_1}$	2.2574	549.238	9.94 02	1.08 03	1.17 03	1.26 03	1.34 03
81	Tl M$_{\alpha_1}$	2.2706	546.045	9.80 02	1.06 03	1.15 03	1.24 03	1.32 03
80	Hg M$_\beta$	2.2825	543.199	9.67 02	1.05 03	1.14 03	1.23 03	1.31 03
42	Mo L$_{\alpha_1}$	2.2932	540.664	9.56 02	1.03 03	1.13 03	1.21 03	1.29 03
16	S K$_{\alpha_1}$	2.3080	537.197	9.40 02	1.02 03	1.11 03	1.19 03	1.27 03
82	Pb M$_{\alpha_1}$	2.3457	528.563	9.03 02	9.77 02	1.06 03	1.14 03	1.22 03
81	Tl M$_\beta$	2.3621	524.893	8.87 02	9.60 02	1.05 03	1.12 03	1.29 03
42	Mo L$_{\beta_1}$	2.3948	517.726	8.57 02	9.28 02	1.01 03	1.09 03	1.16 03
83	Bi M$_{\alpha_1}$	2.4226	511.785	8.33 02	9.01 02	9.83 02	1.06 03	1.13 03
43	Tl L$_{\alpha_1}$	2.4240	511.490	8.31 02	9.00 02	9.81 02	1.05 03	1.12 03
82	Pb M$_\beta$	2.4427	507.574	8.15 02	8.83 02	9.63 02	1.03 03	1.10 03
16	S K$_\beta$	2.4640	503.186	7.98 02	8.64 02	9.42 02	1.01 03	1.08 03
83	Bi M$_{\beta_1}$	2.5255	490.933	7.50 02	8.12 02	8.87 02	9.52 02	1.01 03
43	Tc L$_{\beta_1}$	2.5368	488.746	7.42 02	8.03 02	8.77 02	9.41 02	1.00 03
44	Ru L$_{\alpha_1}$	2.5586	484.582	7.26 02	7.86 02	8.58 02	9.21 02	9.82 02
17	Cl K$_{\alpha_1}$	2.6224	472.792	6.82 02	7.39 02	8.08 02	8.66 02	9.23 02
44	Ru L$_{\beta_1}$	2.6832	462.079	6.44 02	6.98 02	7.63 02	8.18 02	8.72 02
45	Rh L$_{\alpha_1}$	2.6967	459.766	6.36 02	6.89 02	7.54 02	8.08 02	8.61 02
17	Cl K$_\beta$	2.8156	440.350	5.71 02	6.18 02	6.77 02	7.25 02	7.73 02
45	Rh L$_{\beta_1}$	2.8344	437.430	5.61 02	6.08 02	6.66 02	7.13 02	7.60 02
46	Pd L$_{\alpha_1}$	2.8386	436.782	5.59 02	6.06 02	6.64 02	7.11 02	7.57 02
18	Ar K$_{\alpha_1}$	2.9577	419.194	5.04 02	5.47 02	5.99 02	6.41 02	6.83 02
47	Ag L$_{\alpha_1}$	2.9843	415.458	4.93 02	5.35 02	5.86 02	6.27 02	6.68 02
46	Pd L$_{\beta_1}$	2.9902	414.638	4.91 02	5.32 02	5.83 02	6.24 02	6.65 02
90	Th M$_{\alpha_1}$	2.9961	413.821	4.88 02	5.29 02	5.81 02	6.21 02	6.62 02
91	Pa M$_{\alpha_1}$	3.0823	402.248	4.55 02	4.93 02	5.41 02	5.79 02	6.17 02
48	Cd L$_{\alpha_1}$	3.1337	395.651	4.36 02	4.73 02	5.19 02	5.56 02	5.92 02
90	Th M$_\beta$	3.1458	394.129	4.32 02	4.68 02	5.14 02	5.50 02	5.86 02
47	Ag L$_{\beta_1}$	3.1509	393.491	4.30 02	4.66 02	5.12 02	5.48 02	5.84 02
92	U M$_{\alpha_1}$	3.1708	391.021	4.24 02	4.59 02	5.04 02	5.40 02	5.74 02
18	Ar K$_\beta$	3.1905	388.607	1.52 03	4.52 02	4.97 02	5.32 02	5.66 02
91	Pa M$_\beta$	3.2397	382.705	1.45 03	4.35 02	4.78 02	5.12 02	5.44 02
49	In L$_{\alpha_1}$	3.2869	377.210	1.40 03	4.20 02	4.61 02	4.94 02	5.25 02
19	K K$_{\alpha_1}$	3.3138	374.148	1.37 03	4.11 02	4.52 02	4.84 02	5.15 02
48	Cd L$_{\beta_1}$	3.3166	373.832	<u>1.36 03</u>	4.10 02	4.51 02	4.83 02	5.13 02
92	U M$_\beta$	3.3360	371.658	1.88 03	<u>4.04 02</u>	4.45 02	4.76 02	5.06 02
50	Sn L$_{\alpha_1}$	3.4440	360.003	1.72 03	1.31 03	4.11 02	4.40 02	4.67 02
49	In L$_{\beta_1}$	3.4872	355.543	1.66 03	<u>1.26 03</u>	<u>3.98 02</u>	4.27 02	4.53 02
19	K K$_\beta$	3.5896	345.401	<u>1.53 03</u>	1.63 03	1.23 03	3.97 02	4.21 02
51	Sb L$_{\alpha_1}$	3.6047	343.854	1.82 03	1.61 03	1.21 03	3.93 02	4.17 02
50	Sn L$_{\beta_1}$	3.6628	338.498	1.74 03	1.54 03	1.16 03	3.78 02	4.01 02
20	Ca K$_{\alpha_1}$	3.6917	335.848	1.70 03	1.51 03	<u>1.14 03</u>	<u>3.71 02</u>	3.93 02

Z	transition	E [keV]	λ [pm]	$_{46}$Pd	$_{47}$Ag	$_{48}$Cd	$_{49}$In	$_{50}$Sn
52	Te L$_{\alpha_1}$	3.7693	328.934	1.61 03	1.43 03	1.50 03	1.13 03	3.73 02
51	Sb L$_{\beta_1}$	3.8436	322.575	1.52 03	1.62 03	1.43 03	1.07 03	3.55 02
53	I L$_{\alpha_1}$	3.9377	314.867	1.42 03	1.52 03	1.34 03	1.01 03	1.07 03
20	Ca K$_\beta$	4.0127	308.981	1.35 03	1.44 03	1.27 03	1.34 03	1.02 03
52	Te L$_{\beta_1}$	4.0296	307.686	1.34 03	1.42 03	1.50 03	1.32 03	1.01 03
21	Sc K$_{\alpha_1}$	4.0906	303.097	1.28 03	1.37 03	1.44 03	1.27 03	9.66 02
54	Xe L$_{\alpha_1}$	4.1099	301.674	1.27 03	1.35 03	1.42 03	1.25 03	9.53 02
53	I L$_{\beta_1}$	4.2207	293.755	1.18 03	1.25 03	1.32 03	1.16 03	2.02 03
55	Cs L$_{\alpha_1}$	4.2865	289.245	1.13 03	1.20 03	1.26 03	1.34 03	3.15 03
21	Sc K$_{\beta_1}$	4.4605	277.962	1.01 03	1.08 03	1.13 03	1.20 03	9.90 03
56	Sc K$_{\beta_1}$	4.4663	277.601	1.01 03	1.07 03	1.13 03	1.19 03	1.27 03
22	Ti K$_{\alpha_1}$	4.5108	274.863	9.78 02	1.04 03	1.10 03	1.16 03	1.24 03
55	Cs L$_{\beta_1}$	4.6198	268.377	9.16 02	9.78 02	1.03 03	1.09 03	1.16 03
57	La L$_{\alpha_1}$	4.6510	266.577	8.99 02	9.60 02	1.01 03	1.07 03	1.14 03
56	Ba L$_{\beta_1}$	4.8273	256.841	8.11 02	8.67 02	9.09 02	9.65 02	1.03 03
58	Ce L$_{\alpha_1}$	4.8402	256.157	8.05 02	8.61 02	9.03 02	9.58 02	1.02 03
22	Ti K$_{\beta_{1,3}}$	4.9318	251.399	7.64 02	8.17 02	8.57 02	9.10 02	9.69 02
23	V K$_{\alpha_1}$	4.9522	250.363	7.56 02	8.08 02	8.47 02	8.99 02	9.59 02
59	Pr L$_{\alpha_1}$	5.0337	246.310	7.22 02	7.73 02	8.10 02	8.60 02	9.16 02
57	La L$_{\beta_1}$	5.0421	245.899	7.19 02	7.69 02	8.06 02	8.56 02	9.12 02
60	Nd L$_{\alpha_1}$	5.2304	237.047	6.51 02	6.96 02	7.29 02	7.74 02	8.24 02
58	Ce L$_{\beta_1}$	5.2622	235.614	6.40 02	6.85 02	7.17 02	7.61 02	8.10 02
24	Cr K$_{\alpha_1}$	5.4147	228.978	5.92 02	6.34 02	6.63 02	7.03 02	7.49 02
23	V K$_{\beta_{1,3}}$	5.4273	228.447	5.88 02	6.30 02	6.59 02	6.99 02	7.44 02
61	Pm L$_{\alpha_1}$	5.4325	228.228	5.87 02	6.28 02	6.58 02	6.97 02	7.42 02
59	Pr L$_{\beta_1}$	5.4889	225.883	5.70 02	6.11 02	6.39 02	6.77 02	7.21 02
62	Sm L$_{\alpha_1}$	5.6361	219.984	5.30 02	5.68 02	5.95 02	6.30 02	6.70 02
60	Nd L$_{\beta_1}$	5.7216	216.696	5.09 02	5.45 02	5.71 02	6.04 02	6.42 02
63	Eu L$_{\alpha_1}$	5.8457	212.096	4.80 02	5.14 02	5.38 02	5.69 02	6.05 02
25	Mn K$_{\alpha_1}$	5.8988	210.187	4.68 02	5.02 02	5.25 02	5.55 02	5.90 02
24	Cr K$_{\beta_{1,3}}$	5.9467	208.494	4.58 02	4.91 02	5.13 02	5.43 02	5.77 02
61	Pm L$_{\beta_1}$	5.9610	207.993	4.55 02	4.88 02	5.10 02	5.39 02	5.73 02
64	Gd L$_{\alpha_1}$	6.0572	204.690	4.36 02	4.67 02	4.88 02	5.16 02	5.49 02
62	Sm L$_{\beta_1}$	6.2051	199.811	4.08 02	4.37 02	4.57 02	4.84 02	5.14 02
65	Tb L$_{\alpha_1}$	6.2728	197.655	3.96 02	4.24 02	4.43 02	4.70 02	4.99 02
26	Fe K$_{\alpha_1}$	6.4038	193.611	3.74 02	4.01 02	4.19 02	4.44 02	4.71 02
63	Eu L$_{\beta_1}$	6.4564	192.034	3.66 02	3.92 02	4.10 02	4.34 02	4.61 02
25	Mn K$_{\beta_{1,3}}$	6.4905	191.025	3.61 02	3.86 02	4.04 02	4.28 02	4.54 02
66	Dy L$_{\alpha_1}$	6.4952	190.887	3.60 02	3.86 02	4.03 02	4.27 02	4.54 02
64	Gd L$_{\beta_1}$	6.7132	184.688	3.29 02	3.52 02	3.68 02	3.91 02	4.14 02
67	Ho L$_{\alpha_1}$	6.7198	184.507	3.28 02	3.52 02	3.67 02	3.90 02	4.13 02
27	Co K$_{\alpha_1}$	6.9303	178.903	3.02 02	3.23 02	3.37 02	3.58 02	3.80 02
68	Er L$_{\alpha_1}$	6.9487	178.429	2.99 02	3.21 02	3.35 02	3.56 02	3.77 02
65	Tb L$_{\beta_1}$	6.9780	177.680	2.96 02	3.17 02	3.31 02	3.52 02	3.73 02

Z	transition	E [keV]	λ [pm]	$_{46}$Pd	$_{47}$Ag	$_{48}$Cd	$_{49}$In	$_{50}$Sn
26	Fe K$_{\beta_{1,3}}$	7.0580	175.666	2.87 02	3.07 02	3.21 02	3.41 02	3.61 02
69	Tm L$_{\alpha_1}$	7.1799	172.683	2.74 02	2.93 02	3.06 02	3.26 02	3.45 02
66	Dy L$_{\beta_1}$	7.2477	171.068	2.67 02	2.86 02	2.98 02	3.17 02	3.36 02
70	Yb L$_{\alpha_1}$	7.4156	167.195	2.51 02	2.69 02	2.80 02	2.98 02	3.16 02
28	Ni K$_{\alpha_1}$	7.4782	165.795	2.45 02	2.62 02	2.74 02	2.92 02	3.09 02
67	Ho L$_{\beta_1}$	7.5253	164.757	2.41 02	2.58 02	2.69 02	2.87 02	3.03 02
27	Co K$_{\beta_{1,3}}$	7.6494	162.084	2.30 02	2.47 02	2.57 02	2.74 02	2.90 02
71	Lu L$_{\alpha_1}$	7.6555	161.955	2.30 02	2.46 02	2.57 02	2.74 02	2.90 02
68	Er L$_{\beta_1}$	7.8109	158.733	2.18 02	2.33 02	2.43 02	2.59 02	2.74 02
72	Hf L$_{\alpha_1}$	7.8990	156.963	2.11 02	2.26 02	2.36 02	2.51 02	2.66 02
29	Cu K$_{\alpha_1}$	8.0478	154.060	2.00 02	2.15 02	2.24 02	2.39 02	2.55 02
69	Tm L$_{\beta_1}$	8.1010	153.049	1.97 02	2.11 02	2.20 02	2.35 02	2.48 02
73	Ta L$_{\alpha_1}$	8.1461	152.201	1.94 02	2.08 02	2.17 02	2.31 02	2.44 02
28	Ni K$_{\beta_{1,3}}$	8.2647	150.017	1.86 02	2.00 02	2.08 02	2.22 02	2.35 02
74	W L$_{\alpha_1}$	8.3976	147.643	1.79 02	1.91 02	1.99 02	2.13 02	2.25 02
70	Yb L$_{\beta_1}$	8.4018	147.569	1.78 02	1.91 02	1.99 02	2.13 02	2.25 02
30	Zn K$_{\alpha_1}$	8.6389	143.519	1.65 02	1.77 02	1.85 02	1.97 02	2.08 02
75	Re L$_{\alpha_1}$	8.6525	143.294	1.65 02	1.77 02	1.84 02	1.96 02	2.07 02
71	Lu L$_{\beta_1}$	8.7090	142.364	1.62 02	1.74 02	1.81 02	1.93 02	2.04 02
29	Cu K$_{\beta_1}$	8.9053	139.226	1.52 02	1.63 02	1.70 02	1.82 02	1.92 02
76	Os L$_{\alpha_1}$	8.9117	139.126	1.52 02	1.63 02	1.70 02	1.81 02	1.91 02
72	Hf L$_{\beta_1}$	9.0227	137.414	1.47 02	1.58 02	1.64 02	1.75 02	1.85 02
77	Ir L$_{\alpha_1}$	9.1751	135.132	1.40 02	1.51 02	1.57 02	1.68 02	1.77 02
31	Ga K$_{\alpha_1}$	9.2517	134.013	1.37 02	1.47 02	1.53 02	1.64 02	1.73 02
73	Ta L$_{\beta_1}$	9.3431	132.702	1.33 02	1.44 02	1.49 02	1.60 02	1.68 02
78	Pt L$_{\alpha_1}$	9.4423	131.308	1.30 02	1.40 02	1.45 02	1.55 02	1.63 02
30	Zn K$_{\beta_{1,3}}$	9.5720	129.529	1.25 02	1.34 02	1.39 02	1.49 02	1.57 02
74	W L$_{\beta_1}$	9.6724	128.184	1.21 02	1.31 02	1.36 02	1.45 02	1.53 02
79	Au L$_{\alpha_1}$	9.7133	127.644	1.20 02	1.29 02	1.34 02	1.44 02	1.51 02
32	Ge K$_{\alpha_1}$	9.8864	125.409	1.14 02	1.23 02	1.28 02	1.37 02	1.44 02
80	Hg L$_{\alpha_1}$	9.9888	124.124	1.11 02	1.20 02	1.24 02	1.33 02	1.40 02
75	Re L$_{\beta_1}$	10.010	123.861	1.11 02	1.19 02	1.23 02	1.32 02	1.39 02
31	Ga K$_{\beta_1}$	10.264	120.796	1.03 02	1.11 02	1.15 02	1.24 02	1.30 02
81	Tl L$_{\alpha_1}$	10.269	120.757	1.03 02	1.11 02	1.15 02	1.24 02	1.30 02
76	Os L$_{\beta_1}$	10.355	119.734	1.01 02	1.09 02	1.13 02	1.21 02	1.27 02
33	As K$_{\alpha_1}$	10.544	117.588	9.61 01	1.04 02	1.07 02	1.15 02	1.21 02
82	Pb L$_{\alpha_1}$	10.552	117.499	9.59 01	1.03 02	1.07 02	1.15 02	1.21 02
77	Ir L$_{\beta_1}$	10.708	115.787	9.22 01	9.93 01	1.03 02	1.10 02	1.16 02
83	Bi L$_{\alpha_1}$	10.839	114.388	8.92 01	9.61 01	9.95 01	1.07 02	1.12 02
32	Ge K$_{\beta_1}$	10.982	112.898	8.61 01	9.28 01	9.61 01	1.03 02	1.08 02
78	Pt L$_{\beta_1}$	11.071	111.990	8.43 01	9.08 01	9.40 01	1.01 02	1.06 01
84	Po L$_{\alpha_1}$	11.131	111.387	8.30 01	8.99 01	9.26 01	9.94 01	1.04 02
34	Se K$_{\alpha_1}$	11.222	110.484	8.12 01	8.75 01	9.06 01	9.71 01	1.02 01
85	At L$_{\alpha_1}$	11.427	108.501	7.74 01	8.33 01	8.63 01	9.26 01	9.72 01

Z	transition	E [keV]	λ [pm]	$_{46}$Pd	$_{47}$Ag	$_{48}$Cd	$_{49}$In	$_{50}$Sn
79	Au L_{β_1}	11.442	108.359	7.71 01	8.30 01	8.60 01	9.23 01	9.68 01
33	As K_{β_1}	11.726	105.735	7.21.01	7.77 01	8.05 01	8.68 01	9.06 01
86	Rn L_{α_1}	11.727	105.726	7.21 01	7.77 01	8.05 01	8.63 01	9.06 01
80	Hg L_{β_1}	11.823	104.867	7.05 01	7.60 01	7.87 01	8.44 01	8.86 01
35	Br K_{α_1}	11.924	103.979	6.89 01	7.43 01	7.69 01	8.25 01	8.66 01
87	Fr L_{α_1}	12.031	103.054	6.73 01	7.25 01	7.51 01	8.06 01	8.45 01
81	Tl L_{β_1}	12.213	101.519	6.46 01	6.96 01	7.21 01	7.74 01	8.11 01
88	Ra L_{α_1}	12.340	100.474	6.28 01	6.77 01	7.01 01	7.52 01	7.89 01
34	Se K_{β_1}	12.496	99.219	6.07 01	6.55 01	6.78 01	7.27 01	7.63 01
82	Pb L_{β_1}	12.614	98.291	5.92 01	6.38 01	6.61 01	7.09 01	7.43 01
36	Kr K_{α_1}	12.649	98.019	5.88 01	6.33 01	6.56 01	7.04 01	7.38 01
89	Ac L_{α_1}	12.652	97.996	5.87 01	6.33 01	6.55 01	7.03 01	7.37 01
90	Th L_{α_1}	12.969	95.601	5.49 01	5.92 01	6.13 01	6.58 01	6.90 01
83	Bi L_{β_1}	13.024	95.197	5.43 01	5.85 01	6.06 01	6.50 01	6.82 01
91	Pa L_{α_1}	13.291	93.285	5.14 01	5.54 01	5.74 01	6.16 01	6.45 01
35	Br K_{β_1}	13.291	93.285	5.14 01	5.54 01	5.74 01	6.16 01	6.45 01
37	Rb K_{α_1}	13.395	92.560	5.03 01	5.43 01	5.62 01	6.03 01	6.32 01
84	Po L_{β_1}	13.447	92.202	4.98 01	5.37 01	5.56 01	5.97 01	6.25 01
92	U L_{α_1}	13.615	91.065	4.82 01	5.19 01	5.37 01	5.77 01	6.04 01
85	At L_{β_1}	13.876	89.352	4.58 01	4.93 01	5.11 01	5.48 01	5.74 01
36	Kr K_{β_1}	14.112	87.857	4.37 01	4.71 01	4.88 01	5.24 01	5.48 01
38	Sr K_{α_1}	14.165	87.529	4.33 01	4.61 01	4.83 01	5.19 01	5.43 01
86	Rn L_{β_1}	14.316	86.606	4.21 01	4.53 01	4.69 01	5.04 01	5.27 01
87	Fr L_{β_1}	14.770	83.943	3.87 01	4.17 01	4.31 01	4.63 01	4.85 01
39	Y K_{α_1}	14.958	82.888	3.74 01	4.03 01	4.17 01	4.48 01	4.68 01
37	Rb K_{β_1}	14.961	82.872	3.73 01	4.03 01	4.17 01	4.47 01	4.68 01
88	Ra L_{β_1}	15.236	81.376	3.56 01	3.83 01	3.97 01	4.26 01	4.46 01
89	Ac L_{β_1}	15.713	78.906	3.28 01	3.53 01	3.65 01	3.92 01	4.10 01
40	Zr K_{α_1}	15.775	78.596	3.24 01	3.50 01	3.62 01	3.88 01	4.06 01
38	Sr K_{β_1}	15.836	78.293	3.21 01	3.46 01	3.56 01	3.84 01	4.02 01
90	Th L_{β_1}	16.202	76.524	3.02 01	3.26 01	3.37 01	3.62 01	3.78 01
41	Nb K_{α_1}	16.615	74.622	2.82 01	3.05 01	3.15 01	3.38 01	3.53 01
91	Pa L_{β_1}	16.702	74.233	2.78 01	3.00 01	3.10 01	3.33 01	3.49 01
39	Y K_{β_1}	16.738	74.074	2.77 01	2.99 01	3.09 01	3.32 01	3.47 01
92	U L_{β_1}	17.220	72.000	2.57 01	2.77 01	2.86 01	3.07 01	3.21 01
42	Mo K_{α_1}	17.479	70.933	2.47 01	2.66 01	2.75 01	2.95 01	3.09 01
40	Zr K_{β_1}	17.668	70.175	2.40 01	2.59 01	2.67 01	2.87 01	3.00 01
43	Tc K_{α_1}	18.367	67.504	2.16 01	2.33 01	2.41 01	2.59 01	2.70 01
41	Nb K_{β_1}	18.623	66.576	2.08 01	2.25 01	2.32 01	2.49 01	2.60 01
44	Ru K_{α_1}	19.279	64.311	1.90 01	2.05 01	2.12 01	2.27 01	2.37 01
42	Mo K_{β_1}	19.608	63.231	1.81 01	1.96 01	2.02 01	2.17 01	2.27 01
45	Rh K_{α_1}	20.216	61.330	1.67 01	1.81 01	1.86 01	2.00 01	2.09 01
43	Tc K_{β_1}	20.619	60.131	1.59 01	1.72 01	1.77 01	1.90 01	1.98 01
46	Pd K_{α_1}	21.177	58.547	1.48 01	1.60 01	1.65 01	1.77 01	1.85 01

Z	transition	E [keV]	λ [pm]	$_{46}$Pd	$_{47}$Ag	$_{48}$Cd	$_{49}$In	$_{50}$Sn
44	Ru K$_{\beta_1}$	21.657	57.249	1.39 01	1.51 01	1.55 01	1.67 01	1.74 01
47	Ag K$_{\alpha_1}$	22.163	55.942	1.31 01	1.42 01	1.46 01	1.57 01	1.64 01
45	Rh K$_{\beta_1}$	22.724	54.561	1.23 01	1.33 01	1.37 01	1.47 01	1.53 01
48	Cd K$_{\alpha_1}$	23.174	53.501	1.17 01	1.26 01	1.30 01	1.40 01	1.46 01
46	Pd K$_{\beta_1}$	23.819	52.053	1.09 01	1.17 01	1.21 01	1.30 01	1.35 01
49	In K$_{\alpha_1}$	24.210	51.212	1.04 01	1.13 01	1.16 01	1.25 01	1.30 01
47	Ag K$_{\beta_1}$	24.942	49.709	5.74 01	1.04 01	1.07 01	1.15 01	1.20 01
50	Sn K$_{\alpha_1}$	25.271	49.062	5.54 01	1.01 01	1.03 01	1.11 01	1.16 01
48	Cd K$_{\beta_1}$	26.096	47.511	5.10 01	5.36 01	9.51 00	1.02 01	1.06 01
51	Sb K$_{\alpha_1}$	26.359	47.037	4.96 01	5.22 01	9.26 00	9.97 00	1.04 01
49	In K$_{\beta_1}$	27.276	45.455	4.54 01	4.78 01	4.97 01	9.11 00	9.47 00
52	Te K$_{\alpha_1}$	27.472	45.131	4.45 01	4.69 01	4.88 01	8.94 00	9.29 00
50	Sn K$_{\beta_1}$	28.486	43.524	4.05 01	4.26 01	4.44 01	4.94 01	8.45 00
53	I K$_{\alpha_1}$	28.612	43.333	4.00 01	4.21 01	4.39 01	4.88 01	8.35 00
51	Sb K$_{\beta_1}$	29.726	41.709	3.62 01	3.81 01	3.97 01	4.40 01	4.27 01
54	Xe K$_{\alpha_1}$	29.779	41.635	3.60 01	3.79 01	3.95 01	4.38 01	4.25 01
55	Cs K$_{\alpha_1}$	30.973	40.030	3.24 01	3.42 01	3.56 01	3.93 01	3.84 01
52	Te K$_{\beta_1}$	30.996	40.000	3.23 01	3.41 01	3.55 01	3.92 01	3.83 01
56	Ba K$_{\alpha_1}$	32.194	38.511	2.92 01	3.08 01	3.21 01	3.53 01	3.47 01
53	I K$_{\beta_1}$	32.295	38.391	2.89 01	3.06 01	3.18 01	3.50 01	3.44 01
57	La K$_{\alpha_1}$	33.442	37.074	2.63 01	2.79 01	2.90 01	3.18 01	3.14 01
54	Xe K$_{\beta_1}$	33.624	36.874	2.60 01	2.75 01	2.86 01	3.14 01	3.10 01
58	Ce K$_{\alpha_1}$	34.279	36.169	2.46 01	2.61 01	2.72 01	2.97 01	2.95 01
55	Cs K$_{\beta_1}$	34.987	35.437	2.33 01	2.47 01	2.57 01	2.81 01	2.79 01
59	Pr K$_{\alpha_1}$	36.026	34.415	2.15 01	2.29 01	2.38 01	2.59 01	2.59 01
56	Ba K$_{\beta_1}$	36.378	34.082	2.10 01	2.23 01	2.32 01	2.53 01	2.52 01
60	Nd K$_{\alpha_1}$	36.847	33.648	2.03 01	2.15 01	2.24 01	2.44 01	2.44 01
57	La K$_{\beta_1}$	37.801	32.799	1.89 01	2.01 01	2.09 01	2.27 01	2.28 01
61	Pm K$_{\alpha_1}$	38.725	32.016	1.77 01	1.89 01	1.96 01	2.13 01	2.14 01
58	Ce K$_{\beta_1}$	39.257	31.582	1.71 01	1.82 01	1.89 01	2.05 01	2.07 01
62	Sm K$_{\alpha_1}$	40.118	30.905	1.61 01	1.72 01	1.79 01	1.93 01	1.95 01
59	Pr K$_{\beta_1}$	40.748	30.427	1.54 01	1.65 01	1.71 01	1.85 01	1.87 01
63	Eu K$_{\alpha_1}$	41.542	29.845	1.46 01	1.56 01	1.62 01	1.75 01	1.78 01
60	Nd K$_{\beta_1}$	42.271	29.331	1.40 01	1.49 01	1.55 01	1.67 01	1.70 01
64	Gd K$_{\alpha_1}$	42.996	28.836	1.33 01	1.42 01	1.48 01	1.59 01	1.62 01
61	Pm K$_{\beta_1}$	43.826	28.290	1.27 01	1.35 01	1.40 01	1.51 01	1.54 01
65	Tb K$_{\alpha_1}$	44.482	27.873	1.22 01	1.30 01	1.35 01	1.45 01	1.48 01
62	Sm K$_{\beta_1}$	45.413	27.301	1.15 01	1.23 01	1.27 01	1.37 01	1.40 01
66	Dy K$_{\alpha_1}$	45.998	26.954	1.11 01	1.18 01	1.23 01	1.32 01	1.35 01
63	Eu K$_{\beta_1}$	47.038	26.358	1.04 01	1.11 01	1.16 01	1.24 01	1.27 01
67	Ho K$_{\alpha_1}$	47.547	26.076	1.01 01	1.08 01	1.12 01	1.21 01	1.24 01
64	Gd K$_{\beta_1}$	48.697	25.460	9.49 00	1.01 01	1.05 01	1.13 01	1.16 01
68	Er K$_{\alpha_1}$	49.128	25.237	9.26 00	9.90 00	1.03 01	1.10 01	1.13 01
65	Tb K$_{\beta_1}$	50.382	24.609	8.65 00	9.24 00	9.60 00	1.03 01	1.06 01

Z	transition	E [keV]	λ [pm]	$_{46}$Pd	$_{47}$Ag	$_{48}$Cd	$_{49}$In	$_{50}$Sn
69	Tm K$_{\alpha_1}$	50.742	24.434	8.48 00	9.07 00	9.42 00	1.01 01	1.04 01
66	Dy K$_{\beta_1}$	52.119	23.788	7.88 00	8.43 00	8.75 00	9.36 00	9.66 00
70	Yb K$_{\alpha_1}$	52.389	23.666	7.77 00	8.31 00	8.63 00	9.22 00	9.52 00
67	Ho K$_{\beta_1}$	53.877	23.012	7.20 00	7.70 00	7.99 00	8.54 00	8.83 00
71	Lu K$_{\alpha_1}$	54.070	22.930	7.13 00	7.63 00	7.91 00	8.45 00	8.74 00
68	Er K$_{\beta_1}$	55.681	22.267	6.58 00	7.04 00	7.30 00	7.80 00	8.08 00
72	Hf K$_{\alpha_1}$	55.790	22.223	6.55 00	7.01 00	7.26 00	7.76 00	8.03 00
69	Tm K$_{\beta_1}$	57.517	21.556	6.02 00	6.45 00	6.68 00	7.13 00	7.40 00
73	Ta K$_{\alpha_1}$	57.532	21.550	6.02 00	6.45 00	6.67 00	7.13 00	7.39 00
74	W K$_{\alpha_1}$	59.318	20.901	5.54 00	5.93 00	6.14 00	6.55 00	6.80 00
70	Yb K$_{\beta_1}$	59.370	20.883	5.52 00	5.92 00	6.12 00	6.53 00	6.79 00
75	Re K$_{\alpha_1}$	61.140	20.278	5.10 00	5.46 00	5.65 00	6.03 00	6.27 00
71	Lu K$_{\beta_1}$	61.283	20.231	5.07 00	5.43 00	5.62 00	5.99 00	6.23 00
76	Os K$_{\alpha_1}$	63.001	19.679	4.70 00	5.04 00	5.21 00	5.55 00	5.77 00
72	Hf K$_{\beta_1}$	63.234	19.607	4.65 00	4.99 00	5.16 00	5.49 00	5.71 00
77	Ir K$_{\alpha_1}$	64.896	19.105	4.33 00	4.65 00	4.81 00	5.11 00	5.32 00
73	Ta K$_{\beta_1}$	65.223	19.009	4.27 00	4.59 00	4.74 00	5.04 00	5.25 00
78	Pt K$_{\alpha_1}$	66.832	18.551	4.00 00	4.29 00	4.44 00	4.72 00	4.91 00
74	W K$_{\beta_1}$	67.244	18.438	3.93 00	4.22 00	4.36 00	4.64 00	4.83 00
79	Au K$_{\alpha_1}$	68.804	18.020	3.70 00	3.97 00	4.10 00	4.55 00	4.54 00
75	Re K$_{\beta_1}$	69.310	17.888	3.62 00	3.89 00	4.02 00	4.27 00	4.45 00
80	Hg K$_{\alpha_1}$	70.819	17.507	3.42 00	3.67 00	3.79 00	4.02 00	4.19 00
76	Os K$_{\beta_1}$	71.413	17.361	3.34 00	3.59 00	3.71 00	3.93 00	4.10 00
81	Tl K$_{\alpha_1}$	72.872	17.014	3.16 00	3.39 00	3.51 00	3.72 00	3.88 00
77	Ir K$_{\beta_1}$	73.561	16.854	3.08 00	3.31 00	3.42 00	3.62 00	3.78 00
82	Pb K$_{\alpha_1}$	74.969	16.538	2.93 00	3.14 00	3.25 00	3.44 00	3.59 00
78	Pt K$_{\beta_1}$	75.748	16.368	2.84 00	3.06 00	3.16 00	3.34 00	3.49 00
83	Bi K$_{\alpha_1}$	77.108	16.079	2.71 00	2.91 00	3.01 00	3.18 00	3.32 00
79	Au K$_{\beta_1}$	77.948	15.906	2.63 00	2.83 00	2.92 00	3.09 00	3.22 00
84	Po K$_{\alpha_1}$	79.290	15.636	2.51 00	2.70 00	2.79 00	2.95 00	3.08 00
80	Hg K$_{\alpha_1}$	80.253	15.449	2.43 00	2.61 00	2.70 00	2.85 00	2.98 00
85	At K$_{\beta_1}$	81.520	15.209	2.33 00	2.51 00	2.59 00	2.73 00	2.86 00
81	Tl K$_{\beta_1}$	82.576	15.014	2.25 00	2.42 00	2.50 00	2.64 00	2.76 00
86	Rn K$_{\alpha_1}$	83.780	14.798	2.17 00	2.33 00	2.41 00	2.54 00	2.65 00
82	Pb K$_{\beta_1}$	84.936	14.597	2.09 00	2.25 00	2.32 00	2.45 00	2.56 00
87	Fr K$_{\alpha_1}$	86.100	14.400	2.02 00	2.17 00	2.24 00	2.36 00	2.47 00
83	Bi K$_{\beta_1}$	87.343	14.195	1.94 00	2.09 00	2.16 00	2.27 00	2.37 00
88	Ra K$_{\alpha_1}$	88.470	14.014	1.88 00	2.02 00	2.08 00	2.20 00	2.29 00
84	Po K$_{\beta_1}$	89.800	13.806	1.81 00	1.94 00	2.00 00	2.11 00	2.20 00
89	Ac K$_{\alpha_1}$	90.884	13.642	1.75 00	1.88 00	1.94 00	2.04 00	2.13 00
85	At K$_{\beta_1}$	92.300	13.432	1.68 00	1.81 00	1.86 00	1.96 00	2.05 00
90	Th K$_{\alpha_1}$	93.350	13.281	1.63 00	1.75 00	1.81 00	1.90 00	1.98 00
86	Rn K$_{\beta_1}$	94.870	13.068	1.56 00	1.68 00	1.73 00	1.82 00	1.90 00
91	Pa K$_{\alpha_1}$	95.868	12.932	1.52 00	1.63 00	1.68 00	1.77 00	1.85 00

Z	transition	E [keV]	λ [pm]	$_{46}$Pd	$_{47}$Ag	$_{48}$Cd	$_{49}$In	$_{50}$Sn
87	Fr K$_{\beta_1}$	97.470	12.720	1.45 00	1.57 00	1.61 00	1.70 00	1.77 00
92	U K$_{\alpha_1}$	98.439	12.595	1.42 00	1.52 00	1.57 00	1.65 00	1.72 00
88	Ra K$_{\beta_1}$	100.130	12.382	1.35 00	1.46 00	1.50 00	1.58 00	1.64 00
89	Ac K$_{\beta_1}$	102.850	12.054	1.27 00	1.36 00	1.40 00	1.48 00	1.54 00
90	Th K$_{\beta_1}$	105.610	11.739	1.19 00	1.28 00	1.32 00	1.38 00	1.44 00
91	Pa K$_{\beta_1}$	108.430	11.434	1.12 00	1.20 00	1.23 00	1.29 00	1.34 00
92	U K$_{\beta_1}$	111.300	11.139	1.05 00	1.12 00	1.16 00	1.21 00	1.26 00

Z	transition	E [keV]	λ [pm]	$_{51}$Sb	$_{52}$Te	$_{53}$I	$_{54}$Xe	$_{55}$Cs
4	Be K$_\alpha$	0.1085	11427.207	5.14 04	6.79 04	6.97 04	1.08 05	1.24 05
38	Sr M$_\xi$	0.1140	10875.895	3.63 04	4.83 04	5.79 04	8.95 04	1.17 05
39	Y M$_\xi$	0.1328	9339.235	7.77 03	1.13 04	1.36 04	2.94 04	5.90 04
16	S L$_l$	0.1487	8337.942	4.87 03	5.06 03	5.21 03	1.10 04	2.30 04
40	Zr M$_\xi$	0.1511	8205.506	4.79 03	4.73 03	5.03 03	9.55 03	2.00 04
41	Nb M$_\xi$	0.1717	7221.037	6.60 03	6.48 03	5.57 03	4.44 03	6.84 03
5	B K$_\alpha$	0.1833	6764.059	6.85 03	6.76 03	5.88 03	4.10 03	4.73 03
42	Mo M$_\xi$	0.1926	6473.445	7.07 03	7.00 03	7.20 03	4.29 03	4.16 03
6	C K$_\alpha$	0.2770	4476.000	6.60 03	6.67 03	7.09 03	6.75 03	6.51 03
47	Ag M$_\xi$	0.3117	3977.709	6.01 03	6.11 03	6.55 03	5.74 03	6.16 03
7	N K$_\alpha$	0.3924	3159.664	5.02 03	5.15 03	5.61 03	5.70 03	6.12 03
22	Ti L$_l$	0.3953	3136.484	4.98 03	5.11 03	5.58 03	5.70 03	6.12 03
22	Ti L$_\alpha$	0.4522	2741.822	4.19 03	4.34 03	4.75 03	5.09 03	5.34 03
23	V L$_\alpha$	0.5113	2424.901	3.54 03	3.69 03	4.04 03	4.34 03	4.59 03
8	O K$_\alpha$	0.5249	2362.072	3.42 03	3.57 03	3.91 03	4.20 03	4.44 03
25	Mn L$_l$	0.5563	2228.747	2.34 04	3.28 03	3.60 03	3.87 03	4.10 03
24	Cr L$_\alpha$	0.5728	2164.549	2.18 04	3.95 03	3.44 03	3.70 03	3.93 03
25	Mn L$_\alpha$	0.6374	1945.171	1.70 04	2.25 04	2.04 04	3.15 03	3.35 03
9	F K$_\alpha$	0.6768	1831.932	1.47 04	1.64 04	1.72 04	2.73 04	3.03 03
26	Fe L$_\alpha$	0.7050	1758.655	1.35 04	1.39 04	1.56 04	2.28 04	2.85 03
27	Co L$_\alpha$	0.7162	1597.335	1.10 04	1.14 04	1.24 04	1.34 04	1.54 04
28	Ni L$_\alpha$	0.8515	1456.080	1.13 04	1.14 04	1.02 04	1.08 04	1.15 04
29	Cu L$_\alpha$	0.9297	1336.044	9.51 03	9.72 03	1.07 04	8.92 03	9.42 03
30	Zn L$_\alpha$	1.0117	1215.513	1.08 04	1.13 04	1.12 04	1.08 04	1.16 04
11	Na K$_\alpha$	1.0410	1191.020	1.01 04	1.05 04	1.04 04	1.00 04	1.08 04
11	Na K$_\beta$	1.0711	1157.550	9.35 03	9.78 03	9.70 03	9.36 03	1.11 04
12	Mg K$_\alpha$	1.2536	989.033	6.28 03	6.57 03	7.15 03	6.95 03	8.14 03
33	As L$_\alpha$	1.2820	967.123	5.93 03	6.20 03	6.76 03	6.58 03	7.68 03
12	Mg K$_\beta$	1.3022	952.121	5.70 03	5.96 03	6.49 03	6.32 03	7.38 03
33	As L$_{\beta_1}$	1.3170	941.421	5.54 03	5.80 03	6.31 03	6.15 03	7.17 03
66	Dy M$_\beta$	1.3250	935.737	5.45 03	5.71 03	6.21 03	6.06 03	7.06 03
67	Ho M$_\alpha$	1.3480	919.771	5.22 03	5.46 03	5.95 03	5.80 03	6.75 03
34	Se L$_\alpha$	1.3791	899.029	4.93 03	5.16 03	5.61 03	5.48 03	6.36 03
67	Ho M$_\beta$	1.3830	896.494	4.89 03	5.12 03	5.57 03	5.44 03	6.32 03

Z	transition	E [keV]	λ [pm]	$_{51}$Sb	$_{52}$Te	$_{53}$I	$_{54}$Xe	$_{55}$Cs
68	Er M$_\alpha$	1.4060	881.829	4.69 03	4.91 03	5.35 03	5.22 03	6.05 03
34	Se L$_\beta$	1.4192	873.627	4.58 03	4.80 03	5.22 03	5.10 03	5.91 03
68	Er M$_\beta$	1.4430	859.218	4.39 03	4.60 03	5.01 03	4.90 03	5.66 03
69	Tm M$_\alpha$	1.4620	848.051	4.25 03	4.45 03	4.84 03	4.74 03	5.47 03
35	Br L$_\alpha$	1.4804	837.511	4.12 03	4.31 03	4.69 03	4.59 03	5.30 03
13	Al K$_{\alpha_1}$	1.4867	833.962	4.07 03	4.26 03	4.64 03	4.54 03	5.24 03
69	Tm M$_\beta$	1.5030	824.918	3.96 03	4.15 03	4.51 03	4.42 03	5.09 03
70	Yb M$_\alpha$	1.5214	814.941	3.84 03	4.02 03	4.38 03	4.29 03	4.94 03
35	Br L$_{\beta_1}$	1.5259	812.538	3.81 03	3.99 03	4.34 03	4.26 03	4.90 03
13	Al K$_\beta$	1.5574	796.103	3.62 03	3.79 03	4.13 03	4.05 03	4.65 03
70	Yb M$_\beta$	1.5675	790.974	3.56 03	3.73 03	4.06 03	3.98 03	4.58 03
71	Lu M$_\alpha$	1.5813	784.071	3.48 03	3.65 03	3.97 03	3.90 03	4.48 03
36	Kr L$_\alpha$	1.5860	781.747	3.46 03	3.62 03	3.94 03	3.87 03	4.44 03
71	Lu M$_\beta$	1.6312	760.085	3.22 03	3.37 03	3.67 03	3.61 03	4.14 03
36	Kr L$_{\beta_1}$	1.6366	757.577	3.19 03	3.34 03	3.64 03	3.58 03	4.10 03
72	Hf M$_{\alpha_1}$	1.6446	753.892	3.16 03	3.30 03	3.59 03	3.53 03	4.05 03
37	Rb L$_{\alpha_1}$	1.6941	731.864	2.93 03	3.06 03	3.33 03	3.28 03	3.76 03
72	Hf M$_\beta$	1.6976	730.355	2.91 03	3.05 03	3.32 03	3.26 03	3.74 03
73	Ta M$_{\alpha_1}$	1.8096	725.229	2.86 03	2.99 03	3.26 03	3.21 03	3.67 03
14	Si K$_{\alpha_1}$	1.7400	712.558	2.74 03	2.86 03	3.11 03	3.07 03	3.51 03
37	Rb L$_{\beta_1}$	1.7522	707.597	2.69 03	2.81 03	3.06 03	3.02 03	3.45 03
73	Ta M$_\beta$	1.7655	702.266	2.64 03	2.76 03	3.00 03	2.96 03	3.38 03
74	W M$_{\alpha_1}$	1.7754	698.350	2.60 03	2.72 03	2.96 03	2.92 03	3.34 03
38	Sr L$_{\beta_1}$	1.8066	686.290	2.49 03	2.60 03	2.83 03	2.80 03	3.19 03
74	W M$_\beta$	1.8349	675.705	2.39 03	2.50 03	2.72 03	2.69 03	3.07 03
14	Si K$_\beta$	1.8359	675.337	2.39 03	2.50 03	2.72 03	2.69 03	3.06 03
75	Re M$_{\alpha_1}$	1.8420	673.100	2.37 03	2.48 03	2.70 03	2.66 03	3.04 03
38	Sr L$_{\beta_1}$	1.8717	662.420	2.28 03	2.38 03	2.59 03	2.56 03	2.92 03
75	Re M$_\beta$	1.9061	650.465	2.17 03	2.27 03	2.47 03	2.45 03	2.79 03
76	Os M$_{\alpha_1}$	1.9102	649.069	2.16 03	2.26 03	2.46 03	2.43 03	2.77 03
39	Y L$_{\alpha_1}$	1.9226	644.882	2.13 03	2.22 03	2.42 03	2.39 03	2.72 03
76	Os M$_\beta$	1.9783	626.725	1.98 03	2.07 03	2.25 03	2.23 03	2.53 03
77	Ir M$_{\alpha_1}$	1.9799	626.219	1.97 03	2.06 03	2.24 03	2.22 03	2.53 03
39	Y L$_\beta$	1.9958	621.230	1.93 03	2.02 03	2.20 03	2.18 03	2.48 03
15	P K$_{\alpha_1}$	2.0137	615.708	1.89 03	1.98 03	2.15 03	2.13 03	2.42 03
40	Zr L$_{\alpha_1}$	2.0424	607.056	1.83 03	1.91 03	2.07 03	2.06 03	2.34 03
78	Pt M$_{\alpha_1}$	2.0505	604.658	1.81 03	1.89 03	2.05 03	2.04 03	2.31 03
77	Ir M$_\beta$	2.0535	603.775	1.80 03	1.88 03	2.05 03	2.03 03	2.30 03
79	Au M$_{\alpha_1}$	2.1229	584.036	1.66 03	1.73 03	1.88 03	1.87 03	2.12 03
40	Zr L$_{\beta_{1,3}}$	2.1244	583.624	1.65 03	1.73 03	1.88 03	1.87 03	2.11 03
78	Pt M$_\beta$	2.1273	582.828	1.65 03	1.72 03	1.87 03	1.86 03	2.11 03
15	P K$_{\beta_{1,3}}$	2.1390	579.640	1.62 03	1.70 03	1.85 03	1.84 03	2.08 03
41	Nb L$_{\alpha_1}$	2.1659	572.441	1.52 03	1.59 03	1.73 03	1.72 03	1.94 03
80	Hg M$_{\alpha_1}$	2.1953	564.775	1.52 03	1.59 03	1.73 03	1.72 03	1.94 03

Z	transition	E [keV]	λ [pm]	$_{51}$Sb	$_{52}$Te	$_{53}$I	$_{54}$Xe	$_{55}$Cs
79	Au M$_\beta$	2.2046	562.393	1.51 03	1.57 03	1.71 03	1.70 03	1.92 03
41	Nb L$_{\beta_1}$	2.2574	549.238	1.42 03	1.48 03	1.61 03	1.61 03	1.81 03
81	Tl M$_{\alpha_1}$	2.2706	546.045	1.40 03	1.46 03	1.59 03	1.58 03	1.78 03
80	Hg M$_\beta$	2.2825	543.199	1.38 03	1.44 03	1.57 03	1.56 03	1.76 03
42	Mo L$_{\alpha_1}$	2.2932	540.664	1.36 03	1.42 03	1.55 03	1.55 03	1.74 03
16	S K$_{\alpha_1}$	2.3080	537.197	1.34 03	1.40 03	1.52 03	1.52 03	1.71 03
82	Pb M$_{\alpha_1}$	2.3457	528.563	1.29 03	1.34 03	1.46 03	1.46 03	1.64 03
81	Tl M$_\beta$	2.3621	524.893	1.27 03	1.32 03	1.44 03	1.44 03	1.61 03
42	Mo L$_{\beta_1}$	2.3948	517.726	1.22 03	1.28 03	1.39 03	1.39 03	1.56 03
83	Bi M$_{\alpha_1}$	2.4226	511.785	1.19 03	1.24 03	1.35 03	1.35 03	1.51 03
43	Tc L$_{\alpha_1}$	2.4240	511.490	1.19 03	1.24 03	1.34 03	1.35 03	1.48 03
82	Pb M$_\beta$	2.4427	507.574	1.16 03	1.21 03	1.32 03	1.32 03	1.48 03
16	S K$_\beta$	2.4640	503.186	1.14 03	1.19 03	1.29 03	1.29 03	1.45 03
83	Bi M$_{\beta_1}$	2.5255	490.933	1.07 03	1.12 03	1.21 03	1.22 03	1.36 03
43	Tc L$_{\beta_1}$	2.5368	488.746	1.06 03	1.10 03	1.20 03	1.20 03	1.34 03
44	Ru L$_{\alpha_1}$	2.5586	484.582	1.03 03	1.08 03	1.17 03	1.18 03	1.31 03
17	Cl K$_{\alpha_1}$	2.6224	472.792	9.72 02	1.01 03	1.10 03	1.11 03	1.23 03
44	Ru L$_{\beta_1}$	2.6832	462.079	9.18 02	9.57 02	1.04 03	1.05 03	1.16 03
45	Rh L$_{\alpha_1}$	2.6967	459.766	9.06 02	9.45 02	1.03 03	1.03 03	1.15 03
17	Cl K$_\beta$	2.8156	440.350	8.13 02	8.48 02	9.20 02	9.29 02	1.03 03
45	Rh L$_{\beta_1}$	2.8344	437.430	7.99 02	8.34 02	9.04 02	9.14 02	1.01 03
46	Pd L$_{\alpha_1}$	2.8386	436.782	7.96 02	8.30 02	9.01 02	9.10 02	1.01 03
18	Ar K$_{\alpha_1}$	2.9577	419.194	7.18 02	7.49 02	8.12 02	8.22 02	9.07 02
47	Ag L$_{\alpha_1}$	2.9843	415.458	7.02 02	7.32 02	7.94 02	8.04 02	8.86 02
46	Pd L$_{\beta_1}$	2.9902	414.638	6.98 02	7.28 02	7.90 02	8.00 02	8.82 02
90	Tb M$_{\alpha_1}$	2.9961	413.821	6.95 02	7.25 02	7.86 02	7.96 02	8.77 02
91	Pa M$_{\alpha_1}$	3.0823	402.248	6.47 02	6.75 02	7.32 02	7.42 02	8.17 02
48	Cd L$_{\alpha_1}$	3.1337	395.651	6.21 02	6.47 02	7.02 02	7.13 02	7.83 02
90	Th M$_\beta$	3.1458	394.129	6.15 02	6.41 02	6.95 02	7.06 02	7.75 02
47	Ag L$_{\beta_1}$	3.1509	393.491	6.12 02	6.38 02	6.92 02	7.03 02	7.72 02
92	U M$_{\alpha_1}$	3.1708	391.021	6.02 02	6.28 02	6.81 02	6.92 02	7.60 02
18	Ar K$_\beta$	3.1905	388.607	5.93 02	6.18 02	6.71 02	6.82 02	7.48 02
91	Pa M$_\beta$	3.2397	382.705	5.71 02	5.95 02	6.46 02	6.56 02	7.20 02
49	In L$_{\alpha_1}$	3.2869	377.210	5.50 02	5.74 02	6.23 02	6.33 02	6.94 02
19	K K$_{\alpha_1}$	3.3138	374.148	5.39 02	5.62 02	6.10 02	6.21 02	6.80 02
48	Cd L$_{\beta_1}$	3.3166	373.832	5.38 02	5.61 02	6.09 02	6.19 02	6.78 02
92	U M$_\beta$	3.3360	371.658	5.30 02	5.53 02	6.00 02	6.10 02	6.68 02
50	Sn L$_{\alpha_1}$	3.4440	360.003	4.89 02	5.10 02	5.54 02	5.64 02	6.16 02
49	In L$_{\beta_1}$	3.4872	355.543	4.74 02	4.94 02	5.37 02	5.47 02	5.97 02
19	K K$_\beta$	3.5896	345.401	4.41 02	4.59 02	4.99 02	5.09 02	5.55 02
51	Sb L$_{\alpha_1}$	3.6047	343.954	4.36 02	4.55 02	4.94 02	5.04 02	5.49 02
50	Sn L$_{\beta_1}$	3.6628	338.498	4.19 02	4.37 02	4.74 02	4.85 02	5.27 02
20	Ca K$_{\alpha_1}$	3.6917	335.848	4.11 02	4.28 02	4.65 02	4.75 02	5.17 02
52	Te L$_{\alpha_1}$	3.7693	328.934	3.90 02	4.06 02	4.41 02	4.52 02	4.90 02

Z	transition	E [keV]	λ [pm]	$_{51}$Sb	$_{52}$Te	$_{53}$I	$_{54}$Xe	$_{55}$Cs
51	Sb L_{β_1}	3.8436	322.575	3.71 02	3.87 02	4.20 02	4.30 02	4.67 02
53	I L_{α_1}	3.9377	314.867	3.49 02	3.64 02	3.95 02	4.05 02	4.39 02
20	Ca K_β	4.0127	308.981	3.33 02	3.47 02	3.77 02	3.87 02	4.18 02
52	Te L_{β_1}	4.0296	307.686	3.29 02	3.43 02	3.73 02	3.83 02	4.14 02
21	Sc K_{α_1}	4.0906	303.097	3.17 02	3.31 02	3.59 02	3.69 02	3.99 02
54	Xe L_{α_1}	4.1099	301.674	<u>3.13 02</u>	3.27 02	3.55 02	3.65 02	3.94 02
53	I L_{β_1}	4.2207	293.755	8.97 02	3.06 02	3.32 02	3.42 02	3.68 02
55	Cs L_{α_1}	4.2865	289.245	<u>8.60 02</u>	<u>2.94 02</u>	3.20 02	3.29 02	3.54 02
21	Sc K_β	4.4605	277.962	1.08 03	8.04 02	2.89 02	2.98 02	3.20 02
56	Ba L_{α_1}	4.4663	277.601	1.08 03	8.01 02	2.88 02	2.97 02	3.19 02
22	Ti K_{α_1}	4.5108	274.863	1.05 03	<u>7.81 02</u>	<u>2.81 02</u>	2.90 02	3.11 02
55	Cs L_{β_1}	4.6198	268.377	9.83 02	1.02 03	7.96 02	2.74 02	2.93 02
57	La L_{α_1}	4.6510	266.577	<u>9.65 02</u>	1.01 03	7.82 02	<u>2.69 02</u>	2.88 02
56	Ba L_{β_1}	4.8275	256.841	1.04 03	9.09 02	7.07 02	7.19 02	2.62 02
58	Ce L_{α_1}	4.8402	256.157	1.04 03	9.03 02	<u>7.02 02</u>	7.14 02	2.60 02
22	Ti $K_{\beta_{1,3}}$	4.9318	251.399	9.86 02	<u>8.58 02</u>	9.31 02	6.77 02	2.48 02
23	V K_{α_1}	4.9522	250.363	9.75 02	1.02 03	9.21 02	6.69 02	<u>2.46 02</u>
59	Pr L_{α_1}	5.0337	246.310	9.33 02	9.73 02	8.81 02	6.41 02	6.99 02
57	La L_{β_1}	5.0421	245.899	9.28 02	9.69 02	<u>8.77 02</u>	<u>6.38 02</u>	6.95 02
60	Nd L_{α_1}	5.2304	237.047	8.40 02	8.77 02	9.52 02	8.07 02	6.29 02
58	Ce L_{β_1}	5.2622	235.614	8.27 02	8.63 02	9.36 02	7.04 02	<u>6.19 02</u>
24	Cr K_{α_1}	5.4147	228.978	7.65 02	7.98 02	8.66 02	7.34 02	7.97 02
23	V $K_{\beta_{1,3}}$	5.4273	228.447	7.60 02	7.93 02	8.61 02	7.29 02	7.92 02
61	Pm L_{α_1}	5.4325	228.228	7.58 02	7.91 02	8.58 02	<u>7.28 02</u>	7.90 02
59	Pr L_{β_1}	5.4889	225.883	7.37 02	7.69 02	8.35 02	8.52 02	7.68 02
62	Sm L_{α_1}	5.6361	219.984	6.86 02	7.16 02	7.77 02	7.93 02	<u>7.15 02</u>
60	Nd L_{β_1}	5.7216	216.696	6.59 02	6.87 02	7.45 02	7.61 02	8.26 02
63	Eu L_{α_1}	5.8457	212.096	6.21 02	6.48 02	7.03 02	7.19 02	7.79 02
25	Mn K_{α_1}	5.8988	210.187	6.06 02	6.33 02	6.86 02	7.01 02	7.60 02
24	Cr $K_{\beta_{1,3}}$	5.9467	208.494	5.93 02	6.19 02	6.71 02	6.86 02	7.43 02
61	Pm L_{β_1}	5.9610	207.993	5.89 02	6.15 02	6.67 02	6.82 02	7.38 02
64	Gd L_{α_1}	6.0572	204.690	5.64 02	5.89 02	6.38 02	6.53 02	7.07 02
62	Sm L_{β_1}	6.2051	199.811	5.29 02	5.52 02	5.98 02	6.12 02	6.62 02
65	Tb L_{α_1}	6.2728	197.655	5.14 02	5.36 02	5.81 02	5.94 02	6.43 02
26	Fe K_{α_1}	6.4038	193.611	4.86 02	5.07 02	5.49 02	5.62 02	6.08 02
63	Eu L_{β_1}	6.4564	192.034	4.76 02	4.96 02	5.37 02	5.49 02	5.95 02
25	Mn $K_{\beta_{1,3}}$	6.4905	191.025	4.69 02	4.89 02	5.30 02	5.42 02	5.86 02
66	Dy L_{α_1}	6.4952	190.887	4.68 02	4.88 02	5.29 02	5.41 02	5.85 02
64	Gd L_{β_1}	6.7132	184.688	4.28 02	4.47 02	4.83 02	4.94 02	5.35 02
67	Ho L_{α_1}	6.7198	184.507	4.27 02	4.45 02	4.82 02	4.93 02	5.34 02
27	Co K_{α_1}	6.9303	178.903	3.93 02	4.10 02	4.44 02	4.54 02	4.91 02
68	Er L_{α_1}	6.9487	178.429	3.90 02	4.07 02	4.40 02	4.50 02	4.87 02
65	Tb L_{β_1}	6.9780	177.680	3.86 02	4.02 02	4.35 02	4.45 02	4.82 02
26	Fe $K_{\beta_{1,3}}$	7.0580	175.666	3.74 02	3.90 02	4.22 02	4.32 02	4.67 02

Z	transition	E [keV]	λ [pm]	$_{51}$Sb	$_{52}$Te	$_{53}$I	$_{54}$Xe	$_{55}$Cs
69	Tm L$_{\alpha_1}$	7.1799	172.683	3.58 02	3.73 02	4.03 02	4.12 02	4.46 02
66	Dy L$_{\beta_1}$	7.2477	171.068	3.49 02	3.63 02	3.93 02	4.02 02	4.33 02
70	Yb L$_{\alpha_1}$	7.4158	167.195	3.28 02	3.42 02	3.69 02	3.78 02	4.09 02
28	Ni K$_{\alpha_1}$	7.4782	165.795	3.21 02	3.34 02	3.61 02	3.69 02	4.00 02
67	Ho L$_{\beta_1}$	7.5253	164.757	3.15 02	3.28 02	3.55 02	3.63 02	3.93 02
27	Co K$_{\beta_{1,3}}$	7.6494	162.084	3.02 02	3.14 02	3.40 02	3.47 02	3.76 02
71	Lu L$_{\alpha_1}$	7.6555	161.955	3.01 02	3.14 02	3.39 02	3.47 02	3.75 02
68	Er L$_{\beta_1}$	7.8109	158.733	2.85 02	2.97 02	3.21 02	3.28 02	3.55 02
72	Hf L$_{\alpha_1}$	7.8990	156.963	2.77 02	2.88 02	3.11 02	3.19 02	3.45 02
29	Cu K$_{\alpha_1}$	8.0478	154.060	2.63 02	2.74 02	2.96 02	3.03 02	3.28 02
69	Tm L$_{\beta_1}$	8.1010	153.049	2.59 02	2.69 02	2.91 02	2.98 02	3.22 02
73	Ta L$_{\alpha_1}$	8.1461	152.201	2.55 02	2.65 02	2.87 02	2.93 02	3.17 02
28	Ni K$_{\beta_{1,3}}$	8.2647	150.017	2.45 02	2.55 02	2.76 02	2.82 02	3.05 02
74	W L$_{\alpha_1}$	8.3976	147.643	2.35 02	2.44 02	2.64 02	2.70 02	2.92 02
70	Yb L$_{\beta_1}$	8.4018	147.569	2.35 02	2.44 02	2.64 02	2.70 02	2.92 02
30	Zn K$_{\alpha_1}$	8.6389	143.519	2.18 02	2.27 02	2.45 02	2.50 02	2.71 02
75	Re L$_{\alpha_1}$	8.6525	143.294	2.17 02	1.26 02	2.43 02	2.49 02	2.70 02
71	Lu L$_{\beta_1}$	8.7090	142.364	2.13 02	2.22 02	2.39 02	2.45 02	2.65 02
29	Cu K$_{\beta_1}$	8.9053	139.226	2.01 02	2.09 02	2.25 02	2.31 02	2.50 02
76	Os L$_{\alpha_1}$	8.9117	139.126	2.00 02	2.08 02	2.25 02	2.30 02	2.49 02
72	Hf L$_{\beta_1}$	9.0227	137.414	1.94 02	2.02 02	2.17 02	2.23 02	2.41 02
77	Ir L$_{\alpha_1}$	9.1751	135.132	1.85 02	1.93 02	2.08 02	2.13 02	2.30 02
31	Ga K$_{\alpha_1}$	9.2517	134.013	1.81 02	1.88 02	2.03 02	2.08 02	2.25 02
73	Ta L$_{\beta_1}$	9.3431	132.702	1.76 02	1.84 02	1.98 02	2.03 02	2.19 02
78	Pt L$_{\alpha_1}$	9.4423	131.308	1.71 02	1.78 02	1.92 02	1.97 02	2.13 02
30	Zn K$_{\beta_{1,3}}$	9.5720	129.529	1.65 02	1.72 02	1.85 02	1.90 02	2.05 02
74	W L$_{\beta_1}$	9.6724	128.184	1.61 02	1.67 02	1.80 02	1.85 02	2.00 02
79	Au L$_{\alpha_1}$	9.7133	127.644	1.59 02	1.65 02	1.78 02	1.83 02	1.97 02
32	Ge K$_{\alpha_1}$	9.8864	125.409	1.52 02	1.58 02	1.70 02	1.74 02	1.88 02
80	Hg L$_{\alpha_1}$	9.9888	124.124	1.47 02	1.53 02	1.65 02	1.69 02	1.83 02
75	Re L$_{\beta_1}$	10.010	123.861	1.47 02	1.53 02	1.64 02	1.68 02	1.82 02
31	Ga K$_{\beta_1}$	10.264	120.796	1.37 02	1.43 02	1.54 02	1.57 02	1.70 02
81	Tl L$_{\alpha_1}$	10.269	120.737	1.37 02	1.42 02	1.53 02	1.57 02	1.70 02
76	Os L$_{\beta_1}$	10.355	119.734	1.34 02	1.39 02	1.50 02	1.54 02	1.66 02
33	As K$_{\alpha_1}$	10.544	117.588	1.28 02	1.33 02	1.43 02	1.47 02	1.58 02
82	Pb L$_{\alpha_1}$	10.552	117.499	1.27 02	1.33 02	1.43 02	1.46 02	1.58 02
77	Ir L$_{\beta_1}$	10.708	115.787	1.22 02	1.27 02	1.37 02	1.41 02	1.58 02
83	Bi L$_{\alpha_1}$	10.839	114.388	1.19 02	1.23 02	1.33 02	1.36 02	1.47 02
32	Ge K$_{\beta_1}$	10.982	112.898	1.15 02	1.19 02	1.28 02	1.32 02	1.42 02
78	Pt L$_{\beta_1}$	11.071	111.990	1.12 02	1.17 02	1.25 02	1.29 02	1.39 02
84	Po L$_{\alpha_1}$	11.131	111.387	1.10 02	1.15 02	1.24 02	1.27 02	1.37 02
34	Se K$_{\alpha_1}$	11.222	110.484	1.08 02	1.12 02	1.21 02	1.24 02	1.34 02
85	At L$_{\alpha_1}$	11.427	108.501	1.03 02	1.07 02	1.15 02	1.18 02	1.28 02
79	Au L$_{\beta_1}$	11.442	108.359	1.03 02	1.07 02	1.15 02	1.18 02	1.27 02

Z	transition	E [keV]	λ [pm]	$_{51}$Sb	$_{52}$Te	$_{53}$I	$_{54}$Xe	$_{55}$Cs
33	As K$_{\beta_1}$	11.726	105.735	9.62 01	1.00 02	1.07 02	1.10 02	1.19 02
86	Rn L$_{\alpha_1}$	11.727	105.726	9.62 01	1.00 02	1.07 02	1.10 02	1.19 02
80	Hg L$_{\beta_1}$	11.823	104.867	9.41 01	9.79 01	1.05 02	1.08 02	1.17 02
35	Br K$_{\alpha_1}$	11.924	103.979	9.20 01	9.57 01	1.03 02	1.06 02	1.14 02
87	Fr L$_{\alpha_1}$	12.031	103.054	8.98 01	9.34 01	1.00 02	1.03 02	1.11 02
81	Tl L$_{\beta_1}$	12.213	101.519	8.63 01	8.98 01	9.64 01	9.91 01	1.07 02
88	Ra L$_{\alpha_1}$	12.340	100.474	8.40 01	8.73 01	9.37 01	9.64 01	1.04 02
34	Se K$_{\beta_1}$	12.496	99.219	8.12 01	8.45 01	9.06 01	9.33 01	1.01 02
82	Pb L$_{\beta_1}$	12.614	98.291	7.92 01	8.24 01	8.84 01	9.09 01	9.80 01
36	Kr K$_{\alpha_1}$	12.649	98.019	7.86 01	8.18 01	8.77 01	9.03 01	9.73 01
89	Ac L$_{\alpha_1}$	12.652	97.996	7.86 01	8.17 01	8.76 01	9.02 01	9.72 01
90	Th L$_{\alpha_1}$	12.969	95.601	7.36 01	7.65 01	8.20 01	8.45 01	9.10 01
83	Bi L$_{\beta_1}$	13.024	95.197	7.28 01	7.56 01	8.11 01	8.35 01	9.00 01
91	Pa L$_{\alpha_1}$	13.291	93.285	6.89 01	7.17 01	7.68 01	7.91 01	8.52 01
35	Br K$_{\beta_1}$	13.291	93.285	6.89 01	7.17 01	7.68 01	7.91 01	8.52 01
37	Rb K$_{\alpha_1}$	13.395	92.560	6.75 01	7.02 01	7.52 01	7.75 01	8.34 01
84	Po L$_{\beta_1}$	13.447	92.202	6.68 01	6.95 01	7.44 01	7.67 01	8.26 01
92	U L$_{\alpha_1}$	13.615	91.065	6.47 01	6.72 01	7.20 01	7.42 01	7.99 01
85	At L$_{\beta_1}$	13.876	89.352	6.15 01	6.39 01	6.84 01	7.06 01	7.59 01
36	Kr K$_{\beta_1}$	14.112	87.857	5.88 01	6.11 01	6.54 01	6.75 01	7.26 01
38	Sr K$_{\alpha_1}$	14.165	87.529	5.82 01	6.05 01	6.47 01	6.68 01	7.18 01
86	Rn L$_{\beta_1}$	14.316	86.606	5.66 01	5.88 01	6.29 01	6.49 01	6.98 01
87	Fr L$_{\beta_1}$	14.770	83.943	5.21 01	5.41 01	5.79 01	5.97 01	6.42 01
39	Y K$_{\alpha_1}$	14.958	82.888	5.03 01	5.23 01	5.59 01	5.78 01	6.21 01
37	Rb K$_{\beta_1}$	14.961	82.872	5.03 01	5.23 01	5.59 01	5.77 01	6.21 01
88	Ra L$_{\beta_1}$	15.236	81.376	4.79 01	4.98 01	5.33 01	5.50 01	5.91 01
89	Ac L$_{\beta_1}$	15.713	78.906	4.42 01	4.59 01	4.91 01	5.07 01	5.45 01
40	Zr K$_{\alpha_1}$	15.775	78.596	4.37 01	4.54 01	4.86 01	5.01 01	5.39 01
38	Sr K$_{\beta_1}$	15.836	78.293	4.33 01	4.50 01	4.81 01	4.96 01	5.33 01
90	Th L$_{\beta_1}$	16.202	76.524	4.08 01	4.23 01	4.53 01	4.67 01	5.02 01
41	Nb K$_{\alpha_1}$	16.615	74.622	3.81 01	3.96 01	4.24 01	4.37 01	4.69 01
91	Pa L$_{\beta_1}$	16.702	74.233	3.76 01	3.90 01	4.18 01	4.31 01	4.63 01
39	Y K$_{\beta_1}$	16.738	74.074	3.74 01	3.88 01	4.15 01	4.28 01	4.60 01
92	U L$_{\beta_1}$	17.220	72.000	3.47 01	3.60 01	3.85 01	3.97 01	4.27 01
42	Mo K$_{\alpha_1}$	17.479	70.933	3.33 01	3.46 01	3.70 01	3.82 01	4.10 01
40	Zr K$_{\beta_1}$	17.668	70.175	3.24 01	3.36 01	3.60 01	3.71 01	3.98 01
43	Tc K$_{\alpha_1}$	18.367	67.504	2.92 01	3.03 01	3.25 01	3.34 01	3.59 01
41	Nb K$_{\beta_1}$	18.623	66.576	2.82 01	2.93 01	3.13 01	3.22 01	3.46 01
44	Ru K$_{\alpha_1}$	19.279	64.311	2.57 01	2.67 01	2.86 01	2.94 01	3.16 01
42	Mo K$_{\beta_1}$	19.608	63.231	2.46 01	2.55 01	2.73 01	2.81 01	3.02 01
45	Rh K$_{\alpha_1}$	20.216	61.330	2.27 01	2.35 01	2.52 01	2.59 01	2.78 01
43	Tc K$_{\beta_1}$	20.619	60.131	2.16 01	2.24 01	2.39 01	2.46 01	2.64 01
46	Pd K$_{\alpha_1}$	21.177	58.547	2.01 01	2.09 01	2.23 01	2.30 01	2.46 01
44	Ru K$_{\beta_1}$	21.657	57.249	1.90 01	1.97 01	2.10 01	2.17 01	2.32 01

Z	transition	E [keV]	λ [pm]	$_{51}$Sb	$_{52}$Te	$_{53}$I	$_{54}$Xe	$_{55}$Cs
47	Ag K$_{\alpha_1}$	22.163	55.942	1.79 01	1.85 01	1.98 01	2.04 01	2.19 01
45	Rh K$_{\beta_1}$	22.724	54.561	1.67 01	1.73 01	1.85 01	1.91 01	2.05 01
48	Cd K$_{\alpha_1}$	23.174	53.501	1.59 01	1.65 01	1.76 01	1.82 01	1.95 01
46	Pd K$_{\beta_1}$	23.819	52.053	1.48 01	1.53 01	1.64 01	1.69 01	1.81 01
49	In K$_{\alpha_1}$	24.210	51.212	1.42 01	1.47 01	1.57 01	1.62 01	1.73 01
47	Ag K$_{\beta_1}$	24.942	49.709	1.31 01	1.36 01	1.45 01	1.50 01	1.60 01
50	Sn K$_{\alpha_1}$	25.271	49.062	1.27 01	1.31 01	1.40 01	1.45 01	1.55 01
48	Cd K$_{\beta_1}$	26.096	47.511	1.17 01	1.21 01	1.29 01	1.33 01	1.42 01
51	Sb K$_{\alpha_1}$	26.359	47.037	1.14 01	1.18 01	1.25 01	1.30 01	1.39 01
49	In K$_{\beta_1}$	27.276	45.455	1.04 01	1.08 01	1.15 01	1.19 01	1.27 01
52	Te K$_{\alpha_1}$	27.472	45.131	1.02 01	1.06 01	1.13 01	1.17 01	1.24 01
50	Sn K$_{\beta_1}$	28.486	43.524	9.29 00	9.62 00	1.02 01	1.06 01	1.13 01
53	I K$_{\alpha_1}$	28.612	43.333	9.18 00	9.51 00	1.01 01	1.05 01	1.12 01
51	Sb K$_{\beta_1}$	29.726	41.709	8.31 00	8.61 00	9.18 00	9.49 00	1.01 01
54	Xe K$_{\alpha_1}$	29.779	41.635	8.27 00	8.57 00	9.14 00	9.45 00	1.01 01
55	Cs K$_{\alpha_1}$	30.973	40.030	4.26 01	7.74 00	8.26 00	8.53 00	9.10 00
52	Te K$_{\beta_1}$	30.996	40.000	4.25 01	7.72 00	8.25 00	8.51 00	9.09 00
56	Ba K$_{\alpha_1}$	32.194	38.511	3.84 01	3.94 01	7.48 00	7.72 00	8.25 00
53	I K$_{\beta_1}$	32.295	38.391	3.81 01	3.90 01	7.42 00	7.65 00	8.18 00
57	La K$_{\alpha_1}$	33.442	37.074	3.46 01	3.55 01	3.83 01	6.99 00	7.48 00
54	Xe K$_{\beta_1}$	33.624	36.874	3.41 01	3.50 01	3.77 01	6.90 00	7.38 00
58	Ce K$_{\alpha_1}$	34.279	36.169	3.24 01	3.33 01	3.58 01	6.56 00	7.03 00
55	Cs K$_{\beta_1}$	34.987	35.437	3.06 01	3.15 01	3.39 01	3.41 01	6.67 00
59	Pr K$_{\alpha_1}$	36.026	34.415	2.83 01	2.91 01	3.13 01	3.15 01	3.14 01
56	Ba K$_{\beta_1}$	36.378	34.082	2.76 01	2.84 01	3.04 01	3.07 01	3.35 01
60	Nd K$_{\alpha_1}$	36.847	33.648	2.66 01	2.74 01	2.94 01	2.97 01	3.23 01
57	La K$_{\beta_1}$	37.801	32.799	2.48 01	2.56 01	2.74 01	2.77 01	3.02 01
61	Pm K$_{\alpha_1}$	38.725	32.016	2.33 01	2.40 01	2.57 01	2.60 01	2.82 01
58	Ce K$_{\beta_1}$	39.257	31.582	2.24 01	2.31 01	2.47 01	2.50 01	2.72 01
62	Sm K$_{\alpha_1}$	40.118	30.905	2.12 01	2.18 01	2.33 01	2.36 01	2.56 01
59	Pr K$_{\beta_1}$	40.748	30.427	2.03 01	2.09 01	2.23 01	2.26 01	2.45 01
63	Eu K$_{\alpha_1}$	41.542	29.845	1.92 01	1.98 01	2.12 01	2.15 01	2.33 01
60	Nd K$_{\beta_1}$	42.271	29.331	1.83 01	1.89 01	2.02 01	2.05 01	2.22 01
64	Gd K$_{\alpha_1}$	42.996	28.836	1.75 01	1.81 01	1.93 01	1.96 01	2.12 01
61	Pm K$_{\beta_1}$	43.826	28.290	1.66 01	1.72 01	1.83 01	1.86 01	2.01 01
65	Tb K$_{\alpha_1}$	44.482	27.873	1.59 01	1.65 01	1.75 01	1.78 01	1.93 01
62	Sm K$_{\beta_1}$	45.413	27.301	1.51 01	1.56 01	1.66 01	1.69 01	1.82 01
66	Dy K$_{\alpha_1}$	45.998	26.954	1.45 01	1.50 01	1.60 01	1.63 01	1.76 01
63	Eu K$_{\beta_1}$	47.038	26.358	1.37 01	1.42 01	1.50 01	1.53 01	1.65 01
67	Ho K$_{\alpha_1}$	47.547	26.076	1.33 01	1.37 01	1.46 01	1.49 01	1.60 01
64	Gd K$_{\beta_1}$	48.697	25.460	1.24 01	1.29 01	1.37 01	1.39 01	1.50 01
68	Er K$_{\alpha_1}$	49.128	25.237	1.21 01	1.26 01	1.33 01	1.36 01	1.46 01
65	Tb K$_{\beta_1}$	50.382	24.609	1.13 01	1.17 01	1.25 01	1.27 01	1.37 01
69	Tm K$_{\alpha_1}$	50.742	24.434	1.11 01	1.15 01	1.22 01	1.25 01	1.34 01

Z	transition	E [keV]	λ [pm]	$_{51}$Sb	$_{52}$Te	$_{53}$I	$_{54}$Xe	$_{55}$Cs
66	Dy K$_{\beta_1}$	52.119	23.788	1.03 01	1.07 01	1.13 01	1.16 01	1.24 01
70	Yb K$_{\alpha_1}$	52.389	23.666	1.02 01	1.05 01	1.12 01	1.14 01	1.23 01
67	Ho K$_{\beta_1}$	53.877	23.012	9.41 00	9.77 00	1.03 01	1.06 01	1.14 01
71	Lu K$_{\alpha_1}$	54.070	22.930	9.31 00	9.67 00	1.02 01	1.05 01	1.12 01
68	Er K$_{\beta_1}$	55.681	22.167	8.58 00	8.92 00	9.45 00	9.68 00	1.04 01
72	Hf K$_{\alpha_1}$	55.790	22.223	8.54 00	8.88 00	9.40 00	9.62 00	1.03 01
69	Tm K$_{\beta_1}$	57.517	21.556	7.84 00	8.16 00	8.64 00	8.85 00	9.47 00
73	Ta K$_{\alpha_1}$	57.532	21.550	7.84 00	8.16 00	8.63 00	8.85 00	9.47 00
74	W K$_{\alpha_1}$	59.318	20.901	7.20 00	7.50 00	7.94 00	8.14 00	8.70 00
70	Yb K$_{\beta_1}$	59.370	20.883	7.18 00	7.48 00	7.92 00	8.12 00	8.68 00
75	Re K$_{\alpha_1}$	61.140	20.278	6.63 00	6.90 00	7.30 00	7.49 00	8.00 00
71	Lu K$_{\beta_1}$	61.283	20.231	6.58 00	6.86 00	7.25 00	7.45 00	7.95 00
76	Os K$_{\alpha_1}$	63.001	19.679	6.10 00	6.36 00	6.72 00	6.90 00	7.37 00
72	Hf K$_{\beta_1}$	63.234	19.607	6.04 00	6.29 00	6.65 00	6.83 00	7.29 00
77	Ir K$_{\alpha_1}$	64.896	19.105	5.63 00	5.86 00	6.20 00	6.37 00	6.79 00
73	Ta K$_{\beta_1}$	65.223	19.009	5.55 00	5.78 00	6.11 00	6.28 00	6.70 00
78	Pt K$_{\alpha_1}$	66.832	18.551	5.19 00	5.41 00	5.71 00	5.87 00	6.26 00
74	W K$_{\beta_1}$	67.244	18.438	5.11 00	5.32 00	5.62 00	5.78 00	6.16 00
79	Au K$_{\alpha_1}$	68.804	18.020	4.80 00	4.99 00	5.27 00	5.42 00	5.78 00
75	Re K$_{\beta_1}$	69.310	17.888	4.70 00	4.89 00	5.17 00	5.32 00	5.67 00
80	Hg K$_{\alpha_1}$	70.819	17.507	4.43 00	4.61 00	4.87 00	5.01 00	5.34 00
76	Os K$_{\beta_1}$	71.413	17.361	4.33 00	4.51 00	4.76 00	4.90 00	5.22 00
81	Tl K$_{\alpha_1}$	72.872	17.014	4.10 00	4.26 00	4.50 00	4.64 00	4.94 00
77	Ir K$_{\beta_1}$	73.561	16.854	4.00 00	4.15 00	4.39 00	4.52 00	4.81 00
82	Pb K$_{\alpha_1}$	74.969	16.538	3.79 00	3.94 00	4.17 00	4.29 00	4.57 00
78	Pt K$_{\beta_1}$	75.748	16.368	3.69 00	3.83 00	4.05 00	4.17 00	4.44 00
83	Bi K$_{\alpha_1}$	77.108	16.079	3.51 00	3.65 00	3.86 00	3.97 00	4.23 00
79	Au K$_{\beta_1}$	77.948	15.906	3.41 00	3.54 00	3.74 00	3.86 00	4.10 00
84	Po K$_{\alpha_1}$	79.290	15.636	3.25 00	3.38 00	3.57 00	3.68 00	3.91 00
80	Hg K$_{\beta_1}$	80.253	15.449	3.15 00	3.27 00	3.45 00	3.56 00	3.79 00
85	At K$_{\alpha_1}$	81.520	15.209	3.02 00	3.14 00	3.31 00	3.41 00	3.63 00
81	Tl K$_{\beta_1}$	82.576	15.014	2.92 00	3.03 00	3.20 00	3.30 00	3.51 00
86	Rn K$_{\alpha_1}$	83.780	14.798	2.80 00	2.91 00	3.08 00	3.17 00	3.37 00
82	Pb K$_{\beta_1}$	84.936	14.597	2.70 00	2.81 00	2.96 00	3.06 00	3.25 00
87	Fr K$_{\alpha_1}$	86.100	14.400	2.61 00	2.71 00	2.86 00	2.95 00	3.13 00
83	Bi K$_{\beta_1}$	87.343	14.195	2.51 00	2.60 00	2.75 00	2.84 00	3.01 00
88	Ra K$_{\alpha_1}$	88.470	14.014	2.42 00	2.52 00	2.66 00	2.74 00	2.91 00
84	Po K$_{\beta_1}$	89.800	13.806	2.33 00	2.42 00	2.55 00	2.63 00	2.79 00
89	Ac K$_{\alpha_1}$	90.884	13.642	2.25 00	2.34 00	2.47 00	2.55 00	2.71 00
85	At K$_{\beta_1}$	92.300	13.432	2.16 00	2.24 00	2.37 00	2.45 00	2.59 00
90	Th K$_{\alpha_1}$	93.350	13.281	2.10 00	2.18 00	2.30 00	2.37 00	2.52 00
86	Rn K$_{\beta_1}$	94.870	13.068	2.01 00	2.08 00	2.20 00	2.27 00	2.41 00
91	Pa K$_{\alpha_1}$	95.868	12.932	1.95 00	2.03 00	2.14 00	2.21 00	2.34 00
87	Fr K$_{\beta_1}$	97.470	12.720	1.87 00	1.94 00	2.04 00	2.11 00	2.24 00

Z	transition	E [keV]	λ [pm]	$_{51}$Sb	$_{52}$Te	$_{53}$I	$_{54}$Xe	$_{55}$Cs
92	U K$_{\alpha_1}$	98.439	12.595	1.82 00	1.89 00	1.99 00	2.06 00	2.18 00
88	Ra K$_{\beta_1}$	100.130	12.382	1.74 00	1.80 00	1.90 00	1.97 00	2.08 00
89	Ac K$_{\beta_1}$	102.850	12.054	1.62 00	1.68 00	1.78 00	1.84 00	1.94 00
90	Th K$_{\beta_1}$	105.610	11.739	1.51 00	1.57 00	1.66 00	1.72 00	1.82 00
91	Pa K$_{\beta_1}$	108.430	11.434	1.42 00	1.47 00	1.55 00	1.60 00	1.70 00
92	U K$_{\beta_1}$	111.300	11.139	1.33 00	1.38 00	1.45 00	1.50 00	1.59 00

Z	transition	E [keV]	λ [pm]	$_{56}$Ba	$_{57}$La	$_{58}$Ce	$_{59}$Pr	$_{60}$Nd
4	Be K$_\alpha$	0.1085	11427.207	1.32 05	3.73 04	1.18 04	1.31 04	1.15 04
38	Sr M$_\xi$	0.1140	10875.895	1.24 05	6.33 04	1.59 04	1.87 04	1.10 04
39	Y M$_\xi$	0.1328	9339.235	3.40 04	5.57 04	6.11 04	1.23 05	8.10 04
16	S L$_l$	0.1487	8337.942	1.26 04	2.24 04	2.51 04	5.45 04	4.99 04
49	Zr M$_\xi$	0.1511	8205.506	1.13 04	1.96 04	2.22 04	4.99 04	4.58 04
41	Nb M$_\xi$	0.1717	7221.037	5.89 03	6.02 03	1.09 04	2.04 04	2.18 04
5	B K$_\alpha$	0.1833	6764.059	4.94 03	3.73 03	8.25 03	1.44 04	1.51 04
42	Mo M$_\xi$	0.1926	6473.445	4.33 03	2.99 03	7.23 03	1.13 04	1.15 04
6	C K$_\alpha$	0.2776	4476.000	4.45 03	6.63 03	6.48 03	8.07 03	6.37 03
47	Ag M$_\xi$	0.3117	3977.709	5.35 03	6.75 03	7.42 03	1.11 04	8.35 03
7	N K$_\alpha$	0.3924	3159.664	5.73 03	6.72 03	6.86 03	8.69 03	8.03 03
22	Ti L$_l$	0.3953	3136.484	5.71 03	6.30 03	6.83 03	8.56 03	7.97 03
22	Ti L$_\alpha$	0.4522	2741.822	5.22 03	5.56 03	6.11 03	6.77 03	6.84 03
23	V L$_\alpha$	0.5113	2424.901	4.63 03	4.83 03	5.38 03	5.66 03	5.88 03
8	O K$_\alpha$	0.5249	2362.072	4.56 03	4.69 03	5.21 03	5.50 03	5.69 03
25	Mn L$_l$	0.5563	2228.747	4.21 03	4.35 03	4.81 03	5.08 03	5.26 03
24	Cr L$_\alpha$	0.5728	2164.549	4.05 03	4.18 03	4.62 03	4.88 03	5.04 03
25	Mn L$_\alpha$	0.6374	1945.171	3.46 03	3.60 03	3.95 03	4.17 03	4.31 03
9	F K$_\alpha$	0.6768	1831.932	3.16 03	3.28 03	3.58 03	3.78 03	3.91 03
26	Fe L$_\alpha$	0.7050	1758.655	2.96 03	3.10 03	3.38 03	3.57 03	3.69 03
27	Co L$_\alpha$	0.7762	1597.335	2.53 03	2.65 03	2.88 03	3.04 03	3.14 03
28	Ni L$_\alpha$	0.8515	1456.080	1.16 04	1.23 04	2.46 03	2.59 03	2.68 03
29	Cu L$_\alpha$	0.9297	1336.044	9.74 03	1.02 04	1.15 04	2.22 03	2.29 03
30	Zn L$_\alpha$	1.0117	1215.513	8.10 03	8.44 03	9.39 03	1.01 04	8.68 04
11	Na K$_\alpha$	1.0410	1191.020	7.62 03	7.93 03	8.80 03	9.41 03	8.28 04
11	Na K$_\beta$	1.0711	1157.550	1.04 04	7.22 03	7.82 03	8.39 03	7.98 04
12	Mg K$_\alpha$	1.2536	989.033	7.64 03	8.18 03	7.92 01	8.59 00	7.42 03
33	As L$_\alpha$	1.2820	967.123	7.22 03	7.72 03	8.23 03	8.11 03	7.01 03
12	Mg K$_\beta$	1.3022	952.121	7.62 03	7.42 03	7.91 03	7.78 03	8.06 03
33	As L$_{\beta_1}$	1.3170	941.421	7.40 03	7.21 03	7.69 03	7.56 03	7.83 03
66	Dy M$_\beta$	1.3250	935.737	7.29 03	7.10 03	7.57 03	7.44 03	7.71 03
67	Ho M$_\alpha$	1.3480	919.771	6.98 03	6.80 03	7.25 03	7.83 03	7.39 03
34	Se L$_\alpha$	1.3791	899.029	6.58 03	7.04 03	6.84 03	7.38 03	6.98 03
67	Ho M$_\beta$	1.3830	896.494	6.53 03	6.99 03	6.80 03	7.32 03	6.93 03
68	Er M$_\alpha$	1.4060	881.829	6.26 03	6.70 03	6.52 03	7.02 03	7.31 03

Z	transition	E [keV]	λ [pm]	$_{56}$Ba	$_{57}$La	$_{58}$Ce	$_{59}$Pr	$_{60}$Nd
34	Se L$_{\beta_1}$	1.4192	873.627	6.11 03	6.54 03	<u>6.37 03</u>	6.85 03	7.13 03
68	Er M$_\beta$	1.4430	859.218	5.86 03	6.27 03	6.69 03	6.56 03	6.83 03
69	Tm M$_\alpha$	1.4620	848.051	5.67 03	6.07 03	6.47 03	6.35 03	6.60 03
35	Br L$_\alpha$	1.4804	837.511	5.49 03	5.87 03	6.27 03	6.15 03	6.39 03
13	Al K$_{\alpha_1}$	1.4867	833.962	5.43 03	5.81 03	6.20 03	6.08 02	6.32 03
69	Tm M$_\beta$	1.5030	824.918	5.28 03	5.65 03	6.03 03	<u>5.92 03</u>	6.19 03
70	Yb M$_\alpha$	1.5214	814.941	5.12 03	5.48 03	5.85 03	6.30 03	5.96 03
35	Br L$_{\beta_1}$	1.5259	812.538	5.08 03	5.43 03	5.80 03	6.25 03	5.92 03
13	Al K$_\beta$	1.5574	796.103	4.82 03	5.15 03	5.51 03	5.93 03	5.63 03
70	Yb M$_\beta$	1.5675	790.974	4.74 03	5.07 03	5.42 03	5.84 03	<u>5.53 03</u>
71	Lu M$_\alpha$	1.5813	784.071	4.64 03	4.96 03	5.30 03	5.71 03	5.95 03
36	Kr L$_\alpha$	1.5860	781.747	4.61 03	4.92 03	5.26 03	5.66 03	5.91 03
71	Lu M$_\beta$	1.6312	760.085	4.29 03	4.57 03	4.89 03	5.27 03	5.50 03
36	Kr L$_{\beta_1}$	1.6366	757.577	4.25 03	4.53 03	4.85 03	5.22 03	5.45 03
72	Hf M$_{\alpha_1}$	1.6446	753.892	4.20 03	4.48 03	4.79 03	5.16 03	5.38 03
37	Rb L$_{\alpha_1}$	1.6941	731.864	3.90 03	4.15 03	4.44 03	4.78 03	4.99 03
72	Hf M$_\beta$	1.6976	730.355	3.88 03	4.13 03	4.42 03	4.76 03	4.96 03
73	Ta M$_{\alpha_1}$	1.7096	725.229	3.81 03	4.05 03	4.34 03	4.67 03	4.87 03
14	Si K$_{\alpha_1}$	1.7400	712.558	3.64 03	3.87 03	4.15 03	4.46 03	4.66 03
37	Rb L$_{\beta_1}$	1.7522	707.597	3.58 03	3.80 03	4.08 03	4.39 03	4.58 03
73	Ta M$_\beta$	1.7655	702.266	3.51 03	3.73 03	4.00 03	4.30 03	4.49 03
74	W M$_{\alpha_1}$	1.7754	698.350	3.46 03	3.67 03	3.94 03	4.24 03	4.42 03
38	Sr L$_{\alpha_1}$	1.8066	686.290	3.31 03	3.51 03	3.77 03	4.05 03	4.23 03
74	W M$_\beta$	1.8349	675.705	3.18 03	3.37 03	3.62 03	3.90 03	4.07 03
14	Si K$_\beta$	1.8359	675.337	3.18 03	3.37 03	3.62 03	3.89 03	4.06 03
75	Re M$_{\alpha_1}$	1.8420	673.100	3.15 03	3.34 03	3.59 03	3.86 03	4.03 03
38	Sr L$_{\beta_1}$	1.8717	662.420	3.02 03	3.21 03	3.45 03	3.70 03	3.86 03
75	Re M$_\beta$	1.9061	650.465	2.89 03	3.06 03	3.29 03	3.53 03	5.69 03
76	Os M$_{\alpha_1}$	1.9102	649.069	2.87 03	3.04 03	3.27 03	3.51 03	3.67 03
39	Y L$_{\alpha_1}$	1.9226	644.882	2.83 03	2.99 03	3.22 03	3.46 03	3.61 03
76	Os M$_\beta$	1.9783	626.725	2.63 03	2.78 03	2.99 03	3.21 03	3.35 03
77	Ir M$_{\alpha_1}$	1.9799	626.219	2.62 03	2.77 03	2.99 03	3.21 03	3.35 03
39	Y L$_\beta$	1.9958	621.230	2.57 03	2.72 03	2.93 03	3.14 03	3.28 03
15	P K$_{\alpha_1}$	2.0137	615.708	2.51 03	2.66 03	2.86 03	3.07 03	3.20 03
40	Zr L$_{\alpha_1}$	2.0424	607.056	2.42 03	2.56 03	2.76 03	2.96 03	3.09 03
78	Pt M$_{\alpha_1}$	2.0505	604.658	2.40 03	2.54 03	2.73 03	2.93 03	3.06 03
77	Ir M$_\beta$	2.0535	603.775	2.39 03	2.53 03	2.72 03	2.92 03	3.05 03
79	Au M$_{\alpha_1}$	2.1229	584.036	2.20 03	2.32 03	2.50 03	2.68 03	2.80 03
40	Zr L$_{\beta_1}$	2.1244	583.624	2.19 03	2.32 03	2.50 03	2.68 03	2.80 03
78	Pt M$_\beta$	2.1273	582.828	2.19 03	2.31 03	2.49 03	2.67 03	2.79 03
15	P K$_{\beta_{1,3}}$	2.1390	579.640	2.16 03	2.28 03	2.45 03	2.63 03	2.75 03
41	Nb L$_{\alpha_1}$	2.1659	572.441	2.09 03	2.21 03	2.38 03	2.55 03	2.66 03
80	Hg M$_{\alpha_1}$	2.1953	564.775	2.02 03	2.14 03	2.29 03	2.46 03	2.57 03
79	Au M$_\beta$	2.2046	562.393	2.00 03	2.11 03	2.27 03	2.43 03	2.54 03

Z	transition	E [keV]	λ [pm]	$_{56}$Ba	$_{57}$La	$_{58}$Ce	$_{59}$Pr	$_{60}$Nd
41	Nb L$_{\beta_1}$	2.2574	549.238	1.88 03	1.99 03	2.14 03	2.29 03	2.40 03
81	Tl M$_{\alpha_1}$	2.2706	546.045	1.85 03	1.96 03	2.11 03	2.26 03	2.36 03
80	Hg M$_\beta$	2.2825	543.199	1.83 03	1.94 03	2.08 03	2.23 03	2.33 03
42	Mo L$_{\alpha_1}$	2.2932	540.664	1.81 03	1.91 03	2.05 03	2.20 03	2.30 03
16	S K$_{\alpha_1}$	2.3080	537.197	1.78 03	1.88 03	2.02 03	2.16 03	2.26 03
82	Pb M$_{\alpha_1}$	2.3457	528.563	1.70 03	1.81 03	1.94 03	2.07 03	2.17 03
81	Tl M$_\beta$	2.3621	524.893	1.67 03	1.78 03	1.90 03	2.04 03	2.13 03
42	Mo L$_{\beta_1}$	2.3948	517.726	1.62 03	1.71 03	1.84 03	1.97 03	2.06 03
83	Bi M$_{\alpha_1}$	2.4226	511.785	1.57 03	1.67 03	1.79 03	1.91 03	2.00 03
43	Tc L$_{\alpha_1}$	2.4240	511.490	1.57 03	1.66 03	1.78 03	1.91 03	2.00 03
82	Pb M$_\beta$	2.4427	507.574	1.54 03	1.63 03	1.75 03	1.87 03	1.96 03
16	S K$_\beta$	2.4640	503.186	1.50 03	1.60 03	1.71 03	1.83 03	1.92 03
83	Bi M$_{\beta_1}$	2.5255	490.933	1.41 03	1.50 03	1.61 03	1.72 03	1.80 03
43	Tc L$_{\beta_1}$	2.5368	488.746	1.40 03	1.48 03	1.59 03	1.70 03	1.78 03
44	Ru L$_{\alpha_1}$	2.5586	484.582	1.37 03	1.45 03	1.55 03	1.66 03	1.74 03
17	Cl K$_{\alpha_1}$	2.6224	472.792	1.28 03	1.36 03	1.46 03	1.56 03	1.64 03
44	Ru L$_{\beta_1}$	2.6832	462.079	1.21 03	1.29 03	1.38 03	1.47 03	1.54 03
45	Rh L$_{\alpha_1}$	2.6967	459.766	1.20 03	1.27 03	1.36 03	1.45 03	1.52 03
17	Cl K$_\beta$	2.8156	440.350	1.07 03	1.14 03	1.22 03	1.30 03	1.36 03
45	Rh L$_{\beta_1}$	2.8344	437.430	1.05 03	1.12 03	1.20 03	1.28 03	1.34 03
46	Pd L$_{\alpha_1}$	2.8386	436.782	1.05 03	1.12 03	1.19 03	1.27 03	1.34 03
18	Ar K$_{\alpha_1}$	2.9577	419.194	9.46 02	1.01 03	1.07 03	1.14 03	1.20 03
47	Ag L$_{\alpha_1}$	2.9843	415.458	9.24 02	9.84 02	1.05 03	1.12 03	1.18 03
46	Pd L$_{\beta_1}$	2.9902	414.638	9.20 02	9.79 02	1.04 03	1.11 03	1.17 03
90	Th M$_{\alpha_1}$	2.9961	413.821	9.15 02	9.74 02	1.04 03	1.11 03	1.16 03
91	Pa M$_{\alpha_1}$	3.0823	402.248	8.51 02	9.07 02	9.67 02	1.03 03	1.08 03
48	Cd L$_{\alpha_1}$	3.1337	395.651	8.16 02	8.70 02	9.27 02	9.87 02	1.04 03
90	Th M$_\beta$	3.1458	394.129	8.08 02	8.61 02	9.18 02	9.77 02	1.03 03
47	Ag L$_{\beta_1}$	3.1509	393.491	8.05 02	8.58 02	9.14 02	9.73 02	1.02 03
92	U M$_{\alpha_1}$	3.1708	391.021	7.92 02	8.44 02	9.00 02	9.58 02	1.01 03
18	Ar K$_\beta$	3.1905	388.607	7.80 02	8.31 02	8.86 02	9.43 02	9.93 02
91	Pa M$_\beta$	3.2397	382.705	7.50 02	7.99 02	8.52 02	9.07 02	9.55 02
49	In L$_{\alpha_1}$	3.2869	377.210	7.23 02	7.70 02	8.21 02	8.74 02	9.21 02
19	K K$_{\alpha_1}$	3.3138	374.148	7.08 02	7.55 02	8.04 02	8.56 02	9.02 02
48	Cd L$_{\beta_1}$	3.3166	373.832	7.06 02	7.53 02	8.03 02	8.54 02	9.00 02
92	U M$_\beta$	3.3360	371.658	6.96 02	7.42 02	7.91 02	8.41 02	8.87 02
50	Sn L$_{\alpha_1}$	3.4440	360.003	6.42 02	6.84 02	7.29 02	7.76 02	8.18 02
49	In L$_{\beta_1}$	3.4872	355.543	6.22 02	6.63 02	7.07 02	7.52 02	7.92 02
19	K K$_\beta$	3.5896	345.401	5.78 02	6.16 02	6.56 02	6.98 02	7.36 02
51	Sb L$_{\alpha_1}$	3.6047	343.954	4.71 02	6.10 02	6.50 02	6.91 02	7.28 02
50	Sn L$_{\beta_1}$	3.6628	338.498	5.49 02	5.85 02	6.24 02	6.63 02	6.99 02
20	Ca K$_{\alpha_1}$	3.6917	335.848	5.38 02	5.74 02	6.11 02	6.50 02	6.86 02
52	Te L$_{\alpha_1}$	3.7693	328.934	5.10 02	5.44 02	5.80 02	6.17 02	6.50 02
51	Sb L$_{\beta_1}$	3.8436	322.575	4.85 02	5.18 02	5.52 02	5.87 02	6.19 02

Z	transition	E [keV]	λ [pm]	$_{56}$Ba	$_{57}$La	$_{58}$Ce	$_{59}$Pr	$_{60}$Nd
53	I L$_{\alpha_1}$	3.9377	314.867	4.56 02	4.87 02	5.19 02	5.52 02	5.82 02
20	Ca K$_\beta$	4.0127	308.981	4.35 02	4.65 02	4.95 02	5.26 02	5.55 02
52	Te L$_{\beta_1}$	4.0296	307.686	4.31 02	4.60 02	4.89 02	5.20 02	5.49 02
21	Sc K$_{\alpha_1}$	4.0906	303.097	4.15 02	4.43 02	4.71 02	5.01 02	5.28 02
54	Xe L$_{\alpha_1}$	4.1099	301.674	4.10 02	4.37 02	4.66 02	4.95 02	5.22 02
53	I L$_{\beta_1}$	4.2207	293.755	3.83 02	4.09 02	4.35 02	4.62 02	4.88 02
55	Cs L$_{\alpha_1}$	4.2865	289.245	3.69 02	3.93 02	4.19 02	4.44 02	4.69 02
21	Sc K$_\beta$	4.4605	277.962	3.34 02	3.56 02	3.79 02	4.01 02	4.24 02
56	Ba L$_{\alpha_1}$	4.4663	277.601	3.33 02	3.55 02	3.78 02	4.00 02	4.23 02
22	Ti K$_{\alpha_1}$	4.5108	274.863	3.25 02	3.46 02	3.68 02	3.90 02	4.12 02
55	Cs L$_{\beta_1}$	4.6198	268.377	3.06 02	3.26 02	3.47 02	3.67 02	3.88 02
57	La L$_{\alpha_1}$	4.6510	266.577	3.02 02	3.20 02	3.41 02	3.61 02	3.81 02
56	Ba L$_{\beta_1}$	4.8275	256.841	2.74 02	2.92 02	3.10 02	3.28 02	3.47 02
58	Ce L$_{\alpha_1}$	4.8402	256.157	2.72 02	2.90 02	3.06 02	3.26 02	3.45 02
22	Ti K$_{\beta_{1,3}}$	4.9318	251.399	2.60 02	2.76 02	2.94 02	3.11 02	3.29 02
23	V K$_{\alpha_1}$	4.9522	250.363	2.57 02	2.74 03	2.91 02	3.07 02	3.25 02
59	Pr L$_{\alpha_1}$	5.0337	246.310	2.47 02	2.63 02	2.79 02	2.95 02	3.12 02
57	La L$_{\beta_1}$	5.0421	245.899	2.46 02	2.62 02	2.78 02	2.94 02	3.11 02
60	Nd L$_{\alpha_1}$	5.2304	237.047	<u>2.24 02</u>	2.38 02	2.54 02	2.68 02	2.83 02
58	Ce L$_{\beta_1}$	5.2622	235.614	6.44 02	2.35 02	2.50 02	2.64 02	2.79 02
24	Cr K$_{\alpha_1}$	5.4147	228.978	5.97 02	2.18 02	2.32 02	2.45 02	2.60 02
23	V K$_{\beta_{1,3}}$	5.4273	228.447	5.93 02	2.17 02	2.31 02	2.44 02	2.58 02
61	Pm L$_{\alpha_1}$	5.4325	228.228	5.92 02	<u>2.17 02</u>	2.30 02	2.43 02	2.58 02
59	Pr L$_{\beta_1}$	5.4889	225.883	<u>5.75 02</u>	6.14 02	2.24 02	2.37 02	2.51 02
62	Sm L$_{\alpha_1}$	5.6361	219.984	7.45 02	5.72 02	2.10 02	2.21 02	2.35 02
60	Nd L$_{\beta_1}$	5.7216	216.696	7.15 02	5.49 02	<u>2.02 02</u>	2.13 02	2.26 02
63	Eu L$_{\alpha_1}$	5.8457	212.096	6.75 02	<u>5.18 02</u>	5.48 02	2.02 02	2.14 02
25	Mn K$_{\alpha_1}$	5.8988	210.187	6.58 02	7.04 02	5.35 02	1.97 02	2.10 02
24	Cr K$_{\beta_{1,3}}$	5.9467	208.494	6.44 02	6.89 02	5.24 02	1.93 02	2.05 02
61	Pm L$_{\beta_1}$	5.9610	207.993	<u>6.40 02</u>	6.84 02	5.20 02	<u>1.92 02</u>	2.04 02
64	Gd L$_{\alpha_1}$	6.0572	204.690	7.35 02	6.55 02	<u>4.99 02</u>	5.29 02	1.96 02
62	Sm L$_{\beta_1}$	6.2051	199.811	6.89 02	<u>6.14 02</u>	6.50 02	4.96 02	<u>1.84 02</u>
65	Tb L$_{\alpha_1}$	6.2728	197.655	6.69 02	7.18 02	6.32 02	4.82 02	5.08 02
26	Fe K$_{\alpha_1}$	6.4038	193.611	6.32 02	6.78 02	5.98 02	<u>4.56 02</u>	4.80 02
63	Eu L$_{\beta_1}$	6.4564	192.034	6.19 02	6.63 02	5.85 02	6.20 02	4.70 02
25	Mn K$_{\beta_{1,3}}$	6.4905	191.025	6.10 02	6.54 02	5.77 02	6.11 02	4.63 02
66	Dy L$_{\alpha_1}$	6.4952	190.887	6.09 02	6.52 02	<u>5.76 02</u>	6.10 02	4.62 02
64	Gd L$_{\beta_1}$	6.7132	184.688	5.57 02	5.96 02	6.31 02	5.57 02	4.23 02
67	Ho L$_{\alpha_1}$	6.7198	184.507	5.55 02	5.94 02	6.29 02	<u>5.55 02</u>	<u>4.22 02</u>
27	Co K$_{\alpha_1}$	6.9303	178.903	5.11 02	5.46 02	5.79 02	6.13 02	5.39 02
68	Er L$_{\alpha_1}$	6.9487	178.429	5.07 02	5.42 02	5.75 02	6.09 02	5.35 02
65	Tb L$_{\beta_1}$	6.9780	177.680	5.01 02	5.36 02	5.68 02	6.02 02	5.30 02
26	Fe K$_{\beta_{1,3}}$	7.0580	175.666	4.86 02	5.20 02	5.51 02	5.84 02	<u>5.14 02</u>
69	Tm L$_{\alpha_1}$	7.1799	172.683	4.64 02	4.96 02	5.26 02	5.57 02	5.89 02

Z	transition	E [keV]	λ [pm]	$_{56}$Ba	$_{57}$La	$_{58}$Ce	$_{59}$Pr	$_{60}$Nd
66	Dy L$_{\beta_1}$	7.2477	171.068	4.52 02	4.83 02	5.13 02	5.43 02	5.74 02
70	Yb L$_{\alpha_1}$	7.4156	167.195	4.25 02	4.54 02	4.82 02	5.11 02	5.39 02
28	Ni K$_{\alpha_1}$	7.4782	165.795	4.16 02	4.44 02	4.71 02	4.99 02	5.27 02
67	Ho L$_{\beta_1}$	7.5253	164.757	4.09 02	4.36 02	4.63 02	4.91 02	5.18 02
27	Co K$_{\beta_{1,3}}$	7.6494	162.084	3.91 02	4.17 02	4.43 02	4.70 02	4.95 02
71	Lu L$_{\alpha_1}$	7.6555	161.955	3.90 02	4.16 02	4.42 02	4.68 02	4.94 02
68	Er L$_{\beta_1}$	7.8109	158.733	3.70 02	3.94 02	4.19 02	4.44 02	4.68 02
72	Hf L$_{\alpha_1}$	7.8990	156.963	3.59 02	3.82 02	4.06 02	4.30 02	4.54 02
29	Cu K$_{\alpha_1}$	8.0478	154.060	3.41 02	3.63 02	3.86 02	4.09 02	4.31 02
69	Tm L$_{\beta_1}$	8.1010	153.049	3.35 02	3.57 02	3.79 02	4.02 02	4.24 02
73	Ta L$_{\alpha_1}$	8.1461	152.201	3.30 02	3.51 02	3.74 02	3.96 02	4.17 02
28	Ni K$_{\beta_{1,3}}$	8.2647	150.017	3.17 02	3.38 02	3.59 02	3.81 02	4.01 02
74	W L$_{\alpha_1}$	8.3976	147.643	3.04 02	3.23 02	3.44 02	3.65 02	3.84 02
70	Yb L$_{\beta_1}$	8.4018	147.569	3.04 02	3.23 02	3.44 02	3.64 02	3.84 02
30	Zn K$_{\alpha_1}$	8.6389	143.519	2.82 02	2.99 02	3.19 02	3.38 02	3.56 02
75	Re L$_{\alpha_1}$	8.6525	143.294	2.80 02	2.98 02	3.17 02	3.36 02	3.54 02
71	Lu L$_{\beta_1}$	8.7090	142.364	2.76 02	2.93 02	3.12 02	3.30 02	3.48 02
29	Cu K$_{\beta_1}$	8.9053	139.226	2.59 02	2.76 02	2.94 02	3.11 02	3.27 02
76	Os L$_{\alpha_1}$	8.9117	139.126	2.59 02	2.75 02	2.93 02	3.11 02	3.27 02
72	Hf L$_{\beta_1}$	9.0227	137.414	2.51 02	2.66 02	2.83 02	3.00 02	3.16 02
77	Ir L$_{\alpha_1}$	9.1751	135.132	2.39 02	2.54 02	2.71 02	2.87 02	3.02 02
31	Ga K$_{\alpha_1}$	9.2517	134.013	2.34 02	2.49 02	2.65 02	2.81 02	2.95 02
73	Ta L$_{\beta_1}$	9.3431	132.702	2.28 02	2.42 02	2.58 02	2.73 02	2.87 02
78	Pt L$_{\alpha_1}$	9.4423	131.308	2.22 02	2.35 02	2.51 02	2.66 02	2.79 02
30	Zn K$_{\beta_{1,3}}$	9.5720	129.529	2.14 02	2.27 02	2.41 02	2.56 02	2.69 02
74	W L$_{\beta_1}$	9.6724	128.184	2.08 02	2.20 02	2.35 02	2.49 02	2.61 02
79	Au L$_{\alpha_1}$	9.7133	127.644	2.05 02	2.18 02	2.32 02	2.46 02	2.58 02
32	Ge K$_{\alpha_1}$	9.8864	125.409	1.96 02	2.08 02	2.21 02	2.35 02	2.46 02
80	Hg L$_{\alpha_1}$	9.9888	124.124	1.90 02	2.02 02	2.15 02	2.28 02	2.40 02
75	Re L$_{\beta_1}$	10.010	123.861	1.89 02	2.01 02	2.14 02	2.27 02	2.38 02
31	Ga K$_{\beta_1}$	10.264	120.796	1.77 02	1.88 02	2.00 02	2.12 02	2.23 02
81	Tl L$_{\alpha_1}$	10.269	120.737	1.77 02	1.87 02	2.00 02	2.12 02	2.22 02
76	Os L$_{\beta_1}$	10.355	119.734	1.73 02	1.83 02	1.95 02	2.07 02	2.17 02
33	As K$_{\alpha_1}$	10.544	117.588	1.65 02	1.74 02	1.86 02	1.97 02	2.07 02
82	Pb L$_{\alpha_1}$	10.552	117.499	1.64 02	1.74 02	1.86 02	1.97 02	2.07 02
77	Ir L$_{\beta_1}$	10.708	115.787	1.58 02	1.67 02	1.78 02	1.89 02	1.98 02
83	Bi L$_{\alpha_1}$	10.839	114.388	1.53 02	1.62 02	1.73 02	1.83 02	1.92 02
32	Ge K$_{\beta_1}$	10.982	112.898	1.47 02	1.56 02	1.67 02	1.77 02	1.85 02
78	Pt L$_{\beta_1}$	11.071	111.990	1.44 02	1.53 02	1.63 02	1.73 02	1.81 02
84	Po L$_{\alpha_1}$	11.131	111.387	1.42 02	1.51 02	1.61 02	1.71 02	1.79 02
34	Se K$_{\alpha_1}$	11.222	110.484	1.39 02	1.47 02	1.57 02	1.67 02	1.75 02
85	At L$_{\alpha_1}$	11.427	108.501	1.32 02	1.40 02	1.50 02	1.59 02	1.66 02
79	Au L$_{\beta_1}$	11.442	108.359	1.32 02	1.40 02	1.49 02	1.58 02	1.66 02
33	As K$_{\beta_1}$	11.726	105.735	1.24 02	1.31 02	1.40 02	1.48 02	1.55 02

Z	transition	E [keV]	λ [pm]	$_{56}$Ba	$_{57}$La	$_{58}$Ce	$_{59}$Pr	$_{60}$Nd
86	Rn L_{α_1}	11.727	105.726	1.24 02	1.31 02	1.40 02	1.48 02	1.55 02
80	Hg L_{β_1}	11.823	104.867	1.21 02	1.28 02	1.37 02	1.45 02	1.52 02
35	Br K_{α_1}	11.924	103.979	1.18 02	1.25 02	1.34 02	1.42 02	1.48 02
87	Fr L_{α_1}	12.031	103.054	1.15 02	1.22 02	1.31 02	1.38 02	1.45 02
81	Tl L_{β_1}	12.213	101.519	1.11 02	1.17 02	1.25 02	1.33 02	1.39 02
88	Ra L_{α_1}	12.340	100.474	1.08 02	1.14 02	1.22 02	1.29 02	1.35 02
34	Se K_{β_1}	12.496	99.219	1.04 02	1.10 02	1.18 02	1.25 02	1.31 02
82	Pb L_{β_1}	12.614	98.291	1.01 02	1.07 02	1.15 02	1.22 02	1.27 02
36	Kr K_{α_1}	12.649	98.019	1.01 02	1.07 02	1.14 02	1.21 02	1.26 02
89	Ac L_{α_1}	12.652	97.996	1.01 02	1.06 02	1.14 02	1.21 02	1.26 02
90	Th L_{α_1}	12.969	95.601	9.41 01	9.96 01	1.07 02	1.13 02	1.18 02
83	Bi L_{β_1}	13.024	95.197	9.31 01	9.85 01	1.06 02	1.12 02	1.17 02
91	Pa L_{α_1}	13.291	93.285	8.81 01	9.32 01	9.99 01	1.06 02	1.11 02
35	Br K_{β_1}	13.291	93.285	8.81 01	9.32 01	9.99 01	1.06 02	1.11 02
37	Rb K_{α_1}	13.395	92.560	8.63 01	9.13 01	9.78 01	1.04 02	1.08 02
84	Po L_{β_1}	13.447	92.202	8.54 01	9.03 01	9.68 01	1.03 02	1.07 02
92	U L_{α_1}	13.615	91.065	8.26 01	8.73 01	9.37 01	9.92 01	1.04 02
85	At L_{β_1}	13.876	89.352	7.85 01	8.30 01	8.90 01	9.43 01	9.84 01
36	Kr K_{β_1}	14.112	87.857	7.50 01	7.93 01	8.51 01	9.01 01	9.41 01
38	Sr K_{α_1}	14.165	87.529	7.42 01	7.85 01	8.42 01	8.92 01	9.31 01
86	Rn L_{β_1}	14.316	86.606	7.21 01	7.62 01	8.18 01	8.67 01	9.05 01
87	Fr L_{β_1}	14.770	83.943	6.63 01	7.01 01	7.53 01	7.97 01	8.31 01
39	Y K_{α_1}	14.958	82.888	6.41 01	6.77 01	7.28 01	7.71 01	8.03 01
37	Rb K_{β_1}	14.961	82.872	6.40 01	6.77 01	7.27 01	7.70 01	8.03 01
88	Ra L_{β_1}	15.236	81.376	6.10 01	6.45 01	6.93 01	7.34 01	7.65 01
89	Ac L_{β_1}	15.713	78.906	5.63 01	5.94 01	6.38 01	6.76 01	7.04 01
40	Zr K_{α_1}	15.775	78.596	5.57 01	5.87 01	6.31 01	6.69 01	6.97 01
38	Sr K_{β_1}	15.836	78.293	5.51 01	5.81 01	6.25 01	6.62 01	6.89 01
90	Th L_{β_1}	16.202	76.524	5.19 01	5.47 01	5.88 01	6.22 01	6.48 01
41	Nb K_{α_1}	16.615	74.622	4.86 01	5.11 01	5.49 01	5.82 01	6.06 01
91	Pa L_{β_1}	16.702	74.233	4.79 01	5.04 01	5.42 01	5.74 01	5.98 01
39	Y K_{β_1}	16.738	74.074	4.76 01	5.01 01	5.39 01	5.71 01	5.94 01
92	U L_{β_1}	17.220	72.000	4.42 01	4.65 01	4.99 01	5.29 01	5.51 01
42	Mo K_{α_1}	17.479	70.933	4.25 01	4.46 01	4.80 01	5.08 01	5.29 01
40	Zr K_{β_1}	17.668	70.175	4.13 01	4.34 01	4.66 01	4.94 01	5.14 01
43	Tc K_{α_1}	18.367	67.504	3.73 01	3.91 01	4.20 01	4.45 01	4.63 01
41	Nb K_{β_1}	18.623	66.576	3.59 01	3.77 01	4.05 01	4.29 01	4.46 01
44	Ru K_{α_1}	19.279	64.311	3.28 01	3.43 01	3.69 01	3.91 01	4.07 01
42	Mo K_{β_1}	19.608	63.231	3.14 01	3.28 01	3.53 01	3.74 01	3.89 01
45	Rh K_{α_1}	20.216	61.330	2.89 01	3.03 01	3.25 01	3.44 01	3.58 01
43	Tc K_{β_1}	20.619	60.131	2.75 01	2.87 01	3.09 01	3.27 01	3.40 01
46	Pd K_{α_1}	21.177	58.547	2.56 01	2.68 01	2.88 01	3.05 01	3.17 01
44	Ru K_{β_1}	21.657	57.249	2.42 01	2.52 01	2.72 01	2.87 01	2.99 01
47	Ag K_{α_1}	22.163	55.942	2.27 01	2.37 01	2.56 01	2.70 01	2.81 01

Z	transition	E [keV]	λ [pm]	$_{56}$Ba	$_{57}$La	$_{58}$Ce	$_{59}$Pr	$_{60}$Nd
45	Rh K$_{\beta_1}$	22.724	54.561	2.13 01	2.22 01	2.39 01	2.53 01	2.63 01
48	Cd K$_{\alpha_1}$	23.174	53.501	2.02 01	2.11 01	2.27 01	2.40 01	2.50 01
46	Pd K$_{\beta_1}$	23.819	52.053	1.88 01	1.96 01	2.11 01	2.24 01	2.32 01
49	In K$_{\alpha_1}$	24.210	51.212	1.80 01	1.88 01	2.03 01	2.14 01	2.22 01
47	Ag K$_{\beta_1}$	24.942	49.709	1.67 01	1.74 01	1.87 01	1.98 01	2.06 01
50	Sn K$_{\alpha_1}$	25.271	49.062	1.61 01	1.68 01	1.81 01	1.91 01	1.99 01
48	Cd K$_{\beta_1}$	26.096	47.511	1.48 01	1.54 01	1.66 01	1.76 01	1.82 01
51	Sb K$_{\alpha_1}$	26.359	47.037	1.44 01	1.50 01	1.62 01	1.71 01	1.78 01
49	In K$_{\beta_1}$	27.276	45.455	1.32 01	1.37 01	1.48 01	1.57 01	1.62 01
52	Te K$_{\alpha_1}$	27.472	45.131	1.29 01	1.35 01	1.45 01	1.54 01	1.59 01
50	Sn K$_{\beta_1}$	28.486	43.524	1.18 01	1.22 01	1.32 01	1.40 01	1.45 01
53	I K$_{\alpha_1}$	28.612	43.333	1.16 01	1.21 01	1.30 01	1.38 01	1.43 01
51	Sb K$_{\beta_1}$	29.726	41.709	1.05 01	1.09 01	1.18 01	1.25 01	1.29 01
54	Xe K$_{\alpha_1}$	29.779	41.635	1.05 01	1.09 01	1.17 01	1.24 01	1.29 01
55	Cs K$_{\alpha_1}$	30.973	40.030	9.47 00	9.81 00	1.06 01	1.12 01	1.16 01
52	Te K$_{\beta_1}$	30.996	40.000	9.45 00	9.79 00	1.06 01	1.12 01	1.16 01
56	Ba K$_{\alpha_1}$	32.194	38.511	8.59 00	8.86 00	9.58 00	1.01 01	1.05 01
53	I K$_{\beta_1}$	32.295	38.391	8.52 00	8.79 00	9.50 00	1.01 01	1.04 01
57	La K$_{\alpha_1}$	33.442	37.074	7.78 00	8.02 00	8.67 00	9.18 00	9.49 00
54	Xe K$_{\beta_1}$	33.624	36.874	7.67 00	7.91 00	8.55 00	9.05 00	9.35 00
58	Ce K$_{\alpha_1}$	34.279	36.169	7.29 00	7.52 00	8.13 00	8.61 00	8.89 00
55	Cs K$_{\beta_1}$	34.987	35.437	6.91 00	7.14 00	7.71 00	8.16 00	8.43 00
59	Pr K$_{\alpha_1}$	36.026	34.415	6.40 00	6.62 00	7.15 00	7.56 00	7.81 00
56	Ba K$_{\beta_1}$	36.378	34.082	6.24 00	6.45 00	6.97 00	7.38 00	7.61 00
60	Nd K$_{\alpha_1}$	36.847	33.648	<u>6.04 00</u>	6.24 00	6.75 00	7.14 00	7.36 00
57	La K$_{\beta_1}$	37.801	32.799	3.11 01	5.84 00	6.32 00	6.68 00	6.89 00
61	Pm K$_{\alpha_1}$	38.725	32.016	2.91 01	<u>5.49 00</u>	5.93 00	6.27 00	6.48 00
58	Ce K$_{\beta_1}$	39.257	31.582	2.80 01	2.91 01	5.73 00	6.06 00	6.26 00
62	Sm K$_{\alpha_1}$	40.118	30.905	2.64 01	2.74 01	<u>5.42 00</u>	5.73 00	5.93 00
59	Pr K$_{\beta_1}$	40.748	30.427	2.53 01	2.63 01	2.88 01	5.50 00	5.69 00
63	Eu K$_{\alpha_1}$	41.542	29.845	2.40 01	2.50 01	2.73 01	<u>5.23 00</u>	5.42 00
60	Nd K$_{\beta_1}$	42.271	29.331	2.29 01	2.38 01	2.60 01	2.75 01	5.19 00
64	Gd K$_{\alpha_1}$	42.996	28.836	2.18 01	2.27 01	2.48 01	2.63 01	<u>4.97 00</u>
61	Pm K$_{\beta_1}$	43.826	28.290	2.07 01	2.16 01	2.35 01	2.49 01	2.57 01
65	Tb K$_{\alpha_1}$	44.482	27.873	1.99 01	2.07 01	2.26 01	2.39 01	2.46 01
62	Sm K$_{\beta_1}$	45.413	27.301	1.88 01	1.96 01	2.13 01	2.26 01	2.33 01
66	Dy K$_{\alpha_1}$	45.998	26.954	1.81 01	1.89 01	2.06 01	2.18 01	2.25 01
63	Eu K$_{\beta_1}$	47.038	26.358	1.70 01	1.78 01	1.93 01	2.05 01	2.11 01
67	Ho K$_{\alpha_1}$	47.547	26.076	1.65 01	1.73 01	1.87 01	1.99 01	2.05 01
64	Gd K$_{\beta_1}$	48.697	25.460	1.55 01	1.62 01	1.75 01	1.86 01	1.92 01
68	Er K$_{\alpha_1}$	49.128	25.237	1.51 01	1.58 01	1.71 01	1.81 01	1.87 01
65	Tb K$_{\beta_1}$	50.382	24.609	1.41 01	1.48 01	1.59 01	1.69 01	1.75 01
69	Tm K$_{\alpha_1}$	50.742	24.434	1.38 01	1.45 01	1.56 01	1.66 01	1.71 01
66	Dy K$_{\beta_1}$	52.119	23.788	1.28 01	1.35 01	1.45 01	1.54 01	1.59 01

Z	transition	E [keV]	λ [pm]	$_{56}$Ba	$_{57}$La	$_{58}$Ce	$_{59}$Pr	$_{60}$Nd
70	Yb K$_{\alpha_1}$	52.389	23.666	1.26 01	1.33 01	1.43 01	1.52 01	1.57 01
67	Ho K$_{\beta_1}$	53.877	23.012	1.17 01	1.23 01	1.32 01	1.40 01	1.45 01
71	Lu K$_{\alpha_1}$	54.070	22.930	1.16 01	1.22 01	1.31 01	1.39 01	1.44 01
68	Er K$_{\beta_1}$	55.681	22.167	1.07 01	1.13 01	1.21 01	1.28 01	1.32 01
72	Hf K$_{\alpha_1}$	55.790	22.223	1.06 01	1.12 01	1.20 01	1.27 01	1.32 01
69	Tm K$_{\beta_1}$	57.517	21.556	9.76 00	1.03 01	1.10 01	1.17 01	1.21 01
73	Ta K$_{\alpha_1}$	57.532	21.550	9.75 00	1.03 01	1.10 01	1.17 01	1.21 01
74	W K$_{\alpha_1}$	59.318	20.901	8.96 00	9.48 00	1.01 01	1.07 01	1.11 01
70	Yb K$_{\beta_1}$	59.370	20.883	8.94 00	9.46 00	1.01 01	1.07 01	1.11 01
75	Re K$_{\alpha_1}$	61.140	20.278	8.25 00	8.73 00	9.30 00	9.86 00	1.02 01
71	Lu K$_{\beta_1}$	61.283	20.231	8.19 00	8.68 00	9.24 00	9.80 00	1.02 01
76	Os K$_{\alpha_1}$	63.001	19.679	7.59 00	8.05 00	8.56 00	9.08 00	9.42 00
72	Hf K$_{\beta_1}$	63.234	19.607	7.52 00	7.97 00	8.48 00	8.99 00	9.34 00
77	Ir K$_{\alpha_1}$	64.896	19.105	7.00 00	7.42 00	7.89 00	8.37 00	8.68 00
73	Ta K$_{\beta_1}$	65.223	19.009	6.90 00	7.32 00	7.78 00	8.25 00	8.56 00
78	Pt K$_{\alpha_1}$	66.832	18.551	6.46 00	6.85 00	7.28 00	7.72 00	8.01 00
74	W K$_{\beta_1}$	67.244	18.438	6.35 00	6.74 00	7.15 00	7.59 00	7.87 00
79	Au K$_{\alpha_1}$	68.804	18.020	5.96 00	6.33 00	6.71 00	7.12 00	7.39 00
75	Re K$_{\beta_1}$	69.310	17.888	5.84 00	6.20 00	6.58 00	6.98 00	7.25 00
80	Hg K$_{\alpha_1}$	70.819	17.507	5.51 00	5.85 00	6.20 00	6.58 00	6.83 00
76	Os K$_{\beta_1}$	71.413	17.361	5.38 00	5.72 00	6.06 00	6.43 00	6.67 00
81	Tl K$_{\alpha_1}$	72.872	17.014	5.09 00	5.41 00	5.73 00	6.08 00	6.31 00
77	Ir K$_{\beta_1}$	73.561	16.854	4.96 00	5.28 00	5.58 00	5.92 00	6.15 00
82	Pb K$_{\alpha_1}$	74.969	16.538	4.71 00	5.01 00	5.30 00	5.62 00	5.84 00
78	Pt K$_{\beta_1}$	75.748	16.368	4.58 00	4.87 00	5.15 00	5.46 00	5.68 00
83	Bi K$_{\alpha_1}$	77.108	16.079	4.36 00	4.64 00	4.90 00	5.20 00	5.41 00
79	Au K$_{\beta_1}$	77.948	15.906	4.23 00	4.51 00	4.76 00	5.05 00	5.25 00
84	Po K$_{\alpha_1}$	79.290	15.636	4.04 00	4.30 00	4.54 00	4.82 00	5.01 00
80	Hg K$_{\beta_1}$	80.253	15.449	3.91 00	4.16 00	4.39 00	4.66 00	4.84 00
85	At K$_{\alpha_1}$	81.520	15.209	3.75 00	3.99 00	4.21 00	4.47 00	4.64 00
81	Tl K$_{\beta_1}$	82.576	15.014	3.62 00	3.85 00	4.07 00	4.31 00	4.48 00
86	Rn K$_{\alpha_1}$	83.780	14.798	3.48 00	3.71 00	3.92 00	4.14 00	4.31 00
82	Pb K$_{\beta_1}$	84.936	14.597	3.35 00	3.57 00	3.78 00	3.99 00	4.15 00
87	Fr K$_{\alpha_1}$	86.100	14.400	3.23 00	3.44 00	3.65 00	3.85 00	4.00 00
83	Bi K$_{\beta_1}$	87.343	14.195	3.11 00	3.31 00	3.51 00	3.70 00	3.85 00
88	Ra K$_{\alpha_1}$	88.470	14.014	3.00 00	3.20 00	3.40 00	3.57 00	3.71 00
84	Po K$_{\beta_1}$	89.800	13.806	2.89 00	3.08 00	3.27 00	3.43 00	3.57 00
89	Ac K$_{\alpha_1}$	90.884	13.642	2.79 00	2.98 00	3.16 00	3.32 00	3.45 00
85	At K$_{\beta_1}$	92.300	13.432	2.68 00	2.86 00	3.04 00	3.18 00	3.31 00
90	Th K$_{\alpha_1}$	93.350	13.281	2.60 00	2.77 00	2.95 00	3.08 00	3.21 00
86	Rn K$_{\beta_1}$	94.870	13.068	2.49 00	2.65 00	2.83 00	2.95 00	3.07 00
91	Pa K$_{\alpha_1}$	95.868	12.932	2.42 00	2.58 00	2.75 00	2.87 00	2.98 00
87	Fr K$_{\beta_1}$	97.470	12.720	2.31 00	2.47 00	2.63 00	2.74 00	2.85 00
92	U K$_{\alpha_1}$	98.439	12.595	2.25 00	2.40 00	2.56 00	2.67 00	2.78 00

Z	transition	E [keV]	λ [pm]	$_{56}$Ba	$_{57}$La	$_{58}$Ce	$_{59}$Pr	$_{60}$Nd
88	Ra K$_{\beta_1}$	100.130	12.382	2.15 00	2.29 00	2.45 00	2.55 00	2.65 00
89	Ac K$_{\alpha_1}$	102.850	12.054	2.01 00	2.14 00	2.29 00	2.38 00	2.47 00
90	Th K$_{\beta_1}$	105.610	11.739	1.88 00	2.00 00	2.13 00	2.22 00	2.31 00
91	Pa K$_{\beta_1}$	108.430	11.434	1.75 00	1.87 00	1.99 00	2.07 00	2.16 00
92	U K$_{\beta_1}$	111.300	11.139	1.64 00	1.75 00	1.86 00	1.94 00	2.02 00

Z	transition	E [keV]	λ [pm]	$_{61}$Pm	$_{62}$Sm	$_{63}$Eu	$_{64}$Gd	$_{65}$Tb
4	Be K$_\alpha$	0.1085	11427.207	1.23 04	1.81 04	1.50 04	1.94 04	2.05 04
38	Sr M$_\xi$	0.1140	10875.895	<u>1.59 04</u>	<u>1.70 04</u>	1.52 04	1.80 04	2.00 04
39	Y M$_\xi$	0.1328	9339.235	5.79 04	4.79 04	<u>2.76 04</u>	<u>1.34 04</u>	<u>1.44 04</u>
16	S L$_l$	0.1487	8337.942	6.70 04	9.84 04	6.06 04	2.29 05	1.47 04
40	Zr M$_\xi$	0.1511	8205.506	6.17 04	8.81 04	5.49 04	8.54 04	8.11 04
41	Nb M$_\xi$	0.1717	7221.037	3.10 04	3.68 04	3.74 04	2.91 04	3.34 04
5	B K$_\alpha$	0.1833	6764.059	2.35 04	2.74 04	3.13 04	2.30 04	3.11 04
42	Mo M$_\xi$	0.1926	6473.445	<u>1.91 04</u>	<u>2.33 04</u>	<u>2.59 04</u>	<u>2.37 04</u>	2.69 04
6	C K$_\alpha$	0.2770	4476.000	8.78 03	1.11 04	<u>1.24 04</u>	<u>1.20 04</u>	<u>1.60 04</u>
47	Ag M$_\xi$	0.3117	3977.709	8.73 03	<u>1.05 04</u>	<u>1.11 04</u>	<u>1.11 04</u>	<u>1.40 04</u>
7	N K$_\alpha$	0.3924	3159.664	8.49 03	9.41 03	9.58 03	9.55 03	<u>1.02 04</u>
22	Ti L$_l$	0.3953	3136.484	8.44 03	9.35 03	9.51 03	9.47 03	1.01 04
22	Ti L$_\alpha$	0.4522	2741.822	7.22 03	7.67 03	8.06 03	8.04 03	8.23 03
23	V L$_\alpha$	0.5113	2424.901	6.17 03	6.53 03	6.86 03	6.87 03	7.01 03
8	O K$_\alpha$	0.5249	2362.072	5.96 03	6.31 03	6.63 03	6.64 03	6.82 03
25	Mn L$_l$	0.5563	2228.747	5.50 03	5.80 03	6.10 03	6.13 03	6.35 03
24	Cr L$_\alpha$	0.5728	2164.549	5.27 03	5.55 03	5.84 03	5.88 03	6.13 03
25	Mn L$_\alpha$	0.6374	1945.171	4.50 03	4.73 03	4.97 03	5.02 03	5.38 03
9	F K$_\alpha$	0.6768	1831.932	4.08 03	4.28 03	4.50 03	4.55 03	5.00 03
26	Fe L$_\alpha$	0.7050	1758.655	3.85 03	4.04 03	4.24 03	4.29 03	4.70 03
27	Co L$_\alpha$	0.7762	1597.335	3.28 03	3.45 03	3.62 03	3.66 03	4.00 03
28	Ni L$_\alpha$	0.8815	1456.080	2.80 03	2.93 03	3.08 03	3.12 03	3.39 03
29	Cu L$_\alpha$	0.9297	1336.044	2.39 03	2.51 03	2.63 03	2.67 03	2.90 03
30	Zn L$_\alpha$	1.0117	1225.513	<u>2.05 03</u>	2.15 03	2.25 03	2.25 03	2.29 03
11	Na K$_\alpha$	1.0410	1191.020	<u>7.92 03</u>	2.03 03	2.13 03	2.13 03	2.17 03
11	Na K$_\beta$	1.0711	1157.550	9.12 04	<u>1.75 03</u>	<u>1.92 03</u>	<u>1.92 03</u>	<u>1.96 03</u>
12	Mg K$_\alpha$	1.2536	989.033	7.83 03	8.16 03	8.69 03	8.98 03	<u>6.32 03</u>
33	As L$_\alpha$	1.2820	967.123	7.39 03	7.71 03	8.21 03	8.48 03	8.93 03
12	Mg K$_\beta$	1.3022	952.121	7.10 03	7.40 03	7.88 03	8.15 03	8.58 03
33	As L$_{\beta_1}$	1.3170	941.421	6.90 03	7.19 03	7.66 03	7.91 03	8.34 03
66	Dy M$_\beta$	1.3250	935.737	6.79 03	7.08 03	7.54 03	7.79 03	8.21 03
67	Ho M$_\alpha$	1.3480	919.771	<u>6.50 03</u>	6.78 03	7.21 03	7.46 03	7.86 03
34	Se L$_\alpha$	1.3791	899.029	7.35 03	6.39 03	6.80 03	7.03 03	7.42 03
67	Ho M$_\beta$	1.3830	896.494	7.30 03	6.34 03	6.75 03	6.98 03	7.36 03
68	Er M$_\alpha$	1.4060	881.829	6.99 03	6.08 03	6.48 03	6.69 03	7.06 03
34	Se L$_{\beta_1}$	1.4192	873.627	6.83 03	<u>5.94 03</u>	6.32 03	6.53 03	6.90 03

Z	transition	E [keV]	λ [pm]	$_{61}$Pm	$_{62}$Sm	$_{63}$Eu	$_{64}$Gd	$_{65}$Tb
68	Er M_β	1.4430	859.218	6.54 03	6.85 03	6.06 03	6.26 03	6.61 03
69	Tm M_α	1.4620	848.051	<u>6.33 03</u>	6.62 03	5.86 03	6.05 03	6.39 03
35	Br L_α	1.4804	837.511	6.73 03	6.42 03	<u>5.67 03</u>	5.86 03	6.19 03
13	Al K_{α_1}	1.4867	833.962	6.66 03	6.35 03	6.74 03	5.80 03	6.13 03
69	Tm M_β	1.5030	824.918	6.48 03	6.18 03	6.55 03	5.64 03	5.96 03
70	Yb M_α	1.5214	814.941	6.29 03	5.97 03	6.35 03	5.46 03	5.77 03
35	Br L_{β_1}	1.5259	812.538	6.24 03	<u>5.93 03</u>	6.30 03	<u>5.42 03</u>	5.73 03
13	Al K_β	1.5574	796.103	5.93 03	6.20 03	5.98 03	6.18 03	5.43 03
70	Yb M_β	1.5675	790.974	5.83 03	6.09 03	5.89 03	6.08 03	5.33 03
71	Lu M_α	1.5813	784.071	5.70 03	5.96 03	5.76 03	5.94 03	5.21 03
36	Kr L_α	1.5860	781.747	5.66 03	5.91 03	<u>5.72 03</u>	5.90 03	<u>5.17 03</u>
71	Lu M_β	1.6312	760.085	5.27 03	5.50 03	5.82 03	5.48 03	5.77 03
36	Kr L_{β_1}	1.6366	757.577	5.23 03	5.45 03	5.77 03	5.44 03	5.72 03
72	Hf M_{α_1}	1.6446	753.892	<u>5.16 03</u>	5.38 03	5.70 03	<u>5.37 03</u>	5.65 03
37	Rb L_{α_1}	1.6941	731.864	5.25 03	4.98 03	5.28 03	5.46 03	5.24 03
72	Hf M_β	1.6976	730.355	5.22 03	4.95 03	5.25 03	5.44 03	5.21 03
73	Ta M_{α_1}	1.7096	725.229	5.13 03	<u>4.86 03</u>	5.16 03	5.34 03	5.12 03
14	Si K_{α_1}	1.7400	712.558	4.90 03	5.12 03	4.93 03	5.10 03	4.89 03
37	Rb L_{β_1}	1.7522	707.597	4.81 03	5.03 03	4.84 03	5.01 03	4.81 03
73	Ta M_β	1.7655	702.266	4.72 03	4.93 03	4.75 03	4.91 03	<u>4.72 03</u>
74	W M_{α_1}	1.7754	698.350	4.65 03	4.86 03	<u>4.68 03</u>	4.84 03	5.10 03
38	Sr L_{α_1}	1.8069	686.290	4.45 03	4.65 03	4.95 03	4.63 03	4.88 03
74	W M_β	1.8349	675.705	4.28 03	4.47 03	4.75 03	4.45 04	4.70 03
14	Si K_β	1.8359	675.337	4.27 03	4.46 03	4.75 03	4.44 03	4.69 03
75	Re M_{α_1}	1.8420	673.100	4.23 03	4.43 03	4.71 03	4.41 03	4.65 03
38	Sr L_{β_1}	1.8717	662.420	4.06 03	4.25 03	4.52 03	<u>4.23 03</u>	4.47 03
75	Re M_β	1.9061	650.465	3.88 03	4.06 03	4.31 03	4.45 03	4.27 03
76	Os M_{α_1}	1.9102	649.069	3.86 03	4.04 03	4.29 03	4.42 03	4.25 03
39	Y L_{α_1}	1.9226	644.882	3.79 03	3.97 03	4.22 03	4.35 03	<u>4.18 03</u>
76	Os M_β	1.9783	626.725	3.53 03	3.69 03	3.92 03	4.05 03	4.26 03
77	Ir M_{α_1}	1.9799	626.219	3.52 03	3.68 03	3.91 03	4.05 03	4.25 03
39	Y L_{β_1}	1.9958	621.230	3.45 03	3.61 03	3.83 03	3.97 03	4.16 03
15	P K_{α_1}	2.0137	615.708	3.37 03	3.53 03	3.74 03	3.88 03	4.06 03
40	Zr L_{α_1}	2.0424	607.056	3.25 03	3.40 03	3.61 03	3.74 03	3.92 03
78	Pt M_{α_1}	2.0505	604.658	3.22 03	3.37 03	3.57 03	3.70 03	3.88 03
77	Ir M_β	2.0535	603.775	3.21 03	3.36 03	3.56 03	3.69 03	3.86 03
79	Au M_{α_1}	2.1229	584.036	2.94 03	3.08 03	3.27 03	3.38 03	3.55 03
40	Zr L_{β_1}	2.1244	583.624	2.94 03	3.08 03	3.26 03	3.38 03	3.54 03
78	Pt M_β	2.1273	582.828	2.93 03	3.07 03	3.25 03	3.37 03	3.53 03
15	P $K_{\beta_{1,3}}$	2.1390	579.640	2.89 03	3.02 03	3.21 03	3.32 03	3.48 03
41	Nb L_{α_1}	2.1659	572.441	2.80 03	2.93 03	3.10 03	3.21 03	3.37 03
80	Hg M_{α_1}	2.1953	564.775	2.70 03	2.83 03	3.00 03	3.10 03	3.26 03
79	Au M_β	2.2046	562.393	2.67 03	2.80 03	2.97 03	3.07 03	3.22 03
41	Nb L_{β_1}	2.2574	549.238	2.52 03	2.64 03	2.79 03	2.89 03	3.03 03

Z	transition	E [keV]	λ [pm]	$_{61}$Pm	$_{62}$Sm	$_{63}$Eu	$_{64}$Gd	$_{65}$Tb
81	Tl M$_{\alpha_1}$	2.2706	546.045	2.48 03	2.60 03	2.75 03	2.85 03	2.99 03
80	Hg M$_\beta$	2.2825	543.199	2.45 03	2.56 03	2.71 03	2.81 03	2.95 03
42	Mo L$_{\alpha_1}$	2.2932	540.664	2.42 03	2.53 03	2.68 03	2.77 03	2.91 03
16	S K$_{\alpha_1}$	2.3080	537.197	2.38 03	2.49 03	2.64 03	2.73 03	2.87 03
82	Pb M$_{\alpha_1}$	2.3457	528.563	2.28 03	2.39 03	2.53 03	2.62 03	2.75 03
81	Tl M$_\beta$	2.3621	524.893	2.24 03	2.35 03	2.49 03	2.57 03	2.70 03
42	Mo L$_{\beta_1}$	2.3948	517.726	2.16 03	2.27 03	2.40 03	2.48 03	2.61 03
83	Bi M$_{\alpha_1}$	2.4226	511.785	2.10 03	2.20 03	2.33 03	2.41 03	2.53 03
43	Tc L$_{\alpha_1}$	2.4240	511.490	2.10 03	2.20 03	2.33 03	2.41 03	2.53 03
82	Pb M$_\beta$	2.4427	507.574	2.06 03	2.16 03	2.28 03	2.36 03	2.48 03
16	S K$_\beta$	2.4640	503.186	2.01 03	2.11 03	2.23 03	2.31 03	2.42 03
83	Bi M$_{\beta_1}$	2.5255	490.933	1.89 03	1.98 03	2.09 03	2.16 03	2.28 03
43	Tc L$_{\beta_1}$	2.5368	488.746	1.87 03	1.96 03	2.07 03	2.14 03	2.25 03
44	Ru L$_{\alpha_1}$	2.5586	484.582	1.83 03	1.92 03	2.03 03	2.09 03	2.20 03
17	Cl K$_\beta$	2.6224	472.792	1.72 03	1.80 03	1.90 03	1.97 03	2.07 03
44	Ru L$_{\beta_1}$	2.6332	462.079	1.62 03	1.70 03	1.79 03	1.85 03	1.95 03
45	Rh L$_{\alpha_1}$	2.6967	459.766	1.60 03	1.68 03	1.77 03	1.83 03	1.93 03
17	Cl K$_\beta$	2.8156	440.350	1.43 03	1.50 03	1.59 03	1.64 03	1.72 03
45	Rh L$_{\beta_1}$	2.8344	437.430	1.41 03	1.48 03	1.56 03	1.61 03	1.70 03
46	Pd L$_{\alpha_1}$	2.8386	436.782	1.40 03	1.47 03	1.55 03	1.60 03	1.69 03
18	Ar K$_{\alpha_1}$	2.9577	419.194	1.26 03	1.32 03	1.40 03	1.44 03	1.52 03
47	Ag L$_{\alpha_1}$	2.9843	415.458	1.23 03	1.30 03	1.37 03	1.41 03	1.49 03
46	Pd L$_{\beta_1}$	2.9902	414.638	1.23 03	1.29 03	1.36 03	1.40 03	1.48 03
90	Th M$_{\alpha_1}$	2.9961	413.821	1.22 03	1.28 03	1.35 03	1.39 03	1.47 03
91	Pa M$_{\alpha_1}$	3.0823	402.248	1.14 03	1.19 03	1.26 03	1.30 03	1.37 03
48	Cd L$_{\alpha_1}$	3.1337	395.651	1.09 03	1.14 03	1.21 03	1.24 03	1.31 03
90	Th M$_\beta$	3.1458	394.129	1.08 03	1.13 03	1.19 03	1.23 03	1.30 03
47	Ag L$_{\beta_1}$	3.1509	393.491	1.07 03	1.13 03	1.19 03	1.23 03	1.29 03
92	U M$_{\alpha_1}$	3.1708	391.021	1.06 03	1.11 03	1.17 03	1.21 03	1.27 03
18	Ar K$_\beta$	3.1905	388.607	1.04 03	1.09 03	1.15 03	1.19 03	1.25 03
91	Pa M$_\beta$	3.2397	382.705	1.00 03	1.05 03	1.11 03	1.14 03	1.21 03
49	In L$_{\alpha_1}$	3.2869	377.210	9.65 02	1.01 03	1.07 03	1.10 03	1.16 03
19	K K$_{\alpha_1}$	3.3138	374.148	9.45 02	9.91 02	1.04 03	1.08 03	1.14 03
48	Cd L$_{\beta_1}$	3.3166	373.832	9.43 02	9.89 02	1.04 03	1.08 03	1.14 03
92	U M$_\beta$	3.3360	371.658	9.29 02	9.74 02	1.03 03	1.06 03	1.12 03
50	Sn L$_{\alpha_1}$	3.4440	360.003	8.57 02	8.98 02	9.47 02	9.77 02	1.03 03
49	In L$_{\beta_1}$	3.4872	355.543	8.30 02	8.70 02	9.17 02	9.46 02	1.00 03
19	K K$_\beta$	3.5896	345.401	7.71 02	8.08 02	8.52 02	8.79 02	9.29 02
51	Sb L$_{\alpha_1}$	3.6047	343.954	7.63 02	8.00 02	8.43 02	8.69 02	9.19 02
50	Sn L$_{\beta_1}$	3.6628	338.498	7.33 02	7.68 02	8.09 02	8.35 02	8.82 02
20	Ca K$_{\alpha_1}$	3.6917	335.848	7.18 02	7.52 02	7.93 02	8.18 02	8.65 02
52	Te L$_{\alpha_1}$	3.7693	328.934	6.81 02	7.13 02	7.52 02	7.76 02	8.20 02
51	Sb L$_{\beta_1}$	3.8436	322.575	6.48 02	6.79 02	7.15 02	7.38 02	7.80 02
53	I L$_{\alpha_1}$	3.9377	314.867	6.10 02	6.38 02	6.72 02	6.94 02	7.34 02

Z	transition	E [keV]	λ [pm]	$_{61}$Pm	$_{62}$Sm	$_{63}$Eu	$_{64}$Gd	$_{65}$Tb
20	Ca K$_\beta$	4.0127	308.981	5.81 02	6.08 02	6.41 02	6.61 02	6.99 02
52	Te L$_{\beta_1}$	4.0296	307.686	5.75 02	6.02 02	6.34 02	6.54 02	6.92 02
21	Sc K$_{\alpha_1}$	4.0906	303.097	5.54 02	5.79 02	6.10 02	6.29 02	6.66 02
54	Xe L$_{\alpha_1}$	4.1099	301.674	5.47 02	5.72 02	6.03 02	6.22 02	6.58 02
53	I L$_{\beta_1}$	4.2207	293.755	5.11 02	5.35 02	5.64 02	5.81 02	6.14 02
55	Cs L$_{\alpha_1}$	4.2865	289.245	4.92 02	5.15 02	5.42 02	5.59 02	5.90 02
21	Sc K$_\beta$	4.4605	277.962	4.45 02	4.66 02	4.90 02	5.05 02	5.33 02
56	Ba L$_{\alpha_1}$	4.4663	277.601	4.43 02	4.64 02	4.89 02	5.03 02	5.31 02
22	Ti K$_{\alpha_1}$	4.5108	274.863	4.32 02	4.53 02	4.77 02	4.91 02	5.18 02
55	Cs L$_{\beta_1}$	4.6198	268.377	4.07 02	4.26 02	4.49 02	4.62 02	4.87 02
57	La L$_{\alpha_1}$	4.6510	266.577	4.00 02	4.19 02	4.41 02	4.54 02	4.79 02
56	Ba L$_{\beta_1}$	4.8273	256.841	3.64 02	3.82 02	4.02 02	4.13 02	4.35 02
58	Ce L$_{\alpha_1}$	4.8402	256.157	3.62 02	3.79 02	3.99 02	4.10 02	4.32 02
22	Ti K$_{\beta_{1,3}}$	4.9318	251.399	3.45 02	3.62 02	3.80 02	3.91 02	4.12 02
23	V K$_{\alpha_1}$	4.9522	250.363	3.41 02	3.58 02	3.76 02	3.87 02	4.08 02
59	Pr L$_{\alpha_1}$	5.0337	246.310	3.27 02	3.43 02	3.61 02	3.71 02	3.91 02
57	La L$_{\beta_1}$	5.0421	245.899	3.26 02	3.42 02	3.60 02	3.70 02	3.90 02
60	Nd L$_{\alpha_1}$	5.2304	237.047	2.97 02	3.12 02	3.28 02	3.37 02	3.55 02
58	Ce L$_{\beta_1}$	5.2622	235.614	2.93 02	3.07 02	3.23 02	3.32 02	3.50 02
24	Cr K$_{\alpha_1}$	5.4147	228.978	2.72 02	2.86 02	3.00 02	3.09 02	3.25 02
23	V K$_{\beta_{1,3}}$	5.4273	228.447	2.71 02	2.84 02	2.98 02	3.07 02	3.23 02
61	Pm L$_{\alpha_1}$	5.4325	228.228	2.70 02	2.83 02	2.98 02	3.06 02	3.23 02
59	Pr L$_{\beta_1}$	5.4889	225.883	2.63 02	2.76 02	2.90 02	2.98 02	3.14 02
62	Sm L$_{\alpha_1}$	5.6361	219.984	2.46 02	2.58 02	2.71 02	2.79 02	2.94 02
60	Nd L$_{\beta_1}$	5.7216	216.696	2.37 02	2.49 02	2.61 02	2.68 02	2.83 02
63	Eu L$_{\alpha_1}$	5.8457	212.096	2.24 02	2.36 02	2.47 02	2.54 02	2.68 02
25	Mn K$_{\alpha_1}$	5.8988	210.187	2.19 02	2.30 02	2.42 02	2.48 02	2.62 02
24	Cr K$_{\beta_{1,3}}$	5.9467	208.494	2.15 02	2.26 02	2.37 02	2.43 02	2.57 02
61	Pm L$_{\beta_1}$	5.9610	207.993	2.14 02	2.24 02	2.35 02	2.42 02	2.55 02
64	Gd L$_{\alpha_1}$	6.0572	204.690	2.05 02	2.16 02	2.26 02	2.32 02	2.45 02
62	Sm L$_{\beta_1}$	6.2051	199.811	1.93 02	2.03 02	2.13 02	2.18 02	2.31 02
65	Tb L$_{\alpha_1}$	6.2728	197.655	1.88 02	1.97 02	2.07 02	2.13 02	2.24 02
26	Fe K$_{\alpha_1}$	6.4038	193.611	1.78 02	1.87 02	1.97 02	2.02 02	2.13 02
63	Eu L$_{\beta_1}$	6.4564	192.034	<u>1.75 02</u>	1.84 02	1.93 02	1.98 02	2.09 02
25	Mn K$_{\beta_{1,3}}$	6.4905	191.025	4.81 02	1.81 02	1.90 02	1.95 02	2.06 02
66	Dy L$_{\alpha_1}$	6.4952	190.887	4.80 02	1.81 02	1.90 02	1.95 02	2.06 02
64	Gd L$_{\beta_1}$	6.7132	184.688	4.40 02	<u>1.66 02</u>	1.75 02	1.79 02	1.89 02
67	Ho L$_{\alpha_1}$	6.7198	184.507	4.39 02	4.60 02	1.74 02	1.79 02	1.89 02
27	Co K$_{\alpha_1}$	6.9303	178.903	4.04 02	4.23 02	1.61 02	1.65 02	1.75 02
68	Er L$_{\alpha_1}$	6.9487	178.429	4.01 02	4.20 02	<u>1.60 02</u>	1.64 02	1.73 02
65	Tb L$_{\beta_1}$	6.9780	177.680	3.97 02	4.16 02	4.36 02	1.62 02	1.72 02
26	Fe K$_{\beta_{1,3}}$	7.0580	175.666	<u>5.36 02</u>	4.03 02	4.23 02	1.58 02	1.67 02
69	Tm L$_{\alpha_1}$	7.1799	172.683	5.12 02	3.85 02	4.04 02	<u>1.51 02</u>	1.60 02
66	Dy L$_{\beta_1}$	7.2477	171.068	4.99 02	<u>3.75 02</u>	3.94 02	4.05 02	1.56 02

Z	transition	E [keV]	λ [pm]	$_{61}$Pm	$_{62}$Sm	$_{63}$Eu	$_{64}$Gd	$_{65}$Tb
70	Yb L$_{\alpha_1}$	7.4156	167.195	<u>4.69 02</u>	4.93 02	3.71 02	3.81 02	1.47 02
28	Ni K$_{\alpha_1}$	7.4782	165.795	5.51 02	4.81 02	3.62 02	3.73 02	<u>1.44 02</u>
67	Ho L$_{\beta_1}$	7.5253	164.757	5.42 02	4.73 02	<u>3.56 02</u>	3.67 02	3.89 02
27	Co K$_{\beta_{1,3}}$	7.6494	162.084	5.18 02	4.51 02	4.74 02	3.51 02	3.72 02
71	Lu L$_{\alpha_1}$	7.6555	161.955	5.17 02	<u>4.50 02</u>	4.73 02	3.50 02	3.71 02
68	Er L$_{\beta_1}$	7.8109	158.733	4.90 02	5.12 02	4.48 02	3.32 02	3.51 02
72	Hf L$_{\alpha_1}$	7.8990	156.963	4.75 02	4.97 02	4.35 02	<u>3.22 02</u>	3.40 02
29	Cu K$_{\alpha_1}$	8.0478	154.060	4.51 02	4.73 02	<u>4.13 02</u>	4.26 02	3.23 02
69	Tm L$_{\beta_1}$	8.1010	153.049	4.43 02	4.65 02	4.87 02	4.19 02	3.17 02
73	Ta L$_{\alpha_1}$	8.1461	152.201	4.37 02	4.58 02	4.80 02	4.12 02	<u>3.12 02</u>
28	Ni K$_{\beta_{1,3}}$	8.2647	150.017	4.20 02	4.40 02	4.62 02	<u>3.97 02</u>	4.19 02
74	W L$_{\alpha_1}$	8.3976	147.643	4.02 02	4.21 02	4.42 02	4.56 02	4.01 02
70	Yb L$_{\beta_1}$	8.4018	147.018	4.02 02	4.21 02	4.41 02	4.56 02	4.01 02
30	Zn K$_{\alpha_1}$	8.6389	143.519	3.72 02	3.90 02	4.09 02	4.22 02	3.71 02
75	Re L$_{\alpha_1}$	8.6525	143.294	3.71 02	3.88 02	4.08 02	4.21 02	<u>3.70 02</u>
71	Lu L$_{\alpha_1}$	8.7090	142.364	3.64 02	3.82 02	4.01 02	4.13 02	4.36 02
29	Cu K$_{\beta_1}$	8.9053	139.226	3.43 02	3.59 02	3.77 02	3.89 02	4.10 02
76	Os L$_{\alpha_1}$	8.9117	139.126	3.42 02	3.58 02	3.76 02	3.88 02	4.09 02
72	Hf L$_{\beta_1}$	9.0227	137.414	3.31 02	3.47 02	3.64 02	3.75 02	3.96 02
77	Ir L$_{\alpha_1}$	9.1751	135.132	3.16 02	3.31 02	3.48 02	3.58 02	3.78 02
31	Ga K$_{\alpha_1}$	9.2517	134.013	3.09 02	3.24 02	3.40 02	3.50 02	3.70 02
73	Ta L$_{\beta_1}$	9.3431	132.702	3.01 02	3.15 02	3.31 02	3.41 02	3.60 02
78	Pt L$_{\alpha_1}$	9.4423	131.308	2.92 02	3.06 02	3.22 02	3.31 02	3.50 02
30	Zn K$_{\beta_{1,3}}$	9.5720	129.529	2.82 02	2.95 02	3.10 02	3.19 02	3.37 02
74	W L$_{\beta_1}$	9.6724	128.184	2.74 02	2.87 02	3.01 02	3.10 02	3.27 02
79	Au L$_{\alpha_1}$	9.7133	127.644	2.71 02	2.84 02	2.98 02	3.07 02	3.24 02
32	Ge K$_{\alpha_1}$	9.8864	125.409	2.58 02	2.70 02	2.84 02	2.92 02	3.08 02
80	Hg L$_{\alpha_1}$	9.9888	124.124	2.51 02	2.63 02	2.76 02	2.84 02	3.00 02
75	Re L$_{\beta_1}$	10.010	123.861	2.49 02	2.61 02	2.75 02	2.83 02	2.98 02
31	Ga K$_{\beta_1}$	10.264	120.796	2.33 02	2.44 02	2.57 02	2.64 02	2.79 02
81	Tl L$_{\alpha_1}$	10.269	120.737	2.33 02	2.44 02	2.56 02	2.64 02	2.78 02
76	Os L$_{\beta_1}$	10.355	119.734	2.27 02	2.38 02	2.51 02	2.58 02	2.72 02
33	As K$_{\alpha_1}$	10.544	117.588	2.17 02	2.27 02	2.39 02	2.46 02	2.59 02
82	Pb L$_{\alpha_1}$	10.552	117.499	2.16 02	2.27 02	2.38 02	2.45 02	2.59 02
77	Ir L$_{\beta_1}$	10.708	115.787	2.08 02	2.18 02	2.29 02	2.36 02	2.48 02
83	Bi L$_{\alpha_1}$	10.839	114.388	2.01 02	2.11 02	2.21 02	2.28 02	2.40 02
32	Ge K$_{\beta_1}$	10.982	112.898	1.94 02	2.03 02	2.14 02	2.20 02	2.32 02
78	Pt L$_{\beta_1}$	11.071	111.990	1.90 02	1.99 02	2.09 02	2.15 02	2.27 02
84	Po L$_{\alpha_1}$	11.131	111.387	1.87 02	1.96 02	2.06 02	2.12 02	2.24 02
34	Se K$_{\alpha_1}$	11.222	110.484	1.83 02	1.92 02	2.02 02	2.07 02	2.19 02
85	At L$_{\alpha_1}$	11.427	108.501	1.74 02	1.83 02	1.92 02	1.98 02	2.08 02
79	Au L$_{\beta_1}$	11.442	108.359	1.74 02	1.82 02	1.91 02	1.97 02	2.08 02
33	As K$_{\beta_1}$	11.726	105.735	1.63 02	1.70 02	1.79 02	1.84 02	1.94 02
86	Rn L$_{\alpha_1}$	11.727	105.726	1.63 02	1.70 02	1.79 02	1.84 02	1.94 02

Z	transition	E [keV]	λ [pm]	$_{61}$Pm	$_{62}$Sm	$_{63}$Eu	$_{64}$Gd	$_{65}$Tb
80	Hg L$_{\beta_1}$	11.823	104.867	1.59 02	1.67 02	1.75 02	1.80 02	1.90 02
35	Br K$_{\alpha_1}$	11.924	103.979	1.55 02	1.63 02	1.71 02	1.76 02	1.86 02
87	Fr L$_{\alpha_1}$	12.031	103.054	1.52 02	1.59 02	1.67 02	1.72 02	1.81 02
81	Tl L$_{\beta_1}$	12.213	101.519	1.46 02	1.53 02	1.60 02	1.65 02	1.74 02
88	Ra L$_{\alpha_1}$	12.340	100.474	1.42 02	1.49 02	1.56 02	1.60 02	1.69 02
34	Se K$_{\beta_1}$	12.496	99.219	1.37 02	1.44 02	1.51 02	1.55 02	1.63 02
82	Pb L$_{\beta_1}$	12.614	98.291	1.33 02	1.40 02	1.47 02	1.51 02	1.59 02
36	Kr K$_{\alpha_1}$	12.649	98.019	1.32 02	1.39 02	1.46 02	1.50 02	1.58 02
89	Ac L$_{\alpha_1}$	12.652	97.996	1.32 02	1.39 02	1.46 02	1.50 02	1.58 02
90	Th L$_{\alpha_1}$	12.969	95.601	1.24 02	1.30 02	1.36 02	1.40 02	1.48 02
83	Bi L$_{\beta_1}$	13.024	95.197	1.22 02	1.28 02	1.35 02	1.39 02	1.46 02
91	Pa L$_{\alpha_1}$	13.291	93.285	1.16 02	1.22 02	1.28 02	1.31 02	1.38 02
35	Br K$_{\beta_1}$	13.291	93.285	1.16 02	1.22 02	1.28 02	1.31 02	1.38 02
37	Rb K$_{\alpha_1}$	13.395	92.560	1.13 02	1.19 02	1.25 02	1.28 02	1.35 02
84	Po L$_{\beta_1}$	13.447	92.202	1.12 02	1.18 02	1.24 02	1.27 02	1.34 02
92	U L$_{\alpha_1}$	13.615	91.065	1.09 02	1.14 02	1.19 02	1.23 02	1.30 02
85	At L$_{\beta_1}$	13.876	89.352	1.03 02	1.08 02	1.14 02	1.17 02	1.23 02
36	Kr K$_{\beta_1}$	14.112	87.857	9.86 01	1.03 02	1.08 02	1.12 02	1.18 02
38	Sr K$_{\alpha_1}$	14.165	87.529	9.76 01	1.02 02	1.07 02	1.10 02	1.16 02
86	Rn L$_{\beta_1}$	14.316	86.606	9.48 01	9.95 01	1.04 02	1.07 02	1.13 02
87	Fr L$_{\beta_1}$	14.770	83.943	8.71 01	9.15 01	9.59 01	9.86 01	1.04 02
39	Y K$_{\alpha_1}$	14.958	82.888	8.42 01	8.84 01	9.26 01	9.53 01	1.00 02
31	Rb K$_{\alpha_1}$	14.961	82.872	8.42 01	8.83 01	9.26 01	9.53 01	1.00 02
88	Ra L$_{\beta_1}$	15.236	81.376	8.02 01	8.41 01	8.82 01	9.07 01	9.56 01
89	Ac L$_{\beta_1}$	15.713	78.906	7.38 01	7.74 01	8.11 01	8.35 01	8.81 01
40	Zr K$_{\alpha_1}$	15.775	78.596	7.31 01	7.66 01	8.03 01	8.26 01	8.71 01
38	Sr K$_{\beta_1}$	15.836	78.293	7.23 01	7.58 01	7.95 01	8.18 01	8.62 01
90	Th L$_{\beta_1}$	16.202	76.524	6.80 01	7.13 01	7.47 01	7.69 01	8.11 01
41	Nb K$_{\alpha_1}$	16.615	74.622	6.36 01	6.66 01	6.98 01	7.19 01	7.59 01
91	Pa L$_{\beta_1}$	16.702	74.233	6.27 01	6.57 01	6.89 01	7.09 01	7.48 01
39	Y K$_{\beta_1}$	16.738	74.074	6.24 01	6.53 01	6.85 01	7.05 01	7.44 01
92	U L$_{\beta_1}$	17.220	72.000	5.78 01	6.05 01	6.34 01	6.53 01	6.90 01
42	Mo K$_{\alpha_1}$	17.479	70.933	5.55 01	5.81 01	6.09 01	6.27 01	6.63 01
40	Zr K$_{\beta_1}$	17.668	70.175	5.40 01	5.65 01	5.92 01	6.09 01	6.44 01
43	Tc K$_{\alpha_1}$	18.367	67.504	4.87 01	5.09 01	5.33 01	5.49 01	5.80 01
41	Nb K$_{\beta_1}$	18.623	66.576	4.69 01	4.90 01	5.14 01	5.29 01	5.59 01
44	Ru K$_{\alpha_1}$	19.279	64.311	4.28 01	4.47 01	4.68 01	4.82 01	5.10 01
42	Mo K$_{\beta_1}$	19.608	63.231	4.09 01	4.27 01	4.47 01	4.60 01	4.87 01
45	Rh K$_{\alpha_1}$	20.216	61.330	3.77 01	3.93 01	4.12 01	4.24 01	4.49 01
43	Tc K$_{\beta_1}$	20.619	60.131	3.58 01	3.73 01	3.91 01	4.03 01	4.26 01
46	Pd K$_{\alpha_1}$	21.177	58.547	3.33 01	3.48 01	3.65 01	3.75 01	3.97 01
44	Ru K$_{\beta_1}$	21.657	57.249	3.14 01	3.28 01	3.44 01	3.54 01	3.74 01
47	Ag K$_{\alpha_1}$	22.163	55.942	2.95 01	3.08 01	3.23 01	3.33 01	3.52 01
45	Rh K$_{\beta_1}$	22.724	54.561	2.77 01	2.88 01	3.03 01	3.11 01	3.29 01

Z	transition	E [keV]	λ [pm]	$_{61}$Pm	$_{62}$Sm	$_{63}$Eu	$_{64}$Gd	$_{65}$Tb
48	Cd K_{α_1}	23.174	53.501	2.63 01	2.74 01	2.87 01	2.95 01	3.12 01
46	Pd K_{β_1}	23.819	52.053	2.44 01	2.55 01	2.67 01	2.75 01	2.90 01
49	In K_{α_1}	24.210	51.2l2	2.34 01	2.44 01	2.56 01	2.63 01	2.78 01
47	Ag K_{β_1}	24.942	49.709	2.16 01	2.25 01	2.37 01	2.43 01	2.57 01
50	Sn K_{α_1}	25.271	49.062	2.09 01	2.18 01	2.29 01	2.35 01	2.48 01
48	Cd K_{β_1}	26.096	47.511	1.92 01	2.00 01	2.10 01	2.16 01	2.28 01
51	Sb K_{α_1}	26.359	47.037	1.87 01	1.95 01	2.05 01	2.10 01	2.22 01
49	In K_{β_1}	27.276	45.455	1.71 01	1.78 01	1.87 01	1.92 01	2.02 01
52	Te K_{α_1}	27.472	45.131	1.68 01	1.75 01	1.83 01	1.88 01	1.98 01
50	Sn K_{β_1}	28.486	43.524	1.52 01	1.59 01	1.67 01	1.71 01	1.80 01
53	I K_{α_1}	28.6l2	43.333	1.50 01	1.57 01	1.65 01	1.69 01	1.78 01
51	Sb K_{β_1}	29.726	41.709	1.36 01	1.42 01	1.49 01	1.53 01	1.61 01
54	Xe K_{α_1}	29.779	41.635	1.35 01	1.41 01	1.48 01	1.52 01	1.60 01
55	Cs K_{α_1}	30.973	40.030	1.22 01	1.27 01	1.34 01	1.37 01	1.44 01
52	Te K_{β_1}	30.996	40.000	1.22 01	1.27 01	1.34 01	1.37 01	1.44 01
56	Ba K_{α_1}	32.194	38.511	1.11 01	1.15 01	1.21 01	1.24 01	1.30 01
53	I K_{β_1}	32.295	38.391	1.10 01	1.14 01	1.20 01	1.23 01	1.29 01
57	La K_{α_1}	33.442	37.074	1.00 01	1.04 01	1.09 01	1.l2 01	1.18 01
54	Xe K_{β_1}	33.624	36.874	9.87 00	1.03 01	1.08 01	1.10 01	1.16 01
58	Ce K_{α_1}	34.279	36.169	9.39 00	9.76 00	1.03 01	1.05 01	1.11 01
55	Cs K_{β_1}	34.987	35.437	8.90 00	9.26 00	9.73 00	9.96 00	1.05 01
69	Pr K_{α_1}	36.026	34.415	8.25 00	8.58 00	9.01 00	9.22 00	9.72 00
56	Ba K_{β_1}	36.378	34.082	8.05 00	8.36 00	8.79 00	8.99 00	9.47 00
60	Nd K_{α_1}	36.847	33.648	7.78 00	8.09 00	8.50 00	8.69 00	9.16 00
57	La K_{β_1}	37.801	32.799	7.28 00	7.57 00	7.95 00	8.13 00	8.57 00
61	Pm K_{β_1}	38.725	32.016	6.84 00	7.11 00	7.46 00	7.63 00	8.04 00
58	Ce K_{β_1}	39.257	31.582	6.59 00	6.86 00	7.20 00	7.36 00	7.76 00
62	Sm K_{α_1}	40.118	30.905	6.22 00	6.49 00	6.80 00	6.96 00	7.33 00
59	Pr K_{β_1}	40.748	30.427	5.98 00	6.23 00	6.54 00	6.68 00	7.04 00
63	Eu K_{α_1}	41.542	29.845	5.69 00	5.93 00	6.22 00	6.36 00	6.70 00
60	Nd K_{β_1}	42.271	29.331	5.44 00	5.68 00	5.95 00	6.08 00	6.40 00
64	Gd K_{α_1}	42.996	28.836	5.21 00	5.43 00	5.69 00	5.82 00	6.13 00
61	Pm K_{β_1}	43.826	28.290	4.96 00	5.17 00	5.42 00	5.54 00	5.83 00
65	Tb K_{α_1}	44.482	27.873	<u>4.78 00</u>	4.98 00	5.21 00	5.33 00	5.61 00
62	Sm K_{β_1}	45.413	27.301	2.45 01	4.72 00	4.94 00	5.06 00	5.32 00
66	Dy K_{α_1}	45.998	26.954	2.36 01	<u>4.57 00</u>	4.78 00	4.89 00	5.15 00
63	Eu K_{β_1}	47.038	26.358	2.22 01	2.30 01	4.51 00	4.62 00	4.87 00
67	Ho K_{α_1}	47.547	26.076	2.15 01	2.23 01	<u>4.39 00</u>	4.49 00	4.74 00
64	Gd K_{β_1}	48.697	25.460	2.02 01	2.09 01	2.20 01	4.22 00	4.46 00
68	Er K_{α_1}	49.l28	25.237	1.97 01	2.04 01	2.14 01	<u>4.13 00</u>	4.36 00
65	Tb K_{β_1}	50.382	24.609	1.83 01	1.90 01	2.00 01	2.00 01	4.09 00
69	Tm K_{α_1}	50.742	24.434	1.80 01	1.86 01	1.96 01	1.96 01	<u>4.01 00</u>
66	Dy K_{β_1}	52.119	23.788	1.67 01	1.73 01	1.82 01	1.82 01	1.97 01
70	Yb K_{α_1}	52.389	23.666	1.64 01	1.71 01	1.79 01	1.80 01	1.94 01

Z	transition	E [keV]	λ [pm]	$_{61}$Pm	$_{62}$Sm	$_{63}$Eu	$_{64}$Gd	$_{65}$Tb
67	Ho K$_{\beta_1}$	53.877	23.012	1.52 01	1.58 01	1.66 01	1.66 01	1.79 01
71	Lu K$_{\alpha_1}$	54.707	22.930	1.51 01	1.56 01	1.64 01	1.65 01	1.77 01
68	Er K$_{\beta_1}$	55.681	22.267	1.39 01	1.44 01	1.51 01	1.52 01	1.63 01
72	Hf K$_{\alpha_1}$	55.790	22.223	1.38 01	1.43 01	1.50 01	1.52 01	1.63 01
69	Tm K$_{\beta_1}$	57.517	21.556	1.27 01	1.32 01	1.38 01	1.40 01	1.49 01
73	Ta K$_{\alpha_1}$	57.532	21.550	1.27 01	1.31 01	1.38 01	1.39 01	1.49 01
74	W K$_{\alpha_1}$	59.318	20.901	1.16 01	1.21 01	1.27 01	1.28 01	1.37 01
70	Yb K$_{\beta_1}$	59.370	20.883	1.16 01	1.20 01	1.27 01	1.28 01	1.37 01
75	Re K$_{\alpha_1}$	61.140	20.278	1.07 01	1.11 01	1.17 01	1.18 01	1.26 01
71	Lu K$_{\beta_1}$	61.283	20.231	1.06 01	1.10 01	1.16 01	1.18 01	1.25 01
76	Os K$_{\alpha_1}$	63.001	19.679	9.84 00	1.02 01	1.07 01	1.09 01	1.16 01
72	Hf K$_{\beta_1}$	63.234	19.607	9.74 00	1.01 01	1.06 01	1.08 01	1.15 01
77	Ir K$_{\alpha_1}$	64.896	19.105	9.07 00	9.42 00	9.89 00	1.01 01	1.07 01
73	Ta K$_{\beta_1}$	65.223	19.009	8.94 00	9.29 00	9.75 00	9.93 00	1.05 01
78	Pt K$_{\alpha_1}$	66.632	18.551	8.36 00	8.69 00	9.12 00	9.29 00	9.85 00
74	W K$_{\beta_1}$	67.244	18.438	8.22 00	8.54 00	8.97 00	9.14 00	9.68 00
79	Au K$_{\alpha_1}$	68.804	18.020	7.71 00	8.02 00	8.42 00	8.59 00	9.08 00
75	Re K$_{\beta_1}$	69.310	17.888	7.56 00	7.86 00	8.25 00	8.42 00	8.90 00
80	Hg K$_{\alpha_1}$	70.819	17.507	7.12 00	7.40 00	7.77 00	7.94 00	8.39 00
76	Os K$_{\beta_1}$	71.413	17.361	6.96 00	7.23 00	7.59 00	7.76 00	8.20 00
81	Tl K$_{\alpha_1}$	72.872	17.014	6.58 00	6.84 00	7.18 00	7.35 00	7.75 00
77	Ir K$_{\beta_1}$	73.561	16.854	6.41 00	6.67 00	7.00 00	7.16 00	7.55 00
82	Pb K$_{\alpha_1}$	74.969	16.538	6.08 00	6.33 00	6.64 00	6.81 00	7.17 00
78	Pt K$_{\beta_1}$	75.748	16.368	5.91 00	6.15 00	6.45 00	6.62 00	6.96 00
83	Bi K$_{\alpha_1}$	77.108	16.079	5.63 00	5.85 00	6.14 00	6.31 00	6.63 00
79	Au K$_{\beta_1}$	77.948	15.906	5.46 00	5.68 00	5.96 00	6.12 00	6.43 00
84	Po K$_{\alpha_1}$	79.290	15.636	5.21 00	5.42 00	5.69 00	5.85 00	6.14 00
80	Hg K$_{\beta_1}$	80.253	15.449	5.04 00	5.24 00	5.50 00	5.66 00	5.94 00
85	At K$_{\alpha_1}$	81.520	15.209	4.83 00	5.02 00	5.27 00	5.42 00	5.69 00
81	Tl K$_{\beta_1}$	82.576	15.014	4.66 00	4.85 00	5.09 00	5.24 00	5.49 00
86	Rn K$_{\alpha_1}$	83.780	14.798	4.48 00	4.66 00	4.89 00	5.04 00	5.28 00
82	Pb K$_{\beta_1}$	84.936	14.597	4.32 00	4.49 00	4.72 00	4.86 00	5.08 00
87	Fr K$_{\alpha_1}$	86.100	14.400	4.16 00	4.33 00	4.54 00	4.68 00	4.90 00
83	Bi K$_{\beta_1}$	87.343	14.195	4.01 00	4.17 00	4.37 00	4.50 00	4.71 00
88	Ra K$_{\alpha_1}$	88.470	14.014	3.87 00	4.02 00	4.22 00	4.35 00	4.55 00
84	Po K$_{\beta_1}$	89.800	13.806	3.72 00	3.86 00	4.05 00	4.18 00	4.36 00
89	Ac K$_{\alpha_1}$	90.884	13.642	3.60 00	3.74 00	3.92 00	4.05 00	4.22 00
85	At K$_{\beta_1}$	92.300	13.432	3.45 00	3.59 00	3.76 00	3.88 00	4.05 00
90	Th K$_{\alpha_1}$	93.350	13.281	3.35 00	3.48 00	3.65 00	3.77 00	3.93 00
86	Rn K$_{\beta_1}$	94.870	13.068	3.20 00	3.33 00	3.49 00	3.61 00	3.76 00
91	Pa K$_{\alpha_1}$	95.868	12.932	3.11 00	3.23 00	3.40 00	3.51 00	3.65 00
87	Fr K$_{\beta_1}$	97.470	12.720	2.98 00	3.09 00	3.25 00	3.35 00	3.49 00
92	U K$_{\alpha_1}$	98.439	12.595	2.90 00	3.01 00	3.16 00	3.26 00	3.39 00
88	Ra K$_{\beta_1}$	100.130	12.382	2.77 00	2.87 00	3.02 00	3.12 00	3.24 00

Z	transition	E [keV]	λ [pm]	$_{61}$Pm	$_{62}$Sm	$_{63}$Eu	$_{64}$Gd	$_{65}$Tb
89	Ac K$_{\beta_1}$	102.850	12.054	2.58 00	2.68 00	2.81 00	2.91 00	3.02 00
90	Th K$_{\beta_1}$	105.610	11.739	2.41 00	2.50 00	2.63 00	2.71 00	2.82 00
91	Pa K$_{\beta_1}$	108.430	11.434	2.25 00	2.34 00	2.45 00	2.53 00	2.63 00
92	U K$_{\beta_1}$	111.300	11.139	2.10 00	2.18 00	2.29 00	2.37 00	2.46 00

Z	transition	E [keV]	λ [pm]	$_{66}$Dy	$_{67}$Ho	$_{68}$Er	$_{69}$Tm	$_{70}$Yb
4	Be K$_\alpha$	0.1085	11427.207	2.92 04	2.27 04	2.29 04	2.64 04	2.69 04
38	Sr M$_\xi$	0.1140	10875.895	2.72 04	2.23 04	2.18 04	2.50 04	2.64 04
39	Y M$_\xi$	0.1328	9339.235	2.12 04	2.06 04	2.06 04	2.19 04	2.37 04
16	S L$_l$	0.1487	8337.942	2.26 04	1.39 04	1.91 04	2.03 04	2.40 04
40	Zr M$_\xi$	0.1511	8205.506	3.76 04	1.34 04	1.81 04	2.00 04	2.40 04
41	Nb M$_\xi$	0.1717	7221.037	4.16 04	5.42 04	4.97 04	2.48 04	2.02 04
5	B K$_\alpha$	0.1833	6764.059	3.21 04	3.06 04	2.80 04	2.95 04	2.37 04
42	Mo M$_\xi$	0.1926	6473.445	2.75 04	2.68 04	2.19 04	2.42 04	2.26 04
6	C K$_\alpha$	0.2770	4476.000	1.74 04	1.60 04	1.75 04	1.71 04	1.77 04
47	Ag M$_\xi$	0.3117	3977.709	1.65 04	1.51 04	1.59 04	1.50 04	1.63 04
7	N K$_\alpha$	0.3924	3159.664	1.19 04	1.20 04	1.22 04	1.25 04	1.38 04
22	Ti L$_l$	0.3953	3136.484	1.18 04	1.19 04	1.21 04	1.24 04	1.37 04
22	Ti L$_\alpha$	0.4522	2741.822	9.99 03	1.03 04	1.05 04	1.12 04	1.16 04
23	V L$_\alpha$	0.5113	2424.901	8.18 03	8.66 03	9.21 03	9.85 03	1.03 04
8	O K$_\alpha$	0.5249	2362.072	7.88 03	8.34 03	8.92 03	9.56 03	1.00 04
25	Mn L$_l$	0.5563	2228.747	7.24 03	7.65 03	8.19 03	8.77 03	9.18 03
24	Cr L$_\alpha$	0.5728	2164.549	6.92 03	7.31 03	7.83 03	8.38 03	8.78 03
25	Mn L$_\alpha$	0.6374	1945.171	5.86 03	6.19 03	6.62 03	7.09 03	7.43 03
9	F K$_\alpha$	0.6768	1831.932	5.28 03	5.59 03	5.95 03	6.38 03	6.68 03
26	Fe L$_\alpha$	0.7050	1758.655	4.97 03	5.26 03	5.59 03	6.00 03	6.28 03
27	Co L$_\alpha$	0.7762	1597.335	4.22 03	4.44 03	4.73 03	5.06 03	5.31 03
28	Ni L$_\alpha$	0.8515	1456.086	3.57 03	3.76 03	4.00 03	4.27 03	4.48 03
29	Cu L$_\alpha$	0.9297	1336.044	3.04 03	3.20 03	3.40 03	3.63 03	3.80 03
30	Zn L$_\alpha$	1.0117	1225.513	2.59 03	2.73 03	2.89 03	3.09 03	3.28 03
11	Na K$_\alpha$	1.0410	1191.020	2.45 03	2.61 03	2.81 03	2.92 03	2.73 03
11	Na K$_\beta$	1.0711	1157.550	2.36 03	2.38 03	2.56 03	2.71 03	2.44 03
12	Mg K$_\alpha$	1.2536	989.033	2.32 03	2.34 03	2.44 03	2.52 03	2.53 03
33	As L$_\alpha$	1.2820	967.123	2.21 03	2.22 03	2.31 03	2.39 03	2.38 03
12	Mg K$_\beta$	1.3022	952.121	6.03 03	2.14 03	2.23 03	2.29 03	2.29 03
33	As L$_{\beta_1}$	1.3170	941.421	5.87 03	2.08 03	2.17 03	2.23 03	2.22 03
66	Dy M$_\beta$	1.3250	935.737	5.97 03	2.05 03	2.14 03	2.20 03	2.18 03
67	Ho M$_\alpha$	1.3480	919.771	8.28 03	1.97 03	2.05 03	2.11 03	2.08 03
34	Se L$_\alpha$	1.3791	899.029	7.81 03	5.46 03	1.94 03	1.99 03	1.96 03
67	Ho M$_\beta$	1.3830	896.494	7.76 03	5.42 03	1.93 03	1.98 03	1.95 03
68	Er M$_\alpha$	1.4060	881.829	7.44 03	7.79 03	1.86 03	1.90 03	1.86 03
34	Se L$_{\beta_1}$	1.4192	873.627	7.26 03	7.60 03	5.40 03	1.85 03	1.82 03
68	Er M$_\beta$	1.4430	859.218	6.96 03	7.28 03	5.17 03	1.78 03	1.74 03

Z	transition	E [keV]	λ [pm]	$_{66}$Dy	$_{67}$Ho	$_{68}$Er	$_{69}$Tm	$_{70}$Yb
69	Tm M_α	1.4620	848.051	6.73 03	7.03 03	7.48 03	<u>1.72 03</u>	1.68 03
35	Br L_α	1.4804	837.511	6.52 03	6.80 03	7.25 03	5.06 03	1.62 03
13	Al K_{α_1}	1.4867	833.962	6.45 03	6.73 03	7.18 03	5.00 03	1.60 03
69	Tm M_β	1.5030	824.918	6.27 03	6.54 03	6.99 03	<u>4.86 03</u>	1.56 03
70	Yb M_α	1.5214	814.941	6.08 03	6.34 03	6.77 03	7.05 03	1.51 03
35	Br L_{β_1}	1.5259	812.538	6.03 03	6.29 03	6.71 03	7.00 03	<u>1.50 03</u>
13	Al K_β	1.5574	796.103	5.72 03	5.97 03	6.37 03	6.64 03	4.56 03
70	Yb M_β	1.5675	790.974	5.63 03	5.87 03	6.26 03	6.53 03	<u>4.48 03</u>
71	Lu M_α	1.5813	784.071	5.50 03	5.74 03	6.12 03	6.39 03	6.59 03
36	Kr L_α	1.5860	781.747	5.46 03	5.70 03	6.07 03	6.34 03	6.54 03
71	Lu M_β	1.6312	760.085	5.08 03	5.31 03	5.64 03	5.90 03	6.08 03
36	Kr L_{β_1}	1.6366	757.577	5.04 03	5.26 03	5.59 03	5.85 03	6.03 03
72	Hf M_{α_1}	1.6446	753.892	<u>4.98 03</u>	5.20 03	5.52 03	5.77 03	5.95 03
37	Rb L_{α_1}	1.6941	731.864	5.52 03	4.82 03	5.11 03	5.35 03	5.51 03
72	Hf M_β	1.6976	730.355	5.49 03	4.79 03	5.08 03	5.32 03	5.48 03
73	Ta M_{α_1}	1.7096	725.229	5.39 03	4.71 03	4.99 03	5.22 03	5.38 03
14	Si K_{α_1}	1.7400	712.558	5.15 03	<u>4.50 03</u>	4.76 03	4.99 03	5.14 03
37	Rb L_{β_1}	1.7522	707.597	5.06 03	5.28 03	4.68 03	4.90 03	5.05 03
73	Ta M_β	1.7655	702.266	4.96 03	5.18 03	4.59 03	4.81 03	4.59 03
74	W M_{α_1}	1.7754	698.350	4.89 03	5.11 03	4.52 03	4.74 03	4.88 03
38	Sr L_{α_1}	1.8066	686.290	4.67 03	4.88 03	<u>4.32 03</u>	4.53 03	4.67 03
74	W M_β	1.8349	675.705	4.49 03	4.69 03	4.99 03	4.36 03	4.48 03
14	Si K_β	1.8359	675.337	<u>4.49 03</u>	4.69 03	4.98 03	4.35 03	4.48 03
75	Re M_{α_1}	1.8420	673.100	4.45 03	4.65 03	4.94 03	4.31 03	4.44 03
38	Sr L_{β_1}	1.8717	662.420	4.70 03	4.46 03	4.74 03	<u>4.14 03</u>	4.26 03
75	Re M_β	1.9061	650.465	4.48 03	4.26 03	4.52 03	4.71 03	4.06 03
76	Os M_{α_1}	1.9102	649.069	4.46 03	4.23 03	4.50 03	4.69 03	4.04 03
39	Y L_{α_1}	1.9226	644.882	4.39 03	<u>4.17 03</u>	4.42 03	4.61 03	<u>3.97 03</u>
76	Os M_β	1.9783	626.725	4.08 03	4.24 03	4.11 03	4.29 03	4.43 03
77	Ir M_{α_1}	1.9799	626.219	4.07 03	4.24 03	4.10 03	4.28 03	4.42 03
39	Y L_{β_1}	1.9958	621.230	3.99 03	4.15 03	<u>4.02 03</u>	4.19 03	4.34 03
15	P K_{α_1}	2.0137	615.708	3.90 03	4.06 03	4.31 03	4.10 03	4.24 03
40	Zr L_{α_1}	2.0424	607.056	<u>3.76 03</u>	3.91 03	4.16 03	3.94 03	4.09 03
78	Pt M_{α_1}	2.0505	604.658	4.10 03	3.88 03	4.12 03	3.90 03	4.05 03
77	Ir M_β	2.0535	603.775	4.08 03	3.86 03	4.10 03	<u>3.89 03</u>	4.03 03
79	Au M_{α_1}	2.1229	584.036	3.74 03	3.55 03	3.76 03	3.94 03	3.70 03
40	Zr L_{β_1}	2.1244	583.624	3.74 03	3.55 03	3.75 03	3.93 03	3.69 03
78	Pt M_β	2.1273	582.828	3.72 03	<u>3.53 03</u>	3.74 03	3.92 03	3 68 03
15	P $K_{\beta_{1.3}}$	2.1390	579.640	3.67 03	3.81 03	3.69 03	3.86 03	3.63 03
41	Nb L_{α_1}	2.1659	572.441	3.56 03	3.69 03	3.57 03	3.74 03	<u>3.51 03</u>
80	Hg M_{α_1}	2.1953	564.775	3.43 03	3.57 03	3.45 03	3.61 03	3.73 03
79	Au M_β	2.2046	362.393	3.40 03	3.53 03	<u>3.41 03</u>	3.57 03	3.69 03
41	Nb L_{β_1}	2.2574	549.238	3.20 03	3.32 03	3.55 03	3.36 03	3.47 03
81	Tl M_{α_1}	2.2706	546.045	3.15 03	3.27 03	3.48 03	3.31 03	3.42 03

Z	transition	E [keV]	λ [pm]	$_{66}$Dy	$_{67}$Ho	$_{68}$Er	$_{69}$Tm	$_{70}$Yb
80	Hg M$_\beta$	2.2825	543.199	3.11 03	3.23 03	3.43 03	3.26 03	3.37 03
42	Mo L$_{\alpha_1}$	2.2932	540.664	3.07 03	3.19 03	3.39 03	3.22 03	3.33 03
16	S L$_{\alpha_1}$	2.3080	537.197	3.02 03	3.14 03	3.33 03	3.48 03	3.28 03
82	Pb M$_{\alpha_1}$	2.3457	528.563	2.89 03	3.01 03	3.20 03	3.34 03	3.14 03
81	Tl M$_\beta$	2.3621	524.893	2.84 03	2.96 03	3.14 03	3.28 03	3.09 03
42	Mo L$_{\beta_1}$	2.3948	517.726	2.74 03	2.85 03	3.03 03	3.17 03	2.98 03
83	Bi M$_{\alpha_1}$	2.4226	511.785	2.66 03	2.17 03	2.94 03	3.07 03	3.18 03
43	Tc L$_{\alpha_1}$	2.4240	511.490	2.66 03	2.77 03	2.94 03	3.07 03	3.17 03
82	Pb M$_\beta$	2.4427	507.574	2.61 03	2.71 03	2.88 03	3.01 03	3.11 03
16	S K$_\beta$	2.4640	503.186	2.55 03	2.65 03	2.81 03	2.94 03	3.04 03
83	Bi M$_{\beta_1}$	2.5255	490.933	2.39 03	2.49 03	2.64 03	2.76 03	2.86 03
43	Tc L$_{\beta_1}$	2.5368	488.746	2.37 03	2.46 03	2.61 03	2.73 03	2.82 03
44	Ru L$_{\alpha_1}$	2.5586	484.582	2.31 03	2.41 03	2.55 03	2.67 03	2.76 03
17	Cl K$_{\alpha_1}$	2.6224	472.792	2.17 03	2.26 03	2.39 03	2.51 03	2.59 03
44	Ru L$_{\beta_1}$	2.6832	462.079	2.05 03	2.13 03	2.26 03	2.36 03	2.44 03
45	Rh L$_{\alpha_1}$	2.6967	459.766	2.02 03	2.11 03	2.23 03	2.33 03	2.41 03
17	Cl K$_\beta$	2.8156	440.350	1.81 03	1.89 03	1.99 03	2.09 03	2.16 03
45	Rh L$_{\beta_1}$	2.8344	437.430	1.78 03	1.85 03	1.96 03	2.05 03	2.12 03
46	Pd L$_{\alpha_1}$	2.8386	436.782	1.77 03	1.85 03	1.95 03	2.04 03	2.11 03
18	Ar K$_{\alpha_1}$	2.9577	419.194	1.59 03	1.66 03	1.75 03	1.84 03	1.90 03
47	Ag L$_{\alpha_1}$	2.9843	415.458	1.56 03	1.62 03	1.71 03	1.80 03	1.86 03
46	Pd L$_{\beta_1}$	2.9902	414.638	1.55 03	1.62 03	1.70 03	1.79 03	1.85 03
90	Th M$_{\alpha_1}$	2.9961	413.821	1.54 03	1.61 03	1.69 03	1.78 03	1.84 03
91	Pa M$_{\alpha_1}$	3.0823	402.248	1.43 03	1.50 03	1.57 03	1.65 03	1.71 03
48	Cd L$_{\alpha_1}$	3.1337	395.651	1.37 03	1.43 03	1.51 03	1.58 03	1.64 03
90	Th M$_\beta$	3.1458	394.129	1.36 03	1.42 03	1.49 03	1.57 03	1.62 03
47	Ag L$_{\beta_1}$	3.1509	393.491	1.35 03	1.41 03	1.49 03	1.56 03	1.62 03
92	U M$_{\alpha_1}$	3.1708	391.021	1.33 03	1.39 03	1.46 03	1.54 03	1.59 03
18	Ar K$_\beta$	3.1905	388.607	1.31 03	1.37 03	1.44 03	1.51 03	1.56 03
91	Pa M$_\beta$	3.2397	382.705	1.26 03	1.32 03	1.38 03	1.45 03	1.50 03
49	In L$_{\alpha_1}$	3.2869	377.210	1.21 03	1.27 03	1.33 03	1.40 03	1.45 03
19	K K$_{\alpha_1}$	3.3138	374.148	1.19 03	1.24 03	1.31 03	1.37 03	1.42 03
48	Cd L$_{\beta_1}$	3.3166	373.832	1.19 03	1.24 03	1.30 03	1.37 03	1.42 03
92	U M$_\beta$	3.3360	371.658	1.17 03	1.22 03	1.28 03	1.35 03	1.39 03
50	Sn L$_{\alpha_1}$	3.4440	360.003	1.08 03	1.13 03	1.18 03	1.24 03	1.28 03
49	In L$_{\beta_1}$	3.4872	355.543	1.04 03	1.09 03	1.14 03	1.20 03	1.24 03
19	K K$_\beta$	3.5896	345.401	9.69 02	1.01 03	1.06 03	1.12 03	1.15 03
51	Sb L$_{\alpha_1}$	3.6047	343.954	9.58 02	1.00 03	1.05 03	1.10 03	1.14 03
50	Sn L$_{\beta_1}$	3.6628	338.498	9.20 02	9.61 02	1.01 03	1.06 03	1.10 03
20	Ca K$_{\alpha_1}$	3.6917	335.848	9.01 02	9.42 02	9.88 02	1.04 03	1.07 03
52	Te L$_{\alpha_1}$	3.7693	328.934	8.55 02	8.93 02	9.36 02	9.85 02	1.02 03
51	Sb L$_{\beta_1}$	3.8436	322.575	8.13 02	8.49 02	8.90 02	9.36 02	9.68 02
53	I L$_{\alpha_1}$	3.9377	314.867	7.64 02	7.98 02	8.36 02	8.80 02	9.10 02
20	Ca K$_\beta$	4.0127	308.981	7.28 02	7.61 02	7.97 02	8.38 02	8.66 02

Z	transition	E [keV]	λ [pm]	$_{66}$Dy	$_{67}$Ho	$_{68}$Er	$_{69}$Tm	$_{70}$Yb
52	Te L$_{\beta_1}$	4.0296	307.686	7.20 02	7.53 02	7.88 02	8.29 02	8.57 02
21	Sc K$_{\alpha_1}$	4.0906	303.097	6.93 02	7.24 02	7.58 02	7.98 02	8.25 02
54	Xe L$_{\alpha_1}$	4.1099	301.674	6.85 02	7.16 02	7.49 02	7.88 02	8.15 02
53	I L$_{\beta_1}$	4.2207	293.755	6.40 02	6.69 02	1.00 02	7.37 02	7.62 02
55	Cs L$_{\alpha_1}$	4.2865	289.245	6.15 02	6.43 02	6.73 02	7.08 02	7.32 02
21	Sc K$_{\beta}$	4.4605	277.962	5.56 02	5.81 02	6.08 02	6.40 02	6.61 02
56	Ba L$_{\alpha_1}$	4.4663	277.601	5.54 02	5.79 02	6.06 02	6.37 02	6.59 02
22	Ti K$_{\alpha_1}$	4.5108	274.863	5.40 02	5.65 02	5.90 02	6.21 02	6.43 02
55	Cs L$_{\beta_1}$	4.6198	268.377	5.08 02	5.32 02	5.55 02	5.85 02	6.05 02
57	La L$_{\alpha_1}$	4.6510	266.577	4.99 02	5.22 02	5.46 02	5.75 02	5.94 02
56	Ba L$_{\beta_1}$	4.8273	256.841	4.54 02	4.75 02	4.96 02	5.23 02	5.41 02
58	Ce L$_{\alpha_1}$	4.8402	256.157	4.51 02	4.72 02	4.93 02	5.19 02	5.31 02
22	Ti K$_{\beta_{1,3}}$	4.9318	251.399	4.30 02	4.50 02	4.70 02	4.95 02	5.37 02
23	V K$_{\alpha_1}$	4.9522	250.363	4.25 02	4.45 02	4.65 02	4.90 02	5.06 02
59	Pr L$_{\alpha_1}$	5.0337	246.310	4.08 02	4.27 02	4.46 02	4.70 02	4.86 02
57	La L$_{\beta_1}$	5.0421	245.899	4.06 02	4.25 02	4.44 02	4.68 02	4.84 02
60	Nd L$_{\alpha_1}$	5.2304	237.047	3.70 02	3.87 02	4.04 02	4.26 02	4.40 02
58	Ce L$_{\beta_1}$	5.2622	235.614	3.64 02	3.81 02	3.98 02	4.19 02	4.34 02
24	Cr K$_{\alpha_1}$	5.4147	228.978	3.39 02	3.55 02	3.70 02	3.90 02	4.03 02
23	V K$_{\beta_{1,3}}$	5.4273	228.447	3.37 02	3.52 02	3.67 02	3.87 02	4.01 02
61	Pm L$_{\alpha_1}$	5.4325	228.228	3.36 02	3.52 02	3.66 02	3.86 02	4.00 02
59	Pr L$_{\beta_1}$	5.4889	225.883	3.27 02	3.42 02	3.57 02	3.76 02	3.89 02
62	Sm L$_{\alpha_1}$	5.6361	219.984	3.06 02	3.20 02	3.33 02	3.52 02	3.64 02
60	Nd L$_{\beta_1}$	5.7216	216.696	2.94 02	3.08 02	3.21 02	3.38 02	3.50 02
63	Eu L$_{\alpha_1}$	5.8457	212.096	2.78 02	2.92 02	3.03 02	3.20 02	3.31 02
25	Mn K$_{\alpha_1}$	5.8988	210.187	2.72 02	2.85 02	2.96 02	3.13 02	3.24 02
24	Cr K$_{\beta_{1,3}}$	5.9467	208.494	2.67 02	2.79 02	2.90 02	3.06 02	3.17 02
61	Pm L$_{\beta_1}$	5.9610	207.993	2.65 02	2.77 02	2.89 02	3.05 02	3.15 02
64	Gd L$_{\alpha_1}$	6.0572	204.690	2.54 02	2.66 02	2.77 02	2.92 02	3.03 02
62	Sm L$_{\beta_1}$	6.2051	199.811	2.39 02	2.51 02	2.61 02	2.75 02	2.85 02
65	Tb L$_{\alpha_1}$	6.2728	197.655	2.33 02	2.44 02	2.53 02	2.68 02	2.77 02
26	Fe K$_{\alpha_1}$	6.4038	193.611	2.21 02	2.31 02	2.40 02	2.54 02	2.63 02
63	Eu L$_{\beta_1}$	6.4564	192.034	2.16 02	2.27 02	2.35 02	2.49 02	2.57 02
25	Mn K$_{\beta_{1,3}}$	6.4905	191.025	2.14 02	2.24 02	2.32 02	2.45 02	2.54 02
66	Dy L$_{\alpha_1}$	6.4952	190.887	2.13 02	2.23 02	2.32 02	2.45 02	2.53 02
64	Gd L$_{\beta_1}$	6.7132	184.688	1.96 02	2.05 02	2.13 02	2.25 02	2.33 02
67	Ho L$_{\alpha_1}$	6.7198	184.507	1.96 02	2.05 02	2.13 02	2.25 02	2.33 02
27	Co K$_{\alpha_1}$	6.9303	178.903	1.81 02	1.90 02	1.97 02	2.08 02	2.15 02
68	Er L$_{\alpha_1}$	6.9487	178.429	1.80 02	1.88 02	1.95 02	2.06 02	2.14 02
65	Tb L$_{\beta_1}$	6.9780	177.680	1.78 02	1.86 02	1.93 02	2.04 02	2.11 02
26	Fe K$_{\beta_{1,3}}$	7.0580	175.666	1.73 02	1.81 02	1.88 02	1.98 02	2.05 02
69	Tm L$_{\alpha_1}$	7.1799	172.683	1.65 02	1.73 02	1.80 02	1.90 02	1.97 02
66	Dy L$_{\beta_1}$	7.2477	171.068	1.61 02	1.69 02	1.75 02	1.85 02	1.92 02
70	Yb L$_{\alpha_1}$	7.4156	167.195	1.52 02	1.60 02	1.66 02	1.75 02	1.81 02

Z	transition	E [keV]	λ [pm]	$_{66}$Dy	$_{67}$Ho	$_{68}$Er	$_{69}$Tm	$_{70}$Yb
28	Ni K$_{\alpha_1}$	7.4782	165.795	1.49 02	1.56 02	1.62 02	1.71 02	1.77 02
67	Ho L$_{\beta_1}$	7.5253	164.757	1.47 02	1.54 02	1.59 02	1.69 02	1.74 02
27	Co K$_{\beta_{1,3}}$	7.6494	162.084	1.41 02	1.48 02	1.53 02	1.62 02	1.67 02
71	Lu L$_{\alpha_1}$	7.6555	161.955	<u>1.41 02</u>	1.47 02	1.53 02	1.61 02	1.67 02
68	Er L$_{\beta_1}$	7.8109	158.733	3.63 02	1.40 02	1.45 02	1.53 02	1.59 02
72	Hf L$_{\alpha_1}$	7.8990	156.963	3.52 02	1.36 02	1.41 02	1.49 02	1.54 02
29	Cu K$_{\alpha_1}$	8.0478	154.060	3.35 02	<u>1.30 02</u>	1.34 02	1.42 02	1.47 02
69	Tm L$_{\beta_1}$	8.1010	153.049	3.29 02	3.42 02	1.32 02	1.40 02	1.45 02
73	Ta L$_{\alpha_1}$	8.1461	152.201	3.24 02	3.37 02	1.30 02	1.38 02	1.43 02
28	Ni K$_{\beta_{1,3}}$	8.2647	150.017	3.11 02	3.24 02	<u>1.26 02</u>	1.33 02	1.38 02
74	W L$_{\alpha_1}$	8.3976	147.643	2.98 02	3.10 02	3.23 02	1.28 02	1.32 02
70	Yb L$_{\beta_1}$	8.4018	147.569	<u>2.98 02</u>	3.10 02	3.23 02	1.28 02	1.32 02
30	Zn K$_{\alpha_1}$	8.6389	143.519	3.86 02	2.87 02	2.99 02	<u>1.19 02</u>	1.23 02
75	Re L$_{\alpha_1}$	8.6525	143.294	3.84 02	2.86 02	2.98 02	3.14 02	1.23 02
71	Lu L$_{\beta_1}$	8.7090	142.364	3.77 02	2.81 02	2.93 02	3.09 02	1.21 02
29	Cu K$_{\beta_1}$	8.9053	139.226	3.55 02	2.65 02	2.76 02	2.90 02	1.14 02
76	Os L$_{\alpha_1}$	8.9117	139.126	<u>2.64 02</u>	3.54 02	2.75 02	2.90 02	<u>1.14 02</u>
72	Hf L$_{\beta_1}$	9.0227	137.414	<u>3.42 02</u>	3.57 02	2.66 02	2.80 02	2.91 02
77	Ir L$_{\alpha_1}$	9.1751	135.132	3.92 02	3.41 02	2.54 02	2.68 02	2.78 02
31	Ga K$_{\alpha_1}$	9.2517	134.013	3.83 02	3.33 02	<u>2.49 02</u>	2.62 02	2.72 02
73	Ta L$_{\beta_1}$	9.3431	132.702	3.73 02	<u>3.24 02</u>	3.37 02	2.55 02	2.65 02
78	Pt L$_{\alpha_1}$	9.4423	131.308	3.63 02	3.78 02	3.28 02	2.48 02	2.57 02
30	Zn K$_{\beta_{1,3}}$	9.5720	129.529	3.49 02	3.64 02	3.16 02	<u>3.39 02</u>	2.48 02
74	W L$_{\beta_1}$	9.6724	128.184	3.40 02	3.54 02	3.07 02	3.24 02	2.41 02
79	Au L$_{\alpha_1}$	9.7133	127.644	3.36 02	3.50 02	<u>3.04 02</u>	3.20 02	2.38 02
32	Ge K$_{\alpha_1}$	9.8864	125.409	3.20 02	3.33 02	3.47 02	3.05 02	<u>2.27 02</u>
80	Hg L$_{\alpha_1}$	9.9888	124.124	3.11 02	3.24 02	3.37 02	2.97 02	3.08 02
75	Re L$_{\beta_1}$	10.010	123.861	3.09 02	3.22 02	3.35 02	<u>2.95 02</u>	3.06 02
31	Ga K$_{\beta_1}$	10.264	120.796	2.89 02	3.01 02	3.14 02	3.30 02	2.86 02
81	Tl L$_{\alpha_1}$	10.269	120.737	2.89 02	3.01 02	3.13 02	3.30 02	2.86 02
76	Os L$_{\beta_1}$	10.355	119.734	2.82 02	2.94 02	3.06 02	3.22 02	<u>2.79 02</u>
33	As K$_{\alpha_1}$	10.544	117.588	2.69 02	2.80 02	2.92 02	3.07 02	3.18 02
82	Pb L$_{\alpha_1}$	10.552	117.499	2.68 02	2.79 02	2.91 02	3.06 02	3.17 02
77	Ir L$_{\beta_1}$	10.708	115.787	2.58 02	2.68 02	2.80 02	2.94 02	3.05 02
83	Bi L$_{\alpha_1}$	10.839	114.388	2.49 02	2.60 02	2.71 02	2.85 02	2.95 02
32	Ge K$_{\beta_1}$	10.982	112.898	2.41 02	2.51 02	2.61 02	2.75 02	2.85 02
78	Pt L$_{\beta_1}$	11.071	111.990	2.35 02	2.45 02	2.56 02	2.69 02	2.79 02
84	Po L$_{\alpha_1}$	11.131	111.387	2.32 02	2.42 02	2.52 02	2.65 02	2.75 02
34	Se K$_{\alpha_1}$	11.222	110.484	2.27 02	2.37 02	2.46 02	2.59 02	2.69 02
85	At L$_{\alpha_1}$	11.427	108.501	2.16 02	2.25 02	2.35 02	2.47 02	2.56 02
79	Au L$_{\beta_1}$	11.442	108.359	2.15 02	2.24 02	2.34 02	2.46 02	2.55 02
33	As K$_{\beta_1}$	11.726	105.735	2.01 02	2.10 02	2.19 02	2.30 02	2.39 02
86	Rn L$_{\alpha_1}$	11.727	105.726	2.01 02	2.10 02	2.19 02	2.30 02	2.39 02
80	Hg L$_{\beta_1}$	11.823	104.867	1.97 02	2.05 02	2.14 02	2.25 02	2.33 02

Z	transition	E [keV]	λ [pm]	₆₆Dy	₆₇Ho	₆₈Er	₆₉Tm	₇₀Yb
35	Br K$_{\alpha_1}$	11.924	103.979	1.93 02	2.01 02	2.09 02	2.20 02	2.28 02
87	Fr L$_{\alpha_1}$	12.031	103.054	1.88 02	1.96 02	2.04 02	2.15 02	2.23 02
81	Tl L$_{\beta_1}$	12.213	101.519	1.80 02	1.88 02	1.96 02	2.06 02	2.14 02
88	Ra L$_{\alpha_1}$	12.340	100.474	1.75 02	1.83 02	1.91 02	2.01 02	2.08 02
34	Se K$_{\beta_1}$	12.496	99.219	1.70 02	1.77 02	1.84 02	1.94 02	2.01 02
82	Pb L$_{\beta_1}$	12.614	98.291	1.65 02	1.73 02	1.80 02	1.89 02	1.96 02
36	Kr K$_{\alpha_1}$	12.649	98.019	1.64 02	1.71 02	1.78 02	1.88 02	1.94 02
89	Ac L$_{\alpha_1}$	12.652	97.996	1.64 02	1.71 02	1.78 02	1.88 02	1.94 02
90	Th L$_{\alpha_1}$	12.969	95.601	1.53 02	1.60 02	1.67 02	1.75 02	1.82 02
83	Bi L$_{\beta_1}$	13.024	95.197	1.52 02	1.58 02	1.65 02	1.73 02	1.80 02
91	Pa L$_{\alpha_1}$	13.291	93.285	1.43 02	1.50 02	1.56 02	1.64 02	1.70 02
35	Br K$_{\beta_1}$	13.291	93.285	1.43 02	1.50 02	1.56 02	1.64 02	1.70 02
37	Rb K$_{\alpha_1}$	13.395	92.560	1.41 02	1.47 02	1.53 02	1.61 02	1.66 02
84	Po L$_{\beta_1}$	13.447	92.202	1.39 02	1.45 02	1.51 02	1.59 02	1.65 02
92	U L$_{\alpha_1}$	13.615	91.065	1.34 02	1.40 02	1.46 02	1.54 02	1.59 02
85	At L$_{\beta_1}$	13.876	89.352	1.28 02	1.33 02	1.39 02	1.46 02	1.51 02
36	Kr K$_{\beta_1}$	14.112	87.857	1.22 02	1.27 02	1.33 02	1.40 02	1.45 02
38	Sr K$_{\alpha_1}$	14.165	87.529	1.21 02	1.26 02	1.32 02	1.38 02	1.43 02
86	Rn L$_{\beta_1}$	14.316	86.606	1.17 02	1.23 02	1.28 02	1.34 02	1.39 02
87	Fr L$_{\beta_1}$	14.770	83.943	1.08 02	1.13 02	1.18 02	1.23 02	1.28 02
39	Y K$_{\alpha_1}$	14.958	82.888	1.04 02	1.09 02	1.14 02	1.19 02	1.23 02
37	Rb K$_{\beta_1}$	14.961	82.872	1.04 02	1.09 02	1.14 02	1.19 02	1.23 02
88	Ra L$_{\beta_1}$	15.236	81.376	9.92 01	1.04 02	1.08 02	1.13 02	1.17 02
89	Ac L$_{\beta_1}$	15.713	78.906	9.13 01	9.55 01	9.95 01	1.04 02	1.08 02
40	Zr K$_{\alpha_1}$	15.775	78.596	9.03 01	9.45 01	9.85 01	1.03 02	1.07 02
38	Sr K$_{\beta_1}$	15.836	78.293	8.94 01	9.35 01	9.75 01	1.02 02	1.06 02
90	Th L$_{\beta_1}$	16.202	76.524	8.41 01	8.79 01	9.17 01	9.62 01	9.95 01
41	Nb K$_{\alpha_1}$	16.615	74.622	7.86 01	8.22 01	8.57 01	8.99 01	9.30 01
91	Pa L$_{\beta_1}$	16.702	74.233	7.75 01	8.11 01	8.46 01	8.86 01	9.17 01
39	Y K$_{\beta_1}$	16.738	74.074	7.71 01	8.06 01	8.41 01	8.81 01	9.11 01
92	U L$_{\beta_1}$	17.220	72.000	7.14 01	7.47 01	7.79 01	8.16 01	8.44 01
42	Mo K$_{\alpha_1}$	17.479	70.933	6.86 01	7.18 01	7.49 01	7.84 01	8.11 01
40	Zr K$_{\beta_1}$	17.668	70.175	6.67 01	6.97 01	7.27 01	7.62 01	7.88 01
43	Tc K$_{\alpha_1}$	18.367	67.504	6.01 01	6.29 01	6.56 01	6.86 01	7.09 01
41	Nb K$_{\beta_1}$	18.623	66.576	5.79 01	6.06 01	6.32 01	6.61 01	6.83 01
44	Ru K$_{\alpha_1}$	19.279	64.311	5.28 01	5.52 01	5.76 01	6.02 01	6.22 01
42	Mo K$_{\beta_1}$	19.608	63.231	5.04 01	5.28 01	5.51 01	5.75 01	5.95 01
45	Rh K$_{\alpha_1}$	20.216	61.330	4.65 01	4.86 01	5.07 01	5.30 01	5.48 01
43	Tc K$_{\beta_1}$	20.619	60.131	4.41 01	4.61 01	4.81 01	5.03 01	5.20 01
46	Pd K$_{\alpha_1}$	21.177	58.547	4.11 01	4.30 01	4.48 01	4.69 01	4.84 01
44	Ru K$_{\beta_1}$	21.657	57.249	3.87 01	4.05 01	4.22 01	4.42 01	4.56 01
47	Ag K$_{\alpha_1}$	22.163	55.942	3.64 01	3.81 01	3.97 01	4.15 01	4.29 01
45	Rh K$_{\beta_1}$	22.724	54.561	3.40 01	3.56 01	3.72 01	3.89 01	4.02 01
48	Cd K$_{\alpha_1}$	23.174	53.501	3.23 01	3.38 01	3.53 01	3.69 01	3.81 01

Z	transition	E [keV]	λ [pm]	$_{66}$Dy	$_{67}$Ho	$_{68}$Er	$_{69}$Tm	$_{70}$Yb
46	Pd K$_{\beta_1}$	23.819	52.053	3.00 01	3.15 01	3.28 01	3.43 01	3.54 01
49	In K$_{\alpha_1}$	24.210	51.212	2.88 01	3.01 01	3.14 01	3.29 01	3.39 01
47	Ag K$_{\beta_1}$	24.942	49.709	2.66 01	2.78 01	2.90 01	3.04 01	3.14 01
50	Sn K$_{\alpha_1}$	25.271	49.067	2.57 01	2.69 01	2.80 01	2.93 01	3.03 01
48	Cd K$_{\beta_1}$	26.096	47.511	2.36 01	2.47 01	2.57 01	2.69 01	2.78 01
51	Sb K$_{\alpha_1}$	29.359	47.037	2.29 01	2.40 01	2.50 01	2.62 01	2.71 01
49	In K$_{\beta_1}$	27.276	45.455	2.10 01	2.20 01	2.29 01	2.40 01	2.47 01
52	Te K$_{\alpha_1}$	27.472	45.131	2.06 01	2.15 01	2.24 01	2.35 01	2.43 01
50	Sn K$_{\beta_1}$	28.486	43.524	1.87 01	1.96 01	2.04 01	2.13 01	2.20 01
53	I K$_{\alpha_1}$	28.612	43.333	1.85 01	1.95 01	2.01 01	2.11 01	2.18 01
51	Sb K$_{\beta_1}$	29.726	41.709	1.67 01	1.75 01	1.82 01	1.91 01	1.97 01
54	Xe K$_{\alpha_1}$	29.779	41.635	1.66 01	1.74 01	1.81 01	1.90 01	1.96 01
55	Cs K$_{\alpha_1}$	30.996	40.030	1.50 01	1.57 01	1.63 01	1.71 01	1.77 01
52	Te K$_{\beta_1}$	30.996	40.000	1.49 01	1.57 01	1.63 01	1.71 01	1.76 01
56	Ba K$_{\alpha_1}$	32.194	38.511	1.35 01	1.42 01	1.48 01	1.55 01	1.60 01
53	I K$_{\beta_1}$	32.295	38.391	1.34 01	1.41 01	1.46 01	1.53 01	1.58 01
57	La K$_{\alpha_1}$	33.442	37.074	1.22 01	1.28 01	1.34 01	1.40 01	1.44 01
54	Xe K$_{\beta_1}$	33.624	36.874	1.21 01	1.27 01	1.32 01	1.38 01	1.42 01
58	Ce K$_{\alpha_1}$	34.279	36.169	1.15 01	1.20 01	1.25 01	1.31 01	1.35 01
55	Cs K$_{\beta_1}$	34.987	35.437	1.09 01	1.14 01	1.19 01	1.24 01	1.28 01
59	Pr K$_{\alpha_1}$	36.026	34.415	1.01 01	1.06 01	1.10 01	1.15 01	1.19 01
56	Ba K$_{\beta_1}$	36.378	34.082	9.83 00	1.03 01	1.07 01	1.12 01	1.16 01
60	Nd K$_{\alpha_1}$	36.847	33.648	9.51 00	9.96 00	1.04 01	1.09 01	1.12 01
57	La K$_{\beta_1}$	37.801	32.799	8.89 00	9.32 00	9.69 00	1.02 01	1.05 01
61	Pm K$_{\alpha_1}$	38.725	32.016	8.35 00	8.75 00	9.09 00	9.55 00	9.81 00
58	Ce K$_{\beta_1}$	39.257	31.582	8.06 00	8.44 00	8.77 00	9.21 00	9.47 00
62	Sm K$_{\alpha_1}$	40.257	30.905	7.61 00	7.97 00	8.29 00	8.70 00	8.94 00
59	Pr K$_{\beta_1}$	40.748	30.427	7.31 00	7.66 00	7.97 00	8.36 00	8.59 00
63	Eu K$_{\alpha_1}$	41.542	29.845	6.95 00	7.29 00	7.58 00	7.95 00	8.17 00
60	Nd K$_{\beta_1}$	42.271	29.331	6.65 00	6.97 00	7.25 00	7.60 00	7.81 00
64	Gd K$_{\alpha_1}$	42.996	28.836	6.36 00	6.67 00	6.94 00	7.27 00	7.47 00
61	Pm K$_{\beta_1}$	43.826	28.290	6.05 00	6.35 00	6.61 00	6.92 00	7.11 00
65	Tb K$_{\alpha_1}$	44.482	27.873	5.82 00	6.11 00	6.36 00	6.66 00	6.86 00
62	Sm K$_{\beta_1}$	45.413	27.301	5.52 00	5.80 00	6.04 00	6.31 00	6.48 00
66	Dy K$_{\alpha_1}$	45.998	26.954	5.34 00	5.61 00	5.84 00	6.11 00	6.27 00
63	Eu K$_{\beta_1}$	47.038	26.358	5.04 00	5.29 00	5.52 00	5.76 00	5.91 00
67	Ho K$_{\alpha_1}$	47.547	26.076	4.90 00	5.15 00	5.37 00	5.60 00	5.75 00
64	Gd K$_{\beta_1}$	48.697	25.460	4.60 00	4.84 00	5.05 00	5.27 00	5.40 00
68	Er K$_{\alpha_1}$	49.128	25.237	4.50 00	4.74 00	4.93 00	5.15 00	5.28 00
65	Tb K$_{\beta_1}$	50.382	24.609	4.22 00	4.44 00	4.63 00	4.83 00	4.95 00
69	Tm K$_{\alpha_1}$	50.742	24.434	4.14 00	4.36 00	4.54 00	4.74 00	4.86 00
66	Dy K$_{\beta_1}$	52.119	23.788	3.88 00	4.08 00	4.25 00	4.43 00	4.55 00
70	Yb K$_{\alpha_1}$	52.389	23.666	<u>3.83 00</u>	4.03 00	4.19 00	4.37 00	4.49 00
67	Ho K$_{\beta_1}$	53.877	23.012	1.86 01	3.75 00	3.90 00	4.07 00	4.18 00

Z	transition	E [keV]	λ [pm]	$_{66}$Dy	$_{67}$Ho	$_{68}$Er	$_{69}$Tm	$_{70}$Yb
71	Lu K$_{\alpha_1}$	54.070	22.930	1.84 01	3.72 00	3.87 00	4.04 00	4.14 00
68	Er K$_{\beta_1}$	55.681	22.267	1.69 01	1.76 01	3.59 00	3.75 00	3.85 00
72	Hf K$_{\alpha_1}$	55.790	22.223	1.68 01	1.75 01	3.57 00	3.73 00	3.83 00
69	Tm K$_{\beta_1}$	57.517	21.556	1.55 01	1.61 01	1.68 00	3.45 00	3.54 00
73	Ta K$_{\alpha_1}$	57.53Z	21.550	1.55 01	1.61.01	1.68 00	3.45 00	3.54 00
74	W K$_{\alpha_1}$	59.318	20.901	1.42 01	1.48 01	1.54 01	3.19 00	3.28 00
70	Yb K$_{\beta_1}$	59.370	20.883	1.42 01	1.47 01	1.54 01	3.19 00	3.27 00
75	Re K$_{\alpha_1}$	61.140	20.278	1.30 01	1.36 01	1.42 01	1.48 01	3.04 00
71	Lu K$_{\beta_1}$	61.283	20.231	1.30 01	1.35 01	1.41 01	1.47 01	3.02 00
76	Os K$_{\alpha_1}$	63.001	19.679	1.20 01	1.25 01	1.30 01	1.36 01	1.38 01
72	Hf K$_{\beta_1}$	63.234	19.607	1.19 01	1.24 01	1.29 01	1.35 01	1.36 01
77	Ir K$_{\alpha_1}$	64.896	19.105	1.11 01	1.15 01	1.20 01	1.25 01	1.27 01
73	Ta K$_{\beta_1}$	65.223	19.009	1.09 01	1.14 01	1.18 01	1.24 01	1.25 01
78	Pt K$_{\alpha_1}$	66.832	18.551	1.02 01	1.06 01	1.11 01	1.16 01	1.17 01
74	W K$_{\beta_1}$	67.244	18.438	1.00 01	1.04 01	1.09 01	1.14 01	1.15 01
79	Au K$_{\alpha_1}$	68.804	18.020	9.40 00	9.81 00	1.02 01	1.07 01	1.08 01
75	Re K$_{\beta_1}$	69.310	17.888	9.21 00	9.61 00	1.00 01	1.04 01	1.06 01
80	Hg K$_{\alpha_1}$	70.819	17.507	8.68 00	9.05 00	9.43 00	9.84 00	1.00 01
76	Os K$_{\beta_1}$	71.413	17.361	8.48 00	8.85 00	9.22 00	9.62 00	9.80 00
81	Tl K$_{\alpha_1}$	72.872	17.014	8.02 00	8.37 00	8.71 00	9.09 00	9.27 00
77	Ir K$_{\beta_1}$	73.561	16.854	7.81 00	8.15 00	8.49 00	8.86 00	9.04 00
8Z	Pb K$_{\alpha_1}$	74.969	16.538	7.51 00	7.74 00	8.06 00	8.41 00	8.59 00
78	Pt K$_{\beta_1}$	75.748	16.368	7.20 00	7.52 00	7.83 00	8.17 00	8.35 00
83	Bi K$_{\alpha_1}$	77.108	16.079	6.85 00	7.16 00	7.45 00	7.78 00	7.96 00
79	Au K$_{\beta_1}$	77.948	15.906	6.65 00	6.95 00	7.Z4 00	7.55 00	7.73 00
84	Po K$_{\alpha_1}$	79.290	15.636	6.34 00	6.63 00	6.90 00	7.20 00	7.38 00
80	Hg K$_{\beta_1}$	80.253	15.449	6.14 00	6.41 00	6.68 00	6.96 00	7.14 00
85	At K$_{\alpha_1}$	81.520	15.209	5.88 00	6.14 00	6.40 00	6.67 00	6.85 00
81	Tl K$_{\beta_1}$	82.576	15.014	5.68 00	5.93 00	6.18 00	6.44 00	6.61 00
86	Rn K$_{\alpha_1}$	83.780	14.798	5.46 00	5.70 00	5.94 00	6.19 00	6.36 00
82	Pb K$_{\beta_1}$	84.936	14.597	5.26 00	5.49 00	5.72 00	5.96 00	6.12 00
87	Fr K$_{\alpha_1}$	86.100	14.400	5.07 00	5.29 00	5.52 00	5.74 00	5.90 00
83	Bi K$_{\beta_1}$	87.343	14.195	4.87 00	5.09 00	5.30 00	5.52 00	5.68 00
88	Ra K$_{\alpha_1}$	88.470	14.014	4.70 00	4.91 00	5.12 00	5.33 00	5.48 00
84	Po K$_{\beta_1}$	89.800	13.806	4.52 00	4.72 00	4.92 00	5.12 00	5.27 00
89	Ac K$_{\alpha_1}$	90.884	13.642	4.37 00	4.56 00	4.76 00	4.95 00	5.10 00
85	At K$_{\beta_1}$	92.300	13.432	4.19 00	4.38 00	4.57 00	4.75 00	4.89 00
90	Th K$_{\alpha_1}$	93.350	13.281	4.06 00	4.24 00	4.43 00	4.60 00	4.74 00
86	Rn K$_{\beta_1}$	94.870	13.068	3.89 00	4.06 00	4.24 00	4.41 00	4.54 00
91	Pa K$_{\alpha_1}$	95.868	12.932	3.78 00	3.95 00	4.12 00	4.28 00	4.41 00
87	Fr K$_{\beta_1}$	97.470	12.720	3.61 00	3.77 00	3.94 00	4.09 00	4.22 00
92	U K$_{\alpha_1}$	98.439	12.595	3.52 00	3.67 00	3.83 00	3.98 00	4.10 00
88	Ra K$_{\beta_1}$	100.130	12.382	3.36 00	3.50 00	3.66 00	3.80 00	3.92 00
89	Ac K$_{\beta_1}$	102.850	12.054	3.13 00	3.27 00	3.41 00	3.54 00	3.65 00

Z	transition	E [keV]	λ [pm]	$_{66}$Dy	$_{67}$Ho	$_{68}$Er	$_{69}$Tm	$_{70}$Yb
90	Th K$_{\beta_1}$	105.610	11.739	2.92 00	3.05 00	3.18 00	3.31 00	3.41 00
91	Pa K$_{\beta_1}$	108.430	11.434	2.72 00	2.84 00	2.97 00	3.08 00	3.18 00
92	U K$_{\beta_1}$	111.300	11.139	2.54 00	2.65 00	2.77 00	2.88 00	2.97 00

Z	transition	E [keV]	λ [pm]	$_{71}$Lu	$_{72}$Hf	$_{73}$Ta	$_{74}$W	$_{75}$Re
4	Be K$_\alpha$	0.1085	11427.207	2.21 04	2.08 04	2.23 04	2.04 04	1.60 04
38	Sr M$_\xi$	0.1140	10875.895	2.22 04	2.07 04	2.21 04	2.01 04	1.63 04
39	Y M$_\xi$	0.1328	9339.235	2.28 04	2.12 04	2.15 04	1.93 04	1.72 04
16	S L$_l$	0.1487	8337.942	2.32 04	2.22 04	2.13 04	1.83 04	1.75 04
40	Zr M$_\xi$	0.1511	8205.506	2.32 04	2.23 04	2.13 04	1.85 04	1.75 04
41	Nb M$_\xi$	0.1717	7221.037	2.23 04	2.18 04	2.08 04	1.93 04	1.73 04
5	B K$_\alpha$	0.1833	6764.059	2.19 04	2.16 04	2.08 04	1.97 04	1.80 04
42	Mo M$_\xi$	0.1926	6473.445	2.16 04	2.15 04	2.09 04	2.00 04	1.86 04
6	C K$_\alpha$	0.2770	4476.000	1.74 04	1.80 04	1.84 04	1.88 04	1.87 04
47	Ag M$_\xi$	0.3117	3977.709	1.61 04	1.61 04	1.65 04	1.69 04	1.70 04
7	N K$_\alpha$	0.3924	3159.664	1.39 04	1.29 04	1.34 04	1.39 04	1.41 04
22	Ti L$_l$	0.3953	3136.484	1.38 04	1.28 04	1.33 04	1.38 04	1.40 04
22	Ti L$_\alpha$	0.4522	2741.822	1.17 04	1.22 04	1.26 04	1.30 04	1.21 04
23	V L$_\alpha$	0.5113	2424.901	1.05 04	1.05 04	1.09 04	1.13 04	1.14 04
8	O K$_\alpha$	0.5249	2362.072	1.03 04	1.01 04	1.06 04	1.10 04	1.13 04
25	Mn L$_l$	0.5563	2228.747	9.43 03	9.90 03	9.75 03	1.02 04	1.04 04
24	Cr L$_\alpha$	0.5728	2164.549	9.02 03	9.46 03	9.93 03	9.74 03	1.00 04
25	Mn L$_\alpha$	0.6374	1945.171	7.65 03	8.01 03	8.40 03	8.78 03	9.16 03
9	F K$_\alpha$	0.6768	1831.932	6.90 03	7.23 03	7.59 03	7.93 03	8.23 03
26	Fe L$_\alpha$	0.7050	1758.655	6.48 03	6.80 03	7.14 03	7.46 03	7.72 03
27	Co L$_\alpha$	0.7762	1597.335	5.48 03	5.74 03	6.05 03	6.34 03	6.57 03
28	Ni L$_\alpha$	0.8515	1456.080	4.62 03	4.86 03	5.13 03	5.36 03	5.57 03
29	Cu L$_\alpha$	0.9297	1336.044	3.93 03	4.14 03	4.37 03	4.56 03	4.75 03
30	Zn L$_\alpha$	1.0117	1225.513	3.35 03	3.53 03	3.71 03	3.88 03	4.06 03
11	Na K$_\alpha$	1.0410	1191.020	3.16 03	3.33 03	3.51 03	3.67 03	3.84 03
11	Na K$_\beta$	1.0711	1157.550	2.84 03	2.93 03	3.16 03	3.35 03	3.48 03
12	Mg K$_\alpha$	1.2536	989.033	2.66 03	2.75 03	2.87 03	3.00 03	3.13 03
33	As L$_\alpha$	1.2820	967.123	2.50 03	2.59 03	2.70 03	2.83 03	2.95 03
12	Mg K$_\beta$	1.3022	952.121	2.40 03	2.48 03	2.59 03	2.72 03	2.83 03
33	As L$_\beta$	1.3170	441.421	2.33 03	2.41 03	2.51 03	2.64 03	2.75 03
66	Dy M$_\beta$	1.3250	935.737	2.29 03	2.37 03	2.47 03	2.60 03	2.71 03
67	Ho M$_\alpha$	1.3480	919.771	2.19 03	2.27 03	2.36 03	2.48 03	2.59 03
34	Se L$_\alpha$	1.3791	899.029	2.06 03	2.13 03	2.22 03	2.34 03	2.44 03
67	Ho M$_\beta$	1.3830	896.494	2.04 03	2.12 03	2.21 03	2.32 03	2.42 03
68	Er M$_\alpha$	1.4060	881.829	1.96 03	2.03 03	2.11 03	2.23 03	2.32 03
34	Se L$_\beta$	1.4192	873.627	1.91 03	1.98 03	2.06 03	2.17 03	2.27 03
68	Er M$_\beta$	1.4430	859.218	1.82 03	1.89 03	1.97 03	2.08 03	2.17 03

Z	transition	E [keV]	λ [pm]	$_{71}$Lu	$_{72}$Hf	$_{73}$Ta	$_{74}$W	$_{75}$Re
69	Tm M_α	1.4620	848.051	1.76 03	1.82 03	1.90 03	2.01 03	2.10 03
35	Br L_α	1.4804	837.511	1.70 03	1.76 03	1.84 03	1.95 03	2.03 03
13	Al K_{α_1}	1.4867	833.962	1.68 03	1.74 03	1.82 03	1.92 03	2.01 03
69	Tm M_β	1.5030	824.918	1.64 03	1.69 03	1.76 03	1.87 03	1.95 03
70	Yb M_α	1.5214	814.941	1.58 03	1.64 03	1.71 03	1.81 03	1.89 03
35	Br L_{β_1}	1.5259	812.538	1.57 03	1.63 03	1.69 03	1.80 03	1.88 03
13	Al K_β	1.5574	796.103	1.49 03	1.54 03	1.61 03	1.71 03	1.78 03
70	Yb M_β	1.5675	790.974	1.46 03	1.51 03	1.58 03	1.68 03	1.75 03
71	Lu M_α	1.5813	784.071	1.43 03	1.48 03	1.54 03	1.64 03	1.71 03
36	Kr L_α	1.5860	781.747	<u>1.42 03</u>	1.47 03	1.53 03	1.63 03	1.70 03
71	Lu M_β	1.6312	760.085	4.28 03	1.36 03	1.42 03	1.51 03	1.58 03
36	Kr L_{β_1}	1.6366	757.577	4.25 03	1.35 03	1.41 03	1.50 03	1.57 03
72	Hf M_{α_1}	1.6446	753.892	<u>6.28 03</u>	<u>1.33 03</u>	1.39 03	1.48 03	1.55 03
37	Rb L_{α_1}	1.6941	731.864	5.81 03	4.05 03	1.28 03	1.37 03	1.43 03
72	Hf M_β	1.6976	730.355	5.78 03	4.02 03	1.28 03	1.36 03	1.43 03
73	Ta M_{α_1}	1.7096	725.229	5.68 03	<u>3.95 03</u>	<u>1.25 03</u>	1.34 03	1.40 03
14	Si K_{α_1}	1.7400	712.558	5.43 03	5.66 03	4.03 03	1.28 03	1.34 03
37	Rb L_{β_1}	1.7522	707.597	5.33 03	5.56 03	3.96 03	1.26 03	1.31 03
73	Ta M_β	1.7655	702.266	5.23 03	5.45 03	3.88 03	1.23 03	1.29 03
74	W M_{α_1}	1.7754	698.350	5.15 03	5.38 03	<u>3.82 03</u>	1.21 03	1.27 03
38	Sr L_{α_1}	1.8066	686.290	4.92 03	5.14 03	5.48 03	<u>1.16 03</u>	1.21 03
74	W M_β	1.8349	675.705	4.73 03	4.94 03	5.26 03	3.59 03	1.17 03
14	Si K_β	1.8359	675.337	4.72 03	4.93 03	5.25 03	3.59 03	1.16 03
75	Re M_{α_1}	1.8420	673.100	4.68 03	4.89 03	5.21 03	3.56 03	1.15 03
38	Sr L_{β_1}	1.8717	662.420	4.49 03	4.69 03	4.99 03	<u>3.41 03</u>	<u>1.11 03</u>
75	Re M_β	1.9061	650.465	4.29 03	4.47 03	4.76 03	4.88 03	3.39 03
76	Os M_{α_1}	1.9102	649.069	4.26 03	4.45 03	4.73 03	4.86 03	3.37 03
39	Y L_{α_1}	1.9226	644.882	4.19 03	4.37 03	4.65 03	4.78 03	<u>3.31 03</u>
76	Os M_β	1.9783	626.725	3.89 03	4.06 03	4.32 03	4.45 03	4.62 03
77	Ir M_{α_1}	1.9799	626.219	3.89 03	4.05 03	4.31 03	4.44 03	4.61 03
39	Y L_{β_1}	1.9958	621.230	3.81 03	3.97 03	4.22 03	4.35 03	4.52 03
15	P K_{α_1}	2.0137	615.708	<u>3.73 03</u>	3.88 03	4.12 03	4.25 03	4.42 03
40	Zr L_{α_1}	2.0424	607.056	4.30 03	3.74 03	3.97 03	4.09 03	4.26 03
78	Pt M_{α_1}	2.0505	604.658	4.26 03	3.70 03	3.93 03	4.05 03	4.21 03
77	Ir M_β	2.0535	603.775	4.24 03	<u>3.68 03</u>	3.91 03	4.04 03	4.20 03
79	Au M_{α_1}	2.1229	584.036	3.89 03	4.07 03	3.59 03	3.70 03	3.85 03
40	Zr L_{β_1}	2.1244	583.624	3.89 03	4.07 03	3.58 03	3.69 03	3.84 03
78	Pt M_β	2.1273	582.828	3.87 03	4.05 03	3.57 03	3.68 03	3.83 03
15	P $K_{\beta_{1,3}}$	2.1390	579.640	3.82 03	3.99 03	3.52 03	3.63 03	3.77 03
41	Nb L_{α_1}	2.1659	572.441	3.70 03	3.87 03	<u>3.40 03</u>	3.51 03	3.65 03
80	Hg M_{α_1}	2.1953	564.775	3.57 03	3.73 03	3.73 03	3.39 03	3.52 03
79	Au M_β	2.2046	562.393	3.53 03	3.69 03	3.91 03	3.35 03	3.49 03
41	Nb L_{β_1}	2.2574	549.238	<u>3.32 03</u>	3.47 03	3.67 03	3.15 03	3.28 03
81	Tl M_{α_1}	2.2706	546.045	3.59 00	3.42 03	3.62 03	<u>3.10 03</u>	3.23 03

Z	transition	E [keV]	λ [pm]	$_{71}$Lu	$_{72}$Hf	$_{73}$Ta	$_{74}$W	$_{75}$Re
80	Hg M$_\beta$	2.2825	543.199	3.54 03	3.37 03	3.57 03	3.66 03	3.18 03
42	Mo L$_{\alpha_1}$	2.2932	540.664	3.50 03	3.33 03	3.52 03	3.62 03	3.14 03
16	S K$_{\alpha_1}$	2.3080	537.197	3.44 03	3.27 03	3.46 03	3.56 03	3.09 03
82	Pb M$_{\alpha_1}$	2.3457	528.563	3.30 03	3.14 03	3.32 03	3.41 03	2.96 03
81	Tl M$_\beta$	2.3621	524.893	3.24 03	<u>3.08 03</u>	3.26 03	3.35 03	<u>2.91 03</u>
42	Mo L$_{\beta_1}$	2.3948	517.726	3.13 03	3.27 03	3.14 03	3.24 03	3.36 03
83	Bi M$_{\alpha_1}$	2.4226	511.785	3.04 03	3.17 03	3.05 03	3.14 03	3.26 03
43	Tc L$_{\alpha_1}$	2.4240	511.490	3.04 03	3.17 03	3.04 03	3.14 03	3.25 03
82	Pb M$_\beta$	2.4427	507.574	2.98 03	3.10 03	2.98 03	3.07 03	3.19 03
16	S K$_\beta$	2.4640	503.186	<u>2.91 03</u>	3.03 03	<u>2.91 03</u>	3.01 03	3.12 03
83	Bi M$_{\beta_1}$	2.5255	490.933	3.01 03	2.85 03	3.00 03	2.82 03	2.93 03
43	Tc L$_{\beta_1}$	2.5368	488.746	2.97 03	2.81 03	2.97 03	2.79 03	2.90 03
44	Ru L$_{\alpha_1}$	2.5586	484.582	2.91 03	<u>2.75 03</u>	2.90 03	<u>2.73 03</u>	2.83 03
17	Cl K$_{\alpha_1}$	2.6224	472.792	2.73 03	2.84 03	2.72 03	2.81 03	<u>2.66 03</u>
44	Ru L$_{\beta_1}$	2.6832	462.079	2.57 03	2.68 03	2.56 03	2.65 03	2.76 03
45	Rh L$_{\alpha_1}$	2.6967	459.766	2.54 03	2.64 03	<u>2.53 03</u>	2.62 03	2.72 03
17	Cl K$_\beta$	2.8156	440.350	2.27 03	2.36 03	2.49 03	<u>2.34 03</u>	2.43 03
45	Rh L$_{\beta_1}$	2.8344	437.430	2.23 03	2.32 03	2.44 03	2.53 03	2.39 03
46	Pd L$_{\alpha_1}$	2.8386	436.782	2.22 03	2.31 03	2.43 03	2.52 03	<u>2.38 03</u>
18	Ar K$_{\alpha_1}$	2.9577	419.194	2.00 03	2.08 03	2.19 03	2.26 03	2.36 03
47	Ag L$_{\alpha_1}$	2.9843	415.458	1.95 03	2.03 03	2.14 03	2.21 03	2.30 03
46	Pd L$_{\beta_1}$	2.9902	414.638	1.94 03	2.02 03	2.12 03	2.20 03	2.29 03
90	Th M$_{\alpha_1}$	2.9961	413.821	1.93 03	2.01 03	2.11 03	2.19 03	2.28 03
91	Pa M$_{\alpha_1}$	3.0823	402.248	1.79 03	1.87 03	1.96 03	2.03 03	2.12 03
48	Cd L$_{\alpha_1}$	3.1337	395.651	1.72 03	1.79 03	1.88 03	1.95 03	2.03 03
90	Th M$_\beta$	3.1458	394.129	1.70 03	1.77 03	1.86 03	1.93 03	2.01 03
47	Ag L$_{\beta_1}$	3.1509	393.491	1.70 03	1.76 03	1.85 03	1.92 03	2.00 03
92	U M$_{\alpha_1}$	3.1708	391.021	1.67 03	1.74 03	1.82 03	1.89 03	1.97 03
18	Ar K$_\beta$	3.1905	388.607	1.64 03	1.71 03	1.79 03	1.86 03	1.94 03
91	Pa M$_\beta$	3.2397	382.705	1.58 03	1.64 03	1.72 03	1.79 03	1.86 03
49	In L$_{\alpha_1}$	3.2869	377.210	1.52 03	1.58 03	1.66 03	1.72 03	1.79 03
19	K K$_{\alpha_1}$	3.3136	374.148	1.49 03	1.55 03	1.63 03	1.69 03	1.75 03
48	Cd L$_{\beta_1}$	3.3166	373.832	1.49 03	1.55 03	1.62 03	1.68 03	1.75 03
92	U M$_\beta$	3.3360	371.658	1.46 03	1.52 03	1.60 03	1.66 03	1.72 03
50	Sn L$_{\alpha_1}$	3.4440	360.003	1.35 03	1.40 03	1.47 03	1.53 03	1.59 03
49	In L$_{\beta_1}$	3.4872	355.543	1.30 03	1.36 03	1.42 03	1.48 03	1.54 03
19	K K$_\beta$	3.5896	345.401	1.21 03	1.26 03	1.32 03	1.37 03	1.43 03
51	Sb L$_{\alpha_1}$	3.6047	343.954	1.20 03	1.25 03	1.31 03	1.36 03	1.41 03
50	Sn L$_{\beta_1}$	3.6628	338.498	1.15 03	1.20 03	1.25 03	1.30 03	1.35 03
20	Ca K$_{\alpha_1}$	3.6917	335.848	1.13 03	1.17 03	1.23 03	1.27 03	1.33 03
52	Te L$_{\alpha_1}$	3.7693	328.934	1.07 03	1.11 03	1.16 03	1.21 03	1.26 03
51	Sb L$_{\beta_1}$	3.8436	322.575	1.01 03	1.05 03	1.10 03	1.15 03	1.19 03
53	I L$_{\alpha_1}$	3.9377	314.867	9.53 02	9.91 02	1.04 03	1.08 03	1.12 03
20	Ca K$_\beta$	4.0127	308.981	9.08 02	9.44 02	9.87 02	1.03 03	1.07 03

Z	transition	E [keV]	λ [pm]	$_{71}$Lu	$_{72}$Hf	$_{73}$Ta	$_{74}$W	$_{75}$Re
52	Te L_{β_1}	4.0296	307.686	8.98 02	9.34 02	9.76 02	1.02 03	1.06 03
21	Sc K_{α_1}	4.0906	303.097	8.64 02	8.98 02	9.39 02	9.77 02	1.02 03
54	Xe L_{α_1}	4.1099	301.674	8.54 02	8.87 02	9.27 02	9.65 02	1.00 03
53	I L_{β_1}	4.2207	293.755	7.98 02	8.29 02	8.65 02	9.00 02	9.37 02
55	Cs L_{α_1}	4.2865	289.245	7.67 02	7.97 02	8.31 02	8.65 02	9.01 02
21	Sc K_β	4.4605	277.962	6.92 02	7.19 02	7.49 02	7.80 02	8.12 02
56	Ba L_{α_1}	4.4663	277.601	6.90 02	7.17 02	7.46 02	7.77 02	8.10 02
22	Ti K_{α_1}	4.5108	274.863	6.73 02	6.99 02	7.27 02	7.58 02	7.89 02
55	Cs L_{β_1}	4.6198	268.377	6.33 02	6.57 02	6.83 02	7.12 02	7.42 02
57	La L_{α_1}	4.6510	266.577	6.22 02	6.46 02	6.71 02	7.00 02	7.29 02
56	Ba L_{β_1}	4.8273	256.841	5.65 02	5.87 02	6.09 02	6.35 02	6.62 02
58	Ce L_{α_1}	4.8402	256.157	5.61 02	5.83 02	6.05 02	6.31 02	6.58 02
22	Ti $K_{\beta_{1,3}}$	4.9318	251.399	5.35 02	5.56 02	5.76 02	6.01 02	6.27 02
23	V K_{α_1}	4.9522	250.363	5.29 02	5.50 02	5.70 02	5.95 02	6.20 02
59	Pr L_{α_1}	5.0337	246.310	5.07 02	5.27 02	5.46 02	5.70 02	5.94 02
57	La L_{β_1}	5.0421	245.899	5.05 02	5.25 02	5.44 02	5.68 02	5.92 02
60	Nd L_{α_1}	5.2304	237.047	4.60 02	4.78 02	4.94 02	5.17 02	5.39 02
58	Ce L_{β_1}	5.2622	235.614	4.53 02	4.70 02	4.87 02	5.09 02	5.30 02
24	Cr K_{α_1}	5.4147	228.978	4.21 02	4.37 02	4.52 02	4.73 02	4.93 02
23	V $K_{\beta_{1,3}}$	5.4273	228.447	4.18 02	4.34 02	4.49 02	4.70 02	4.90 02
61	Pm L_{α_1}	5.4325	228.228	4.17 02	4.33 02	4.48 02	4.69 02	4.89 02
59	Pr L_{β_1}	5.4889	225.883	4.06 02	4.22 02	4.36 02	4.57 02	4.76 02
62	Sm L_{α_1}	5.6361	219.984	3.80 02	3.94 02	4.07 02	4.27 02	4.44 02
60	Nd L_{β_1}	5.7216	216.696	3.65 02	3.79 02	3.92 02	4.11 02	4.28 02
63	Eu L_{α_1}	5.8457	212.096	3.46 02	3.59 02	3.70 02	3.89 02	4.05 02
25	Mn K_{α_1}	5.8988	210.187	3.38 02	3.51 02	3.62 02	3.80 02	3.95 02
24	Cr $K_{\beta_{1,3}}$	5.9467	208.494	3.31 02	3.44 02	3.54 02	3.72 02	3.87 02
61	Pm L_{β_1}	5.9610	207.993	3.29 02	3.41 02	3.52 02	3.70 02	3.85 02
64	Gd L_{α_1}	6.0572	204.690	3.16 02	3.28 02	3.38 02	3.55 02	3.69 02
62	Sm L_{β_1}	6.2051	199.811	2.97 02	3.08 02	3.18 02	3.34 02	3.47 02
65	Tb L_{α_1}	6.2728	197.655	2.89 02	3.00 02	3.09 02	3.24 02	3.38 02
26	Fe K_{α_1}	6.4038	193.611	2.74 02	2.84 02	2.93 02	3.08 02	3.20 02
63	Eu L_{β_1}	6.4564	192.034	2.68 02	2.78 02	2.87 02	3.01 02	3.14 02
25	Mn $K_{\beta_{1,3}}$	6.4905	191.025	2.65 02	2.75 02	2.83 02	2.97 02	3.10 02
66	Dy L_{α_1}	6.4952	190.887	2.64 02	2.74 02	2.83 02	2.97 02	3.09 02
64	Gd L_{β_1}	6.7132	184.688	2.43 02	2.52 02	2.60 02	2.72 02	2.84 02
67	Ho L_{α_1}	6.7198	184.507	2.42 02	2.51 02	2.59 02	2.72 02	2.83 02
27	Co K_{α_1}	6.9303	178.903	2.24 02	2.32 02	2.39 02	2.51 02	2.62 02
68	Er L_{α_1}	6.9487	178.429	2.23 02	2.31 02	2.38 02	2.49 02	2.60 02
65	Tb L_{β_1}	6.9780	177.680	2.20 02	2.28 02	2.35 02	2.47 02	2.57 02
26	Fe $K_{\beta_{1,3}}$	7.0580	175.666	2.14 02	2.22 02	2.28 02	2.40 02	2.50 02
69	Tm L_{α_1}	7.1799	172.683	2.05 02	2.12 02	2.19 02	2.29 02	2.39 02
66	Dy L_{β_1}	7.2477	171.068	2.00 02	2.07 02	2.13 02	2.24 02	2.34 02
70	Yb L_{α_1}	7.4156	167.195	1.89 02	1.95 02	2.01 02	2.11 02	2.21 02

Z	transition	E [keV]	λ [pm]	$_{71}$Lu	$_{72}$Hf	$_{73}$Ta	$_{74}$W	$_{75}$Re
28	Ni K$_{\alpha_1}$	7.4782	165.795	1.85 02	1.91 02	1.97 02	2.06 02	2.16 02
67	Ho L$_{\beta_1}$	7.5253	164.757	1.82 02	1.88 02	1.94 02	2.03 02	2.12 02
27	Co K$_{\beta_{1,3}}$	7.6494	162.084	1.74 02	1.80 02	1.86 02	1.95 02	2.04 02
71	Lu L$_{\alpha_1}$	7.6555	161.955	1.74 02	1.80 02	1.86 02	1.94 02	2.03 02
68	Er L$_{\beta_1}$	7.8190	158.733	1.65 02	1.71 02	1.76 02	1.85 02	1.93 02
72	Hf L$_{\alpha_1}$	7.8990	156.963	1.61 02	1.66 02	1.71 02	1.79 02	1.88 02
29	Cu K$_{\alpha_1}$	8.0478	154.060	1.53 02	1.59 02	1.63 02	1.71 02	1.79 02
69	Tm L$_{\beta_1}$	8.1010	153.049	1.51 02	1.56 02	1.60 02	1.68 02	1.76 02
73	Ta L$_{\alpha_1}$	8.1461	152.201	1.49 02	1.54 02	1.58 02	1.66 02	1.74 02
28	Ni K$_{\beta_{1,3}}$	8.2647	150.017	1.43 02	1.48 02	1.52 02	1.60 02	1.67 02
74	W L$_{\alpha_1}$	8.3976	147.643	1.38 02	1.42 02	1.46 02	1.54 02	1.61 02
70	Yb L$_{\beta_1}$	8.4018	147.569	1.37 02	1.42 02	1.46 02	1.53 02	1.60 02
30	Zn K$_{\alpha_1}$	8.6389	143.519	1.28 02	1.33 02	1.36 02	1.43 02	1.50 02
75	Re L$_{\alpha_1}$	8.6525	143.294	1.27 02	1.32 02	1.36 02	1.42 02	1.49 02
71	Lu L$_{\beta_1}$	8.7090	142.364	1.25 02	1.30 02	1.33 02	1.40 02	1.47 02
29	Cu K$_{\beta_1}$	8.9053	139.226	1.18 02	1.23 02	1.26 02	1.32 02	1.38 02
76	Os L$_{\alpha_1}$	8.9117	139.126	1.18 02	1.23 02	1.26 02	1.32 02	1.38 02
72	Hf L$_{\beta_1}$	9.0227	137.414	1.15 02	1.19 02	1.22 02	1.28 02	1.34 02
77	Ir L$_{\alpha_1}$	9.1751	135.132	1.10 02	1.14 02	1.17 02	1.23 02	1.28 02
31	Ga K$_{\alpha_1}$	9.2517	134.013	2.84 02	1.11 02	1.14 02	1.20 02	1.26 02
73	Ta L$_{\beta_1}$	9.3431	132.702	2.77 02	1.09 02	1.11 02	1.17 02	1.23 02
78	Pt L$_{\alpha_1}$	9.4423	131.308	2.69 02	<u>1.06 02</u>	1.08 02	1.14 02	1.19 02
30	Zn K$_{\beta_{1,3}}$	9.5720	129.529	2.59 02	2.70 02	1.05 02	1.10 02	1.15 02
74	W L$_{\beta_1}$	9.6724	128.184	2.52 02	2.63 02	1.02 02	1.07 02	1.12 02
79	Au L$_{\alpha_1}$	9.7133	127.644	2.49 02	2.60 02	<u>1.01 02</u>	1.06 02	1.11 02
32	Ge K$_{\alpha_1}$	9.8864	125.409	2.37 02	2.48 02	2.45 02	1.02 02	1.06 02
80	Hg L$_{\alpha_1}$	9.9888	124.124	2.31 02	2.41 02	2.39 02	9.89 01	1.03 02
75	Re L$_{\beta_1}$	10.010	123.861	2.30 02	2.40 02	2.37 02	<u>9.83 01</u>	1.03 02
31	Ga K$_{\beta_1}$	10.264	120.796	2.15 02	2.24 02	2.22 02	2.43 02	9.65 01
81	Tl L$_{\alpha_1}$	10.269	120.737	<u>2.14 02</u>	2.24 02	2.22 02	2.43 02	9.64 01
76	Os L$_{\beta_1}$	10.355	119.734	2.92 02	2.19 02	2.17 02	2.37 02	<u>6.44 01</u>
33	As K$_{\alpha_1}$	10.544	117.588	2.78 02	2.08 02	2.07 02	2.26 02	2.32 02
82	Pb L$_{\alpha_1}$	10.552	117.499	2.77 02	2.08 02	2.06 02	2.26 02	2.31 02
77	Ir L$_{\beta_1}$	10.708	115.787	2.66 02	<u>2.00 02</u>	1.98 02	2.17 02	2.22 02
83	Bi L$_{\alpha_1}$	10.839	114.388	<u>2.58 02</u>	2.69 02	1.92 02	2.10 02	2.15 02
32	Ge K$_{\beta_1}$	10.982	112.898	2.98 02	2.59 02	1.85 02	2.02 02	2.08 02
78	Pt L$_{\beta_1}$	11.071	111.990	2.91 02	2.54 02	1.81 02	1.98 02	2.03 02
84	Po L$_{\alpha_1}$	11.131	111.387	2.87 02	2.50 02	<u>1.79 02</u>	1.95 02	2.00 02
34	Se K$_{\alpha_1}$	11.222	110.484	2.81 02	<u>2.45 02</u>	2.43 02	1.91 02	1.96 02
85	At L$_{\alpha_1}$	11.427	108.501	2.67 02	2.79 02	2.32 02	1.82 02	1.86 02
79	Au L$_{\beta_1}$	11.442	108.359	2.66 02	2.78 02	<u>2.31 02</u>	<u>1.81 02</u>	1.86 02
33	As K$_{\beta_1}$	11.726	105.735	2.49 02	2.60 02	2.59 02	2.36 02	1.74 02
86	Rn L$_{\alpha_1}$	11.727	105.726	2.49 02	2.60 02	2.59 02	2.36 02	1.74 02
80	Hg L$_{\beta_1}$	11.823	104.867	2.44 02	2.54 02	2.53 02	2.31 02	1.70 02

Z	transition	E [keV]	λ [pm]	$_{71}$Lu	$_{72}$Hf	$_{73}$Ta	$_{74}$W	$_{75}$Re
35	Br K$_{\alpha_1}$	11.924	103.979	2.38 02	2.48 02	2.47 02	2.25 02	1.66 02
87	Fr L$_{\alpha_1}$	12.031	103.054	2.33 02	2.42 02	2.42 02	2.20 02	2.26 02
81	Tl L$_{\beta_1}$	12.213	101.519	2.23 02	2.33 02	2.32 02	2.53 02	2.17 02
88	Ra L$_{\alpha_1}$	12.340	100.474	2.17 02	2.26 02	2.26 02	2.46 02	2.11 02
34	Se K$_{\beta_1}$	12.496	99.219	2.10 02	2.19 02	2.18 02	2.37 02	2.04 02
82	Pb L$_{\beta_1}$	12.614	98.291	2.05 02	2.13 02	2.13 02	2.32 02	2.38 02
36	Kr K$_{\alpha_1}$	12.649	98.019	2.03 02	2.12 02	2.11 02	2.30 02	2.36 02
89	Ac L$_{\alpha_1}$	12.652	97.996	2.03 02	2.12 02	2.11 02	2.30 02	2.36 02
90	Th L$_{\alpha_1}$	12.969	95.601	1.90 02	1.98 02	1.98 02	2.15 02	2.21 02
83	Bi L$_{\beta_1}$	13.024	95.197	1.88 02	1.96 02	1.95 02	2.12 02	2.18 02
91	Pa L$_{\alpha_1}$	13.291	93.285	1.78 02	1.85 02	1.85 02	2.01 02	2.07 02
35	Br K$_{\beta_1}$	13.291	93.285	1.78 02	1.85 02	1.85 02	2.01 02	2.07 02
37	Rb K$_{\alpha_1}$	13.395	92.560	1.74 02	1.81 02	1.81 02	1.97 02	2.02 02
84	Po L$_{\beta_1}$	13.447	92.202	1.72 02	1.79 02	1.79 02	1.95 02	2.00 02
92	U L$_{\alpha_1}$	13.615	91.065	1.66 02	1.73 02	1.73 02	1.88 02	1.94 02
85	At L$_{\beta_1}$	13.876	89.352	1.58 02	1.65 02	1.65 02	1.79 02	1.84 02
36	Kr K$_{\beta_1}$	14.112	87.857	1.51 02	1.5702	1.58 02	1.71 02	1.76 02
38	Sr K$_{\alpha_1}$	14.165	87.529	1.50 02	1.56 02	1.56 02	1.69 02	1.74 02
86	Rn L$_{\beta_1}$	14.316	86.606	1.45 02	1.51 02	1.52 02	1.64 02	1.69 02
87	Fr L$_{\beta_1}$	14.770	83.943	1.34 02	1.39 02	1.39 02	1.51 02	1.55 02
39	Y K$_{\alpha_1}$	14.958	82.888	1.29 02	1.34 02	1.35 02	1.46 02	1.50 02
37	Rb K$_{\beta_1}$	14.961	82.872	1.29 02	1.34 02	1.35 02	1.46 02	1.50 02
88	Ra L$_{\beta_1}$	15.236	81.376	1.23 02	1.28 02	1.28 02	1.42 02	1.43 02
89	Ac L$_{\beta_1}$	15.713	78.906	1.13 02	1.18 02	1.18 02	1.36 02	1.32 02
40	Zr K$_{\alpha_1}$	15.775	78.596	1.12 02	1.16 02	1.17 02	1.36 02	1.30 02
38	Sr K$_{\beta_1}$	15.836	78.293	1.11 02	1.15 02	1.16 02	1.35 02	1.29 02
90	Th L$_{\beta_1}$	16.202	76.524	1.04 02	1.08 02	1.09 02	1.31 02	1.21 02
41	Nb K$_{\alpha_1}$	16.615	74.622	9.73 01	1.01 02	1.02 02	1.27 02	1.13 02
91	Pa L$_{\beta_1}$	16.702	74.233	9.60 01	9.98 01	1.01 02	1.26 02	1.12 02
39	Y K$_{\beta_1}$	16.738	74.074	9.54 01	9.92 01	9.99 01	1.25 02	1.11 02
92	U L$_{\beta_1}$	17.220	72.000	8.84 01	9.19 01	9.27 01	1.21 02	1.03 02
42	Mo K$_{\alpha_1}$	17.479	70.933	8.50 01	8.83 01	8.91 01	1.18 02	9.89 01
40	Zr K$_{\beta_1}$	17.668	70.175	8.25 01	8.58 01	8.65 01	1.17 02	9.61 01
43	Tc K$_{\alpha_1}$	18.367	67.504	7.44 01	7.72 01	7.81 01	1.11 02	8.66 01
41	Nb K$_{\beta_1}$	18.623	66.576	7.17 01	7.44 01	7.52 01	1.09 02	8.34 01
44	Ru K$_{\alpha_1}$	19.279	64.311	6.53 01	6.78 01	6.86 01	1.04 02	7.60 01
42	Mo K$_{\beta_1}$	19.608	63.231	6.24 01	6.48 01	6.56 01	1.02 02	7.26 01
45	Rh K$_{\alpha_1}$	20.216	61.330	5.75 01	5.97 01	6.05 01	9.54 01	6.69 01
43	Tc K$_{\beta_1}$	20.619	60.131	5.46 01	5.66 01	5.74 01	8.88 01	6.35 01
46	Pd K$_{\alpha_1}$	21.177	58.547	5.08 01	5.27 01	5.35 01	8.05 01	5.91 01
47	Ag K$_{\alpha_1}$	22.163	55.942	4.50 01	4.67 01	4.75 01	6.82 01	5.24 01
44	Ru K$_{\beta_1}$	21.657	57.249	4.79 01	4.79 01	5.04 01	7.42 01	5.57 01
45	Rh K$_{\beta_1}$	22.724	54.561	4.21 01	4.37 01	4.44 01	6.22 01	4.90 01
48	Cd K$_{\alpha_1}$	23.174	53.501	4.00 01	4.15 01	4.22 01	5.79 01	4.65 01

Z	transition	E [keV]	λ [pm]	$_{71}$Lu	$_{72}$Hf	$_{73}$Ta	$_{74}$W	$_{75}$Re
46	Pd K$_{\beta_1}$	23.819	52.053	3.71 01	3.85 01	3.92 01	5.24 01	4.32 01
49	In K$_{\alpha_1}$	24.210	51.212	3.56 01	3.69 01	3.76 01	4.93 01	4.14 01
47	Ag K$_{\beta_1}$	24.942	49.709	3.28 01	3.41 01	3.47 01	4.42 01	3.82 01
50	Sn K$_{\alpha_1}$	25.271	49.062	3.17 01	3.29 01	3.36 01	4.22 01	3.69 01
48	Cd K$_{\beta_1}$	26.096	47.511	2.91 01	3.02 01	3.08 01	3.75 01	3.39 01
51	Sb K$_{\alpha_1}$	26.359	47.037	2.83 01	2.94 01	3.00 01	3.61 01	3.30 01
49	In K$_{\beta_1}$	27.276	45.455	2.59 01	2.68 01	2.74 01	3.19 01	3.01 01
52	Te K$_{\alpha_1}$	27.472	45.131	2.54 01	2.63 01	2.69 01	3.11 01	2.95 01
50	Sn K$_{\beta_1}$	28.486	43.524	2.30 01	2.39 01	2.45 01	2.72 01	2.68 01
53	I K$_{\alpha_1}$	28.612	43.333	2.28 01	2.36 01	2.42 01	2.68 01	2.65 01
51	Sb K$_{\beta_1}$	29.726	41.709	2.06 01	2.13 01	2.19 01	2.33 01	2.39 01
54	Xe K$_{\alpha_1}$	29.779	41.635	2.05 01	2.12 01	2.18 01	2.31 01	2.38 01
55	Cs K$_{\alpha_1}$	30.973	40.030	1.85 01	1.91 01	1.96 01	2.07 01	2.15 01
52	Te K$_{\beta_1}$	30.996	40.000	1.84 01	1.91 01	1.96 01	2.06 01	2.14 01
56	Ba K$_{\alpha_1}$	32.194	38.511	1.67 01	1.73 01	1.78 01	1.87 01	1.94 01
53	I K$_{\beta_1}$	32.295	38.391	1.65 01	1.71 01	1.76 01	1.85 01	1.92 01
57	La K$_{\alpha_1}$	33.442	37.074	1.51 01	1.56 01	1.61 01	1.69 01	1.76 01
54	Xe K$_{\beta_1}$	33.624	36.874	1.49 01	1.54 01	1.59 01	1.66 01	1.73 01
58	Ce K$_{\alpha_1}$	34.279	36.169	1.41 01	1.46 01	1.51 01	1.58 01	1.64 01
55	Cs K$_{\beta_1}$	34.987	35.437	1.34 01	1.39 01	1.43 01	1.50 01	1.56 01
59	Pr K$_{\alpha_1}$	36.026	34.415	1.24 01	1.28 01	1.32 01	1.39 01	1.44 01
56	Ba K$_{\beta_1}$	36.378	34.082	1.21 01	1.25 01	1.29 01	1.35 01	1.41 01
60	Nd K$_{\alpha_1}$	36.847	33.648	1.17 01	1.21 01	1.25 01	1.31 01	1.36 01
57	La K$_{\beta_1}$	37.801	32.799	1.09 01	1.13 01	1.17 01	1.22 01	1.27 01
61	Pm K$_{\alpha_1}$	38.725	32.016	1.03 01	1.06 01	1.10 01	1.15 01	1.19 01
58	Ce K$_{\beta_1}$	39.257	31.582	9.91 00	1.02 01	1.06 01	1.11 01	1.15 01
62	Sm K$_{\alpha_1}$	40.118	30.905	9.36 00	9.68 00	1.00 01	1.04 01	1.09 01
59	Pr K$_{\beta_1}$	40.748	30.427	8.99 00	9.29 00	9.62 00	1.00 01	1.04 01
63	Eu K$_{\alpha_1}$	41.542	29.845	8.54 00	8.84 00	9.15 00	9.35 00	9.94 00
60	Nd K$_{\beta_1}$	42.271	29.331	8.16 00	8.45 00	8.75 00	9.11 00	9.50 00
64	Gd K$_{\alpha_1}$	42.996	28.836	7.81 00	8.08 00	8.38 00	8.72 00	9.09 00
61	Pm K$_{\beta_1}$	43.826	28.290	7.43 00	7.69 00	7.98 00	8.30 00	8.65 00
65	Tb K$_{\alpha_1}$	44.482	27.873	7.15 00	7.40 00	7.68 00	7.98 00	8.32 00
62	Sm K$_{\beta_1}$	45.413	27.301	6.77 00	7.02 00	7.28 00	7.57 00	7.89 00
66	Dy K$_{\alpha_1}$	45.998	26.954	6.55 00	6.79 00	7.05 00	7.32 00	7.63 00
63	Eu K$_{\beta_1}$	47.038	26.358	6.18 00	6.40 00	6.66 00	6.91 00	7.20 00
67	Ho K$_{\alpha_1}$	47.547	26.076	6.01 00	6.23 00	6.47 00	6.72 00	7.00 00
64	Gd K$_{\beta_1}$	48.697	25.460	5.64 00	5.85 00	6.09 00	6.31 00	6.58 00
68	Er K$_{\alpha_1}$	49.128	25.237	5.51 00	5.72 00	5.95 00	6.17 00	6.43 00
65	Tb K$_{\beta_1}$	50.382	24.609	5.17 00	5.36 00	5.58 00	5.78 00	6.03 00
69	Tm K$_{\alpha_1}$	50.742	24.434	5.08 00	5.26 00	5.48 00	5.68 00	5.92 00
66	Dy K$_{\beta_1}$	52.119	23.788	4.75 00	4.91 00	5.13 00	5.30 00	5.52 00
70	Yb K$_{\alpha_1}$	52.389	23.666	4.69 00	4.84 00	5.06 00	5.23 00	5.45 00
67	Ho K$_{\beta_1}$	53.877	23.012	4.37 00	4.51 00	4.71 00	4.87 00	5.07 00

Z	transition	E [keV]	λ [pm]	$_{71}$Lu	$_{72}$Hf	$_{73}$Ta	$_{74}$W	$_{75}$Re
71	Lu K$_{\alpha_1}$	54.070	22.930	4.34 00	4.46 00	4.67 00	4.82 00	5.03 00
68	Er K$_{\beta_1}$	55.681	22.267	4.02 00	4.14 00	4.34 00	4.47 00	4.66 00
72	Hf K$_{\alpha_1}$	55.790	22.223	4.00 00	4.12 00	4.32 00	4.45 00	4.64 00
69	Tm K$_{\beta_1}$	57.517	21.556	3.70 00	3.82 00	4.00 00	4.11 00	4.29 00
73	Ta K$_{\alpha_1}$	57.532	21.550	3.70 00	3.82 00	3.99 00	4.11 00	4.29 00
74	W K$_{\alpha_1}$	59.318	20.901	3.42 00	3.54 00	3.70 00	3.80 00	3.96 00
70	Yb K$_{\beta_1}$	59.370	20.883	3.41 00	3.53 00	3.69 00	3.79 00	3.95 00
75	Re K$_{\alpha_1}$	61.140	20.278	3.17 00	3.28 00	3.43 00	3.53 00	3.69 00
71	Lu K$_{\beta_1}$	61.283	20.231	3.15 00	3.26 00	3.41 00	3.51 00	3.67 00
76	Os K$_{\alpha_1}$	63.001	19.679	2.94 00	3.04 00	3.18 00	3.27 00	3.42 00
72	Hf K$_{\beta_1}$	63.234	19.607	<u>2.91 00</u>	3.01 00	3.15 00	3.24 00	3.39 00
77	Ir K$_{\alpha_1}$	64.896	19.105	1.35 01	2.82 00	2.95 00	3.03 00	3.17 00
73	Ta K$_{\beta_1}$	65.223	19.009	1.33 01	<u>2.78 00</u>	2.91 00	2.99 00	3.13 00
78	Pt K$_{\alpha_1}$	66.832	18.551	1.24 01	1.25 01	2.74 00	2.81 00	2.94 00
74	W K$_{\beta_1}$	67.244	18.438	1.22 01	1.23 01	<u>2.70 00</u>	2.77 00	2.90 00
79	Au K$_{\alpha_1}$	68.804	18.020	1.15 01	1.16 01	1.17 01	2.61 00	2.74 00
75	Re K$_{\beta_1}$	69.310	17.888	1.12 01	1.13 01	1.14 01	<u>2.57 00</u>	2.69 00
80	Hg K$_{\alpha_1}$	70.819	17.507	1.06 01	1.07 01	1.08 01	1.11 01	2.54 00
76	Os K$_{\beta_1}$	71.413	17.361	1.03 01	1.05 01	1.06 01	1.09 01	<u>2.49 00</u>
81	Tl K$_{\alpha_1}$	72.872	17.014	9.76 00	9.89 00	1.00 01	1.03 01	1.12 01
77	Ir K$_{\beta_1}$	73.561	16.854	9.51 00	9.64 00	9.75 00	1.01 01	1.10 01
82	Pb K$_{\alpha_1}$	74.969	16.538	9.03 00	9.15 00	9.27 00	9.56 00	1.04 01
78	Pt K$_{\beta_1}$	75.748	16.368	8.66 00	8.89 00	9.01 00	9.31 00	1.01 01
83	Bi K$_{\alpha_1}$	77.108	16.079	8.35 00	8.47 00	8.59 00	8.88 00	9.63 00
79	Au K$_{\beta_1}$	77.948	15.906	8.10 00	8.22 00	8.35 00	8.63 00	9.34 00
84	Po K$_{\alpha_1}$	79.290	15.636	7.73 00	7.85 00	7.98 00	8.25 00	8.92 00
80	Hg K$_{\beta_1}$	80.253	15.449	7.47 00	7.59 00	7.72 00	7.99 00	8.62 00
85	At K$_{\alpha_1}$	81.520	15.209	7.16 00	7.28 00	7.41 00	7.67 00	8.26 00
81	Tl K$_{\beta_1}$	82.576	15.014	6.91 00	7.03 00	7.16 00	7.42 00	7.98 00
86	Rn K$_{\alpha_1}$	83.780	14.798	6.64 00	6.76 00	6.88 00	7.14 00	7.67 00
82	Pb K$_{\beta_1}$	84.936	14.597	6.40 00	6.51 00	6.64 00	6.89 00	7.38 00
87	Fr K$_{\alpha_1}$	86.100	14.400	6.17 00	6.28 00	6.40 00	6.65 00	7.11 00
83	Bi K$_{\beta_1}$	87.343	14.195	5.93 00	6.04 00	6.16 00	6.40 00	6.84 00
88	Ra K$_{\alpha_1}$	88.470	14.014	5.73 00	5.83 00	5.95 00	6.19 00	6.60 00
84	Po K$_{\beta_1}$	89.800	13.806	5.50 00	5.60 00	5.72 00	5.95 00	6.34 00
89	Ac K$_{\alpha_1}$	90.884	13.642	5.32 00	5.42 00	5.54 00	5.77 00	6.13 00
85	At K$_{\beta_1}$	92.300	13.432	5.10 00	4.20 00	5.32 00	5.54 00	5.88 00
90	Th K$_{\alpha_1}$	93.350	13.281	4.94 00	5.04 00	5.16 00	5.38 00	5.70 00
86	Rn K$_{\beta_1}$	94.870	13.068	4.73 00	4.83 00	4.94 00	5.15 00	5.45 00
91	Pa K$_{\alpha_1}$	95.868	12.932	4.60 00	4.69 00	4.80 00	5.01 00	5.30 00
87	Fr K$_{\beta_1}$	97.470	12.720	4.39 00	4.48 00	4.60 00	4.80 00	5.06 00
92	U K$_{\alpha_1}$	98.439	12.595	4.28 00	4.37 00	4.48 00	4.68 00	4.93 00
88	Ra K$_{\beta_1}$	100.130	12.382	4.08 00	4.17 00	4.28 00	4.47 00	4.71 00
89	Ac K$_{\beta_1}$	102.850	12.054	3.80 00	3.89 00	3.99 00	4.17 00	4.38 00

Z	transition	E [keV]	λ [pm]	$_{71}$Lu	$_{72}$Hf	$_{73}$Ta	$_{74}$W	$_{75}$Re
90	Rh K$_{\beta_1}$	105.610	11.739	3.55 00	3.63 00	3.73 00	3.89 00	4.09 00
91	Pa K$_{\beta_1}$	108.430	11.434	3.31 00	3.39 00	3.48 00	3.64 00	3.82 00
92	U K$_{\beta_1}$	111.300	11.139	3.09 00	3.16 00	3.25 00	3.40 00	3.56 00

Z	transition	E [keV]	λ [pm]	$_{76}$Os	$_{77}$Ir	$_{78}$Pt	$_{79}$Au	$_{80}$Hg
4	Be K$_\alpha$	0.1085	11427.207	1.67 04	1.71 04	1.54 04	1.71 04	3.63 04
38	Sr M$_\xi$	0.1140	10875.895	1.71 04	1.70 04	1.38 04	1.40 04	<u>3.03 04</u>
39	Y M$_\xi$	0.1328	9339.235	1.87 04	1.53 04	1.17 04	8.67 03	1.58 04
16	S L$_l$	0.1487	8337.942	1.78 04	1.62 04	1.23 04	7.98 03	1.06 04
40	Zr M$_\xi$	0.1511	8205.506	1.77 04	1.62 04	1.24 04	8.02 03	1.03 04
41	Nb M$_\xi$	0.1717	7221.037	1.67 04	1.58 04	1.38 04	9.37 03	9.40 03
5	B K$_\alpha$	0.1833	6764.059	1.68 04	1.61 04	1.47 04	1.04 04	1.01 04
42	Mo M$_\xi$	0.1926	6473.445	<u>1.69 04</u>	1.63 04	1.54 04	1.13 04	1.09 04
6	C K$_\alpha$	0.2770	4476.000	<u>1.71 04</u>	<u>1.64 04</u>	1.61 04	1.52 04	1.68 04
47	Ag M$_\xi$	0.3117	3977.709	1.60 04	<u>1.73 04</u>	<u>1.58 04</u>	<u>1.53 04</u>	<u>1.67 04</u>
7	N K$_\alpha$	0.3924	3159.664	1.41 04	1.47 04	1.52 04	1.54 04	1.59 04
22	Ti L$_l$	0.3953	3136.484	1.41 04	1.46 04	1.51 04	1.54 04	1.59 04
22	Ti L$_\alpha$	0.4522	2741.822	<u>1.22 04</u>	<u>1.27 04</u>	1.33 04	1.37 04	1.43 04
23	V L$_\alpha$	0.5113	2424.901	1.15 04	1.19 04	<u>1.17 04</u>	1.21 04	1.28 04
8	O K$_\alpha$	0.5249	2362.072	<u>1.13 04</u>	1.17 04	1.13 04	<u>1.18 04</u>	1.24 04
25	Mn L$_l$	0.5563	2228.747	1.06 04	1.10 04	1.15 04	1.09 04	<u>1.16 04</u>
24	Cr L$_\alpha$	0.5728	2164.549	1.02 04	<u>1.07 04</u>	<u>1.12 04</u>	1.05 04	1.11 04
25	Mn L$_\alpha$	0.6374	1945.171	<u>8.79 03</u>	9.27 03	9.77 03	<u>1.02 04</u>	1.05 04
9	F K$_\alpha$	0.6768	1831.932	8.44 03	<u>8.43 03</u>	8.89 03	9.29 03	<u>9.77 03</u>
26	Fe L$_\alpha$	0.7050	1758.655	7.92 03	8.40 03	<u>8.40 03</u>	<u>8.79 03</u>	<u>9.29 03</u>
27	Co L$_\alpha$	0.7762	1597.335	6.76 03	7.16 03	7.59 03	7.95 03	<u>7.94 03</u>
28	Ni L$_\alpha$	0.8515	1456.080	5.76 03	6.11 03	6.45 03	6.76 03	7.19 03
29	Cu L$_\alpha$	0.9297	1336.044	4.92 03	5.23 03	5.51 03	5.79 03	6.12 03
30	Zn L$_\alpha$	1.0117	1225.513	4.19 03	4.46 03	4.71 03	4.97 03	5.23 03
11	Na K$_\alpha$	1.0410	1191.020	3.97 03	4.22 03	4.45 03	4.70 03	4.94 03
11	Na K$_\beta$	1.0711	1157.550	3.72 03	4.01 03	4.12 03	4.18 03	4.32 03
12	Mg K$_\alpha$	1.2536	989.033	3.25 03	3.41 03	3.55 03	3.67 03	3.88 03
33	As L$_\alpha$	1.2820	967.123	3.06 03	3.22 03	3.36 03	3.47 03	3.67 03
12	Mg K$_\beta$	1.3022	952.121	2.94 03	3.09 03	3.23 03	3.34 03	3.53 03
33	As L$_{\beta_1}$	1.3170	941.421	2.86 03	3.01 03	3.14 03	3.25 03	3.43 03
66	Dy M$_\beta$	1.3250	935.737	2.81 03	2.96 03	3.09 03	3.20 03	3.38 03
67	Ho M$_\alpha$	1.3480	919.771	2.69 03	2.83 03	2.96 03	3.07 03	3.24 03
34	Se L$_\alpha$	1.3791	899.029	2.54 03	2.68 03	2.79 03	2.90 03	3.07 03
67	Ho M$_\beta$	1.3830	896.494	2.52 03	2.66 03	2.77 03	2.88 03	3.05 03
68	Er M$_\alpha$	1.4060	881.829	2.41 03	2.55 03	2.66 03	2.76 03	2.92 03
34	Se L$_{\beta_1}$	1.4192	873.627	2.36 03	2.49 03	2.60 03	2.70 03	2.86 03
68	Er M$_\beta$	1.4430	859.218	2.26 03	2.39 03	2.49 03	2.59 03	2.74 03
69	Tm M$_\alpha$	1.4620	848.051	2.18 03	2.31 03	2.41 03	2.51 03	2.66 03

Z	transition	E [keV]	λ [pm]	$_{76}$Os	$_{77}$Ir	$_{78}$Pt	$_{79}$Au	$_{80}$Hg
35	Br L$_\alpha$	1.4804	837.511	2.11 03	2.24 03	2.34 03	2.44 03	2.58 03
13	Al K$_{\alpha_1}$	1.4867	833.962	2.09 03	2.21 03	2.31 03	2.41 03	2.55 03
69	Tm M$_\beta$	1.5030	824.918	2.03 03	2.15 03	2.25 03	2.35 03	2.48 03
70	Yb M$_\alpha$	1.5214	814.941	1.97 03	2.09 03	2.18 03	2.28 03	2.41 03
35	Br L$_{\beta_1}$	1.5259	812.538	1.95 03	2.07 03	2.17 03	2.26 03	2.39 03
13	Al K$_\beta$	1.5574	796.103	1.85 03	1.97 03	2.06 03	2.15 03	2.27 03
70	Yb M$_\beta$	1.5675	790.974	1.82 03	1.93 03	2.03 03	2.12 03	2.24 03
71	Lu M$_\alpha$	1.5813	784.071	1.78 03	1.89 03	1.98 03	2.07 03	2.19 03
36	Kr L$_\alpha$	1.5860	781.747	1.77 03	1.88 03	1.97 03	2.06 03	2.18 03
71	Lu M$_\beta$	1.6312	760.085	1.65 03	1.75 03	1.83 03	1.92 03	2.03 03
36	Kr L$_{\beta_1}$	1.6366	757.577	1.63 03	1.73 03	1.82 03	1.90 03	2.01 03
72	Hf M$_\beta$	1.6446	753.892	1.61 03	1.71 03	1.80 03	1.88 03	1.99 03
37	Rb L$_{\alpha_1}$	1.6941	731.864	1.49 03	1.59 03	1.67 03	1.75 03	1.85 03
72	Hf M$_\beta$	1.6976	730.355	1.49 03	1.58 03	1.66 03	1.74 03	1.84 03
73	Ta K$_{\alpha_1}$	1.7096	725.229	1.46 03	1.55 03	1.63 03	1.71 03	1.81 03
14	Si K$_{\alpha_1}$	1.7400	712.558	1.39 03	1.48 03	1.56 03	1.64 03	1.73 03
37	Rb L$_{\beta_1}$	1.7522	707.597	1.37 03	1.46 03	1.53 03	1.61 03	1.70 03
73	Ta M$_\beta$	1.7655	702.266	1.34 03	1.43 03	1.51 03	1.58 03	1.67 03
74	W M$_{\alpha_1}$	1.7754	698.350	1.32 03	1.41 03	1.48 03	1.56 03	1.65 03
38	Sr L$_{\alpha_1}$	1.8066	686.290	1.27 03	1.35 03	1.42 03	1.49 03	1.58 03
74	W M$_\beta$	1.8349	675.705	1.22 03	1.30 03	1.37 03	1.44 03	1.52 03
14	Si K$_\beta$	1.8359	675.337	1.21 03	1.30 03	1.37 03	1.44 03	1.52 03
75	Re M$_{\alpha_1}$	1.8420	673.100	1.20 03	1.29 03	1.35 03	1.43 03	1.51 03
38	Sr L$_{\beta_1}$	1.8717	662.420	1.16 03	1.23 03	1.30 03	1.37 03	1.45 03
75	Re M$_\beta$	1.9061	650.465	1.10 03	1.18 03	1.24 03	1.31 03	1.39 03
76	Os M$_{\alpha_1}$	1.9102	649.069	1.10 03	1.17 03	1.24 03	1.30 03	1.38 03
39	Y L$_{\alpha_1}$	1.9226	644.882	<u>1.08 03</u>	1.15 03	1.22 03	1.28 03	1.36 03
76	Os M$_\beta$	1.9783	626.725	3.18 03	1.07 03	1.13 03	1.20 03	1.26 03
77	Ir M$_{\alpha_1}$	1.9799	626.219	3.17 03	1.07 03	1.13 03	1.19 03	1.26 03
39	Y L$_{\beta_1}$	1.9958	621.230	3.10 03	1.05 03	1.11 03	1.17 03	1.24 03
15	P K$_{\alpha_1}$	2.0137	615.708	<u>3.03 03</u>	<u>1.03 03</u>	1.08 03	1.15 03	1.21 03
40	Zr L$_{\alpha_1}$	2.0424	607.056	4.37 03	3.05 03	1.05 03	1.11 03	1.17 03
78	Pt M$_{\alpha_1}$	2.0505	604.658	4.32 03	3.01 03	1.04 03	1.10 03	1.16 03
77	Ir M$_\beta$	2.0535	603.775	4.31 03	<u>3.00 03</u>	<u>1.03 03</u>	1.09 03	1.15 03
79	Au M$_{\alpha_1}$	2.1229	584.036	3.95 03	4.13 03	2.93 03	1.01 03	1.06 03
40	Zr L$_{\beta_1}$	2.1244	583.624	3.95 03	4.12 03	2.92 03	1.01 03	1.06 03
78	Pt M$_\beta$	2.1273	582.828	3.93 03	4.11 03	2.91 03	1.00 03	1.06 03
15	P K$_{\beta_{1,3}}$	2.1390	579.640	3.88 03	4.05 03	2.87 03	9.89 02	1.04 03
41	Nb L$_{\alpha_1}$	2.1659	572.441	3.76 03	3.92 03	2.78 03	9.59 03	1.01 03
80	Hg M$_{\alpha_1}$	2.1953	564.775	3.63 03	3.79 03	<u>2.68 03</u>	9.28 02	9.80 02
79	Au M$_\beta$	2.2046	562.393	3.59 03	3.75 03	3.97 03	<u>9.19 02</u>	9.70 02
41	Nb L$_{\beta_1}$	2.2574	549.238	3.38 03	3.52 03	3.73 03	2.46 03	9.16 02
81	Tl M$_{\alpha_1}$	2.2706	546.045	3.33 03	3.47 03	3.68 03	2.43 03	9.03 02
80	Hg M$_\beta$	2.2825	543.199	3.28 03	3.42 03	3.63 03	<u>2.39 03</u>	8.91 02

Z	transition	E [keV]	λ [pm]	$_{76}$Os	$_{77}$Ir	$_{78}$Pt	$_{79}$Au	$_{80}$Hg
42	Mo L$_{\alpha_1}$	2.2932	540.664	3.24 03	3.38 03	3.58 03	3.54 03	<u>8.81 02</u>
16	S K$_{\alpha_1}$	2.3080	537.197	3.19 03	3.33 03	3.52 03	3.48 03	2.46 03
82	Pb M$_{\alpha_1}$	2.3457	528.563	3.06 03	3.19 03	3.38 03	3.34 03	2.36 03
81	Tl M$_\beta$	2.3621	524.893	3.00 03	3.13 03	3.32 03	3.28 03	<u>2.32 03</u>
42	Mo L$_{\beta_1}$	2.3948	517.726	2.90 03	3.02 03	3.20 03	3.17 03	3.36 03
83	Bi M$_{\alpha_1}$	2.4226	511.785	2.81 03	2.93 03	3.10 03	3.08 03	3.26 03
43	Tc L$_{\alpha_1}$	2.4240	511.490	2.81 03	2.93 03	3.10 03	3.07 03	3.25 03
82	Pb M$_\beta$	2.4427	507.574	<u>2.75 03</u>	2.87 03	3.04 03	3.01 03	3.19 03
16	S K$_\beta$	2.4640	503.186	3.24 03	2.81 03	2.97 03	2.94 03	3.12 03
83	Bi M$_{\beta_1}$	2.5255	490.933	3.04 03	2.63 03	2.79 03	2.76 03	2.92 03
43	Tc L$_{\beta_1}$	2.5368	488.746	3.00 03	<u>2.60 03</u>	2.75 03	2.73 03	2 89 03
44	Ru L$_{\alpha_1}$	2.5586	484.582	2.93 03	3.05 03	2.69 03	2.67 03	2.83 03
17	Cl K$_{\alpha_1}$	2.6224	472.792	2.75 03	2.86 03	<u>2.53 03</u>	2.51 03	2.65 03
44	Ru L$_{\beta_1}$	2.6832	462.079	2.59 03	2.70 03	2.85 03	2.37 03	2.50 03
45	Rh L$_{\alpha_1}$	2.6967	459.766	<u>2.56 03</u>	2.66 03	2.81 03	<u>2.34 03</u>	2.47 03
17	Cl K$_\beta$	2.8156	440.350	2.51 03	2.38 03	2.51 03	2.51 03	2.21 03
45	Rh L$_{\beta_1}$	2.8344	437.430	2.47 03	2.34 03	2.47 03	2.47 03	2.17 03
46	Pd L$_{\alpha_1}$	2.8386	436.782	2.46 03	<u>2.33 03</u>	2.46 03	2.46 03	<u>2.16 03</u>
18	Ar K$_{\alpha_1}$	2.9577	419.194	2.21 03	2.30 03	2.21 03	2.21 03	2.33 03
47	Ag L$_{\alpha_1}$	2.9843	415.458	2.16 03	2.25 03	2.16 03	2.16 03	2.28 03
46	Pd L$_{\beta_1}$	2.9902	414.638	2.15 03	2.24 03	2.15 03	2.15 03	2.26 03
90	Th M$_{\alpha_1}$	2.9961	413.821	<u>2.14 03</u>	2.23 03	<u>2.14 03</u>	2.14 03	2.25 03
91	Pa M$_{\alpha_1}$	3.0823	402.248	2.19 03	2.07 03	2.18 03	1.99 03	2.09 03
48	Cd L$_{\alpha_1}$	3.1337	395.651	2.09 03	1.98 03	2.09 03	1.91 03	2.01 03
90	Th M$_\beta$	3.1458	394.129	2.07 03	1.97 03	2.07 03	<u>1.89 03</u>	1.99 03
47	Ag L$_{\beta_1}$	3.1509	393.491	2.06 03	1.96 03	2.06 03	2.07 03	1.98 03
92	U M$_{\alpha_1}$	3.1708	391.021	2.03 03	<u>1.93 03</u>	2.03 03	2.04 03	1.95 03
18	Ar K$_\beta$	3.1905	388.607	2.00 03	2.08 03	1.99 03	2.01 03	1.92 03
91	Pa M$_\beta$	3.2397	382.705	1.92 03	2.00 03	1.92 03	1.93 03	1.84 03
49	In L$_{\alpha_1}$	3.2869	377.69	1.85 03	1.93 03	<u>1.84 03</u>	1.86 03	1.95 03
19	K K$_{\alpha_1}$	3.3138	374.148	1.81 03	1.89 03	1.98 03	1.82 03	1.91 03
48	Cd L$_{\beta_1}$	3.3166	373.832	1.81 03	1.89 03	1.98 03	1.82 03	1.91 03
92	U M$_\beta$	3.3360	371.658	1.78 03	1.86 03	1.95 03	<u>1.79 03</u>	1.88 03
50	Sn L$_{\alpha_1}$	3.4440	360.003	1.64 03	1.71 03	1.80 03	1.82 03	1.73 03
49	In L$_{\beta_1}$	3.4872	355.543	1.59 03	1.66 03	1.74 03	1.76 03	<u>1.67 03</u>
19	K K$_\beta$	3.5896	345.401	1.47 03	1.54 03	1.61 03	1.63 03	1.71 03
51	Sb L$_{\alpha_1}$	3.6047	343.954	1.46 03	1.52 03	1.59 03	1.62 03	1.69 03
50	Sn L$_{\beta_1}$	3.6628	338.498	1.40 03	1.46 03	1.53 03	1.55 03	1.62 03
20	Ca K$_{\alpha_1}$	3.6917	335.848	1.37 03	1.43 03	1.50 03	1.52 03	1.59 03
52	Te L$_{\alpha_1}$	3.7693	328.934	1.30 03	1.35 03	1.42 03	1.44 03	1.51 03
51	Sb L$_{\beta_1}$	3.8436	322.575	1.24 03	1.29 03	1.35 03	1.37 03	1.43 03
53	I L$_{\alpha_1}$	3.9377	314.867	1.16 03	1.21 03	1.27 03	1.29 03	1.35 03
20	Ca K$_\beta$	4.0127	308.981	1.11 03	1.15 03	1.21 03	1.23 03	1.28 03
52	Te L$_{\beta_1}$	4.0296	307.686	1.09 03	1.14 03	1.19 03	1.22 03	1.27 03

Z	transition	E [keV]	λ [pm]	$_{76}$Os	$_{77}$Ir	$_{78}$Pt	$_{79}$Au	$_{80}$Hg
21	Sc K$_{\alpha_1}$	4.0906	303.097	1.05 03	1.10 03	1.15 03	1.17 03	1.22 03
34	Xe L$_{\alpha_1}$	4.1099	301.674	1.04 03	1.08 03	1.13 03	1.16 03	1.21 03
53	I L$_{\beta_1}$	4.2207	293.755	9.70 02	1.01 03	1.06 03	1.08 03	1.13 03
55	Cs L$_{\alpha_1}$	4.2865	289.245	9.32 02	9.71 02	1.02 03	1.04 03	1.08 03
21	Sc K$_\beta$	4.4605	277.962	8.41 02	8.76 02	9.15 02	9.38 02	9.78 02
56	Ba L$_{\alpha_1}$	4.4663	277.601	8.39 02	8.73 03	9.12 03	9.34 02	9.75 02
22	Ti K$_{\alpha_1}$	4.5108	274.863	8.17 02	8.51 02	8.89 02	9.11 02	9.50 02
55	Cs L$_{\beta_1}$	4.6198	268.377	2.69 02	8.00 02	8.35 02	8.57 02	8.94 02
57	La L$_{\alpha_1}$	4.6510	266.577	7.55 02	7.86 02	8.20 02	8.42 02	8.78 02
56	Ba L$_{\beta_1}$	4.8273	256.841	6.86 02	7.14 02	7.44 02	7.66 02	7.98 02
58	Ce L$_{\alpha_1}$	4.8402	256.157	6.82 02	7.09 02	7.39 02	7.61 02	7.93 02
22	Ti K$_{\beta_{1,3}}$	4.9318	251.399	6.49 02	6.75 02	7.04 02	7.25 02	7.56 02
23	V K$_{\alpha_1}$	4.9522	250.363	6.43 02	6.68 02	6.96 02	7.18 02	7.48 02
59	Pr L$_{\alpha_1}$	5.0337	246.310	6.16 02	6.40 02	6.67 02	6.88 02	7.17 02
57	La L$_{\beta_1}$	5.0421	245.899	6.14 02	6.38 02	6.65 02	6.86 02	7.14 02
60	Nd L$_{\alpha_1}$	5.2304	237.047	5.58 02	5.80 02	6.04 02	6.24 02	6.50 02
58	Ce L$_{\beta_1}$	5.2622	235.614	5.50 02	5.71 02	5.95 02	6.15 02	6.40 02
24	Cr K$_{\alpha_1}$	5.4147	228.978	5.11 02	5.31 02	5.52 02	5.72 02	5.94 02
23	V K$_{\beta_{1,3}}$	5.4273	228.447	5.08 02	5.27 02	5.49 02	5.68 02	5.91 02
61	Pm L$_{\alpha_1}$	5.4325	228.228	5.07 02	5.26 02	5.47 02	5.67 02	5.89 02
59	Pr L$_{\beta_1}$	5.4889	225.883	4.93 02	5.12 02	5.33 02	5.52 02	5.74 02
62	Sm L$_{\alpha_1}$	5.6361	219.984	4.61 02	4.79 02	4.98 02	5.16 02	5.36 02
60	Nd L$_{\beta_1}$	5.7216	216.696	4.44 02	4.60 02	4.78 02	4.97 02	5.16 02
63	Eu L$_{\alpha_1}$	5.8457	212.096	4.20 02	4.36 02	4.53 02	4.71 02	4.88 02
25	Mn K$_{\alpha_1}$	5.8986	210.187	4.10 02	4.26 02	4.42 02	4.60 02	4.77 02
24	Cr K$_{\beta_{1,3}}$	5.9467	208.494	4.02 02	4.17 02	4.33 02	4.50 02	4.67 02
61	Pm L$_{\beta_1}$	5.9610	207.993	3.99 02	4.14 02	4.30 02	4.48 02	4.64 02
64	Gd L$_{\alpha_1}$	6.0572	204.690	3.83 02	3.98 02	4.13 02	4.30 02	4.45 02
62	Sm L$_{\beta_1}$	6.2051	199.811	3.60 02	3.74 02	3.88 02	4.05 02	4.19 02
65	Tb L$_{\alpha_1}$	6.2728	197.655	3.50 02	3.64 02	3.77 02	3.94 02	4.07 02
26	Fe K$_{\alpha_1}$	6.4038	193.611	3.32 02	3.45 02	3.57 02	3.74 02	3.87 02
63	Eu L$_{\beta_1}$	6.4564	192.034	3.25 02	3.38 02	3.50 02	3.66 02	3.79 02
25	Mn K$_{\beta_{1,3}}$	6.4905	191.025	3.21 02	3.33 02	3.45 02	3.61 02	3.74 02
66	Dy L$_{\alpha_1}$	6.4952	190.887	3.20 02	3.33 02	3.45 02	3.61 02	3.73 02
64	Gd L$_{\beta_1}$	6.7132	184.688	2.94 02	3.06 02	3.16 02	3.32 02	3.43 02
67	Ho L$_{\alpha_1}$	6.7198	184.507	2.94 02	3.05 02	3.16 02	3.31 02	3.42 02
27	Co K$_{\alpha_1}$	6.9303	178.903	2.71 02	2.82 02	2.92 02	3.06 02	3.16 02
68	Er L$_{\alpha_1}$	6.9487	178.429	2.69 02	2.80 02	2.90 02	3.04 02	3.14 02
65	Tb L$_{\beta_1}$	6.9780	177.680	2.67 02	2.77 02	2.86 02	3.01 02	3.11 02
26	Fe K$_{\beta_{1,3}}$	7.0580	175.666	2.59 02	2.69 02	2.78 02	2.92 02	3.02 02
69	Tm L$_{\alpha_1}$	7.1799	172.683	2.48 02	2.58 02	2.66 02	2.80 02	2.89 02
66	Dy L$_{\beta_1}$	7.2477	171.068	2.42 02	2.52 02	2.60 02	2.73 02	2.82 02
70	Yb L$_{\alpha_1}$	7.4156	167.195	2.28 02	2.38 02	2.45 02	2.58 02	2.66 02
28	Ni K$_{\alpha_1}$	7.4782	165.795	2.23 02	2.32 02	2.40 02	2.53 02	2.61 02

Z	transition	E [keV]	λ [pm]	$_{76}$Os	$_{77}$Ir	$_{78}$Pt	$_{79}$Au	$_{80}$Hg
67	Ho L$_{\beta_1}$	7.5253	164.757	2.20 02	2.29 02	2.36 02	2.49 02	2.56 02
27	Co K$_{\beta_{1,3}}$	7.6494	162.084	2.11 02	2.19 02	2.26 02	2.39 02	2.46 02
71	Lu L$_{\alpha_1}$	7.6555	161.955	2.10 02	2.19 02	2.25 02	2.38 02	2.45 02
68	Er L$_{\beta_1}$	7.8109	158.733	2.00 02	2.08 02	2.14 02	2.26 02	2.33 02
72	Hf L$_{\alpha_1}$	7.8990	156.963	1.94 02	2.02 02	2.08 02	2.20 02	2.27 02
29	Cu K$_{\alpha_1}$	8.0478	154.060	1.85 02	1.93 02	1.98 02	2.10 02	2.16 02
69	Tm L$_{\beta_1}$	8.1010	153.049	1.82 02	1.90 02	1.95 02	2.07 02	2.13 02
73	Ta L$_{\alpha_1}$	8.1461	152.201	1.79 02	1.87 02	1.92 02	2.04 02	2.10 02
28	Ni K$_{\beta_{1,3}}$	8.2647	150.017	1.73 02	1.80 02	1.85 02	1.96 02	2.02 02
74	W L$_{\alpha_1}$	8.3976	147.643	1.66 02	1.73 02	1.78 02	1.89 02	1.94 02
70	Yb L$_{\beta_1}$	8.4018	147.569	1.66 02	1.73 02	1.78 02	1.88 02	1.94 02
30	Zn K$_{\alpha_1}$	8.6389	143.519	1.55 02	1.61 02	1.65 02	1.76 02	1.81 02
75	Re L$_{\alpha_1}$	8.6525	143.294	1.54 02	1.61 02	1.65 02	1.75 02	1.80 02
71	Lu L$_{\beta_1}$	8.7090	142.364	1.51 02	1.58 02	1.62 02	1.72 02	1.77 02
29	Cu K$_{\beta_1}$	8.9053	139.226	1.43 02	1.49 02	1.53 02	1.63 02	1.67 02
76	Os L$_{\alpha_1}$	8.9117	139.126	1.43 02	1.49 02	1.53 02	1.63 02	1.67 02
72	Hf L$_{\beta_1}$	9.0227	137.414	1.38 02	1.44 02	1.48 02	1.58 02	1.62 02
77	Ir L$_{\alpha_1}$	9.1751	135.132	1.33 02	1.38 02	1.42 02	1.51 02	1.55 02
31	Ga K$_{\alpha_1}$	9.2517	134.013	1.30 02	1.35 02	1.39 02	1.48 02	1.52 02
73	Ta L$_{\beta_1}$	9.3431	132.702	1.27 02	1.32 02	1.35 02	1.44 02	1.48 02
78	Pt L$_{\alpha_1}$	9.4423	131.308	1.23 02	1.29 02	1.32 02	1.41 02	1.44 02
30	Zn K$_{\beta_{1,3}}$	9.5720	129.529	1.19 02	1.24 02	1.27 02	1.36 02	1.39 02
74	W L$_{\beta_1}$	9.6724	128.184	1.16 02	1.21 02	1.24 02	1.32 02	1.36 02
79	Au L$_{\alpha_1}$	9.7133	127.644	1.15 02	1.20 02	1.22 02	1.31 02	1.34 02
32	Ge K$_{\alpha_1}$	9.8864	125.409	1.10 02	1.14 02	1.17 02	1.25 02	1.28 02
80	Hg L$_{\alpha_1}$	9.9886	124.124	1.07 02	1.12 02	1.14 02	1.22 02	1.25 02
75	Re L$_{\beta_1}$	10.010	123.861	1.06 02	1.11 02	1.13 02	1.21 02	1.24 02
31	Ga K$_{\beta_1}$	10.264	120.796	9.99 01	1.04 02	1.06 02	1.14 02	1.17 02
81	Tl L$_{\alpha_1}$	10.269	120.737	9.98 01	1.04 02	1.06 02	1.14 02	1.17 02
76	Os L$_{\beta_1}$	10.355	119.734	9.77 01	1.02 02	1.04 02	1.12 02	1.14 02
33	As K$_{\alpha_1}$	10.544	117.588	9.33 01	9.74 01	9.94 01	1.07 02	1.09 02
82	Pb L$_{\alpha_1}$	10.552	117.499	9.32 01	9.72 01	9.92 01	1.06 02	1.09 02
77	Ir L$_{\beta_1}$	10.708	115.787	8.98 01	9.37 01	9.55 01	1.03 02	1.05 02
83	Bi L$_{\alpha_1}$	10.839	114.388	<u>8.71 01</u>	9.09 01	9.27 01	9.96 01	1.02 02
32	Ge K$_{\beta_1}$	10.982	112.898	2.13 02	8.79 01	8.96 01	9.64 01	9.85 01
78	Pt L$_{\beta_1}$	11.071	111.990	2.09 02	8.62 01	8.78 01	9.45 01	9.65 01
84	Po L$_{\alpha_1}$	11.131	111.387	2.06 02	<u>8.50 01</u>	8.66 01	9.32 01	9.52 01
34	Se K$_{\alpha_1}$	11.222	110.484	2.01 02	2.09 02	8.49 01	9.13 01	9.33 01
85	At L$_{\alpha_1}$	11.427	108.501	1.92 02	1.99 02	8.11 01	8.73 01	8.92 01
79	Au L$_{\beta_1}$	11.442	108.359	1.91 02	1.99 02	<u>8.08 01</u>	8.70 01	8.89 01
33	As K$_{\beta_1}$	11.726	105.735	1.79 02	1.86 02	1.93 02	8.19 01	8.36 01
86	Rn L$_{\alpha_1}$	11.727	105.726	1.79 02	1.86 02	1.93 02	8.19 01	8.36 01
80	Hg L$_{\beta_1}$	11.823	104.867	1.75 02	1.82 02	1.88 02	<u>8.02 01</u>	8.19 01
35	Br K$_{\alpha_1}$	11.924	103.979	1.71 02	1.78 02	1.84 02	1.86 02	8.01 01

Z	transition	E [keV]	λ [pm]	$_{76}$Os	$_{77}$Ir	$_{78}$Pt	$_{79}$Au	$_{80}$Hg
87	Fr K$_{\alpha_1}$	12.031	103.054	1.67 02	1.74 02	1.80 02	1.82 02	7.84 01
81	Tl L$_{\beta_1}$	12.213	101.519	1.60 02	1.67 02	1.73 02	1.75 02	<u>7.55 01</u>
88	Ra L$_{\alpha_1}$	12.340	100.474	<u>1.56 02</u>	1.62 02	1.68 02	1.70 02	1.79 02
34	Se K$_{\beta_1}$	12.496	99.219	2.10 02	1.57 02	1.62 02	1.64 02	1.73 02
82	Pb L$_{\beta_1}$	12.614	98.291	2.04 02	1.53 02	1.58 02	1.60 02	1.69 02
36	Kr K$_{\alpha_1}$	12.649	98.019	2.03 02	1.52 02	1.57 02	1.59 02	1.68 02
89	Ac L$_{\alpha_1}$	12.652	97.996	<u>2.03 02</u>	<u>1.52 02</u>	1.57 02	1.59 02	1.67 02
90	Th L$_{\alpha_1}$	12.969	95.601	2.90 02	1.97 02	1.47 02	1.49 02	1.57 02
83	Bi L$_{\beta_1}$	13.024	95.197	2.24 02	1.95 02	<u>1.45 02</u>	1.47 02	1.55 02
91	Pa L$_{\alpha_1}$	13.291	93.285	2.12 02	1.85 02	1.92 02	1.40 02	1.47 02
35	Br K$_{\beta_1}$	13.291	93.285	2.12 02	1.85 02	1.92 02	1.40 02	1.47 02
37	Rb K$_{\alpha_1}$	13.395	92.560	2.08 02	<u>1.81 02</u>	1.88 02	1.37 02	1.44 02
84	Po L$_{\beta_1}$	13.447	92.202	2.06 02	2.14 02	1.86 02	1.35 02	1.42 02
92	U L$_{\alpha_1}$	13.615	91.065	1.99 02	2.07 02	1.79 02	<u>1.31 02</u>	1.38 02
85	At L$_{\beta_1}$	13.876	89.352	1.89 02	1.97 02	<u>1.71 02</u>	1.73 02	1.31 02
36	Kr K$_{\beta_1}$	14.112	87.857	1.81 02	1.88 02	1.95 02	1.65 02	1.25 02
38	Sr K$_{\alpha_1}$	14.165	87.529	1.79 02	1.86 02	1.93 02	1.64 02	<u>1.24 02</u>
86	Rn L$_{\beta_1}$	14.316	86.606	1.74 02	1.81 02	1.88 02	<u>1.59 02</u>	1.67 02
87	Fr L$_{\beta_1}$	14.770	83.943	1.60 02	1.66 02	1.72 02	1.75 02	<u>1.54 02</u>
39	Y K$_{\alpha_1}$	14.958	82.888	1.54 02	1.61 02	1.67 02	1.69 02	1.78 02
37	Rb K$_{\beta_1}$	14.961	82.872	1.54 02	1.61 02	1.67 02	1.69 02	1.78 02
88	Ra L$_{\beta_1}$	15.236	81.376	1.47 02	1.53 02	1.59 02	1.61 02	1.69 02
89	Ac L$_{\beta_1}$	15.713	78.906	1.35 02	1.40 02	1.46 02	1.49 02	1.56 02
40	Zr K$_{\alpha_1}$	15.775	78.596	1.34 02	1.39 02	1.45 02	1.47 02	1.54 02
38	Sr K$_{\beta_1}$	15.836	78.293	1.33 02	1.37 02	1.43 02	1.46 02	1.53 02
90	Th L$_{\beta_1}$	16.202	76.524	1.25 02	1.29 02	1.35 02	1.37 02	1.44 02
41	Nb K$_{\alpha_1}$	16.615	74.622	1.17 02	1.20 02	1.26 02	1.28 02	1.34 02
91	Pa L$_{\beta_1}$	16.702	74.233	1.15 02	1.19 02	1.24 02	1.26 02	1.33 02
39	Y K$_{\beta_1}$	16.738	74.074	1.14 02	1.18 02	1.23 02	1.26 02	1.32 02
92	U L$_{\beta_1}$	17.220	72.000	1.06 02	1.09 02	1.14 02	1.17 02	1.22 02
42	Mo K$_{\alpha_1}$	17.479	70.933	1.02 02	1.05 02	1.10 02	1.12 02	1.17 02
40	Zr K$_{\beta_1}$	17.668	70.175	9.88 01	1.01 02	1.07 02	1.09 02	1.14 02
43	Tc K$_{\alpha_1}$	18.367	67.504	8.91 01	9.12 01	9.62 01	9.82 01	1.03 02
41	Nb K$_{\beta_1}$	18.623	66.576	8.58 01	8.77 01	9.27 01	9.46 01	9.91 01
44	Ru K$_{\alpha_1}$	19.279	64.311	7.82 01	7.97 01	8.45 01	8.63 01	9.04 01
42	Mo K$_{\beta_1}$	19.608	63.231	7.48 01	7.61 01	8.07 01	8.25 01	8.64 01
45	Rh K$_{\alpha_1}$	20.216	61.330	6.89 01	7.01 01	7.44 01	7.61 01	7.97 01
43	Tc K$_{\beta_1}$	20.619	60.131	6.54 01	6.66 01	7.06 01	7.22 01	7.56 01
46	Pd K$_{\alpha_1}$	21.177	58.547	6.09 01	6.21 01	6.57 01	6.73 01	7.04 01
44	Ru K$_{\beta_1}$	21.657	57.249	5.74 01	5.86 01	6.19 01	6.35 01	6.64 01
47	Ag K$_{\alpha_1}$	22.163	55.942	5.40 01	5.52 01	5.82 01	5.97 01	6.25 01
45	Rh K$_{\beta_1}$	22.724	54.561	5.05 01	5.17 01	5.45 01	5.59 01	5.85 01
48	Cd K$_{\alpha_1}$	23.174	53.501	4.79 01	4.91 01	5.17 01	5.31 01	5.55 01
46	Pd K$_{\beta_1}$	23.819	52.053	4.46 01	4.58 01	4.81 01	4.94 01	5.16 01

Z	transition	E [keV]	λ [pm]	$_{76}$Os	$_{77}$Ir	$_{78}$Pt	$_{79}$Au	$_{80}$Hg
49	In K_{α_1}	24.210	51.212	4.27 01	4.39 01	4.60 01	4.73 01	4.94 01
47	Ag K_{β_1}	24.942	49.709	3.94 01	4.06 01	4.25 01	4.73 01	4.57 01
50	Sn K_{α_1}	25.271	49.062	3.81 01	3.92 01	4.11 01	4.22 01	4.41 01
48	Cd K_{β_1}	26.096	47.511	3.50 01	3.61 01	3.77 01	3.88 01	4.05 01
51	Sb K_{α_1}	26.359	47.037	3.40 01	3.52 01	3.67 01	3.78 01	3.95 01
49	In K_{β_1}	27.276	45.455	3.11 01	3.22 01	3.35 01	3.45 01	3.61 01
52	Te K_{α_1}	27.472	45.131	3.05 01	3.16 01	3.29 01	3.39 01	3.54 01
50	Sn K_{β_1}	28.486	43.524	2.77 01	2.87 01	2.98 01	3.08 01	3.22 01
53	I K_{α_1}	28.612	43.333	2.74 01	2.84 01	2.95 01	3.04 01	3.18 01
51	Sb K_{β_1}	29.726	41.709	2.47 01	2.57 01	2.66 01	2.75 01	2.87 01
54	Xe K_{α_1}	29.779	41.635	2.46 01	2.56 01	2.65 01	2.74 01	2.86 01
55	Cs K_{α_1}	30.973	40.030	2.22 01	2.31 01	2.39 01	2.47 01	2.58 01
52	Te K_{β_1}	30.996	40.000	2.21 01	2.31 01	2.39 01	2.47 01	2.57 01
56	Ba K_{α_1}	32.194	38.511	2.00 01	2.09 01	2.16 01	2.23 01	2.33 01
53	I K_{β_1}	32.295	38.391	1.99 01	2.07 01	2.14 01	2.22 01	2.31 01
57	La K_{α_1}	33.442	37.074	1.81 01	1.89 01	1.95 01	2.02 01	2.11 01
54	Xe K_{β_1}	33.624	36.874	1.79 01	1.86 01	1.93 01	1.99 01	2.08 01
58	Ce K_{α_1}	34.279	36.169	1.70 01	1.77 01	1.83 01	1.90 01	1.98 01
55	Cs K_{β_1}	34.987	35.437	1.61 01	1.68 01	1.73 01	1.80 01	1.88 01
59	Pr K_{β_1}	36.026	34.415	1.49 01	1.55 01	1.61 01	1.67 01	1.74 01
56	Ba K_{β_1}	36.378	34.082	1.45 01	1.52 01	1.57 01	1.63 01	1.69 01
60	Nd K_{α_1}	36.847	33.648	1.41 01	1.47 01	1.51 01	1.57 01	1.64 01
57	La K_{β_1}	37.801	32.799	1.31 01	1.37 01	1.41 01	1.47 01	1.53 01
61	Pm K_{α_1}	38.725	32.016	1.23 01	1.29 01	1.33 01	1.38 01	1.44 01
58	Ce K_{β_1}	39.257	31.582	1.19 01	1.24 01	1.28 01	1.33 01	1.39 01
62	Sm K_{α_1}	40.118	30.905	1.12 01	1.17 01	1.21 01	1.26 01	1.31 01
59	Pr K_{β_1}	40.748	30.427	1.08 01	1.13 01	1.16 01	1.21 01	1.26 01
63	Eu K_{α_1}	41.542	29.845	1.03 01	1.07 01	1.11 01	1.15 01	1.20 01
60	Nd K_{β_1}	42.271	29.331	9.81 00	1.02 01	1.06 01	1.10 01	1.15 01
64	Gd K_{α_1}	42.996	28.836	9.39 00	9.79 00	1.01 01	1.05 01	1.10 01
61	Pm K_{β_1}	43.826	28.290	8.93 00	9.31 00	9.61 00	1.00 01	1.04 01
65	Tb K_{α_1}	44.482	27.873	8.59 00	8.96 00	9.25 00	9.64 00	1.00 01
62	Sm K_{β_1}	45.413	27.301	8.14 00	8.49 00	8.76 00	9.14 00	9.51 00
66	Dy K_{α_1}	45.998	26.954	7.87 00	8.21 00	8.48 00	8.84 00	9.20 00
63	Eu K_{β_1}	47.038	26.358	7.43 00	7.75 00	8.00 00	8.34 00	8.68 00
67	Ho K_{α_1}	47.547	26.076	7.22 00	7.53 00	7.78 00	8.12 00	8.45 00
64	Gd K_{β_1}	48.697	25.460	6.78 00	7.08 00	7.31 00	7.63 00	7.94 00
68	Er K_{α_1}	49.128	25.237	6.63 00	6.92 00	7.14 00	7.46 00	7.76 00
65	Tb K_{β_1}	50.382	24.609	6.21 00	6.48 00	6.69 00	6.99 00	7.27 00
69	Tm K_{α_1}	50.742	24.434	6.10 00	6.37 00	6.57 00	6.86 00	7.14 00
66	Dy K_{β_1}	52.119	23.788	5.70 00	5.95 00	6.14 00	6.42 00	6.67 00
70	Yb K_{α_1}	52.389	23.666	5.63 00	5.87 00	6.06 00	6.33 00	6.58 00
67	Ho K_{β_1}	53.877	23.012	5.24 00	5.47 00	5.64 00	5.90 00	6.13 00
71	Lu K_{α_1}	54.070	22.930	5.19 00	5.42 00	5.59 00	5.85 00	6.07 00

Z	transition	E [keV]	λ [pm]	$_{76}$Os	$_{77}$Ir	$_{78}$Pt	$_{79}$Au	$_{80}$Hg
68	Er K$_{\beta_1}$	55.681	22.267	4.82 00	5.03 00	5.19 00	5.43 00	5.63 00
72	Hf K$_{\alpha_1}$	55.790	22.223	4.80 00	5.01 00	5.16 00	5.40 00	5.60 00
69	Tm K$_{\beta_1}$	57.517	21.556	4.44 00	4.64 00	4.78 00	5.00 00	5.18 00
73	Ta K$_{\alpha_1}$	57.532	21.550	4.44 00	4.63 00	4.77 00	5.00 00	5.18 00
74	W K$_{\alpha_1}$	59.318	20.901	4.11 00	4.29 00	4.42 00	4.63 00	4.79 00
70	Yb K$_{\beta_1}$	59.370	20.883	4.10 00	4.28 00	4.41 00	4.62 00	4.78 00
75	Re K$_{\alpha_1}$	61.140	20.278	3.80 00	3.97 00	4.09 00	4.29 00	4.44 00
71	Lu K$_{\beta_1}$	61.283	20.231	3.78 00	3.95 00	4.07 00	4.26 00	4.41 00
76	Os K$_{\alpha_1}$	63.001	19.679	3.51 00	3.68 00	3.80 00	3.97 00	4.11 00
72	Hf K$_{\beta_1}$	63.234	19.607	3.48 00	3.65 00	3.76 00	3.94 00	4.08 00
77	Ir K$_{\alpha_1}$	64.896	19.105	3.26 00	3.41 00	3.52 00	3.69 00	3.82 00
73	Ta K$_{\beta_1}$	65.223	19.009	3.22 00	3.37 00	3.48 00	3.64 00	3.77 00
78	Pt K$_{\alpha_1}$	66.832	18.551	3.03 00	3.17 00	3.27 00	3.42 00	3.54 00
74	W K$_{\beta_1}$	67.244	18.438	2.99 00	3.12 00	3.22 00	3.37 00	3.49 00
79	Au K$_{\alpha_1}$	68.804	18.020	2.82 00	2.95 00	3.04 00	3.18 00	3.29 00
75	Re K$_{\beta_1}$	69.310	17.888	2.77 00	2.90 00	2.99 00	3.12 00	3.23 00
80	Hg K$_{\alpha_1}$	70.819	17.507	2.63 00	2.74 00	2.83 00	2.96 00	3.06 00
76	Os K$_{\beta_1}$	71.413	17.361	2.57 00	2.69 00	2.77 00	2.90 00	3.00 00
81	Tl K$_{\alpha_1}$	72.872	17.014	2.45 00	2.55 00	2.63 00	2.76 00	2.85 00
77	Ir K$_{\beta_1}$	73.561	16.854	<u>2.39 00</u>	2.49 00	2.57 00	2.70 00	2.79 00
82	Pb K$_{\alpha_1}$	74.969	16.538	1.07 01	2.38 00	2.45 00	2.58 00	2.66 00
78	Pt K$_{\beta_1}$	75.748	16.368	1.04 01	<u>2.32 00</u>	2.39 00	2.51 00	2.59 00
83	Bi K$_{\alpha_1}$	77.108	16.079	9.91 00	1.03 01	2.29 00	2.40 00	2.48 00
79	Au K$_{\beta_1}$	77.948	15.906	9.62 00	9.98 00	<u>2.23 00</u>	2.34 00	2.42 00
84	Po K$_{\alpha_1}$	79.290	15.636	9.18 00	9.53 00	9.75 00	2.24 00	2.32 00
80	Hg K$_{\beta_1}$	80.253	15.449	8.88 00	9.22 00	9.42 00	<u>2.18 00</u>	2.25 00
85	At K$_{\alpha_1}$	81.520	15.209	8.51 00	8.84 00	9.02 00	8.85 00	2.16 00
81	Tl K$_{\beta_1}$	82.576	15.014	8.22 00	8.53 00	8.70 00	8.57 00	<u>2.10 00</u>
86	Rn K$_{\alpha_1}$	83.780	14.798	7.90 00	8.20 00	8.36 00	8.25 00	8.98 00
82	Pb K$_{\beta_1}$	84.936	14.597	7.61 00	7.90 00	8.04 00	7.97 00	8.65 00
87	Fr K$_{\alpha_1}$	86.100	14.400	7.33 00	7.61 00	7.74 00	7.69 00	8.33 00
83	Bi K$_{\beta_1}$	87.343	14.195	7.05 00	7.31 00	7.44 00	7.41 00	8.01 00
88	Ra K$_{\alpha_1}$	88.470	14.014	6.81 00	7.06 00	7.18 00	7.17 00	7.73 00
84	Po K$_{\beta_1}$	89.800	13.806	6.54 00	6.78 00	6.88 00	6.90 00	7.42 00
89	Ac K$_{\alpha_1}$	90.884	13.642	6.33 00	6.56 00	6.65 00	6.69 00	7.18 00
85	At K$_{\beta_1}$	92.300	13.432	6.07 00	6.28 00	6.37 00	6.43 00	6.88 00
90	Th K$_{\alpha_1}$	93.350	13.281	5.88 00	6.09 00	6.17 00	6.24 00	6.67 00
86	Rn K$_{\beta_1}$	94.870	13.068	5.63 00	5.83 00	5.90 00	5.99 00	6.36 00
91	Pa K$_{\alpha_1}$	95.868	12.932	5.47 00	5.66 00	5.73 00	5.83 00	6.20 00
87	Fr K$_{\beta_1}$	97.470	12.720	5.23 00	5.41 00	5.47 00	5.59 00	5.93 00
92	U K$_{\alpha_1}$	98.439	12.595	5.09 00	5.27 00	5.32 00	5.45 00	5.77 00
88	Ra K$_{\beta_1}$	100.130	12.382	4.86 00	5.03 00	5.08 00	5.21 00	5.50 00
89	Ac K$_{\beta_1}$	102.850	12.054	4.53 00	4.68 00	4.73 00	4.86 00	5.12 00
90	Th K$_{\beta_1}$	105.610	11.739	4.22 00	4.37 00	4.41 00	4.54 00	4.77 00

Z	transition	E [keV]	λ [pm]	$_{76}$Os	$_{77}$Ir	$_{78}$Pt	$_{79}$Au	$_{80}$Hg
91	Pa K$_{\beta_1}$	108.430	11.434	3.94 00	4.08 00	4.11 00	4.25 00	4.45 00
92	U K$_{\beta_1}$	l11.300	11.139	3.67 00	3.80 00	3.84 00	3.97 00	4.15 00

Z	transition	E [keV]	λ [pm]	$_{81}$Tl	$_{82}$Pb	$_{83}$Bi	$_{84}$Po	$_{85}$At
4	Be K$_\alpha$	0.1085	11427.207	1.84 04	1.77 04	2.99 04	3.94 04	4.91 04
38	Sr M$_\xi$	0.1140	10875.895	<u>1.60 04</u>	1.52 04	<u>2.53 04</u>	3.43 04	4.25 04
39	Y M$_\xi$	0.1328	9339.235	<u>1.11 04</u>	<u>8.67 03</u>	1.54 04	2.09 04	2.73 04
16	S L$_l$	0.1487	8337.942	9.42 03	6.71 03	9.79 03	1.45 04	1.84 04
40	Zr M$_\xi$	0.1511	8205.506	9.26 03	6.57 03	<u>9.16 03</u>	1.36 04	1.75 04
41	Nb M$_\xi$	0.1717	7221.037	8.61 03	5.90 03	6.64 03	8.05 03	1.09 04
5	B K$_\alpha$	0.1833	6764.059	8.82 03	6.03 03	6.11 03	6.91 03	8.23 03
42	Mo M$_\xi$	0.1926	6473.445	9.28 03	6.29 03	5.90 03	6.40 03	6.98 03
6	C K$_\alpha$	0.2770	4476.000	1.47 04	1.32 04	1.18 04	8.99 03	6.75 03
47	Ag M$_\xi$	0.3117	3977.709	<u>1.50 04</u>	1.40 04	1.35 04	1.25 04	1.02 04
7	N K$_\alpha$	0.3924	3159.664	1.43 04	1.37 04	1.40 04	1.41 04	1.33 04
22	Ti L$_l$	0.3953	3136.484	<u>1.49 04</u>	<u>1.37 04</u>	<u>1.40 04</u>	1.41 04	1.35 04
22	Ti L$_\alpha$	0.4522	2741.822	1.37 04	1.37 04	<u>1.34 04</u>	<u>1.37 04</u>	1.31 04
23	V L$_\alpha$	0.5113	2424.901	1.25 04	1.27 04	1.28 04	1.38 04	1.26 04
8	O K$_\alpha$	0.5249	2362.072	1.22 04	1.25 04	1.27 04	1.35 04	<u>1.25 04</u>
25	Mn L$_l$	0.5563	2228.747	1.14 04	1.17 04	1.19 04	1.26 04	1.23 04
24	Cr L$_\alpha$	0.5728	2164.549	<u>1.10 04</u>	1.13 04	1.16 04	1.21 04	1.22 04
25	Mn L$_\alpha$	0.6374	1945.171	9.62 03	<u>9.87 03</u>	1.01 04	1.06 04	1.12 04
9	F K$_\alpha$	0.6768	1831.932	9.69 03	9.08 03	<u>9.33 03</u>	<u>9.75 03</u>	1.03 04
26	Fe L$_\alpha$	0.7050	1758.655	<u>9.33 03</u>	<u>9.59 03</u>	8.81 03	9.26 03	<u>9.78 03</u>
27	Co L$_\alpha$	0.7762	1597.335	<u>8.01 03</u>	8.29 03	<u>8.53 03</u>	<u>8.95 03</u>	8.44 03
28	Ni L$_\alpha$	0.8515	1456.080	7.31 03	<u>7.13 03</u>	7.39 03	7.75 03	<u>8.19 03</u>
29	Cu L$_\alpha$	0.9297	1336.044	6.28 03	6.47 03	<u>6.38 03</u>	<u>6.68 03</u>	7.07 03
30	Zn L$_\alpha$	1.0117	1225.513	5.46 03	5.56 03	5.77 03	6.06 03	6.10 03
11	Na K$_\alpha$	1.0410	1191.020	5.15 03	5.26 03	5.46 03	5.74 03	5.80 03
11	Na K$_\beta$	1.0711	1157.550	4.72 03	4.83 03	5.01 03	5.29 03	5.32 03
12	Mg K$_\alpha$	1.2536	989.033	4.03 03	4.18 03	4.42 03	4.68 03	5.16 03
33	As L$_\alpha$	1.2820	967.123	3.82 03	3.96 03	4.19 03	4.44 03	4.89 03
12	Mg K$_\beta$	1.3022	952.121	3.68 03	3.81 03	4.04 03	4.28 03	4.70 03
33	As L$_{\beta_1}$	1.3117	941.421	3.58 03	3.71 03	3.95 03	4.17 03	4.57 03
66	Dy M$_\beta$	1.3225	935.737	3.53 03	3.66 03	3.87 03	4.11 03	4.51 03
67	Ho M$_\alpha$	1.3480	919.771	3.38 03	3.51 03	3.72 03	3.95 03	4.32 03
34	Se L$_\alpha$	1.3791	899.029	3.20 03	3.32 03	3.52 03	3.74 03	4.09 03
67	Ho M$_\beta$	1.3830	896.494	3.18 03	3.30 03	3.49 03	5.72 03	4.06 03
68	Er M$_\alpha$	1.4060	881.829	3.05 03	3.17 03	3.36 03	3.57 03	3.90 03
34	Se L$_{\beta_1}$	1.4192	873.627	2.98 03	3.10 03	3.28 03	3.50 03	3.81 03
68	Er M$_\beta$	1.4430	859.218	2.87 03	2.97 03	3.16 03	3.36 03	3.66 03
69	Tm M$_\alpha$	1.4620	848.051	2.78 03	2.88 03	3.06 03	3.26 03	3.54 03

Z	transition	E [keV]	λ [pm]	$_{81}$Tl	$_{82}$Pb	$_{83}$Bi	$_{84}$Po	$_{85}$At
35	Br L_α	1.4804	837.511	2.69 03	2.80 03	2.97 03	3.17 03	3.44.03
13	Al K_{α_1}	1.4867	833.962	2.67 03	2.77 03	2.94 03	3.13 03	3.40 03
69	Tm M_β	1.5030	824.918	2.60 03	2.70 03	2.86 03	3.05 03	3.31 03
70	Yb M_α	1.5214	814.941	2.52 03	2.62 03	2.78 03	2.97 03	3.21 03
35	Br L_{β_1}	1.5259	812.538	2.50 03	2.60 03	2.76 03	2.95 03	2.19 03
13	Al K_β	1.5574	796.103	2.38 03	2.47 03	2.63 03	2.81 03	3.04 03
70	Yb M_β	1.5675	790.974	2.34 03	2.44 03	2.59 03	2.77 03	2.99 03
71	Lu M_α	1.5813	784.071	2.30 03	2.39 03	2.53 03	2.71 03	2.93 03
36	Kr L_α	1.5860	781.747	2.28 03	2.37 03	2.52 03	2.69 03	2.91 03
71	Lu M_β	1.6312	760.085	2.13 03	2.21 03	2.35 03	2.52 03	2.72 03
36	Kr L_{β_1}	1.6366	757.577	2.11 03	2.20 03	2.33 03	2.50 03	2.69 03
72	Hf M_{α_1}	1.6446	753.892	2.09 03	2.17 03	2.31 03	2.47 03	2.66 03
37	Rb L_{α_1}	1.6941	731.864	1.94 03	2.02 03	2.15 03	2.30 03	2.48 03
72	Hf M_β	1.6976	730.355	1.93 03	2.01 03	2.14 03	2.29 03	2.47 03
73	Ta M_{α_1}	1.7096	725.229	1.90 03	1.98 03	2.10 03	2.25 03	2.42 03
14	Si K_{α_1}	1.7400	712.558	1.82 03	1.89 03	2.02 03	2.16 03	2.32 03
37	Rb L_{β_1}	1.7522	707.597	1.79 03	1.86 03	1.98 03	2.13 03	2.28 03
73	Ta M_β	1.7655	702.266	1.76 03	1.83 03	1.95 03	2.09 03	2.24 03
74	W M_{α_1}	1.7754	698.350	1.73 03	1.81 03	1.92 03	2.06 03	2.21 03
38	Sr L_{α_1}	1.8066	686.290	1.66 03	1.73 03	1.84 03	1.98 03	2.12 03
74	W M_β	1.8349	675.705	1.60 03	1.67 03	1.77 03	1.91 03	2.04 03
14	Si K_β	1.8359	675.337	1.60 03	1.67 03	1.77 03	1.91 03	2.04 03
75	Re M_{α_1}	1.8420	673.100	1.59 03	1.65 03	1.76 03	1.89 03	2.02 03
38	Sr L_{β_1}	1.8717	662.420	1.53 03	1.59 03	1.69 03	1.82 03	1.95 03
75	Re M_β	1.9061	650.465	1.46 03	1.52 03	1.62 03	1.74 03	1.86 03
76	Os M_{α_1}	1.9102	649.069	1.45 03	1.51 03	1.61 03	1.74 03	1.85 03
39	Y L_{α_1}	1.9226	644.882	1.43 03	1.49 03	1.59 03	1.71 03	1.82 03
76	Os M_β	1.9783	626.725	1.33 03	1.39 03	1.48 03	1.60 03	1.70 03
77	Ir M_{α_1}	1.9799	626.219	1.33 03	1.39 03	1.48 03	1.59 03	1.70 03
39	Y L_{β_1}	1.9958	621.230	1.31 03	1.36 03	1.45 03	1.57 03	1.67 03
15	P K_{α_1}	2.0137	615.708	1.28 03	1.33 03	1.42 03	1.53 03	1.63 03
40	Zr L_{α_1}	2.0424	607.056	1.24 03	1.29 03	1.37 03	1.48 03	1.58 03
78	Pt M_{α_1}	2.0505	604.658	1.22 03	1.28 03	1.36 03	1.47 03	1.56 03
77	Ir M_{α_1}	2.0535	603.775	1.22 03	1.27 03	1.36 03	1.46 03	1.56 03
79	Au M_{α_1}	2.1229	584.036	1.13 03	1.17 03	1.25 03	1.35 03	1.43 03
40	Zr L_{β_1}	2.1244	584.624	1.12 03	1.17 03	1.25 03	1.35 03	1.43 03
78	Pt M_β	2.1273	582.828	1.12 03	1.17 03	1.25 03	1.35 03	1.43 03
15	P $K_{\beta_{1,3}}$	2.1390	579.640	1.10 03	1.15 03	1.23 03	1.33 03	1.41 03
41	Nb L_{α_1}	2.1659	572.441	1.07 03	1.12 03	1.19 03	1.29 03	1.37 03
80	Hg M_{α_1}	2.1953	564.775	1.04 03	1.08 03	1.16 03	1.25 03	1.32 03
79	Au M_β	2.2046	562.393	1.03 03	1.07 03	1.14 03	1.24 03	1.31 03
41	Nb L_{β_1}	2.2574	549.238	9.70 02	1.01 03	1.08 03	1.17 03	1.24 03
81	Tl M_{α_1}	2.2706	546.045	9.56 02	9.99 02	1.07 03	1.16 03	1.22 03
80	Hg M_β	2.2825	543.199	9.44 02	9.87 02	1.05 03	1.14 03	1.20 03

Z	transition	E [keV]	λ [pm]	$_{81}$Tl	$_{82}$Pb	$_{83}$Bi	$_{84}$Po	$_{85}$At
42	Mo L$_{\alpha_1}$	2.2932	540.664	9.34 02	9.76 02	1.04 03	1.13 03	1.19 03
16	S K$_{\alpha_1}$	2.3080	537.197	9.19 02	9.61 02	1.03 03	1.11 03	1.17 03
82	Pb M$_{\alpha_1}$	2.3457	528.563	8.84 02	9.24 02	9.87 02	1.07 03	1.13 03
81	Tl M$_\beta$	2.3621	524.893	8.69 02	9.09 02	9.71 02	1.05 03	1.11 03
42	Mo L$_{\beta_1}$	2.3948	517.726	2.30 03	8.79 02	9.40 02	1.02 03	1.07 03
83	Bi M$_{\alpha_1}$	2.4226	511.785	2.24 03	8.55 02	9.14 02	9.94 02	1.04 03
43	Tc L$_{\alpha_1}$	2.4240	511.490	2.23 03	8.54 02	9.13 02	9.93 02	1.04 03
82	Pb M$_\beta$	2.4427	507.574	2.19 03	8.38 02	8.96 02	9.75 02	1.02 03
16	S K$_\beta$	2.4640	503.186	2.14 03	8.21 02	8.78 02	9.56 02	1.00 03
83	Bi M$_{\beta_1}$	2.5255	490.933	3.00 03	2.08 03	8.28 02	9.02 02	9.43 02
43	Tc L$_{\beta_1}$	2.5368	488.746	2.96 03	2.05 03	8.19 02	8.93 02	9.33 02
44	Ru L$_{\alpha_1}$	2.5586	484.582	2.90 03	2.01 03	8.03 02	8.75 02	9.16 02
17	Cl K$_{\alpha_1}$	2.6224	472.792	2.72 03	2.83 03	1.95 03	8.26 02	8.61 02
44	Ru L$_{\beta_1}$	2.6832	462.079	2.57 03	2.67 03	1.84 03	1.92 03	8.14 02
45	Rh L$_{\alpha_1}$	2.6967	459.766	2.54 03	2.63 03	2.72 03	1.89 03	8.04 02
17	Cl K$_\beta$	2.8156	440.350	2.27 03	2.35 03	2.44 03	2.54 03	1.78 03
45	Rh L$_{\beta_1}$	2.8344	437.430	2.23 03	2.31 03	2.40 03	2.49 03	1.75 03
46	Pd L$_{\alpha_1}$	2.8386	436.782	2.22 03	2.31 03	2.39 03	2.48 03	1.74 03
18	Ar K$_{\alpha_1}$	2.9577	419.194	2.40 03	2.07 03	2.15 03	2.24 03	2.34 03
47	Ag L$_{\alpha_1}$	2.9843	415.458	2.35 03	2.02 03	2.10 03	2.19 03	2.29 03
46	Pd L$_{\alpha_1}$	2.9902	414.638	2.33 03	2.01 03	2.09 03	2.17 03	2.28 03
90	Tb M$_{\alpha_1}$	2.9961	413.821	2.32 03	2.00 03	2.08 03	2.16 03	2.27 03
91	Pa M$_{\alpha_1}$	3.0823	402.248	2.16 03	2.23 03	1.93 03	2.01 03	2.11 03
48	Cd L$_{\alpha_1}$	3.1337	395.651	2.07 03	2.14 03	1.85 03	1.93 03	2.02 03
90	Th M$_\beta$	3.1458	394.129	2.05 03	2.12 03	1.84 03	1.91 03	2.00 03
47	Ag L$_{\beta_1}$	3.1509	393.491	2.04 03	2.11 03	1.83 03	1.90 03	1.99 03
92	U M$_{\alpha_1}$	3.1708	391.021	2.01 03	2.08 03	1.80 03	1.87 03	1.96 03
18	Ar K$_\beta$	3.1905	388.607	1.97 03	2.04 03	2.12 03	1.84 03	1.93 03
91	Pa M$_\beta$	3.2397	382.705	1.90 03	1.96 03	2.04 03	1.77 03	1.85 03
49	In L$_{\alpha_1}$	3.2869	377.210	1.83 03	1.89 03	1.97 03	1.71 03	1.79 03
19	K K$_{\alpha_1}$	3.3138	374.148	1.79 03	1.85 03	1.92 03	1.89 03	1.75 03
48	Cd L$_{\beta_1}$	3.3166	373.832	1.79 03	1.85 03	1.92 03	1.88 03	1.75 03
92	U M$_\beta$	3.3360	371.658	1.76 03	1.82 03	1.89 03	1.86 03	1.72 03
50	Sn L$_{\alpha_1}$	3.4440	360.003	1.78 03	1.68 03	1.74 03	1.77 03	1.90 03
49	In L$_{\beta_1}$	3.4872	355.543	1.73 03	1.63 03	1.69 03	1.73 03	1.84 03
19	K K$_\beta$	3.5896	345.401	1.60 03	1.66 03	1.57 03	1.63 03	1.71 03
51	Sb L$_{\alpha_1}$	3.6047	343.954	1.59 03	1.64 03	1.55 03	1.62 03	1.69 03
50	Sn L$_{\beta_1}$	3.6628	338.498	1.52 03	1.58 03	1.49 03	1.56 03	1.62 03
20	Ca K$_{\alpha_1}$	3.6917	335.848	1.49 03	1.55 03	1.46 03	1.53 03	1.59 03
52	Te L$_{\alpha_1}$	3.7693	328.934	1.55 03	1.47 03	1.52 03	1.46 03	1.51 03
51	Sb L$_{\beta_1}$	3.8436	322.575	1.48 03	1.39 03	1.45 03	1.38 03	1.44 03
53	I L$_{\alpha_1}$	3.9377	314.867	1.39 03	1.44 03	1.36 03	1.41 03	1.35 03
20	Ca K$_\beta$	4.0127	308.981	1.32 03	1.37 03	1.42 03	1.35 03	1.41 03
52	Te L$_{\beta_1}$	4.0296	307.686	1.31 03	1.35 03	1.41 03	1.33 03	1.40 03

Z	transition	E [keV]	λ [pm]	$_{81}$Tl	$_{82}$Pb	$_{83}$Bi	$_{84}$Po	$_{85}$At
21	Sc K$_{\alpha_1}$	4.0906	303.097	1.26 03	1.30 03	1.36 03	1.28 03	1.34 03
54	Xe L$_{\alpha_1}$	4.1099	301.674	1.24 03	1.29 03	1.34 03	<u>1.27 03</u>	1.33 03
53	I L$_{\beta_1}$	4.2207	293.755	1.16 03	1.20 03	1.25 03	1.30 03	1.24 03
55	Cs L$_{\alpha_1}$	4.2865	289.245	1.12 03	1.16 03	1.20 03	1.25 03	<u>1.19 03</u>
21	Sc K$_\beta$	4.4605	277.962	1.01 03	1.04 03	1.09 03	1.13 03	1.18 03
56	Ba L$_{\alpha_1}$	4.4663	277.601	1.00 03	1.04 03	1.08 03	1.13 03	1.18 03
22	Ti K$_{\alpha_1}$	4.5108	274.863	9.79 02	1.01 03	1.05 03	1.10 03	1.15 03
55	Cs L$_{\beta_1}$	4.6198	268.377	9.21 02	9.54 02	9.92 02	1.03 03	1.08 03
57	La L$_{\alpha_1}$	4.6510	266.577	9.05 02	9.38 02	9.75 02	1.01 03	1.06 03
56	Ba L$_{\beta_1}$	4.8273	256.841	8.22 02	8.53 02	8.86 02	9.23 02	9.67 02
58	Ce L$_{\beta_1}$	4.8402	256.157	8.17 02	8.47 02	8.80 02	9.16 02	9.60 02
22	Ti K$_{\beta_{1,3}}$	4.9318	251.399	7.78 02	8.07 02	8.39 02	8.73 02	9.15 02
23	V K$_{\alpha_1}$	4.9522	250.363	7.70 02	7.99 02	8.30 02	8.64 02	9.05 02
59	Pr L$_{\alpha_1}$	5.0337	246.310	7.39 02	7.66 02	7.96 02	8.29 02	8.68 02
57	La L$_{\beta_1}$	5.0421	245.899	7.36 02	7.63 02	7.93 02	8.25 02	8.65 02
60	Nd L$_{\alpha_1}$	5.2304	237.047	6.70 02	6.95 02	7.22 02	7.52 02	7.88 02
58	Ce L$_{\beta_1}$	5.2622	235.614	6.59 02	6.85 02	7.11 02	7.41 02	7.76 02
24	Cr K$_{\alpha_1}$	5.4147	228.978	6.13 02	6.37 02	6.62 02	6.89 02	7.22 02
23	V K$_{\beta_{1,3}}$	5.4273	228.447	6.09 02	6.33 02	6.58 02	6.85 02	7.18 02
61	Pm L$_{\alpha_1}$	5.4325	228.228	6.08 02	6.32 02	6.56 02	6.83 02	7.16 02
59	Pr L$_{\beta_1}$	5.4889	225.883	5.92 02	6.15 02	6.39 02	6.65 02	6.97 02
62	Sm L$_{\alpha_1}$	5.6361	219.984	5.53 02	5.76 02	5.98 02	6.22 02	6.52 02
60	Nd L$_{\beta_1}$	5.7216	216.696	5.32 02	5.54 02	5.75 02	5.99 02	6.28 02
63	Eu L$_{\alpha_1}$	5.8457	212.096	5.04 02	5.25 02	5.45 02	5.67 02	5.94 02
25	Mn K$_{\alpha_1}$	5.8988	210.187	4.92 02	5.13 02	5.33 02	5.54 02	5.81 02
24	Cr K$_{\beta_{1,3}}$	5.9467	208.494	4.82 02	5.02 02	5.22 02	5.43 02	5.69 02
61	Pm L$_{\beta_1}$	5.9610	207.993	4.79 02	4.99 02	5.19 02	5.39 02	5.66 02
64	Gd L$_{\alpha_1}$	6.0572	204.690	4.60 02	4.80 02	4.98 02	5.18 02	5.43 02
62	Sm L$_{\beta_1}$	6.2051	199.811	4.33 02	4.51 02	4.68 02	4.87 02	5.11 02
65	Tb L$_{\alpha_1}$	6.2728	197.655	4.21 02	4.39 02	4.55 02	4.74 02	4.97 02
26	Fe K$_{\alpha_1}$	6.4038	193.611	3.99 02	4.16 02	4.32 02	4.49 02	4.71 02
63	Eu L$_{\beta_1}$	6.4564	192.034	3.91 02	4.07 02	4.23 02	4.40 02	4.62 02
25	Mn K$_{\beta_{1,3}}$	6.4905	191.025	3.86 02	4.02 02	4.17 02	4.34 02	4.55 02
66	Dy L$_{\alpha_1}$	6.4952	190.887	3.85 02	4.01 02	4.17 02	4.34 02	4.55 02
64	Gd L$_{\beta_1}$	6.7132	184.688	3.54 02	3.69 02	3.83 02	3.99 02	4.18 02
67	Ho L$_{\alpha_1}$	6.7198	184.507	3.53 02	3.68 02	3.82 02	3.98 02	4.17 02
27	Co K$_{\alpha_1}$	6.9303	178.903	3.26 02	3.40 02	3.53 02	3.68 02	3.85 02
68	Er L$_{\alpha_1}$	6.9487	178.429	3.24 02	3.38 02	3.51 02	3.65 02	3.83 02
65	Tb L$_{\beta_1}$	6.9780	177.680	3.21 02	3.34 02	3.47 02	3.61 02	3.79 02
26	Fe K$_{\beta_{1,3}}$	7.0580	175.666	3.12 02	3.25 02	3.37 02	3.51 02	3.68 02
69	Tm L$_{\alpha_1}$	7.1799	172.683	2.98 02	3.11 02	3.23 02	3.36 02	3.52 02
66	Dy L$_{\beta_1}$	7.2477	171.068	2.91 02	3.03 02	3.15 02	3.28 02	3.44 02
70	Yb L$_{\alpha_1}$	7.4156	167.195	2.75 02	2.86 02	2.97 02	3.09 02	3.24 02
28	Ni K$_{\alpha_1}$	7.4782	165.795	2.69 02	2.80 02	2.91 02	3.03 02	3.17 02

Z	transition	E [keV]	λ [pm]	$_{81}$Tl	$_{82}$Pb	$_{83}$Bi	$_{84}$Po	$_{85}$At
67	Ho L$_{\beta_1}$	7.5253	164.157	2.65 02	2.76 02	2.86 02	2.98 02	3.12 02
27	Co K$_{\beta_{1,3}}$	7.6494	162.084	2.54 02	2.64 02	2.75 02	2.86 02	3.00 02
71	Lu L$_{\alpha_1}$	7.6555	161.955	2.53 02	2.64 02	2.74 02	2.85 02	2.99 02
68	Er L$_{\beta_1}$	7.8109	158.733	2.41 02	2.51 02	2.60 02	2.71 02	2.84 02
72	Hf L$_{\alpha_1}$	7.8990	156.963	2.34 02	2.44 02	2.53 02	2.63 02	2.76 02
29	Cu K$_{\alpha_1}$	8.0478	154.060	2.23 02	2.32 02	2.41 02	2.51 02	2.63 02
69	Tm L$_{\beta_1}$	8.1010	153.049	2.19 02	2.28 02	2.37 02	2.47 02	2.59 02
73	Ta L$_{\alpha_1}$	8.1461	152.201	2.16 02	2.25 02	2.34 02	2.44 02	2.55 02
28	Ni K$_{\beta_{1,3}}$	8.2647	150.017	2.08 02	2.17 02	2.26 02	2.35 02	2.46 02
74	W L$_{\alpha_1}$	8.3976	147.643	2.00 02	2.09 02	2.17 02	2.26 02	2.36 02
70	Yb L$_{\beta_1}$	8.4018	147.569	2.00 02	2.08 02	2.16 02	2.25 02	2.36 02
30	Zn K$_{\alpha_1}$	8.6389	143.519	1.86 02	1.94 02	2.02 02	2.10 02	2.20 02
75	Re L$_{\alpha_1}$	8.6525	143.294	1.86 02	1.93 02	2.01 02	2.09 02	2.19 02
71	Lu L$_{\beta_1}$	8.7090	142.364	1.83 02	1.90 02	1.98 02	2.06 02	2.16 02
29	Cu K$_{\beta_1}$	8.9053	139.226	1.73 02	1.80 02	1.87 02	1.94 02	2.04 02
76	Os L$_{\alpha_1}$	8.9117	139.126	1.72 02	1.80 02	1.87 02	1.94 02	2.03 02
72	Hf L$_{\beta_1}$	9.0227	137.414	1.67 02	1.74 02	1.81 02	1.88 02	1.97 02
77	Ir L$_{\alpha_1}$	9.1751	135.132	1.60 02	1.67 02	1.73 02	1.80 02	1.89 02
31	Ga K$_{\alpha_1}$	9.2517	134.013	1.57 02	1.63 02	1.70 02	1.77 02	1.85 02
73	Ta L$_{\beta_1}$	9.3431	132.702	1.53 02	1.59 02	1.66 02	1.72 02	1.81 02
78	Pt L$_{\alpha_1}$	9.4423	131.308	1.49 02	1.55 02	1.61 02	1.68 02	1.76 02
30	Zb K$_{\beta_{1,3}}$	9.5720	129.529	1.44 02	1.50 02	1.56 02	1.62 02	1.70 02
74	W L$_{\beta_1}$	9.6724	128.184	1.40 02	1.46 02	1.52 02	1.58 02	1.65 02
79	Au L$_{\alpha_1}$	9.7133	127.644	1.39 02	1.45 02	1.50 02	1.56 02	1.64 02
32	Ge K$_{\alpha_1}$	9.8864	125.409	1.33 02	1.38 02	1.44 02	1.49 02	1.56 02
80	Hg L$_{\alpha_1}$	9.9888	124.124	1.29 02	1.35 02	1.40 02	1.46 02	1.52 02
75	Re L$_{\beta_1}$	10.010	123.861	1.28 02	1.34 02	1.39 02	1.45 02	1.52 02
31	Ga K$_{\beta_1}$	10.264	120.796	1.21 02	1.26 02	1.31 02	1.36 02	1.42 02
81	Tl L$_{\alpha_1}$	10.269	120.737	1.20 02	1.26 02	1.31 02	1.36 02	1.42 02
76	Os L$_{\beta_1}$	10.355	119.743	1.18 02	1.23 02	1.28 02	1.33 02	1.39 02
33	As K$_{\alpha_1}$	10.544	117.588	1.13 02	1.18 02	1.22 02	1.27 02	1.33 02
82	Pb L$_{\alpha_1}$	10.552	117.499	1.12 02	1.17 02	1.22 02	1.27 02	1.33 02
77	Ir L$_{\beta_1}$	10.708	115.787	1.08 02	1.13 02	1.17 02	1.22 02	1.28 02
83	Bi L$_{\alpha_1}$	10.839	114.388	1.05 02	1.10 02	1.14 02	1.19 02	1.24 02
32	Ge K$_{\beta_1}$	10.982	112.898	1.02 02	1.06 02	1.10 02	1.15 02	1.20 02
78	Pt L$_{\beta_1}$	11.071	111.990	9.96 01	1.04 02	1.08 02	1.12 02	1.18 02
84	Po L$_{\alpha_1}$	11.131	111.387	9.83 01	1.03 02	1.07 02	1.11 02	1.16 02
34	Se K$_{\alpha_1}$	11.222	110.484	9.63 01	1.01 02	1.04 02	1.09 02	1.14 02
85	At L$_{\alpha_1}$	11.427	108.501	9.20 01	9.60 01	9.97 01	1.04 02	1.09 02
79	Au L$_{\beta_1}$	11.442	108.359	9.17 01	9.57 01	9.94 01	1.03 02	1.09 02
33	As K$_{\beta_1}$	11.726	105.735	8.62 01	9.00 01	9.35 01	9.73 01	1.02 02
86	Rn L$_{\alpha_1}$	11.727	105.726	8.62 01	9.00 01	9.34 01	9.73 01	1.02 02
80	Hg L$_{\beta_1}$	11.823	104.867	8.44 01	8.81 01	9.16 01	9.53 01	1.00 02
35	Br K$_{\alpha_1}$	11.924	103.979	8.26 01	8.63 01	8.96 01	9.33 01	9.79 01

Z	transition	E [keV]	λ [pm]	$_{81}$Tl	$_{82}$Pb	$_{83}$Bi	$_{84}$Po	$_{85}$At
87	Fr L_{α_1}	12.031	103.054	8.08 01	8.44 01	8.76 01	9.12 01	9.57 01
81	Tl L_{β_1}	12.213	101.519	7.78 01	8.12 01	8.44 01	8.79 01	9.22 01
88	Ra L_{α_1}	12.340	100.474	7.58 01	7.92 01	8.22 01	8.56 01	8.98 01
34	Se K_{β_1}	12.496	99.219	7.34 01	7.67 01	7.97 01	8.29 01	8.70 01
82	Pb L_{β_1}	12.614	98.291	7.17 01	7.49 01	7.78 01	8.10 01	8.50 01
36	Kr K_{α_1}	12.649	98.019	7.12 01	7.44 01	7.73 01	8.04 01	8.44 01
89	Ac L_{α_1}	12.652	97.996	7.11 01	7.43 01	7.72 01	8.04 01	8.44 01
90	Th L_{α_1}	12.969	95.601	1.61 02	6.98 01	7.25 01	7.56 01	7.93 01
83	Bi L_{β_1}	13.024	95.197	1.59 02	6.91 01	7.18 01	7.48 01	7.85 01
91	Pa L_{α_1}	13.291	93.285	1.51 02	1.62 02	6.82 01	7.10 01	7.46 01
35	Br K_{β_1}	13.291	93.285	1.51 02	1.62 02	6.82 01	7.10 01	7.46 01
37	Rb K_{α_1}	13.395	92.560	1.47 02	1.59 02	6.69 01	6.97 01	7.31 01
84	Po L_{β_1}	13.447	92.202	1.46 02	1.57 02	1.54 02	6.90 01	7.24 01
92	U L_{α_1}	13.615	91.065	1.41 02	1.52 02	1.49 02	6.69 01	7.02 01
85	At L_{β_1}	13.876	89.352	1.34 02	1.44 02	1.41 02	1.48 02	6.70 01
36	Kr K_{β_1}	14.112	87.857	1.28 02	1.38 02	1.35 02	1.41 02	6.42 01
38	Sr K_{α_1}	14.165	87.529	1.27 02	1.37 02	1.34 02	1.40 02	6.36 01
86	Rn L_{β_1}	14.316	86.606	1.24 02	1.33 02	1.30 02	1.36 02	1.43 02
87	Fr L_{β_1}	14.770	83.943	1.58 02	1.22 02	1.20 02	1.25 02	1.31 02
39	Y K_{α_1}	14.958	82.888	1.53 02	1.18 02	1.16 02	1.21 02	1.27 02
37	Rb K_{β_1}	14.961	82.872	1.53 02	1.18 02	1.16 02	1.21 02	1.27 02
88	Ra L_{β_1}	15.236	81.376	1.45 02	1.56 02	1.11 02	1.16 02	1.21 02
89	Ac L_{β_1}	15.713	78.906	1.60 02	1.44 02	1.42 02	1.07 02	1.12 02
40	Zr K_{α_1}	15.775	78.596	1.58 02	1.42 02	1.40 02	1.06 02	1.11 02
38	Sr K_{β_1}	15.836	78.293	1.57 02	1.41 02	1.39 02	1.04 02	1.09 02
90	Th L_{β_1}	16.202	76.524	1.48 02	1.58 02	1.31 02	9.84 01	1.03 02
41	Nb K_{α_1}	16.615	74.622	1.38 02	1.48 02	1.46 02	1.28 02	9.65 01
91	Pa L_{β_1}	16.702	74.233	1.36 02	1.46 02	1.44 02	1.26 02	9.52 01
39	Y K_{β_1}	16.738	74.074	1.35 02	1.45 02	1.43 02	1.25 02	9.47 01
92	U L_{β_1}	17.220	72.000	1.25 02	1.34 02	1.33 02	1.39 02	1.22 02
42	Mo K_{α_1}	17.479	70.933	1.21 02	1.29 02	1.28 02	1.34 02	1.17 02
40	Zr K_{β_1}	17.668	70.175	1.17 02	1.25 02	1.24 02	1.30 02	1.36 02
43	Tc K_{α_1}	18.367	67.504	1.06 02	1.13 02	1.12 02	1.17 02	1.23 02
41	Nb K_{β_1}	18.623	66.576	1.02 02	1.09 02	1.08 02	1.13 02	1.18 02
44	Ru K_{α_1}	19.279	64.311	9.29 01	9.89 01	9.88 01	1.03 02	1.08 02
42	Mo K_{β_1}	19.608	63.231	8.88 01	9.44 01	9.44 01	9.88 01	1.03 02
45	Rh K_{α_1}	20.216	61.330	8.19 01	8.70 01	8.72 01	9.12 01	9.54 01
43	Tc K_{β_1}	20.619	60.131	7.78 01	8.25 01	8.28 01	8.66 01	9.05 01
46	Pd K_{α_1}	21.177	58.547	7.25 01	7.68 01	7.72 01	8.07 01	8.44 01
44	Ru K_{β_1}	21.657	57.249	6.83 01	7.23 01	7.28 01	7.61 01	7.96 01
47	Ag K_{α_1}	22.163	55.942	6.43 01	6.80 01	6.86 01	7.17 01	7.49 01
45	Rh K_{β_1}	22.724	54.561	6.02 01	6.36 01	6.42 01	6.71 01	7.02 01
48	Cd K_{α_1}	23.174	53.501	5.71 01	6.04 01	6.10 01	6.38 01	6.66 01
46	Pd K_{β_1}	23.819	52.053	5.31 01	5.61 01	5.68 01	5.93 01	6.20 01

Z	transition	E [keV]	λ [pm]	$_{81}$Tl	$_{82}$Pb	$_{83}$Bi	$_{84}$Po	$_{85}$At
49	In K$_{\alpha_1}$	24.210	51.212	5.09 01	5.37 01	5.45 01	5.68 01	5.94 01
47	Ag K$_{\beta_1}$	24.942	49.709	4.70 01	4.96 01	5.04 01	5.26 01	5.49 01
50	Sn K$_{\alpha_1}$	25.271	49.062	4.54 01	4.79 01	4.87 01	5.08 01	5.31 01
48	Cd K$_{\beta_1}$	26.096	47.511	4.17 01	4.39 01	4.48 01	4.67 01	4.88 01
51	Sb K$_{\alpha_1}$	26.359	47.037	4.06 01	4.28 01	4.36 01	4.55 01	4.75 01
49	In K$_{\beta_1}$	27.276	45.455	3.71 01	3.90 01	3.99 01	4.16 01	4.34 01
52	Te K$_{\alpha_1}$	27.472	45.131	3.64 01	3.83 01	3.92 01	4.08 01	4.26 01
50	Sn K$_{\beta_1}$	28.486	43.524	3.31 01	3.47 01	3.56 01	3.71 01	3.88 01
53	I K$_{\alpha_1}$	28.612	43.333	3.27 01	3.43 01	3.52 01	3.67 01	3.83 01
51	Sb K$_{\beta_1}$	29.726	41.709	2.96 01	3.10 01	3.19 01	3.32 01	3.47 01
54	Xe K$_{\alpha_1}$	29.779	41.635	2.95 01	3.08 01	3.17 01	3.30 01	3.45 01
55	Cs K$_{\alpha_1}$	30.973	40.030	2.66 01	2.78 01	2.86 01	2.98 01	3.12 01
52	Te K$_{\beta_1}$	30.996	40.000	2.65 01	2.77 01	2.86 01	2.98 01	3.11 01
56	Ba K$_{\alpha_1}$	32.194	38.511	2.40 01	2.51 01	2.59 01	2.70 01	2.82 01
53	I K$_{\beta_1}$	32.295	38.391	2.38 01	2.49 01	2.57 01	2.67 01	2.80 01
57	La K$_{\alpha_1}$	33.442	37.074	2.17 01	2.27 01	2.35 01	2.44 01	2.55 01
54	Xe K$_{\beta_1}$	33.624	36.874	2.14 01	2.24 01	2.31 01	2.41 01	2.52 01
58	Ce K$_{\alpha_1}$	34.279	36.169	2.04 01	2.13 01	2.20 01	2.29 01	2.40 01
55	Cs K$_{\beta_1}$	34.987	35.437	1.93 01	2.02 01	2.09 01	2.17 01	2.27 01
59	Pr K$_{\alpha_1}$	36.026	34.415	1.79 01	1.87 01	1.93 01	2.01 01	2.11 01
56	Ba K$_{\beta_1}$	36.378	34.082	1.74 01	1.82 01	1.89 01	1.96 01	2.05 01
60	Nd K$_{\alpha_1}$	36.847	33.648	1.69 01	1.76 01	1.82 01	1.90 01	1.99 01
57	La K$_{\beta_1}$	37.801	32.799	1.58 01	1.64 01	1.71 01	1.77 01	1.86 01
61	Pm K$_{\alpha_1}$	38.725	32.016	1.48 01	1.54 01	1.60 01	1.67 01	1.75 01
58	Ce K$_{\beta_1}$	39.257	31.582	1.43 01	1.49 01	1.55 01	1.61 01	1.69 01
62	Sm K$_{\alpha_1}$	40.118	30.905	1.35 01	1.40 01	1.46 01	1.52 01	1.59 01
59	Pr K$_{\beta_1}$	40.748	30.427	1.30 01	1.35 01	1.40 01	1.46 01	1.53 01
63	Eu K$_{\alpha_1}$	41.542	29.845	1.24 01	1.28 01	1.34 01	1.39 01	1.46 01
60	Nd K$_{\beta_1}$	42.271	29.331	1.18 01	1.22 01	1.28 01	1.33 01	1.39 01
64	Gd K$_{\alpha_1}$	42.996	28.836	1.13 01	1.17 01	1.22 01	1.27 01	1.33 01
61	Pm K$_{\beta_1}$	43.826	28.290	1.08 01	1.11 01	1.16 01	1.21 01	1.27 01
65	Tb K$_{\alpha_1}$	44.482	27.873	1.04 01	1.07 01	1.12 01	1.17 01	1.22 01
62	Sm K$_{\beta_1}$	45.413	27.301	9.82 00	1.01 01	1.06 01	1.11 01	1.16 01
66	Dy K$_{\alpha_1}$	45.998	26.954	9.50 00	9.81 00	1.03 01	1.07 01	1.12 01
63	Eu K$_{\beta_1}$	47.038	26.358	8.97 00	9.25 00	9.71 00	1.01 01	1.06 01
67	Ho K$_{\alpha_1}$	47.547	26.076	8.72 00	8.99 00	9.45 00	9.82 00	1.03 01
64	Gd K$_{\beta_1}$	48.697	25.460	8.20 00	8.44 00	8.88 00	9.24 00	9.67 00
68	Er K$_{\alpha_1}$	49.128	25.237	8.02 00	8.25 00	8.68 00	9.03 00	9.45 00
65	Tb K$_{\beta_1}$	50.382	24.609	7.51 00	7.72 00	8.14 00	8.47 00	8.86 00
69	Tm K$_{\alpha_1}$	50.742	24.434	7.38 00	7.58 00	7.99 00	8.31 00	8.70 00
66	Dy K$_{\beta_1}$	52.119	23.788	6.89 00	7.07 00	7.47 00	7.76 00	8.13 00
70	Yb K$_{\alpha_1}$	52.389	23.666	6.80 00	6.98 00	7.37 00	7.66 00	8.02 00
67	Ho K$_{\beta_1}$	53.877	23.012	6.33 00	6.49 00	6.86 00	7.13 00	7.47 00
71	Lu K$_{\alpha_1}$	54.070	22.930	6.27 00	6.43 00	6.80 00	7.07 00	7.40 00

Z	transition	E [keV]	λ [pm]	$_{81}$Tl	$_{82}$Pb	$_{83}$Bi	$_{84}$Po	$_{85}$At
68	Er K$_{\beta_1}$	55.681	22.267	5.82 00	5.96 00	6.31 00	6.56 00	6.87 00
72	Hf K$_{\alpha_1}$	55.790	22.223	5.79 00	5.93 00	6.28 00	6.53 00	6.83 00
69	Tm K$_{\beta_1}$	57.517	21.556	5.35 00	5.48 00	5.81 00	6.04 00	6.32 00
73	Ta K$_{\alpha_1}$	57.532	21.550	5.35 00	5.48 00	5.81 00	6.03 00	6.32 00
74	W K$_{\alpha_1}$	59.318	20.901	4.95 00	5.06 00	5.37 00	5.58 00	5.85 00
70	Yb K$_{\beta_1}$	59.370	20.883	4.94 00	5.05 00	5.36 00	5.57 00	5.83 00
75	Re K$_{\alpha_1}$	61.140	20.278	4.58 00	4.68 00	4.98 00	5.17 00	5.42 00
71	Lu K$_{\beta_1}$	61.283	20.231	4.55 00	4.66 00	4.95 00	5.14 00	5.39 00
76	Os K$_{\alpha_1}$	63.001	19.679	4.25 00	4.34 00	4.62 00	4.80 00	5.03 00
72	Hf K$_{\beta_1}$	63.234	19.607	4.21 00	4.30 00	4.58 00	4.76 00	4.98 00
77	Ir K$_{\alpha_1}$	64.896	19.105	3.94 00	4.03 00	4.29 00	4.46 00	4.67 00
73	Ta K$_{\beta_1}$	65.223	19.009	3.89 00	3.98 00	4.24 00	4.40 00	4.61 00
78	Pt K$_{\alpha_1}$	66.832	18.551	3.66 00	3.74 00	3.99 00	4.14 00	4.34 00
74	W K$_{\beta_1}$	67.244	18.438	3.61 00	3.68 00	3.93 00	4.08 00	4.27 00
79	Au K$_{\alpha_1}$	68.804	18.020	3.40 00	3.48 00	3.71 00	3.86 00	4.04 00
75	Re K$_{\beta_1}$	69.210	17.888	3.34 00	3.41 00	3.65 00	3.79 00	3.96 00
80	Hg K$_{\alpha_1}$	70.819	17.507	3.17 00	3.23 00	3.46 00	3.59 00	3.76 00
76	Os K$_{\beta_1}$	71.413	17.361	3.10 00	3.16 00	3.39 00	3.51 00	3.68 00
81	Tl K$_{\alpha_1}$	72.872	17.014	2.95 00	3.01 00	3.22 00	3.34 00	3.50 00
77	Ir K$_{\beta_1}$	73.561	16.854	2.88 00	2.94 00	3.15 00	3.26 00	3.42 00
82	Pb K$_{\alpha_1}$	74.969	16.538	2.75 00	2.80 00	3.00 00	3.11 00	3.26 00
78	Pt K$_{\beta_1}$	75.748	16.368	2.68 00	2.73 00	2.92 00	3.03 00	3.18 00
83	Bi K$_{\alpha_1}$	77.108	16.079	2.56 00	2.61 00	2.80 00	2.90 00	3.04 00
79	Au K$_{\beta_1}$	77.948	15.906	2.50 00	2.54 00	2.73 00	2.83 00	2.96 00
84	Po K$_{\alpha_1}$	79.290	15.636	2.39 00	2.43 00	2.61 00	2.71 00	2.84 00
80	Hg K$_{\beta_1}$	80.253	15.449	2.32 00	2.36 00	2.54 00	2.63 00	2.75 00
85	At K$_{\alpha_1}$	81.520	15.209	2.24 00	2.27 00	2.44 00	2.53 00	2.65 00
81	Tl K$_{\beta_1}$	82.576	15.014	2.17 00	2.20 00	2.37 00	2.45 00	2.57 00
86	Rn K$_{\alpha_1}$	83.780	14.798	2.09 00	2.12 00	2.29 00	2.37 00	2.48 00
82	Pb K$_{\beta_1}$	84.936	14.597	<u>2.02 00</u>	2.05 00	2.21 00	2.29 00	2.40 00
87	Fr K$_{\alpha_1}$	86.100	14.400	8.56 00	1.99 00	2.14 00	2.22 00	2.32 00
83	Bi K$_{\beta_1}$	87.343	14.195	8.23 00	<u>1.92 00</u>	2.07 00	2.14 00	2.24 00
88	Ra K$_{\alpha_1}$	88.470	14.014	7.94 00	7.63 00	2.00 00	2.08 00	2.17 00
84	Po K$_{\beta_1}$	89.800	13.806	7.62 00	7.34 00	<u>1.93 00</u>	2.00 00	2.09 00
89	Ac K$_{\alpha_1}$	90.884	13.642	7.37 00	7.12 00	8.01 00	1.94 00	2.03 00
85	At K$_{\beta_1}$	92.300	13.432	7.06 00	6.85 00	7.68 00	<u>1.87 00</u>	1.96 00
90	Th K$_{\alpha_1}$	93.350	13.281	6.85 00	6.65 00	7.45 00	7.69 00	1.91 00
86	Rn K$_{\beta_1}$	94.870	13.068	6.55 00	6.38 00	7.12 00	7.36 00	<u>1.83 00</u>
91	Pa K$_{\alpha_1}$	95.868	12.932	6.36 00	6.22 00	6.92 00	7.16 00	7.49 00
87	Fr K$_{\beta_1}$	97.470	12.720	6.07 00	5.96 00	6.62 00	6.85 00	7.16 00
92	U K$_{\alpha_1}$	98.439	12.595	5.91 00	5.81 00	6.44 00	6.67 00	6.97 00
88	Ra K$_{\beta_1}$	100.130	12.382	5.64 00	5.56 00	6.15 00	6.37 00	6.66 00
89	Ac K$_{\beta_1}$	102.850	12.054	5.25 00	5.20 00	5.72 00	5.94 00	6.20 00
90	Th K$_{\beta_1}$	105.610	11.739	4.89 00	4.86 00	5.33 00	5.54 00	5.78 00

Z	transition	E [keV]	λ [pm]	$_{81}$Tl	$_{82}$Pb	$_{83}$Bi	$_{84}$Po	$_{85}$At
91	Pa K$_{\beta_1}$	108.430	11.434	4.56 00	4.54 00	4.97 00	5.16 00	5.40 00
92	U K$_{\beta_1}$	111.300	11.139	4.26 00	4.25 00	4.63 00	4.82 00	5.04 00

Z	transition	E [keV]	λ [pm]	$_{86}$Rn	$_{87}$Fr	$_{88}$Ra	$_{89}$Ac	$_{90}$Th
4	Be K$_\alpha$	0.1085	11427.207	5.61 04	5.61 04	5.94 04	5.91 04	6.84 04
38	Sr M$_\xi$	0.1140	10875.895	5.03 04	5.10 04	5.40 04	5.40 04	5.46 04
39	Y M$_\xi$	0.1328	9339.235	3.22 04	3.35 04	3.53 04	3.97 04	3.06 04
16	S L$_l$	0.1487	8337.942	2.25 04	2.38 04	2.49 04	2.90 04	1.96 04
40	Zr M$_\xi$	0.1511	8205.506	2.13 04	2.25 04	2.37 04	2.78 04	1.85 04
41	Nb M$_\xi$	0.1717	7221.037	1.38 04	1.47 04	1.61 04	1.89 04	1.12 04
5	B K$_\alpha$	0.1833	6764.059	1.10 04	1.19 04	1.30 04	1.52 04	8.67 03
42	Mo M$_\xi$	0.1926	6473.445	9.24 03	1.01 04	1.11 04	1.29 04	7.15 03
6	C K$_\alpha$	0.2770	4476.000	5.29 03	5.04 03	4.54 03	4.54 03	6.28 03
47	Ag M$_\xi$	0.3117	3977.709	5.91 03	5.83 03	4.88 03	4.64 03	7.25 03
7	N K$_\alpha$	0.3924	3159.664	1.23 04	1.11 04	9.20 03	7.99 03	8.81 03
22	Ti L$_l$	0.3953	3136.484	1.24 04	1.14 04	9.60 03	8.32 03	8.83 03
22	Ti L$_\alpha$	0.4522	2741.822	1.25 04	1.22 04	1.14 04	1.05 04	9.33 03
23	V L$_\alpha$	0.5113	2424.901	1.19 04	1.19 04	1.18 04	1.15 04	1.09 04
8	O K$_\alpha$	0.5249	2362.072	1.18 04	1.19 04	1.19 04	1.17 04	1.13 04
25	Mn L$_l$	0.5563	2228.747	1.17 04	1.13 04	1.13 04	1.12 04	1.09 04
24	Cr L$_\alpha$	0.5728	2164.549	1.18 04	1.10 04	1.10 04	1.10 04	1.07 04
25	Mn L$_\alpha$	0.6374	1945.171	1.11 04	1.16 04	1.08 04	9.89 03	9.64 03
9	F K$_\alpha$	0.6768	1831.932	1.02 04	1.06 04	1.06 04	1.03 04	1.04 04
26	Fe L$_\alpha$	0.7050	1758.655	9.75 03	1.01 04	1.03 04	1.04 04	9.98 03
27	Co L$_\alpha$	0.7762	1597.335	8.45 03	8.79 03	9.00 03	9.12 03	8.98 03
28	Ni L$_\alpha$	0.8515	1456.080	8.32 03	7.58 03	7.78 03	7.91 03	7.89 03
29	Cu L$_\alpha$	0.9297	1336.044	7.06 03	7.35 03	6.73 03	6.87 03	6.87 03
30	Zn L$_\alpha$	1.0117	1225.513	8.07 03	6.36 03	6.57 03	6.67 03	5.96 03
11	Na K$_\alpha$	1.0410	1191.020	7.52 03	6.04 03	6.24 03	6.34 03	6.31 03
11	Na K$_\beta$	1.0711	1157.550	7.02 03	5.38 03	5.92 03	5.94 03	6.12 03
12	Mg K$_\alpha$	1.2536	989.033	5.25 03	5.58 03	5.98 03	5.96 03	5.97 03
33	As L$_\alpha$	1.2820	967.123	4.97 03	5.29 03	5.66 03	6.18 03	5.63 03
12	Mg K$_\beta$	1.3022	952.121	4.79 03	5.09 03	5.44 03	5.94 03	5.41 03
33	As L$_{\beta_1}$	1.3170	941.421	4.66 03	4.96 03	5.29 03	5.78 03	5.26 03
66	Dy M$_\beta$	1.3250	955.737	4.59 03	4.88 03	5.21 03	5.69 03	5.18 03
67	Ho M$_\alpha$	1.3480	919.771	4.40 03	4.69 03	5.00 03	5.45 03	5.46 03
34	Se L$_\alpha$	1.3791	899.029	4.16 03	4.43 03	4.72 03	5.14 03	5.15 03
67	Ho M$_\beta$	1.3830	896.494	4.13 03	4.40 03	4.69 03	5.11 03	5.11 03
68	Er M$_\alpha$	1.4060	881.829	3.97 03	4.23 03	4.50 03	4.90 03	4.90 03
34	Se L$_{\beta_1}$	1.4192	873.627	3.88 03	4.14 03	4.40 03	4.79 03	4.78 03
68	Er M$_\beta$	1.4430	859.218	3.73 03	3.97 03	4.22 03	4.59 03	4.58 03
69	Tm M$_\alpha$	1.4620	848.051	3.61 03	3.85 03	4.09 03	4.44 03	4.43 03
35	Br L$_\alpha$	1.4804	837.511	3.50 03	3.74 03	3.96 03	4.31 03	4.29 03

Z	transition	E [keV]	λ [pm]	$_{86}$Rn	$_{87}$Fr	$_{88}$Ra	$_{89}$Ac	$_{90}$Th
13	Al K$_{\alpha_1}$	1.4867	833.962	3.47 03	3.70 03	3.92 03	4.26 03	4.26 03
69	Tm M$_\beta$	1.5030	824.918	3.37 03	3.60 03	3.82 03	4.14 03	4.13 03
70	Yb M$_\alpha$	1.5214	814.941	3.28 03	3.50 03	3.70 03	4.02 03	4.00 03
35	Br L$_{\beta_1}$	1.5259	812.638	3.25 03	3.47 03	3.68 03	3.99 03	3.97 03
13	Al K$_\beta$	1.5574	796.103	3.09 03	3.30 03	3.50 03	3.79 03	3.77 03
70	Yb M$_\beta$	1.5675	790.974	3.04 03	3.25 03	3.44 03	3.73 03	3.71 03
71	Lu M$_\alpha$	1.5813	784.071	2.98 03	3.18 03	3.37 03	3.64 03	3.63 03
36	Kr L$_\alpha$	1.5860	781.747	2.96 03	3.16 03	3.34 03	3.62 03	3.60 03
71	Lu M$_\beta$	1.6312	760.085	2.76 03	2.95 03	3.12 03	3.37 03	3.35 03
36	Kr L$_{\beta_1}$	1.6366	757.577	2.74 03	2.93 03	3.09 03	3.34 03	3.32 03
72	Hf M$_{\alpha_1}$	1.6446	753.892	2.71 03	2.89 03	3.06 03	3.30 03	3.28 03
37	Rb L$_{\alpha_1}$	1.6941	731.864	2.52 03	2.69 03	2.84 03	3.06 03	3.04 03
72	Hf M$_\beta$	1.6976	730.355	2.50 03	2.68 03	2.82 03	3.05 03	3.03 03
73	Ta M$_{\alpha_1}$	1.7096	725.229	2.46 03	2.63 03	2.78 03	2.99 03	2.97 03
14	Si K$_{\alpha_1}$	1.7400	712.558	2.36 03	2.52 03	2.66 03	2.86 03	2.84 03
37	Rb L$_{\beta_1}$	1.7522	707.597	2.32 03	2.48 03	2.61 03	2.81 03	2.79 03
73	Ta M$_\beta$	1.7655	702.266	2.28 03	2.44 03	2.56 03	2.76 03	2.74 03
74	W M$_{\alpha_1}$	1.7754	698.350	2.24 03	2.40 03	2.53 03	2.72 03	2.70 03
38	Sr L$_{\alpha_1}$	1.8066	686.290	2.15 03	2.30 03	2.42 03	2.60 03	2.59 03
74	W M$_\beta$	1.8349	675.705	2.07 03	2.22 03	2.33 03	2.50 03	2.48 03
14	Si K$_\beta$	1.8359	675.337	2.07 03	2.21 03	2.33 03	2.50 03	2.48 03
75	Re M$_{\alpha_1}$	1.8420	673.100	2.05 03	2.20 03	2.31 03	2.48 03	2.46 03
38	Sr L$_{\beta_1}$	1.8717	662.420	1.97 03	2.11 03	2.22 03	2.38 03	2.36 03
75	Re M$_\beta$	1.9061	650.465	1.89 03	2.02 03	2.12 03	2.27 03	2.26 03
76	Os M$_{\alpha_1}$	1.9102	649.069	1.88 03	2.01 03	2.11 03	2.26 03	2.24 03
39	Y L$_{\alpha_1}$	1.9226	644.882	1.85 03	1.98 03	2.08 03	2.22 03	2.21 03
76	Os M$_\beta$	1.9783	626.725	1.72 03	1.85 03	1.93 03	2.07 03	2.05 03
77	Ir M$_{\alpha_1}$	1.9799	626.219	1.72 03	1.84 03	1.93 03	2.06 03	2.05 03
39	Y L$_{\beta_1}$	1.9958	621.230	1.69 03	1.81 03	1.89 03	2.02 03	2.01 03
15	P K$_{\alpha_1}$	2.0137	615.708	1.65 03	1.77 03	1.85 03	1.98 03	1.96 03
40	Zr L$_{\alpha_1}$	2.0424	607.056	1.59 03	1.71 03	1.79 03	1.91 03	1.89 03
78	Pt M$_{\alpha_1}$	2.0505	604.658	1.58 03	1.69 03	1.77 03	1.89 03	1.87 03
77	Ir M$_\beta$	2.0535	603.775	1.57 03	1.69 03	1.77 03	1.88 03	1.87 03
79	Au M$_{\alpha_1}$	2.1229	584.036	1.45 03	1.56 03	1.63 03	1.73 03	1.72 03
40	Zr L$_{\beta_1}$	2.1244	583.624	1.45 03	1.55 03	1.62 03	1.73 03	1.71 03
78	Pt M$_\beta$	2.1273	582.828	1.44 03	1.55 03	1.62 03	1.72 03	1.71 03
15	P K$_{\beta_{1,3}}$	2.1390	579.640	1.42 03	1.53 03	1.60 03	1.70 03	1.68 03
41	Nb L$_{\alpha_1}$	2.1659	572.441	1.38 03	1.48 03	1.55 03	1.65 03	1.63 03
80	Hg M$_{\alpha_1}$	2.1953	564.775	1.34 03	1.44 03	1.50 03	1.59 03	1.58 03
79	Au M$_\beta$	2.2046	562.393	1.32 03	1.42 03	1.48 03	1.57 03	1.56 03
41	Nb L$_{\beta_1}$	2.2574	549.238	1.25 03	1.34 03	1.40 03	1.48 03	1.47 03
81	Tl M$_{\alpha_1}$	2.2706	546.045	1.23 03	1.32 03	1.38 03	1.46 03	1.45 03
80	Hg M$_\beta$	2.2825	543.199	1.22 03	1.31 03	1.36 03	1.44 03	1.43 03
42	Mo L$_{\alpha_1}$	2.2932	540.664	1.20 03	1.29 03	1.35 03	1.43 03	1.41 03

Z	transition	E [keV]	λ [pm]	$_{86}$Rn	$_{87}$Fr	$_{88}$Ra	$_{89}$Ac	$_{90}$Th
16	S K$_{\alpha_1}$	2.3080	537.197	1.18 03	1.27 03	1.32 03	1.40 03	1.39 03
82	Pb M$_{\alpha_1}$	2.3457	528.563	1.14 03	1.22 03	1.27 03	1.35 03	1.33 03
81	Tl M$_\beta$	2.3621	524.893	1.12 03	1.20 03	1.25 03	1.32 03	1.31 03
42	Mo L$_{\beta_1}$	2.3948	517.726	1.08 03	1.16 03	1.21 03	1.28 03	1.26 03
83	Bi M$_{\alpha_1}$	2.4226	511.785	1.05 03	1.13 03	1.18 03	1.24 03	1.23 03
43	Tl L$_{\alpha_1}$	2.4240	511.490	1.05 03	1.13 03	1.17 03	1.24 03	1.23 03
82	Pb M$_\beta$	2.4427	507.574	1.03 03	1.11 03	1.15 03	1.22 03	1.20 03
16	S K$_\beta$	2.4640	503.186	1.01 03	1.09 03	1.13 03	1.19 03	1.18 03
83	Bi M$_{\beta_1}$	2.5255	490.933	9.51 02	1.02 03	1.06 03	1.12 03	1.11 03
43	Tc L$_{\beta_1}$	2.5368	488.746	9.40 02	1.01 03	1.05 03	1.11 03	1.09 03
44	Ru L$_{\alpha_1}$	2.5586	484.582	9.21 02	9.92 02	1.03 03	1.08 03	1.07 03
17	Cl K$_{\alpha_1}$	2.6224	472.792	8.67 02	9.35 02	9.68 02	1.02 03	1.00 03
44	Ru L$_{\beta_1}$	2.6832	462.079	8.20 02	8.84 02	9.15 02	9.62 02	9.48 02
45	Rh L$_{\alpha_1}$	2.6967	459.766	8.10 02	8.74 02	9.04 02	9.50 02	9.36 02
17	Cl K$_\beta$	2.8156	440.350	7.30 02	7.87 02	8.13 02	8.52 02	8.39 02
45	Rh L$_{\beta_1}$	2.8344	437.430	7.18 02	7.75 02	7.99 02	8.38 02	8.25 02
46	Pd L$_{\alpha_1}$	2.8386	436.782	<u>7.15 02</u>	7.72 02	7.97 02	8.35 02	8.22 02
18	Ar K$_{\alpha_1}$	2.9577	419.194	1.55 03	6.99 02	7.20 02	7.53 02	7.41 02
47	Ag L$_{\alpha_1}$	2.9843	415.458	1.51 03	6.84 02	7.04 02	7.37 02	7.24 02
46	Pd L$_{\beta_1}$	2.9902	414.638	1.51 03	6.81 02	7.01 02	7.33 02	7.20 02
90	Th M$_{\alpha_1}$	2.9961	413.821	<u>1.50 03</u>	<u>6.77 02</u>	6.98 02	7.29 02	7.17 02
91	Pa M$_{\alpha_1}$	3.0823	402.248	2.09 03	1.44 03	<u>6.51 02</u>	6.80 02	6.68 02
48	Cd L$_{\alpha_1}$	3.1337	395.651	2.00 03	<u>1.38 03</u>	1.42 03	6.52 02	6.41 02
90	Th M$_\beta$	3.1458	394.129	1.98 03	2.04 03	1.41 03	6.46 02	6.35 02
47	Ag L$_{\beta_1}$	3.1509	393.491	1.97 03	2.04 03	1.40 03	6.43 02	6.32 02
92	U M$_{\alpha_1}$	3.1708	391.021	1.94 03	2.00 03	1.38 03	6.33 02	6.22 02
18	Ar K$_\beta$	3.1905	388.607	1.91 03	1.97 03	1.36 03	<u>6.24 02</u>	6.13 02
91	Pa M$_\beta$	3.2397	382.705	1.84 03	1.90 03	<u>1.31 03</u>	1.36 03	5.90 02
49	In L$_{\alpha_1}$	3.2869	377.210	1.77 03	1.83 03	1.88 03	1.31 03	5.69 02
19	K K$_{\alpha_1}$	3.3138	374.148	1.74 03	1.79 03	1.85 03	1.28 03	5.57 02
48	Cd L$_{\beta_1}$	3.3166	373.832	1.73 03	1.79 03	1.84 03	1.28 03	<u>5.56 02</u>
92	U M$_\beta$	3.3360	371.658	1.71 03	1.76 03	1.81 03	<u>1.26 03</u>	1.22 03
50	Sn L$_{\alpha_1}$	3.4440	360.003	1.57 03	1.62 03	1.67 03	1.74 03	1.13 03
49	In L$_{\beta_1}$	3.4872	355.543	<u>1.52 03</u>	1.57 03	1.62 03	1.69 03	<u>1.09 03</u>
19	K K$_\beta$	3.5896	345.401	1.69 01	1.46 03	1.51 03	1.57 03	1.51 03
51	Sb L$_{\alpha_1}$	3.6047	343.954	1.68 03	1.44 03	1.49 03	1.55 03	1.50 03
50	Sn L$_{\beta_1}$	3.6628	338.498	1.61 03	<u>1.39 03</u>	1.43 03	1.49 03	1 44 03
20	Ca K$_{\alpha_1}$	3.6917	335.848	1.58 03	1.63 03	1.40 03	1.46 03	1.41 03
52	Te L$_{\alpha_1}$	3.7693	328.934	1.49 03	1.54 03	<u>1.33 03</u>	1.38 03	1.34 03
51	Sb L$_{\beta_1}$	3.8436	322.575	1.42 03	1.47 03	1.52 03	<u>1.32 03</u>	1.27 03
53	I L$_{\alpha_1}$	3.9377	314.867	1.34 03	1.38 03	1.43 03	1.48 03	1.20 03
20	Ca K$_\beta$	4.0127	308.981	1.27 03	1.32 03	1.36 03	1.42 03	1.14 03
52	Te L$_{\beta_1}$	4.0296	307.686	1.26 03	1.30 03	1.34 03	1.40 03	<u>1.13 03</u>
21	Sc K$_{\alpha_1}$	4.0906	303.097	1.21 03	1.26 03	1.29 03	1.35 03	1.30 03

Z	transition	E [keV]	λ [pm]	$_{86}$Rn	$_{87}$Fr	$_{88}$Ra	$_{89}$Ac	$_{90}$Th
54	Xe L$_{\alpha_1}$	4.1099	301.674	1.20 03	1.24 03	1.28 03	1.33 03	1.29 03
53	I L$_{\beta_1}$	4.2207	293.755	1.23 03	1.16 03	1.19 03	1.24 03	1.20 03
55	Cs L$_{\alpha_1}$	4.2865	289.245	1.18 03	1.11 03	1.15 03	1.19 03	1.15 03
21	Sc K$_{\beta_1}$	4.4605	277.962	1.07 03	1.11 03	1.04 03	1.08 03	1.04 03
56	Ba L$_{\alpha_1}$	4.4663	277.601	1.06 03	1.10 03	1.03 03	1.08 03	1.04 03
22	Ti K$_{\alpha_1}$	4.5108	274.863	1.14 03	1.07 03	1.11 03	1.05 03	1.01 03
55	Cs L$_{\beta_1}$	4.6198	268.377	1.07 03	1.01 03	1.04 03	9.88 02	9.54 02
57	La L$_{\alpha_1}$	4.6510	266.577	1.05 03	9.95 02	1.03 03	9.71 02	9.38 02
56	Ba L$_{\beta_1}$	4.8273	256.841	9.59 02	9.94 02	1.03 03	9.71 02	8.53 02
58	Ce L$_{\alpha_1}$	4.8402	256.157	9.52 02	9.88 02	1.02 03	9.65 02	9.33 02
22	Ti K$_{\beta_{1,3}}$	4.9318	251.399	9.08 02	9.42 02	9.71 02	9.20 02	8.90 02
23	V K$_{\alpha_1}$	4.9522	250.363	8.98 02	9.32 02	9.61 02	9.11 02	8.80 02
59	Pr L$_{\alpha_1}$	5.0337	246.310	8.61 02	8.95 02	9.22 02	9.58 02	8.44 02
57	La L$_{\beta_1}$	5.0421	245.899	8.58 02	8.91 02	9.18 02	9.54 02	8.41 02
60	Nd L$_{\alpha_1}$	5.2304	237.047	7.81 02	8.12 02	8.37 02	8.69 02	8.42 02
58	Ce L$_{\beta_1}$	5.2622	235.614	7.69 02	7.99 02	8.24 02	8.56 02	8.29 02
24	Cr K$_{\alpha_1}$	5.4147	228.978	7.15 02	7.43 02	7.67 02	1.96 02	7.71 02
23	V K$_{\beta_{1,3}}$	5.4273	228.447	7.11 02	7.39 02	7.62 02	7.91 02	7.66 02
61	Pm L$_{\alpha_1}$	5.4325	228.228	7.09 02	7.37 02	7.60 02	7.89 02	7.64 02
59	Pr L$_{\beta_1}$	5.4889	225.883	6.91 02	7.18 02	7.41 02	7.69 02	7.45 02
62	Sm L$_{\alpha_1}$	5.6361	219.984	6.46 02	6.71 02	6.93 02	7.19 02	6.96 02
60	Nd L$_{\beta_1}$	5.7216	216.696	6.22 02	6.46 02	6.67 02	6.92 02	6.70 02
63	Eu L$_{\alpha_1}$	5.8457	212.096	5.89 02	6.12 02	6.32 02	6.55 02	6.35 02
25	Mn K$_{\alpha_1}$	5.8988	210.187	5.75 02	5.98 02	6.17 02	6.40 02	6.20 02
24	Cr K$_{\beta_{1,3}}$	5.9467	208.494	5.63 02	5.86 02	6.05 02	6.27 02	6.08 02
61	Pm L$_{\beta_1}$	5.9610	207.993	5.60 02	5.82 02	6.01 02	6.23 02	6.04 02
64	Gd L$_{\alpha_1}$	6.0572	204.690	5.38 02	5.59 02	5.77 02	5.98 02	5.80 02
62	Sm L$_{\beta_1}$	6.2051	199.811	5.06 02	5.26 02	5.43 02	5.63 02	5.46 02
65	Tb L$_{\alpha_1}$	6.2728	197.655	4.92 02	5.11 02	5.26 02	5.48 02	5.31 02
26	Fe K$_{\alpha_1}$	6.4038	193.611	4.67 02	4.85 02	5.01 02	5.20 02	5.04 02
63	Eu L$_{\beta_1}$	6.4564	192.034	4.57 02	4.75 02	4.91 02	5.10 02	4.94 02
25	Mn K$_{\beta_{1,3}}$	6.4905	191.025	4.51 02	4.69 02	4.85 02	5.03 02	4.87 02
66	Dy L$_{\alpha_1}$	6.4952	190.887	4.50 02	4.68 02	4.84 02	5.02 02	4.86 02
64	Gd L$_{\beta_1}$	6.7132	184.688	4.14 02	4.30 02	4.45 02	4.62 02	4.47 02
67	Ho L$_{\alpha_1}$	6.7198	184.507	4.13 02	4.29 02	4.44 02	4.61 02	4.46 02
27	Co K$_{\alpha_1}$	6.9303	178.903	3.82 02	3.97 02	4.10 02	4.26 02	4.13 02
68	Er L$_{\alpha_1}$	6.9487	178.429	3.79 02	3.94 02	4.08 02	4.24 02	4.10 02
65	Tb L$_{\beta_1}$	6.9780	177.680	3.75 02	3.90 02	4.03 02	4.19 02	4.06 02
26	Fe K$_{\beta_{1,3}}$	7.0580	175.666	3.64 02	3.79 02	3.92 02	4.07 02	3.94 02
69	Tm L$_{\alpha_1}$	7.1799	172.683	3.49 02	3.63 02	3.75 02	3.90 02	3.77 02
66	Dy L$_{\beta_1}$	7.2477	171.068	3.40 02	3.54 02	3.66 02	3.81 02	3.69 02
70	Yb L$_{\alpha_1}$	7.4156	167.195	3.21 02	3.34 02	3.46 02	3.60 02	3.48 02
28	Ni K$_{\alpha_1}$	7.4782	165.795	3.14 02	3.27 02	3.38 02	3.52 02	3.41 02
67	Ho L$_{\beta_1}$	7.5253	164.757	3.09 02	3.22 02	3.33 02	3.47 02	3.35 02

Z	transition	E [keV]	λ [pm]	$_{86}$Rn	$_{87}$Fr	$_{88}$Ra	$_{89}$Ac	$_{90}$Th
27	Co K$_{\beta_{1,3}}$	7.6494	162.084	2.97 02	3.09 02	3.19 02	3.33 02	3.22 02
71	Lu L$_{\alpha_1}$	7.6555	161.955	2.96 02	3.08 02	3.19 02	3.32 02	3.21 02
68	Er L$_{\beta_1}$	7.8109	158.733	2.81 02	2.93 02	3.03 02	3.16 02	3.05 02
72	Hf L$_{\alpha_1}$	7.8990	156.963	2.73 02	2.85 02	2.94 02	3.07 02	2.97 02
29	Cu K$_{\alpha_1}$	8.0478	154.060	2.61 02	2.71 02	2.81 02	2.93 02	2.83 02
69	Tm L$_{\beta_1}$	8.1010	153.049	2.57 02	2.67 02	2.76 02	2.86 02	2.78 02
73	Ta L$_{\alpha_1}$	8.1461	152.201	2.53 02	2.63 02	2.72 02	2.84 02	2.74 02
28	Ni K$_{\beta_{1,3}}$	8.2647	150.017	2.44 02	2.54 02	2.63 02	2.74 02	2.65 02
74	W L$_{\alpha_1}$	8.3976	147.643	2.34 02	2.44 02	2.52 02	2.63 02	2.54 02
70	Yb L$_{\beta_1}$	8.4018	147.569	2.34 02	2.44 02	2.52 02	2.63 02	2.54 02
30	Zn K$_{\alpha_1}$	8.6389	143.519	2.18 02	2.27 02	2.35 02	2.45 02	2.37 02
75	Re L$_{\alpha_1}$	8.6525	143.294	2.17 02	2.26 02	2.34 02	2.44 02	2.36 02
71	Lu L$_{\beta_1}$	8.7090	142.364	2.14 02	2.23 02	2.30 02	2.40 02	2.32 02
29	Cu K$_{\beta_1}$	8.9053	139.226	2.02 02	2.11 02	2.18 02	2.27 02	2.19 02
76	Os L$_{\alpha_1}$	8.9117	139.126	2.02 02	2.10 02	2.17 02	2.26 02	2.19 02
72	Hf L$_{\beta_1}$	9.0227	137.414	1.95 02	2.04 02	2.11 02	2.19 02	2.12 02
77	Ir L$_{\alpha_1}$	9.1751	135.132	1.87 02	1.95 02	2.02 02	2.10 02	1.02 02
31	Ga K$_{\alpha_1}$	9.2517	134.013	1.83 02	1.91 02	1.98 02	2.06 02	1.99 02
73	Ta L$_{\beta_1}$	9.3431	132.702	1.79 02	1.87 02	1.93 02	2.01 02	1.94 02
78	Pt L$_{\alpha_1}$	9.4423	131.308	1.74 02	1.82 02	1.88 02	1.96 02	1.89 02
30	Zn K$_{\beta_{1,3}}$	9.5720	129.529	1.68 02	1.76 02	1.81 02	1.89 02	1.83 02
74	W L$_{\beta_1}$	9.6724	128.184	1.64 02	1.71 02	1.77 02	1.84 02	1.78 02
79	Au L$_{\alpha_1}$	9.7133	127.644	1.62 02	1.69 02	1.75 02	1.82 02	1.76 02
32	Ge K$_{\alpha_1}$	9.8864	125.409	1.55 02	1.62 02	1.67 02	1.74 02	1.68 02
80	Hg L$_{\alpha_1}$	9.9888	124.124	1.51 02	1.58 02	1.63 02	1.70 02	1.64 02
75	Re L$_{\beta_1}$	10.010	123.861	1.50 02	1.57 02	1.62 02	1.69 02	1.63 02
31	Ga K$_{\beta_1}$	10.264	120.796	1.41 02	1.48 02	1.52 02	1.59 02	1.53 02
81	Tl L$_{\alpha_1}$	10.269	120.737	1.41 02	1.47 02	1.52 02	1.58 02	1.53 02
76	Os L$_{\beta_1}$	10.355	119.734	1.38 02	1.44 02	1.49 02	1.55 02	1.50 02
33	As K$_{\alpha_1}$	10.544	117.588	1.32 02	1.38 02	1.42 02	1.48 02	1.43 02
82	Pb L$_{\alpha_1}$	10.552	117.499	1.32 02	1.38 02	1.42 02	1.48 02	1.43 02
77	Ir L$_{\beta_1}$	10.708	115.787	1.27 02	1.33 02	1.37 02	1.43 02	1.38 02
83	Bi L$_{\alpha_1}$	10.839	114.388	1.23 02	1.29 02	1.33 02	1.38 02	1.34 02
32	Ge K$_{\beta_1}$	10.982	112.898	1.19 02	1.25 02	1.29 02	1.34 02	1.29 02
78	Pt L$_{\beta_1}$	11.071	111.990	1.17 02	1.22 02	1.26 02	1.31 02	1.27 02
84	Po L$_{\alpha_1}$	11.131	111.387	1.15 02	1.21 02	1.24 02	1.30 02	1.25 02
34	Se K$_{\alpha_1}$	11.222	110.484	1.13 02	1.18 02	1.22 02	1.27 02	1.23 02
85	At L$_{\alpha_1}$	11.427	108.501	1.08 02	1.13 02	1.17 02	1.21 02	1.17 02
79	Au L$_{\beta_1}$	11.442	108.359	1.08 02	1.13 02	1.16 02	1.21 02	1.17 02
33	As K$_{\beta_1}$	11.726	105.735	1.01 02	1.06 02	1.09 02	1.14 02	1.10 02
86	Rn L$_{\alpha_1}$	11.727	105.726	1.01 02	1.06 02	1.09 02	1.14 02	1.10 02
80	Hg L$_{\beta_1}$	11.823	104.867	9.91 01	1.04 02	1.07 02	1.11 02	1.08 02
35	Br K$_{\alpha_1}$	11.924	103.979	9.70 01	1.02 02	1.05 02	1.09 02	1.05 02
87	Fr L$_{\alpha_1}$	12.031	103.054	9.49 01	9.95 01	1.02 02	1.07 02	1.03 02

Z	transition	E [keV]	λ [pm]	$_{86}$Rn	$_{87}$Fr	$_{88}$Ra	$_{89}$Ac	$_{90}$Th
81	Tl L$_{\beta_1}$	12.213	101.519	9.14 01	9.58 01	9.87 01	1.03 02	9.93 01
88	Ra L$_{\alpha_1}$	12.340	100.474	8.91 01	9.34 01	9.62 01	1.00 02	9.67 01
34	Se K$_{\beta_1}$	12.496	99.219	8.63 01	9.05 01	9.32 01	9.71 01	9.37 01
82	Pb L$_{\beta_1}$	12.614	98.291	8.43 01	8.85 01	9.11 01	9.48 01	9.16 01
36	Kr K$_{\alpha_1}$	12.649	98.019	8.37 01	8.79 01	9.05 01	9.42 01	9.09 01
89	Ac L$_{\alpha_1}$	12.652	97.996	8.37 01	8.78 01	9.04 01	9.41 01	9.09 01
90	Th L$_{\alpha_1}$	12.969	95.601	7.87 01	8.26 01	8.50 01	8.85 01	8.54 01
83	Bi L$_{\beta_1}$	13.024	95.197	7.78 01	8.17 01	8.41 01	8.76 01	8.45 01
91	Pa L$_{\alpha_1}$	13.291	93.285	7.40 01	7.77 01	8.00 01	8.32 01	8.04 01
35	Br K$_{\beta_1}$	13.291	93.285	7.40 01	7.77 01	8.00 01	8.32 01	8.04 01
37	Rb K$_{\alpha_1}$	13.395	92.560	7.26 01	7.62 01	7.84 01	8.16 01	7.88 01
84	Po L$_{\beta_1}$	13.447	92.202	7.19 01	7.55 01	7.77 01	8.08 01	7.81 01
92	U L$_{\alpha_1}$	13.615	91.065	6.97 01	7.32 01	7.53 01	7.84 01	7.57 01
85	At L$_{\beta_1}$	13.876	89.352	6.65 01	6.98 01	7.18 01	7.48 01	7.22 01
36	Kr K$_{\beta_1}$	14.112	87.857	6.37 01	6.69 01	6.89 01	7.17 01	6.92 01
38	Sr K$_{\alpha_1}$	14.165	87.529	6.31 01	6.63 01	6.82 01	7.10 01	6.85 01
86	Rn L$_{\beta_1}$	14.316	86.606	<u>6.15 01</u>	6.46 01	6.64 01	6.92 01	6.68 01
87	Fr L$_{\beta_1}$	14.770	83.943	1.30 02	5.98 01	6.15 01	6.40 01	6.17 01
39	Y K$_{\alpha_1}$	14.958	82.888	1.26 02	5.79 01	5.96 01	6.20 01	5.98 01
37	Rb K$_{\beta_1}$	14.961	82.872	1.26 02	<u>5.79 01</u>	5.95 01	6.20 01	5.98 01
88	Ra L$_{\beta_1}$	15.236	81.376	1.20 02	1.25 02	<u>5.69 01</u>	5.92 01	5.72 01
89	Ac L$_{\beta_1}$	15.713	78.906	1.11 02	1.15 02	1.19 02	5.49 01	5.30 01
40	Zr K$_{\alpha_1}$	15.775	78.596	1.10 02	1.14 02	1.18 02	5.44 01	5.24 01
38	Sr K$_{\beta_1}$	15.836	78.293	1.08 02	1.13 02	1.16 02	<u>5.39 01</u>	5.19 01
90	Th L$_{\beta_1}$	16.202	76.524	1.02 02	1.06 02	1.10 02	1.14 02	<u>4.91 01</u>
41	Nb K$_{\alpha_1}$	16.615	74.622	9.56 01	9.94 01	1.03 02	1.07 02	1.04 02
91	Pa L$_{\beta_1}$	16.702	74.233	9.43 01	9.80 01	1.01 02	1.06 02	1.03 02
39	Y K$_{\beta_1}$	16.738	74.074	9.38 01	9.75 01	1.01 02	1.05 02	1.02 02
92	U L$_{\beta_1}$	17.220	72.002	<u>8.70 01</u>	9.05 01	9.36 01	9.75 01	9.48 01
42	Mo K$_{\alpha_1}$	17.479	70.933	1.16 02	8.70 01	9.00 01	9.38 01	9.12 01
40	Zr K$_{\beta_1}$	17.668	70.175	<u>1.13 02</u>	<u>8.45 01</u>	8.75 01	9.12 01	8.87 01
43	Tc K$_{\alpha_1}$	18.367	67.504	1.22 02	1.06 02	<u>7.90 01</u>	8.24 01	8.01 01
41	Nb K$_{\beta_1}$	18.623	66.576	1.17 02	<u>1.02 02</u>	<u>1.06 02</u>	<u>7.94 01</u>	7.73 01
44	Ru K$_{\alpha_1}$	19.279	64.311	1.07 02	1.12 02	1.15 02	1.00 02	7.06 01
42	Mo K$_{\beta_1}$	19.608	63.231	1.02 02	1.07 02	1.10 02	<u>9.60 01</u>	<u>6.75 01</u>
45	Rh K$_{\alpha_1}$	20.216	61.330	9.44 01	9.85 01	1.02 02	1.06 02	<u>8.62 01</u>
43	Tc K$_{\beta_1}$	20.619	60.131	8.97 01	9.35 01	9.64 01	1.00 02	9.75 01
46	Pd K$_{\alpha_1}$	21.177	58.547	8.36 01	8.71 01	8.99 01	9.37 01	9.09 01
44	Ru K$_{\beta_1}$	21.657	57.249	7.89 01	8.21 01	8.47 01	8.83 01	8.57 01
47	Ag K$_{\alpha_1}$	22.163	55.942	7.43 01	7.73 01	7.97 01	8.32 01	8.07 01
45	Rh K$_{\beta_1}$	22.724	54.561	6.96 01	7.24 01	7.46 01	7.79 01	7.55 01
48	Cd K$_{\alpha_1}$	23.174	53.501	6.61 01	6.87 01	7.09 01	7.40 01	7.17 01
46	Pd K$_{\beta_1}$	23.819	52.053	6.15 01	6.39 01	6.60 01	6.89 01	6.67 01
49	In K$_{\alpha_1}$	24.210	51.212	5.89 01	6.13 01	6.32 01	6.60 01	6.39 01

Z	transition	E [keV]	λ [pm]	$_{86}$Rn	$_{87}$Fr	$_{88}$Ra	$_{89}$Ac	$_{90}$Th
47	Ag K$_{\beta_1}$	24.942	49.709	5.45 01	5.66 01	5.84 01	6.11 01	5.91 01
50	Sn K$_{\alpha_1}$	25.271	49.062	5.27 01	5.47 01	5.65 01	5.90 01	5.71 01
48	Cd K$_{\beta_1}$	26.096	47.511	4.84 01	5.03 01	5.19 01	5.42 01	5.25 01
51	Sb K$_{\alpha_1}$	26.359	47.037	4.72 01	4.90 01	5.05 01	5.28 01	5.11 01
49	In K$_{\beta_1}$	27.276	45.455	4.31 01	4.48 01	4.62 01	4.83 01	4.67 01
52	Te K$_{\alpha_1}$	27.472	45.131	4.23 01	4.39 01	4.53 01	4.74 01	4.58 01
50	Sn K$_{\beta_1}$	28.486	43.524	3.85 01	3.99 01	4.12 01	4.31 01	4.16 01
53	I K$_{\alpha_1}$	28.612	43.333	3.81 01	3.95 01	4.07 01	4.26 01	4.12 01
51	Sb K$_{\beta_1}$	29.726	41.709	3.45 01	3.57 01	3.68 01	3.86 01	3.72 01
54	Xe K$_{\alpha_1}$	29.779	41.635	3.43 01	3.55 01	3.67 01	3.84 01	3.70 01
55	Cs K$_{\alpha_1}$	30.973	40.030	3.10 01	3.21 01	3.31 01	3.47 01	3.34 01
52	Te K$_{\beta_1}$	30.996	40.000	3.09 01	3.20 01	3.31 01	3.46 01	3.34 01
56	Ba K$_{\alpha_1}$	32.194	38.511	2.80 01	2.90 01	3.00 01	3.13 01	3.02 01
53	I K$_{\beta_1}$	32.295	38.391	2.78 01	2.88 01	2.97 01	3.11 01	3.00 01
57	La K$_{\alpha_1}$	33.442	37.074	2.53 01	2.63 01	2.72 01	2.84 01	2.74 01
54	Xe K$_{\beta_1}$	33.624	36.874	2.50 01	2.60 01	2.68 01	2.80 01	2.70 01
58	Ce K$_{\alpha_1}$	34.279	36.169	2.38 01	2.47 01	2.55 01	2.66 01	2.57 01
55	Cs K$_{\beta_1}$	34.987	35.437	2.25 01	2.34 01	2.42 01	2.52 01	2.44 01
59	Pr K$_{\alpha_1}$	36.026	34.415	2.09 01	2.17 01	2.24 01	2.34 01	2.26 01
56	Ba K$_{\beta_1}$	36.378	34.082	2.04 01	2.12 01	2.19 01	2.28 01	2.20 01
60	Nd K$_{\alpha_1}$	36.847	33.648	1.97 01	2.05 01	2.11 01	2.20 01	2.13 01
57	La K$_{\beta_1}$	37.801	32.799	1.84 01	1.92 01	1.98 01	2.06 01	1.99 01
61	Pm K$_{\alpha_1}$	38.725	32.016	1.73 01	1.80 01	1.86 01	1.94 01	1.87 01
58	Ce K$_{\beta_1}$	39.257	31.582	1.67 01	1.74 01	1.80 01	1.87 01	1.81 01
62	Sm K$_{\alpha_1}$	40.118	30.905	1.58 01	1.65 01	1.70 01	1.77 01	1.71 01
59	Pr K$_{\beta_1}$	40.748	30.427	1.52 01	1.58 01	1.63 01	1.70 01	1.64 01
63	Eu K$_{\alpha_1}$	41.542	29.845	1.44 01	1.50 01	1.55 01	1.61 01	1.56 01
60	Nd K$_{\beta_1}$	42.271	29.331	1.38 01	1.44 01	1.48 01	1.54 01	1.49 01
64	Gd K$_{\alpha_1}$	42.996	28.836	1.32 01	1.38 01	1.42 01	1.48 01	1.43 01
61	Pm K$_{\beta_1}$	43.826	28.290	1.26 01	1.31 01	1.35 01	1.41 01	1.36 01
65	Tb K$_{\alpha_1}$	44.482	27.873	1.21 01	1.26 01	1.30 01	1.35 01	1.31 01
62	Sm K$_{\beta_1}$	45.413	27.301	1.15 01	1.20 01	1.23 01	1.28 01	1.24 01
66	Dy K$_{\alpha_1}$	45.998	26.954	1.11 01	1.16 01	1.19 01	1.24 01	1.20 01
63	Eu K$_{\beta_1}$	47.038	26.358	1.05 01	1.09 01	1.13 01	1.17 01	1.13 01
67	Ho K$_{\alpha_1}$	47.547	26.076	1.02 01	1.06 01	1.09 01	1.14 01	1.10 01
64	Gd K$_{\beta_1}$	48.697	25.460	9.58 00	1.00 01	1.03 01	1.07 01	1.04 01
68	Er K$_{\alpha_1}$	49.128	25.237	9.37 00	9.78 00	1.01 01	1.05 01	1.01 01
65	Tb K$_{\beta_1}$	50.382	24.609	8.87 00	9.17 00	9.43 00	9.81 00	9.49 00
69	Tm K$_{\alpha_1}$	50.742	24.434	8.63 00	9.00 00	9.26 00	9.63 00	9.31 00
66	Dy K$_{\beta_1}$	52.119	23.788	8.06 00	8.40 00	8.65 00	9.00 00	8.70 00
70	Yb K$_{\alpha_1}$	52.389	23.666	7.96 00	8.29 00	8.54 00	8.88 00	8.58 00
67	Ho K$_{\beta_1}$	53.877	23.012	7.41 00	7.72 00	7.95 00	8.27 00	7.99 00
71	Lu K$_{\alpha_1}$	54.070	22.930	7.34 00	7.65 00	7.88 00	8.20 00	7.92 00
68	Er K$_{\beta_1}$	55.681	22.267	6.82 00	7.10 00	7.31 00	7.60 00	7.34 00

Z	transition	E [keV]	λ [pm]	$_{86}$Rn	$_{87}$Fr	$_{88}$Ra	$_{89}$Ac	$_{90}$Th
72	Hf K$_{\alpha_1}$	55.790	22.223	6.78 00	7.06 00	7.28 00	7.57 00	7.31 00
69	Tm K$_{\beta_1}$	57.517	21.556	6.28 00	6.53 00	6.73 00	7.00 00	6.76 00
73	Ta K$_{\alpha_1}$	57.532	21.550	6.28 00	6.53 00	6.73 00	7.00 00	6.75 00
74	W K$_{\alpha_1}$	59.318	20.901	5.81 00	6.04 00	6.23 00	6.47 00	6.24 00
70	Yb K$_{\beta_1}$	59.370	20.883	5.80 00	6.02 00	6.21 00	6.46 00	6.23 00
75	Re K$_{\alpha_1}$	61.140	20.278	5.38 00	5.59 00	5.77 00	6.00 00	5.78 00
71	Lu K$_{\beta_1}$	61.283	20.231	5.35 00	5.56 00	5.73 00	5.96 00	5.75 00
76	Os K$_{\alpha_1}$	63.001	19.679	4.99 00	5.19 00	5.35 00	5.56 00	5.36 00
72	Hf K$_{\beta_1}$	63.234	19.607	4.95 00	5.14 00	5.30 00	5.51 00	5.31 00
77	Ir K$_{\alpha_1}$	64.896	19.105	4.63 00	4.82 00	4.96 00	5.16 00	4.97 00
73	Ta K$_{\beta_1}$	65.223	19.009	4.58 00	4.76 00	4.90 00	5.10 00	4.91 00
78	Pt K$_{\alpha_1}$	66.832	18.551	4.30 00	4.48 00	4.61 00	4.79 00	4.61 00
74	W K$_{\beta_1}$	67.244	18.438	4.24 00	4.41 00	4.54 00	4.72 00	4.54 00
79	Au K$_{\alpha_1}$	68.804	18.020	4.00 00	4.17 00	4.28 00	4.46 00	4.28 00
75	Re K$_{\beta_1}$	69.310	17.888	3.93 00	4.09 00	4.21 00	4.38 00	4.20 00
80	Hg K$_{\alpha_1}$	70.819	17.507	3.72 00	3.88 00	3.98 00	4.14 00	3.98 00
76	Os K$_{\beta_1}$	71.413	17.361	3.64 00	3.80 00	3.90 00	4.06 00	3.90 00
81	Tl K$_{\alpha_1}$	72.872	17.014	3.46 00	3.61 00	3.71 00	3.86 00	3.70 00
77	Ir K$_{\beta_1}$	73.561	16.854	3.38 00	3.53 00	3.62 00	3.77 00	3.61 00
82	Pb K$_{\alpha_1}$	74.969	16.538	3.23 00	3.37 00	3.45 00	3.59 00	3.45 00
78	Pt K$_{\beta_1}$	75.748	16.368	3.14 00	3.28 00	3.36 00	3.50 00	3.36 00
83	Bi K$_{\alpha_1}$	77.108	16.079	3.01 00	3.14 00	3.22 00	3.35 00	3.21 00
79	Au K$_{\beta_1}$	77.948	15.906	2.92 00	3.05 00	3.13 00	3.26 00	3.12 00
84	Po K$_{\alpha_1}$	79.290	15.636	2.80 00	2.93 00	3.00 00	3.12 00	2.99 00
80	Hg K$_{\alpha_1}$	80.253	15.449	2.72 00	2.84 00	2.91 00	3.03 00	2.90 00
85	At K$_{\beta_1}$	81.520	15.209	2.62 00	2.73 00	2.80 00	2.91 00	2.79 00
81	Tl K$_{\beta_1}$	82.576	15.014	2.54 00	2.65 00	2.71 00	2.82 00	2.70 00
86	Rn K$_{\alpha_1}$	83.780	14.798	2.45 00	2.55 00	2.62 00	2.73 00	2.61 00
82	Pb K$_{\beta_1}$	84.936	14.597	2.37 00	2.47 00	2.53 00	2.64 00	2.52 00
87	Fr K$_{\alpha_1}$	86.100	14.400	2.29 00	2.39 00	2.45 00	2.55 00	2.44 00
83	Bi K$_{\beta_1}$	87.343	14.195	2.21 00	2.30 00	2.37 00	2.46 00	2.36 00
88	Ra K$_{\alpha_1}$	88.470	14.014	2.14 00	2.23 00	2.30 00	2.39 00	2.29 00
84	Po K$_{\beta_1}$	89.800	13.806	2.07 00	2.15 00	2.21 00	2.30 00	2.20 00
89	Ac K$_{\alpha_1}$	90.884	13.642	2.01 00	2.09 00	2.15 00	2.23 00	2.14 00
85	At K$_{\beta_1}$	92.300	13.432	1.94 00	2.02 00	2.07 00	2.15 00	2.06 00
90	Th K$_{\alpha_1}$	93.350	13.281	1.88 00	1.96 00	2.02 00	2.09 00	2.01 00
86	Rn K$_{\beta_1}$	94.870	13.068	1.81 00	1.89 00	1.94 00	2.01 00	1.93 00
91	Pa K$_{\alpha_1}$	95.868	12.932	1.77 00	1.84 00	1.89 00	1.96 00	1.88 00
87	Fr K$_{\beta_1}$	97.470	12.720	<u>1.70 00</u>	1.77 00	1.82 00	1.89 00	1.81 00
92	U K$_{\alpha_1}$	98.439	12.595	6.88 00	1.73 00	1.77 00	1.84 00	1.76 00
88	Ra K$_{\beta_1}$	100.130	12.382	6.57 00	<u>1.66 00</u>	1.70 00	1.77 00	1.69 00
89	Ac K$_{\beta_1}$	102.850	12.054	6.12 00	6.28 00	<u>1.60 00</u>	1.81 00	1.59 00
90	Th K$_{\beta_1}$	105.610	11.739	5.71 00	5.86 00	6.02 00	<u>1.86 00</u>	1.49 00
91	Pa K$_{\beta_1}$	108.430	11.434	5.33 00	5.47 00	5.62 00	6.05 00	<u>1.40 00</u>

Z	transition	E [keV]	λ [pm]	$_{86}$Rn	$_{87}$Fr	$_{88}$Ra	$_{89}$Ac	$_{90}$Th
92	U K$_{\beta_1}$	111.300	11.139	4.97 00	5.10 00	5.25 00	5.87 00	4.85 00

Z	transition	E [keV]	λ [pm]	$_{91}$Pa	$_{92}$U	$_{93}$Np	$_{94}$Pu
4	Be K$_\alpha$	0.1085	11427.207	1.16 05	7.64 04	<u>7.87 04</u>	4.55 04
38	Sr M$_\xi$	0.1140	10875.895	7.17 04	1.01 05	9.04 04	<u>5.39 04</u>
39	Y M$_\xi$	0.1328	9339.235	3.26 04	3.48 04	5.27 04	6.93 04
16	S L$_l$	0.1487	8337.942	2.08 04	2.22 04	2.99 04	4.43 04
40	Zr M$_\xi$	0.1511	8205.506	1.95 04	2.06 04	2.78 04	4.11 04
41	Nb M$_\xi$	0.1717	7227.037	1.17 04	1.18 04	1.58 04	2.28 04
5	B K$_\alpha$	0.1833	6764.059	9.00 03	9.02 03	1.19 04	1.72 04
42	Mo M$_\xi$	0.1926	6473.445	<u>7.39 03</u>	<u>7.38 03</u>	9.60 03	<u>1.41 04</u>
6	C K$_\alpha$	0.2770	4476.000	<u>6.47 03</u>	6.44 03	<u>5.18 03</u>	4.88 03
47	Ag M$_\xi$	0.3117	3977.709	<u>7.47 03</u>	<u>7.43 03</u>	5.80 03	<u>4.86 03</u>
7	N K$_\alpha$	0.3924	3159.664	9.08 03	9.03 03	7.56 03	5.99 03
22	Ti L$_l$	0.3953	3136.484	9.11 03	9.06 03	<u>7.62 03</u>	<u>6.06 03</u>
22	Ti L$_\alpha$	0.4522	2741.822	1.02 04	9.94 03	8.79 03	7.42 03
23	V L$_\alpha$	0.5113	2424.901	1.13 04	1.09 04	9.85 03	8.33 03
8	O K$_\alpha$	0.5249	2362.072	1.15 04	1.11 04	9.83 03	8.53 03
25	Mn L$_l$	0.5563	2228.747	1.13 04	1.13 04	1.04 04	8.98 03
24	Cr L$_\alpha$	0.5728	2164.549	1.12 04	1.11 04	1.07 04	9.22 03
25	Mn L$_\alpha$	0.6374	1945.171	1.03 04	1.02 04	1.05 04	1.08 04
9	F K$_\alpha$	0.6768	1831.932	9.53 03	9.49 03	9.87 03	1.03 04
26	Fe L$_\alpha$	0.7050	1758.655	<u>9.11 03</u>	<u>9.09 03</u>	<u>9.45 03</u>	9.93 03
27	Co L$_\alpha$	0.7762	1597.335	<u>9.22 03</u>	<u>8.90 03</u>	<u>8.28 03</u>	<u>8.70 03</u>
28	Ni L$_\alpha$	0.8515	1456.080	8.40 03	8.42 03	8.66 03	8.40 03
29	Cu L$_\alpha$	0.9297	1336.044	<u>7.33 03</u>	7.36 03	7.70 03	8.07 03
30	Zn L$_\alpha$	1.0117	1225.513	6.35 03	6.40 03	6.73 03	7.04 03
11	Na K$_\alpha$	1.0410	1191.020	6.04 03	<u>6.09 03</u>	6.40 03	6.69 03
11	Na K$_\beta$	1.0711	1157.550	<u>5.38 03</u>	6.52 03	<u>5.82 03</u>	<u>6.71 03</u>
12	Mg K$_\alpha$	1.2536	989.033	6.82 03	<u>6.42 03</u>	6.71 03	6.84 03
33	As L$_\alpha$	1.2820	967.123	6.44 03	6.68 03	6.35 03	6.49 03
12	Mg K$_\beta$	1.3022	952.121	6.19 03	6.42 03	6.11 03	6.25 03
33	As L$_{\beta_1}$	1.3170	941.421	6.01 03	6.24 03	5.94 03	6.09 03
66	Dy M$_\beta$	1.3250	935.737	5.92 03	6.14 03	<u>5.85 03</u>	6.00 03
67	Ho M$_\alpha$	1.3480	919.771	5.66 03	5.88 03	6.17 03	<u>5.76 03</u>
34	Se L$_\alpha$	1.3791	899.029	5.34 03	5.55 03	5.84 03	6.00 03
67	Ho M$_\beta$	1.3830	896.494	<u>5.31 03</u>	5.51 03	5.80 03	5.96 03
68	Er M$_\alpha$	1.4060	881.829	5.59 03	5.28 03	5.57 03	5.73 03
34	Se L$_{\beta_1}$	1.4192	873.627	<u>5.46 03</u>	<u>5.16 03</u>	5.44 03	5.60 03
68	Er M$_\beta$	1.4430	859.218	5.23 03	5.45 03	5.22 03	5.38 03
69	Tm M$_\alpha$	1.4620	848.051	5.06 03	5.27 03	5.06 03	5.21 03
35	Br L$_\alpha$	1.4804	837.511	4.91 03	5.10 03	4.91 03	5.06 03

Z	transition	E [keV]	λ [pm]	$_{91}$Pa	$_{92}$U	$_{93}$Np	$_{94}$Pu
13	Al K$_{\beta_1}$	1.4867	833.962	4.85 03	5.05 03	<u>4.86 03</u>	5.01 03
69	Tm M$_\beta$	1.5030	824.918	4.72 03	4.91 03	5.19 03	4.88 03
70	Yb M$_\alpha$	1.5214	814.941	4.58 03	4.76 03	5.04 03	4.74 03
35	Br L$_{\beta_1}$	1.5259	812.538	4.34 03	4.72 03	5.00 03	4.71 03
13	Al K$_\beta$	1.5574	796.103	4.31 03	4.49 03	4.76 03	<u>4.49 03</u>
70	Yb M$_\beta$	1.5675	790.974	4.24 03	4.41 03	4.68 03	4.86 03
71	Lu M$_\alpha$	1.5813	784.071	4.15 03	4.32 03	4.58 03	4.76 03
36	Kr L$_\alpha$	1.5860	781.747	4.12 03	4.28 03	4.55 03	4.73 03
71	Lu M$_\beta$	1.6312	760.085	3.83 03	3.99 03	4.24 03	4.42 03
36	Kr L$_{\beta_1}$	1.6366	757.577	3.80 03	3.96 03	4.21 03	4.39 03
72	Hf M$_{\alpha_1}$	1.6446	753.892	3.75 03	3.91 03	4.16 03	4.34 03
37	Rb L$_{\alpha_1}$	1.6941	731.864	3.48 03	3.63 03	3.86 03	4.04 03
72	Hf M$_\beta$	1.6976	730.355	3.46 03	3.61 03	3.85 03	4.02 03
73	Ta M$_{\alpha_1}$	1.7096	725.229	3.40 03	3.54 03	3.78 03	3.95 03
14	Si K$_{\alpha_1}$	1.7400	712.558	3.25 03	3.39 03	3.62 03	3.79 03
37	Rb L$_{\beta_1}$	1.7522	707.597	3.19 03	3.33 03	3.56 03	3.72 03
73	Ta M$_\beta$	1.7655	702.266	3.13 03	3.27 03	3.49 03	3.66 03
74	W M$_{\alpha_1}$	1.7754	698.350	3.09 03	3.22 03	3.44 03	3.61 03
38	Sr L$_{\alpha_1}$	1.8066	686.290	2.95 03	3.08 03	3.30 03	3.46 03
74	W M$_\beta$	1.8349	675.705	2.84 03	2.96 03	3.17 03	3.34 03
14	Si K$_\beta$	1.8359	675.337	2.83 03	2.96 03	3.17 03	3.33 03
75	Re M$_{\alpha_1}$	1.8420	673.100	2.81 03	2.93 03	3.14 03	3.30 03
38	Sr L$_{\beta_1}$	1.8717	662.420	2.70 03	2.82 03	3.02 03	3.18 03
75	Re M$_\beta$	1.9061	650.465	2.58 03	2.69 03	2.89 03	3.04 03
76	Os M$_{\alpha_1}$	1.9102	649.069	2.56 03	2.68 03	2.87 03	3.03 03
39	Y L$_{\alpha_1}$	1.9226	644.882	2.52 03	2.63 03	2.83 03	2.98 03
76	Os M$_\beta$	1.9783	626.725	2.34 03	2.45 03	2.64 03	2.78 03
77	Ir M$_{\alpha_1}$	1.9799	626.219	2.34 03	2.44 03	2.63 03	2.78 03
39	Y L$_{\beta_1}$	1.9958	621.230	2.29 03	2.40 03	2.58 03	2.73 03
15	P K$_{\alpha_1}$	2.0137	615.708	2.24 03	2.34 03	2.52 03	2.67 03
40	Zr L$_{\alpha_1}$	2.0424	607.056	2.16 03	2.26 03	2.44 03	2.58 03
78	Pt M$_{\alpha_1}$	2.0505	604.658	2.14 03	2.24 03	2.42 03	2.56 03
77	Ir M$_\beta$	2.0535	603.775	2.13 03	2.23 03	2.41 03	2.55 03
79	Au M$_{\alpha_1}$	2.1229	584.036	1.96 03	2.05 03	2.22 03	2.36 03
40	Zr L$_{\beta_1}$	2.1244	583.624	1.96 03	2.05 03	2.22 03	2.35 03
78	Pt M$_\beta$	2.1275	582.828	1.95 03	2.04 03	2.21 03	2.34 03
15	P K$_{\beta_{1,3}}$	2.1390	579.640	1.92 03	2.01 03	2.18 03	2.31 03
41	Nb L$_{\alpha_1}$	2.1659	572.441	1.86 03	1.95 03	2.11 03	2.25 03
80	Hg M$_{\alpha_1}$	2.1953	564.775	1.80 03	1.88 03	2.05 03	2.18 03
79	Au M$_\beta$	2.2046	562.393	1.78 03	1.86 03	2.02 03	2.16 03
41	Nb L$_{\beta_1}$	2.5274	549.238	1.68 03	1.76 03	1.91 03	2.04 03
81	Tl M$_{\alpha_1}$	2.2706	546.045	1.65 03	1.73 03	1.88 03	2.01 03
80	Hg M$_\beta$	2.2825	543.199	1.63 03	1.71 03	1.86 03	1.99 03
42	Mo L$_{\alpha_1}$	2.2932	540.664	1.61 03	1.69 03	1.84 03	1.96 03

Z	transition	E [keV]	λ [pm]	$_{91}$Pa	$_{92}$U	$_{93}$Np	$_{94}$Pu
16	S K$_{\alpha_1}$	2.3080	537.197	1.59 03	1.66 03	1.81 03	1.93 03
82	Pb M$_{\alpha_1}$	2.3457	528.563	1.52 03	1.59 03	1.74 03	1.86 03
81	Tl M$_\beta$	2.3621	524.893	1.49 03	1.57 03	1.71 03	1.83 03
42	Mo L$_{\beta_1}$	2.3948	517.726	1.44 03	1.51 03	1.65 03	1.77 03
83	Bi M$_{\alpha_1}$	2.4226	511.785	1.40 03	1.47 03	1.61 03	1.72 03
43	Tc L$_{\alpha_1}$	2.4240	511.490	1.40 03	1.47 03	1.61 03	1.72 03
82	Pb M$_\beta$	2.4427	507.574	1.37 03	1.44 03	1.56 03	1.69 03
16	S K$_\beta$	2.4640	503.186	1.34 03	1.41 03	1.54 03	1.66 03
83	Bi M$_{\beta_1}$	2.5255	490.933	1.26 03	1.32 03	1.45 03	1.56 03
43	Tc L$_{\beta_1}$	2.5368	488.746	1.25 03	1.31 03	1.44 03	1.55 03
44	Ru L$_{\alpha_1}$	2.5586	484.582	1.22 03	1.28 03	1.41 03	1.52 03
17	Cl K$_{\alpha_1}$	2.6224	472.792	1.15 03	1.20 03	1.33 03	1.43 03
44	Ru L$_{\beta_1}$	2.6832	462.079	1.08 03	1.14 03	1.25 03	1.35 03
45	Rh L$_{\alpha_1}$	2.6967	459.766	1.07 03	1.12 03	1.24 03	1.34 03
17	Cl K$_\beta$	2.8156	440.350	9.58 02	1.01 03	1.12 03	1.21 03
45	Rh L$_{\beta_1}$	2.8344	437.430	9.42 02	9.90 02	1.10 03	1.19 03
46	Pd L$_{\alpha_1}$	2.8386	436.782	9.39 02	9.86 02	1.09 03	1.19 03
18	Ar K$_{\alpha_1}$	2.9577	419.194	8.46 02	8.89 02	9.89 02	1.08 03
47	Ag L$_{\alpha_1}$	2.9843	415.458	8.27 02	8.69 02	9.68 02	1.05 03
46	Pd L$_{\beta_1}$	2.9902	414.638	8.23 02	8.65 02	9.63 02	1.05 03
90	Th M$_{\alpha_1}$	2.9961	413.821	8.19 02	8.60 02	9.58 02	1.04 03
91	Pa M$_{\alpha_1}$	3.0823	402.248	7.62 02	8.01 02	8.95 02	9.76 02
48	Cd L$_{\alpha_1}$	3.1337	395.651	7.31 02	7.69 02	8.59 02	9.39 02
90	Th M$_\beta$	3.1458	394.129	7.24 02	7.61 02	8.51 02	9.31 02
47	Ag L$_{\beta_1}$	3.1509	393.491	7.21 02	7.58 02	8.48 02	9.27 02
92	U M$_{\alpha_1}$	3.1708	391.021	7.09 02	7.47 02	8.35 02	9.13 02
18	Ar K$_\beta$	3.1905	388.607	6.98 02	7.35 02	8.23 02	9.00 02
91	Pa M$_\beta$	3.2397	382.705	6.72 02	7.07 02	7.93 02	8.68 02
49	In L$_{\alpha_1}$	3.2869	377.210	6.48 02	6.82 02	7.65 02	8.39 02
19	K K$_{\alpha_1}$	3.3138	374.148	6.34 02	6.68 02	7.50 02	8.23 02
48	Cd L$_{\beta_1}$	3.3166	373.832	6.33 02	6.67 02	7.49 02	8.22 02
92	U M$_\beta$	3.3360	371.658	<u>6.24 02</u>	6.57 02	7.38 02	8.10 02
50	Sn L$_{\alpha_1}$	3.4440	360.003	1.25 03	6.07 02	6.83 02	7.52 02
49	In L$_{\beta_1}$	3.4872	355.543	1.21 03	<u>5.88 02</u>	6.63 02	7.30 02
19	K K$_\beta$	3.5896	345.401	1.13 03	1.12 03	6.18 02	6.82 02
51	Sb L$_{\alpha_1}$	3.6047	343.954	<u>1.11 03</u>	1.11 03	6.12 02	6.75 02
50	Sn L$_{\beta_1}$	3.6628	338.498	1.60 03	1.06 03	<u>5.88 02</u>	6.51 02
20	Ca K$_{\alpha_1}$	3.6917	335.848	1.57 03	<u>1.04 03</u>	1.12 03	6.39 02
52	Te L$_{\beta_1}$	3.7693	328.934	1.49 03	1.88 03	1.06 03	<u>6.08 02</u>
51	Sb L$_{\beta_1}$	3.8436	322.575	1.41 03	1.65 03	<u>1.01 03</u>	1.04 03
53	I L$_{\alpha_1}$	3.9377	314.867	1.33 03	1.41 03	1.42 03	<u>9.75 02</u>
20	Ca K$_\beta$	4.0127	308.981	1.27 03	1.26 03	1.35 03	1.39 03
52	Te L$_{\beta_1}$	4.0296	307.686	1.25 03	1.25 03	1.34 03	1.37 03
21	Sc K$_{\alpha_1}$	4.0906	303.097	1.21 03	1.20 03	1.29 03	1.32 03

Z	transition	E [keV]	λ [pm]	$_{91}$Pa	$_{92}$U	$_{93}$Np	$_{94}$Pu
54	Xe L$_{\alpha_1}$	4.1099	301.674	<u>1.19 03</u>	1.19 03	1.27 03	1.31 03
53	I L$_{\beta_1}$	4.2207	293.755	1.34 03	1.11 03	1.19 03	1.22 03
55	Cs L$_{\alpha_1}$	4.2865	289.245	1.28 03	<u>1.07 03</u>	<u>1.14 03</u>	1.17 03
21	Sc K$_\beta$	4.4605	277.962	1.16 03	1.16 03	1.24 03	1.06 03
56	Ba L$_{\alpha_1}$	4.4663	277.601	1.16 03	1.15 03	1.23 03	1.06 03
22	Ti K$_{\alpha_1}$	4.5108	274.863	1.13 03	1.12 03	1.20 03	<u>1.03 03</u>
55	Cs L$_{\beta_1}$	4.6198	268.377	1.06 03	1.06 03	1.13 03	1.16 03
57	La L$_{\alpha_1}$	4.6510	266.577	1.04 03	1.04 03	1.11 03	1.14 03
56	Ba L$_{\beta_1}$	4.8273	256.841	9.49 02	9.46 02	1.01 03	1.03 03
58	Ce L$_{\alpha_1}$	4.8402	256.157	9.43 02	9.39 02	1.00 03	1.03 03
22	Ti K$_{\beta_{1,3}}$	4.9318	251.399	8.90 02	8.96 02	9.58 02	9.80 02
23	V K$_{\alpha_1}$	4.9522	250.363	<u>8.90 02</u>	8.87 02	9.48 02	9.69 02
59	Pr L$_{\alpha_1}$	5.0337	246.310	9.38 02	8.50 02	9.09 02	9.30 02
57	La L$_{\beta_1}$	5.0421	245.899	9.34 02	<u>8.47 02</u>	9.05 02	9.26 02
60	Nd L$_{\alpha_1}$	5.2304	237.047	8.50 02	8.48 02	8.25 02	8.43 02
58	Ce L$_{\beta_1}$	5.2622	235.614	<u>8.37 02</u>	8.35 02	<u>8.12 02</u>	8.30 02
24	Cr K$_{\alpha_1}$	5.4147	228.978	8.56 02	7.77 02	8.29 02	7.71 02
23	V K$_{\beta_{1,3}}$	5.4273	228.447	8.51 02	7.73 02	8.24 02	7.66 02
61	Pm L$_{\alpha_1}$	5.4325	228.228	8.49 02	7.71 02	8.22 02	7.65 02
59	Pr L$_{\beta_1}$	5.4889	225.883	8.27 02	<u>7.51 02</u>	8.01 03	<u>7.45 02</u>
62	Sm L$_{\alpha_1}$	5.6361	219.984	7.73 02	7.71 02	7.48 02	7.65 02
60	Nd L$_{\beta_1}$	5.7216	216.696	7.44 02	7.42 02	<u>7.20 02</u>	7.37 02
63	Eu L$_{\alpha_1}$	5.8457	212.096	7.05 02	7.03 02	7.49 02	6.98 02
25	Mn K$_{\alpha_1}$	5.8988	210.187	6.89 02	6.87 02	7.32 02	<u>6.82 02</u>
24	Cr K$_{\beta_{1,3}}$	5.9467	208.494	6.75 02	6.73 02	7.17 02	7.33 02
61	Pm L$_{\beta_1}$	5.9610	207.993	6.71 02	6.69 02	7.13 02	7.28 02
64	Gd L$_{\alpha_1}$	6.0572	204.690	6.44 02	6.42 02	6.85 02	7.00 02
62	Sm L$_{\beta_1}$	6.2051	199.811	6.06 02	6.04 02	6.44 02	6.58 02
65	Tb L$_{\alpha_1}$	6.2728	197.655	5.89 02	5.88 02	6.26 02	6.40 02
26	Fe K$_{\alpha_1}$	6.4038	193.611	5.59 02	5.58 02	5.94 02	6.07 02
63	Eu L$_{\beta_1}$	6.4564	192.034	5.48 02	5.47 02	5.82 02	5.94 02
25	Mn K$_{\beta_{1,3}}$	6.4905	191.025	5.40 02	5.40 02	5.75 02	5.86 02
66	Dy L$_{\alpha_1}$	6.4952	190.887	5.39 02	5.39 02	5.73 02	5.85 02
64	Gd L$_{\beta_1}$	6.7132	184.688	4.96 02	4.96 02	5.27 02	5.38 02
67	Ho L$_{\alpha_1}$	6.7198	184.507	4.95 02	4.95 02	5.26 02	5.37 02
27	Co K$_{\alpha_1}$	6.9303	178.903	4.58 02	4.58 02	4.86 02	4.96 02
68	Er L$_{\alpha_1}$	6.9487	178.429	4.55 02	4.55 02	4.83 02	4.92 02
65	Tb L$_{\beta_1}$	6.9780	177.680	4.50 02	4.50 02	4.78 02	4.87 02
26	Fe K$_{\beta_{1,3}}$	7.0580	175.666	4.37 02	4.37 02	4.64 02	4.73 02
69	Tm L$_{\alpha_1}$	7.1799	172.683	4.18 02	4.19 02	4.45 02	4.53 02
66	Dy L$_{\beta_1}$	7.2477	171.068	4.09 02	4.09 02	4.34 02	4.42 02
70	Yb L$_{\alpha_1}$	7.4156	167.195	3.85 02	3.86 02	4.10 02	4.17 02
28	Ni K$_{\alpha_1}$	7.4782	165.795	3.77 02	3.78 02	4.01 02	4.08 02
67	Ho L$_{\beta_1}$	7.5253	164.757	3.71 02	3.72 02	3.95 02	4.02 02

Z	transition	E [keV]	λ [pm]	$_{91}$Pa	$_{92}$U	$_{93}$Np	$_{94}$Pu
27	Co K$_{\beta_{1,3}}$	7.6494	162.084	3.56 02	3.57 02	3.79 02	3.85 02
71	Lu L$_{\alpha_1}$	7.6555	161.955	3.56 02	3.56 02	3.78 02	3.84 02
68	Er L$_{\beta_1}$	7.8109	158.733	3.38 02	3.39 02	3.59 02	3.65 02
72	Hf L$_{\alpha_1}$	7.8990	156.963	3.28 02	3.29 02	3.49 02	3.55 02
29	Cu K$_{\alpha_1}$	8.0478	154.060	3.13 02	3.14 02	3.33 02	3.38 02
69	Tm L$_{\beta_1}$	8.1010	153.049	3.08 02	3.09 02	3.27 02	3.33 02
73	Ta L$_{\alpha_1}$	8.1461	152.201	3.04 02	3.05 02	3.23 02	3.28 02
28	Ni K$_{\beta_{1,3}}$	8.2647	150.017	2.93 02	2.94 02	3.11 02	3.16 02
74	W L$_{\alpha_1}$	8.3976	147.643	2.81 02	2.82 02	2.99 02	3.04 02
70	Yb L$_{\beta_1}$	8.4018	147.569	2.81 02	2.82 02	2.98 02	3.03 02
30	Zn K$_{\alpha_1}$	8.6389	143.519	2.62 02	2.63 02	2.78 02	2.83 02
75	Re L$_{\alpha_1}$	8.6525	143.294	2.61 02	2.62 02	2.77 02	2.82 02
71	Lu L$_{\beta_1}$	8.7090	142.364	2.57 02	2.58 02	2.72 02	2.77 02
29	Cu K$_{\beta_1}$	8.9053	139.226	2.43 02	2.44 02	2.57 02	2.62 02
76	Os L$_{\alpha_1}$	8.9117	139.126	2.42 02	2.43 02	2.57 02	2.61 02
72	Hf L$_{\beta_1}$	9.0227	137.414	2.35 02	2.36 02	2.49 02	2.53 02
77	Ir L$_{\alpha_1}$	9.1751	135.132	2.25 02	2.26 02	2.38 02	2.43 02
31	Ga K$_{\alpha_1}$	9.2517	134.013	2.20 02	2.21 02	2.34 02	2.38 02
73	Ta L$_{\beta_1}$	9.3431	132.702	2.15 02	2.16 02	2.28 02	2.32 02
78	Pt L$_{\alpha_1}$	9.4423	131.308	2.09 02	2.10 02	2.22 02	2.26 02
30	Zn K$_{\beta_{1,3}}$	9.5720	129.529	2.02 02	2.03 02	2.14 02	2.18 02
74	W L$_{\beta_1}$	9.6724	128.184	1.97 02	1.98 02	2.09 02	2.13 02
79	Au L$_{\alpha_1}$	9.7133	127.644	1.95 02	1.96 02	2.06 02	2.10 02
32	Ge K$_{\alpha_1}$	9.8864	125.409	1.86 02	1.87 02	1.97 02	2.01 02
80	Hg L$_{\alpha_1}$	9.9888	124.124	1.82 02	1.83 02	1.92 02	1.96 02
75	Re L$_{\beta_1}$	10.010	123.861	1.81 02	1.82 02	1.91 02	1.95 02
31	Ga K$_{\beta_1}$	10.264	120.796	1.70 02	1.71 02	1.79 02	1.83 02
81	Tl L$_{\alpha_1}$	10.269	120.737	1.70 02	1.71 02	1.79 02	1.83 02
76	Os L$_{\beta_1}$	10.355	119.734	1.66 02	1.67 02	1.76 02	1.79 02
33	As K$_{\alpha_1}$	10.544	117.588	1.59 02	1.60 02	1.68 02	1.71 02
82	Pb L$_{\alpha_1}$	10.552	117.499	1.58 02	1.59 02	1.67 02	1.71 02
77	Ir L$_{\beta_1}$	10.708	115.787	1.53 02	1.54 02	1.61 02	1.64 02
83	Bi L$_{\alpha_1}$	10.839	114.388	1.48 02	1.49 02	1.57 02	1.60 02
32	Ge K$_{\beta_1}$	10.982	112.898	1.43 02	1.44 02	1.52 02	1.54 02
78	Pt L$_{\beta_1}$	11.071	111.990	1.41 02	1.41 02	1.48 02	1.49 02
84	Po L$_{\alpha_1}$	11.131	111.387	1.39 02	1.40 02	1.46 02	1.46 02
34	Se K$_{\alpha_1}$	11.222	110.484	1.36 02	1.37 02	1.44 02	1.46 02
85	At L$_{\alpha_1}$	11.427	108.501	1.30 02	1.31 02	1.37 02	1.40 02
79	Au L$_{\beta_1}$	11.442	108.359	1.30 02	1.30 02	1.37 02	1.39 02
33	As K$_{\beta_1}$	11.726	105.735	1.22 02	1.23 02	1.29 02	1.31 02
86	Rn L$_{\alpha_1}$	11.727	105.726	1.22 02	1.23 02	1.29 02	1.31 02
80	Hg L$_{\beta_1}$	11.823	104.867	1.19 02	1.20 02	1.26 02	1.28 02
35	Br K$_{\alpha_1}$	11.924	103.979	1.17 02	1.18 02	1.23 02	1.25 02
87	Fr L$_{\alpha_1}$	12.031	103.054	1.14 02	1.15 02	1.21 02	1.23 02

Z	transition	E [keV]	λ [pm]	$_{91}$Pa	$_{92}$U	$_{93}$Np	$_{94}$Pu
81	Tl L$_{\beta_1}$	12.213	101.519	1.10 02	1.11 02	1.16 02	1.18 02
88	Ra L$_{\alpha_1}$	12.340	100.474	1.07 02	1.08 02	1.13 02	1.15 02
34	Se K$_{\beta_1}$	12.496	99.219	1.04 02	1.05 02	1.10 02	1.12 02
82	Pb L$_{\beta_1}$	12.614	98.291	1.02 02	1.02 02	1.07 02	1.09 02
36	Kr K$_{\alpha_1}$	12.649	98.019	1.01 02	1.02 02	1.06 02	1.08 02
89	Ac L$_{\alpha_1}$	12.652	97.996	1.01 02	1.02 02	1.06 02	1.08 02
90	Th L$_{\alpha_1}$	12.969	95.601	9.48 01	9.55 01	9.99 01	1.02 02
83	Bi L$_{\beta_1}$	13.024	95.197	9.38 01	9.45 01	9.88 01	1.01 02
91	Pa L$_{\alpha_1}$	13.291	93.285	8.92 01	8.99 01	9.39 01	9.55 01
35	Br K$_{\beta_1}$	13.291	93.285	8.92 01	8.99 01	9.39 01	9.55 01
37	Rb K$_{\alpha_1}$	13.395	92.560	8.74 01	8.81 01	9.21 01	9.36 01
84	Po L$_{\beta_1}$	13.447	92.202	8.66 01	8.73 01	9.12 01	9.27 01
92	U L$_{\alpha_1}$	13.615	91.065	8.40 01	8.46 01	8.84 01	8.99 01
85	At L$_{\beta_1}$	13.876	89.352	8.01 01	8.07 01	8.43 01	8.57 01
36	Kr K$_{\beta_1}$	14.112	87.857	7.68 01	7.74 01	8.08 01	8.21 01
38	Sr K$_{\alpha_1}$	14.165	87.529	7.61 01	7.67 01	8.01 01	8.14 01
86	Rn L$_{\beta_1}$	14.316	86.606	7.41 01	7.47 01	7.80 01	7.92 01
87	Fr L$_{\beta_1}$	14.770	83.943	6.85 01	6.91 01	7.21 01	7.32 01
39	Y K$_{\alpha_1}$	14.958	82.888	6.64 01	6.70 01	6.99 01	7.09 01
37	Rb K$_{\beta_1}$	14.961	82.872	6.64 01	6.70 01	6.98 01	7.09 01
88	Ra L$_{\beta_1}$	15.236	81.376	6.35 01	6.40 01	6.67 01	6.78 01
89	Ac L$_{\beta_1}$	15.713	78.906	5.88 01	5.93 01	6.18 01	6.28 01
40	Zr K$_{\alpha_1}$	15.775	78.596	5.82 01	5.87 01	6.12 01	6.22 01
38	Sr K$_{\beta_1}$	15.836	78.293	5.77 01	5.82 01	6.06 01	6.16 01
90	Th L$_{\beta_1}$	16.202	76.524	5.45 01	5.49 01	5.72 01	5.82 01
41	Nb K$_{\alpha_1}$	16.615	74.622	5.12 01	5.16 01	5.38 01	5.47 01
91	Pa L$_{\beta_1}$	16.702	74.233	<u>5.05 01</u>	5.09 01	5.31 01	5.40 01
39	Y K$_{\beta_1}$	16.738	74.074	1.14 02	<u>5.07 01</u>	5.28 01	5.37 01
92	U L$_{\beta_1}$	17.220	72.000	1.05 02	1.05 02	4.92 01	5.01 01
42	Mo K$_{\alpha_1}$	17.479	70.933	1.01 02	1.02 02	<u>4.74 01</u>	4.83 01
40	Zr K$_{\beta_1}$	17.668	70.175	9.85 01	9.86 01	1.05 02	<u>4.70 01</u>
43	Tc K$_{\alpha_1}$	18.367	67.504	8.90 01	8.91 01	9.51 01	9.69 01
41	Nb K$_{\beta_1}$	18.623	66.576	8.58 01	8.60 01	9.17 01	9.34 01
44	Ru K$_{\alpha_1}$	19.279	64.311	7.84 01	7.85 01	8.37 01	8.53 01
42	Mo K$_{\beta_1}$	19.608	63.231	7.50 01	7.51 01	8.00 01	8.16 01
45	Rh K$_{\alpha_1}$	20.216	61.330	<u>6.92 01</u>	6.94 01	7.39 01	7.53 01
43	Tc K$_{\beta_1}$	20.619	60.131	<u>9.10 01</u>	<u>6.59 01</u>	7.01 01	7.15 01
46	Pd K$_{\alpha_1}$	21.177	58.547	1.01 02	8.51 01	<u>6.53 01</u>	6.67 01
44	Ru K$_{\beta_1}$	21.657	57.249	9.54 01	<u>8.02 01</u>	8.53 01	6.29 01
47	Ag K$_{\alpha_1}$	22.163	55.942	8.98 01	9.02 01	<u>8.02 01</u>	<u>5.92 01</u>
45	Rh K$_{\beta_1}$	22.724	54.561	8.40 01	8.44 01	8.99 01	<u>7.67 01</u>
48	Cd K$_{\alpha_1}$	23.274	53.501	7.98 01	8.02 01	8.53 01	8.69 01
46	Pd K$_{\beta_1}$	23.819	52.053	7.42 01	7.46 01	7.94 01	8.08 01
49	In K$_{\alpha_1}$	24.210	51.212	7.11 01	7.15 01	7.60 01	7.74 01

Z	transition	E [keV]	λ [pm]	$_{91}$Pa	$_{92}$U	$_{93}$Np	$_{94}$Pu
47	Ag K$_{\beta_1}$	24.942	49.709	6.57 01	6.61 01	7.03 01	7.16 01
50	Sn K$_{\alpha_1}$	25.271	49.062	6.35 01	6.39 01	6.79 01	6.92 01
48	Cd K$_{\beta_1}$	26.096	47.511	5.84 01	5.87 01	6.24 01	6.36 01
51	Sb K$_{\alpha_1}$	26.359	47.037	5.68 01	5.72 01	6.08 01	6.19 01
49	In K$_{\beta_1}$	27.276	45.455	5.19 01	5.23 01	5.55 01	5.66 01
52	Te K$_{\alpha_1}$	27.472	45.131	5.10 01	5.13 01	5.45 01	5.55 01
50	Sn K$_{\beta_1}$	28.486	43.524	4.63 01	4.67 01	4.95 01	5.05 01
53	I K$_{\alpha_1}$	28.612	43.333	4.58 01	4.61 01	4.89 01	4.99 01
51	Sb K$_{\beta_1}$	29.726	41.709	4.19 01	4.17 01	4.43 01	4.51 01
54	Xe K$_{\alpha_1}$	29.779	41.635	4.12 01	4.15 01	4.40 01	4.49 01
55	Cs K$_{\alpha_1}$	30.973	40.030	3.72 01	3.75 01	3.97 01	4.05 01
52	Te K$_{\beta_1}$	30.996	40.000	3.71 01	3.74 01	3.97 01	4.04 01
56	Ba K$_{\alpha_1}$	32.194	38.511	3.36 01	3.39 01	3.59 01	3.66 01
53	I K$_{\beta_1}$	32.295	38.391	3.34 01	3.36 01	3.56 01	3.63 01
57	La K$_{\alpha_1}$	33.442	37.074	3.05 01	3.07 01	3.25 01	3.31 01
54	Xe K$_{\alpha_1}$	33.624	36.874	3.00 01	3.03 01	3.21 01	3.27 01
58	Ce K$_{\alpha_1}$	34.279	36.169	2.86 01	2.88 01	3.05 01	3.11 01
55	Cs K$_{\beta_1}$	34.987	35.437	2.71 01	2.73 01	2.89 01	2.94 01
59	Pr K$_{\alpha_1}$	36.026	34.415	2.51 01	2.53 01	2.68 01	2.73 01
56	Ba K$_{\beta_1}$	36.378	34.082	2.45 01	2.47 01	2.61 01	2.66 01
60	Nd K$_{\alpha_1}$	36.847	33.648	2.37 01	2.39 01	2.52 01	2.57 01
57	La K$_{\beta_1}$	37.801	32.799	2.22 01	2.24 01	2.36 01	2.40 01
61	Pm K$_{\alpha_1}$	38.725	32.016	2.08 01	2.10 01	2.22 01	2.26 01
58	Ce K$_{\beta_1}$	39.257	31.582	2.01 01	2.03 01	2.14 01	2.18 01
62	Sm K$_{\alpha_1}$	40.118	30.905	1.90 01	1.92 01	2.02 01	2.06 01
59	Pr K$_{\beta_1}$	40.748	30.427	1.82 01	1.84 01	1.94 01	1.98 01
63	Eu K$_{\alpha_1}$	41.542	29.845	1.74 01	1.75 01	1.85 01	1.88 01
60	Nd K$_{\beta_1}$	42.271	29.331	1.66 01	1.67 01	1.76 01	1.80 01
64	Gd K$_{\alpha_1}$	42.996	28.836	1.59 01	1.60 01	1.69 01	1.72 01
61	Pm K$_{\beta_1}$	43.826	28.290	1.51 01	1.53 01	1.61 01	1.64 01
65	Tb K$_{\alpha_1}$	44.482	27.873	1.45 01	1.47 01	1.55 01	1.57 01
62	Sm K$_{\beta_1}$	45.413	27.301	1.38 01	1.39 01	1.46 01	1.49 01
66	Dy K$_{\alpha_1}$	45.998	26.954	1.33 01	1.35 01	1.42 01	1.44 01
63	Eu K$_{\beta_1}$	47.038	26.358	1.26 01	1.27 01	1.34 01	1.36 01
67	Ho K$_{\alpha_1}$	47.547	26.076	1.22 01	1.24 01	1.30 01	1.32 01
64	Gd K$_{\beta_1}$	48.697	25.460	1.15 01	1.16 01	1.22 01	1.24 01
68	Er K$_{\alpha_1}$	49.128	25.237	1.12 01	1.14 01	1.19 01	1.22 01
65	Tb K$_{\beta_1}$	50.382	24.609	1.05 01	1.07 01	1.12 01	1.14 01
69	Tm K$_{\alpha_1}$	50.742	24.434	1.03 01	1.05 01	1.10 01	1.12 01
66	Dy K$_{\beta_1}$	52.119	23.788	9.66 00	9.79 00	1.03 01	1.04 01
70	Yb K$_{\alpha_1}$	52.389	23.666	9.53 00	9.66 00	1.01 01	1.03 01
67	Ho K$_{\beta_1}$	53.877	23.012	8.87 00	8.99 00	9.42 00	9.59 00
71	Lu K$_{\alpha_1}$	54.070	22.930	8.79 00	8.91 00	9.33 00	9.50 00
68	Er K$_{\beta_1}$	55.681	22.267	8.15 00	8.27 00	8.65 00	8.81 00

Z	transition	E [keV]	λ [pm]	$_{91}$Pa	$_{92}$U	$_{93}$Np	$_{94}$Pu
72	Hf K$_{\alpha_1}$	55.790	22.223	8.11 00	8.22 00	8.61 00	8.77 00
69	Tm K$_{\beta_1}$	57.517	21.556	7.50 00	7.61 00	7.96 00	8.10 00
73	Ta K$_{\alpha_1}$	57.532	21.550	7.49 00	7.60 00	7.95 00	8.10 00
74	W K$_{\alpha_1}$	59.318	20.901	6.93 00	7.03 00	7.35 00	7.49 00
70	Yb K$_{\beta_1}$	59.370	20.883	6.91 00	7.02 00	7.34 00	7.47 00
75	Re K$_{\alpha_1}$	61.140	20.278	6.41 00	6.52 00	6.81 00	6.93 00
71	Lu K$_{\beta_1}$	61.283	20.231	6.38 00	6.48 00	6.77 00	6.89 00
76	Os K$_{\alpha_1}$	63.001	19.679	5.95 00	6.04 00	6.31 00	6.42 00
72	Hf K$_{\beta_1}$	63.234	19.607	5.89 00	5.99 00	6.25 00	6.36 00
77	Ir K$_{\alpha_1}$	64.896	19.105	5.52 00	5.61 00	5.86 00	5.96 00
73	Ta K$_{\beta_1}$	65.223	19.009	5.45 00	5.54 00	5.78 00	5.88 00
78	Pt K$_{\alpha_1}$	66.832	18.551	5.13 00	5.21 00	5.44 00	5.53 00
74	W K$_{\beta_1}$	67.244	18.438	5.05 00	5.13 00	5.36 00	5.45 00
79	Au K$_{\alpha_1}$	68.804	18.020	4.76 00	4.84 00	5.05 00	5.14 00
75	Re K$_{\beta_1}$	69.310	17.888	4.68 00	4.75 00	4.96 00	5.05 00
80	Hg K$_{\alpha_1}$	70.819	17.507	4.43 00	4.50 00	4.70 00	4.78 00
76	Os K$_{\beta_1}$	71.413	17.361	4.34 00	4.41 00	4.60 00	4.68 00
81	Tl K$_{\alpha_1}$	72.872	17.014	4.12 00	4.19 00	4.37 00	4.44 00
77	Ir K$_{\beta_1}$	73.561	16.854	4.03 00	4.09 00	4.27 00	4.34 00
82	Pb K$_{\alpha_1}$	74.969	16.538	3.84 00	3.90 00	4.07 00	4.14 00
78	Pt K$_{\beta_1}$	75.748	16.368	3.74 00	3.80 00	3.97 00	4.03 00
83	Bi K$_{\alpha_1}$	77.108	16.079	3.58 00	3.64 00	3.79 00	3.85 00
79	Au K$_{\beta_1}$	77.948	15.906	3.48 00	3.54 00	3.69 00	3.75 00
84	Po K$_{\alpha_1}$	79.290	15.636	3.33 00	3.39 00	3.53 00	3.59 00
80	Hg K$_{\beta_1}$	80.253	15.449	3.23 00	3.29 00	3.43 00	3.48 00
85	At K$_{\alpha_1}$	81.520	15.209	3.11 00	3.16 00	3.30 00	3.35 00
81	Tl K$_{\beta_1}$	82.576	15.014	3.01 00	3.07 00	3.19 00	3.24 00
86	Rn K$_{\alpha_1}$	83.780	14.798	2.91 00	2.96 00	3.08 00	3.13 00
82	Pb K$_{\beta_1}$	84.936	14.597	2.81 00	2.86 00	2.98 00	3.03 00
87	Fr K$_{\alpha_1}$	86.100	14.400	2.72 00	2.77 00	2.88 00	2.93 00
83	Bi K$_{\beta_1}$	87.343	14.195	2.63 00	2.67 00	2.78 00	2.82 00
88	Ra K$_{\alpha_1}$	88.470	14.014	2.54 00	2.59 00	2.69 00	2.74 00
84	Po K$_{\beta_1}$	89.800	13.806	2.45 00	2.49 00	2.59 00	2.64 00
89	Ac K$_{\alpha_1}$	90.884	13.642	2.38 00	2.42 00	2.52 00	2.56 00
85	At K$_{\beta_1}$	92.300	13.432	2.29 00	2.33 00	2.42 00	2.46 00
90	Th K$_{\alpha_1}$	93.350	13.281	2.23 00	2.27 00	2.36 00	2.40 00
86	Rn K$_{\beta_1}$	94.870	13.068	2.14 00	2.18 00	2.26 00	2.30 00
91	Pa K$_{\alpha_1}$	95.868	12.932	2.09 00	2.12 00	2.21 00	2.24 00
87	Fr K$_{\beta_1}$	97.470	12.720	2.01 00	2.04 00	2.12 00	2.15 00
92	U K$_{\alpha_1}$	98.439	12.595	1.96 00	1.99 00	2.07 00	2.10 00
88	Ra K$_{\beta_1}$	100.130	12.382	1.88 00	1.91 00	1.98 00	2.01 00
89	Ac K$_{\beta_1}$	102.850	12.054	1.76 00	1.79 00	1.86 00	1.89 00
90	Th K$_{\beta_1}$	105.610	11.739	1.65 00	1.68 00	1.74 00	1.77 00
91	Pa K$_{\beta_1}$	108.430	11.434	1.55 00	1.58 00	1.63 00	1.66 00

Z	transition	E [keV]	λ [pm]	$_{91}$Pa	$_{92}$U	$_{93}$Np	$_{94}$Pu
92	U K$_{\beta_1}$	111.300	11.139	1.46 00	1.48 00	1.54 00	1.56 00

11 Fit Parameters for the Calculation of Mass Attenuation Coefficients

In the following fit parameters for the approximation of mass attenuation coefficients μ for the elements up to plutonium (Z=94) are given for the energy region $1\,\text{keV} \le E \le 150\,\text{keV}$.

As basis mass attenuation coefficients determined by [547] derived from a wide varity of literature values are used in the approximation. The fit is provided with a χ^2-approximation for the function

$$\ln \mu = x_0 + x_1 \ln E + x_2 (\ln E)^2 + x_3 (\ln E)^3$$

As criteria for the evaluation of the used parameter sets the approximation quality is given by a mean relative error quotient (RFQ)

$$RFQ = \frac{1}{m} \sum_{i=1}^{m} \frac{|\mu_{th}(i) - \mu_{ex}(i)|}{\Delta\mu_{ex}(i)}$$

Here μ_{th} describes the approximation results, μ_{ex} the original values and and $\Delta\mu_{ex}$ the error of the original values. For instance $RFQ = 0.1$ means that the deviations of the fit values do not exceed 10% of the experimental errors, i.e. they are an order of magnitude smaller as the the experimental errors.

In the column "region" intervals between the corresponding absorption edges are indicated. A characterizes the start value, E the end value and Z are intermediate values for reaching a higher approximation quality.

Z	region		energy interval [keV]			x_0	x_1	x_2	x_3	RFQ
1	H	(A-Z)	1.000	-	3.000	1.9782	-3.1810	0.2663	0.4646	0.0001
1	H	(Z-Z)	3.000	-	8.000	2.3064	-4.7411	2.3680	-0.4032	0.0098
1	H	(Z-Z)	8.000	-	40.000	-1.1768	0.3192	-0.1256	0.9120	0.0540
1	H	(Z-E)	40.000	-	150.000	-1.5268	0.3818	-0.0725	0.0009	0.0124
2	He	(A-Z)	1.000	-	4.000	4.1342	-0.1842	-0.1900	0.2179	0.0252
2	He	(Z-Z)	4.000	-	10.000	6.2952	-7.0916	2.0715	-0.1909	0.0233
2	He	(Z-Z)	10.000	-	40.000	2.1555	-2.9261	0.7622	-0.0689	0.0455
2	He	(Z-E)	40.000	-	150.000	-1.9594	0.2798	-0.0624	0.0009	0.0412
3	Li	(A-Z)	1.000	-	4.000	5.4603	-3.1208	-0.1262	0.0759	0.0115
3	Li	(Z-Z)	4.000	-	10.000	4.9815	-1.9013	-1.1537	0.3625	0.0257
3	Li	(Z-Z)	10.000	-	40.000	10.6173	-10.1751	2.8249	-0.2666	0.0331
3	Li	(Z-E)	40.000	-	150.000	-2.6383	0.6895	-0.1623	0.0087	0.0466
4	Be	(A-Z)	1.000	-	5.000	6.4310	-2.9919	-0.1353	0.0474	0.0319

Z		region	energy interval [keV]			x_0	x_1	x_2	x_3	RFQ
4	Be	(Z-Z)	5.000	-	15.000	6.2244	-2.0962	-0.9718	0.2721	0.0525
4	Be	(Z-Z)	15.000	-	50.000	14.0212	-11.8160	2.9831	-0.2557	0.0630
4	Be	(Z-E)	50.000	-	150.000	-2.6112	0.7384	-0.1809	0.0105	0.0182
5	B	(A-Z)	1.000	-	4.000	7.1423	-2.9402	-0.1013	0.0216	0.0115
5	B	(Z-Z)	4.000	-	10.000	6.5962	-1.8275	-0.8584	0.1942	0.0191
5	B	(Z-Z)	10.000	-	40.000	15.6224	-11.4874	2.4828	-0.1742	0.0707
5	B	(Z-E)	40.000	-	150.000	-1.3145	0.2090	-0.1264	0.0108	0.0926
6	C	(A-Z)	1.000	-	4.000	7.7102	-2.8547	-0.0778	0.0003	0.0068
6	C	(Z-Z)	4.000	-	10.000	9.0028	-5.1288	1.2123	-0.2324	0.0540
6	C	(Z-Z)	10.000	-	40.000	10.0877	-4.8492	0.1742	0.0769	0.1210
6	C	(Z-E)	40.000	-	150.000	4.6797	-3.5526	0.6729	-0.0459	0.0833
7	N	(A-Z)	1.000	-	4.000	8.1495	-2.7883	-0.1042	0.0107	0.0215
7	N	(Z-Z)	4.000	-	10.000	6.4505	0.1488	-1.7796	0.3270	0.0889
7	N	(Z-Z)	10.000	-	40.000	8.6328	-2.3069	-0.7958	0.1840	0.0701
7	N	(Z-E)	40.000	-	150.000	9.8129	-6.6764	1.3102	-0.0894	0.0996
8	O	(A-Z)	1.000	-	5.000	8.4664	-2.6619	-0.1544	0.0220	0.0028
8	O	(Z-Z)	5.000	-	15.000	8.1187	-2.0347	-0.5296	0.0948	0.0068
8	O	(Z-Z)	15.000	-	50.000	9.9595	-2.9911	-0.5859	0.1532	0.0999
8	O	(Z-E)	50.000	-	150.000	6.0915	-3.8800	0.6363	-0.0367	0.0854
9	F	(A-Z)	1.000	-	5.000	8.7139	-2.6522	-0.1305	0.0137	0.0069
9	F	(Z-Z)	5.000	-	15.000	8.4276	-2.1423	-0.4321	0.0730	0.0089
9	F	(Z-Z)	15.000	-	50.000	10.3174	-2.9096	-0.6130	0.1494	0.0781
9	F	(Z-E)	50.000	-	150.000	19.7311	-12.6064	2.4979	-0.1695	0.0343
10	Ne	(A-Z)	1.000	-	5.000	8.9794	-2.5658	-0.1526	0.0176	0.0126
10	Ne	(Z-Z)	5.000	-	15.000	8.9435	-2.4401	-0.2552	0.0413	0.0126
10	Ne	(Z-Z)	15.000	-	50.000	8.6594	-1.0223	-1.1564	0.1952	0.0710
10	Ne	(Z-E)	50.000	-	150.000	19.8683	-12.0722	2.2690	-0.1457	0.0983
11	Na	(A-K)	1.000	-	1.072	6.492	-2.5589	0.2069	-1.9866	0.0002
11	Na	(K-Z)	1.072	-	5.000	9.1937	-2.5599	-0.1406	0.0144	0.0043
11	Na	(Z-Z)	5.000	-	20.000	8.7473	-1.8442	-0.5181	0.0797	0.0182
11	Na	(Z-Z)	20.000	-	60.000	11.5477	-3.2117	-0.5236	0.1297	0.0121
11	Na	(Z-E)	60.000	-	150.000	2.3765	-0.0597	-0.4497	0.0572	0.0819
12	Mg	(A-K)	1.000	-	1.305	6.8628	-2.8442	-0.2259	0.5522	0.0038
12	Mg	(K-Z)	1.305	-	5.000	9.3765	-2.4337	-0.2069	0.0307	0.0079
12	Mg	(Z-Z)	5.000	-	20.000	8.9194	-1.7667	-0.5141	0.0736	0.0108
12	Mg	(Z-Z)	20.000	-	60.000	12.0705	-3.4060	-0.4459	0.1164	0.0781
12	Mg	(Z-E)	60.000	-	150.000	4.9596	-1.1838	-0.3026	0.0525	0.0917
13	Al	(A-K)	1.000	-	1.560	7.0813	-2.7215	-0.0598	0.0746	0.0053
13	Al	(K-Z)	1.560	-	6.000	9.4955	-2.3690	-0.1814	0.0149	0.0177
13	Al	(Z-Z)	6.000	-	20.000	10.1688	-3.1658	0.1124	-0.0174	0.0629
13	Al	(Z-Z)	20.000	-	60.000	9.3874	-0.9191	-1.1208	0.1732	0.0885
13	Al	(Z-E)	60.000	-	150.000	9.8438	-3.7963	0.1608	0.0251	0.0921

Z		region	energy interval [keV]			x_0	x_1	x_2	x_3	RFQ
14	Si	(A-K)	1.000	-	1.839	7.3707	-2.6915	-0.0623	0.1091	0.0032
14	Si	(K-Z)	1.839	-	6.000	9.7569	-2.4209	-0.1610	0.0127	0.0221
14	Si	(Z-Z)	6.000	-	20.000	10.1325	-2.8635	-0.0031	-0.0027	0.0596
14	Si	(Z-Z)	20.000	-	60.000	9.2002	-0.7004	-1.1267	0.1662	0.0755
14	Si	(Z-E)	60.000	-	150.000	11.5491	-4.4530	0.2330	0.0237	0.0694
15	P	(A-K)	1.000	-	2.146	7.5666	-2.7296	0.2250	-0.1832	0.0070
15	P	(K-Z)	2.146	-	6.000	9.7201	-2.0315	-0.4399	0.0835	0.0379
15	P	(Z-Z)	6.000	-	20.000	8.9597	-1.1517	-0.7184	0.0965	0.0475
15	P	(Z-Z)	20.000	-	60.000	8.0900	0.2728	-1.3420	0.1783	0.0849
15	P	(Z-E)	60.000	-	150.000	14.3809	-5.6628	0.3836	0.0193	0.0941
16	S	(A-K)	1.000	-	2.472	7.8080	-2.6465	-0.0884	0.0815	0.0023
16	S	(K-Z)	2.472	-	8.000	9.9950	-2.2813	-0.2215	0.0267	0.0175
16	S	(Z-Z)	8.000	-	30.000	10.2549	-2.4798	-0.1847	0.0264	0.0629
16	S	(Z-Z)	30.000	-	60.000	12.0817	-2.5465	-0.6256	0.1l54	0.0263
16	S	(Z-E)	60.000	-	150.000	13.0883	-4.2778	-0.0018	0.0517	0.0891
17	Cl	(A-K)	1.000	-	2.822	7.9625	-2.6348	-0.0578	0.0489	0.0007
17	Cl	(K-Z)	2.822	-	6.000	10.0532	-2.2392	-0.2124	0.0208	0.0087
17	Cl	(Z-Z)	6.000	-	20.000	10.7233	-3.1023	0.1483	-0.0278	0.0642
17	Cl	(Z-Z)	20.000	-	60.000	9.8514	-1.1646	-0.8421	0.1195	0.0999
17	Cl	(Z-E)	60.000	-	150.000	13.8834	-4.2897	-0.0846	0.0622	0.0780
18	Ar	(A-K)	1.000	-	3.203	8.0980	-2.6887	0.0264	-0.0101	0.0191
18	Ar	(K-Z)	3.203	-	8.000	10.0898	-2.0921	-0.3072	0.0411	0.0129
18	Ar	(Z-Z)	8.000	-	30.000	10.7941	-2.8889	-0.0016	0.0003	0.0579
18	Ar	(Z-Z)	30.000	-	60.000	6.7230	1.6632	-1.6418	0.1926	0.0234
18	Ar	(Z-E)	60.000	-	150.000	8.4368	-0.3376	-1.0070	0.1319	0.0915
19	K	(A-K)	1.000	-	3.607	8.3328	-2.6534	-0.0031	0.0055	0.0101
19	K	(K-Z)	3.607	-	8.000	9.3967	-0.5435	-1.2175	0.2209	0.0405
19	K	(Z-Z)	8.000	-	30.000	11.0888	-3.0349	0.0677	-0.0089	0.0584
19	K	(Z-Z)	30.000	-	60.000	10.7944	-1.5088	-0.7587	0.1097	0.0310
19	K	(Z-E)	60.000	-	150.000	10.0274	-0.9735	-0.9134	0.1267	0.0595
20	Ca	(A-K)	1.000	-	4.038	8.5272	-2.6739	0.0182	-0.0047	0.0063
20	Ca	(K-Z)	4.038	-	10.000	11.3251	-3.7263	0.6452	-0.1344	0.0698
20	Ca	(Z-Z)	10.000	-	40.000	8.7788	-0.3549	-0.8821	0.1019	0.0177
20	Ca	(Z-E)	40.000	-	150.000	13.8980	-3.0968	-0.5092	0.1003	0.0420
21	Sc	(A-K)	1.000	-	4.493	8.6210	-2.6716	-0.0055	0.0083	0.0129
21	Sc	(K-Z)	4.493	-	10.000	10.3056	-1.9903	-0.2839	0.0314	0.0146
21	Sc	(Z-Z)	10.000	-	40.000	11.0462	-2.7062	-0.0551	0.0066	0.0935
21	Sc	(Z-E)	40.000	-	150.000	16.2561	-4.6667	-0.1399	0.0701	0.0917
22	Ti	(A-K)	1.000	-	4.966	8.7678	-2.6858	-0.0053	0.0059	0.0133
22	Ti	(K-Z)	4.966	-	10.000	11.0905	-3.2374	0.4194	-0.0935	0.0276
22	Ti	(Z-Z)	10.000	-	40.000	8.5484	-0.0558	-0.9349	0.1027	0.0300
22	Ti	(Z-E)	40.000	-	150.000	14.2169	-3.0953	-0.5048	0.0967	0.0798
23	V	(A-K)	1.000	-	5.465	8.8785	-2.7030	-0.0020	0.0058	0.0087

Z		region	energy interval [keV]			x_0	x_1	x_2	x_3	RFQ
23	V	(K-Z)	5.465	-	15.000	10.2481	-1.7272	-0.3719	0.0419	0.0072
23	V	(Z-Z)	15.000	-	50.000	7.1580	1.2422	-1.3076	0.1381	0.0251
23	V	(Z-E)	50.000	-	150.000	14.5009	-3.1092	-0.5148	0.0971	0.0364
24	Cr	(A-K)	1.000	-	5.989	9.0185	-2.6816	-0.0077	0.0067	0.0138
24	Cr	(K-Z)	5.989	-	15.000	10.6126	-2.1103	-0.1628	0.0063	0.0068
24	Cr	(Z-Z)	15.000	-	50.000	11.1070	-2.2894	-0.2216	0.0276	0.0664
24	Cr	(Z-E)	50.000	-	150.000	15.2180	-3.4169	-0.4524	0.0918	0.0633
25	Mn	(A-Z)	1.000	-	3.000	9.1407	-2.6878	-0.0360	0.0274	0.0001
25	Mn	(Z-K)	3.000	-	6.539	8.5102	-1.3948	-0.8804	0.1990	0.0547
25	Mn	(K-Z)	6.539	-	30.000	10.0986	-2.4196	-0.0759	0.0011	0.0395
25	Mn	(Z-E)	30.000	-	150.000	2.5203	5.2040	-2.3705	0.2323	0.0288
26	Fe	(A-Z)	1.000	-	3.000	9.2954	-2.7152	0.0259	-0.0140	0.0001
26	Fe	(Z-K)	3.000	-	7.112	9.0445	-2.1903	-0.3479	0.0803	0.0512
26	Fe	(K-Z)	7.112	-	20.000	11.0983	-2.5335	0.0219	-0.0175	0.0274
26	Fe	(Z-Z)	20.000	-	60.000	11.9428	-2.7324	-0.1196	0.0206	0.0829
26	Fe	(Z-E)	60.000	-	150.000	19.9935	-6.5028	0.2670	0.0338	0.0788
27	Co	(A-Z)	1.000	-	3.000	9.3805	-2.6890	-0.0192	0.0130	0.0000
27	Co	(Z-K)	3.000	-	7.709	9.7769	-3.4972	0.5196	-0.1059	0.0459
27	Co	(K-Z)	7.709	-	30.000	12.1201	-3.4434	0.3086	-0.0460	0.0876
27	Co	(Z-E)	30.000	-	150.000	4.5706	3.7659	-1.9835	0.1968	0.0907
28	Ni	(A-L_1)	1.000	-	1.008	9.3582	-3.3702	-0.6828	1.9903	0.0001
28	Ni	(L_1-Z)	1.008	-	4.000	9.5351	-2.6954	-0.0188	0.0095	0.0005
28	Ni	(Z-K)	4.000	-	8.333	9.5553	-2.7222	-0.0087	0.0088	0.0275
28	Ni	(K-Z)	8.333	-	30.000	10.4996	-1.6968	-0.2844	0.0224	0.0048
28	Ni	(Z-Z)	30.000	-	60.000	10.3381	-1.0648	-0.6105	0.0678	0.0358
28	Ni	(Z-E)	60.000	-	150.000	16.3188	-3.7065	-0.3750	0.0807	0.0304
29	Cu	(A-L_1)	1.000	-	1.097	9.4191	-2.7178	-0.2751	1.9938	0.0006
29	Cu	(L_1-Z)	1.097	-	4.000	9.5966	-2.6880	-0.0230	0.0128	0.0134
29	Cu	(Z-K)	4.000	-	8.979	9.6409	-2.7600	0.0200	0.0025	0.0103
29	Cu	(K-Z)	8.979	-	40.000	11.9847	-3.1930	0.2245	-0.0339	0.0861
29	Cu	(Z-E)	40.000	-	150.000	-0.2807	7.2613	-2.7670	0.2532	0.0152
30	Zn	(A-L_3)	1.000	-	1.020	7.2896	-2.5712	-2.2958	1.7779	0.0005
30	Zn	(L_3-L_2)	1.020	-	1.043	9.1259	-1.4875	-18.7907	1.9493	0.0174
30	Zn	(L_2-L_1)	1.043	-	1.194	9.4789	-2.6315	-0.4980	1.5028	0.0000
30	Zn	(L_1-Z)	1.194	-	3.000	9.6608	-2.6751	0.0045	-0.0028	0.0001
30	Zn	(Z-Z)	3.000	-	6.000	8.2635	0.3119	-2.0894	0.4813	0.0634
30	Zn	(Z-K)	6.000	-	9.659	9.6758	-2.7573	0.0630	-0.0099	0.0052
30	Zn	(K-Z)	9.659	-	30.000	11.1939	-2.2636	-0.0842	-0.0010	0.0203
30	Zn	(Z-Z)	30.000	-	60.000	10.2106	-0.9735	-0.5938	0.0623	0.0084
30	Zn	(Z-E)	60.000	-	150.000	15.7552	-3.3403	-0.4139	0.0788	0.0434
31	Ga	(A-L_3)	1.000	-	1.115	7.4645	-2.6863	0.3120	-1.7796	0.0001
31	Ga	(L_3-L_2)	1.115	-	1.142	9.0885	0.0396	-11.3943	0.3901	0.0117
31	Ga	(L_2-L_1)	1.142	-	1.298	9.5753	-2.5098	-0.9070	1.4514	0.0014

Z		region	energy interval [keV]			x_0	x_1	x_2	x_3	RFQ
31	Ga	$(L_1\text{-}Z)$	1.298	-	4.000	9.7686	-2.7069	0.0262	-0.0130	0.0147
31	Ga	$(Z\text{-}Z)$	4.000	-	8.000	9.3986	-2.0430	-0.3877	0.0787	0.0143
31	Ga	$(Z\text{-}K)$	8.000	-	10.367	11.6092	-5.3305	1.2432	-0.1912	0.0067
31	Ga	$(K\text{-}Z)$	10.367	-	30.000	11.8195	-2.9182	0.1527	-0.0288	0.0523
31	Ga	$(Z\text{-}Z)$	30.000	-	60.000	11.4986	-2.1214	-0.2470	0.0280	0.0200
31	Ga	$(Z\text{-}E)$	60.000	-	150.000	16.3221	-3.8192	-0.2737	0.0656	0.0477
32	Ge	$(A\text{-}L_3)$	1.000	-	1.217	7.6171	-2.7234	0.2652	-0.9012	0.0001
32	Ge	$(L_3\text{-}L_2)$	1.217	-	1.248	9.0383	0.3633	-7.6143	0.7052	0.0083
32	Ge	$(L_2\text{-}L_1)$	1.248	-	1.414	9.6128	-1.9610	-2.2135	1.9640	0.0145
32	Ge	$(L_1\text{-}Z)$	1.414	-	4.000	9.8792	-2.7502	0.0544	-0.0220	0.0107
32	Ge	$(Z\text{-}Z)$	4.000	-	8.000	9.5619	-2.1781	-0.3110	0.0627	0.0198
32	Ge	$(Z\text{-}K)$	8.000	-	11.104	8.7746	-1.3133	-0.6016	0.0900	0.0017
32	Ge	$(K\text{-}Z)$	11.104	-	30.000	11.9055	-2.9585	0.1597	-0.0278	0.0022
32	Ge	$(Z\text{-}Z)$	30.000	-	60.000	15.5513	-5.3158	0.6044	-0.0474	0.0291
32	Ge	$(Z\text{-}E)$	60.000	-	150.000	12.9936	-1.6251	-0.7379	0.0976	0.0287
33	As	$(A\text{-}L_3)$	1.000	-	1.323	7.9506	-2.7470	-0.5879	1.3641	0.0001
33	As	$(L_3\text{-}L_2)$	1.323	-	1.359	8.9623	0.7495	-6.8013	1.9119	0.0067
33	As	$(L_2\text{-}L_1)$	1.359	-	1.527	9.5021	-0.9888	-3.1732	1.4856	0.0420
33	As	$(L_1\text{-}Z)$	1.527	-	4.000	9.9611	-2.7329	0.0446	-0.0182	0.0171
33	As	$(Z\text{-}Z)$	4.000	-	8.000	9.7538	-2.3614	-0.1972	0.0406	0.0126
33	As	$(Z\text{-}K)$	8.000	-	11.867	9.6364	-2.1907	-0.2751	0.0516	0.0015
33	As	$(K\text{-}Z)$	11.867	-	30.000	11.5655	-2.4693	-0.0162	-0.0071	0.0049
33	As	$(Z\text{-}Z)$	30.000	-	60.000	10.1267	-0.8326	-0.6034	0.0606	0.0139
33	As	$(Z\text{-}E)$	60.000	-	150.000	15.3071	-3.1505	-0.3813	0.0692	0.0401
34	Se	$(A\text{-}L_3)$	1.000	-	1.436	8.0004	-2.8090	0.5305	-0.9489	0.0000
34	Se	$(L_3\text{-}L_2)$	1.436	-	1.476	8.9452	0.5368	-5.3790	1.9360	0.0044
34	Se	$(L_2\text{-}L_1)$	1.476	-	1.654	9.7118	-1.9588	-1.3616	0.7777	0.0039
34	Se	$(L_1\text{-}Z)$	1.654	-	4.000	10.0350	-2.7447	0.0468	-0.0154	0.0000
34	Se	$(Z\text{-}Z)$	4.000	-	8.000	9.5858	-1.9591	-0.4190	0.0801	0.0210
34	Se	$(Z\text{-}K)$	8.000	-	12.658	10.5207	-3.4155	0.3258	-0.0452	0.0051
34	Se	$(K\text{-}Z)$	12.658	-	30.000	11.8807	-2.8087	0.1256	-0.0259	0.0285
34	Se	$(Z\text{-}Z)$	30.000	-	60.000	14.4591	-4.3931	0.3835	-0.0302	0.0289
34	Se	$(Z\text{-}E)$	60.000	-	150.000	15.7503	-3.4680	-0.2902	0.0604	0.0625
35	Br	$(A\text{-}L_1)$	1.000	-	1.550	8.0906	-2.7307	0.2669	-0.4900	0.0037
35	Br	$(L_3\text{-}L_2)$	1.550	-	1.596	8.9866	0.0823	-3.1087	-0.0700	0.0053
35	Br	$(L_2\text{-}L_1)$	1.596	-	1.782	9.6068	-1.1087	-2.2838	0.9600	0.0178
35	Br	$(L_1\text{-}Z)$	1.782	-	4.000	10.0001	-2.2845	-0.4057	0.1224	0.0586
35	Br	$(Z\text{-}Z)$	4.000	-	8.000	9.7457	-1.9876	-0.4435	0.0919	0.0016
35	Br	$(Z\text{-}K)$	8.000	-	13.474	10.8304	-3.6154	0.3843	-0.0504	0.0036
35	Br	$(K\text{-}Z)$	13.474	-	30.000	12.6008	-3.4185	0.3145	-0.0445	0.0116
35	Br	$(Z\text{-}Z)$	30.000	-	60.000	10.9451	-1.4278	-0.4258	0.0432	0.0093
35	Br	$(Z\text{-}E)$	60.000	-	150.000	14.6365	-2.7417	-0.4286	0.0684	0.0416
36	Kr	$(A\text{-}L_3)$	1.000	-	1.675	8.1208	-2.7939	-0.0638	0.0750	0.0016
36	Kr	$(L_3\text{-}L_2)$	1.675	-	1.727	8.9019	0.5459	-3.9760	1.1005	0.0038

Z		region	energy interval [keV]			x_0	x_1	x_2	x_3	RFQ
36	Kr	(L_2-L_1)	1.727	-	1.921	9.8965	-2.1654	-0.8819	0.4816	0.0004
36	Kr	(L_1-Z)	1.921	-	4.000	10.2300	-2.8234	0.1054	-0.0272	0.0024
36	Kr	$(Z-Z)$	4.000	-	8.000	11.3386	-4.7284	1.1704	-0.2200	0.0272
36	Kr	$(Z-K)$	8.000	-	14.326	10.4267	-2.9754	0.0936	-0.0061	0.0024
36	Kr	$(K-Z)$	14.326	-	30.000	11.7381	-2.5830	0.0577	-0.0179	0.0042
36	Kr	$(Z-Z)$	30.000	-	60.000	12.9129	-3.1124	0.0629	-0.0035	0.0163
36	Kr	$(Z-E)$	60.000	-	150.000	14.6896	-2.7964	-0.3989	0.0646	0.0400
37	Rb	$(A-L_3)$	1.000	-	1.804	8.2365	-2.7404	-0.0266	0.0269	0.0012
37	Rb	(L_3-L_2)	1.804	-	1.864	8.4177	1.7635	-3.4675	-0.2660	0.0077
37	Rb	(L_2-L_1)	1.864	-	2.065	9.1775	0.3407	-2.6778	0.3811	0.0483
37	Rb	(L_1-Z)	2.065	-	5.000	10.2320	-2.4776	-0.2139	0.0587	0.0210
37	Rb	$(Z-Z)$	5.000	-	10.000	10.3345	-2.7594	0.0164	0.0000	0.0009
37	Rb	$(Z-K)$	10.000	-	15.200	11.3555	-3.9683	0.4856	-0.0594	0.0143
37	Rb	$(K-Z)$	15.200	-	50.000	12.3734	-2.9529	0.1164	-0.0173	0.0662
37	Rb	$(Z-E)$	50.000	-	150.000	-18.0997	19.1559	-5.2626	0.4220	0.0905
38	Sr	$(A-L_3)$	1.000	-	1.940	8.3128	-2.6796	-0.0188	0.0479	0.0023
38	Sr	(L_3-L_2)	1.940	-	2.007	8.5191	1.1731	-2.5369	-0.3452	0.0085
38	Sr	(L_2-L_1)	2.007	-	2.216	8.9900	0.4682	-1.8685	-0.2538	0.0531
38	Sr	(L_1-Z)	2.216	-	6.000	10.2413	-2.3015	-0.3493	0.0922	0.0558
38	Sr	$(Z-K)$	6.000	-	16.105	10.9351	-3.4293	0.2978	-0.0384	0.0273
38	Sr	$(K-Z)$	16.105	-	50.000	13.0020	-3.4546	0.2651	-0.0317	0.0506
38	Sr	$(Z-E)$	50.000	-	150.000	6.1503	2.7472	-1.5623	0.1447	0.0467
39	Y	$(A-L_3)$	1.000	-	2.080	8.3976	-2.7023	0.3289	-0.2654	0.0057
39	Y	(L_3-L_2)	2.080	-	2.156	7.9206	1.7188	-0.8111	-1.9381	0.0148
39	Y	(L_2-L_1)	2.156	-	2.373	8.1759	2.4112	-2.9200	-0.1926	0.0693
39	Y	(L_1-Z)	2.373	-	6.000	10.1608	-1.9038	-0.6467	0.1640	0.0653
39	Y	$(Z-K)$	6.000	-	17.039	11.0502	-3.4428	0.2940	-0.0365	0.0273
39	Y	$(K-Z)$	17.039	-	50.000	12.1706	-2.6753	0.0445	-0.0109	0.0298
39	Y	$(Z-E)$	50.000	-	150.000	6.8944	2.2211	-1.4234	0.1325	0.0284
40	Zr	$(A-L_3)$	1.000	-	2.222	8.4748	-2.4989	-0.2381	0.2127	0.0022
40	Zr	(L_3-L_2)	2.222	-	2.307	8.1826	1.1585	-0.9829	-1.1550	0.0119
40	Zr	(L_2-L_1)	2.307	-	2.532	7.5080	2.7469	-1.1922	-1.4523	0.0969
40	Zr	(L_1-Z)	2.532	-	6.000	10.1378	-1.6694	-0.8045	0.1960	0.0531
40	Zr	$(Z-K)$	6.000	-	17.998	12.4367	-5.1732	1.0461	-0.1448	0.0859
40	Zr	$(K-Z)$	17.998	-	50.000	13.0407	-3.6016	0.3604	-0.0444	0.0253
40	Zr	$(Z-E)$	50.000	-	150.000	13.0798	-2.0389	-0.4339	0.0560	0.0977
41	Nb	$(A-L_3)$	1.000	-	2.371	8.5671	-2.5212	-0.1394	0.1073	0.0043
41	Nb	(L_3-L_2)	2.371	-	2.465	7.4570	2.3185	-0.9358	-1.4842	0.0150
41	Nb	(L_2-L_1)	2.465	-	2.698	7.0690	3.2746	-1.1729	-1.4126	0.0878
41	Nb	(L_1-Z)	2.698	-	6.000	10.8229	-3.0135	0.1762	-0.0365	0.0063
41	Nb	$(Z-K)$	6.000	-	18.986	11.2570	-3.4786	0.3038	-0.0381	0.0440
41	Nb	$(K-Z)$	18.986	-	50.000	11.7544	-2.4658	0.0364	-0.0128	0.0634
41	Nb	$(Z-E)$	50.000	-	150.000	14.1355	-2.7633	-0.2539	0.0411	0.0975
42	Mo	$(A-L_3)$	1.000	-	2.520	8.6304	-2.4787	-0.0956	0.0789	0.0027

Z		region	energy interval [keV]			x_0	x_1	x_2	x_3	RFQ
42	Mo	(L$_3$-L$_2$)	2.520	-	2.625	7.0084	2.4378	-0.0571	-1.9058	0.0202
42	Mo	(L$_2$-L$_1$)	2.625	-	2.866	6.9328	3.3392	-1.2735	-1.1580	0.0820
42	Mo	(L$_1$-Z)	2.688	-	6.000	9.4995	0.0170	-1.9689	0.4575	0.0682
42	Mo	(Z - K)	6.000	-	20.000	10.0894	- 1.8523	- 0.3995	0.0606	0.0397
42	Mo	(K - Z)	20.000	-	50.000	12.1106	- 2.6805	0.0907	- 0.0178	0.0146
42	Mo	(Z - E)	50.000	-	150.000	14.5610	- 3.1598	- 0.1305	0.0293	0.0996
43	Te	(A - L$_3$)	1.000	-	2.677	8.7052	- 2.4793	- 0.0488	0.0380	0.0005
43	Te	(L$_3$ - L$_2$)	2.677	-	2.793	5.6897	4.4375	- 0.6096	- 1.9693	0.0262
43	Te	(L$_2$ - L$_1$)	2.793	-	3.043	6.0985	3.8517	- 0.4652	- 1.6261	0.0900
43	Te	(L$_1$ - Z)	3.043	-	8.000	10.7115	- 2.5542	- 0.1263	0.0292	0.0158
43	Te	(Z - K)	8.000	-	21.044	11.3508	- 3.4189	0.2491	- 0.0268	0.0247
43	Te	(K - Z)	21.044	-	50.000	12.4623	- 2.7472	0.0708	- 0.0134	0.0203
43	Te	(Z - E)	50.000	-	150.000	14.6450	- 3.1128	- 0.1547	0.0318	0.0968
44	Ru	(A - L$_3$)	1.000	-	2.838	8.7880	- 2.4568	- 0.0887	0.0618	0.0000
44	Ru	(L$_3$ - L$_2$)	2.838	-	2.967	5.7471	3.7999	- 0.1450	- 1.8113	0.0261
44	Ru	(L$_2$ - L$_1$)	2.967	-	3.224	5.8512	3.5253	0.3099	- 1.8215	0.0840
44	Ru	(L$_1$ - Z)	3.244	-	8.000	10.8251	- 2.6269	- 0.0848	0.0215	0.0037
44	Ru	(Z - K)	8.000	-	22.117	11.3784	- 3.3006	0.1957	- 0.0191	0.0139
44	Ru	(K - Z)	22.117	-	50.000	12.1186	- 2.3774	- 0.0401	- 0.0024	0.0101
44	Ru	(Z - E)	50.000	-	150.000	13.9733	- 2.6528	- 0.2470	0.0376	0.0986
45	Rh	(A - L$_3$)	1.000	-	3.004	8.8730	- 2.4547	- 0.0794	0.0515	0.0052
45	Rh	(L$_3$ - L$_2$)	3.004	-	3.146	5.3732	3.7081	0.4929	- 1.9925	0.0295
45	Rh	(L$_2$ - L$_1$)	3.146	-	3.412	5.9269	3.0395	0.5162	- 1.6507	0.0785
45	Rh	(L$_1$ - Z)	3.412	-	8.000	10.8270	- 2.5523	- 0.1159	0.0265	0.0087
45	Rh	(Z - K)	8.000	-	23.220	10.8729	- 2.6569	- 0.0424	0.0103	0.0070
45	Rh	(K - Z)	23.220	-	50.000	13.2033	- 3.3939	0.2756	- 0.0336	0.0146
45	Rh	(Z - E)	50.000	-	150.000	14.2128	- 2.8316	- 0.1910	0.0321	0.0844
46	Pd	(A - L$_3$)	1.000	-	3.173	8.9560	- 2.5077	- 0.0467	0.0320	0.0037
46	Pd	(L$_3$ - L$_2$)	3.173	-	3.330	5.7408	2.8657	0.7919	- 1.7979	0.0289
46	Pd	(L$_2$ - L$_1$)	3.330	-	3.604	5.2425	3.2498	1.2778	- 1.9897	0.0853
46	Pd	(L$_1$ - Z)	3.604	-	8.000	10.0105	- 0.8353	- 1.1811	0.2393	0.0523
46	Pd	(Z - K)	8.000	-	24.350	11.4971	- 3.2507	0.1641	- 0.0144	0.0244
46	Pd	(K - Z)	24.350	-	50.000	10.1581	- 0.8806	- 0.4099	0.0290	0.0154
46	Pd	(Z - E)	50.000	-	150.000	15.5733	- 3.8277	0.0578	0.0117	0.0996
47	Ag	(A - L$_3$)	1.000	-	3.351	9.0326	- 2.5060	- 0.0428	0.0295	0.0068
47	Ag	(L$_3$ - L$_2$)	3.351	-	3.524	5.3753	2.8292	1.1597	- 1.8350	0.0317
47	Ag	(L$_2$ - L$_1$)	3.524	-	3.806	5.1524	3.0160	1.3542	- 1.8300	0.0760
47	Ag	(L$_1$ - Z)	3.806	-	8.000	11.7920	- 4.0185	0.7503	- 0.1440	0.0127
47	Ag	(Z - K)	8.000	-	25.514	10.5418	- 2.0832	- 0.2742	0.0402	0.0077
47	Ag	(K - Z)	25.514	-	50.000	12.0455	- 2.4762	0.0451	- 0.0134	0.0103
47	Ag	(Z - E)	50.000	-	150.000	14.3593	- 3.0833	- 0.0811	0.0199	0.0786
48	Cd	(A - L$_3$)	1.000	-	3.538	9.0994	- 2.5181	0.0316	- 0.0083	0.0054
48	Cd	(L$_3$ - L$_2$)	3.538	-	3.727	6.0494	2.6346	- 0.1435	- 0.9912	0.0239
48	Cd	(L$_2$ - L$_1$)	3.727	-	4.018	5.4120	2.6713	1.0627	- 1.5034	0.0443

Z	region	energy interval [keV]			x_0	x_1	x_2	x_3	RFQ
48	Cd (L_1 - Z)	4.018	-	10.000	10.8848	- 2.2773	- 0.2834	0.0558	0.0337
48	Cd (Z - K)	10.000	-	26.711	11.3372	- 2.9255	0.0366	0.0019	0.0164
48	Cd (K - Z)	26.711	-	50.000	12.2756	- 2.6613	0.1046	- 0.0198	0.0024
48	Cd (Z - E)	50.000	-	150.000	13.3383	- 2.3392	- 0.2527	0.0328	0.0647
49	In (A - L_3)	1.000	-	3.730	9.1745	- 2.4986	- 0.0155	0.0136	0.0084
49	In (L_3 - L_2)	3.730	-	3.938	4.6858	3.1672	1.1040	- 1.6264	0.0357
49	In (L_2 - L_1)	3.938	-	4.238	5.2469	2.6032	1.1355	- 1.4380	0.0442
49	In (L_1 - Z)	4.238	-	10.000	11.5516	- 3.2887	0.2619	- 0.0393	0.0183
49	In (Z - Z)	10.000	-	27.940	11.1550	- 2.6686	- 0.0503	0.0118	0.0177
49	In (K - Z)	27.940	-	50.000	12.2142	- 2.1809	- 0.1077	0.0053	0.0072
49	In (Z - E)	50.000	-	150.000	14.8200	- 3.2239	- 0.0672	0.0196	0.0799
50	Sn (A - L_3)	1.000	-	3.929	9.2447	- 2.5274	0.0309	- 0.0101	0.0024
50	Sn (L_3 - L_2)	3.929	-	4.156	4.4550	3.1534	1.2659	- 1.6221	0.0208
50	Sn (L_2 - L_1)	4.156	-	4.465	- 17.4613	4.5683	10.1212	- 0.8360	0.0738
50	Sn (L_1-Z)	4.465	-	10.000	11.7937	-3.5661	0.4096	-0.0669	0.0210
50	Sn (Z-K)	10.000	-	29.200	11.4504	-2.9132	0.0318	0.0022	0.0033
50	Sn (K-Z)	29.200	-	50.000	8.0188	0.6741	-0.7560	0.0550	0.0018
50	Sn (Z-E)	50.000	-	150.000	13.3316	-2.3920	-0.2083	0.0267	0.0945
51	Sb (A-L_1)	1.000	-	4.132	9.3174	-2.5407	0.0154	-0.0035	0.0016
51	Sb (L_3L_2)	4.132	-	4.380	4.6074	2.9941	0.8513	-1.3004	0.0332
51	Sb (L_2-L_1)	4.380	-	4.698	4.1099	2.6527	1.9840	-1.6528	0.0553
51	Sb (L_1-Z)	4.698	-	10.000	11.3479	-2.9036	0.0877	-0.0112	0.0176
51	Sb (Z-K)	10.000	-	30.491	12.0975	-3.6329	0.3083	-0.0308	0.0390
51	Sb (K-E)	30.491	-	150.000	6.6584	2.2150	-1.2520	0.1049	0.0240
52	Te (A-M_1)	1.000	-	1.006	9.2660	-2.1697	-14.4403	-0.2082	0.0003
52	Te (M_1-L_3)	1.006	-	4.341	9.3626	-2.5336	-0.0041	0.0057	0.0026
52	Te (L_3-L_2)	4.341	-	4.612	4.4479	2.4939	1.4273	-1.3984	0.0324
52	Te (L_2-L_1)	4.612	-	4.939	4.0017	2.8362	1.6123	-1.4472	0.0442
52	Te (L_1-Z)	4.939	-	8.000	10.4272	-1.2645	-0.8336	0.1597	0.0035
52	Te (Z-Z)	8.000	-	20.000	12.3993	-3.9093	0.4055	-0.0423	0.0362
52	Te (Z-K)	20.000	-	31.814	10.2316	-1.8040	-0.2769	0.0314	0.0045
52	Te (K-Z)	31.814	-	60.000	18.4605	-7.3073	1.3022	-0.1222	0.0188
52	Te (Z-E)	60.000	-	150.000	14.1699	-2.7452	-0.1557	0.0243	0.0294
53	I (A-M_1)	1.000	-	1.072	9.3530	-2.5140	-0.1147	1.2114	0.0001
53	I (M_1-L_3)	1.072	-	4.557	9.4469	-2.5229	-0.0281	0.0168	0.0032
53	I (L_3-L_2)	4.557	-	4.852	3.9060	2.6347	1.7478	-1.4928	0.0453
53	I (L_2-L_1)	4.852	-	5.188	4.6273	2.4737	1.1050	-1.1201	0.0419
53	I (L_1-Z)	5.188	-	15.000	11.2064	-2.4911	-0.1186	-0.0215	0.0102
53	I (Z-K)	15.000	-	33.170	14.2305	-5.4626	0.8549	-0.0851	0.0448
53	I (K-E)	33.170	-	150.000	9.0371	0.5999	-0.8734	0.0755	0.0703
54	Xe (A-M_1)	1.000	-	1.145	9.3148	-2.4859	0.1507	-0.7450	0.0004
54	Xe (M_1-L_3)	1.145	-	4.782	9.4109	-2.4961	0.0001	0.0057	0.0016
54	Xe (L_3-L_2)	4.782	-	5.104	3.8948	2.9121	1.1453	-1.2150	0.0811
54	Xe (L_2-L_1)	5.104	-	5.453	3.9269	2.6069	1.4800	-1.2355	0.0455

Z		region	energy interval [keV]			x_0	x_1	x_2	x_3	RFQ
54	Xe	(L_1-Z)	5.453	-	15.000	10.5553	-1.5455	-0.5579	0.0892	0.0218
54	Xe	(Z-K)	15.000	-	34.561	12.6045	-3.8397	0.3272	-0.0281	0.0133
54	Xe	(K-E)	34.561	-	150.000	10.6930	-0.6824	-0.5465	0.0484	0.0959
55	Cs	(A-M_2)	1.000	-	1.065	9.3933	-2.6005	0.1471	-1.8188	0.0001
55	Cs	(M_2-M_1)	1.065	-	1.217	9.4866	-2.5082	-0.5225	1.3419	0.0007
55	Cs	(M_1-L_3)	1.217	-	5.012	9.5850	-2.5738	0.0138	0.0003	0.0112
55	Cs	(L_3-L_2)	5.012	-	5.359	4.2679	1.9321	1.7485	-1.2815	0.0456
55	Cs	(L_2-L_1)	5.359	-	5.714	5.5332	1.5832	0.8578	-0.8244	0.0308
55	Cs	(L_1-Z)	5.714	-	15.000	12.0792	-3.5662	0.3730	-0.0522	0.0115
55	Cs	(Z-K)	15.000	-	35.985	11.7163	-2.8509	-0.0077	0.0092	0.0200
55	Cs	(K-E)	35.985	-	150.000	10.9351	-0.6925	-0.5651	0.0510	0.0655
56	Ba	(A-M_3)	1.000	-	1.062	9.2404	-2.5158	-0.0007	0.0714	0.0000
56	Ba	(M_3-M_2)	1.062	-	1.137	9.4176	-2.3986	-1.2255	1.9731	0.0034
56	Ba	(M_2-M_1)	1.137	-	1.293	9.5221	-2.5663	-0.0118	0.0203	0.0000
56	Ba	(M_1-L_3)	1.293	-	5.247	9.6028	-2.5075	-0.0541	0.0236	0.0141
56	Ba	(L_3-L_2)	5.247	-	5.624	5.3060	1.4118	1.0368	-0.8829	0.0369
56	Ba	(L_2-L_1)	5.624	-	5.989	3.8759	2.4923	1.2997	-1.0558	0.0375
56	Ba	(L_1-Z)	5.989	-	15.000	11.3684	-2.5566	-0.0708	0.0114	0.0107
56	Ba	(Z-K)	15.000	-	37.441	11.1306	-2.4020	-0.1040	0.0146	0.0308
56	Ba	(K-E)	37.441	-	150.000	3.4590	4.6056	-1.8028	0.1467	0.0581
57	La	(A-M_3)	1.000	-	1.123	9.3069	-2.5538	0.1362	-0.7707	0.0003
57	La	(M_3-M_2)	1.123	-	1.204	9.4253	-1.6200	-3.5397	2.0000	0.0142
57	La	(M_2-M_1)	1.204	-	1.361	9.5593	-2.2921	-0.7937	0.7621	0.0038
57	La	(M_1-L_3)	1.361	-	5.483	9.6959	-2.6205	0.0626	-0.0127	0.0145
57	La	(L_3-L_2)	5.483	-	5.891	3.9366	1.8751	1.7707	-1.1841	0.0504
57	La	(L_2-L_1)	5.891	-	6.266	4.0727	2.2066	1.2597	-0.9664	0.0377
57	La	(L_1-Z)	6.266	-	15.000	7.3851	3.0259	-2.6009	0.3879	0.0888
57	La	(Z-K)	15.000	-	38.925	12.2515	-3.3149	0.1553	-0.0103	0.0182
57	La	(K-E)	38.925	-	150.000	10.5652	-0.5177	-0.5764	0.0500	0.0496
58	Ce	(A-M_3)	1.000	-	1.185	9.3738	-2.5259	-0.2547	0.9791	0.0009
58	Ce	(M_3-M_2)	1.185	-	1.273	9.4866	-1.7818	-2.4306	1.5002	0.0080
58	Ce	(M_2-M_1)	1.273	-	1.435	9.5321	-1.6093	-2.3182	1.7180	0.0116
58	Ce	(M_1-L_3)	1.435	-	5.723	9.7436	-2.5453	-0.0133	0.0086	0.0041
58	Ce	(L_3-L_2)	5.723	-	6.164	5.5252	1.1567	0.8986	-0.7376	0.0454
58	Ce	(L_2-L_1)	6.164	-	6.549	4.5437	1.6777	1.2789	-0.8862	0.0272
58	Ce	(L_2-Z)	6.549	-	20.000	12.7591	-4.1441	0.5811	-0.0763	0.0548
58	Ce	(Z-K)	20.000	-	40.444	10.3546	-1.5028	-0.4002	0.0465	0.0241
58	Ce	(K-E)	40.444	-	150.000	14.0724	-2.6825	-0.1224	0.0181	0.0802
59	Pr	(A-M_3)	1.000	-	1.242	9.4554	-2.5473	-0.3936	1.1832	0.0027
59	Pr	(M_3-M_2)	1.242	-	1.337	9.5537	-1.8634	-1.4255	-0.0866	0.0090
59	Pr	(M_2-M_1)	1.337	-	1.511	9.6696	-2.1461	-0.9043	0.6072	0.0157
59	Pr	(M_1-L_3)	1.511	-	5.964	9.8217	-2.5508	-0.0222	0.0108	0.0056
59	Pr	(L_3-L_2)	5.964	-	6.440	3.8452	2.0842	1.3231	-0.9616	0.0424
59	Pr	(L_2-L_1)	6.440	-	6.835	4.1439	2.2146	0.9625	-0.8006	0.0282

Z	region	energy interval [keV]			x_0	x_1	x_2	x_3	RFQ
59	Pr	(L$_1$-Z)	6.835	- 20.000	9.7092	-0.2407	-1.0302	0.1424	0.0949
59	Pr	(Z-K)	20.000	- 41.991	11.0929	-2.2057	-0.1618	0.0199	0.0084
59	Pr	(K-E)	41.991	- 150.000	14.2005	-2.7592	-0.0977	0.0156	0.0775
60	Nd	(A-M$_3$)	1.000	- 1.297	9.4871	-2.4752	-0.7717	1.9691	0.0076
60	Nd	(M$_3$-M$_2$)	1.297	- 1.403	9.4802	-1.4780	-0.9006	-1.8336	0.0175
60	Nd	(M$_2$-M$_1$)	1.403	- 1.575	9.7399	-2.1427	-1.5789	1.7697	0.0025
60	Nd	(M$_1$-L$_3$)	1.575	- 6.208	9.8465	-2.5065	-0.0492	0.0182	0.0122
60	Nd	(L$_3$-L$_2$)	6.208	- 6.722	4.0361	2.1000	1.0099	-0.8185	0.0502
60	Nd	(L$_2$-L$_1$)	6.722	- 7.126	4.4545	2.0763	0.6939	-0.6592	0.0237
60	Nd	(L$_1$-Z)	7.126	- 20.000	11.8737	-2.8396	0.0256	0.0004	0.0034
60	Nd	(Z-K)	20.000	- 43.569	12.3547	-3.1421	0.0700	0.0014	0.0235
60	Nd	(K-E)	43.569	- 150.000	14.5998	-3.0371	-0.0294	0.0102	0.0765
61	Pm	(A-M$_5$)	1.000	- 1.027	7.9456	-2.5779	-0.8929	1.4862	0.0006
61	Pm	(M$_5$-M$_4$)	1.027	- 1.052	9.1121	-1.1433	-19.1893	1.9954	0.0122
61	Pm	(M$_4$-M$_3$)	1.052	- 1.357	9.5374	-2.4090	-0.9640	1.7894	0.0064
61	Pm	(M$_3$-M$_2$)	1.357	- 1.471	9.4237	-0.9347	-1.7579	-1.1562	0.0214
61	Pm	(M$_2$-M$_1$)	1.471	- 1.650	9.6629	-1.9700	-0.3162	-0.4551	0.0106
61	Pm	(M$_1$-L$_3$)	1.650	- 6.459	9.8827	-2.4608	-0.0943	0.0308	0.0162
61	Pm	(L$_3$-L$_2$)	6.459	- 7.013	3.9098	2.2622	0.8218	-0.7399	0.0513
61	Pm	(L$_2$-L$_1$)	7.013	- 7.428	5.1210	2.1069	0.0720	-0.4328	0.0183
61	Pm	(L$_1$Z)	7.428	- 20.000	12.3990	-3.4150	0.2513	-0.0285	0.0227
61	Pm	(Z-K)	20.000	- 45.185	9.8314	-0.9552	-0.5445	0.0585	0.0207
61	Pm	(K-E)	45.185	- 150.000	9.0332	0.8800	-0.9336	0.0792	0.0071
62	Sm	(A-M$_5$)	1.000	- 1.080	7.9890	-2.5513	0.1692	-1.4509	0.0001
62	Sm	(M$_5$-M$_4$)	1.080	- 1.106	9.1157	-1.0361	-8.8759	-0.2525	0.0053
62	Sm	(M$_4$-M$_3$)	1.106	- 1.420	9.5736	-2.3346	-1.1343	1.6798	0.0047
62	Sm	(M$_3$-M$_2$)	1.420	- 1.541	9.3232	-0.4697	-1.7818	-1.5837	0.0113
62	Sm	(M$_2$-M$_1$)	1.541	- 1.723	9.1099	0.3375	-2.1798	-1.1442	0.0498
62	Sm	(M$_1$-L$_3$)	1.723	- 6.716	9.9472	-2.5252	-0.0326	0.0136	0.0073
62	Sm	(L$_3$-L$_2$)	6.716	- 7.312	4.1253	1.4228	1.4286	-0.8520	0.0523
62	Sm	(L$_2$-L$_1$)	7.312	- 7.737	4.7163	1.4542	1.0376	-0.6955	0.0206
62	Sm	(L$_1$-Z)	7.737	- 20.000	13.4511	-4.6790	0.7756	-0.1002	0.0445
62	Sm	(Z-K)	20.000	- 46.835	11.7341	-2.5341	-0.1006	0.0171	0.0093
62	Sm	(K-E)	46.835	- 150.000	14.5273	-2.8798	-0.0741	0.0140	0.0374
63	Eu	(A-M$_5$)	1.000	- 1.131	8.0839	-2.5651	0.1530	-0.8281	0.0002
63	Eu	(M$_5$-M$_4$)	1.131	- 1.161	9.1440	-0.8965	-6.9778	1.8816	0.0047
63	Eu	(M$_4$-M$_3$)	1.161	- 1.481	9.6221	-2.2148	-1.3673	1.6716	0.0053
63	Eu	(M$_3$-M$_1$)	1.481	- 1.614	9.0206	0.8131	-2.6472	-1.8309	0.0427
63	Eu	(M$_2$-M$_1$)	1.164	- 1.800	9.0117	0.3294	-1.1522	-1.9549	0.0566
63	Eu	(M$_1$-L$_3$)	1.800	- 6.977	10.2478	-3.1928	0.5164	-0.1278	0.0701
63	Eu	(L$_3$-L$_2$)	6.977	- 7.617	4.4416	1.7009	0.7827	-0.6308	0.0510
63	Eu	(L$_2$-L$_1$)	7.617	- 8.052	5.4785	1.4737	0.3263	-0.4353	0.0096
63	Eu	(L$_1$-Z)	8.052	- 20.000	11.6638	-2.4501	-0.1158	0.0172	0.0009
63	Eu	(Z-K)	20.000	- 48.519	11.2837	-2.2085	-0.1612	0.0196	0.0068

Z	region	energy interval [keV]			x_0	x_1	x_2	x_3	RFQ
63	Eu (K-E)	48.519	-	150.000	15.4859	-3.4938	0.0638	0.0037	0.0355
64	Gd (A-M$_5$)	1.000	-	1.185	8.1194	-2.5972	0.4712	-1.8537	0.0010
64	Gd (M$_5$-M$_4$)	1.185	-	1.217	9.0485	-0.0817	-6.3660	-1.6010	0.0057
64	Gd (M$_4$-M$_3$)	1.217	-	1.544	9.6304	-2.0529	-1.5667	1.5084	0.0111
64	Gd (M$_3$-M$_2$)	1.544	-	1.688	9.2077	-0.1062	-1.6031	-1.3572	0.0314
64	Gd (M$_2$-M$_1$)	1.688	-	1.881	9.1307	0.4928	-3.1382	0.5763	0.0279
64	Gd (M$_1$-L$_3$)	1.881	-	7.243	10.2597	-3.1106	0.4323	-0.1038	0.0579
64	Gd (L$_3$-L$_2$)	7.243	-	7.930	3.7662	1.5979	1.2934	-0.7726	0.0621
64	Gd (L$_2$-L$_1$)	7.930	-	8.376	4.6610	1.4038	0.8733	-0.5877	0.0310
64	Gd (L$_1$-Z)	8.376	-	30.000	12.2557	-3.0499	0.0931	-0.0066	0.0158
64	Gd (Z-K)	30.000	-	50.239	11.1809	-1.9778	-0.2598	0.0318	0.0156
64	Gd (K-E)	50.239	-	150.000	13.4555	-2.3399	-0.1493	0.0161	0.0462
65	Tb (A-M$_5$)	1.000	-	1.241	8.2012	-2.5064	-0.6517	1.9990	0.0022
65	Tb (M$_5$-M$_4$)	1.241	-	1.275	8.9511	0.5583	-5.9453	-1.8341	0.0059
65	Tb (M$_4$-M$_3$)	1.275	-	1.611	9.6340	-1.7657	-1.9755	1.5488	0.0111
65	Tb (M$_3$-M$_2$)	1.611	-	1.768	9.8144	-2.0677	-0.8421	0.5000	0.0005
65	Tb (M$_2$-M$_1$)	1.7680	-	1.968	8.8869	0.3723	-0.8104	-1.5861	0.0483
65	Tb (M$_1$-L$_3$)	1.968	-	7.514	10.0056	-2.2986	-0.2163	0.0574	0.0221
65	Tb (L$_3$-L$_2$)	7.514	-	8.252	3.6506	2.1018	0.8433	-0.6530	0.0337
65	Tb (L$_2$-L$_1$)	8.252	-	8.708	5.6772	1.0655	0.4668	-0.4217	0.0072
65	Tb (L$_1$-Z)	8.708	-	30.000	12.5733	-3.3553	0.2091	-0.0210	0.0071
65	Tb (Z-K)	30.000	-	51.996	6.2395	2.1216	-1.3801	0.1338	0.0059
65	Tb (K-E)	51.996	-	150.000	15.4617	-3.3636	0.0228	0.0075	0.0218
66	Dy (A-M$_5$)	1.000	-	1.295	8.2276	-2.0896	-0.3317	0.8606	0.0016
66	Dy (M$_5$-M$_4$)	1.295	-	1.333	8.8333	1.1469	-5.6479	-1.9994	0.0068
66	Dy (M$_4$-M$_3$)	1.333	-	1.676	9.7548	-2.3080	-0.6523	0.5496	0.0012
66	Dy (M$_3$-M$_2$)	1.676	-	1.842	8.8893	0.8264	-1.8052	-1.4236	0.0418
66	Dy (M$_2$-M$_1$)	1.842	-	2.047	8.4057	1.5546	-1.1074	-1.9903	0.0595
66	Dy (M$_1$-L$_3$)	2.047	-	7.790	9.9387	-2.0059	-0.4408	0.1090	0.0622
66	Dy (L$_3$-L$_2$)	7.790	-	8.581	3.3237	1.9619	1.0809	-0.6946	0.0295
66	Dy (L$_2$-L$_1$)	8.581	-	9.046	5.5669	1.4584	0.1338	-0.3370	0.0064
66	Dy (L$_1$-Z)	9.046	-	30.000	13.9082	-4.7605	0.7091	-0.0796	0.0254
66	Dy (Z-K)	30.000	-	53.788	13.9449	-4.1637	0.3347	-0.0220	0.0160
66	Dy (K-E)	53.788	-	150.000	14.6521	-2.7983	-0.1023	0.0166	0.0139
67	Ho (A-M$_5$)	1.000	-	1.351	8.2890	-2.2810	-0.7035	1.5671	0.0037
67	Ho (M$_5$-M$_4$)	1.351	-	1.392	8.9070	0.9942	-6.4900	1.5211	0.0046
67	Ho (M$_4$-M$_3$)	1.392	-	1.741	9.7230	-1.6583	-2.3720	1.9727	0.0094
67	Ho (M$_3$-M$_2$)	1.741	-	1.923	8.9579	0.4733	-1.3468	-1.2943	0.0376
67	Ho (M$_2$-M$_1$)	1.923	-	2.128	8.9211	0.6522	-1.9830	-0.2737	0.0345
67	Ho (M$_1$-L$_3$)	2.128	-	8.071	10.2257	-2.6222	0.0311	-0.0029	0.0070
67	Ho (L$_3$-L$_2$)	8.071	-	8.918	3.1045	1.5514	1.5317	-0.7887	0.0358
67	Ho (L$_2$- L$_1$)	8.918	-	9.394	5.3639	1.3066	0.3698	-0.3898	0.0061
67	Ho (L$_1$-Z)	9.394	-	30.000	8.9539	0.6763	-1.2373	0.1499	0.0444
67	Ho (Z-K)	30.000	-	55.618	8.4957	0.3220	-0.8845	0.0884	0.0018

Z	region	energy interval [keV]			x_0	x_1	x_2	x_3	RFQ
67	Ho	(K-E)	55.618	- 150.000	14.1838	-2.5410	-0.1412	0.0182	0.0166
68	Er	(A-M_5)	1.000	- 1.409	8.3371	-2.3386	-0.3364	0.6463	0.0028
68	Er	(M_5-M_4)	1.409	- 1.453	8.9223	0.7120	-4.7027	0.0045	0.0049
68	Er	(M_4-M_3)	1.453	- 1.812	9.8312	-2.1944	-0.6519	0.3152	0.0073
68	Er	(M_3-M_2)	1.812	- 2.006	8.9778	0.0335	-0.1572	-1.9056	0.0377
68	Er	(M_2-M_1)	2.006	- 2.207	8.5170	1.3561	-1.5472	-0.9947	0.0382
68	Er	(M_3-L_3)	2.207	- 8.358	10.4162	-2.9019	0.2122	-0.0427	0.0201
68	Er	(L_3-L_2)	8.358	- 9.264	4.6503	1.1506	0.8233	-0.5241	0.0305
68	Er	(L_2-L_1)	9.264	- 9.751	4.9585	1.1055	0.7027	-0.4585	0.0070
68	Er	(L_1-Z)	9.751	- 30.000	12.9862	-3.6908	0.3340	-0.0361	0.0058
68	Er	(Z-K)	30.000	- 57.486	11.7352	-2.3923	-0.1198	0.0168	0.0101
68	Er	(K-E)	57.486	- 150.000	14.8398	-2.9161	-0.0651	0.0130	0.0035
69	Tm	(A-M_5)	1.000	- 1.468	8.3940	-2.4715	-0.0899	0.1571	0.0009
69	Tm	(M_5-M_4)	1.468	- 1.515	8.6800	2.1825	-7.3185	1.9692	0.0053
69	Tm	(M_4-M_3)	1.515	- 1.885	9.6344	-0.8797	-3.0653	1.8233	0.0101
69	Tm	(M_3-M_2)	1.885	- 1.090	8.5982	1.0734	-0.8795	-1.7430	0.0397
69	Tm	(M_2-M_1)	2.090	- 2.307	7.8412	2.3803	-1.0392	-1.7940	0.0572
69	Tm	(M_1-L_3)	2.307	- 8.648	9.8155	-1.4656	-0.7853	0.1777	0.0845
69	Tm	(L_3-L_2)	8.648	- 9.617	3.8583	1.3547	1.0722	-0.6000	0.0364
69	Tm	(L_2-L_1)	9.617	- 10.120	6.1203	1.0690	0.0262	-0.2482	0.0060
69	Tm	(L_1-Z)	10.120	- 30.000	11.9036	-2.4258	-0.1293	0.0195	0.0048
69	Tm	(Z-K)	30.000	- 59.390	11.4639	-2.0542	-0.2306	0.0283	0.0079
69	Tm	(K-E)	59.390	- 150.000	12.7282	-1.3950	-0.4204	0.0405	0.0080
70	Yb	(A-M_5)	1.000	- 1.528	8.4398	-2.6976	0.1815	-0.2992	0.0019
70	Yb	(M_5-M_4)	1.528	- 1.576	8.7215	1.6965	-6.0510	1.6176	0.0042
70	Yb	(M_4-M_3)	1.576	- 1.950	9.7050	-1.1229	-2.5820	1.4993	0.0041
70	Yb	(M_3-M_2)	1.950	- 2.173	8.3881	1.4467	-0.9697	-1.6628	0.0443
70	Yb	(M_2-M_1)	2.173	- 2.398	7.5542	2.8698	-1.3012	-1.6121	0.0571
70	Yb	(M_1-L_3)	2.398	- 8.944	9.8491	-1.4985	-0.7432	0.1648	0.0795
70	Yb	(L_3-L_2)	8.944	- 9.978	3.7058	1.4098	1.0454	-0.5819	0.0382
70	Yb	(L_2-L_1)	9.978	- 10.490	4.7482	1.2419	0.6036	-0.4163	0.0053
70	Yb	(L_1-Z)	10.490	- 40.000	12.3349	-2.8377	0.0129	0.0031	0.0071
70	Yb	(Z-K)	40.000	- 61.332	13.2287	-4.0627	0.4569	-0.0451	0.0449
70	Yb	(K-E)	61.332	- 150.000	13.7249	-2.6206	-0.0294	0.0035	0.0594
71	Lu	(A-M_5)	1.000	- 1.589	8.4881	-2.7298	0.3525	-0.4864	0.0059
71	Lu	(M_5-M_4)	1.589	- 1.639	8.8323	0.4260	-2.2973	-1.0904	0.0041
71	Lu	(M_4-M_3)	1.639	- 2.024	9.7562	-1.2409	-2.1829	1.1799	0.0111
71	Lu	(M_3-M_2)	2.024	- 2.264	8.2814	1.5665	-0.8791	-1.6121	0.0637
71	Lu	(M_2-M_1)	2.264	- 2.491	7.6775	2.0422	-0.1824	-1.9012	0.0519
71	Lu	(M_1-L_3)	2.491	- 9.244	10.0119	-1.7543	-0.5628	0.1236	0.0610
71	Lu	(L_3-L_2)	9.244	- 10.350	3.0380	1.1678	1.5828	-0.7107	0.0448
71	Lu	(L_2-L_1)	10.350	- 10.870	5.3106	1.2262	0.2752	-0.3136	0.0052
71	Lu	(L_1-Z)	10.870	- 40.000	12.6201	-3.0974	0.1049	-0.0075	0.0138
71	Lu	(Z-K)	40.000	- 63.314	14.1783	-3.7707	0.1303	0.0040	0.0146

Z		region	energy interval [keV]			x_0	x_1	x_2	x_3	RFQ
71	Lu	(K-E)	63.314	-	150.000	8.4190	1.5286	-1.0683	0.0881	0.0007
72	Hf	(A-M_5)	1.000	-	1.662	8.5213	-2.5843	-0.4152	0.5078	0.0082
72	Hf	(M_5-M_4)	1.662	-	1.716	8.4272	1.9602	-3.7797	-0.7100	0.0056
72	Hf	(M_4-M_3)	1.716	-	2.108	9.4864	-0.0434	-3.5986	1.6631	0.0135
72	Hf	(M_3-M_2)	2.108	-	2.365	9.8785	-1.1825	-1.6893	0.6592	0.0015
72	Hf	(M_2-M_1)	2.365	-	2.601	8.0269	1.2064	0.1113	-1.6122	0.0430
72	Hf	(M_1-L_3)	2.601	-	9.561	9.9367	-1.5376	-0.6916	0.1474	0.0643
72	Hf	(L_3-L_2)	9.561	-	10.740	4.2933	1.2074	0.7685	-0.4639	0.0361
72	Hf	(L_2-L_1)	10.740	-	11.270	5.0249	0.8478	0.6998	-0.4008	0.0056
72	Hf	(L_1-Z)	11.270	-	40.000	12.5993	-3.0169	0.0735	-0.0039	0.0047
72	Hf	(Z-K)	40.000	-	65.351	11.8020	-2.1440	-0.2273	0.0294	0.0139
72	Hf	(K-E)	65.351	-	150.000	15.5544	-3.3483	0.0364	0.0054	0.0152
73	Ta	(A-M_5)	1.000	-	1.735	8.5674	-2.6619	-0.1084	0.1571	0.0008
73	Ta	(M_5-M_4)	1.735	-	1.793	8.5264	2.0445	-5.0430	1.1161	0.0038
73	Ta	(M_4-M_3)	1.793	-	2.194	9.7710	-0.9322	-2.4311	1.1562	0.0055
73	Ta	(M_3-M_2)	2.194	-	2.496	7.6417	2.4174	-0.5288	-1.9378	0.0843
73	Ta	(M_2-M_1)	2.469	-	2.708	6.6767	3.5525	-0.4876	-1.9343	0.0569
73	Ta	(M_1-L_3)	2.708	-	9.881	10.3569	-2.2642	-0.2385	0.0543	0.0180
73	Ta	(L_3-L_2)	9.881	-	11.140	3.5809	1.3024	0.9870	-0.5195	0.0702
73	Ta	(L_2-L_1)	11.140	-	11.680	5.3426	1.3754	0.0276	-0.2359	0.0084
73	Ta	(L_1-Z)	11.680	-	40.000	13.0324	-3.4986	0.2425	-0.0223	0.0319
73	Ta	(Z-K)	40.000	-	67.417	6.8629	1.4261	-1.0822	0.0979	0.0146
73	Ta	(K-E)	67.417	-	150.000	13.5053	-2.2047	-0.1743	0.0185	0.0312
74	W	(A-M_5)	1.000	-	1.809	8.5951	-2.5999	-0.0269	0.0411	0.0002
74	W	(M_5-M_4)	1.809	-	1.872	8.3348	1.9062	-3.0862	-0.7454	0.0053
74	W	(M_4-M_3)	1.872	-	2.281	9.4965	-0.1459	-2.9086	1.1123	0.0062
74	W	(M_3-M_2)	2.281	-	2.575	7.5097	2.5824	-0.7720	-1.6262	0.0767
74	W	(M_2-M_1)	2.575	-	2.820	6.9223	2.6003	0.3149	-1.9823	0.0470
74	W	(M_1-Z)	2.820	-	5.000	10.2358	-1.9098	-0.5082	0.1240	0.0047
74	W	(Z-L_3)	5.000	-	10.210	10.7006	-2.8514	0.1242	-0.0167	0.0117
74	W	(L_3-L_2)	10.210	-	11.540	5.1408	1.1859	0.2574	-0.3013	0.0580
74	W	(L_2-L_1)	11.540	-	12.100	4.5680	1.2794	0.4989	-0.3537	0.0097
74	W	(L_1-Z)	12.100	-	40.000	13.3350	-3.6541	0.2798	-0.0261	0.0299
74	W	(Z-K)	40.000	-	69.525	9.5662	-0.5331	-0.5958	0.0570	0.0127
74	W	(K-E)	69.525	-	150.000	14.8098	-3.2874	0.1169	-0.0058	0.0010
75	Re	(A-M_5)	1.000	-	1.883	8.6360	-2.6185	0.0768	-0.0639	0.0027
75	Re	(M_5-M_4)	1.883	-	1.949	8.0899	2.1546	-2.5205	-1.1292	0.0062
75	Re	(M_4-M_3)	1.949	-	2.367	9.4851	-0.1904	-2.5576	0.8547	0.0143
75	Re	(M_3-M_2)	2.367	-	2.682	10.2818	-2.2878	-0.3224	0.1218	0.0003
75	Re	(M_2-M_1)	2.682	-	2.932	5.8305	4.3263	-0.2854	-1.9830	0.0567
75	Re	(M_1-Z)	2.932	-	5.000	9.9206	-1.1033	-1.1145	0.2750	0.0032
75	Re	(Z-L_3)	5.000	-	10.540	10.8777	-3.0641	0.2318	-0.0340	0.0016
75	Re	(L_3-L_2)	10.540	-	11.960	3.6130	1.6127	0.6357	-0.4207	0.0857
75	Re	(L_2-L_1)	11.960	-	12.530	5.2877	1.1646	0.1786	-0.2515	0.0087

Z	region	energy interval [keV]			x_0	x_1	x_2	x_3	RFQ
75	Re (L_1-Z)	12.530	-	40.000	13.7825	-4.0824	0.4223	-0.0414	0.0510
75	Re (Z-K)	40.000	-	71.677	11.3209	-1.7495	-0.3110	0.0351	0.0198
75	Re (K-E)	71.677	-	150.000	10.9239	-0.0358	-0.7232	0.0628	0.0106
76	Os (A-M_5)	1.000	-	1.960	8.6700	-2.5858	-0.0045	0.0177	0.0006
76	Os (M_5-M_4)	1.960	-	2.031	7.9680	1.9268	-1.2519	-1.9996	0.0118
76	Os (M_4-M_3)	2.031	-	2.457	9.2255	0.3421	-2.6086	0.6568	0.0413
76	Os (M_3-M_2)	2.457	-	2.792	7.4311	1.9533	0.3656	-1.8192	0.0906
76	Os (M_2-M_1)	2.792	-	3.049	6.3238	3.6196	-0.6323	-1.4114	0.0409
76	Os (M_1-L_3)	3.049	-	10.870	11.4828	-4.2004	0.9403	-0.1740	0.0836
76	Os (L_3-L_2)	10.870	-	12.390	4.0487	0.9898	0.8705	-0.4403	0.0901
76	Os (L_2-L_1)	12.390	-	12.970	5.4816	1.1543	0.0633	-0.4146	0.0073
76	Os (L_1-Z)	12.970	-	40.000	13.0615	-3.3530	0.1865	-0.0160	0.0212
76	Os (Z-K)	40.000	-	73.871	12.1420	-2.3163	-0.1757	0.0243	0.0251
76	Os (K-E)	73.871	-	150.000	13.1369	-1.5944	-0.3558	0.0342	0.0077
77	Ir (A-M_5)	1.000	-	2.040	8.7073	-2.4256	-0.4360	0.4002	0.0011
77	Ir (M_5-M_4)	2.040	-	2.116	7.4683	3.0388	-1.8972	-1.7542	0.0079
77	Ir (M_4-M_3)	2.116	-	2.551	9.5520	-0.1421	-2.7316	1.0059	0.0067
77	Ir (M_3-M_2)	2.551	-	2.909	7.1260	2.1426	0.5753	-1.9626	0.0941
77	Ir (M_2-M_1)	2.909	-	3.174	6.5227	3.3467	-0.8774	-1.0787	0.0383
77	Ir (M_1-L_3)	3.174	-	11.220	10.6967	-2.6543	0.0144	0.0042	0.0097
77	Ir (L_3-L_2)	11.220	-	12.820	3.8819	0.9864	0.9201	-0.4461	0.0871
77	Ir (L_2-L_1)	12.820	-	13.420	5.2535	1.1189	0.1860	-0.2423	0.0072
77	Ir (L_1-Z)	13.420	-	40.000	9.8503	-0.2058	-0.8201	0.0904	0.0440
77	Ir (Z-K)	40.000	-	76.111	9.0606	-0.0513	-0.7218	0.0681	0.0274
77	Ir (K-E)	76.111	-	150.000	14.1877	-2.1002	-0.2791	0.0309	0.0178
78	Pt (A-M_5)	1.000	-	2.122	8.7428	-2.4546	-0.2198	0.2033	0.0079
78	Pt (M_5-M_4)	2.122	-	2.202	7.5064	2.5570	-1.0577	-1.9927	0.0172
78	Pt (M_4-M_3)	2.202	-	2.645	9.6857	-0.3393	-2.5819	0.9769	0.0027
78	Pt (M_3-M_2)	2.645	-	3.027	9.9743	-0.9938	-1.5567	0.4979	0.0007
78	Pt (M_2-M_1)	3.027	-	3.296	5.8218	3.6263	0.0646	-1.6094	0.0884
78	Pt (M_1-L_3)	3.296	-	11.560	10.9199	-2.9406	0.1688	-0.0253	0.0459
78	Pt (L_3-L_2)	11.560	-	13.270	7.7858	-1.2287	0.7749	-0.2813	0.0570
78	Pt (L_2-L_1)	13.270	-	13.880	5.6050	0.8743	0.1828	-0.2215	0.0070
78	Pt (L_1-Z)	13.880	-	40.000	12.3408	-2.5866	-0.0561	0.0093	0.0088
78	Pt (Z-K)	40.000	-	78.395	10.1108	-0.7947	-0.5409	0.0534	0.0222
78	Pt (K-E)	78.395	-	150.000	18.6026	-4.2330	0.0340	0.0184	0.0356
79	Au (A-M_5)	1.000	-	2.206	8.7627	-2.4581	-0.0132	0.0240	0.0001
79	Au (M_5-M_4)	2.206	-	2.291	7.3864	2.3266	-0.5924	-1.9979	0.0173
79	Au (M_4-M_3)	2.291	-	2.743	8.1960	3.6777	-6.0933	1.9549	0.0504
79	Au (M_3-M_2)	2.743	-	3.148	9.8010	-0.6735	-1.7473	0.5383	0.0006
79	Au (M_2-M_1)	3.148	-	3.425	6.0567	2.5110	1.0666	-1.7936	0.0843
79	Au (M_1-L_3)	3.425	-	11.920	10.7303	-2.6474	0.0313	0.0002	0.0101
79	Au (L_3-L_2)	11.920	-	13.730	7.9872	-1.3658	0.7495	-0.2616	0.0555
79	Au (L_2-L_1)	13.730	-	14.350	5.5844	1.0368	0.0316	-0.1856	0.0062

Z		region	energy interval [keV]			x_0	x_1	x_2	x_3	RFQ
79	Au	$(L_1\text{-}Z)$	14.350	-	40.000	13.6639	-3.8695	0.3553	-0.0336	0.0318
79	Au	$(Z\text{-}K)$	40.000	-	80.725	13.2048	-3.0630	0.0168	0.0081	0.0166
79	Au	$(K\text{-}E)$	80.725	-	150.000	14.4173	-3.1183	0.1017	-0.0070	0.0018
80	Hg	$(A\text{-}M_5)$	1.000	-	2.295	8.8180	-2.4450	-0.0449	0.0413	0.0004
80	Hg	$(M_5\text{-}M_4)$	2.295	-	2.385	7.5778	1.7129	-0.2039	-1.8122	0.0159
80	Hg	$(M_4\text{-}M_3)$	2.385	-	2.847	8.8752	2.1401	-4.9395	1.7139	0.0042
80	Hg	$(M_3\text{-}M_2)$	2.847	-	3.279	9.0320	-0.5169	-0.0463	-0.5190	0.0774
80	Hg	$(M_2\text{-}M_1)$	3.279	-	3.562	5.1647	3.3511	1.2501	-1.9880	0.0926
80	Hg	$(M_1\text{-}L_3)$	3.562	-	12.280	10.1549	-1.5785	-0.5602	0.1041	0.0826
80	Hg	$(L_3\text{-}L_2)$	12.280	-	14.210	4.3910	0.9309	0.5922	-0.3331	0.0940
80	Hg	$(L_2\text{-}L_1)$	14.210	-	14.840	5.7522	0.9882	-0.0241	-0.1641	0.0056
80	Hg	$(L_1\text{-}Z)$	14.840	-	40.000	13.3851	-3.5353	0.2448	-0.0218	0.0135
80	Hg	$(Z\text{-}K)$	40.000	-	83.103	14.5749	-4.0551	0.2655	-0.0129	0.0158
80	Hg	$(K\text{-}E)$	83.103	-	150.000	10.1597	0.2525	-0.7143	0.0567	0.0292
81	Tl	$(A\text{-}M_5)$	1.000	-	2.389	8.8514	-2.4356	0.0166	-0.0041	0.0001
81	Tl	$(M_5\text{-}M_4)$	2.389	-	2.485	8.1298	1.3354	-1.7138	-0.3031	0.0089
81	Tl	$(M_4\text{-}M_3)$	2.485	-	2.957	9.2124	0.7781	-3.1143	1.0019	0.0120
81	Tl	$(M_3\text{-}M_2)$	2.957	-	3.416	9.0353	1.4133	-3.4387	0.9872	0.0031
81	Tl	$(M_2\text{-}M_1)$	3.416	-	3.704	5.3967	2.8085	1.3429	-1.8179	0.0812
81	Tl	$(M_1\text{-}L_3)$	3.704	-	12.660	10.7305	-2.5116	-0.0447	0.0120	0.0096
81	Tl	$(L_3\text{-}L_2)$	12.660	-	14.700	7.9422	-1.3945	0.7661	-0.2567	0.0618
81	Tl	$(L_2\text{-}L_1)$	14.700	-	15.350	5.1800	1.0977	0.1062	-0.1969	0.0012
81	Tl	$(L_1\text{-}Z)$	15.350	-	50.000	13.0091	-3.1289	0.1083	-0.0065	0.0229
81	Tl	$(Z\text{-}K)$	50.000	-	85.531	10.8688	-1.1900	-0.4625	0.0485	0.0045
81	Tl	$(K\text{-}E)$	85.531	-	150.000	14.1531	-2.0711	-0.2669	0.0285	0.0401
82	Pb	$(A\text{-}M_5)$	1.000	-	2.484	8.8831	-2.4199	-0.0103	0.0147	0.0002
82	Pb	$(M_5\text{-}M_4)$	2.484	-	2.586	6.5124	3.5770	-1.0610	-1.6053	0.0199
82	Pb	$(M_4\text{-}M_3)$	2.586	-	3.066	8.5834	1.4163	-2.5067	0.3662	0.0692
82	Pb	$(M_3\text{-}M_2)$	3.066	-	3.554	10.1077	-1.6121	-0.5666	0.0946	0.0110
82	Pb	$(M_2\text{-}M_1)$	3.554	-	3.851	4.4274	3.6184	1.5364	-1.9864	0.0896
82	Pb	$(M_1\text{-}L_3)$	3.851	-	13.040	10.8552	-2.6676	0.0428	-0.0033	0.0182
82	Pb	$(L_3\text{-}L_2)$	13.040	-	15.200	4.2254	0.9844	0.5749	-0.3195	0.0863
82	Pb	$(L_2\text{-}L_1)$	15.200	-	15.860	4.7628	1.1163	0.2552	-0.2300	0.0066
82	Pb	$(L_1\text{-}Z)$	15.860	-	50.000	16.0619	-5.1840	0.8946	-0.0846	0.0613
82	Pb	$(Z\text{-}K)$	50.000	-	88.005	11.3867	-1.9259	-0.1904	0.0187	0.0529
82	Pb	$(K\text{-}E)$	88.005	-	150.000	17.0377	-5.0123	0.5601	-0.0421	0.0266
83	Bi	$(A\text{-}M_5)$	1.000	-	2.580	8.9370	-2.3919	-0.0269	0.0261	0.0016
83	Bi	$(M_5\text{-}M_4)$	2.580	-	2.688	6.8200	3.0449	-1.2698	-1.1118	0.0168
83	Bi	$(M_4\text{-}M_3)$	2.688	-	3.177	9.7299	-0.5657	-1.8310	0.5621	0.0042
83	Bi	$(M_3\text{-}M_2)$	3.177	-	3.696	10.5567	-2.3821	-0.1438	0.0385	0.0001
83	Bi	$(M_2\text{-}M_1)$	3.696	-	3.999	5.4331	2.1121	1.8302	-1.7673	0.0733
83	Bi	$(M_1\text{-}L_3)$	3.999	-	13.420	11.2679	-3.2649	0.3513	-0.0550	0.0561
83	Bi	$(L_3\text{-}L_2)$	13.420	-	15.710	6.8805	0.1985	-0.0250	-0.1251	0.0648
83	Bi	$(L_2\text{-}L_1)$	15.710	-	16.390	5.0070	0.8437	0.2997	-0.2226	0.0062

Z		region	energy interval [keV]			x_0	x_1	x_2	x_3	RFQ
83	Bi	(L$_1$-Z)	16.390	-	50.000	12.7042	-2.8675	0.0476	-0.0016	0.0172
83	Bi	(Z-K)	50.000	-	90.526	15.8718	-4.8038	0.4169	-0.0224	0.0036
83	Bi	(K-E)	90.526	-	150.000	12.6776	-1.4380	-0.3252	0.0273	0.0227
84	Po	(A-M$_5$)	1.000	-	2.683	8.9836	-2.3304	-0.0774	0.0571	0.0021
84	Po	(M$_5$-M$_4$)	2.683	-	2.798	6.4388	2.7821	0.1798	-1.8733	0.0050
84	Po	(M$_4$-M$_3$)	2.798	-	3.302	6.3298	2.6131	0.8639	-1.9215	0.0842
84	Po	(M$_3$-M$_2$)	3.302	-	3.854	18.0493	3.2895	-8.9704	0.4205	0.0904
84	Po	(M$_2$-M$_1$)	3.854	-	4.149	5.0547	2.7873	1.2276	-1.5240	0.0133
84	Po	(M$_1$-L$_3$)	4.149	-	13.810	11.5715	-3.6496	0.5337	-0.0831	0.0204
84	Po	(L$_3$-L$_2$)	13.810	-	16.240	3.2953	1.3728	0.5820	-0.3264	0.0602
84	Po	(L$_2$-L$_1$)	16.240	-	16.940	6.1859	0.8442	-0.1633	-0.1090	0.0019
84	Po	(L$_1$-Z)	16.940	-	50.000	12.6968	-2.8050	0.0243	0.0009	0.0097
84	Po	(Z-K)	50.000	-	93.105	11.6249	-1.7412	-0.3100	0.0348	0.0029
84	Po	(K-E)	93.105	-	150.000	17.5452	-4.3004	0.2309	-0.0080	0.0175
85	At	(A-N$_1$)	1.000	-	1.042	9.0052	-2.4319	0.0839	0.0536	0.0001
85	At	(N$_1$-M$_5$)	1.042	-	2.787	9.1031	-2.4720	0.0696	-0.0278	0.0010
85	At	(M$_5$-M$_4$)	2.787	-	2.909	6.1181	2.9948	0.3974	-1.9476	0.0053
85	At	(M$_4$-M$_3$)	2.909	-	3.426	5.4753	4.2409	-0.2804	-1.5574	0.0745
85	At	(M$_3$-M$_2$)	3.426	-	4.008	5.2814	2.4683	1.8618	-1.9232	0.0681
85	At	(M$_2$-M$_1$)	4.008	-	4.317	3.9462	3.6625	1.4370	-1.6991	0.0164
85	At	(M$_1$-L$_3$)	4.317	-	14.210	11.2186	-3.0176	0.2095	-0.0290	0.0088
85	At	(L$_3$-L$_2$)	14.210	-	16.780	3.1554	0.9114	0.9424	-0.3872	0.0667
85	At	(L$_2$-L$_1$)	16.780	-	17.490	4.9529	0.9565	0.2054	-0.1967	0.0026
85	At	(L$_1$-Z)	17.490	-	50.000	12.8751	-2.9089	0.0511	-0.0013	0.0043
85	At	(Z-K)	50.000	-	95.730	8.3958	0.5414	-0.8388	0.0756	0.0039
85	At	(K-E)	95.730	-	150.000	17.6643	-4.0904	0.1335	0.0025	0.0255
86	Rn	(A-N$_1$)	1.000	-	1.097	9.0242	-2.4372	-0.2767	2.0000	0.0002
86	Rn	(N$_1$-M$_5$)	1.097	-	2.892	9.1150	-2.4068	-0.0812	0.0501	0.0005
86	Rn	(M$_5$-M$_4$)	2.892	-	3.022	6.0427	2.6401	0.8134	-1.9719	0.0055
86	Rn	(M$_4$-M$_3$)	3.022	-	3.538	5.5959	3.3291	0.6270	-1.7487	0.0771
86	Rn	(M$_3$-M$_2$)	3.538	-	4.159	5.5988	2.0205	1.6971	-1.6879	0.0663
86	Rn	(M$_2$-M$_1$)	4.159	-	4.482	4.5574	3.2139	0.8626	-1.2919	0.0149
86	Rn	(M$_1$-L$_3$)	4.482	-	14.620	11.3462	-3.2083	0.2953	-0.0415	0.0115
86	Rn	(L$_3$-L$_2$)	14.620	-	17.340	2.7433	0.9855	0.9933	-0.3961	0.0598
86	Rn	(L$_2$-L$_1$)	17.340	-	18.050	5.5441	0.9124	-0.0160	-0.1396	0.0022
86	Rn	(L$_1$-Z)	18.050	-	50.000	12.5728	-2.6768	-0.0087	0.0037	0.0059
86	Rn	(Z-K)	50.000	-	98.404	8.9563	0.1030	-0.7255	0.0657	0.0089
86	Rn	(K-E)	98.404	-	150.000	17.9733	-4.1920	0.1337	0.0039	0.0306
87	Fr	(A-N$_1$)	1.000	-	1.153	9.0830	-2.4278	-0.4277	1.9492	0.0006
87	Fr	(N$_1$-M$_5$)	1.153	-	3.000	9.1709	-2.3793	-0.0925	0.0526	0.0019
87	Fr	(M$_5$-M$_4$)	3.000	-	3.136	6.0927	2.5681	0.5391	-1.6778	0.0056
87	Fr	(M$_4$-M$_3$)	3.136	-	3.663	5.4196	3.0245	1.2201	-1.9123	0.0775
87	Fr	(M$_3$-M$_2$)	3.663	-	4.327	4.9705	2.4495	1.8119	-1.7417	0.0824
87	Fr	(M$_2$-M$_1$)	4.327	-	4.652	3.9973	2.3753	2.5125	-1.8409	0.0168

Z	region	energy interval [keV]		x_0	x_1	x_2	x_3	RFQ
87	Fr (M_1-L_3)	4.652	- 15.030	10.8178	-2.3790	-0.1028	0.0216	0.0072
87	Fr (L_3-L_2)	15.030	- 17.910	3.2763	0.8609	0.8474	-0.3506	0.0715
87	Fr (L_2-L_1)	17.910	- 18.640	5.2366	1.1380	-0.0720	-0.1329	0.0022
87	Fr (L_1-Z)	18.640	- 50.000	13.2895	-3.2107	0.1280	-0.0076	0.0093
87	Fr (Z-K)	50.000	- 101.137	15.1037	-4.1785	0.2731	-0.0117	0.0049
87	Fr (K-E)	101.137	- 150.000	12.2880	-3.7350	0.6815	-0.0782	0.0773
88	Ra (A-N_2)	1.000	- 1.058	9.0632	-2.4649	0.1006	-1.0430	0.0000
88	Ra (N_2-N_1)	1.058	- 1.208	9.1561	-2.3977	-0.7471	1.9598	0.0005
88	Ra (N_1-M_5)	1.208	- 3.105	9.2308	-2.2921	-0.3513	0.1919	0.0120
88	Ra (M_5-M_4)	3.105	- 3.248	5.4458	2.9809	1.0240	-1.9631	0.0065
88	Ra (M_4-M_2)	3.248	- 3.792	4.9417	3.4502	1.1808	-1.8899	0.0791
88	Ra (M_3-M_2)	3.792	- 4.490	3.5257	3.7050	1.9703	-1.9566	0.0944
88	Ra (M_2-M_1)	4.490	- 4.822	3.4963	3.0460	2.0582	-1.6809	0.0161
88	Ra (M_1-L_3)	4.822	- 15.440	10.6674	-2.1265	-0.2163	0.0382	0.0023
88	Ra (L_3-L_2)	15.440	- 18.480	4.4475	0.7919	0.4184	-0.2405	0.0582
88	Ra (L_2-L_1)	18.480	- 19.240	5.8144	0.8504	-0.1101	-0.1080	0.0019
88	Ra (L_1-Z)	19.240	- 50.000	14.4841	-4.2441	0.4328	-0.0374	0.0053
88	Ra (Z-K)	50.000	- 103.922	7.0634	1.5549	-1.0813	0.0946	0.0078
88	Ra (K-E)	103.922	- 150.000	13.8481	-3.5601	0.4086	-0.0429	0.0317
89	Ac (A-N_2)	1.000	- 1.080	9.1675	-2.5324	-0.1460	1.2882	0.0001
89	Ac (N_2-N_1)	1.080	- 1.269	9.2557	-2.3762	-0.9460	1.9872	0.0008
89	Ac (N_1-M_5)	1.269	- 3.219	9.5836	-2.0851	-0.7356	0.3613	0.0215
89	Ac (M_5-M_4)	3.219	- 3.370	5.1124	3.2877	1.0103	-1.9440	0.0063
89	Ac (M_4-M_3)	3.370	- 3.909	3.5064	5.2927	0.7410	-1.9651	0.0771
89	Ac (M_3-M_2)	3.909	- 4.656	3.9707	3.7128	2.1539	-1.9858	0.0898
89	Ac (M_2-M_1)	4.656	- 5.002	4.5660	1.8007	2.1285	-1.4849	0.0112
89	Ac (M_1-L_3)	5.002	- 15.870	11.9028	-3.8585	0.6012	-0.0877	0.0296
89	Ac (L_3-L_2)	15.870	- 19.080	1.6025	1.2623	1.1298	-0.4231	0.0859
89	Ac (L_2-L_1)	19.080	- 19.840	4.3149	0.8091	0.4361	-0.2284	0.0024
89	Ac (L_1-Z)	19.840	- 50.000	12.6134	-2.6429	-0.0090	0.0028	0.0048
89	Ac (Z-K)	50.000	- 106.755	12.3829	-2.2208	-0.1827	0.0234	0.0064
89	Ac (K-E)	106.755	- 150.000	18.7491	-4.0299	-0.0260	0.0240	0.0372
90	Th (A-N_2)	1.000	- 1.168	9.1745	-2.5644	0.0412	-0.1713	0.0001
90	Th (N_2-N_1)	1.168	- 1.330	9.2535	-2.3967	-0.3168	-0.0718	0.0034
90	Th (N_1-M_5)	1.330	- 3.332	9.1820	-1.5537	-1.5583	0.7260	0.0493
90	Th (M_5-M_4)	3.332	- 3.491	5.0926	3.7532	-0.1316	-1.3248	0.0121
90	Th (M_4-M_3)	3.491	- 4.046	8.5703	1.9926	-3.4207	0.8590	0.0028
90	Th (M_3-M_2)	4.046	- 4.830	8.7286	-1.5768	1.4753	-0.8096	0.0742
90	Th (M_2-M_1)	4.830	- 5.182	4.7568	2.3052	1.0399	-1.0561	0.0181
90	Th (M_1-L_3)	5.182	- 16.300	8.8568	0.4149	-1.3780	0.2117	0.0559
90	Th (L_3-L_2)	16.300	- 19.690	3.3392	0.697	1.0533	-0.3628	0.0827
90	Th (L_2-L_1)	19.690	- 20.470	4.4448	0.7388	0.3801	-0.2077	0.0034
90	Th (L_1-Z)	20.470	- 60.000	14.4006	-4.1120	0.3821	-0.0316	0.0062
90	Th (Z-K)	60.000	- 109.651	14.032	-3.1320	-0.0287	0.0156	0.0076
90	Th (K-E)	109.671	- 150.000	3.2554	0.2486	0.1526	-0.0596	0.0775

Z		region	energy interval [keV]			x_0	x_1	x_2	x_3	RFQ
91	Pa	(A-N$_3$)	1.000	-	1.007	9.2114	-2.6287	-1.6485	1.4556	0.0001
91	Pa	(N$_3$-N$_2$)	1.007	-	1.224	9.3063	-2.5049	-0.6161	1.8790	0.0016
91	Pa	(N$_2$-N$_1$)	1.224	-	1.387	9.3127	-1.7001	-2.4198	2.0000	0.0072
91	Pa	(N$_1$)-M$_5$)	1.387	-	3.442	9.3134	-1.6808	-1.2178	0.5251	0.0422
91	Pa	(M$_5$-M$_4$)	3.442	-	3.611	5.7215	2.0966	1.2750	-1.6564	0.0127
91	Pa	(M$_4$-M$_3$)	3.611	-	4.174	8.3896	-0.5017	0.7361	-0.7328	0.0512
91	Pa	(M$_3$-M$_2$)	4.174	-	5.001	8.7013	1.6679	-2.7375	0.5929	0.0023
91	Pa	(M$_2$-M$_1$)	5.001	-	5.367	3.5604	3.4532	0.9311	-1.1205	0.0255
91	Pa	(M$_1$-L$_3$)	5.367	-	16.730	11.3747	-2.9608	0.1657	-0.0193	0.0074
91	Pa	(L$_3$-L$_2$)	16.730	-	20.310	3.6406	0.4908	0.8603	-0.3187	0.0670
91	Pa	(L$_2$-L$_1$)	20.310	-	21.100	5.5294	0.6823	0.0665	-0.1332	0.0018
91	Pa	(L$_1$-Z)	21.000	-	60.000	12.9583	-2.7945	0.0119	0.0028	0.0057
91	Pa	(Z-K)	60.000	-	112.601	10.1596	-0.5984	-0.5624	0.0528	0.0004
91	Pa	(K-E)	112.601	-	150.000	2.2338	0.2668	0.3174	-0.0840	0.0829
92	U	(A-N$_3$)	1.000	-	1.045	9.5470	-2.5807	0.1166	-1.8471	0.0000
92	U	(N$_3$-N$_2$)	1.045	-	1.273	9.3417	-2.5432	0.0071	-0.0178	0.0000
92	U	(N$_2$-N$_1$)	1.273	-	1.441	9.3188	-1.5582	-2.4428	1.7206	0.0069
92	U	(N$_1$-M$_5$)	1.441	-	3.552	9.4051	-1.9333	-0.8273	0.3508	0.0259
92	U	(M$_5$-M$_4$)	3.552	-	3.728	5.4039	2.3579	1.1532	-1.5713	0.0118
92	U	(M$_4$-M$_3$)	3.728	-	4.303	5.7876	6.8471	-5.9127	1.2137	0.0255
92	U	(M$_3$-M$_2$)	4.303	-	5.182	9.7225	-0.5194	-1.2004	0.2364	0.0022
92	U	(M$_2$-M$_1$)	5.182	-	5.548	3.8366	2.4499	1.6154	-1.2295	0.0217
92	U	(M$_1$-L$_3$)	5.548	-	17.170	10.6134	-1.9518	-0.2748	0.0444	0.0139
92	U	(L$_3$-L$_2$)	17.170	-	20.950	2.6254	1.0995	0.7710	-0.3186	0.0766
92	U	(L$_2$-L$_1$)	20.950	-	21.760	4.9207	0.5395	0.3513	-0.1897	0.0025
92	U	(L$_1$-Z)	21.760	-	60.000	12.3788	-2.3008	-0.1282	0.0162	0.0043
92	U	(Z-K)	60.000	-	115.606	11.8025	-2.2610	-0.0595	0.0054	0.0510
92	U	(K-E)	115.606	-	150.000	1.3964	0.1975	0.4216	-0.0956	0.0671
93	Np	(A-N$_3$)	1.000	-	1.087	9.2730	-2.4764	0.2425	-1.9341	0.0001
93	Np	(N$_3$-N$_2$)	1.087	-	1.328	9.3640	-2.3908	-0.4263	0.7644	0.0007
93	Np	(N$_2$-N$_1$)	1.328	-	1.501	9.2907	-1.3789	-1.9384	0.7984	0.0092
93	Np	(N$_1$-M$_5$)	1.501	-	3.666	9.8943	-1.3652	-1.3703	0.5310	0.0562
93	Np	(M$_5$-M$_4$)	3.666	-	3.850	5.0613	3.1982	0.3872	-1.2914	0.0063
93	Np	(M$_4$-M$_3$)	3.850	-	4.435	4.2490	2.9176	1.9994	-1.8432	0.0584
93	Np	(M$_3$-M$_2$)	4.450	-	5.366	4.6045	2.2281	1.5497	-1.2855	0.0888
93	Np	(M$_2$-M$_1$)	5.366	-	5.723	3.6366	2.2623	1.9554	-1.3106	0.0l10
93	Np	(M$_1$-L$_3$)	5.723	-	17.610	12.2157	-4.0102	0.6311	-0.0876	0.0151
93	Np	(L$_3$-L$_2$)	17.610	-	21.600	4.3965	0.8191	0.3394	-0.2068	0.0661
93	Np	(L$_2$-L$_1$)	21.600	-	22.430	4.8l15	0.7126	0.2654	-0.1742	0.0021
93	Np	(L$_1$) -Z)	22.430	-	60.000	13.6235	-3.2826	0.1455	-0.0096	0.0052
93	Np	(Z-K)	60.000	-	118.678	12.7094	-2.3155	-0.1656	0.0220	0.0117
93	Np	(K-E)	118.678	-	150.000	2.1165	0.3549	0.2849	-0.0796	0.0537
94	Pu	(A-N$_3$)	1.000	-	1.115	9.2737	-2.4060	0.3346	-1.7145	0.0004
94	Pu	(N$_3$-N$_2$)	1.l15	-	1.372	9.3648	-2.3354	-0.1797	0.2825	0.0003
94	Pu	(N$_2$-N$_1$)	1.372	-	1.559	9.2698	-1.2279	-1.9715	0.7776	0.0122

Z	region	energy interval [keV]			x_0	x_1	x_2	x_3	RFQ
94	Pu (N_1-M_5)	1.559	-	3.778	9.3546	-1.5567	-1.0173	0.3923	0.0404
94	Pu (M_5-M_4)	3.778	-	3.973	4.7611	2.5478	1.6490	-1.7353	0.0076
94	Pu (M_4-M_3)	3.973	-	4.557	3.0704	3.9887	2.0397	-1.9824	0.0600
94	Pu (M_3-M_2)	4.557	-	5.541	3.7786	3.1475	1.2162	-1.2289	0.0997
94	Pu (M_2-M_1)	5.541	-	5.933	4.5288	2.0840	1.0985	-0.9236	0.0097
94	Pu (M_1-L_3)	5.933	-	18.060	10.8522	-2.1104	-0.2224	0.0377	0.0037
94	Pu (L_3-L_2)	18.060	-	22.270	3.6184	1.0461	0.4187	-0.2288	0.0855
94	Pu (L_2-L_1)	22.270	-	23.100	3.8336	1.0191	0.3533	-0.2009	0.0023
94	Pu (L_1-Z)	23.100	-	60.000	2.9682	5.6257	-2.3200	0.2167	0.0405
94	Pu (Z-K)	60.000	-	121.818	14.1820	-3.2943	0.0538	0.0056	0.0057
94	Pu (K-E)	121.818	-	150.000	2.4739	0.1807	0.2651	-0.0719	0.0381

12 Atomic Weights and Isotope Masses

The symbols used in the following table have the meaning:

w   Elements for which the variation of the isotopic abundance in conventional material probes make worse a precise determinationof the atomic mass, i.e. the value for A is valid only for arbitrary "normal" probes.

x   Elements for which geological probes with anomalous isotopic abundance are known.

y   Elements for which a variation of the isotopic abuncances is known for commercial probes

z   Elements for which the isotopic abundance can be spezified for the longest lived isotops.

*   Characterization of the longest lived isotops

The atomic weights relate to $\frac{1}{12} m(^{12}C) = 1$ amu $= 1.660\,5.38\,86\,(28) \cdot 10^{-27}$ kg $= 931.50$ MeV

Used references:

[222]
[441]

Z	symbol	element	A[amu]	symbol and mass number of the isotope	isotope mass [amu]
0	n	neutron	-	-	1.00866
1	H	hydrogen	1.0079^w	^{1}H hydrogen	1.007825
				^{2}H deuterium	2.01410
				^{3}H tritium	3.01605
2	He	helium	4.00260^x	^{3}He helium-3	3.01603
				^{4}He helium-4	4.00260
3	Li	lithium	$6.941^{w,x,y}$	^{6}Li	6.01512
				^{7}Li	7.01600
4	Be	beryllium	9.01218	^{9}Be	9.01218
5	B	boron	$10.811^{w,y}$	^{10}B	10.0129
				^{11}B	11.0093
6	C	carbon	12.01115^w	^{12}C	12
				^{13}C	13.00335
7	N	nitrogen	14.0067	^{14}N	14.00307

Z	symbol	element	A[amu]	symbol and mass number of the isotope	isotope mass [amu]
				^{15}N	15.00011
8	O	oxygen	15.9994^w	^{16}O	15.99491
				^{17}O	16.99913
				^{18}O	17.99916
9	F	fluorine	18.998403	^{19}F	18.99840
10	Ne	neon	20.179^y	^{20}Ne	19.99244
				^{22}Ne	21.99138
11	Na	sodium	22.98977	23Na	22.98977
12	Mg	magnesium	24.305^x	^{24}Mg	23.9850
				^{25}Mg	24.9858
				^{26}Mg	25.9826
13	Al	aluminium	26.98154	^{27}Al	26.9815
14	Si	silicon	28.0855	^{28}Si	27.9769
				^{29}Si	28.9765
15	P	phosphorus	30.97376	^{31}P	30.97376
16	S	sulphur	32.064^w	^{32}S	31.9721
				^{34}S	33.9679
17	Cl	chlorine	35.453	^{35}Cl	34.9689
				37Cl	36.9659
18	Ar	argon	39.948^{w,x}	^{36}Ar	35.9675
				^{40}Ar	39.9624
19	K	potassium	39.0983	^{39}K	38.9637
				^{41}K	40.9618
20	Ca	calcium	40.08^x	^{40}Ca	39.9626
				^{42}Ca	41.9586
				^{44}Ca	43.9555
21	Sc	scandium	44.9559	^{45}Sc	44.95591
22	Ti	titanium	47.9	^{48}Ti	47.94795
23	V	vanadium	50.9415	^{51}V	50.94396
24	Cr	chromium	51.996	^{52}Cr	51.9405
				^{53}Cr	52.9407
25	Mn	manganese	54.9380	^{55}Mn	54.93805
26	Fe	iron	55.847	^{54}Fe	53.9396
				^{56}Fe	55.9349
				^{57}Fe	56.9354
27	Co	cobalt	58.9332	^{59}Co	58.03320
28	Ni	nickel	58.70	^{58}Ni	57.9353
				^{60}Ni	59.9308
29	Cu	copper	63.546^w	^{63}Cu	62.9296
				^{65}Cu	64.9278
30	Zn	zinc	65.37	^{64}Zn	63.9291
				^{66}Zn	65.9260
				^{68}Zn	67.9248

Z	symbol	element	A[amu]	symbol and mass number of the isotope	isotope mass [amu]
31	Ga	gallium	69.72	^{69}Ga	68.9256
				^{71}Ga	70.9247
32	Ge	germanium	72.59	^{70}Ge	69.9242
				^{72}Ge	71.9221
				^{74}Ge	73.9212
33	As	arsenic	74.9216	^{75}As	74.92160
34	Se	selenium	78.96	^{78}Se	77.9173
				^{80}Se	79.9165
35	Br	bromine	79.904	^{79}Br	78.9183
				^{81}Br	80.9163
36	Kr	krypton	83.80^{x,y}	^{82}Kr	81.9135
				^{83}Kr	82.9141
				^{84}Kr	83.9115
				^{85}Kr	84.9125
37	Rb	rubidium	85.467^{x}	^{85}Rb	84.9118
				^{87}Rb	86.9092
38	Sr	strontium	87.62^{x}	^{86}Sr	85.9093
				^{87}Sr	86.9089
				^{88}Sr	87.9056
39	Y	yttrium	88.9059	^{89}Y	88.90586
40	Zr	zirconium	91.22^{x}	^{90}Zr	89.9047
				^{91}Zr	90.9056
				^{94}Zr	93.9063
41	Nb	niobium	92.9064	^{93}Nb	92.90638
42	Mo	molybdenum	95.94	^{92}Mo	91.9068
				^{95}Mo	94.9058
				^{96}Mo	95.9047
				^{98}Mo	97.9054
43	Tc	technetium	[97]	^{97}Tc*	96.9064
				^{98}Tc*	97.9072
				^{99}Tc*	98.9062
44	Ru	ruthenium	101.07^{x}	^{101}Ru	100.9056
				^{102}Ru	101.9043
				^{104}Ru	103.9054
45	Rh	rhodium	102.9055	^{103}Rh	102.90550
46	Pd	palladium	106.4^{x}	^{104}Pd	103.9040
				^{105}Pd	104.9051
				^{106}Pd	105.9035
				^{108}Pd	107.9039
				^{110}Pd	109.9052
47	Ag	silver	107.868^{x}	^{107}Ag	106.9051
				^{109}Ag	108.9048
48	Cd	cadmium	112.41^{x}	^{112}Cd	111.9028

Z	symbol	element	A[amu]	symbol and mass number of the isotope	isotope mass [amu]
				^{114}Cd	113.9034
49	In	indium	114.82^x	^{113}In	112.9041
				^{115}In	114.9039
50	Sn	tin	118.69	^{116}Sn	115.9017
				^{118}Sn	117.9016
				^{120}Sn	119.9022
51	Sb	antimony	121.75	^{121}Sb	120.9038
				^{123}Sb	122.9042
52	Te	tellurium	127.6^x	^{126}Te	125.9033
				^{128}Te	127.9045
				^{130}Te	129.9062
53	I	iodine	126.9045	^{127}I	126.90448
54	Xe	xenon	131.30^{x,y}	^{129}Xe	128.9048
				^{131}Xe	130.9051
				^{132}Xe	131.9041
55	Cs	caesium	132.9054	^{133}Cs	132.90543
56	Ba	barium	137.33^x	^{137}Ba	136.9058
				^{138}Ba	137.9052
57	La	lanthanum	138.9055^x	^{139}La	138.90636
58	Ce	cerium	140.12^x	^{140}Ce	139.9054
				^{142}Ce	141.9092
59	Pr	praseodymium	140.9077	^{141}Pr	140.90766
60	Nd	neodymium	144.24^x	^{142}Nd	141.9071
				^{143}Nd	142.9098
				^{144}Nd	143.9101
				^{146}Nd	145.9131
61	Pm	promethium	[145]	^{145}Pm*	144.9145
				^{146}Pm*	145.9175
				^{147}Pm*	146.9190
62	Sm	samarium	150.4^x	^{147}Sm	146.9149
				^{148}Sm	147.9148
				^{149}Sm	148.9172
				^{152}Sm	151.9197
				^{154}Sm	153.9222
63	Eu	europium	151.96^x	^{151}Eu	150.937
				^{153}Eu	152.935
64	Gd	gadolinium	157.25^x	^{155}Gd	154.9226
				^{156}Gd	155.9221
				^{157}Gd	156.9240
				^{158}Gd	157.9241
				^{160}Gd	159.9271
65	Tb	terbium	158.9254	^{159}Tb	158.92535
66	Dy	dysprosium	162.5	^{161}Dy	160.9269

Z	symbol	element	A[amu]	symbol and mass number of the isotope	isotope mass [amu]
				^{162}Dy	161.9268
				^{163}Dy	162.9287
				^{164}Dy	163.9292
67	Ho	holmium	164.9304	^{165}Ho	164.93033
68	Er	erbium	167.26	^{166}Er	165.9303
				^{167}Er	166.9321
				^{168}Er	167.9324
				^{170}Er	169.9355
69	Tm	thulium	168.9342	^{169}Tm	168.93425
70	Yb	ytterbium	173.04	^{171}Yb	170.9363
				^{172}Yb	171.9364
				^{173}Yb	172.9389
				^{174}Yb	173.9389
				^{176}Yb	175.9426
71	Lu	lutetium	174.967	^{175}Lu	174.9408
				^{176}Lu	175.9427
72	Hf	hafnium	178.49	^{177}Hf	176.9432
				^{178}Hf	177.9437
				^{179}Hf	178.9458
				^{180}Hf	179.9466
73	Ta	tantalum	180.9479	^{181}Ta	180.94801
74	W	tungsten	183.85	^{182}W	181.9482
				^{183}W	182.9502
				^{184}W	183.9510
				^{186}W	185.9544
75	Re	rhenium	186.2	^{185}Re	184.9530
				^{187}Re	186.9558
76	Os	osmium	190.2^{x}	^{188}Os	187.9559
				^{189}Os	188.9582
				^{190}Os	189.9585
				^{192}Os	191.9615
77	Ir	iridium	192.22	^{191}Ir	190.9606
				^{193}Ir	192.9629
78	Pt	platinum	195.09	^{194}Pt	193.9627
				^{195}Pt	194.9648
				^{196}Pt	195.9649
				^{198}Pt	197.9679
79	Au	gold	196.9665	^{197}Au	196.96656
80	Hg	mercury	200.59	^{198}Hg	197.9668
				^{199}Hg	198.9683
				^{200}Hg	199.9683
				^{201}Hg	200.9703

Z	symbol	element	A[amu]	symbol and mass number of the isotope	isotope mass [amu]
				^{202}Hg	201.9706
81	Tl	thallium	204.37	^{203}Tl	202.9723
				^{205}Tl	204.9744
82	Pb	lead	207.2^{w,x}	^{206}Pb	205.9744
				^{207}Pb	206.9759
				^{208}Pb	207.9766
83	Bi	bismuth	208.9804	^{209}Bi	208.98039
84	Po	polonium	[209]	^{208}Po*	207.9812
				^{209}Po*	208.9824
85	At	astatine	[210]	^{210}At*	209.9871
86	Rn	radon	[222]	^{222}Rn*	222.0176
87	Fr	francium	[223]	^{223}Fr*	223.0197
88	Ra	radium	226.0254^{x,z}	^{226}Ra*	226.02541
				^{228}Ra*	228.0311
89	Ac	actinium	227.028^{z}	^{227}Ac*	227.02775
90	Th	thorium	232.038^{x,z}	^{232}Th*	232.03805
				^{229}Th*	229.0318
				^{230}Th*	230.0331
91	Pa	protactinium	231.0359^{z}	^{231}Pa*	231.03588
92	U	uranium	238.029^{x,y}	^{232}U*	232.03714
				^{233}U*	233.03963
				^{234}U*	234.04095
				^{235}U*	235.04393
				^{236}U*	236.04556
				^{238}U*	238.05079
93	Np	neptunium	237.0482^{z}	^{235}Np*	235.04406
				^{237}Np*	237.04817
94	Pu	plutonium	[244]	^{238}Pu*	238.04956
				^{239}Pu*	239.05216
				^{240}Pu*	240.05381
				^{241}Pu*	241.05685
				^{242}Pu*	242.05874
				^{244}Pu*	244.06420
95	Am	americium	[243]	^{241}Am*	241.0568
				^{243}Am*	243.0614
96	Cm	curium	[247]	^{243}Cm*	243.0614
				^{244}Cm*	244.0627
				^{245}Cm*	245.0655
				^{246}Cm*	246.0672
				^{247}Cm*	247.0703
				^{248}Cm*	248.0723
				^{250}Cm*	250.0784
97	Bk	berkelium	[247]	^{247}Bk*	247.0703

Z	symbol	element	A[amu]	symbol and mass number of the isotope	isotope mass [amu]
				^{249}Bk*	249.0750
98	Cf	californium	[251]	^{248}Cf*	248.0722
				^{249}Cf*	249.0748
				^{250}Cf*	250.0748
				^{251}Cf*	251.0796
				^{252}Cf*	252.0816
99	Es	einsteinium	[252]	^{253}Es*	253.0848
				^{254}Es*	254.0880
100	Fm	fermium	[257]	^{252}Fm*	252.0825
				^{253}Fm*	253.0852
				^{257}Fm*	257.0951
101	Md	mendelevium	[258]	-	-
102	No	nobelium	[259]	^{252}No*	252.0880
				^{256}No*	256.0943
				^{259}No*	259.1009
103	Lr	lawrencium	[260]	-	-

13 Parameters of Stable Isotopes

Explanation of the footnotes:

$^1)$ $T_{1/2}(\text{EE}) = 6 \cdot 10^{15}$ a $^5)$ $T_{1/2}(\alpha) = 3.25 \cdot 10^4$ a
$^2)$ $T_{1/2}(\beta^-) = 4.8 \cdot 10^{10}$ a $^6)$ $T_{1/2}(\alpha) = 2.6 \cdot 10^5$ a
$^3)$ $T_{1/2}(\beta^-, \text{EE}) = 3.6 \cdot 10^{10}$ a $^7)$ $T_{1/2}(\alpha) = 7.04 \cdot 10^8$ a
$^4)$ $T_{1/2}(\alpha) = 1.4 \cdot 10^{10}$ a $^8)$ $T_{1/2}(\alpha) = 4.5 \cdot 10^9$ a

The minus-sign after the nuclear spin denotes negative parity.

References:

[191]   [192]
[222]   [238]
[315]   [441]

Z	isotope	frequency [%]	nuclear spin	magnetic moment μ_K
1	^{1}H	99.985	1/2	2.79285
	^{2}H	0.015	1	0.85744
2	^{3}He	0.00014	1/2	-2.12762
	^{4}He	99.99986	0	0
3	^{6}Li	7.59	1	0.82205
	^{7}Li	92.41	3/2-	3.2564
4	^{9}Be	100	3/2-	-1.178
5	^{10}B	19.9	3	1.8006
	^{11}B	80.1	3/2-	2.6886
6	^{12}C	98.93	0	0
	^{13}C	1.07	1/2-	0.7024
7	^{14}N	99.636	1	0.40376
	^{15}N	0.364	1/2-	-0.2832
8	^{16}O	99.757	0	0
	^{17}O	0.038	5/2	-1.8938
	^{18}O	0.205	0	0

Z	isotope	frequency [%]	nuclear spin	magnetic moment μ_K
9	^{19}F	100	1/2	2.6289
10	^{20}Ne	90.48	0	0
	^{21}Ne	0.27	3/2	-0.6618
	^{22}Ne	9.25	0	0
11	^{23}Na	100	3/2	2.2175
12	^{24}Mg	78.99	0	0
	^{25}Mg	10.00	5/2	-0.8554
	^{26}Mg	11.01	0	0
13	^{27}Al	100	5/2	3.6415
14	^{28}Si	92.2	0	0
	^{29}Si	4.7	1/2	-0.5553
	^{30}Si	3.1	0	0
15	^{31}P	100	1/2	1.1316
16	^{32}S	95.0	0	0
	^{33}S	0.75	3/2	0.644
	^{34}S	4.21	0	0
17	^{35}Cl	75.78	3/2	0.82187
	^{37}Cl	24.22	3/2	0.68412
18	^{36}Ar	0.34	0	0
	^{38}Ar	0.06	0	0
	^{40}Ar	99.60	0	0
19	^{39}K	93.26	3/2	0.3915
	^{40}K	0.01	4-	-1.2981
	^{41}K	6.73	3/2	0.21487
20	^{40}Ca	96.941	0	0
	^{42}Ca	0.647	0	0
	^{43}Ca	0.135	7/2-	-1.317
	^{44}Ca	2.086	0	0
	^{46}Ca	0.004	0	0
	^{48}Ca	0.187	0	0
21	^{45}Sc	100	7/2-	4.7565
22	^{46}Ti	8.25	0	0
	^{47}Ti	7.44	5/2-	-0.7885
	^{48}Ti	73.72	0	0
	^{49}Ti	5.41	7/2-	-1.1042
	^{50}Ti	5.18	0	0
23	^{50}V[1])	0.25	6	3.3475
	^{51}V	99.75	7/2-	5.151
24	^{50}Cr	4.345	0	0
	^{52}Cr	83.789	0	0
	^{53}Cr	9.501	3/2-	-0.475

Z	isotope	frequency [%]	nuclear spin	magnetic moment μ_K
	^{54}Cr	2.365	0	0
25	^{55}Mn	100	5/2-	3.47
26	^{54}Fe	5.845	0	0
	^{56}Fe	91.754	0	0
	^{57}Fe	2.119	1/2-	0.0907
	^{58}Fe	0.282	0	0
27	^{59}Co	100	7/2-	4.6
28	^{58}Ni	68.08	0	0
	^{60}Ni	26.22	0	0
	^{61}Ni	1.14	3/2-	-0.750
	^{62}Ni	3.63	0	0
	^{64}Ni	0.93	0	0
29	^{63}Cu	69.15	3/2-	2.223
	^{65}Cu	30.85	3/2-	2.382
30	^{64}Zn	48.27	0	0
	^{66}Zn	27.98	0	0
	^{67}Zn	4.10	5/2-	0.876
	^{68}Zn	19.02	0	0
	^{70}Zn	0.63	0	0
31	^{69}Ga	60.1	3/2-	2.017
	^{71}Ga	39.9	3/2-	2.562
32	^{70}Ge	20.37	0	0
	^{72}Ge	27.38	0	0
	^{73}Ge	7.76	9/2	-0.8795
	^{74}Ge	36.66	0	0
	^{76}Ge	7.83	0	0
33	^{75}As	100	3/2-	1.439
34	^{74}Se	0.89	0	0
	^{76}Se	9.37	0	0
	^{77}Se	7.63	1/2-	0.534
	^{78}Se	23.77	0	0
	^{80}Se	49.61	0	0
	^{82}Se	8.73	0	0
35	^{79}Br	50.69	3/2-	2.1064
	^{81}Br	49.31	3/2-	2.271
36	^{78}Kr	0.36	0	0
	^{80}Kr	2.29	0	0
	^{82}Kr	11.59	0	0
	^{83}Kr	11.50	9/2	-0.9707
	^{84}Kr	56.99	0	0
	^{85}Kr	17.27	0	0
37	^{85}Rb	72.17	5/2-	1.353

Z	isotope	frequency [%]	nuclear spin	magnetic moment μ_K
	$^{87}Rb^{2)}$	27.83	3/2-	2.751
38	^{84}Sr	0.56	0	0
	^{86}Sr	9.86	0	0
	^{87}Sr	7.00	9/2	-1.093
	^{88}Sr	82.58	0	0
39	^{89}Y	100	1/2-	-0.1374
40	^{90}Zr	51.45	0	0
	^{91}Zr	11.22	5/2	-1.304
	^{92}Zr	17.15	0	0
	^{94}Zr	17.38	0	0
	^{96}Zr	2.80	0	0
41	^{93}Nb	100	9/2	6.171
42	^{92}Mo	14.77	0	0
	^{94}Mo	9.23	0	0
	^{95}Mo	15.90	5/2	-0.914
	^{96}Mo	16.67	0	0
	^{97}Mo	9.56	5/2	-0.933
	^{98}Mo	24.20	0	0
	^{100}Mo	9.67	0	0
44	^{96}Ru	5.54	0	0
	^{98}Ru	1.87	0	0
	^{99}Ru	12.76	5/2	-0.641
	^{100}Ru	12.60	0	0
	^{101}Ru	17.06	5/2	-0.719
	^{102}Ru	31.55	0	0
	^{104}Ru	18.62	0	0
45	^{103}Rh	100	1/2-	-0.088
46	^{102}Pd	1.02	0	0
	^{104}Pd	11.14	0	0
	^{105}Pd	22.33	5/2	-0.642
	^{106}Pd	27.33	0	0
	^{108}Pd	26.46	0	0
	^{110}Pd	11.72	0	0
47	^{107}Ag	51.84	1/2-	-0.1137
	^{109}Ag	48.16	1/2-	-0.1307
48	^{106}Cd	1.25	0	0
	^{108}Cd	0.89	0	0
	^{110}Cd	12.49	0	0
	^{111}Cd	12.80	1/2	-0.595
	^{112}Cd	24.13	0	0
	^{113}Cd	12.22	1/2	0.622
	^{114}Cd	28.73	0	0
	^{116}Cd	7.49	0	0

Z	isotope	frequency [%]	nuclear spin	magnetic moment μ_K
49	^{113}In	4.29	9/2	5.529
	^{115}In	95.71	9/2	5.541
50	^{112}Sn	0.97	0	0
	^{114}Sn	0.66	0	0
	^{115}Sn	0.34	1/2	-0.919
	^{116}Sn	14.54	0	0
	^{117}Sn	7.68	1/2	-1.001
	^{118}Sn	24.22	0	0
	^{119}Sn	8.59	1/2	-1.047
	^{120}Sn	32.58	0	0
	^{122}Sn	4.63	0	0
	^{124}Sn	5.79	0	0
51	^{121}Sb	57.21	5/2	3.363
	^{123}Sb	42.79	7/2	2.550
52	^{120}Te	0.09	0	0
	^{122}Te	2.55	0	0
	^{123}Te	0.89	1/2	-0.737
	^{124}Te	4.74	0	0
	^{125}Te	7.07	1/2	-0.888
	^{126}Te	18.84	0	0
	^{128}Te	31.74	0	0
	^{130}Te	34.08	0	0
53	^{127}I	100	5/2	2.813
54	^{124}Xe	0.10	0	0
	^{126}Xe	0.09	0	0
	^{128}Xe	1.91	0	0
	^{129}Xe	26.40	1/2	-0.778
	^{130}Xe	4.07	0	0
	^{131}Xe	21.23	3/2	0.692
	^{132}Xe	26.91	0	0
	^{134}Xe	10.44	0	0
	^{136}Xe	8.85	0	0
55	^{133}Cs	100	7/2	2.582
56	^{130}Ba	0.11	0	0
	^{132}Ba	0.10	0	0
	^{134}Ba	2.42	0	0
	^{135}Ba	6.59	3/2	0.838
	^{136}Ba	7.85	0	0
	^{137}Ba	11.23	3/2	0.937
	^{138}Ba	71.70	0	0
57	^{138}La	0.09	5	3.714
	^{139}La	99.91	7/2	2.783
58	^{136}Ce	0.19	0	0
	^{138}Ce	0.25	0	0

Z	isotope	frequency [%]	nuclear spin	magnetic moment μ_K
	^{140}Ce	88.45	0	0
	^{142}Ce	11.11	0	0
59	^{141}Pr	100	5/2	4.275
60	^{142}Nd	27.2	0	0
	^{143}Nd	12.2	7/2-	-1.07
	^{144}Nd	23.8	0	0
	^{145}Nd	8.3	7/2-	-0.66
	^{146}Nd	17.2	0	0
	^{148}Nd	5.7	0	0
	^{150}Nd	5.6	0	0
62	^{144}Sm	3.07	0	0
	^{147}Sm	14.99	7/2-	-0.815
	^{148}Sm	11.24	0	0
	^{149}Sm	13.82	7/2-	-0.672
	^{150}Sm	7.38	0	0
	^{152}Sm	26.75	0	0
	^{154}Sm	22.75	0	0
63	^{151}Eu	47.81	5/2	3.472
	^{153}Eu	52.19	5/2	1.533
64	^{152}Gd	0.20	0	0
	^{154}Gd	2.18	0	0
	^{155}Gd	14.80	3/2-	-0.259
	^{156}Gd	20.47	0	0
	^{157}Gd	15.65	3/2-	-0.340
	^{158}Gd	24.84	0	0
	^{160}Gd	21.86	0	0
65	^{159}Tb	100	3/2	2.014
66	^{156}Dy	0.05	0	0
	^{158}Dy	0.09	0	0
	^{160}Dy	2.38	0	0
	^{161}Dy	18.88	5/2	-0.480
	^{162}Dy	25.47	0	0
	^{163}Dy	24.88	5/2-	0.673
	^{164}Dy	28.25	0	0
67	^{165}Ho	100	7/2-	4.17
68	^{162}Er	0.14	0	0
	^{164}Er	1.60	0	0
	^{166}Er	33.50	0	0
	^{167}Er	22.87	7/2	-0.564
	^{168}Er	26.98	0	0
68	^{170}Er	14.91	0	0
69	^{169}Tm	100	1/2	-0.232
70	^{168}Yb	0.13	0	0

Z	isotope	frequency [%]	nuclear spin	magnetic moment μ_K
	^{170}Yb	3.04	0	0
	^{171}Yb	14.28	1/2-	0.49
	^{172}Yb	21.83	0	0
	^{173}Yb	16.13	5/2-	-0.68
	^{174}Yb	31.83	0	0
	^{176}Yb	12.76	0	0
71	^{175}Lu	97.41	7/2	2.233
	^{176}Lu[3])	2.59	7-	3.169
72	^{174}Hf	0.16	0	0
	^{176}Hf	5.26	0	0
	^{177}Hf	18.60	7/2-	0.794
	^{178}Hf	27.28	0	0
	^{179}Hf	13.62	9/2	-0.641
	^{180}Hf	35.08	0	0
73	^{180}Ta	0.01	9-	4.82
	^{181}Ta	99.99	7/2	2.37
74	^{180}W	0.12	0	0
	^{182}W	26.50	0	0
	^{183}W	14.31	1/2-	0.118
	^{184}W	30.64	0	0
	^{186}W	28.43	0	0
75	^{185}Re	37.40	5/2	3.187
	^{187}Re	62.60	5/2	3.220
76	^{184}Os	0.02	0	
	^{186}Os	1.59	0	0
	^{187}Os	1.96	1/2-	0.065
	^{188}Os	13.24	0	0
	^{189}Os	16.15	3/2	0.660
	^{190}Os	26.26	0	0
	^{192}Os	40.78	0	0
77	^{191}Ir	37.3	3/2	0.151
	^{193}Ir	62.7	3/2	0.164
78	^{190}Pt	0.01	0	0
	^{192}Pt	0.78	0	0
	^{194}Pt	32.97	0	0
	^{195}Pt	33.83	1/2-	0.610
	^{196}Pt	25.24	0	0
	^{198}Pt	7.17	0	0
79	^{197}Au	100	3/2	0.146
80	^{196}Hg	0.15	0	0
	^{198}Hg	9.97	0	0
	^{199}Hg	16.87	1/2-	0.506
	^{200}Hg	23.10	0	0

Z	isotope	frequency [%]	nuclear spin	magnetic moment μ_K
	^{201}Hg	13.18	3/2-	-0.560
	^{202}Hg	29.86	0	0
	^{204}Hg	6.87	0	0
81	^{203}Tl	29.52	1/2	1.622
	^{205}Tl	70.48	1/2	1.638
82	^{204}Pb	1.4	0	0
	^{206}Pb	24.1	0	0
	^{207}Pb	22.1	1/2-	0.593
	^{208}Pb	52.4	0	0
83	^{209}Bi	100	9/2-	4.111
90	^{232}Th[4]	100	0	0
91	^{231}Pa[5]	100	3/2-	2.01
92	^{234}U[6]	0.005	0	0
	^{235}U[7]	0.720	7/2-	-0.38
	^{238}U[8]	99.274	0	0

14 Parameters of Long-Live Radioactive Isotopes

Meaning of the symbols:

I	nuclear spin ("-" if negative parity)	$T_{1/2}$	half-life time
μ	magnetic nuclear moment	EC	electron capture
μ_K	nuclear magneton	$\alpha, \beta^{\pm}$	α or $\beta\pm$ decay
sf	spontaneous fission		

References

[441]
[191]
[192]
[222]
[238]
[315]

Z	symbol	I	μ/μ_K	$T_{1/2}$	decay channel
0	^{1}n	1/2	-1.913	10.6 min	β^-
1	^{3}H	1/2	2.979	12.33 a	β^-
4	^{10}Be	0	-	$2.7 \cdot 10^6$ a	β^-
6	^{11}C	3/2-	1.0	0.34 h	β^+, EC
	^{14}C	0	0	5700 a	β^-
7	^{13}N	1/2-	±0.322	0.17 h	β^+
9	^{18}F	1	0.8	1.83 h	β^+, EC
11	^{22}Na	3	1.75	2.6 a	β^+, EC
	^{24}Na	4	1.69	15.0 h	β^-
12	^{28}Mg	0	-	21.4 h	β^-
13	^{26}Al	5	2.8	$7.4 \cdot 10^5$ a	β^+, EC
14	^{31}Si	3/2	-	2.6 h	β^-
	^{32}Si	0	-	650 a	β^+
15	^{32}P	1	-0.252	14.3 d	β^-
	^{33}P	1/2	-	24.8 d	β^-
16	^{35}S	3/2	1.0	87 d	β^-

Z	symbol	I	μ/μ_K	$T_{1/2}$	decay channel
17	^{36}Cl	0	1.285	$3.0 \cdot 10^5$ a	β^-, EC
	^{38}Cl	2-	2.0	0.62 h	β^-
	^{39}Cl	3/2	-	1 h	β^-
18	^{37}Ar	3/2	0.9	35 d	EC
	^{39}Ar	7/2-	-1.3	270 a	β^-
	^{41}Ar	7/2-	-	1.83 h	β^-
19	^{40}K	4-	-1.30	$1.3 \cdot 10^9$ a	β^-, EC
	^{42}K	2-	-1.14	12.4 h	β^-
	^{43}K	3/2	±0.16	22 h	β^-
	^{44}K	2-	-	0.37 h	β^-
	^{45}K	3/2	±0.17	0.3 h	β^-
20	^{41}Ca	7/2-	-1.595	$8 \cdot 10^4$ a	EC
	^{45}Ca	7/2-	-	160 d	β^-
	^{47}Ca	7/2-	-	4.5 d	β^-
21	^{43}Sc	7/2-	4.6	3.9 h	β^+, EC
	^{44}Sc	2	2.6	3.9 h	β^+, EC
	^{46}Sc	4	3.0	84 d	β^-
	^{47}Sc	7/2-	5.3	3.4 d	β^-
	^{48}Sc	6	-	44 h	β^-
	^{49}Sc	7/2-	-	0.95 h	β^-
22	^{44}Ti	0	-	48 a	EC
	^{45}Ti	7/2-	0.09	3.1 h	β^+, EC
23	^{47}V	3/2-	-	0.52 h	β^+
	^{48}V	4	±1.6	16 d	β^+, EC
	^{49}V	7/2-	±4.5	1 a	EC
	^{50}V	6	3.347	$6 \cdot 10^{15}$ a	EC
24	^{49}Cr	5/2-	±0.48	0.70 h	β^+
	^{51}Cr	7/2-	±0.93	27.7 d	EC
25	^{51}Mn	5/2-	±3.57	0.75 h	β^+
	^{52}Mn	6	3.06	5.6 d	β^+, EC
	^{53}Mn	7/2-	±5.02	$3.7 \cdot 10^6$ a	EC
	^{56}Mn	3	3.23	2.58 h	β^-
26	^{52}Fe	0	-	8.5 h	β^+, EC
	^{55}Fe	3/2-	-	2.9 a	EC
	^{59}Fe	3/2-	±1.1	46 d	β^-
27	^{55}Co	7/2-	±4.5	18 h	β^+, EC
	^{56}Co	4	3.83	77 d	EC, β^+
	^{57}Co	7/2-	4.72	270 d	EC
	^{58}Co	2	4.04	71 d	EC, β^+
	^{60}Co	5	3.80	5.3 a	β^-
	^{61}Co	7/2-	-	1.7 h	β^-
	^{62}Co	2	-	0.23 h	β^-

Z	symbol	I	μ/μ_K	$T_{1/2}$	decay channel
28	^{56}Ni	0	-	6.1 d	EC
	^{57}Ni	3/2-	-	1.5 d	EC, β^+
	^{59}Ni	3/2-	-	$8\cdot10^4$a	EC
	^{63}Ni	1/2-	-	92 a	β^-
	^{65}Ni	5/2-	-	2.56 h	β^-
29	^{60}Cu	2	1.22	0.38 h	β^+, EC
	^{61}Cu	3/2-	2.13	3.3 h	β^+, EC
	^{64}Cu	1	-0.22	12.9 h	EC, β^-, β^+
	^{67}Cu	3/2-	-	59 h	β^-
30	^{62}Zn	0	-	9.1 h	EC, β^+
	^{63}Zn	3/2-	-0.282	0.64 h	β^+, EC
	^{65}Zn	5/2-	0.77	240 d	EC, β^+
	^{69}Zn	1/2-	-	1 h	β^-
	^{72}Zn	0	-	46 h	β^-
31	^{66}Ga	0	-	9.4 h	β^+, EC
	^{67}Ga	3/2-	1.85	3.3 d	EC
	^{68}Ga	1	±0.012	1.1 d	β^+, EC
	^{70}Ga	1	-	0.35 h	β^-
	^{72}Ga	3-	-0.132	14.1 h	β^-
	^{73}Ga	3/2-	-	5 h	β^-
32	^{66}Ge	0	-	2.4 h	EC, β^+
	^{67}Ge	1/2-	-	0.3 h	β^+, EC
	^{68}Ge	0	-	280 d	EC
	^{69}Ge	5/2-	-	40 h	EC, β^+
	^{71}Ge	1/2-	0.55	11 d	EC
	^{77}Ge	1/2-	-	11 h	β^-
	^{78}Ge	0	-	1.5 h	β^-
33	^{70}As	4	-	0.9 h	EC, β^+
	^{71}As	5/2-	-	2.7 d	EC, β^+
	^{72}As	2-	±2.2	1.1 d	EC, β^+
	^{73}As	3/2-	-	80 d	EC
	^{74}As	2-	-	18 d	EC, β^-, β^+
	^{76}As	2-	-0.91	1.1 d	β^-
	^{77}As	3/2-	-	1.6 d	β^-
	^{78}As	2-	-	1.5 h	β^-
34	^{70}Se	0	-	0.7 h	β^+
	^{72}Se	0	-	8.7 d	EC
	^{73}Se	9/2	-	7 h	β^+, EC
	^{75}Se	5/2	-	120 d	EC
	^{79}Se	7/2	-1.02	$6.5\cdot10^4$a	β^-
	^{81}Se	1/2-	-	0.3 h	β^-
	^{83}Se	9/2	-	0.4 h	β^-
35	^{75}Br	3/2-	-	1.7 h	β^+, EC

Z	symbol	I	μ/μ_K	$T_{1/2}$	decay channel
	^{76}Br	1-	±0.55	16 h	β^+, EC
	^{77}Br	3/2-	-	2.4 d	β^+, EC
	^{80}Br	1	±0.514	0.3 h	β^-, EC, β^+
	^{82}Br	5-	1.63	1.47 d	β^-
	^{83}Br	3/2-	-	2.4 h	β^-
	^{84}Br	2-	-	0.53 h	β^-
36	^{74}Kr	0	-	0.2 h	β^+
	^{76}Kr	0	-	15 h	EC
	^{77}Kr	5/2	-	1.2 h	β^+, EC
	^{79}Kr	1/2-	-	35 h	EC, β^+
	^{81}Kr	7/2	-	$2 \cdot 10^5$ a	EC
	^{85}Kr	9/2	±1.00	10.7 a	β^-
	^{87}Kr	5/2	-	1.3 h	β^-
	^{88}Kr	0	-	2.8 h	β^-
37	^{81}Rb	3/2-	2.4	4.6 h	EC, β^+
	^{83}Rb	5/2-	1.4	86 d	EC
	^{84}Rb	2-	-1.3	33 d	EC, β^+
	^{86}Rb	2-	-1.69	18.8 d	β^-
	^{87}Rb	3/2-	2.75	$4.8 \cdot 10^{10}$ a	β^-
	^{88}Rb	2-	±0.51	0.30 h	β^-
	^{89}Rb	3/2-	-	0.25 h	β^-
38	^{82}Sr	-	-	1.0 d	EC
	^{83}Sr	7/2	-	1.4 d	EC, β^+
	^{85}Sr	9/2	-	64 d	EC
	^{89}Sr	5/2	-	53 d	β^-
	^{90}Sr	0	-	28 a	β^-
	^{91}Sr	5/2	-	9.7 h	β^-
	^{92}Sr	0	-	2.7 h	β^-
39	^{84}Y	5-	-	0.7 h	β^+
	^{85}Y	1/2-	-	5 h	β^+, EC
	^{86}Y	4-	-	15 h	EC, β^+
	^{87}Y	1/2-	-	3.3 d	EC
	^{88}Y	4-	-	110 d	EC
	^{90}Y	2-	-1.63	2.7 d	β^-
	^{91}Y	1/2-	±0.164	59 d	β^-
	^{92}Y	2-	-	3.5 h	β^-
	^{93}Y	1/2-	-	10 h	β^-
	^{94}Y	2-	-	0.3 h	β^-
	^{95}Y	1/2-	-	0.2 h	β^-
40	^{85}Zr	7/2	-	1.4 h	β^+, EC
	^{86}Zr	0	-	16 h	EC
	^{87}Zr	9/2	-	1.5 h	β^+, EC
	^{88}Zr	0	-	85 d	EC
	^{89}Zr	9/2	-	3.3 d	EC, β^+

Z	symbol	I	μ/μ_K	$T_{1/2}$	decay channel
	^{93}Zr	5/2	-	$1.5\cdot10^6$a	β^-
	^{95}Zr	5/2	-	65 d	β^-
41	^{88}Nb	8	-	0.3 h	β^+
	^{89}Nb	1/2-	-	2 h	β^+, EC
	^{90}Nb	4-	-	14.6 h	β^+, EC
	^{91}Nb	9/2-	-	8 a	EC
	^{92}Nb	7	-	10 d	EC
	^{94}Nb	6	-	$2\cdot10^4$a	β^-
	^{96}Nb	6	-	1 d	β^-
	^{97}Nb	9/2	-	1.2 h	β^-
	^{98}Nb	1	-	0.8 h	β^-
	^{100}Nb	-	-	0.2 h	β^-
42	^{90}Mo	0	-	5.7 h	EC, β^+
	^{91}Mo	9/2-	-	0.26 h	β^+
	^{93}Mo	5/2	-	$3\cdot10^3$a	EC
	^{99}Mo	1/2	-	2.8 d	β^-
	^{101}Mo	1/2	-	0.24 h	β^-
	^{102}Mo	0	-	0.2 h	β^-
43	^{93}Tc	9/2	-	2.8 h	EC, β^+
	^{94}Tc	7	-	4.9 h	EC, β^+
	^{95}Tc	9/2	-	20 h	EC
	^{96}Tc	7	±5.4	4.3 d	EC
	^{97}Tc	9/2	-	$2.6\cdot10^6$a	EC
	^{98}Tc	6	-	$1.5\cdot10^6$a	β^-
	^{99}Tc	9/2	5.68	$2.1\cdot10^5$a	β^-
	^{101}Tc	9/2	-	0.23 h	β^-
	^{104}Tc	-	-	0.3 h	β^-
44	^{94}Ru	0	-	1 h	EC
	^{95}Ru	5/2	-	1.6 h	EC, β^+
	^{97}Ru	5/2	-	2.9 d	EC
	^{103}Ru	3/2	-	40 d	β^-
	^{105}Ru	3/2	-	4.4 h	β^-
	^{106}Ru	0	-	1.0 a	β^-
45	^{97}Rh	9/2	-	0.6 h	β^+
	^{99}Rh	1/2-	-	4.6 h	EC, β^+
	^{100}Rh	5	-	21 h	EC, β^+
	^{101}Rh	1/2-	-	3 a	EC
	^{102}Rh	-	4.1	2.9 a	β^+, β^-
	^{105}Rh	7/2	-	1.5 d	β^-
	^{107}Rh	7/2	-	0.36 h	β^-
	^{109}Rh	7/2	-	1 h	β^-
46	^{98}Pd	0	-	0.3 h	EC
	^{99}Pd	5/2	-	0.36 h	β^+
	^{100}Pd	0	-	3.7 d	EC

Z	symbol	I	μ/μ_K	$T_{1/2}$	decay channel
	^{101}Pd	5/2	-	8.4 h	EC, β^+
	^{103}Pd	5/2	-	17 d	EC
	^{107}Pd	5/2	-	$7 \cdot 10^6$ a	β^-
	^{109}Pd	5/2	-	13.5 h	β^-
	^{111}Pd	5/2	-	0.37 h	β^-
	^{112}Pd	0	-	21 h	β^-
47	^{103}Ag	7/2	4.5	1.1 h	EC, β^+
	^{104}Ag	5	4.0	1.1 h	β^+, EC
	^{105}Ag	1/2-	±0.10	41 d	EC
	^{106}Ag	1	2.9	0.4 h	β^+, EC
	^{111}Ag	1/2-	-0.15	7.5 d	β^-
	^{112}Ag	2-	±0.055	3.1 h	β^-
	^{113}Ag	1/2-	±0.16	5.3 h	β^-
	^{115}Ag	1/2-	-	0.35 h	β^-
48	^{103}Cd	5/2	-	0.2 h	β^+
	^{104}Cd	0	-	0.95 h	EC
	^{105}Cd	5/2	-0.74	0.91 h	EC, β^+
	^{107}Cd	5/2	-0.615	6.5 h	EC, β^+
	^{109}Cd	5/2	-0.83	1.3 a	EC
	^{115}Cd	1/2	-0.65	2.2 d	β^-
	^{117}Cd	1/2	-	3 h	β^-
	^{118}Cd	0	-	0.8 h	β^-
49	^{107}In	9/2	-	0.5 h	β^+, EC
	^{108}In	3	-	1 h	EC, β^+
	^{109}In	9/2	5.5	4.2 h	EC, β^+
	^{110}In	2	4.36	1.1 h	β^+, EC
	^{111}In	9/2	5.5	2.8 d	EC
	^{112}In	1	2.8	0.24 h	β^-, EC, β^+
	^{117}In	9/2	-	0.8 h	β^-
50	^{109}Sn	7/2	-	0.30 h	EC, β^+
	^{110}Sn	0	-	4 h	EC
	^{111}Sn	7/2	-	0.6 h	EC, β^+
	^{113}Sn	1/2	±0.88	115 d	EC
	^{121}Sn	3/2	±0.70	1.1 d	β^-
	^{123}Sn	11/2-	-	0.7 h	β^-
	^{125}Sn	11/2-	-	9.6 d	β^-
	^{126}Sn	0	-	10^5 a	β^-
	^{128}Sn	0	-	1 h	β^-
51	^{115}Sb	5/2	3.46	0.5 h	EC, β^+
	^{116}Sb	3	-	0.25 h	β^+, EC
	^{117}Sb	5/2	3.4	2.8 h	EC, β^+
	^{119}Sb	5/2	3.45	38 h	EC
	^{120}Sb	1	±2.3	0.27 h	β^+, EC
	^{122}Sb	2-	11.90	2.7 d	β^-, EC

Z	symbol	I	μ/μ_K	$T_{1/2}$	decay channel
	^{124}Sb	3-	±1.2	60 d	β^-
	^{125}Sb	7/2	±2.6	2.7 a	β^-
	^{126}Sb	8-	±1.3	12 d	β^-
	^{127}Sb	7/2	±2.6	2.9 a	β^-
	^{128}Sb	8-	±1.3	9 h	β^-
	^{129}Sb	7/2-	-	3.4 h	β^-
	^{131}Sb	7/2	-	0.4 h	β^-
52	^{116}Te	0	-	2.5 h	EC
	^{117}Te	1/2	-	1.1 h	EC, β^+
	^{118}Te	0	-	6 d	EC, β^+
	^{119}Te	1/2	±0.25	16 h	EC, β^+
	^{121}Te	1/2	-	17 d	EC
	^{127}Te	3/2	±0.66	9.4 h	β^-
	^{129}Te	3/2	±0.66	1.1 h	β^-
	^{131}Te	3/2	-	0.4 h	β^-
	^{132}Te	0	-	1.3 h	β^-
53	^{118}I	2-	-	0.2 h	β^+, EC
	^{120}I	2-	-	1.4 h	EC, β^+
	^{121}I	5/2	-	2.1 h	EC, β^+
	^{123}I	5/2	-	13 h	EC
	^{124}I	2-	-	4.2 d	EC, β^+
	^{125}I	5/2	3	60 d	EC
	^{126}I	2-	-	13 d	EC, β^-, β^+
	^{128}I	1	-	0.42 h	β^-, EC
	^{129}I	7/2	2.62	$1.6 \cdot 10^7$ a	β^-
	^{130}I	5	-	12.3 h	β^-
	^{131}I	7/2	2.74	8.0 d	β^-
	^{132}I	4	±3.09	2.3 h	β^-
	^{133}I	7/2	2.86	21 h	β^-
	^{134}I	4	-	0.87 h	β^-
	^{135}I	7/2	-	6.7 h	β^-
54	^{121}Xe	5/2	-	0.65 h	β^+
	^{122}Xe	0	-	20 h	EC
	^{123}Xe	1/2	-	2.1 h	EC, β^+
	^{125}Xe	1/2	-	17 h	EC
	^{125}Xe	1/2	-	36.4 d	EC
	^{133}Xe	3/2	-	5.3 d	β^-
	^{135}Xe	3/2	-	9.1 h	β^-
	^{138}Xe	0	-	0.3 h	β^-
55	^{125}Cs	1/2	1.41	0.8 h	β^+, EC
	^{127}Cs	1/2	1.46	6.2 h	EC, β^+
	^{129}Cs	1/2	1.48	32 h	EC
	^{130}Cs	1	1.4	0.5 h	β^+, EC, β^-
	^{131}Cs	5/2	3.54	9.7 d	EC

Z	symbol	I	μ/μ_K	$T_{1/2}$	decay channel
	^{132}Cs	2-	2.22	6.5 d	EC
	^{134}Cs	4	2.99	2.06 a	β^-
	^{135}Cs	7/2	2.73	$3 \cdot 10^6$a	β^-
	^{136}Cs	5	3.7	13 d	β^-
	^{137}Cs	7/2	2.84	30 a	β^-
	^{138}Cs	3-	0.5	0.54 h	β^-
56	^{126}Ba	0	-	1.6 h	EC
	^{128}Ba	0	-	2.4 d	EC
	^{129}Ba	1/2	-	2.6 h	EC, β^+
	^{131}Ba	1/2	-	12 d	EC
	^{133}Ba	1/2	-	10 a	EC
	^{139}Ba	7/2-	-	1.4 h	β^-
	^{140}Ba	0	-	12.8 d	β^-
	^{141}Ba	3/2-	-	0.3 h	β^-
57	^{131}La	3/2	-	1.0 h	EC, β^+
	^{132}La	2-	-	5 h	β^+
	^{133}La	5/2	-	4 h	EC, β^+
	^{135}La	5/2	-	19 h	EC
	^{137}La	7/2	2.69	$6 \cdot 10^4$a	EC
	^{140}La	3-	0.73	40 h	β^-
	^{141}La	7/2	-	3.9 h	β^-
	^{142}La	2-	-	1.5 h	β^-
	^{143}La	7/2-	-	0.2 h	β^-
58	^{130}Ce	0	-	0.5 h	EC, β^+
	^{132}Ce	0	-	4.2 h	EC, β^+
	^{133}Ce	9/2-	-	5.4 h	EC, β^+
	^{134}Ce	0	-	72 h	EC
	^{135}Ce	1/2	-	17 h	EC
	^{137}Ce	3/2	±0.9	9 h	EC
	^{139}Ce	3/2	0.9	137 d	EC
	^{141}Ce	7/2-	1.0	33 d	β^-
	^{143}Ce	3/2-	-	33 h	β^-
	^{144}Ce	0	-	258 d	β^-
59	^{134}Pr	2	-	0.3 h	β^+
	^{135}Pr	3/2	-	0.4 h	EC, β^+
	^{136}Pr	2	-	0.23 h	EC, β^+
	^{137}Pr	5/2	-	1.3 h	EC, β^+
	^{138}Pr	1	-	2.1 h	EC, β^+
	^{139}Pr	5/2	-	4.5 h	EC, β^+
	^{142}Pr	2-	-0.23	19 h	β^-
	^{143}Pr	7/2	-	13.6 h	β^-
	^{144}Pr	0-	-	0.29 h	β^-
	^{145}Pr	7/2	-	6.0 h	β^-
	^{146}Pr	2-	-	0.4 h	β^-

Z	symbol	I	μ/μ_K	$T_{1/2}$	decay channel
	^{147}Pr	3/2	-	0.2 h	β^-
60	^{137}Nd	1/2	-	0.9 h	β^+
	^{140}Nd	0	-	3.3 d	EC, β
	^{141}Nd	3/2	-	2.4 h	EC, β
	^{147}Nd	5/2-	±0.55	11 d	β^-
	^{149}Nd	5/2-	±0.35	1.7 h	β^-
	^{151}Nd	-	-	0.2 h	β^-
61	^{141}Pm	5/2	-	0.3 h	β^+, EC
	^{143}Pm	5/2	±3.8	265 d	EC
	^{144}Pm	5-	±1.7	350 d	EC
	^{145}Pm	5/2	-	18 a	EC
	^{146}Pm	3-	-	2 a	EC, β^-
	^{147}Pm	7/2	2.6	2.6 a	β^-
	^{148}Pm	1-	2.1	5.4 d	β^-
	^{149}Pm	7/2	±3.3	53 h	β^-
	^{150}Pm	-	-	3.7 h	β^-
	^{151}Pm	5/2	±1.8	28 h	β^-
62	^{142}Sm	0	-	1.2 h	EC
	^{145}Sm	7/2-	±0.9	340 d	EC
	^{151}Sm	5/2-	-	80 a	β^-
	^{153}Sm	3/2	-0.022	47 h	β^-
	^{155}Sm	3/2-	-	0.4 h	β^-
	^{156}Sm	0	-	9.4 h	β^-
63	^{145}Eu	5/2	-	6 d	EC, β^+
	^{146}Eu	4-	-	4.6 d	EC, β^+
	^{147}Eu	5/2	-	22 d	EC
	^{148}Eu	5-	-	58 d	EC
	^{149}Eu	5/2	-	110 d	EC
	^{150}Eu	5-	-	13 h	β^-
	^{152}Eu	3-	-1.94	13 a	EC, β^-
	^{154}Eu	3-	±2.0	8.5 a	β^-
	^{155}Eu	5/2	2.0	5 a	β^-
	^{156}Eu	1	1.97	15 d	β^-
	^{157}Eu	-	-	15 h	β^-
	^{158}Eu	-	-	0.9 h	β^-
	^{159}Eu	-	-	0.3 h	β^-
64	^{147}Gd	7/2-	-	1.4 d	EC
	^{148}Gd	0	-	90 d	EC
	^{149}Gd	7/2-	-	10 d	EC
	^{150}Gd	0	-	$2 \cdot 10^6$a	α
	^{151}Gd	7/2-	-	120 d	EC
	^{153}Gd	3/2-	-	240 d	EC
	^{159}Gd	3/2 -	±0.4	18 h	β^-
65	^{147}Tb	5/2	-	0.4 h	EC, β^+

Z	symbol	I	μ/μ_K	$T_{1/2}$	decay channel
	^{148}Tb	2-	-	1.1 h	EC, β^+
	^{149}Tb	1/2	-	4.1 h	EC, α
	^{150}Tb	2-	-	3.1 h	β^+, α, EC
	^{151}Tb	1/2	-	18 h	EC
	^{152}Tb	2-	-	17 h	EC, β^+
	^{153}Tb	5/2	-	55 h	EC
	^{154}Tb	0-	-	21 h	EC
	^{155}Tb	3/2	-	5 d	EC
	^{156}Tb	3-	± 1.4	5 d	EC
	^{157}Tb	3/2	± 2.0	150 a	EC
	^{158}Tb	3-	± 1.75	150 a	EC, β^-
	^{160}Tb	3-	± 1.70	72 d	β^-
	^{161}Tb	3/2	-	7 d	β^-
	^{162}Tb	-	-	2 h	β^-
	^{163}Tb	3/2	-	7 h	β^-
66	^{151}Dy	7/2-	-	0.3 h	EC, β^+, α
	^{152}Dy	0	-	2.4 h	EC, β^+, α
	^{153}Dy	7/2-	± 0.7	6 h	EC, α
	^{155}Dy	3/2-	± 0.3	10 h	EC, β^+
	^{157}Dy	3/2-	± 0.3	8.1 h	EC
	^{159}Dy	3/2-	-	140 d	EC
	^{165}Dy	7/2	± 0.5	2.3 h	β^-
	^{166}Dy	0	0	3.3 d	β^-
67	^{155}Ho	5/2	-	0.3 h	EC, β^+
	^{156}Ho	5	-	1 h	EC, β^+
	^{157}Ho	7/2-	-	0.3 h	β^+
	^{159}Ho	7/2-	-	0.6 h	EC
	^{160}Ho	5	-	0.4 h	EC
	^{161}Ho	7/2-	-	3 h	EC
	^{162}Ho	1	-	0.2 h	EC
	^{166}Ho	0-	0	1.1 d	β^-
	^{167}Ho	7/2-	-	3 h	β^-
68	^{157}Er	3/2-	-	0.4 h	β^+
	^{158}Er	0	-	2.3 h	EC, β^+
	^{159}Er	3/2-	-	0.6 h	EC, β^+
	^{160}Er	0	-	2.2 d	EC
	^{161}Er	3/2-	-0.37	3.2 h	EC
	^{163}Er	5/2-	0.56	1.4 h	EC
	^{165}Er	5/2-	± 0.66	10 h	EC
	^{169}Er	1/2-	0.51	9.4 d	β^-
	^{172}Er	-	-	2 d	β^-
69	^{161}Tm	7/2	-	0.5 h	EC, β^+
	^{162}Tm	1-	-	0.4 h	β^+, EC
	^{163}Tm	1/2	± 0.08	1.8 h	β^+, EC

Z	symbol	I	μ/μ_K	$T_{1/2}$	decay channel
	^{165}Tm	1/2	±0.14	30 h	EC
	^{166}Tm	2	±0.09	7.7 h	EC, β^+
	^{167}Tm	1/2	-0.20	9.3 d	EC
	^{168}Tm	3	-	90 d	EC, β^+
	^{170}Tm	1-	±0.248	130 d	β^-
	^{171}Tm	1/2	±0.23	1.9 a	β^-
	^{172}Tm	2-	-	64 h	β^-
	^{173}Tm	1/2	-	8 h	β^-
70	^{164}Yb	0	-	1.3 h	EC
	^{166}Yb	0	-	2.4 d	EC
	^{167}Yb	5/2-	-	0.3 h	EC
	^{169}Yb	7/2	±0.6	32 d	EC
	^{175}Yb	7/2-	±0.6	4 d	β^-
	^{177}Yb	9/2	-	1.9 h	β^-
71	^{167}Lu	7/2	-	0.9 h	EC, β^+
	^{169}Lu	7/2	-	1.5 d	EC, β^+
	^{170}Lu	0	0	2 d	β^+, EC
	^{171}Lu	7/2	-	8 d	EC
	^{172}Lu	4-	-	6.7 d	EC
	^{173}Lu	7/2	-	1.4 a	EC
	^{174}Lu	1-	-	4 a	EC
	^{176}Lu	7-	3.2	$3.6 \cdot 10^{10}$ a	β^-, EC
	^{177}Lu	7/2	2.24	6.7 d	β^-
	^{179}Lu	7/2	-	4.6 h	β^-
72	^{168}Hf	0	-	0.4 h	EC
	^{169}Hf	5/2	-	1.5 h	EC, β^+
	^{170}Hf	0	-	10 h	EC
	^{171}Hf	7/2	-	12 h	EC
	^{172}Hf	0	-	5 a	EC
	^{173}Hf	1/2-	-	1.0 d	EC
	^{175}Hf	5/2-	-	70 d	EC
	^{181}Hf	1/2-	-	42 d	β^-
	^{182}Hf	0	-	10^7 a	β^-
	^{183}Hf	3/2-	-	1.0 h	β^-
73	^{172}Ta	3-	-	0.5 h	β^+, EC
	^{173}Ta	5/2-	-	3 h	EC
	^{174}Ta	3	-	1.2 h	EC, β^+
	^{175}Ta	7/2	-	10 h	EC
	^{176}Ta	1-	-	8.0 h	EC
	^{177}Ta	7/2-	-	2.4 d	EC
	^{178}Ta	1	-	2.1 h	EC, β^+
	^{179}Ta	7/2	-	1.6 a	EC
	^{182}Ta	3-	±2.6	115 d	β^-
	^{183}Ta	7/2	-	5 d	β^-

Z	symbol	I	μ/μ_K	$T_{1/2}$	decay channel
	^{184}Ta	5-	-	9 h	β^-
	^{185}Ta	7/2	-	0.8 h	β^-
74	^{173}W	-	-	0.3 h	EC
	^{174}W	0	-	0.5 h	EC
	^{175}W	1/2-	-	0.6 h	EC
	^{176}W	0	-	2 h	EC
	^{178}W	0	-	22 d	EC, β^+
	^{179}W	7/2-	-	0.6 h	EC
	^{181}W	9/2	-	130 d	EC
	^{185}W	3/2-	-	75 d	β^-
	^{187}W	3/2-	-	1.0 d	β^-
	^{188}W	0	-	2.9 d	β^-
75	^{177}Re	5/2-	-	0.3 h	β^+
	^{178}Re	3-	-	0.3 h	β^+
	^{179}Re	5/2	-	0.3 h	EC
	^{181}Re	1-	-	18 h	EC
	^{182}Re	7	-	2.7 d	EC
	^{183}Re	5/2	±3.1	71 d	EC
	^{184}Re	3-	±2.5	38 d	EC, β^-
	^{186}Re	1-	1.74	3.7 d	β^-, EC
	^{188}Re	1-	1.79	17 h	β^-
76	^{180}Os	0	-	0.36 h	EC
	^{181}Os	7/2-	-	2 h	EC
	^{182}Os	0	-	21 h	EC
	^{183}Os	9/2	-	14 h	EC
	^{185}Os	1/2-	-	95 d	EC
	^{191}Os	9/2-	-	15 d	β^-
	^{193}Os	3/2-	-	1.3 d	β^-
	^{194}Os	0-	-	6 a	β^-
77	^{182}Ir	-	-	0.25 h	EC
	^{183}Ir	-	-	0.9 h	EC
	^{184}Ir	5-	-	3.2 h	EC, β^+
	^{185}Ir	5/2-	-	15 h	EC
	^{186}Ir	5	-	15 h	EC, β^+
	^{187}Ir	3/2-	-	12 h	EC
	^{188}Ir	2-	-	1.7 d	EC
	^{189}Ir	3/2	-	12 d	EC
	^{190}Ir	4	-	11 d	EC
	^{192}Ir	4-	1.9	74 d	β^-, EC
	^{194}Ir	1-	±0.4	19 h	β^-
	^{195}Ir	3/2	-	3 h	β^-
	^{196}Ir	0-	-	2 h	β^-
78	^{185}Pt	9/2	-	1 h	EC
	^{186}Pt	0	-	2.6 h	EC

Z	symbol	I	μ/μ_K	$T_{1/2}$	decay channel
	^{187}Pt	3/2	-	2.5 h	EC
	^{188}Pt	0	-	10 d	EC
	^{189}Pt	3/2-	-	11 h	EC
	^{191}Pt	3/2-	-	3.0 d	EC
	^{197}Pt	1/2-	±0.5	18 h	β^-
	^{199}Pt	5/2-	-	0.5 h	β^-
79	^{186}Au	3-	-	0.2 h	EC
	^{189}Au	1/2	-	0.5 h	EC
	^{190}Au	1-	±0.07	0.6 h	EC, β^+
	^{191}Au	3/2	±0.14	3.2 h	EC
	^{192}Au	1-	±0.008	5 h	EC, β^+
	^{193}Au	3/2	±0.14	17 h	EC
	^{194}Au	1-	±0.07	1.7 d	EC, β^+
	^{195}Au	3/2	±0.15	183 d	EC
	^{196}Au	2-	±0.6	6.2 d	EC, β^+
	^{198}Au	2-	0.59	2.7 d	β^-
	^{199}Au	3/2	0.27	3.1 d	β^-
	^{200}Au	1-	-	0.8 h	β^-
	^{201}Au	3/2	-	0.4 h	β^-
80	^{190}Hg	0	-	0.4 h	EC
	^{191}Hg	3/2-	-	0.9 h	EC
	^{192}Hg	0	-	5 h	EC, β^+
	^{193}Hg	3/2-	-0.63	4 h	EC
	^{194}Hg	0	-	1.2 a	EC
	^{195}Hg	1/2-	0.54	10 h	EC
	^{197}Hg	1/2-	0.527	2.7 d	EC
	^{203}Hg	5/2-	0.85	1.9 d	β^-
81	^{193}Tl	1/2	-	0.4 h	EC
	^{194}Tl	2-	-	0.5 h	EC
	^{195}Tl	1/2	1.6	1.2 h	EC, β^+
	^{196}Tl	2-	±0.07	1.8 h	EC
	^{197}Tl	1/2	1.6	2.8 h	EC
	^{198}Tl	2-	±0.001	5.3 h	EC
	^{199}Tl	1/2-	1.6	7.4 h	EC
	^{200}Tl	2-	±0.04	1.1 d	EC
	^{201}Tl	1/2	1.6	3.0 d	EC
	^{202}Tl	2-	±0.06	12 d	EC
	^{204}Tl	2-	±0.09	3.8 d	β^-, EC
82	^{195}Pb	-	-	0.6 h	EC
	^{197}Pb	3/2-	-	1 h	EC
	^{198}Pb	0	-	2.4 h	EC
	^{199}Pb	5/2-	-	1.5 h	EC
	^{200}Pb	0	-	20 h	EC
	^{201}Pb	5/2-	-	9 h	EC

Z	symbol	I	μ/μ_K	$T_{1/2}$	decay channel
	^{202}Pb	0	-	$3 \cdot 10^5$ a	EC
	^{203}Pb	5/2-	-	52 h	EC
	^{205}Pb	5/2-	-	$3 \cdot 10^7$ a	EC
	^{209}Pb	9/2	-	3.3 h	β^-
	^{210}Pb	0	-	22 a	β^-
	^{211}Pb	9/2	-	0.6 h	β^-
	^{214}Pb	0	-	0.44 h	β^-
83	^{199}Bi	9/2-	-	0.41 h	EC
	^{201}Bi	9/2-	-	1.8 h	EC
	^{202}Bi	5	-	1.6 h	EC
	^{203}Bi	9/2-	4.6	11.8 h	EC
	^{204}Bi	6	4.3	11.2 h	EC
	^{205}Bi	9/2-	4.2	15 d	EC
	^{206}Bi	6	4.6	6.2 d	EC
	^{207}Bi	9/2-	-	30 a	EC
	^{208}Bi	5	-	$4 \cdot 10^5$ a	EC
	^{210}Bi	1-	-0.045	5 d	β^-
	^{212}Bi	1-	-	1 h	β^-, α
	^{213}Bi	9/2-	-	0.8 h	β^-, α
	^{214}Bi	-	-	0.3 h	β^-
84	^{201}Po	3/2-	-	0.25 h	EC, α
	^{202}Po	0	0	0.7 h	EC, α
	^{203}Po	5/2-	-	0.6 h	EC, α
	^{204}Po	0	0	3.6 h	EC, α
	^{205}Po	5/2-	0.3	1.8 h	EC
	^{206}Po	0	0	9 d	EC, α
	^{207}Po	5/2-	0.3	6 h	EC
	^{208}Po	0	-	2.93 a	α
	^{209}Po	1/2-	0.8	102 a	α
	^{210}Po	0	0	138 d	α
85	^{205}At	9/2-	-	0.4 h	EC, α
	^{206}At	5	-	0.5 h	α, EC
	^{207}At	9/2-	-	1.8 h	EC, α
	^{208}At	6	-	1.6 h	EC, α
	^{209}At	6	-	5.5 h	EC, α
	^{210}At	5	-	8.3 h	EC
	^{211}At	9/2-	-	7.2 h	α, EC
86	^{208}Rn	0	-	0.4 h	EC, α
	^{209}Rn	5/2-	-	0.5 h	EC, α
	^{210}Rn	0	-	2.4 h	α, EC
	^{211}Rn	1/2-	-	16 h	EC, α
	^{212}Rn	0	-	0.4 h	α
	^{221}Rn	7/2	-	0.4 h	β^-, α
	^{222}Rn	0	-	3.82 d	α

Z	symbol	I	μ/μ_K	$T_{1/2}$	decay channel
87	^{212}Fr	5	-	0.32 h	EC, α
	^{222}Fr	2-	-	0.25 h	β^-
	^{223}Fr	3/2	-	0.35 h	β^-
88	^{223}Ra	3/2	-	11.4 d	α
	^{224}Ra	0	-	3.65 d	α
	^{225}Ra	3/2	-	15 d	β^-
	^{226}Ra	0	-	1620 a	α
	^{227}Ra	3/2	-	0.7 h	β^-
	^{228}Ra	0	-	6.7 a	β^-
	^{230}Ra	0	-	1 h	β^-
89	^{224}Ac	0-	-	3 h	EC, α
	^{225}Ac	3/2	-	10 d	α
	^{226}Ac	1-	-	1 d	β^-, EC
	^{227}Ac	3/2-	1.1	22 a	β^-, α
	^{228}Ac	3	-	6.1 h	β^-
	^{229}Ac	3/2	-	1.1 h	β^-
90	^{226}Th	0	-	0.5 h	α
	^{227}Th	3/2	-	18.7 d	α
	^{228}Th	0	-	1.9 a	α
	^{229}Th	5/2	0.5	7300 a	α
	^{230}Th	0	-	$8\cdot10^4$ a	α
	^{231}Th	5/2	-	25.5 h	β^-
	^{232}Th	0	0	$1.4\cdot10^{10}$ a	α
	^{233}Th	1/2	-	0.37 h	β^-
	^{234}Th	0	-	24 d	β^-
91	^{227}Pa	5/2-	-	0.64 h	α, EC
	^{228}Pa	3	-	22 h	EC, α
	^{229}Pa	5/2	-	1.5 d	EC, α
	^{230}Pa	2-	-	17 d	EC, β^-
	^{231}Pa	3/2-	±2.0	$3.3\cdot10^4$ a	α
	^{232}Pa	2-	-	1.3 d	β^-
	^{233}Pa	3/2-	3.5	27 d	β^-
	^{234}Pa	4	-	6.7 h	β^-
	^{235}Pa	3/2-	-	0.4 h	β^-
	^{236}Pa	1-	-	0.2 h	β^-
	^{237}Pa	1/2	-	0.65 h	β^-
92	^{229}U	3/2	-	1 h	EC, α
	^{230}U	0	-	20 d	α
	^{232}U	0	-	70 a	α
	^{233}U	5/2	0.6	$1.59\cdot10^5$ a	α
	^{234}U	0	0	$2.6\cdot10^5$ a	α
	^{235}U	7/2-	-0.3	$7.04\cdot10^8$ a	α
	^{236}U	0	-	$2.4\cdot10^7$ a	α
	^{237}U	1/2	-	6.7 d	β^-

Z	symbol	I	μ/μ_K	$T_{1/2}$	decay channel
	^{238}U	0	0	$4.5 \cdot 10^9$a	α
	^{239}U	5/2	-	0.4 h	β^-
	^{240}U	0	-	14 h	β^-
93	^{231}Np	5/2	-	0.9 h	EC, α
	^{232}Np	4-	-	0.2 h	EC
	^{233}Np	5/2	-	0.6 h	EC
	^{234}Np	0	-	4.4 d	EC
	^{235}Np	5/2	-	1.1 a	EC
	^{236}Np	6-	-	20 h	EC, β^-
	^{237}Np	5/2	3.1	$2.1 \cdot 10^6$a	α
	^{238}Np	2	-	2.1 d	β^-
	^{239}Np	5/2	0.3	2.35 d	β^-
	^{240}Np	5	-	1.1 h	β^-
	^{241}Np	5/2	-	0.3 h	α, β^-
94	^{232}Pu	0	-	0.6 h	EC, α
	^{233}Pu	-	-	0.3 h	EC
	^{234}Pu	0	-	9 h	EC, α
	^{235}Pu	5/2	-	0.4 h	EC
	^{236}Pu	0	-	2.8 a	α
	^{237}Pu	7/2-	-	46 d	EC
	^{238}Pu	0	-	89 a	α
	^{239}Pu	1/2	0.20	$2.411 \cdot 10^4$a	α
	^{240}Pu	0	-	6600 a	α
	^{241}Pu	5/2	-0.71	14 a	β^-
	^{242}Pu	0	-	$3.8 \cdot 10^5$a	α
	^{243}Pu	7/2	-	5.0 h	β^-
	^{244}Pu	0	-	$8 \cdot 10^7$a	α
	^{245}Pu	9/2-	-	10 h	β^-
	^{246}Pu	0	-	11 d	β^-
95	^{237}Am	5/2-	-	1.4 h	EC
	^{238}Am	1	-	1.9 h	EC
	^{239}Am	5/2-	-	12 h	EC
	^{240}Am	3-	-	2.1 d	EC
	^{241}Am	5/2-	1.6	433 a	α
	^{242}Am	1-	0.39	16 h	β^-, EC
	^{243}Am	5/2-	1.6	$7.4 \cdot 10^3$a	α
	^{244}Am	-	-	10 h	β^-
	^{245}Am	5/2	-	2 h	β^-
	^{246}Am	7-	-	0.4 h	β^-
	^{247}Am	-	-	0.4 h	β^-
96	^{238}Cm	0	-	2.5 h	EC, α
	^{239}Cm	-	-	3 h	EC
	^{240}Cm	0	-	27 d	α
	^{241}Cm	1/2	-	36 d	EC, α

Z	symbol	I	μ/μ_K	$T_{1/2}$	decay channel
	^{242}Cm	0	0	162 d	α
	^{243}Cm	5/2	± 0.4	28 a	α
	^{244}Cm	0	-	18 a	α
	^{245}Cm	7/2	± 0.5	$8 \cdot 10^3$ a	α
	^{246}Cm	0	-	$3 \cdot 10^3$ a	α
	^{247}Cm	9/2-	± 0.4	$1.6 \cdot 10^7$ a	α
	^{248}Cm	0	-	$5 \cdot 10^5$ a	α, sf
	^{249}Cm	1/2	-	1.1 h	β^-
	^{250}Cm	0	-	$2 \cdot 10^4$ a	sf
97	^{243}Bk	3/2-	-	4.5 h	EC
	^{244}Bk	4-	-	4.4 h	EC
	^{245}Bk	3/2-	-	5.0 d	EC
	^{246}Bk	2-	-	1.8 d	EC
	^{247}Bk	3/2-	-	58.3 d	α
	^{248}Bk	1-	-	20 h	β^-, EC
	^{249}Bk	7/2	± 2	310 d	β^-
	^{250}Bk	2-	-	2 h	β^-
	^{251}Bk	3/2-	-	1 h	β^-
98	^{244}Cf	0	-	0.3 h	α
	^{245}Cf	-	-	35 h	α, sf
	^{247}Cf	7/2	-	2.5 h	EC
	^{248}Cf	0	-	1 a	α
	^{249}Cf	9/2-	-	400 a	α
	^{250}Cf	0	-	13 a	α, sf
	^{251}Cf	1/2	-	898 a	α
	^{252}Cf	0	-	2.6 a	α, sf
	^{253}Cf	7/2	-	18 d	β^-
	^{254}Cf	0	-	60 d	sf
99	^{248}Es	-	-	0.4 h	EC
	^{249}Es	7/2	-	2 h	EC
	^{250}Es	6	-	8 h	EC
	^{251}Es	3/2-	-	1.5 d	EC
	^{252}Es	5-	-	140 d	α
	^{253}Es	7/2	4.1	20.5 d	α
	^{254}Es	7	-	270 d	α
	^{255}Es	7/2	-	40 d	β^-, α
100	^{250}Fm	0	-	0.5 h	α, EC
	^{251}Fm	9/2-	-	7 h	EC, α
	^{252}Fm	0	-	1 d	α, sf
	^{253}Fm	1/2	-	3 d	EC, α
	^{254}Fm	0	-	3.2 h	α
	^{255}Fm	7/2	-	20 h	α, sf
	^{256}Fm	0	-	3 h	sf, α
	^{257}Fm	9/2	-	90 d	α, sf

Z	symbol	I	μ/μ_K	$T_{1/2}$	decay channel
101	^{255}Md	7/2-	-	0.6 h	EC, α
	^{256}Md	-	-	1.3 h	EC, α
	^{257}Md	7/2-	-	3 h	α
	^{258}Md	8-	-	50 d	α

15 Mean X-Ray Transition Energies

The table summarizes mean X-ray transition energies $\bar{E}$ on the basis of line energies given in [50] by using X-ray emission rates I calculated in [488].

The calculation uses the relation

$$\bar{E} = \frac{\sum E_i I_i}{\sum I_i}.$$

For the M series, intensities given in [116] and [335] were used.

Mean X-ray transition energies are given for line combinations for transitions with identical initial vacancy states.

Li	Lithium	Z = 3
transition		$\bar{E}$ [eV]
K_α		51.3

Be	Beryllium	Z = 4
transition		$\bar{E}$ [eV]
K_α		108.5

B	Boron	Z = 5
transition		$\bar{E}$ [eV]
K_α		183.3

C	Carbon	Z = 6
transition		$\bar{E}$ [eV]
K_α		277.0

N	Nitrogen	Z = 7
transition		$\bar{E}$ [eV]
K_α		392.4

O	Oxygen	Z = 8
transition		$\bar{E}$ [eV]
K_α		524.9

F	Fluorine	Z = 9
transition		$\bar{E}$ [eV]
K_α		676.8

Ne	Neon	Z = 10
transition		$\bar{E}$ [eV]
K_α		848.6

Na	Sodium	Z = 11
transition		$\bar{E}$ [eV]
$K_{\alpha_1}\ K_{\alpha_2}$		1041.0

Mg	Magnesium	Z = 12
transition		$\bar{E}$ [eV]
$K_{\alpha_1}\ K_{\alpha_2}$		1253.6

Al   **Aluminium**   Z = 13

transition	$\bar{E}$ [eV]
K_{α_1} K_{α_2}	1486.6
K_β	1557.4

Si   **Silicon**   Z = 14

transition	$\bar{E}$ [eV]
K_{α_1} K_{α_2}	1739.8
K_β	1835.9

P   **Phosphorus**   Z = 15

transition	$\bar{E}$ [eV]
K_{α_1} K_{α_2}	2013.4
K_{β_1} K_{β_3}	2139.0

S   **Sulfur**   Z = 16

transition	$\bar{E}$ [eV]
K_{α_1} K_{α_2}	2307.4

Cl   **Chlorine**   Z = 17

transition	$\bar{E}$ [eV]
K_{α_1} K_{α_2}	2621.6
K_β	2815.6

Ar   **Argon**   Z = 18

transition	$\bar{E}$ [eV]
K_{α_1} K_{α_2}	2957.0
K_{β_1} K_{β_3}	3190.5

K   **Potassium**   Z = 19

transition	$\bar{E}$ [eV]
K_{α_1} K_{α_2}	3312.9
K_{β_1} K_{β_3}	3589.6

Ca   **Calcium**   Z = 20

transition	$\bar{E}$ [eV]
K_{α_1} K_{α_2}	3690.5
K_{β_1} K_{β_3}	4012.7

Sc   **Scandium**   Z = 21

transition	$\bar{E}$ [eV]
K_{α_1} K_{α_2}	4089.1
K_{β_1} K_{β_3}	4460.5
L_{β_3} K_{β_4}	467.8
L_{α_1} L_{α_2}	395.4

Ti   **Titanium**   Z = 22

transition	$\bar{E}$ [eV]
K_{α_1} K_{α_2}	4508.8
K_{β_1} K_{β_3}	4931.8
L_{α_1} L_{α_2}	452.2

V   **Vanadium**   Z = 23

transition	$\bar{E}$ [eV]
K_{α_1} K_{α_2}	4949.6
K_{β_1} K_{β_3}	5427.3
L_{β_3} K_{β_4}	585.0
L_{α_1} L_{α_2}	511.3

Cr   **Chromium**   Z = 24

transition	$\bar{E}$ [eV]
K_{α_1} K_{α_2}	5411.6
K_{β_1} K_{β_3}	5946.7
L_{β_3} K_{β_4}	654.0
L_{α_1} L_{α_2}	572.8

Mn   **Manganese**   Z=25

transition	$\bar{E}$ [eV]
K_{α_1} K_{α_2}	5895.1
K_{β_1} K_{β_3}	6490.5
L_{β_3} K_{β_4}	721.0
L_{α_1} L_{α_2}	637.4

Fe   Iron   Z=26

transition	$\bar{E}$ [eV]
$K_{\alpha_1}\ K_{\alpha_2}$	6399.4
$K_{\beta_1}\ K_{\beta_3}$	7058.0
$L_{\beta_3}\ K_{\beta_4}$	792.0
$L_{\alpha_1}\ L_{\alpha_2}$	705.0

Co   Cobalt   Z = 27

transition	$\bar{E}$ [eV]
$K_{\alpha_1}\ K_{\alpha_2}$	6925.2
$K_{\beta_1}\ K_{\beta_3}$	7649.4
$L_{\beta_3}\ K_{\beta_4}$	870.0
$L_{\alpha_1}\ L_{\alpha_2}$	776.2

Ni   Nickel   Z = 28

transition	$\bar{E}$ [eV]
$K_{\alpha_1}\ K_{\alpha_2}$	7472.3
$K_{\beta_1}\ K_{\beta_3}$	8264.7
$L_{\beta_3}\ K_{\beta_4}$	941.0
$L_{\alpha_1}\ L_{\alpha_2}$	851.5

Cu   Copper   Z = 29

transition	$\bar{E}$ [eV]
$K_{\alpha_1}\ K_{\alpha_2}$	8041.0
$K_{\beta_1}\ K_{\beta_3}$	8904.5
$L_{\beta_3}\ K_{\beta_4}$	1022.8
$L_{\alpha_1}\ L_{\alpha_2}$	922.7

Zn   Zinc   Z = 30

transition	$\bar{E}$ [eV]
$K_{\alpha_1}\ K_{\alpha_2}$	8631.1
$K_{\beta_1}\ K_{\beta_3}$	9572.0
$L_{\beta_3}\ K_{\beta_4}$	1107.0
$L_{\alpha_1}\ L_{\alpha_2}$	1011.7

Ga   Gallium   Z=31

transition	$\bar{E}$ [eV]
$K_{\alpha_1}\ K_{\alpha_2}$	9242.6
$K_{\beta_1}\ K_{\beta_3}$	10262.6
$L_{\beta_3}\ K_{\beta_4}$	1197.0
$L_{\alpha_1}\ L_{\alpha_2}$	1097.9

Ge   Germanium   Z=32

transition	$\bar{E}$ [eV]
$K_{\alpha_1}\ K_{\alpha_2}$	9875.8
$K_{\beta_1}\ K_{\beta_3}$	10980.6
$L_{\beta_3}\ K_{\beta_4}$	1291.3
$L_{\alpha_1}\ L_{\alpha_2}$	1188.0

As   Arsenic   Z = 33

transition	$\bar{E}$ [eV]
$K_{\alpha_1}\ K_{\alpha_2}$	10531.8
$K_{\beta_1}\ K_{\beta_3}$	11724.0
$L_{\beta_3}\ K_{\beta_4}$	1388.4
$L_{\alpha_1}\ L_{\alpha_2}$	1282.0

Se   Selenium   Z = 34

transition	$\bar{E}$ [eV]
$K_{\alpha_1}\ K_{\alpha_2}$	11209.1
$K_{\beta_1}\ K_{\beta_3}$	12494.0
$L_{\alpha_1}\ L_{\alpha_2}$	1379.1

Br   Bromine   Z = 35

transition	$\bar{E}$ [eV]
$K_{\alpha_1}\ K_{\alpha_2}$	11908.3
$K_{\beta_1}\ K_{\beta_3}$	13289.0
$L_{\alpha_1}\ L_{\alpha_2}$	1480.4

Kr   Krypton   Z = 36

transition	$\bar{E}$ [eV]
K_{α_1} K_{α_2}	12631.6
K_{β_1} K_{β_3}	14109.3
L_{β_3} L_{β_4}	11703.4
L_{α_1} L_{α_2}	1586.0

Rb   Rubidium   Z = 37

transition	$\bar{E}$ [eV]
K_{α_1} K_{α_2}	13375.0
K_{β_1} K_{β_3}	14958.0
L_{β_3} L_{β_4}	1823.4
L_{γ_2} L_{γ_3}	2050.7
L_{α_1} L_{α_2}	1693.9

Sr   Strontium   Z = 38

transition	$\bar{E}$ [eV]
K_{α_1} K_{α_2}	14142.0
K_{β_1} K_{β_3}	15832.0
L_{β_3} L_{β_4}	1943.3
L_{γ_2} L_{γ_3}	2196.5
L_{α_1} L_{α_2}	1806.4

Y   Yttrium   Z = 39

transition	$\bar{E}$ [eV]
K_{α_1} K_{α_2}	14932.3
K_{β_1} K_{β_3}	16733.7
L_{β_3} L_{β_4}	2067.8
L_{γ_2} L_{γ_3}	2346.8
L_{α_1} L_{α_2}	1922.4

Zr   Zirconium   Z = 40

transition	$\bar{E}$ [eV]
K_{α_1} K_{α_2}	15746.2
K_{β_1} K_{β_3}	17663.1
K'_{β_2}	17970.0
L_{β_3} L_{β_4}	2196.0
L_{γ_2} L_{γ_3}	2502.9
L_{α_1} L_{α_2}	2042.1
L_{β_2} $L_{\beta_{15}}$	2219.4

Nb   Niobium   Z = 41

transition	$\bar{E}$ [eV]
K_{α_1} K_{α_2}	16582.7
K_{β_1} K_{β_3}	18617.0
K'_{β_2}	18953.0
L_{β_3} L_{β_4}	2329.2
L_{γ_2} L_{γ_3}	2663.8
L_{α_1} L_{α_2}	2165.6
L_{β_2} $L_{\beta_{15}}$	2367.0

Mo   Molybdenum   Z = 42

transition	$\bar{E}$ [eV]
K_{α_1} K_{α_2}	17442.9
K_{β_1} K_{β_3}	19601.9
K'_{β_2}	19994.7
L_{β_3} L_{β_4}	2466.6
L_{γ_2} L_{γ_3}	2830.6
L_{α_1} L_{α_2}	2292.9
L_{β_2} $L_{\beta_{15}}$	2518.3

Tc   Technetium   Z = 43

transition	$\bar{E}$ [eV]
K_{α_1} K_{α_2}	18327.0
K_{β_1} K_{β_3}	20612.0

Ru   Ruthenium   Z = 44

transition	$\bar{E}$ [eV]
K_{α_1} K_{α_2}	19234.5
K_{β_1} K_{β_3}	21649.5
K'_{β_2}	22074.1
L_{β_3} L_{β_4}	2755.1
L_{γ_2} L_{γ_3}	3180.9
L_{α_1} L_{α_2}	2558.2
L_{β_2} $L_{\beta_{15}}$	2836.0

Rh   Rhodium   Z = 45

transition	$\bar{E}$ [eV]
K_{α_1} K_{α_2}	20166.9
K_{β_1} K_{β_3}	22715.5
K'_{β_2}	23171.4
L_{β_3} L_{β_4}	2906.4
L_{γ_2} L_{γ_3}	3364.0
L_{α_1} L_{α_2}	2696.2
L_{β_2} $L_{\beta_{15}}$	3001.3

Pd   Palladium   Z = 46

transition	$\bar{E}$ [eV]
K_{α_1} K_{α_2}	21122.6
K_{β_1} K_{β_3}	23809.5
K'_{β_2}	24299.2
L_{β_3} L_{β_4}	3062.7
L_{γ_2} L_{γ_3}	3553.3
L_{α_1} L_{α_2}	2838.1
L_{β_2} $L_{\beta_{15}}$	3171.8

Ag   Silver   Z = 47

transition	$\bar{E}$ [eV]
K_{α_1} K_{α_2}	22103.1
K_{β_1} K_{β_3}	24931.8
K'_{β_2}	25456.3
L_{β_3} L_{β_4}	3222.9
L_{γ_2} L_{γ_3}	3747.3
L_{α_1} L_{α_2}	2983.7
L_{β_2} $L_{\beta_{15}}$	3347.8

Cd   Cadmium   Z = 48

transition	$\bar{E}$ [eV]
K_{α_1} K_{α_2}	23108.0
K_{β_1} K_{β_3}	26084.1
L_{β_3} L_{β_4}	3388.6
L_{γ_2} L_{γ_3}	3951.3
L_{α_1} L_{α_2}	3133.0
L_{β_2} $L_{\beta_{15}}$	3528.1

In   Indium   Z = 49

transition	$\bar{E}$ [eV]
K_{β_1} K_{β_3}	27263.1
K'_{β_2}	27861.5
L_{β_3} L_{β_4}	3558.8
L_{γ_2} L_{γ_3}	4160.5
L_{α_1} L_{α_2}	3286.1
L_{β_2} $L_{\beta_{15}}$	3713.8

Sn   Tin   Z = 50

transition	$\bar{E}$ [eV]
K_{β_1} K_{β_3}	28471.7
K'_{β_2} $KO_{2,3}$	29112.2
L_{β_3} L_{β_4}	3734.1
L_{γ_2} L_{γ_3}	4376.8
L_{α_1} L_{α_2}	3443.1
L_{β_2} $L_{\beta_{15}}$	3904.9

Sb   Antimony   Z = 51

transition	$\bar{E}$ [eV]
K_{β_1} K_{β_3}	29710.0
K'_{β_2} $KO_{2,3}$	30396.1
L_{β_3} L_{β_4}	3915.0
L_{γ_2} L_{γ_3}	4599.9
L_{α_1} L_{α_2}	3603.7
L_{β_2} $L_{\beta_{15}}$	4100.8

Te   Tellurium   Z = 52

transition	$\bar{E}$ [eV]
K_{β_1} K_{β_3}	30978.3
L_{β_3} L_{β_4}	4100.8
L_{γ_2} L_{γ_3}	4829.0
L_{γ_1} L_{γ_5}	4565.1
L_{α_1} L_{α_2}	3768.2
L_{β_2} $L_{\beta_{15}}$	4301.7

I   Iodine   Z = 53

transition	$\bar{E}$ [eV]
$K_{\beta_1}\ K_{\beta_3}$	32276.0
$L_{\beta_3}\ L_{\beta_4}$	4291.8
$L_{\gamma_2}\ L_{\gamma_3}$	5065.7
$L_{\alpha_1}\ L_{\alpha_2}$	3936.5
$L_{\beta_2}\ L_{\beta_{15}}$	4507.5

Xe   Xenon   Z = 54

transition	$\bar{E}$ [eV]
$K_{\beta_1}\ K_{\beta_3}$	33602.9

Cs   Cesium   Z = 55

transition	$\bar{E}$ [eV]
$K_{\beta_1}\ K_{\beta_3}$	34963.8
$L_{\beta_3}\ L_{\beta_4}$	4690.4
$L_{\gamma_2}\ L_{\gamma_3}$	5548.6
$L_{\alpha_1}\ L_{\alpha_2}$	4285.0
$L_{\beta_2}\ L_{\beta_{15}}$	4935.9

Ba   Barium   Z = 56

transition	$\bar{E}$ [eV]
$K_{\beta_1}\ K_{\beta_3}$	36352.8
$L_{\beta_3}\ L_{\beta_4}$	4897.5
$L_{\gamma_2}\ L_{\gamma_3}\ L_{\gamma_4}$	5818.4
$L_{\alpha_1}\ L_{\alpha_2}$	4464.7
$L_{\beta_2}\ L_{\beta_{15}}$	5165.5

La   Lanthanum   Z = 57

transition	$\bar{E}$ [eV]
$K_{\beta_1}\ K_{\beta_3}$	37773.4
$K'_{\beta_2}\ KO_{2,3}$	38754.6
$L_{\beta_3}\ L_{\beta_4}$	5111.3
$L_{\gamma_2}\ L_{\gamma_3}$	6068.6
$L_{\alpha_1}\ L_{\alpha_2}$	4649.3
$L_{\beta_2}\ L_{\beta_{15}}$	5383.5
$M_{\alpha_1}\ M_{\alpha_2}$	833.0

Ce   Cerium   Z = 58

transition	$\bar{E}$ [eV]
$K_{\beta_1}\ K_{\beta_3}$	39227.4
$K'_{\beta_2}\ KO_{2,3}$	40257.9
$L_{\beta_3}\ L_{\beta_4}$	5329.9
$L_{\gamma_2}\ L_{\gamma_3}$	6339.9
$L_{\alpha_1}\ L_{\alpha_2}$	4838.5
$L_{\beta_2}\ L_{\beta_{15}}$	5613.4
$M_{\alpha_1}\ M_{\alpha_2}$	883.0

Pr   Praseodymium   Z = 59

transition	$\bar{E}$ [eV]
$K_{\beta_1}\ K_{\beta_3}$	40715.6
$L_{\beta_3}\ L_{\beta_4}$	5554.4
$L_{\gamma_2}\ L_{\gamma_3}$	6609.0
$L_{\alpha_1}\ L_{\alpha_2}$	5031.6
$L_{\beta_2}\ L_{\beta_{15}}$	5850.0
$M_{\alpha_1}\ M_{\alpha_2}$	929.2

Nd   Neodymium   Z = 60

transition	$\bar{E}$ [eV]
$K_{\beta_1}\ K_{\beta_3}$	42235.6
$L_{\beta_3}\ L_{\beta_4}$	5786.1
$L_{\gamma_2}\ L_{\gamma_3}$	6894.5
$L_{\alpha_1}\ L_{\alpha_2}$	5228.1
$L_{\beta_2}\ L_{\beta_{15}}$	6089.4
$M_{\alpha_1}\ M_{\alpha_2}$	978.0

Pm   Promethium   Z = 61

transition	$\bar{E}$ [eV]
$K_{\beta_1}\ K_{\beta_3}$	43787.5
$L_{\alpha_1}\ L_{\alpha_2}$	5429.9
$L_{\beta_2}\ L_{\beta_{15}}$	6339.0

Sm — Samarium — Z = 62

transition	$\bar{E}$ [eV]
K_{β_1} K_{β_3}	45370.8
L_{β_3} L_{β_4}	6268.5
L_{γ_2} L_{γ_3}	7478.4
L_{α_1} L_{α_2}	5633.3
L_{β_2} $L_{\beta_{15}}$	6587.0
M_{α_1} M_{α_2}	1081.0

Eu — Europium — Z = 63

transition	$\bar{E}$ [eV]
K_{β_1} K_{β_3}	46992.3
L_{β_3} L_{β_4}	6399.0
L_{γ_2} L_{γ_3}	7784.7
L_{α_1} L_{α_2}	5842.7
L_{β_2} $L_{\beta_{15}}$	6843.2
M_{γ_1} M_{γ_2}	1346.0
M_{α_1} M_{α_2}	1131.0

Gd — Gadolinium — Z = 64

transition	$\bar{E}$ [eV]
K_{β_1} K_{β_3}	48648.6
L_{β_3} L_{β_4}	6771.8
L_{γ_2} L_{γ_3}	8097.8
L_{α_1} L_{α_2}	6053.9
L_{β_2} $L_{\beta_{15}}$	7110.3
M_{γ_1} M_{γ_2}	1402.0
M_{α_1} M_{α_2}	1185.0

Tb — Terbium — Z = 65

transition	$\bar{E}$ [eV]
K_{β_1} K_{β_3}	50329.9
L_{β_3} L_{β_4}	7031.3
L_{γ_2} L_{γ_3}	8412.9
L_{α_1} L_{α_2}	6269.3
L_{β_2} $L_{\beta_{15}}$	7366.7
M_{α_1} M_{α_2}	1240.0

Dy — Dysprosium — Z = 66

transition	$\bar{E}$ [eV]
K_{β_1} K_{β_3}	52063.8
L_{β_3} L_{β_4}	7300.7
L_{γ_2} L_{γ_3}	8737.2
L_{α_1} L_{α_2}	6491.4
L_{β_2} $L_{\beta_{15}}$	7635.7
M_{α_1} M_{α_2}	1293.0

Ho — Holmium — Z = 67

transition	$\bar{E}$ [eV]
K_{β_1} K_{β_3}	53820.4
L_{β_3} L_{β_4}	7575.7
L_{γ_2} L_{γ_3}	9072.2
L_{α_1} L_{α_2}	6715.7
L_{β_2} $L_{\beta_{15}}$	7911.0
M_{α_1} M_{α_2}	1348.0

Er — Erbium — Z = 68

transition	$\bar{E}$ [eV]
L_{β_3} L_{β_4}	7857.0
L_{γ_2} L_{γ_3}	9412.0
L_{α_1} L_{α_2}	6944.0
L_{β_2} $L_{\beta_{15}}$	8189.0
M_{α_1} M_{α_2}	1406.0

Tm — Thulium — Z = 69

transition	$\bar{E}$ [eV]
L_{γ_2} L_{γ_3}	9758.6
L_{α_1} L_{α_2}	7175.1
L_{β_2} $L_{\beta_{15}}$	8468.0
M_{α_1} M_{α_2}	1402.0

Yb — Ytterbium — Z = 70

transition	$\bar{E}$ [eV]
L_{γ_2} L_{γ_3}	10120.8
L_{α_1} L_{α_2}	7410.7
L_{β_2} $L_{\beta_{15}}$	8758.8
M_{α_1} M_{α_2}	1521.4

Lu	Lutetium	Z = 71
transition		$\bar{E}$ [eV]
$L_{\gamma_2}\ L_{\gamma_3}$		10489.5
$L_{\alpha_1}\ L_{\alpha_2}$		7650.4
$L_{\beta_2}\ L_{\beta_{15}}$		9048.0
$M_{\alpha_1}\ M_{\alpha_2}$		1581.3

Hf	Hafnium	Z = 72
transition		$\bar{E}$ [eV]
$L_{\gamma_2}\ L_{\gamma_3}$		10866.8
$L'_{\gamma_4}\ L_{\gamma_4}$		11237.4
$L_{\alpha_1}\ L_{\alpha_2}$		7893.5
$L_{\beta_2}\ L_{\beta_{15}}$		9346.3
$M_{\alpha_1}\ M_{\alpha_2}$		1644.6

Ta	Tantalum	Z = 73
transition		$\bar{E}$ [eV]
K'_{β_2}		66994.5
$L_{\gamma_2}\ L_{\gamma_3}$		11252.0
$L'_{\gamma_4}\ L_{\gamma_4}$		11641.5
$L_{\alpha_1}\ L_{\alpha_2}$		8140.2
$L_{\beta_2}\ L_{\beta_{15}}$		9650.6
$M_{\gamma_1}\ M_{\gamma_2}$		1957.5
$M_{\alpha_1}\ M_{\alpha_2}$		1709.6

W	Tungsten	Z = 74
transition		$\bar{E}$ [eV]
K'_{β_2}		69080.9
$L_{\gamma_2}\ L_{\gamma_3}$		11645.6
$L'_{\gamma_4}\ L_{\gamma_4}$		12058.6
$L_{\alpha_1}\ L_{\alpha_2}$		8391.2
$L_{\beta_2}\ L_{\beta_{15}}$		9960.1
$M_{\gamma_1}\ M_{\gamma_2}$		2028.0
$M_{\alpha_1}\ M_{\alpha_2}$		1775.3

Re	Rhenium	Z = 75
transition		$\bar{E}$ [eV]
K'_{β_2}		71208.2
$L_{\gamma_2}\ L_{\gamma_3}$		12050.8
$L_{\alpha_1}\ L_{\alpha_2}$		8645.8
$L_{\beta_2}\ L_{\beta_{15}}$		10273.9

Os	Osmium	Z = 76
transition		$\bar{E}$ [eV]
$L_{\gamma_2}\ L_{\gamma_3}$		12466.0
$L'_{\gamma_4}\ L_{\gamma_4}$		12917.0
$L_{\alpha_1}\ L_{\alpha_2}$		8904.5
$L_{\beta_2}\ L_{\beta_{15}}$		10597.3
$M_{\gamma_1}\ M_{\gamma_2}$		2174.0
$M_{\alpha_1}\ M_{\alpha_2}$		1910.2

Ir	Iridium	Z = 77
transition		$\bar{E}$ [eV]
$L_{\gamma_2}\ L_{\gamma_3}$		12889.0
$L'_{\gamma_4}\ L_{\gamma_4}$		13362.6
$L_{\alpha_1}\ L_{\alpha_2}$		9167.3
$L_{\beta_2}\ L_{\beta_{15}}$		10918.4
$M_{\gamma_1}\ M_{\gamma_2}$		2246.0

Pt	Platinum	Z = 78
transition		$\bar{E}$ [eV]
$L_{\gamma_2}\ L_{\gamma_3}$		13320.7
$L'_{\gamma_4}\ L_{\gamma_4}$		13822.1
$L_{\alpha_1}\ L_{\alpha_2}$		9434.1
$M_{\gamma_1}\ M_{\gamma_2}$		2322.5

Au	Gold	Z = 79
transition		$\bar{E}$ [eV]
$L_{\gamma_2}\ L_{\gamma_3}$		13764.9
$L'_{\gamma_4}\ L_{\gamma_4}$		14290.8
$L_{\alpha_1}\ L_{\alpha_2}$		9704.6
$L_{\beta_2}\ L_{\beta_{15}}$		11583.2
$M_{\gamma_1}\ M_{\gamma_2}$		2409.8
$M_{\alpha_1}\ M_{\alpha_2}$		2122.7

Hg — Mercury — Z = 80

transition	$\bar{E}$ [eV]
$L_{\gamma_2}\ L_{\gamma_3}$	14218.7
$L'_{\gamma_4}\ L_{\gamma_4}$	14768.4
$L_{\alpha_1}\ L_{\alpha_2}$	9979.5
$L_{\beta_2}\ L_{\beta_{15}}$	11922.0

Tl — Thallium — Z = 81

transition	$\bar{E}$ [eV]
$L_{\gamma_2}\ L_{\gamma_3}$	14686.3
$L'_{\gamma_4}\ L_{\gamma_4}$	15261.0
$L_{\alpha_1}\ L_{\alpha_2}$	10252.2
$L_{\beta_2}\ L_{\beta_{15}}$	12269.9
$M_{\gamma_1}\ M_{\gamma_2}$	2559.5

Pb — Lead — Z = 82

transition	$\bar{E}$ [eV]
$L_{\gamma_2}\ L_{\gamma_3}$	15164.6
$L'_{\gamma_4}\ L_{\gamma_4}$	15765.5
$L_{\alpha_1}\ L_{\alpha_2}$	10541.6
$L_{\beta_2}\ L_{\beta_{15}}$	12620.8
$M_{\gamma_1}\ M_{\gamma_2}$	2650.1
$M_{\alpha_1}\ M_{\alpha_2}$	2345.4

Bi — Bismuth — Z = 83

transition	$\bar{E}$ [eV]
$L_{\gamma_2}\ L_{\gamma_3}$	15651.1
$L'_{\gamma_4}\ L_{\gamma_4}$	16283.3
$L_{\alpha_1}\ L_{\alpha_2}$	10828.0
$L_{\beta_2}\ L_{\beta_{15}}$	12977.5
$M_{\gamma_1}\ M_{\gamma_2}$	2723.5

Po — Polonium — Z = 84

transition	$\bar{E}$ [eV]
$L_{\gamma_2}\ L_{\gamma_3}$	16149.3
$L_{\alpha_1}\ L_{\alpha_2}$	11119.3
$L_{\beta_2}\ L_{\beta_{15}}$	13337.4

At — Astatine — Z = 85

transition	$\bar{E}$ [eV]
$L_{\alpha_1}\ L_{\alpha_2}$	11414.5

Rn — Radon — Z = 86

transition	$\bar{E}$ [eV]
$L_{\alpha_1}\ L_{\alpha_2}$	11713.8

Fr — Francium — Z = 87

transition	$\bar{E}$ [eV]
$L_{\alpha_1}\ L_{\alpha_2}$	12017.1

Ra — Radium — Z = 88

transition	$\bar{E}$ [eV]
$L_{\gamma_2}\ L_{\gamma_3}$	18271.3
$L'_{\gamma_4}\ L_{\gamma_4}$	19060.9
$L_{\alpha_1}\ L_{\alpha_2}$	12325.3
$L_{\beta_2}\ L_{\beta_{15}}$	14837.8

Ac — Actinium — Z = 89

transition	$\bar{E}$ [eV]
$L_{\alpha_1}\ L_{\alpha_2}$	12636.6

Th — Thorium — Z = 90

transition	$\bar{E}$ [eV]
$L_{\beta_9}\ L_{\beta_{10}}$	17075.6
$L'_{\gamma_4}\ L_{\gamma_4}$	20267.6
$L_{\alpha_1}\ L_{\alpha_2}$	12952.8
$L_{\beta_2}\ L_{\beta_{15}}$	15620.4
$M_{\gamma_1}\ M_{\gamma_2}$	3352.5

Pa   Protactinium   Z = 91

transition	$\bar{E}$ [eV]
L_{β_9} $L_{\beta_{10}}$	17596.7
L'_{γ_4} L_{γ_4}	20882.0
L_{α_1} L_{α_2}	13273.7
M_{γ_1} M_{γ_2}	3447.9
M_{α_1} M_{α_2}	3081.7

U   Uranium   Z = 92

transition	$\bar{E}$ [eV]
L_{β_9} $L_{\beta_{10}}$	18135.2
L'_{γ_4} L_{γ_4}	21530.1
L_{α_1} L_{α_2}	13597.0
L_{β_2} $L_{\beta_{15}}$	16423.9
M_{γ_1} M_{γ_2}	3558.1
M_{α_1} M_{α_2}	3170.3

Np   Neptunium   Z = 93

transition	$\bar{E}$ [eV]
L_{β_9} $L_{\beta_{10}}$	18686.9
L'_{γ_4} L_{γ_4}	22171.1
L_{α_1} L_{α_2}	13925.1
L_{β_2} $L_{\beta_{15}}$	16835.5

Pu   Plutonium   Z = 94

transition	$\bar{E}$ [eV]
L_{β_9} $L_{\beta_{10}}$	19284.0
L'_{γ_4} L_{γ_4}	22856.4
L_{α_1} L_{α_2}	14258.7
L_{β_2} $L_{\beta_{15}}$	17250.4

Am   Americium   Z = 95

transition	$\bar{E}$ [eV]
L_{β_2} $L_{\beta_{15}}$	17671.5

References

A

1. T. Åberg: Phys. Rev. A **4**, 1735 (1971)
2. T. Åberg: *Proc. Int. Conf. Inner Shell Ioniz. Phenomena and Future Appl.* , Atlanta, US At. Energy Comm. Rep. No CONF-720404 (Oak Ridge, Tennesee 1973), p 1509
3. H. Ågren et al.: UIIP-944 Report, Uppsala University, Institute of Physics, Dec. 1976
4. B. K. Agarwal: *X-Ray Spectroscopy*, 2nd edition (Springer, Berlin Heidelberg New York 1991)
5. I. Ahmad, F. T. Porter, M. S. Freedman, R. F. Barnes, R. K. Sjoblom, F. Wagner Jr, J. Milsted, P. R. Fields: Phys. Rev. C **3**, 390 (1971)
6. H. Aiginger, P. Wobrauschek, C. Brauner: Nuclear Instruments and Methods **120**, 541 (1974)
7. K. Alder et al.: Rev. Mod. Phys. **28**, 432 (1956)
8. K. Alder, G. Baur, U. Raff: Helvetica Physica Acta **45**, 765 (1972)
9. K. L. Allawadhi, B. S. Sood, R. Mittal, N. Singh, J. K. Sharma: X-Ray Spectrometry **25**, 233 (1996)
10. S. K. Allinson: Phys. Rev. **44**, 63 (1933)
11. M. Ya. Amusia, V. M. Balmistrov, B. A. Zon et al.: *Polarization Bremsstrahlung from Particles and Atoms* (Nauka, Moscow 1987; in Russian)
12. M. Ya. Amusia, V. N. Buimistrov, V. N. Tsitovich, B. A. Zon: *Polarization Radiation*, ed by V. N. Tsitovich and I. M. Oiringel (Plenum Press 1993)
13. M. Ya. Amusia: Many-Body Theory of Atomic Structure and Processes, in: *Atomic, Molecular and Optical Physics Handbook*, ed by Gordon W. F. Drake (American Institute of Physics, Woodbury New York 1996), p 299
14. G. Angloher, P. Hettl, M. Huber, J. Jochum, F. Feilizsch, R. L. Mössbauer: J. Appl. Phys. **89**, 1425 (2001)
15. A. Anders: Phys. Rev. E **55**, 969 (1997)
16. K. V. Anisovic, N. I. Komjak: PTE (in Russian) **1**, 228 (1981)
17. E. Arndt et al.: Physics Letters **83A**, 164 (1981)
18. A. Assmann, M. Wendt: Spectrochimica Acta B **58**, 711 (2003)
19. A. Assmann, J. Dellith, M. Wendt: Microchimica Acta **155**, 87 (2006)
20. J. Auerhammer, H. Genz, A. Richter: Journ. de Physique, Coll. C **9**, 621 (1987)
21. J. Auerhammer, H. Genz, A. Richter: Journ. Phys. D **7**, 301 (1988)
22. P. Auger: C. R. Acad. Sci. **178**, 929 (1924)
23. P. Auger: C. R. Acad. Sci. **180**, 65 (1925)
24. P. Auger: J. Phys. Rad. **6**, 205 (1925)
25. P. Auger: C. R. Acad. Sci. **182**, 773 and 1215 (1926)
26. P. Auger: Ann. Phys. **6**, 183 (1926)
27. N. B. Avdonina, R. H. Pratt: J. Quant. Spectrosc. Radiat. Transfer **50**, 349 (1993)

B

28. J. J. Balmer: Ann. Physik **25**, 80 (1885)
29. F. A. Babushkin: Acta Phys. Polonia **XXV**, 749 (1964)

30. F. A. Babushkin: Optics and Spectroscopy **XIX**, 3 (1965)
31. F. A. Babushkin: Acta Phys. Polonia **XXXI** 459 (1967)
32. P. S. Bagus: Physical Review **139**, A619 (1965)
33. K. M. Balakrishna, N. Govinda Nayak, N. Lingappa, K. Siddappa: J. Phys. B: At. Mol. Opt. Phys. B **27**, 715 (1994)
34. W. Bambynek: Z. Phys. **206**, 662 (1967)
35. W. Bambynek et al.: *Proc. Int. Conf. on Electron Capture and Higher Order Processes in Nuclear Decay* (Budapest 1968) p 253
36. W. Bambynek, D. Reher: Z. Phys. **214**, 374 (1968)
37. W. Bambynek, D. Reher: Z. Phys. **228**, 49 (1970)
38. W. Bambynek, B. Crasemann, R. W. Fink, H.-U. Freund, H. Mark, C. D. Swift, R. E. Price, V. Venugopala Rao: Review of Modern Physics **44**, 716 (1972)
39. W. Bambynek: New Evaluation of K-Fluorescence Yields, in: *Proc. Int. Conf. on X-Ray and Inner-Shell Processes in Atoms, Molecules and Solids* (Leipzig 1984), Post-Deadline Abstracts, p 1
40. P. Bailey, C. Castelli, M. Cross, P. van Essen, A. Holland, F. Jansen, P. de Korte, D. Lumb, K. M. Carthy, P. Pool, P. Verhoeve: XRA 90/05, University of Leicester, 1990
41. L. E. Baily, J. B. Swedlund: Phys. Rev. **158** 6 (1967)
42. C. G. Barkla: Philos. Trans. R. Soc. (London) **204A**, 467 (1905)
43. G. Barreau, H. G. Börner, T. v. Egedy, R. W. Hoff: Zeitschrift für Physik **308**, 209 (1982)
44. K. Bartschat: Phys. Rep. **180** 1 (1989)
45. E. Baydas, N. Ekinci, E. Büyükkasap, Y. Sahin: Spectrochimica Acta B **53**, 151 (1998)
46. J. A. Bearden: Phys. Rev. **37**, 1210 (1931)
47. J. A. Bearden: Phys. Rev. **48**, 385 (1935)
48. J. A. Bearden, T. M. Snyder: Phys. Rev. **59**, 162 (1941)
49. J. A. Bearden et al.: Phys. Rev. **135**, A899 (1964)
50. J. A. Bearden: Review of Modern Physics **39**, 78 (1967)
51. J. A. Bearden, A. F. Burr: Review of Modern Physics **39**, 125 (1967)
52. W. W. Beeman, H. Friedman: Phys. Rev. **56**, 392 (1939)
53. W. Beer: IPF-SP-004 (Fribourg 1974)
54. P. Beiersdorfer et al.: Rev. Sci. Instruments **61** (1990)
55. K. L. Bell et al.: J. Phys. B **3**, 959 (1970)
56. S. Bemis: Physics Today, September 1973, 17
57. P. F. Bernath: *Spectra of Atoms and Molecules* (Oxford University Press, New York Oxford 1995)
58. H. Bethe: Ann. d. Phys. **5**, 325 (1930)
59. H. A. Bethe and J. Ashkin: *Experimental Nuclear Physics*, Vol. I, New York 1953
60. H. A. Bethe, W. Heitler: Proc. Royal Soc. A. **146**, 83 (1934)
61. H. F. Beyer; GSI-Report 79-6 (Gesellschaft für Schwerionenforschung, Darmstadt 1979)
62. H. F. Beyer, H.-J. Kluge, V. P. Shevelko: *X-Ray Radiation of Highly Charged Ions* (Springer Berlin Heidelberg New York 1997)
63. K. Bethge, G. Gruber: *Physik der Atome und Moleküle* (VCH, Weinheim und Basel 1990)
64. C. P. Bhalla: Nuclear Instruments and Methods **90**, 149 (1970)
65. C. P. Bhalla: J. Phys. B **3**, 916 (1970)
66. C. P. Bhalla: Phys. Rev. A **2**, 2573 (1970)
67. C. P. Bhalla: Phys. Rev. B **3**, L9 (1970)
68. C. P. Bhalla, M. Hein: Phys. Rev. Letters **30**, 39 (1973)
69. C. Bhan, S. N. Chaturvedi, N. Nath: X-Ray Spectrometry **15**, 217 (1986)
70. L. Bischoff, J. von Borany, H. Morgenstern, B. Schmidt, D. Schubert: ZfK-579 (Dresden-Rossendorf 1986)
71. S. Bloch: Ann. d. Phys. **38**, 559 (1912)
72. M. A. Blochin: *Physik der Röntgenstrahlen* (Verlag der Technik, Berlin 1957)
73. M. Blochin, I. Nikiforov: Bull. Acad. Sci. USSR **28**, 780 (1964)
74. M. A. Blochin, U. G. Schweitzer; *X-Ray Spectral Handbook* (Nauka, Moscow 1982; in Russian)
75. http://www.bmsc.wasgington.edu/scatter/AS_periodic.html

76. F. Boehm: Isotope Shifts, Chemical Shifts and Hyperfine Interactions of Atomic K X-Rays, in: *Atomic Inner-Shell Processes*, Vol. I (Academic Press, New York a. o. 1975) p 411
77. N. Bohr: Philos. Mag. **26**, 1 (1913)
78. G. L. Borchert: Z. Naturforsch. **31A**, 102 (1976)
79. G. L. Borchert et al.: Physics Letters **66A**, 347 (1978)
80. G. L. Borchert, P. G. Hansen, B. Johnson, H. L. Ravn, J. P. Desclaux: in: *Atomic Masses and Fundamental Constants 6*, edited by J. Nolen and W. Benenson (Plenum Press, New York 1980) p 189
81. P. Boyer, J. L. Barat: Nuclear Physics **A115**, 521 (1968)
82. W. H. Bragg, W. L. Bragg: Proc. Roy. Soc. **88A**, 428 (1913)
83. F. S. Brackett: Astrophys. J. **56**, 154 (1922)
84. M. Breinig et al.: Phys. Rev. **A22**, 520 (1980)
85. J. S. Briggs, K. Taulbjerg: Theory of Inelastic Atom-Atom-Collisions, in: *Structures and Collisions of Ions and Atoms* (Springer, Berlin Heidelberg New York 1978) p 105
86. G. Brogren: Ark. Phys. **8**, 391 (1954)
87. N. Broll: X-Ray Spectrometry **15**, 271 (1986)
88. C. F. Bunge: Phys. Rev. A **14**, 1965 (1967)
89. A. Burges: *Proc. of the 3rd Int. Conf. on the Physics of Electronic and Atomic Collisions*, ed by M. R. C. McDowell (North Holland Publ. Comp. , 1964) p 237
90. E. H. S. Burhop: *The Auger Effect and Other Radiationless Transitions* (Cambridge Press, Cambridge 1952), p 50
91. E. H. S. Burhop, W. N. Asaad: in: *Advances in Atomic and Molecular Physics*, Vol. 8, ed by D. R. Bates (Academic Press, New York 1972)

C

92. J. L. Campbell, P. L. Mc Ghee: Journal de Physique **48**, 597 (1987)
93. J. L. Campbell, T. Papp: X-Ray Spectrometry **24**, 307 (1995)
94. J. L. Campbell, T. Papp: Atomic Data and Nuclear Data Tables **77**, 1 (2001)
95. M. Carlen, J. C. Dousse, M. Gasser, J. Hoszowska, J. Kern, Ch. Rheme, P. Rymuza, Z. Sujkowski: Z. Phys. D **23**, 71 (1992)
96. T. A. Carlson 1966: Phys. Rev. **151**, 41 (1966)
97. T. A. Carlson et al.: Phys. Rev. **169**, 27 (1968)
98. T. A. Carlson, C. W. Nestor, Jr. , F. B. Malik, T. C. Tucker: Nuclear Physics A **135**, 57 (1969)
99. T. A. Carlson, C. W. Nestor, Jr. , N. Wassermann, J. D. McDowell: Atomic Data **2**, 63 (1970)
100. T. A. Carlson, C. W. Nestor Jr: Phys. Rev. A **8**, 2887 (1973)
101. Y. Cauchois; Compt. Rend. **194**, 362 (1932) and **195**, 228 (1932)
102. Y. Cauchois: Comp. Rend. **199**, 857 (1934)
103. Y. Cauchois, C. Bonelle: X-Ray Diffraction Spectrometry, in: *Atomic Inner-Shell Processes*, Vol. II (Academic Press, New York a. o. 1975)
104. Y. Cauchois, C. Senemaud: *Wavelengths of X-Ray Emission Lines and Absorption Edges* (Pergamon Press, Oxford 1978)
105. G. Charpak, R. Bouclier, T. Bressani, J. Favier, C. Zupancic: Nuclear Instruments and Methods **62**, 235 (1968)
106. G. Charpac, R. Bouclier, M. Steiner: Nuclear Instruments and Methods **80**, 13 (1970)
107. M. H. Chen et al.: Phys. Rev. A **4**, 1 (1971)
108. M. H. Chen et al.: Phys. Rev. A **19**, 2253 (1979)
109. M. H. Chen, B. Craseman, H. Mark: Atomic Data and Nuclear Data Tables **24**, 13 (1979)
110. M. H. Chen, B. Craseman, H. Mark: Phys. Rev. A **24**, 13 (1979)
111. M. H. Chen, B. Craseman, H. Mark: Phys. Rev. A **21**, 449 (1980)
112. M. H. Chen, B. Craseman, H. Mark: Phys. Rev. A **21**, 436 (1980)
113. M. H. Chen, B. Craseman, H. Mark: Phys. Rev. A **21**, 436 and 449 (1981)
114. M. H. Chen, B. Craseman, H. Mark: Phys. Rev. A **24**, 177 (1981)
115. M. H. Chen, B. Craseman, H. Mark: Phys. Rev. A **27**, 2989 (1983)

116. M. H. Chen, B. Craseman: Phys. Rev. A **30**, 170 (1984)
117. Y-Y-Chu, M. L. Perlman, P. F. Dittner, C. E. Bermis: Phys. Rev. A **5**, 67 (1972)
118. F. N. Chukhowski et al.: J. Appl. Phys. **77**, 1843 and 1849 (1995)
119. P. H. Citrin et al.: Phys. Rev. B **10**, 1762 (1974)
120. W. M. Coates: Phys. Rev. **46** 542 (1934)
121. D. C. Cohen: X-Ray Spectrometry **16**, 237 (1987)
122. D. D. Cohen: Nuclear Instruments and Methods A **267**, 492 (1988)
123. A. Compton, W. Duane: Proc. Nat. Acad. Sci. **11**, 598 (1925)
124. E. U. Condon, G. H. Shortley: *The Theory of Atomic Spectra* (Cambridge Univ. Press, Cambridge 1953)
125. D. Coster: Phil. Mag. **43**, 1070 and 1088 (1922)
126. R. D. Cowan: *The Theory of Atomic Structure and Spectra* (University of California Press, Berkeley a. o. 1981)
127. D. H. Crandall: Physica Scripta **23**, 153 (1981)
128. B. Craseman et al.: Phys. Rev. A **4**, 2161 (1971)
129. D. T. Cromer, J. T. Waber: Atomic Scaterring Factors for X-Rays, in: *International Tables for X-Ray Crystallography*, Vol. III (Kynoch Press, Birmingham 1974) p 237

D

130. A. Dalgarno: in: *At. Phys. Proc. Int. Conf.* (Plenum Press, New York 1969) p 161
131. A. Dalgarno: Nuclear Instruments and Methods **110**, 183 (1973)
132. Davis, Purks: Proc. Nat. Acad. Sci. **13**, 419 (1927)
133. C. F. G. Delaney, E. C. Finch: *Radiation Detectors* (Clarendon Press, Oxford 1992)
134. J. Dellith, M. Wendth: Microchim. Acta **145**, 25 (2004)
135. G. B. Deodhar, P. P. Varma, R. B. Singh: J. Phys. B **1**, 997 (1968)
136. G. B. Deodhar, P. P. Varma, R. B. Singh: Can. J. Phys. **47**, 341 (1969)
137. G. B. Deodhar, P. P. Varma: J. Phys. B **2**, 410 (1969)
138. G. B. Deodhar, P. P. Varma: Physica A **43**, 209 (1969)
139. J. P. Desclaux: J. Phys. B **4**, 631 (1971)
140. J. P. Desclaux: Computer Physics Communications **9**, 31 (1975)
141. R. D. Deslattes: Review of Scientific Instruments **38**, 616 (1967)
142. R. D. Deslattes et al.: Visible to Gamma-Ray Wavelength Ratio, in: *Atomic Masses and Fundamental Constants* (Plenum Press, London 1975) p 48
143. R. D. Deslattes, E. G. Kessler, W. C. Sauder, A. Henins: Annals of Physics, **37** (1980) 378
144. R. D. Deslattes, E. G. Kessler Jr: Experimental Evaluation of Inner-Vacancy Levels Energies for Comparison with Theory, in: *Inner-Shell Physics*, ed by B. Crasemann (Plenum Publ. Corp., New York 1985)
145. R. D. Deslattes, E. G. Kessler, Jr., P. Indelicato, L. deBilly, E. Lindroth, J. Anton, J. S. Coursey, D. J. Schwab, K. Olsen, R. A. Dragoset: *X-Ray Transition Energies*, Version 1. 0; [Online] available: http://physics. nist. gov/XrayTrans [2004,November 3]. National Institute of Standards and Technology, Gaithersburg, MD; originally published as: R. D. Deslattes, E. G. Kessler, Jr., P. Indelicato, L. deBilly, E. Lindroth, J. Anton: Review of Modern Physics **75**, 35 (2003)
146. M. Deutsch, M. Hart: Phys. Rev. B **26**, 5558 (1982)
147. M. Deutsch, G. Hölzer, J. Härtwig, J. Wolf, M. Fritsch, E. Förster: Phys. Rev. A **51**, 283 (1995)
148. M. Deutsch, O. Gang, G. Hölzer, J. Härtwig, J. Wolf, M. Fritsch, E. Förster: Phys. Rev. A **52**. 3661 (1995)
149. P. F. Dittner, C. E. Bemis, Jr, D. C. Hensley, R. J. Silva, C. D. Goodman: Phys. Rev. Letters **26**, 1037 (1971)
150. P. F. Dittner, C. E. Bemis, Jr: Phys. Rev. A **5**, 481 (1972)
151. K. T. Dolder et al.: Proc. Roy. Soc. A **264**, 367 (1961)
152. E. D. Donets, V. P. Ovsyannikov: Preprint P7-80-404 (JINR, Dubna 1980)
153. E. D. Donets: ECAJA **13**, 941 (1982)
154. J. C. Dousse, J. Kern: Physics Letters **59A**, 159 (1976)
155. R. Durak, Y. Sahin: Nuclear Instruments and Methods in Physics Research B **124**, 1 (1997)
156. *Atomic, Molecular and Optical Handbook*, ed by G. W. F. Drake (AIP, Woodbury, New York 1996)

157. B. S. Dselepov, S. A. Schestopalova: *Yaderno-Spektroskopicheskye Normaly* (in Russian) (Atomizdat, Moskau 1980)
158. J. W. M. DuMont: Review of Scientific Instruments **18**, 626 (1947)
159. K. G. Dyall, I. P. Grant: J. Phys. B **17**, 1281 (1984)
160. K. G. Dyall, I. P. Grant, C. T. Johnson, F. A. Parpia, E. P. Plummer: Computer Physics Communications **55**, 425 (1989)
161. N. A. Dyson: *X-Rays in Atomic and Nuclear Physics* (Longman Group Limited, London 1973)

E

162. I. Edamoto: Rep. Res. Inst. Tohoku Univ. A **2**, 561 (1950)
163. W. Ehrenberg, H. Mark: Z. Physik **42**, 807 (1927)
164. W. Ehrenberg, G. Susich: Z. Physik **42**, 823 (1927)
165. A. Einstein: Physik. Z. **18**, 121 (1917)
166. C. Enns (editor): *Cryogenic Particle Detection*, (Springer, Heidelberg a.o. 2005)
167. P. Erman, Z. Sujkowski: Ark. Fys. **20**, 209 (1961)
168. M. Ertugrul: Nuclear Instruments and Physics in Nuclear Research B **124**, 475 (1997)

F

169. D. J. Fabian: *Soft X-Ray Band Spectra* (Academic Press, New York 1969)
170. U. Fano, J. W. Cooper: Review of Modern Physics **40**, 441 (1968)
171. U. Fano, J. W. Cooper: Review of Modern Physics **41**, 724 (1969)
172. E. Fermi, E. Segre: Mem. Accad. Italia **4**, 131 (1933)
173. V. Fisher, v. Bernhstam, H. Golten, Y. Maron: *Electron Impact Excitation Cross-Sections for Allowed Transitions in Atoms*, 1995
     http://xxx. lanl. gov/abs/atom-ph/9509006
174. V. Fock: Z. Physik **61**, 126 (1930)
175. V. Fock, M. J. Petrashen: Z. Sowjetunion **6**, 368 (1934)
176. E. Förster, W. Z. Chang, M. Dirksmöller: *Int. Conf. on Applications of Laser Plasma Radiation II*, 12-14 July 1995, San Diego, California; in: SPIE Proceedings vol. **2523**, Bellingham, 1995, p 140
177. C. Foin et al.: Nuclear Physics A **113**, 241 (1968)
178. M. Frank, C. A. Mears, S. E. Labov, F. Azgui, M. A. Lindeman, L. J. Hiller, H. Netel, A. Barfknecht: Nuclear Instruments and Methods A **370**, 41 (1996)
179. M. Frank, S. Friedrich, J. Höhne, J. Jochum: *Cryogenic Detectors and their Application to XPF*, (Draft, 2002)
180. M. S. Freedman, I. Ahmad, F. T. Porter, R. K. Sjoblom, R. F. Barnes, J. Lerner, P. R. Fields: Phys. Rev. C **15**, 760 (1977)
181. H. M. Freund, R. W. Fink: Phys. Rev. **178**, 1952 (1969)
182. H. M. Freund et al.: Nuclear Physics A **138**, 200 (1969)
183. H. M. Freund: X-Ray Spectrometry **4**, 90 (1975)
184. H. Friedrich: *Theoretische Atomphysik* (Springer, Berlin Heidelberg New York 1990)
185. S. Friedrich, J. B. leGrand, L. J. Hiller, J. Kipp, M. Frank, S. E. Labov, S. P. Cramer, A. T. Barfknecht: IEEE Transactions on Applied Superconductivity **9** 3330 (1999)
186. L. Friens, F. M. Bonson: J. Phys. B **1**, 1123 (1968)
187. M. Fritsch: *Hochauflösende Röntgenspektroskopie an 3d-Übergangsmetallen*, Doctor Thesis, Physikalisch-- Astronomisch-Technikwissenschaftliche Fakultät der Friedrich-Schiller-Universität Jena, Jena (1998)
188. S. Fritzsche: Journal of Electron Spectroscopy and Related Phenomena **114-116**, 1155 (2001)
189. S. Fritzsche: Physica Scripta **T100**, 46 (2002)
190. C. Froese-Fischer: *The Hartree-Fock Method for Atoms* (John Wiley, New York 1977)
191. G. H. Fuller, V. W. Cohen: Nuclear Data Tables A **5**, 433 (1969)
192. G. H. Fuller: J. Phys. Chem. Ref. Data **5**, 835 (1976)
193. B. Furmann, D. Stefanska, A. Krzykowski, A. Jarosz, A. Kajoch: Z. Phys. D **37**, 289 (1996)

G

194. A. H. Gabriel: Mon. Not. R. Astron. Soc. **160**, 99 (1972)
195. R. H. Garstang. in: *Atomic and Molecular Processes,* ed by D. R. Bates (Academic Press, New York 1972) p 1
196. R. H. Garstang: in: *Topics in Modern Physics,* ed by W. E. Britin and H. Odabasi (Colo. Ass. Press, Boulder 1971) p 153
197. V. Gerling et al.: Z. Phys. **246**, 376 (1971)
198. R. C. Gibbs, D. T. Wilber, H. E. White: Phys. Rev. **29**, 790 (1927)
199. T. P. Gill: *The Doppler Effect* (Logo Press, London 1965)
200. G. Gilmore, J. Hemmingway: *Practical Gamma-Ray Spectrometry* (John Wiley and Sons, Chichester a. o. 1995)
201. R. Glocker: *Materialprüfung und Röntgenstrahlen* (Springer, Berlin Heidelberg New York 1971)
202. B. G. Gokhale: Ann. Phys. (Paris) **7**, 852 (1952)
203. B. G. Gokhale et al.: Phys. Rev. Letters **18**, 957 (1967)
204. B. G. Gokhale, S. N. Shukla: J. Phys. B **2**, 282 (1969)
205. B. G. Gokhale, S. N. Shukla: J. Phys. B **3**, 1392 (1970)
206. B. G. Gokhale, S. N. Shukla: J. Phy. B **3**, 438 (1970)
207. B. G. Gokhale, S. N. Shukla: J. Phys. B **3**, 1175 (1970)
208. Y. Gohshi, Y. Hukao, K. Hori: Spectrochimica Acta **27B**, 135 (1982)
209. Y. Goshi, H. Kamada, K. Kohra, T. Utaka, T. Arai: Applied Spectroscopy **36**, 171 (1982)
210. S. Goudsmit: Phys. Rev. **43**, 636 (1933)
211. R. L. Graham et al.: Can. J. Phys. **39**, 1058 (1961)
212. R. L. Graham et al.: Can. J. Phys. **39**, 1086 (1961)
213. I. P. Grant: Adv. Phys. **19**, 747 (1970)
214. I. P. Grant et al.: Computer Physics Communications **21**, 207 (1980)
215. H. H. Grother et al.: Z. Phys. **225**, 293 (1969)
216. M. Gryzinski: Phys. Rev. A **138**, 336 (1965)
217. O. Gunarsson, K. Schönhammer: Phys. Rev. Letters **41**, 1608 (1978)
218. S. N. Gupta, A. N. Gandhi, V. P. Vijayavargiya: J. Phys. B **9**, L481 (1976)
219. N. G. Gusev, P. P. Dmitriev: *Quantum Radiation of Radioactive Nuclides* (Atomizdat, Moscow 1977)

H

220. E. E. Haller: IEEE Transactions on Nuclear Science **NS-29**, 1109 (1982)
221. W. Hammer: Z. Phys. **216**, 355 (1968)
222. *Handbook of Chemistry and Physics,* ed by D. R. Lide (CRC Press, Boca Raton a. o. 2004)
223. J. Hanins, J. A. Bearden: Phys. Rev. **4A**, 890 (1964)
224. W. Hanke, J. Wernisch, C. Pöhn: X-Ray Spectrometry **14**, 43 (1985)
225. J. S. Hansen et al.: Bull. Americ. Phys. Soc. **16**, 578 (1971)
226. P. G. Hansen et al.: in: *Inner-Shell and X-Ray Physics of Atoms and Solids* (Plenum Press, New York 1981) p 163
227. P. G. Hansen, B. Jonson, G. L. Borchert, O. W. B. Schult: Mechanisms for Energy Shifts of Atomic K X-Rays, in: *Atomic Inner-Shell Physics,* ed by B. Crasemann (Plenum Publ. Corporation, New York 1985)
228. D. R. Hartree: *Calculation of Atomic Structures* (John Wiley, New York 1957)
229. J. Härtwig, G. Hölzer, J. Wolf, E. Förster: J. Appl. Cryst. **26**, 539 (1993)
230. J. Heintze: Z. Phys. **143**, 153 (1955)
231. W. Heitler: *The Quantum Theory of Radiation* (Clarendon Press, Oxford 1954)
232. A. Henins: Ruled Grating Measurement of the AL $K_{\alpha_{1,2}}$ Wavelength, in: *Precision Measurement and Fundamental Constants,* NBS 343, Maryland, August 1971, p 255
233. B. L. Henke, E. S. Ebisu: *Advances in X-Ray Analysis,* vol. 17 (Plenum Press, New York 1974) p 639
234. B. L. Henke et al.: Atomic Data and Nuclear Data Tables **27**, 1 (1982)
235. G. von Hevesy: *Chemical Analysis by X-Rays and its Applications* (New York 1932)
236. W. R. Hindmarsh: *Atomic Spectra* (Pergamon Press Ltd. , London 1967)

237. W. R. Hindmarsh: *Atomspektren* (Akademie Verlag, Berlin 1972)
238. N. E. Holden: *Isotopic Composition of the Elements and Their Variation in Nature: a Preliminary Report*, National Nuclear Data Center, Brookhaven National Laboratory, USA, March 1977
239. J. M. Hollander, M. D. Holtz, T. Novakov, R. L. Graham: Ark. Fys. **28**, 375 (1965)
240. C. P. Holmes, V. O. Kostroun: Bull. Americ. Phys. Soc. **15**, 561 (1970)
241. G. Hölzer, M. Fritzsch, M. Deutsch, J. Härtwig, E. Förster: Phys. Rev. A **56**, 4554 (1997)
242. L. D. Horakeri, B. Hanumaiah, S. R. Thontadarys: X-Ray Spectrometry **26**, 69 (1997)
243. J. Hoszowska, J.-Cl. Dousse, J. Kern, Ch. Rheme: Nuclear Instruments and Methods in Physics Research A **376**, 129 (1996)
244. C. M. Hsiung, B. Crasemann, H. Mark: Phys. Rev. A **24**, 177 (1981)
245. C. M. Hsiung, B. Crasemann, H. Mark: Phys. Rev. A **27**, 2989 (1983)
246. http://www.seilnacht.com/Lexikon/85Astat.htm
247. http://www.liverpool.K12.ny.uk/HyperChart/chemicalprops/
248. http://www.uniterra.de/rutherford/
249. K. N. Huang et al.: Atomic Data and Nuclear Data Tables **18**, 243 (1976)
250. D. J. Hubbard, E. Pantos: Nuclear Instruments and Methods **208**, 319 (1983)
251. J. H. Hubbell: *Photon Cross-Sections, Attenuation Coefficients and Energy Absorption Coefficients from 10 keV to 100 GeV*, NSRDS-NBS 29, NIST, Washington D.C., 1969
252. J. H. Hubbell: *Bibliography and Current Status of K, L, and Higher Shell Fluorescence Yields for Computations of Photon Energy-Absorption Coefficients*, NIST, Center for Radiation Research, NISTIR 89-4144, Gaithersburg, 1989
253. J. H. Hubbell, P. N. Trehan, N. Singh, B. Chand, D. Mehta, M. L. Garg, R. R. Garg, S. Singh, S. Puri: J. Phys. Chem. Ref. Data **23**, 339 (1994)
254. F. Hund: *Linienspektren und Periodisches System der Elemente* (Julius Springer Verlag, Berlin 1927)
255. P. Hungerford et al.: Z. Naturforsch. **A36**, 919 (1981)

I

256. O. Icelli, S. Erzeneoglu: Nuclear Instruments and Methods in Physics Research B **197**, 39 (2002)
257. P. Indelicato, E. Lindroth: Phys. Rev. A **46**, 2426 (1992)
258. E. Ingelstam: Acta Regia Soc. Sci. Uppsala **10**, 148 (1936)
259. K. D. Irwin, G. C. Hilton, J. M. Martinis, S. Deiker, N. Bergren, S. W. Nam, D. A. Rudnman, D. A. Wollman: Nuclear Instruments and Methods A **444**, 184 (2000)
260. IUPAC Commision on Atomic Weights and Isotopic Abundances: *Atomic Weigts of the Elements 1995*, Pure Appl. Chem. **68**, 2339 (1996)

J

261. H. Jacubowicz, D. L. Moores: Comm. At. Phys. **9**, 55 (1980)
262. R. W. James: *The Optical Principles of the Diffraction of X-Rays* (Bell, London 1948)
263. R. Jenkins: X-Ray Analysis **2**, 207 (1973)
264. R. Jenkins, R. Manne, J. Robin, C. Senemaud: Pure Appl. Chem. **63**, 735 (1991)
265. W. Jitschin, G. Grosse, P. Röhl: Phys. Rev. A **39**, 103 (1989)
266. W. Jitschin, R. Stötzel, T. Papp, M. Sarkar: Phys. Rev. A **59**, 3408 (1999)
267. H. H. Johann: Z. Phys. **69**, 185 (1931)
268. T. Johansson: Z. Phys. **82**, 507 (1933)
269. R. C. Jopson et al.: Phys. Rev. **381A**, 133 (1964)

K

270. E. Kartunen et al.: Nuclear Physics A **131**, 343 (1969)
271. S. K. Kataria, R. Govil, A. Saxena, H. N. Bajpei: X-Ray Spectrometry **15**, 49 (1986)
272. R. C. Kauffman et al.: Nuclear Physics Letters **31**, 621 (1973)
273. H. P. Kelly: Many-Body Pertubation Approaches to the Calculation of Transition Probabilities, in: *Atomic Inner-Shell Processes*, Vol. I, ed by B. Crasemann (Academic Press, New York a. o. 1975) p 331
274. R. L. Kelly, D. E. Harrison: At. Data Nucl. Data Tables **19**, 301 (1977)
275. O. Keski-Rahkonen, M. O. Krause: Phys. Rev. A **15**, 959 (1977)
276. E. G. Kessler Jr et al.: Phys. Rev. A **19**, 215 (1979)
277. E. G. Kessler Jr., R. D. Deslattes, W. Schwitz, L. Jacobs, O. Renner: Phys. Rev. A **26**, 2696 (1982)
278. A. E. Kingston: J. Phys. B **1**, 559 (1968)
279. I. K. Kikoin et al.: *Handbook of Physical Quantities* (in Russian), ed by I. K. Kikoin (Atomizdat, Moscow 1976)
280. Y. K. Kim: Phys. Rev. **154**, 1 (1967)
281. D. Kim, C. Tath, C. P. Barty: Phys. Rev. A **59**, R4129 (1999)
282. Y. S. Kim, R. H. Pratt: Phys. Rev. A **27**, 2913 (1983)
283. W. H. King: *Isotope Shifts in Atomic Spectra* (Plenum Press, New York 19849
284. *Handbook of X-Ray Techniques*, ed by W. W. Kluev (Mashinostroenije, Moscow 1980; in Russian)
285. V. Kment, A. Kuhn: *Technik des Messens radioaktiver Strahlung* (Geest & Portig, Leipzig 1963)
286. J. W. Knowles: Nuclear Instruments and Methods **162**, 677 (1979)
287. H. W. Koch, J. W. Motz: Review of Modern Physics **31**, 920 (1959)
288. A. A. Konstantinov et al.: Izv. Akad. Nauk SSSR, Ser. Fizika **28**, 107 (1964)
289. A. V. Korol, A. G. Lyalin, A. V. Solovyov: J. Phys. B **28**, 4947 (1995)
290. A. V. Korol, A. G. Lyalin, A. V. Solovyov: Phys. Rev. A **52**, 1 (1995)
291. A. V. Korol, A. G. Lyalin, A. S. Shukalov, A. V. Solovyov: JETP **82**, 631 (1996)
292. A. V. Korol, A. G. Lyalin, A. S. Shukalov, A. V. Solovyov: Journal of Electron Spectroscopy and Related Phenomena **79**, 323 (1996)
293. V. D. Kostroun et al.: Phys. Rev. A **3**, 553 (1971)
294. S. Kraft, J. Stümpel, P. Becker, U. Kuetgens: Rev. Sci. Instrum. **67**, 681 (1996)
295. S. Kraft, P. Verhoeve, A. Peacock, N. Rando, D. J. Goldie, R. Hart, D. Glowacka, F. Scholze, G. Ulm: Journal of Applied Physics **86**, 7189 (1999)
296. H. A. Kramers: Philos. Mag. **46**, 836 (1923)
297. H. W. Kraner: *Radiation Damage in Silicon Detectors*, Brookhaven National Laboratory Preprint, BNL 33692, New York, 1983
298. M. O. Krause, C. W. Nestor Jr: Physica Scripta **16**, 285 (1977)
299. M. O. Krause: *Creation of Multiple Vacancies*, in: Proc. Int. Conf. Inn. Shell Ioniz. Phenomena and Future Applications, Atlanta, US At. Energy Comm. Rep. No. CONF-720404, Oak Ridge, Tennessee, 1973, p 1586
300. M. O. Krause: J. Phys. Chem. Ref. Data **8**, 307 (1979)
301. M. O. Krause. J. H. Oliver: J. Phys. Chem. Ref. Data **8**, 329 (1979)
302. D. Küchler: Diploma Thesis, Institut für Kern- und Teilchenphysik, TU Dresden, Dresden (1994)
303. A. Kumar, B. N. Roy: Physics Letters **66A**, 362 (1978)
304. A. Kumar, S. Puri, D. Mehta, M. L. Garg, N. Singh: X-Ray Spectrometry **31**, 103 (2002)

L

305. J. L. Labar, C. J. Salter: *Electron Probe Quantitation*, ed by K. F. J. Heinrich and D. E. Newbury (Plenum Press, New York 1991) p 223
306. M. Lamoureux, N. B. Avdonina: Phys. Rev. E **55**, 912 (1997)
307. L. D. Landau, E. M. Lifschitz: *Quantenmechanik*, Lehrbuch der Theoretischen Physik, Band III (Akademie-Verlag, Berlin 19669

308. B. Lai et al.: Nuclear Instruments and Methods in Physics Research A **266**, 544 (1988)

309. A. Langenberg, J. van Eck: Journal of Physics B **12**, 1331 (1979)

310. G. Lapicki: Journ. Phys. Chem. Reference Data **18**, 111 (1989)

311. V. S. Lisitsa: *Atoms in Plasmas* (Springer, Berlin Heidelberg New York 1994)

312. F. P. Larkins: J. Physics B **4**, L29 (1971)

313. D. Layzer, R. H. Garstang: Ann. Rev. Astron. Astrophys. **6**, 449 (1968)

314. J. M. Leclercy: Phys. Rev. A **1**, 1358 (1970)

315. *Tables of Isotopes*, ed by C. M. Lederer and V. S. Shirley (John Wiley and Sons, New York 1978)

316. P. L. Lee et al.: Phys. Rev. A **9**, 614 (1973)

317. P. Lee, S. Salem: Phys. Rev. **10**, 2027 (1974)

318. J. B. leGrand, C. A. Mears, L. J. Hiller, M. Frank, S. E. Labov, H. Netel, D. Bochow, S. Friedrich, M. A. Lindeman, A. T. Barfknecht: Applied Physics Letters **73**, 1295 (1998)

319. U. Lehnert: Diploma Thesis, Fachrichtung Physik, Technische Universität Dresden, Dresden (1992)

320. H. J. Leisi et al.: Helv. Phys. Acta **34**, 161 (1961)

321. P. Lenard: Wied. Ann. **51**, 225 (1894)

322. P. Lenard: Ann. Physik **15**, 485 (1904)

323. W. R. Leo: *Techniques for Nuclear and Particle Physics Experiments: A How-to Approach* (Springer, Berlin Heidelberg New York 1988)

324. W. S. Letochow: *Laserspektroskopie* (Akademie Verlag, Berlin 1977)

325. L. Ley et al.: Phys. Rev. B **8**, 2392 (1973)

326. D. A. Liberman et al.: Computer Physics Communications **2**, 107 (1971)

327. A. A. Lind, O. Lundquist: Ark. Mat. Phys. **18**, 3 (1924)

328. W. Lotz et al.: Zeitschrift f. Physik **206**, 205 (1967)

329. C. C. Lu: Nucl. Phys. A **175**, 289 (1971)

330. G. K. Lum, C. E. Wiegand, E. G. Kessler Jr., R. D. Deslattes, L. Jacobs, W. Schwitz, R. Seki: Phys. Rev. D **23**, 2522 (1981)

331. G. Lutz: *Semiconductor Radiation Detectors: Device Physics* (Springer, Berlin Heidelberg New York 1999)

332. T. Lyman: Astrophys. J. **23**, 181 (1906) and Nature **93**, 241 (1914)

M

333. T. D. Märk, G. H. Dunn: *Electron Impact Ionization* (Springer, Berlin Heidelberg New York 1985)

334. J. B. Mann, J. T. Waber: Atomic Data **5**, 201 (1973)

335. S. T. Manson, D. J. Kennedy: Atomic Data and Nuclear Data Tables **14**, 111 (1974)

336. H. Maria, J. Dalmasso, G. Ardisson, A. Hachern: X-Ray Spectrometry **11** 79 (1982)

337. A. A. Markowicz: X-Ray Physics, in: *Handbook of X-Ray Spectrometry*, ed by R. E. Van Grieken and A. A. Markowicz (Marcel Dekker Inc., New York a. o. 1993)

338. W. C. Martin, L. Hagan, J. Reader, J. Sugar: J. Phys. Chem. Ref. Data **3**, 771 (1974)

339. N. Maskil, M. Deutsch: Phys. Rev. A **37**, 2947 (1988)

340. H. S. W. Massey, D. R. Bates: Rep. Progr. Phys. **9**, 62 (1942)

341. H. S. W. Massey, E. H. S. Burhop: Proc. Cambr. Phil. Soc. **32**, 461 (1936) and Proc. Roy. Soc. A **153** 661 (1936)

342. *MaTeck's Periodic Table of the Elements*, Juelich (2004); http://www. mateck. com

343. J. H. McCrary et al.: Phys. Rev. A **4**, 1745 (1971)

344. J. C. McGeorge et al.: Nuclear Physics A **151**, 526 (1970)

345. J. C. McGeorge et al.: Phys. Rev. A **4**, 1317 (1971)

346. E. J. McGuire: Phys. Rev. **185**, 1 (1969)

347. E. J. McGuire: Phys. Rev. A **2**, 273 (1970)

348. E. J. McGuire: Phys. Rev. A **3**, 587 (1971)

349. E. J. McGuire: Phys. Rev. A **5**, 1043 (1971)

350. E. J. McGuire: Phys. Rev. A **6**, 851 (1972)

351. B. J. McKenzie, I. P. Grant, P. H. Norrington: Computer Physics Communications, **21**, 233 (1980)

352. W. H. McMaster et al.: *Compilation of X-Ray Cross-Sections*, UCRL-50174, Sect. II, Rev. 1, Lawrence Radiation Laboratory, Livermore (1969)
353. W. Mehlhorn: Physical Letters A **26**, 166 (1968)
354. W. Meiling: *Kernphysikalische Elektronik* (Akademie-Verlag, Berlin 1975)
355. A. Meisel, W. Nefedov: Z. Chem. **1**, 337 (1961)
356. A. Meisel, G. Leonhard, R. Szargan: *Röntgenspektren und chemische Bindung* (Geest & Portig, Leipzig 1977)
357. L. Meitner: Z. Phys. **9**, 131 (1922)
358. A. G. Miller: Phys. Rev. A **13**, 2153 (1976)
359. J. E. Minor, E. C. Lingafelter: J. Amer. Chem. Soc. **71**, 1145 (1949)
360. M. Mizushima: *Quantum Mechanics of Atomic Spectra and Atomic Structure* (Benjamin Press, New York 1970)
361. S. Mohan et al.: Phys. Rev. C **1**, 254 (1970)
362. P. J. Mohr, N. B. Taylor: http://physics. nist. gov/constants
363. P. H. Mokler, F. Folkmann: X-Ray Production in Heavy Ion-Atom Collisions, in: *Structure and Collisions of Ions and Atoms* (Springer, Berlin Heidelberg New York 1978) p 214
364. E. Monnand, A. Moussa: Nuclear Physics **25**, 292 (1961)
365. T. Mooney, E. Kindroth, P. Indelicato, E. G. Kessler Jr, R. D. Deslattes: Phys. Rev. A **45**, 1531 (1992)
366. private communication from T. M. Mooney 1996 as cited in [145]
367. C. E. Moore: *Ionization Potentials and Ionization Limits Derived from the Analysis of Optical Spectra*, NSRDS-NBS 34, U.S. Govt. Printing Off. , Washington, D.C. (1970)
368. D. L. Moores: *Electronic and Atomic Collisions*, in: Papers and Progress Reports, ICPEAC XII, ed by S. Datz (North Holland Publ. Comp. , Gatlinburg 1981)
369. S. Morita: Japanese Journal of Applied Physics **22**, 1030 (1983)
370. H. G. Moseley: Philos. Mag. **26**, 1024 (1913)
371. H. G. Moseley: Philos. Mag. **27**, 703 (1914)

N

372. J. C. Nall et al.: Phys. Rev. **118**, 1278 (1960)
373. Y. Neeman, Y. Kirsh: *Die Teilchenjäger* (Springer, Berlin Heidelberg New York 1995)
374. G. C. Nelson, B. G. Saunders, W. John: Phys. Rev. **188**, 4 (1969)
375. G. C. Nelson et al.: Phys. Rev. **187**, 1 (1969)
376. G. C. Nelson et al.: Atomic Data **1**, 377 (1970)
377. G. C. Nelson, B. G. Saunders: Phys. Rev. **188**, 108 (1971)
378. R. C. Newman: Rep. Prog. Phys. **45**, 1163 (1982)
379. A. N. Nigam, K. B. Garg: Naturwissenschaften **54**, 641 (1967)
380. A. N. Nigam, Q. S. Kapoor: Indian J. Pure Appl. Phys. **6**, 644 (1968)
381. A. N. Nigam, P. P. Varma, R. B. Singh: Can. J. Phys. **47**, 341 (1969)
382. A. N. Nigam, K. B. Garg: Physica (Amsterdam) **45**, 203 (1969)
383. A. N. Nigam, K. B. Garg: J. Phys. B **2**, 419 (1969)
384. A. N. Nigam, K. B. Garg: Chem. Phys. Lett. **3**, 398 (1969)
385. A. N. Nigam, Q. S. Kapoor: Chem. Phys. Lett. **4**, 639 (1970)
386. A. N. Nigam, Q. S. Kapoor: Chem. Phys. Lett. **20**, 219 (1973)
387. A. N. Nigam, Q. S. Kapoor: J. Phy. B **6**, 2464 (1973)
388. A. N. Nigam, R. B. Mathur, R. Jain: J. Phys. B **7**, 2489 (1974)
389. A. N. Nigam, R. B. Mathur: Chem. Phys. Lett. **33**, 579 (1975)
390. A. N. Nigam, R. B. Mathur, B. G. Gokhale: Phys. Rev. A **13**, 1756 (1976)
391. A. N. Nigam, R. B. Mathur: J. Phys. B **9**, 2613 (1976)
392. A. N. Nigam, R. B. Mathur: Ltt. Nuovo Cimento Soc. Ital. Fis. **17**, 421 (1976)
393. A. N. Nigam, R. B. Mathur: Phys. Lett. **73A**, 159 (1979)
394. A. N. Nigam, R. B. Mathur: Physica B & C **100B+c**, 279 (1980)
395. A. Nigavekar, S. Bergwall: J. Phys. B **2**, 507 (1969)
396. http://physics. nist. gov/PUBS/AtSpec/node38. html
397. J. Nordgreen et al.: Physica Scripta **16**, 280 (1977)
398. C. Nordling, S. Hagström: Z. Phys. **178**, 418 (1964)

O

399. M. Ohno, G. Wendin: Physica Scripta **16**, 299 (1977)
400. M. Ohno: J. Phys. C: Solid State Physics **17**, 1437 (1984)
401. M. Ohno: Physica Scripta **34**, 146 (1986)
402. M. Ohno, R. E. LaVilla: Phys. Rev. A **38**, 3479 (1988)
403. M. Ohno, R. E. LaVilla: Phys. Rev. B **39**, 8845 (1989)
404. M. Ohno, R. E. La Villa: Phys. Rev. A **45**, 4713 (1992)
405. E. Öz, H. Erdogan, M. Ertuğrul: X-Ray Spectrometry **28**, 198 (1999)
406. K. Onoue, T. Suzuki: Jpn. J. Appl. Phys. **17**, 439 (1978)
407. V. P. Ovsyannikov, G. Zschornack: Review of Scientific Instruments **70**, 2646 (1999)

P

408. J. Pajor, A. Moljik: Comt. Rend. , B **264**, 550 (1967)
409. J. Pahor et al.: Nuclear Physics A **109**, 62 (1967)
410. J. Pahor et al.: Z. Phys. **221**, 490 (1969)
411. J. Pahor et al.: Z. Phys. **230**, 287 (1970)
412. J. Pahor: Review of Modern Physics **44**, 758 (1972)
413. J. M. Palms et al.: Phys. Rev. C **2**, 592 (1970)
414. J. J. Park, P. Christmas: Can. J. Phys. **45**, 2621 (1967)
415. F. A. Parpia, C. Froese Fischer, I. P. Grant: Computer Physics Communications **94**, 249 (1996)
416. L. G. Parratt: Phys. Rev. **50**, 1 (1936)
417. L. G. Parratt: Phys. Rev. **54**, 99 (1938)
418. L. G. Parratt: Review of Modern Physics **31**, 616 (1959)
419. F. Paschen: Ann. Physik **27**, 537 (1908)
420. W. Pauli: Naturwiss. **12**, 741 (1924)
421. W. Pauli: Z. Phys. **31**, 765 (1925)
422. H.-H. Perkampus: *Lexikon Spektroskopie* (VCH Verlagsgesellschaft mbH, Weinheim o. a. 1993)
423. S. T. Perkins, D. E. Cullen, M. H. Chen, J. H. Hubbell, J. Rathkopf, J. H. Scofield: *Tables and Graphs of Atomic Subshell Relaxation Data Derived from the LLNL Evaluated Atomic Data Library Z=1-100*, Lawrence Livermore National Laboratory Report, UCRL 50400, vol. 30, Livermore (1991)
424. B. Perny, J.-Cl. Dousse, M. Gasser, J. Kern, R. Lanners, Ch. Rheme, W. Schwitz: Nuclear Instruments and Methods in Physics Research A **267**, 120 (1987)
425. V. M. Pessa, E. Suoninen, T. Valkonen: Phys. Fenn. **8**, 71 (1973)
426. V. M. Pessa: X-Ray Spectrometry **2**, 169 (1973)
427. M. Petel, H. Houtermans: *Standardisation of Radionuclids* (IAEA, Vienna 1968) p 301
428. A. H. Pfund: J. Opt. Soc. Am. **9**, 193 (1924)
429. Z. G. Pinsker: *Dynamical Scattering of X-Rays in Crystals* (Springer, Berlin Heidelberg New York 1978)
430. J. K. Pious, K. M. Balakrishna, N. Lingpapa, K. Sidappa: Journ. Phys. B: At. Mol. Opt. Phys. **25**, 1155 (1992)
431. F. T. Porter, M. S. Freedman: Phys. Rev. Letters **27**, 293 (1971)
432. F. T. Porter, I. Ahmad, M. S. Freedman, J. Milstedt, A. M. Friedman: Phys. Rev. C **10**, 803 (1974)
433. F. T. Porter, M. S. Freedman: J. Phys. Chem. Ref. Data **7**, 1267 (1978)
434. C. J. Powell: Rev. Mod. Phys. **48**, 33 (1976)
435. R. E. Price et al.: Phys. Rev. **176** 3 (1968)
436. S. Puri, D. Mehta, B. Chand, N. Singh, P. N. Trehan: X-Ray Spectrometry **22**, 358 (1993)
437. P. Putila-Mäntylä, M. Ohno, G. Graeffe: Journal of Physics B **16**, 3503 (1983)

R

438. P.-A. Raboud, J.-C. Dousse: EAS-20, 20. Arbeitsbericht *Energiereiche atomare Stösse*, ed by B. Fricke et al. (GSI, Darmstadt 1999) p 70
439. P.-A. Raboud, J.-C. Dousse, J. Hoszowska, I. Savoy: Phys. Rev. A **61**, 012507 (1999)
440. P.-A. Raboud, M. Berset, J.-C. Dousse, Y.-P. Maillard: Phys. Rev. A **65**, 022512 (2002)
441. A. A. Radzig, B. M. Smirnov: *Handbook of Atomic and Molecular Physics* (Atomizdat, Moscow 1980; in Russian)
442. A. A. Radzig, B. M-Smirnov: *Reference Data on Atoms, Molecules and Ions*, Springer Ser. Chem. Phys. , vol. 13 (Springer, Berlin Heidelberg New York 1985)
443. S. Rai, G. B. Deodhar: Acta Phys. Polon. A **48**, 835 (1975)
444. S. Raj, H. C. Padhi, M. Polasik: Nuclear Instruments and Methods in Physics Research B **160**, 443 (2000)
445. A. Rani, N. Nath, S. N. Chaturvedi: X-Ray Spectrometry **18**, 77 (1989)
446. D. V. Rao, R. Cesareo, G. E. Gigante: Radiat. Phys. Chem. **46**, 317 (1997)
447. L. Rebohle, U. Lehnert, G. Zschornack: X-Ray Spectrometry **25**, 295 (1996)
448. I. Reiche: *Emissionsraten strahlender Elektronenübergänge in freien Atomen unter Beachtung der Elektronenrelaxation*, Doctor Thesis, TU Dresden, Fakultät für Naturwissenschaften und Mathematik, Dresden (1992)
449. J. J. Reidy: Curved Crystal Spectrometers and Use for Gamma-Ray Energy and Intensity Measurements, in: *The Electromagnetic Interaction in Nuclear Spectroscopy* (North Holland Publ. Corp. , 1975)
450. O. Renner, M. Kopecky, J. S. Wark, H. He, E. Förster: Review of Scientific Instruments **66**, 3234 (1995)
451. R. Revel, C. Den Auwer, C. Madic, F. David, B. Fourest, S. Hubert, J.-F. De Lu, L. R. Morss: Inorg. Chem. **38**, 4139 (1999)
452. P. Rice-Evans: *Spark, Streamer, Proportional and Drift Chambers* (Richelieu Press, London 1974)
453. P. K. Richtmyer, S. W. Barnes: Phys. Rev. **46**, 35 (1934)
454. W. Ritz: Phys. Z. **9**, 521 (1908)
455. H. Robinson: Proc. Roy. Soc. **104**, 455 (1923)
456. W. C. Röntgen; Science **3**, 227 and 726 (1896)
457. C. C. J. Roothaan: Review of Modern Physics **23**, 69 (1951)
458. C. C. J. Roothaan, P. S. Bayus: Meth. Comput. Phys. **2**, 47 (1963)
459. H. E. Roscoe, A. Harden: *A New View of the Origin of Daltons Theory* (Macmillan, 1896)
460. H. R. Rosner, C. P. Bhalla: Z. Phys. **231**, 347 (1970)
461. E. P. Ross: Phys. Rev. **100**, 1267 (1955)
462. W. Rubinson, K. P. Gopinathan: Phys. Rev. **170**, 170 and 696 (1968)
463. B. Rupp: *X-Ray Absorption Edge Energies*, UCRL-MI-125269. October 1999; http://www-structure. llnl. gov/X-Ray/elements. html
464. H. N. Russell, F. A. Saunders: Astrophys. J. **61**, 38 (1925)
465. H. N. Russell: Phys. Rev. **29**, 782 (1927)
466. H. N. Russell, A. G. Shenstone, L. A. Turner: Phys. Rev. **33**, 900 (1929)
467. E. Rutherford: Phil. Mag. **21**, 669 (1911)
468. J. R. Rydberg: Kgl. Svenska Vetensk. Akad. Handl. **23**, No. 11, Sect. 15 and 16 (1889)
469. R. W. Ryon, J. D. Zahrt: Polarized Beam X-Ray Fluorescence, in: *Handbook of X-Ray Spectrometry* (Marcel Dekker Inc. , New York a. o. 1993) p 491

S

470. S. I. Salem, R. J. Wimmer: Phys. Rev. A **2**, 1121 (1970)
471. S. I. Salem, C. W. Schultz: Atomic Data **3**, 215 (1971)
472. S. I. Salem et al.: Atomic Data and Nuclear Data Tables **14**, 91 (1974)
473. S. I. Salem, P. L. Lee: Atomic Data and Nuclear Data Tables **18**, 233 (1976)
474. A. Salop: Phys. Rev. A **14**, 2095 (1976)
475. E. Salzborn: Electron-Impact Ionization of Ions, in: *Physics of Ion–Ion and Electron–Ion Collisions*, ed by F. Brouillard and J. W. Mc Gowan (Plenum Press Publ. Comp. , New York 1983)

476. H. J. Sanchez, M. Rubio, R. D. Perez, E. Burattini: X-Ray Spectrometry **23**, 267 (1994)
477. H. J. Sanchez, R. D. Perez, M. Rubio, G. Castellano: X-Ray Spectrometry **24**, 221 (1995)
478. M. Sanchez del Rio et al.: Physica Scripta **55**, 735 (1997)
479. W. C. Sander et al.: Phys. Lett. **63A**, 313 (1977)
480. P. R. Sarode: X-Ray Spectrometry **22**, 138 (1993)
481. A. Scheffel. A. Assmann, J. Delith, M. Wendt: Microchimica Acta **155**, 269 (2006)
482. E. W. Schpolski: *Atomphysik*, Vol. II (Deutscher Verlag der Wissenschaften, Berlin 1969)
483. G. E. R. Schulze: *Metallphysik* (Akademie Verlag, Berlin 1974)
484. J. Schweppe, R. D. Deslattes, T. Mooney, C. J. Powell: Electron Spectrosc. Relat. Phenom **67**, 463 (1994)
485. pivate communication from J. Schweppe as cited in [145]
486. J. H. Scofield: Phys. Rev. **179**, 9 (1969)
487. J. H. Scofield: Phys. Rev. A **9**, 1041 (1974)
488. J. H. Scofield: Atomic Data and Nuclear Data Tables **14**, 121 (1974)
489. J. H. Scofield: Radiative Transitions, in: *Atomic Inner-Shell Processes*, Vol. I, ed by B. Crasemann (Academic Press, New York a. o. 1975) p 265
490. M. J. Seaton: Proc. Phys. Soc. **79**, 1105 (1962)
491. E. C. Seltzer: Phys. Rev. **188**, 1916 (1969)
492. P. B. Semmes, R. A. Braga, J. C. Griffin, R. W. Fink: Phys. Rev. C **35**, 749 (1987)
493. K. D. Sevier: Atomic Data and Nuclear Data Tables **24**, 323 (1979)
494. V. P. Shevelko: *Atoms and Their Spectroscopic Properties* (Springer, Berlin Heidelberg New York 1997)
495. D. A. Shirley et al.: Phys. Rev. B **15**, 544 (1977)
496. B. W. Shore, D. H. Menzel: *Principles of Atomic Spectra* (Wiley, New York 1968)
497. B. D. Shrivastava, R. K. Jain, V. S. Dubey: J. Phys. B **8**, 2948 (1975)
498. B. D. Shrivastava, R. K. Jain, V. S. Dubey: Phys. Lett. **59A**, 323 (1976)
499. B. D. Shrivastava, R. K. Jain, V. S. Dubey: Physica B & C **84 B+C**, 281 (1976)
500. B. D. Shrivastava, R. K. Jain, V. S. Dubey: Can. J. Phys. **55**, 1521 (1977)
501. B. D. Shrivastava, G. D. Gupta, S. K. Joshi: X-Ray Spectrometry **21**, 21 (1992)
502. K. Siegbahn, W. Stenstrom: Physik. Zeitschrift **17**
503. *ESCA - Atomic, Molecular and Solid State Structure Studied by Means of Electron Spectroscopy* (Almquvist and Wiksells, Uppsala 1967)
504. O. Sinanaoglu: Nuclear Instruments and Methods **110**, 193 (1973)
505. S. Singh, D. Mehta, R. R. Garg, S. Kumar, M. L. Garg, N. Singh, P. C. Mangal, J. H. Hubbell, P. N. Trehan: Nuclear Instruments and Methods in Physics Research B **51**, 5 (1990)
506. J. C. Slater: Phys. Rev. **81**, 385 (1951)
507. J. C. Slater: *The Quantum Theory of Atomic Structure*, vols. I and II (Mc Graw Hill, New York 1960)
508. I. I. Sobelman: *Atomic Spectra and Radiative Transitions*, Springer Ser. Atom. Plas. vol. 12 (Springer, Berlin Heidelberg New York 19929
509. I. I. Sobelman: *Introduction into the Theory of Atomic Spectra* (Nauka, Moskau 1977; in Russian)
510. F. R. S. Soddy: *The Story of Atomic Energy* (Nova Atlantis, London 1949)
511. Ö. Söğüt, E. Büyükkasap, A. Kücükönder, M. Ertuğrul, Ö. Simsek: Appl. Spectrosc. Rev. *30*, 175 (1995)
512. Ö. Söğüt, E. Büyükkasap, A. Kücükönder, M. Ertuğrul, O. Doğan, H. Erdoğan, Ö. Simsek: X-Ray Spectrometry **31**, 62 (2002)
513. Ö. Söğüt, E. Büyükkasap, A. Kücükönder, M. Ertuğrul, H. Erdoğan: X-Ray Spectrometry **31**, 71 (2002)
514. A. Sommerfeld: Proc. Nat. Acad. Sci. USA **15**, 393 (1929)
515. A. Sommerfeld: Ann. d. Phys. **11**, 257 (1931)
516. A. Sommerfeld: *Atombau und Spektrallinien* (Vieweg und Sohn, Berlin 1944)
517. S. L. Sorenson, R. Carr, S. J. Schaphorst, S. B. Whitfield, B. Crasemann: Pysical Review A **39**, 6241 (1989)
518. S. L. Sorenson, S. J. Schaphorst, S. B. Whitfield, B. Crasemann, R. Carr: Pysical Review A **44**, 350 (1991)
519. H. Sorum: J. Phys. F **17**, 417 (1987)
520. C. J. Sparks Jr. , B. S. Borie, J. B. Hastings: Nuclear Instruments and Methods **172**, 237 (1980)
521. J. Stark: Sitzber. Preuss. Akad. Wiss. Berlin (1913) p. 932

522. W. Stenstrom: Ann. Physik **57**, 343 (1918)
523. D. J. Stephanson: Phys. Rev. **178**, 1997 (1969)
524. M. Stobbe: Ann. Phys. **7**, 661 (1930)
525. R. Stoetzel, U. Werner, M. Sarker, W. Jitschin: Journ. Phys. B: At. Mol. Opt. Phys. **25**, 2295 (1992)
526. W. Stolz: *Messung ionisierender Strahlung* (Akademie Verlag, Berlin 1985)
527. W. Stolz: *Radioaktivität* (B. G. Teubner, Stuttgart 2003)
528. *Taschenbuch der Physik*, ed by H. Stöcker (Harri Deutsch, Thun and Frankfurt am Main 1998)
529. E. Storm, H. I. Israel. Nuclear Data Tables A **7**, 565 (1970)
530. J. Sugar: J. Opt. Soc. Am. **65**, 1366 (1975)
531. O. I. Sumbaev et al.: JETF **50**, 859 (1965)
532. O. I. Sumbaev, A. F. Mezentsev, V. I. Marushenko, A. S. Rylnikov, G. A. Ivanov. Sov. J. Nucl. Phys. **9**, 529 (1969)
533. O. I. Sumbaev: in: *Modern Physics in Chemistry*, ed by E. Fluck and V. I. Goldanskii, vol. I (Academic Press, London 1976) p 31
534. S. Svanberg. *Atomic and Molecular Spectroscopy* (Springer, Berlin Heidelberg New York 1991)
535. S. Svenson et al.: Physica Scripta **14**, 141 (1976)

T

536. N. Tawara: Atomic Data and Nuclear Data Tables **36**, 167 (1987)
537. J. G. V. Taylor, J. S. Merritt: *Role of Atomic Electrons in Nuclear Transitions* (Nuclear Energy Information Center, Warsaw 1963) p 465
538. J. J. Thompson: Phil. Mag. **23**, 449 (1912)
539. G. L. Trigg: *Crucial Experiments in Modern Physics* (Van Nostrand Reinhold Company, New York 1971)
540. M. E. Troughton: *Stand. of Radionucl.* (IAEA, Vienna 1967) p 32
541. K. Tsutsumi, H. Nakamori: J. Phys. Soc. Jpn. **25**, 1418 (1968)
542. K. Tsutsumi, H. Nakamori: in: *X-Ray Spectra and Electronic Structure of Matter*, ed by A. Faessler and G. Wiech (Fotodruck Frank OHG, München 1973)

U

543. W. Uchai, C. W. Nestor Jr, S. Raman, C. R. Vane: Atomic Data and Nuclear Data Tables **34**, 201 (1986)
544. J. H. Underwood Optics and Detctors, in: *X-Ray Data Booklet*, LBNL/PUB-490 Rev. 2 (LBNL, Berkeley 2001); see also http://xdb. lbl. gov (permanently updated content of the booklet)
545. F. Ullmann, G. Zschornack: *Zur Physik der Ionisation in der Dresden EBIT* (Leybold Vacuum Dresden GmbH, Dresden 2004)

V

546. R. V. Vedrinski, S. A. Prosandeev, Yu. A. Teterin: Teoretisheskaya i Eksperimentalnaya Chimia (in Russian) **16**, 620 (1980)
547. W. M. J. Veigele: Atomic Data Tables **5**, 51 (1973)
548. P. Venegopala Rao, B. Craseman: Phys. Rev. **139**, 1926 (1965)
549. P. Venegopala Rao et al.: Phys. Rev. **178**, 1997 (1969)
550. P. Venogopala Rao et al.: Phys. Rev. A **3**, 1568 (1971)
551. P. Verhoeve, N. Rando, A. Peacock, A. Dordrecht, B. G. Taylor, D. J. Goldie: Applied Physics Letters **72**, 3359 (1998)
552. W. Voigt: Sitzber. Akad. Wiss. (München, 1912) p 603
553. L. von Hamos: Naturw. , **20**, 705 (1932)
554. L. von Hamos: Ann. Phys. **17**, 716 (1933)
555. Z. Vylov et al.: *Radiation Spectra of Radioactive Nuclides* (Fan, Tashkent 1980; in Russian)

W

556. E. E. Wainstein: *Atomic and Molecular X-Ray Spectra in Chemical Compounds ans Alloys* (Academy of Science USSR, Moscow and Leningrad 1950)
557. D. l. Walters, C. P. Bhalla: Phys. Rev. A **3**, 273 and 519 (1971)
558. T. Watanbe: Phys. Rev. **139**, 1747 (1965)
559. R. L. Watson et al.: Phys. Rev. A **10**, 1230 (1974)
560. R. L. Watson et al.: Phys. Rev. B **13**, 2358 (1976)
561. A. W. Weiss: Nuclear Instruments and Methods **90**, 121 (1970)
562. V. Weisskopf, E. Wigner: Z. Phys. **63**, 54 and 65 (1930)
563. V. Weisskopf: Phys. Z. **34**, 1 (1933)
564. M. Wendt: Mikrochim. Acta **139**, 195 (2002)
565. M. Wendt, J. Dellith: Microchim. Acta **145**, 261 (2004)
566. G. Wentzel: Z. Phys. **43**, 524 (1927)
567. U. Werner, W. Jitschin: Phys. Rev. A **38**, 4009 (1988)
568. W. L. Wieseand: in: *A Physicist's Desk Reference* (American Institute of Physics, New York 1994)
569. K. Wiesemann: *Einführung in die Gaselektronik* (B. G. Teubner, Stuttgart 1976)
570. A. R. Williams, C. P. Long: Phys. Rev. A **3**, 519 (1971)
571. D. B. Wittry, S. Sun: J. Appl. Phys. **68**, 387 (1990)
572. D. B. Wittry: Proc. of the Int. Congr. X-Ray Optics and Microanalysis, Manchester 1992, Inst. Phys. Conf. Ser. , No. 130 (1993) p 535
573. D. A. Wollman, S. W. Nam, D. E. Newbury, G. C. Hilton, K. D. Irwin, N. F. Bergren, S. Deiker, D. A. Rudman, J. M. Martinis. Nuclear Instruments and Methods A **444**, 145 (2000)
574. R. E. Wood et al.: Phys. Rev. **187**, 1497 (1969)
575. F. J. Wuilleumier: J. Phys. (Paris) **32**, C4-88 (1971)

X

576. J. Q. Xu, E. Rosato: Journ. de Physique, Coll. C9 **48**, 661 (1987)
577. J. Q. Xu, E. Rosato: Phys. Rev. A **6**, 1946 (1988)
578. J. Q. Xu: Phys. Rev. A **43**, 4771 (1991)
579. J. Q. Xu, X. J. Xu: J. Phys. B: At. Mol. Opt. Phys. **25**, 695 (1992)

Y

580. T. Yamazaki, J. M. Hollander: Nuclear Physics **84**, 505 (1966)
581. S. M. Yonger: Phys. Rev. A **23**, 1138 (1981) and A **24**, 1278 (1981)

Z

582. J. D. Zahrt, R. W. Ryon: Adv. X-Ray Anal. **29**, 435 (1986)
583. A. Zararsiz, E. Aygün: J. Radioanal. Nucl. Chem. Articles **129**, 367 (1989)
584. P. Zeeman: Phil. Mag. **43**, 226 (1897) and **44**, 55 and 255 (1897)
585. G. Zschornack, G. Müller, G. Musiol: Nuclear Instruments and Methods **200**, 481 (1982)
586. G. Zschornack et al.: **P10-83-75** (JINR, Dubna 1983)
587. G. Zschornack, G. Musiol, W. Wagner: **ZfK-574** (FZ Rossendorf, Dresden-Rossendorf 1986)
588. L. Zülicke: *Quantenchemie*, Vol. 1 (Deutscher Verlag der Wissenschaften, Berlin 1973)

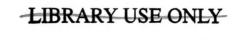

Physical constants, units and conversion factors[1]

Avogadro constant N_A	$6.022\,1415\,(10) \times 10^{23}\ \text{mol}^{-1}$
Faraday constant $F = N_A\,e$	$96\,485.3383(83)\ \text{C mol}^{-1}$
Bohr radius a_0	$0.529\,177\,210\,3\,(18) \times 10^{-10}\ \text{m}$
Bohr magneton μ_B	$927.400\,949(80) \times 10^{-26}\ \text{J T}^{-1}$ $5.788\,381\,804(39) \times 10^{-5}\ \text{eV T}^{-1}$
classical electron radius $r_e = \alpha^2 a_o$	$2.817\,940\,325(28) \times 10^{-15}\ \text{m}$
electron Compton wavelength λ_C	$2.426\,310\,238(16) \times 10^{-12}\ \text{m}$
electron mass m_e	$9.109\,3826(16) \times 10^{-31}\ \text{kg}$
electron mass energy equivalent $m_e c^2$	$0.510\,998\,918(44)\ \text{MeV}$
elementary charge e	$1.602\,176\,462(63) \times 10^{-19}\ \text{C}$
electric constant $\varepsilon_0 = 1/\mu_0 c^2$	$8.854\,187\,817\ldots \times 10^{-12}\ \text{F m}^{-1}\ (\text{exact})$
fine-structure constant $\alpha = e^2/(4\pi\varepsilon_0\,\hbar c)$	$7.297\,352\,568(24) \times 10^{-3}$
inverse fine-structure constant α^{-1}	$137.035\,999\,11(46)$
Hartree energy $E_h = 2\,R_\infty hc$	$27.211\,3845(23)\ \text{eV}$
magnetic constant μ_0	$4\pi \times 10^{-7}\ \text{N A}^{-2}$ $12.566\,370\,614\ldots \times 10^{-7}\ \text{N A}^{-2}\ (\text{exact})$
molar gas constant R	$8.314\,472(15)\ \text{J mol}^{-1}\ \text{K}^{-1}$
nuclear magneton $\mu_N = e\hbar/2m_p$	$5.050\,783\,43(43) \times 10^{-27}\ \text{J T}^{-1}$ $3.152\,451\,259(21) \times 10^{-8}\ \text{eV T}^{-1}$
photon wavelength Å	$1 \times 10^{-10}\text{m}$
Planck constant h	$6.626\,0693(11) \times 10^{-34}\ \text{J s}$ $4.135\,667\,43(35) \times 10^{-15}\ \text{eV s}$
Planck constant $\hbar = h/2\pi$	$1.054\,571\,68(18) \times 10^{-34}\ \text{J s}$ $6.582\,119\,15(56) \times 10^{-16}\ \text{eV s}$

---

[1] CODATA recommended values based on the 2002 adjustment [357], also extensively summarized in [220]. Fundamental constants were published by the TasK Group on Fundamental Constants of the Committee for Science and Technology (CODATA) of the International Council of Scientific Unions (CSU).
see also: http://physics.nist.gov/cuu/constants/index.html